Physical Constants

Name	Symbol	SI Units	cgs Units
Avogadro's number	N	$6.022137 \times 10^{23}/mol$	$6.022137 \times 10^{23}/mol$
Boltzmann constant	k	$1.38066 \times 10^{-23} J/K$	$1.38066 \times 10^{-16} erg/K$
Curie	Ci	$3.7 \times 10^{10} d/s$	$3.7 \times 10^{10} d/s$
Electron charge	e	1.602177×10^{-19} coulomb[†]	4.80321×10^{-10} esu
Faraday constant	$\mathscr{F}$	$96485 J/V \cdot mol$	$9.6485 \times 10^{11} erg/V \cdot mol$
Gas constant*	R	$8.31451 J/K \cdot mol$	$8.31451 \times 10^{7} erg/K \cdot mol$
Gravity acceleration	g	$9.80665 m/s^2$	$980.665 cm/s^2$
Light speed (vacuum)	c	$2.99792 \times 10^{8} m/s$	$2.99792 \times 10^{10} cm/s$
Planck's constant	h	$6.626075 \times 10^{-34} J \cdot s$	$6.626075 \times 10^{-27} erg \cdot s$

*Other values of R: 1.9872 cal/K $\cdot$ mol $= 0.082$ liter $\cdot$ atm/K $\cdot$ mol

[†]1 coulomb $= 1$ J/V

Conversion Factors

Energy: 1 Joule $= 10^7$ ergs $= 0.239$ cal
      1 cal $= 4.184$ Joule
Length: 1 nm $= 10$ Å $= 1 \times 10^{-7}$ cm
Mass: 1 kg $= 1000$ g $= 2.2$ lb
     1 lb $= 453.6$ g

Pressure: 1 atm $= 760$ torr $= 14.696$ psi
     1 torr $= 1$ mm Hg
Temperature: K $= °C + 273$
     C $= (5/9)(°F - 32)$
Volume: 1 liter $= 1 \times 10^{-3}$ m^3 $= 1000$ cm^3

Useful Equations

The Henderson–Hasselbalch Equation
$$pH = pK_a + \log([A^-]/[HA])$$
The Michaelis–Menten Equation
$$v = V_{max}[S]/(K_m + [S])$$
Temperature Dependence of the Equilibrium Constant
$$\Delta H° = -Rd(\ln K_{eq})/d(1/T)$$
Free Energy Change under Non-Standard-State Conditions
$$\Delta G = \Delta G° + RT \ln ([C][D]/[A][B])$$

Free Energy Change and Standard Reduction Potential
$$\Delta G°' = -n\mathscr{F}\Delta\mathscr{E}_o'$$
Reduction Potentials in a Redox Reaction
$$\Delta\mathscr{E}_o' = \mathscr{E}_o'(\text{acceptor}) - \mathscr{E}_o'(\text{donor})$$
The Proton-Motive Force
$$\Delta p = \Delta\Psi - (2.3RT/\mathscr{F})\Delta pH$$
Passive Diffusion of a Charged Species
$$\Delta G = G_2 - G_1 = RT \ln(C_2/C_1) + Z\mathscr{F}\Delta\Psi$$

Principles of

Biochemistry

WITH A HUMAN FOCUS

Principles of Biochemistry
WITH A HUMAN FOCUS

REGINALD H. GARRETT

CHARLES M. GRISHAM

University of Virginia

BROOKS/COLE

THOMSON LEARNING

Australia • Canada • Mexico • Singapore • Spain
United Kingdom • United States

BROOKS/COLE
THOMSON LEARNING

Chemistry Editor: John Vondeling
Developmental Editor: Sandi Kiselica
Marketing Strategist: Pauline Mula
Production Manager: Charlene Catlett Squibb
Project Editor: Bonnie Boehme
Art Director: Paul Fry
Copy Editor: Cathy Dexter
Illustrator: Dartmouth

Printed in the United States of America
2 3 4 5 6 7 05 04 03 02 01

For more information about our products, contact us at:
Thomson Learning Academic Resource Center
1-800-423-0563

For permission to use material from this text, contact us by:
Phone: 1-800-730-2214
Fax: 1-800-730-2215
Web: http://www.thomsonrights.com

Library of Congress Catalog Number: 2001086902
PRINCIPLES OF BIOCHEMISTRY, WITH A HUMAN FOCUS, First Edition
ISBN: 0-03-097369-4

Cover Printer: Lehigh Press
Compositor: TechBooks
Printer: Von Hoffmann Press
Cover Image: Protein folding: The process by which a polypeptide chain folds into a compact yet complex, stable, functional conformation can be modeled as a funnel-shaped energy landscape of many possible paths, all of which culminate in a unique, native folded state.

Asia
Thomson Learning
60 Albert Street, #15-01
Albert Complex
Singapore 189969

Australia
Nelson Thomson Learning
102 Dodds Street
South Melbourne, Victoria 3205
Australia

Canada
Nelson Thomson Learning
1120 Birchmount Road
Toronto, Ontario M1K 5G4
Canada

Europe/Middle East/Africa
Thomson Learning
Berkshire House
168-173 High Holborn
London WC1 V7AA
United Kingdom

Latin America
Thomson Learning
Seneca, 53
Colonia Polanco
11560 Mexico D.F.
Mexico

Spain
Paraninfo Thomson Learning
Calle/Magallanes, 25
28015 Madrid, Spain

Synopsis of Icon and Color Use in Illustrations

The following symbols and colors are used in this text to help in illustrating structures, reactions, and biochemical principles.

Elements:

= Oxygen   = Nitrogen   = Phosphorus   = Sulfur   = Carbon   = Chlorine

Small molecules and groups, which are common reactants or products in many biochemical reactions, are symbolized by the following icons:

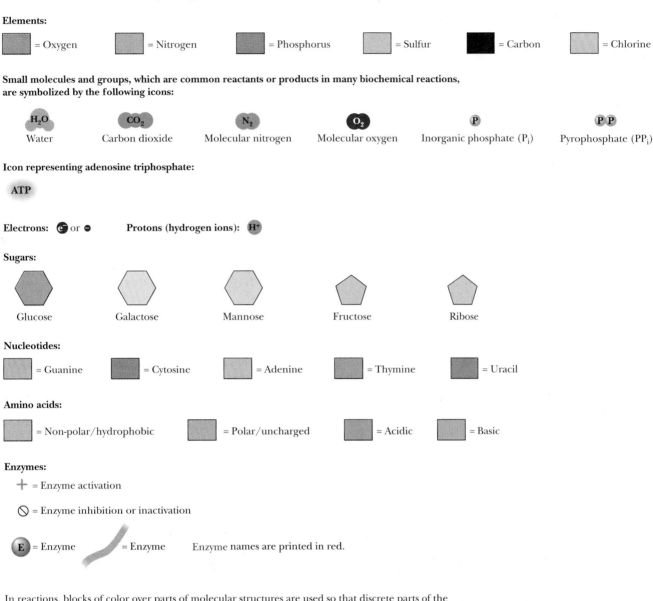

H_2O	CO_2	N_2	O_2	P	P P
Water	Carbon dioxide	Molecular nitrogen	Molecular oxygen	Inorganic phosphate (P_i)	Pyrophosphate (PP_i)

Icon representing adenosine triphosphate:

ATP

Electrons: e^- or ●   **Protons (hydrogen ions):** H^+

Sugars:

Glucose   Galactose   Mannose   Fructose   Ribose

Nucleotides:

= Guanine   = Cytosine   = Adenine   = Thymine   = Uracil

Amino acids:

= Non-polar/hydrophobic   = Polar/uncharged   = Acidic   = Basic

Enzymes:

+ = Enzyme activation

⊘ = Enzyme inhibition or inactivation

E = Enzyme   = Enzyme   Enzyme names are printed in red.

In reactions, blocks of color over parts of molecular structures are used so that discrete parts of the reaction can be easily followed from one intermediate to another, making it easy to see where the reactants originate and how the products are produced.

Some examples:

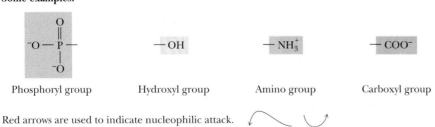

Phosphoryl group   Hydroxyl group   Amino group   Carboxyl group

Red arrows are used to indicate nucleophilic attack.

These colors are internally consistent within reactions and are generally consistent within the scope of a chapter or treatment of a particular topic.

ABOUT THE AUTHORS

Reginald H. Garrett was educated in the Baltimore city public schools and at the Johns Hopkins University, where he received his Ph.D. in biology in 1968. Since that time, he has been at the University of Virginia, where he is currently professor of biology. He is the author of *Biochemistry*, second edition (Harcourt College Publishers), and numerous papers and review articles on biochemical, genetic, and molecular biological aspects of inorganic nitrogen metabolism. Since 1964, his research interests have centered on the pathway of nitrate assimilation in filamentous fungi. His investigations have contributed substantially to our understanding of the enzymology, genetics, and regulation of this major pathway of biological nitrogen acquisition. His research has been supported by grants from the National Institutes of Health, the National Science Foundation, and private industry. He is a former Fulbright Scholar and has been a Visiting Scholar at the University of Cambridge on two sabbatical occasions; during the second, he was the Thomas Jefferson Visiting Fellow in Downing College. He has taught biochemistry at the University of Virginia for 33 years. He is a member of the American Society for Biochemistry and Molecular Biology.

Charles M. Grisham was born and raised in Minneapolis, Minnesota, and educated at Benilde High School. He received his B.S. in chemistry from the Illinois Institute of Technology in 1969 and his Ph.D. in chemistry from the University of Minnesota in 1973. Following a postdoctoral appointment at the Institute for Cancer Research in Philadelphia, he joined the faculty of the University of Virginia, where he is professor of chemistry. He is the author of *Biochemistry*, second edition (Harcourt College Publishers), and numerous papers and review articles on active transport of sodium, potassium, and calcium in mammalian systems, on protein kinase C, and on the applications of NMR and EPR spectroscopy to the study of biological systems. He has also authored the Harcourt *Interactive Biochemistry CD-ROM and Workbook*, a tutorial CD for students. His work has been supported by the National Institutes of Health, the National Science Foundation, the Muscular Dystrophy Association of America, the Research Corporation, the American Heart Association, and the American Chemical Society. He is a Research Career Development Awardee of the National Institutes of Health, and in 1983 and 1984 he was a Visiting Scientist at the Aarhus University Institute of Physiology, Aarhus, Denmark. In 1999, he was Knapp Professor of Chemistry at the University of San Diego. He has taught biochemistry and physical chemistry at the University of Virginia for 27 years. He is a member of the American Society for Biochemistry and Molecular Biology.

Left to right: Clancy, Reginald Garrett, Jazmine, Jasper, Charles Grisham, and Jatszi. *(Rosemary Jurbala Grisham)*

PREFACE

Scientific understanding of the molecular nature of life is growing at an astounding rate. Significantly, society is the prime beneficiary of this increased understanding. Cures for diseases, better public health, remediations for environmental pollution, and the development of cheaper and safer natural products are just a few practical benefits of this knowledge.

In addition, this expansion of information fuels, in the words of Thomas Jefferson, *"the illimitable freedom of the human mind."* Scientists can use the tools of biochemistry and molecular biology to explore all aspects of an organism—from basic questions about its chemical composition, through inquiries into the complexities of its metabolism, its differentiation and development, to analysis of its evolution and even its behavior. *New procedures based on the results of these explorations lie at the heart of the many modern medical miracles.* Biochemistry is a science whose boundaries now encompass all aspects of biology, from molecules to cells, to organisms, to ecology, and *to all aspects of health care.*

As the explication of natural phenomena rests more and more on biochemistry, its inclusion in undergraduate and graduate curricula in biology, chemistry, and the health sciences becomes imperative. And the challenge to authors and instructors is a formidable one: how to familiarize students with the essential features of modern biochemistry in an introductory course or textbook. Fortunately, the increased scope of knowledge allows scientists to make generalizations connecting the biochemical properties of living systems with the character of their constituent molecules. As a consequence, these generalizations, validated by repetitive examples, emerge in time as principles of biochemistry, principles that are useful in discerning and describing new relationships between diverse biomolecular functions and in predicting the mechanisms underlying newly discovered biomolecular processes. This biochemistry textbook is designed to communicate the fundamental principles governing the structure, function, and interactions of biological molecules to students encountering biochemistry for the first time. We aim to bring an appreciation of the science of biochemistry to a broad audience that includes undergraduates majoring in biology, chemistry, or premedical programs, as well as medical students and graduate students in the various health sciences for whom biochemistry is an important route to understanding human physiology. *To make this subject matter more relevant and interesting to all readers, we emphasize, where appropriate, the biochemistry of humans.*

We are both biochemists, but one of us is in a biology department, and the other is in a chemistry department. Undoubtedly, our approaches to biochemistry are influenced by the academic perspectives of our respective disciplines. We believe, however, that our collaboration on this textbook represents a melding of our perspectives that will provide new dimensions of appreciation and understanding for all students.

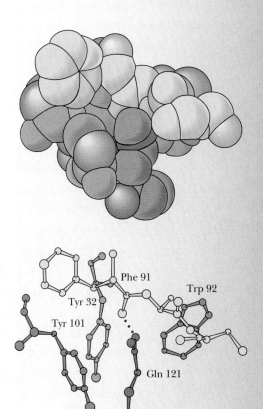

Pedagogical Features

- Up-to-date coverage gives students the latest information on active research.

- *Human Biochemistry* essays emphasize the central role of basic biochemistry in medicine and the health sciences.
- *A Deeper Look* essays expand on the text, highlighting selected topics or experimental observations.
- *Critical Developments in Biochemistry* essays emphasize recent and historical advances in the field.
- The four-color art program complements the text and aids in the students' ability to visualize biochemistry as a three-dimensional science.
- Photographs, chosen specifically for this text, show biochemistry as it applies to everyday life, to laboratory settings, and to cutting-edge research.
- The text and art use consistently colored icons for specific molecules, such as ATP, NADH, and coenzyme A. For a listing of all the icons, see page v.
- In-chapter examples and end-of-chapter problems give students practice in testing their mastery of the material.
- Up-to-date references at the end of each chapter make it easy for students to find additional information about each topic.
- CD-ROM icons point out the information contained on *Interactive Biochemistry CD-ROM and Workbook* by Charles Grisham.

Features and Organization

The organizational approach we have taken in this textbook is traditional in that it builds from the simple to the complex. Part 1, **Molecular Components of Cells,** creates a continuum between biochemistry and the prerequisite course in organic chemistry that should precede it. This section starts with the basics—the structure and hierarchical organization of biomolecular structures in cells (Chapter 1: *Chemistry Is the Logic of Biological Phenomena*) and the central role of water as the solvent of life (Chapter 2: *Water: The Medium of Life*). Chapter 3: *Thermodynamics of Biological Systems* presents fundamental thermodynamic relationships useful to understanding the energetics of cellular metabolism. The thermodynamics of protein folding is illustrated as an apt example for the relative contributions of enthalpy and entropy changes to free energy changes in biological systems. This chapter also discusses the particular chemical features of ATP and other biomolecules that make these substances effective as cellular energy carriers.

Chapter 4: *Amino Acids and Polypeptides* begins our survey of biomolecules by describing the structure and chemistry of amino acids and the fundamental properties of polypeptides, the polymers formed from amino acids. Chapter 5: *Proteins: Secondary, Tertiary, and Quaternary Structure* is devoted to a fresh and comprehensive description of the molecular anatomy of proteins, illustrated by a gallery of ribbon diagrams and space-filling computer graphics that capture the diversity of these structurally most sophisticated of all the biological macromolecules. Particular emphasis is given to the intriguing question of how proteins spontaneously fold in order to achieve the complex three-dimensional architecture that is the basis of their function.

Chapter 6: *Lipids, Membranes, and Transport* describes the structure and chemistry of lipids, allowing an in-depth exploration of the structure, chemistry, and function of membranes, the noncovalent structures that define cellular boundaries and compartments. This chapter also includes descriptions of the various structural classes of integral membrane proteins and the structural chemistry of lipid-anchored membrane proteins. The chapter also addresses the problem of membrane transport: how cells acquire nutrients and dispose of waste despite their enclosure within otherwise impermeable envelopes. Chapter 7: *Carbohydrates and Cell Surfaces* describes the structure and chemistry

(Loretto Chapel, Santa Fe, NM/© Sarbo)

of carbohydrates and the role of carbohydrate polymers as energy storage forms and as structural elements, including the complex proteoglycan structures that comprise bacterial cell wells and those that adorn animal cell surfaces.

Chapter 8: *Nucleotides and Nucleic Acids* describes the chemistry of nucleotides and the organization of these units into the polymeric macromolecules, ribonucleic acid and deoxyribonucleic acid (RNA and DNA). The procedures used to elucidate the sequential order of nucleotides in DNA and thus access the genetic information encoded therein are presented in an easily understandable way. This chapter is highlighted by an explanation of the structural order in the various forms of the DNA double helix (the ABZs of DNA structure) and the newly emerging views we have of the tertiary structure of RNA molecules. Chapter 9: *Recombinant DNA: Cloning and Creation of Chimeric Genes* is provided at this point in the text to familiarize students with the basic concepts and applications of this methodology. This technology is one of the newest tools available to biochemical research, and its enormous success in illuminating the relationships between genetic information, biomolecular structure, and biomolecular function is unparalleled. This chapter culminates in a discussion of the methodologies currently being considered for human gene therapy—the treatment of disease through introduction of engineered genes into humans.

Some instructors might prefer to present lipids and carbohydrates after a discussion of nucleotides and nucleic acids, and these chapters have been written so that this alternative sequence of presentation will go smoothly. The chapters of Part 1, like all the chapters in this text, feature **Text Boxes** of three kinds: **Human Biochemistry, A Deeper Look,** and **Critical Developments in Biochemistry.** These text boxes focus on aspects of medical biochemistry or delve a little deeper into selected topics or experimental observations, drawing the student into a greater appreciation for the relevance, logic, and historical context of significant biochemical advances. A prominent example is the *Human Biochemistry* box in Chapter 5 entitled "Diseases of Protein Folding," in which we illuminate the emerging paradigm of protein folding as a process in which impairments are manifested as disease states.

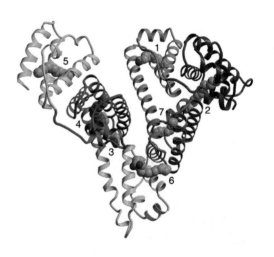

Part 2, **Protein Dynamics,** begins with a quantitative but easily grasped discussion of the kinetics of enzyme-catalyzed reactions (Chapter 10: *Enzymes: Their Kinetics, Specificity, and Regulation*). Also described are the exciting new discoveries of catalytic roles for certain RNA molecules (*ribozymes*) and the purposeful design of antibodies with catalytic properties (*abzymes*). This chapter also reviews the mechanisms by which enzymes—and hence metabolic processes—are regulated. Highlights of this chapter include a close scrutiny of structure–function relationships in glycogen phosphorylase and its regulation by allosteric mechanisms and reversible phosphorylation. The atomic structure of this paradigm of allosteric enzymes has been solved by X-ray crystallography, and the details of its structure provide fundamental insights into the catalytic and regulatory properties of enzymes. Chapter 11: *Mechanisms of Enzyme Action* presents the basic physicochemical principles of transition-state stabilization that underlie the catalytic power of enzymes. It also examines a gallery of enzyme mechanisms, including the well-studied serine proteases and aspartate proteases. This latter class includes clinically important HIV aspartate protease (HIV is the causative agent in AIDS).

Chapter 12: *Proteins of the Blood* may appear to be an unusual chapter to include in a basic textbook of biochemistry. However, discussion of blood proteins provides an excellent opportunity to touch upon some of the many functions that proteins display, while at the same time providing examples of fundamental biochemical phenomena, such as protein : ligand interactions, both general and specific, including allosteric regulation; protein processing (fibrinogen → fibrin); the biochemistry of the immune system; and the relationships between lipid transport and human cardiovascular health. Included here are transport achieved through protein:ligand interactions (both nonspecific, as in serum albumin or lipoprotein complexes, and specific, as in hemoglobin), antigen recognition through protein:ligand interactions (as in IgG), and catalysis through protein:substrate interactions (with formation of the blood clot as the example).

Chapter 13: *Molecular Motors* collects in one chapter a discussion of molecular motors—protein complexes that transduce chemical energy into the mechanics of movement. Included here is a discussion of the structure and function of muscle, as well as other systems of movement, such as the movement of cilia, vesicle and organelle transport within cells by kinesin and dyenin, and the novel rotating mechanism by which flagella impart motion to cells.

Part 3, **Metabolism and Its Regulation,** encompassing Chapters 14 through 22, describes the metabolic pathways that orchestrate the synthetic and degradative chemistry of life. Emphasis is placed on the chemical logic of intermediary metabolism. Chapter 14: *The Organization of Metabolism* points out the basic similarities in metabolism that unite all forms of life and gives a survey of nutrition and the underlying principles of metabolism, with particular emphasis on the role of vitamins as coenzymes. The fundamental aspects of catabolic metabolism are described in Chapter 15: *Glycolysis,* Chapter 16: *The Tricarboxylic Acid Cycle,* and Chapter 17: *Electron Transport and Oxidative Phosphorylation.* The contemporary view that energy coupling in mitochondria has an non-integral stoichiometry, yielding a P/O ratio of 2.5 rather than the traditionally held value of 3.0, is a pertinent feature of Chapter 17.

The photosynthetic processes that provide the energy and fundamental carbohydrate synthesis upon which virtually all life depends are described in Chapter 18: *Photosynthesis.* Focal points of this chapter include the exciting new description of the molecular structure of photosynthetic reaction centers and the mechanism and regulation of ribulose bisphosphate carboxylase. Chapter 19: *Gluconeogenesis, Glycogen Metabolism, and the Pentose Phosphate Pathway* includes a detailed discussion of carbohydrate metabolism beyond glycolysis and stresses the interconnections and interrelationships between the different pathways of carbohydrate conversion. Chapter 20: *Lipid Metabolism* begins with the details of fatty acid oxidation and liberation of the abundant energy stored in fats. The chapter then moves on to considerations of lipid synthesis, beginning with the novel properties of acetyl-CoA carboxylase, which catalyzes the principal regulated step in fatty acid biosynthesis. The chapter concludes with the metabolic pathways for synthesis of cholesterol and the biologically active lipids (eicosanoids and steroid hormones).

(© Thomas D. Mangelsen)

Chapter 21: *Amino Acid and Nucleotide Metabolism* examines the pathways for degradation and biosynthesis of amino acids and then profiles purine and pyrimidine metabolism. Highlighted within this chapter are the biochemical basis for genetic defects in nucleotide metabolism and the potential for pharmacological intervention to control unwanted cell proliferation, such as occurs in cancer or microbial infections. Chapter 22: *Metabolic Integration and Organ Specialization* is unique among textbook chapters in defining the essentially unidirectional nature of metabolic pathways and the stoichiometric role of ATP in driving vital processes that are thermodynamically unfavorable. This chapter also reveals the interlocking logic of metabolic pathways and the metabolic relationships between the various major organs of the human body.

Part 4, **Information Transfer,** addresses the storage and transmission of genetic information in organisms, as well as mechanisms by which organisms interpret and respond to chemical and physical information coming from the environment. The role of DNA molecules as the repository of inheritable information is presented in Chapter 23: *DNA: Replication, Recombination, and Repair,* along with the latest discoveries unraveling the molecular mechanisms un-

Preface | **xi**

derlying the enzymology of DNA replication. Sections on DNA replication and DNA repair treat the biochemistry involved in the maintenance and the replication of genetic information for transmission to daughter cells and accent the exciting new awareness that replication, recombination, and repair are interrelated aspects more appropriately treated together as DNA metabolism. Chapter 24: *Transcription and the Regulation of Gene Expression* then characterizes the means by which DNA-encoded information is expressed through synthesis of RNA and how expression of this information is regulated. Highlights of this chapter include the molecular structure and mechanism of RNA polymerase and the DNA-binding transcription factors that modulate its activity. The new attention given to nucleosomes as general repressors of transcription and the prerequisite for chromatin rearrangements in order to activate transcription is given prominent treatment, with emphasis on the roles of histone acetylation/deacylation and chromatin remodeling in these processes.

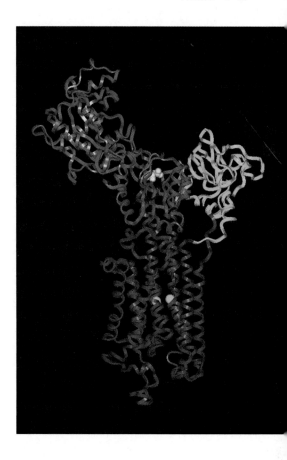

Chapter 25: *Protein Synthesis and Degradation* discusses the genetic code by which triplets of bases (codons) in mRNA specify particular amino acids in proteins and describes the molecular events that underlie the "second" genetic code—how aminoacyl-tRNA synthetases uniquely recognize their specific tRNA acceptors. This chapter also presents the structure and function of ribosomes, with emphasis on new, detailed information on ribosome structure and the interesting realization that 23SrRNA is the peptidyl transferase enzyme responsible for peptide bond formation. This chapter also reviews the necessity for molecular chaperones in the proper folding of proteins, and the emerging importance of proteasome-mediated protein degradation as a means to regulate cellular levels of specific proteins.

Chapter 26: *Signal Transduction: The Relay of Metabolic and Environmental Information* pulls together an up-to-date perspective on the rapidly changing fields of cellular signaling and stresses the information transfer aspects involved in the interpretation of environmental information, with coverage of hormone action, signal transduction cascades, membrane receptors, oncogenes, tumor suppressor genes, sensory transduction and neurotransmission, and the biochemistry of neurological disorders.

Acknowledgments

We are indebted to the many experts in biochemistry and molecular biology who carefully reviewed the manuscript at several stages for their outstanding and invaluable advice on how to construct an effective textbook. Their names are cited in the *List of Reviewers* at the end of this Preface.

We wish to warmly and gratefully acknowledge many other people who assisted and encouraged us in this endeavor. These include the highly skilled and dedicated professionals at Harcourt College Publishers. The superior quality of this group, exemplified in the books they have produced, is the living legacy of their leader and our publisher, John Vondeling. John recruited us to this monumental task to complete his personal pantheon of chemistry texts, and over the years, he has relentlessly urged us on. His threats, admonishments, and entreaties, laced with wisdom drawn from his vast experience in the publishing world, have been generally heeded by us. We knew well that John expected no more; he was not impressed by capitulation to all of his whims. He understood fully that excellence stems from the tension between different views, and he nourished such tensions as a way of bringing out the best in people. A relationship with John was a tempestuous blessing. He was a jolly good fellow; that nobody can deny.

Sandi Kiselica, our Developmental Editor, is a biochemist in her own right. Her fascination with our shared discipline has given her a particular interest in our book and a singular purpose: to keep us focused on the matters at hand,

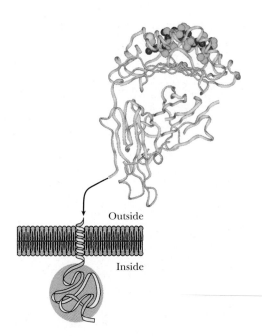

Outside

Inside

the urgencies of the schedule, and the limits of scale in a textbook's dimensions. The dint of her efforts is a major factor in the fruition of this project. We also applaud the unsung but absolutely indispensable contributions by those colleagues whose efforts transformed a rough manuscript into this final product: Bonnie Boehme, our Project Editor; Charlene Catlett Squibb, the Production Manager; Paul Fry, the Art Director and Cover Designer; and Pauline Mula, who is more than the Marketing Strategist—she is our friend and confidante. If this book has visual appeal and editorial grace, it is due to them. The beautiful illustrations that not only decorate this text but explain its contents are a testament to the creative and tasteful work of the team at Dartmouth Graphics and to the legacy of John Woolsey and Patrick Lane at J/B Woolsey Associates. We are thankful to our many colleagues who provided original art and graphic images for this work, particularly Professor Jane Richardson of Duke University, who gave us numerous original line drawings of the protein ribbon structures, and Michal Sabat and Mindy Whaley, who prepared many of the molecular graphics displayed herein. We owe a very special thank-you to Rosemary Jurbala Grisham, devoted spouse of Charles and wonderfully tolerant friend of Reg, who works tirelessly as our cheerleader and our photograph acquisitions specialist; in appreciation for her many contributions, spoken and unspoken, we dedicate this book to her. Also to be acknowledged with love and pride are our children, Jeffrey, Randal, and Robert Garrett, and David, Emily, and Andrew Grisham, as well as Clancy, a golden retriever of epic patience and perspicuity, and Jatszi, Jazmine, and Jasper, three Hungarian pulis whose unseen eyes view life with an energetic curiosity we all should emulate. With the publication of this new version of our text, we celebrate and commemorate the role of our mentors in bringing biochemistry to life for us— Alvin Nason, Kenneth R. Schug, William D. McElroy, Ronald E. Barnett, Maurice J. Bessman, Albert S. Mildvan, Ludwig Brand, and Rufus Lumry.

REGINALD H. GARRETT
Charlottesville, VA

CHARLES M. GRISHAM
Ivy, VA
February 2001

LIST OF REVIEWERS

We are very grateful to these reviewers for their careful reading of the manuscript. Their suggestions have proven invaluable.

David Brautigan, *University of Virginia Health System*

Jerry Brown, *University of Colorado Health Sciences Center*

Judith Campbell, *California Institute of Technology*

Jack Correia, *University of Mississippi Medical Center*

Donald Creighton, *University of Maryland, Baltimore County*

Christina Goode, *California State University, Fullerton*

Michael Griswold, *Washington State University*

Russ Hille, *Ohio State University, University Medical Center*

Diane Hills, *University of Osteopathic Medicine and Health Sciences, Iowa*

Larry Jackson, *Montana State University*

Tuajuanda Jordan-Stark, *Xavier University of Louisiana*

Gary Kunkel, *Texas A&M University*

Sidney Kushner, *University of Georgia*

Robert Lawther, *University of South Carolina*

Dennis Lohr, *Arizona State University*

Robert Mackin, *Creighton University*

Robin Miskimis, *University of South Dakota, School of Medicine*

Michael K. Reddy, *University of Wisconsin—Milwaukee*

Mary (Sam) Rigler, *California Polytechnic State University*

Joel Schildbach, *The Johns Hopkins University*

Jonathan Scholey, *University of California, Davis*

John Scocca, *The Johns Hopkins University*

Roger Sloboda, *Dartmouth College*

John Stewart, *University of Rochester, School of Medicine*

David Teller, *University of Washington*

Steven Theg, *University of California, Davis*

John Wilson, *Michigan State University*

Patrick Woster, *Wayne State University*

DEDICATION

We dedicate this book to two people who have
given us love and acceptance without
question and without end.

—WILLIAM W. GARRETT II   —ROSEMARY JURBALA GRISHAM

CONTENTS IN BRIEF

CONTENTS

5S rRNA

Dom V

Dom II

Dom I

Dom IV

Dom VI

Dom III

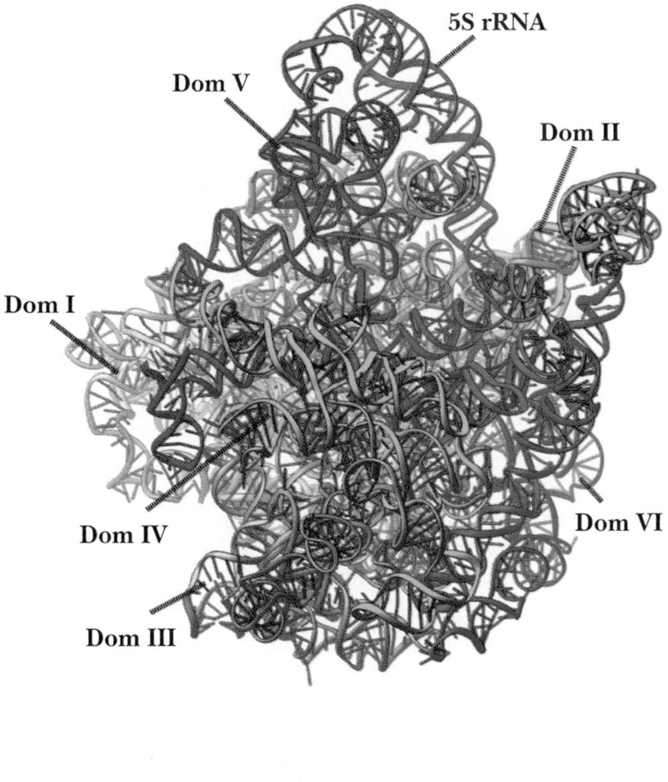

PART 2 PROTEIN DYNAMICS 319

(a) A dimeric protein can exist in either of two conformational states at equilibrium.

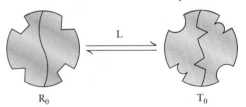

$$L = \frac{T_0}{R_0} \qquad \text{L is large. } (T_0 >> R_0)$$

(b) Substrate binding shifts equilibrium in favor of R.

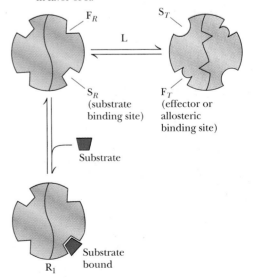

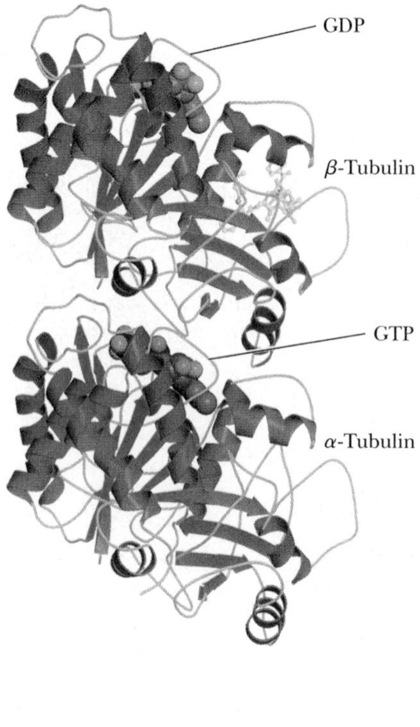

GDP

β-Tubulin

GTP

α-Tubulin

PART 3 METABOLISM AND ITS REGULATION   437

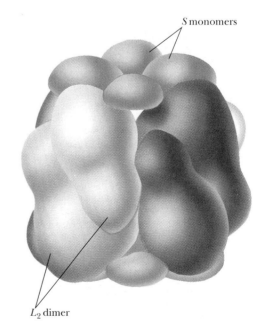

S monomers

L_2 dimer

(Frank Lane/Parfitt/Tony Stone Images)

Chapter 21 Amino Acid and Nucleotide Metabolism   657

Ammonium Formation and Amino Acid Metabolism   658

Nucleotide Metabolism   680

PART 4   INFORMATION TRANSFER   721

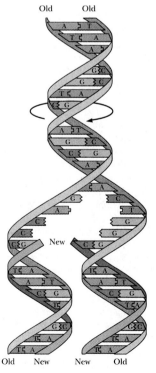

Emerging progeny DNA

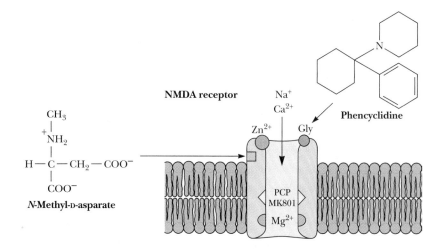

LIST OF BOXES

CRITICAL DEVELOPMENTS IN BIOCHEMISTRY

A DEEPER LOOK

SUPPORT PACKAGE

SUPPLEMENTS FOR STUDENTS

Study Guide by David Jemiolo (Vassar College) and Steven Theg (University of California, Davis) contains chapter summaries, detailed solutions to all end-of-chapter problems, additional problems with answers, self-tests, important definitions, and illustrations of major metabolic pathways.

Interactive Biochemistry CD-ROM and Workbook, developed by Charles Grisham, University of Virginia, includes three-dimensional protein graphics, virtual reality animations, and hundreds of interactive Java-based exercises. The presence of this icon throughout the text indicates links to the CD-ROM. Available for a fee from the publisher.

Student Lecture Outline. The same set of images available for professors as overhead transparency acetates is available in printed form for students to take notes during lectures.

Brooks/Cole Web Site for *Principles of Biochemistry* developed by Barrie Kitto and James Caras (University of Texas) can be accessed through the following address: www.brookscole.com/chemistry_d/ Students can watch animations of biochemical processes, test themselves with hundreds of multiple-choice questions, view and manipulate biomolecular structures, browse the Protein Data Bank for almost any structure found in the text, read about new and cutting-edge research, review basic concepts, and browse a collection of hundreds of biochemistry Web links organized by topic.

SUPPLEMENTS FOR INSTRUCTORS

Overhead Transparency Acetates. A set of 200 of the most pedagogically important images from the text.

Instructor's Resource CD-ROM. A teaching tool containing all the images and tables from the book.

PowerPoint™ Lectures. The Web site contains PowerPoint™ lecture images to accompany this text.

Test Bank by Larry Jackson (Montana State University). It includes multiple-choice questions for each chapter for professors to use as tests, quizzes, or homework assignments. The bank of questions is available in printed and computerized form for Macintosh® and Windows™ users.

MOLECULAR COMPONENTS OF CELLS

All life depends on water. All organisms are aqueous chemical systems. "Swamp Animals and Birds on the River Gambia," c. 1912 by Harry Hamilton Johnston (1858–1927).

(*Royal Geographical Society, London/The Bridgeman Art Library*)

Chemistry Is the Logic of Biological Phenomena

. . . everything that living things do can be understood in terms of the jigglings and wigglings of atoms.

Richard P. Feynman
Lectures on Physics
Addison-Wesley Publishing Company, 1963

"Noah's Ark" by Andre Normil (Superstock)

Molecules are lifeless. Yet, in appropriate complexity and number, molecules compose living things. These living systems are distinct from the inanimate world because they have certain extraordinary properties. They can grow, move, perform the incredible chemistry of metabolism, respond to stimuli from the environment, and, most significantly, replicate themselves with exceptional fidelity. The complex structure and behavior of living organisms veils the basic truth that their molecular constitution can be described and understood. The chemistry of the living cell resembles the chemistry of organic reactions. Indeed, cellular constituents, or **biomolecules,** must conform to the chemical and physical principles that govern all matter. Despite the spectacular diversity of life, the intricacy of biological structures, and the complexity of vital mechanisms, life functions are ultimately interpretable in chemical terms. *Chemistry is the logic of biological phenomena.*

1.1 Distinctive Properties of Living Systems

First, the most obvious quality of **living organisms** is that they are *complicated and highly organized* (Figure 1.1). For example, organisms large enough to be seen with the naked eye are composed of many **cells,** typically of many types. In turn, these cells possess subcellular structures, or **organelles,** which are complex assemblies of very large polymeric molecules, or **macromolecules.** These macromolecules themselves show an exquisite degree of organization in their intricate three-dimensional architecture, even though they are composed of simple sets of chemical building blocks, such as sugars and amino acids. Indeed, the complex three-dimensional structure, or **conformation,** characteristic of macromolecules is a consequence of interactions between their building block components, according to their individual chemical properties.

Second, *biological structures serve functional purposes.* That is, biological structures have a role in terms of the organism's existence. From parts of organisms, such as limbs and organs, down to the chemical agents of metabolism, such as enzymes and metabolic intermediates, a purpose can be given for each component. Indeed, it is this functional characteristic of biological structures that separates the science of biology from studies of the inanimate world such as chemistry, physics, and geology. In biology, it is always appropriate to ask why observed structures, organizations, or patterns exist, that is, to ask what functional role they serve within the organism.

Third, *living systems are actively engaged in energy transformations.* The maintenance of the highly organized structure and activity of living systems is dependent upon their ability to extract energy from the environment. The ultimate source of energy is the sun. Solar energy flows from *photosynthetic organisms* (those organisms able to capture light energy by the process of photosynthesis) through food chains to herbivores and ultimately to carnivorous predators at the apex of the food pyramid (Figure 1.2). The biosphere is thus a system through which energy flows. Organisms capture some of this energy, be it from photosynthesis or the metabolism of food, by forming special energized biomolecules, of which **ATP** and **NADPH** are the two most prominent examples (Figure 1.3). ATP and NADPH are energized biomolecules because they represent chemically useful forms of stored energy. When these molecules react

(a)

(b)

Figure 1.1 **(a)** Mandrill (*Mandrillus sphinx*), a baboon native to West Africa. **(b)** Tropical orchid (*Bulbophyllum blumei*), New Guinea. *(a, Tony Angemayer/Photo Researchers, Inc.; b, Thomas C. Boydon/Marie Selby Botanical Gardens)*

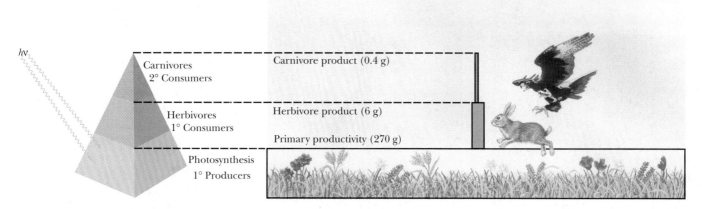

Productivity per square meter of a Tennessee field

Figure 1.2 The food pyramid: photosynthetic organisms at the base capture light energy. Herbivores and carnivores derive their energy ultimately from these primary producers.

Figure 1.3 ATP and NADPH—two biochemically important energy-rich compounds.

with other molecules in the cell, the energy released can be used to drive unfavorable processes. That is, ATP, NADPH, and related compounds are the power sources that drive the energy-requiring activities of the cell, including biosynthesis, movement, osmotic work against concentration gradients, and, in special instances, light emission (bioluminescence). Only upon death does an

Figure 1.4 Organisms resemble their parents.
(a) The Garretts: Jeffrey, Jackson, Reginald, Randal, little Reggie, and Robert. **(b)** Orangutan with infant.
(c) The Grishams: Charles, Emily, David, Rosemary, and Andrew. *(a, Kristin Koontz Garrett; b, Randal Harrison Garrett; c, Charles Grisham)*

(a)

(b)

(c)

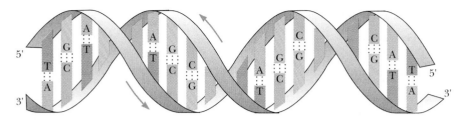

Figure 1.5 The DNA double helix. Two complementary polynucleotide chains running in opposite directions can pair through hydrogen bonding between their nitrogenous bases. Their complementary nucleotide sequences give rise to structural complementarity.

organism reach equilibrium with its inanimate environment. *The living state is characterized by the flow of energy through the organism.* At the expense of this energy flow, the organism can maintain its intricate order and activity far removed from equilibrium with its surroundings, yet exist in a state of apparent constancy over time. This state of apparent constancy, or so-called **steady state,** is actually a very dynamic condition: energy and material are consumed by the organism and used to maintain its stability and order. In contrast, inanimate matter, as exemplified by the universe in totality, is moving to a condition of increasing disorder, or, in thermodynamic terms, maximum entropy.

Fourth, *living systems have a remarkable capacity for self-replication.* Generation after generation, organisms reproduce virtually identical copies of themselves. This self-replication can proceed by a variety of mechanisms, ranging from simple division in bacteria to sexual reproduction in animals and plants, but in every case, it is characterized by an astounding degree of fidelity (Figure 1.4). Indeed, if the accuracy of self-replication were significantly greater, the evolution of organisms would be hampered. This is so because evolution depends upon natural selection operating on individual organisms that vary slightly in their fitness for the environment. The fidelity of self-replication resides ultimately in the chemical nature of the genetic material. This substance consists of polymeric chains of deoxyribonucleic acid, or **DNA,** which are structurally complementary to one another (Figure 1.5). These molecules can generate copies of themselves in a rigorously executed polymerization process that ensures a faithful reproduction of the original DNA strands. In contrast, the molecules of the inanimate world lack this capacity to replicate. A crude mechanism of replication, or specification of unique chemical structure according to some blueprint, must have existed at life's origin. This primordial system no doubt shared the property of **structural complementarity** (see page 15) with the highly evolved patterns of replication prevailing today.

1.2 Biomolecules: The Molecules of Life

The elemental composition of living matter differs markedly from the relative abundance of elements in the earth's crust (Table 1.1). Hydrogen, oxygen, carbon, and nitrogen constitute more than 99% of the atoms in the human body, with most of the H and O occurring as H_2O. Oxygen, silicon, aluminum, and iron are the most abundant atoms in the earth's crust, with hydrogen, carbon, and nitrogen being relatively rare (less than 0.2% each). Nitrogen as dinitrogen (N_2) is the predominant gas in the atmosphere, and carbon dioxide (CO_2) is present at a level of 0.05%, a small but critical amount. Oxygen is also abundant in the atmosphere and in the oceans. What property unites H, O, C, and

Table 1.1   Composition of the Earth's Crust, Seawater, and the Human Body*

Earth's Crust		Seawater		Human Body†	
Element	**%**	**Compound**	**mM**	**Element**	**%**
O	47	Cl^-	548	H	63
Si	28	Na^+	470	O	25.5
Al	7.9	Mg^{2+}	54	C	9.5
Fe	4.5	SO_4^{2-}	28	N	1.4
Ca	3.5	Ca^{2+}	10	Ca	0.31
Na	2.5	K^+	10	P	0.22
K	2.5	HCO_3^-	2.3	Cl	0.08
Mg	2.2	NO_3^-	0.01	K	0.06
Ti	0.46	HPO_4^{2-}	<0.001	S	0.05
H	0.22			Na	0.03
C	0.19			Mg	0.01

*Figures for the earth's crust and the human body are presented as percentages of the total number of atoms; seawater data are millimoles per liter. Figures for the earth's crust do *not* include water, while figures for the human body do.
†Trace elements found in the human body serving essential biological functions include Mn, Fe, Co, Cu, Zn, Mo, I, Ni, and Se.

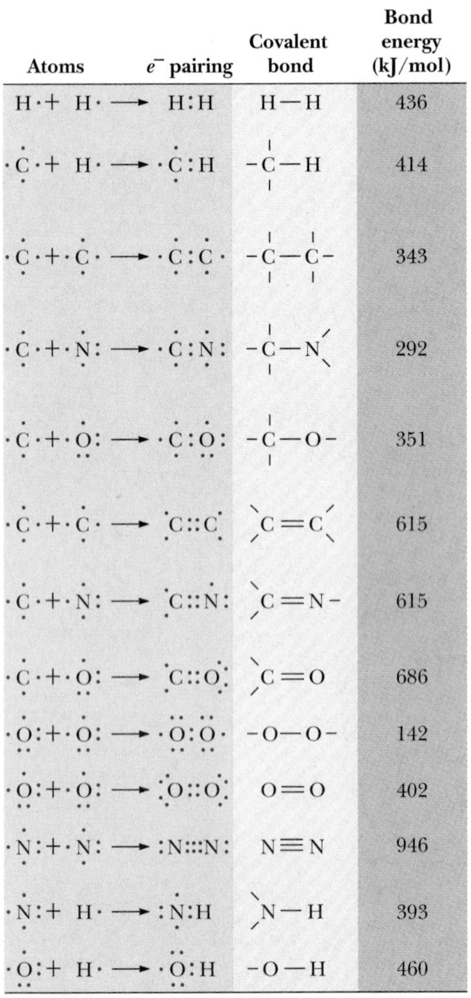

N and renders these atoms so suitable to the chemistry of life? It is their ability to form covalent bonds by electron-pair sharing. Furthermore, H, C, N, and O are among the lightest elements of the periodic table capable of forming such bonds (Figure 1.6). Since the strength of covalent bonds is inversely proportional to the atomic weights of the atoms involved, H, C, N, and O form the strongest covalent bonds. Two other covalent bond–forming elements, phosphorus (as phosphate $[-PO_4^{2-}]$ derivatives) and sulfur, also play important roles in biomolecules.

Biomolecules Are Carbon Compounds

All biomolecules contain carbon. The prevalence of C is due to its unparalleled versatility in forming stable covalent bonds by electron-pair sharing. Carbon can form as many as four such bonds by sharing each of the four electrons in its outer shell with electrons contributed by other atoms. Atoms commonly found in covalent linkage to C are C itself, H, O, and N. Hydrogen can form one such bond by contributing its single electron to the formation of an electron pair. Oxygen, with two unpaired electrons in its outer shell, can participate in two covalent bonds, and nitrogen, which has three unshared electrons, can form three such covalent bonds. Furthermore, C, N, and O can share two electron pairs to form double bonds with one another within biomolecules, a property that enhances their chemical versatility. Carbon and nitrogen can even share three electron pairs to form triple bonds.

Two properties of carbon covalent bonds deserve particular attention. One is the ability of carbon to form covalent bonds with itself. The other is the tetrahedral nature of the four covalent bonds when a carbon atom forms only single bonds. Together these properties hold the potential for an incredible variety of linear, branched, and cyclic compounds of C. This diversity is multiplied further by the possibilities for including N, O, and H atoms in these compounds (Figure 1.7). We can therefore envision the ability of C to generate complex structures in three dimensions. These structures, by virtue of appropriately included N, O, and H atoms, can display unique chemistries suitable to the living state. Thus, we may ask: is there any pattern or underlying organization that brings order to this astounding potentiality?

Atoms	e^- pairing	Covalent bond	Bond energy (kJ/mol)
$H\cdot + H\cdot \longrightarrow$	H:H	H—H	436
$\cdot C\cdot + H\cdot \longrightarrow$	$\cdot C$:H	$-\overset{\mid}{\underset{\mid}{C}}-H$	414
$\cdot C\cdot + \cdot C\cdot \longrightarrow$	$\cdot C$:$C\cdot$	$-\overset{\mid}{\underset{\mid}{C}}-\overset{\mid}{\underset{\mid}{C}}-$	343
$\cdot C\cdot + \cdot N$: $\longrightarrow$	$\cdot C$:N:	$-\overset{\mid}{\underset{\mid}{C}}-N\overset{\diagup}{\diagdown}$	292
$\cdot C\cdot + \cdot O$: $\longrightarrow$	$\cdot C$:O:	$-\overset{\mid}{\underset{\mid}{C}}-O-$	351
$\cdot C\cdot + \cdot C\cdot \longrightarrow$	C::C	$\overset{\diagdown}{\diagup}C=C\overset{\diagup}{\diagdown}$	615
$\cdot C\cdot + \cdot N$: $\longrightarrow$	C::N:	$\overset{\diagdown}{\diagup}C=N-$	615
$\cdot C\cdot + \cdot O$: $\longrightarrow$	C::O:	$\overset{\diagdown}{\diagup}C=O$	686
$\cdot O$: $+ \cdot O$: $\longrightarrow$	:O:O:	$-O-O-$	142
$\cdot O$: $+ \cdot O$: $\longrightarrow$	:O::O:	$O=O$	402
$\cdot N$: $+ \cdot N$: $\longrightarrow$	:N:::N:	$N\equiv N$	946
$\cdot N$: $+ H\cdot \longrightarrow$	:N:H	$\overset{\diagdown}{\diagup}N-H$	393
$\cdot O$: $+ H\cdot \longrightarrow$	$\cdot O$:H	$-O-H$	460

Figure 1.6   Covalent bond formation by e^- pair sharing.

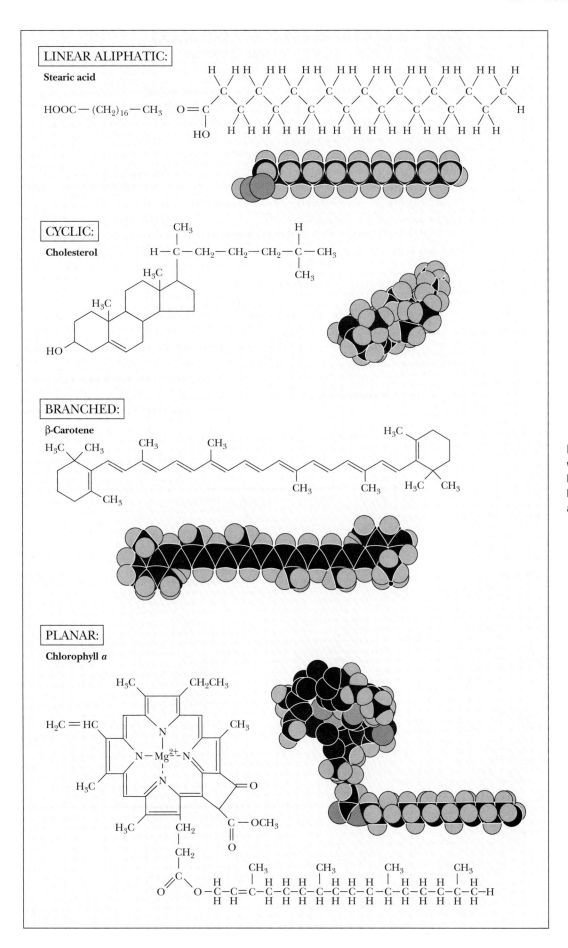

LINEAR ALIPHATIC:

Stearic acid

$$HOOC - (CH_2)_{16} - CH_3$$

CYCLIC:

Cholesterol

BRANCHED:

β-Carotene

PLANAR:

Chlorophyll *a*

Figure 1.7 Examples of the versatility of C—C bonds in building complex structures: linear aliphatic, cyclic, branched, and planar.

<table>
<tr><td>**1.3**</td><td>**A Biomolecular Hierarchy: Simple Molecules Are the Units for Building Complex Structures**</td></tr>
</table>

Examination of the chemical composition of cells reveals a dazzling variety of organic compounds covering a wide range of molecular dimensions (Table 1.2). As this complexity is sorted out and biomolecules are classified according to the similarity of their sizes and chemical properties, an organizational pattern emerges. The molecular constituents of living matter do not reflect randomly the infinite possibilities for combining carbon, hydrogen, oxygen, and nitrogen atoms. Instead, only a limited set of the many possibilities is found, and these collections share certain properties essential to the establishment and maintenance of the living state. We will first characterize the hierarchical relationships of biomolecules and then briefly consider what properties they possess that have rendered them so appropriate for the condition of life.

Metabolites and Macromolecules

The major precursors for the formation of biomolecules are water, carbon dioxide, and three inorganic nitrogen compounds—ammonium (NH_4^+), nitrate (NO_3^-), and dinitrogen (N_2). Metabolic processes assimilate and transform these inorganic precursors through ever more complex levels of biomolecular order (Figure 1.8). In the first step, precursors are converted to **metabolites,** simple organic compounds that are intermediates in cellular energy transformation and in the biosynthesis of various sets of **building blocks:** amino acids,

---

◼ Table 1.2   Biomolecular Dimensions

The dimensions of mass* and length for biomolecules are given typically in daltons and nanometers,[†] respectively. One dalton (D) is the mass of one hydrogen atom, 1.66×10^{-24} g. One nanometer (nm) is 10^{-9} m, or 10 Å (angstroms).

Biomolecule	Length (long dimension, nm)	Mass Daltons	Mass Picograms
Water	0.3	18	
Alanine	0.5	89	
Glucose	0.7	180	
Phospholipid	3.5	750	
Ribonuclease (a small protein)	4	12,600	
Immunoglobulin G (IgG)	14	150,000	
Myosin (a large muscle protein)	160	470,000	
Ribosome (bacteria)	18	2,520,000	
Bacteriophage ϕx174 (a very small bacterial virus)	25	4,700,000	
Pyruvate dehydrogenase complex (a multienzyme complex)	60	7,000,000	
Tobacco mosaic virus (a plant virus)	300	40,000,000	6.68×10^{-5}
Mitochondrion (liver)	1,500		1.5
Escherichia coli cell	2,000		2
Chloroplast (spinach leaf)	8,000		60
Liver cell	20,000		8,000

*Molecular mass is expressed in units of daltons (D) or kilodaltons (kD) in this book; alternatively, the dimensionless term *molecular weight*, symbolized by M_r and defined as the ratio of the mass of a molecule to 1 dalton of mass, is used.

[†]Prefixes used for powers of 10 are

10^6	mega M	10^{-3}	milli m	
10^3	kilo k	10^{-6}	micro μ	
10^{-1}	deci d	10^{-9}	nano n	
10^{-2}	centi c	10^{-12}	pico p	
		10^{-15}	femto f	

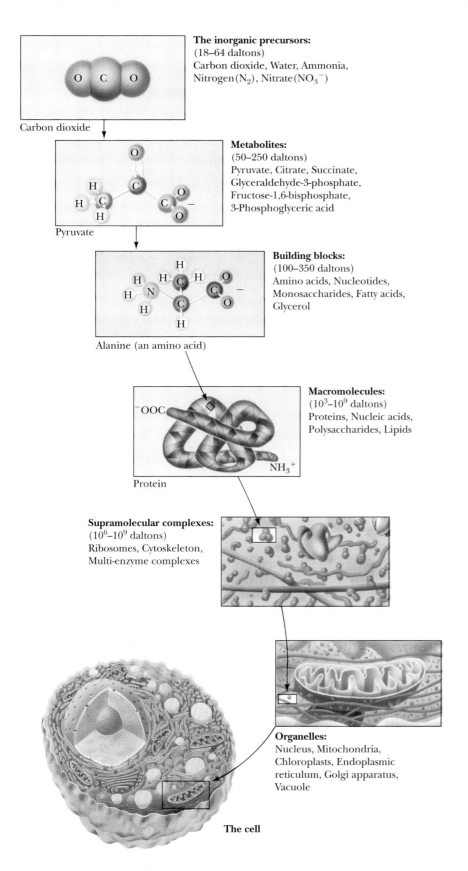

Figure 1.8 Molecular organization in the cell is a hierarchy.

The inorganic precursors:
(18–64 daltons)
Carbon dioxide, Water, Ammonia,
Nitrogen (N_2), Nitrate (NO_3^-)

Carbon dioxide

Metabolites:
(50–250 daltons)
Pyruvate, Citrate, Succinate,
Glyceraldehyde-3-phosphate,
Fructose-1,6-bisphosphate,
3-Phosphoglyceric acid

Pyruvate

Building blocks:
(100–350 daltons)
Amino acids, Nucleotides,
Monosaccharides, Fatty acids,
Glycerol

Alanine (an amino acid)

Macromolecules:
$(10^3–10^9$ daltons)
Proteins, Nucleic acids,
Polysaccharides, Lipids

Protein

Supramolecular complexes:
$(10^6–10^9$ daltons)
Ribosomes, Cytoskeleton,
Multi-enzyme complexes

Organelles:
Nucleus, Mitochondria,
Chloroplasts, Endoplasmic
reticulum, Golgi apparatus,
Vacuole

The cell

sugars, nucleotides, fatty acids, and glycerol. By covalent linkage of these building blocks, the **macromolecules** are constructed: proteins, polysaccharides, polynucleotides (DNA and RNA), and lipids. (Strictly speaking, lipids contain relatively few building blocks and are therefore not really "macromolecules"; however, lipids are important contributors to higher levels of complexity.)

Interactions among macromolecules lead to the next level of structural organization, **supramolecular complexes.** Here, various members of one or more of the classes of macromolecules come together to form specific assemblies serving important subcellular functions. Examples of these supramolecular assemblies are multifunctional enzyme complexes, ribosomes, chromosomes, and cytoskeletal elements. For example, a eukaryotic ribosome (see Section 1.5) contains four different RNA molecules and at least 70 unique proteins. These supramolecular assemblies are an interesting contrast to their individual components because their structural integrity is maintained by noncovalent forces, not by covalent bonds. These noncovalent forces include hydrogen bonds, ionic attractions, van der Waals forces, and hydrophobic interactions between macromolecules. Such forces maintain these supramolecular assemblies in a highly ordered functional state. Although noncovalent forces are weak (less than 40 kJ/mol), they are numerous in these assemblies and thus can collectively maintain the essential architecture of the supramolecular complex under conditions of temperature, pH, and ionic strength that are consistent with cell life.

Organelles

The next higher rung in the hierarchical ladder is occupied by the **organelles,** entities of considerable dimensions compared with the cell itself. Organelles are found only in **eukaryotic cells,** that is, the cells of "higher" organisms (eukaryotic cells are described in Section 1.5). Several kinds, such as mitochondria (and the chloroplasts of plant cells), evolved from bacteria that gained entry to the cytoplasm of early eukaryotic cells. Organelles share two attributes: they are cellular inclusions, usually membrane-bounded, and are dedicated to important cellular tasks. Organelles include the nucleus, mitochondria, endoplasmic reticulum, Golgi apparatus, and vacuoles, as well as other relatively small cellular inclusions, such as lysosomes. The **nucleus** is the repository of genetic information as contained within the linear sequences of nucleotides in the DNA of chromosomes. **Mitochondria** are the "power plants" of cells by virtue of their ability to carry out the energy-releasing aerobic metabolism of carbohydrates and fatty acids, capturing the energy in metabolically useful forms such as ATP. In plants, chloroplasts are the organelles that endow cells with the ability to carry out photosynthesis. They are the biological agents for harvesting light energy and transforming it into metabolically useful chemical forms.

Membranes

Membranes define the boundaries of cells and organelles. As such, they are not easily classified as supramolecular assemblies or organelles, although they share the properties of both. Membranes resemble supramolecular complexes in their construction, since they are complexes of proteins and lipids maintained by noncovalent forces. **Hydrophobic interactions** are particularly important in maintaining membrane structure. These interactions derive from the tendency of nonpolar molecules to come together when they are excluded by the polar solvent, water. The spontaneous assembly of membranes in the aqueous environment where life arose and exists is the natural result of their hydrophobic ("water-fearing") character. Hydrophobic interactions are the creative means of membrane formation and the driving force that presumably established the boundary of the first cell. The membranes of organelles, such as nuclei, mitochondria, and chloroplasts, differ from one another, with each having a unique protein and lipid composition suited to the organelle's function. Furthermore, the creation of discrete volumes or **compartments** within cells is not only an inevitable consequence of the presence of membranes but usually an essential condition for proper organellar function.

The Unit of Life Is the Cell

The **cell** is characterized as the unit of life, the smallest entity capable of displaying the attributes associated uniquely with the living state: growth, metabolism, stimulus response, and replication. In the previous discussions, we explicitly narrowed the infinity of chemical complexity potentially available to organic life, and we previewed an organizational arrangement, moving from simple to complex, that provides interesting insights into the functional and structural plan of the cell. Nevertheless, we find no obvious explanation within these features for the living characteristics of cells. Can we find other themes represented within biomolecules that are explicitly chemical yet anticipate or illuminate the living condition?

1.4	Properties of Biomolecules Reflect Their Fitness to the Living Condition

If we consider what attributes of biomolecules render them so fit as components of growing, replicating systems, several biologically relevant themes of structure and organization emerge. Furthermore, as we study biochemistry, we will see that these themes serve as principles of biochemistry. Prominent among them is *the necessity for information and energy in the maintenance of the living state*. Some biomolecules must have the capacity to contain the information or "recipe" of life. Other biomolecules must have the capacity to translate this information so that the blueprint is transformed into the functional, organized structures essential to life. An orderly mechanism for abstracting energy from the environment must also exist in order to obtain the energy needed to drive the exquisitely dynamic, intricate processes that collectively constitute life. What properties of biomolecules endow them with the potential for such remarkable qualities?

Biological Macromolecules and Their Building Blocks Have a "Sense" or Directionality

The macromolecules of cells are built of units—amino acids in proteins, nucleotides in nucleic acids, and carbohydrates in polysaccharides—that have **structural polarity.** That is, these molecules are not symmetrical, and so they can be thought of as having a "head" and a "tail." Polymerization of these units to form macromolecules occurs by head-to-tail linear connections. Because of this, the polymer will also have a head and a tail, and hence, the macromolecule has a "sense" or direction to its structure (Figure 1.9).

Biological Macromolecules Are Informational

Because biological macromolecules have a sense to their structure, their component building blocks, when read along the length of the molecule, have the capacity to specify information in the same manner that the letters of the alphabet can form words when arranged in a linear sequence (Figure 1.10). Not all biological macromolecules are rich in information. Polysaccharides are often composed of the same sugar unit repeated over and over, as in cellulose or starch, which are polymers of many glucose units. On the other hand, proteins and polynucleotides are typically composed of building blocks arranged in no obvious repetitive way; that is, their sequences are unique, like the letters and punctuation that form this descriptive sentence. In these unique sequences lies meaning. To discern the meaning, however, requires some mechanism for recognition.

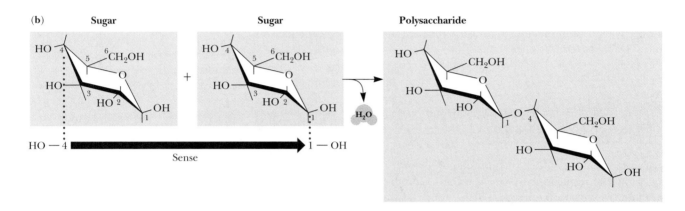

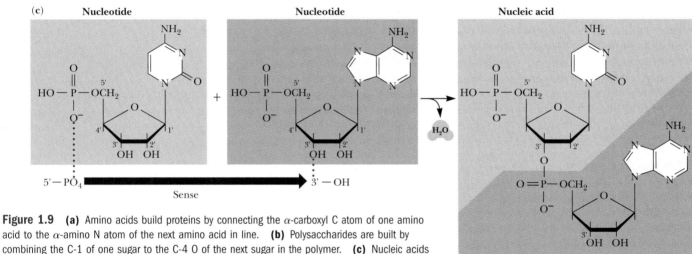

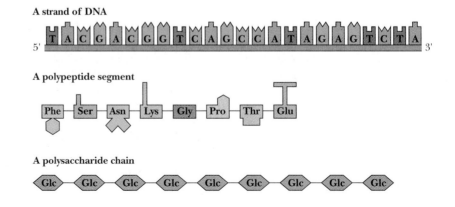

Figure 1.9 **(a)** Amino acids build proteins by connecting the α-carboxyl C atom of one amino acid to the α-amino N atom of the next amino acid in line. **(b)** Polysaccharides are built by combining the C-1 of one sugar to the C-4 O of the next sugar in the polymer. **(c)** Nucleic acids are polymers of nucleotides linked by bonds between the 3′-OH of the ribose ring of one nucleotide to the 5′-PO_4 of its neighboring nucleotide. All three of these polymerization processes involve bond formations accompanied by the elimination of water.

Figure 1.10 The sequence of monomeric units in a biological polymer has the potential to contain information if the diversity and order of the units is not overly simple or repetitive. Nucleic acids and proteins are information-rich molecules; polysaccharides are not.

Biomolecules Have Characteristic Three-Dimensional Architecture

The structure of any molecule is a unique and specific aspect of its identity. Molecular structure reaches its pinnacle in the intricate complexity of biological macromolecules, particularly the proteins. Although proteins are linear sequences of covalently linked amino acids, the course of the protein chain can turn, fold, and coil in the three dimensions of space to establish a specific, highly ordered architecture that is an identifying characteristic of the given protein molecule (Figure 1.11).

Weak Forces Maintain Biological Structure and Determine Biomolecular Interactions

Covalent bonds hold atoms together so that molecules are formed. **Weak chemical forces,** which include hydrogen bonds, van der Waals forces, ionic bonds, and hydrophobic interactions, are intramolecular or intermolecular attractions between atoms. None of these forces, which typically range from 4 to 30 kJ/mol, are strong enough to bind free atoms together (Table 1.3). The average kinetic energy of molecules at 25°C is 2.5 kJ/mol, so the energy of weak forces is only several times greater than the dissociating tendency due to the thermal motion of molecules. These weak forces are the basis of noncovalent interactions that are constantly forming and breaking at physiological temperature, unless by cumulative number they give stability to the structures generated by their collective action. These weak forces merit further discussion because their properties profoundly influence the nature of the biological structures they build.

Van der Waals Interactions

Van der Waals attractive forces are the result of induced electrical interactions between closely approaching atoms or molecules as their negatively charged electron clouds fluctuate instantaneously in time. These fluctuations allow attractions to occur between the positively charged nuclei and the electron density of the incoming atom. The strength of van der Waals attractive forces varies inversely with the sixth power of the distance between atoms. However, when two atoms approach each other so closely that their electron clouds interpenetrate, **van der Waals repulsive forces** limit the approach of the two atoms. The

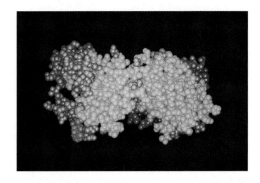

Figure 1.11 Three-dimensional representation of a protein molecule, the antigen-binding domain of immunoglobulin G (IgG). Immunoglobulin G is a major type of circulating antibody.

■ **Table 1.3** **Weak Chemical Forces and Their Relative Strengths and Distances**

Force	Strength (kJ/mol)	Distance (nm)	Description
Van der Waals interactions	0.4–4.0	0.2	Strength depends on the relative size of the atoms or molecules and the distance between them. The size factor determines the area of contact between two molecules: the greater the area, the stronger the interaction. Attractive force is inversely proportional to the sixth power of the distance, r, separating two atoms or molecules: $F \approx 1/r^6$.
Hydrogen bonds	12–30	0.3	Relative strength is proportional to the polarity of the H bond donor and H bond acceptor. More polar atoms form stronger H bonds.
Ionic bonds	20	0.25	Strength also depends on the relative polarity of the interacting charged species. Some ionic bonds are also H bonds: $-NH_3^+ \cdots {}^-OOC-$
Hydrophobic interactions	<40	–	Force is a complex phenomenon determined by the degree to which the structure of water is disordered as discrete hydrophobic molecules or molecular regions coalesce.

Table 1.4 **Radii of the Common Atoms of Biomolecules**

Atom	Van der Waals Radius, nm	Covalent Radius, nm	Atom Represented to Scale
H	0.1	0.037	
C	0.17	0.077	
N	0.15	0.070	
O	0.14	0.066	
P	0.19	0.096	
S	0.185	0.104	
Half-thickness of an aromatic ring	0.17	—	

strength of van der Waals repulsive forces varies inversely with the twelfth power of the distance between atoms. The balance between the attractive and repulsive aspects of van der Waals interactions determines the **van der Waals contact distance,** which is the interatomic distance that results if only van der Waals forces hold two atoms together. The optimal distance of contact between two atoms can be found by adding their van der Waals radii (Table 1.4). Van der Waals attractions operate only over a limited interatomic distance and are an effective bonding interaction at physiological temperatures only when a number of atoms in a molecule can interact with several atoms in a neighboring molecule. For this to occur, the atoms on interacting molecules must pack together neatly. That is, their molecular surfaces must possess a degree of structural complementarity (Figure 1.12).

(a) (b)

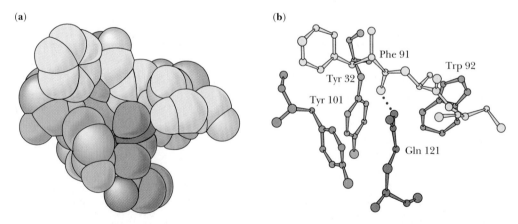

Figure 1.12 Van der Waals packing is enhanced in molecules which are structurally complementary. Gln[121] represents a surface projection on the protein lysozyme. This projection fits nicely within a pocket (formed by Tyr[101], Phe[91], and Trp[92]) in the antigen-binding domain of an antibody raised against lysozyme. (See also Figure 1.15.) **(a)** A space-filling representation. **(b)** A ball-and-stick model. *(From* Science *233:751 [1986], Figure 5)*

Hydrogen Bonds

Hydrogen bonds form between a hydrogen atom covalently bonded to an electronegative atom (such as oxygen or nitrogen) and a second electronegative atom that serves as the hydrogen bond acceptor. Several important biological examples are given in Figure 1.13. Hydrogen bonds, at a strength of 12 to 30 kJ/mol, are stronger than van der Waals forces and have an additional property: H bonds tend to be highly directional, forming straight bonds between donor, hydrogen, and acceptor atoms. Hydrogen bonds are also more specific than van der Waals interactions because they require the presence of complementary hydrogen donor and acceptor groups.

Ionic Interactions

Ionic bonds are the result of attractive electrostatic interactions between oppositely charged polar functions, such as negative carboxyl groups and positive amino groups (Figure 1.14). These electrostatic forces average about 20 kJ/mol in aqueous solutions. Because the electrical charge is typically radially distributed, these bonds may lack the directionality of hydrogen bonds or the precise fit of van der Waals interactions. Nevertheless, since the opposite charges are restricted to sterically defined positions, ionic bonds can impart a high degree of structural specificity. Repulsive electrostatic interactions also occur as a result of repulsion between like charges.

Hydrophobic Interactions

Hydrophobic interactions are due to the strong tendency of water to exclude nonpolar groups or molecules (see Chapter 2). Hydrophobic interactions arise not so much because of any intrinsic affinity of nonpolar substances for one another (although van der Waals forces do promote the weak bonding of nonpolar substances), but because water molecules prefer the stronger interactions that they share with one another. Since the strongest chemical interaction that is possible between two molecules actually determines what occurs, the preferential hydrogen bonding interactions between polar water molecules excludes nonpolar groups. It is this exclusion that drives the tendency of nonpolar substances to cluster in aqueous solution. Thus, nonpolar regions of biological macromolecules are often buried in the molecule's interior to exclude them from the aqueous surroundings. The formation of oil droplets as hydrophobic nonpolar lipid molecules coalesce in the presence of water is an approximation of this phenomenon. These tendencies have important consequences in the creation and maintenance of the macromolecular structures and supramolecular assemblies of living cells.

Structural Complementarity Determines Biomolecular Interactions

Structural complementarity is the means of recognition in biomolecular interactions. The complicated and highly organized patterns of life are dependent upon the ability of biomolecules to recognize and interact with one another in very specific ways. Such interactions are fundamental to metabolism, growth, replication, and other vital processes. The interaction of one molecule with another—a protein with a metabolite, for example—can be most precise if the structure of one is complementary to the structure of the other, as in two connecting pieces of a puzzle or, in the more popular analogy for macromolecules and their **ligands** (see page 16), a lock and its key (Figure 1.15). This principle of structural complementarity is the very essence of biomolecular recognition. *Structural complementarity is the significant clue to understanding the functional properties of biological systems.* Biological systems from the macromolecular level to the cellular level operate via specific molecular recognition

(a) H bonds Bonded atoms	Approximate bond length*
O—H---O	0.27 nm
O—H---O⁻	0.26 nm
O—H---N	0.29 nm
N—H---O	0.30 nm
⁺N—H---O	0.29 nm
N—H---N	0.31 nm

*Lengths given are distances from the atom covalently linked to the H to the atom H-bonded to the hydrogen:

$$O-H---O$$
$$\mid\leftarrow 0.27\ \text{nm}\rightarrow\mid$$

(b) **Functional groups which are important H bond donors and acceptors:**

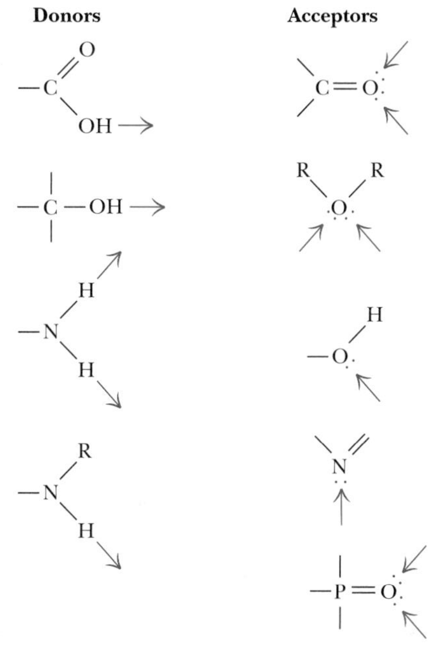

Figure 1.13 Biologically important H bonds and functional groups that are important H-bond donors and acceptors.

ligand something that binds; a molecule that is bound to another molecule; from the Latin *ligare*, meaning "to bind."

Magnesium ATP

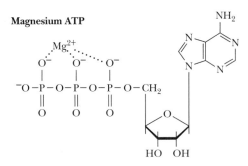

Intramolecular ionic bonds between oppositely charged groups on amino acid residues in a protein

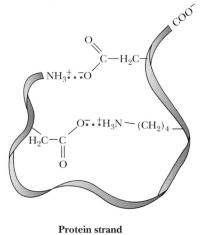

Protein strand

Histone–DNA complexes in chromosomes

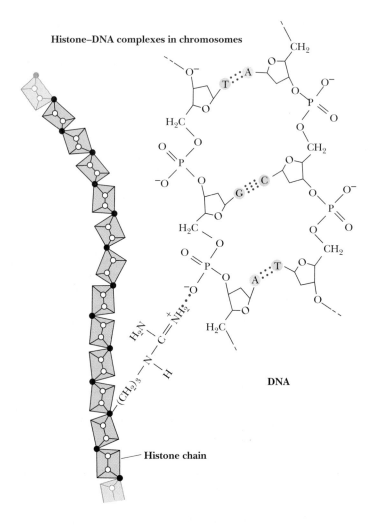

DNA

Histone chain

Figure 1.14 Ionic bonds in biological molecules.

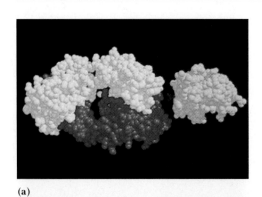

(a)

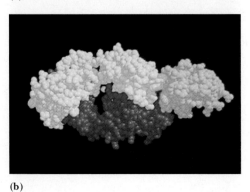

(b)

Puzzle

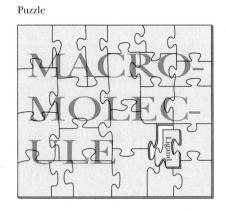

Lock and key

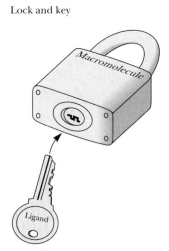

Figure 1.15 Structural complementarity: the pieces of a puzzle, the lock and its key, the biological macromolecule and its ligand—an antigen–antibody complex. **(a)** The antigen on the right (green) is a small protein, lysozyme, from hen egg white. The part of the antibody molecule (IgG), shown on the left in blue and yellow, includes the antigen-binding domain. **(b)** This domain has a pocket that is structurally complementary to a surface projection (Gln[121], shown in red between antigen and antigen-binding domain) on the antigen. (See also Figure 1.12.) *(Photos courtesy of Professor Simon E. V. Philips)*

mechanisms based on structural complementarity: a protein recognizes its specific metabolite, a strand of DNA recognizes its complementary strand, sperm recognize an egg. All these interactions involve structural complementarity between molecules.

Biomolecular Recognition Is Mediated by Weak Chemical Forces

The biomolecular recognition events that occur through structural complementarity are mediated by the weak chemical forces previously discussed. It is important to realize that, since these interactions are sufficiently weak, they are readily reversible under physiological conditions. Consequently, biomolecular interactions tend to be transient; rigid, static lattices of biomolecules that might paralyze cellular activities are not formed. Instead, there is a dynamic interplay between metabolites and macromolecules, hormones and receptors, and all the other participants instrumental to life processes. This interplay is initiated upon specific recognition between complementary molecules and ultimately culminates in unique physiological activities. Thus, *biological function is achieved through mechanisms based on structural complementarity and weak chemical interactions.*

This principle of structural complementarity extends to higher interactions essential to the establishment of the living condition. For example, the formation of supramolecular complexes occurs because of recognition and interaction between their various macromolecular components, as governed by the weak forces formed between them. If a sufficient number of weak bonds can be formed, as in macromolecules complementary in structure to one another, larger structures will assemble spontaneously. The tendency for nonpolar molecules and parts of molecules to come together through hydrophobic interactions also promotes the formation of supramolecular assemblies. Very complex subcellular structures are actually spontaneously formed in an assembly process that is driven by weak forces accumulated through structural complementarity.

Weak Forces Restrict Organisms to a Narrow Range of Environmental Conditions

The central role of weak forces in biomolecular interactions restricts living systems to a narrow range of environmental conditions. Biological macromolecules are functionally active only within a narrow range of environmental conditions, such as temperature, ionic strength, and relative acidity. Extremes of these conditions disrupt the weak forces essential to maintaining the intricate structure of macromolecules. The loss of structural order in these complex macromolecules, so-called **denaturation,** is accompanied by loss of function (Figure 1.16). As a consequence, cells cannot tolerate reactions in which large

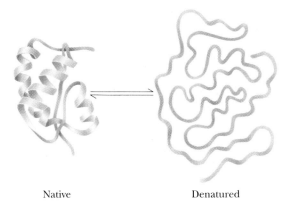

Native   Denatured

Figure 1.16 Denaturation and renaturation of the intricate structure of a protein.

amounts of energy are released. Nor can they generate a large energy burst to drive energy-requiring processes. Instead, such transformations take place via sequential series of chemical reactions whose overall effect achieves dramatic energy changes, even though any given reaction in the series proceeds with only modest input or release of energy (Figure 1.17). These sequences of reactions are organized either to provide for the release of useful energy to the cell from the breakdown of food or to take such energy and use it to drive the synthesis of biomolecules essential to the living state. Collectively, these reaction sequences constitute cellular **metabolism:** the ordered reaction pathways by which cellular chemistry proceeds and biological energy transformations are accomplished.

Enzymes

The sensitivity of cellular constituents to environmental extremes places another constraint on the reactions of metabolism. The rate at which cellular reactions proceed is a very important factor in maintenance of the living state. However, the common ways chemists accelerate reactions are not available to cells; the temperature cannot be increased, acid or base cannot be added, the

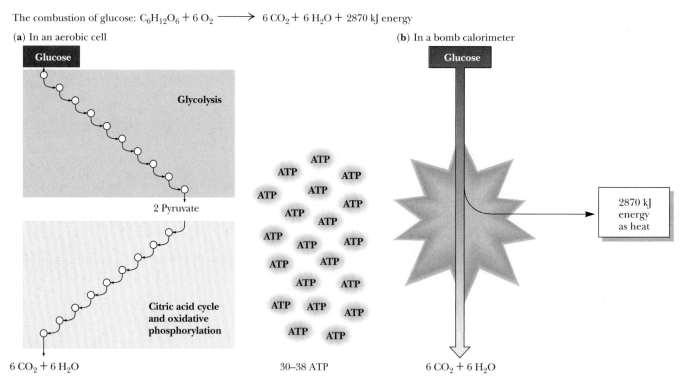

The combustion of glucose: $C_6H_{12}O_6 + 6 O_2 \longrightarrow 6 CO_2 + 6 H_2O + 2870$ kJ energy

(a) In an aerobic cell

Glucose

Glycolysis

2 Pyruvate

Citric acid cycle and oxidative phosphorylation

$6 CO_2 + 6 H_2O$

30–38 ATP

(b) In a bomb calorimeter

Glucose

2870 kJ energy as heat

$6 CO_2 + 6 H_2O$

Figure 1.17 Metabolism is the organized release or capture of small amounts of energy in processes whose overall change in energy is large. **(a)** The combustion of glucose by cells is a major pathway of energy production, with the energy captured appearing as 30 to 38 equivalents of ATP, the principal energy-rich chemical of cells. The ten reactions of glycolysis, the nine reactions of the citric acid cycle, and the successive linked reactions of oxidative phosphorylation release the energy of glucose in a stepwise fashion and the small "packets" of energy appear in ATP. **(b)** Combustion of glucose in a bomb calorimeter results in an uncontrolled, explosive release of energy in its least useful form, heat.

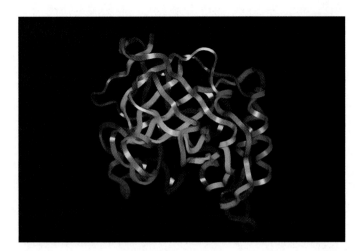

Figure 1.18 Carbonic anhydrase, a representative enzyme, and the reaction it catalyzes. Dissolved carbon dioxide is slowly hydrated by water to form bicarbonate ion and H^+:

$$CO_2 + H_2O \longrightarrow HCO_3^- + H^+$$

At 20°C, the rate constant for this uncatalyzed reaction is 0.03/sec. In the presence of carbonic anhydrase, the rate constant for this reaction, k_{cat}, is 10^6/sec. Thus, carbonic anhydrase accelerates the rate of this reaction 3.3×10^7 times. Carbonic anhydrase is a 29-kD protein.

pressure cannot be raised, and concentrations cannot be markedly changed. Instead, biomolecular catalysts mediate cellular reactions by accelerating their rates many orders of magnitude and by ensuring the selectivity or specificity of the substances undergoing reaction. These catalysts are called **enzymes,** and virtually every metabolic reaction is served by an enzyme whose sole biological purpose is to catalyze its specific reaction (Figure 1.18).

Metabolic Regulation Is Achieved by Controlling the Activity of Enzymes

Thousands of reactions mediated by an equal number of enzymes are occurring at any given instant within the cell. Since metabolism has many branch points, cycles, and interconnections, as a glance at a metabolic pathway map will reveal (Figure 1.19), the need for metabolic regulation is obvious. All of these reactions, many of which are at apparent cross-purposes in the cell, must be fine-tuned and integrated so that metabolism and, in turn, life proceed harmoniously. This metabolic regulation is achieved through controls on enzyme activity so that the rates of cellular reactions are appropriate to cellular requirements.

Despite the organized pattern of metabolism and the thousands of enzymes required, cellular reactions nevertheless conform to the same thermodynamic principles that govern any chemical reaction. Enzymes have no influence over energy changes (the thermodynamic component) in their reactions; they only influence reaction rate. Thus, cells are systems that take in food, release waste, and carry out complex degradative and biosynthetic reactions essential to their survival while operating under conditions of essentially constant temperature and pressure and maintaining a constant internal environment (**homeostasis**) with no outwardly apparent changes. *Cells are open thermodynamic systems exchanging matter and energy with their environment and functioning as highly regulated isothermal chemical engines.*

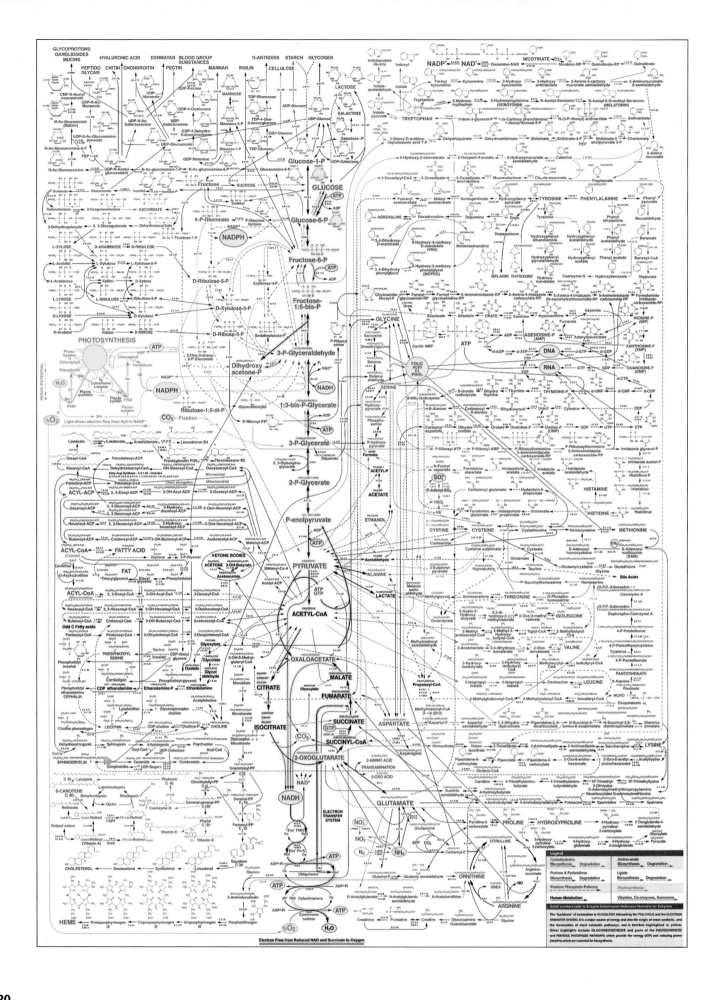

◄**Figure 1.19** Reproduction of a metabolic map. *(Figure courtesy of D. E. Nicholson, University of Leeds, and the International Union of Biochemistry and Molecular Biology)*

1.5 Organization and Structure of Cells

All living cells fall into one of two broad categories—**prokaryotic** or **eukaryotic.** The distinction is based on whether or not the cell has a nucleus. Prokaryotes are single-celled organisms that lack nuclei and other organelles; the word is derived from *pro*, meaning "prior to," and *karyote*, meaning "nucleus." In conventional biological classification schemes, prokaryotes are grouped together as members of the kingdom Monera, represented by bacteria and cyanobacteria (formerly called blue-green algae). The other four living kingdoms are all eukaryotes—the single-celled Protists, such as amoebae, and all multicellular life forms, including the Fungi, Plant, and Animal kingdoms. Eukaryotic cells have true nuclei and other organelles such as mitochondria, with the prefix *eu*, meaning "true."

Early Evolution of Cells

Until recently, most biologists accepted the idea that eukaryotes evolved from the simpler prokaryotes in some linear progression from simple to complex over the course of geological time. Contemporary evidence favors the view that present-day organisms are better grouped into three classes or lineages: eukaryotes and two prokaryotic groups, the **eubacteria** and the **archaea** (formerly designated as **archaebacteria**). All are believed to have evolved approximately 3.5 billion years ago from a common ancestral form called the **progenote.** It is now understood that eukaryotic cells are, in reality, composite cells derived from various prokaryotic contributions. Thus, the dichotomy between prokaryotic cells and eukaryotic cells, though useful, is an artificial distinction.

Despite the great diversity in form and function, cells and organisms share a common biochemistry. This well-established understanding has been substantiated by **whole genome sequencing** (the determination of the complete nucleotide sequence within the DNA of an organism). As an example, the recently sequenced genome of the archaean *Methanococcus jannaschii* shows 44% similarity to known genes in eubacteria and eukaryotes, yet 45% of its genes are new to science.

Structural Organization of Prokaryotic Cells

Among prokaryotes (the simplest cells), most known species are eubacteria and they form a widely spread group. Certain of them are pathogenic to humans. In contrast, no human illness has been attributed to a member of the archaea. The archaea are remarkable because they can be found in unusual environments where other cells cannot survive. Archaea include the **thermoacidophiles** (heat- and acid-loving bacteria) of hot springs, the **halophiles** (salt-loving bacteria) of salt lakes and ponds, and the **methanogens** (bacteria that generate methane from CO_2 and H_2). Prokaryotes are typically very small, on the order of several microns in length, and are usually surrounded by a rigid **cell wall** that protects the cell and gives it its shape. The characteristic structural organization of a prokaryotic cell is depicted in Figure 1.20.

Prokaryotic cells have only a single membrane, the **plasma membrane** or **cell membrane.** Since they have no other membranes, prokaryotic cells contain no nucleus or organelles. Nevertheless, they possess a distinct **nuclear area**

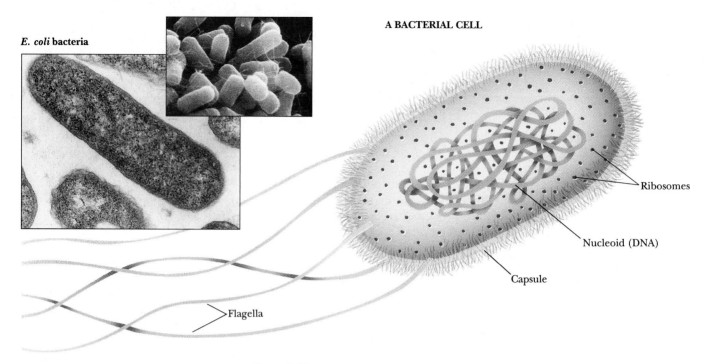

E. coli bacteria

A BACTERIAL CELL

Ribosomes

Nucleoid (DNA)

Capsule

Flagella

Figure 1.20   This bacterium is *Escherichia coli*, a member of the coliform group of bacteria that colonize the intestinal tract of humans. *E. coli* have rather simple nutritional requirements. They grow and multiply quite well if provided with a simple carbohydrate source of energy (such as glucose), ammonium ions as a source of nitrogen, and a few mineral salts. The simple nutrition of this "lower" organism means that its biosynthetic capacities must be quite advanced. When growing at 37°C on a rich organic medium, *E. coli* cells divide every 20 minutes. Subcellular features are indicated, including the cell wall, plasma membrane, nuclear region, ribosomes, storage granules, and cytosol (see Table 1.5). *(Photo, Martin Rotker/Phototake, Inc.; inset photo, David M. Phillips/The Population Council/Science Source/Photo Researchers, Inc.)*

where a single circular chromosome is localized, and some have an internal membranous structure called a **mesosome** that is derived from and is continuous with the cell membrane. Reactions of cellular respiration are localized on these membranes. In photosynthetic prokaryotes such as the **cyanobacteria,** flat, sheetlike, membranous structures called **lamellae** are formed from cell membrane infoldings. These lamellae are the sites of photosynthetic activity, but in prokaryotes, they are not contained within **plastids,** the organelles of photosynthesis found in higher plant cells. Prokaryotic cells also lack a cytoskeleton; the cell wall maintains their structure. Some bacteria have **flagella,** single, long filaments used for motility. Prokaryotes largely reproduce by asexual division, although sexual exchanges can occur. Table 1.5 lists the major features of prokaryotic cells. Table 1.6 lists some of the many human illnesses caused by prokaryotes.

Structural Organization of Eukaryotic Cells

In comparison with prokaryotic cells, eukaryotic cells are much greater in size, typically having cell volumes 10^3 to 10^4 times larger. Also, they are much more complex. These two features require that eukaryotic cells partition their diverse metabolic processes into organized compartments, with each compartment dedicated to particular functions. A system of internal membranes accomplishes this partitioning. A typical animal cell is shown in Figure 1.21.

Table 1.5 **Major Features of Prokaryotic Cells**

Structure	Molecular Composition	Function
Cell wall	Peptidoglycan: a rigid framework of polysaccharide cross-linked by short peptide chains. Some bacteria possess a lipopolysaccharide- and protein-rich outer membrane.	Mechanical support, shape, and protection against swelling in hypotonic media. The cell wall is a porous nonselective barrier allowing most small molecules to pass.
Cell membrane	The cell membrane is composed of about 45% lipid and 55% protein. The lipids form a bilayer that has a continuous nonpolar hydrophobic phase in which the proteins are embedded.	The cell membrane is a highly selective permeability barrier that controls the entry of most substances into the cell. Important enzymes in the generation of cellular energy are located in the membrane.
Nuclear area or nucleoid	The genetic material is a single, tightly coiled DNA molecule 2 nm in diameter but over 1 mm in length (molecular mass of E. coli DNA is 3×10^9 daltons; 4.64×10^6 nucleotide pairs).	DNA is the blueprint of the cell, the repository of the cell's genetic information. During cell division, each strand of the double-stranded DNA molecule is replicated to yield two double-helical daughter molecules. Messenger RNA (mRNA) is transcribed from DNA to direct the synthesis of cellular proteins.
Ribosomes	Bacterial cells contain about 15,000 ribosomes. Each is composed of a small (30S) subunit and a large (50S) subunit. The mass of a single ribosome is 2.3×10^6 daltons. It consists of 65% RNA and 35% protein.	Ribosomes are the sites of protein synthesis. The mRNA binds to ribosomes, and the mRNA nucleotide sequence specifies the protein that is synthesized.
Storage granules	Bacteria contain granules that represent storage forms of polymerized metabolites such as sugars or β-hydroxybutyric acid.	When needed as metabolic fuel, the monomeric units of the polymer are liberated and degraded by energy-yielding pathways in the cell.
Cytosol	Despite its amorphous appearance, the cytosol is now recognized to be an organized gelatinous compartment that is 20% protein by weight and rich in the organic molecules that are the intermediates in metabolism.	The cytosol is the site of intermediary metabolism, the interconnecting sets of chemical reactions by which cells generate energy and form the precursors necessary for biosynthesis of macromolecules essential to cell growth and function.

Eukaryotic cells possess a discrete, membrane-bounded **nucleus,** the principal repository of the cell's genetic material, which is distributed among a few or many **chromosomes.** During cell division, equivalent copies of this genetic material must be passed to both daughter cells through the duplication and orderly partitioning of the chromosomes by the process known as **mitosis.** Like

Table 1.6 **Some Human Diseases Caused by Prokaryotes**

Disease	Causative Organism
Tetanus (lockjaw)	Clostridium tetani
Botulism	Clostridium botulinum
Whooping cough	Bordetella pertussis
Cholera	Vibrio cholerae
Bubonic plague	Yersinia pestis
Diphtheria	Corynebacterium diphtheriae
Tuberculosis	Mycobacterium tuberculosis
Leprosy	Mycobacterium leprae
Gonorrhea	Neisseria gonorrhoeae

Rough endoplasmic
reticulum (plant and animal)

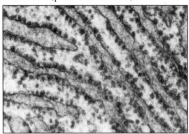

Smooth endoplasmic
reticulum (plant and animal)

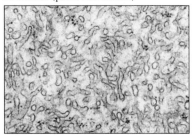

Mitochondrion
(plant and animal)

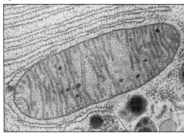

AN ANIMAL CELL

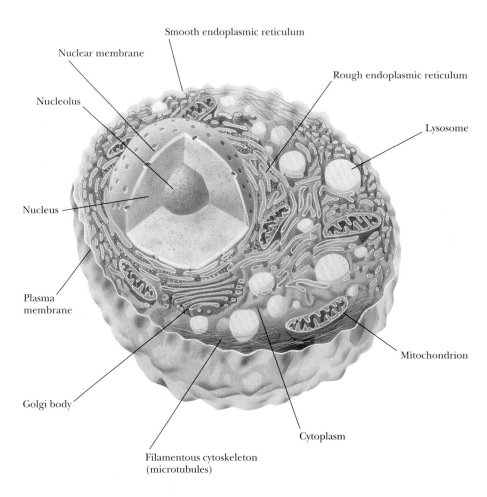

Figure 1.21 This figure diagrams a rat liver cell, a typical higher animal cell in which the characteristic features of animal cells are evident, such as the nucleus, nucleolus, mitochondria, Golgi bodies, peroxisomes, lysosomes, and endoplasmic reticulum (ER). Microtubules and the network of filaments constituting the cytoskeleton are also depicted. *(Photos: top, Dwight R. Kuhn/Visuals Unlimited; middle, D. W. Fawcett/Visuals Unlimited; bottom, Keith Porter/Photo Researchers, Inc.)*

prokaryotic cells, eukaryotic cells are surrounded by a plasma membrane. Unlike prokaryotic cells, eukaryotic cells are rich in internal membranes that are differentiated into specialized structures such as the **endoplasmic reticulum (ER)** and the **Golgi apparatus.** Membranes also surround certain organelles (**mitochondria** and **chloroplasts,** for example) and various vesicles, including **vacuoles, lysosomes,** and **peroxisomes.** The common purpose of these membranous partitionings is the creation of cellular compartments that have specific, organized metabolic functions, such as the mitochondrion's role as the principal site of cellular energy production. Eukaryotic cells also have a **cytoskeleton** composed of arrays of filaments that give the cell its shape and its capacity to move. Some eukaryotic cells also have long projections on their surface—cilia or flagella—which provide propulsion. Table 1.7 lists the major features of a typical animal cell.

Table 1.7   Major Features of a Typical Animal Cell

Structure	Molecular Composition	Function
Extracellular matrix	The surfaces of animal cells are covered with a flexible and sticky layer of complex carbohydrates, proteins, and lipids.	This complex coating is cell-specific, serves in cell–cell recognition and communication, creates cell adhesion, and provides a protective outer layer.
Cell membrane (plasma membrane)	Roughly 50:50 lipid:protein as a 5-nm-thick continuous sheet of lipid bilayer in which a variety of proteins are embedded.	The plasma membrane is a selectively permeable outer boundary of the cell, containing specific systems—pumps, channels, transporters—for the exchange of nutrients and other materials with the environment. Important enzymes are also located here.
Nucleus	The nucleus is separated from the cytosol by a double membrane, the nuclear envelope. The DNA is complexed with basic proteins (histones) to form chromatin fibers, the material from which chromosomes are made. A distinct RNA-rich region, the nucleolus, is the site of ribosome assembly.	The nucleus is the repository of genetic information encoded in DNA and organized into chromosomes. During mitosis, the chromosomes are replicated and transmitted to the daughter cells. The genetic information of DNA is transcribed into RNA in the nucleus and passes into the cytosol, where it is translated into protein by ribosomes.
Mitochondria	Mitochondria are organelles surrounded by two membranes that differ markedly in their protein and lipid composition. The inner membrane and its interior volume, the matrix, contain many important enzymes of energy metabolism. Mitochondria are about the size of bacteria, $\approx 1\ \mu m$. Mitochondria collectively occupy about one-fifth of the cell volume.	Mitochondria are the power plants of eukaryotic cells where carbohydrates, fats, and amino acids are oxidized to CO_2 and H_2O. The energy released is trapped as high-energy phosphate bonds in ATP.
Golgi apparatus	A system of flattened membrane-bounded vesicles often stacked into a complex. Numerous small vesicles are found peripheral to the Golgi and contain secretory material packaged by the Golgi.	Involved in the packaging and processing of macromolecules for secretion and for delivery to other cellular compartments.
Endoplasmic reticulum (ER) and ribosomes	Flattened sacs, tubes, and sheets of internal membrane extending throughout the cytoplasm of the cell and enclosing a large interconnecting series of volumes called *cisternae*. The ER membrane is continuous with the outer membrane of the nuclear envelope. Portions of the sheetlike areas of the ER are studded with ribosomes, giving rise to *rough ER*. Eukaryotic ribosomes are larger than prokaryotic ribosomes.	The endoplasmic reticulum is a labyrinthine organelle where both membrane proteins and lipids are synthesized. Proteins made by the ribosomes of the rough ER pass through the outer ER membrane into the cisternae and can be transported via the Golgi to the periphery of the cell. Other ribosomes unassociated with the ER carry on protein synthesis in the cytosol.
Lysosomes	Lysosomes are vesicles 0.2–0.5 μm in diameter, bounded by a single membrane. They contain hydrolytic enzymes such as proteases and nucleases which, if set free, could degrade essential cell constituents. They are formed by budding from the Golgi apparatus.	Lysosomes function in intracellular digestion of materials entering the cell via phagocytosis or pinocytosis. They also function in the controlled degradation of cellular components.
Peroxisomes	Like lysosomes, peroxisomes are 0.2–0.5 μm single-membrane-bounded vesicles. They contain a variety of oxidative enzymes that use molecular oxygen and generate peroxides. They are formed by budding from the smooth ER.	Peroxisomes act to oxidize certain nutrients, such as amino acids. In doing so, they form potentially toxic hydrogen peroxide, H_2O_2, and then decompose it to H_2O and O_2 by way of the peroxide-cleaving enzyme catalase.
Cytoskeleton	The cytoskeleton is composed of a network of protein filaments: actin filaments (or microfilaments), 7 nm in diameter; intermediate filaments, 8–10 nm; and microtubules, 25 nm. These filaments interact in establishing the structure and functions of the cytoskeleton. This interacting network of protein filaments gives structure and organization to the cytoplasm.	The cytoskeleton determines the shape of the cell and gives it its ability to move. It also mediates the internal movements that occur in the cytoplasm, such as the migration of organelles and mitotic movements of chromosomes. The propulsion instruments of cells—cilia and flagella—are constructed of microtubules.

1.6 Viruses Are Supramolecular Assemblies Acting As Cell Parasites

Viruses are supramolecular complexes of nucleic acid, either DNA or RNA, encapsulated in a protein coat and, in some instances, surrounded by a membrane envelope (Figure 1.22). The bits of nucleic acid in viruses are, in reality, mobile elements of genetic information. The protein coat serves to protect the nucleic acid and allows it to gain entry to the cells that are its specific hosts. Viruses for virtually every kind of cell are known. Viruses infecting bacteria are called **bacteriophages** ("bacteria eaters"); different viruses infect animal cells and plant cells. Once the nucleic acid of a virus gains access to its specific host, it typically takes over the metabolic machinery of the host cell, diverting it to the production of virus particles. The host metabolic functions are subjugated to the synthesis of viral nucleic acid and proteins. Mature virus particles arise by encapsulating the nucleic acid within a protein coat called the **capsid.** Viruses are thus supramolecular assemblies that act as parasites of cells (Figure 1.23).

Usually, viruses cause the lysis of the cells they infect. It is their cytolytic properties that are the basis of viral disease. In certain circumstances, the viral genetic elements may integrate into the host chromosome and become quiescent. Such a state is termed **lysogeny.** Typically, damage to the host cell activates the replicative capacities of the quiescent viral nucleic acid, leading to viral propagation and release. Some viruses are implicated in transforming cells into a cancerous state, that is, in converting their hosts to an unregulated state of cell division and proliferation. Since all viruses are heavily dependent on their host for the production of viral progeny, viruses must have arisen after cells were established in the course of evolution. Presumably, the first viruses were fragments of nucleic acid that developed the ability to replicate independently of the chromosome and then acquired the necessary genes enabling protection, autonomy, and transfer between cells. Table 1.8 lists some of the human diseases caused by viruses.

Figure 1.22 Viruses are genetic elements enclosed in a protein coat. Viruses are not free-living, but can reproduce only within cells. Viruses show an almost absolute specificity for their particular host cells, infecting and multiplying only within those cells. Viruses are known for virtually every kind of cell. Shown here are examples of **(a)** a bacterial virus, bacteriophage T₄; **(b)** an animal virus, adenovirus (inset at greater magnification); and **(c)** a plant virus, tobacco mosaic virus. *(a, M. Wurtz/ Biozeetrum/University of Basel/SPL/Photo Researchers, Inc.; b, Dr. Thomas Broker/Phototake, NYC; inset, CNRI/SPL/Photo Researchers, Inc.; c, Biology Media/Photo Researchers, Inc.)*

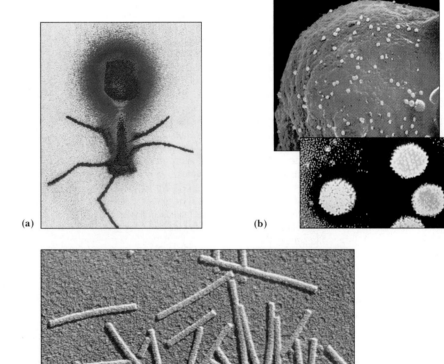

(a)

(b)

(c)

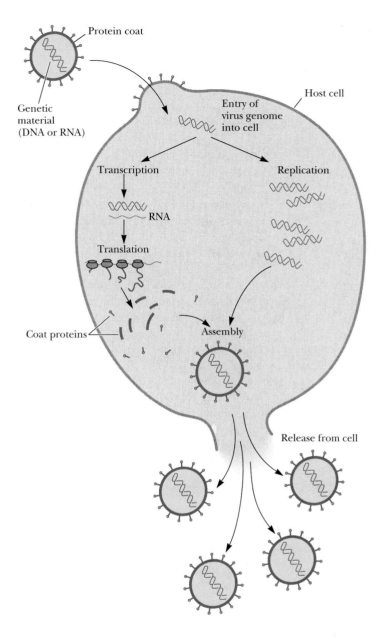

Protein coat

Genetic material (DNA or RNA)

Entry of virus genome into cell

Host cell

Transcription

Replication

RNA

Translation

Coat proteins

Assembly

Release from cell

Figure 1.23 The virus life cycle. Viruses are mobile bits of genetic information encapsulated in a protein coat. The genetic material may be either DNA or RNA. Once this genetic material gains entry to its host cell, it takes over the host machinery for macromolecular synthesis and subverts it to the synthesis of viral-specific nucleic acids and proteins. These virus components are then assembled into mature virus particles, which are released from the cell. Often, this parasitic cycle of virus infection leads to cell death and disease.

■ **Table 1.8   Some Human Diseases Caused by Viruses**

Disease	Causative Agent
DNA viruses	
Measles	*Rubeola virus*
Mumps	*Paramyxovirus*
Hepatitis A, B	*Hepatitis viruses*
Influenza (flu)	*Influenza virus*
Polio	*Poliomyelitis virus*
Smallpox	*Poxvirus*
Dengue fever	*Arbovirus*
RNA viruses	
Common cold	*Rhinoviruses (picornaviruses)*
Chickenpox	*Herpes zoster*
Shingles	*Herpes zoster*
Rabies	*Rabies virus*
AIDS	*HIV*
(acquired immunodeficiency syndrome)	*(human immunodeficiency virus)*

PROBLEMS

1. The nutritional requirements of *Escherichia coli* cells are far simpler than those of humans, yet the macromolecules found in bacteria are about as complex as those of animals. Since bacteria can make all their essential biomolecules while subsisting on a simpler diet, do you think bacteria may have more biosynthetic capacity and hence more metabolic complexity than animals? Organize your thoughts on this question, pro and con, into a rational argument.

2. Without consulting chapter figures, sketch the characteristic prokaryotic and animal cell types and label their pertinent organelle and membrane systems.

3. *Escherichia coli* cells are about 2 μm (microns) long and 0.8 μm in diameter.

   a. How many *E. coli* cells laid end to end would fit across the diameter of a pinhead? (Assume a pinhead diameter of 0.5 mm.)

   b. What is the volume of an *E. coli* cell? (Assume it is a cylinder. The volume of a cylinder is given by V = $\pi r^2 h$, where π = 3.14.)

   c. What is the surface area of an *E. coli* cell? What is the surface-to-volume ratio of an *E. coli* cell?

   d. Glucose, a major energy-yielding nutrient, is present in bacterial cells at a concentration of about 1 mM. How many glucose molecules are contained in a typical *E. coli* cell? (Recall that Avogadro's number = 6.023 $\times$ 10^{23}.)

   e. A number of regulatory proteins are present in *E. coli* at only one or two molecules per cell. If we assume that an *E. coli* cell contains just one molecule of a particular protein, what is the molar concentration of this protein in the cell?

   f. An *E. coli* cell contains about 15,000 ribosomes, which carry out protein synthesis. Assuming ribosomes are spherical and have a diameter of 20 nm (nanometers), what fraction of the *E. coli* cell volume is occupied by ribosomes?

   g. The *E. coli* chromosome is a single DNA molecule whose mass is about 3 $\times$ 10^9 daltons. This macromolecule is actually a linear array of nucleotide pairs. The average molecular weight of a nucleotide pair is 660 and each pair imparts 0.34 nm to the length of the DNA molecule. What is the total length of the *E. coli* chromosome? How does this length compare with the overall dimensions of an *E. coli* cell? How many nucleotide pairs does this DNA contain?

   h. The average *E. coli* protein is a linear chain of 360 amino acids. If three nucleotide pairs in a gene encode one amino acid in a protein, how many different proteins can the *E. coli* chromosome encode? (The answer to this question is a reasonable approximation of the maximum number of different kinds of proteins that can be expected in bacteria.)

4. Assume that mitochondria are cylinders 1.5 μm in length and 0.6 μm in diameter.

   a. What is the volume of a single mitochondrion?

   b. Oxaloacetate is an intermediate in the citric acid cycle, an important metabolic pathway localized in the mitochondria of eukaryotic cells. The concentration of oxaloacetate in mitochondria is about 0.03 μM. How many molecules of oxaloacetate are in a single mitochondrion?

5. Assume that liver cells are cuboidal in shape, 20 μm on a side.

   a. How many liver cells laid end to end would fit across the diameter of a pinhead? (Assume a pinhead diameter of 0.5 mm.)

   b. What is the volume of a liver cell? (Assume it is a cube.)

   c. What is the surface area of a liver cell? What is the surface-to-volume ratio of a liver cell? How does this compare with the surface-to-volume ratio of an *E. coli* cell? (Compare this answer with that of problem 3c.) What problems must cells with low surface-to-volume ratios confront that don't affect cells with high surface-to-volume ratios?

   d. A human liver cell contains two sets of 23 chromosomes, both roughly equivalent in information content. The total mass of DNA contained in these 46 enormous DNA molecules is 4 $\times$ 10^{12} daltons. Since each nucleotide pair contributes 660 daltons to the mass of DNA and 0.34 nm to the length of DNA, what is the total number of nucleotide pairs and the complete length of the DNA in a liver cell? How does this length compare with the overall dimensions of a liver cell?

   e. The maximal information in each set of liver cell chromosomes should be related to the number of nucleotide pairs in the chromosome set's DNA. This number can be obtained by dividing the total number of nucleotide pairs calculated above by 2. What is this value? If this information is expressed in proteins that average 400 amino acids in length and three nucleotide pairs encode one amino acid in a protein, how many different kinds of proteins might a liver cell be able to produce? (In reality, liver cells express at most about 30,000 different proteins. Thus, a large discrepancy exists between the theoretical information content of DNA in these cells and the amount of information actually expressed.)

6. Biomolecules interact with one another through molecular surfaces that are structurally complementary. How can various proteins interact with molecules as different as simple ions, hydrophobic lipids, polar but uncharged carbohydrates, or even nucleic acids?

7. What structural features allow biological polymers to be informational macromolecules? Is it possible for polysaccharides to be informational macromolecules?

8. Why is it important that weak forces, not strong forces, mediate biomolecular recognition interactions?

9. Why does the central role of weak forces in biomolecular interactions restrict living systems to a rather narrow range of environmental conditions?

10. Describe what is meant by the phrase "cells are steady-state systems."

FURTHER READING

Alberts, B., et al., 1998. *Essential Cell: An Introduction to the Molecular Biology of the Cell.* New York: Garland Press.

Goodsell, D. S., 1991. Inside a living cell. *Trends in Biochemical Sciences* **16**:203–206.

Koonin, E. V., et al., 1996. Sequencing and analysis of bacterial genomes. *Current Biology* **6**:404–416.

Lloyd, C., ed., 1986. Cell organization. *Trends in Biochemical Sciences* **11**:437–485.

Lodish, H., et al., 2000. *Molecular Cell Biology,* 4th ed. New York: W. H. Freeman and Co.

Pace, N. R., 1996. New perspective on the natural microbial world: Molecular microbial ecology. *ASM News* **62**:463–470.

Service, R. F., 1997. Microbiologists explore life's rich, hidden kingdoms. *Science* **275**:1740–1742.

Solomon, E. P., Berg, L. R., and Martin, D. W., 1999. *Biology,* 5th ed. Philadelphia: Saunders College Publishing.

Tobin, A. J., and Morel, R. E., 1997. *Asking About Cells.* Philadelphia: Saunders College Publishing.

Wald, G., 1964. The origins of life. *Proceedings of the National Academy of Science, U.S.A.* **52**:595–611.

Watson, J. D., Hopkins, N. H., Roberts, J. W., et al., 1987. *Molecular Biology of the Gene,* 4th ed. Menlo Park, Calif.: Benjamin Cummings Publishing Co.

Woese, C. R., 1996. Phylogenetic trees: Whither microbiology? *Current Biology* **6**:1060–1063.

Water: The Medium of Life

If there is magic on this planet, it is contained in water.

Loren Eisley
(inscribed on the wall of the National Aquarium in Baltimore, Maryland)

Outline

Some of the magic: students and teacher view a coral crab in Graham's Harbour, San Salvador Island, the Bahamas. (Lara Call)

Water is a major chemical component of the earth's surface. It is indispensable to life. Indeed, it is the only liquid that most organisms ever encounter. We alternately take it for granted because of its ubiquity and bland nature or marvel at its many unusual and fascinating properties. At the center of this fascination is the role of water as both the medium and the continuum of life. Life originated, evolved, and thrives in the seas. Organisms invaded and occupied terrestrial and aerial niches, but none gained true independence from water. Typically, organisms are constituted of 70% to 90% water. Indeed, normal metabolic activity can occur only when cells are at least 65% H_2O. This dependency of life on water is not a simple matter, but it can be grasped through a consideration of the unusual chemical and physical properties of H_2O. Subsequent chapters will establish that water and its ionization products, hydrogen ions and hydroxide ions, are critical determinants of the structure and function of proteins, nucleic acids, and membranes. In yet another essential biological phenomenon, water is an indirect participant—a difference in concentration of hydrogen ions on opposite sides of a membrane represents an energized condition essential to biological mechanisms of energy transformation. First, let's review the remarkable properties of water.

2.1 | Properties of Water

Unusual Properties

In comparison with chemical compounds of similar atomic organization and molecular size, water displays rather aberrant properties. For example, compare water, the hydride of oxygen, with hydrides of oxygen's nearest neighbors in the periodic table—namely, ammonia (NH_3) and hydrogen fluoride (HF)—or with the hydride of its nearest congener, sulfur (H_2S). Water has a substantially higher boiling point, melting point, heat of vaporization, and surface tension. Indeed, all of these physical properties are anomalously high for a substance of this molecular weight that is neither metallic or ionic. These properties suggest that intermolecular forces of attraction between H_2O molecules are high. Thus, the internal cohesion of this substance is high. Furthermore, water has an unusually high dielectric constant, its maximum density is found in the liquid (not the solid) state, and it has a negative volume of melting (that is, the solid form, ice, occupies more space than the liquid form, water). It is truly remarkable that so many eccentric properties should occur together in a single substance. As chemists, we expect to find an explanation for these apparent anomalies in the structure of water. The clue to understanding its intermolecular attractions is found in its atomic constitution. Indeed, *the fact crucial to understanding water is its unrivaled ability to form hydrogen bonds.*

The Structure of Water

The two H atoms of H_2O are linked covalently to oxygen, each sharing an electron pair, to give a nonlinear arrangement (Figure 2.1). This "bent" structure of the H_2O molecule is of enormous significance to its properties. If H_2O were linear, it would be a nonpolar substance. In the bent configuration, however, the electronegative oxygen atom and the two hydrogen atoms form a dipole that renders the molecule distinctly polar. Furthermore, this structure is ideally suited to hydrogen bond formation. Water can serve as both an H donor and an H acceptor in H-bond formation. The potential to form four H bonds per water molecule is the source of the strong intermolecular attractions that endow this substance with its anomalously high boiling point, melting point, heat of vaporization, and surface tension. In ordinary ice, the common crystalline form of water, each H_2O molecule has four nearest neighbors to which it is hydrogen-bonded: each H atom donates an H bond to the oxygen of a neighbor while the O atom serves as an H-bond acceptor from H atoms bound to two different water molecules (Figure 2.2). A local tetrahedral symmetry results.

Hydrogen bonding in water is cooperative. That is, an H-bonded water molecule serving as an acceptor is a better H-bond donor than an unbonded molecule (and an H-bonded H_2O molecule serving as an H-bond donor becomes a better H-bond acceptor). Thus, participation in H bonding by H_2O molecules is a phenomenon of mutual reinforcement. The H bonds between neighboring molecules are weak (23 kJ/mol each) relative to the H—O covalent bonds (420 kJ/mol). As a consequence, the hydrogen atoms are situated

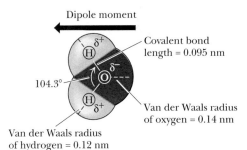

Figure 2.1 The structure of water. Two lobes of negative charge formed by the lone-pair electrons of the oxygen atom lie above and below the plane of the diagram. This electron density contributes substantially to the large dipole moment and polarizability of the water molecule. The dipole moment of water corresponds to the O—H bonds having 33% ionic character. Note that the H—O—H angle is 104.3°, *not* 109°, the angular value found in molecules with tetrahedral symmetry, such as CH_4. Many of the important properties of water derive from this angular value, such as the decreased density of its crystalline state, ice. (The dipole moment in this figure points in the direction of negative to positive, the convention used by physicists and physical chemists; organic chemists draw it pointing in the opposite direction.)

Figure 2.2   The structure of normal ice. The hydrogen bonds in ice form a three-dimensional network. The smallest number of H_2O molecules in any closed circuit of H-bonded molecules is six, so that this structure bears the name *hexagonal ice*. Covalent bonds are represented as sticks, whereas hydrogen bonds are shown as dashed lines. The directional preference of H bonds leads to a rather open lattice structure for crystalline water and, consequently, a low density for the solid state. The distance between neighboring oxygen atoms linked by a hydrogen bond is 0.274 nm. Since the covalent H—O bond is 0.095 nm, the H—O hydrogen bond length in ice is 0.18 nm.

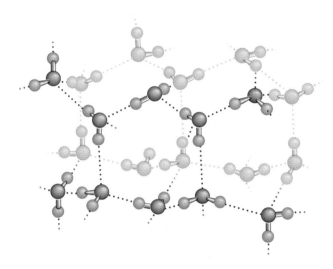

asymmetrically between the two oxygen atoms along the O—O axis. There is never any ambiguity about which O atom the H atom is chemically bound to, nor, naturally, to which O it is H-bonded.

The Structure of Ice

In ice, the hydrogen bonds form a space-filling, three-dimensional network. These bonds are directional and they are straight; that is, the H atom lies on the direct line between the two O atoms. This linearity and directionality means that the resultant H bonds are strong. In addition, the directional preference of the H bonds leads to an open lattice structure. For example, if the water molecules are approximated as rigid spheres centered at the positions of the O atoms in the lattice, then the observed density of ice is actually only 57% of that expected for a tightly packed arrangement of such spheres. The H bonds in ice hold the water molecules apart. Melting involves breaking some of the H bonds which maintain the crystal structure of ice so that the molecules of water (now liquid) can actually pack closer together. Thus, the density of ice is slightly less than the density of water. Ice floats, a property of great importance to aquatic organisms in cold climates.

In liquid water, the rigidity of ice is replaced by fluidity, and the crystalline periodicity of ice gives way to spatial homogeneity. The H_2O molecules in liquid water form a random, H-bonded network, with each molecule having an average of 4.4 close neighbors situated within a center-to-center distance of 0.284 nm (2.84 Å). At least half of the hydrogen bonds have nonideal orientations (that is, they are not perfectly straight); consequently, liquid H_2O lacks the regular latticelike structure of ice. The space around an O atom is not defined by the presence of four hydrogens but can be occupied by other water molecules randomly oriented so that the local environment, over time, is essentially uniform. Nevertheless, the heat of melting for ice is but a small fraction (13%) of the heat of sublimation for ice (the energy needed to go from the solid to the vapor state). This fact indicates that the majority of H bonds between H_2O molecules survive the transition from solid to liquid. At 10°C, 2.3 H bonds per H_2O molecule remain, and the tetrahedral bond order persists even though substantial disorder is now present.

Molecular Interactions in Liquid Water

In the liquid state, H_2O molecules are connected by H-bond paths running in every direction, spanning the whole sample. Each water molecule experiences an average state of H bonding to its neighbors via a fluid network of H bonds.

Figure 2.3 The fluid network of H bonds linking water molecules in the liquid state. It is revealing to note that, in 10 psec, a photon of light (which travels at about 3×10^8 m/sec) would move a distance of only 0.003 m. ▶

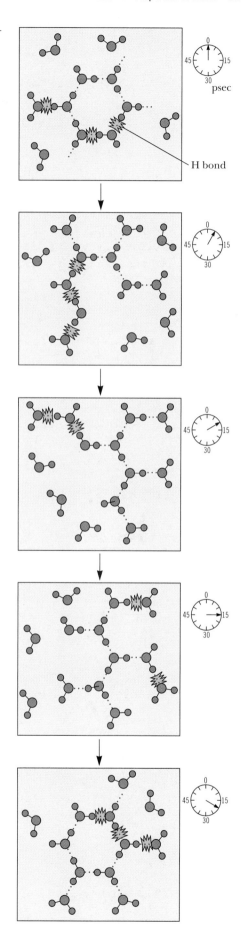

The average lifetime of an H-bonded connection between two H_2O molecules in water is 9.5 psec (picoseconds, where 1 psec $= 10^{-12}$ sec). Thus, about every 10 psec, the average H_2O molecule moves, reorients, and interacts with new neighbors, as illustrated in Figure 2.3. Therefore, pure liquid water consists of H_2O molecules held in a random, three-dimensional network which has a local preference for tetrahedral geometry but contains a large number of strained or broken hydrogen bonds. The presence of strain creates a dynamic situation in which H_2O molecules can switch H-bond allegiances; fluidity ensues.

Solvent Properties

Because of its highly polar nature, water is an excellent solvent for ionic substances such as salts; nonionic but polar substances such as sugars, simple alcohols, and amines; and carbonyl-containing molecules such as aldehydes and ketones. Although the electrostatic attractions between the positive and negative ions in the crystal lattice of a salt are very strong, water readily dissolves salts. For example, sodium chloride is dissolved because dipole water molecules participate in strong electrostatic interactions with the Na^+ and Cl^- ions, leading to the formation of **hydration shells** surrounding these ions (Figure 2.4). Although hydration shells are stable structures, they are also dynamic. Each water molecule in the inner hydration shell around a Na^+ ion is replaced on average every 2 to 4 nsec (nanoseconds, where 1 nsec $= 10^{-9}$ sec) by another H_2O. Consequently, a water molecule is trapped only several hundred

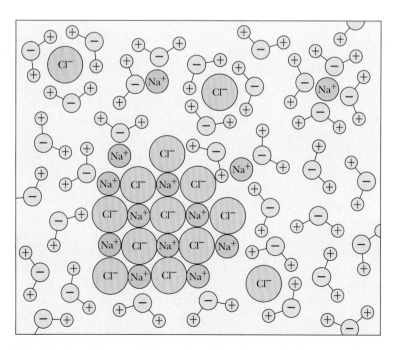

Figure 2.4 Hydration shells surrounding ions in solution. Water molecules orient so that the electrical charge on the ion is sequestered by the water dipole. For positive ions (cations), the partially negative oxygen atom of H_2O is toward the ion in solution. Negatively charged ions (anions) attract the partially positive hydrogen atoms of water in creating their hydration shells.

Table 2.1 Dielectric Constants* of Some Common Solvents at 25°C

Solvent	Dielectric Constant (D)
Water	78.5
Methyl alcohol	32.6
Ethyl alcohol	24.3
Acetone	20.7
Acetic acid	6.2
Chloroform	5.0
Benzene	2.3
Hexane	1.9

*The dielectric constant is also referred to as *relative permittivity* by physical chemists.

times longer by the electrostatic force field of an ion than it is by the H-bonded network of water. (Recall that the average lifetime of H bonds between water molecules is about 10 psec.)

Water Has a High Dielectric Constant

The attractions between the water molecules interacting with, or **hydrating,** ions are much greater than the tendency of oppositely charged ions to attract one another. The ability of water to surround these ions in dipole–dipole interactions and diminish their attraction for each other is a measure of its **dielectric constant, D.**[1] Indeed, ionization in solution depends on the dielectric constant of the solvent; otherwise, the strongly attracted positive and negative ions would unite to form neutral molecules. Table 2.1 lists the dielectric constants of some common liquids. Note that the dielectric constant for water is more than twice that of methanol and more than 40 times that of hexane.

Water Forms H Bonds with Polar Solutes

In the case of nonionic but polar compounds such as sugars, the excellent solvent properties of water are attributed to its ability to readily form hydrogen bonds with the polar functional groups on these compounds, such as hydroxyls, amines, and carbonyls. These polar interactions between solvent and solute are stronger than the intermolecular attractions between solute molecules caused by van der Waals forces and weaker hydrogen bonding. Thus, the solute molecules readily dissolve in water.

Hydrophobic Interactions

The behavior of water toward nonpolar solutes is different from the interactions just discussed. Nonpolar solutes (or nonpolar functional groups on biological macromolecules) do not readily H-bond to H_2O, and as a result, such compounds tend to be only sparingly soluble in water. The process of dissolving such substances is accompanied by significant reorganization of the water surrounding the solute so that the response of the solvent water to such solutes can be equated to "structure making." Since nonpolar solutes must occupy space, the random H-bond network of water must reorganize to accommodate them. At the same time, the water molecules participate in as many H-bonded

---

[1]The strength of the dielectric constant of a medium is related to the force, F, experienced between two ions of opposite charge separated by a distance, r, as given in the relationship $F = e_1 e_2 / Dr^2$, where e_1 and e_2 are the charges on the two ions.

interactions with one another as the temperature will permit. Consequently, the H-bonded water network rearranges toward formation of a local cagelike (**clathrate**) structure surrounding each solute molecule (Figure 2.5). A major consequence of this rearrangement is that the molecules of H_2O participating in the cage layer have markedly reduced orientational options. Water molecules tend to straddle the nonpolar solute such that most of their H-bond donor ($O—H\rightarrow$) and H-bond acceptor ($\rightarrow:O$) functions are not directed toward the nonpolar solute. This "straddling" means no water H-bonding capacity is lost because no H-bond donor or acceptor of the H_2O is directed toward the caged solute. The water molecules forming these clathrates are involved in highly ordered structures. That is, clathrate formation is accompanied by significant ordering of structure.

Under these conditions, nonpolar solute molecules experience a net attraction for one another that is called **hydrophobic interaction.** The basis of this interaction is that when two nonpolar solutes meet, their joint solvation cage involves less overall ordering of the water molecules than do their separate cages, and the attraction is an entropy-driven process, that is, a net decrease in order among the H_2O molecules. To be specific, hydrophobic interactions between nonpolar solutes are maintained not so much by direct interactions between the inert solutes themselves as by the stability achieved when the water cages coalesce and reorganize. Because interactions between nonpolar solute molecules and the water surrounding them are of uncertain stoichiometry and do not share the equality of atom-to-atom participation implicit in chemical bonding, the term *hydrophobic interaction* is more correct than the misleading expression *hydrophobic bond.*

Amphiphilic Molecules

Compounds containing both strongly polar and strongly nonpolar groups are called **amphiphilic molecules** (from the Greek *amphi,* meaning "both," and *philos,* meaning "loving"), also referred to as **amphipathic molecules** (from the Greek *pathos,* meaning "passion, suffering"). Salts of fatty acids are a typical example that has biological relevance. They have a long nonpolar hydrocarbon tail and a strongly polar carboxyl head group, as in the sodium salt of palmitic

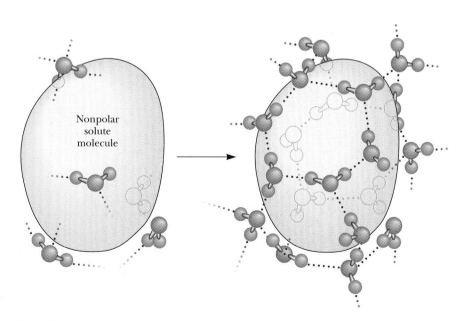

Figure 2.5 Formation of a clathrate structure by water molecules surrounding a hydrophobic solute.

Figure 2.6 An amphiphilic molecule: sodium palmitate. Amphiphilic molecules are frequently symbolized by a ball and zig-zag line structure, ●〜〜 , where the ball represents the hydrophilic polar head and the zig-zag represents the nonpolar hydrophobic hydrocarbon tail.

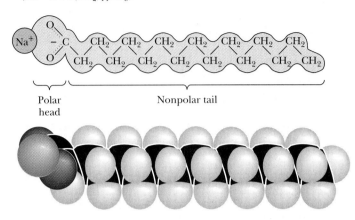

The sodium salt of palmitic acid: Sodium palmitate
$(Na^{+-}OOC(CH_2)_{14}CH_3)$

Polar head   Nonpolar tail

acid (Figure 2.6). Their behavior in aqueous solution reflects the combination of the contrasting polar and nonpolar nature of these substances. The ionic carboxylate function hydrates readily, whereas the long hydrophobic tail is intrinsically insoluble. Thus, sodium palmitate, a soap, shows little tendency to form a true ionic solution in water. Nevertheless, sodium palmitate and other amphiphilic molecules readily disperse in water because the hydrocarbon tails of these substances are joined together in hydrophobic interactions as their polar carboxylate functions are hydrated in typical hydrophilic fashion. Such clusters of amphipathic molecules are termed **micelles.** Figure 2.7 depicts their structure. Of enormous biological significance is the contrasting solute behavior of the two ends of amphipathic molecules upon introduction into aqueous solutions. The polar ends express their hydrophilicity in ionic interactions with the solvent, whereas their nonpolar counterparts are excluded from the water into a hydrophobic domain constituted from the hydrocarbon tails of many like molecules. It is exactly this behavior that accounts for the formation of membranes, the structures that define the limits and compartments of cells (see Chapter 6).

Figure 2.7 Micelle formation by amphiphilic molecules in aqueous solution. Negatively charged carboxylate head groups orient to the micelle surface and interact with the polar H_2O molecules via H bonding. The nonpolar hydrocarbon tails cluster in the interior of the spherical micelle, driven by hydrophobic interactions and the formation of favorable van der Waals interactions. Because of their negatively charged surfaces, neighboring micelles repel one another and thereby maintain a relative stability in solution.

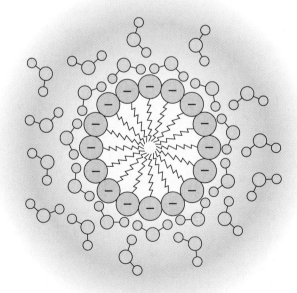

Influence of Solutes on Water Properties

The presence of dissolved substances disturbs the structure of liquid water so that its properties change. The dynamic hydrogen-bonding pattern of water must now accommodate the intruding substance. The net effect is that solutes, regardless of whether they are polar or nonpolar, fix nearby water molecules in a more ordered array. Ions, by the establishment of hydration shells through interactions with the water dipoles, create local order. Hydrophobic effects, for different reasons, make structures within water. To put it another way, by limiting the orientations that neighboring water molecules can assume, solutes give order to the solvent and diminish the dynamic interplay between H_2O molecules that occurs in pure water.

Colligative Properties of Water

This influence of the solute on water is reflected in a set of characteristic changes in behavior that are termed **colligative properties,** or properties related by a common principle. These alterations in solvent properties are related in that they all depend only on the number of solute particles per unit volume of solvent and not on the chemical nature of the solute. These effects include freezing point depression, boiling point elevation, vapor pressure lowering, and osmotic pressure effects. For example, one mol of an ideal solute dissolved in 1000 grams of water (a 1 m, or molal, solution) at 1 atm pressure depresses the freezing point by 1.86°C, raises the boiling point by 0.543°C, lowers the vapor pressure in a temperature-dependent manner, and yields a solution whose osmotic pressure relative to pure water is 22.4 atm. In effect, by imposing local order on the water molecules, solutes make it more difficult for water to assume its crystalline lattice (freeze) or escape into the atmosphere (boil or vaporize). Furthermore, when a solution (such as the 1 m solution discussed here) is separated from a volume of pure water by a semipermeable membrane, the solution will draw water molecules across this barrier. The water molecules are moving from a region of higher effective concentration (pure H_2O) to a region of lower effective concentration (the solution). This movement of water into the solution dilutes the effects of the solute that is present. The osmotic force exerted by each mole of solute is so strong that it requires the imposition of 22.4 atm of pressure in order to be negated (Figure 2.8). Osmotic pressure from high concentrations of dissolved solutes is a serious problem for cells. Bacterial and plant cells have strong, rigid cell walls in order to contain these pressures. In contrast, animal cells are bathed in extracellular fluids of comparable osmolarity, so no net osmotic gradient exists. Also, to minimize the osmotic pressure created by the contents of their cytosol, cells tend to store

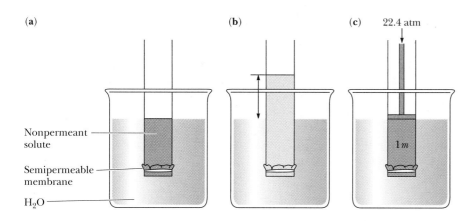

(a) (b) (c) 22.4 atm

Nonpermeant solute

Semipermeable membrane

H_2O

1 m

Figure 2.8 The osmotic pressure of a 1 molal (*m*) solution is equal to 22.4 atmospheres of pressure. **(a)** If a nonpermeant solute is separated from pure water by a semipermeable membrane through which H_2O passes freely, **(b)** water molecules enter the solution (osmosis) and the height of the solution column in the tube rises. The pressure necessary to push water back through the membrane at a rate exactly equaled by the water influx is the osmotic pressure of the solution. **(c)** For a 1 *m* solution, this force is equal to 22.4 atm of pressure. Osmotic pressure is directly proportional to the concentration of the nonpermeant solute.

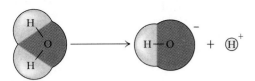

Figure 2.9   The ionization of water.

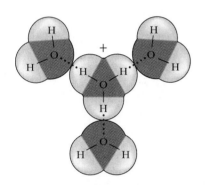

Figure 2.10   The hydration of H_3O^+. Solid lines denote covalent bonds; dashed lines represent the H bonds formed between the hydronium ion and its waters of hydration.

substances such as amino acids or sugars in polymeric form. A molecule of glycogen or starch containing a thousand glucose units will exert only 1/1000 the osmotic pressure that 1000 free glucose molecules would.

Ionization of Water

Water shows a small but finite tendency to form ions. This tendency is demonstrated by the electrical conductivity of pure water, a property which clearly establishes the presence of charged species (ions). Water ionizes because the larger, strongly electronegative oxygen atom strips the electron from one of its hydrogen atoms, leaving the proton to dissociate (Figure 2.9):

$$H-O-H \rightarrow H^+ + OH^-$$

Two ions are thus formed, **protons** or **hydrogen ions,** H^+, and **hydroxyl ions,** OH^-. Free protons are immediately hydrated to form **hydronium ions,** H_3O^+:

$$H^+ + H_2O \rightarrow H_3O^+$$

Indeed, since most hydrogen atoms in liquid water are hydrogen-bonded to a neighboring water molecule, this protonic hydration is an instantaneous process and the ion products of water are H_3O^+ and OH^-:

The amount of H_3O^+ or OH^- in 1 L (liter) of pure water at 25°C is 1×10^{-7} moles; the concentrations are equal, since the dissociation is stoichiometric.

While it is important to keep in mind that the hydronium ion, or hydrated hydrogen ion, represents the true state in solution, the convention is to speak of hydrogen ion concentrations in aqueous solution, even though such "naked" protons are virtually nonexistent. Indeed, H_3O^+ itself attracts a hydration shell by H bonding to adjacent water molecules to form an $H_9O_4^+$ species (Figure 2.10) and even more highly hydrated forms. Similarly, the hydroxyl ion, like all other highly charged species, is also hydrated.

K_w, the Ion Product of Water

The dissociation of water into hydrogen ions and hydroxyl ions occurs to the extent that 10^{-7} mol of H^+ and 10^{-7} mol of OH^- are present at equilibrium in 1 L of water at 25°C.

$$H_2O \rightarrow H^+ + OH^-$$

The equilibrium constant for this process is

$$K_{eq} = \frac{[H^+][OH^-]}{[H_2O]}$$

where brackets denote concentrations in moles per liter. Since the concentration of H_2O in 1 L of pure water is equal to the number of grams in a liter divided by the gram molecular weight of H_2O, or 1000/18, the molar concentration of H_2O in pure water is 55.5 M. The decrease in H_2O concentration as a result of ion formation ($[H^+], [OH^-] = 10^{-7}\ M$) is negligible in comparison and thus its influence on the overall concentration of H_2O can be ignored. Thus,

$$K_{eq} = \frac{(10^{-7})(10^{-7})}{55.5} = 1.8 \times 10^{-16}$$

Since the concentration of H_2O in pure water is essentially constant, a new constant, K_w, the **ion product of water,** can be written as

$$K_w = 55.5 \, K_{eq} = 10^{-14} = [H^+][OH^-]$$

This equation has the virtue of revealing the reciprocal relationship between the H^+ and OH^- concentrations of aqueous solutions. If a solution is acidic—that is, of significant $[H^+]$—then the ion product of water dictates that the OH^- concentration is correspondingly less. For example, if $[H^+]$ is 10^{-2} M, $[OH^-]$ must be 10^{-12} M ($K_w = 10^{-14} = [10^{-2}][OH^-]$; $[OH^-] = 10^{-12}$ M). Similarly, in an alkaline, or basic, solution where $[OH^-]$ is great, $[H^+]$ is low.

2.2 pH

To avoid the cumbersome use of negative exponents to express concentrations which range over 14 orders of magnitude, Sørensen, a Danish biochemist, devised the **pH scale** by defining **pH** as *the negative logarithm of the hydrogen ion concentration:*

$$pH = -\log_{10}[H^+]$$

Table 2.2 gives the pH scale. Note again the reciprocal relationship between $[H^+]$ and $[OH^-]$. Also because the pH scale is based on negative logarithms, low pH values represent the highest H^+ concentrations (and the lowest $[OH^-]$, as K_w specifies). Note also that

$$pK_w = pH + pOH = 14$$

The pH scale is widely used in biological applications because hydrogen ion concentrations in biological fluids are very low, around 10^{-7} M or 0.0000001 M, a value more easily represented as pH 7. The pH of blood plasma, for example, is 7.4, or 0.00000004 M H^+. Certain disease conditions may lower the plasma pH level to 6.8 or less, a situation which may result in death. At pH 6.8, the H^+ concentration is 0.00000016 M, four times greater than at pH 7.4.

Table 2.2 pH Scale

The hydrogen ion and hydroxyl ion concentrations are given in moles per liter at 25°C.

pH	$[H^+]$		$[OH^-]$	
0	(10^0)	1.0	0.00000000000001	(10^{-14})
1	(10^{-1})	0.1	0.0000000000001	(10^{-13})
2	(10^{-2})	0.01	0.000000000001	(10^{-12})
3	(10^{-3})	0.001	0.00000000001	(10^{-11})
4	(10^{-4})	0.0001	0.0000000001	(10^{-10})
5	(10^{-5})	0.00001	0.000000001	(10^{-9})
6	(10^{-6})	0.000001	0.00000001	(10^{-8})
7	$\mathbf{(10^{-7})}$	**0.0000001**	**0.0000001**	$\mathbf{(10^{-7})}$
8	(10^{-8})	0.00000001	0.000001	(10^{-6})
9	(10^{-9})	0.000000001	0.00001	(10^{-5})
10	(10^{-10})	0.0000000001	0.0001	(10^{-4})
11	(10^{-11})	0.00000000001	0.001	(10^{-3})
12	(10^{-12})	0.000000000001	0.01	(10^{-2})
13	(10^{-13})	0.0000000000001	0.1	(10^{-1})
14	(10^{-14})	0.00000000000001	1.0	(10^0)

Table 2.3 The pH of Various Common Fluids

Fluid	pH
Household lye	13.6
Bleach	12.6
Household ammonia	11.4
Milk of magnesia	10.3
Baking soda	8.4
Seawater	8.0
Pancreatic fluid	7.8–8.0
Blood plasma	7.4
Intracellular fluids	
Liver	6.9
Muscle	6.1
Saliva	6.6
Urine	5–8
Boric acid	5.0
Beer	4.5
Orange juice	4.3
Grapefruit juice	3.2
Vinegar	2.9
Soft drinks	2.8
Lemon juice	2.3
Gastric juice	1.2–3.0
Battery acid	0.35

At pH 7, $[H^+] = [OH^-]$. That is, there is no excess acidity or basicity. The point of **neutrality** is at pH 7, and solutions having a pH of 7 are said to be **at neutral pH.** The pH of various fluids of biological origin or relevance is given in Table 2.3. Since the pH scale is a logarithmic scale, two solutions whose pH values differ by one pH unit have a 10-fold difference in $[H^+]$. For example, grapefruit juice at pH 3.2 contains more than 12 times as much H^+ as orange juice at pH 4.3.

Dissociation of Strong Electrolytes

Substances that are almost completely dissociated to form ions in solution are called **strong electrolytes.** The term **electrolyte** describes substances capable of generating ions in solution and thereby causing an increase in the electrical conductivity of the solution. Many salts (such as NaCl or K_2SO_4) fit this category, as do strong acids (such as HCl) and strong bases (such as NaOH). Recall from general chemistry that acids are proton donors and bases are proton acceptors. In effect, the dissociation of a strong acid such as HCl in water can be treated as a proton transfer reaction between the acid HCl and the base H_2O to give the **conjugate acid** H_3O^+ and the **conjugate base** Cl^-:

$$HCl + H_2O \rightarrow H_3O^+ + Cl^-$$

The equilibrium constant for this reaction is

$$K = \frac{[H_3O^+][Cl^-]}{[H_2O][HCl]}$$

Customarily, since the term $[H_2O]$ is essentially constant in dilute aqueous solutions, it is incorporated into the equilibrium constant K to give a new term, K_a, the acid dissociation constant (where $K_a = K[H_2O]$). Also, the term H_3O^+ is often replaced by H^+, such that

$$K_a = \frac{[H^+][Cl^-]}{[HCl]}$$

For HCl, the value of K_a is exceedingly large because the concentration of HCl in aqueous solution is vanishingly small. Because this is so, the pH of HCl solutions is readily calculated from the amount of HCl used to make the solution:

$$[H^+] \text{ in solution} = [HCl] \text{ added to solution}$$

Thus, a 1 M solution of HCl has a pH $= 0$; a 0.01 M HCl solution has a pH $= 2$. Similarly, a 0.1 M NaOH solution has a pH of 13. (Since $[OH^-] = 0.1\ M$, $[H^+]$ must be $10^{-13}\ M$.)

Viewing the dissociation of strong electrolytes another way, we see that the ions show little affinity for one another. For example, in HCl in water, Cl^- has very little affinity for H^+

$$HCl \rightarrow H^+ + Cl^-$$

and in NaOH solutions, Na^+ has little affinity for OH^-. The dissociation of these substances in water is effectively complete.

Dissociation of Weak Electrolytes

Substances with only a slight tendency to dissociate to form ions in solution are called **weak electrolytes.** Acetic acid, CH_3COOH, is a good example:

$$CH_3COOH + H_2O \rightleftharpoons CH_3COO^- + H_3O^+$$

The acid dissociation constant K_a for acetic acid is 1.74×10^{-5}:

$$K_a = \frac{[H^+][CH_3COO^-]}{[CH_3COOH]} = 1.74 \times 10^{-5}$$

The term K_a is also called an **ionization constant** because it states the extent to which a substance forms ions in water. The relatively low value of K_a for acetic acid reveals that the un-ionized form, CH_3COOH, predominates over H^+ and CH_3COO^- in aqueous solutions of acetic acid. Viewed another way, CH_3COO^-, the acetate ion, has a high affinity for H^+.

▼ EXAMPLE

What is the pH of a 0.1 M solution of acetic acid? Or, to restate the question, what is the final pH when 0.1 mol of acetic acid (HAc) is added to water and the volume of the solution is adjusted to equal 1 L?

ANSWER

The dissociation of HAc in water can be written simply as

$$HAc \rightleftharpoons H^+ + Ac^-$$

where Ac^- represents the acetate ion, CH_3COO^-. In solution, some amount x of HAc dissociates, generating x amount of Ac^- and an equal amount x of H^+. Ionic equilibria characteristically are established very rapidly. At equilibrium, the concentration of HAc $+$ Ac^- must equal 0.1 M. So, [HAc] can be represented as $(0.1 - x)$, and $[Ac^-]$ and $[H^+]$ then both equal x molar. From $1.74 \times 10^{-5} = ([H^+][Ac^-])/[HAc]$, we get $1.74 \times 10^{-5} = x^2/(0.1 - x)$. The solution to quadratic equations of this form $(ax^2 + bx + c = 0)$ is $x = (-b \pm (b^2 - 4ac)^{1/2})/2a$. However, the calculation of x can be simplified by noting that, since K_a is quite small, $x \ll 0.1\ M$. Therefore, K_a is essentially equal to $x^2/0.1$. This simplification yields $x^2 = 1.74 \times 10^{-6}$, or $x = 1.32 \times 10^{-3}\ M$ and pH $= 2.88$.

The Henderson–Hasselbalch Equation

Consider the ionization of some weak acid, HA, occurring with an acid dissociation constant, K_a. Then,

$$HA \rightleftharpoons H^+ + A^-$$

and

$$K_a = \frac{[H^+][A^-]}{[HA]}$$

Rearranging this expression in terms of the parameter of interest, $[H^+]$, we have

$$[H^+] = \frac{K_a[HA]}{[A^-]}$$

Taking the logarithm of both sides gives

$$\log[H^+] = \log K_a + \log\left(\frac{[HA]}{[A^-]}\right)$$

If we change the signs and define $pK_a = -\log K_a$, we have

$$pH = pK_a - \log\left(\frac{[HA]}{[A^-]}\right)$$

or

$$\mathbf{pH = pK_a + \log\left(\frac{[A^-]}{[HA]}\right)}$$

This relationship is known as the **Henderson-Hasselbalch equation.** Thus, the pH of a solution can be calculated, provided K_a and the concentrations of the weak acid, HA, and its conjugate base, A^-, are known. Note particularly that when $[HA] = [A^-]$, $pH = pK_a$. For example, if equal volumes of 0.1 M HAc and 0.1 M sodium acetate are mixed,

$$pH = pK_a = 4.76$$
$$pK_a = -\log K_a = -\log_{10}(1.74 \times 10^{-5}) = 4.76$$

(Sodium acetate, the sodium salt of acetic acid, is a strong electrolyte and dissociates completely in water to yield Na^+ and Ac^-).

The Henderson–Hasselbalch equation provides a general solution to the quantitative treatment of acid–base equilibria in biological systems. Table 2.4 gives the acid dissociation constants and pK_a values for some weak electrolytes of biochemical interest.

▼ **EXAMPLE**

What is the pH when 100 mL of 0.1 N NaOH is added to 150 mL of 0.2 M HAc if pK_a for acetic acid = 4.76?

ANSWER

100 mL 0.1 N NaOH = 0.01 mol OH^-, which neutralizes 0.01 mol of HAc, giving an equivalent amount of Ac^-:

$$OH^- + HAc \rightarrow Ac^- + H_2O$$

Note that 0.02 mol of the original 0.03 mol of HAc remains essentially undissociated. The final volume is 250 mL.

$$pH = pK_a + \log_{10}\frac{[Ac^-]}{[HAc]} = 4.76 + \log\frac{(0.01 \ mol)}{(0.02 \ mol)}$$

$$pH = 4.76 - \log_{10}2 = 4.46$$

If 150 mL of 0.2 M HAc had merely been diluted with 100 mL of water, this would leave 250 mL of a 0.12 M HAc solution. The pH would be given by

$$K_a = \frac{[H^+][Ac^-]}{[HAc]} = \frac{x^2}{0.12 \ M} = 1.74 \times 10^{-5} \ M$$

$$x = 1.44 \times 10^{-3} \ M = [H^+]$$

$$pH = 2.84$$

Clearly, the presence of sodium hydroxide has mostly neutralized the acidity of the acetic acid through the formation of acetate ion.

Titration Curves

Titration is the analytical method used to determine the amount of acid in a solution. A measured volume of the acid solution is titrated by slowly adding a solution of base, typically NaOH, of known concentration. As incremental amounts of NaOH are added, the pH of the solution is determined and a plot of the pH of the solution versus the amount of OH^- added yields a **titration curve**. The titration curve for acetic acid is shown in Figure 2.11. In considering the progress of this titration, keep in mind two important equilibria:

1. $HAc \rightleftharpoons H^+ + Ac^-$ $K_a = 1.74 \times 10^{-5}$
2. $H^+ + OH^- \rightleftharpoons H_2O$ $K = [H_2O]/K_w = 5.55 \times 10^{15}$

As the titration begins, mostly HAc is present, plus some H^+ and Ac^- in amounts that can be calculated (see the Example on page 41). Addition of a solution

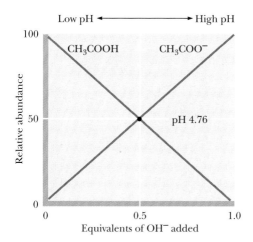

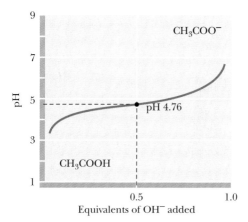

Figure 2.11 The titration curve for acetic acid. Note that the titration curve is relatively flat at pH values near the pK_a; in other words, the pH changes relatively little as OH^- is added in this region of the titration curve.

Table 2.4 **Acid Dissociation Constants and pK$_a$ Values for Some Weak Electrolytes (at 25°C)**

Acid	K_a (M)	pK_a
HCOOH (formic acid)	1.78×10^{-4}	3.75
CH$_3$COOH (acetic acid)	1.74×10^{-5}	4.76
CH$_3$CH$_2$COOH (propionic acid)	1.35×10^{-5}	4.87
CH$_3$CHOHCOOH (lactic acid)	1.38×10^{-4}	3.86
HOOCCH$_2$CH$_2$COOH (succinic acid) pK_1*	6.16×10^{-5}	4.21
HOOCCH$_2$CH$_2$COO$^-$ (succinic acid) pK_2	2.34×10^{-6}	5.63
H$_3$PO$_4$ (phosphoric acid) pK_1	7.08×10^{-3}	2.15
H$_2$PO$_4^-$ (phosphoric acid) pK_2	6.31×10^{-8}	7.20
HPO$_4^{2-}$ (phosphoric acid) pK_3	3.98×10^{-13}	12.40
C$_3$N$_2$H$_3^+$ (imidazole)	1.02×10^{-7}	6.99
C$_6$O$_2$N$_3$H$_{11}^+$ (histidine–imidazole group) pK_R†	9.12×10^{-7}	6.04
H$_2$CO$_3$ (carbonic acid) pK_1	1.70×10^{-4}	3.77
HCO$_3^-$ (bicarbonate) pK_2	5.75×10^{-11}	10.24
(HOCH$_2$)$_3$CNH$_3^+$ (*tris*-hydroxymethyl aminomethane)	8.32×10^{-9}	8.07
NH$_4^+$ (ammonium)	5.62×10^{-10}	9.25
CH$_3$NH$_4^+$ (methylammonium)	2.46×10^{-11}	10.62

*These pK values listed as pK_1, pK_2, or pK_3 are in actuality pK_a values for the respective dissociations. This simplification in notation is used throughout this book.
†pK_R refers to the imidazole ionization of histidine.

Data from *CRC Handbook of Biochemistry*. Chemical Rubber Co., 1968.

of NaOH allows hydroxide ion to neutralize any H^+ present. Note that reaction (2) as written is strongly favored; its apparent equilibrium constant is greater than 10^{15}! As H^+ is neutralized, more HAc will dissociate to H^+ and Ac^-. As further NaOH is added, the pH gradually increases as Ac^- accumulates at the expense of diminishing HAc and the neutralization of H^+. At the point where half of the HAc has been neutralized—that is, where 0.5 equivalent of OH^- has been added—the concentrations of HAc and Ac^- are equal and pH = pK_a for HAc. Thus, we have an experimental method for determining the pK_a values of weak electrolytes. These pK_a values lie at the midpoint of their respective titration curves. After all of the acid has been neutralized (that is, when one equivalent of base has been added), the pH rises exponentially.

The shapes of the titration curves of weak electrolytes are identical, as Figure 2.12 reveals. Note, however, that the midpoints of the different curves vary in a way that characterizes the particular electrolytes. The pK_a for acetic acid is 4.76, the pK_a for imidazole is 6.99, and that for ammonium is 9.25. These pK_a values are directly related to the dissociation constants of these substances, or, viewed the other way, to the relative affinities of the conjugate bases for protons. NH_3 has a high affinity for protons compared with Ac^-; NH_4^+ is a poor acid in comparison with HAc.

Phosphoric Acid Has Three Dissociable H^+

Figure 2.13 shows the titration curve for phosphoric acid, H_3PO_4. This substance is a *polyprotic acid*, meaning it has more than one dissociable proton. Indeed, it has three, and thus three equivalents of OH^- are required to neutralize it, as Figure 2.13 shows. Note that the three dissociable H^+ are lost in discrete steps, each dissociation showing a characteristic pK_a. pK_1 occurs at pH = 2.15, and the concentrations of the acid H_3PO_4 and the conjugate base, $H_2PO_4^-$, are equal. As the next dissociation is approached, $H_2PO_4^-$ is treated as the acid and HPO_4^{2-} is its conjugate base. Their concentrations are equal

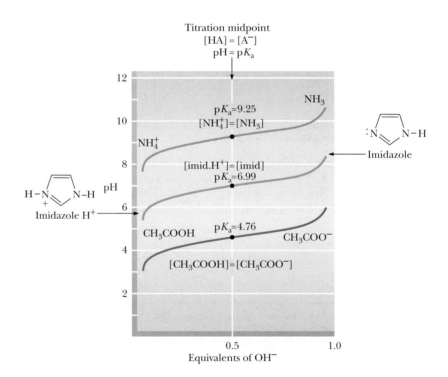

Figure 2.12 The titration curves of several weak electrolytes: acetic acid, imidazole, and ammonium. Note that the shape of these different curves is identical. Only their position along the pH scale is displaced, in accordance with their respective affinities for H^+ ions, as reflected in their differing pK_a values.

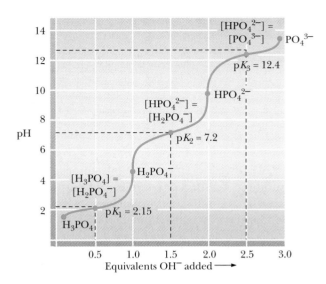

Figure 2.13 The titration curve for phosphoric acid. The chemical formulas show the prevailing ionic species present at various pH values. Phosphoric acid (H_3PO_4) has three titratable hydrogens and therefore three midpoints are seen: at 25°C, these midpoints are at pH 2.15 (pK_1), pH 7.20 (pK_2), and pH 12.4 (pK_3).

at pH 7.20, so $pK_2 = 7.20$. (Note that at this point, 1.5 equivalents of OH^- have been added.) As more OH^- is added, the last dissociable hydrogen is titrated, and pK_3 occurs at pH = 12.4, where $[HPO_4^{2-}] = [PO_4^{3-}]$.

A biologically important point is revealed by the basic shape of the titration curves of weak electrolytes: in the region of the pK_a, pH remains relatively unaffected as increments of OH^- (or H^+) are added. The weak acid and its conjugate base are acting as a buffer.

2.3 Buffers

Buffers are solutions that tend to resist changes in their pH as acid or base is added. Typically, a buffer system is composed of a weak acid *and* its conjugate base. A solution of a weak acid that has a pH near its pK_a by definition contains an amount of the conjugate base nearly equivalent to the weak acid. Note that in this region, the titration curve is relatively flat (Figure 2.14). Addition of H^+ then has little effect because it is absorbed by the following reaction:

$$H^+ + A^- \rightarrow HA$$

Similarly, added OH^- is consumed by the process

$$OH^- + HA \rightarrow A^- + H_2O$$

The pH then remains relatively constant. The components of a buffer system are chosen such that the pK_a of the weak acid is close to the pH of interest. It is at the pK_a that the buffer system shows its greatest buffering capacity. At pH values more than 1 pH unit from the pK_a, buffer systems become ineffective because the concentration of one of the components is too low to absorb the influx of H^+ or OH^-. The molarity of a buffer is defined as the *sum* of the concentrations of the acid and conjugate base forms.

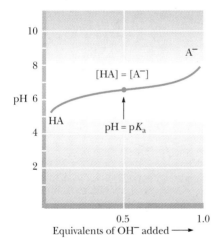

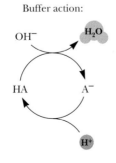

Figure 2.14 Buffer systems consist of a weak acid, HA, and its conjugate base, A^-. The pH varies only slightly in the region of the titration curve where $[HA] = [A^-]$. The unshaded area denotes this area of greatest buffering capacity. Buffer action: when HA and A^- are both available in sufficient concentration, the solution can absorb input of either H^+ or OH^- and pH is maintained essentially constant.

(a)

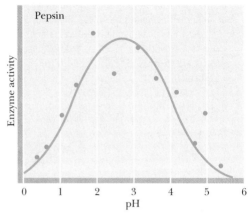

(b)

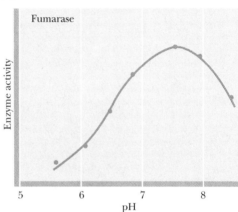

(c)

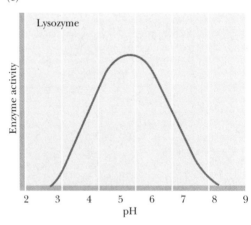

Figure 2.15 Enzymatic activity *versus* pH. The activity of enzymes is very sensitive to pH. The pH optimum of an enzyme is one of its most important characteristics. **(a)** Pepsin is a protein-digesting enzyme active in the gastric fluid of the stomach. **(b)** Fumarase is a mitochondrial enzyme participating in the citric acid cycle (this graph is the pH *versus* activity profile for fumarase operating in the malate $\rightarrow$ fumarate $+$ H_2O direction). **(c)** Lysozyme digests the cell walls of bacteria; it is found in tears.

Maintenance of pH is vital to all cells. Cellular processes such as metabolism are dependent on the activities of enzymes, and, in turn, enzyme activity is markedly influenced by pH, as the graphs in Figure 2.15 show. Consequently, changes in pH would be very disruptive to metabolism for reasons that will become apparent in later chapters. Organisms have a variety of mechanisms to keep the pH of their intra- and extracellular fluids essentially constant, but the primary protection against harmful pH changes is provided by buffer systems. The buffer systems selected reflect both the need for a pK_a value near pH 7 and the compatibility of the buffer components with the metabolic machinery of cells. Two buffer systems act to maintain intracellular pH essentially constant: the $HPO_4^{2-}/H_2PO_4^-$ system and the histidine system. The pH of the extracellular fluid that bathes the cells and tissues of animals is maintained by the bicarbonate/carbonic acid (HCO_3^-/H_2CO_3) system.

The Phosphate System

The **phosphate system** serves to buffer the intracellular fluid of cells at physiological pH because pK_2 lies near this pH value. The intracellular pH of most cells is maintained in the range between 6.9 and 7.4. Phosphate is an abundant anion in cells, both in inorganic form and as an important substituent of organic molecules that serve as metabolites or macromolecular precursors. In both organic and inorganic forms, phosphate, with its characteristic pK_2, is represented by ionized states at physiological pH that can either donate H^+ or accept H^+ to buffer any changes in pH, as the titration curve for H_3PO_4 shown in Figure 2.13 reveals. For example, if the total cellular concentration of phosphate is 20 mM (millimolar) and the pH is 7.4, the distribution of the major phosphate species is given by

$$pH = pK_2 + \log_{10}\left(\frac{[HPO_4^{2-}]}{[H_2PO_4^-]}\right)$$

$$7.4 = 7.20 + \log_{10}\left(\frac{[HPO_4^{2-}]}{[H_2PO_4^-]}\right)$$

$$\frac{[HPO_4^{2-}]}{[H_2PO_4^-]} = 1.58$$

Thus, if $[HPO_4^{2-}] + [H_2PO_4^-] = 20$ mM, then

$$[HPO_4^{2-}] = 12.25 \text{ m}M \text{ and } [H_2PO_4^-] = 7.75 \text{ m}M$$

The Histidine System

Histidine is one of the 20 naturally occurring amino acids commonly found in proteins (see Chapter 4). It possesses as part of its structure an imidazole group, a five-membered heterocyclic ring possessing two nitrogen atoms. The pK_a for dissociation of the imidazole hydrogen of histidine is 6.04.

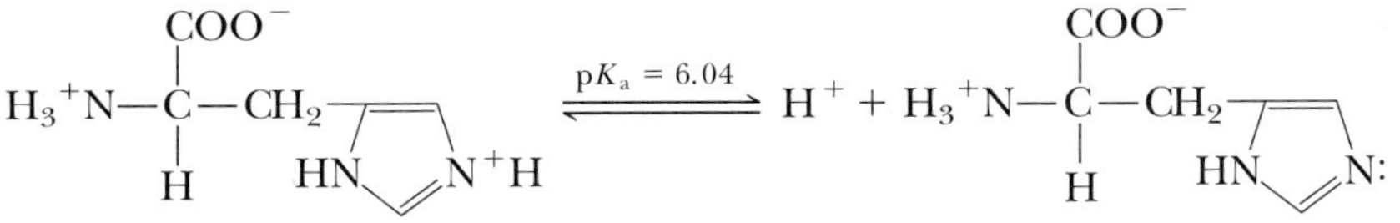

In cells, histidine occurs as the free amino acid, as a constituent of proteins, and as part of dipeptides in combination with other amino acids. Because the concentration of free histidine is low and its imidazole pK_a is more than one pH unit removed from prevailing intracellular pH, its role in intracellular buffering is minor. However, protein-bound and dipeptide histidine may be the dominant buffering system in some cells. In combination with other amino acids, as in proteins or dipeptides, the imidazole pK_a may increase substantially. For example, the imidazole pK_a is 7.04 in **anserine,** a dipeptide containing β-alanine and histidine (Figure 2.16). Thus, this pK_a is near physiological pH, and some histidine peptides are well suited for buffering at physiological pH.

The Bicarbonate Buffer System of Blood Plasma

The important buffer system of blood plasma is the bicarbonate/carbonic acid couple:

$$H_2CO_3 \rightleftharpoons H^+ + HCO_3^-$$

The relevant pK_a is pK_1 for carbonic acid, which has a value of 3.80 at 37°C, far removed from the normal pH for blood plasma (pH 7.4). At pH 7.4, the concentration of H_2CO_3 is a minuscule fraction of the HCO_3^- concentration, and thus the plasma would appear to be poorly protected against an influx of OH^- ions.

$$pH = 7.4 = 3.8 + \log_{10}\left(\frac{[HCO_3^-]}{[H_2CO_3]}\right)$$

$$\frac{[HCO_3^-]}{[H_2CO_3]} = 3981$$

For example, if $[HCO_3^-] = 4\ mM$, $[H_2CO_3]$ is barely $1\ \mu M\,(10^{-6}\ M)$, and an equivalent amount of OH^- would swamp the buffer system, causing a dangerous rise in the plasma pH. How then can this bicarbonate system function effectively? The bicarbonate buffer system works well because the critical concentration of H_2CO_3 is maintained relatively constant through equilibrium with

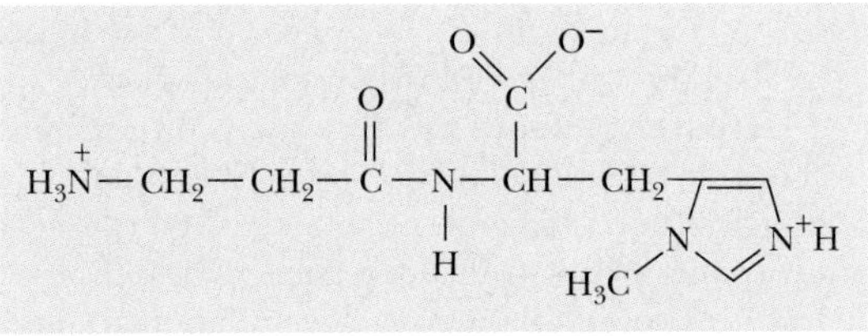

Figure 2.16 Anserine (N-β-alanyl-3-methyl-L-histidine) is an important dipeptide buffer in the maintenance of intracellular pH in some tissues. The structure shown is the predominant ionic species at pH 7. $pK_1(COOH) = 2.64$; pK_2 (imidazole$-N^+H$) = 7.04; $pK_3(NH_3^+) = 9.49$.

A DEEPER LOOK

How the Bicarbonate Buffer System Works

Gaseous carbon dioxide from the lungs and tissues is dissolved in the plasma, symbolized as $CO_2(d)$, and hydrated to form H_2CO_3:

$$CO_2(g) \rightleftharpoons CO_2(d) \qquad (1)$$

$$CO_2(d) + H_2O \rightleftharpoons H_2CO_3 \qquad (2)$$

$$H_2CO_3 \rightleftharpoons H^+ + HCO_3^- \qquad (3)$$

Thus, the concentration of H_2CO_3 is itself buffered by the available pools of CO_2. The hydration of CO_2 (reaction [2]) is rapidly catalyzed by the enzyme *carbonic anhydrase.*

Under the conditions of temperature and ionic strength prevailing in mammalian body fluids, the equilibrium for reaction (2) lies far to the left, such that over 300 CO_2 molecules are present in solution for every molecule of H_2CO_3. Since dissolved CO_2 and H_2CO_3 are in equilibrium, the proper expression for H_2CO_3 availability is $([CO_2(d)] + [H_2CO_3])$, the so-called total carbonic acid pool, consisting primarily of $CO_2(d)$.

An expression for the ionization of H_2CO_3 in the presence of dissolved CO_2 can be obtained from K_h, the equilibrium constant for reaction (2), the hydration of CO_2, and from K_1, the first dissociation constant for H_2CO_3 (reaction [3]):

$$K_h = \frac{[H_2CO_3]}{[CO_2(d)]}$$

Thus,

$$[H_2CO_3] = K_h[CO_2(d)]$$

Putting this value for $[H_2CO_3]$ into the expression for the first dissociation of H_2CO_3 gives

$$K_1 = \frac{[H^+][HCO_3^-]}{[H_2CO_3]} = \frac{[H^+][HCO_3^-]}{K_h[CO_2(d)]}$$

Therefore, the overall equilibrium constant for the ionization of H_2CO_3 in equilibrium with $CO_2(d)$ is given by

$$K_1 K_h = \frac{[H^+][HCO_3^-]}{[CO_2(d)]}$$

and $K_1 K_h$, the product of two constants, can be defined as a new equilibrium constant, $K_{overall}$. The value of K_h is 0.003 at 37°C and K_a, the ionization constant for H_2CO_3, is $10^{-3.57} = 0.000269$. Therefore,

$$K_{overall} = (0.000269)(0.0031) = 8.07 \times 10^{-7}$$
$$pK_{overall} = 6.1$$

which yields the following Henderson–Hasselbalch relationship:

$$pH = pK_{overall} + \log\left(\frac{[HCO_3^-]}{[CO_2(d)]}\right)$$

Although the prevailing blood pH of 7.4 is more than 1 pH unit away from $pK_{overall}$, the bicarbonate system is still an effective buffer. That is, at blood pH, the concentration of $CO_2(d)$ (the acid component of the buffer) is less than 10% of $[HCO_3^-]$, the conjugate base component. One might imagine that this buffer component could be overwhelmed by relatively small amounts of alkali, with consequent disastrous rises in blood pH. However, the concentration of $CO_2(d)$ is stabilized by its equilibrium with $CO_2(g)$. Gaseous CO_2 buffers any losses from the total carbonic acid pool by entering solution as $CO_2(d)$, and blood pH is effectively maintained. Thus, the bicarbonate/CO_2 buffer system is an *open system.* The natural presence of CO_2 gas at a partial pressure of 40 mm Hg in the alveoli of the lungs, and the equilibrium

$$CO_2(g) \rightleftharpoons CO_2(d)$$

keeps the concentration of $CO_2(d)$ in the neighborhood of 1.2 mM. Plasma $[HCO_3^-]$ is slightly less than 15 mM under such conditions, and blood pH = 7.4.

HUMAN BIOCHEMISTRY

Blood pH and Respiration

Hyperventilation, defined as a breathing rate more rapid than necessary for normal CO_2 elimination from the body, can result in an inappropriately low $[CO_2(g)]$ in the blood. Central nervous system disorders such as meningitis, encephalitis, or cerebral hemorrhage, as well as a number of drug- or hormone-induced physiological changes, can lead to hyperventilation. As $[CO_2(g)]$ drops due to excessive exhalation, $[H_2CO_3]$ in the blood plasma falls, followed by decline in $[H^+]$ and $[HCO_3^-]$ in the blood plasma. Blood pH rises within 20 sec of the onset of hyperventilation, becoming maximal within 15 min. $[H^+]$ can change from its normal value of 40 nM

(pH = 7.4) to 18 nM (pH = 7.74). This rise in plasma pH (increase in alkalinity) is termed **respiratory alkalosis.**

Hypoventilation is the opposite of hyperventilation and is characterized by an inability to excrete CO_2 rapidly enough to meet physiological needs. Hypoventilation can be caused by narcotics, sedatives, anesthetics, and depressant drugs; diseases of the lung also lead to hypoventilation. Hypoventilation results in **respiratory acidosis** as $CO_2(g)$ accumulates, giving rise to H_2CO_3, which dissociates to form $[H^+]$ and $[HCO_3^-]$.

dissolved CO_2 produced in the tissues and available as a gaseous CO_2 reservoir in the lungs.[2]

The remarkable properties of water render it particularly suited to its unique role in living processes and the environment, and its presence in abundance favors the existence of life. Let's examine water's physical and chemical properties to see the extent to which they provide conditions advantageous to organisms.

As a *solvent*, water is powerful yet innocuous. No other chemically inert solvent compares to water for the substances it can dissolve. Also, it is very important to life that water is a "poor" solvent for nonpolar substances. Thus, through hydrophobic interactions, lipids coalesce, membranes form, boundaries are created, and the cellular nature of life is established. Because of its high dielectric constant, water is a medium for ionization. Ions enrich the living environment by enhancing the variety of chemical species and introducing an important class of chemical reactions. They provide electrical properties to solutions and, therefore, to organisms. Aqueous solutions are the prime source of ions.

The *thermal properties* of water are especially relevant to its environmental fitness. It has great power as a buffer resisting thermal (temperature) change. Its heat capacity, or specific heat ($4.1840 \, J/g/°C$), is remarkably high; it is 10 times greater than that of iron, 5 times greater than that of quartz or salt, and twice as great as that of hexane. Its heat of fusion is $335 \, J/g$. Thus, at $0°C$, it takes a loss of 335 J to change the state of 1 g of H_2O from liquid to solid. Its heat of vaporization, $2.24 \, kJ/g$, is exceptionally high. These thermal properties mean that it takes substantial amounts of heat to alter the temperature, and especially the *state*, of water. Water's thermal properties allow it to buffer the climate through processes such as condensation, evaporation, melting, and freezing. Furthermore, these properties allow very effective temperature regulation in living organisms. For example, heat generated within the organism as a result of metabolism can be efficiently eliminated by evaporation or conduction. The thermal conductivity of water is very high in comparison with other liquids. The anomalous expansion of water as it cools to temperatures near its freezing point is particularly significant to the environment. As water cools, H bonding increases because the thermal motions of the molecules are lessened. Hydrogen bonding tends to separate the water molecules (Figure 2.2), and thus the density of water decreases. These changes in density mean that, at temperatures below $4°C$, cool water rises and, most important, ice freezes on the surfaces of bodies of water, forming an insulating layer protecting the liquid water underneath.

Water has the highest *surface tension* (75 dyn/cm) of all common liquids (except mercury). Together, surface tension and density determine how high a liquid will rise in a capillary system. Capillary movement of water plays a prominent role in the life of plants. Finally, consider *osmosis*, the bulk movement of water from a dilute aqueous solution to a more concentrated one across a semipermeable boundary. Such bulk movements determine the shape and form of living things.

Water is truly a crucial determinant of the fitness of the environment. In a very real sense, organisms are aqueous systems in a watery world.

---

[2]Well-fed human adults exhale about 1 kilogram of carbon dioxide daily. Imagine the excretory problem if CO_2 were not a volatile gas!

PROBLEMS

1. Calculate the pH of
   a. $5 \times 10^{-4}\,M\,\text{HCl}$
   b. $7 \times 10^{-5}\,M\,\text{NaOH}$
   c. $2 \times 10^{-6}\,M\,\text{HCl}$
   d. $3 \times 10^{-2}\,M\,\text{KOH}$
   e. $4 \times 10^{-5}\,M\,\text{HCl}$
   f. $6 \times 10^{-9}\,M\,\text{HCl}$

2. Calculate from the values in Table 2.3:
   a. the $[\text{H}^+]$ in vinegar
   b. the $[\text{H}^+]$ in saliva
   c. the $[\text{H}^+]$ in household ammonia
   d. the $[\text{OH}^-]$ in milk of magnesia
   e. the $[\text{OH}^-]$ in beer
   f. the $[\text{H}^+]$ inside a liver cell

3. Measurement of the pH of a 0.02 M solution of an acid gave a value of 4.6.
   a. What is $[\text{H}^+]$ in this solution?
   b. Calculate the dissociation constant K_a and pK_a for this acid.

4. The K_a for formic acid is $1.78 \times 10^{-4}\,M$.
   a. What is the pH of a 0.1 M solution of formic acid?

   b. 150 mL of 0.1 M NaOH is added to 200 mL of 0.1 M formic acid, and water is added to give volume of 1 L. What is the pH of the final solution?

5. Given 0.1 M solutions of acetic acid and sodium acetate, describe the preparation of 1 L of 0.1 M acetate buffer at pH 5.4.

6. If the internal pH of muscle cells is 6.8, what is the $[\text{HPO}_4^{-2}]/[\text{H}_2\text{PO}_4^{-}]$ ratio in this cell?

7. What are the approximate fractional concentrations of the following phosphate species at pH values of 2, 3, 6, 8, 10, and 13?
   a. H_3PO_4 b. $\text{H}_2\text{PO}_4^{-}$ c. HPO_4^{2-} d. PO_4^{3-}

8. a. If 50 mL of 0.01 M HCl is added to 100 mL of 0.05 M phosphate buffer, pH 7.2, what is the resulting pH? What are the concentrations of $\text{H}_2\text{PO}_4^{-}$ and HPO_4^{2-} in the final solution?
   b. If 50 mL of 0.01 M NaOH is added to 100 mL of 0.05 M phosphate buffer, pH 7.2, what is the resulting pH? What are the concentrations of $\text{H}_2\text{PO}_4^{-}$ and HPO_4^{2-} in this final solution?

FURTHER READING

Cooper, T. G., 1977. *The Tools of Biochemistry.* Chap. 1. New York: John Wiley & Sons.

Darvey, I. G., and Ralston, G. B., 1993. Titration curves—misshapen or mislabeled? *Trends in Biochemical Sciences* **18**:69–71.

Edsall, J. T., and Wyman, J., 1958. Carbon dioxide and carbonic acid. *Biophysical Chemistry*, Vol. 1, Chap. 10. New York: Academic Press.

Franks, F., ed., 1982. *The Biophysics of Water.* New York: John Wiley & Sons.

Henderson, L. J., 1913. *The Fitness of the Environment.* New York: Macmillan Co. (republished 1970, Gloucester, Mass.: P. Smith).

Hille, B. 1992. *Ionic Channels of Excitable Membranes*, 2nd ed., Chap. 10. Sunderland, Mass.: Sinauer Associates.

Masoro, E. J., and Siegel, P. D., 1971. *Acid-Base Regulation: Its Physiology and Pathophysiology.* Philadelphia: W. B. Saunders Co.

Segel, I. H., 1976. *Biochemical Calculations*, 2nd ed., Chap. 1. New York: John Wiley & Sons.

Stillinger, F. H., 1980. Water revisited. *Science* **209**:451–457.

Thermodynamics of Biological Systems

Sun emblem of Louis XIV on a gate at Versailles. The sun is the prime source of energy for life, and thermodynamics is the gateway to understanding metabolism. (Giraudon/Art Research, New York)

A theory is the more impressive the greater is the simplicity of its premises, the more different are the kinds of things it relates and the more extended is its range of applicability. Therefore, the deep impression which classical thermodynamics made upon me. It is the only physical theory of universal content which I am convinced, that within the framework of applicability of its basic concepts, will never be overthrown.

ALBERT EINSTEIN

Outline

The activities of living things require energy. Movement, growth, synthesis of biomolecules, and the transport of ions and molecules across membranes all demand energy input. All organisms must acquire energy from their surroundings and must utilize that energy efficiently to carry out life processes. To study such bioenergetic phenomena requires familiarity with **thermodynamics,** a collection of laws and principles describing the flows and interchanges of heat, energy, and matter in systems of interest. Thermodynamics also allows us to determine whether or not chemical processes and reactions occur spontaneously. Students who make the effort to understand thermodynamic properties soon appreciate the power and practical value of thermodynamic reasoning.

Even the most complicated aspects of thermodynamics are based ultimately on three rather simple and straightforward laws. These laws and their extensions sometimes run counter to our intuition. However, once truly understood, the basic principles of thermodynamics become powerful devices for sorting out complicated chemical and biochemical problems. At this milestone in our scientific development, thermodynamic thinking becomes an enjoyable and satisfying activity.

Several basic thermodynamic principles are presented in this chapter, including the analysis of heat flow, entropy production, and free energy functions and the relationship between entropy and information. In addition, some ancillary concepts are considered, including the concept of standard states, the effect of pH on standard-state free energies, the effect of concentration on the net free energy change of a reaction, and the importance of coupled processes in living things. The chapter concludes with a discussion of ATP and other energy-rich compounds.

3.1 Basic Thermodynamic Concepts

In any consideration of thermodynamics, a distinction must be made between the system and the surroundings. The **system** is that portion of the universe with which we are concerned, whereas the **surroundings** include everything else in the universe (Figure 3.1). The nature of the system must also be specified. There are three basic systems: isolated, closed, and open. An **isolated system** cannot exchange matter or energy with its surroundings. A **closed system** may exchange energy, but not matter, with the surroundings. An **open system** may exchange matter, energy, or both with the surroundings. Living things are typically open systems that exchange matter (nutrients and waste products) and heat (from metabolism, for example) with their surroundings. Since organisms operate under conditions where temperatures are not changing (or changing slowly), we can also think of them as isothermal energy conversion systems.

isothermal condition of constant temperature

The First Law: Heat, Work, Internal Energy, and Enthalpy

It was realized early in the development of thermodynamics that heat could be converted into other forms of energy, and moreover that all forms of energy could ultimately be converted to some other form. The **first law of thermodynamics** states that the total energy of an isolated system is conserved. Thermodynamicists have formulated two mathematical functions for keeping track of heat transfers and work expenditures in thermodynamic systems. The **in-**

Isolated system:
No exchange of matter or energy

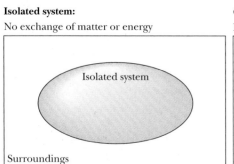

Closed system:
Energy exchange may occur

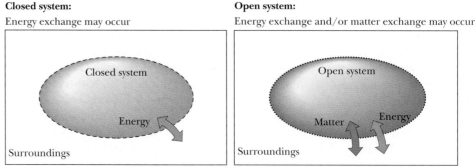

Open system:
Energy exchange and/or matter exchange may occur

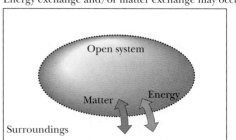

Figure 3.1 The characteristics of isolated, closed, and open systems. Isolated systems exchange neither matter nor energy with their surroundings. Closed systems may exchange energy, but not matter, with their surroundings. Open systems may exchange either matter or energy with the surroundings.

ternal energy, designated E or U, is defined as *the sum of the heat (q) absorbed by the system from the surroundings plus the work (w) done on the system by the surroundings:*

$$\Delta E = E_2 - E_1 = q + w \qquad (3.1)$$

The **enthalpy,** designated H, is defined in turn by

$$H = E + PV \qquad (3.2)$$

In biological systems, where pressure (P) and volume (V) are constant, work and the product PV terms drop out of these equations, and the enthalpy change in most processes is given by

$$\Delta H = \Delta E = q \qquad (3.3)$$

Thus, the enthalpy change is equal to the heat transferred for a typical biochemical process. In what units should H, E, and q be measured? The **calorie,** abbreviated **cal,** and **kilocalorie (kcal),** have been traditional choices of chemists and biochemists, but the SI unit, the **joule,** is now recommended.

Note that H and E depend only on the present state of a system and hence are referred to as **state functions.** Such functions do not depend on how the system got to a given state and thus are said to be **independent of path.**

In order to compare the thermodynamic parameters of different reactions, it is convenient to define a *standard state.* For solutes in a solution, the standard state is normally unit activity (often simplified to 1 M concentration). Enthalpy, internal energy, and other thermodynamic quantities are often given or determined for standard-state conditions and are then denoted by a superscript degree sign ($\degree$), as in $\Delta H\degree$, $\Delta E\degree$, and so on.

Enthalpy changes for biochemical processes can be determined experimentally by measuring the heat absorbed (or given off) by the process in a *calorimeter* (Figure 3.2). Alternatively, for any process A $\rightleftharpoons$ B at equilibrium, the standard-state enthalpy change for the process can be determined from the temperature dependence of the equilibrium constant:

$$\Delta H = -R\frac{d(\ln K_{eq})}{d(1/T)} \qquad (3.4)$$

Here, R is the gas constant, defined as $R = 8.314 \, \text{J/mol} \cdot \text{K}$. A plot of $R(\ln K_{eq})$ versus $1/T$ is called a **van't Hoff plot.**

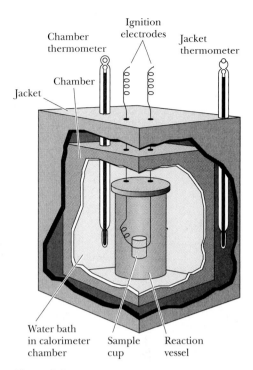

Figure 3.2 Diagram of a calorimeter. The reaction vessel is completely submerged in a water bath. The heat evolved by a reaction is determined by measuring the rise in temperature of the water bath.

▼ **EXAMPLE**

In a study[1] of the temperature-induced reversible denaturation of the protein chymotrypsinogen,

$$\text{Native state (N)} \rightleftharpoons \text{denatured state (D)}$$

$$K_{eq} = \frac{[\text{D}]}{[\text{N}]}$$

John F. Brandts measured the equilibrium constants for the unfolding of the protein (denaturation) over a range of pH and temperatures. The data for pH 3 are

T(K):	324.4	326.1	327.5	329.0	330.7	332.0	333.8
K_{eq}:	0.041	0.12	0.27	0.68	1.9	5.0	21

[1]Brandts, J. F., 1964. The thermodynamics of protein denaturation. I. The denaturation of chymotrypsinogen. *Journal of the American Chemical Society* **86:**4291–4301.

■ Table 3.1 **Thermodynamic Parameters for Protein Denaturation***

Protein (and conditions)	$\Delta H°$ kJ/mol	$\Delta S°$ kJ/mol · K	$\Delta G°$ kJ/mol	ΔC_p kJ/mol · K
Chymotrypsinogen (pH 3, 25°C)	164	0.440	31	10.9
β-Lactoglobulin (5 *M* urea, pH 3, 25°C)	−88	−0.300	2.5	9.0
Myoglobin (pH 9, 25°C)	180	0.400	57	5.9
Ribonuclease (pH 2.5, 30°C)	240	0.780	3.8	8.4

*Adapted from Cantor, C., and Schimmel, P., 1980. *Biophysical Chemistry*. San Francisco: W. H. Freeman, and Tanford, C., 1968. Protein denaturation. *Advances in Protein Chemistry* **23:**121–282.

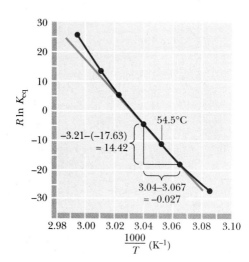

Figure 3.3 The enthalpy change, $\Delta H°$, for a reaction can be determined from the slope of a plot of $R \ln K_{eq}$ versus $1/T$. To illustrate the method, the values of the data points on either side of the 327.5 K (54.5°C) data point have been used to calculate $\Delta H°$ at 54.5°C. Regression analysis would normally be preferable. *(Adapted from Brandts, J. F., 1964. The thermodynamics of protein denaturation: I. The denaturation of chymotrypsinogen.* Journal of the American Chemical Society **86:**4291–4301.*)*

A plot of $R(\ln K_{eq})$ versus $1/T$ (a van't Hoff plot) is shown in Figure 3.3. $\Delta H°$ for the denaturation process at any temperature is the negative of the slope of the plot at that temperature. As shown, $\Delta H°$ at 54.5°C (327.5 K) is

$$\Delta H° = \frac{-[-3.2 - (-17.6)]}{[(3.04 - 3.067) \times 10^{-3}]} = +533 \text{ kJ/mol}$$

What does this value of $\Delta H°$ mean for the unfolding of the protein? Positive values of $\Delta H°$ would be expected for the breaking of hydrogen bonds as well as for the exposure of hydrophobic groups from the interior of the native, folded protein during the unfolding process. Such events would raise the energy of the protein–water solution. The magnitude of this enthalpy change (533 kJ/mol) at 54.5°C is large, compared with similar values of $\Delta H°$ for other proteins and for this same protein at 25°C (see Table 3.1). If we consider only this positive enthalpy change for the unfolding process, the native, folded state is strongly favored. As we shall see, however, other parameters must be taken into account.

The Second Law and Entropy: An Orderly Way of Thinking About Disorder

The **second law of thermodynamics** has been described and expressed in many different ways, including the following:

1. Systems tend to proceed from ordered (*low entropy* or low probability) states to disordered (*high entropy* or high probability) states.
2. The entropy of the system plus surroundings is unchanged by *reversible processes;* the entropy of the system plus surroundings increases for *irreversible processes.*
3. All naturally occurring processes proceed toward **equilibrium,** that is, to a state of minimum potential energy.

Several of these statements of the second law invoke the concept of **entropy,** which is a measure of disorder and randomness in the system (or the surroundings). An organized or ordered state is a low-entropy state, whereas a disordered state is a high-entropy state. All else being equal, reactions involving large, positive entropy changes, ΔS, are more likely to occur than reactions for which ΔS is not large and positive.

A DEEPER LOOK

Entropy, Information, and the Importance of "Negentropy"

When a thermodynamic system undergoes an increase in entropy, it becomes more disordered. On the other hand, a decrease in entropy reflects an increase in order. A more ordered system is more highly organized and possesses a greater information content. To appreciate the implications of decreasing the entropy of a system, consider the random collection of letters in the figure. This disorganized array of letters possesses no inherent information content, and nothing can be learned by its perusal. On the other hand, this particular array of letters can be systematically arranged to construct the first sentence of the Einstein quotation that opened this chapter: "A theory is the more impressive the greater is the simplicity of its premises, the more different are the kinds of things it relates and the more extended is its range of applicability."

Arranged in this way, this same collection of 151 letters possesses enormous information content—the profound words of a great scientist. Just as it would have required significant effort to rearrange these 151 letters in this way, so large amounts of energy are required to construct and maintain living organisms. Energy input is required to produce information-rich, organized structures such as proteins and nucleic acids. Information content can be thought of as *negative entropy*. In 1945 Erwin Schrödinger took time out from his studies of quantum mechanics to publish a delightful book entitled *What Is Life?* (See Further Reading, page 67.) In it, Schrödinger coined the term *negentropy* to describe the negative entropy changes that confer organization and information content to living organisms. Schrödinger pointed out that organisms must "acquire negentropy" to sustain life.

Entropy can be defined in several quantitative ways. If W is the number of ways to arrange the components of a system without changing the internal energy or enthalpy (that is, the number of possible states at a given temperature, pressure, and amount of material), then the entropy is given by

$$S = k \ln W \qquad (3.5)$$

where k is Boltzmann's constant ($k = 1.38 \times 10^{-23}$ J/K). This definition is useful for statistical calculations (it is, in fact, a foundation of statistical thermodynamics). However, a more common definition relates entropy to the heat transferred in a process:

$$dS_{reversible} = \frac{dq}{T} \qquad (3.6)$$

where $dS_{reversible}$ is the entropy change of the system in a reversible[2] process, q is the heat transferred, and T is the temperature at which the heat transfer occurs.

The Third Law: Why Is "Absolute Zero" So Important?

The **third law of thermodynamics** states that the entropy of any crystalline, perfectly ordered substance must approach zero as the temperature approaches 0 K, and at $T = 0$ K *entropy is exactly zero*. Based on this, it is possible to establish a *quantitative, absolute* **entropy scale** for any substance as

---

[2]A reversible process is one that can be reversed by an infinitesimal modification of a variable.

$$S = \int_0^T C_P \, d\ln T \tag{3.7}$$

where C_P is the *heat capacity* at constant pressure. The heat capacity of any substance is the amount of heat one mole of it can store as the temperature of that substance is raised by one degree. For a constant pressure process, this is described mathematically as

$$C_P = \frac{dH}{dT} \tag{3.8}$$

If the heat capacity can be evaluated at all temperatures between 0 K and the temperature of interest, an absolute entropy can be calculated. For biological processes, *entropy changes* are more useful than absolute entropies. The entropy change for a process can be calculated if the enthalpy change <u>and</u> a quantity known as the **free energy change** are known.

Free Energy: A Hypothetical But Useful Device

An important question for chemists, and particularly for biochemists, is, "Will the reaction proceed in the direction written?" J. Willard Gibbs, one of the founders of thermodynamics, realized that the answer to this question lay in a comparison of the enthalpy change and the entropy change for a reaction at a given temperature. The **Gibbs free energy,** G, is defined as

$$G = H - TS \tag{3.9}$$

For any process A $\rightarrow$ B at constant pressure and temperature, the free energy change is given by

$$\Delta G = \Delta H - T\Delta S \tag{3.10}$$

If ΔG is equal to 0, the process is at *equilibrium,* and there is no net flow either in the forward or reverse direction. When $\Delta G = 0$, $\Delta S = \Delta H/T$, and the enthalpic and entropic changes are exactly balanced. Any process with a nonzero ΔG proceeds spontaneously to a final state of lower free energy. If ΔG is negative, the process proceeds spontaneously in the direction written. If ΔG is positive, the reaction or process proceeds spontaneously in the reverse direction. (The sign and value of ΔG do not allow us to determine *how fast* the process will go.) If the process has a negative ΔG, it is said to be **exergonic,** whereas processes with positive ΔG values are **endergonic.**

The Standard-State Free Energy Change

The free energy change, ΔG, for any reaction depends upon the nature of the reactants and products, but it is also affected by the conditions of the reaction, including temperature, pressure, pH, and the concentrations of the reactants and products. As explained earlier, it is useful to define a standard state for such processes. If the free energy change for a reaction is sensitive to solution conditions, what is the particular significance of the standard-state free energy change? To answer this question, consider a reaction between two reactants A and B to produce the products C and D.

$$A + B \rightleftharpoons C + D \tag{3.11}$$

The free energy change for non–standard-state concentrations is given by

$$\Delta G = \Delta G^\circ + RT \ln\frac{[C][D]}{[A][B]} \tag{3.12}$$

At equilibrium, $\Delta G = 0$ and $\dfrac{[C][D]}{[A][B]} = K_{eq}$. We then have

$$\Delta G^\circ = -RT \ln K_{eq} \qquad (3.13)$$

or, in base 10 logarithms,

$$\Delta G^\circ = -2.3RT \log_{10} K_{eq} \qquad (3.14)$$

This can be rearranged to

$$K_{eq} = -10^{-\Delta G^\circ / 2.3RT} \qquad (3.15)$$

In any of these forms, this relationship allows the standard-state free energy change for any process to be determined if the equilibrium constant is known. More importantly, it states that *the equilibrium established for a reaction in solution is a function of the standard-state free energy change for the process.* That is, ΔG° is another way of writing an equilibrium constant.

▼ **EXAMPLE**

The equilibrium constants determined by Brandts at several temperatures for the denaturation of chymotrypsinogen (see previous Example) can be used to calculate the free energy changes for the denaturation process. For example, the equilibrium constant at 54.5°C is 0.27, so

$$\Delta G^\circ = -(8.314 \,\text{J/mol} \cdot \text{K})(327.5 \,\text{K}) \ln (0.27)$$

$$\Delta G^\circ = -(2.72 \,\text{kJ/mol}) \ln (0.27)$$

$$\Delta G^\circ = 3.56 \,\text{kJ/mol}$$

The positive sign of ΔG° means that the unfolding process is unfavorable; that is, the stable form of the protein at 54.5°C is the folded form. On the other hand, the relatively small magnitude of ΔG° means that the folded form is only slightly favored. Figure 3.4 shows the dependence of ΔG° on temperature for the denaturation data at pH 3 (from the data given in the Example on page 53).

Having calculated both ΔH° and ΔG° for the denaturation of chymotrypsinogen, we can also calculate ΔS°, using Equation (3.10):

$$\Delta S^\circ = -\frac{(\Delta G^\circ - \Delta H^\circ)}{T} \qquad (3.16)$$

At 54.5°C (327.5 K),

$$\Delta S^\circ = -(3560 - 533{,}000 \,\text{J/mol})/327.5 \,\text{K}$$

$$\Delta S^\circ = 1620 \,\text{J/mol} \cdot \text{K}$$

Figure 3.5 presents the dependence of ΔS° on temperature for chymotrypsinogen denaturation at pH 3. A positive ΔS° indicates that the protein solution has become more disordered as the protein unfolds. Comparison of the value of 1.62 kJ/mol · K with the values of ΔS° in Table 3.1 shows that the present value (for chymotrypsinogen at 54.5°C) is quite large. The physical significance of the thermodynamic parameters for the unfolding of chymotrypsinogen becomes clear in the next section.

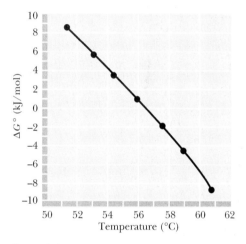

Figure 3.4 The dependence of ΔG° on temperature for the denaturation of chymotrypsinogen. *(Adapted from Brandts, J. F., 1964. The thermodynamics of protein denaturation: I. The denaturation of chymotrypsinogen. Journal of the American Chemical Society* **86:**4291–4301.)

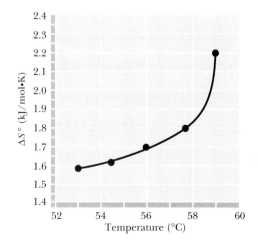

Figure 3.5 The dependence of ΔS° on temperature for the denaturation of chymotrypsinogen. *(Adapted from Brandts, J. F., 1964. The thermodynamics of protein denaturation: I. The denaturation of chymotrypsinogen. Journal of the American Chemical Society* **86:**4291–4301.)

3.2 The Physical Significance of Thermodynamic Properties

What can thermodynamic parameters tell us about biochemical events? The best answer to this question is that a single parameter (ΔH or ΔS, for example) is not very meaningful. A positive $\Delta H°$ for the unfolding of a protein might reflect *either* the breaking of hydrogen bonds within the protein or the exposure of hydrophobic groups to water (Figure 3.6). However, *comparison of several thermodynamic parameters can provide meaningful insights about a process.* For example, the transfer of Na^+ and Cl^- ions from the gas phase to aqueous solution involves a very large negative $\Delta H°$ (thus a very favorable stabilization of the ions) and a comparatively small $\Delta S°$ (Table 3.2). The negative entropy term reflects the ordering of water molecules in the hydration shells of the Na^+ and Cl^- ions. This unfavorable effect is more than offset by the large heat of hydration, which makes the hydration of ions a very favorable process overall. The negative entropy change for the dissociation of acetic acid in water also reflects the ordering of water molecules in the ion hydration shells. In this case, however, the enthalpy change is much smaller in magnitude. As a result, $\Delta G°$ for dissociation of acetic acid in water is positive, and acetic acid is thus a weak (largely undissociated) acid.

The transfer of a nonpolar hydrocarbon molecule from its pure liquid to water is an appropriate model for the exposure of protein hydrophobic groups to solvent when a protein unfolds. The transfer of toluene from liquid toluene to water involves a negative $\Delta S°$, a positive $\Delta G°$, and a $\Delta H°$ that is small compared with $\Delta G°$ (a pattern similar to that observed for the dissociation of acetic acid). *What distinguishes these two very different processes is the change in heat capacity* (see Table 3.2). A positive heat capacity change for a process indicates that the molecules have acquired new ways to move (and thus to store heat energy). A negative ΔC_P means that the process has resulted in less freedom of motion for the molecules involved. ΔC_P is negative for the dissociation of acetic acid and positive for the transfer of toluene to water. The explanation is that *both* polar and nonpolar molecules induce organization of nearby water molecules, but in different ways. The water molecules near a nonpolar solute become ordered, but the hydrogen bonds formed by water molecules near nonpolar solutes rearrange more rapidly than the hydrogen bonds of pure water.

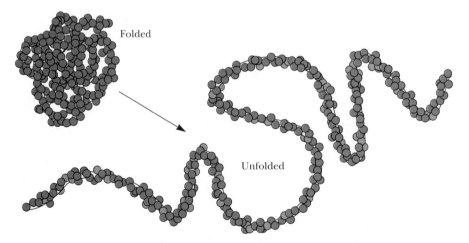

Figure 3.6 Unfolding of a soluble protein exposes significant numbers of nonpolar groups to water, forcing order on the solvent and resulting in a negative $\Delta S°$ for the unfolding process. Yellow spheres represent nonpolar groups; blue spheres are polar and/or charged groups.

Table 3.2 **Thermodynamic Parameters for Several Simple Processes***

Process	$\Delta H°$ kJ/mol	$\Delta S°$ kJ/mol · K	$\Delta G°$ kJ/mol	ΔC_P kJ/mol · K
Hydration of ions[†] $Na^+(g) + Cl^-(g) \longrightarrow Na^+(aq) + Cl^-(aq)$	−760.0	−0.185	−705.0	
Dissociation of ions in solution[‡] $H_2O + CH_3COOH \longrightarrow H_3O^+ + CH_3COO^-$	−10.3	−0.126	27.26	−0.143
Transfer of hydrocarbon from pure liquid to water[‡] Toluene (in pure toluene) $\longrightarrow$ toluene (aqueous)	1.72	−0.071	22.7	0.265

*All data collected for 25°C.
[†]Berry, R. S., Rice, S. A., and Ross, J., 1980. *Physical Chemistry*. New York: John Wiley & Sons.
[‡]Tanford, C., 1980. *The Hydrophobic Effect*. New York: John Wiley & Sons.

On the other hand, the hydrogen bonds formed between water molecules near an ion are less labile (rearrange more slowly) than they would be in pure water. This means that ΔC_P should be negative for the dissociation of ions in solution, as observed for acetic acid (see Table 3.2).

3.3 The Effect of pH on Standard-State Free Energies

For biochemical reactions in which hydrogen ions (H^+) are consumed or produced, the usual definition of the standard state is awkward. Standard state for the H^+ ion is 1 M, which corresponds to pH 0. At this pH, nearly all enzymes would be denatured, and biological reactions could not occur. It makes more sense to use free energies and equilibrium constants determined at pH 7. Biochemists have thus adopted a modified standard state, designated with prime (') symbols, as in $\Delta G°'$, K_{eq}', $\Delta H°'$, and so on. For values determined in this way, a standard state of $10^{-7}\,M\,H^+$ and unit activity (1 M for solutions, 1 atm for gases, the solvent (H_2O), and pure solids defined as unit activity) for all other components (in the ionic forms that exist at pH 7) is assumed. The two standard states can be related easily. For a reaction in which H^+ is produced,

$$A \longrightarrow B^- + H^+ \tag{3.17}$$

the relation of the equilibrium constants for the two standard states is

$$K_{eq}' = K_{eq}[H^+] \tag{3.18}$$

and $\Delta G°'$ is given by

$$\Delta G°' = \Delta G° + RT \ln [H^+] \tag{3.19}$$

For a reaction in which H^+ is consumed,

$$A^- + H^+ \longrightarrow B \tag{3.20}$$

the equilibrium constants are related by

$$K_{eq}' = \frac{K_{eq}}{[H^+]} \tag{3.21}$$

and $\Delta G°'$ is given by

$$\Delta G°' = \Delta G° + RT \ln \left(\frac{1}{[H^+]}\right) = \Delta G° - RT \ln [H^+] \tag{3.22}$$

3.4 The Important Effect of Concentration on Net Free Energy Changes

Equation (3.12) shows that the free energy change for a reaction can be very different from the standard-state value if the concentrations of reactants and products differ significantly from unit activity (1 M for solutions). The effects can often be dramatic. Consider the hydrolysis of phosphocreatine:

$$\text{Phosphocreatine} + H_2O \longrightarrow \text{creatine} + P_i \qquad \textbf{(3.23)}$$

This reaction is strongly exergonic and $\Delta G°$ at 37°C is -42.8 kJ/mol. Physiological concentrations of phosphocreatine, creatine, and inorganic phosphate are normally between 1 mM and 10 mM. Assuming 1 mM concentrations and using Equation (3.12), the ΔG for the hydrolysis of phosphocreatine is

$$\Delta G = -42.8 \text{ kJ/mol} + (8.314 \text{ J/mol} \cdot \text{K})(310 \text{ K}) \ln\frac{[0.001][0.001]}{[0.001]} \qquad \textbf{(3.24)}$$

$$\Delta G = -60.5 \text{ kJ/mol} \qquad \textbf{(3.25)}$$

At 37°C, the difference in free energy change for the reaction at standard-state and 1 mM concentrations is thus approximately -17.7 kJ/mol.

3.5 The Importance of Coupled Processes in Living Things

Many of the reactions necessary to keep cells and organisms alive must run against their **thermodynamic potential,** that is, in the direction of positive ΔG. Among these are the synthesis of adenosine triphosphate and other high-energy molecules and the creation of ion gradients in all mammalian cells. These processes are driven in the thermodynamically unfavorable direction via coupling with highly favorable processes. Many such *coupled processes* are discussed later in this text. They are crucially important in intermediary metabolism, oxidative phosphorylation, and membrane transport, as we shall see.

We can predict whether pairs of coupled reactions will proceed spontaneously by simply summing the free energy changes for each reaction. For example, consider the reaction from glycolysis (discussed in Chapter 15) involving the conversion of phospho(enol)pyruvate (PEP) to pyruvate (Figure 3.7). The hydrolysis of PEP is energetically very favorable, and it is used to drive phosphorylation of ADP to form ATP, a process that is energetically unfavorable. Using values of ΔG that would be typical for a human erythrocyte:

$$\text{PEP} + H_2O \longrightarrow \text{pyruvate} + P_i \qquad \Delta G = -78 \text{ kJ/mol} \qquad \textbf{(3.26)}$$

$$\text{ADP} + P_i \longrightarrow \text{ATP} + H_2O \qquad \Delta G = +55 \text{ kJ/mol} \qquad \textbf{(3.27)}$$

$$\text{PEP} + \text{ADP} \longrightarrow \text{pyruvate} + \text{ATP} \qquad \text{Total } \Delta G = -23 \text{ kJ/mol} \qquad \textbf{(3.28)}$$

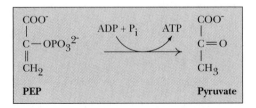

Figure 3.7 The pyruvate kinase reaction.

The net reaction catalyzed by this enzyme depends upon coupling between the two reactions shown in Equations (3.26) and (3.27) to produce the net reaction shown in Equation (3.28) with a net negative $\Delta G^{\circ\prime}$. Many other examples of coupled reactions are considered in our discussions of intermediary metabolism (Part 3). In addition, many of the complex biochemical systems discussed in the later chapters of this text involve reactions and processes with positive $\Delta G^{\circ\prime}$ values that are driven forward by coupling to reactions with a negative $\Delta G^{\circ\prime}$.

3.6 | The High-Energy Biomolecules

Virtually all life on earth depends on energy from the sun. Among life forms, there is a hierarchy of energetics: certain organisms capture solar energy directly, whereas others derive their energy from this group in subsequent processes. Organisms that absorb light energy directly are called **phototrophic organisms.** These organisms store solar energy in the form of various organic molecules. Organisms that feed on these latter molecules, releasing the stored energy in a series of oxidative reactions, are called **chemotrophic organisms.** Despite these differences, both types of organisms share common mechanisms for generating a useful form of chemical energy. Once captured in chemical form, energy can be released in controlled exergonic reactions to drive a variety of life processes (which require energy). A small family of universal biomolecules mediates the flow of energy from exergonic reactions to the energy-requiring processes of life. These molecules are the *reduced coenzymes* and the *high-energy phosphate compounds.* Phosphate compounds are considered high-energy if they exhibit large negative free energies of hydrolysis (that is, if $\Delta G^{\circ\prime}$ is more negative than -25 kJ/mol).

Table 3.3 lists the most important members of the high-energy phosphate compounds. Such molecules include *phosphoric anhydrides* (ATP, ADP), an *enol phosphate* (PEP), an *acyl phosphate* (acetyl phosphate), and a *guanidino phosphate* (creatine phosphate). Also included are *thioesters,* such as acetyl-CoA, which do not contain phosphorus but which have a high free energy of hydrolysis. As noted earlier in this chapter, the exact amount of chemical free energy available from the hydrolysis of such compounds depends on concentration, pH, temperature, and so on, but the $\Delta G^{\circ\prime}$ values for hydrolysis of these substances are substantially more negative than for most other metabolic species. Two important points: first, high-energy phosphate compounds are not long-term energy storage substances. They are transient forms of stored energy, meant to carry energy from point to point, from one enzyme system to another, in the minute-to-minute existence of the cell. (As we shall see in subsequent chapters, other molecules bear the responsibility for long-term storage of energy supplies.) Second, the term *high-energy compound* should not be construed to imply that these molecules are unstable and hydrolyze or decompose unpredictably. ATP, for example, is quite a stable molecule. A substantial activation energy must be delivered to ATP to hydrolyze the terminal, or γ, phosphate group. In fact, as shown in Figure 3.8, the *activation energy* that must be absorbed by the molecule to break the $O—P_\gamma$ bond is normally 200 to 400 kJ/mol, which is substantially larger than the net 30.5 kJ/mol released in the hydrolysis reaction. Biochemists are much more concerned with the *net release* of 30.5 kJ/mol than with the activation energy for the reaction (because suitable enzymes cope with the latter). The net release of large quantities of free energy distinguishes the high-energy phosphoric anhydrides from their "low-energy" ester cousins, such as glucose-6-phosphate (see Table 3.3).

(Text continued on page 64)

▪ **Table 3.3 Free Energies of Hydrolysis of Some High-Energy Compounds***

Compound (and Hydrolysis Product)	$\Delta G^{\circ\prime}$ (kJ/mol)	Structure
Phosphoenolpyruvate (pyruvate + P_i)	−62.2	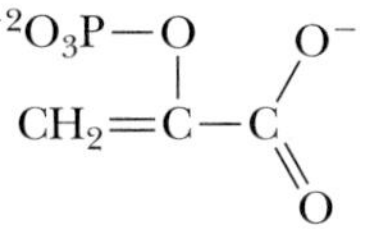
1,3-Bisphosphoglycerate (3-phosphoglycerate + P_i)	−49.6	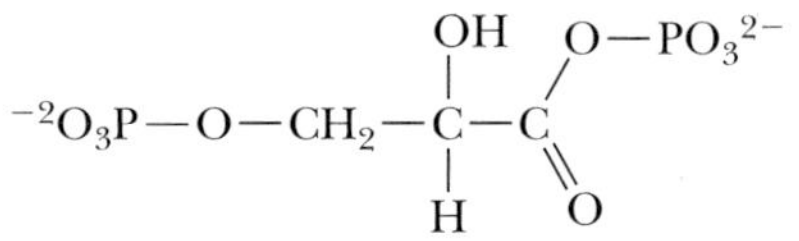
Creatine phosphate (creatine + P_i)	−43.3	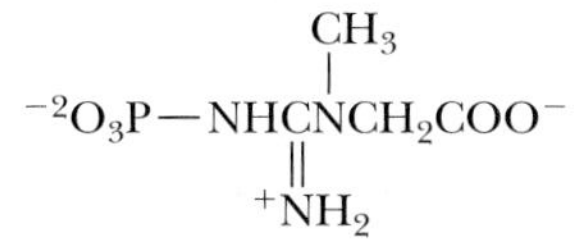
Acetyl phosphate (acetate + P_i)	−43.3	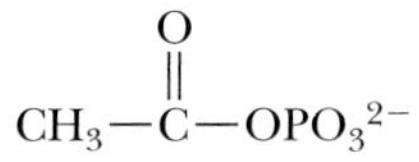
Adenosine-5′-triphosphate (ADP + P_i), excess Mg^{2+}	**−30.5**	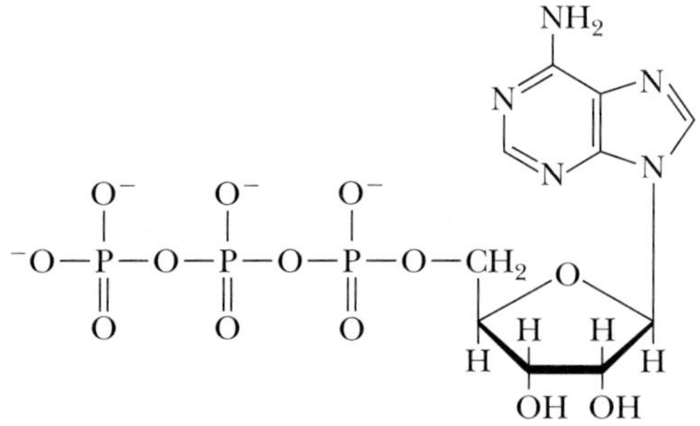
Adenosine-5′-diphosphate (AMP + P_i)	−35.7	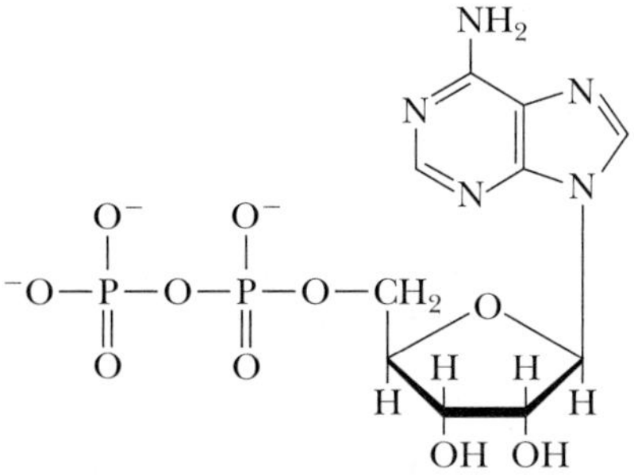
Pyrophosphate (P_i + P_i) in 5 mM Mg^{2+}	−33.6	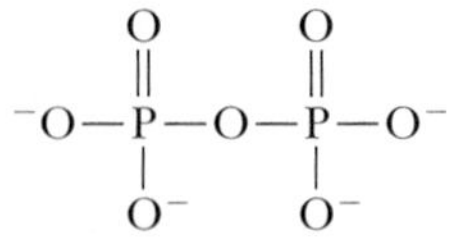
Adenosine-5′-triphosphate (AMP + PP_i), excess Mg^{2+}	−32.3	(See ATP structure above)
Acetyl-coenzyme A (acetate + CoA)	−31.5	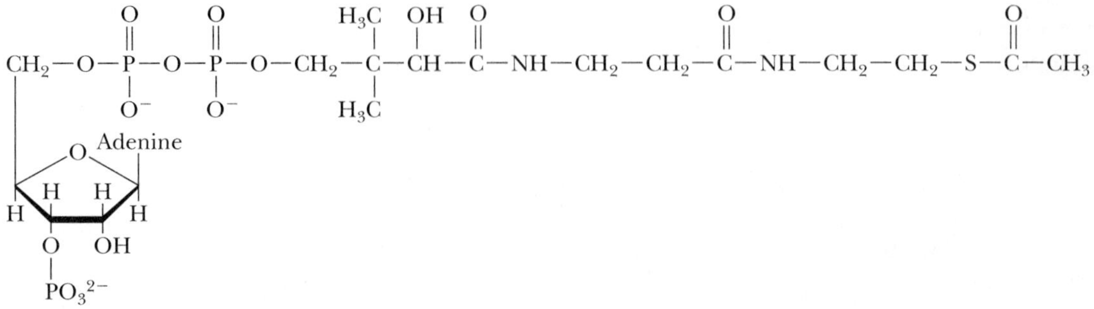

Table 3.3 Free Energies of Hydrolysis of Some High-Energy Compounds* *(Continued)*

Compound (and Hydrolysis Product)	$\Delta G^{\circ\prime}$ (kJ/mol)	Structure
Lower-Energy Phosphate Compounds		
Glucose-1-P (glucose + P_i)	−21.0	
Glucose-6-P (glucose + P_i)	−13.9	
Adenosine-5′-monophosphate (adenosine + P_i)	−9.2	

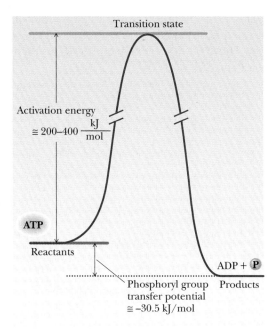

*Adapted primarily from *Handbook of Biochemistry and Molecular Biology*, 1976, 3rd ed. In *Physical and Chemical Data*, G. Fasman, ed., Vol. 1, pp. 296–304. Boca Raton, FL: CRC Press.

Figure 3.8 The activation energies for phosphoryl group-transfer reactions (200 to 400 kJ/mol) are substantially larger than the free energy of hydrolysis of ATP (−30.5 kJ/mol).

3.7 ATP Is an Intermediate Energy-Shuttle Molecule

Given the central importance of ATP as a high-energy phosphate in biology, students are sometimes surprised to find that ATP holds an intermediate place in the rank of high-energy phosphates. PEP, 1,3-BPG, phosphocreatine, acetyl phosphate, and pyrophosphate all exhibit higher values of $\Delta G^{\circ\prime}$. This is not a biological anomaly. ATP is uniquely situated between the very high energy phosphates synthesized in the breakdown of fuel molecules and the numerous lower-energy acceptor molecules that are phosphorylated in the course of further metabolic reactions. ADP can accept both phosphates and energy from the higher-energy phosphates, and the ATP thus formed can donate both phosphates and energy to the lower-energy molecules of metabolism. The ATP/ADP pair is an intermediately placed acceptor/donor system among high-energy phosphates. In this context, ATP functions as a very versatile but intermediate energy-shuttle device that interacts with many different energy-coupling enzymes of metabolism.

ATP Contains Phosphoric Acid Anhydride Linkages

ATP contains two *pyrophosphoryl* or *phosphoric acid anhydride* linkages, as shown in Figure 3.9. Other common biomolecules possessing phosphoric acid anhydride linkages include ADP, GTP, GDP and the other nucleoside di- and triphosphates, sugar nucleotides such as UDP-glucose, and inorganic pyrophosphate itself. All exhibit large negative free energies of hydrolysis, as shown in Table 3.3. The chemical reasons for the large negative $\Delta G^{\circ\prime}$ values for the hydrolysis reactions include destabilization of the reactant due to bond strain caused by electrostatic repulsion, stabilization of the products by ionization and resonance, and entropy factors due to hydrolysis and subsequent ionization.

A Comparison of the Free Energy of Hydrolysis of ATP, ADP, and AMP

The free energies of hydrolysis of ATP and ADP are much larger than that for AMP, because AMP is a phosphate ester (not an anhydride). AMP possesses only a single phosphoryl group and is not markedly different from its hydrolysis product, inorganic phosphate, in terms of electrostatic repulsion and resonance stabilization. The $\Delta G^{\circ\prime}$ for hydrolysis of AMP is much smaller than the corresponding values for ATP and ADP.

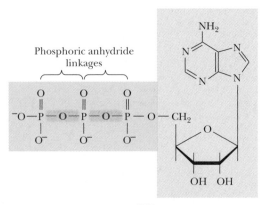

Figure 3.9 The triphosphate chain of ATP contains two pyrophosphate linkages, both of which release large amounts of energy upon hydrolysis.

ATP
(adenosine-5'-triphosphate)

Effects of Concentration and Metal Ions on the Free Energy of Hydrolysis of ATP

The concentrations of ATP, ADP, and other high-energy metabolites in living cells never even begin to approach the standard state of 1 M. In most cells, the concentrations of these species are more typically 1 to 5 mM or even less. Earlier, we described the effect of concentration on equilibrium constants and free energies in the form of Equation (3.12). For the present case, we can rewrite this as

$$\Delta G = \Delta G° + RT \ln \frac{[\Sigma ADP][\Sigma P_i]}{[\Sigma ATP]} \qquad (3.36)$$

where the terms in brackets represent the sum (Σ) of the concentrations of all the ionic forms of ATP, ADP, and P_i.

It is clear that changes in the concentrations of these species can have large effects on ΔG. The concentrations of ATP, ADP, and P_i may, of course, vary rather independently in real biological environments, but if, for the sake of some model calculations, we assume that all three concentrations are equal, then the effect of concentration on ΔG is as shown in Figure 3.10. The free energy of hydrolysis of ATP, which is -35.7 kJ/mol at 1 M, becomes -49.4 kJ/mol at 5 mM (that is, the concentration for which pC $= -2.3$ in Figure 3.10). At 1 mM ATP, ADP, and P_i, the free energy change becomes even more negative at -53.6 kJ/mol.

Does the "concentration effect" change ATP's position in the energy hierarchy (in Table 3.3)? Not really. All the other high- and low-energy phosphates experience roughly similar changes in concentration under physiological conditions and thus similar changes in their free energies of hydrolysis. The roles of the very high energy phosphates (PEP, 1,3-bisphosphoglycerate, and creatine phosphate) in the synthesis and maintenance of ATP in the cell are considered in our discussions of metabolic pathways. In the meantime, several of the problems at the end of this chapter address some of the more interesting cases.

There is one last consideration for the hydrolysis of ATP in biological systems. Most biological environments contain substantial amounts of divalent and monovalent metal ions, including Mg^{2+}, Ca^{2+}, Na^+, K^+, and so on. Metal ions can affect the equilibrium constant for ATP hydrolysis and the associated free energy change. A widely used "consensus value" for $\Delta G°'$ of ATP in biological systems is **-30.5 kJ/mol** (Table 3.3). This value, cited in the 1976 *Handbook of Biochemistry and Molecular Biology* (3rd ed., *Physical and Chemical Data*, Vol. 1, pp. 296–304, Boca Raton Fla.: CRC Press), was determined in the presence of "excess Mg^{2+}." *This is the value we use for metabolic calculations in the balance of this text.*

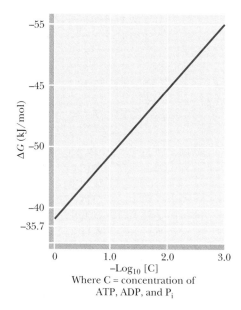

Figure 3.10 The free energy of hydrolysis of ATP as a function of concentration at 38°C, pH 7.0. The plot follows the relationship described in Equation (3.36), with the concentrations [C] of ATP, ADP, and P_i assumed to be equal.

3.8 The Daily Human Requirement for ATP

We can end this discussion of ATP and the other important high-energy compounds in biology by discussing the daily metabolic consumption of ATP by humans. An approximate calculation gives a somewhat surprising and impressive result. Assume that the average adult human consumes approximately 11,700 kJ (2800 kcal, that is, 2800 Calories) per day. Assume also that the metabolic pathways leading to ATP synthesis operate at a thermodynamic efficiency of approximately 50%. Thus, of the 11,700 kJ a person consumes as food, about 5860 kJ end up in the form of synthesized ATP. As indicated earlier, the hydrolysis of 1 mole of ATP yields approximately 50 kJ of free energy under cellular conditions. This means that the body cycles through 5860/50 = 117 moles of ATP each day. The disodium salt of ATP has a molecular weight of 551 g/mol, so that an average person hydrolyzes about

$$(117 \text{ moles})(551 \text{ g/mole}) = 64{,}467 \text{ g of ATP per day}$$

The average adult human, with a typical weight of 70 kg or so, thus consumes approximately 65 kg of ATP per day, an amount nearly equal to his or her own body weight! Fortunately, we have a highly efficient recycling system for ATP/ADP utilization. The energy released from food is stored transiently in the form of ATP. Once ATP energy is used and ADP and phosphate are released, our bodies recycle it to ATP through intermediary metabolism, so that it may be reused. The typical 70-kg body contains only about 50 grams of ATP/ADP total. Therefore, each ATP molecule in our bodies must be recycled nearly 1300 times each day! Were it not for this fact, at current commercial prices of about $10 per gram, our ATP "habit" would cost approximately $650,000 per day! In these terms, the ability of biochemistry to sustain the marvelous activity and vigor of organisms gains our respect and fascination.

PROBLEMS

1. An enzymatic hydrolysis of fructose-1-P

$$\text{Fructose-1-P} + H_2O \rightleftharpoons \text{fructose} + P_i$$

was allowed to proceed to equilibrium at 25°C. The original concentration of fructose-1-P was 0.2 M, but when the system had reached equilibrium, the concentration of fructose-1-P was only 6.52×10^{-5} M. Calculate the equilibrium constant for this reaction and the free energy of hydrolysis of fructose-1-P.

2. The equilibrium constant for some process A $\rightleftharpoons$ B is 0.5 at 20°C and 10 at 30°C. Assuming that $\Delta H°$ is independent of temperature, calculate $\Delta H°$ for this reaction. Determine $\Delta G°$ and $\Delta S°$ at 20°C and at 30°C. Why is it important in this problem to assume that $\Delta H°$ is independent of temperature?

3. The standard-state free energy of hydrolysis for acetyl phosphate is $\Delta G° = -42.3$ kJ/mol.

$$\text{Acetyl-P} + H_2O \longrightarrow \text{acetate} + P_i$$

Calculate the free energy change for acetyl phosphate hydrolysis in a solution of 2 mM acetate, 2 mM phosphate, and 3 nM acetyl phosphate.

4. Define a state function. Name three thermodynamic quantities that are state functions and three that are not.

5. ATP hydrolysis at pH 7 is accompanied by release of a hydrogen ion to the medium.

$$ATP^{4-} + H_2O \rightleftharpoons ADP^{3-} + HPO_4^{2-} + H^+$$

If the $\Delta G°'$ for this reaction is -30.5 kJ/mol, what is $\Delta G°$ (that is, the free energy change for the same reaction with all components, including H^+, at a standard state of 1 M)?

6. For the process A $\rightleftharpoons$ B, $K_{eq}(AB)$ is 0.02 at 37°C. For the process B $\rightleftharpoons$ C, $K_{eq}(BC) = 1000$ at 37°C.
   a. Determine $K_{eq}(AC)$, the equilibrium constant for the overall process A $\rightleftharpoons$ C, from $K_{eq}(AB)$ and $K_{eq}(BC)$.
   b. Determine standard-state free energy changes for all three processes, and use $\Delta G°(AC)$ to determine $K_{eq}(AC)$. Make sure that this value agrees with that determined in part **a** of this problem.

7. Draw all possible resonance structures for creatine phosphate and discuss their possible effects on resonance stabilization of the molecule.

8. Write the equilibrium constant, K_{eq}, for the hydrolysis of creatine phosphate and calculate a value for K_{eq} at 25°C from the value of $\Delta G°'$ in Table 3.3.

9. Imagine that creatine phosphate, rather than ATP, is the universal energy carrier molecule in the human body. Repeat the calculation presented in Section 3.8, calculating the weight of creatine phosphate that would need to be consumed each day by a typical adult human if creatine phosphate could not be recycled. If recycling of creatine phosphate were possible, and if the typical adult human body contained 20 grams of creatine phosphate, how many times would each creatine phosphate molecule need to be turned over or recycled each day? Repeat the calculation assuming that glucose-6-phosphate is the universal energy carrier, and that the body contains 20 grams of glucose-6-phosphate.

10. Calculate the free energy of hydrolysis of ATP in a rat liver cell in which the ATP, ADP, and P_i concentrations are 3.4, 1.3, and 4.8 mM, respectively.

11. Hexokinase catalyzes the phosphorylation of glucose from ATP, yielding glucose-6-P and ADP. Using the values of Table 3.3, calculate the standard-state free energy change and equilibrium constant for the hexokinase reaction.

12. Would you expect the free energy of hydrolysis of acetoacetyl-coenzyme A (see diagram) to be greater than, equal to, or less than that of acetyl-coenzyme A? Provide a chemical rationale for your answer.

$$CH_3-\overset{\overset{\displaystyle O}{\|}}{C}-CH_2-\overset{\overset{\displaystyle O}{\|}}{C}-SCoA$$

13. Consider carbamoyl phosphate, a precursor in the biosynthesis of pyrimidines:

$$H_3\overset{+}{N}\quad\overset{\overset{\displaystyle O}{\|}}{C}\quad O-PO_3{}^{2-}$$

Based on the discussion of high-energy phosphates in this chapter, would you expect carbamoyl phosphate to possess a high free energy of hydrolysis? Provide a chemical rationale for your answer.

FURTHER READING

Alberty, R. A., 1968. Effect of pH and metal ion concentration on the equilibrium hydrolysis of adenosine triphosphate to adenosine diphosphate. *Journal of Biological Chemistry* **243**:1337–1343.

Alberty, R. A., 1969. Standard Gibbs free energy, enthalpy, and entropy changes as a function of pH and pMg for reactions involving adenosine phosphates. *Journal of Biological Chemistry* **244**:3290–3302.

Brandts, J. F., 1964. The thermodynamics of protein denaturation: I. The denaturation of chymotrypsinogen. *Journal of the American Chemical Society* **86**:4291–4301.

Cantor, C. R., and Schimmel, P. R., 1980. *Biophysical Chemistry*. San Francisco: W. H. Freeman.

Dickerson, R. E., 1969. *Molecular Thermodynamics*. New York: Benjamin Co.

Edsall, J. T., and Gutfreund, H., 1983. *Biothermodynamics: The Study of Biochemical Processes at Equilibrium*. New York: John Wiley & Sons.

Edsall, J. T., and Wyman, J. 1958. *Biophysical Chemistry*. New York: Academic Press.

Gwynn, R. W., and Veech, R. L., 1973. The equilibrium constants of the adenosine triphosphate hydrolysis and the adenosine triphosphate-citrate lyase reactions. *Journal of Biological Chemistry* **248**:6966–6972.

Klotz, I. M., 1967. *Energy Changes in Biochemical Reactions*. New York: Academic Press.

Lehninger, A. L., 1972. *Bioenergetics*. 2nd ed. New York: Benjamin Co.

Morris, J. G., 1968. *A Biologist's Physical Chemistry*. Reading, Mass.: Addison-Wesley.

Patton, A. R., 1965. *Biochemical Energetics and Kinetics*. Philadelphia: W. B. Saunders.

Schrödinger, E., 1945. *What Is Life?* New York: Macmillan.

Segel, I. H., 1976. *Biochemical Calculations*. 2nd ed. New York: John Wiley & Sons.

Tanford, C., 1980. *The Hydrophobic Effect*. 2nd ed. New York: John Wiley & Sons.

4

Amino Acids and Polypeptides

. . . by small and simple things are great things brought to pass.

Alma 37.6, *The Book of Mormon*

Outline

Although helices are uncommon in man-made architecture, they are a common structural theme in biological macromolecules—proteins, nucleic acids, and even polysaccharides. (Loretto Chapel, Santa Fe, NM/© Sarbo)

Proteins are the indispensable agents of biological function, and **amino acids** are the building blocks of proteins. The stunning diversity of the thousands of proteins found in nature arises from the intrinsic properties of only 20 commonly occurring amino acids.

Proteins are the most abundant class of biomolecules, constituting more than 50% of the dry weight of cells. This abundance reflects the central role of proteins in virtually all aspects of cell structure and function. The extraordinary diversity of cellular activity is possible only because of the versatility displayed by proteins, each of which is specifically tailored to its biological role. The pattern by which each is tailored resides within the genetic information of cells, encoded in a specific sequence of nucleotide bases in DNA. Each such segment of encoded information defines a gene, and expression of the gene leads to synthesis of the specific protein encoded by it, endowing the cell with the functions unique to that particular protein. Proteins are the agents of biological function; they are also the expressions of genetic information.

Amino Acids: Building Blocks of Proteins

The special chemical features of amino acids include (1) the capacity to polymerize, (2) novel acid–base properties, (3) varied structure and chemical functionality in the amino acid side chains, and (4) chirality. This chapter describes each of these properties and lays a foundation for discussions of protein structure.

Structure of a Typical Amino Acid

The structure of a single typical amino acid is shown in Figure 4.1. Central to this structure is the tetrahedral alpha (α) carbon (C$_\alpha$), which is covalently linked to both an amino group and a carboxyl group. Also bonded to this α-carbon is a hydrogen and a variable side chain. It is the side chain, the so-called R group, that gives each amino acid its identity. The detailed acid–base properties of amino acids are discussed in the following sections. It is sufficient for now to realize that, in neutral solution (pH 7), the carboxyl group exists as —COO$^-$ and the amino group as —NH$_3$$^+$. Since the resulting amino acid contains one positive and one negative charge, it is a neutral molecule called a **zwitterion.** Amino acids are also *chiral* molecules. With four different groups attached to it, the α-carbon is said to be *asymmetric.* The two possible configurations for the α-carbon constitute nonidentical mirror image isomers or *enantiomers.* Details of amino acid stereochemistry are discussed in Section 4.4.

Amino Acids Can Join via Peptide Bonds

The crucial feature of amino acids that allows them to polymerize to form peptides and proteins is the existence of their two identifying chemical groups: the amino —NH$_3$$^+$ and carboxyl —COO$^-$ groups, as shown in Figure 4.2. The amino and carboxyl groups of amino acids can react in a head-to-tail fashion, eliminating a water molecule and forming a covalent amide linkage, which, in the case of peptides and proteins, is typically referred to as a **peptide bond.** The equilibrium for this reaction in aqueous solution favors peptide bond hydrolysis. For this reason, biological systems as well as peptide chemists in the laboratory must carry out peptide bond formation in an indirect manner or with energy input.

Repetition of the reaction shown in Figure 4.2 produces **polypeptides** and **proteins.** The remarkable properties of proteins, which we shall discover and come to appreciate later, all depend in one way or another on the unique properties and chemical diversity of the 20 common amino acids found in proteins.

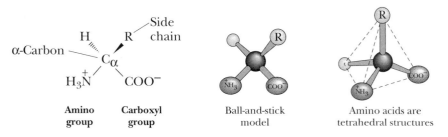

Figure 4.1 Anatomy of an amino acid. Except for proline and its derivatives, all of the amino acids commonly found in proteins possess this type of structure.

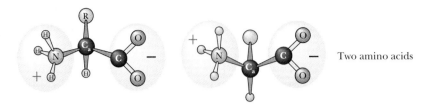

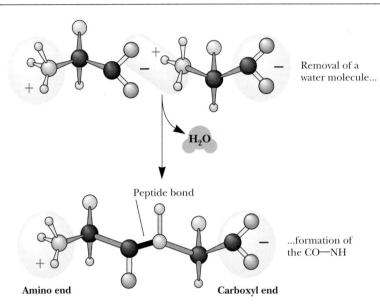

Figure 4.2 The α-COOH and α-NH$_3^+$ groups of two amino acids can react with the resulting loss of a water molecule to form a covalent amide bond. *(Irving Geis)*

Common Amino Acids

See *Interactive Biochemistry CD-ROM and Workbook,* pages 4–5

The structures and abbreviations for the 20 amino acids commonly found in proteins are shown in Figure 4.3. All the amino acids except proline have both free α-amino and free α-carboxyl groups (see Figure 4.1). There are several ways to classify the common amino acids. The most useful of these classifications is based on the polarity of the side chains. Thus, the structures shown in Figure 4.3 are grouped into the following categories: (1) nonpolar (hydrophobic) amino acids, (2) neutral (uncharged) but polar amino acids, (3) acidic amino acids (which have a net negative charge at pH 7.0), and (4) basic amino acids (which have a net positive charge at neutral pH). In later chapters, the importance of this classification system for predicting protein properties will become clear. Also shown in Figure 4.3 are the three-letter and one-letter codes used to represent the amino acids. These codes are useful when displaying and comparing amino acid sequences of proteins in shorthand form. (Note that several of the one-letter abbreviations are phonetic in origin: arginine = "Rginine" = R; phenylalanine = "Fenylalanine" = F; aspartic acid = "asparDic" = D.)

Nonpolar Amino Acids

The nonpolar or hydrophobic amino acids (Figure 4.3a) include all those with alkyl chain R groups (alanine, valine, leucine, and isoleucine), as well as proline (with its unusual cyclic structure), methionine (one of the two sulfur-containing amino acids), and two aromatic amino acids, phenylalanine and tryptophan. Tryptophan is sometimes considered a borderline member of this group, since it can interact favorably with water via the N—H moiety of the indole ring. Proline, strictly speaking, is not an amino acid but rather an α-imino acid.

Polar, Uncharged Amino Acids

The polar, uncharged amino acids (Figure 4.3b) except for glycine contain R groups that can form hydrogen bonds with water. Thus, these amino acids are usually more soluble in water than the nonpolar amino acids. The amide groups of asparagine and glutamine; the hydroxyl groups of tyrosine, threonine, and serine; and the sulfhydryl group of cysteine are all good hydrogen bond–forming moieties. Glycine, the simplest amino acid, has only a single hydrogen for an R group, and this hydrogen is not a good hydrogen bond former. Glycine's solubility properties are mainly influenced by its polar amino and carboxyl groups, and thus glycine is best considered as a member of the polar, uncharged group. It should be noted that tyrosine has significant nonpolar characteristics due to its aromatic ring and could arguably be placed in the nonpolar group (Figure 4.3a). However, with a pK_a of 10.1, tyrosine's phenolic hydroxyl is a charged, polar entity at high pH.

Acidic Amino Acids

There are two acidic amino acids—aspartic acid and glutamic acid—whose R groups contain a carboxyl group (Figure 4.3c). These side-chain carboxyl groups are weaker acids than the α-COOH group, but are sufficiently acidic to exist as carboxylate groups (—COO⁻) at neutral pH. Aspartic acid and glutamic acid thus have a net negative charge at pH 7. These forms are appropriately referred to as aspartate and glutamate. These negatively charged amino acids play several important roles in proteins. Many proteins that bind metal ions for structural or functional purposes possess metal binding sites containing one or more aspartate and glutamate side chains. The acid–base chemistry of such groups is considered in detail in Section 4.2.

Basic Amino Acids

Three of the common amino acids have side chains with net positive charges at neutral pH: histidine, arginine, and lysine (Figure 4.3d). The side chains of

HUMAN BIOCHEMISTRY

Eosinophilia–Myalgia Syndrome and the L-Tryptophan Disaster

The amino acid L-tryptophan is the biosynthetic precursor of serotonin, a brain chemical involved in regulating sleep, mood, and appetite. With this rationale, the health-food industry marketed L-tryptophan in tablet, capsule, and powder form during the 1970s and 1980s. The practice of supplementing one's diet with L-tryptophan became particularly popular in the late 1980s. However, in 1989, an outbreak of eosinophilia–myalgia syndrome (EMS) occurred among L-tryptophan users. This disease, ordinarily very rare, is characterized by severe muscle and joint pain, weakness, swelling of the arms and legs, fever, skin rash, and an increase of eosinophils (white blood cells that bind a dye called eosin) in the

blood. During the next year, more than 1500 cases and 28 deaths were reported to the U.S. Centers for Disease Control and Prevention. The actual toll was probably more than 5000 people affected, many of whom have suffered for years and remain disabled. When a link between EMS and L-tryptophan became apparent, the FDA banned the import and sale of L-tryptophan in 1990. Since it can still be obtained illegally, new cases of EMS continue to occur.

Extensive research into L-tryptophan–induced EMS has not been able to identify clearly the underlying causes for the disease, but tryptophan-induced metabolic abnormalities are suspected.

(Text continued on page 74)

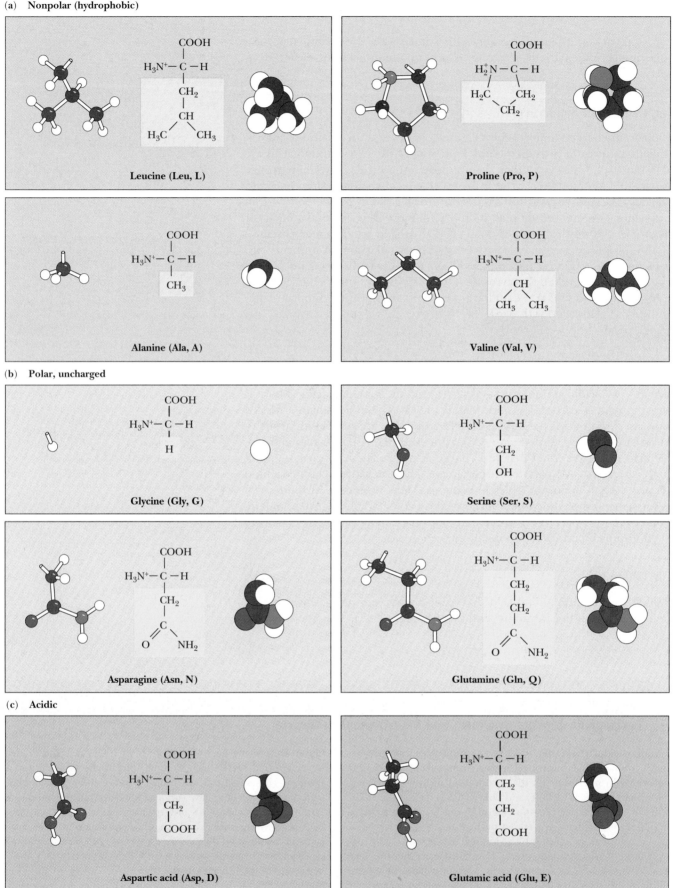

Leucine (Leu, L)

Proline (Pro, P)

Alanine (Ala, A)

Valine (Val, V)

(b) Polar, uncharged

Glycine (Gly, G)

Serine (Ser, S)

Asparagine (Asn, N)

Glutamine (Gln, Q)

(c) Acidic

Aspartic acid (Asp, D)

Glutamic acid (Glu, E)

See pages 4–5

Figure 4.3 The 20 amino acids that are the building blocks of most proteins can be classified as **(a)** nonpolar (hydrophobic), **(b)** polar, neutral, **(c)** acidic, or *(Continued)*

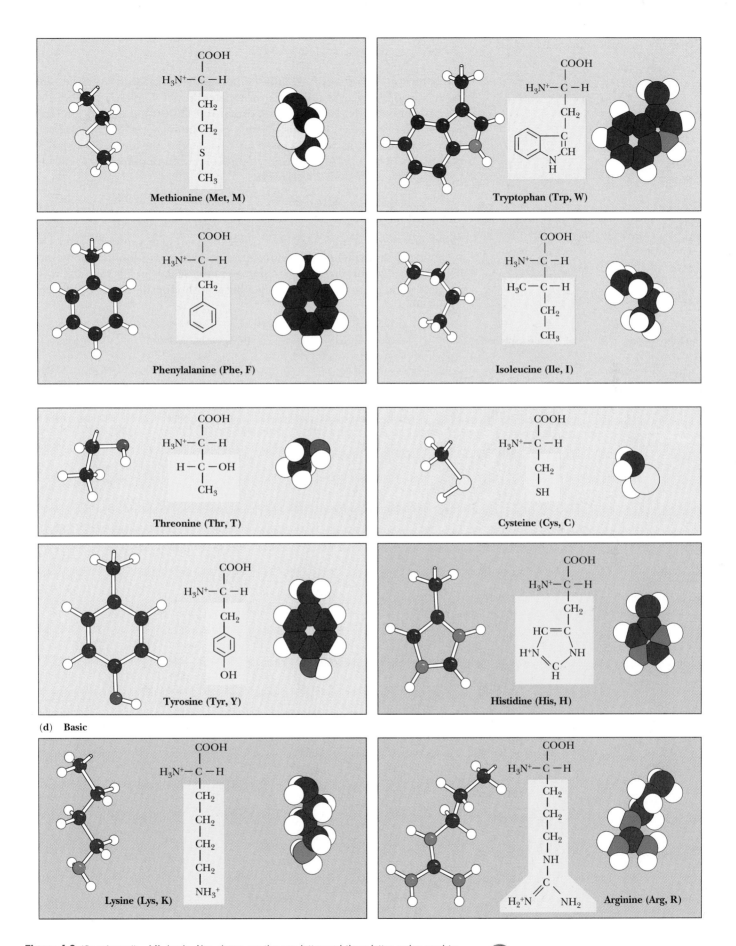

Figure 4.3 *(Continued)* **(d)** basic. Also shown are the one-letter and three-letter codes used to denote amino acids. For each amino acid, the ball-and-stick (*left*) and space-filling (*right*) models show only the side chain. *(Irving Geis)*

See pages 4–5

the latter two amino acids are fully protonated at pH 7, but histidine, with a side-chain pK_a of 6.0, is only 10% protonated at pH 7. With a pK_a near neutrality, histidine side chains play important roles as proton donors and acceptors in many enzyme reactions. Histidine-containing peptides are important biological buffers, as was discussed in Chapter 2. Arginine and lysine side chains, which are protonated under physiological conditions, participate in electrostatic interactions in proteins.

Other Amino Acids

There are several amino acids that occur only rarely in proteins (Figure 4.4), including **hydroxylysine** and **hydroxyproline,** which are found mainly in the collagen and gelatin proteins, and **pyroglutamic acid,** which is found in a light-driven proton-pumping protein called bacteriorhodopsin, discussed elsewhere in this book.

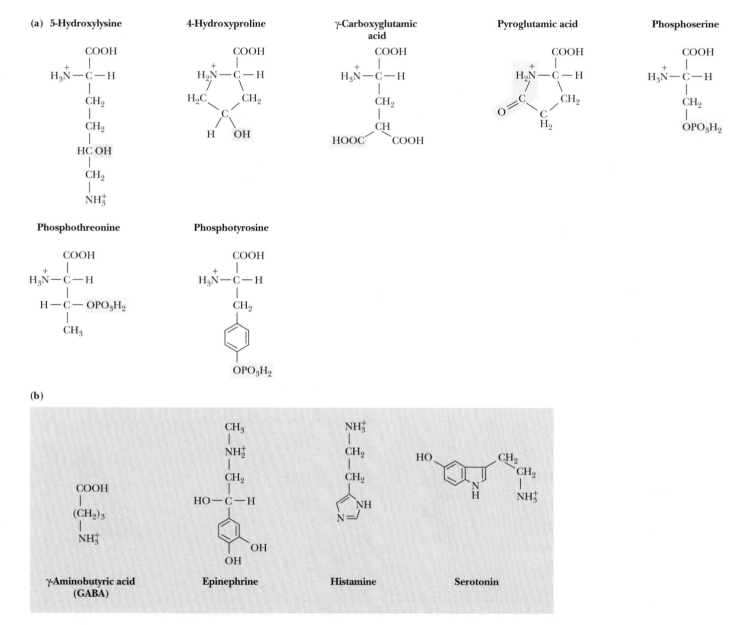

Figure 4.4 The structures of **(a)** amino acids that are less common but are found in certain proteins, and **(b)** an amino acid (GABA) and three amino acid derivatives that all act as neurotransmitters in neural tissue.

Amino Acids Are Weak Polyprotic Acids

From a chemical point of view, the common amino acids are all weak polyprotic acids. The ionizable groups are not strongly dissociating ones, and the degree of dissociation thus depends on the pH of the medium. All the amino acids contain at least two dissociable hydrogens.

Consider the acid–base behavior of glycine, the simplest amino acid. At low pH, both the amino and carboxyl groups are protonated and the molecule has a net positive charge. If the counterion in solution is a chloride ion, this form is referred to as **glycine hydrochloride.** If the pH is increased, the carboxyl group is the first to dissociate, yielding a neutral zwitterionic species (Figure 4.5). Further increase in pH eventually results in dissociation of the amino group to yield the negatively charged **glycinate.** If we denote these three forms as Gly^+, Gly^0, and Gly^-, we can write the first dissociation of Gly^+ as

$$Gly^+ + H_2O \rightleftharpoons Gly^0 + H_3O^+$$

and the dissociation constant K_1 as

$$K_1 = \frac{[Gly^0][H_3O^+]}{[Gly^+]}$$

Values for K_1 for the common amino acids are typically 0.4 to 1.0×10^{-2} M, so that typical values of pK_1 center around values of 2.0 to 2.4 (see Table 4.1). In a similar manner, we can write the second dissociation reaction as

$$Gly^0 + H_2O \rightleftharpoons Gly^- + H_3O^+$$

and the dissociation constant K_2 as

$$K_2 = \frac{[Gly^-][H_3O^+]}{[Gly^0]}$$

Typical values for pK_2 are in the range of 9.0 to 9.8. At physiological pH, the α-carboxyl group of a simple amino acid (with no ionizable side chains) is completely dissociated, whereas the α-amino group has not really begun its dissociation. The titration curve for such an amino acid is shown in Figure 4.6.

Note that the dissociation constants of both the α-carboxyl and α-amino groups are affected by the presence of the other group. The adjacent α-amino group makes the α-COOH group more acidic (that is, it lowers the pK_a) so that it gives up a proton more readily than simple alkyl carboxylic acids. Thus, the pK_1 of 2.0 to 2.1 for α-carboxyl groups of amino acids is substantially lower than that of acetic acid ($pK_a = 4.76$), for example. What is the chemical basis for the low pK_a of the α-COOH group of amino acids? The α-NH$_3^+$ (ammonium) group is strongly electron-withdrawing, and the positive charge of the amino group exerts a strong field effect and stabilizes the carboxylate anion.

HUMAN BIOCHEMISTRY

Some Amino Acids and Derivatives Are Neurotransmitters and Hormones

Amino acids in general show no particular biological activity. On the other hand, certain amino acids (and their derivatives) are potent neuroactive agents, acting either as neurotransmitters or as hormones. Notable examples (see Figure 4.4) include glutamate, aspartate, γ-**aminobutyric acid** (or **GABA,** which is produced by the decarboxylation of glutamic acid), as well as **histamine, serotonin,** and **epinephrine,** derived from histidine, tryptophan, and tyrosine, respectively.

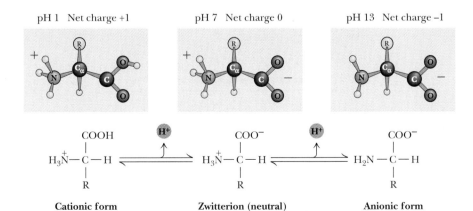

pH 1   Net charge +1   pH 7   Net charge 0   pH 13   Net charge −1

Cationic form   **Zwitterion (neutral)**   **Anionic form**

Figure 4.5 The ionic forms of the amino acids, shown without consideration of any ionizations on the side chain. The cationic form is the low pH form, and the titration of the cationic species with base will yield the zwitterion and finally the anionic form. *(Irving Geis)*

Table 4.1 pKₐ Values of Common Amino Acids

Amino Acid	α-COOH pK_a	α-NH$_3^+$ pK_a	R group pK_a
Alanine	2.4	9.7	
Arginine	2.2	9.0	12.5
Asparagine	2.0	8.8	
Aspartic acid	2.1	9.8	3.9
Cysteine	1.7	10.8	8.3
Glutamic acid	2.2	9.7	4.3
Glutamine	2.2	9.1	
Glycine	2.3	9.6	
Histidine	1.8	9.2	6.0
Isoleucine	2.4	9.7	
Leucine	2.4	9.6	
Lysine	2.2	9.0	10.5
Methionine	2.3	9.2	
Phenylalanine	1.8	9.1	
Proline	2.1	10.6	
Serine	2.2	9.2	~13
Threonine	2.6	10.4	~13
Tryptophan	2.4	9.4	
Tyrosine	2.2	9.1	10.1
Valine	2.3	9.6	

Ionization of Side Chains

As we have seen, the side chains of several of the amino acids also contain dissociable groups. Thus, aspartic and glutamic acids contain an additional carboxyl function, while lysine possesses an aliphatic amino function. Histidine contains an ionizable imidazolium proton, while arginine carries a guanidinium function. Typical values for the pK_a values of these groups are shown in Table 4.1. The β-carboxyl group of aspartic acid and the γ-carboxyl side chain of glutamic acid exhibit pK_a values intermediate to the α-COOH on the one hand and typical of aliphatic carboxyl groups on the other hand. In a similar fash-

Figure 4.6 Titration of glycine, a simple amino acid.

See pages 6–16

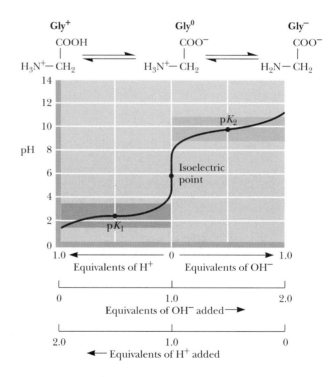

Figure 4.7 Titrations of glutamic acid and lysine.

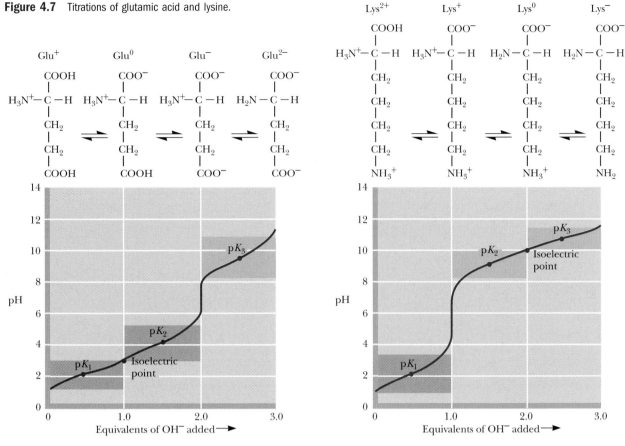

See pages 6–16

ion, the ε-amino group of lysine exhibits a pK_a that is higher than the α-amino group but similar to that for a typical aliphatic amino group. These intermediate values for side-chain pK_a values reflect the slightly diminished effect of the α-carbon dissociable groups that lie several carbons removed from the side-chain functional groups. Figure 4.7 shows typical titration curves for glutamic acid and lysine, along with the ionic species that predominate at various points in the titration. The only other side-chain groups that exhibit any significant degree of dissociation are the *para*-OH group of tyrosine and the —SH group of cysteine. The pK_a of the cysteine sulfhydryl is 8.3, so that it is approximately 8% to 10% dissociated at pH 7. The tyrosine *para*-OH group is a very weakly acidic group, with a pK_a of about 10.1.

4.3 Reactions of Amino Acids

The amino groups and the carboxyl groups of amino acids undergo all the simple reactions common to such functional groups. The side chains of amino acids exhibit specific chemical reactivities, depending on the nature of the functional groups, and it is the characteristic behavior of the side chain that governs the reactivity of amino acids incorporated into proteins. The biological functions of proteins depend in turn on the behavior and reactivity of specific R groups.

In recent years, biochemists have developed an arsenal of reactions that are relatively specific to the side chains of particular amino acids. These reactions can be used to identify functional amino acids at the active sites of enzymes or to label proteins with appropriate reagents for further study. Cysteine residues in proteins, for example, react with one another to form disulfide species and also react with iodoacetic acid to yield S-carboxymethyl cysteine derivatives, as shown in Figure 4.8. The side-chain amino group of lysine reacts with aldehydes and ketones to form Schiff bases, whereas the hydroxyl groups of serine, threonine, and tyrosine can all be phosphorylated.

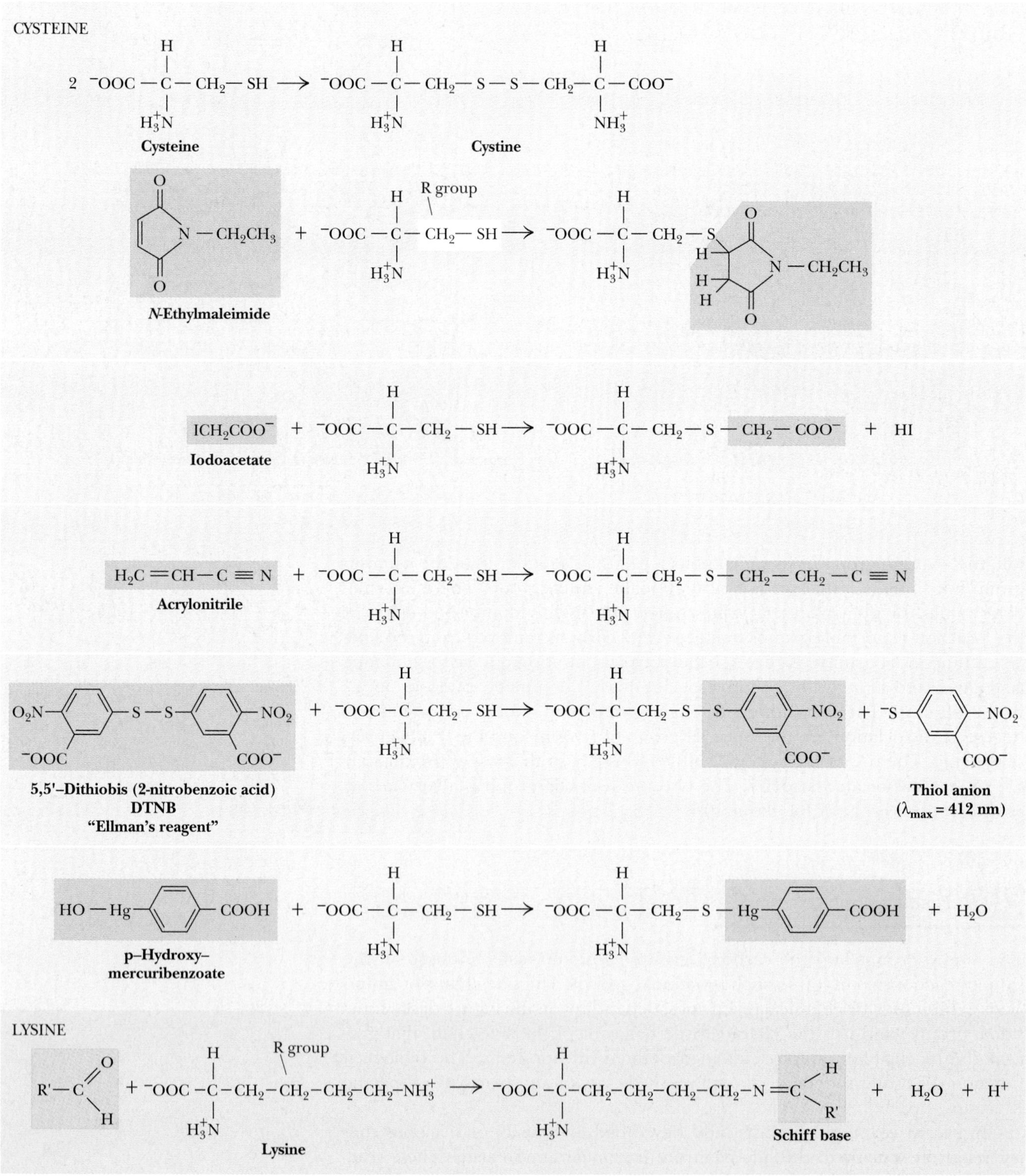

Figure 4.8 Some reactions of amino acid side-chain functional groups.

CRITICAL DEVELOPMENTS IN BIOCHEMISTRY

Green Fluorescent Protein—The "Light Fantastic" from Jellyfish to Gene Expression

Aquorea victoria, a species of jellyfish found in the northwest Pacific Ocean, contains a **green fluorescent protein (GFP)** that works together with another protein, aequorin, to provide a defense mechanism for the jellyfish. When the jellyfish is disturbed physically, aequorin produces a blue light. This light energy is captured by GFP, which then emits a bright green flash that presumably blinds or startles the attacker. Remarkably, the fluorescence of GFP occurs without the assistance of a **prosthetic group**—a "helper molecule" that would mediate GFP's fluorescence. Instead, the light-transducing capability of GFP is the result of a reaction between three amino acids in the protein itself. As shown below, adjacent **serine, tyrosine,** and **glycine** in the sequence of the protein react to form the pigment complex—termed a **chromophore.** No enzymes are required; the reaction is autocatalytic.

Because the light-transducing talents of GFP depend only on the protein itself (see upper photo, chromophore highlighted), GFP has quickly become a darling of genetic engineering laboratories. The promoter of any gene whose cellular expression is of interest can be fused to the DNA sequence coding for GFP. Telltale green fluorescence tells the researcher when this fused gene has been expressed (see lower photo and also Chapter 9).

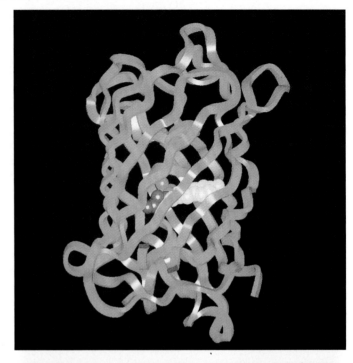

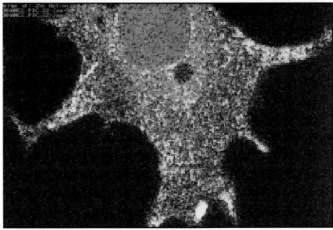

Phe-**Ser-Tyr-Gly**-Val-Gln $\xrightarrow{O_2}$
64   69

Autocatalytic oxidation of GFP amino acids leads to the chromophore shown above. The green fluorescence requires further interactions of the chromophore with other parts of the protein.

  See page 44

Boxer, S.G., 1997. Another green revolution. *Nature* **383:**484–485.

4.4 Optical Activity and Stereochemistry of Amino Acids

Amino Acids Are Chiral Molecules

Except for glycine, all of the amino acids isolated from proteins have four different groups attached to the α-carbon atom. In such a case, the α-carbon is said to be **asymmetric** or **chiral** (from the Greek *cheir,* meaning "hand"), and the two possible configurations for the α-carbon constitute nonsuperimposable mirror image isomers, or **enantiomers** (Figure 4.9). Enantiomeric molecules

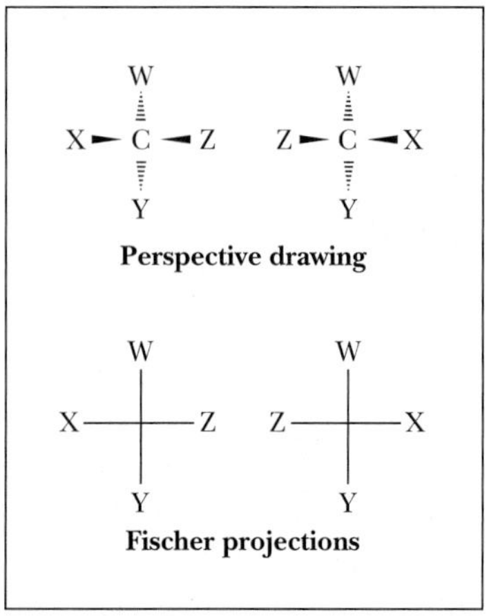

Figure 4.9 Enantiomeric molecules based on a chiral carbon atom. Enantiomers are nonsuperimposable mirror images of each other.

display a special property called **optical activity**—the ability to rotate the plane of polarization of plane-polarized light. Clockwise rotation of incident light is referred to as **dextrorotatory** behavior, and counterclockwise rotation is called **levorotatory** behavior. The magnitude and direction of the optical rotation depend on the nature of the amino acid side chain. Some protein-derived amino acids at a given pH are dextrorotatory and others are levorotatory, even though all of them are of the L-configuration. The direction of optical rotation can be specified in the name by using a (+) for dextrorotatory compounds and a (−) for levorotatory compounds, as in L(+)-leucine.

Nomenclature for Chiral Molecules

The discoveries of optical activity and enantiomeric structures made it important to develop suitable nomenclature for chiral molecules. Two systems are in common use today: the so-called D,L system and the (R,S) system.

In the **D,L-system** of nomenclature, the (+) and (−) isomers of glyceraldehyde are denoted as **D-glyceraldehyde** and **L-glyceraldehyde,** respectively (Figure 4.10). Absolute configurations of all other carbon-based molecules are referenced to D- and L-glyceraldehyde. When sufficient care is taken to avoid racemization of the amino acids during hydrolysis of proteins, it is found that all of the amino acids derived from natural proteins are of the L-configuration. Amino acids of the D-configuration are nonetheless found in nature, especially as components of certain peptide antibiotics, such as valinomycin, gramicidin, and actinomycin D, and in the cell walls of certain microorganisms.

CRITICAL DEVELOPMENTS IN BIOCHEMISTRY

Rules for Description of Chiral Centers in the (R,S) System

Naming a chiral center in the (R,S) system is accomplished by viewing the molecule from the chiral center to the atom with the lowest priority. If the other three atoms facing the viewer then decrease in priority in a clockwise direction, the center is said to have the (R) configuration (where R is from the Latin *rectus,* meaning "right"). If the three atoms in question decrease in priority in a counterclockwise fashion, the chiral center is of the (S) configuration (where S is from the Latin *sinistrus,* meaning "left"). If two of the atoms coordinated to a chiral center are identical, the atoms

bound to these two are considered for priorities. For such purposes, the priorities of certain functional groups found in amino acids and related molecules are in the following order:

$$SH > OH > NH_2 > COOH > CHO > CH_2OH > CH_3$$

From this, it is clear that D-glyceraldehyde is (R)-glyceraldehyde, and L-alanine is (S)-alanine (see figure). Interestingly, the α-carbon configuration of all the L-amino acids *except for cysteine* is (S). Cysteine, by virtue of its thiol group, is in fact (R)-cysteine.

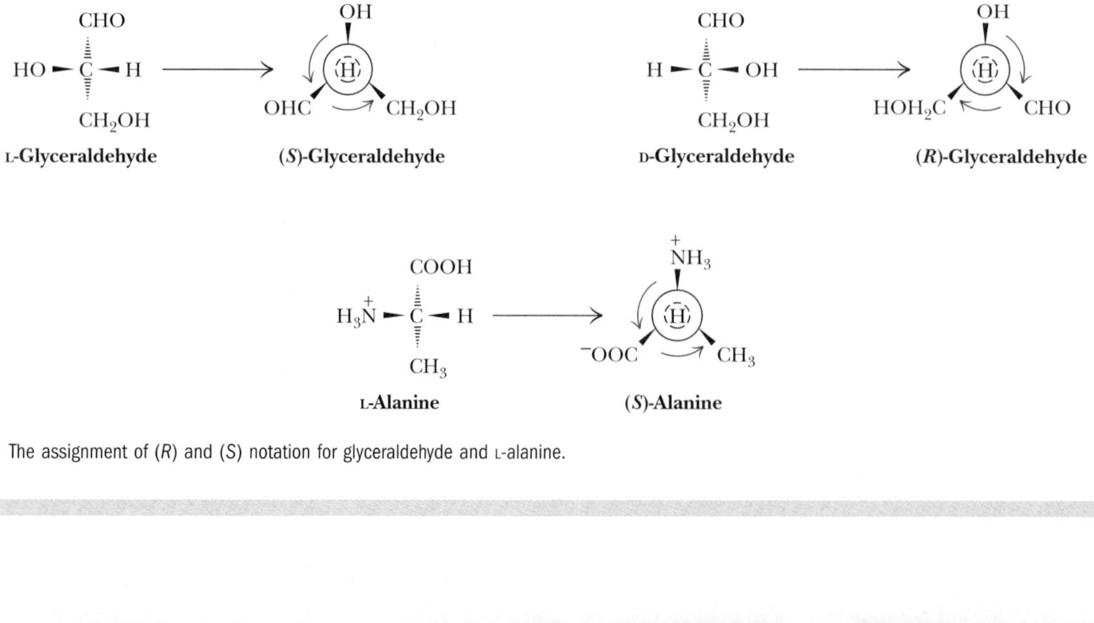

The assignment of (R) and (S) notation for glyceraldehyde and L-alanine.

A DEEPER LOOK

The Murchison Meteorite—Discovery of Extraterrestrial Handedness

The predominance of L-amino acids in biological systems is one of life's most intriguing features. Prebiotic syntheses of amino acids would be expected to produce equal amounts of L- and D-enantiomers. Some kind of enantiomeric selection process must have intervened to select L-amino acids over their D-counterparts as the constituents of proteins. Was it random chance that chose L- over D-isomers?

Analysis of carbon compounds—even amino acids—from extraterrestrial sources might provide deeper insights into this mystery. John Cronin and Sandra Pizzarello have examined the enantiomeric distribution of unusual amino acids obtained from the Murchison meteorite, which struck the earth on September 28, 1969, near Murchison, Australia. (By selecting unusual amino acids

for their studies, Cronin and Pizzarello ensured that they were examining materials that were native to the meteorite and not earth-derived contaminants.) Four α-dialkyl amino acids—α-methylisoleucine, α-methylalloisoleucine, α-methylnorvaline, and isovaline—were found to have an L-enantiomeric excess of 2% to 9%.

This may be the first demonstration that a natural L-enantiomer enrichment occurs in certain cosmological environments. Could these observations be relevant to the emergence of L-enantiomers as the dominant amino acids on the earth? And, if so, could there be life elsewhere in the universe that is based upon the same amino acid handedness?

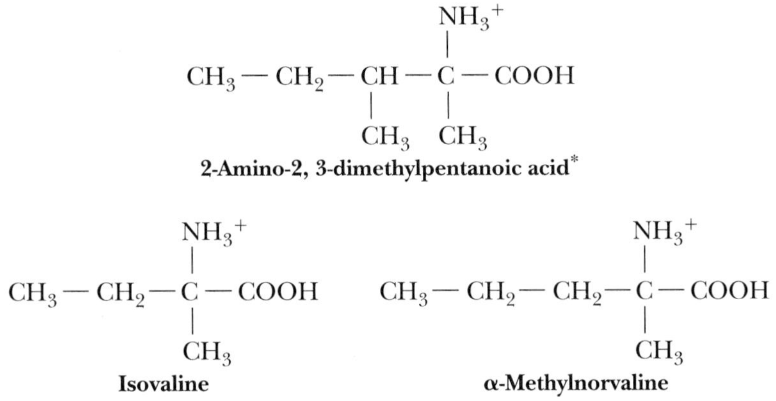

$$CH_3 - CH_2 - CH - \overset{\overset{\displaystyle NH_3^+}{|}}{C} - COOH$$
$$\underset{CH_3 \quad CH_3}{}$$

2-Amino-2, 3-dimethylpentanoic acid*

$$CH_3 - CH_2 - \overset{\overset{\displaystyle NH_3^+}{|}}{C} - COOH$$
$$\underset{CH_3}{}$$
Isovaline

$$CH_3 - CH_2 - CH_2 - \overset{\overset{\displaystyle NH_3^+}{|}}{C} - COOH$$
$$\underset{CH_3}{}$$
α-Methylnorvaline

Amino acids found in the Murchison meteorite.

*The four stereoisomers of this amino acid include the D- and L-forms of α-methylisoleucine and α-methylalloisoleucine.
Cronin, J. R., and Pizzarello, S., 1997. Enantiomeric excesses in meteoritic amino acids. *Science* **275**:951–955.

In spite of its widespread acceptance, there are problems with the D,L system of nomenclature. For example, this system can be ambiguous for molecules with two or more chiral centers. To address such problems, the (**R,S**) **system** of nomenclature for chiral molecules was proposed in 1956 by Robert Cahn, Sir Christopher Ingold, and Vladimir Prelog. In this more versatile system, priorities are assigned to each of the groups attached to a chiral center on the basis of atomic number, atoms with higher atomic numbers having higher priorities (see *Critical Developments in Biochemistry*, page 80).

The newer (R,S) system of nomenclature is superior to the older D,L system in one important way. The configuration of molecules with more than one chiral center can be more easily, completely, and unambiguously described with (R,S) notation. There are several amino acids, including isoleucine, threonine, hydroxyproline, and hydroxylysine, which have two chiral centers. In the (R,S) system, L-threonine is (2S,3R)-threonine.

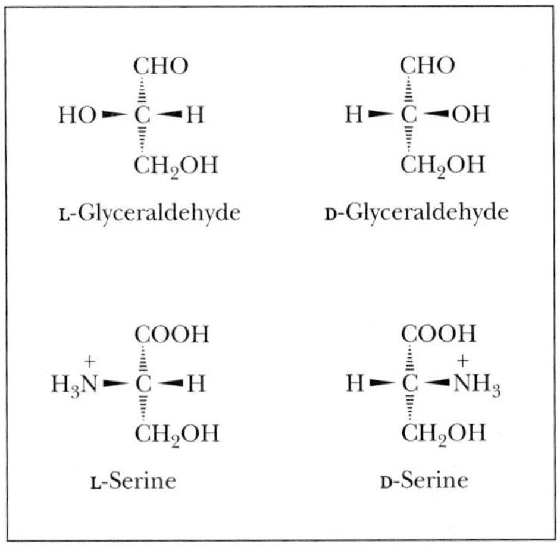

Figure 4.10 The configuration of the common L-amino acids can be related to the configuration of L(−)-glyceraldehyde as shown. These drawings are known as Fischer projections. The horizontal lines of the Fischer projections are meant to indicate bonds coming out of the page from the central carbon, while vertical lines represent bonds extending behind the page from the central carbon atom.

CHO
HO — C — H
CH₂OH
L-Glyceraldehyde

CHO
H — C — OH
CH₂OH
D-Glyceraldehyde

COOH
H₃N⁺ — C — H
CH₂OH
L-Serine

COOH
H — C — NH₃⁺
CH₂OH
D-Serine

Chromatographic Methods

A wide variety of methods is available for the separation and analysis of amino acids (and other biological molecules and macromolecules). All of these methods take advantage of the relative differences in the physical and chemical characteristics of amino acids, particularly ionization behavior and solubility characteristics. Separations of amino acids are usually based on **partition** properties (the tendency to associate with one solvent or phase over another) or on **electrical charge.** In all of the partition methods discussed here, the molecules of interest are allowed (or forced) to flow through a medium consisting of two phases: solid—liquid, liquid—liquid, or gas—liquid. The molecules partition, or distribute themselves, between the two phases in a manner based on their particular properties and their consequent preference for associating with one or the other phase.

In 1903, a separation technique based on repeated partitioning between phases was developed by Mikhail Tswett for the separation of plant pigments (carotenes and chlorophylls). Due to the colorful nature of the pigments thus separated, Tswett called his technique **chromatography.** This term is now applied to a wide variety of separation methods, regardless of whether the products are colored or not. The success of all chromatography techniques depends on the repeated microscopic partitioning of a solute mixture between the available phases. The more frequently this partitioning can be made to occur within a given time span or over a given volume, the more efficient is the resulting separation. Chromatographic methods have advanced rapidly in recent years, due in part to the development of sophisticated new solid-phase materials. Methods important for amino acid separations include ion exchange chromatography, gas chromatography (GC), and high-performance liquid chromatography (HPLC).

Ion Exchange Chromatography

The separation of amino acids and other solutes is often achieved by means of **ion exchange chromatography,** in which the molecule of interest is *exchanged* for another ion on and off of a charged solid support. In a typical procedure, solutes in a liquid phase, usually water, are passed through columns filled with a porous solid phase, usually a bed of synthetic resin particles, containing charged groups. Resins containing positive charges attract negatively charged solutes and are referred to as *anion exchangers.* Solid supports possessing negative charges attract positively charged species and are referred to as *cation exchangers.* The bare charges on such solid phases must be counterbalanced by oppositely charged ions in solution ("counterions"). Washing a cation exchange resin, such as Dowex-50, which has strongly acidic phenyl-SO_3^- groups, with a NaCl solution results in the formation of the so-called sodium form of the resin (see Figure 4.11). When the mixture whose separation is desired is added to the column, the positively charged solute molecules displace the Na^+ ions and bind to the resin. A gradient of an appropriate salt is then applied to the column, and the solute molecules are competitively (and sequentially) displaced (eluted) from the column by the rising concentration of cations in the gradient, in an order that is inversely proportional to their affinities for the column. The separation of mixtures of amino acids on such columns is shown in Figures 4.12 and 4.13; the latter is taken from a classic 1958 paper by Stanford Moore, Darrel Spackman, and William Stein.

Cation exchange bead before adding sample

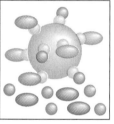

(a)

Add mixture of Asp, Ser, Lys

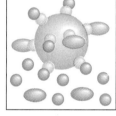

(b)

Add Na^+ (NaCl)

Increase $[Na^+]$

(c) Asp, the least positively charged amino acid, is eluted first

(d) Serine is eluted next

Increase $[Na^+]$

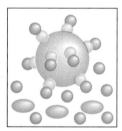

(e) Lysine, the most positively charged amino acid, is eluted last

Figure 4.11 Operation of a cation exchange column, separating a mixture of Asp, Ser, and Lys. **(a)** The cation exchange resin in the beginning, Na^+ form. **(b)** A mixture of Asp, Ser, and Lys is added to the column containing the resin. **(c)** A gradient of the eluting salt (e.g., NaCl) is added to the column. Asp, the least positively charged amino acid, is eluted first. **(d)** As the salt concentration increases, Ser is eluted. **(e)** As the salt concentration is increased further, Lys, the most positively charged of the three amino acids, is eluted last.

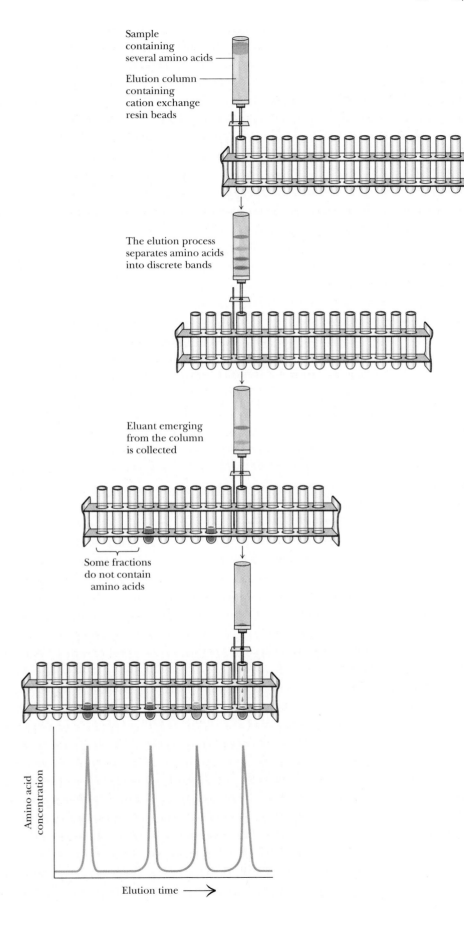

Sample containing several amino acids

Elution column containing cation exchange resin beads

The elution process separates amino acids into discrete bands

Eluant emerging from the column is collected

Some fractions do not contain amino acids

Amino acid concentration

Elution time

Figure 4.12 The separation of amino acids on a cation exchange column.

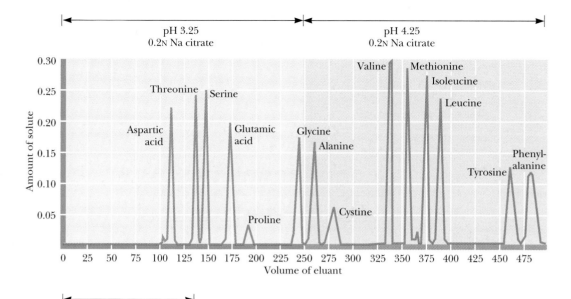

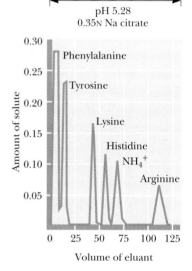

Figure 4.13 Chromatographic fractionation of a synthetic mixture of amino acids on ion exchange columns using Amberlite IR-120, a sulfonated polystyrene resin similar to Dowex-50. A typical separation of the common amino acids is shown. The events occurring in this separation are essentially those depicted in Figures 4.11 and 4.12. The amino acids are applied to the column at low pH (4.25), under which conditions the acidic amino acids (aspartate and glutamate, among others) are weakly bound and the basic amino acids, such as arginine and lysine, are tightly bound. Sodium citrate solutions, at two different concentrations and three different values of pH, are used to elute the amino acids gradually from the column. A second column with different buffer conditions is used to resolve the basic amino acids. *(Adapted from Moore, S., Spackman, D., and Stein, W., 1958. Chromatography of amino acids on sulfonated polystyrene resins.* Analytical Chemistry **30**:1185–1190.)

Figure 4.14 Gradient separation of common PTH-amino acids which absorb UV light. Absorbance was monitored at 269 nm. PTH peaks are identified by single-letter notation for amino acid residues and by other abbreviations. D, Asp; CMC, carboxymethyl Cys; E, Glu; N, Asn; S, Ser; Q, Gln; H, His; T, Thr; G, Gly; R, Arg; MO_2, Met sulfoxide; A, Ala; Y, Tyr; M, Met; V, Val; P, Pro; W, Trp; K, Lys; F, Phe; I, Ile; L, Leu. See Figure 4.22 for PTH derivatization. *(Adapted from Persson, B., and Eaker, D., 1990. An optimized procedure for the separation of amino acid phenylthiohydantoins by reversed-phase HPLC.* Journal of Biochemical and Biophysical Methods **21**:341–350.)

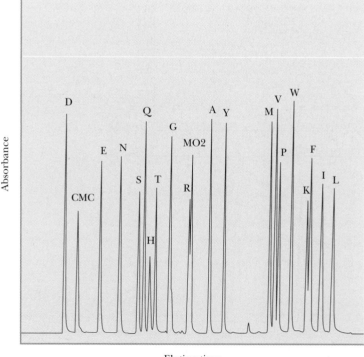

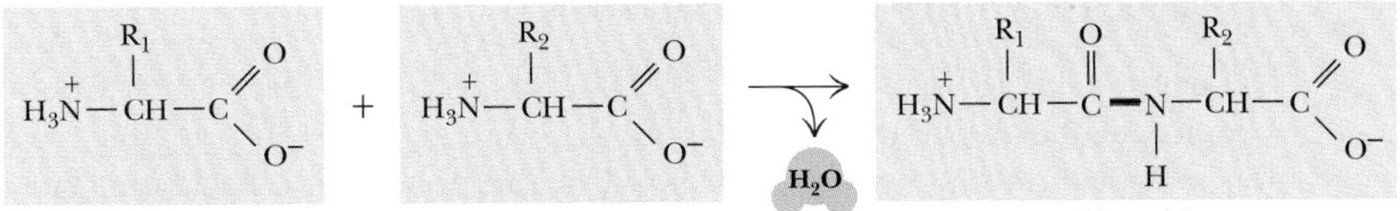

Figure 4.15 Peptide formation is the creation of an amide bond between the carboxyl group of one amino acid and the amino group of another amino acid. R_1 and R_2 represent the R groups of two different amino acids.

A typical HPLC chromatogram using precolumn modification of amino acids to form phenylthiohydantoin (PTH) derivatives is shown in Figure 4.14. HPLC is the chromatographic technique of choice for most modern biochemists. The very high resolution, excellent sensitivity, and high speed of this technique usually outweigh the disadvantage of relatively low capacity.

4.6 Proteins Are Linear Polymers of Amino Acids

Chemically, proteins are unbranched polymers of amino acids linked head to tail, from carboxyl group to amino group, through formation of covalent **peptide bonds,** a type of amide linkage (Figure 4.15).

Peptide bond formation results in the release of H_2O. The peptide "backbone" of a protein consists of the repeated sequence $-N-C_\alpha-C_o-$, where the N represents the amide nitrogen, the C_α is the α-carbon atom of an amino acid in the polymer chain, and the final C is the carbonyl carbon of the amino acid, which in turn is linked to the amide N of the next amino acid down the line. The geometry of the peptide backbone is shown in Figure 4.16. Note that the carbonyl oxygen and the amide hydrogen are *trans* to each other in this figure. This conformation is favored energetically since it results in less steric hindrance between nonbonded atoms in neighboring amino acids. Since the α-carbon atom of the amino acid is an asymmetric center (in all amino acids except glycine), the polypeptide chain is inherently asymmetric. Only L-amino acids are found in proteins.

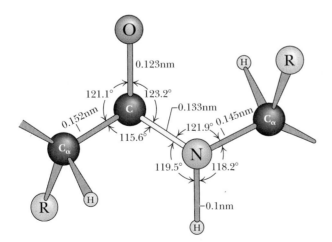

Figure 4.16 The peptide bond is shown in its usual *trans* conformation of carbonyl O and amide H. The C_α atoms are the α-carbons of two adjacent amino acids joined in peptide linkage. The dimensions and angles are the average values observed by crystallographic analysis of amino acids and small peptides. The peptide bond is the light gray bond between C and N. (*Adapted from Ramachandran, G. N., et al., 1974.* Biochimica Biophysica Acta **359**:298–302.)

The Peptide Bond Has Partial Double-Bond Character

The peptide linkage is usually portrayed by a single bond between the carbonyl carbon and the amide nitrogen (Figure 4.17a). Therefore, in principle, rotation may occur about any covalent bond in the polypeptide backbone, since all three kinds of bonds ($N-C_\alpha$, $C_\alpha-C_o$, and the C_o-N peptide bond) are single bonds. In this representation, the C and N atoms of the peptide grouping are both in planar sp^2 hybridization and the C and O atoms are linked by a π bond, leaving the nitrogen with a lone pair of electrons in a $2p$ orbital. However, another resonance form for the peptide bond is feasible in which the C and N atoms participate in a π bond, leaving a lone e^- pair on the oxygen (Figure 4.17b). This structure, because it has a double bond, clearly prevents free rotation about the C_o-N peptide bond. The real nature of the peptide bond lies somewhere between these extremes; that is, it has partial double-bond character, as represented by the intermediate form shown in Figure 4.17c.

(a)

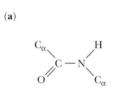

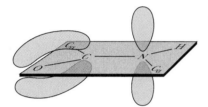

A pure double bond between C and O would permit free rotation around the C—N bond.

(b)

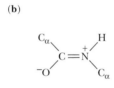

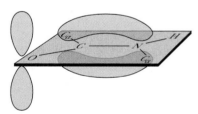

The other extreme would prohibit C—N bond rotation but would place too great a charge on O and N.

Figure 4.17 The partial double-bond character of the peptide bond. Resonance interactions among the carbon, oxygen, and nitrogen atoms of the peptide group can be represented by two resonance extremes (a and b). **(a)** The usual way the peptide atoms are drawn. **(b)** In an equally feasible form, the peptide bond is now a double bond; the amide N bears a positive charge and the carbonyl O has a negative charge. **(c)** The actual peptide bond is best described as a resonance hybrid of the forms in (a) and (b). Significantly, all of the atoms associated with the peptide group are coplanar, rotation about C_o-N is restricted, and the peptide is distinctly polar. *(Irving Geis)*

(c)

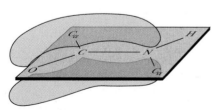

The true electron density is intermediate. The barrier to C—N bond rotation of about 88 kJ/mol is enough to keep the amide group planar.

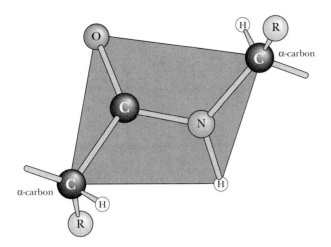

Figure 4.18 The coplanar relationship of the atoms in the amide group is highlighted as an imaginary shaded plane lying between two successive α-carbon atoms in the peptide backbone.

 See pages 17–19

Peptide bond resonance has several important consequences. First, it restricts free rotation around the peptide bond and leaves the peptide backbone with only two degrees of freedom per amino acid group: rotation around the $N—C_\alpha$ bond and rotation around the $C_\alpha—C_o$ bond. Second, the six atoms composing the peptide bond group tend to be coplanar, forming the so-called **amide plane** of the polypeptide backbone (Figure 4.18). Third, the $C_o—N$ bond length is 0.133 nm, which is shorter than normal C—N bond lengths (for example, the $C_\alpha—N$ bond of 0.145 nm) but longer than typical C=N bonds (0.125 nm). The peptide bond is estimated to have 40% double-bond character.

The Polypeptide Backbone Is Relatively Polar

Peptide bond resonance also causes the peptide backbone to be relatively polar. As shown in Figure 4.17b, the amide nitrogen represents a protonated or positively charged form, and the carbonyl oxygen becomes a negatively charged atom in the double-bonded resonance state. In actuality, the hybrid state of the partially double-bonded peptide arrangement gives a net positive charge of 0.28 on the amide N and an equivalent net negative charge of 0.28 on the carbonyl O. The presence of these partial charges means that the peptide bond has a permanent dipole. Nevertheless, the peptide backbone is relatively unreactive chemically, and protons are gained or lost by the peptide groups only at extreme pH conditions.

Peptide Classification

Peptide is the name assigned to short polymers of amino acids. Peptides are classified by the number of amino acid units in the chain. Each unit is called an **amino acid residue,** the word *residue* denoting what is left after the release of H_2O when an amino acid forms a peptide link upon joining the peptide chain. **Dipeptides** have two amino acid residues, tripeptides have three, tetrapeptides four, and so on. Since after about 12 residues this terminology becomes cumbersome, peptide chains of more than 12 and less than about 20 amino acid residues are referred to as **oligopeptides,** and, when the chain exceeds several dozen amino acids in length, the term **polypeptide** is used. The distinctions in this terminology are not precise.

Proteins Are Composed of One or More Polypeptide Chains

The terms *polypeptide* and *protein* are used interchangeably in discussing single polypeptide chains. The term **protein** broadly defines molecules composed of one or more polypeptide chains. Proteins having only one polypeptide chain are **monomeric proteins.** Proteins composed of more than one polypeptide chain are **multimeric proteins.** Multimeric proteins may contain only one kind of polypeptide, in which case they are **homomultimeric,** or they may be composed of several different kinds of polypeptide chains, in which instance they are **heteromultimeric.** Multimeric proteins are usually designated by Greek letters and subscripts to denote their polypeptide composition. Thus, an α_2-type protein is a dimer of identical polypeptide subunits, or a **homodimer.** Hemoglobin (Table 4.2) consists of four polypeptides of two different kinds; it is an $\alpha_2\beta_2$ heteromultimer.

Polypeptide chains typically range in length from about 100 amino acids to 1800, the number found in each of the two polypeptide chains of myosin, a contractile protein of muscle. However, titin, another muscle protein, has nearly 27,000 amino acid residues. The average molecular weight of polypeptide chains in eukaryotic cells is about 31,700, corresponding to about 270 amino acid residues. Table 4.2 is a representative list of proteins according to size. The molecular weights (M_r) of proteins can be estimated by a number of physical or chemical methods such as electrophoresis[1] or ultracentrifugation. Precise determinations of protein molecular masses are best obtained by simple calculations based on knowledge of their amino acid sequence. No simple generalizations correlate the size of proteins with their functions. For instance, the same function may be fulfilled in different cells by proteins of different molecular weight. The *Escherichia coli* enzyme responsible for glutamine synthesis (a protein known as *glutamine synthetase*) has a molecular weight of 600,000, whereas the analogous enzyme in brain tissue has a molecular weight of just 380,000.

Acid Hydrolysis of Proteins

Peptide bonds of proteins are hydrolyzed by either strong acid or strong base. Because acid hydrolysis proceeds without racemization and with less destruction of certain amino acids (Ser, Thr, Arg, and Cys) than alkaline treatment, it is the method of choice in analysis of the amino acid composition of proteins and polypeptides. Typically, samples of a protein are hydrolyzed with 6 N HCl at 110° C for 24, 48, and 72 hr in sealed glass vials. Tryptophan is destroyed by acid and must be estimated by other means to determine its contribution to the total amino acid composition. Another complication arises because the β- and γ-amide linkages in asparagine (Asn) and glutamine (Gln) are acid labile. The amide nitrogen is released as free ammonia, and all of the Asn and Gln residues of the protein become aspartic acid (Asp) and glutamic acid (Glu), respectively. The amount of ammonia released during acid hydrolysis gives an estimate of the total number of Asn and Gln residues in the original protein, but not the amounts of either.

Amino Acid Analysis of Proteins

The complex amino acid mixture in the hydrolysate obtained after digestion of a protein for one to three days in 6 M HCl can be resolved by chromatographic

---

[1]Electrophoresis is a method for separating molecules based on their mobility in an electric field. Generally, electrophoresis is carried out in a porous support matrix such as polyacrylamide or agarose, which retards the movement of molecules according to their dimensions relative to the size of the pores in the matrix. For molecules whose charge-to-size ratio is roughly constant (such as proteins complexed with the detergent SDS [sodium dodecyl sulfate]), the relative mobility will be directly proportional to the size of the molecule.

Table 4.2 Size of Protein Molecules*

Protein	M_r	Number of Residues per Chain		Subunit Organization
Insulin (bovine)	5,733	21	(A)	$\alpha\beta$
		30	(B)	
Cytochrome c (equine)	12,500	104		α_1
Ribonuclease A (bovine pancreas)	12,640	124		α_1
Lysozyme (egg white)	13,930	129		α_1
Myoglobin (horse)	16,980	153		α_1
Chymotrypsin (bovine pancreas)	22,600	13	(α)	$\alpha\beta\gamma$
		132	(β)	
		97	(γ)	
Hemoglobin (human)	64,500	141	(α)	$\alpha_2\beta_2$
		146	(β)	
Serum albumin (human)	68,500	550		α_1
Hexokinase (yeast)	96,000	200		α_4
γ-Globulin (horse)	149,900	214	(α)	$\alpha_2\beta_2$
		446	(β)	
Glutamate dehydrogenase (liver)	332,694	500		α_6
Myosin (rabbit)	470,000	1800	(heavy, h)	$h_2\alpha_1\alpha'_2\beta_2$
		190	(α)	
		149	(α')	
		160	(β)	
Ribulose bisphosphate carboxylase (spinach)	560,000	475	(α)	$\alpha_8\beta_8$
		123	(β)	
Glutamine synthetase (E. coli)	600,000	468		α_{12}

Insulin

Cytochrome c

Ribonuclease

Lysozyme

Myoglobin

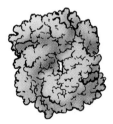

Hemoglobin

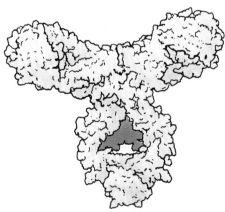

Immunoglobulin

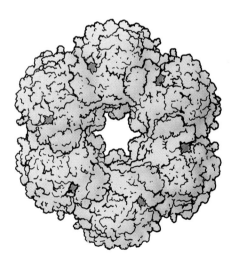

Glutamine synthetase

*Illustrations of selected proteins listed in Table 4.2 are drawn to constant scale. (*Adapted from Goodsell and Olson, 1993.* Trends in Biochemical Sciences **18**:65–68.)

■ **Table 4.3** **Amino Acid Composition of Some Selected Proteins**

Values expressed are percent representation of each amino acid.

Amino Acid	Proteins*				
	RNase	ADH	Mb	Histone H3	Collagen
Ala	6.9	7.5	9.8	13.3	11.7
Arg	3.7	3.2	1.7	13.3	4.9
Asn	7.6	2.1	2.0	0.7	1.0
Asp	4.1	4.5	5.0	3.0	3.0
Cys	6.7	3.7	0	1.5	0
Gln	6.5	2.1	3.5	5.9	2.6
Glu	4.2	5.6	8.7	5.2	4.5
Gly	3.7	10.2	9.0	5.2	32.7
His	3.7	1.9	7.0	1.5	0.3
Ile	3.1	6.4	5.1	5.2	0.8
Leu	1.7	6.7	11.6	8.9	2.1
Lys	7.7	8.0	13.0	9.6	3.6
Met	3.7	2.4	1.5	1.5	0.7
Phe	2.4	4.8	4.6	3.0	1.2
Pro	4.5	5.3	2.5	4.4	22.5
Ser	12.2	7.0	3.9	3.7	3.8
Thr	6.7	6.4	3.5	7.4	1.5
Trp	0	0.5	1.3	0	0
Tyr	4.0	1.1	1.3	2.2	0.5
Val	7.1	10.4	4.8	4.4	1.7
Acidic	8.4	10.2	13.7	8.1	7.5
Basic	15.0	13.1	21.8	24.4	8.8
Aromatic	6.4	6.4	7.2	5.2	1.7
Hydrophobic	18.0	30.7	27.6	23.0	6.5

*Proteins are as follows:

RNase: Bovine ribonuclease A, an enzyme; 124 amino acid residues. Note that RNase lacks tryptophan.
ADH: Horse liver alcohol dehydrogenase, an enzyme; dimer of identical 374 amino acid polypeptide chains. The amino acid composition of ADH is reasonably representative of the norm for water-soluble proteins.
Mb: Sperm whale myoglobin, an oxygen-binding protein; 153 amino acid residues. Note that Mb lacks cysteine.
Histone H3: Histones are DNA-binding proteins found in chromosomes: 135 amino acid residues. Note the very basic nature of this protein due to its abundance of Arg and Lys residues. It also lacks tryptophan.
Collagen: Collagen is an extracellular structural protein: 1052 amino acid residues. Collagen has an unusual amino acid composition: it is about one-third glycine and is rich in proline. Note that it also lacks Cys and Trp and is deficient in aromatic amino acid residues in general.

methods, which separate the component amino acids. The chromatographic separation and analysis is completely automated in instruments called **amino acid analyzers,** which can perform the determination on as little as 1 nmol of protein in less than 1 hr.

Table 4.3 gives the amino acid composition of several selected proteins: ribonuclease A, alcohol dehydrogenase, myoglobin, histone H3, and collagen. Each of the 20 naturally occurring amino acids is usually represented at least once in a polypeptide chain. However, some small proteins may not have a representative of every amino acid. Note that ribonuclease A (12.6 kD, 124 amino acid residues) does not contain any tryptophan. Amino acids virtually never occur in equimolar ratios in proteins, indicating that proteins are not composed of repeating arrays of amino acids. There are a few exceptions to this rule. Collagen, for example, contains large proportions of glycine and proline, and much of its structure is composed of (Gly-*x*-Pro) repeating units, where *x* is any amino acid. Other proteins show unusual abundances of various amino acids. For example, histones are rich in positively charged amino acids such as arginine and lysine. Histones are a class of proteins found associated with the anionic phosphate groups of eukaryotic DNA.

Amino acid analysis itself does not directly give the number of residues of each amino acid in a polypeptide, but it does give amounts from which the percentages or ratios of the various amino acids can be obtained (See Table 4.3). If the molecular weight *and* the exact amount of the protein analyzed are known (or the number of amino acid residues per molecule is known), the molar ratios of amino acids in the protein can be calculated. Amino acid analysis provides no information on the order or sequence of amino acid residues in the polypeptide chain. Since the polypeptide chain is unbranched, it has only two ends, an amino-terminal or **N-terminal end** and a carboxyl-terminal or **C-terminal end.**

Ultraviolet Absorption Can Quantitate Some Amino Acids

One of the most important and exciting advances in modern biochemistry has been the application of **spectroscopic methods,** which measure the absorption and emission of energy of different frequencies by molecules and atoms. Many details of the structure and chemistry of the amino acids and proteins have been elucidated or at least confirmed by spectroscopic measurements. None of the amino acids absorb light in the visible region of the electromagnetic spectrum. Several of the amino acids, however, do absorb **ultraviolet** radiation, and all absorb in the **infrared** region. Only the aromatic amino acids phenylalanine, tyrosine, and tryptophan exhibit significant ultraviolet absorption above 250 nm, as shown in Figure 4.19. These strong absorptions are the basis for spectroscopic determinations of protein concentration.

The Sequence of Amino Acids in Proteins

The unique characteristic of each protein is the distinctive sequence of amino acid residues in its polypeptide chain(s). Indeed, it is the **amino acid sequence** of proteins that is encoded by the nucleotide sequence of DNA. This amino acid sequence, then, is a form of genetic information. By convention, the amino acid sequence is read from the N-terminal end of the polypeptide chain through to the C-terminal end. As an example, every molecule of

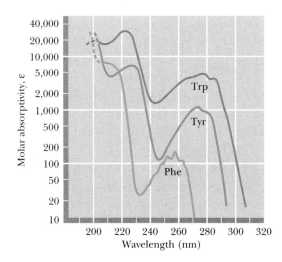

Figure 4.19 The ultraviolet absorption spectra of the aromatic amino acids at pH 6. *(From Wetlaufer, D. B., 1962. Ultraviolet spectra of proteins and amino acids. Advances in Protein Chemistry* **17:***303–390.)*

Figure 4.20 Bovine pancreatic ribonuclease A contains 124 amino acid residues, none of which are tryptophan. Four intrachain disulfide bridges (S—S) form cross-links in this polypeptide between Cys[26] and Cys[84], Cys[40] and Cys[95], Cys[58] and Cys[110], and Cys[65] and Cys[72]. These disulfides are depicted by yellow bars.

See page 54

ribonuclease A from bovine pancreas has the same amino acid sequence, beginning with N-terminal lysine at position 1 and ending with C-terminal valine at position 124 (Figure 4.20). Given the possibility of any of the 20 amino acids at each position, the number of unique amino acid sequences is astronomically large. The astounding sequence variation possible within polypeptide chains provides a key insight into the incredible functional diversity of protein molecules in biological systems, which is discussed shortly.

A DEEPER LOOK

The Virtually Limitless Number of Different Amino Acid Sequences

For a chain of n residues, there are 20^n possible sequence arrangements. To portray this, consider the number of tripeptides possible if there were only three different amino acids, A, B, and C (tripeptide = 3 = n; $n^3 = 3^3 = 27$):

AAA	BBB	CCC
AAB	BBA	CCA
AAC	BBC	CCB
ABA	BAB	CBC
ACA	BCB	CAC
ABC	BAA	CBA
ACB	BCC	CAB
ABB	BAC	CBB
ACC	BCA	CAA

For a polypeptide chain of 100 residues in length, a rather modest size, the number of possible sequences is 20^{100} or, since $20 = 10^{1.3}$, 10^{130} unique possibilities. These numbers are more than astronomical! Since an average protein molecule of 100 residues would have a mass of 13,800 daltons (average molecular mass of an amino acid residue = 138), 10^{130} such molecules would have a mass of 1.38×10^{134} daltons. The mass of the observable universe is estimated to be 10^{80} proton masses (about 10^{80} daltons). Thus, the universe lacks enough material to make just one molecule of each possible polypeptide sequence for a protein only 100 residues in length.

A DEEPER LOOK

Estimating Protein Concentrations by Dye Binding

Several protocols for protein estimation enjoy prevalent usage in biochemical laboratories. Perhaps the most common is the **Bradford assay,** a rapid and reliable technique that uses a dye called *Coomassie Brilliant Blue G-250.* This dye undergoes a change in its color upon noncovalent binding to proteins. The binding is quantitative and relatively unaffected by variations in the protein's amino acid composition. The color change is easily measured by a spectrophotometer.

Coomassie Brilliant Blue G-250

4.7 Reactions of Peptides and Proteins

The chemical properties of peptides and proteins are most easily considered in terms of the chemistry of their component functional groups. That is, they possess reactive amino and carboxyl termini and they display reactions characteristic of the chemistry of the R groups on their amino acids. These reactions are familiar to us from Section 4.3 and from the study of organic chemistry and need not be repeated here.

4.8 Purification of Protein Mixtures

Cells contain thousands of different proteins. A major problem for protein chemists is to purify a chosen protein so that they can study its specific properties in the absence of other proteins. Traditionally, proteins have been separated and purified on the basis of their two prominent physical properties: size and electrical charge.

Separation Methods

Separation methods based on size include size exclusion chromatography, ultrafiltration, and ultracentrifugation. The ionic properties of peptides and proteins are determined principally by their complement of amino acid side chains. Furthermore, the ionization of these groups is pH-dependent.

A variety of procedures exploit electrical charge as a means of discriminating between proteins, including ion exchange chromatography (see Section 4.5), electrophoresis, and solubility. Proteins tend to be least soluble at their **isoelectric point,** the pH value at which the sum of their positive and negative electrical charges is zero. At this pH, electrostatic repulsion between protein molecules is minimal and they are more likely to coalesce and precipitate out of solution. Ionic strength also profoundly influences protein solubility. Most globular proteins tend to be more soluble at intermediate (10 mM) ionic strength and less so at low (1 mM) or high (1 M) ionic strength.

Although the side chains of most nonpolar amino acids in proteins are usually buried in the interior of the protein away from contact with the aqueous solvent, a portion of them is exposed at the protein's surface, giving it a partially hydrophobic character. *Hydrophobic interaction chromatography* is a protein purification technique that exploits this hydrophobicity.

4.9 The Primary Structure of a Protein: Determining the Amino Acid Sequence

In 1953, Frederick Sanger of Cambridge University in England reported the amino acid sequences of the two polypeptide chains composing the protein insulin (Figure 4.21). Not only was this a remarkable achievement in analytical chemistry but it helped to demystify speculation about the chemical nature of proteins. Sanger's results clearly established that all of the molecules of a given protein have a fixed amino acid composition, a defined amino acid sequence, and therefore an invariant molecular weight. In short, proteins are well defined chemically. Today, the amino acid sequences of hundreds of thousands of different proteins are known. While many sequences have been determined from application of the principles first established by Sanger, most are now deduced from knowledge of the nucleotide sequence of the gene that encodes the protein.

Protein Sequencing Strategy

The usual strategy for determining the amino acid sequence of a protein involves a series of steps, including:

1. The N-terminal and C-terminal residues are identified.
2. Each polypeptide chain is cleaved into smaller fragments, and the amino acid composition and sequence of each fragment are determined.
3. Step 2 is repeated, using a different cleavage procedure to generate a different and therefore overlapping set of peptide fragments.
4. The overall amino acid sequence of the protein is reconstructed from the sequences in the overlapping fragments.

Each of these steps is discussed in greater detail in the following sections.

Step 1. Identification of the N- and C-Terminal Residues

End-group analysis reveals several things. First, it identifies the N- and C-terminal residues in the polypeptide chain. Second, it can be a clue to the number of ends in the protein. That is, if the protein consists of two or more different polypeptide chains, then more than one end group may be discovered, alerting the investigator to the presence of multiple polypeptides.

A. N-Terminal Analysis The amino acid residing at the N-terminal end of a protein can be identified in a number of ways, some of which are destructive to the protein sample. One method, Edman degradation, has become the procedure of choice because it allows the sequential identification of a series of residues beginning at the N-terminus.

Edman Degradation. In weakly basic solutions, phenylisothiocyanate, or **Edman's reagent** (phenyl-N=C=S), will combine with the free amino terminus of a protein (Figure 4.22), which can be excised from the end of the polypeptide chain and recovered as a phenylthiohydantoin (PTH) derivative. This PTH derivative can be identified by chromatographic methods.

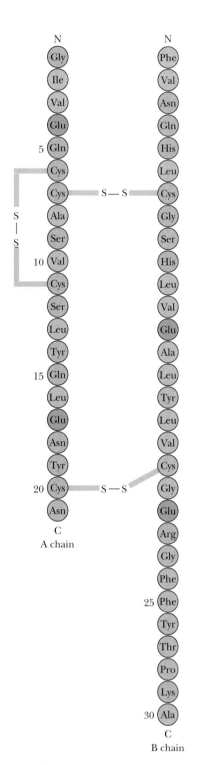

Figure 4.21 The hormone insulin consists of two polypeptide chains, A and B, held together by two disulfide cross-bridges (S—S). The A chain has 21 amino acid residues and an intrachain disulfide; the B polypeptide contains 30 amino acids. The sequence shown is for bovine insulin.

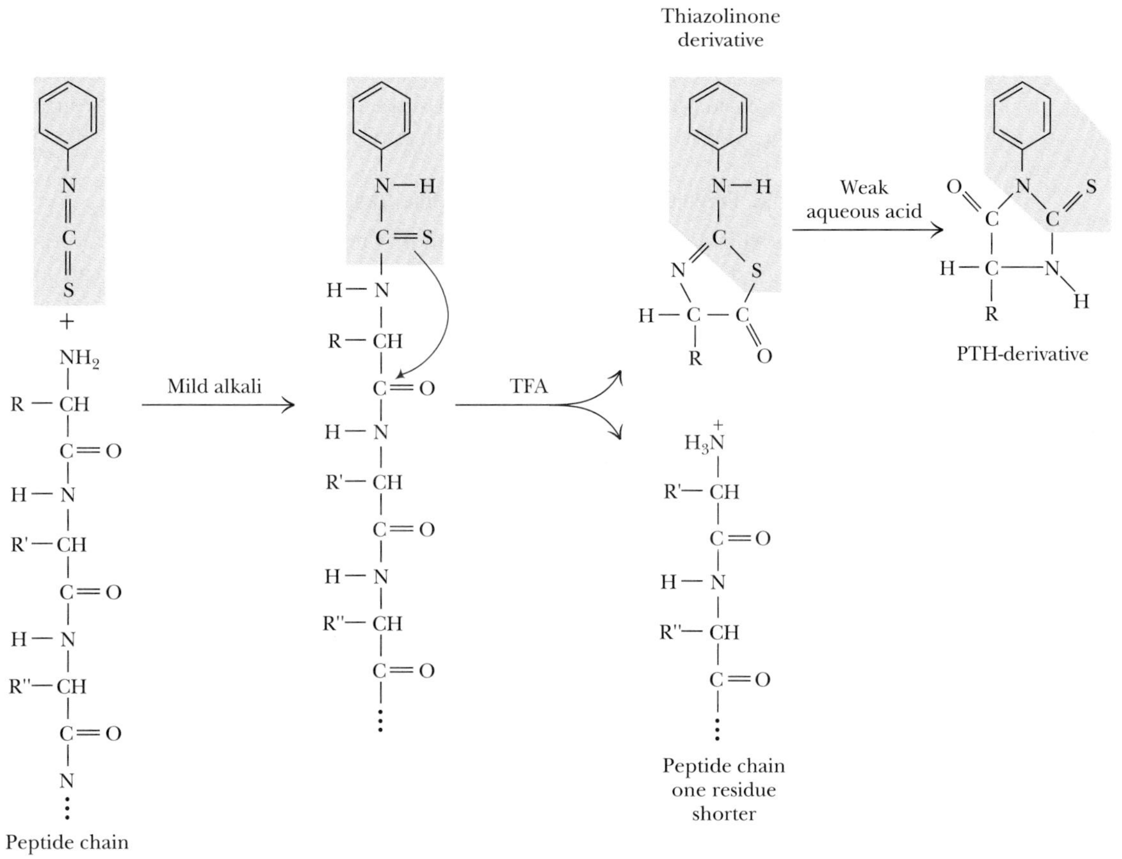

Figure 4.22 N-Terminal analysis using Edman's reagent, phenylisothiocyanate. Phenylisothiocyanate will combine with the N-terminus of a peptide under mildly alkaline conditions to form a phenylthiocarbamoyl substitution. Upon treatment with TFA (trifluoroacetic acid), this cyclizes to release the N-terminal amino acid residue, but the other peptide bonds are not hydrolyzed. Organic extraction and treatment with aqueous acid yields the amino acid as a phenylthiohydantoin (PTH) derivative.

Importantly, in this procedure, the rest of the polypeptide chain remains intact and can be subjected to further rounds of Edman degradation to identify successive amino acid residues in the chain. Often, the carboxyl terminus of the polypeptide under analysis is coupled to an insoluble matrix, allowing the polypeptide to be easily recovered by filtration following each round of Edman reaction. Thus, Edman reaction not only identifies the N-terminus of proteins but can also reveal additional information regarding sequence. Instruments automated to carry out the reaction cycles of the procedure, so-called Edman sequenators, can yield the sequence of the first 30 to 60 amino acid residues in a polypeptide under appropriate conditions.

B. C-Terminal Analysis For identification of the C-terminal residue of polypeptides, an enzymatic approach is commonly used.

Enzymatic Analysis with Carboxypeptidases. Carboxypeptidases are enzymes that cleave amino acid residues from the C-termini of polypeptides in a successive fashion. Four carboxypeptidases are in general use: A, B, C, and Y. *Carboxypeptidase A* (from bovine pancreas) works well in hydrolyzing the C-terminal peptide bond of all residues except proline, arginine, and lysine. The analogous enzyme from hog pancreas, *carboxypeptidase B*, is effective only when Arg or Lys are the C-terminal residues. Thus, a mixture of carboxypeptidases

A and B will liberate any C-terminal amino acid except proline. *Carboxypeptidase C* from citrus leaves and *carboxypeptidase Y* from yeast act on any C-terminal residue. Since the nature of the amino acid residue at the end often determines the rate at which it is cleaved and since these enzymes remove residues successively, care must be taken in interpreting results.

Steps 2 and 3. Fragmentation of the Polypeptide Chain

The aim at this step is to produce fragments amenable to sequence analysis. The cleavage methods employed are usually enzymatic, but proteins can also be fragmented by specific or nonspecific chemical means (such as partial acid hydrolysis). Proteolytic enzymes offer an advantage in that they may hydrolyze only specific peptide bonds, and this specificity immediately gives information about the peptide products. As a first approximation, fragments produced upon cleavage should be small enough to yield their sequences through end-group analysis and Edman degradation, yet not so small that an overabundance of products must be resolved before analysis.

A. Trypsin The digestive enzyme *trypsin* is the most commonly used reagent for specific proteolysis. Trypsin is specific in hydrolyzing only peptide bonds in which the carbonyl function is contributed by an arginine or a lysine residue. That is, trypsin cleaves on the C-side of Arg or Lys, generating a set of peptide fragments having Arg or Lys at their C-termini. The number of smaller peptides resulting from trypsin action is equal to the total number of Arg and Lys residues in the protein *plus* one—the protein's C-terminal peptide fragment (Figure 4.23).

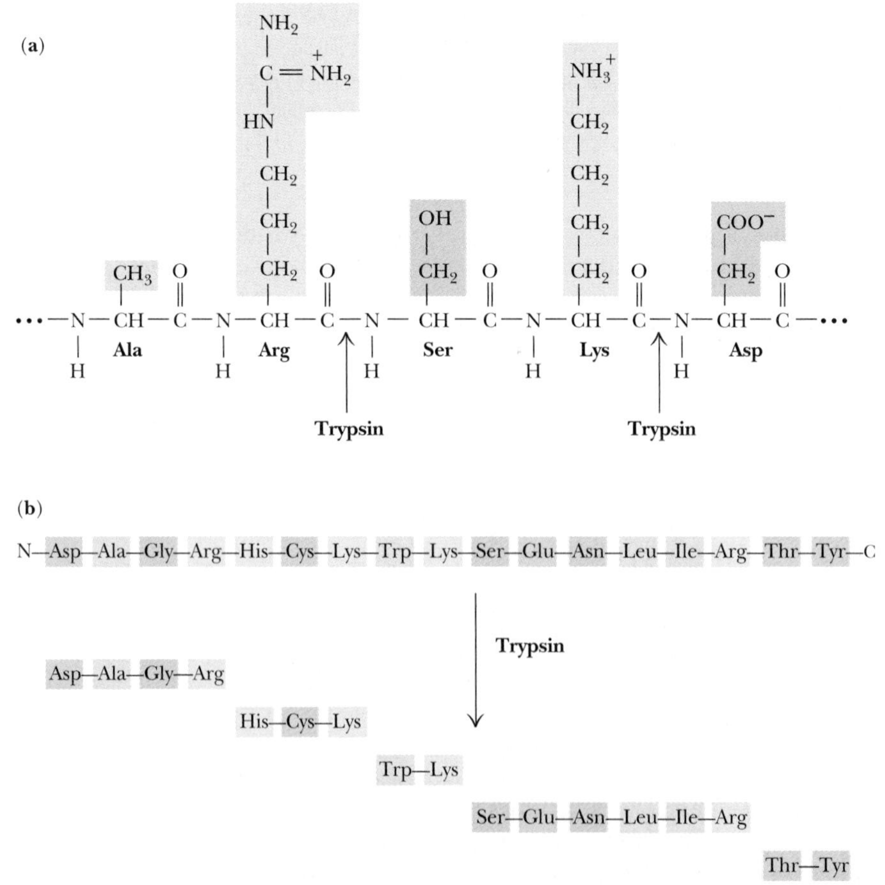

Figure 4.23 Trypsin is a proteolytic enzyme, or *protease*, that specifically cleaves only those peptide bonds in which arginine or lysine contributes the carbonyl function. The products of the reaction are a mixture of peptide fragments with C-terminal Arg or Lys residues *and* a single peptide derived from the polypeptide's C-terminal end.

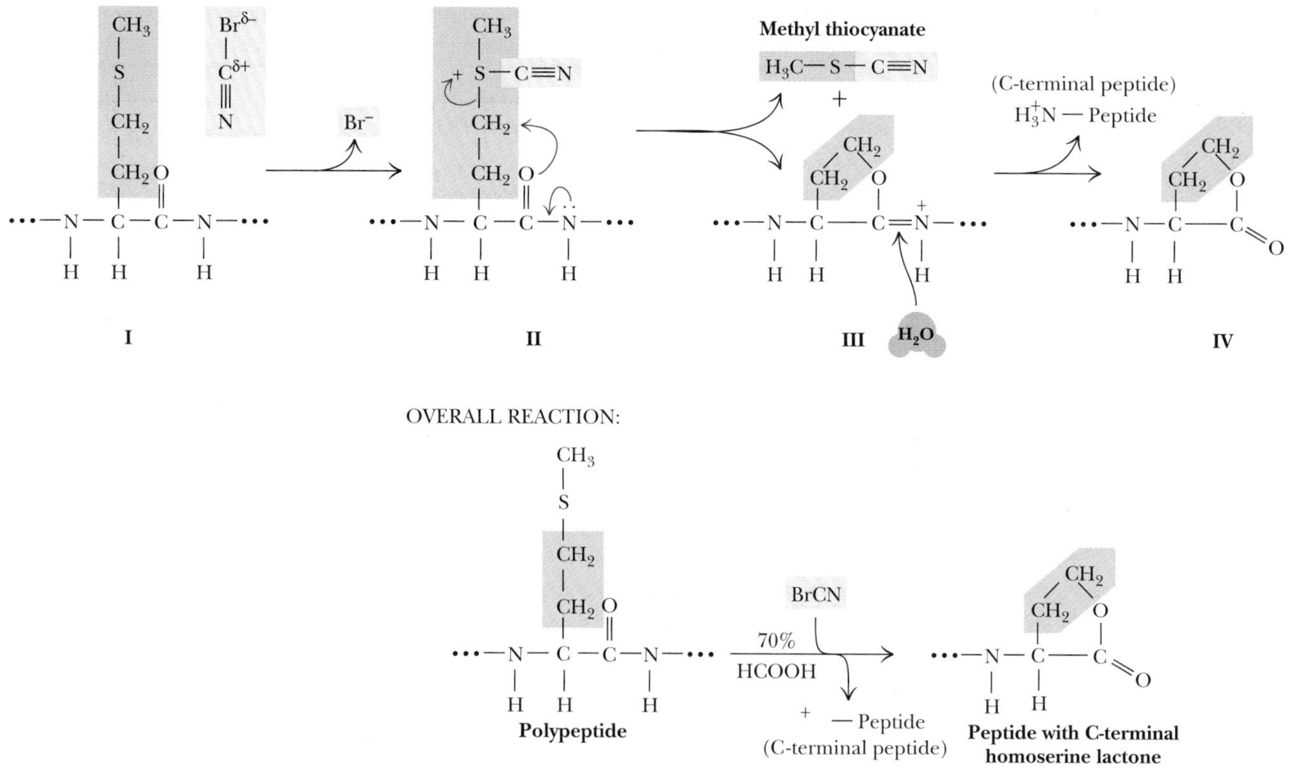

CH$_3$ | S | CH$_2$ | CH$_2$ O | $\cdots$—N—C—C—N—$\cdots$ | H H H

Br$^{\delta-}$ | C$^{\delta+}$ ||| N

Br$^-$

I

CH$_3$ | $^+$S—C≡N | CH$_2$ | CH$_2$ O | $\cdots$—N—C—C—N—$\cdots$ | H H H

II

Methyl thiocyanate

H$_3$C—S—C≡N

+

CH$_2$ | CH$_2$ O | $\cdots$—N—C—C≡N$^+$—$\cdots$ | H H H

H$_2$O

III

(C-terminal peptide)

H$_3^+$N—Peptide

CH$_2$ | CH$_2$ O | $\cdots$—N—C—C | H H O

IV

OVERALL REACTION:

CH$_3$ | S | CH$_2$ | CH$_2$ O | $\cdots$—N—C—C—N—$\cdots$ | H H H
Polypeptide

BrCN

70% HCOOH

+ —Peptide (C-terminal peptide)

CH$_2$ | CH$_2$ O | $\cdots$—N—C—C | H H O
Peptide with C-terminal homoserine lactone

Figure 4.24 Cyanogen bromide (CNBr) is a highly selective reagent for cleavage of peptides only at methionine residues. The reaction occurs in 70% formic acid via nucleophilic attack of the Met S atom on the —C≡N carbon atom, with displacement of Br$^-$ (I). The cyano intermediate undergoes nucleophilic attack by the Met carbonyl oxygen atom on the R group (II), resulting in formation of the cyclic derivative (III), which is unstable in aqueous solution. Hydrolysis ensues, producing cleavage of the Met peptide bond and release of peptide fragments with C-terminal homoserine lactone residues where Met residues once were (IV). One peptide will not have a C-terminal homoserine lactone: the original C-terminal end of the polypeptide.

B. Chymotrypsin *Chymotrypsin* shows a strong preference for hydrolyzing peptide bonds formed by the carboxyl groups of the aromatic amino acids, phenylalanine, tyrosine, and tryptophan. Since chymotrypsin produces a very different set of products than trypsin, treatment of samples of a protein with these two enzymes will generate fragments whose sequences overlap. Resolution of the order of amino acid residues in the fragments yields the amino acid sequence in the original protein.

C. Other Endopeptidases A number of other *endopeptidases* (proteases that cleave peptide bonds within the interior of a polypeptide chain) are occasionally used in sequence investigations. These include *clostripain*, which acts only at Arg residues, and *staphylococcal protease*, which acts at the acidic residues, Asp and Glu. Other enzymes which do not display a high degree of specificity are handy for digesting large tryptic or chymotryptic fragments. *Pepsin, papain, subtilisin,* and *elastase* are some examples. Papain is the active ingredient in meat tenderizer and in soft contact lens cleaner as well as in some laundry detergents. The abundance of papain in papaya, and a similar protease (bromelain) in pineapple, causes the hydrolysis of gelatin and prevents the preparation of Jell-O containing either of these fresh fruits. Cooking these fruits thermally denatures their proteolytic enzymes so that they can be used in gelatin desserts.

D. Cyanogen Bromide Several highly specific chemical methods of proteolysis are available, the most widely used being *cyanogen bromide (CNBr)* cleavage. CNBr acts upon methionine residues (Figure 4.24). Met reacts with CNBr, and undergoes a rapid intramolecular rearrangement to form a cyclic iminolactone. Water readily hydrolyzes this iminolactone, cleaving the polypeptide and generating peptide fragments having C-terminal homoserine lactone residues at the former Met positions. Table 4.4 summarizes the various procedures described here for polypeptide cleavage. These methods are only a partial list of

Table 4.4 Specificity of Representative Polypeptide Cleavage Procedures Used in Sequence Analysis

Method	Susceptible Residue(s)
Proteolytic enzymes	
Trypsin	Arg or Lys
Chymotrypsin	Phe, Trp, or Tyr; Leu
Clostripain	Arg
Staphylococcal protease	Asp or Glu
Chemical methods	
Cyanogen bromide	Met
NH_2OH	Asn-Gly bonds
pH 2.5, 40°C	Asp-Pro bonds

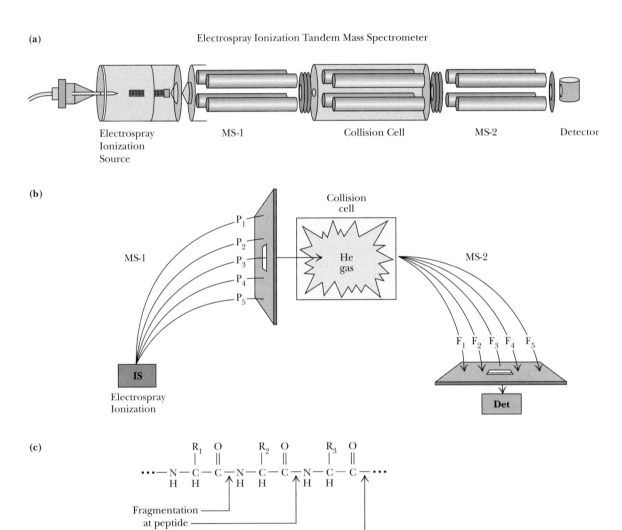

Figure 4.25 Tandem mass spectrometry. **(a)** Configuration used in tandem MS. **(b)** Schematic description of tandem MS: tandem MS involves electrospray ionization of a protein digest (IS in this figure), followed by selection of a single peptide ion mass for collision with inert gas molecules (He) and mass analysis of the fragment ions resulting from the collisions. **(c)** Fragmentation usually occurs at peptide bonds, as indicated. *(I: Adapted from Yates, J. R., 1996. Methods in Enzymology* **271**:*351– 376; II: Adapted from Gillece-Castro, B. L., and Stults, J. T., 1996. Methods in Enzymology* **271**:*427–447.)*

the arsenal of reactions available to protein chemists. Peptide sequencing today is most commonly done by Edman degradation of relatively large peptides.

Sequence Determination by Mass Spectrometric Methods

Mass spectrometry has come into vogue for the sequence determination of peptides of 15 or fewer residues. This technique has great advantages of sensitivity and accuracy. Protein is introduced into the mass spectrometer and fragmented in an electron beam to produce a complex set of cationic species, which are then identified solely by their mass to charge ratio. An enormous advantage of mass spectrometry is that the peptides need not be pure. Tandem arrangement of two or more mass spectrometers (*tandem mass spectrometry*) allows one instrument to be used to separate peptides, a second to fragment them by ion bombardment, and a third to analyze the masses of the fragments (Figure 4.25). In such devices, oligopeptides and proteins up to 13 kD in molecular weight can be analyzed directly, and only 1 μg of the protein is consumed in the analysis.

Step 4. Reconstruction of the Overall Amino Acid Sequence

The sequences obtained for the sets of fragments derived from two or more cleavage procedures are now compared, with the objective being to find overlaps that establish continuity of the overall amino acid sequence of the polypeptide chain. The strategy is illustrated by the example shown in Figure 4.26. Peptides generated from specific hydrolysis of the polypeptide can be aligned to reveal the overall amino acid sequence. Such comparisons are necessary in eliminating errors and validating the accuracy of the sequences determined for the individual fragments.

Figure 4.26 Summary of the sequence analysis of catrocollastatin-C, a 23.6-kD protein found in the venom of the western diamondback rattlesnake *Crotalus atrox*. Sequences shown are given in the one-letter amino acid code. The overall amino acid sequence (216 amino acid residues long) as deduced from the overlapping sequences of peptide fragments is shown on the lines headed **CAT-C**. The other lines report the various sequences used to obtain the overlaps. These sequences were obtained from **(a) N-term.:** Edman degradation of the intact protein in an automated Edman sequenator; **(b) M:** proteolytic fragments generated by CNBr cleavage, followed by Edman sequencing of the individual fragments (numbers denote fragments M1 through M5); **(c) K:** proteolytic fragments (K3 through K6) from endopeptidase Lys-C cleavage, followed by Edman sequencing; **(d) E:** proteolytic fragments from *Staphylococcus* protease (E13 through E15) digestion of catrocollastatin sequenced in the Edman sequenator. *(Adapted from Shimokawa, K., et al., 1997.* Archives of Biochemistry and Biophysics ***343**:35–43.)

Sequence Databases

Today, most protein sequence information is derived from translating the nucleotide sequences of genes into codons and, thus, amino acid sequences (see Chapter 8). Sequencing the order of nucleotides in cloned genes is a more rapid, efficient, and informative process than determining the amino acid sequences of proteins. A number of electronic databases containing continuously updated sequence information are readily accessible by personal computer. Prominent among these are PIR (Protein Identification Resource Protein Sequence Database), GenBank (Genetic Sequence Data Bank), and EMBL (European Molecular Biology Laboratory Data Library), and those found at the National Center for Biotechnology Information Web site.

4.10   Nature of Amino Acid Sequences

Table 4.5 lists the relative frequencies of the amino acids in various proteins. It is very unusual for a globular protein to have an amino acid composition that deviates substantially from these values. Apparently, these abundances reflect a distribution of amino acid polarities that is optimal for protein stability in an aqueous medium. Membrane proteins have relatively more hydrophobic and fewer ionic amino acids, a condition consistent with their location. Fibrous proteins may show compositions that are atypical with respect to these norms, indicating an underlying relationship between the composition and the structure of these proteins.

Proteins have unique amino acid sequences, and it is this uniqueness of sequence that ultimately gives each protein its own particular personality. Because the number of possible amino acid sequences in a protein is astronomically large, the probability that two proteins will, by chance, have similar amino

Table 4.5   Frequency of Occurrence of Amino Acid Residues in Proteins

Amino Acid		M_r*	Occurrence in Proteins (%)[†]
Alanine	Ala A	71.1	9.0
Arginine	Arg R	156.2	4.7
Asparagine	Asn N	114.1	4.4
Aspartic acid	Asp D	115.1	5.5
Cysteine	Cys C	103.1	2.8
Glutamine	Gln Q	128.1	3.9
Glutamic acid	Glu E	129.1	6.2
Glycine	Gly G	57.1	7.5
Histidine	His H	137.2	2.1
Isoleucine	Ile I	113.2	4.6
Leucine	Leu L	113.2	7.5
Lysine	Lys K	128.2	7.0
Methionine	Met M	131.2	1.7
Phenylalanine	Phe F	147.2	3.5
Proline	Pro P	97.1	4.6
Serine	Ser S	87.1	7.1
Threonine	Thr T	101.1	6.0
Tryptophan	Trp W	186.2	1.1
Tyrosine	Tyr Y	163.2	3.5
Valine	Val V	99.1	6.9

*Molecular weight of amino acid *minus* that of water.
[†]Frequency of occurrence of each amino acid residue in the polypeptide chains of 207 unrelated proteins of known sequence. Values from Klapper, M. H., 1977. *Biochemical and Biophysical Research Communications* **78:**1018–1024.

Figure 4.27 Cytochrome *c* is a small protein consisting of a single polypeptide chain of 104 ▶ residues in terrestrial vertebrates, 103 or 104 in fishes, 107 in insects, 107 to 109 in fungi and yeasts, and 111 or 112 in green plants. Analysis of the sequence of cytochrome *c* from more than 40 different species reveals that 28 residues are invariant. These invariant residues are scattered irregularly along the polypeptide chain, except for a cluster between residues 70 and 80. All cytochrome *c* polypeptide chains have a cysteine residue at position 17, and all but one have another Cys at position 14. These Cys residues serve to link the heme prosthetic group of cytochrome *c* to the protein, a role explaining their invariable presence.

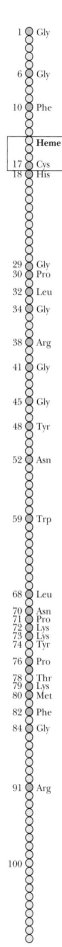

acid sequences is negligible. Consequently, sequence similarities between proteins imply evolutionary relatedness.

Homologous Proteins from Different Organisms Have Homologous Amino Acid Sequences

Proteins sharing a significant degree of sequence similarity are said to be **homologous.** Proteins that perform the same function in different organisms are also referred to as homologous. For example, the oxygen transport protein, hemoglobin, serves a similar role and has a similar structure in all vertebrates. The study of the amino acid sequences of homologous proteins from different organisms provides very strong evidence for their evolutionary origin within a common ancestor. Homologous proteins characteristically have polypeptide chains that are nearly identical in length, and their sequences share identity in direct correlation to the relatedness of the species from which they are derived.

Cytochrome *c*

The electron transport protein, cytochrome *c*, found in the mitochondria of all eukaryotic organisms, provides the best-studied example of homology. The polypeptide chain of cytochrome *c* from most species contains slightly more than 100 amino acids and has a molecular weight of about 12,500. Amino acid sequencing of cytochrome *c* from more than 40 different species has revealed that there are 28 positions in the polypeptide chain where the same amino acid residues are always found (Figure 4.27). These **invariant residues** apparently serve roles crucial to the biological function of this protein, and thus substitutions of other amino acids at these positions cannot be tolerated.

Furthermore, as shown in Figure 4.28, the number of amino acid differences between two cytochrome *c* sequences is proportional to the phylogenetic

	Chimpanzee	Sheep	Rattlesnake	Carp	Snail	Moth	Yeast	Cauliflower	Parsnip
Human	0	10	14	18	29	31	44	44	43
Chimpanzee		10	14	18	29	31	44	44	43
Sheep			20	11	24	27	44	46	46
Rattlesnake				26	28	33	47	45	43
Carp					26	26	44	47	46
Garden snail						28	48	51	50
Tobacco hornworm moth							44	44	41
Baker's yeast (iso-1)								47	47
Cauliflower									13

Figure 4.28 The number of amino acid differences among the cytochrome *c* sequences of various organisms can be compared. The numbers bear a direct relationship to the degree of relatedness between the organisms. Each of these species has a cytochrome *c* of at least 104 residues, so any given pair of species has more than half their residues in common. *(Adapted from Creighton, 1984. Proteins: Structure and Molecular Properties. San Francisco: W. H. Freeman and Co.)*

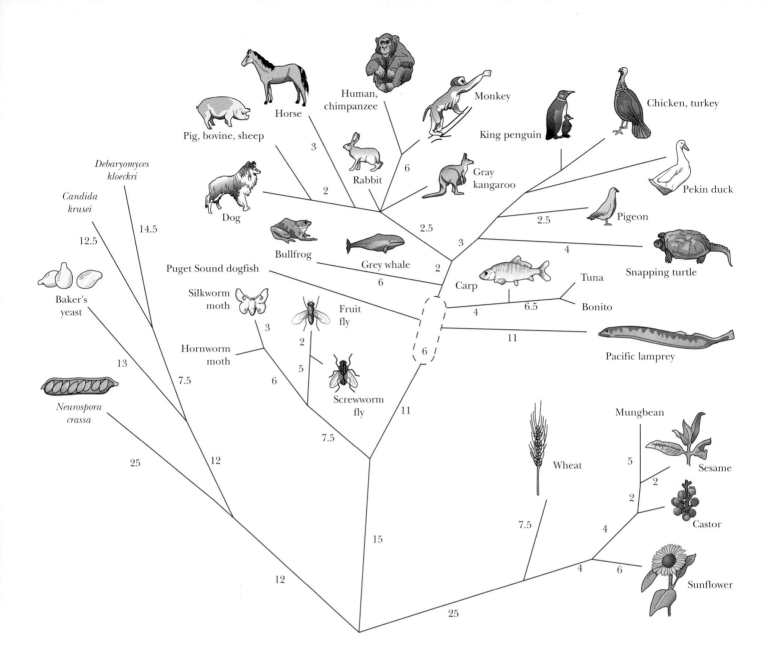

Ancestral cytochrome *c*

1										10										20						

−Pro−Ala −Gly−Asp− ? −Lys−Lys −Gly −Ala −Lys −Ile −Phe −Lys −Thr− ? −Cys−Ala −Gln−Cys −His−Thr−Val−Glu− ? − Gly−Gly − ? −

Human cytochrome *c*

Gly−Asp−Val −Glu−Lys −Gly −Lys −Lys −Ile −Phe −Ile −Met−Lys −Cys−Ser −Gln−Cys −His−Thr−Val−Glu−Lys −Gly−Gly−Lys −

								30											40									50

−His−Lys −Val−Gly−Pro−Asn−Leu−His−Gly−Leu−Phe−Gly −Arg−Lys − ? −Gly−Gln−Ala − ? −Gly−Tyr −Ser −Tyr−Thr−Asp−

−His−Lys −Thr−Gly−Pro−Asn−Leu−His−Gly−Leu−Phe−Gly −Arg−Lys −Thr−Gly−Gln−Ala −Pro −Gly−Tyr −Ser −Tyr−Thr−Ala−

								60											70									

−Ala −Asn−Lys−Asn−Lys −Gly − ? − ? −Trp− ? −Glu−Asn−Thr−Leu−Phe−Glu−Tyr−Leu−Glu−Asn−Pro−Lys −Lys −Tyr−Ile −

−Ala −Asn−Lys−Asn−Lys −Gly − Ile − Ile −Trp−Gly −Glu−Asp−Thr−Leu−Met−Gln−Tyr−Leu−Glu−Asn−Pro − Lys − Lys −Tyr −Pro−

								80											90									100

−Pro−Gly−Thr−Lys−Met− ? −Phe− ? −Gly−Leu−Lys−Lys − ? − ? −Asp−Arg−Ala−Asp−Leu−Ile−Ala−Tyr−Leu−Lys − ? −

−Pro−Gly−Thr−Lys−Met− Ile −Phe−Val−Gly −Ile −Lys −Lys −Lys −Glu−Glu−Arg−Ala−Asp−Leu−Ile−Ala−Tyr−Leu−Lys−Lys−

−Ala−Thr−Ala

−Ala−Thr−Asn−Glu

◀ **Figure 4.29** This phylogenetic tree depicts the evolutionary relationships among organisms as determined by the similarity of their cytochrome *c* amino acid sequences. The numbers along the branches give the amino acid changes between a species and a hypothetical progenitor. Note that extant species are located only at the tips of branches. Below, the sequence of human cytochrome *c* is compared with an inferred ancestral sequence represented by the base of the tree. Uncertainties are denoted by question marks. *(Adapted from Creighton, 1984. Proteins: Structure and Molecular Properties. San Francisco: W. H. Freeman and Co.)*

difference between the species from which they are derived. The cytochrome *c* in humans and in chimpanzees is identical; human and another mammalian (sheep) cytochrome *c* differ at 10 residues. The human cytochrome *c* sequence has 14 variant residues from a reptile sequence (rattlesnake), 18 from a fish (carp), 29 from a mollusc (snail), 31 from an insect (moth), and more than 40 from yeast or higher plants (cauliflower).

The Phylogenetic Tree for Cytochrome *c*

Figure 4.29 displays a **phylogenetic tree** (a diagram illustrating the evolutionary relationships among a group of organisms) constructed from the sequences of cytochrome *c*. The tips of the branches are occupied by contemporary species whose sequences have been determined. The tree has been deduced by computer analysis of these sequences to find the minimum number of mutational changes connecting the branches. Such analysis ultimately suggests a primordial cytochrome *c* sequence lying at the base of the tree. Evolutionary trees constructed in this manner—that is, solely on the basis of amino acid differences occurring in the primary sequence of one selected protein—show remarkable fidelity to phylogenetic relationships derived from more classic approaches and have given rise to the field of *molecular evolution.*

Related Proteins Share a Common Evolutionary Origin

Amino acid sequence analysis reveals that proteins with related functions often show a high degree of sequence similarity. Such findings suggest a common ancestry for these proteins.

Oxygen-Binding Heme Proteins

The oxygen-binding heme protein of muscle, **myoglobin,** consists of a single polypeptide chain of 153 residues. **Hemoglobin,** the oxygen transport protein of erythrocytes, is a tetramer composed of two **α-chains** (141 residues each) and two **β-chains** (146 residues each). These globin polypeptides—myoglobin, α-globin, and β-globin—share a strong degree of sequence homology (Figure 4.30). Myoglobin and the α-globin chain show 38 amino acid identities, whereas α-globin and β-globin have 64 residues in common. The relatedness suggests an evolutionary sequence of events in which chance mutations led to amino acid substitutions and divergence in primary structure. The ancestral myoglobin gene diverged first, after duplication of a primordial globin gene had given rise to its progenitor and an ancestral hemoglobin gene (Figure 4.31). Subsequently, the ancestral hemoglobin gene duplicated to generate the progenitors of the present-day α-globin and β-globin genes. The ability to bind O_2 via a heme prosthetic group is retained by all three of these polypeptides.

Serine Proteases

While the globins provide an example of gene duplication giving rise to a set of proteins in which the biological function has been highly conserved, other

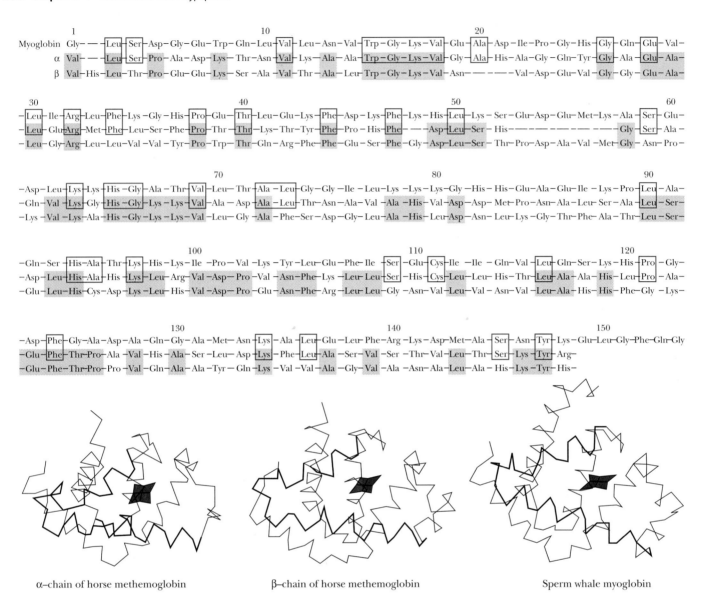

α–chain of horse methemoglobin   β–chain of horse methemoglobin   Sperm whale myoglobin

Figure 4.30 Inspection of the amino acid sequences of the globin chains of human hemoglobin and myoglobin reveals a strong degree of homology. The α-globin and β-globin chains share 64 residues of their approximately 140 residues in common. Myoglobin and the α-globin chain have 38 amino acid sequence identities. This homology is further reflected in these proteins' tertiary structure. *(Irving Geis)*

sets of proteins united by strong sequence homology show more divergent biological functions. Trypsin, chymotrypsin (see Section 4.9), and elastase are members of a class of proteolytic enzymes called **serine proteases** because of the central role played by specific serine residues in their catalytic activity. **Thrombin,** an essential enzyme in blood clotting, is also a serine protease. These enzymes show sufficient sequence homology to conclude that they arose via

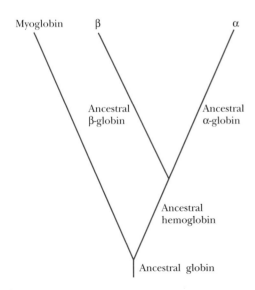

Figure 4.31 This evolutionary tree is inferred from the homology between the amino acid sequences of the α-globin, β-globin, and myoglobin chains. Duplication of an ancestral globin gene allowed the divergence of the myoglobin and ancestral hemoglobin genes. Another gene duplication event subsequently gave rise to ancestral α and β forms, as indicated. Gene duplication is an important evolutionary force in creating diversity.

duplication of a progenitor serine protease gene, even though their substrate preferences are now quite different.

Mutant Proteins

Given a large population of individuals, a considerable number of sequence variants can be found for a protein. These variants are a consequence of **mutations** in a gene (base substitutions in DNA) that have arisen naturally within the population. Gene mutations lead to mutant forms of the protein in which the amino acid sequence is altered at one or more positions. Many of these mutant forms will be "neutral" in that the functional properties of the protein will be unaffected by the amino acid substitution. Others may be nonfunctional (if loss of function is not lethal to the individual), and yet others may display a range of aberrations between these two extremes. The severity of the effects on function will depend on the nature of the amino acid substitution and its role in the protein. These conclusions are exemplified by the more than 300 human hemoglobin variants that have been discovered to date. Some of these are listed in Table 4.6.

A variety of effects on the hemoglobin molecule are seen in these mutants, including alterations in oxygen affinity, heme affinity, stability, solubility, and subunit interactions between the α-globin and β-globin polypeptide chains. Some variants show no apparent changes, while others, such as HbS, sickle-cell hemoglobin (see Chapter 12), result in serious illness. This diversity of response substantiates the conclusion that various amino acid residues play widely different roles in the function of a protein.

4.11 Some Proteins Have Chemical Groups Other than Amino Acids

Many proteins consist of only amino acids and contain no other chemical groups. The enzyme ribonuclease and the contractile protein actin are two such examples. Such proteins are called **simple proteins.** However, many other

■ Table 4.6 **Some Pathological Sequence Variants of Human Hemoglobin**

Abnormal Hemoglobin*	Normal Residue and Position	Substitution
Alpha chain		
Torino	Phenylalanine 43	Valine
M$_{Boston}$	Histidine 58	Tyrosine
Chesapeake	Arginine 92	Leucine
G$_{Georgia}$	Proline 95	Leucine
Tarrant	Aspartate 126	Asparagine
Suresnes	Arginine 141	Histidine
Beta chain		
S	Glutamate 6	Valine
Riverdale-Bronx	Glycine 24	Arginine
Genova	Leucine 28	Proline
Zurich	Histidine 63	Arginine
M$_{Milwaukee}$	Valine 67	Glutamate
M$_{Hyde Park}$	Histidine 92	Tyrosine
Yoshizuka	Asparagine 108	Aspartate
Hiroshima	Histidine 146	Aspartate

*Hemoglobin variants are often given the geographical name of their origin.

Adapted from Dickerson, R. E., and Geis, I., 1984. *Hemoglobin: Structure, Function, Evolution and Pathology.* Menlo Park, Calif.: Benjamin Cummings Publishing Co.

proteins contain various chemical constituents as an integral part of their structure. These proteins are termed **conjugated proteins** (Table 4.7). If the non-protein part is crucial to the protein's function, it is referred to as a **prosthetic group.** Conjugated proteins are typically classified according to the chemical nature of their non–amino acid component; a representative selection of them is given here and in Table 4.7. (Note that comparisons of Tables 4.7 and 4.8 reveal two distinctly different ways of considering the nature of proteins—function versus chemistry.)

Glycoproteins. Glycoproteins are proteins that contain carbohydrate. Proteins destined for extracellular location are characteristically glycoproteins. For example, fibronectin and proteoglycans are important components of the

Table 4.7 Representative Conjugated Proteins

Class	Prosthetic Group	Percent by Weight (approx.)
Glycoproteins contain carbohydrate		
Fibronectin		
γ-Globulin		
Proteoglycan		
Lipoproteins contain lipid		
Blood plasma lipoproteins:		
High density lipoprotein (HDL) (α-lipoprotein)	Triacylglycerols, phospholipids, cholesterol	75
Low density lipoprotein (LDL) (β-lipoprotein)	Triacylglycerols, phospholipids, cholesterol	67
Nucleoprotein complexes contain nucleic acid		
Ribosomes	RNA	50–60
Tobacco mosaic virus	RNA	5
Adenovirus	DNA	
HIV-1 (AIDS virus)	RNA	
Phosphoproteins contain phosphate		
Casein	Phosphate groups	
Glycogen phosphorylase *a*	Phosphate groups	
Metalloproteins contain metal atoms		
Ferritin	Iron	35
Alcohol dehydrogenase	Zinc	
Cytochrome oxidase	Copper and iron	
Nitrogenase	Molybdenum and iron	
Pyruvate carboxylase	Manganese	
Hemoproteins contain heme		
Hemoglobin		
Cytochrome *c*		
Catalase		
Nitrate reductase		
Ammonium oxidase		
Flavoproteins contain flavin		
Succinate dehydrogenase	FAD	
NADH dehydrogenase	FMN	
Dihydroorotate dehydrogenase	FAD and FMN	
Sulfite reductase	FAD and FMN	

extracellular matrix that surrounds the cells of most tissues in multicellular organisms. Immunoglobulin G molecules are the principal antibody species found circulating free in the blood plasma. Many membrane proteins are glycosylated on their extracellular segments.

Lipoproteins. Blood plasma lipoproteins are prominent examples of the class of proteins conjugated with lipid. The plasma lipoproteins function primarily in the transport of lipids to sites of active membrane synthesis. Serum levels of *low density lipoproteins* (*LDLs*) are often used as a clinical index of susceptibility to vascular disease.

Nucleoproteins. Nucleoprotein conjugates have many roles in the storage and transmission of genetic information. Ribosomes are the sites of protein synthesis. Virus particles and even chromosomes are protein–nucleic acid complexes.

Phosphoproteins. These proteins have phosphate groups esterified to the hydroxyls of serine, threonine, or tyrosine residues. Casein, the major protein of milk, contains many phosphates and serves to bring essential phosphorus to the growing infant. Many key steps in metabolism are regulated between states of activity or inactivity, depending on the presence or absence of phosphate groups on proteins, as we shall see in Chapter 10. Glycogen phosphorylase *a* is one well-studied example.

Metalloproteins. Metalloproteins are either metal storage forms, as in the case of ferritin, or enzymes in which the metal atom participates in a catalytically important manner. We will encounter many examples throughout this book of the vital metabolic functions served by metalloenzymes.

Hemoproteins. These proteins are actually a subclass of metalloproteins because their prosthetic group is **heme,** the name given to iron protoporphyrin IX (Figure 4.32). Since heme-containing proteins enjoy so many prominent biological functions, they are considered a class by themselves.

Flavoproteins. *Flavin* is an essential substance for the activity of a number of important oxidoreductases. We discuss the chemistry of the flavin derivatives, FMN and FAD, in Chapter 14, where we consider vitamins and coenzymes.

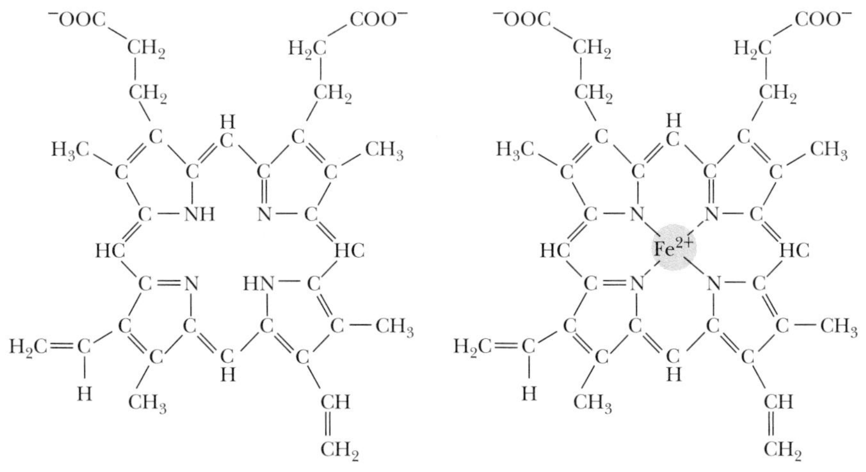

Protoporphyrin IX

Heme
(Fe-protoporphyrin IX)

Figure 4.32 Heme consists of protoporphyrin IX and an iron atom. Protoporphyrin, a highly conjugated system of double bonds, is composed of four 5-membered heterocyclic rings (pyrroles) fused together to form a tetrapyrrole macrocycle. The specific isomeric arrangement of methyl, vinyl, and propionate side chains shown is protoporphyrin IX. Coordination of an atom of ferrous iron (Fe^{2+}) by the four pyrrole nitrogen atoms yields heme.

4.12 The Many Biological Functions of Proteins

Proteins are the agents of biological function. Virtually every cellular activity is dependent on one or more particular proteins. Thus, a convenient way to classify the enormous number of proteins is by the biological roles they fill. Table 4.8 summarizes the classification of proteins by function and gives examples of representative members of each class.

By far the largest class of proteins is **enzymes.** More than 2000 different enzymes are listed in *Enzyme Nomenclature,* the standard reference volume on enzyme classification. **Enzymes** are catalysts that accelerate the rates of biological reactions. Each enzyme is very specific in its function and acts only in a particular metabolic reaction. Virtually every step in metabolism is catalyzed by an enzyme.

Table 4.8 Biological Functions of Proteins and Some Representative Examples

Functional Class	Examples
Enzymes	Ribonuclease Trypsin Phosphofructokinase Alcohol dehydrogenase Catalase
Regulatory proteins	Insulin Somatotropin Thyrotropin *lac* repressor NF-1 (nuclear factor-1) Catabolite activator protein
Transport proteins	Hemoglobin Serum albumin Glucose transporter
Storage proteins	Ovalbumin Casein Zein Phaseolin Ferritin
Contractile and motile proteins	Actin Myosin Tubulin Dynein Kinesin
Structural proteins	α-Keratins Collagen Elastin Proteoglycans
Protective proteins	Immunoglobulins Thrombin Fibrinogen Snake and bee venom proteins Diphtheria toxin Ricin
Exotic proteins	Antifreeze proteins Monellin Resilin Glue proteins

A number of proteins do not perform any obvious chemical transformation, but nevertheless can regulate the ability of other proteins to carry out their physiological functions. Such proteins are referred to as **regulatory proteins.** A well-known example is *insulin,* the hormone regulating glucose metabolism in animals. Other hormones that are also proteins include the pituitary *somatotropin* (21 kD) and *thyrotropin* (28 kD), which stimulates the thyroid gland. Another group of regulatory proteins is involved in the regulation of gene expression. These proteins characteristically act by binding to DNA sequences that are adjacent to coding regions of genes, either activating or inhibiting the transcription of genetic information into RNA. Examples include **repressors,** which, because they block transcription, are considered negative control elements. A prokaryotic representative is *lac repressor* (37 kD), which controls expression of the enzyme system responsible for the metabolism of lactose (milk sugar); a mammalian example is *NF-1* (*nuclear factor-1,* 60 kD) which inhibits transcription of the gene encoding the β-globin polypeptide chain of hemoglobin. The *E. coli* catabolite gene activator protein (**CAP**) (44 kD)is a positively acting regulatory protein. Under appropriate metabolic conditions, CAP binds to specific sites along the *E. coli* chromosome and increases the rate of transcription of adjacent genes. The mammalian *AP1* is a heterodimeric transcription factor composed of polypeptides from the *Jun* and *Fos* families of gene regulatory proteins.

A third class of proteins comprises the **transport proteins.** These proteins function to transport specific substances from one place to another. One type of transport is exemplified by the transport of oxygen from the lungs to the tissues by *hemoglobin* (Figure 4.33) or by the transport of fatty acids from adipose tissue to various organs by the blood protein *serum albumin.* A very different type is the transport of metabolites across permeability barriers such as cell membranes, as mediated by specific membrane proteins. These *membrane transport proteins* take up metabolite molecules on one side of a membrane, transport them across the membrane, and release them on the other side(see Figure 4.33).

Proteins whose biological function is to provide a reservoir of an essential nutrient are called **storage proteins.** Since proteins are amino acid polymers and since nitrogen is commonly a limiting nutrient for growth, organisms have exploited proteins as a means to provide sufficient nitrogen in times of need. Examples include *ovalbumin,* the protein of egg white, which provides the developing bird embryo with a source of nitrogen; *casein,* the most abundant protein of milk and the major nitrogen source for mammalian infants; and proteins such as *zein* and *phaseolin* in the seeds of higher plants. Proteins can also serve to store nutrients other than the more obvious elements composing amino acids (N, C, H, O, and S). As an example, *ferritin* is a protein found in animal tissues that serves in iron storage.

Certain proteins endow cells with unique capabilities for movement. Cell division, muscle contraction, and cell motility represent some of the ways in which cells execute motion. The **contractile** and **motile proteins** underlying these motions include *actin* and *myosin,* the filamentous proteins forming the contractile systems of cells, and *tubulin,* the major component of microtubules (the filaments involved in the mitotic spindle of cell division as well as in flagella and cilia). Another class of proteins involved in movement includes *dynein* and *kinesin,* so-called **motor proteins** that propel intracellular movement of vesicles, granules, and organelles along established cytoskeletal "tracks."

An apparently passive but very important role of proteins is their function in creating and maintaining biological structures. **Structural proteins** provide strength and protection to cells and tissues. Monomeric units of structural proteins typically polymerize to generate long fibers (as in hair) or protective sheets

Figure 4.33 Two basic types of biological transport are **(a)** transport within or between different cells or tissues and **(b)** transport into or out of cells. Proteins function in both of these phenomena. For example, the protein hemoglobin transports oxygen from the lungs to actively respiring tissues. Transport proteins of the other type are localized in cellular membranes where they function in the uptake of specific nutrients, such as glucose (shown here) and amino acids, or the export of metabolites and waste products.

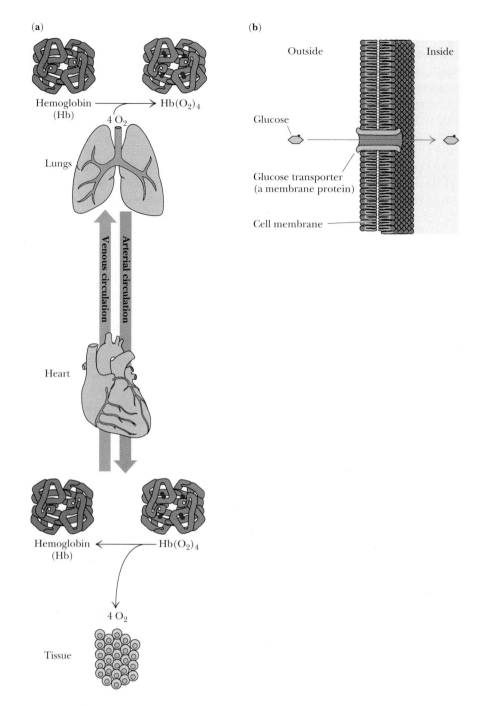

of fibrous arrays, as in cowhide (leather). *α-Keratins* are insoluble fibrous proteins making up hair, horns, and fingernails. *Collagen,* another insoluble fibrous protein, is found in bone, connective tissue, tendons, cartilage, and hide, where it forms inelastic fibrils of great strength. One-third of the total protein in a vertebrate animal is collagen. A structural protein having elastic properties is, appropriately, *elastin,* a component of ligaments. An important protective barrier for animal cells is the *ground substance,* an extracellular matrix containing *proteoglycans,* covalent protein–polysaccharide complexes that cushion and lubricate.

In contrast to the passive protective nature of some structural proteins, another group can be more aptly classified as **protective** or **exploitive proteins** because of their biologically active role in cell defense, protection, or exploitation. Prominent among the protective proteins are the *immunoglobulins* or *antibodies* produced by the lymphocytes of vertebrates. Another group of protective proteins are the blood-clotting proteins, *thrombin* and *fibrinogen,* which prevent the loss of blood when the circulatory system is damaged. Arctic and Antarctic fishes have *antifreeze proteins* to protect their blood against freezing in the below-zero temperatures of high-latitude seas. In addition, various proteins serve defensive or exploitive roles for organisms, including the lytic and neurotoxic proteins of snake and bee venoms and toxic plant proteins, such as *ricin,* whose apparent purpose is to thwart predation by herbivores. Another class of exploitive proteins includes the toxins produced by bacteria, such as diphtheria toxin or cholera toxin.

Some proteins display rather exotic functions that do not quite fit the previous classifications. *Monellin,* a protein found in an African plant, has a very sweet taste and is being considered as an artificial sweetener for human consumption. *Resilin,* a protein having exceptional elastic properties, is found in the hinges of insect wings. Certain marine organisms such as mussels secrete *glue proteins,* allowing them to attach firmly to hard surfaces. It is worth repeating that the great diversity of function in proteins, as reflected in this survey, is attained using just 20 amino acids.

PROBLEMS

1. Without consulting chapter figures, draw Fischer projection formulas for glycine, aspartate, leucine, isoleucine, methionine, and threonine.

2. Without reference to the text, give the one-letter and three-letter abbreviations for asparagine, arginine, cysteine, lysine, proline, tyrosine, and tryptophan.

3. Write equations for the ionic dissociations of alanine, glutamate, histidine, lysine, and phenylalanine.

4. How is the pK_a of the α-NH$_3^+$ group affected by the presence of the α-COOH on an amino acid?

5. Draw an appropriate titration curve for aspartic acid, labeling the axes and indicating the equivalence points and the pK_a values.

6. Calculate the pH at which the γ-carboxyl group of glutamic acid is two-thirds dissociated.

7. Calculate the pH at which the ε-amino group of lysine is 20% dissociated.

8. Calculate the pH of a 0.3 *M* solution of (a) leucine hydrochloride, (b) sodium leucinate, and (c) isoelectric leucine.

9. Describe the stereochemical aspects of the structure of cystine, the structure that is a disulfide-linked pair of cysteines.

10. Describe the expected elution pattern for a mixture of aspartate, histidine, isoleucine, valine, and arginine on a column of Dowex-50.

11. Assign (*R,S*) nomenclature to all the possible isomers of threonine.

12. The element molybdenum (atomic weight 95.95) constitutes 0.08% of the weight of nitrate reductase. If the molecular weight of nitrate reductase is 240,000, what is its likely quaternary structure?

13. Amino acid analysis of an oligopeptide seven residues long gave

Asp   Leu   Lys   Met   Phe   Tyr

The following facts were observed:

a. Trypsin treatment had no apparent effect.

b. The phenylthiohydantoin released by Edman degradation was

c. Brief chymotrypsin treatment yielded several products, including a dipeptide and a tetrapeptide. The amino acid composition of the tetrapeptide was Leu, Lys, and Met.

d. Cyanogen bromide treatment yielded a dipeptide, a tetrapeptide, and free Lys.

What is the amino acid sequence of this heptapeptide?

14. Amino acid analysis of another heptapeptide gave

Asp   Glu   Leu   Lys   Met   Tyr   Trp   NH$_4^+$

The following facts were observed:

a. Trypsin had no effect.

b. The phenylthiohydantoin released by Edman degradation was

c. Brief chymotrypsin treatment yielded several products, including a dipeptide and a tetrapeptide. The amino acid composition of the tetrapeptide was Glx, Leu, Lys, and Met.

d. Cyanogen bromide treatment yielded a tetrapeptide that had a net positive charge at pH 7 and a tripeptide that had a zero net charge at pH 7.

What is the amino acid sequence of this heptapeptide?

15. Amino acid analysis of a decapeptide revealed the presence of the following products:

$$NH_4^+ \quad Asp \quad Glu \quad Tyr \quad Arg \quad Met \quad Pro \quad Lys \quad Ser \quad Phe$$

The following facts were observed:

a. Neither carboxypeptidase A nor B treatment of the decapeptide had any effect.

b. Trypsin treatment yielded two tetrapeptides and free Lys.

c. Clostripain treatment yielded a tetrapeptide and a hexapeptide.

d. Cyanogen bromide treatment yielded an octapeptide and a dipeptide of sequence NP (using the one-letter codes).

e. Chymotrypsin treatment yielded two tripeptides and a tetrapeptide. The N-terminal chymotryptic peptide had a net charge of -1 at neutral pH and a net charge of -3 at pH 12.

f. One cycle of Edman degradation gave the PTH derivative

What is the amino acid sequence of this decapeptide?

FURTHER READING

Barker, R., 1971. *Organic Chemistry of Biological Compounds*, Chap. 4. Englewood Cliffs, N.J.: Prentice-Hall.

Barrett, G. C., ed., 1985. *Chemistry and Biochemistry of the Amino Acids.* New York: Chapman and Hall.

Bovey, F. A., and Tiers, G. V. D., 1959. Proton N.S.R. spectroscopy. V. Studies of amino acids and peptides in trifluoroacetic acid. *Journal of the American Chemical Society* **81**:2870–2878.

Cahn, R. S., 1964. An introduction to the sequence rule. *Journal of Chemical Education* **41**:116–125.

Creighton, T. E., 1983. *Proteins: Structure and Molecular Properties.* San Francisco: W. H. Freeman and Co.

Creighton, T. E., 1997. *Protein Function—A Practical Approach*, 2nd ed. Oxford: IRL Press at Oxford University Press.

Deutscher, M. P., ed., 1990. Guide to Protein Purification. *Methods in Enzymology*, Vol. 182. San Diego: Academic Press.

Goodsell, D. S., and Olson, A. J., 1993. Soluble proteins: Size, shape and function. *Trends in Biochemical Sciences* **18**:65–68.

Greenstein, J. P., and Winitz, M., 1961. *Chemistry of the Amino Acids.* New York: John Wiley & Sons.

Heiser, T., 1990. Amino acid chromatography: The "best" technique for student labs. *Journal of Chemical Education* **67**:964–966.

Hsieh, Y. L., et al., 1996. Automated analytical system for the examination of protein primary structure. *Analytical Chemistry* **68**:455–462. An analytical system is described in which a protein is purified by affinity chromatography, digested with trypsin, and its peptides separated by HPLC and analyzed by tandem MS in order to determine its amino acid sequence.

Hunt, D. F., et al., 1987. Tandem quadrupole Fourier transform mass spectrometry of oligopeptides and small proteins. *Proceedings of the National Academy of Sciences, U.S.A.* **84**:620–624.

Johnstone, R. A. W., and Rose M. E., 1996. *Mass Spectrometry for Chemists and Biochemists*, 2nd ed. Cambridge, England: Cambridge University Press.

Karger, B. L., and Hancock, W. S., eds. 1996. *Methods in Enzymology* **271**, Section III: *Protein Structure Analysis by Mass Spectrometry.* J. R. Yates, *Peptide Characterization by Mass Spectrometry*, B. L. Gillece-Castro and J. T. Stults. New York: Academic Press.

Kauffman, G. B., and Priebe, P. M., 1990. The Emil Fischer–William Ramsey Friendship. *Journal of Chemical Education* **67**:93–101.

Mabbott, G., 1990. Qualitative amino acid analysis of small peptides by GC/MS. *Journal of Chemical Education* **67**:441–445.

Mann, M., and Wilm, M., 1995. Electrospray mass spectrometry for protein characterization. *Trends in Biochemical Sciences* **20**:219–224. A review of the basic application of mass spectrometric methods to the analysis of protein sequence and structure.

Moore, S., Spackman, D., and Stein, W. H., 1958. Chromatography of amino acids on sulfonated polystyrene resins. *Analytical Chemistry* **30**:1185–1190.

Roberts, G. C. K., and Jardetzky, O., 1970. Nuclear magnetic resonance spectroscopy of amino acids, peptides and proteins. *Advances in Protein Chemistry* **24**:447–545.

Segel, I. H., 1976. *Biochemical Calculations*, 2nd ed. New York: John Wiley & Sons.

Von Heijne, G., 1987. *Sequence Analysis in Molecular Biology: Treasure Trove or Trivial Pursuit?* San Diego: Academic Press.

Proteins: Secondary, Tertiary, and Quaternary Structure

CHAPTER

5

Like the Greek sea god Proteus, who could assume different forms, proteins act through changes in conformation. Proteins (from the Greek proteios, *meaning "primary") are the primary agents of biological function. ("Proteus, Old Man of the Sea," Roman period mosaic, from Thessalonika, 1st century A.D. National Archaeological Museum, Athens/Ancient Art and Architecture Collection Ltd./Bridgeman Art Library, London/New York)*

Growing in size and complexity
Living things, masses of atoms, DNA, protein
Dancing a pattern ever more intricate.
Out of the cradle onto the dry land
Here it is standing
Atoms with consciousness
Matter with curiosity.
Stands at the sea
Wonders at wondering
I
A universe of atoms
An atom in the universe.

Richard P. Feynman (1918–1988)
From "The Value of Science," in Edward Hutchings, Jr., ed.,
Frontiers of Science: A Survey. New York: Basic Books, 1958.

Outline

5.1 Architecture of Protein Molecules

5.2 Role of the Amino Acid Sequence in Protein Structure

5.3 Secondary Structure in Proteins

5.4 Protein Folding and Tertiary Structure

5.5 Subunit Interactions and Quaternary Structure

Nearly all biological processes involve the specialized functions of one or more protein molecules. Essentially all of the information required to initiate, conduct, and regulate each of these functions must be contained in the structure of proteins themselves. The previous chapter described the details of protein primary structure. However, proteins do not normally exist as fully extended polypeptide chains but rather as compact, folded structures, and the function of a given protein is rarely if ever dependent only on the amino acid sequence. Instead, the ability of a particular protein to carry out its function is determined by its overall three-dimensional shape or conformation. This native, folded structure of the protein is dictated by several factors: (1) interactions with solvent molecules (normally water), (2) the pH and ionic composition of the solvent, and, most important, (3) the sequence of the protein.

The first two of these effects are intuitively reasonable, but the third, the role of the primary structure, may not be. In ways that are just now beginning to be understood, the primary structure facilitates the development of short-range interactions among adjacent parts of the sequence and also long-range interactions among distant parts of the sequence. Although the resulting overall structure of the complete protein molecule may at first look like a disorganized and random arrangement, it is in nearly all cases a delicate and sophisticated balance of numerous forces that combine to determine the protein's unique conformation. This chapter considers the details of protein structure and the forces that maintain these structures.

5.1 Architecture of Protein Molecules

As a first approximation, proteins can be assigned to one of three great classes on the basis of shape: fibrous, globular, or membrane (Figure 5.1). **Fibrous proteins** tend to have relatively simple, regular, linear structures. These proteins often serve structural roles in cells. Typically, they are insoluble in water or in dilute salt solutions. In contrast, **globular proteins** are roughly spherical in shape. The polypeptide chain is compactly folded so that most of the hydrophobic amino acid side chains are in the interior of the molecule and most of the hydrophilic side chains are on the outside exposed to the solvent, water. Consequently, globular proteins are usually very soluble in aqueous solutions. Most soluble proteins of the cell, such as the cytosolic enzymes, are globular in shape. **Membrane proteins** are found in association with the various membrane systems of cells. For interaction with the nonpolar phase within

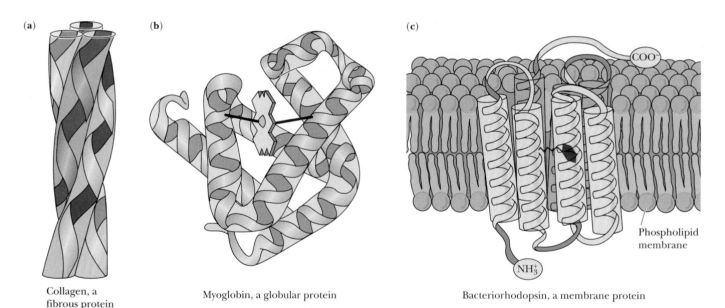

(a) Collagen, a fibrous protein

(b) Myoglobin, a globular protein

(c) Bacteriorhodopsin, a membrane protein

See *Interactive Biochemistry CD-ROM and Workbook,* pages 30, 49, 60

Figure 5.1 **(a)** Proteins having structural roles in cells are typically fibrous and often water-insoluble. Collagen is a good example. Collagen is composed of three polypeptide chains that intertwine. **(b)** Soluble proteins serving metabolic functions can be characterized as compactly folded globular molecules, such as myoglobin. The folding pattern puts hydrophilic amino acid side chains on the outside and buries hydrophobic side chains in the interior, making the protein highly water soluble. **(c)** Membrane proteins fold so that hydrophobic amino acid side chains are exposed in their membrane-associated regions. The portions of membrane proteins extending into or exposed at the aqueous environments are hydrophilic in character, like soluble proteins. Bacteriorhodopsin is a typical membrane protein; it binds the light-absorbing pigment, *cis*-retinal, shown here in red. (*a, b, Irving Geis*)

α-Helix
Only the N — C_α — C backbone is represented. The vertical line is the helix axis.

β-Strand
The N — C_α — C_O backbone as well as the C_β of R groups are represented here. Note that the amide planes are perpendicular to the page.

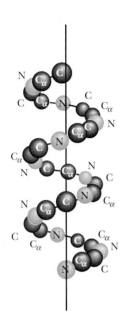

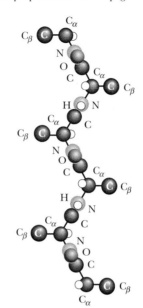

Figure 5.2 Two structural motifs that arrange the primary structure of proteins into a higher level of organization predominate in proteins: the α-helix and the β-pleated strand. Atomic representations of these secondary structures are shown here, along with the symbols used by structural chemists to represent them: the flat, helical ribbon for the α-helix and the flat, wide arrow for the β-structures. Both of these structures owe their stability to the formation of hydrogen bonds between N — H and O ═ C functions along the polypeptide backbone.

"Shorthand" α-helix   "Shorthand" β-strand

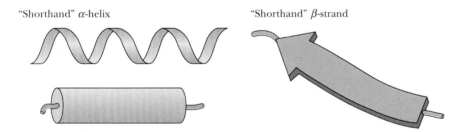

membranes, membrane proteins have hydrophobic amino acid side chains oriented outward. As such, membrane proteins are insoluble in aqueous solutions but can be solubilized in solutions of detergents. Membrane proteins characteristically have fewer hydrophilic amino acids than cytosolic proteins.

The Levels of Protein Structure

The architecture of protein molecules is quite complex. Nevertheless, this complexity can be understood by defining various levels of structural organization.

Primary Structure

The amino acid sequence is the **primary (1°) structure** of a protein, such as that shown in Figure 4.20.

Secondary Structure

Through hydrogen bonding interactions between nearby amino acid residues, the polypeptide chain can arrange itself into characteristic patterns, such as helical or pleated segments. These segments constitute structural conformities, so-called **regular structures,** that extend along one dimension, like the coils of a spring. Such architectural features of a protein are designated as **secondary (2°) structures** (Figure 5.2). Secondary structures are just one of the higher levels of structure that represent the three-dimensional arrangement of the polypeptide in space.

Tertiary Structure

When the polypeptide chains of protein molecules bend and fold in order to assume a more compact three-dimensional shape, the **tertiary (3°) level of structure** is generated (Figure 5.3). By virtue of their tertiary structure, proteins adopt a globular shape, especially those proteins commonly existing in solution in the aqueous compartments of cells. A globular conformation gives the lowest surface-to-volume ratio, shielding much of the protein from interaction with the solvent.

Quaternary Structure

Many proteins consist of two or more interacting polypeptide chains of characteristic tertiary structure, each of which is commonly referred to as a **sub-**

(a) Chymotrypsin primary structure

H₂N–CGVPAIQPVL$_{10}$SGL[SR]IVNGE$_{20}$EAVPGSWPWQ$_{30}$VSLQDKTGFH$_{40}$GGSLINEN$_{50}$WVVTAAHCGV$_{60}$TTSDVVVAGE$_{70}$FDQGSSSEKI$_{80}$QKLKIA KVFK$_{90}$NSKYNSLTIN$_{100}$NDITLLKLST$_{110}$AASFSQTVSA$_{120}$VCLPSASDDF$_{130}$AAGTTCVTTG$_{140}$WGLTRY[TN]AN$_{150}$LPSDRLQQASL$_{160}$PLLSNTNCK K$_{170}$YWGTKIKDAM$_{180}$ICAGASGVSS$_{190}$CMGDSGGPLV$_{200}$CKKNGAWTLV$_{210}$GIVSWGSSTC$_{220}$STSTPGVYAR$_{230}$VTALVNWVQQ$_{240}$TLAAN–COOH

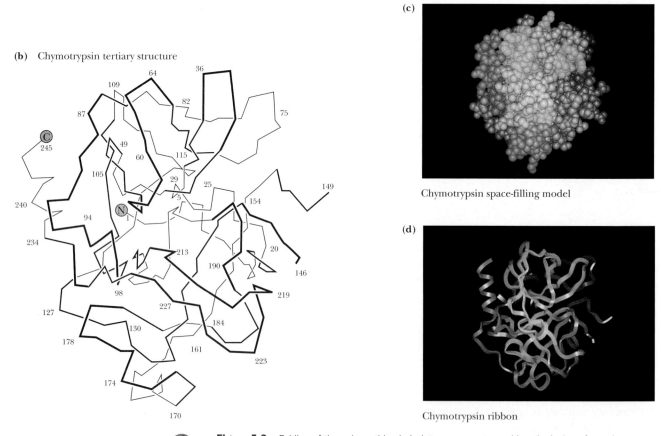

(b) Chymotrypsin tertiary structure

(c)

Chymotrypsin space-filling model

(d)

Chymotrypsin ribbon

See page 37

Figure 5.3 Folding of the polypeptide chain into a compact, roughly spherical conformation creates the tertiary level of protein structure. **(a)** The primary structure and **(b)** a representation of the tertiary structure of chymotrypsin, a proteolytic enzyme, are shown here. The tertiary representation on the left shows the course of the chymotrypsin folding pattern by successive numbering of the amino acids in its sequence. (Residues 14 and 15 and 147 and 148 are missing because these residues are proteolytically removed when chymotrypsin is formed from its larger precursor, chymotrypsinogen.) The ribbon diagram depicts the three-dimensional track of the polypeptide in space.

unit of the protein. Subunit organization in proteins constitutes another level in the hierarchy of protein structure, defined as the protein's **quaternary (4°) structure** (Figure 5.4). Quaternary structure is determined by the kinds of subunits within a protein molecule, the number of each, and the ways in which they interact with one another.

Whereas the primary structure of a protein is determined by the covalently linked amino acid residues in the polypeptide backbone, the secondary and higher orders of structure are determined principally by noncovalent forces such as hydrogen bonds, ionic bonds, and van der Waals and hydrophobic interactions.

Protein Conformation

The overall three-dimensional architecture of a protein is generally referred to as its **conformation.** This term is not to be confused with **configuration,** which denotes the geometric possibilities for a particular set of atoms (Figure 5.5). In going from one configuration to another, covalent bonds must be broken and rearranged. In contrast, the conformational possibilities of a molecule are achieved without breaking any covalent bonds. In proteins, rotations about each of the single bonds along the peptide backbone have the potential to alter the course of the polypeptide chain in three-dimensional space. These rotational possibilities create many possible orientations for the protein chain, referred to as its *conformational possibilities.* Among the great number of conformations a given protein might theoretically adopt, only a very few are favored energetically under physiological conditions. At this time, little is understood regarding the rules that direct the folding of protein chains into energetically favorable conformations.

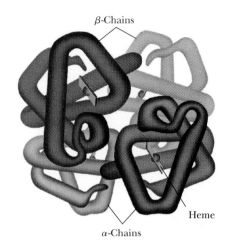

Figure 5.4 Hemoglobin, which consists of two α and two β polypeptide chains, is an example of the quaternary level of protein structure. In this drawing, the β-chains are the two uppermost polypeptides and the two α-chains are the lower half of the molecule. The two closest chains (darkest colored) are the β_2-chain (*upper left*) and the α_1-chain (*lower right*). The heme groups of the four globin chains are represented by rectangles with spheres (the heme iron atom). Note the symmetry of this macromolecular arrangement.

 See page 45

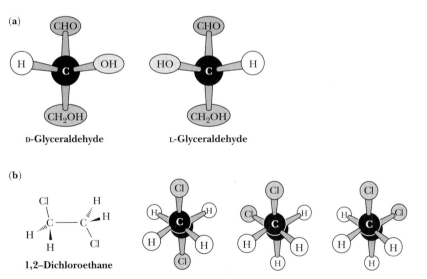

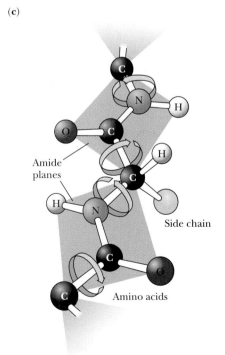

Figure 5.5 Configuration and conformation are *not* synonymous. **(a)** Rearrangements between configurational alternatives of a molecule can be achieved only by breaking and remaking bonds, as in the transformation between the D- and L-configurations of glyceraldehyde. No possible rotational reorientation of bonds linking the atoms of D-glyceraldehyde will yield geometric identity with L-glyceraldehyde, even though they are mirror images of each other. **(b)** The intrinsic free rotation about single covalent bonds creates a great variety of three-dimensional conformations, even for relatively simple molecules. Consider 1,2-dichloroethane. Viewed end-on in a Newman projection, three principal rotational orientations or conformations predominate. Steric repulsion between eclipsed and partially eclipsed conformations keeps the possibilities at a reasonable number. **(c)** Imagine the conformational possibilities for a protein in which two out of every three bonds along its backbone are freely rotating single bonds.

5.2 Role of the Amino Acid Sequence in Protein Structure

Many different forces work together in a delicate balance to determine the overall three-dimensional structure of a protein. These forces operate both within the protein structure itself and also between the protein and the water solvent. How, then, does nature dictate the manner of protein folding to generate the three-dimensional structure that is optimized and balanced by these many forces? *All of the information necessary for folding the peptide chain into its "native" structure is contained in the primary structure (amino acid sequence) of the peptide.* This principle was first appreciated by C. B. Anfinsen and F. White, whose work in the early 1960s dealt with the chemical denaturation and subsequent renaturation of bovine pancreatic ribonuclease. Ribonuclease was first denatured with urea and mercaptoethanol, a treatment that unfolded the protein and cleaved its four covalent disulfide cross-bridges. Subsequent air oxidation permitted random formation of disulfide cross-bridges, most of which were incorrect. Thus, the air-oxidized material showed little enzymatic activity. However, treatment of these inactive preparations with small amounts of mercaptoethanol allowed a reshuffling of the disulfide bonds and permitted formation of significant amounts of active native enzyme. It may be difficult to imagine "uncooking" a hard-boiled egg, but experiments such as these indicate that ovalbumin, the principal egg-white protein, could also be, in theory, first denatured and then renatured and returned to its native state by a suitable sequence of steps. In such experiments, the only road map for the protein—that is, the only "set of instructions" it has—is that directed by its primary structure, the linear sequence of its amino acid residues.

Just how proteins recognize and interpret the information that is stored in the polypeptide sequence is not well understood yet. It may be assumed that certain loci along the peptide chain act as nucleation points, which initiate folding processes that eventually lead to the correct structures. Regardless of how this process operates, it must take the protein correctly to the final native structure without getting trapped in a partially folded state which, though stable, may be different from the native state itself. A long-range goal of many researchers in the protein structure field is the prediction of three-dimensional conformations from the amino acid sequence. As the details of secondary and tertiary structure are described in this chapter, the complexity and immensity of such a prediction will be more fully appreciated. This gap is perhaps the greatest uncharted frontier remaining in molecular biology.

5.3 Secondary Structure in Proteins

Any discussion of protein folding and structure must begin with the peptide bond, the fundamental structural unit in all proteins. As we saw in Chapter 4, the resonance structures experienced by a peptide bond constrain the oxygen, carbon, nitrogen, and hydrogen atoms of the peptide group, as well as the adjacent α-carbons, to all lie in a plane.

Consequences of the Amide Plane

The planarity of the peptide bond means that there are only two degrees of freedom per residue for the peptide chain. Rotation is allowed about the bond linking the α-carbon and the carbon of the peptide bond and also about the bond linking the nitrogen of the peptide bond and the adjacent α-carbon. As shown in Figure 5.6, each α-carbon is the joining point for two planes defined

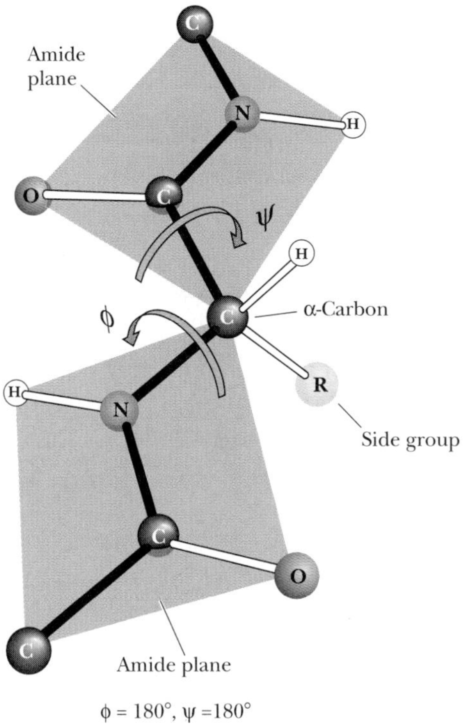

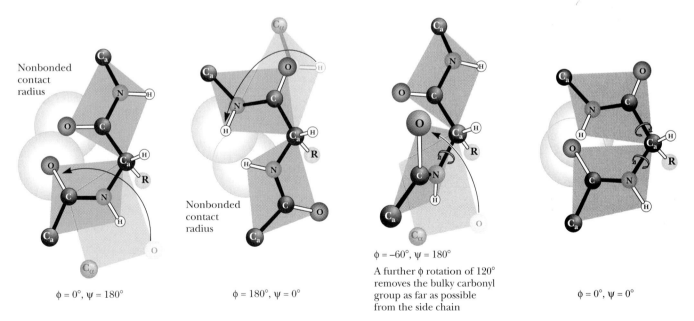

Figure 5.6 The amide or peptide bond planes are joined by the tetrahedral bonds of the α-carbon. The rotation parameters are ϕ and ψ. The conformation shown corresponds to $\phi = 180°$ and $\psi = 180°$. Note that positive values of ϕ and ψ correspond to clockwise rotation as viewed from C_α. Starting from $0°$, a rotation of $180°$ in the clockwise direction ($+180°$) is equivalent to a rotation of $180°$ in the counterclockwise direction ($-180°$). (*Irving Geis*)

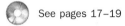 See pages 17–19

by peptide bonds. The angle about the C_α—N bond is denoted by the Greek letter ϕ (phi) and that about the C_α—C_O is denoted by ψ (psi). For either of these bond angles, a value of $0°$ corresponds to an orientation with the amide plane bisecting the H—C_α—R (side-chain) plane and a *cis* conformation of the main chain around the rotating bond in question (Figure 5.7). In any case,

$\phi = 0°$, $\psi = 180°$

$\phi = 180°$, $\psi = 0°$

$\phi = -60°$, $\psi = 180°$

A further ϕ rotation of $120°$ removes the bulky carbonyl group as far as possible from the side chain

$\phi = 0°$, $\psi = 0°$

Figure 5.7 Many of the possible conformations about an α-carbon between two peptide planes are forbidden because of steric crowding. Several noteworthy examples are shown here.

Note: The formal IUPAC-IUB Commission on Biochemical Nomenclature convention for the definition of the torsion angles ϕ and ψ in a polypeptide chain (*Biochemistry* **9**:3471–3479, 1970) is different from that used here, where the C_α atom serves as the point of reference for both rotations, but the result is the same. (*Irving Geis*)

the entire path of the peptide backbone in a protein is known if the ϕ and ψ rotation angles are all specified. Some values of ϕ and ψ are not allowed due to steric interference between nonbonded atoms. As shown in Figure 5.7, values of $\phi = 180°$ and $\psi = 0°$ are not allowed because of the forbidden overlap of the N—H hydrogens. Similarly, the set $\phi = 0°$ and $\psi = 180°$ is forbidden because of unfavorable overlap between the carbonyl oxygens.

G. N. Ramachandran and his coworkers in Madras, India, first showed that it was convenient to plot ϕ values against ψ values to show the distribution of allowed values in a protein or in a family of proteins. A typical **Ramachandran plot** is shown in Figure 5.8. Note the clustering of ϕ and ψ values in a few "allowed" regions of the plot. Most combinations of ϕ and ψ are sterically forbidden, and the corresponding regions of the Ramachandran plot are sparsely populated. The combinations that are sterically allowed represent the subclasses of structure described in the remainder of this section.

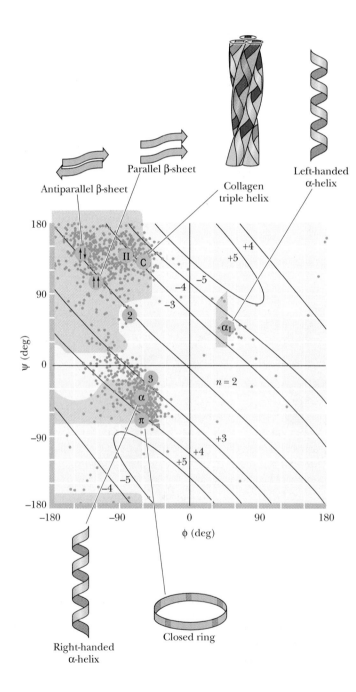

Figure 5.8 A Ramachandran diagram showing the sterically reasonable values of the angles ϕ and ψ. The shaded regions indicate particularly favorable values of these angles. Dots in purple indicate actual angles measured for 1000 residues (excluding glycine, for which a wider range of angles is permitted) in eight proteins. The lines running across the diagram (numbered +5 through 2 and −5 through −3) signify the number of amino acid residues per turn of the helix; "+" means right-handed helices; "−" means left-handed helices. (*From Richardson, J. S., 1981,* Advances in Protein Chemistry *34:167–339.*)

See pages 17–19

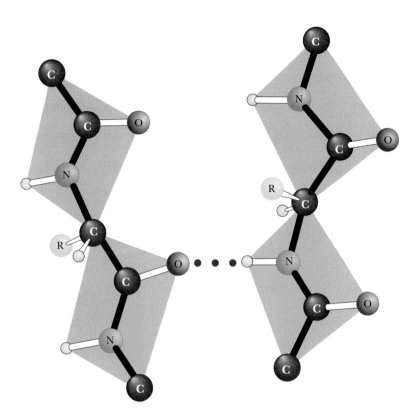

Figure 5.9 A hydrogen bond between the amide proton and carbonyl oxygen of adjacent peptide groups.

The carbonyl oxygen and amide hydrogen of the peptide bond can participate in H bonds either with water molecules in the solvent or with other H-bonding groups in the peptide chain. In nearly all proteins, the carbonyl oxygens and the amide protons of many peptide bonds participate in H bonds that link one peptide group to another, as shown in Figure 5.9. These structures tend to form in cooperative fashion and involve substantial portions of the peptide chain. Structures resulting from these interactions constitute the secondary structure of proteins. When hydrogen bonds form between portions of the peptide chain, two basic types of structure can result: α-helices and β-pleated sheets.

The Alpha-Helix

Evidence for helical structures in proteins was first obtained in the 1930s in studies of fibrous proteins. However, there was little agreement at that time about the exact structure of these helices, primarily because there was also lack of agreement about interatomic distances and bond angles in peptides. In 1951, Linus Pauling, Robert Corey, and their colleagues at the California Institute of Technology (Caltech) summarized a large volume of crystallographic data in a set of dimensions for polypeptide chains. (A summary of data similar to what they reported is shown in Figure 4.16.) With these data in hand, Pauling, Corey, and their colleagues proposed a new model for a helical structure in proteins, which they called the **α-helix.** The report from Caltech was of particular interest to Max Perutz in Cambridge, England, a crystallographer who was also interested in protein structure. By taking into account a critical but previously ignored feature of the X-ray data, Perutz realized that the α-helix existed in keratin, a protein from hair, and also in several other proteins. Since then, the α-helix has proved to be a fundamentally important peptide structure. Several representations of the α-helix are shown in Figure 5.10. One turn of the helix represents 3.6 amino acid residues. (A single turn of the α-helix involves 13 atoms, including the carbonyl O and the amide H of the H bond. For this

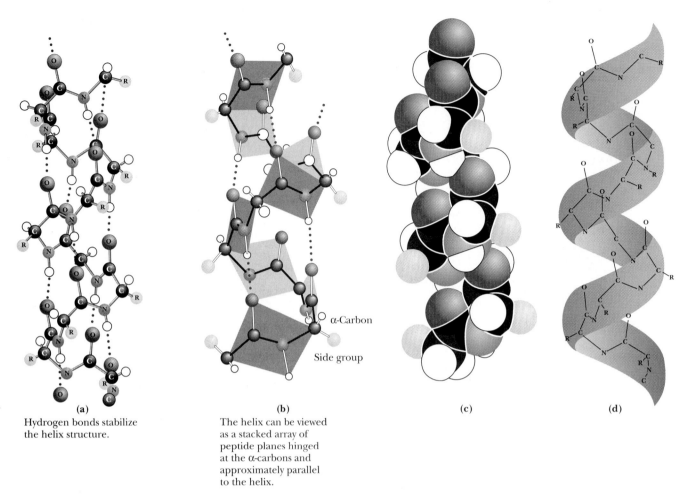

(a)

Hydrogen bonds stabilize the helix structure.

(b)

The helix can be viewed as a stacked array of peptide planes hinged at the α-carbons and approximately parallel to the helix.

α-Carbon

Side group

(c)

(d)

See pages 20–23, 26

Figure 5.10 Four different graphic representations of the α-helix. **(a)** As it originally appeared in Pauling's 1960 *The Nature of the Chemical Bond.* **(b)** Showing the arrangement of peptide planes in the helix. **(c)** A space-filling computer graphic presentation. **(d)** A "ribbon structure" with an inlaid stick figure, showing how the ribbon indicates the path of the polypeptide backbone. (*Irving Geis*)

reason, the α-helix is sometimes referred to as the 3.6_{13} helix.) This is in fact the feature that most confused crystallographers before the Pauling and Corey α-helix model. Crystallographers were so accustomed to finding twofold, three-fold, sixfold, and similar integral axes in simpler molecules that the notion of a nonintegral number of units per turn was never taken seriously before the work of Pauling and Corey.

Each amino acid residue extends 1.5 Å (0.15 nm) along the helix axis. With 3.6 residues per turn, this amounts to 3.6×1.5 Å or 5.4 Å (0.54 nm) of travel along the helix axis per turn. This is referred to as the translation distance or the **pitch** of the helix. If one ignores side chains, the helix is about 6 Å in diameter. The side chains, extending outward from the core structure of the helix, are removed from steric interference with the polypeptide backbone. As can be seen in Figure 5.10, each peptide carbonyl is hydrogen-bonded to the peptide N—H group four residues farther up the chain. Note that all of the H bonds lie parallel to the helix axis and that all of the carbonyl groups are pointing in one direction along the helix axis while the N—H groups are pointing in the opposite direction. Recall that the entire path of the peptide backbone can be known if the ϕ and ψ twist angles are specified for each residue. The α-helix is formed if the values of ϕ are approximately $-60°$ and

CRITICAL DEVELOPMENTS IN BIOCHEMISTRY

In Bed with a Cold, Pauling Stumbles onto the α-Helix and a Nobel Prize*

As high technology continues to transform the modern biochemical laboratory, it is interesting to reflect on Linus Pauling's discovery of the α-helix. It involved only a piece of paper, a pencil, scissors, and a sick Linus Pauling, who had tired of reading detective novels. The story is told in the excellent book *The Eighth Day of Creation* by Horace Freeland Judson (Simon & Schuster, New York, 1979):

> From the spring of 1948 through the spring of 1951 . . . rivalry sputtered and blazed between Pauling's lab and [Sir Lawrence] Bragg's—over protein. The prize was to propose and verify in nature a general three-dimensional structure for the polypeptide chain. Pauling was working up from the simpler structures of components. In January 1948, he went to Oxford as a visiting professor for two terms, to lecture on the chemical bond and on molecular structure and biological specificity. "In Oxford, it was April, I believe, I caught cold. I went to bed, and read detective stories for a day, and got bored, and thought why don't I have a crack at that problem of alpha keratin." Confined, and still fingering the polypeptide chain in his mind, Pauling called for paper, pencil, and straightedge and attempted to reduce the problem to an almost Euclidean purity. "I took a sheet of paper—I still have this sheet of paper—and drew, rather roughly, the way that I thought a polypeptide chain would look if it were spread out into a plane." The repetitious herringbone of the chain he could stretch across the paper as simply as this—

putting in lengths and bond angles from memory. . . . He knew that the peptide bond, at the carbon-to-nitrogen link, was always rigid:

And this meant that the chain could turn corners only at the alpha carbons. . . . "I creased the paper in parallel creases through the alpha carbon atoms, so that I could bend it and make the bonds to the alpha carbons, along the chain, have tetrahedral value. And then I looked to see if I could form hydrogen bonds from one part of the chain to the next." He saw that if he folded the strip like a chain of paper dolls into a helix, and if he got the pitch of the screw right, hydrogen bonds could be shown to form, $N—H \cdots O—C$, three or four knuckles apart along the backbone, holding the helix in shape. After several tries, changing the angle of the parallel creases in order to adjust the pitch of the helix, he found one where the hydrogen bonds would drop into place, connecting the turns, as straight lines of the right length. He had a model.

*The discovery of the α-helix structure was only one of many achievements that led to Pauling's Nobel Prize in chemistry in 1954. The official citation for the prize was "for his research into the nature of the chemical bond and its application to the elucidation of the structure of complex substances."

the values of ψ are in the range of $-45°$ to $-50°$. Figure 5.11 shows the structures of two proteins that contain α-helical segments. The number of residues involved in a given α-helix varies from helix to helix and from protein to protein. On average, there are about 10 residues per helix. Myoglobin, one of the first proteins in which α-helices were observed, has eight stretches of α-helix that form a box to contain the heme prosthetic group. The structures of the α and β subunits of hemoglobin are strikingly similar, with only a few differences at the C- and N-termini and on the surfaces of the structure that contact or interact with the other subunits of this multisubunit protein.

Careful studies of the **polyamino acids,** polymers in which all the amino acids are identical, have shown that certain amino acids tend to occur in α-helices, whereas others are less likely to be found in them. Polyleucine and polyalanine, for example, readily form α-helical structures. In contrast,

Figure 5.11 The three-dimensional structures of two proteins that contain substantial amounts of α-helix in their structures. The helices are represented by the regularly coiled sections of the ribbon drawings. Myohemerythrin is the oxygen-carrying protein in certain invertebrates, including Sipunculids, a phylum of marine worms. (*Jane Richardson*)

See pages 45–50

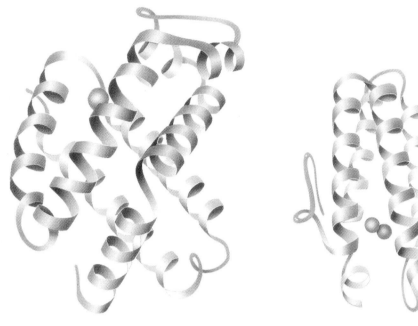

β-Hemoglobin subunit   Myohemerythrin

polyaspartic acid and polyglutamic acid, which are highly negatively charged at pH 7.0, form only random structures because of strong charge repulsion between the R groups along the peptide chain. At pH 1.5 to 2.5, however, where the side chains are protonated and thus uncharged, these latter species spontaneously form α-helical structures. In similar fashion, polylysine is a random coil at pH values below about 11, where repulsion of positive charges prevents helix formation. At pH 12, where polylysine is a neutral peptide chain, it readily forms an α-helix.

The relative tendencies of the common amino acids to stabilize or destabilize α-helices in typical proteins are summarized in Table 5.1. Notably, proline acts as a helix breaker due to its cyclic structure, which fixes the value of ϕ about the C_α — N bond. Helices can be formed from either D- or L-amino acids, but a given helix must be composed entirely of amino acids of one configuration. α-Helices cannot be formed from a mixed copolymer of D- and L-amino acids. An α-helix composed of D-amino acids is left-handed.

Other Helical Structures

There are several other, far less common, types of helices found in proteins. The most common of these is the 3_{10} helix, which contains 3.0 residues per turn (with 10 atoms in the ring formed by making the hydrogen bond three residues up the chain). It normally extends over shorter stretches of sequence than the α-helix.

The Beta-Pleated Sheet

In addition to helices, proteins frequently contain another type of structure stabilized by local, cooperative formation of hydrogen bonds. This is the **β-pleated sheet,** or simply β-sheet. This structure, also postulated by Pauling and Corey in 1951, has been observed in many natural proteins. A β-pleated sheet can be visualized by laying thin, pleated strips of paper side by side to make a "pleated sheet" of paper (Figure 5.12). Each strip of paper can then be pic-

Table 5.1   Helix-Forming and Helix-Breaking Behavior of the Amino Acids

Amino Acid		Helix Behavior*	
A	Ala	H	(I)
C	Cys	Variable	
D	Asp	Variable	
E	Glu	H	
F	Phe	H	
G	Gly	I	(B)
H	His	H	(I)
I	Ile	H	(C)
K	Lys	Variable	
L	Leu	H	
M	Met	H	
N	Asn	C	(I)
P	Pro	B	
Q	Gln	H	(I)
R	Arg	H	(I)
S	Ser	C	(B)
T	Thr	Variable	
V	Val	Variable	
W	Trp	H	(C)
Y	Tyr	H	(C)

*H = helix former; I = indifferent; B = helix breaker; C = random coil; ( ) = secondary tendency

tured as a single peptide strand in which the peptide backbone follows the pleated pattern of the strip, with the α-carbons lying at the folds of the pleats. The pleated sheet can exist in both parallel and antiparallel forms. In the parallel β-pleated sheet, adjacent chains run in the same direction (N → C or C → N). In the antiparallel β-pleated sheet, adjacent strands run in opposite directions.

Each single strand of the β-sheet structure can be pictured as a twofold helix, that is, a helix with two residues per turn. The arrangement of successive amide planes has a pleated appearance due to the tetrahedral nature of the $C_α$ atom. It is important to note that the hydrogen bonds in this structure are interstrand rather than intrastrand. The peptide backbone in the antiparallel β-sheet is in its most extended conformation, whereas optimum formation of H bonds in the parallel pleated sheet results in a slightly less extended conformation. The H bonds thus formed in the parallel β-sheet are bent significantly. The distance between α-carbons (i.e., the "rise per residue" along the strand axis) is 0.347 nm for the antiparallel pleated sheet, but only 0.325 nm for the parallel pleated sheet. Figure 5.13 shows examples of both parallel and antiparallel β-pleated sheets. Note that the side chains in the pleated sheet are oriented perpendicular or normal to the plane of the sheet, extending out from the plane on alternating sides.

Parallel β-sheets tend to be more regular than antiparallel β-sheets. The range of ϕ and ψ angles for the peptide bonds in parallel sheets is much smaller than that for antiparallel sheets. Parallel sheets are typically large structures; those composed of less than five strands are rare. Antiparallel sheets, however, may consist of as few as two strands. Parallel sheets characteristically distribute hydrophobic side chains on both sides of the sheet, while antiparallel sheets are usually arranged with all their hydrophobic residues on one side of the sheet. This often results in an alternation of hydrophilic and hydrophobic residues in the primary structure of peptides involved in antiparallel β-sheets, since alternate side chains project to the same side of the sheet (see Figure 5.12).

Antiparallel pleated sheets are the fundamental structure found in silk, with the polypeptide chains forming the sheets running parallel to the silk fibers. The silk fibers thus formed have properties consistent with those of the

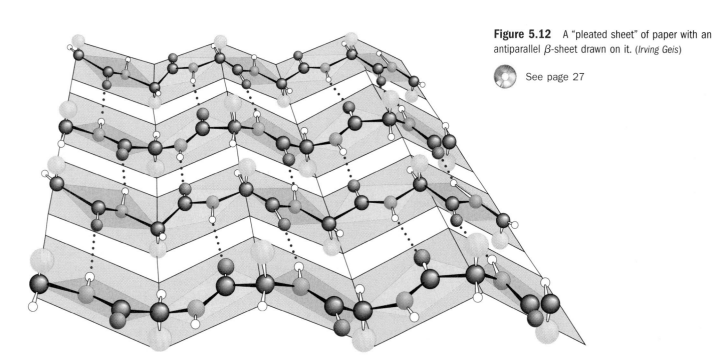

Figure 5.12 A "pleated sheet" of paper with an antiparallel β-sheet drawn on it. (*Irving Geis*)

See page 27

Figure 5.13 The arrangement of hydrogen bonds in **(a)** parallel and **(b)** antiparallel β-pleated sheets.

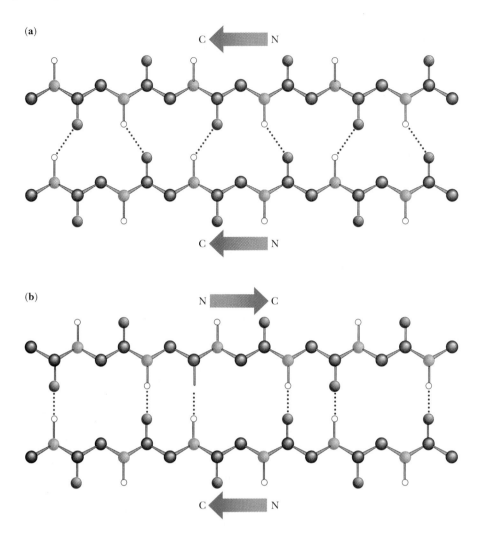

β-sheets that form them. They are quite flexible, but cannot be stretched or extended to any appreciable degree. Antiparallel structures are also observed in many other proteins, including immunoglobulin G, superoxide dismutase from bovine erythrocytes, and concanavalin A. Many proteins, including carbonic anhydrase, egg lysozyme, and glyceraldehyde phosphate dehydrogenase, possess both α-helices and β-pleated sheet structures within a single polypeptide chain.

The Beta-Turn

Most proteins are globular structures. The polypeptide chain must therefore possess the capacity to bend, turn, and reorient itself to produce the required compact, globular structures. A simple structure observed in many proteins is the **β-turn** (also known as the tight turn, reverse turn, or β-bend), in which the peptide chain forms a tight loop with the carbonyl oxygen of one residue hydrogen-bonded with the amide proton of the residue three positions down the chain. This H bond makes the β-turn a relatively stable structure. As shown in Figure 5.14, the β-turn allows the protein to reverse the direction of its peptide chain. This figure shows the two major types of β-turns, but a number of less common types are also found in protein structures. Certain amino acids, such as proline and glycine, occur frequently in β-turn sequences, and the particular conformation of the β-turn sequence depends to some extent on the amino acids composing it. Due to the absence of a side chain, glycine is sterically the most adaptable of the amino acids, and it accommodates conveniently to other steric constraints in the β-turn. Proline, however, has a cyclic structure and a fixed ϕ angle, so, to some extent, it forces the formation of a β-turn, and in many cases this facilitates the turning of a polypeptide chain upon itself. Such bends promote formation of antiparallel β-pleated sheets.

The secondary structures we have described here are all found commonly in proteins in nature. In fact, it is hard to find proteins that do not contain one or more of these structures. The energetic (mostly H-bond) stabilization afforded by α-helices, β-pleated sheets, and β-turns is important to proteins, and they form such structures wherever possible.

5.4 Protein Folding and Tertiary Structure

The folding of a single polypeptide chain in three-dimensional space is referred to as its **tertiary structure.** As discussed in Section 5.2, all of the information needed to fold the protein into its native tertiary structure is contained within the primary structure of the peptide chain itself. With this in mind, it was disappointing to the biochemists of the 1950s when the early protein structures did not reveal the governing principles in any particular detail. It soon became apparent that the proteins knew how they were supposed to fold into tertiary shapes, even if the biochemists did not. Vigorous work in many laboratories has slowly brought important principles to light.

First, secondary structures—helices and sheets—will form whenever possible as a consequence of the formation of large numbers of hydrogen bonds. Second, α-helices and β-sheets often associate and pack close together in the protein. *No protein is stable as a single-layer structure,* for reasons that will become apparent later. There are a few common methods by which such packing occurs. Third, the peptide segments between secondary structures in the protein tend to be short and direct; the peptide does not execute complicated twists and knots as it moves from one region of a secondary structure to another.

A consequence of these three principles is that protein chains are usually folded so that the secondary structures are arranged in one of a few common patterns. For this reason, there are families of proteins that have similar tertiary structure, with little apparent evolutionary or functional relationship among them. Finally, proteins generally fold so as to form the most stable structures possible. The stability of most proteins arises from (1) the formation of

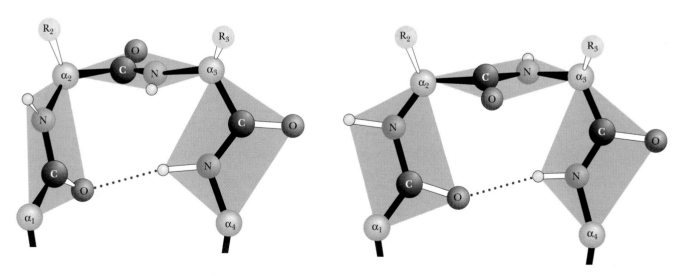

Figure 5.14 The structures of two kinds of β-turns (also called tight turns or β-bends). (*Irving Geis*)

See page 29

large numbers of intramolecular hydrogen bonds and (2) the reduction in the hydrophobic surface area accessible to solvent that occurs upon folding.

Fibrous Proteins

Fibrous proteins contain polypeptide chains organized approximately parallel along a single axis, producing long fibers or large sheets. Such proteins tend to be mechanically strong and resistant to solubilization in water and dilute salt solutions. Fibrous proteins often play a structural role in nature.

α-Keratin

The structure of the **α-keratins** is, as their name suggests, dominated by α-helical segments of polypeptide. The amino acid sequence of α-keratin subunits is composed of central α-helix–rich rod domains about 311 to 314 residues in length, flanked by nonhelical N- and C-terminal domains of varying size and composition (Figure 5.15a). The structure of the central rod domain of a typical α-keratin is shown in Figure 5.15b. It consists of four helical strands arranged as twisted pairs of two-stranded **coiled coils.** X-ray diffraction patterns show that these structures resemble α-helices, but with a pitch of 0.51 nm rather than the expected 0.54 nm. This is consistent with a tilt of each α-helix relative to the long axis of the fiber, as in the two-stranded coiled coil in Figure 5.15.

The primary structure of the central rod segments of α-keratin consists of quasi-repeating seven-residue segments of the form $(a\text{-}b\text{-}c\text{-}d\text{-}e\text{-}f\text{-}g)_n$. These units are not true repeats, but residues a and d are usually nonpolar amino acids. In α-helices, with 3.6 residues per turn, these nonpolar residues are arranged in an inclined row or stripe that twists around the helix axis. These nonpolar residues would make the helix highly unstable if they were exposed to solvent, but the association of hydrophobic strips on two coiled coils to form the two-

(a)

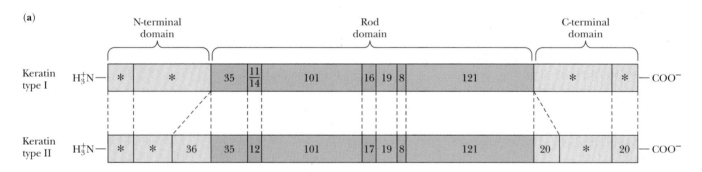

(b)

α-Helix

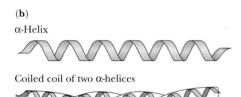

Coiled coil of two α-helices

Protofilament (pair of coiled coils)

Filament (four right-hand twisted protofibrils)

Figure 5.15   (a) Both type I and type II α-keratin molecules have sequences consisting of long, central rod domains with terminal cap domains. The numbers of amino acid residues in each domain are indicated. Asterisks denote domains of variable length.   **(b)** The rod domains form coiled coils consisting of intertwined right-handed α-helices wound around each other in a left-handed twist. Keratin filaments consist of twisted protofibrils (each a bundle of four coiled coils). (*Adapted from Steinert, P., and Parry, D., 1985. Annual Review of Cell Biology **1**:41–65, and Cohlberg, J., 1993. Trends in Biochemical Sciences **18**:360–362.*)

Keratin, Coiled Coils, Permanent Waves, and the Frizzies

The macroscopic structure and appearance of one's hair is a consequence of the interplay between hydrogen bonds (in alpha helices and coiled coils) and covalent disulfide bonds between cysteines. How and where these disulfides form determines the amount of curling in hair and wool fibers. When a hairstylist creates a permanent wave (simply called a "permanent") in a hair salon, disulfides in the hair are first reduced and cleaved, then re-

organized and reoxidized to change the degree of curl or wave. In contrast, a "set" that is created by wetting the hair, rolling it with curlers, and then drying it represents merely a rearrangement of the hydrogen bonds between helices and between fibers. On humid or rainy days, the hydrogen bonds in curled hair may rearrange, and the hair will become "frizzy."

---

stranded protofilament effectively buries the hydrophobic residues and forms a highly stable structure (see Figure 5.15). In addition to these interactions, covalent disulfide bonds form between cysteine residues of adjacent fibers, making the overall structure rigid, inextensible, and insoluble—important properties for structures such as claws, fingernails, hair, and horns in animals.

Fibroin and β-Keratin: β-Sheet Proteins

The **fibroin** proteins found in silk fibers represent another type of fibrous protein. These are composed of stacked antiparallel β-sheets, as shown in Figure 5.16. In the polypeptide sequence of silk proteins, there are large stretches in which every other residue is a glycine. As previously mentioned, the residues of a β-sheet extend alternately above and below the plane of the sheet. As a result, the glycines all end up on one side of the sheet and the other residues (mainly alanines and serines) compose the opposite surface of the sheet. Pairs of β-sheets can then pack snugly together (glycine surface to glycine surface or alanine–serine surface to alanine–serine surface). The β-keratins found in bird feathers are also made up of stacked β-sheets.

Figure 5.16 Silk fibroin consists of a unique stacked array of β-sheets. The primary structure of fibroin molecules consists of long stretches of alternating glycine and alanine or serine residues. When the sheets stack, the more bulky alanine and serine residues on one side of a sheet interdigitate with similar residues on an adjoining sheet. Glycine hydrogens on the alternating faces interdigitate in a similar manner, but with a smaller intersheet spacing. (*Irving Geis*)

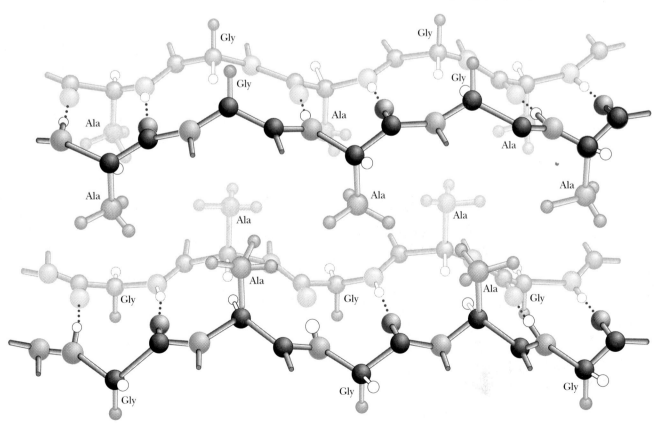

A DEEPER LOOK

The Coiled Coil Motif in Proteins

The **coiled coil** motif was first identified in 1953 by Linus Pauling, Robert Corey, and Francis Crick as the main structural element of fibrous proteins such as keratin and myosin. Since that time, many proteins have been found to contain one or more coiled coil segments or domains. A coiled coil is a bundle of α-helices that are wound into a superhelix. Two, three, or four helical segments may be found in the bundle, and they may be arranged parallel or antiparallel to one another. Coiled coils are characterized by a distinctive and regular packing of side chains in the core of the bundle. This regular meshing of side chains requires that they occupy equivalent positions turn after turn. This is not possible for undistorted α-helices, which have 3.6 residues per turn. The positions of side chains on their surface shift continuously along the helix surface (see figure). However, giving the right-handed α-helix a left-handed twist reduces the number of residues per turn to 3.5, and, because 3.5 times 2 equals 7.0, the positions of the side chains repeat after two turns (seven residues). Thus, a **heptad repeat** pattern in the peptide sequence is diagnostic of a coiled coil structure. The figure shows a sampling of coiled coil structures (highlighted in color) in various proteins.

See pages 57, 58, 90, 91, 95

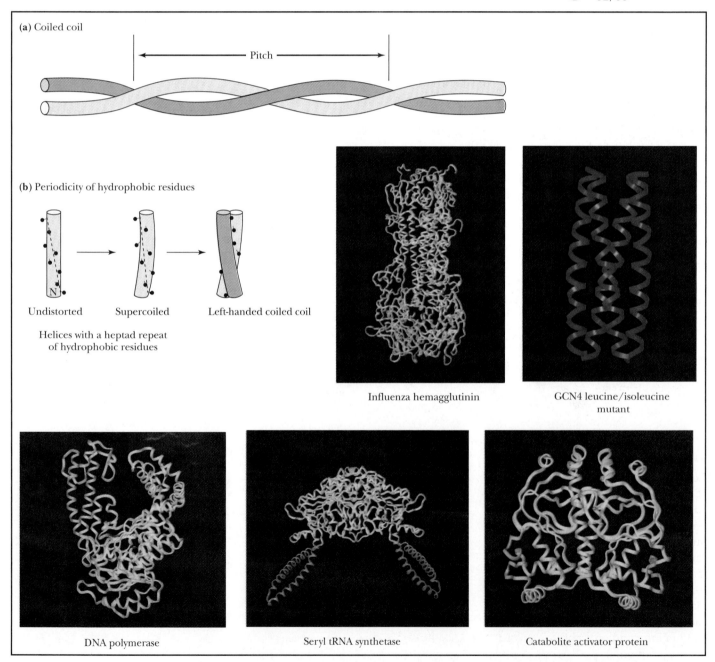

(a) Coiled coil

Pitch

(b) Periodicity of hydrophobic residues

Undistorted   Supercoiled   Left-handed coiled coil

Helices with a heptad repeat
of hydrophobic residues

Influenza hemagglutinin

GCN4 leucine/isoleucine
mutant

DNA polymerase

Seryl tRNA synthetase

Catabolite activator protein

A DEEPER LOOK

Charlotte's Web Revisited: Helix–Sheet Composites in Spider Dragline Silk

E. B. White's endearing story *Charlotte's Web* centers around the web-spinning feats of Charlotte the spider. Though the intricate designs of spiderwebs are eye- (and fly-) catching, it might be argued that the composition of web silk itself is even more remarkable. Spider silk is synthesized in special glands in the spider's abdomen. The silk strands produced by these glands are both strong and elastic. Dragline silk (that from which the spider hangs) has a tensile strength of 200,000 psi (pounds per square inch)—stronger than steel and similar to Kevlar, the synthetic material used in bulletproof vests! This same silk fiber is also flexible enough to withstand strong winds and other natural stresses.

This combination of strength and flexibility derives from the composite nature of spider silk. As keratin protein is extruded from the spider's glands, it encounters shearing forces that break the H

bonds stabilizing keratin α-helices. These regions then form microcrystalline arrays of β-sheets. These microcrystals are surrounded by the keratin strands, which adopt a highly disordered state composed of α-helices and folded structures devoid of helices and sheets.

The β-sheet microcrystals contribute strength, and the disordered array of helix and coil makes the silk strand flexible. The resulting silk strand resembles modern human-engineered composite materials. Certain tennis racquets, for example, consist of fiberglass polymers impregnated with microcrystalline graphite. The fiberglass provides flexibility, and the graphite crystals contribute strength. Modern high technology, for all its sophistication, is merely imitating nature—and Charlotte's web—after all.

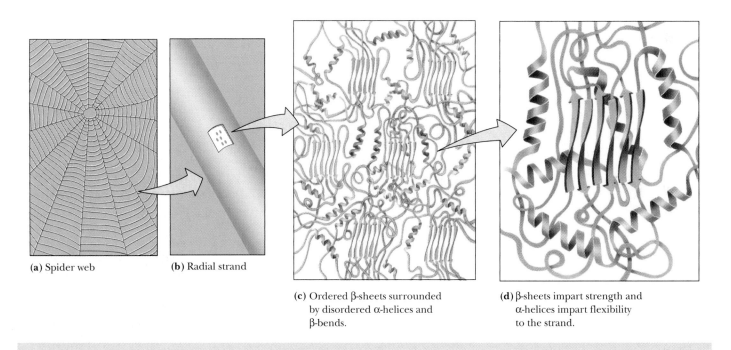

(a) Spider web   (b) Radial strand

(c) Ordered β-sheets surrounded by disordered α-helices and β-bends.

(d) β-sheets impart strength and α-helices impart flexibility to the strand.

Collagen: A Triple Helix

Collagen is a rigid, inextensible fibrous protein that is a principal constituent of connective tissue in animals, including tendons, cartilage, bones, teeth, skin, and blood vessels. The high tensile strength of collagen fibers in these structures makes possible the various activities such as running and jumping that put severe stresses on joints and the skeleton. Broken bones and tendon and cartilage injuries to knees, elbows, and other joints involve tears or hyperextensions of the collagen matrix in these tissues.

The basic structural unit of collagen is **tropocollagen,** which has a molecular weight of 285,000 and consists of three intertwined polypeptide chains, each about 1000 amino acids in length. Tropocollagen molecules are about 300 nm long and only about 1.4 nm in diameter. Several kinds of collagen have been identified. Type I collagen, which is the most common, consists of two identical peptide chains designated α1(I) and one different chain designated

See page 30

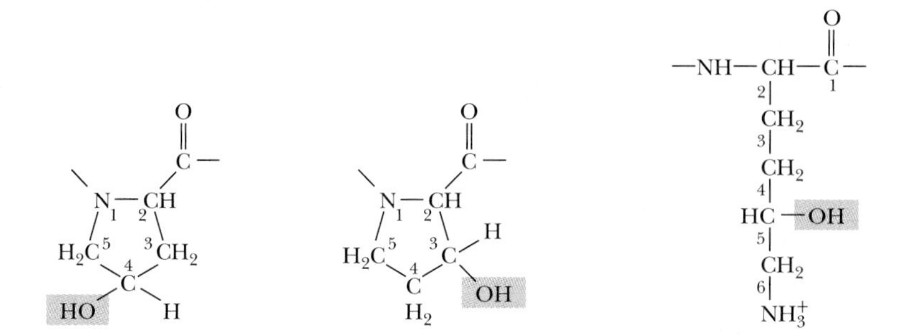

Figure 5.17 The hydroxylated residues typically found in collagen.

4-Hydroxyprolyl residue
(Hyp)

3-Hydroxyprolyl residue

5-Hydroxylysyl residue (Hyl)

$\alpha 2$(I). *Type I collagen* predominates in bones, tendons, and skin. *Type II collagen,* found in cartilage, and *type III collagen,* found in blood vessels, consist of three identical polypeptide chains.

Collagen has an amino acid composition that is crucial to its three-dimensional structure and its characteristic physical properties. Nearly one residue out of three is a glycine, and the proline content is also remarkably high. Three unusual modified amino acids are also found in collagen: 4-hydroxyproline (Hyp), 3-hydroxyproline, and 5-hydroxylysine (Hyl) (Figure 5.17). Proline and Hyp together compose up to 30% of the residues of collagen. Interestingly, these unusual amino acids are formed from normal proline and lysine *after* the collagen polypeptides are synthesized. The modifications are effected by two enzymes: *prolyl hydroxylase* and *lysyl hydroxylase.* The prolyl hydroxylase reaction (Figure 5.18) requires molecular oxygen, α-ketoglutarate, and ascorbic acid (vitamin C) and is activated by Fe^{2+}. The hydroxylation of lysine is similar. These processes are referred to as **post-translational modifi-**

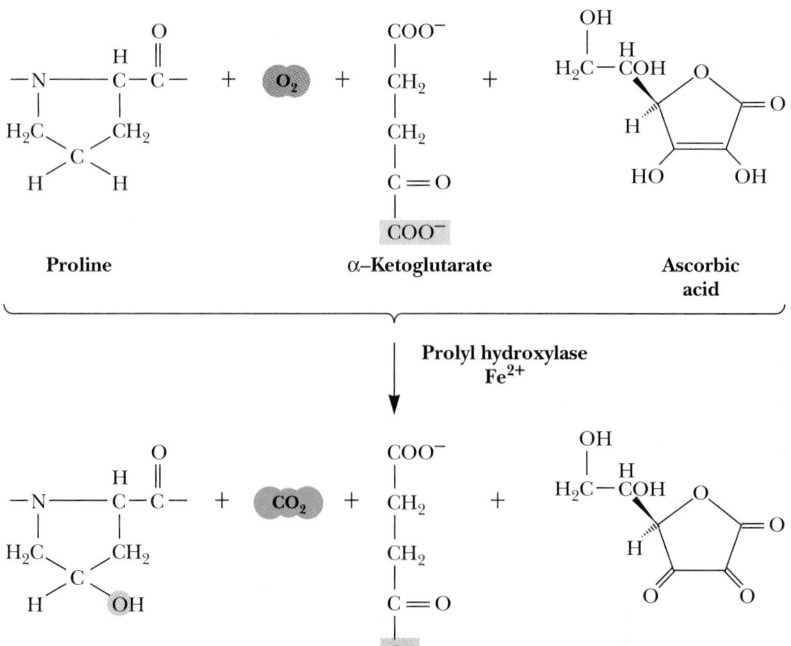

Proline

α–Ketoglutarate

Ascorbic acid

Prolyl hydroxylase
Fe^{2+}

Hydroxyproline

Succinate

Dehydroascorbate

Figure 5.18 Hydroxylation of proline residues is catalyzed by prolyl hydroxylase. The reaction requires O_2, α-ketoglutarate, and ascorbic acid (vitamin C).

Figure 5.19 Poly(Gly-Pro-Pro), a collagenlike right-handed triple helix composed of three left-handed helical chains. (*Adapted from Miller, M. H., and Scheraga, H. A., 1976. Calculation of the structures of collagen models. Role of interchain interactions in determining the triple-helical coiled-coil conformation. I. Poly(glycyl-prolyl-prolyl). Journal of Polymer Science Symposium No. **54**:171–200.*)

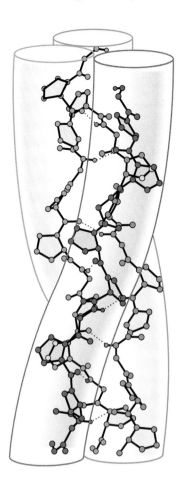

 See page 30

cations, since they occur after genetic information from DNA has been *translated* into newly formed protein.

Because of their high content of glycine, proline, and hydroxyproline, collagen fibers are incapable of forming traditional structures such as α-helices and β-sheets. Instead, collagen polypeptides intertwine to form a right-handed **triple helix,** in which each of the three strands is arranged in a left-handed helical fashion (Figure 5.19). Compared with the α-helix, the collagen helix is much more extended, with a rise per residue along the triple helix axis of 2.9 Å, compared with 1.5 Å for the α-helix. There are about 3.3 residues per turn of each of these helices. *The triple helix is a structure that forms to accommodate the unique composition and sequence of collagen.* Long stretches of the polypeptide sequence are repeats of a Gly-x-y motif, where x is frequently Pro and y is frequently Pro or Hyp. In the triple helix, every third residue faces or contacts the crowded center of the structure. This area is so crowded that only Gly will fit, and thus every third residue must be a Gly (as observed). Moreover, the triple helix is a staggered structure, such that Gly residues from the three strands stack along the center of the triple helix and the Gly from one strand lies adjacent to an x residue from the second strand and to a y from the third. This allows the N—H of each Gly residue to hydrogen bond with the C=O of the adjacent x residue. The triple helix structure is further stabilized and strengthened by the formation of interchain H bonds involving hydroxyproline.

HUMAN BIOCHEMISTRY

Collagen-Related Diseases

Collagen provides an ideal case study of the molecular basis of physiology and disease. For example, the nature and extent of collagen cross-linking depends on the age and function of the tissue. Collagen from young animals is predominantly un–cross-linked and can be extracted in soluble form, whereas collagen from older animals is highly cross-linked and thus insoluble. The loss of flexibility in joints with aging is probably due in part to increased cross-linking of collagen.

Several serious and debilitating diseases involving collagen abnormalities are known. **Lathyrism,** which occurs in animals that regularly consume seeds of *Lathyrus odoratus* (the sweet pea), involves weakening and abnormalities in blood vessels, joints, and bones. These conditions are caused by β-**aminopropionitrile** (see figure), which covalently inactivates lysyl oxidase and leads to greatly reduced intramolecular cross-linking of collagen.

$$N \equiv C - CH_2 - CH_2 - \overset{+}{N}H_3$$

β-aminopropionitrile (present in sweet peas) covalently inactivates lysyl oxidase, preventing intramolecular cross-linking of collagen and causing abnormalities in joints, bones, and blood vessels.

Scurvy results from a dietary vitamin C deficiency and involves the inability to form collagen fibrils properly. This is the result of reduced activity of prolyl hydroxylase, which is vitamin C–dependent, as previously noted. Scurvy leads to lesions in the skin and blood vessels, and, in its advanced stages, it can lead to grotesque disfiguration, bleeding, and eventual death. Although rare in the modern world, it was a disease well known to seafaring explorers in earlier times who did not appreciate the importance of fresh fruits and vegetables in the diet.

A number of rare genetic diseases involve collagen abnormalities, including *Marfan's syndrome* and the *Ehlers–Danlos syndromes*, which result in hyperextensible joints and skin. The formation of *atherosclerotic plaques*, which cause arterial blockages in advanced stages, is due in part to the abnormal formation of collagenous structures in blood vessels.

Globular Proteins

Fibrous proteins, while interesting for their structural properties, represent only a small percentage of the proteins found in nature. **Globular proteins,** so named for their approximately spherical shape, are far more numerous.

Helices and Sheets in Globular Proteins

Globular proteins exist in an enormous variety of three-dimensional structures, but nearly all contain substantial amounts of the α-helices and β-sheets that form the basic structures of the simple fibrous proteins. For example, myoglobin, a small, globular, oxygen-carrying protein of muscle (17 kD, 153 amino acid residues), contains eight α-helical segments, each containing 7 to 26 amino acid residues. These are arranged in an apparently irregular (but invariant) fashion (see Figure 5.1). The space between the helices is tightly packed with (mostly hydrophobic) amino acid side chains. Most of the polar side chains in myoglobin (and in most other globular proteins) face the outside of the protein structure and interact with solvent water. Myoglobin's structure is unusual, since most globular proteins contain relatively smaller amounts of α-helix. A more typical globular protein (Figure 5.20) is ribonuclease A, a small protein (12.6 kD, 124 residues) that contains a few short helices, a small section of antiparallel β-sheet, a few β-turns, and substantial peptide segments without defined secondary structure.

Why should the cores of most globular and membrane proteins consist almost entirely of α-helices and β-sheets? The reason is that the highly polar $N\!-\!H$ and $C\!=\!O$ moieties of the peptide backbone must be neutralized in the hydrophobic core of the protein. The extensively H-bonded nature of α-helices and β-sheets is ideal for this purpose, and these structures effectively stabilize the polar groups of the peptide backbone in the protein core.

Figure 5.20 The three-dimensional structure of bovine ribonuclease A, showing the α-helices as ribbons. (*Jane Richardson*)

See page 48

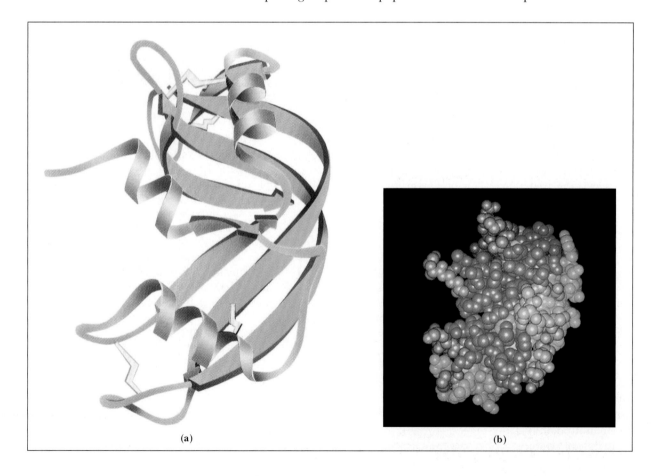

(a)   (b)

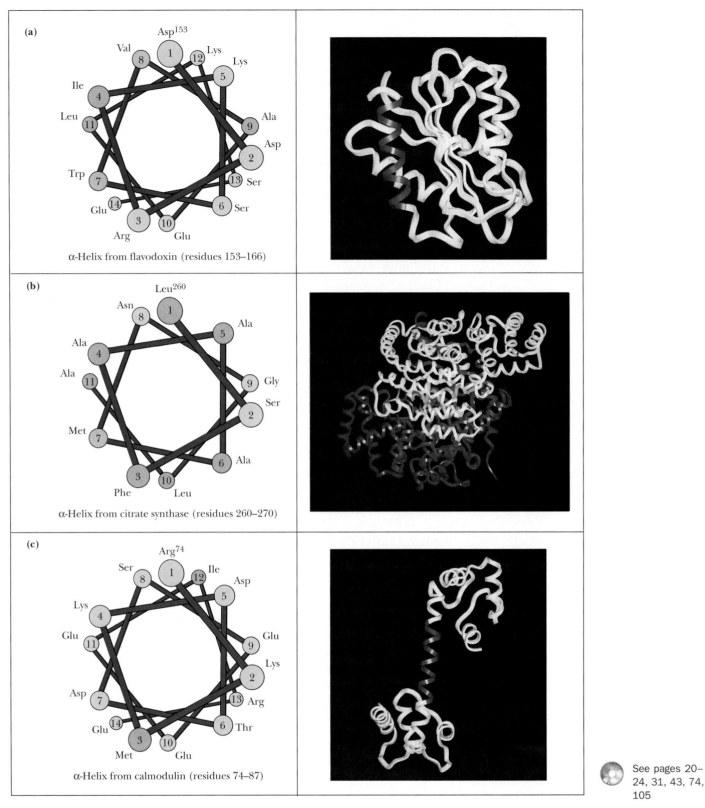

(a)

Asp^{153}

Val — 8

Asp^{153} — 1

Lys — 12

Lys — 5

Ile — 4

Ala — 9

Leu — 11

Asp — 2

Trp — 7

Ser — 13

Glu — 14

Ser — 6

Arg — 3

Glu — 10

α-Helix from flavodoxin (residues 153–166)

(b)

Leu^{260}

Asn — 8

Leu^{260} — 1

Ala — 5

Ala — 4

Gly — 9

Ala — 11

Ser — 2

Met — 7

Ala — 6

Phe — 3

Leu — 10

α-Helix from citrate synthase (residues 260–270)

(c)

Arg^{74}

Ser — 8

Arg^{74} — 1

Ile — 12

Asp — 5

Lys — 4

Glu — 9

Glu — 11

Lys — 2

Asp — 7

Arg — 13

Glu — 14

Thr — 6

Met — 3

Glu — 10

α-Helix from calmodulin (residues 74–87)

See pages 20–24, 31, 43, 74, 105

Figure 5.21 **(a)** The α-helix consisting of residues 153-166 (red) in flavodoxin from *Anabaena* is a surface helix and is amphipathic. **(b)** The two helices (yellow and blue) in the interior of the citrate synthase dimer (residues 260−270) in each monomer) are mostly hydrophobic. **(c)** The exposed helix (residues 74-87—red) of calmodulin is entirely accessible to solvent and consists mainly of polar and charged residues.

In globular protein structures, it is common for one face of an α-helix to be exposed to the water solvent, with the other face toward the hydrophobic interior of the protein. The outward face of such an **amphipathic helix** consists mainly of polar and charged residues, whereas the inward face contains mostly nonpolar, hydrophobic residues. A good example of such a surface helix is that of residues 153 to 166 of **flavodoxin** from *Anabaena* (Figure 5.21).

Note that the **helical wheel presentation** of this helix readily shows that one face contains four hydrophobic residues and that the other is almost entirely polar and charged.

Less commonly, an α-helix can be completely buried in the protein interior or completely exposed to solvent. **Citrate synthase** is a dimeric protein in which α-helical segments form part of the subunit–subunit interface. As shown in Figure 5.21, one of these helices (residues 260 to 270) is highly hydrophobic and contains only two polar residues, as would befit a helix in the protein core. On the other hand, Figure 5.21 also shows the solvent-exposed helix (residues 74 to 87) of **calmodulin,** which consists of 10 charged residues, 2 polar residues, and only 2 nonpolar residues.

Packing Considerations

The secondary and tertiary structures of myoglobin and lysozyme illustrate the importance of packing in tertiary structures. Secondary structures pack closely to one another and also intercalate with (insert between) extended polypeptide chains. If the sum of the van der Waals volumes of a protein's constituent amino acids is divided by the volume occupied by the protein, *packing densities* of 0.72 to 0.77 are typically obtained.[1] This means that, even with close packing, approximately 25% of the total volume of a protein is not occupied by protein atoms. Nearly all of this space is in the form of very small cavities. Cavities the size of water molecules or larger do occasionally occur, but they make up only a small fraction of the total protein volume. It is likely that such cavities provide flexibility for proteins and facilitate conformation changes and a wide range of protein dynamics (to be discussed later).

Ordered, Nonrepetitive Structures

In any protein structure, the segments of the polypeptide chain that cannot be classified as defined secondary structures, such as helices or sheets, have been traditionally referred to as *coil* or *random coil.* Both these terms are misleading. Most of these segments are neither coiled nor random, in any sense of those words. These structures are every bit as highly organized and stable as α-helices or β-sheets. They are just more variable and difficult to describe. These so-called coil structures are strongly influenced by side-chain interactions. Few of these interactions are well understood.

Nonrepetitive but well-defined structures of this type form many important features of enzyme active sites. In some cases, a particular arrangement of "coil" structure providing a specific type of functional site recurs in several functionally related proteins. One example is the central loop portion of the *E–F hand structure* that binds a calcium ion in several calcium-binding proteins, including calmodulin, carp parvalbumin, troponin C, and the intestinal calcium-binding protein. This loop, shown in Figure 5.22, connects two short α-helices. The calcium ion nestles into the pocket formed by this structure.

Flexible, Disordered Segments

In addition to nonrepetitive but well-defined structures, which exist in all proteins, genuinely disordered segments of polypeptide sequence also occur. These sequences either do not show up in electron density maps from X-ray crystallographic studies or give diffuse or ill-defined electron densities. These segments either undergo actual motion in the protein crystals themselves or take on many alternate conformations in different molecules within the protein

---

intercalate to insert between

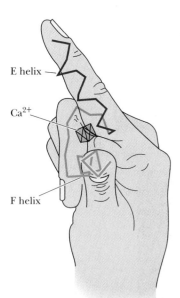

Figure 5.22 A representation of the so-called E–F hand structure, which forms calcium-binding sites in a variety of proteins. The stick drawing shows the peptide backbone of the E–F hand motif. The "E" helix extends along the index finger, a loop traces the approximate arrangement of the curled middle finger, and the "F" helix extends outward along the thumb. A calcium ion (Ca^{2+}) snuggles into the pocket created by the two helices and the loop. Robert Kretsinger and coworkers, who first described this common structural motif, originally assigned letters alphabetically to the helices in parvalbumin, a protein from carp. The E–F hand derives its name from the letters assigned to the helices at one of the Ca^{2+}-binding sites.

See page 105

---

[1]These packing densities are of the same order as hard-sphere packing, e.g., spherical glass beads in a jar.

crystal. Such behavior is quite common for long, charged side chains on the surface of many proteins. For example, 16 of the 19 lysine side chains in myoglobin have uncertain orientations beyond the δ-carbon, and five of these are disordered beyond the β-carbon. Similarly, a majority of the lysine residues are disordered in trypsin, rubredoxin, ribonuclease, and several other proteins. Arginine residues, however, are usually well ordered in protein structures. For the four proteins just mentioned, 70% of the arginine residues are highly ordered, compared with only 26% of the lysines.

Motion in Globular Proteins

Although we have distinguished between well-ordered and disordered segments of the polypeptide chain, it is important to realize that even well-ordered side chains in a protein undergo motion, sometimes quite rapid. These motions should be viewed as momentary oscillations about a single, highly stable conformation. *Proteins are thus best viewed as dynamic structures.* The allowed motions may be motions of individual atoms, groups of atoms, or even whole sections of the protein. Furthermore, they may arise from either thermal energy or specific, triggered conformational changes in the protein. **Atomic fluctuations** such as vibrations typically are random, very fast, and usually occur over small distances (less than 0.5 Å), as shown in Table 5.2. These motions arise from the kinetic energy within the protein and are a function of temperature. These very fast motions can be modeled by molecular dynamics calculations and studied by X-ray diffraction.

A class of slower motions, which may extend over larger distances, is **collective motions.** These are movements of groups of atoms covalently linked in such a way that the group moves as a unit. Such groups range in size from a few atoms to hundreds of atoms. These collective motions also arise from thermal energies in the protein and operate on a time scale of 10^{-12} to 10^{-3} sec. These motions can be studied by nuclear magnetic resonance (NMR) and fluorescence spectroscopy.

Conformational changes involve motions of groups of atoms (individual side chains, for example) or even whole sections of proteins. These motions occur on a time scale of 10^{-9} to 10^{3} sec, and the distances covered can be as large as 1 nm. These motions may occur in response to specific stimuli or arise from specific interactions within the protein, such as hydrogen bonding, electrostatic interactions, and ligand binding. More will be said about conformational changes when enzyme catalysis and regulation are discussed (Chapters 10 and 11).

Table 5.2 Motion and Fluctuations in Proteins

Type of Motion	Spatial Displacement (Å)	Characteristic Time (sec)	Source of Energy
Atomic vibrations	0.01–1	10^{-15}–10^{-11}	Kinetic energy
Collective motions	0.01–5 or more	10^{-12}–10^{-3}	Kinetic energy
1. Fast: Tyr ring flips; methyl group rotations			
2. Slow: hinge bending between domains			
Triggered conformation changes	0.5–10 or more	10^{-9}–10^{3}	Interactions with triggering agent

Adapted from Petsko and Ringe (1984).

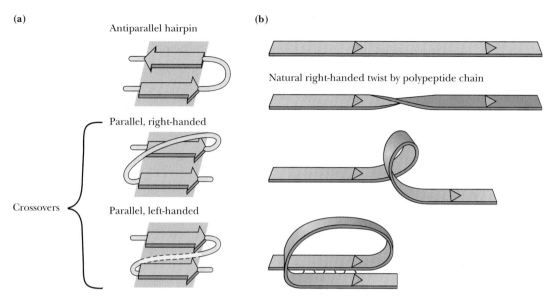

(a)

Antiparallel hairpin

Parallel, right-handed

Crossovers

Parallel, left-handed

(b)

Natural right-handed twist by polypeptide chain

Figure 5.23 The natural right-handed twist exhibited by polypeptide chains, and the variety of structures that arise from this twist.

See page 43

Forces Driving the Folding of Globular Proteins

As already pointed out, the driving force for protein folding and the resulting formation of a tertiary structure is the formation of the most stable structure possible. There are two forces at work here. The peptide chain must both (1) satisfy the constraints inherent in its own structure and (2) fold so as to "bury" the hydrophobic side chains, minimizing their contact with solvent. The polypeptide itself does not usually form simple straight chains. Even in chain segments where helices and sheets are not formed, an extended peptide chain, composed of L-amino acids, has a tendency to twist slightly in a right-handed direction. As shown in Figure 5.23, this tendency is apparently the basis for the formation of a variety of tertiary structures having a right-handed sense. Principal among these are the right-handed twists in arrays of β-sheets and right-handed cross-overs in parallel β-sheet arrays. Right-handed twisted β-sheets are found at the center of a number of proteins and provide an extended, highly stable structural core. Phosphoglycerate mutase, adenylate kinase, and carbonic anhydrase, among others, exist as smoothly twisted planes or saddle-shaped structures. Triose phosphate isomerase, soybean trypsin inhibitor, and domain 1 of pyruvate kinase contain right-handed twisted cylinders or barrel structures at their cores.

Connections between β-strands are of two types—hairpins and cross-overs. **Hairpins,** as shown in Figure 5.23, connect adjacent antiparallel β-strands. **Cross-overs** are necessary to connect adjacent (or nearly adjacent) parallel β-strands. Nearly all cross-over structures are right-handed. In many cross-over structures, the cross-over connection itself contains an α-helical segment. This is referred to as a **βαβ-loop.** As shown in Figure 5.23, the strong tendency in nature to form right-handed cross-overs, the wide occurrence of α-helices in the cross-over connection, and the right-handed twists of β-sheets can all be understood as arising from the tendency of an extended polypeptide chain of L-amino acids to adopt a right-handed twist structure. This is a chiral effect. Proteins composed of D-amino acids would tend to adopt left-handed twist structures.

The second driving force that affects the folding of polypeptide chains is the need to bury the hydrophobic residues of the chain, protecting them from solvent water. From a topological viewpoint, then, all globular proteins must have an "inside" where the hydrophobic core can be arranged and an "outside" toward which the hydrophilic groups must be directed. The sequestration of hydrophobic residues away from water is the dominant force in the arrangement of secondary structures and nonrepetitive peptide segments to form a given tertiary structure. Globular proteins can mainly be classified on the basis of the particular kind of core or backbone structure they use to accomplish this goal. The term *hydrophobic core*, as used here, refers to a region in which hydrophobic side chains cluster together, away from the solvent. Backbone refers to the polypeptide backbone itself, excluding the particular side chains. Globular proteins can be pictured as consisting of "layers" of backbone, with hydrophobic core regions between them. Over half the known globular protein structures have two layers of backbone (separated by one hydrophobic core). Roughly one-third of the known structures are composed of three backbone layers and two hydrophobic cores. There are also a few known four-layer structures and one known five-layer structure. A few structures are not easily classified in this way, but it is remarkable that most proteins fit into one of these classes. Examples of each are presented in Figure 5.24.

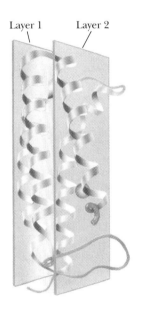

(a) Cytochrome *c'*

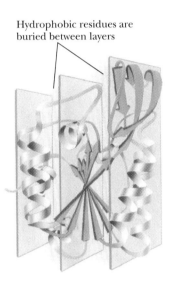

(b) Phosphoglycerate kinase
(Domain 2)

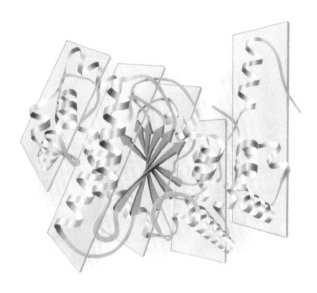

(c) Phosphorylase
(Domain 2)

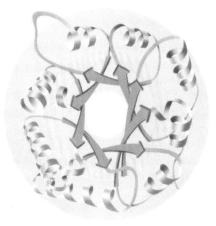

(d) Triose phosphate isomerase

Figure 5.24 Examples of protein domains with different numbers of layers of backbone structure. **(a)** Cytochrome *c'* with two layers of α-helix. **(b)** Domain 2 of phosphoglycerate kinase, composed of a β-sheet layer between two layers of helix, three layers overall. **(c)** An unusual five-layer structure, domain 2 of glycogen phosphorylase, a β-sheet layer sandwiched between four layers of α-helix. **(d)** The concentric "layers" of β-sheet (inside) and α-helix (outside) in triose phosphate isomerase. Hydrophobic residues are buried between these concentric layers in the same manner as in the planar layers of the other proteins. The hydrophobic layers are shaded yellow. (*Jane Richardson*)

See pages 71, 81, 84

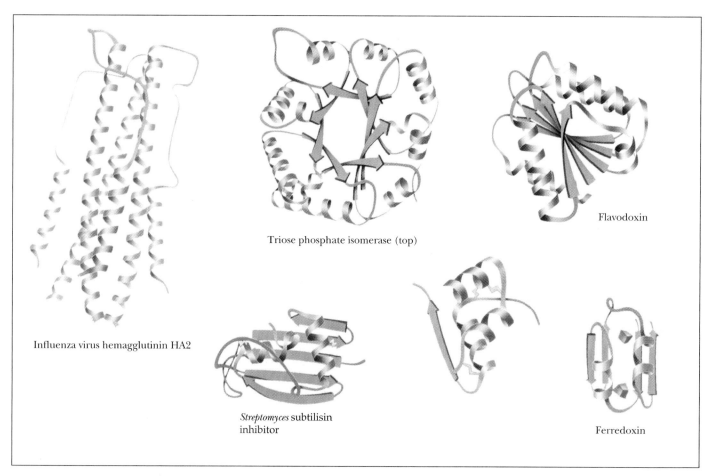

Triose phosphate isomerase (top)

Flavodoxin

Influenza virus hemagglutinin HA2

Streptomyces subtilisin inhibitor

Ferredoxin

Figure 5.25 The four broad categories of protein structure. Influenza virus hemagglutinin HA2 is an *antiparallel α-helix* protein. The simplest way to pack helices is in an antiparallel manner, and most of the proteins in this class consist of bundles of antiparallel helices. Flavodoxin and triose phosphate isomerase are *parallel β-sheet* proteins. Parallel β-sheet arrays distribute hydrophobic side chains on both sides of the sheet, and thus neither side of the sheet can be exposed to solvent. Subtilisin inhibitor from *Streptomyces* is an *antiparallel β-sheet* protein. Hydrophobic residues are found on only one side of an antiparallel β-sheet structure, and one face of an antiparallel β-sheet may thus be exposed to solvent. Insulin and ferredoxin are *disulfide-rich* and *metal-rich* proteins, respectively. These proteins are typically small, and their conformations (and stability) are strongly influenced by their high content of disulfide bridges or bound metal ions. (*Jane Richardson*)

See pages 42, 43, 58, 84

Classification of Globular Proteins

In addition to classification based on layer structure, proteins can be grouped according to the type and arrangement of secondary structure. There are four such broad groups: antiparallel α-helix, parallel or mixed β-sheet, antiparallel β-sheet, and the small metal- and disulfide-rich proteins (Figure 5.25). It is important to note that the similarities of tertiary structure within these groups do not necessarily reflect similar or even related functions.

Protein Modules: Nature's Modular Strategy for Protein Design

Hundreds of thousands of protein sequences are now known. Certain of these protein sequences give rise to distinct structural domains that are used over and over again in modular fashion. These **protein modules** may occur in a wide variety of proteins, often being used for different purposes, or they may be used repeatedly in the same protein. Figure 5.26 shows the tertiary structures of five protein modules, and Figure 5.27 presents several proteins that contain versions of these modules. These modules typically contain about 40 to 100 amino acids and often adopt a stable tertiary structure when isolated from their parent protein. One of the best-known examples of a protein module is the **immunoglobulin module,** which has been found not only in immunoglobulins but also in a wide variety of cell surface proteins, including cell adhesion molecules and growth factor receptors.

How Do Proteins Know How to Fold?

The landmark experiments by Christian Anfinsen on the refolding of ribonuclease showed clearly that the refolding of a pure, denatured protein *in vitro* is a spontaneous process. A corollary result of Anfinsen's work is that the native structures of globular proteins are thermodynamically stable states. But the matter of how a given protein achieves such a stable state is a complex one. Cyrus Levinthal pointed out in 1968 that so many conformations are possible for a typical protein that the protein does not have sufficient time to find its most stable conformational state by sampling all the possible conformations. A simple illustration of "Levinthal's paradox" might go as follows: Consider a protein of 100 amino acids. Assume that there are only two conformational possibilities per amino acid, or $2^{100} = 1.27 \times 10^{30}$ possibilities. Allow 10^{-13} sec (the time of a bond vibration) for the protein to test each conformational possibility in search of the overall energy minimum. The time required for this search is impossibly long:

$$(10^{-13} \text{ sec})(1.27 \times 10^{30}) = 1.27 \times 10^{17} \text{ sec} = 4 \times 10^9 \text{ years}$$

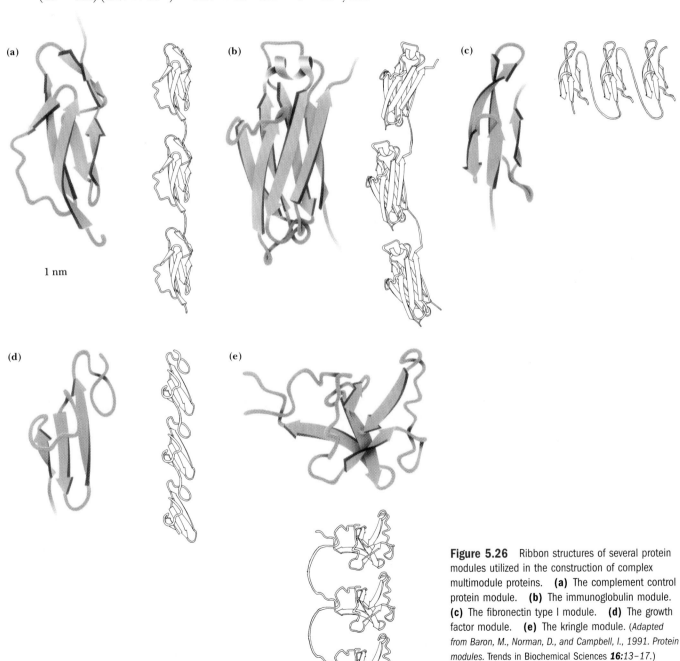

Figure 5.26 Ribbon structures of several protein modules utilized in the construction of complex multimodule proteins. **(a)** The complement control protein module. **(b)** The immunoglobulin module. **(c)** The fibronectin type I module. **(d)** The growth factor module. **(e)** The kringle module. (*Adapted from Baron, M., Norman, D., and Campbell, I., 1991. Protein modules.* Trends in Biochemical Sciences *16:13–17.*)

See page 47

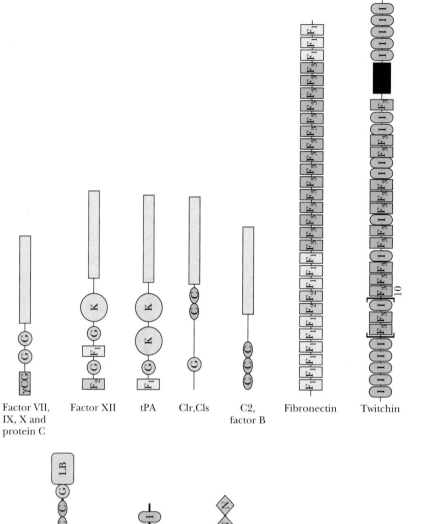

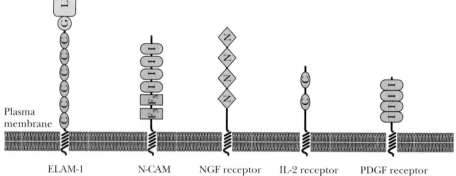

Figure 5.27 A sampling of proteins that consist of mosaics of individual protein modules. The modules shown include: γCG, a module containing γ-carboxyglutamate residues; G, an epidermal growth-factor–like module; K, the "kringle" domain, named for a Danish pastry; C, which is found in complement proteins; F1, F2, and F3, first found in fibronectin; I, the immunoglobulin superfamily domain; N, found in some growth factor receptors; E, a module homologous to the calcium-binding E–F hand domain; and LB, a lectin module found in some cell surface proteins. (*Adapted from Baron, M., Norman, D., and Campbell, I., 1991. Protein modules. Trends in Biochemical Sciences **16**:13–17.*)

Levinthal's paradox led protein chemists to hypothesize that proteins must fold by specific "folding pathways," and much research effort has been devoted to the search for these pathways.

Implicit in the presumption of folding pathways is the existence of intermediate, partially folded conformational states. The notion of intermediate states on the pathway to a tertiary structure raises the possibility that segments of a protein might independently adopt local and well-defined secondary structures (α-helices and β-sheets). The tendency of a peptide segment to prefer a particular secondary structure depends in turn on its amino acid composition and sequence.

Surveys of the frequency with which various residues appear in helices and sheets show (Figure 5.28) that some residues, such as alanine, glutamate, and

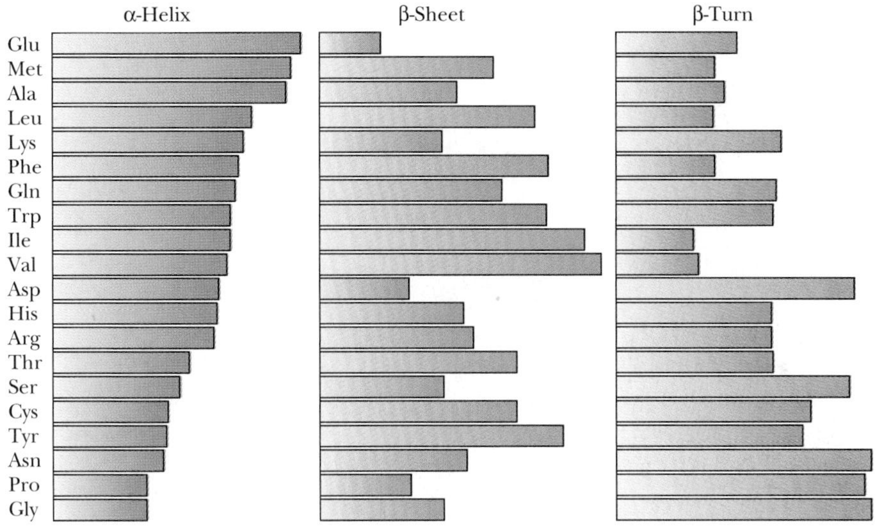

Figure 5.28 Relative frequencies of occurrence of amino acid residues in α-helices, β-sheets, and β-turns in proteins of known structure. (*Adapted from Bell, J. E., and Bell, E. T., 1988.* Proteins and Enzymes. *Englewood Cliffs, N.J.: Prentice-Hall.*)

methionine, occur much more frequently in α-helices than do others. In contrast, glycine and proline are the least likely residues to be found in an α-helix. Likewise, certain residues, including valine, isoleucine, and the aromatic amino acids, are more likely to be found in β-sheets than other residues, and aspartate, glutamate, and proline are much less likely to be found in β-sheets.

Such observations have led to many efforts to predict the occurrence of secondary structure in proteins from knowledge of the amino acid sequence. Such **predictive algorithms** consider the composition of short segments of a polypeptide. If these segments are rich in residues that are found frequently in helices or sheets, then that segment is judged likely to adopt the corresponding secondary structure. The predictive algorithm designed by Peter Chou and Gerald Fasman in 1974 attempted to classify the 20 amino acids for their α-helix–forming and β-sheet–forming propensities. By studying the patterns of occurrence of each of these classes in helices and sheets of proteins with known structures, Chou and Fasman formulated a set of rules to predict the occurrence of helices and sheets in sequences of unknown structure. The Chou–Fasman method has been a useful device for some purposes, but it is able to predict the occurrence of helices and sheets in protein structures only about 50% of the time.

Recent research in protein folding reveals a surprise—*there is, for the typical protein, no single folding pathway.* Rather, there are likely to be many possible pathways.[2] Given so many possible pathways, the folding process can be pictured as a funnel of free energies, also referred to as an **energy landscape** (Figure 5.29). The rim at the top of the funnel represents the many possible unfolded states for a polypeptide chain. Polypeptides fall down the wall of the funnel as contacts made between residues nucleate different folding possibilities. Residue–residue contacts that are similar to those in the final, folded structure are referred to as **native contacts.** Such contacts are made more easily when the involved residues are close in the peptide sequence. The folding

---

[2]Theoretical calculations of the folding process for a globular protein are some of the most complex calculations imaginable. Rising to this challenge, engineers at IBM are designing the world's most powerful computer, nicknamed "Blue Gene." It will contain approximately 1 million processor chips working in parallel, and will operate at a speed of one petaflop (1×10^{15} floating point operations per second). Even for this vastly powerful machine, the complete calculation of a protein's folding process will be a daunting and time-consuming task.

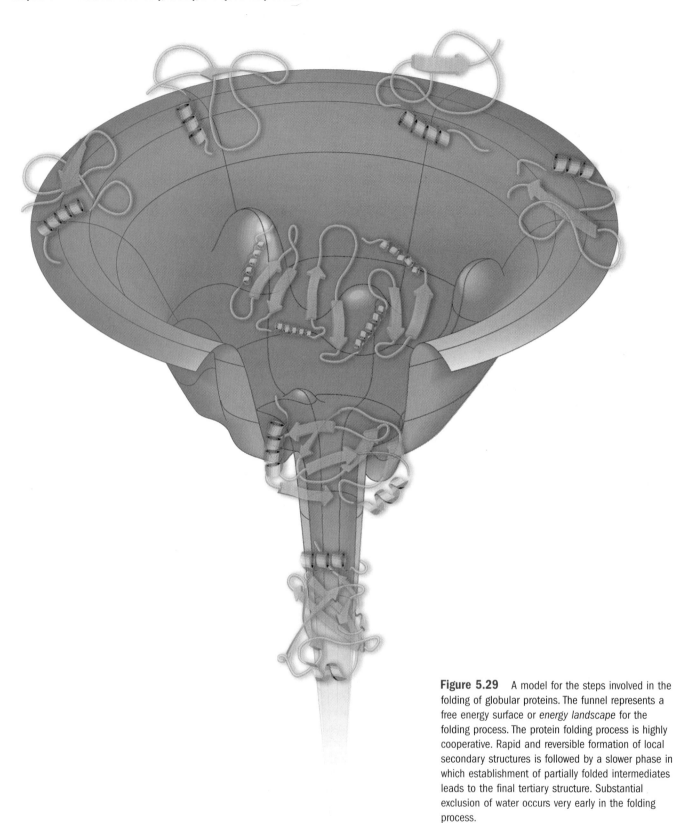

Figure 5.29 A model for the steps involved in the folding of globular proteins. The funnel represents a free energy surface or *energy landscape* for the folding process. The protein folding process is highly cooperative. Rapid and reversible formation of local secondary structures is followed by a slower phase in which establishment of partially folded intermediates leads to the final tertiary structure. Substantial exclusion of water occurs very early in the folding process.

process typically involves a rapid collapse to any one of a family of **semicompact structures.** After this collapse, the polypeptide chain encounters the rate-limiting stage of the folding process, a search through many semicompact states rich in secondary structure for a transition state that leads rapidly to the native, folded, tertiary structure represented by the bottom of the funnel.

CRITICAL DEVELOPMENTS IN BIOCHEMISTRY

Thermodynamics of the Folding Process in Globular Proteins

Section 5.4 considers the noncovalent binding energies that stabilize a protein structure. However, the folding of a protein depends ultimately on the difference in Gibbs free energy (ΔG) between the folded (F) and unfolded (U) states at some temperature T:

$$\Delta G = G_F - G_U = \Delta H - T\Delta S$$
$$= (H_F - H_U) - T(S_F - S_U)$$

In the unfolded state, the peptide chain and its R groups interact with solvent water, and any measurement of the free energy change upon folding must consider contributions to the enthalpy change (ΔH) and the entropy change (ΔS) both for the polypeptide chain and for the solvent:

$$\Delta G_{total} = \Delta H_{chain} + \Delta H_{solvent} - T\Delta S_{chain} - T\Delta S_{solvent}$$

If each of the four terms on the right side of this equation is understood, the thermodynamic basis for protein folding should be clear. A summary of the signs and magnitudes of these quantities for a typical protein is shown in the accompanying figure. The folded protein is a highly ordered structure compared to the unfolded state, so ΔS_{chain} is a negative number and thus $-T\Delta S_{chain}$ is a positive quantity in the equation. The other terms depend on the nature of the particular ensemble of R groups. The nature of ΔH_{chain} depends on both residue–residue interactions and residue–solvent interactions. Nonpolar groups in the folded protein interact mainly with one another via weak van der Waals forces. Interactions between nonpolar groups and water in the unfolded state are stronger because the polar water molecules induce dipoles in the nonpolar groups, producing a significant electrostatic interaction. As a result, ΔH_{chain} is positive for nonpolar groups and favors the unfolded state. $\Delta H_{solvent}$ for nonpolar groups, however, is negative and favors the folded state. This is because folding allows many water molecules to interact (favorably) with one another rather than (less favorably) with the nonpolar side chains. The magnitude of ΔH_{chain} is smaller than that of $\Delta H_{solvent}$, but both these terms are small and usually do not dominate the folding process. However, $\Delta S_{solvent}$ for nonpolar groups is large and

positive and strongly favors the folded state. This is because nonpolar groups force order upon the water solvent in the unfolded state.

For polar side chains, ΔH_{chain} is positive and $\Delta H_{solvent}$ is negative. Because solvent molecules are ordered to some extent around polar groups, $\Delta S_{solvent}$ is small and positive. As shown in the figure, ΔG_{total} for the polar groups of a protein is near zero. Comparison of all the terms considered here makes it clear that *the single largest contribution to the stability of a folded protein is* $\Delta S_{solvent}$ *for the nonpolar residues.*

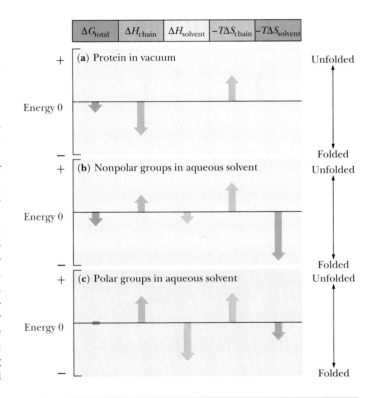

Molecular Chaperones: Proteins That Help Fold Globular Proteins

Christian Anfinsen's experiments demonstrated that proteins can fold spontaneously. As noted (see *Critical Developments in Biochemistry*, above) this folding is driven by the Gibbs free energy difference between the unfolded and folded states. It has also been generally assumed that all the information necessary for the correct folding of a polypeptide chain is contained in the primary structure and requires no additional molecular factors. However, the folding of proteins in the cell is a different matter. The highly concentrated protein matrix in the cell may adversely affect the folding process by causing aggregation of unfolded or partially folded proteins. Also, it may be necessary to accelerate slow steps in the folding process or to suppress or reverse incorrect or premature folding. A family of proteins, known as **molecular chaperones,** appear to be essential for the correct folding of certain polypeptide chains *in vivo*, for their assembly into oligomers, and for preventing inappropriate liaisons with other proteins during their synthesis, folding, and transport. The participation of molecular chaperones in protein folding is discussed in Chapter 25.

5.5 Subunit Interactions and Quaternary Structure

Many proteins exist in nature as **oligomers,** complexes composed of (often symmetric) noncovalent assemblies of two or more monomer subunits. In fact, subunit association is a common feature of macromolecular organization in biology. Most intracellular enzymes are oligomeric and may be composed either of a single type of monomer subunit (**homomultimers**) or of several different kinds of subunits (**heteromultimers).** The simplest case is a protein composed of identical subunits. Liver alcohol dehydrogenase, shown in Figure 5.30, is such a protein. More complicated proteins may have several different subunits in one, two, or more copies. Hemoglobin, for example, contains two each of two different subunits and is referred to as an $\alpha_2\beta_2$-complex. The way in which separate folded monomeric protein subunits associate to form the oligomeric protein constitutes the **quaternary structure** of that protein.

The subunits of an oligomeric protein typically fold into apparently independent globular conformations and then interact with other subunits. The particular surfaces at which protein subunits interact possess both polar and hydrophobic amino acid side chains. The interfaces of mating subunits must

HUMAN BIOCHEMISTRY

Diseases of Protein Folding

A number of human diseases are linked to abnormalities of protein folding. Protein misfolding may cause disease by a variety of mechanisms. For example, misfolding may result in loss of function and the onset of disease. The table below summarizes several other mechanisms and provides an example of each.

Disease	Affected Protein	Mechanism
Alzheimer's disease	β-amyloid peptide (derived from amyloid precursor protein)	Misfolded β-amyloid peptide accumulates in human neural tissue, forming deposits known as neuritic plaques.
Familial amyloidotic polyneuropathy	Transthyretin	Aggregation of unfolded proteins. Nerves and other organs are damaged by deposits of insoluble protein products.
Cancer	p53	p53 prevents cells with damaged DNA from dividing. One class of p53 mutations leads to misfolding; the misfolded protein is unstable and is destroyed.
Creutzfeldt-Jacob disease (human equivalent of mad cow disease)	Prion	Prion protein with an altered conformation (PrP^{sc}) may seed conformational transitions in normal PrP (PrP^c) molecules.
Hereditary emphysema	α_1-antitrypsin	Mutated forms of this protein fold slowly, allowing its target, elastase, to destroy lung tissue.
Cystic fibrosis	CFTR (cystic fibrosis transmembrane conductance regulator)	Folding intermediates of mutant CFTR forms don't dissociate freely from chaperones, preventing the CFTR from reaching its destination in the membrane.

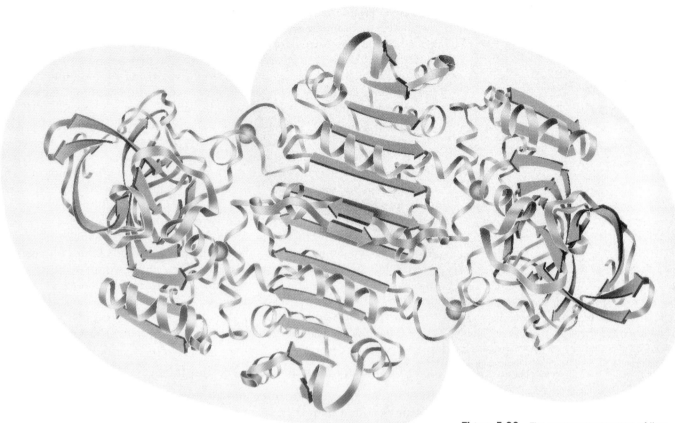

Figure 5.30 The quaternary structure of liver alcohol dehydrogenase. Within each subunit there is a six-stranded parallel sheet. Between the two subunits is a two-stranded antiparallel sheet. (*Jane Richardson*)

 See page 69

therefore possess complementary arrangements of polar and hydrophobic groups. The thermodynamic tendency to bury hydrophobic side chains at the interface between two subunits is akin to what happens when a protein folds so as to bury the majority of its hydrophobic side chains in the interior of the structure.

Forces Driving Quaternary Association

The forces that stabilize quaternary structure have been evaluated for a few proteins. Typical dissociation constants for simple two-subunit associations range from 10^{-8} to 10^{-16} M. These values correspond to free energies of association of about 50 to 100 kJ/mol at 37°C. Dimerization of subunits is accompanied by both favorable and unfavorable energy changes. The favorable interactions include van der Waals interactions, hydrogen bonds, ionic bonds, and hydrophobic interactions. However, considerable entropy loss occurs when subunits interact. When two subunits move as one, three translational degrees of freedom are lost for one subunit, since it is constrained to move with the other one. In addition, many peptide residues at the subunit interface, which were previously free to move on the protein surface, now have their movements restricted by the subunit association. This unfavorable energy of association is in the range of 80 to 120 kJ/mol for temperatures of 25° to 37°C. Thus, to achieve stability, the dimerization of two subunits must involve approximately 130 to 220 kJ/mol of favorable interactions.[3] Van der Waals interactions at protein

---

[3]For example, 130 kJ/mol of favorable interaction minus 80 kJ/mol of unfavorable interaction equals a net free energy of association of 50 kJ/mol.

Figure 5.31 A simplified schematic drawing of an immunoglobulin molecule showing the intramolecular and intermolecular disulfide bridges. (A space-filling model of the same molecule is shown in Figure 1.11.)

See page 47

interfaces are numerous, often running to several hundred for a typical monomer–monomer association. This would account for about 150 to 200 kJ/mol of favorable free energy of association. However, when solvent is removed from the protein surface to form the subunit–subunit contacts, nearly as many van der Waals associations are lost as are made. One subunit is simply trading water molecules for peptide residues in the other subunit. As a result, the energy of subunit association due to van der Waals interactions actually contributes little to the stability of the dimer. Hydrophobic interactions, however, are generally very favorable. For many proteins, the subunit association process effectively buries as much as 20 nm^2 of surface area that had previously been exposed to solvent, resulting in as much as 100 to 200 kJ/mol of favorable hydrophobic interactions. Together with whatever polar interactions occur at the protein–protein interface, this is sufficient to account for the observed stabilization that occurs when two protein subunits associate.

An additional and important factor contributing to the stability of subunit associations for some proteins is the formation of disulfide bonds between different subunits. All antibodies are $\alpha_2\beta_2$-tetramers composed of two heavy chains (53–75 kD) and two relatively light chains (23 kD). In addition to *intrasubunit* disulfide bonds (four per heavy chain, two per light chain), there are two *intersubunit* disulfide bridges holding the two heavy chains together and a disulfide bridge linking each of the two light chains to a heavy chain (Figure 5.31).

Open Quaternary Structures and Polymerization

All of the quaternary structures we have considered to this point are **closed** structures, with a limited capacity to associate. Many proteins in nature associate to form **open** structures, which can polymerize more or less indefinitely, creating structures that are both esthetically attractive and functionally important to the cells or tissue in which they exist. One such protein is **tubulin,** the $\alpha\beta$-dimeric protein that polymerizes into long, tubular structures, which are the structural basis of cilia, flagella, and the cytoskeletal matrix. The *microtubule* thus formed (Figure 5.32) may be viewed as consisting of 13 parallel filaments arising from end-to-end aggregation of the tubulin dimers. The polymerization of 2130 subunits of the coat protein of tobacco mosaic virus to form an enclosure (termed a *capsid*) around the virus's genetic material is another example of extended quaternary structures (see Figure 5.32). The human immunodeficiency virus, or HIV, the causative agent of AIDS, is enveloped by a spherical membrane shell composed of hundreds of coat protein subunits, a large-scale quaternary association.

A DEEPER LOOK

Immunoglobulins—All the Features of Protein Structure Brought Together

The immunoglobulin structure in Figure 5.31 represents the confluence of all the details of protein structure that have been thus far discussed. As for all proteins, the primary structure determines other aspects of structure. There are numerous elements of secondary structure, including β-sheets and tight turns. The tertiary structure consists of 12 distinct domains, and the protein adopts a heterotetrameric quaternary structure. To make matters more interesting, both intrasubunit and intersubunit disulfide linkages act to stabilize the discrete domains and to stabilize the tetramer itself.

One more level of sophistication awaits. As is discussed in Chapter 12, the amino acid sequences of both light and heavy immunoglobulin chains are not constant! Instead, the primary structure of these chains is highly variable in the N-terminal regions (the first 108 residues). Heterogeneity of the amino acid sequence leads to variations in the conformation of these variable regions—variations which account for antibody diversity and the ability of antibodies to recognize and bind a virtually limitless range of antigens. This full potential of antibody–antigen recognition enables organisms to mount immunologic responses to almost any antigen that might challenge them.

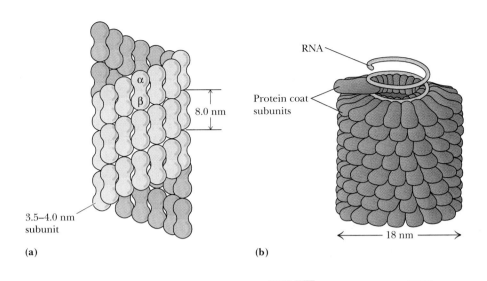

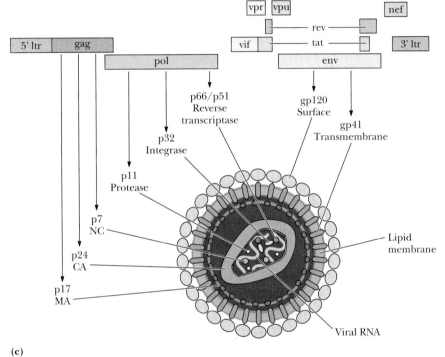

Figure 5.32 **(a)** The structure of a typical microtubule, showing the arrangement of the α- and β-monomers of the tubulin dimer. **(b)** The structure of tobacco mosaic virus. **(c)** The genes and proteins contributing to the structure of the HIV-1 particle.

Structural and Functional Advantages of Quaternary Association

There are several important reasons for protein subunits to associate in oligomeric structures.

Stability

One general benefit of subunit association is a favorable reduction of the protein's surface-to-volume ratio. The surface-to-volume ratio becomes smaller as the radius of any particle or globular object becomes larger. (This is because surface area is a function of the radius squared and volume is a function of the radius cubed.) Since interactions within the protein usually tend to stabilize the protein energetically, more so in fact than interactions of the protein surface with solvent water, it is usually the case that decreased surface-to-volume ratios result in more stable proteins. Subunit association may also serve to shield hydrophobic residues from solvent water.

Genetic Economy and Efficiency

Oligomeric association of protein monomers is genetically economical for an organism. Less DNA is required to code for a monomer that assembles into a homomultimer than to code for a large polypeptide of the same molecular mass. Another way to look at this is to realize that virtually all of the information that determines oligomer assembly and subunit–subunit interaction is contained in the genetic material needed to code for the monomer. For example, HIV-1 protease, an enzyme that is a dimer of identical subunits, performs a catalytic function similar to related cellular protease enzymes that are single polypeptide chains of twice the molecular mass (see Chapter 11).

Bringing Catalytic Sites Together

Many enzymes (see Chapters 10 and 11) derive at least some of their catalytic power from oligomeric associations of monomer subunits. This can happen in several ways. The monomer may not constitute a complete enzyme active site. Formation of the oligomer may bring all the necessary catalytic groups together to form an active enzyme. For example, the active sites of bacterial glutamine synthetase (see Table 4.2) are formed from pairs of adjacent subunits. The dissociated monomers are inactive.

Oligomeric enzymes may also carry out different but related reactions on different subunits. Thus, tryptophan synthase is a tetramer consisting of pairs of different subunits, $\alpha_2 \beta_2$. Purified α-subunits catalyze the following reaction:

$$\text{Indoleglycerol phosphate} \rightarrow \text{Indole} + \text{Glyceraldehyde-3-phosphate}$$

and the β subunits catalyze this reaction:

$$\text{Indole} + \text{L-Serine} \rightarrow \text{L-Tryptophan}$$

When the subunits are associated to form the $\alpha_2 \beta_2$ enzyme, indole, the product of the α-reaction and the reactant for the β-reaction, is passed directly from the α-subunit to the β-subunit and cannot be detected as a free intermediate.

HUMAN BIOCHEMISTRY

Faster-Acting Insulin: Genetic Engineering Solves a Quaternary Structure Protein

Insulin is a peptide hormone secreted by the pancreas that regulates glucose metabolism in the body. Insufficient production of insulin or the failure of insulin to stimulate target sites in liver, muscle, and adipose tissue leads to the serious metabolic disorder known as *diabetes mellitus*. Diabetes afflicts millions of people worldwide. Diabetic individuals typically exhibit high levels of glucose in the blood, but insulin injection therapy allows diabetic individuals to maintain normal levels of blood glucose.

Insulin is composed of two peptide chains covalently linked by disulfide bonds (see Figures 4.21 and 5.25). This "monomer" of insulin is the active form that binds to receptors in target cells. However, in solution, insulin spontaneously forms dimers, which themselves aggregate to form hexamers. The surface of the insulin molecule that self-associates to form hexamers is also the surface that binds to insulin receptors in target cells. Thus, hexamers of insulin are inactive.

Insulin released from the pancreas is monomeric and acts rapidly in target tissues. However, when insulin is administered (by injection) to a diabetic patient, the insulin hexamers dissociate slowly, and the patient's blood glucose levels typically drop slowly (over a period of several hours).

In 1988, G. Dodson showed that insulin could be genetically engineered to prefer the monomeric (active) state. Dodson and his colleagues used recombinant DNA technology to produce insulin with an aspartate residue replacing a proline at the contact interface between adjacent subunits. The negative charge on the Asp side chain creates electrostatic repulsion between subunits and increases the dissociation constant for the hexamer–monomer equilibrium. Injection of this mutant insulin into test animals produced more rapid decreases in blood glucose than did ordinary insulin. The Danish pharmaceutical company Novo is producing this mutant insulin, which may eventually replace ordinary insulin in treatment of diabetes.

Cooperativity

There is another, more important reason for monomer subunits to associate into oligomeric complexes. Most oligomeric enzymes regulate catalytic activity by means of subunit interactions, which may give rise to cooperative phenomena. Multisubunit proteins typically possess multiple binding sites for a given ligand. If the binding of ligand at one site changes the affinity of the protein for ligand at the other binding sites, the binding is said to be **cooperative.** Increases in affinity at subsequent sites represent positive cooperativity, whereas decreases in affinity correspond to negative cooperativity. The points of contact between protein subunits provide a mechanism for communication between the subunits. This in turn provides a way in which the binding of ligand to one subunit can influence the binding behavior at the other subunits. Such cooperative behavior, discussed in greater depth in Chapter 10, is the underlying mechanism for the regulation of many biological processes.

PROBLEMS

1. The central rod domain of a keratin protein is approximately 312 residues in length. What is the length (in Å) of the keratin rod domain? If this same peptide segment were a true α-helix, how long would it be? If the same segment were a β-sheet, what would its length be?

2. A teenager can grow 4 in. in a year during a "growth spurt." Assuming that the increase in height is due to vertical growth of collagen fibers (in bone), calculate the number of collagen helix turns synthesized per minute.

3. Discuss the potential contributions to hydrophobic and van der Waals interactions and ionic and hydrogen bonds of the side chains of Asp, Leu, Tyr, and His in a protein.

4. Figure 5.28 shows that Pro is the amino acid least commonly found in α-helices but most commonly found in β-turns. Discuss the reasons for this behavior.

5. For flavodoxin in Figure 5.25, identify the right-handed cross-overs and the left-handed cross-overs in the parallel β-sheet.

6. Choose any three regions in the Ramachandran plot and discuss the likelihood of observing that combination of ϕ and ψ in a peptide or protein. Defend your answer using suitable molecular models of a peptide.

7. Two polypeptides, A and B, have similar tertiary structures, but A normally exists as a monomer, whereas B exists as a tetramer, B_4. What differences might be expected in the amino acid composition of A versus B?

8. The hemagglutinin protein in influenza virus contains a remarkably long α-helix, with 53 residues.
   a. How long is this α-helix (in nm)?
   b. How many turns does this helix have?
   c. Each residue in an α-helix is involved in two H bonds. How many H bonds are present in this helix?

FURTHER READING

Abeles, R., Frey, P., and Jencks, W., 1992. *Biochemistry*. Boston: Jones and Bartlett.

Baker, D., 2000. A surprising simplicity to protein folding. *Nature* **405:** 39–42.

Cantor, C. R., and Schimmel, P. R., 1980. *Biophysical Chemistry, Part I: The Conformation of Biological Macromolecules*. New York: W. H. Freeman and Co.

Chothia, C., 1984. Principles that determine the structure of proteins. *Annual Review of Biochemistry* **53:**537–572.

Creighton, T. E., 1983. *Proteins: Structure and Molecular Properties*. New York: W. H. Freeman and Co.

Dill, K. A., and Chan, H. S., 1997. From Levinthal to pathways to funnels. *Nature Structural Biology* **4:**10–19.

Dinner, A. R., Sali, A., Smith, L. J., Dobson, C. M., and Karplus, M., 2000. Understanding protein folding via free-energy surfaces from theory and experiment. *Trends in Biochemical Sciences* **25:**331–339.

Doolittle, R. F., 1995. The multiplicity of domains in proteins. *Annual Review of Biochemistry* **64:**287–314.

Hardie, D. G., and Coggins, J. R., eds., 1986. *Multidomain Proteins: Structure and Evolution*. New York: Elsevier.

Judson, H. F., 1979. *The Eighth Day of Creation*. New York: Simon and Schuster.

Klotz, I. M., Langerman, N. R., and Darnell, D. W., 1970. Quaternary structure of proteins. *Annual Review of Biochemistry* **39:**25–62.

Petsko, G. A., and Ringe, D., 1984. Fluctuations in protein structure from X-ray diffraction. *Annual Review of Biophysics and Bioengineering* **13:**331–371.

Radford, S. E., 2000. Protein folding: Progress made and promises ahead. *Trends in Biochemical Sciences* **25:**611–618.

Richardson, J. S., 1981. The anatomy and taxonomy of protein structure. *Advances in Protein Chemistry* **34:**167–339.

Salemme, F. R., 1983. Structural properties of protein β-sheets. *Progress in Biophysics and Molecular Biology* **42:**95–133.

Torchia, D. A., 1984. Solid state NMR studies of protein internal dynamics. *Annual Review of Biophysics and Bioengineering* **13:**125–144.

Wagner, G., Hyberts, S., and Havel, T., 1992. NMR structure determination in solution: A critique and comparison with X-ray crystallography. *Annual Review of Biophysics and Biomolecular Structure* **21:**167–242.

Lipids, Membranes, and Transport

The mighty whales which swim in a sea of water, and have a sea of oil swimming in them.
—Herman Melville, "Extracts," Moby-Dick. *New York: Penguin Books, 1972. (Humpback whale* [Megaptera novaeangliae] *breaching, Cape Cod, MA; photo © Steven Morello/Peter Arnold, Inc.)*

A feast of fat things, a feast of wines on the lees.

Isaiah 25:6

Outline

Lipids are a class of biological molecules defined by low solubility in water and high solubility in nonpolar solvents. As molecules that are largely hydrocarbon in nature, lipids represent highly reduced forms of carbon and, upon oxidation in metabolism, yield large amounts of energy. As such, lipids are thus the molecules of choice for metabolic energy storage.

Lipids are also a principal component of membranes. The lipids found in biological membranes are **amphipathic,** which means they possess both polar and nonpolar groups. The hydrophobic nature of lipid molecules allows membranes to act as effective barriers to more polar molecules. The polar moieties of amphipathic lipids typically lie at the surface of membranes, where they interact with water.

Biological membranes are uniquely organized arrays of lipids and proteins (either of which may be modified with carbohydrate groups). Membranes serve a number of essential cellular functions. They constitute the boundaries of cells and intracellular organelles, and they provide a surface where many important biological reactions and processes occur. Membranes have proteins that mediate and regulate the transport of metabolites, macromolecules, and ions. Membrane proteins interact with the lipids of membranes in a variety of ways. Some proteins associate with membranes via electrostatic interactions with polar groups on the membrane surface, whereas other proteins are embedded to various extents in the hydrophobic core of the membrane. Other proteins are anchored to membranes via covalently bound lipid molecules that associate strongly with the hydrophobic membrane core.

Hormones and many other biological signal molecules and regulatory agents exert their effects via interactions with membranes. Photosynthesis, electron transport, oxidative phosphorylation, muscle contraction, and electrical activity all depend on membranes and membrane proteins.

This chapter discusses the composition, structure, and dynamic processes of biological membranes. We begin with a discussion of the lipid molecules found in living things.

LIPIDS

Lipids can be classified into two great classes: those that contain fatty acids (**complex lipids**) and that don't (**simple lipids).**

6.1 Fatty Acids

A **fatty acid** is composed of a long hydrocarbon chain ("tail") and a terminal carboxyl group (or "head"). The carboxyl group is normally ionized under physiological conditions. Fatty acids occur in large amounts in cells but rarely in the free, uncomplexed state. They typically are esterified to glycerol or other backbone structures. Most of the fatty acids found in nature have an even number of carbon atoms (usually 14 to 24). Certain marine organisms, however, contain substantial amounts of fatty acids with odd numbers of carbon atoms.

Fatty acids are either **saturated** (all carbon–carbon bonds are single bonds) or **unsaturated** (with one or more double bonds in the hydrocarbon chain). If a fatty acid has a single double bond, it is said to be **monounsaturated,** and if it has more than one, it is **polyunsaturated.**

Fatty acids can be named or described in at least three ways, as listed in Table 6.1. For example, a fatty acid composed of an 18-carbon chain with no double bonds can be called by its systematic name (**octadecanoic acid**), its common name (**stearic acid),** or its shorthand notation, in which the number of carbons is followed by a colon and the number of double bonds in the molecule (18:0 for stearic acid). The structures of several fatty acids are given in Figure 6.1. Stearic acid (18:0) and **palmitic acid** (16:0) are the most common saturated fatty acids in nature.

Free rotation around each of the carbon–carbon bonds makes saturated fatty acids extremely flexible molecules. Owing to steric constraints, however, the fully extended conformation (see Figure 6.1) is the most stable for saturated fatty acids. Nonetheless, the degree of stabilization is slight, and (as will be seen) saturated fatty acid chains adopt a variety of conformations.

Unsaturated fatty acids are slightly more abundant in nature than saturated fatty acids, especially in higher plants. The most common unsaturated fatty acid is **oleic acid,** or 18:1(9), with the number in parentheses indicating that the double bond is between carbons 9 and 10. The number of double bonds in an unsaturated fatty acid varies typically from one to four, but in the fatty acids found in most bacteria, this number rarely exceeds one.

Table 6.1 **Common Biological Fatty Acids**

Number of Carbons	Common Name	Systematic Name	Symbol	Structure
Saturated fatty acids				
12	Lauric acid	Dodecanoic acid	12:0	$CH_3(CH_2)_{10}COOH$
14	Myristic acid	Tetradecanoic acid	14:0	$CH_3(CH_2)_{12}COOH$
16	Palmitic acid	Hexadecanoic acid	16:0	$CH_3(CH_2)_{14}COOH$
18	Stearic acid	Octadecanoic acid	18:0	$CH_3(CH_2)_{16}COOH$
20	Arachidic acid	Eicosanoic acid	20:0	$CH_3(CH_2)_{18}COOH$
22	Behenic acid	Docosanoic acid	22:0	$CH_3(CH_2)_{20}COOH$
24	Lignoceric acid	Tetracosanoic acid	24:0	$CH_3(CH_2)_{22}COOH$
Unsaturated fatty acids (all double bonds are *cis*)				
16	Palmitoleic acid	9-Hexadecenoic acid	16:1	$CH_3(CH_2)_5CH = CH(CH_2)_7COOH$
18	Oleic acid	9-Octadecenoic acid	18:1	$CH_3(CH_2)_7CH = CH(CH_2)_7COOH$
18	Linoleic acid	9,12-Octadecadienoic acid	18:2	$CH_3(CH_2)_4(CH = CHCH_2)_2(CH_2)_6COOH$
18	α-Linolenic acid	9,12,15-Octadecatrienoic acid	18:3	$CH_3CH_2(CH = CHCH_2)_3(CH_2)_6COOH$
18	γ-Linolenic acid	6,9,12-Octadecatrienoic acid	18:3	$CH_3(CH_2)_4(CH = CHCH_2)_3(CH_2)_3COOH$
20	Arachidonic acid	5,8,11,14-Eicosatetraenoic acid	20:4	$CH_3(CH_2)_4(CH = CHCH_2)_4(CH_2)_2COOH$
24	Nervonic acid	15-Tetracosenoic acid	24:1	$CH_3(CH_2)_7CH = CH(CH_2)_{13}COOH$

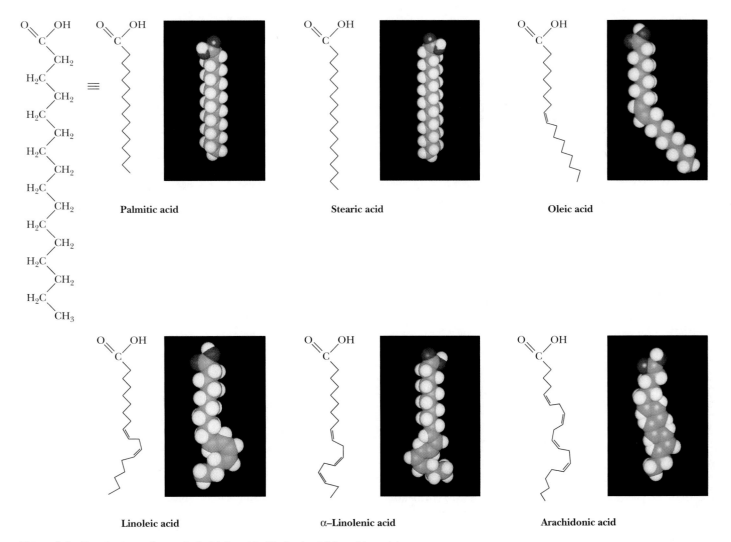

Palmitic acid   Stearic acid   Oleic acid

Linoleic acid   α–Linolenic acid   Arachidonic acid

Figure 6.1 The structures of some typical fatty acids. Most natural fatty acids contain an even number of carbon atoms and any double bonds are nearly always *cis* and rarely conjugated.

A DEEPER LOOK

Fatty Acids in Food: Saturated Versus Unsaturated, *Cis* Versus *Trans*

Fats consumed in the modern human diet vary widely in their fatty acid compositions. The table shown here provides a brief summary. The incidence of cardiovascular disease is correlated with diets high in saturated fatty acids. By contrast, a diet that is relatively higher in unsaturated fatty acids (especially polyunsaturated fatty acids) may reduce the risk of heart attacks and strokes. Corn oil, abundant in the United States and high in (polyunsaturated) linoleic acid, is an attractive dietary choice. Margarine made from corn, safflower, or sunflower oils is much lower in saturated fatty acids than is butter, which is made from milk fat.*

Although vegetable oils usually contain a higher proportion of unsaturated fatty acids than do animal oils and fats, several plant oils are actually high in saturated fats. Palm oil is low in polyunsaturated fatty acids and particularly high in (saturated) palmitic acid (whence the name palmitic). Coconut oil is particularly high in lauric and myristic acids (both saturated) and contains very few unsaturated fatty acids.

Some of the fatty acids found in the diets of developed nations are *trans* fatty acids—fatty acids with one or more double bonds in the *trans* configuration. Some of these derive from dairy fat and ruminant meats, but the bulk are provided by partially hydrogenated vegetable or fish oils. (Margarine, for example, contains substantial amounts of *trans* fatty acids.) Substantial evidence now exists to indicate that *trans* fatty acids may have deleterious health consequences. Numerous studies have shown that *trans* fatty acids raise plasma LDL cholesterol levels when exchanged for *cis*-unsaturated fatty acids in the diet, and that they may also lower HDL cholesterol levels and raise triglyceride levels. The effects of *trans* fatty acids on LDL, HDL, and overall cholesterol levels are similar to those of saturated fatty acids, and diets aimed at reducing the risk of coronary heart disease should be low in both *trans* and saturated fatty acids.

Fatty Acid Compositions of Some Dietary Lipids*

Source	Lauric and Myristic	Palmitic	Stearic	Oleic	Linoleic
Beef	5	24–32	20–25	37–43	2–3
Milk		25	12	33	3
Coconut	74	10	2	7	–
Corn		8–12	3–4	19–49	34–62
Olive		9	2	84	4
Palm		39	4	40	8
Safflower		6	3	13	78
Soybean		9	6	20	52
Sunflower		6	1	21	66

Data from *Merck Index*, 10th ed. Rahway, NJ: Merck and Co., and Wilson et al., 1967, *Principles of Nutrition*, 2nd ed. New York: Wiley.
*Values are percentages of total fatty acids.

Oleic acid
cis double bond

Elaidic acid
trans double bond

*Margarine was invented by a French chemist, H. Mège Mouriès, who won a prize from Napoleon III in 1869 for developing a substitute for butter.

The double bonds found in fatty acids are nearly always in the *cis* configuration. As shown in Figure 6.1, this causes a bend or "kink" in the fatty acid chain. This bend has very important consequences for the structure of biological membranes. Saturated fatty acid chains can pack closely together to form ordered, rigid arrays under certain conditions, but unsaturated fatty acids prevent such close packing and produce flexible, fluid aggregates.

Some fatty acids are not synthesized by mammals and yet are necessary for normal growth and life. These **essential fatty acids** include **linoleic** and **α-linolenic acids.** These must be obtained by humans in their diet (specifically from plant sources). **Arachidonic acid,** which is not found in plants, can only be synthesized by mammals from linoleic acid. At least one function of the essential fatty acids is to serve as a precursor for the synthesis of **eicosanoids,** such as *prostaglandins,* a class of compounds that exert hormonelike effects in many physiological processes (discussed in Chapter 20).

A DEEPER LOOK

Polar Bears Use Triacylglycerols to Survive Long Periods of Fasting

The polar bear is magnificently adapted to thrive in its harsh Arctic environment. Research by Malcolm Ramsey (at the University of Saskatchewan in Canada) and others has shown that polar bears eat only during a few weeks of the year and then fast for periods of eight months or more, consuming no food or water during that time. Eating mainly in the winter, the adult polar bear feeds almost exclusively on seal blubber (largely composed of triacylglycerols), thus building up its own triacylglycerol reserves. Through the Arctic summer, the polar bear maintains normal physical activity, roaming over long distances, but relies entirely on its body fat for sustenance, burning as much as 1 to 1.5 kg of fat per day. It neither urinates nor defecates for extended periods. All the water needed to sustain life is provided from the metabolism of triacylglycerides (because oxidation of fatty acids yields carbon dioxide and water).

Interestingly, the word *Arctic* comes from the ancient Greeks, who understood that the northernmost part of the earth lay under the stars of the constellation Ursa Major, the Great Bear. Although unaware of the polar bear, they called this region *Arktikós*, which means "the country of the great bear."

(© *Thomas D. Mangelsen*)

6.2 Triacylglycerols

A significant number of the fatty acids in plants and animals exist in the form of **triacylglycerols** (also called **triglycerides**). Triacylglycerols are a major energy reserve and the principal neutral derivatives of glycerol found in animals. These molecules consist of a glycerol esterified with three fatty acids (Figure 6.2). If all three fatty acid groups are the same, the molecule is called a simple triacylglycerol. Examples include **tristearoylglycerol** (common name: *tristearin*) and **trioleoylglycerol** (*triolein*). Mixed triacylglycerols contain two or three different fatty acids. Triacylglycerols in animals are found primarily in the adipose tissue (body fat), which serves as a depot or storage site for lipids. Monoacylglycerols and diacylglycerols also exist but are far less common than the triacylglycerols. Most natural plant and animal fat is composed of mixtures of simple and mixed triacylglycerols.

Figure 6.2 Triacylglycerols are formed from glycerol and fatty acids.

Glycerol

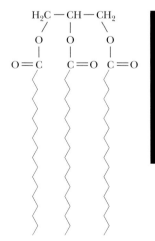

Tristearin
(a simple triacylglycerol)

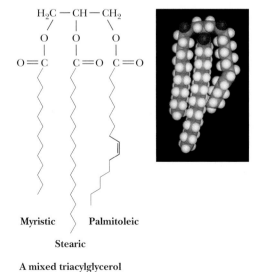

Myristic   Palmitoleic

Stearic

A mixed triacylglycerol

Triacylglycerols are rich in highly reduced carbons and thus yield large amounts of energy in the oxidative reactions of metabolism. Complete oxidation of 1 g of triacylglycerols yields about 38 kJ of energy, whereas proteins and carbohydrates yield only about 17 kJ/g. Also, their hydrophobic nature allows them to aggregate in highly anhydrous forms, whereas polysaccharides and proteins are highly hydrated. For these reasons, triacylglycerols are the molecules of choice for energy storage in animals. Body fat (mainly triacylglycerols) also provides good insulation. Whales and Arctic mammals rely on body fat for both insulation and energy reserves.

6.3 Glycerophospholipids

A 1,2-diacylglycerol that has a phosphate group esterified at carbon atom 3 of the glycerol backbone is a **glycerophospholipid,** also known as a *phosphoglyceride* or a *glycerol phosphatide* (Figure 6.3). These lipids form one of the largest classes of natural lipids and one of the most important. They are essential components of cell membranes and are found in small concentrations in other parts of the cell. It should be noted that all glycerophospholipids are members of the broader class of lipids known as **phospholipids.**

The numbering and nomenclature of glycerophospholipids present a dilemma in that the number 2 carbon of the glycerol backbone is asymmetric. It is possible to name these molecules as either D- or L-isomers. Thus, glycerol phosphate itself can be referred to either as D-glycerol-1-phosphate or as L-glycerol-3-phosphate (Figure 6.4). Instead of naming the glycerol phosphatides in this way, biochemists have adopted the *stereospecific numbering* system, or *sn*-system. In this system, the *pro-(S)* position of a prochiral atom (see *A Deeper Look—Prochirality*) is denoted as the *1-position,* the prochiral atom as the *2-position,* and so on. When this scheme is used, the prefix *sn*-

A DEEPER LOOK

Prochirality

If a tetrahedral center in a molecule has two identical substituents, it is referred to as **prochiral,** since, if either of the like substituents were converted to a different group, the tetrahedral center would then be chiral. Consider glycerol:

Glycerol

The central carbon of glycerol is prochiral, since replacing either of the CH$_2$OH groups would make the central carbon chiral. Nomenclature for prochiral centers is based on the (*R,S*) system (see Chapter 3). To name the otherwise identical substituents of a prochiral center, imagine increasing slightly the priority of one of them (by substituting a deuterium for a hydrogen, for example) as shown:

1-d, 2(*S*)-Glycerol
[(*S*)-configuration at C-2]

The resulting molecule has an (*S*)-configuration about the (now chiral) central carbon atom. The group that contains the deuterium is thus referred to as the *pro-S* group. As a useful exercise, you should confirm that labeling the other CH$_2$OH group with a deuterium produces the (*R*)-configuration at the central carbon, so that this latter CH$_2$OH group is the *pro-R* substituent.

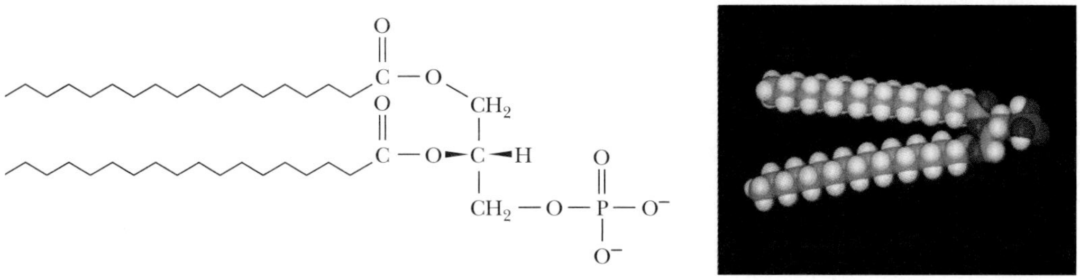

Figure 6.3 Phosphatidic acid, the parent compound for glycerophospholipids.

precedes the molecule name (glycerol phosphate in this case) and distinguishes this nomenclature from other approaches. In this way, the glycerol phosphate in natural phosphoglycerides is named *sn*-glycerol-3-phosphate.

Phosphatidic acid, the parent compound for the glycerol-based phospholipids (see Figure 6.3), consists of *sn*-glycerol-3-phosphate, with fatty acids esterified at the 1- and 2-positions. Phosphatidic acid is found in small amounts in most natural systems and is an important intermediate in the biosynthesis of the more common glycerophospholipids (Figure 6.5). In these compounds, a variety of polar groups are esterified to the phosphoric acid moiety of the molecule. The phosphate, together with such esterified entities, is referred to as a "head" group. **Phosphatides** are formed when a hydroxyl-containing organic function becomes esterified to the phosphate group of phosphatidic acid. Phosphatides with choline or ethanolamine are referred to as **phosphatidylcholine** (known commonly as **lecithin**) or **phosphatidylethanolamine,** respectively. These phosphatides are two of the most common constituents of biological membranes. Other common head groups found in phosphatides include glycerol, serine, and inositol (see Figure 6.5). Another kind of glycerol phosphatide found in many tissues is **diphosphatidylglycerol.** First observed in heart tissue, it is also called **cardiolipin.** In cardiolipin, a phosphatidylglycerol is esterified through the C-1 hydroxyl group of the glycerol moiety of the head group to the phosphoryl group of another phosphatidic acid molecule.

Phosphatides exist in many different varieties, depending on the fatty acids esterified to the glycerol group. As we shall see, the nature of the fatty acids can greatly affect the chemical and physical properties of the phosphatides and the membranes that contain them. In most cases, glycerol phosphatides have a saturated fatty acid at position 1 and an unsaturated fatty acid at position 2 of the glycerol. Thus, **1-stearoyl-2-oleoyl-phosphatidylcholine** (Figure 6.6) is a common constituent in natural membranes, but **1-linoleoyl-2-palmitoylphosphatidylcholine** is not.

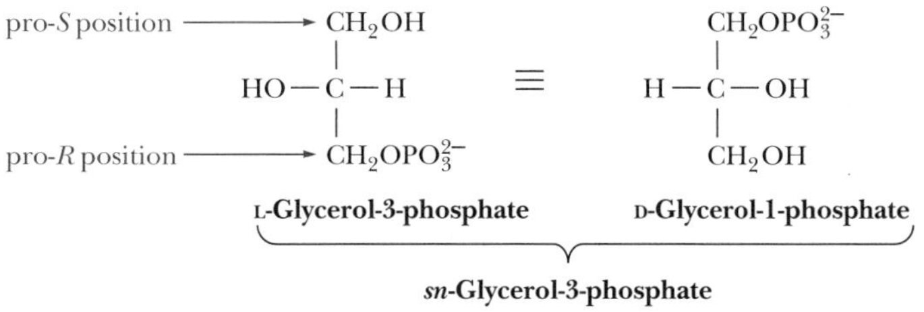

Figure 6.4 The absolute configuration of *sn*-glycerol-3-phosphate. The *pro-(R)* and *pro-(S)* positions of the parent glycerol are also indicated.

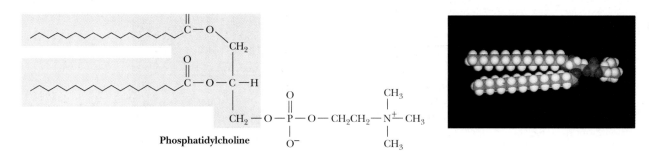

Phosphatidylcholine

GLYCEROLIPIDS WITH OTHER HEAD GROUPS:

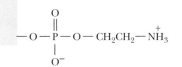

Phosphatidylethanolamine

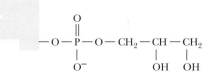

Phosphatidylserine

Phosphatidylglycerol

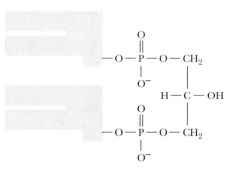

Diphosphatidylglycerol (Cardiolipin)

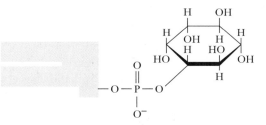

Phosphatidylinositol

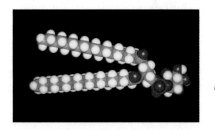

See *Interactive Bio-chemistry CD-ROM and Workbook,* page 119

Figure 6.5 Structures of several glycerophospholipids and space-filling models of phosphatidylcholine, phosphatidylglycerol, and phosphatidylinositol.

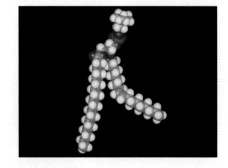

Both structural and functional strategies govern the natural design of the many different kinds of glycerophospholipid head groups and fatty acids. The structural roles of these different glycerophospholipid classes are described later in this chapter. Certain phospholipids, including phosphatidylinositol and phosphatidylcholine, participate in complex cellular signaling events. These roles, appreciated only in recent years, are described in Chapter 26.

◀ **Figure 6.6** A space-filling model of 1-stearoyl-2-oleoyl-phosphatidylcholine.

A DEEPER LOOK

Glycerophospholipid Degradation: One of the Effects of Snake Venoms

The venoms of poisonous snakes contain (among other things) a class of enzymes known as **phospholipases,** enzymes that cause the breakdown of phospholipids. For example, the venoms of the western diamondback rattlesnake (*Crotalus atrox*) and the Indian cobra (*Naja naja*) both contain phospholipase A_2, which catalyzes the hydrolysis of fatty acids at the C-2 position of glycerophospholipids.

The phospholipid breakdown product of this reaction, *lysolecithin,* acts as a detergent and dissolves the membranes of red blood cells, causing them to rupture. Indian cobras kill several thousand people each year.

Western diamondback rattlesnake. (*Tom Bean/CORBIS*)

Indian cobra. (*Joe McDonald/CORBIS*) ▶

Phospholipid

Ether Glycerophospholipids

Ether glycerophospholipids possess an ether linkage instead of an acyl group at the C-1 position of glycerol (Figure 6.7). One of the most versatile biochemical signal molecules found in mammals is **platelet activating factor,** or **PAF,** a unique ether glycerophospholipid (Figure 6.8). The alkyl group at C-1 of PAF is typically a 16-carbon chain, but the acyl group at C-2 is a 2-carbon acetate unit. By virtue of this acetate group, PAF is much more water-soluble

Figure 6.7 A 1-alkyl 2-acyl-phosphatidylethanolamine (an ether glycerophospholipid). ▶

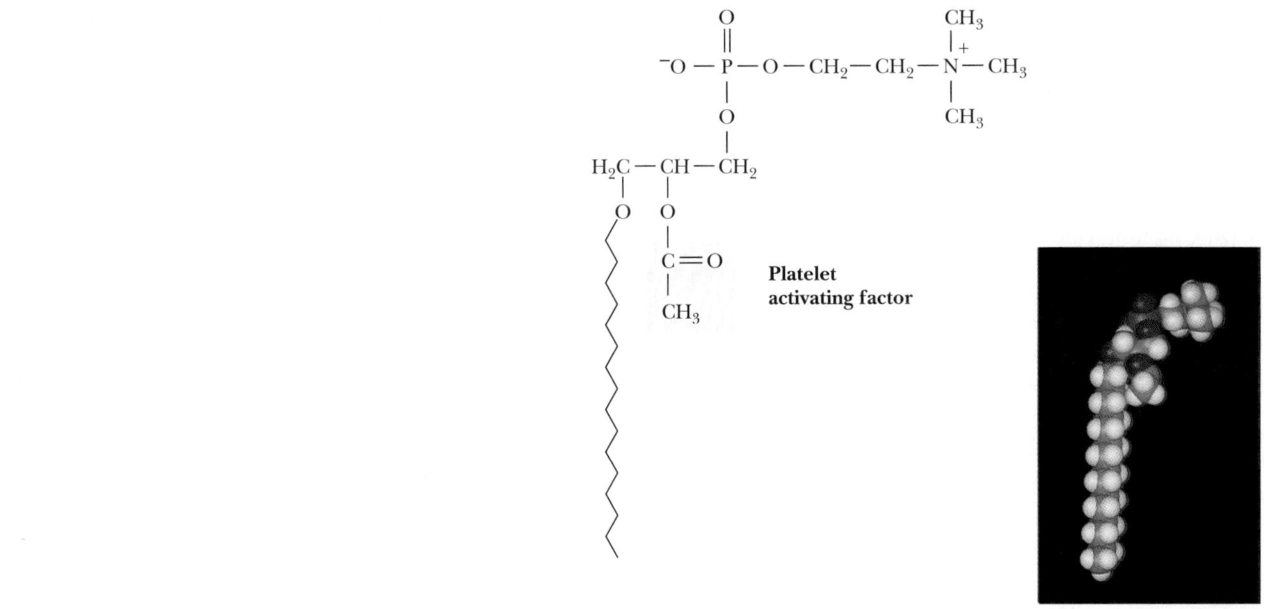

$$
\begin{array}{c}
\text{O} \qquad\qquad\qquad \text{CH}_3 \\
\parallel \qquad\qquad\qquad \mid \; + \\
{}^-\text{O}-\text{P}-\text{O}-\text{CH}_2-\text{CH}_2-\text{N}-\text{CH}_3 \\
\mid \qquad\qquad\qquad\qquad \mid \\
\text{O} \qquad\qquad\qquad\qquad \text{CH}_3 \\
\mid \\
\text{H}_2\text{C}-\text{CH}-\text{CH}_2 \\
\quad\mid \qquad \mid \\
\quad\text{O} \qquad \text{O} \\
\qquad\qquad \mid \\
\qquad\qquad \text{C}=\text{O} \qquad \textbf{Platelet} \\
\qquad\qquad \mid \qquad\qquad \textbf{activating factor} \\
\qquad\qquad \text{CH}_3
\end{array}
$$

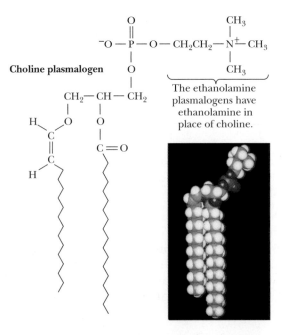

Figure 6.8 The structure of 1-alkyl 2-acetyl-phosphatidylcholine, also known as platelet activating factor or PAF.

than other lipids, allowing PAF to function as a soluble messenger in signal transduction.

   Plasmalogens are ether glycerophospholipids in which the alkyl moiety is *cis-α, β*-unsaturated (Figure 6.9). Common plasmalogen head groups include choline, ethanolamine, and serine. These lipids are referred to as *phosphatidal choline, phosphatidal ethanolamine,* and *phosphatidal serine.*

Figure 6.9 The structure and a space-filling model of a choline plasmalogen.

Platelet Activating Factor: A Potent Glyceroether Mediator

Platelet activating factor (PAF) was first identified by its ability (at low levels) to cause platelet aggregation and dilation of blood vessels, but it is now known to be a potent mediator in inflammation, allergic responses, and shock. PAF effects are observed at tissue concentrations as low as 10^{-12} M. PAF causes a dramatic inflammation of air passages and induces asthmalike symptoms in laboratory animals. **Toxic-shock syndrome** occurs when fragments of destroyed bacteria act as toxins and induce the synthesis of PAF. This results in a drop in blood pressure and a reduced volume of blood pumped by the heart, which leads to shock and, in severe cases, death.

Beneficial effects have also been attributed to PAF. In reproduction, PAF secreted by the fertilized egg is instrumental in the implantation of the egg in the uterine wall. PAF is produced in significant quantities in the lungs of the fetus late in pregnancy and may stimulate the production of fetal lung surfactant, a protein–lipid complex that prevents collapse of the lungs in a newborn infant.

6.4 Sphingolipids

Sphingolipids represent another class of lipids found frequently in biological membranes. An 18-carbon amino alcohol, **sphingosine** (Figure 6.10), forms the backbone of these lipids rather than glycerol. Typically, a fatty acid is joined to a sphingosine via an amide linkage to form a **ceramide. Sphingomyelins** represent a phosphorus-containing subclass of sphingolipids and are especially important in the nervous tissue of higher animals. A **sphingomyelin** is formed by the esterification of a phosphorylcholine or a phosphorylethanolamine to the 1-hydroxy group of a ceramide (Figure 6.11).

There is another class of ceramide-based lipids which, like the sphingomyelins, are important components of muscle and nerve membranes in animals. These are the **glycosphingolipids,** and they consist of a ceramide with one or more sugar residues in a β-glycosidic linkage at the 1-hydroxyl moiety. The neutral glycosphingolipids contain only neutral (uncharged) sugar residues. When a single glucose or galactose is bound in this manner, the molecule is a **cerebroside** (Figure 6.12). Another class of lipids is formed when a sulfate is esterified at the 3-position of the galactose to make a **sulfatide.**

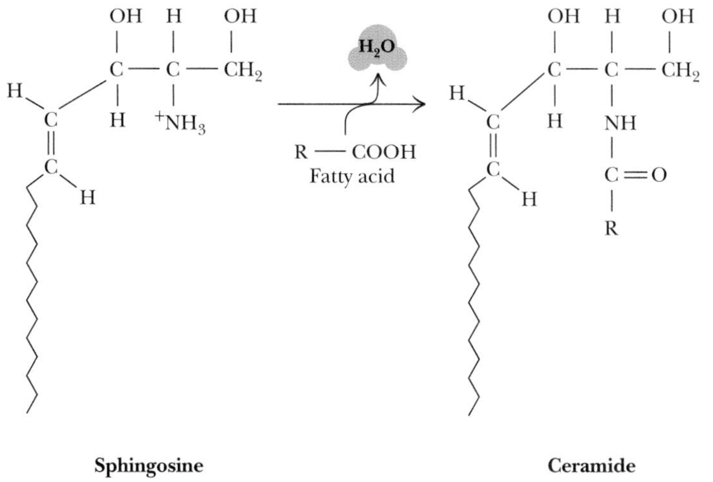

Figure 6.10 Formation of an amide linkage between a fatty acid and sphingosine produces a ceramide.

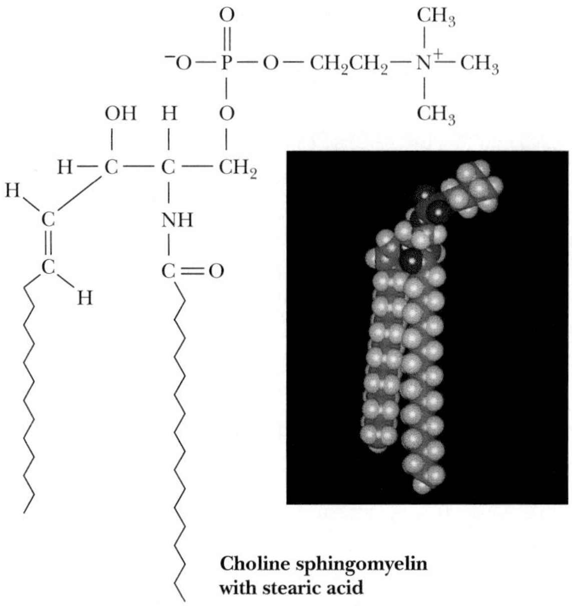

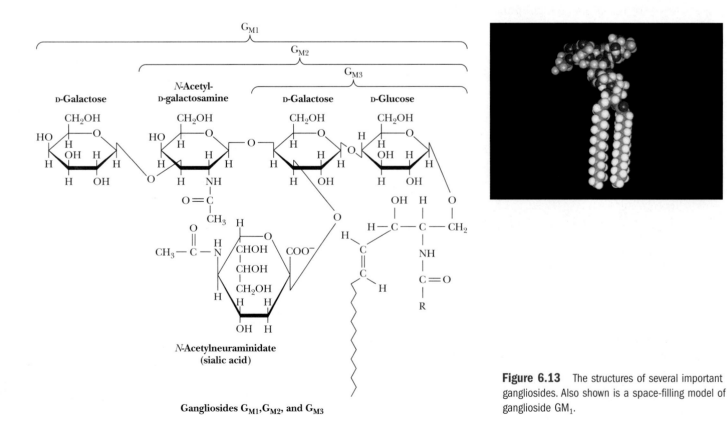

Figure 6.11 A structure and a space-filling model of a choline sphingomyelin formed from stearic acid.

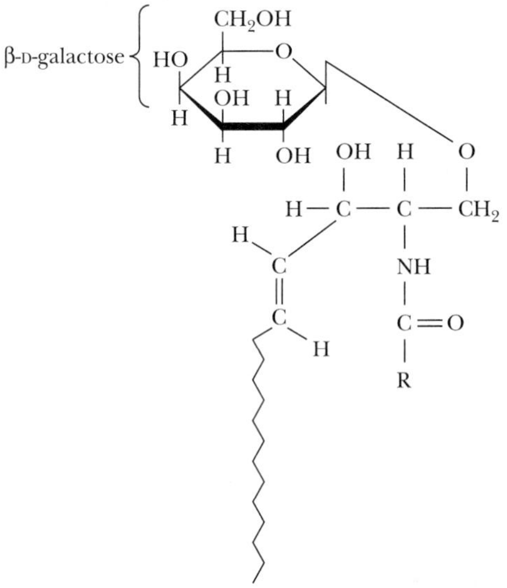

A cerebroside

Figure 6.12 The structure of a cerebroside. Note the sphingosine backbone.

Gangliosides (Figure 6.13) are more complex glycosphingolipids that consist of a ceramide backbone with three or more sugars esterified, one of these being a **sialic acid** such as **N-acetylneuraminic acid.** These latter compounds are referred to as acidic *glycosphingolipids,* and they have a net negative charge at neutral pH.

Figure 6.13 The structures of several important gangliosides. Also shown is a space-filling model of ganglioside GM₁.

Gangliosides G_{M1}, G_{M2}, and G_{M3}

Figure 6.14 An example of a wax. Oleoyl alcohol is esterified to stearic acid in this case. ▶

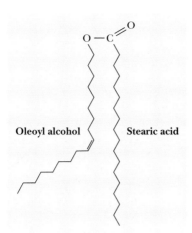

Oleoyl alcohol — Stearic acid

The glycosphingolipids have a number of important cellular functions, despite the fact that they are present only in small amounts in most membranes. Glycosphingolipids at cell surfaces appear to determine, at least in part, certain elements of tissue and organ specificity. Cell–cell recognition and tissue immunity appear to depend upon specific glycosphingolipids. Gangliosides are present in nerve endings and appear to be important in nerve impulse transmission. A number of genetically transmitted diseases involve the accumulation of specific glycosphingolipids due to an absence of the enzymes needed for their degradation. Such is the case for ganglioside GM_2 in the brains of Tay–Sachs disease victims. A rare but fatal disease is characterized by a red spot on the retina, gradual blindness, and loss of weight; victims do not survive beyond early childhood.

<h2>6.5   Waxes</h2>

Waxes are esters of long-chain alcohols with long-chain fatty acids. The resulting molecule can be viewed (in analogy to the glycerolipids) as having a weakly polar head group (the ester moiety itself) and a long, nonpolar tail (the hydrocarbon chains) (Figure 6.14). Fatty acids found in waxes are usually saturated. The alcohols found in waxes may be saturated or unsaturated and may include sterols, such as cholesterol (see page 168). Waxes are highly insoluble due to the weakly polar nature of the ester group. As a result, this class of molecules confers water-repellant character to animal skin, to the leaves of certain plants, and to bird feathers. The glossy surface of a polished apple results from a wax coating. **Carnauba wax,** obtained from the fronds of a species of palm tree in Brazil, is a particularly hard wax used for high-gloss finishes, such as those for automobiles, boats, floors, and shoes. **Lanolin,** a component of wool wax, is used as a base for pharmaceutical and cosmetic products because it is rapidly assimilated by human skin.

A DEEPER LOOK

Moby-Dick and Spermaceti: A Valuable Wax from Whale Oil

When oil from the head of the sperm whale is cooled, spermaceti, a translucent wax with a white, pearly luster, crystallizes from the mixture. Spermaceti, which makes up 11% of whale oil, is composed mainly of the wax **cetyl palmitate:**

$$CH_3(CH_2)_{14}\!-\!\overset{\displaystyle O}{\overset{\displaystyle \|}{C}}\!-\!O\!-\!(CH_2)_{15}CH_3$$

as well as smaller amounts of cetyl alcohol:

$$HO\!-\!(CH_2)_{15}CH_3$$

Spermaceti and cetyl palmitate have been widely used in the making of cosmetics, fragrant soaps, and candles.

In the literary classic *Moby-Dick*, Herman Melville describes Ishmael's impressions of spermaceti when the character muses that the waxes "discharged all their opulence, like fully ripe grapes their wine; as I snuffed that uncontaminated aroma—literally and truly, like the smell of spring violets."*

*Melville, H., 1984. *Moby-Dick.* London: Octopus Books, p. 205. (Adapted from Waddell, T. G., and Sanderlin, R. R., 1986. Chemistry in *Moby-Dick. Journal of Chemical Education* **63:**1019–1020.)

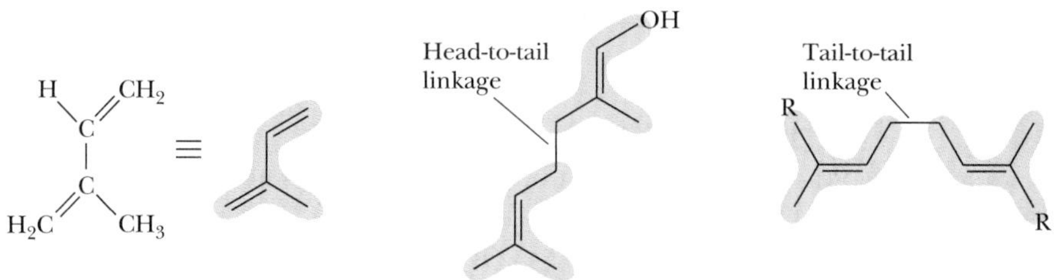

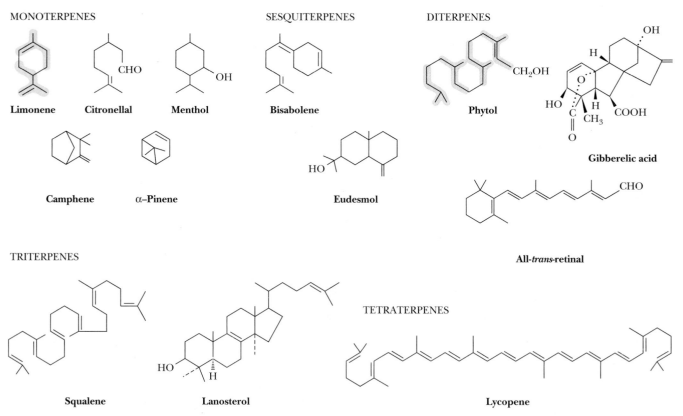

Figure 6.15 The structure of isoprene (2-methyl-1,3-butadiene) and the structure of head-to-tail and tail-to-tail linkages. Isoprene itself is formed by distillation of natural rubber, a linear head-to-tail polymer of isoprene units.

6.6 Terpenes

The **terpenes** are simple lipids because they lack fatty acid components. Terpenes are a class of lipids formed from combinations of two or more molecules of 2-methyl-1,3-butadiene, better known as **isoprene** (a five-carbon unit that is abbreviated C_5). A **monoterpene** (C_{10}) consists of two isoprene units, a

Figure 6.16 Many monoterpenes are readily recognized by their characteristic flavors or odors (limonene in lemons; citronellal in roses, geraniums, and some perfumes; pinene in turpentine; and menthol from peppermint, used in cough drops and nasal inhalers). The diterpenes, which are C_{20} terpenes, include retinal (the essential light-absorbing pigment in rhodopsin, the photoreceptor protein of the eye), phytol (a constituent of chlorophyll), and the gibberellins (potent plant hormones). The triterpene lanosterol is a constituent of wool fat. Lycopene is a carotenoid found in ripe fruit, especially tomatoes.

sesquiterpene (C_{15}) consists of three isoprene units, a **diterpene** (C_{20}) has four isoprene units, and so on. Isoprene units can be linked in terpenes to form straight-chain or cyclic molecules, and the usual method of linking isoprene units is head to tail (Figure 6.15). Monoterpenes occur in all higher plants, while sesquiterpenes and diterpenes are less widely known. Several examples of these classes of terpenes are shown in Figure 6.16. The **triterpenes** are C_{30} terpenes and include **squalene** and **lanosterol,** two of the precursors of cholesterol and other steroids (discussed on page 168). **Tetraterpenes** (C_{40}) are less common but include the carotenoids, a class of colorful photosynthetic pigments. β-Carotene is the precursor of vitamin A, while lycopene, similar to β-carotene but lacking the cyclopentene rings, is a pigment found in tomatoes.

Long-chain polyisoprenoid molecules with a terminal alcohol moiety are called **polyprenols.** The **dolichols,** one class of polyprenols (Figure 6.17), consist of 16 to 22 isoprene units and, in the form of dolichyl phosphates, function to carry carbohydrate units in the biosynthesis of glycoproteins in animals. Polyprenyl groups serve to *anchor* certain proteins to biological membranes (discussed in Section 6.11).

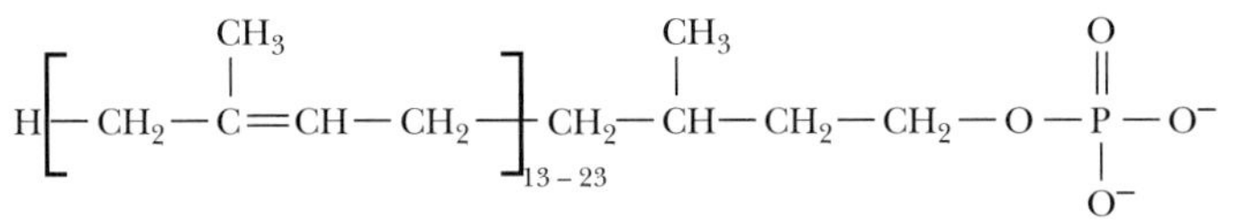

Dolichol phosphate

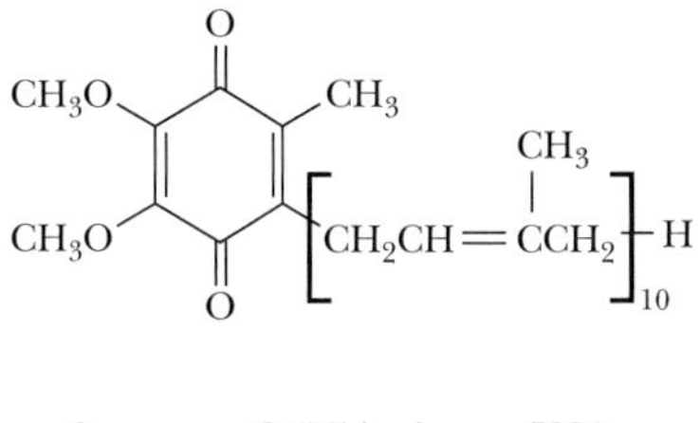

Coenzyme Q (Ubiquinone, UQ)

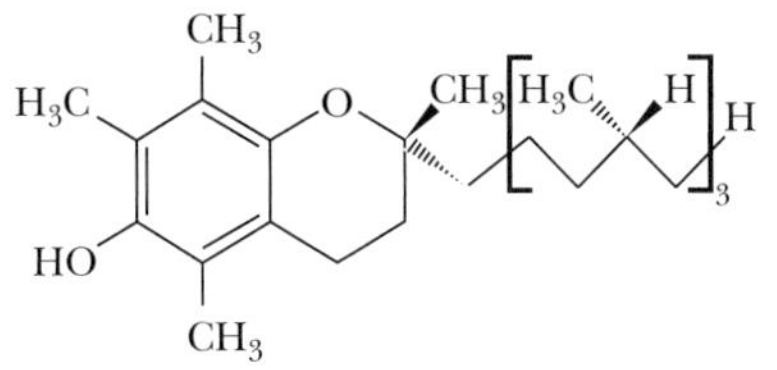

Vitamin E (α-tocopherol)

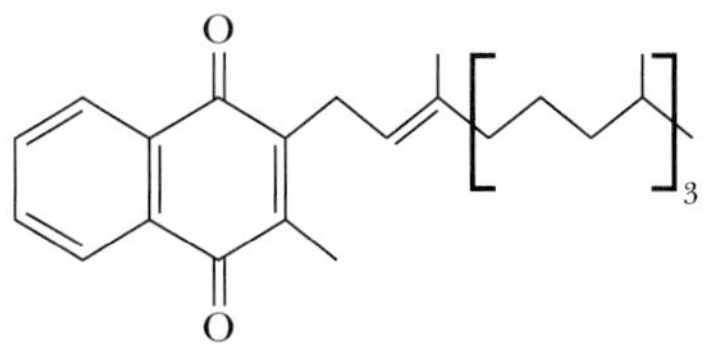

Vitamin K$_1$
(phylloquinone)

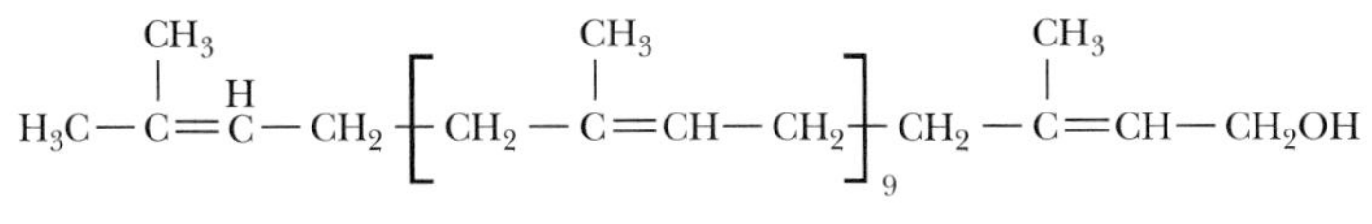

Undecaprenyl alcohol (bactoprenol)

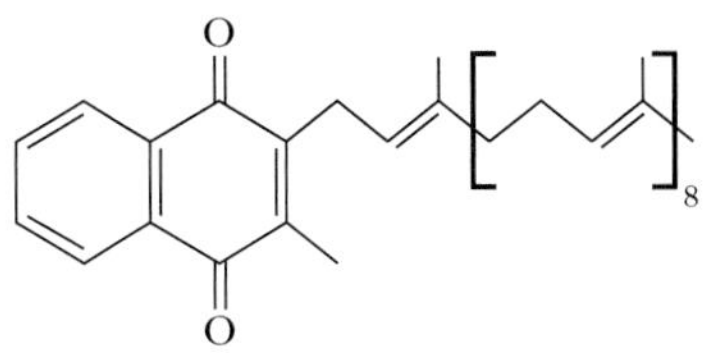

Vitamin K$_2$
(menaquinone)

Figure 6.17 Dolichol phosphate is an initiation point for the synthesis of carbohydrate polymers in animals. The analogous alcohol in bacterial systems, undecaprenol, also known as bactoprenol, consists of 11 isoprene units. Undecaprenyl phosphate delivers sugars from the cytoplasm for the synthesis of cell wall components such as peptidoglycans, lipopolysaccharides, and glycoproteins. Polyprenyl compounds also serve as the side chains of vitamin K, the ubiquinones, plastoquinones, and tocopherols (such as vitamin E).

HUMAN BIOCHEMISTRY

Coumadin or Warfarin—Agent of Life or Death

The isoprene-derived molecule whose structure is shown here is known alternately as **coumadin** or **warfarin.** By the former name, it is a widely prescribed anticoagulant. By the latter name, it is a component of rodent poisons. How can the same chemical species be used for such disparate purposes? The key to both uses lies in its ability to act as an antagonist of vitamin K in the body.

Vitamin K stimulates the carboxylation of glutamate residues on certain proteins, especially certain proteins in the blood-clotting cascade (including **prothrombin, factor VII, factor IX,** and **factor X**), which undergo a Ca^{2+}-dependent conformational change in the course of their biological activity, as well as **protein C** and **protein S,** two regulatory proteins. Carboxylation of these coagulation factors is catalyzed by a carboxylase that requires the reduced form of vitamin K (vitamin KH_2), molecular oxygen, and carbon dioxide (as shown in the figure here). KH_2 is oxidized to vitamin K epoxide, KO, which is recycled to KH_2 by the enzymes **vitamin K epoxide reductase(1)** and **vitamin K reductase(2).** Coumadin/warfarin exerts its anticoagulant effect by inhibiting vitamin K epoxide reductase and possibly also vitamin K reductase. This inhibition depletes vitamin KH_2 and reduces the activity of the carboxylase.

Coumadin/warfarin, given at a typical dosage of 4 to 5 mg per day, prevents the deleterious formation in the bloodstream of small blood clots, and thus reduces the risk of heart attacks and strokes for individuals whose arteries contain sclerotic plaques. Taken in much larger doses—as, for example, in rodent poisons—coumadin/warfarin can cause massive hemorrhages and death.

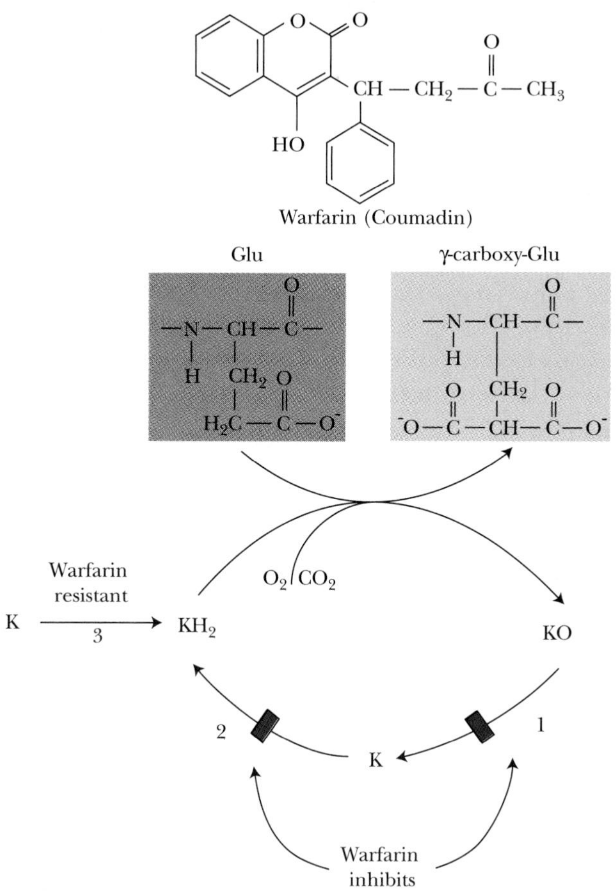

Warfarin (Coumadin)

6.7 | Steroids

A large and important class of terpene-based lipids is the **steroids.** This molecular family, whose members effect an amazing array of cellular functions, is based on a common structural motif of three six-membered rings and one five-membered ring all fused together.

Cholesterol

Cholesterol (Figure 6.18) is the most common steroid in animals and the precursor for all other animal steroids. The numbering system for cholesterol applies to all such molecules. Many steroids contain methyl groups at positions 10 and 13 and an 8- to 10-carbon alkyl side chain at position 17. The polyprenyl nature of this compound is particularly evident in the side chain. Many steroids contain an oxygen at C-3, either a hydroxyl group in sterols or a carbonyl group in other steroids. The carbons at positions 10 and 13 and the alkyl group at position 17 are nearly always oriented on the same side of the steroid nucleus, the β-orientation. Alkyl groups that extend from the other side of the steroid backbone are in an α-orientation.

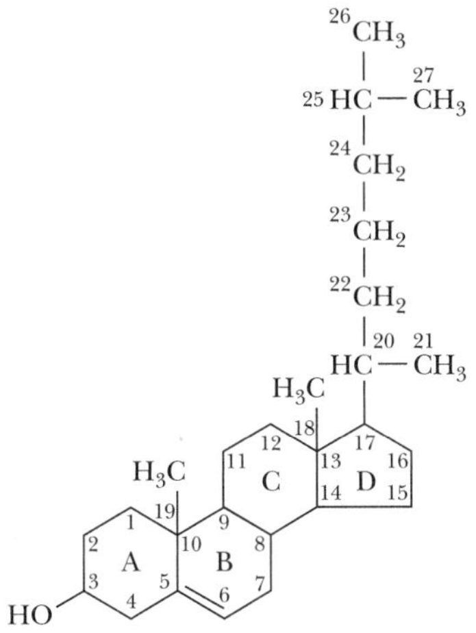

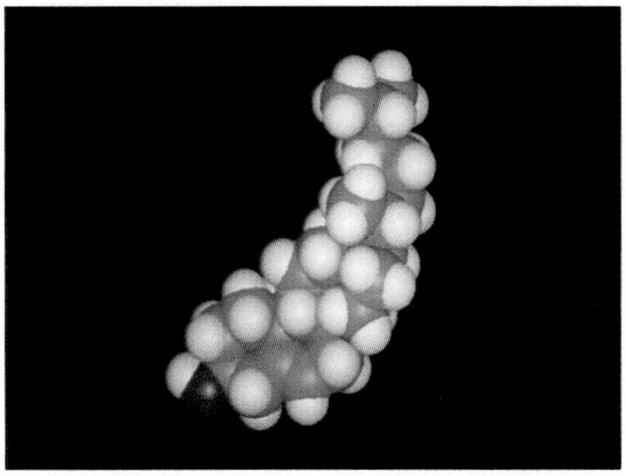

Figure 6.18 The structure of cholesterol, shown with steroid ring designations and carbon numbering.

HUMAN BIOCHEMISTRY

Plant Sterols—Natural Cholesterol Fighters

Dietary guidelines for optimal health call for reducing the intake of cholesterol. One strategy for doing so involves the plant sterols, including sitosterol, stigmasterol, stigmastanol, and campesterol, shown in the figure. Despite their structural similarity to cholesterol, minor isomeric differences and the presence of methyl and ethyl groups in the side chains of these substances result in their poor absorption by intestinal mucosal cells. Interestingly, though plant sterols are not effectively absorbed by the body, they nonetheless are highly effective in blocking the absorption of cholesterol itself by intestinal cells.

Ortho-McNeil Pharmaceutical has developed a sitostanol ester preparation from pine wood products. The company markets this substance as a butter substitute called **Benecol,** with the claim that its use can lead to a 14% decrease in blood cholesterol levels.

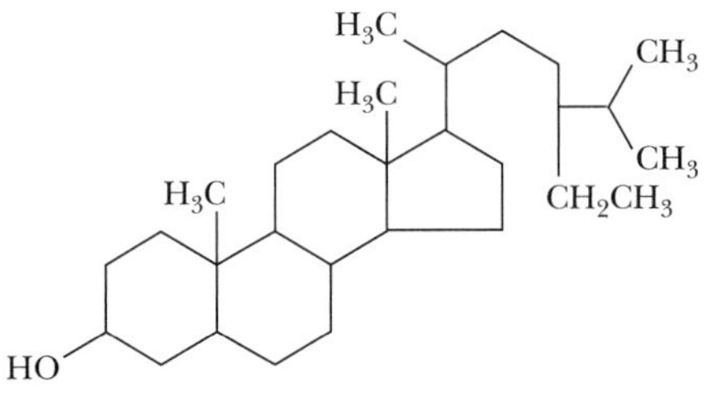

Stigmastanol

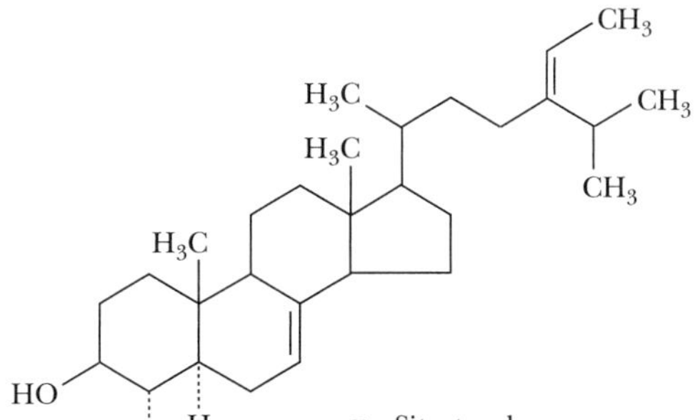

α_1- Sitosterol

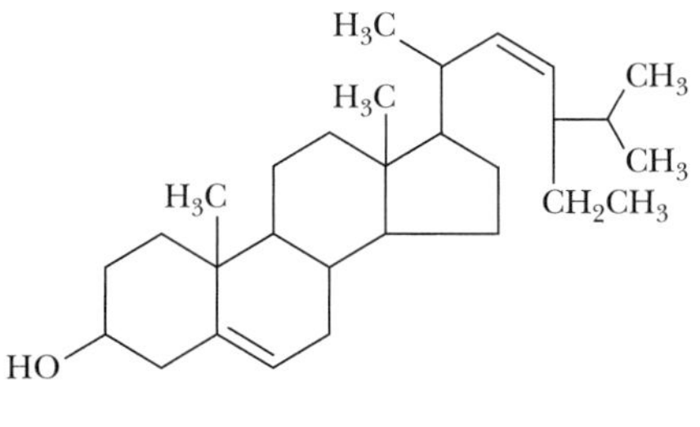

Stigmasterol

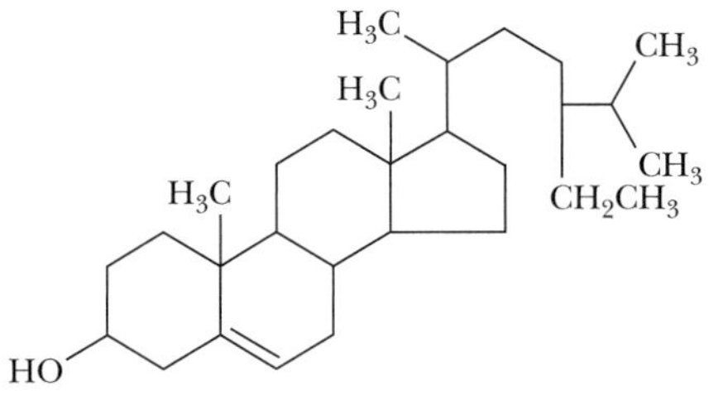

β-Sitosterol

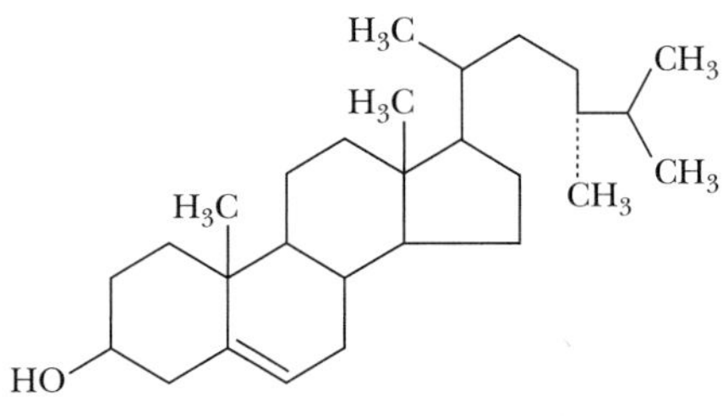

Campesterol

Cholesterol is a principal component of animal cell plasma membranes, and much smaller amounts of cholesterol are found in the membranes of organelles. The relatively rigid fused ring system of cholesterol and the weakly polar alcohol group at the C-3 position have important consequences for the properties of plasma membranes. Cholesterol is also a component of lipoprotein complexes in the blood, and it is one of the constituents of plaques that form on arterial walls in atherosclerosis.

Steroid Hormones

Steroids derived from cholesterol in animals include five families of hormones: androgens, estrogens, progestins, glucocorticoids and mineralocorticoids, and bile acids (Figure 6.19). **Androgens** such as **testosterone** and **estrogens** such as **estradiol** mediate the development of sexual characteristics and sexual function in animals. The **progestins,** such as **progesterone,** participate in control of the menstrual cycle and pregnancy. **Glucocorticoids** (**cortisol,** for example) participate in the control of carbohydrate, protein, and lipid metabolism,

HUMAN BIOCHEMISTRY

17β-Hydroxysteroid Dehydrogenase 3 Deficiency

Testosterone, the principal male sex steroid hormone, is synthesized in five steps from cholesterol, as shown in the figure here. In the last step, five isozymes catalyze the 17β-hydroxysteroid dehydrogenase reactions that interconvert 4-androstenedione and testosterone. Defects in the synthesis or action of testosterone can impair the development of the male phenotype during embryogenesis and cause the disorders of human sexuality termed male

pseudohermaphroditism. Specifically, mutations in isozyme 3 of the 17β-hydroxysteroid dehydrogenase in the fetal testes impair the formation of testosterone and give rise to genetic males with female external genitalia and blind-ending vaginas. Such individuals are typically raised as females but, due to an increase in serum testosterone, they virilize at puberty and develop male hair growth patterns.

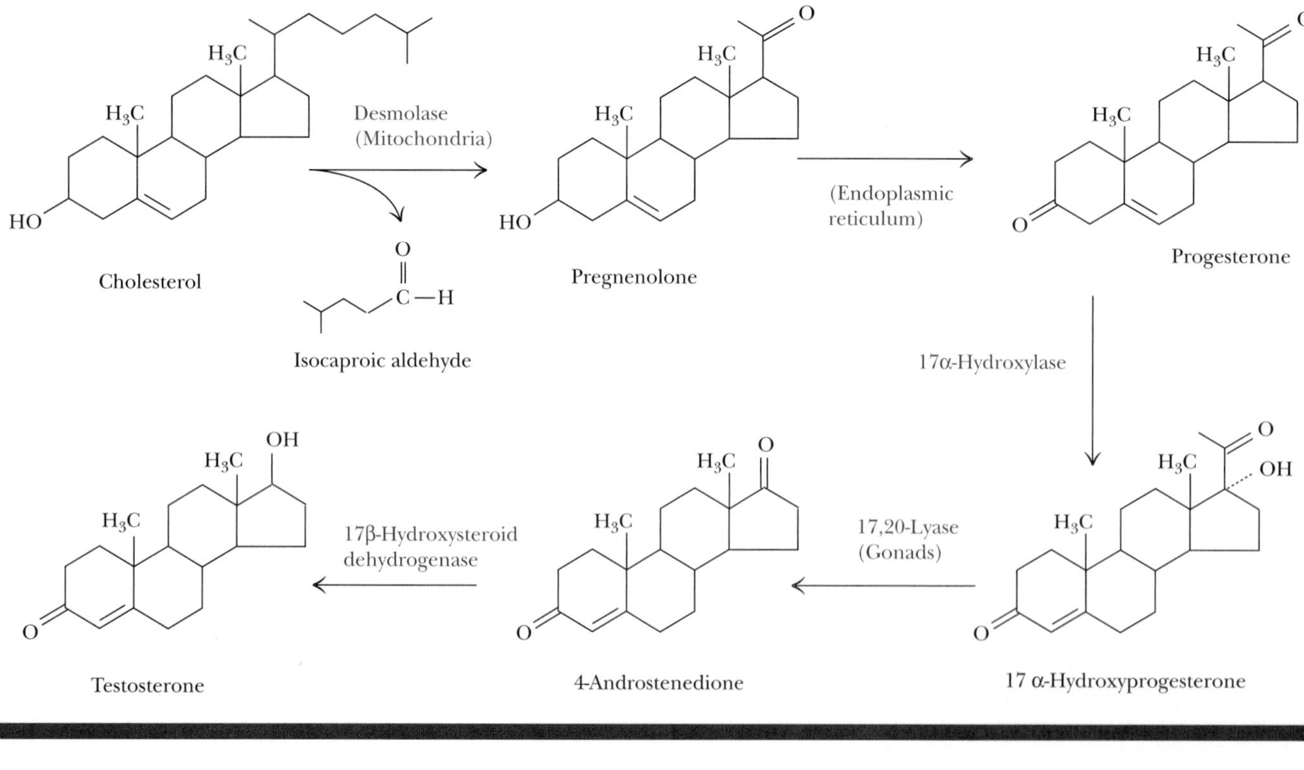

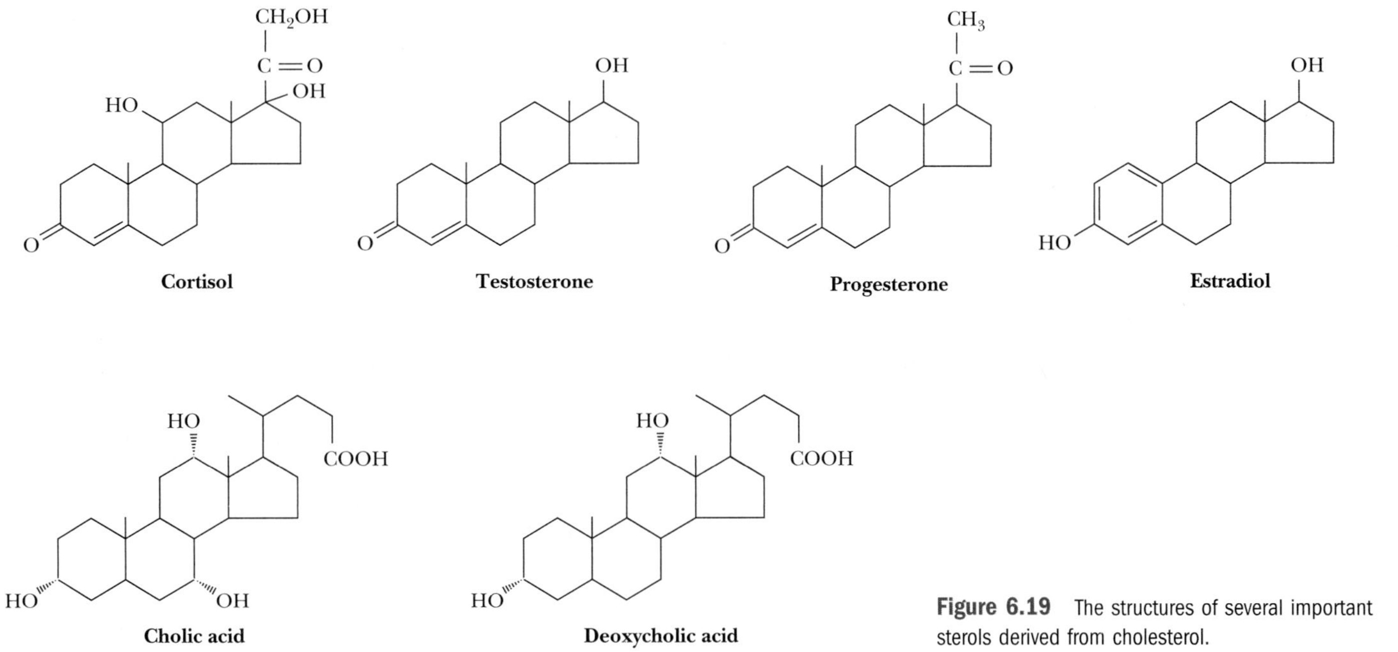

Cortisol   **Testosterone**   **Progesterone**   **Estradiol**

Cholic acid   **Deoxycholic acid**

Figure 6.19 The structures of several important sterols derived from cholesterol.

whereas the **mineralocorticoids** regulate salt (Na^+, K^+, and Cl^-) balances in tissues. The **bile acids** (including **cholic** and **deoxycholic acid**) are detergent molecules secreted in bile from the gallbladder that assist in the absorption of dietary lipids in the intestine.

MEMBRANES

Cells make use of many different types of membranes. All cells have a cytoplasmic membrane, or *plasma membrane,* that functions (in part) to separate the cytoplasm from the surroundings. In the early days of biochemistry, the plasma membrane was not accorded many functions other than this one of partition. We now know that the plasma membrane is also responsible for (1) the exclusion of certain toxic ions and molecules from the cell, (2) the accumulation of cell nutrients, and (3) energy transduction. It functions in (4) cell locomotion, (5) reproduction, (6) signal transduction processes, and (7) interactions with molecules or other cells in the vicinity.

Even the plasma membranes of prokaryotic cells (bacteria) are complex (Figure 6.20). With no intracellular organelles to divide and organize the work, bacteria must carry out all processes either at the plasma membrane or in the cytoplasm itself. Eukaryotic cells, however, contain numerous intracellular organelles that perform specialized tasks. Nucleic acid biosynthesis is handled in the nucleus; mitochondria are the site of electron transport, oxidative phosphorylation, fatty acid oxidation, and the tricarboxylic acid cycle; and secretion of proteins and other substances is handled by the endoplasmic reticulum and the Golgi apparatus. This partitioning of labor is not the only contribution of the membranes in these cells. Many of the processes occurring in these organelles (or in the prokaryotic cell) actively involve membranes. Thus, some of the enzymes involved in nucleic acid metabolism are membrane-associated. The electron transfer chain and its associated system for ATP synthesis are embedded in the mitochondrial membrane. Many enzymes responsible for aspects of lipid biosynthesis are located in the endoplasmic reticulum membrane.

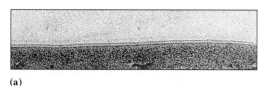

(a)

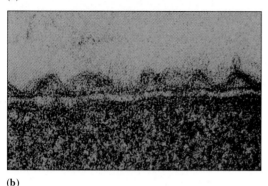

(b)

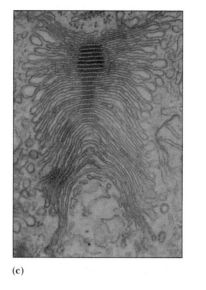

(c)

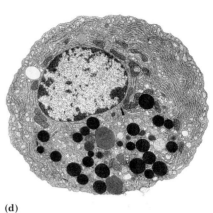

(d)

Figure 6.20 Electron micrographs of several different membrane structures: **(a)** *Menoidium,* a protozoan; **(b)** Gram-negative envelope of *Aquaspirillum serpens;* **(c)** Golgi apparatus; **(d)** pancreatic acinar cell. *(a, T. T. Beveridge/Visuals Unlimited; b, © Cabisco/Visuals Unlimited; c, d, © D. W. Fawcett/Photo Researchers. Inc.)*

6.8 | Amphipathic Lipids Spontaneously Form Aggregate Structures

Monolayers and Micelles

Amphipathic lipids spontaneously form a variety of structures when added to aqueous solution. All these structures form in ways that minimize contact between the hydrophobic lipid chains and the aqueous milieu. For example, when small amounts of a fatty acid are added to an aqueous solution, a monolayer is formed at the air–water interface, with the polar head groups in contact with the water surface and the hydrophobic tails in contact with the air (Figure 6.21). Few lipid molecules are found as monomers in solution.

Further addition of fatty acid eventually results in the formation of micelles. **Micelles** formed from an amphipathic lipid in water position the hydrophobic tails in the center of the lipid aggregation with the polar head groups facing outward. Amphipathic molecules that form micelles are characterized

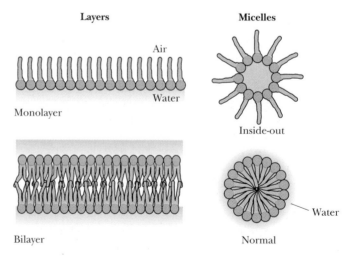

Figure 6.21 Several spontaneously formed lipid structures.

Structure	M_r	CMC	Micelle M_r
Triton X-100 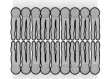	625	0.24 mM	90–95,000
Octyl glucoside	292	25 mM	
$C_{12}E_8$ (Dodecyl octaoxyethylene ether) $C_{12}H_{25}-(OCH_2CH_2)_8-OH$	538	0.071 mM	

Figure 6.22 The structures of some common detergents and their physical properties. Micelles formed by detergents can be quite large. Triton X-100, for example, typically forms micelles with a total molecular mass of 90 to 95 kD. This corresponds to approximately 150 molecules of Triton X-100 per micelle.

by a unique **critical micelle concentration,** or **CMC.** Below the CMC, individual lipid molecules predominate. Nearly all the lipid added above the CMC, however, spontaneously forms micelles. Micelles are the preferred form of aggregation in water for detergents and soaps. Some typical CMC values are listed in Figure 6.22.

Lipid Bilayers

Lipid bilayers consist of back-to-back arrangements of monolayers (see Figure 6.21). Phospholipids prefer to form bilayer structures in aqueous solution because their pairs of fatty acyl chains do not pack well in the interior of a micelle. Phospholipid bilayers form rapidly and spontaneously when phospholipids are added to water, and they are stable structures in aqueous solution. As opposed to micelles, which are small, self-limiting structures of a few hundred molecules, bilayers may form spontaneously over large areas (10^8 nm^2 or more). Since exposure of the edges of the bilayer to solvent is highly unfavorable, extensive bilayers normally wrap around themselves and form closed vesicles (Figure 6.23). The nature and integrity of these vesicle structures are very much dependent on the lipid composition. Phospholipids can form either *unilamellar vesicles* (with a single lipid bilayer) known as *liposomes,* or *multilamellar vesicles.* These latter structures are reminiscent of the layered structure of onions.

Liposomes are highly stable structures that can be subjected to manipulations such as gel filtration chromatography and dialysis. With such methods, it is possible to prepare liposomes with different inside and outside solution compositions. Liposomes can be used as **drug and enzyme delivery systems** in therapeutic applications. For example, liposomes can be used to introduce contrast agents into the animal or human body for diagnostic imaging procedures,

Bilayer

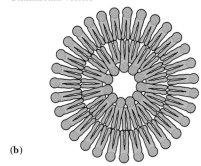

(a)

Unilamellar vesicle

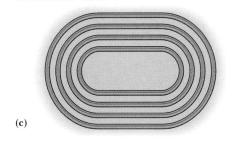

(b)

Multilamellar vesicle

(c)

(d)

Figure 6.23 Drawings of **(a)** a bilayer, **(b)** a unilamellar vesicle, **(c)** a multilamellar vesicle, and **(d)** an electron micrograph of a multilamellar Golgi structure ($\times$ 94,000). (*d, David Phillips/Visuals Unlimited*)

Figure 6.24 A computerized tomography (CT) image of the upper abdomen of a dog, following administration of liposome-encapsulated iodine, a contrast agent that improves the light/dark contrast of objects in the image. The spine is the bright white object at the bottom and the other bright objects on the periphery are ribs. The liver (white) occupies most of the abdominal space. The gallbladder (bulbous object at the center top) and blood vessels appear dark in the image. The liposomal iodine contrast agent has been taken up by Kuppfer cells, which are distributed throughout the liver, except in tumors. The dark object in the lower right is a large tumor. None of these anatomical features would be visible in a CT image in the absence of the liposomal iodine contrast agent. (*Courtesy of Walter Perkins, the Liposome Co., Inc., Princeton, NJ, and Brigham and Women's Hospital, Boston, MA*)

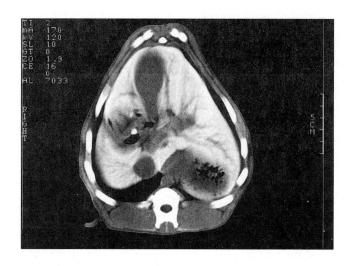

tomography from the Greek *tomos*, section; refers to the visualization of a section of tissue.

including *computerized tomography (CT)* and *magnetic resonance imaging (MRI)* (Figure 6.24). Liposomes can fuse with cells, mixing their contents with the intracellular medium. If methods can be developed to target liposomes to selected cell populations, it may be possible to deliver drugs, therapeutic enzymes, and contrast agents to particular kinds of cells (such as cancer cells).

That vesicles and liposomes form at all is a consequence of the amphipathic nature of the phospholipid molecule. Ionic interactions between the polar head groups and water are maximized, whereas hydrophobic interactions (see Chapter 2) facilitate the association of hydrocarbon chains in the interior of the bilayer. The formation of vesicles results in a favorable increase in the entropy of the solution, since the water molecules are not required to order themselves around the lipid chains.

Lipid bilayers have a polar surface and a nonpolar core. This hydrophobic core provides a substantial barrier to ions and other polar entities. The rates of movement of such species across membranes are thus quite slow. However, this same core also provides a favorable environment for nonpolar molecules and hydrophobic proteins. We will encounter numerous cases of hydrophobic molecules that interact with membranes and regulate biological functions in some way by binding to or embedding themselves in membranes.

6.9 Fluid Mosaic Model for Biological Membranes

In 1972, S. J. Singer and G. L. Nicolson proposed the **fluid mosaic model** for membrane structure, which suggested that membranes are dynamic structures composed of proteins and phospholipids. In this model, the phospholipid bilayer is a *fluid* matrix—in essence, a two-dimensional solvent for proteins. Both lipids and proteins are capable of rotational and lateral movement.

Singer and Nicolson also pointed out that proteins can be associated with the surface of this bilayer or embedded in the bilayer to varying degrees (Figure 6.25). They defined two classes of membrane proteins. The first, called **peripheral proteins** (or **extrinsic proteins**), includes those that do not penetrate the bilayer to any significant degree and are associated with the membrane by virtue of ionic interactions and hydrogen bonds between the membrane surface and the surface of the protein. Peripheral proteins can be dissociated from the membrane by treatment with salt solutions or by changes in pH (treatments that disrupt hydrogen bonds and ionic interactions). **Integral proteins** (or **intrinsic proteins**), in contrast, possess hydrophobic surfaces that can readily penetrate the lipid bilayer itself as well as surfaces that prefer contact with the aqueous medium. These proteins can either insert into the membrane or extend all the way across the membrane and expose themselves to the aqueous solvent on both sides. Because of these intimate associations with membrane lipid, integral proteins can only be removed from the membrane by agents capable of breaking up the hydrophobic interactions within the lipid bilayer itself (such as detergents and organic solvents). The fluid mosaic model has become the paradigm for modern studies of membrane structure and function.

Hydrocarbon Chain Orientation in the Bilayer

An important aspect of membrane structure is the orientation or ordering of lipid molecules in the bilayer. In the bilayers sketched in Figures 6.21 and 6.23, the long axes of the lipid molecules are portrayed as being perpendicular (or normal) to the plane of the bilayer. In fact, the hydrocarbon tails of phospholipids may tilt and bend and adopt a variety of orientations. Typically, the portions of a lipid chain near the membrane surface lie most nearly perpendicular to the membrane plane, and lipid chain ordering decreases toward the end of the chain (toward the middle of the bilayer).

Mobility of Membrane Components

The idea that lipids and proteins could move rapidly in biological membranes was a relatively new one when the fluid mosaic model was proposed. Many of the experiments designed to test this hypothesis involved the use of specially designed probe molecules. The first experiment demonstrating protein lateral movement in the membrane was described by L. Frye and M. Edidin in 1970. In this experiment, human cells and mouse cells were allowed to fuse together. Frye and Edidin used fluorescent antibodies to determine whether integral membrane proteins from the two cell types could move and intermingle in the newly formed, fused cells. The antibodies specific for human cells were labeled

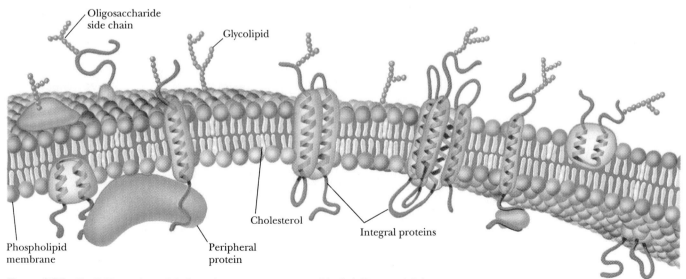

Figure 6.25 The fluid mosaic model of membrane structure proposed by S. J. Singer and G. L. Nicolson. In this model, the lipids and proteins are assumed to be mobile, so that they can move rapidly and laterally in the plane of the membrane. Transverse motion may also occur, but it is much slower.

Human cell   Mouse cell

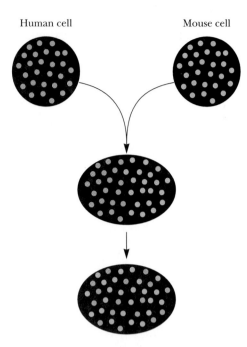

Figure 6.26 The Frye–Edidin experiment. Human cells with membrane antigens for red fluorescent antibodies were mixed and fused with mouse cells having membrane antigens for green fluorescent antibodies. Treatment of the resulting composite cells with red- and green-fluorescent-labeled antibodies revealed a rapid mixing of the membrane antigens in the composite membrane. This experiment demonstrated the lateral mobility of membrane proteins.

with rhodamine, a red fluorescent marker, and the antibodies specific for mouse cells were labeled with fluorescein, a green fluorescent marker. When both types of antibodies were added to newly fused cells, the binding pattern indicated that integral membrane proteins from the two cell types had moved laterally and were dispersed throughout the surface of the fused cell (Figure 6.26). This clearly demonstrated that integral membrane proteins possess significant lateral mobility.

Just how fast can proteins move in a biological membrane? Many membrane proteins can move laterally across a membrane at a rate of a few microns per minute. On the other hand, some integral membrane proteins are much more restricted in their lateral movement, with diffusion rates of about **10 nm/sec** or even slower. These latter proteins are often found to be anchored to the *cytoskeleton* (see Chapter 13), a complex latticelike structure that maintains the cell's shape and assists in the controlled movement of various substances through the cell.

Lipids also undergo rapid lateral motion in membranes. A typical phospholipid can diffuse laterally in a membrane at a linear rate of **several microns per second.** At that rate, a phospholipid could travel from one end of a bacterial cell to the other in less than a second or traverse a typical animal cell in a few minutes. On the other hand, transverse movement of lipids (or proteins) from one face of the bilayer to the other is much slower (and much less likely). For example, it can take as long as several days for half the phospholipids in a bilayer vesicle to "flip" from one side of the bilayer to the other.

Lipids Can Form Clusters in the Membrane

Lipids in model bilayer systems can aggregate to form clusters (Figure 6.27). Such behavior is referred to as a **phase separation,** which arises either spontaneously or as the result of some extraneous influence. Phase separations can be induced in model membranes by divalent cations, which interact with negatively charged moieties on the surface of the bilayer. For example, Ca^{2+} induces phase separations in membranes formed from phosphatidyl-serine (PS) and phosphatidylethanolamine (PE) or from PS, PE, and phosphatidylcholine. Ca^{2+} added to these membranes forms complexes with the negatively charged serine carboxyls, causing the PS to cluster and separate from the other lipids. Such metal-induced lipid phase separations have been shown to regulate the activity of membrane-bound enzymes.

There are other ways in which the lateral organization of lipids in biological membranes can be altered. For example, cholesterol can intercalate between the phospholipid fatty acid chains, its polar hydroxyl group associated with the polar head groups. In this manner, patches of cholesterol and phospholipids can form in an otherwise homogeneous sea of pure phospholipid. These patches or clusters can in turn affect the function of membrane proteins and enzymes.

Protein Aggregates in Membranes

Membrane proteins may also form clusters in the surface of a membrane. This can occur for several reasons. Some proteins must interact intimately with certain other proteins, forming multisubunit complexes that perform specific functions in the membrane. A few integral membrane proteins are known to self-associate in the membrane, forming large multimeric clusters. **Bacteriorhodopsin,** a light-driven proton pump protein, forms such clusters, known as "purple patches," in the membranes of *Halobacterium halobium.* The bacteriorhodopsin protein in these purple patches forms highly ordered, two-dimensional crystals.

The Two Sides of a Membrane Are Different

Membrane proteins and lipids are oriented specifically in the **transverse** direction (from one side of the membrane to the other). This can be appreciated when one considers that many properties of a membrane depend upon its two-sided nature. Properties that are a consequence of membrane "sidedness" include membrane transport, which is driven in one direction only; the effects of hormones at the outsides of cells; and the immunological reactions that occur between cells (necessarily involving only the outside surfaces of the cells). One would surmise that the proteins involved in these and other interactions must be oriented specifically in the membrane (see Section 6.10).

Lipid Orientation in Membranes

The lipids of biological membranes are asymmetrically distributed between the inner and outer monolayers. Figure 6.28 shows the inside–outside distribution of phospholipids observed in the human erythrocyte membrane. Such distributions are important to cells in several ways. The carbohydrate groups of glycolipids (and of glycoproteins) are always on the outside face of plasma membranes where they participate in cell recognition phenomena. Specific lipid distributions may also be important to various integral membrane proteins, which may prefer particular lipid classes in the inner and outer monolayers. The total charge on the inner and outer surfaces of a membrane depends on the distribution of lipids. The resulting charge differences affect the membrane potential, which in turn is known to modulate the activity of certain ion channels and other membrane proteins.

Flippases: Proteins Which Flip Lipids Across the Membrane

Proteins that can "flip" phospholipids from one side of a bilayer to the other have also been identified in several tissues (Figure 6.29). Called **flippases,** these proteins reduce the half-time for phospholipid movement across a membrane from 10 days or more to a few minutes or less. Some of these systems may operate passively, with no required input of energy, but passive transport alone cannot establish or maintain asymmetric transverse lipid distributions. However, rapid phospholipid movement from one monolayer to the other occurs in an *ATP-dependent* manner in erythrocytes. Energy-dependent lipid flippase activity may be responsible for the creation and maintenance of unequal lipid distributions.

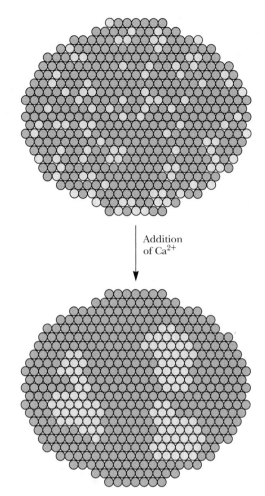

Addition
of Ca^{2+}

Figure 6.27 An illustration of the concept of lateral phase separations in a membrane. Phase separations of phosphatidylserine (yellow circles) can be induced by divalent cations such as Ca^{2+}.

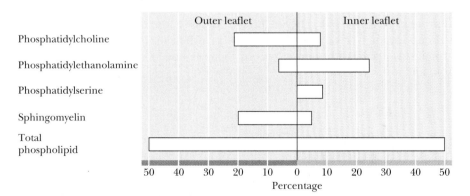

Figure 6.28 Phospholipids are arranged asymmetrically in most membranes, including the human erythrocyte membrane, as shown here. Values are mole percentages. *(After Rothman and Lenard, 1977. Science **194**:1744.)*

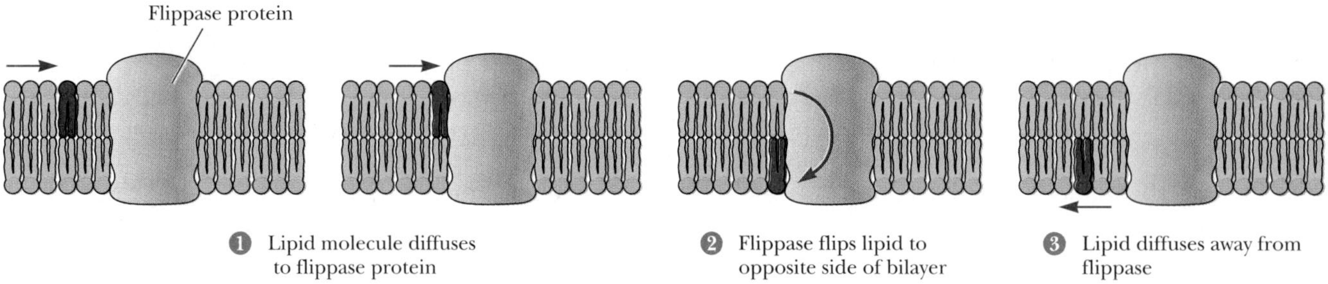

Flippase protein

① Lipid molecule diffuses to flippase protein

② Flippase flips lipid to opposite side of bilayer

③ Lipid diffuses away from flippase

Figure 6.29 Phospholipids can be "flipped" across a bilayer membrane by the action of flippase proteins. When, by normal diffusion through the bilayer, the lipid encounters a flippase, it can be moved quickly to the other face of the bilayer.

Membrane Phase Transitions

Lipids in bilayers undergo radical changes in physical state over characteristic narrow temperature ranges. These changes are in fact true **phase transitions,** and the temperatures at which these changes take place are referred to as **transition temperatures** or **melting temperatures** (T_m). These phase transitions involve substantial changes in the organization and motion of the fatty acyl chains within the bilayer. The bilayer below the phase transition exists in a closely packed gel state, with the fatty acyl chains relatively immobilized in a tightly packed array (Figure 6.30). This leaves the lipid chains in their fully extended conformation. As a result, the surface area per lipid is minimal and the bilayer thickness is maximal. Above the transition temperature, a liquid crystalline state exists in which the mobility of fatty acyl chains is intermediate between that of solid and liquid alkane. In this more fluid, liquid crystalline state, the surface area per lipid increases and the bilayer thickness decreases by 10% to 15%.

6.10 Structure of Membrane Proteins

The lipid bilayer forms the fundamental fabric of all biological membranes. Proteins, in contrast, carry out essentially all of the active functions of membranes, including transport activities, receptor functions, and other related processes. As suggested by Singer and Nicolson, most membrane proteins can be classified as peripheral or integral. The **peripheral proteins** are globular proteins that interact with the membrane mainly through electrostatic and hydrogen-bonding interactions with integral proteins. Although peripheral proteins are not discussed further here, many proteins of this class will be described in the context of other discussions throughout this textbook. **Integral proteins** are those that are firmly embedded in the lipid bilayer. Another class of proteins not anticipated by Singer and Nicolson, the **lipid-anchored proteins,** are important in a variety of functions in different cells and tissues. These proteins associate with membranes by means of a variety of covalently linked lipid anchors.

Figure 6.30 An illustration of the gel-to-liquid crystalline phase transition, which occurs when a membrane is warmed through the transition temperature, T_m. Notice that the surface area must increase and the thickness must decrease as the membrane goes through a phase transition. The mobility of the lipid chains increases dramatically.

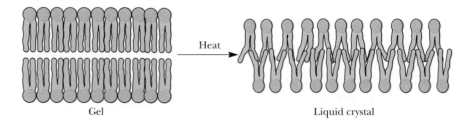

Gel

Heat

Liquid crystal

Integral Membrane Proteins

Despite the diversity of integral membrane proteins, most fall into two general classes. One of these includes proteins attached or anchored to the membrane by only a small hydrophobic segment, such that most of the protein extends out into the water solvent on one or both sides of the membrane. The other class includes those proteins that are more globular in shape and more totally embedded in the membrane, exposing only a small surface to the water solvent outside the membrane.

Glycophorin: A Protein with a Single Transmembrane Segment

In the case of the proteins that are anchored by a small hydrophobic polypeptide segment, that segment often takes the form of a single α-helix. One of the best examples of a membrane protein with such an α-helical structure is **glycophorin.** Most of glycophorin's mass is oriented on the outside surface of the cell, exposed to the aqueous milieu (Figure 6.31). A variety of hydrophilic oligosaccharide units are attached to this extracellular domain. These oligosaccharide groups constitute the ABO and MN blood group antigenic specificities of the red cell. This extracellular portion of the protein also serves as the receptor for the influenza virus. Glycophorin has a total molecular weight of about 31,000 and is approximately 40% protein and 60% carbohydrate. The glycophorin primary structure consists of a segment of 19 hydrophobic amino acid residues with a short hydrophilic sequence on one end and a longer hydrophilic sequence on the other end. The 19-residue sequence is just the right

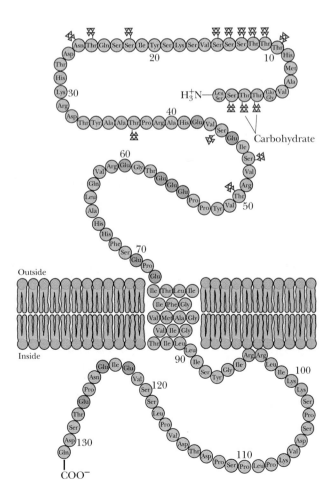

Figure 6.31 Glycophorin A spans the membrane of the human erythrocyte via a single α-helical transmembrane segment. The C-terminus of the peptide, whose sequence is shown here, faces the cytosol of the erythrocyte; the N-terminal domain is extracellular. Points of attachment of carbohydrate groups are indicated.

A DEEPER LOOK

Single TMS Proteins

In addition to glycophorin, numerous other membrane proteins are attached to the membrane by means of a single hydrophobic α-helix, with hydrophilic segments extending into the cytoplasm and the extracellular space. These proteins often function as receptors for extracellular molecules or as recognition sites that allow the immune system to recognize and distinguish the cells of the host organism from invading foreign cells or viruses. The proteins that represent the *major transplantation antigens H2* in mice and *human leukocyte associated (HLA) proteins* in humans are members of this class. Other such proteins include the *surface immunoglobulin receptors* on B lymphocytes and the *spike proteins* of many membrane viruses. The function of many of these proteins depends primarily on their extracellular domain, and thus the segment within the intracellular volume is often a shorter one.

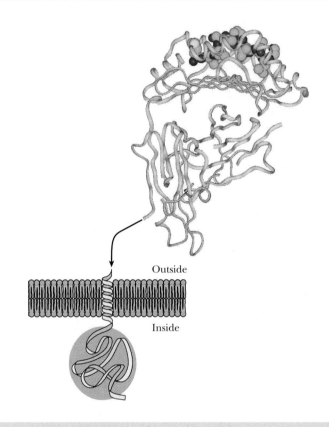

The major histocompatibility antigen HLA-A2 is a membrane-associated protein with a single transmembrane helical segment. The extracellular domain of this protein (in blue and yellow) is shown here complexed to a decapeptide (as a space-filling model) from calreticulin. ▶

---

length to span the cell membrane if it is coiled in the shape of an α-helix. The large hydrophilic sequence includes the amino-terminal residue of the polypeptide chain.

Bacteriorhodopsin: A 7-Transmembrane Segment (7-TMS) Protein

Membrane proteins that take on a more globular shape, instead of the single TMS structure just described, are often involved with transport activities and other functions requiring a substantial portion of the peptide to be embedded in the membrane. In these proteins, the primary sequence may consist of numerous hydrophobic α-helical segments joined by hinge regions so that the protein winds in a zig-zag pattern back and forth across the membrane. A well-characterized example of such a protein is **bacteriorhodopsin,** which clusters in purple patches in the membrane of the bacterium *Halobacterium halobium.* The name *Halobacterium* refers to the fact that this bacterium thrives in solutions having high concentrations of sodium chloride, such as the salt beds of San Francisco Bay. *Halobacterium* carries out a light-driven proton transport by means of bacteriorhodopsin, named in reference to its spectral similarities to rhodopsin in the rod outer segments of the mammalian retina. When this organism is deprived of oxygen for oxidative metabolism, it switches to the capture of energy from sunlight, using this energy to pump protons out of the cell. The proton gradient generated by such light-driven proton pumping represents potential energy, which is exploited elsewhere in the membrane to synthesize ATP.

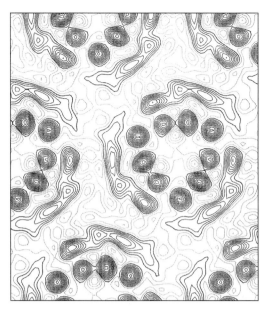

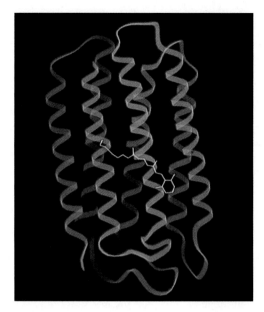

Figure 6.32 An electron density profile illustrating the three centers of threefold symmetry in arrays of bacteriorhodopsin in the purple membrane of *Halobacterium halobium*, together with a computer-generated model showing the seven α-helical transmembrane segments in bacteriorhodopsin. (*Electron density map from Stoeckenius, W., 1980. Purple membrane of halobacteria: A new light-energy converter.* Accounts of Chemical Research **13**:337–344. *Model on right from Henderson, R., 1990. Model for the structure of bacteriorhodopsin based on high-resolution electron cryo-microscopy.* Journal of Molecular Biology **213**:899–929.)

 See page 60

Bacteriorhodopsin clusters in hexagonal arrays (Figure 6.32) in the purple membrane patches of *Halobacterium,* and it was this orderly, repeating arrangement of proteins in the membrane that enabled Nigel Unwin and Richard Henderson in 1975 to determine the bacteriorhodopsin structure. The polypeptide chain crosses the membrane seven times, in seven α-helical segments, with very little of the protein exposed to the aqueous milieu. The bacteriorhodopsin structure has become a model of globular membrane protein structure. The amino acid sequences of many other integral membrane proteins contain numerous hydrophobic sequences which, like those of bacteriorhodopsin, could form α-helical transmembrane segments. For example, the primary structure of the sodium–potassium transport ATPase contains ten hydrophobic segments of length sufficient to span the plasma membrane. By analogy with bacteriorhodopsin, one would expect that these segments form a globular hydrophobic core that anchors the ATPase in the membrane. The helical segments may also account for the transport properties of the enzyme itself.

Porins—A β-Sheet Motif for Membrane Proteins

The β-sheet is another structural motif that provides extensive hydrogen bonding for transmembrane peptide segments. **Porin** proteins found in the outer membranes of bacteria such as *Escherichia coli,* and also in the outer mitochondrial membranes of eukaryotic cells, span their respective membranes with large β-sheets. A good example is **maltoporin,** also known as **LamB protein** or **lambda receptor,** which participates in the entry of maltose and maltodextrins into *E. coli.* Maltoporin is active as a trimer. The 421-residue monomer is an aesthetically pleasing 18-strand β-barrel (Figure 6.33). The β-strands are

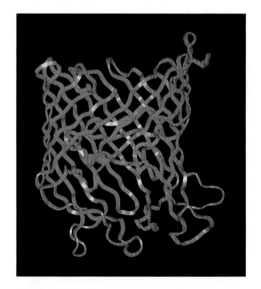

Figure 6.33 The three-dimensional structure of maltoporin from *E. coli.*

 See page 65

HUMAN BIOCHEMISTRY

Treating Allergies at the Cell Membrane

Allergies represent overreactions of the immune system caused by exposure to foreign substances referred to as allergens. The inhalation of allergens, such as pollen, pet dander, and dust, can cause a variety of allergic responses, including itchy eyes, a runny nose, shortness of breath, and wheezing. Allergies can also be caused by food, drugs, dyes, and other chemicals.

The visible symptoms of such an allergic response are caused by the release of **histamine** (see figure) by mast cells, a type of cell found in loose connective tissue. Histamine dilates blood vessels, increases the permeability of capillaries (allowing antibodies to pass from the capillaries to surrounding tissue), and constricts bronchial air passages. Histamine acts by binding to specialized membrane proteins called **histamine H1 receptors.** These integral membrane proteins are 7-TMS proteins with an extracellular amino terminus and a cytoplasmic carboxy terminus. When histamine binds to the extracellular domain of an H1 receptor, the intracellular domain undergoes a conformation change that stimulates a GTP-binding protein, which in turn activates the allergic response in the affected cell.

A variety of highly effective **antihistamine** drugs are available for the treatment of allergy symptoms. These drugs share the property of binding tightly to histamine H1 receptors, without eliciting the same effects as histamine itself. They are referred to as **histamine H1 receptor antagonists** because they prevent the binding of histamine to the receptors. The structures of Allegra (made by Hoechst Marion Roussel, Inc.), Zyrtec (by Pfizer), and Claritin (by Schering-Plough Corp.), are all shown here.

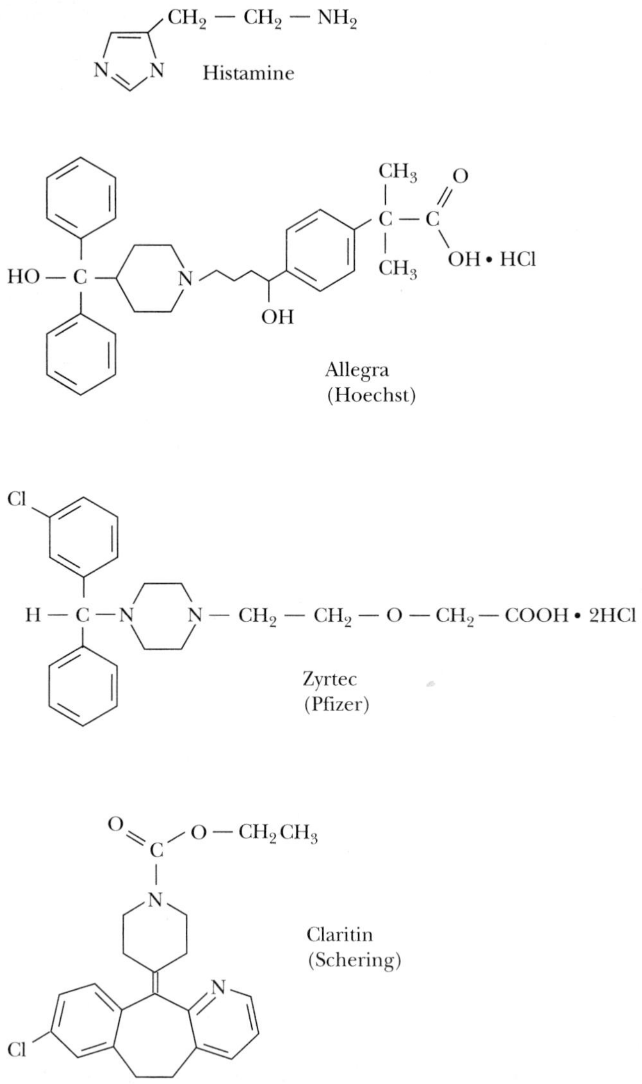

The structures of histamine and three antihistamine drugs. ▶

connected to their nearest neighbors either by long loops or by β-turns (Figure 6.34). The long loops are found at the end of the barrel that is exposed to the cell exterior, whereas the turns are located on the intracellular face of the barrel. Three of the loops fold into the center of the barrel.

The amino acid compositions and sequences of the β-strands in porin proteins are novel. Polar and nonpolar residues alternate along the β-strands, with polar residues facing the central pore or cavity of the barrel and nonpolar residues facing out from the barrel where they can interact with the hydrophobic lipid milieu of the membrane. The smallest diameter of the porin channel is about 5 Å. Thus, a maltodextrin polymer (composed of two or more glucose units) must pass through the porin in an extended conformation (like a spaghetti strand).

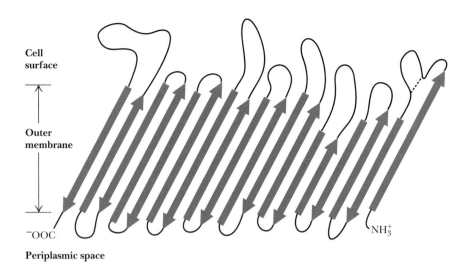

Cell
surface

Outer
membrane

⁻OOC

NH₃⁺

Periplasmic space

Figure 6.34 The arrangement of the peptide chain in maltoporin from *E. coli*.

See page 65

6.11 Lipid-Anchored Membrane Proteins

Certain proteins are covalently linked to lipid molecules. For many of these proteins, covalent attachment of lipid is required for association with a membrane. The lipid moieties can insert into the membrane bilayer, effectively **anchoring** their linked proteins to the membrane. In many cases, attachment to the membrane via the lipid anchor serves to modulate the activity of the protein.

Four different types of lipid-anchoring motifs have been found to date. These are **amide-linked myristoyl** anchors, **thioester-linked fatty acyl** anchors, **thioether-linked prenyl** anchors, and **amide-linked glycosyl phosphatidylinositol**

A DEEPER LOOK

Exterminator Proteins—Biological Pest Control at the Membrane

Control of biological pests, including mosquitoes, houseflies, gnats, and tree-consuming predators like the eastern tent caterpillar, is frequently achieved through the use of microbial membrane proteins. For example, several varieties of *Bacillus thuringiensis* produce proteins that bind to cell membranes in the digestive systems of insects that consume them, creating transmembrane ion channels. Leakage of Na⁺, K⁺, and H⁺ ions through these membranes in the insect gut destroys crucial ion gradients and interferes with digestion of food. Insects that ingest these toxins eventually die of starvation. *B. thuringiensis* toxins account for more than 90% of sales of biological pest control agents.

B. thuringiensis is a common soil bacterium that produces **inclusion bodies,** microcrystalline clusters of many different proteins. These crystalline proteins, called δ-**endotoxins,** are the ion channel toxins that are sold commercially for pest control. Most such endotoxins are **protoxins,** which are inactive until cleaved to smaller, active proteins by proteases in the gut of a susceptible insect. One such crystalline protoxin, lethal to mosquitoes, is a 27-kD protein, which is cleaved to form the active 25-kD toxin in the mosquito. This toxin has no effect on membranes at neutral pH, but at pH 9.5 (the pH of the mosquito gut) the toxin forms cation channels in the gut membranes.

This 25-kD protein is not toxic to tent caterpillars, but a larger, 130-kD protein in the *B. thuringiensis* inclusion bodies is cleaved by a caterpillar gut protease to produce a 55-kD toxin that is active in the caterpillar. Remarkably, the strain of *B. thuringiensis* known as *azawai* produces a protoxin with dual specificity: although cleaved as above in the caterpillar, when the same 130-kD protoxin is consumed by mosquitoes or houseflies, it is cleaved to form a 53-kD protein (15 amino acid residues shorter than the caterpillar toxin) that is toxic to these latter organisms.

Figure 6.35 Certain proteins are anchored to biological membranes by lipid anchors. Particularly common are the *N*-myristoyl- and *S*-palmitoyl-anchoring motifs shown here. Amide-linked *N*-myristoylation always occurs at an N-terminal glycine residue, whereas thioester linkages occur at cysteine residues within the polypeptide chain. G-protein–coupled receptors, with seven transmembrane segments, may contain one (and sometimes two) palmitoyl anchors in thioester linkage to cysteine residues in the C-terminal segment of the protein.

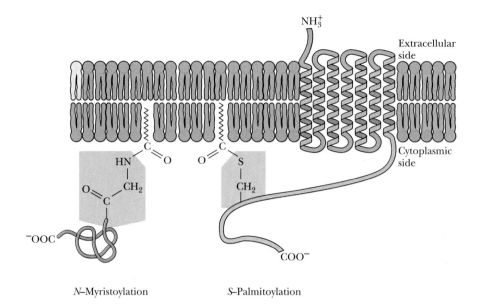

N–Myristoylation   *S*–Palmitoylation

anchors. Each of these anchoring motifs is used by a variety of membrane proteins, but each nonetheless exhibits a characteristic pattern of structural requirements.

Amide-Linked Myristoyl Anchors

Myristic acid may be linked via an amide bond to the α-amino group of the N-terminal glycine residue of selected proteins (Figure 6.35). The reaction is referred to as **N-myristoylation** and is catalyzed by *myristoyl–CoA:protein N-myristoyltransferase*, known simply as **NMT**. N-Myristoyl-anchored proteins include the catalytic subunit of *cAMP-dependent protein kinase*, the *pp60^{src} tyrosine kinase*, the phosphatase known as *calcineurin B*, the α-subunit of *G proteins* (involved in GTP-dependent transmembrane signaling events), and the *gag proteins* of certain retroviruses, including the HIV-1 virus that causes AIDS.

Thioester-Linked Fatty Acyl Anchors

A variety of cellular and viral proteins contain fatty acids covalently bound via ester linkages to the side chains of cysteine and sometimes to serine or threonine residues within a polypeptide chain (see Figure 6.35). This type of fatty acyl chain linkage has a broader fatty acid specificity than *N*-myristoylation. Myristate, palmitate, stearate, and oleate can all be esterified in this way, with the C_{16} and C_{18} chain lengths being most commonly found. Proteins anchored to membranes via fatty acyl thioesters include *G-protein–coupled receptors*, the *surface glycoproteins* of several viruses, and the *transferrin receptor* protein.

Thioether-Linked Prenyl Anchors

As noted in Section 6.6, polyprenyl (or simply prenyl) groups are long-chain polyisoprenoid groups derived from isoprene units. Prenylation of proteins destined for membrane anchoring can involve either **farnesyl** or **geranylgeranyl** groups (Figure 6.36). The addition of a prenyl group typically occurs at the cysteine residue of a carboxy-terminal CAAX sequence of the target protein, where C is cysteine, A is an aliphatic residue, and X can be any amino acid. As shown in Figure 6.37, the result is a thioether-linked farnesyl or geranylgeranyl

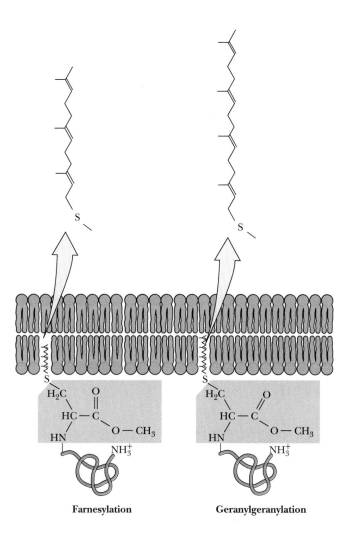

Figure 6.36 Proteins containing the C-terminal sequence CAAX can undergo prenylation reactions that place thioether-linked farnesyl or geranylgeranyl groups at the cysteine side chain. Prenylation is accompanied by removal of the AAX peptide and methylation of the carboxyl group of the cysteine residue, which has become the C-terminal residue.

Farnesylation   **Geranylgeranylation**

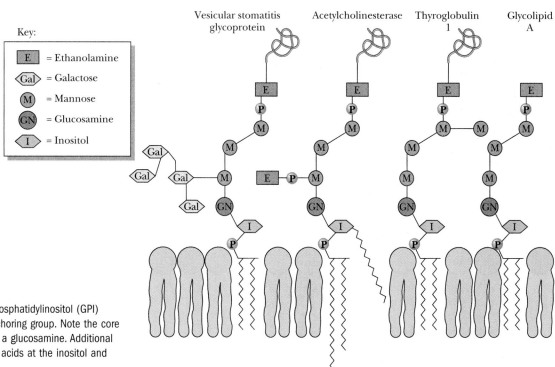

Figure 6.37 The glycosyl phosphatidylinositol (GPI) moiety is an elaborate lipid-anchoring group. Note the core of three mannose residues and a glucosamine. Additional modifications may include fatty acids at the inositol and glycerol —OH groups.

Key:
E = Ethanolamine
Gal = Galactose
M = Mannose
GN = Glucosamine
I = Inositol

Vesicular stomatitis glycoprotein   Acetylcholinesterase   Thyroglobulin 1   Glycolipid A

A Prenyl Protein Protease Is a New Chemotherapy Target

The protein called p21^{ras}, or simply Ras, is a small GTP-binding protein involved in cell signaling pathways that regulate growth and cell division. Mutant forms of Ras cause uncontrolled cell growth, and Ras mutations are involved in one-third of all human cancers. Because the signaling activity of Ras is dependent on prenylation, the prenylation reaction itself, the proteolysis of the -AAX motif, and the methylation of the prenylated Cys residue have all been considered targets for development of new chemotherapy strategies.

Farnesyl transferase catalyzes the prenylation reaction. Farnesyl transferase inhibitors, one of which is shown in the figure, are potent suppressors of tumor growth in mice, but their value in humans has not been established.

Mutations that inhibit prenyl transferases cause defective growth or death of cells, raising questions about the usefulness of prenyl transferase inhibitors in chemotherapy. However, Victor Boyartchuk and his colleagues at the University of California at Berkeley and Acacia Biosciences have shown that the protease that cleaves the -AAX motif from Ras following the prenylation reaction may be a better chemotherapeutic target. They have identified two genes for the prenyl protein protease in the yeast *Saccharomyces cerevisiae* and have shown that deletion of these genes results in loss of proteolytic processing of prenylated proteins, including Ras. Interestingly, normal yeast cells are unaffected by this gene deletion. However, in yeast cells that carry mutant forms of Ras and that display aberrant growth behaviors, deletion of the protease gene restores normal growth patterns. If these remarkable results translate from yeast to human tumor cells, inhibitors of CAAX proteases may be more valuable chemotherapeutic agents than prenyl transferase inhibitors.

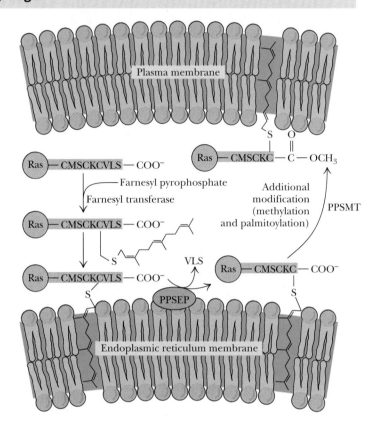

The farnesylation and subsequent processing of the Ras protein. Following farnesylation by the FTase, the carboxy-terminal VLS peptide is removed by methyltransferase, a prenylprotein-specific endoprotease (PPSEP) in the ER, and then a prenylprotein-specific methyltransferase (PPSMT) donates a methyl group from *S*-adenosylmethionine (SAM) to the carboxy-terminal *S*-farnesylated cysteine. Finally, palmitates are added to cysteine residues near the C-terminus of the protein.

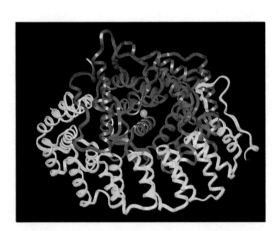

The structure of the farnesyl transferase heterodimer. A novel barrel structure is formed from 12 helical segments in the β-subunit (purple). The α-subunit (yellow) consists largely of seven successive pairs of α-helices that form a series of right-handed antiparallel coiled coils running along the bottom of the structure. These "helical hairpins" are arranged in a double-layered, right-handed superhelix resulting in a cresent-shaped subunit that envelopes part of the subunit.

See page 107

2(S)-{(S)-[2(R)-amino-3-mercapto]propylamino-3(S)-methyl} pentyloxy-3-phenylpropionyl-methioninesulfone methyl ester. This structure is of I-739,749, a farnesyl transferase inhibitor that is a potent tumor growth suppressor.

group. Once the prenylation reaction has occurred, a specific protease cleaves the three carboxy-terminal residues, and the carboxyl group of the now terminal Cys is methylated to produce an ester. All of these modifications appear to be important for subsequent activity of the prenyl-anchored protein. Proteins anchored to membranes via prenyl groups include *yeast mating factors,* the *p21^{ras} protein* (the protein product of the *ras* oncogene), and the *nuclear lamins,* structural components of the lamina underneath the nuclear membrane.

Glycosyl Phosphatidylinositol Anchors

Glycosyl phosphatidylinositol, or **GPI,** groups are structurally more elaborate membrane anchors than fatty acyl or prenyl groups. GPI groups modify the carboxy-terminal amino acid of a target protein via an ethanolamine residue linked to an oligosaccharide, which is linked in turn to the inositol moiety of a phosphatidylinositol (see Figure 6.37). The oligosaccharide typically consists of a conserved tetrasaccharide core of three mannose residues and a glucosamine, which can be altered by modifications of the mannose residues or addition of galactosyl side chains of various sizes, extra phosphoethanolamines, or additional *N*-acetylgalactose or mannosyl residues (see Figure 6.37). The inositol moiety can also be modified by an additional fatty acid, and a variety of fatty acyl groups are found linked to the glycerol group. GPI groups anchor a wide variety of *surface antigens, adhesion molecules,* and *cell surface hydrolases* to plasma membranes in various eukaryotic organisms. GPI anchors have not yet been observed in prokaryotic organisms or plants.

MEMBRANE TRANSPORT

Transport processes are vitally important to all life forms, since all cells exchange materials with their environment. Cells must have ways to bring nutrient molecules into the cell and ways to send waste products and toxic substances out. Also, inorganic electrolytes must be able to pass in and out of cells and across organelle membranes. All cells maintain **concentration gradients** of various metabolites across their plasma membranes and also across the membranes of intracellular organelles. By their very nature, these transmembrane gradients may represent a very large amount of potential energy which cells can exploit. Sodium and potassium ion gradients across the plasma membrane mediate the transmission of nerve impulses and the normal functions of the brain, heart, kidneys, and liver, among other organs. Storage and release of calcium from cellular compartments control muscle contraction, as well as the response of many cells to hormonal signals. High acid concentrations in the stomach are required for the digestion of food. Extremely high hydrogen ion gradients are maintained across the plasma membranes of the mucosal cells lining the stomach in order to maintain high acid levels in the stomach yet protect the cells that constitute the stomach walls from the deleterious effects of such acid.

We shall consider the molecules and mechanisms that mediate these transport activities. In nearly every case, the molecule or ion transported is water-soluble yet moves across the hydrophobic, impermeable lipid membrane at a rate high enough to serve the metabolic and physiologic needs of the cell. In each case the transported species either diffuses through a channel-forming protein or is carried by a carrier protein. Transport proteins are all classed as **integral membrane proteins.**

From a thermodynamic and kinetic perspective, there are only three types of membrane transport processes: *passive diffusion, facilitated diffusion,* and *active transport.* To be thoroughly appreciated, membrane transport phenomena must be considered in terms of thermodynamics. Some of the important kinetic considerations also will be discussed.

6.12 Passive Diffusion

Passive diffusion is the simplest transport process. In passive diffusion, the transported species moves across the membrane in the thermodynamically favored direction without the help of any specific transport system/molecule. For an uncharged molecule, passive diffusion is an entropic process, in which movement of molecules across the membrane proceeds until the concentration of the substance on both sides of the membrane is the same. For an uncharged molecule, the free energy difference between side 1 and side 2 of a membrane (Figure 6.38) is given by

$$\Delta G = G_2 - G_1 = RT \ln \left(\frac{[C_2]}{[C_1]} \right) \tag{6.1}$$

The difference in concentrations, $[C_2] - [C_1]$, is termed the *concentration gradient*, and ΔG here is the *chemical potential difference*.

Passive Diffusion of a Charged Species

For a charged species, the situation is slightly more complicated. In this case, the movement of a molecule across a membrane depends on its **electrochemical potential.** This is given by

$$\Delta G = G_2 - G_1 = RT \ln \left(\frac{[C_2]}{[C_1]} \right) + Z\mathscr{F}\Delta\psi \tag{6.2}$$

where Z is the **charge** on the transported species, $\mathscr{F}$ is **Faraday's constant** (the charge on 1 mole of electrons = 96,485 coulombs/mol = 96,485 joules/volt/mol, since 1 volt = 1 joule/coulomb), and $\Delta\psi$ is the electric potential difference (that is, voltage difference) across the membrane. The second term in the expression thus accounts for the movement of a charge across a potential difference. Note that the effect of this second term on ΔG depends on the magnitude and the sign of both Z and $\Delta\psi$. For example, if, as shown in Figure 6.39, side 2 has a higher potential than side 1 (so that $\Delta\psi$ is positive), for a negatively charged ion the term $Z\mathscr{F}\Delta\psi$ makes a negative contribution to ΔG.

In other words, the negative charge is spontaneously attracted to the more positive potential—and ΔG is negative. In any case, if the sum of the two terms on the right side of Equation (6.2) is a negative number, transport of the ion in question from side 1 to side 2 would occur spontaneously. The driving force for passive transport is the ΔG term for the transported species itself.

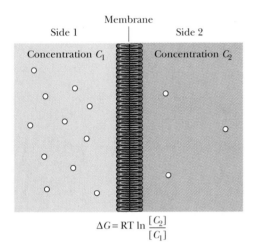

$$\Delta G = RT \ln \frac{[C_2]}{[C_1]}$$

Figure 6.38 Passive diffusion of an uncharged species across a membrane depends only on the concentrations (C_1 and C_2) on the two sides of the membrane.

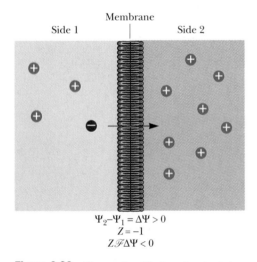

$$\Psi_2 - \Psi_1 = \Delta\Psi > 0$$
$$Z = -1$$
$$Z\mathscr{F}\Delta\Psi < 0$$

Figure 6.39 The passive diffusion of a charged species across a membrane depends upon the concentration and also on the charge of the particle, Z, and the electrical potential difference across the membrane, $\Delta\psi$.

6.13 Facilitated Diffusion

The transport of many substances across simple lipid bilayer membranes via passive diffusion is far too slow to sustain life processes. On the other hand, the transport rates for many ions and small molecules across actual biological membranes are much higher than anticipated from passive diffusion alone. This difference is due to specific proteins embedded in the cell membrane that **facilitate** transport of these species across the membrane. Such proteins capable of effecting **facilitated diffusion** of a variety of solutes are present in essentially all natural membranes. These proteins have two features in common: (a) they

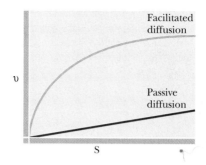

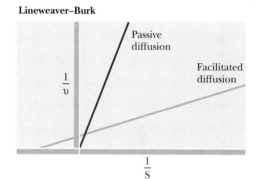
Lineweaver–Burk

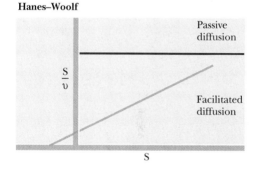
Hanes–Woolf

Figure 6.40 Passive diffusion and facilitated diffusion may be distinguished graphically. The plots for facilitated diffusion are similar to plots of enzyme-catalyzed processes (Chapter 10) and they display saturation behavior.

facilitate net movement of solutes only in the thermodynamically favored direction (that is, $\Delta G < 0$) and (b) they display a measurable affinity and specificity for the transported solute. Consequently, facilitated diffusion rates display **saturation behavior** similar to that observed with substrate binding by enzymes (see Chapter 10). Such behavior provides a simple means for distinguishing between passive diffusion and facilitated diffusion experimentally. The dependence of transport rate on solute concentration takes the form of a rectangular hyperbola (Figure 6.40), so that the transport rate approaches a limiting value, V_{max}, at very high solute concentration. Figure 6.40 also shows the graphical behavior exhibited by simple passive diffusion. Because passive diffusion does not involve formation of a specific solute:protein complex, the plot of rate versus concentration is linear, not hyperbolic.

Glucose Transport in Erythrocytes Occurs by Facilitated Diffusion

Many transport processes in a variety of cells occur by facilitated diffusion. Table 6.2 lists just a few of these. The **glucose transporter** of erythrocytes illustrates many of the important features of facilitated transport systems. Although glucose transport operates variously by passive diffusion, facilitated diffusion, or active transport mechanisms, depending on the particular cell, the **glucose transport system** of erythrocytes (red blood cells) operates exclusively by facilitated diffusion. Typical erythrocytes contain around 500,000 copies of this protein. The active form of this transport protein in the erythrocyte membrane is a trimer. Hydropathy analysis of the amino acid sequence of the erythrocyte glucose transporter provides a model for the structure of the protein (Figure 6.41). In this model, the protein spans the membrane 12 times, with both the N- and C-termini located on the cytoplasmic side. Transmembrane segments 7, 8, and 11 compose a hydrophilic transmembrane channel, with segments 9 and 10 forming a relatively hydrophobic pocket adjacent to the glucose-binding site. The reduced ability of insulin to stimulate glucose transport in diabetic patients can be due to reduced expression of certain glucose transport proteins.

Table 6.2 Facilitated Transport Systems

Permeant	Cell Type	K_m (mM)	V_{max} (mM/min)
D-Glucose	Erythrocyte	4–10	100–500
Chloride	Erythrocyte	25–30	
cAMP	Erythrocyte	0.0047	0.028
Phosphate	Erythrocyte	80	2.8
D-Glucose	Adipocytes	20	
D-Glucose	Yeast	5	
Sugars and amino acids	Tumor cells	0.5–4	2–6
D-Glucose	Rat liver	30	
D-Glucose	*Neurospora crassa*	8.3	46
Choline	Synaptosomes	0.083	
L-Valine	*Arthrobotrys conoides*	0.15–0.75	

Adapted from: Jain, M., and Wagner, R. 1980. *Introduction to Biological Membranes.* New York: John Wiley & Sons.

Figure 6.41 A model for the arrangement of the glucose transport protein in the erythrocyte membrane. Hydropathy analysis is consistent with 12 transmembrane helical segments.

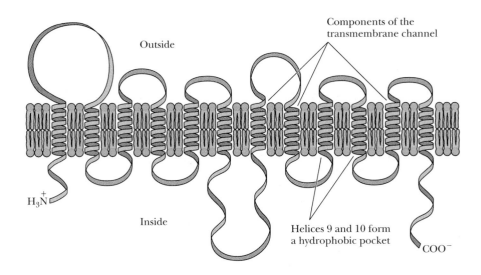

The Anion Transporter of Erythrocytes Also Operates by Facilitated Diffusion

The **anion transport system** is another facilitated diffusion system of the erythrocyte membrane. Chloride and bicarbonate (HCO_3^-) ions are exchanged across the red cell membrane by a 95-kD transmembrane protein. This protein is abundant in the red cell membrane and is represented by **band 3** on SDS electrophoresis gels (see Figure 6.42). The gene for the human erythrocyte anion transporter has been sequenced and hydropathy analysis has yielded a model for the arrangement of the protein in the red cell membrane (Figure 6.43). The model has 14 transmembrane segments, and the sequence includes 3 regions: a hydrophilic, cytoplasmic domain (residues 1 through 403) that interacts with numerous cytoplasmic and membrane proteins; a hydrophobic domain (residues 404 through 882) that comprises the anion transporting channel; and an acidic, C-terminal domain (residues 883 through 911). This transport system facilitates a one-for-one exchange of chloride and bicarbonate, so that the net transport process is electrically neutral. The net direction of anion flow through this protein depends on the sum of the chloride and bicarbonate concentration gradients. Typically, carbon dioxide is collected by red cells in respiring tissues (by means of $Cl^- \rightleftharpoons HCO_3^-$ exchange) and is then carried in the blood to the lungs, where bicarbonate diffuses out of the erythrocytes in exchange for Cl^- ions.

6.14 Active Transport

Passive and facilitated diffusion systems are relatively simple, in the sense that the transported species flow downhill energetically, that is, from high concentration to low concentration. However, other transport processes in biological systems must be *driven* in an energetic sense. In these cases, the transported species move from low concentration to high concentration, and thus the transport requires *energy input*. As such, it is considered an **active transport system.** The most common energy input is **ATP hydrolysis,** with hydrolysis being *tightly coupled* to the transport event. Other energy sources also drive active transport processes, including *light energy* and the *energy stored in ion gradients.*

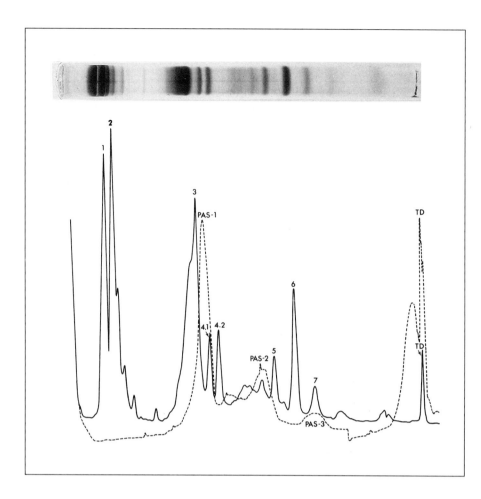

Figure 6.42 SDS-gel electrophoresis of erythrocyte membrane proteins (*top*) and a densitometer tracing of the same gel (*bottom*). Electrophoresis of proteins treated with the detergent sodium dodecyl sulfate (SDS) separates them according to size. The region of the gel between band 4.2 and band 5 is referred to as zone 4.5 or "band 4.5." The bands are numbered from the top of the gel (high molecular weights) to the bottom (low molecular weights). Band 3 is the anion-transporting protein and band 4.5 is the glucose transporter. The dashed line shows the staining of the gel by periodic acid–Schiff's reagent (PAS), which stains carbohydrates. Three "PAS bands" (PAS-1, PAS-2, PAS-3) indicate the positions in the gel of glycoproteins. (*Photo courtesy of Theodore Steck, University of Chicago*)

The original ion gradient is said to arise from a **primary active transport** process, and the transport that depends on the ion gradient for its energy input is referred to as a **secondary active transport process** (see discussion of amino acid and sugar transport on page 198). When transport results in a net movement of electric charge across the membrane, it is referred to as an **electrogenic transport** process. If no net movement of charge occurs during transport, the process is electrically neutral.

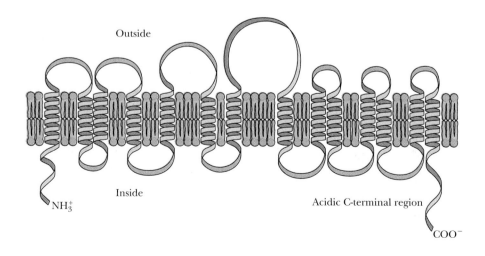

Figure 6.43 Hydropathy plot for the anion transport protein and a model for the arrangement of the protein in the membrane, based on hydropathy analysis.

All Active Transport Systems Are Energy-Coupling Devices

Hydrolysis of ATP is essentially a chemical process, whereas movement of species across a membrane is a mechanical process (that is, movement). An active transport process that depends on ATP hydrolysis thus couples chemical free energy to mechanical (translational) free energy. The bacteriorhodopsin protein in *Halobacterium halobium* couples light energy and mechanical energy. Oxidative phosphorylation (Chapter 17) involves coupling between electron transport, proton translocation, and the capture of chemical energy in the form of ATP synthesis. Similarly, the overall process of photosynthesis amounts to a coupling between captured light energy, proton translocation, and chemical energy stored in ATP.

Transport Processes Driven by ATP

Monovalent Cation Transport: Na⁺,K⁺-ATPase

All animal cells actively extrude Na^+ ions and accumulate K^+ ions. These two transport processes are driven by **Na⁺,K⁺-ATPase,** also known as the **sodium pump,** an integral protein of the plasma membrane. Most animal cells maintain cytosolic concentrations of Na^+ and K^+ of 10 mM and 100 mM, respectively. The extracellular milieu typically contains about 100 to 140 mM Na^+ and 5 to 10 mM K^+. Potassium is required within the cell to activate a variety of processes, whereas high intracellular sodium concentrations are inhibitory. The transmembrane gradients of Na^+ and K^+ and the attendant gradients of Cl^- and other ions provide the means by which neurons communicate. They also serve to regulate cellular volume and shape. Animal cells also depend upon these Na^+ and K^+ gradients to drive transport processes involving amino acids, sugars, nucleotides, and other substances. In fact, maintenance of these Na^+ and K^+ gradients consumes large amounts of energy in animal cells—20% to 40% of total metabolic energy in many cases and up to 70% in neural tissue.

The Na^+- and K^+-dependent ATPase comprises two subunits, an α-subunit of 1016 residues (120 kD) and a 35-kD β-subunit. The sodium pump actively pumps three Na^+ ions out of the cell and two K^+ ions into the cell per ATP hydrolyzed:

$$\text{ATP}^{4-} + \text{H}_2\text{O} + 3\text{Na}^+(\text{inside}) + 2\text{K}^+(\text{outside}) \longrightarrow$$
$$\text{ADP}^{3-} + \text{H}_2\text{PO}_4^- + 3\text{Na}^+(\text{outside}) + 2\text{K}^+(\text{inside}) \tag{6.3}$$

ATP hydrolysis occurs on the cytoplasmic side of the membrane (Figure 6.44), and the net movement of one positive charge outward per cycle makes the sodium pump electrogenic in nature.

Hydropathy analysis of the amino acid sequences of the α- and β-subunits and chemical modification studies have led to a model for the arrangement of the ATPase in the plasma membrane (Figure 6.45). The model describes 10 transmembrane α-helices in the α-subunit, with two large cytoplasmic domains. The larger of these, between transmembrane segments 4 and 5, has been implicated as the ATP-binding domain. The enzyme is covalently phosphorylated at an aspartate residue on the α-subunit in the course of ATP hydrolysis.

A minimal mechanism for Na⁺,K⁺-ATPase postulates that the enzyme cycles between two principal conformations, denoted E_1 and E_2 (Figure 6.46). E_1 has a high affinity for Na^+ and ATP and is rapidly phosphorylated in the presence of Mg^{2+} to form E_1-P, a state which *contains three occluded Na⁺ ions* (occluded in the sense that they are tightly bound and not easily dissociated from the enzyme in this conformation). A conformation change yields E_2-P, a form of the enzyme with relatively low affinity for Na^+ but a high affinity for K^+.

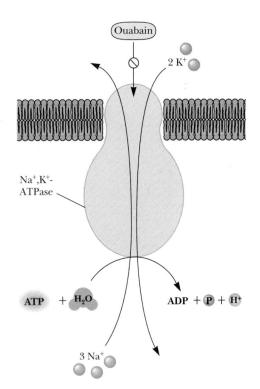

Figure 6.44 A schematic diagram of the Na⁺,K⁺-ATPase in mammalian plasma membrane. ATP hydrolysis occurs on the cytoplasmic side of the membrane, Na⁺ ions are transported out of the cell, and K⁺ ions are transported in. The transport stoichiometry is 3 Na⁺ out and 2 K⁺ in per ATP hydrolyzed. The specific inhibitor ouabain and other cardiac glycosides (see Figure 6.47) inhibit Na⁺,K⁺-ATPase by binding on the extracellular surface of the pump protein.

Figure 6.45 A model for the arrangement of Na$^+$,K$^+$-ATPase in the plasma membrane. The large cytoplasmic domain between transmembrane segments 4 and 5 contains the ATP-binding site and the aspartate residue that is phosphorylated during the catalytic cycle. The β-subunit contains one transmembrane segment and a large extracellular carboxy-terminal segment.

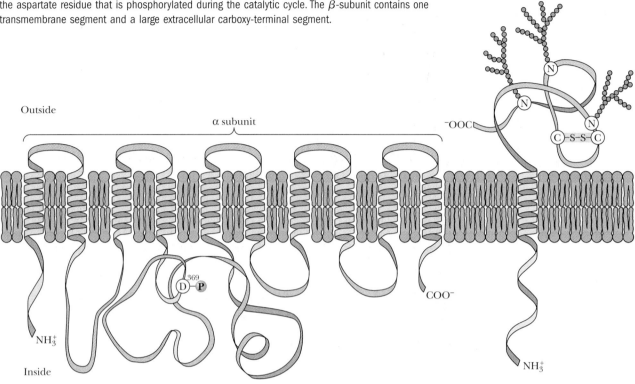

This state presumably releases 3 Na$^+$ ions and binds 2 K$^+$ ions on the outside of the cell. Dephosphorylation leaves E$_2$K$_2$, a form of the enzyme with *two occluded K$^+$ ions*. A conformation change, which appears to be accelerated by the binding of ATP (with a relatively low affinity), releases the bound K$^+$ inside the cell and returns the enzyme to the E$_1$ state. Enzyme forms with occluded cations represent states of the enzyme with cations bound in the transport channel. The alternation between high and low affinities for Na$^+$, K$^+$, and ATP serves to tightly couple the hydrolysis of ATP and ion binding and transport.

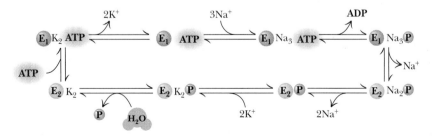

Figure 6.46 A mechanism for Na$^+$,K$^+$-ATPase. The model assumes two principal conformations, E$_1$ and E$_2$. Binding of Na$^+$ ions to E$_1$ is followed by phosphorylation and release of ADP. Na$^+$ ions are transported and released and K$^+$ ions are bound before dephosphorylation of the enzyme. Transport and release of K$^+$ ions complete the cycle.

Na$^+$,K$^+$-ATPase Is Inhibited by Cardiac Glycosides—Potent Drugs from Ancient Times

Plant and animal steroids such as *ouabain* (Figure 6.47) specifically inhibit Na$^+$,K$^+$-ATPase and ion transport. These substances are traditionally referred to as **cardiac glycosides** or **cardiotonic steroids,** both names derived from the potent effects of these molecules on the heart. These molecules all possess a *cis*-configuration of the C—D ring junction, an unsaturated lactone ring (5- or 6-membered) in the β-configuration at C-17, and a β-OH at C-14. There may be one or more sugar residues at C-3. The sugar(s) are not required for inhibition but do contribute to the water solubility of the molecule. Cardiac glycosides bind exclusively to the extracellular surface of Na$^+$,K$^+$-ATPase when it is in the E$_2$-P state, forming a very stable E$_2$-P(cardiac glycoside) complex.

Medical researchers studying high blood pressure have consistently found that people with hypertension have high blood levels of some sort of Na$^+$,K$^+$-ATPase inhibitor. In such patients, inhibition of the sodium pump in the cells lining the blood vessel wall results in accumulation of sodium and calcium in these cells and the narrowing of the vessels to create hypertension. An eight-year study aimed at the isolation and identification of the agent responsible for these effects by researchers at the University of Maryland Medical School and the Upjohn Laboratories in Michigan recently yielded a surprising result. Mass spectrometric analysis of compounds isolated from many hundreds of gallons of blood plasma has revealed that the hypertensive agent is ouabain itself or a closely related molecule!

The cardiac glycosides have a long and colorful history. Many species of plants producing these agents grow in tropical regions and have been used by natives in South America and Africa to prepare poisoned arrows used in fighting and hunting. Zulus in South Africa, for example, have used spears tipped with cardiac glycoside

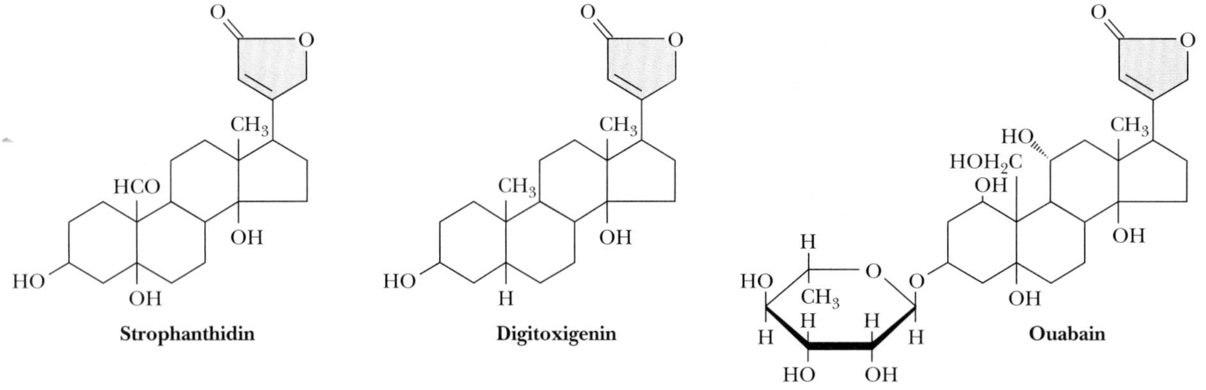

Figure 6.47 The structures of several cardiac glycosides.

Calcium Transport: Ca^{2+}-ATPase

Calcium, an ion acting as a cellular signal in virtually all cells (see Chapter 26), plays a special role in muscles. It is the signal that stimulates muscles to contract (Chapter 13). In the resting state, the levels of Ca^{2+} near the muscle fibers are very low (approximately 0.1 μM), and nearly all of the calcium ion in muscles is sequestered inside a complex network of vesicles called the **sarcoplasmic reticulum,** or **SR** (see Figure 13.11). Nerve impulses induce the sarcoplasmic reticulum membrane to quickly release large amounts of Ca^{2+}, with cytosolic levels rising to approximately 10 μM. At these levels, Ca^{2+} stimulates contraction. Relaxation of the muscle requires that cytosolic Ca^{2+} levels be reduced to their resting levels. This is accomplished by an ATP-driven Ca^{2+} transport protein known as the **Ca^{2+}-ATPase.** This enzyme is the most abundant protein in the SR membrane, accounting for 70% to 80% of the SR protein. Ca^{2+}-ATPase bears many similarities to the Na$^+$,K$^+$-ATPase. It has an α-subunit of the same approximate size, it forms a covalent E-P intermediate during ATP hydrolysis, and its mechanism of ATP hydrolysis and ion transport is similar in many ways to that of the sodium pump.

poisons. The sea onion, found commonly in southern Europe and northern Africa, was used by the Romans and the Egyptians as a cardiac stimulant, diuretic, and expectorant. The Chinese have long used a medicine made from the skins of certain toads for similar purposes. Cardiac glycosides are also found in several species of domestic plants, including the oleander (figure, part a), lily of the valley, foxglove, and milkweed plant. Monarch butterflies (figure, part b) acquire these compounds by feeding on milkweed and then storing the cardiac glycosides in their exoskeletons. Cardiac glycosides deter predation of monarch butterflies by birds, which learn by experience not to feed on monarchs. Viceroy butterflies (figure, part c) mimic monarchs in overall appearance. Although viceroys contain no cardiac glycosides and are edible, they are avoided by birds that mistake them for monarchs.

In modern times, **digitalis** (a preparation of dried leaves prepared from the foxglove, *Digitalis purpurea*) and other purified cardiotonic steroids have been used to increase the contractile force of heart muscle, to slow the rate of beating, and to restore normal function in hearts undergoing fibrillation (a condition in which heart valves do not open and close rhythmically, but rather remain partially open, fluttering in an irregular and ineffective way). Inhibition of the cardiac sodium pump increases the intracellular Na^+ concentration, leading to stimulation of the Na^+-Ca^+ exchanger, which extrudes sodium in exchange for inward movement of calcium. Increased intracellular Ca^{2+} stimulates muscle contraction. Careful use of digitalis drugs has substantial therapeutic benefit for heart patients.

(a) Oleander

(b) Monarch butterfly

(c) Viceroy

(a) Cardiac glycoside inhibitors of Na^+,K^+-ATPase are produced by many plants, including foxglove, lily of the valley, milkweed, and oleander (shown here). **(b)** The monarch butterfly, which concentrates cardiac glycosides in its exoskeleton, is shunned by predatory birds. **(c)** Predators also avoid the viceroy, even though it contains no cardiac glycosides, because it is similar in appearance to the monarch.

The structure of the Ca^{2+}-ATPase (Figure 6.48) includes a transmembrane domain consisting of ten α-helical segments and a large cytoplasmic domain that itself consists of a nucleotide-binding domain, a phosphorylation domain, and an actuator domain. In this structure, two Ca^{2+} ions are buried deep in the membrane-spanning portion of the enzyme.

The Gastric H⁺,K⁺-ATPase

Production of protons is a fundamental activity of cellular metabolism, and proton production plays a special role in the stomach. The highly acidic environment of the stomach is essential for the digestion of food in all animals. The pH of stomach fluid is normally 0.8 to 1. The pH of the parietal cells of the gastric mucosa in mammals is approximately 7.4. This represents a **pH gradient** across the mucosal cell membrane of 6.6, the largest known transmembrane gradient in eukaryotic cells. This enormous gradient must be maintained constantly so that food can be digested in the stomach without damage to the cells and organs adjacent to the stomach. The gradient of H^+ is maintained by a gastric **H⁺,K⁺-ATPase,** which uses the energy of hydrolysis of ATP

Figure 6.48 The structure of Ca^{2+}-ATPase. The transmembrane domain is shown in blue, the nucleotide domain is shown in red, the phosphorylation domain is yellow, and the actuator domain is green.

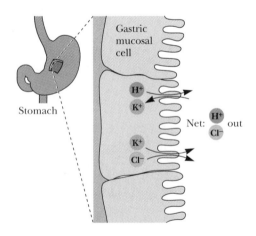

Figure 6.49 The H^+,K^+-ATPase of gastric mucosal cells mediates proton transport into the stomach. Potassium ions are recycled by means of an associated K^+/Cl^- cotransport system. The action of these two pumps results in net transport of H^+ and Cl^- into the stomach.

to pump H^+ out of the mucosal cells and into the stomach interior in exchange for K^+ ions. This transport is electrically neutral, and the K^+ that is transported into the mucosal cell is subsequently pumped back out of the cell together with Cl^- in a second electroneutral process (Figure 6.49). Thus, the net transport effected by these two systems is the movement of HCl into the interior of the stomach. (Only a small amount of K^+ is needed, since it is recycled.) The H^+,K^+-ATPase bears many similarities to the plasma membrane Na^+,K^+-ATPase and the SR Ca^{2+}-ATPase described earlier.

Bone Remodeling by Osteoclast Proton Pumps

Other proton-ATPases exist in eukaryotic and prokaryotic systems. **Vacuolar (V-type) ATPases** are found in vacuoles, lysosomes, endosomes, Golgi, chromaffin granules, and coated vesicles. Various H^+-transporting ATPases occur in yeast and bacteria as well. H^+-transporting ATPases found in **osteoclasts** (multinucleate cells that break down bone during normal bone remodeling) provide a source of circulating calcium for soft tissues such as nerves and muscles. About 5% of bone mass in the human body undergoes remodeling at any given time. Once growth is complete, the body balances formation of new bone tissue by cells called **osteoblasts** with resorption of existing bone matrix by osteoclasts. Osteoclasts possess proton pumps—which are in fact V-type ATPases—on the portion of the plasma membrane that attaches to the bone. This region of the osteoclast membrane is called the ruffled border. The osteoclast attaches to the bone in the manner of a cup turned upside down on a saucer (Figure 6.50), leaving an extracellular space between the bone surface and the cell. The H$^+$-ATPases in the ruffled border pump protons into this space, creating an acidic solution that dissolves the bone mineral matrix. Bone mineral is primarily an inorganic mixture of calcium carbonate and hydroxyapatite (calcium phosphate). In this case, transport of protons out of the osteoclasts lowers the pH of the extracellular space near the bone to about 4, solubilizing the hydroxyapatite.

ATPases That Transport Peptides and Drugs

Species other than protons and inorganic ions are also transported across certain membranes by specialized ATPases. Yeast (*Saccharomyces cerevisiae*) has one such system. Yeasts exist in two haploid mating types, designated **a** and **α**. Each mating type produces a mating factor (**a-factor** or **α-factor,** respectively) and responds to the mating factor of the opposite type. α-Factor is glycosylated in the ER and then secreted from the cell. On the other hand, the a-factor, a 12–amino-acid peptide, is exported from the cell by a 1290-residue protein, which consists of two identical halves joined together—a tandem duplication. Each half contains six putative transmembrane segments arranged in pairs, and a conserved hydrophilic cytoplasmic domain containing a consensus sequence for an ATP-binding site (Figure 6.51). This protein uses the energy of ATP hydrolysis to export a-factor from the cell.

Proteins very similar to the yeast a-factor transporter have been identified in a variety of prokaryotic and eukaryotic cells, and one of these appears to be responsible for the acquisition of **drug resistance** in many human malignancies. Clinical treatment of human cancer often involves chemotherapy, the treatment with one or more drugs that selectively inhibit the growth and proliferation of tumorous tissue. However, the efficacy of a given chemotherapeutic drug often decreases with time, owing to an acquired resistance. Even worse, the acquired resistance to a single drug usually results in a simultaneous resistance to a wide spectrum of drugs with little structural or even functional similarity to the original drug, a phenomenon referred to as **multidrug resistance,** or **MDR.** This perplexing problem has been traced to the induced expression of a 170-kD plasma membrane glycoprotein known as the **P-glycoprotein** or the **MDR ATPase.** Like the yeast a-factor transporter, MDR ATPase is a tandem repeat, each half consisting of a hydrophobic sequence with six transmembrane segments followed by a hydrophilic, cytoplasmic sequence containing a consensus ATP-binding site (see Figure 6.51). The protein uses the energy of ATP hydrolysis to actively transport a wide variety of drugs (Figure 6.52) out of the cell. Ironically, it is probably part of a sophisticated protection system for the cell and the organism. Organic molecules of various types and structures that might diffuse across the plasma membrane are apparently recognized by this protein and actively extruded from the cell. Despite the cancer-fighting nature of chemotherapeutic agents, the MDR ATPase recognizes these agents as cellular intruders and rapidly removes them. It is not yet understood how this large protein can recognize, bind, and transport such a broad group of diverse molecules, but it is known that the yeast a-factor ATPase and the MDR ATPase are just two members of a **superfamily** of transport proteins, many of whose functions are not yet understood.

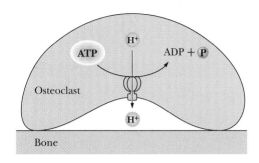

Figure 6.50 Proton pumps cluster on the ruffled border of osteoclast cells and function to pump protons into the space between the cell membrane and the bone surface. High proton concentration in this space dissolves the mineral matrix of the bone.

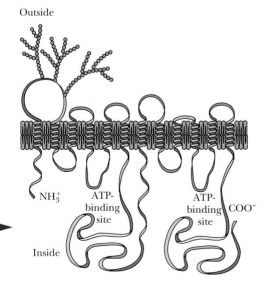

Figure 6.51 A model for the structure of the a-factor transport protein in the yeast plasma membrane. Gene duplication has yielded a protein with two identical halves, each half containing six transmembrane helical segments and an ATP-binding site. Like the yeast a-factor transporter, the multidrug transporter is postulated to have 12 transmembrane helices and 2 ATP-binding sites.

Figure 6.52 Some of the cytotoxic drugs that are transported by the MDR ATPase.

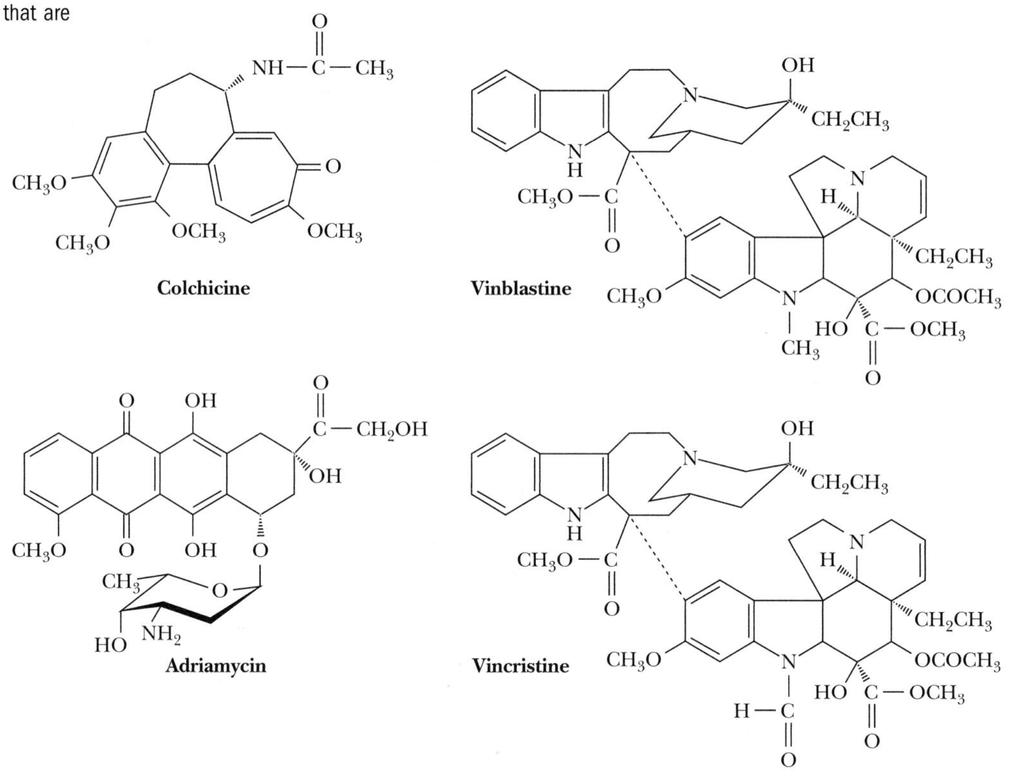

Colchicine

Vinblastine

Adriamycin

Vincristine

6.15 Transport Processes Driven by Ion Gradients

Amino Acid and Sugar Transport

The gradients of H^+, Na^+, and other cations and anions established by ATPases and other energy sources can be used for **secondary active transport** of various substrates. The best-understood systems use Na^+ or H^+ gradients to transport amino acids and sugars in certain cells. Many of these systems operate as **symports,** with the ion and the transported amino acid or sugar moving in the same direction (that is, into the cell). In **antiport** processes, the ion and the

---

Table 6.3 Some Mammalian Amino Acid Transport Systems

System Designation	Ion Dependence	Amino Acids Transported	Cellular Source
A	Na^+	Neutral amino acids	
ASC	Na^+	Neutral amino acids	
L	Na^+-independent	Branched-chain and aromatic amino acids	Ehrlich ascites cells Chinese hamster ovary cells Hepatocytes
N	Na^+	Nitrogen-containing side chains (Gln, Asn, His, etc.)	
y^+	Na^+-independent	Cationic amino acids	
x_{AG}^-	Na^+	Aspartate and glutamate	Hepatocytes
P	Na^+	Proline	Chinese hamster ovary cells

Adapted from: Collarini, E. J., and Oxender, D. L., 1987. Mechanisms of transport of amino acids across membranes. *Annual Review of Nutrition* **7**:75–90.

Cecropin A:

Lys-Trp-Lys-Leu-Phe-Lys-Lys-Ile-Glu-Lys-Val-Gly-Gln-Asn-Ile-Arg-Asp-Gly-Ile-Ile-Lys-Ala-Gly-Pro-Ala-Val-Ala-Val-Val-Gly-Gln-Ala-Thr-Gln-Ile-Ala-Lys-NH₂

Melittin:

Gly-Ile-Gly-Ala-Val-Leu-Lys-Val-Leu-Thr-Thr-Gly-Leu-Pro-Ala-Leu-Ile-Ser-Trp-Ile-Lys-Arg-Lys-Arg-Gln-Gln-NH₂

Magainin 2 amide:

Gly-Ile-Gly-Lys-Phe-Leu-His-Ser-Ala-Lys-Lys-Phe-Gly-Lys-Ala-Phe-Val-Gly-Glu-Ile-Met-Asn-Ser-NH₂

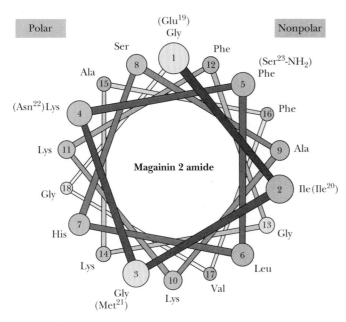

Figure 6.53 The structures of several amphipathic peptide antibiotics. α-Helices formed from these peptides cluster polar residues on one face of the helix, with nonpolar residues at other positions. Magainin peptides, produced by the African clawed frog *Xenopus laevis,* are antibacterial agents that create ion pores in the membranes of bacteria.

 See pages 20–24

other transported species move in opposite directions. (For example, the anion transporter of erythrocytes is an antiport.) **Proton symport** proteins are used by *E. coli* and other bacteria to accumulate lactose, arabinose, ribose, and a variety of amino acids. *E. coli* also possesses Na⁺-symport systems for melibiose as well as for glutamate and other amino acids.

Table 6.3 lists several systems that transport amino acids into mammalian cells. The accumulation of neutral amino acids in the liver by System A represents an important metabolic process. Thus, plasma membrane transport of alanine is the rate-limiting step in hepatic alanine metabolism. This system is normally expressed at low levels in the liver, but substrate deprivation and hormonal activation both stimulate System A expression.

Amphipathic Helices Form Transmembrane Ion Channels

Recently, a variety of natural peptides that form transmembrane channels have been identified and characterized (Figure 6.53). Melittin is a bee venom toxin peptide of 26 residues. The cecropins are peptides induced in *Hyalophora cecropia* (see Figure 6.54) and other related silkworms when challenged by bacterial infections. These peptides are thought to form α-helical aggregates in

Figure 6.54 Adult (*top*) and caterpillar (*bottom*) stages of the cecropia moth, *Hyalophora cecropia.* (*top, Greg Neise/Visuals Unlimited; bottom, Patti Murray/Animals, Animals*)

membranes, creating an ion channel in the center of the aggregate. The unifying feature of these helices is their **amphipathic** character, with polar residues clustered on one face of the helix and nonpolar residues elsewhere. In the membrane, the polar residues face the ion channel, leaving the nonpolar residues elsewhere on the helix to interact with the hydrophobic interior of the lipid bilayer.

6.16 Ionophore Antibiotics

All of the transport systems examined thus far are relatively large proteins. Several small-molecule toxins produced by microorganisms facilitate ion transport across membranes. Due to their relative simplicity, these molecules, the **ionophore antibiotics,** represent paradigms of the **mobile carrier** and the **pore** or **channel** models for membrane transport. Mobile carriers are molecules that form complexes with particular ions and diffuse freely across a lipid membrane (Figure 6.55). Pores or channels, on the other hand, adopt a fixed orientation

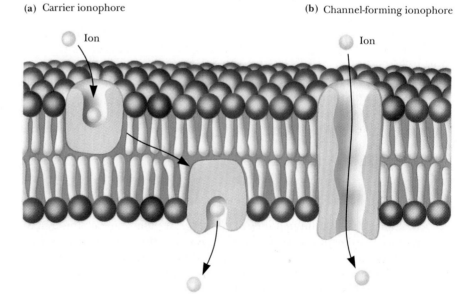

(a) Carrier ionophore

(b) Channel-forming ionophore

Ion

Ion

Figure 6.55 Schematic drawings of mobile carrier and channel ionophores. Carrier ionophores must move from one side of the membrane to the other, acquiring the transported species on one side and releasing it on the other side. Channel ionophores span the entire membrane.

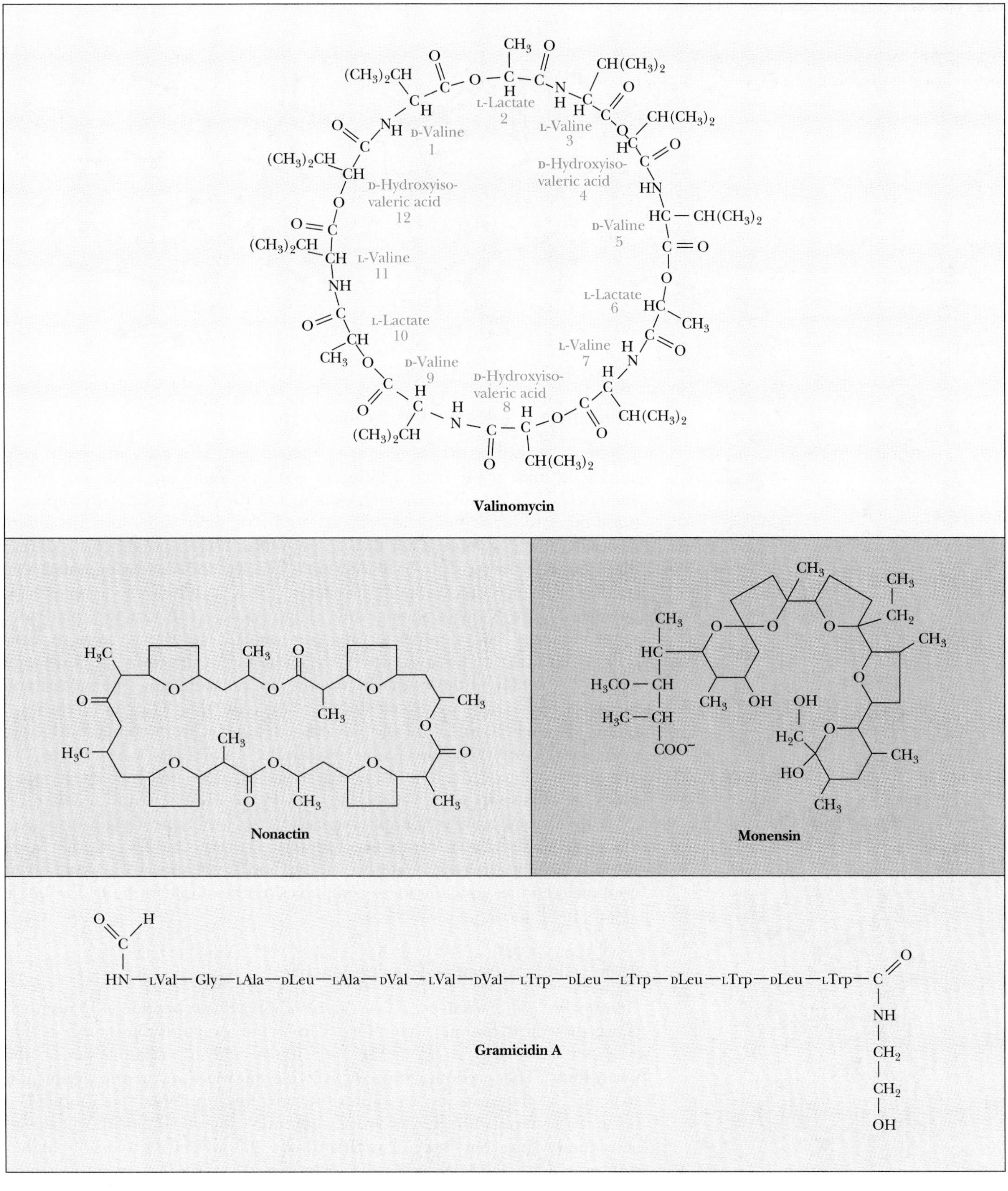

Valinomycin

Nonactin

Monensin

Gramicidin A

HN—ʟVal—Gly—ʟAla—ᴅLeu—ʟAla—ᴅVal—ʟVal—ᴅVal—ʟTrp—ᴅLeu—ʟTrp—ᴅLeu—ʟTrp—ᴅLeu—ʟTrp—C

in a membrane, creating a hole that permits the transmembrane movement of ions. These pores or channels may be formed from monomeric or (more often) multimeric structures in the membrane. Figure 6.56 displays the structures of several of these interesting molecules. As might be anticipated from the variety of structures represented here, these molecules associate with membranes and facilitate transport by different means.

Figure 6.56 Structures of several ionophore antibiotics. Valinomycin consists of three repeats of a four-unit sequence. Because it contains both peptide and ester bonds, it is referred to as a depsipeptide.

Figure 6.57 The structures of the valinomycin-K⁺ complex and of uncomplexed valinomycin.

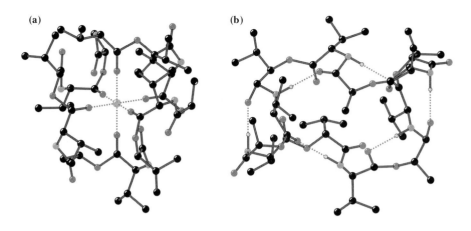

(a)   (b)

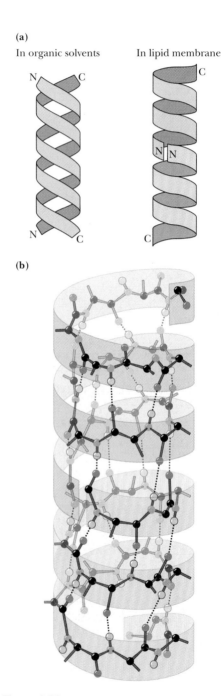

(a)

In organic solvents   In lipid membrane

(b)

Figure 6.58 **(a)** Gramicidin forms a double helix in organic solvents; a helical dimer is the preferred structure in lipid bilayers. The structure is a head-to-head, left-handed helix, with the carboxy termini of the two monomers at the ends of the structure. **(b)** The hydrogen-bonding pattern resembles that of a parallel β-sheet.

Valinomycin Is a Mobile Carrier Ionophore

Valinomycin (isolated from *Streptomyces fulvissimus*) is a cyclic structure containing 12 units made from 4 different residues. Valinomycin consists of the 4-unit sequence (D-valine, L-lactate, L-valine, D-hydroxyisovaleric acid), repeated three times to form the cyclic structure in Figure 6.56. Valinomycin is a **depsipeptide**, that is, a molecule with both peptide and ester bonds. (Considering the 12 units in the structure, valinomycin is called a **dodecadepsipeptide.**) The structures of uncomplexed valinomycin and the K⁺-valinomycin complex have been studied by X-ray crystallography (Figure 6.57). The structure places K⁺ at the center of the valinomycin ring, coordinated with the carbonyl oxygens of the 6 valines. The polar groups of the valinomycin structure are positioned toward the center of the ring, whereas the nonpolar groups (the methyl and isopropyl side chains) are directed outward from the ring. The central carbonyl groups completely surround the K⁺ ion, shielding it from contact with nonpolar solvents or the hydrophobic membrane interior. As a result, the K⁺–valinomycin complex freely diffuses across biological membranes and effects rapid, passive K⁺ transport (up to 10,000 K⁺/sec) in the presence of K⁺ gradients.

Other mobile carrier ionophores include *monensin* and *nonactin* (see Figure 6.56). The unifying feature in all these structures is an inward orientation of polar groups (to coordinate the central ion) and outward orientation of nonpolar residues (making these complexes freely soluble in the hydrophobic membrane interior).

Gramicidin Is a Channel-Forming Ionophore

Gramicidin from *Bacillus brevis* (see Figure 6.56), a linear peptide of 15 residues, is a prototypical channel ionophore. Gramicidin contains alternating L- and D-residues, a formyl group at the N-terminus, and an ethanolamine at the C-terminus. The predominance of hydrophobic residues in the gramicidin structure facilitates its incorporation into membranes. Once incorporated, it permits the rapid diffusion of many different cations. Gramicidin possesses considerably less ionic specificity than does valinomycin, but permits higher transport rates. A single gramicidin channel can transport as many as 10 million K⁺ ions per second. Protons and all alkali cations can diffuse through gramicidin channels, but divalent cations such as Ca²⁺ block the channel.

Gramicidin forms a helical dimer in membranes (Figure 6.58). (An α-helix cannot be formed by gramicidin, because it has both D- and L-amino acid residues.) The helical dimer is a head-to-head or amino terminus–to–amino terminus (N-to-N) dimer oriented perpendicular to the membrane surface, with the formyl groups at the bilayer center and the ethanolamine moieties at the membrane surface. The helix is unusual, with 6.3 residues per turn and a central hole approximately 0.4 nm in diameter. The hydrogen-bonding pattern in this structure, in which N—H groups alternately point up and down the axis of the helix to hydrogen-bond with carbonyl groups, is reminiscent of a β-sheet. For this reason this structure has often been referred to as a β-helix.

PROBLEMS

1. Draw the structures of (a) all the possible triacylglycerols that can be formed from glycerol with stearic and arachidonic acid and (b) all the phosphatidylserine(PS) isomers that can be formed from palmitic and linolenic acids. Which of the PS isomers are not likely to be found in biological membranes?

2. The purple patches of the *Halobacterium halobium* membrane, which contain the protein bacteriorhodopsin, are approximately 75% protein and 25% lipid. If the protein molecular weight is 26,000 and an average phospholipid has a molecular weight of 800, calculate the phospholipid to protein mole ratio.

3. Sucrose gradients for separation of membrane proteins must be able to separate proteins and protein–lipid complexes having a wide range of densities, typically 1.00 to 1.35 g/mL. Consult reference books (such as the *CRC Handbook of Biochemistry*) and plot the density of sucrose solutions versus percent sucrose by volume (g sucrose per 100 mL solution). What would be a suitable range of sucrose concentrations for separation of three membrane-derived protein–lipid complexes with densities of 1.03, 1.07, and 1.08 g/mL?

4. Phospholipid lateral motion in membranes is characterized by a diffusion coefficient of about 1×10^{-8} cm^2/sec. The distance traveled in two dimensions (across the membrane) in a given time is $r = (4Dt)^{1/2}$, where r is the distance traveled in centimeters, D is the diffusion coefficient, and t is the time during which diffusion occurs. Calculate the distance traveled by a phospholipid across a bilayer in 10 msec (milliseconds).

5. Protein lateral motion is much slower than that of lipids because proteins are larger than lipids. Also, some membrane proteins can diffuse freely through the membrane, whereas others are bound or anchored to other protein structures in the membrane. The diffusion constant for the membrane protein fibronectin is approximately 0.7×10^{-12} cm^2/sec, whereas that for rhodopsin is about 3×10^{-9} cm^2/sec.
   a. Calculate the distance traversed by each of these proteins in 10 msec.
   b. What could you surmise about the interactions of these proteins with other membrane components?

6. Discuss the effects on the lipid phase transition of pure dimyristoyl phosphatidylcholine vesicles of added (a) divalent cations, (b) cholesterol, (c) distearoyl phosphatidylserine, (d) dioleoyl phosphatidylcholine, and (e) integral membrane proteins.

7. Calculate the free energy difference at 25°C due to a galactose gradient across a membrane, if the concentration on side 1 is 2 mM and the concentration on side 2 is 10 mM.

8. Consider a phospholipid vesicle containing 10 mM Na$^+$ ions. The vesicle is bathed in a solution that contains 52 mM Na$^+$ ions, and the electrical potential difference across the vesicle membrane $\Delta\psi = \psi_{outside} - \psi_{inside} = -30$ mV. What is the electrochemical potential at 25°C for Na$^+$ ions?

9. Transport of histidine across a cell membrane was measured at several histidine concentrations:

[Histidine], μM	Transport, μmol/min
2.5	42.5
7	119
16	272
31	527
72	1220

Does this transport operate by passive diffusion or by facilitated diffusion?

10. Fructose is present outside a cell at 1 μM concentration. An active transport system in the plasma membrane transports fructose into this cell, using the free energy of ATP hydrolysis to drive fructose uptake. What is the highest intracellular concentration of fructose that this transport system can generate? Assume that one fructose is transported per ATP hydrolyzed, that ATP is hydrolyzed on the intracellular surface of the membrane, and that the concentrations of ATP, ADP, and P$_i$ are 3 mM, 1 mM, and 0.5 mM, respectively. T = 298 K. (*Hint*: Refer to Chapter 3 to read about the effects of concentration on free energy of ATP hydrolysis.)

11. The rate of K$^+$ transport across bilayer membranes reconstituted from dipalmitoylphosphatidylcholine (DPPC) and monensin is approximately the same as that observed across membranes reconstituted from DPPC and cecropin a at 35°C. Would you expect the transport rates across these two membranes to also be similar at 50°C? Explain.

12. In this chapter, we have examined coupled transport systems that rely on ATP hydrolysis, on primary gradients of Na$^+$ or H$^+$, and on phosphotransferase systems. Suppose you have just discovered an unusual strain of bacteria that transports rhamnose across its plasma membrane. Suggest experiments that would test whether it was linked to any of these other transport systems.

FURTHER READING

Blair, H. C., et al., 1989. Osteoclastic bone resorption by a polarized vacuolar proton pump. *Science* **245**:855–857.

Bretscher, M., 1985. The molecules of the cell membrane. *Scientific American* **253**:100–108.

Christensen, B., et al., 1988. Channel-forming properties of cecropins and related model compounds incorporated into planar lipid membranes. *Proceedings of the National Academy of Sciences* **85**:5072–5076.

Collarini, E. J., and Oxender, D., 1987. Mechanisms of transport of amino acids across membranes. *Annual Review of Nutrition* **7**:75–90.

Dawidowicz, E. A., 1987. Dynamics of membrane lipid metabolism and turnover. *Annual Review of Biochemistry* **56**:43–61.

Doering, T. L., Masterson, W. J., Hart, G. W., and Englund, P. T., 1990. Biosynthesis of glycosyl phosphatidylinositol membrane anchors. *Journal of Biological Chemistry* **265**:611–614.

Dowhan, W., 1997. Molecular basis for membrane phospholipid diversity: why are there so many lipids? *Annual Review of Biochemistry* **66**:199–232.

Fasman, G. D., and Gilbert, W. A., 1990. The prediction of transmembrane protein sequences and their conformation: An evaluation. *Trends in Biochemical Sciences* **15**:89–92.

Featherstone, C., 1990. An ATP-driven pump for secretion of yeast mating factor. *Trends in Biochemical Sciences* **15**:169–170.

Frye, C. D., and Edidin, M., 1970. The rapid intermixing of cell surface antigens after formation of mouse-human heterokaryons. *Journal of Cell Science* **7**:319–335.

Garavito, R. M., et al., 1983. X-ray diffraction analysis of matrix porin, an integral membrane protein from *Escherichia coli* outer membrane. *Journal of Molecular Biology* **164**:313–327.

Glomset, J. A., Gelb, M. H., and Farnsworth, C. C., 1990. Prenyl proteins in eukaryotic cells: a new type of membrane anchor. *Trends in Biochemical Sciences* **15**:139–142.

Glynn, I., and Karlish, S., 1990. Occluded ions in active transport. *Annual Review of Biochemistry* **59**:171–205.

Gordon, J. I., Duronio, R. J., Rudnick, D. A., Adams, S. P., and Gokel, G. W., 1991. Protein *N*-myristoylation. *Journal of Biological Chemistry* **266**:8647–8650.

Gould, G. W., and Bell, G. I., 1990. Facilitative glucose transporters: An expanding family. *Trends in Biochemical Sciences* **15**:18–23.

Inesi, G., Sumbilla, C., and Kirtley, M., 1990. Relationships of molecular structure and function in Ca^{2+}-transport ATPase. *Physiological Reviews* **70**:749–759.

Jap, B., and Walian, P. J., 1990. Biophysics of the structure and function of porins. *Quarterly Review of Biophysics* **23**:367–403.

Jay, D., and Cantley, L., 1986. Structural aspects of the red cell anion exchange protein. *Annual Review of Biochemistry* **55**:511–538.

Jennings, M. L., 1989. Topography of membrane proteins. *Annual Review of Biochemistry* **58**:999–1027.

Jennings, M. L., 1989. Structure and function of the red blood cell anion transport protein. *Annual Review of Biophysics and Biophysical Chemistry* **18**:397–430.

Jørgensen, P. L., and Andersen, J. P., 1988. Structural basis for E_1-E_2 conformational transitions in Na^+,K^+-pump and Ca-pump proteins. *Journal of Membrane Biology* **103**:95–120.

Juranka, P. F., Zastawny, R. L., and Ling, V., 1989. P-Glycoprotein: Multidrug-resistance and a superfamily of membrane-associated transport proteins. *The FASEB Journal* **3**:2583–2592.

Kartner, N., and Ling, V., 1989. Multidrug resistance in cancer. *Scientific American* **260**:44–51.

Matthew, M. K., and Balaram, A., 1983. A helix dipole model for alamethicin and related transmembrane channels. *FEBS Letters* **157**:1–5.

Meadow, N. D., Fox, D. K., and Roseman, S., 1990. The bacterial phosphoenolpyruvate:glycose phosphotransferase system. *Annual Review of Biochemistry* **59**:497–542.

Oesterhelt, D., and Tittor, J., 1989. Two pumps, one principle: Light-driven ion transport in *Halobacteria*. *Trends in Biochemical Sciences* **14**:57–61.

Pedersen, P. L., and Carafoli, E., 1987. Ion motive ATPases. *Trends in Biochemical Sciences* **12**:146–150, 186–189.

Prescott, S. M., et al., 2000. Platelet-activating factor and related lipid mediators. *Annual Review of Biochemistry* **69**:419–445.

Seelig, J., and Seelig, A., 1981. Lipid conformation in model membranes and biological membranes. *Quarterly Review of Biophysics* **13**:19–61.

Sefton, B., and Buss, J. E., 1987. The covalent modification of eukaryotic proteins with lipid. *Journal of Cell Biology* **104**:1449–1453.

Singer, S. J., and Nicolson, G. L., 1972. The fluid mosaic model of the structure of cell membranes. *Science* **175**:720–731.

Singer, S. J., and Yaffe, M. P., 1990. Embedded or not? Hydrophobic sequences and membranes. *Trends in Biochemical Sciences* **15**:369–373.

Spudich, J. L., and Bogomolni, R. A., 1988. Sensory rhodopsins of *Halobacteria*. *Annual Review of Biophysics and Biophysical Chemistry* **17**:193–215.

Tanford, C., 1980. *The Hydrophobic Effect: Formation of Micelles and Biological Membranes.* 2nd ed. New York: Wiley-Interscience.

Towler, D. A., Gordon, J. I., Adams, S. P., and Glaser, L., 1988. The biology and enzymology of eukaryotic protein acylation. *Annual Review of Biochemistry* **57**:69–99.

Unwin, N., and Henderson, R., 1984. The structure of proteins in biological membranes. *Scientific American* **250**:78–94.

Wade, D., et al., 1990. All D-amino acid–containing channel-forming antibiotic peptides. *Proceedings of the National Academy of Sciences* **87**:4761–4765.

Wallace, B. A., 1990. Gramicidin channels and pores. *Annual Review of Biophysics and Biophysical Chemistry* **19**:127–157.

Wirtz, K. W. A., 1991. Phospholipid transfer proteins. *Annual Review of Biochemistry* **60**:73–99.

Carbohydrates and Cell Surfaces

"The Discovery of Honey"—Piero di Cosimo (1462) (Courtesy of the Worcester Art Museum)

Sugar in the gourd and honey in the horn,
I never was so happy since the hour I was born.

"Turkey in the Straw" (American folk tune)

Outline

Carbohydrates are the single most abundant class of organic molecules found in nature. The name *carbohydrate* arises from the basic molecular formula $(CH_2O)_n$, which can be rewritten $(C \cdot H_2O)_n$ to show that these substances are hydrates of carbon, where $n = 3$ or more. Carbohydrates constitute a versatile class of molecules. Energy from the sun captured by green plants, algae, and some bacteria during photosynthesis is stored in the form of carbohydrates. In turn, carbohydrates are the metabolic precursors of virtually all other biomolecules. Breakdown of carbohydrates provides the energy that sustains animal life. In addition, carbohydrates are covalently linked with a variety of other molecules. Carbohydrates linked to lipid molecules, or **glycolipids,** are common components of biological membranes. Proteins that have covalently linked carbohydrates are called **glycoproteins.** These two classes of biomolecules, together called **glycoconjugates,** are important components of cell walls and extracellular structures in plants, animals, and bacteria. In addition to the structural roles such molecules play, they also serve in a variety of processes involving *recognition* between cell types or recognition of cellular structures by other molecules. Recognition events are important in normal cell growth, fertilization, transformation of cells, and other processes.

All of these functions are made possible by the characteristic chemical features of carbohydrates: (1) the existence of at least one and often two or more asymmetric centers, (2) the ability to exist either in linear or ring structures, (3) the capacity to form polymeric structures via *glycosidic* bonds, and (4) the potential to form multiple hydrogen bonds with water or other molecules in their environment.

7.1 Carbohydrate Nomenclature

Carbohydrates are generally classified into three groups: **monosaccharides** (and their derivatives), **oligosaccharides,** and **polysaccharides.** The monosaccharides are also called **simple sugars** and have the formula $(CH_2O)_n$. Monosaccharides cannot be broken down into smaller sugars under mild conditions. Oligosaccharides derive their name from the Greek word *oligo,* meaning "few," and consist of from two to ten simple sugar molecules: disaccharides are common in nature, as are trisaccharides; four- to six-sugar-unit oligosaccharides are usually bound covalently to other molecules, including glycoproteins. Polysaccharides, as their name suggests, are polymers of the simple sugars and their derivatives. They may be either linear or branched polymers and may contain hundreds or even thousands of monosaccharide units. Their molecular weights range up to 1 million or more.

7.2 Monosaccharides

Classification

Monosaccharides consist typically of three to seven carbon atoms and are described as either **aldoses** or **ketoses,** depending on whether the molecule contains an aldehyde function or a ketone group. The simplest aldose is glyceraldehyde, and the simplest ketose is dihydroxyacetone (Figure 7.1). These two simple sugars are termed **trioses** because they each contain three carbon atoms. The structures and names of a family of aldoses and ketoses with three, four, five, and six carbons are shown in Figures 7.2 and 7.3. *Hexoses* are the most abundant sugars in nature. Nevertheless, sugars from all these classes are important in metabolism.

Monosaccharides, either aldoses or ketoses, are often given more detailed generic names to describe both the important functional groups and the total number of carbon atoms. Thus, one can refer to *aldotetroses* and *ketotetroses, aldopentoses* and *ketopentoses, aldohexoses* and *ketohexoses,* and so on. Sometimes the ketone-containing monosaccharides are named simply by inserting the letters *-ul-* into the simple generic terms, such as *tetruloses, pentuloses, hexuloses, heptuloses,* and so on. The simplest monosaccharides are water-soluble, and most taste sweet.

Stereochemistry

Aldoses with at least three carbons and ketoses with at least four carbons contain **chiral centers** (Chapter 4). The nomenclature for such molecules must

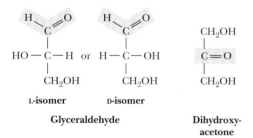

Figure 7.1 Structure of a simple aldose (glyceraldehyde) and a simple ketose (dihydroxyacetone).

specify the **configuration** about each asymmetric center, and drawings of these molecules must be based on a system that clearly specifies these configurations. As noted in Chapter 4, the **Fischer projection** system is used almost universally for this purpose today. The structures shown in Figures 7.2 and 7.3 are Fischer projections. For monosaccharides with two or more asymmetric carbons, the prefixes D and L refer to the configuration of the highest numbered asymmetric carbon (the asymmetric carbon farthest from the carbonyl carbon). A monosaccharide is designated D if the hydroxyl group on the highest numbered asymmetric carbon is drawn to the right in a Fischer projection, as in

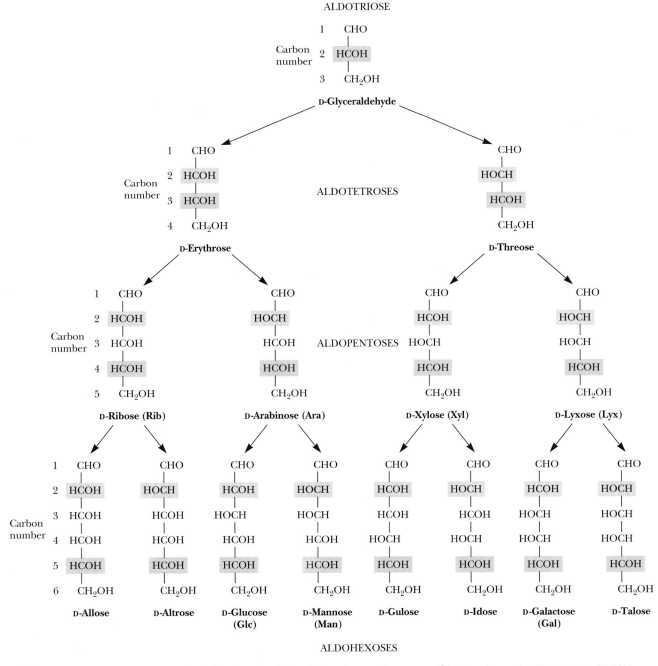

Figure 7.2 The structure and stereochemical relationships of D-aldoses having three to six carbons. The configuration in each case is determined by the highest numbered asymmetric carbon (shown in gray). In each row, the "new" asymmetric carbon is shown in yellow.

 See *Interactive Biochemistry CD-ROM and Workbook,* pages 114–118

Figure 7.3 The structure and stereochemical relationships of D-ketoses having three to six carbons. The configuration in each case is determined by the highest numbered asymmetric carbon (shown in gray). In each row, the "new" asymmetric carbon is shown in yellow.

See pages 114–118

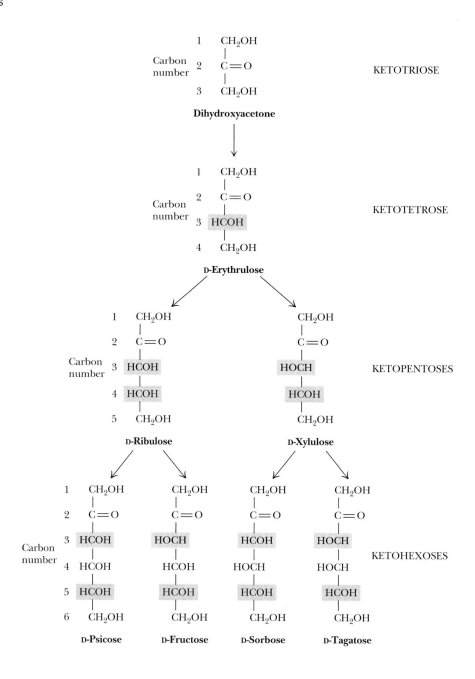

D-glyceraldehyde (see Figure 7.1). Note that the designation D or L merely relates the configuration of a given molecule to that of glyceraldehyde and does not specify the sign of rotation of plane-polarized light. If the sign of optical rotation is to be specified in the name, the Fischer convention of D or L designations may be used along with a + (plus) or − (minus) sign. Thus, D-glucose (see Figure 7.2) may also be called D(+)-glucose, since it is dextrorotatory, whereas D-fructose (see Figure 7.3), which is levorotatory, can also be named D(−)-fructose.

All of the structures shown in Figures 7.2 and 7.3 are D-configurations, and the D-forms of monosaccharides predominate in nature, just as L-amino acids do. These preferences, established in apparently random choices early in evolution, persist uniformly in nature because of the stereospecificity of the enzymes that synthesize and metabolize these small molecules. L-Monosaccharides do exist in nature, serving a few relatively specialized roles. L-Galactose is a constituent of certain polysaccharides, and L-arabinose is a constituent of bacterial cell walls.

According to convention, the D- and L-forms of a monosaccharide are *mirror images* of each other, as shown in Figure 7.4 for fructose. Stereoisomers that are mirror images of each other are called **enantiomers,** or sometimes *enantiomeric pairs.* For molecules that possess two or more chiral centers, more than two stereoisomers can exist. Pairs of isomers that have opposite configurations at one or more of the chiral centers but that are not mirror images of each other are called **diastereomers** or *diastereomeric pairs.* Any two structures in a given row in Figures 7.2 and 7.3 are diastereomeric pairs. Two sugars that differ in configuration at *only one* chiral center are described as **epimers.** For example, D-mannose and D-talose are epimers and D-glucose and D-mannose are epimers, whereas D-glucose and D-talose are *not* epimers but merely diastereomers.

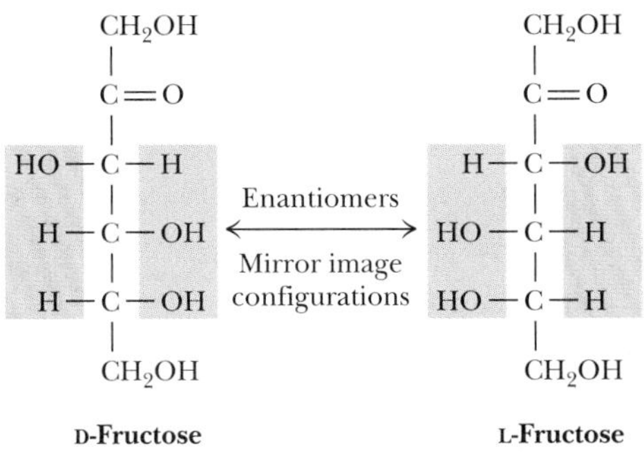

Figure 7.4 D-Fructose and L-fructose, an enantiomeric pair. Note that changing the configuration only at C-5 would change D-fructose to L-sorbose.

Cyclic Structures and Anomeric Forms

Although Fischer projections are useful for presenting the structures of particular monosaccharides and their stereoisomers, they ignore one of the most interesting facets of sugar structure—*the ability to form cyclic structures, with creation of an additional asymmetric center.* Alcohols react readily with aldehydes to form **hemiacetals** (Figure 7.5). The British carbohydrate chemist Sir Norman Haworth showed that the linear form of glucose (and other aldohexoses) could undergo a similar intramolecular reaction to form a *cyclic hemiacetal.* The resulting six-membered oxygen-containing ring is similar to *pyran* and is designated a **pyranose.** The reaction is catalyzed by acid (H^+) or base (OH^-) and is readily reversible.

In a similar manner, ketones can react with alcohols to form **hemiketals.** The analogous intramolecular reaction of a ketose sugar such as fructose yields a *cyclic hemiketal* (Figure 7.6). The five-membered ring thus formed is reminiscent of *furan* and is referred to as a **furanose.** The cyclic pyranose and furanose

β-D-Glucopyranose

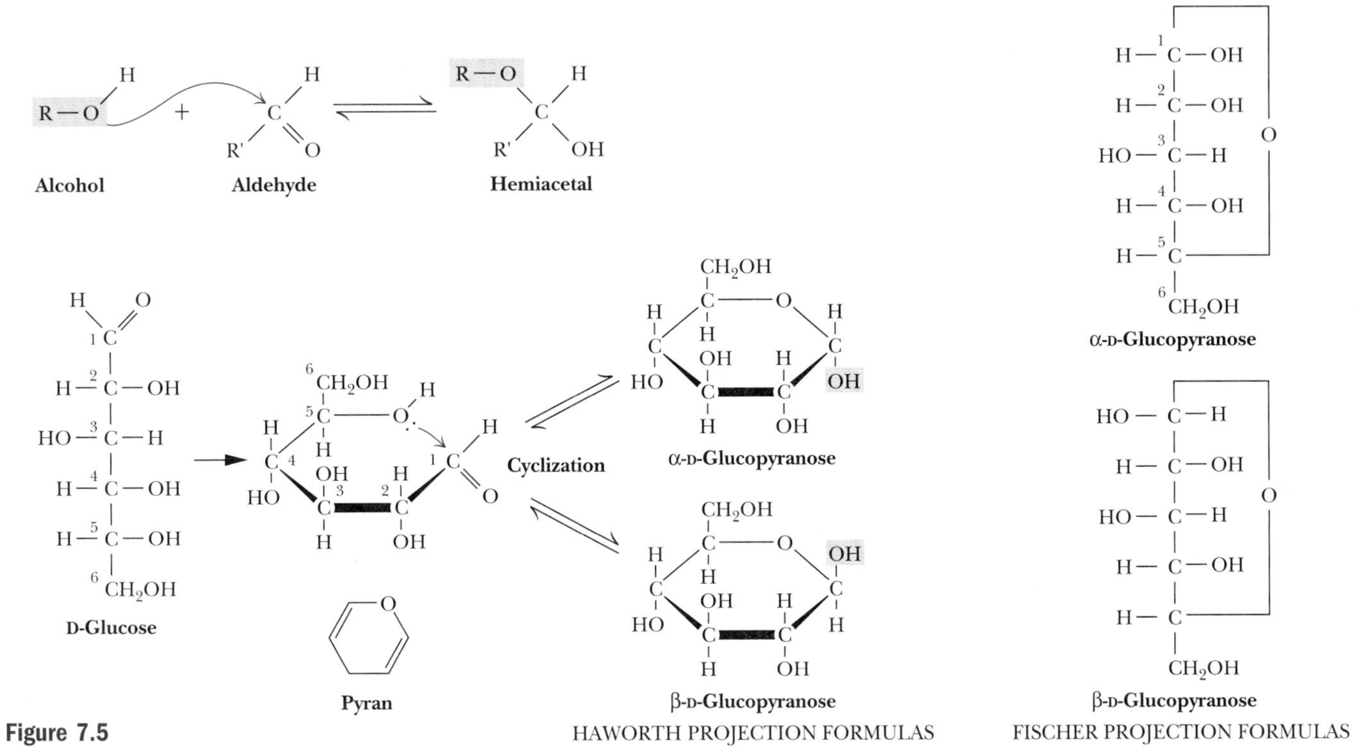

Figure 7.5

HAWORTH PROJECTION FORMULAS

FISCHER PROJECTION FORMULAS

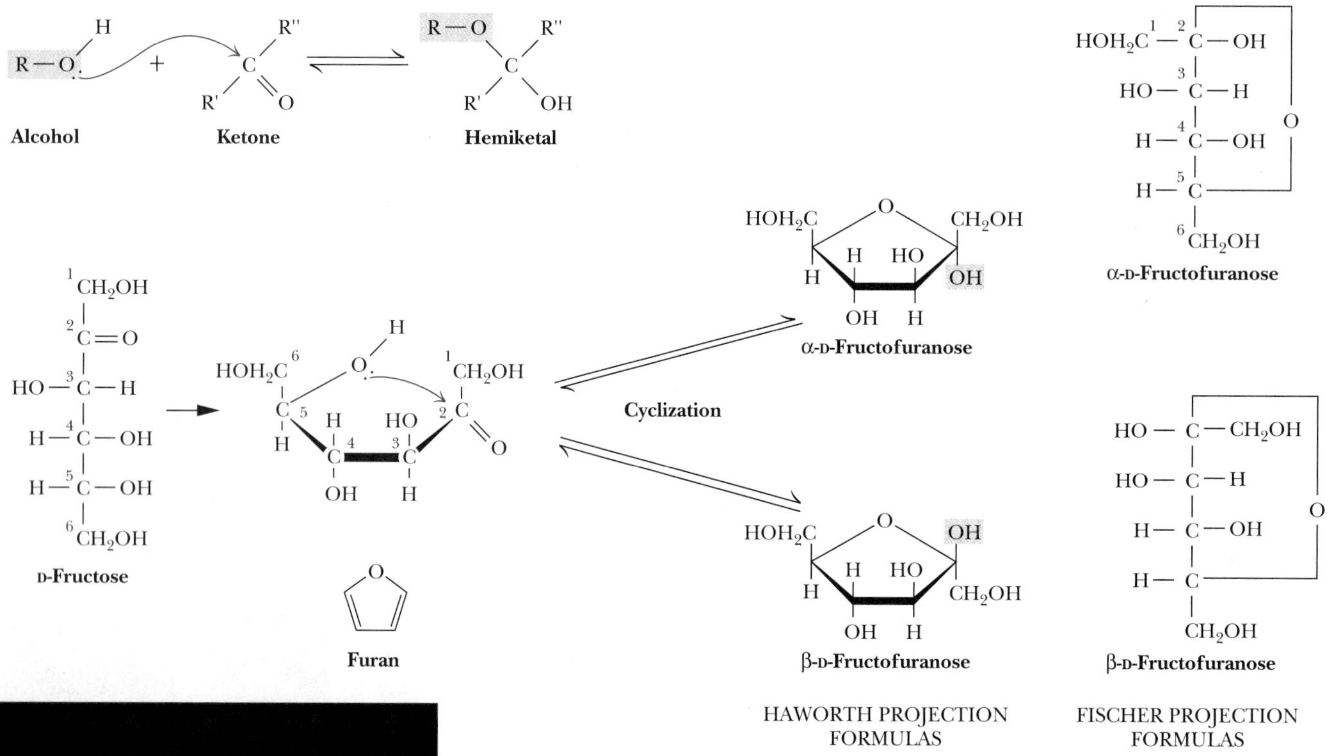

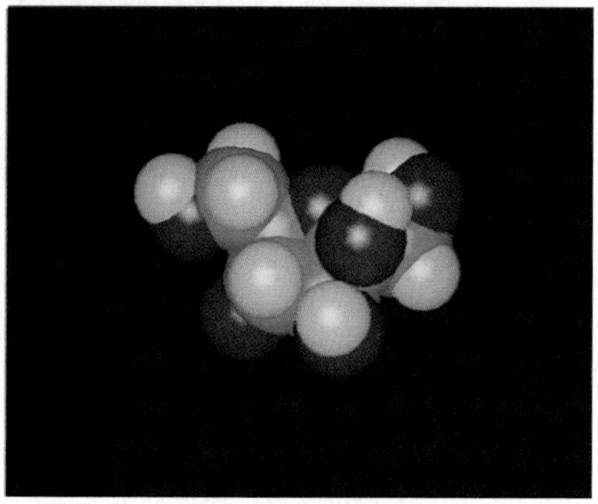

Figure 7.6

forms are the preferred structures for monosaccharides in aqueous solution. At equilibrium, the linear aldehyde or ketone structure is only a minor component of the mixture (generally much less than 1%).

When hemiacetals and hemiketals are formed, the carbon atom that carried the carbonyl function becomes an asymmetric carbon atom. Isomers of monosaccharides that differ only in their configuration about that carbon atom are called **anomers,** designated as α or β, as shown in Figure 7.5, and the carbonyl carbon is thus called the **anomeric carbon.** When the hydroxyl group at the anomeric carbon is on the *same side* of a Fischer projection as the oxygen atom at the highest numbered asymmetric carbon, the configuration at the anomeric carbon is α, as in α-D-glucose. When the anomeric hydroxyl is on the *opposite side* of the Fischer projection, the configuration is β, as in β-D-glucopyranose (see Figure 7.5).

The creation of this asymmetric center upon hemiacetal and hemiketal formation alters the optical rotation properties of monosaccharides, and the original assignment of the α and β notations arose from studies of these properties. Early carbohydrate chemists frequently observed that the optical rotation of glucose (and other sugar) solutions could change with time, a process called **mutarotation.** This indicated that a structural change was occurring. It was eventually found that α-D-glucose has a specific optical rotation, $[\alpha]_D^{20}$, of 112.2°, and that β-D-glucose has a specific optical rotation of 18.7°. Mutarotation involves interconversion of α- and β-forms of the monosaccharide with intermediate formation of the linear aldehyde or ketone, as shown in Figures 7.5 and 7.6.

Haworth Projections

Another of Haworth's lasting contributions to the field of carbohydrate chemistry was his proposal to represent pyranose and furanose structures as hexagonal and pentagonal rings lying perpendicular to the plane of the paper, with

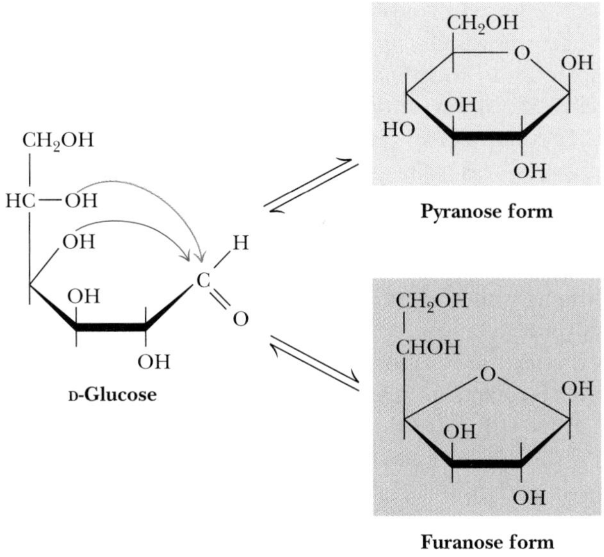

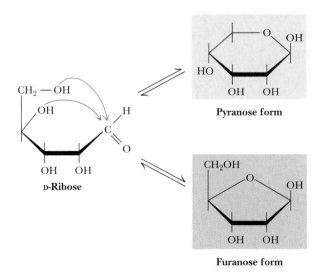

Figure 7.7 D-Glucose can cyclize in two ways, forming either furanose or pyranose structures.

thickened lines indicating the side of the ring closest to the reader. Such **Haworth projections,** which are now widely used to represent saccharide structures (see Figures 7.5 and 7.6), show substituent groups extending either above or below the ring. Substituents drawn to the left in a Fischer projection are drawn above the ring in the corresponding Haworth projection. Substituents drawn to the right in a Fischer projection are below the ring in a Haworth projection. Exceptions to these rules occur in the formation of furanose forms of pentoses and the formation of furanose or pyranose forms of hexoses. In these cases, the structure must be redrawn with a rotation about the carbon whose hydroxyl group is involved in the formation of the cyclic form (Figures 7.7 and 7.8) in order to orient the appropriate hydroxyl group for ring formation. This is merely for illustrative purposes and involves no change in configuration of the saccharide molecule.

The rules previously mentioned for assignment of α- and β-configurations can be readily applied to Haworth projection formulas. For the D-sugars, the anomeric hydroxyl group is below the ring in the α-anomer and above the ring in the β-anomer. For L-sugars, the opposite relationship holds.

Figure 7.8 D-Ribose and other five-carbon saccharides can form either furanose or pyranose structures.

As Figures 7.7 and 7.8 imply, in most monosaccharides there are two or more hydroxyl groups which can react with an aldehyde or ketone at the other end of the molecule to form a hemiacetal or hemiketal. The possibilities for glucose to cyclize are shown in Figure 7.7. D-Ribose, with five carbons, readily forms either five-membered rings (α- or β-D-ribofuranose) or even six-membered rings (α- or β-D-ribopyranose) (see Figure 7.8). In general, aldoses and ketoses with five or more carbons can form *either* furanose or pyranose rings, and the more stable form depends on structural factors. The nature of the substituent groups on the carbonyl and hydroxyl groups and the configuration about the asymmetric carbon will determine whether a given monosaccharide prefers the pyranose or furanose structure. In general, the pyranose form is favored over the furanose ring for aldohexose sugars, although, as we shall see, furanose structures are more stable for ketohexoses.

Although Haworth projections are convenient for display of monosaccharide structures, they do not accurately portray the conformations of pyranose and furanose rings. Given C—C—C tetrahedral bond angles of 109° and C—O—C angles of 111°, neither pyranose nor furanose rings can adopt true planar structures. Instead, they take on puckered conformations, and in the case of pyranose rings, the two favored structures are the **chair conformation** and the **boat conformation,** shown in Figure 7.9. Note that the ring substituents in these structures can be **equatorial,** which means approximately coplanar with the ring, or **axial,** which means parallel to an axis drawn through the ring as shown.

Two general rules dictate the conformation to be adopted by a given saccharide unit. First, bulky substituent groups on such rings are more stable when they occupy equatorial positions rather than axial positions, and second, chair conformations are slightly more stable than boat conformations. For a typical pyranose, such as β-D-glucose, there are two possible chair conformations (see Figure 7.9). Of all the D-aldohexoses, β-D-glucose is the only one that can adopt a conformation with all its bulky groups in an equatorial position. With this advantage of stability, it may come as no surprise that β-D-glucose is the most widely occurring organic group in nature and the central hexose in carbohydrate metabolism.

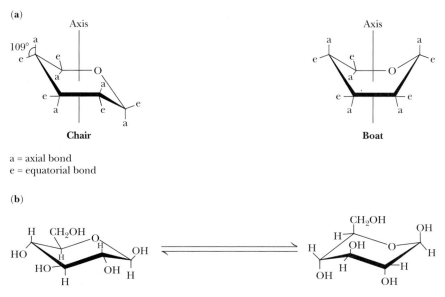

Figure 7.9 **(a)** Chair and boat conformations of a pyranose sugar. **(b)** Two possible chair conformations of β-D-glucose.

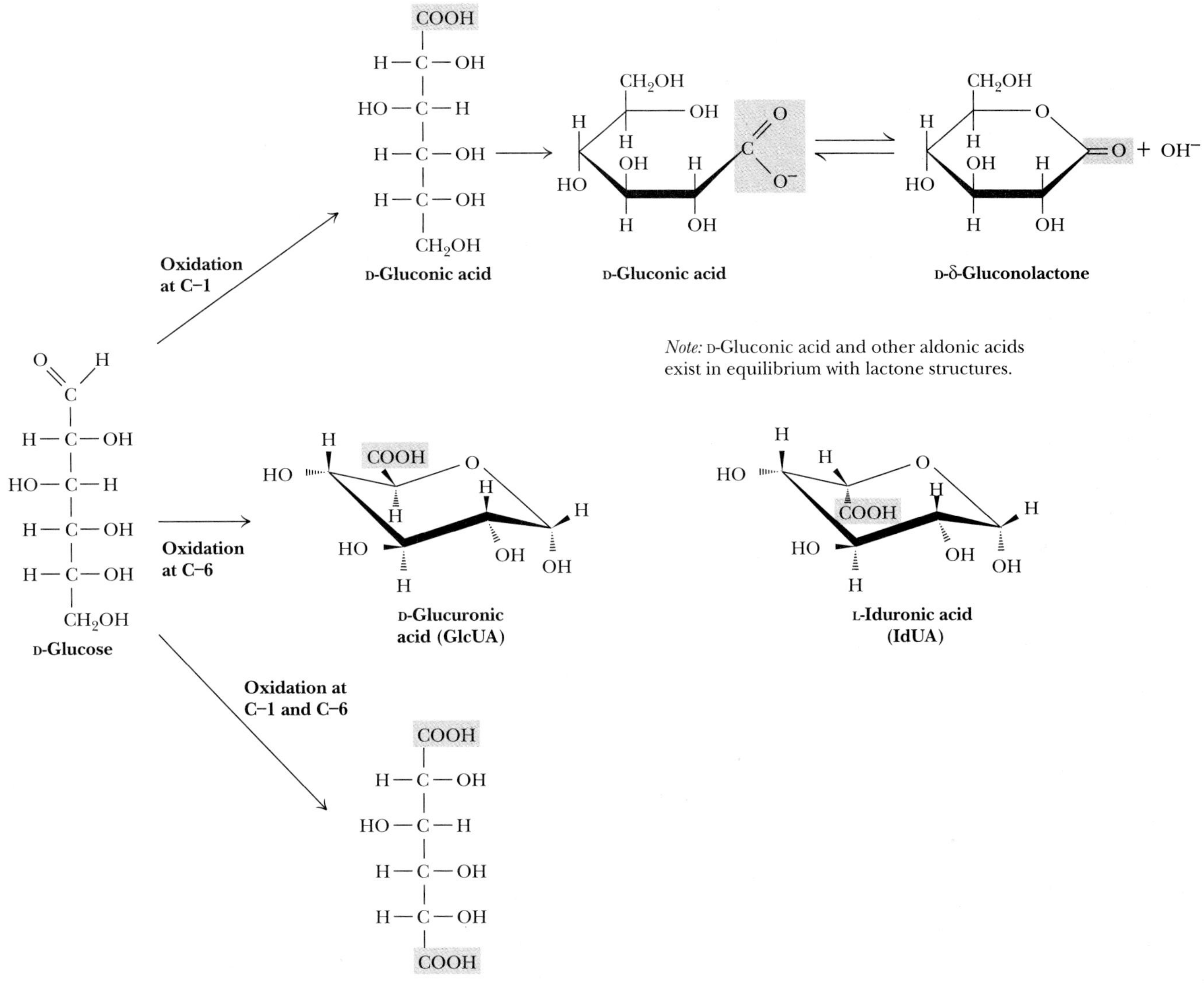

Figure 7.10

Note: D-Gluconic acid and other aldonic acids exist in equilibrium with lactone structures.

Derivatives of Monosaccharides

A variety of chemical and enzymatic reactions produce **derivatives** of the simple sugars. These modifications produce a diverse array of saccharide derivatives. For example, sugars with free anomeric carbon atoms are reasonably good reducing agents and will reduce hydrogen peroxide, ferricyanide, certain metals (Cu^{2+} and Ag^+), and other oxidizing agents. Such reactions convert the sugar to a **sugar acid.** Thus, addition of alkaline $CuSO_4$ (called *Fehling's solution*) to an aldose sugar produces a red cuprous oxide (Cu_2O) precipitate

$$RCHO + 2\,Cu^{2+} + 5\,OH^- \rightarrow RCOO^- + Cu_2O\downarrow + 3\,H_2O$$

Aldehyde Carboxylate

and converts the aldose to an **aldonic acid,** such as **gluconic acid** (Figure 7.10). Formation of a precipitate of red Cu_2O constitutes a positive test for an aldehyde. Carbohydrates that can reduce oxidizing agents in this way are referred to as **reducing sugars.** By quantifying the amount of oxidizing agent reduced by a sugar solution, one can accurately determine the concentration of the sugar. Frequent analysis of reducing sugars in patients with *diabetes mellitus*— a condition that causes high levels of glucose in urine and blood—is an

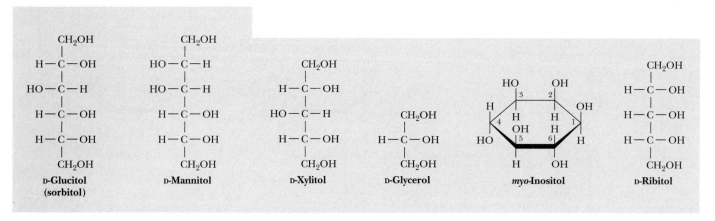

Figure 7.11 Structures of some sugar alcohols.

important part of the diagnosis and treatment of this disease. Over-the-counter kits for the easy and rapid determination of reducing sugars have made this procedure a simple one for diabetics.

Sugar alcohols, another class of sugar derivative, can be prepared by the mild reduction (with NaBH$_4$ or similar agents) of the carbonyl groups of aldoses and ketoses. Sugar alcohols, or **alditols,** are designated by the addition of *-itol* to the name of the parent sugar (Figure 7.11). Alditols are characteristically sweet-tasting, and **sorbitol, mannitol,** and **xylitol** are widely used to sweeten sugarless gum and mints. Sorbitol buildup in the eyes of diabetics is implicated in cataract formation.

The **deoxy sugars** are monosaccharides with one or more hydroxyl groups replaced by hydrogens. 2-Deoxy-D-ribose (Figure 7.12), whose systematic name is 2-deoxy-D-erythropentose, is a constituent of DNA in all living things (see Chapter 8). Deoxy sugars also occur frequently in glycoproteins and polysaccharides.

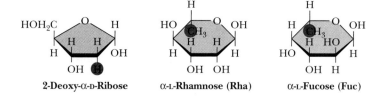

2-Deoxy-α-D-Ribose   **α-L-Rhamnose (Rha)**   **α-L-Fucose (Fuc)**

Figure 7.12 Several deoxy sugars and ouabain, which contains α-L-rhamnose (Rha). Atoms highlighted in red are "deoxy" positions. L-Fucose and L-rhamnose, both 6-deoxy sugars, are components of some cell walls, and rhamnose is a component of ouabain, a highly toxic *cardiac glycoside* found in the bark and root of the ouabaio tree and used by the East African Somalis as an arrow poison. The sugar moiety is not the toxic part of the molecule.

Ouabain

α-D-Glucose-1-phosphate

α-D-Fructose-1,6-bisphosphate

Adenosine-5'-triphosphate

Figure 7.13 Several sugar esters important in metabolism.

Phosphate esters of glucose, fructose, and other monosaccharides are important metabolic intermediates, and the ribose moiety of nucleotides such as ATP and GTP is phosphorylated at the 5'-position (Figure 7.13).

Amino sugars, including D-**glucosamine** and D-**galactosamine** (Figure 7.14), contain an amino group (instead of a hydroxyl group) at the C-2 position. They are found in many oligo- and polysaccharides, including *chitin,* a polysaccharide in the exoskeletons of crustaceans and insects.

Muramic acid and **neuraminic acid,** which are components of the polysaccharides in the cell membranes of higher organisms and also in bacterial cell walls, are glucosamines linked to three-carbon acids at the C-1 or C-3 positions. In muramic acid (from *amine* in bacterial cell wall polysaccharides, and *murus,* which is Latin for "wall"), the hydroxyl group of a lactic acid moiety makes an ether linkage to the C-3 of glucosamine. Neuraminic acid (*amine* isolated from *neural* tissue) forms a C—C bond between the C-1 of *N*-acetylmannosamine and the C-3 of pyruvic acid (Figure 7.15). The *N*-acetyl and *N*-glycolyl derivatives of neuraminic acid are collectively known as **sialic acids** and are distributed widely in bacteria and animal systems.

Acetals, Ketals, and Glycosides

Hemiacetals and hemiketals can react with alcohols in the presence of acid to form **acetals** and **ketals,** as shown in Figure 7.16. This reaction is another example of a *dehydration synthesis* and is similar in this respect to the reactions undergone by amino acids and nucleotides. The pyranose and furanose forms of monosaccharides react with alcohols in this way to form **glycosides** with retention of the α- or β-configuration at the C-1 carbon. The new bond between the anomeric carbon atom and the oxygen atom of the alcohol is called a **glycosidic bond.** Glycosides are named according to the parent monosaccharide. For example, *methyl-β-D-glucoside* (Figure 7.17) can be considered a derivative of β-D-glucose.

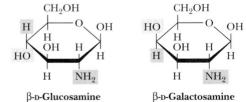

β-D-**Glucosamine**

β-D-**Galactosamine**

Figure 7.14 Structures of D-glucosamine and D-galactosamine.

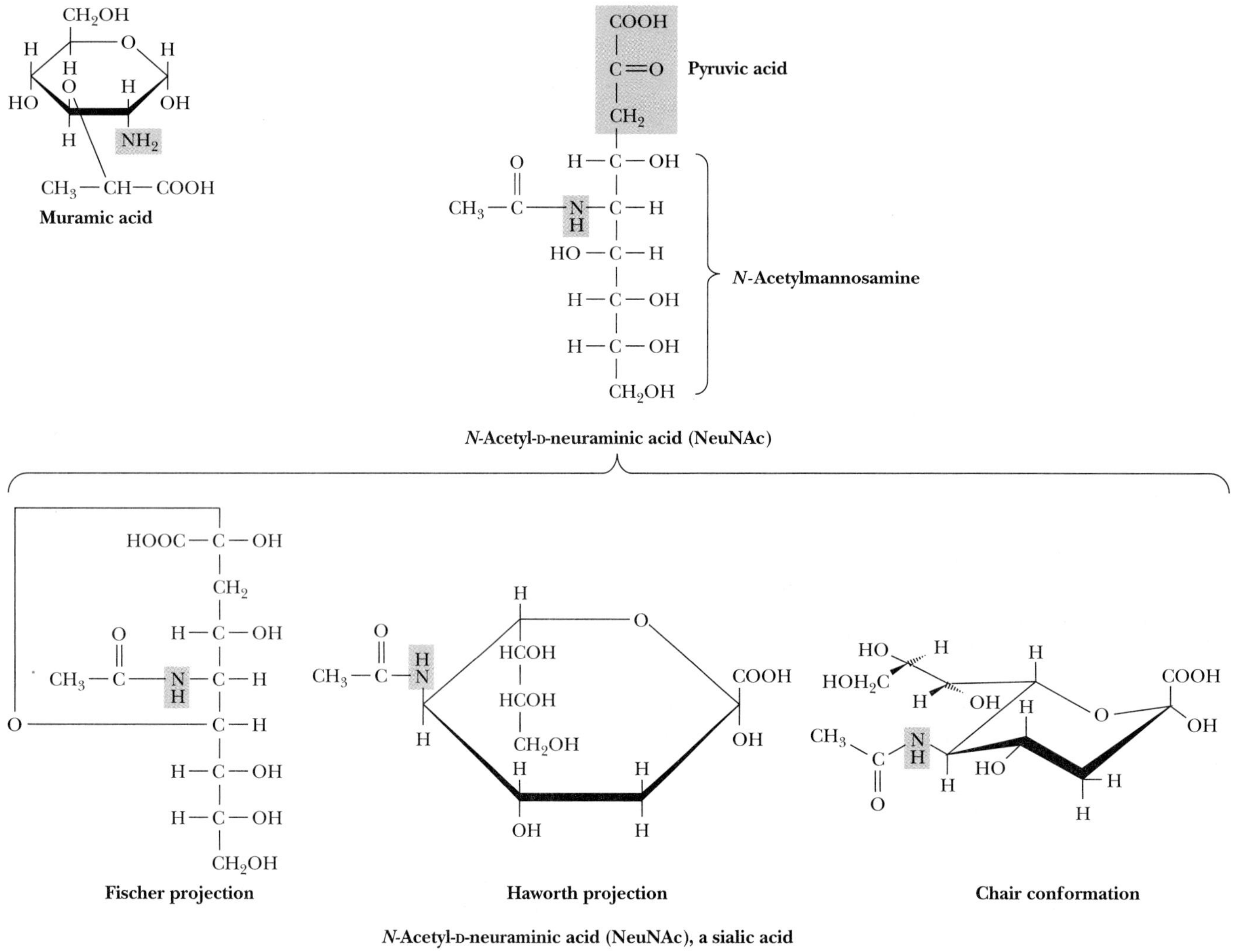

Muramic acid

Pyruvic acid

N-Acetylmannosamine

N-Acetyl-D-neuraminic acid (NeuNAc)

Fischer projection

Haworth projection

Chair conformation

N-Acetyl-D-neuraminic acid (NeuNAc), a sialic acid

Figure 7.15   Structures of muramic acid and neuraminic acid and several depictions of sialic acid.

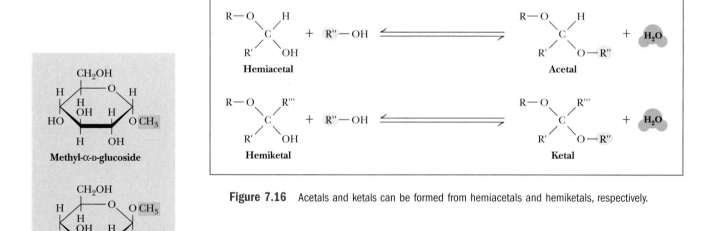

Hemiacetal

Acetal

Hemiketal

Ketal

Figure 7.16   Acetals and ketals can be formed from hemiacetals and hemiketals, respectively.

Methyl-α-D-glucoside

Methyl-β-D-glucoside

◀ **Figure 7.17**   The anomeric forms of methyl-D-glucoside.

Given the relative complexity of oligosaccharides and polysaccharides in higher organisms, it is perhaps surprising that these molecules are formed from relatively few different monosaccharide units. (In this respect, the oligo- and polysaccharides are similar to proteins; both form complicated structures based on a small number of different building blocks.) Monosaccharide units include the hexoses glucose, fructose, mannose, and galactose and the pentoses ribose and xylose.

Disaccharides

The simplest oligosaccharides are the **disaccharides,** which consist of two monosaccharide units linked by a glycosidic bond. As in proteins and nucleic acids, each individual unit in an oligosaccharide is termed a *residue.* The disaccharides shown in Figure 7.18 are all commonly found in nature, with sucrose, maltose, and lactose being the most common. Except for sucrose, each of these structures possesses one free unsubstituted anomeric carbon atom, and thus each of these disaccharides is a reducing sugar. The end of the molecule containing the free anomeric carbon is called the **reducing end,** and the other end is called the **nonreducing end.** In the case of sucrose, both of the anomeric carbon atoms are substituted—that is, neither has a free —OH group. The substituted anomeric carbons cannot be converted to the aldehyde configuration and thus cannot participate in the oxidation–reduction reactions characteristic of reducing sugars. Thus, sucrose is not a reducing sugar.

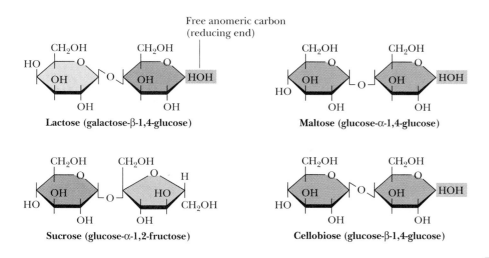
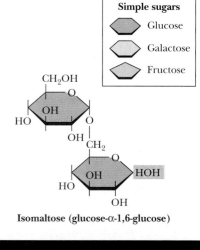

Figure 7.18 The structures of several important disaccharides. Note that the notation —HOH means that the configuration can be either α or β. If the —OH group comes up out of the ring, the configuration is termed β. The configuration is α if the —OH group comes down out of the ring. Also note that sucrose has no free anomeric carbon atoms.

Sucrose

Maltose is a **homodisaccharide,** since it contains only one kind of monosaccharide, namely, glucose. **Maltose** is produced from starch (a polymer of α-D-glucose produced by plants) by the action of amylase enzymes and is a component of malt, a substance obtained by allowing grain (particularly barley) to soften in water and germinate. The enzyme **diastase,** produced during the germination process, catalyzes the hydrolysis of starch to maltose. Maltose is used in beverages (malted milk, for example), and since it is fermented readily by yeast, it is important in the brewing of beer. The glucose units in maltose are **1 → 4 linked,** meaning that the C-1 of one glucose is linked by a glycosidic bond to the C-4 oxygen of the other glucose.

The complete structures of these disaccharides can be specified in shorthand notation by using abbreviations for each monosaccharide, α or β to denote configuration, and appropriate numbers to indicate the nature of the linkage. Thus, maltose is Glcα1-4Glc. Often the glycosidic linkage is written with an arrow, so that maltose would be Glcα1 → 4Glc. Since the linkage carbon on the first sugar is always C-1, a newer trend is to drop the 1- or 1→ and describe a sugar such as maltose as Glcα4Glc. More complete names can also be used, however, so that maltose would be O-α-D-glucopyranosyl-(1 → 4)-D-glucopyranose.

β-D-Lactose (O-β-D-galactopyranosyl-(1 → 4)-D-glucopyranose) (see Figure 7.18) is the principal carbohydrate in milk and is of critical nutritional importance to mammals in the early stages of their lives. It is formed from D-galactose and D-glucose via a β(1 → 4) link, and because it has a free anomeric carbon, it is capable of mutarotation and is a reducing sugar. It is an interesting quirk of nature that lactose cannot be absorbed directly into the bloodstream. It must first be broken down into galactose and glucose by **lactase,** an intestinal enzyme that exists in young, nursing mammals but is not produced in significant quantities in the mature mammal. Most humans, with the ex-

A DEEPER LOOK

Trehalose—A Natural Protectant for Bugs

Insects use an open circulatory system to circulate **hemolymph** (insect blood). Their "blood sugar" is not glucose but rather **trehalose,** an unusual, nonreducing disaccharide (see figure). Trehalose is found typically in organisms that are naturally subject to temperature variations and other environmental stresses—bacterial spores, fungi, yeast, and many insects. (Interestingly, honeybees do not have trehalose in their hemolymph, perhaps because they practice a colonial, rather than solitary, lifestyle. Bee colonies maintain a rather constant temperature of 18°C in winter, protecting the residents from large temperature changes.)

What might explain this correlation between trehalose utilization and environmentally stressful lifestyles? Konrad Bloch[*] suggested that trehalose may act as a natural cryoprotectant. Freezing and thawing of biological tissues frequently causes irreversible structural changes, destroying biological activity. High concentrations of polyhydroxy compounds, such as sucrose and glycerol, can protect biological materials from such damage. Trehalose is par-

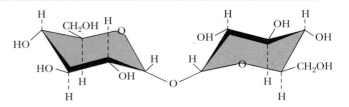

Trehalose, a nonreducing disaccharide.

ticularly well suited for this purpose and has been shown to be superior to other polyhydroxy compounds, especially at low concentrations. Support for this novel idea comes from studies by P. A. Attfield,[†] which show that trehalose levels in the yeast *Saccharomyces cerevisiae* increase significantly during exposure to high salt concentrations or high growth temperatures—the same conditions that elicit the production of heat shock proteins!

[*]Bloch, K., 1994. *Blondes in Venetian Paintings, the Nine-Banded Armadillo, and Other Essays in Biochemistry.* New Haven: Yale University Press.

[†]Attfield, P. A., 1987. Trehalose accumulates in *Saccharomyces cerevisiae* during exposure to agents that induce heat shock responses. *FEBS Letters* **225**:259.

A DEEPER LOOK

Honey—An Ancestral Carbohydrate Treat

Honey, the first sweet known to humankind, is the only sweetening agent that can be stored and used exactly as produced in nature. Bees process the nectar of flowers so that the final product is able to survive long-term storage at ambient temperature. Used as a ceremonial material and medicinal agent in earliest times, honey was not regarded as a food until the era of the Greeks and Romans. Only in modern times have cane and beet sugar surpassed honey as the most frequently used sweetener. What is the chemical nature of this magical, viscous substance?

The bees' processing of honey consists of (1) reducing the water content of the nectar by 30 to 60% to the self-preserving range of 15 to 19%, (2) hydrolyzing the significant amount of sucrose in nectar to glucose and fructose by the action of the enzyme invertase, and (3) producing small amounts of gluconic acid from glucose by the action of the enzyme **glucose oxidase.** Most of the sugar in the final product is glucose and fructose, and the final product is supersaturated with respect to these monosaccharides. Honey actually consists of an emulsion of microscopic glucose hydrate and fructose hydrate crystals in a thick syrup. Sucrose accounts for only about 1% of the sugar in the final product, with fructose at about 38% and glucose at 31% by weight.

The figure shows a ^{13}C nuclear magnetic resonance spectrum of honey from a mixture of wildflowers in southeastern Pennsylvania. Interestingly, five major hexose species contribute to this spectrum. Although most textbooks show fructose exclusively in its furanose form, the predominant form of fructose (67% of total fructose) is β-D-fructopyranose, with the β- and α-fructofuranose forms accounting for 27% and 6% of the fructose, respectively. In polysaccharides, fructose invariably prefers the furanose form, but free fructose (and crystalline fructose) is predominantly β-fructopyranose.

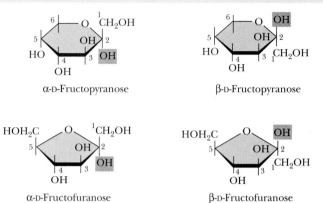

α-D-Fructopyranose   β-D-Fructopyranose

α-D-Fructofuranose   β-D-Fructofuranose

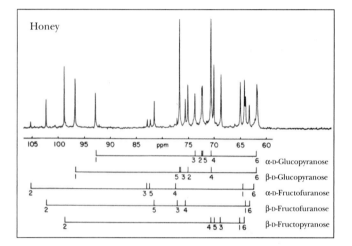

Honey

White, J. W., 1978. Honey. *Advances in Food Research* **24:**287–374.
Prince, R. C., Gunson, D. E., Leigh, J. S., and McDonald, G. G., 1982. The predominant form of fructose is a pyranose, not a furanose ring. *Trends in Biochemical Sciences* **7:**239–240.

 See pages 114–118

ception of certain groups in Africa and northern Europe, produce only low levels of lactase. For most individuals, this is not a problem, but some cannot tolerate lactose and experience intestinal pain and diarrhea upon consumption of milk.

Sucrose, in contrast, is a disaccharide of almost universal appeal and tolerance. Produced by many higher plants and commonly known as table sugar, it is one of the products of photosynthesis and is composed of fructose and glucose. Although sucrose and maltose are important to the human diet, they are not taken up directly in the body. In a manner similar to lactose, they are first hydrolyzed by **sucrase** and **maltase,** respectively, in the human intestine.

Higher Oligosaccharides

In addition to the simple disaccharides, many other oligosaccharides are found in both prokaryotic and eukaryotic organisms, either as naturally occurring substances or as hydrolysis products of natural materials. Figure 7.19 lists a number of simple oligosaccharides, along with descriptions of their origins and interesting features. Several are constituents of the sweet nectars or saps exuded or extracted from plants and trees.

Stachyose is typical of the oligosaccharide components found in substantial quantities in beans, peas, bran, and whole grains. These oligosaccharides are not digested by stomach enzymes but are metabolized readily by bacteria in the intestines. This is the source of the flatulence that often accompanies the consumption of such foods. Commercial products are now available that assist in the digestion of the gas-producing components of these foods. These products contain an enzyme that hydrolyzes the culprit oligosaccharides in the stomach before they become available to intestinal microorganisms.

Another notable glycoside is **amygdalin,** which occurs in bitter almonds and in the kernels or pits of cherries, peaches, and apricots. Hydrolysis of this substance and subsequent oxidation yields **laetrile,** which has been claimed by some to have anticancer properties. There is no scientific evidence for these claims, and the U.S. Food and Drug Administration has never approved laetrile for use in the United States.

Oligosaccharides also occur widely as components (via glycosidic bonds) of *antibiotics* derived from various sources. Figure 7.20 shows the structures of two carbohydrate-containing antibiotics. Some of these antibiotics also show antitumor activity. One of the most important of this type is **bleomycin A₂,** which is used clinically against certain tumors.

Figure 7.19 The structures of some interesting oligosaccharides.

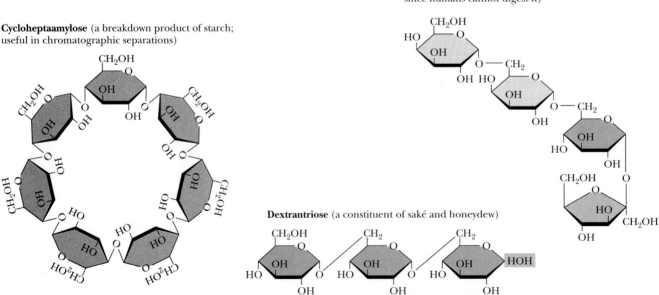

Melezitose (a constituent of honey)

Amygdalin (occurs in seeds of *Rosaceae*, glycoside of bitter almonds, in kernels of cherries, peaches, apricots)

Laetrile (claimed to be an anticancer agent, but there is no scientific evidence for this)

Cycloheptaamylose (a breakdown product of starch; useful in chromatographic separations)

Stachyose (a constituent of many plants: white jasmine, yellow lupine, soybeans, lentils, etc.; causes flatulence since humans cannot digest it)

Dextrantriose (a constituent of saké and honeydew)

Bleomycin A$_2$ (an antitumor agent used clinically against specific tumors)

Streptomycin (a broad spectrum antibiotic)

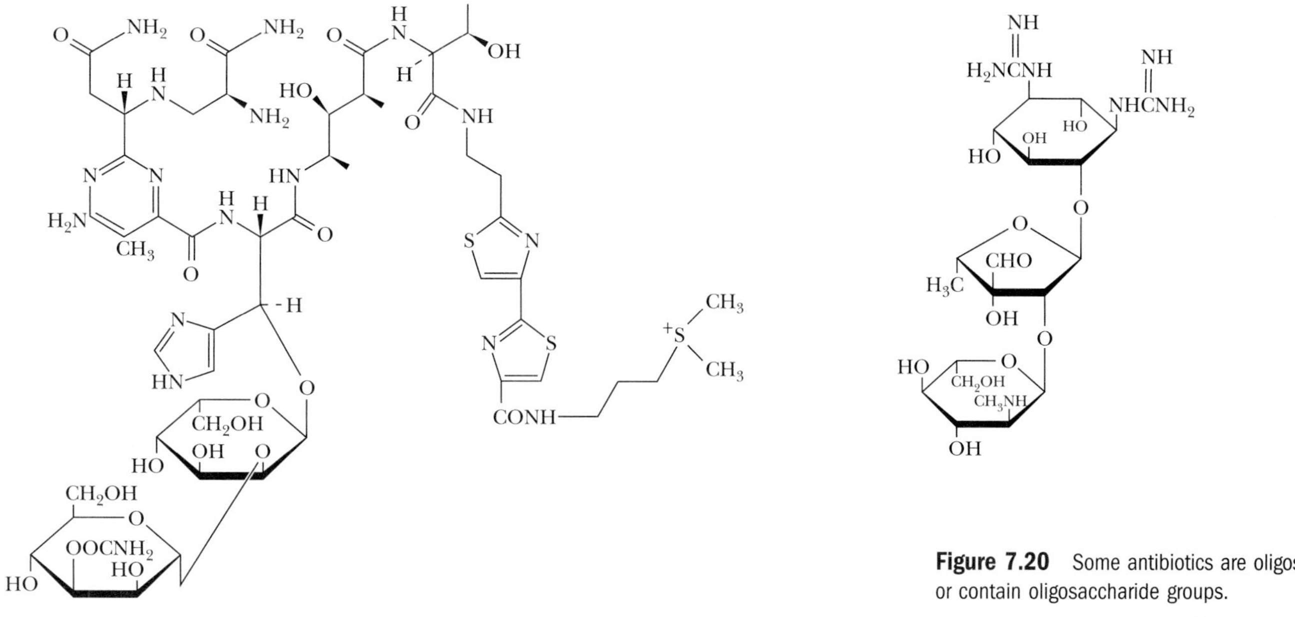

Figure 7.20 Some antibiotics are oligosaccharides or contain oligosaccharide groups.

7.4 Polysaccharides

Structure, Nomenclature, and Function

By far the majority of carbohydrate material in nature occurs in the form of polysaccharides. By our definition, polysaccharides include not only those substances composed solely of glycosidically linked sugar residues but also molecules that contain polymeric saccharide structures linked via covalent bonds to amino acids, peptides, proteins, lipids, and other structures.

Polysaccharides, also called **glycans,** consist of monosaccharides and their derivatives. If a polysaccharide contains only one kind of monosaccharide molecule, it is a **homopolysaccharide,** or **homoglycan,** whereas those containing more than one kind of monosaccharide are **heteropolysaccharides.** Polysaccharides differ not only in the nature of their component monosaccharides but also in the length of their chains and in the amount of chain branching that occurs. Although a given sugar residue has only one anomeric carbon and thus can form only one glycosidic linkage with hydroxyl groups on other molecules, each sugar residue carries several hydroxyls, one or more of which may be an acceptor of glycosyl substituents (Figure 7.21). This ability to form branched structures distinguishes polysaccharides from proteins and nucleic acids, which occur only as linear polymers.

The functions of many individual polysaccharides cannot be assigned uniquely, and some of their functions may not yet be appreciated. Starch, glycogen, and other storage polysaccharides, as readily metabolizable food, provide energy reserves for cells. Chitin and cellulose provide strong support for the skeletons of arthropods and green plants, respectively. Mucopolysaccharides, such as the *hyaluronic acids,* form protective coats on animal cells. In each of these cases, the relevant polysaccharide is either a homopolymer or a polymer of small repeating units. Recent research indicates that oligosaccharides and polysaccharides with more unique structures are involved in much more sophisticated tasks in cells, including a variety of cellular recognition and intercellular communication events, to be discussed later.

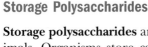

Amylose

Amylopectin

Figure 7.21 Amylose and amylopectin are the two forms of starch. Note that the linear linkages are $\alpha(1 \rightarrow 4)$, but the branches in amylopectin are $\alpha(1 \rightarrow 6)$. Branches in polysaccharides can involve any of the hydroxyl groups on the monosaccharide components. Amylopectin is a highly branched structure, with branches occurring every 12 to 30 residues.

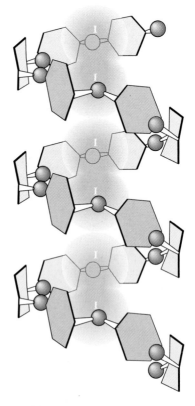

Figure 7.22 Suspensions of amylose in water adopt a helical conformation. Iodine (I_2) can insert into the middle of the amylose helix to give a blue color that is characteristic and diagnostic for starch.

Storage Polysaccharides

Storage polysaccharides are an important carbohydrate form in plants and animals. Organisms store carbohydrates in the form of polysaccharides rather than as monosaccharides in order to lower the osmotic pressure of the sugar reserves. Since osmotic pressures depend only on *numbers of molecules,* the osmotic pressure is greatly reduced by formation of a few polysaccharide molecules out of thousands (or even millions) of monosaccharide units.

Starch

By far the most common storage polysaccharide in plants is **starch,** which exists in two forms: **α-amylose** and **amylopectin,** the structures of which are shown in Figure 7.21. Most forms of starch in nature are 10% to 30% α-amylose and 70% to 90% amylopectin. α-Amylose is composed of linear chains of D-glucose in $\alpha(1 \rightarrow 4)$ linkages. The chains are of varying length, having molecular weights from several thousand to half a million. As can be seen from the structure in Figure 7.21, the chain has a reducing end and a nonreducing end. Although poorly soluble in water, α-amylose forms micelles in which the polysaccharide chain adopts a helical conformation (Figure 7.22). Iodine reacts with α-amylose to give a characteristic blue color, which arises from the insertion of iodine into the middle of the hydrophobic amylose helix.

In contrast to α-amylose, amylopectin, the other component of typical starches, is a highly branched chain of glucose units (see Figure 7.21). Branches occur in these chains every 12 to 30 residues. The linear linkages in amylopectin are $\alpha(1 \rightarrow 4)$, whereas the branch linkages are $\alpha(1 \rightarrow 6)$. As is the case for α-amylose, amylopectin forms micellar suspensions in water; iodine reacts with such suspensions to produce a red-violet color.

In animals, digestion and use of plant starches begins in the mouth with **salivary α-amylase** ($\alpha[1 \rightarrow 4]$-glucan 4-glucanohydrolase), the major enzyme secreted by the salivary glands. Although the capability of making and secreting salivary α-amylases is widespread in the animal world, some animals (such as cats, dogs, birds, and horses) do not secrete them. Salivary α-amylase is an **endoamylase,** which splits $\alpha(1 \rightarrow 4)$ glycosidic linkages only within the chain. Raw starch is not very susceptible to salivary endoamylase. However, when suspensions of starch granules are heated, the granules swell, taking up water and causing the polymers to become more accessible to enzymes. Thus, cooked starch is more digestible.

Glycogen

The major form of storage polysaccharide in animals is **glycogen.** Glycogen is found mainly in the liver (where it may amount to as much as 10% of liver mass) and skeletal muscle (where it accounts for 1% to 2% of muscle mass). Liver glycogen consists of granules containing highly branched molecules, with $\alpha(1 \rightarrow 6)$ branches occurring every 8 to 12 glucose units. Glycogen in these granules has an average molecular weight of several million. Associated with these granules are the enzymes needed to synthesize and degrade glycogen molecules. Like amylopectin, glycogen yields a red-violet color with iodine. Glycogen can be hydrolyzed by both α- and β-amylases, yielding glucose and maltose, respectively, as products and can also be hydrolyzed by **glycogen phosphorylase,** an enzyme present in liver and muscle tissue, to release glucose-1-phosphate.

Molecules of glycogen (or amylopectin) have a single reducing end and many nonreducing ends. (Actually, the number of nonreducing ends is one more than the number of branches.) Enzymes that break down starch and glycogen remove one (or sometimes two) units at a time from the nonreducing ends. Large amounts of energy can thus be provided rapidly by the release (and metabolism) of large numbers of glucose units from the many nonreducing ends of a starch or glycogen molecule.

Structural Polysaccharides

Cellulose

The **structural polysaccharides** have properties dramatically different from those of the storage polysaccharides, even though the compositions of these two classes are similar. The structural polysaccharide **cellulose** is the most abundant natural polymer found in the world. Found in the cell walls of nearly all plants, cellulose is one of the principal components providing physical structure and strength. The wood and bark of trees are insoluble, highly organized structures formed from cellulose and also from lignin (see Figure 21.19). It is awe-inspiring to look at a large tree and realize the amount of weight supported by polymeric structures derived from sugars and organic acids. Cellulose also has its delicate side, however. Cotton, whose woven fibers make some of our most comfortable clothing fabrics, is almost pure cellulose. Derivatives of cellulose have found wide use in our society. **Cellulose acetates** are produced by the action of acetic anhydride on cellulose in the presence of sulfuric acid and can be spun into a variety of fabrics with particular properties. Referred to simply as *acetates,* they have a silky appearance, a luxuriously soft feel, and a deep luster, and are used in dresses, lingerie, linings, and blouses.

Cellulose is a linear homopolymer of D-glucose units, just as in α-amylose. The structural difference, which completely alters the properties of the polymer, is that in cellulose the glucose units are linked by $\beta(1 \rightarrow 4)$-glycosidic bonds, whereas in α-amylose the linkage is $\alpha(1 \rightarrow 4)$. The conformational difference between these two structures is shown in Figure 7.23. The $\alpha(1 \rightarrow 4)$-

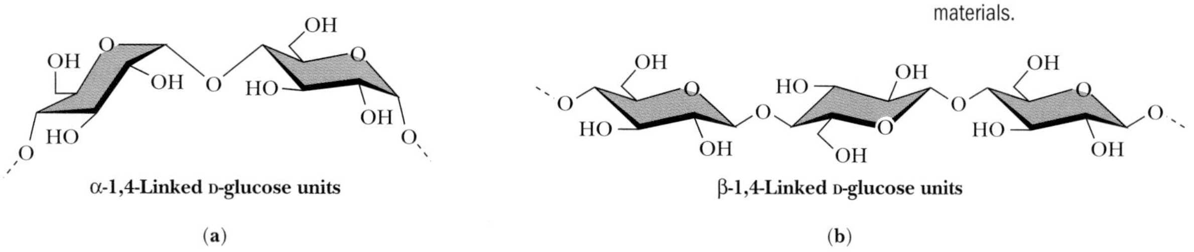

Figure 7.23 **(a)** Amylose, composed exclusively of the relatively bent $\alpha(1 \rightarrow 4)$ linkages, prefers to adopt a helical conformation, whereas **(b)** cellulose, with $\beta(1 \rightarrow 4)$-glycosidic linkages, can adopt a fully extended conformation with alternating 180° flips of the glucose units. The hydrogen bonding inherent in such extended structures is responsible for the great strength of tree trunks and other cellulose-based materials.

α-1,4-Linked D-glucose units

(a)

β-1,4-Linked D-glucose units

(b)

Figure 7.24 The structure of cellulose, showing the hydrogen bonds (blue) between the sheets, which strengthen the structure. Intrachain hydrogen bonds are in red and interchain hydrogen bonds are in green.

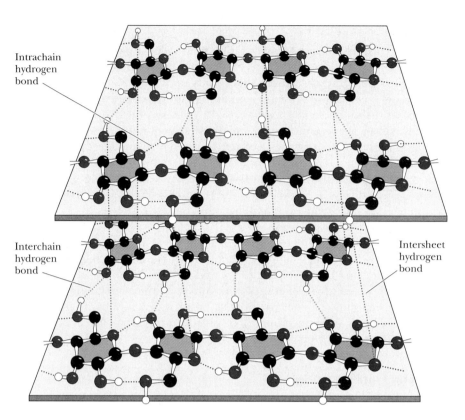

Intrachain hydrogen bond

Interchain hydrogen bond

Intersheet hydrogen bond

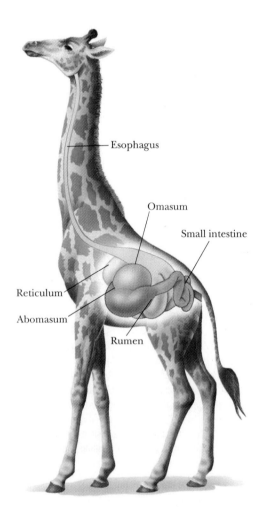

Esophagus

Omasum

Small intestine

Reticulum

Abomasum

Rumen

Figure 7.25 Giraffes, cattle, deer, and camels are ruminant animals that are able to metabolize cellulose, thanks to bacterial cellulase in the rumen, a large first compartment in the stomach of a ruminant.

linkage sites of amylose are naturally bent, conferring a gradual turn to the polymer chain, which results in the helical conformation already described (see Figure 7.22). The most stable conformation about the $\beta(1 \rightarrow 4)$ linkage involves alternating 180° flips of the glucose units along the chain so that the chain adopts a fully extended conformation, referred to as an **extended ribbon.** Juxtaposition of several such chains permits efficient interchain hydrogen bonding, the basis of much of the strength of cellulose.

The structure of one form of cellulose, determined by X-ray and electron diffraction data, is shown in Figure 7.24. The flattened sheets of the chains lie side by side and are joined by hydrogen bonds. These sheets are laid on top of one another in a way that staggers the chains, just as bricks are staggered to give strength and stability to a wall. Cellulose is extremely resistant to hydrolysis, whether by acid or by the digestive tract amylases described earlier. As a result, most animals (including humans) cannot digest cellulose to any significant degree. Ruminant animals, such as cattle, deer, giraffes, and camels, are an exception because bacteria that live in the rumen (Figure 7.25) secrete the enzyme **cellulase,** a β-glucosidase effective in the hydrolysis of cellulose. The resulting glucose is then metabolized in a fermentation process to the benefit of the host animal. Termites and shipworms (*Teredo navalis*) similarly digest cellulose because their digestive tracts also contain bacteria that secrete cellulase.

Chitin

A polysaccharide similar to cellulose, both in its biological function and its primary, secondary, and tertiary structures, is **chitin.** Chitin is present in the cell walls of fungi and is the fundamental material in the exoskeletons of crustaceans, insects, and spiders. The structure of chitin, an extended ribbon, is identical to cellulose, except that the —OH group on each C-2 is replaced by —NHCOCH$_3$, so that the repeating units are *N-acetyl-D-glucosamines* in $\beta(1 \rightarrow 4)$ linkage. Like cellulose (see Figure 7.24), the chains of chitin form extended ribbons (Figure 7.26) and pack side by side in a crystalline, strongly hydrogen-bonded form. One significant difference between cellulose and chitin is in the arrangement of the chains—either **parallel** (all the reducing ends together at one end of a packed bundle) or **antiparallel** (each sheet of chains arranged in opposition to the sheets above and below). Natural cellulose seems to only occur in parallel arrangements. Chitin, however, can occur in three forms, sometimes all in the same organism. α-*Chitin* is an all-parallel arrangement of the chains, whereas β-*chitin* is an antiparallel arrangement. In δ-*chitin*, the structure is thought to involve pairs of parallel sheets separated by single antiparallel sheets.

Chitin is the earth's second most abundant carbohydrate polymer (after cellulose), and its ready availability and abundance offer opportunities for industrial and commercial applications. Chitin-based coatings can extend the shelf life of fruits, and a chitin derivative that binds to iron atoms in meat has been found to slow the reactions that cause rancidity and flavor loss. Without such a coating, the iron in meats activates oxygen from the air, forming

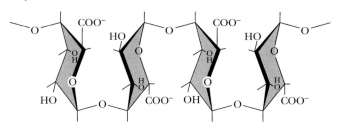

Figure 7.26 Like cellulose, chitin, mannan, and poly (D-mannuronate) form extended ribbons that pack together efficiently, taking advantage of multiple hydrogen bonds. Poly (L-guluronate) forms a buckled ribbon.

reactive free radicals that attack and oxidize polyunsaturated lipids, causing most of the flavor loss associated with rancidity. Chitin-based coatings coordinate the iron atoms, preventing their interaction with oxygen.

Glycosaminoglycans

A class of polysaccharides known as **glycosaminoglycans** is involved in a variety of extracellular (and sometimes intracellular) functions. Glycosaminoglycans consist of linear chains of repeating disaccharides in which one of the monosaccharide units is an amino sugar and one (or both) of the monosaccharide units contains at least one negatively charged sulfate or carboxylate group. The repeating disaccharide structures found commonly in glycosaminoglycans are shown in Figure 7.27. **Heparin,** with the highest net negative charge of the disaccharides shown, is a natural anticoagulant substance. It binds strongly to *antithrombin III* (a protein involved in terminating the clotting process) and inhibits blood clotting. **Hyaluronate** molecules may consist of as many as 25,000 disaccharide units, with molecular weights of up to 10^7. Hyaluronates are important components of the vitreous humor in the eye and of synovial fluid, the lubricant of joints in the body. The **chondroitins** and **keratan sulfate** are found in tendons, cartilage, and other connective tissue. Glycosaminoglycans are fundamental constituents of *proteoglycans* (to be discussed later).

Bacterial Cell Walls

Some of nature's most interesting polysaccharide structures are found in *bacterial cell walls.* Given the strength and rigidity provided by polysaccharide structures, it is not surprising that bacteria use such structures to provide protection for their cellular contents. Bacteria normally exhibit high internal osmotic pressures and frequently encounter variable, often hypotonic, exterior condi-

A DEEPER LOOK

Billiard Balls, Exploding Teeth, and Dynamite—The Colorful History of Cellulose

Although humans cannot digest it, and most people's acquaintance with it is limited to comfortable cotton clothing, cellulose has enjoyed a colorful and varied history of use. In 1838, Théophile Pelouze in France found that paper or cotton could be made explosive if dipped in concentrated nitric acid. Christian Schönbein, a professor of chemistry at the University of Basel in Switzerland, prepared "nitrocotton" in 1845 by dipping cotton in a mixture of nitric and sulfuric acids and then washing the material to remove excess acid. In 1860, Major E. Schultze of the Prussian army used the same material, now called **guncotton,** as a propellant replacement for gunpowder, and its preparation in brass cartridges soon made it popular for this purpose. The only problem was that it was too explosive and could detonate unpredictably in factories where it was produced. The entire town of Faversham, England, was destroyed in such an accident. In 1868, Alfred Nobel mixed guncotton with ether and alcohol, thus creating **nitrocellulose,** and in turn mixed this with nitroglycerine and sawdust to produce **dynamite.** Nobel's income from dynamite and also from his profitable devel-

opment of the Russian oil fields in Baku eventually formed the endowment for the Nobel Prizes.

In 1869, concerned over the precipitous decline (from hunting) of the elephant population in Africa, billiard ball manufacturers Phelan and Collander offered a prize of $10,000 for production of a substitute for ivory. Brothers Isaiah and John Hyatt in Albany, New York, produced such a substitute by mixing guncotton with camphor, then heating and squeezing it to produce **celluloid.** This product found immediate uses well beyond billiard balls. It was easy to shape, was strong and resilient, and it exhibited a high tensile strength. Celluloid was used eventually to make dolls, combs, musical instruments, fountain pens, piano keys, and a variety of other products. The Hyatt brothers eventually formed the Albany Dental Company to make false teeth from celluloid. Because camphor was used in their production, the company advertised that its teeth smelled "clean," but, as reported in the *New York Times* in 1875, the teeth also occasionally exploded!

Portions adapted from Burke, J., 1996. *The Pinball Effect: How Renaissance Water Gardens Made the Carburetor Possible and Other Journeys Through Knowledge.* New York: Little, Brown and Company.

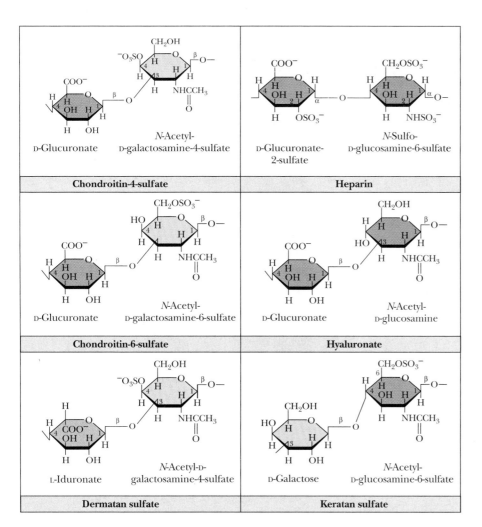

Figure 7.27 Glycosaminoglycans are formed from repeating disaccharide arrays and often occur as components of the proteoglycans.

tions. The rigid cell walls synthesized by bacteria maintain cell shape and size and prevent swelling or shrinkage that would inevitably accompany variations in solution osmotic strength.

Peptidoglycan

Bacteria are conveniently classified as either **Gram-positive** or **Gram-negative** depending on their response to the so-called Gram stain. Despite substantial differences in the various structures surrounding these two types of cell, nearly all bacterial cell walls have a strong, protective peptide–polysaccharide layer called **peptidoglycan.** Gram-positive bacteria have a thick (approximately 25 nm) cell wall consisting of multiple layers of peptidoglycan. This thick cell wall surrounds the bacterial plasma membrane. Gram-negative bacteria, in contrast, have a much thinner (2 to 3 nm) cell wall consisting of a single layer of peptidoglycan sandwiched between the inner and outer lipid bilayer membranes. In either case, peptidoglycan, sometimes called **murein** (from the Latin *murus* for "wall"), is a continuous cross-linked structure—in essence, a single molecule—built around the cell. The structure is shown in Figure 7.28. The backbone is a $\beta(1 \rightarrow 4)$ linked polymer of alternating N-acetylglucosamine and N-acetylmuramic acid units. This part of the structure is similar to chitin, but it is joined to a tetrapeptide, usually L-Ala · D-Glu · L-Lys · D-Ala, in which the L-lysine is linked to the γ-COOH of D-glutamate. The peptide is linked to the N-acetylmuramic acid units via its D-lactate moiety. The ε-amino group of lysine

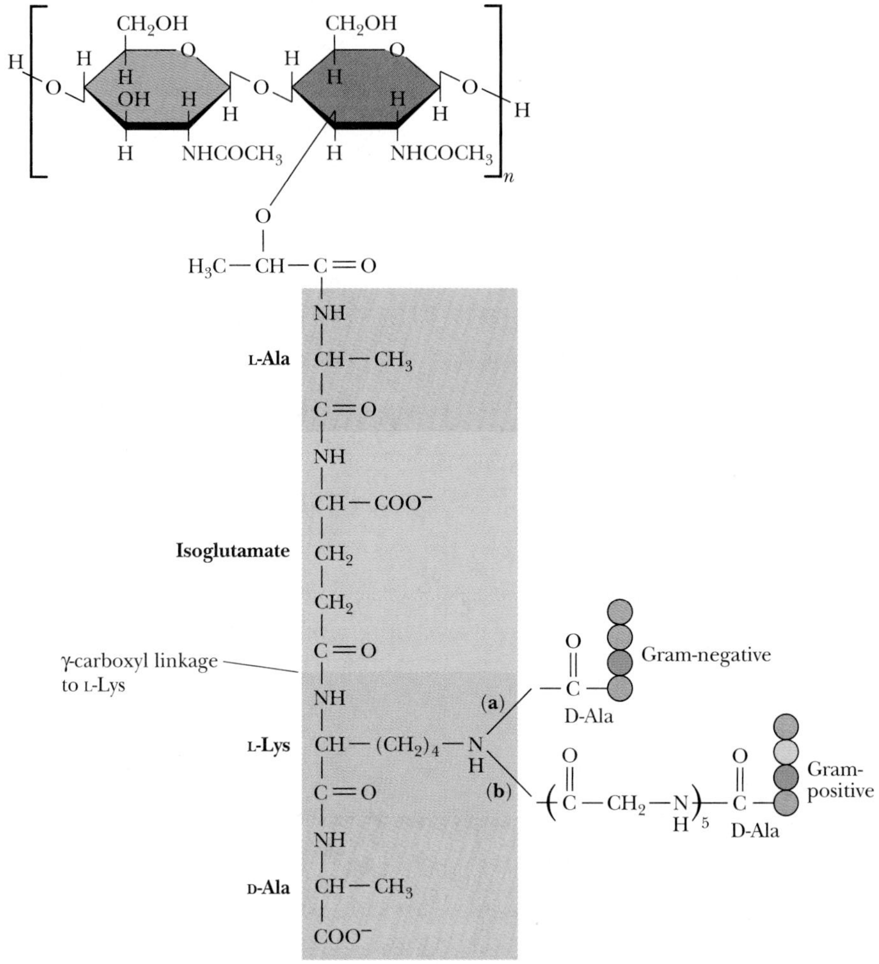

Figure 7.28 The structure of peptidoglycan. The tetrapeptides linking adjacent backbone chains contain an unusual γ-carboxyl linkage.

L-Ala

Isoglutamate

γ-carboxyl linkage to L-Lys

L-Lys

(a)

(b)

D-Ala

Gram-negative

D-Ala

Gram-positive

D-Ala

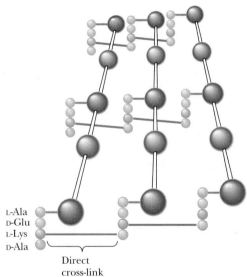

(a) Gram-positive cell wall

(b) Gram-negative cell wall

N-Acetylmuramic acid (NAM)

N-Acetylglucosamine (NAG)

L-Ala
D-Glu
L-Lys
D-Ala

Pentaglycine cross-link

L-Ala
D-Glu
L-Lys
D-Ala

Direct cross-link

Figure 7.29 **(a)** The cross-link in Gram-positive cell walls is a pentaglycine bridge. **(b)** In Gram-negative cell walls, the linkage between the tetrapeptides of adjacent carbohydrate chains in peptidoglycan involves a direct amide bond between the lysine side chain of one tetrapeptide and D-alanine of the other.

in this peptide is linked to the —COOH of D-alanine of an adjacent tetrapeptide. In Gram-negative cell walls, the lysine ε-amino group forms a *direct amide bond* with this D-alanine carboxyl (Figure 7.29). In Gram-positive cell walls, a **pentaglycine chain** bridges the lysine ε-amino group and the D-Ala carboxyl group.

Cell Walls of Gram-Negative Bacteria

In Gram-negative bacteria, the peptidoglycan wall is the rigid framework around which is built an elaborate membrane structure (Figure 7.30). The peptidoglycan layer encloses the *periplasmic space* and is attached to the outer membrane via a group of **hydrophobic proteins.** These proteins, each having 57 amino acid residues, are attached through amide linkages from the sidechains of C-terminal lysines of the proteins to diaminopimelic acid groups on the peptidoglycan. This linkage to the hydrophobic protein replaces one of the D-alanine residues in about 10% of the peptides of the peptidoglycan. On the other end of the hydrophobic protein, the N-terminal residue, a serine, makes a covalent bond to a lipid that is part of the outer membrane.

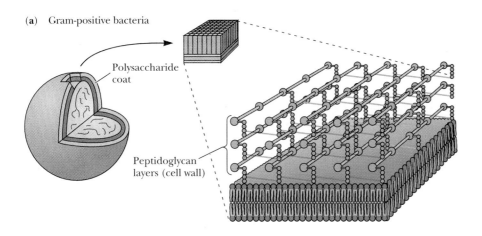

(a) Gram-positive bacteria

Polysaccharide coat

Peptidoglycan layers (cell wall)

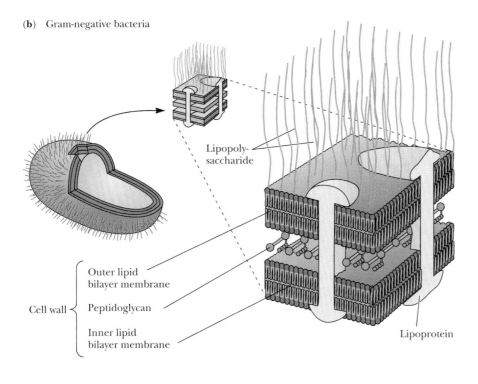

(b) Gram-negative bacteria

Lipopolysaccharide

Cell wall

Outer lipid bilayer membrane

Peptidoglycan

Inner lipid bilayer membrane

Lipoprotein

Figure 7.30 The structures of the cell wall and membrane(s) in Gram-positive **(a)** and Gram-negative **(b)** bacteria. The Gram-positive cell wall is thicker than that in Gram-negative bacteria, compensating for the absence of a second (outer) bilayer membrane.

Selectins, Rolling Leukocytes, and the Inflammatory Response

Human bodies are constantly exposed to a plethora of bacteria, viruses, and other inflammatory substances. To combat these infectious and toxic agents, the body has developed a carefully regulated inflammatory response system. Part of that response is the orderly migration of leukocytes to sites of inflammation. Leukocytes literally roll along the vascular wall and into the tissue site of inflammation. This rolling movement is mediated by reversible adhesive interactions between the leukocytes and the vascular surface.

These interactions involve adhesion proteins called **selectins,** which are found both on the rolling leukocytes and on the endothelial cells of the vascular walls. Selectins have a characteristic domain structure, consisting of an N-terminal extracellular lectin domain (LEC), a single epidermal growth factor (EGF) domain (E), a series of two to nine short consensus repeat (SCR) domains, a single transmembrane segment, and a short cytoplasmic domain. Lectin domains, first characterized in plants, bind carbohydrates with high affinity and specificity. Selectins of three types are known—**E-selectins, L-selectins,** and **P-selectins.** L-Selectin is found on the surfaces of leukocytes, including neutrophils and lymphocytes, and binds to carbohydrate ligands on endothelial cells. The

presence of L-selectin is a necessary component of leukocyte rolling. P-Selectin and E-selectin are located on the vascular endothelium and bind with carbohydrate ligands on leukocytes. Typical neutrophil cells possess 10,000 to 20,000 P-selectin binding sites. Selectins are expressed on the surfaces of their respective cells by exposure to inflammatory signal molecules, such as histamine, hydrogen peroxide, and bacterial endotoxins. P-Selectins, for example, are stored in intracellular granules and are transported to the cell membrane within seconds to minutes of exposure to a triggering agent.

Substantial evidence supports the hypothesis that selectin–carbohydrate ligand interactions modulate the rolling of leukocytes along the vascular wall. Studies with L-selectin–deficient and P-selectin–deficient leukocytes show that L-selectins mediate weaker adherence of the leukocyte to the vascular wall and promote faster rolling along the wall. P-Selectins conversely promote stronger adherence and slower rolling. Thus, leukocyte rolling velocity in the inflammatory response can be modulated by variable exposure of P-selectins and L-selectins at the surfaces of endothelial cells and leukocytes, respectively.

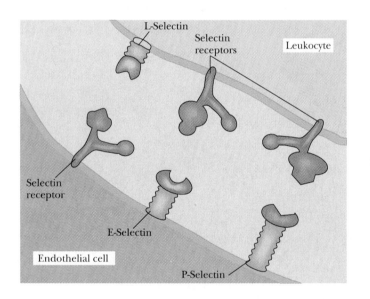

A diagram showing the interactions of selectins with their receptors.

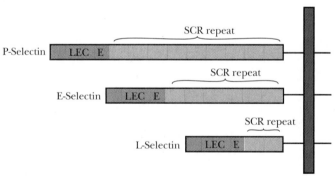

The selectin family of adhesion proteins.

As shown in Figure 7.30, the outer membrane of Gram-negative bacteria is coated with a highly complex **lipopolysaccharide,** which consists of a lipid group (anchored in the outer membrane) joined to a polysaccharide made up of long chains with many different and characteristic repeating structures (Figure 7.31). These many different unique units determine the antigenicity of the bacteria; that is, animal immune systems recognize them as foreign substances and raise antibodies against them. As a group, these **antigenic deter-**

Figure 7.31 Lipopolysaccharide (LPS) coats the outer membrane of Gram-negative bacteria. ▶ The lipid portion of the LPS is embedded in the outer membrane and is linked to a complex polysaccharide.

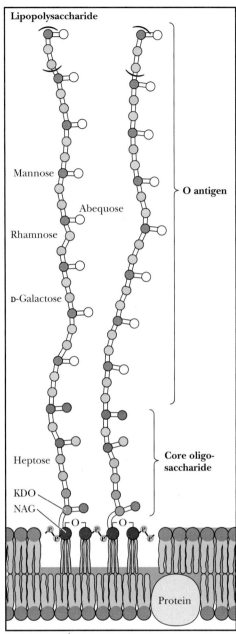

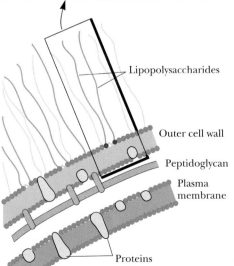

minants are called the **O antigens,** and there are thousands of different ones. The *Salmonella* bacteria alone have well over a thousand known O antigens that have been organized into 17 different groups. The great variation in these O antigen structures apparently plays a role in the recognition of one type of cell by another and in evasion of the host immune system.

Cell Walls of Gram-Positive Bacteria

In Gram-positive bacteria, the cell exterior is less complex than that of Gram-negative cells. Having no outer membrane, Gram-positive cells compensate with a thicker wall. Covalently attached to the peptidoglycan layer are **teichoic acids,** which often account for 50% of the dry weight of the cell wall (Figure 7.32). The teichoic acids are polymers of *ribitol phosphate* or *glycerol phosphate* linked by phosphodiester bonds. In these heteropolysaccharides, the free hydroxyl groups of the ribitol or glycerol are often substituted by glycosidically linked monosaccharides (often glucose or *N*-acetylglucosamine) or disaccharides. D-Alanine is sometimes found in ester linkage to the saccharides. Teichoic acids are not confined to the cell wall itself, and they may be present in the inner membranes of these bacteria. Many teichoic acids are antigenic, and they also serve as the receptors for bacteriophages in some cases.

Animal Cell Surface Polysaccharides

Compared to bacterial cells, which are identical within a given cell type (except for O antigen variations), animal cells display a wondrous diversity of structure, constitution, and function. Although each animal cell contains, in its genetic material, the instructions to replicate the entire organism, each differentiated animal cell carefully controls its composition and behavior within the organism. A great part of each cell's identity begins at the cell surface. This surface uniqueness is critical to each animal cell, since cells spend their entire life span in intimate contact with other cells and must therefore communicate with one another. That cells are able to pass information among themselves is evidenced by numerous experiments. For example, heart myocytes, when grown in culture (in glass dishes) establish *synchrony* when they make contact, so that they "beat" or contract in unison. If removed from the culture and separated, they lose their synchronous behavior, but if allowed to reestablish cell-to-cell contact, they spontaneously restore their synchronous contractions. Kidney cells grown in culture with liver cells seek out and make contact with other kidney cells and shun contact with liver cells.

As these and many other related phenomena show, it is clear that molecular structures on one cell are recognizing and responding to molecules on the adjacent cell or to molecules in the **extracellular matrix,** the complex "soup" of connective proteins and other molecules that exists outside of and among cells. Many of these interactions involve *glycoproteins* on the cell surface and proteoglycans in the extracellular matrix. The "information" held in these special carbohydrate-containing molecules is not encoded directly in the genes (as with proteins), but is determined instead by expression of the appropriate enzymes that assemble carbohydrate units in a characteristic way on these molecules. Also, by virtue of the several hydroxyl linkages that can be formed with each carbohydrate monomer, these structures represent a different form of

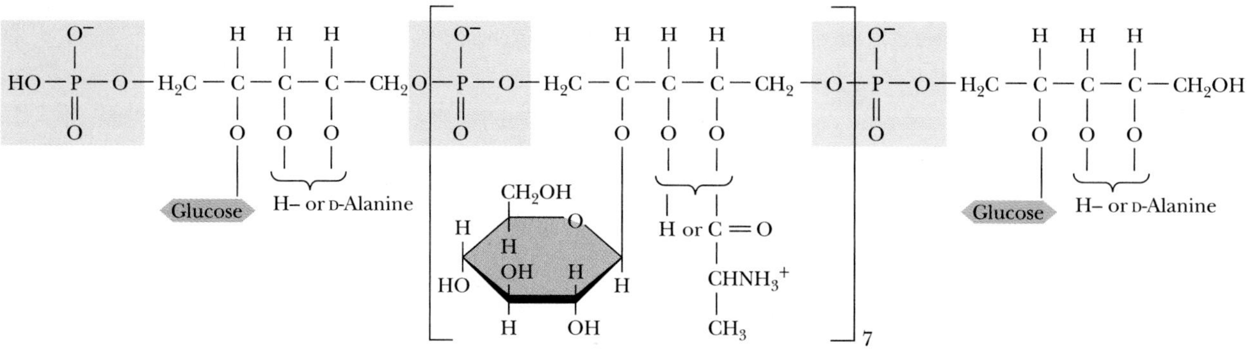

Ribitol teichoic acid from *Bacillus subtilis*

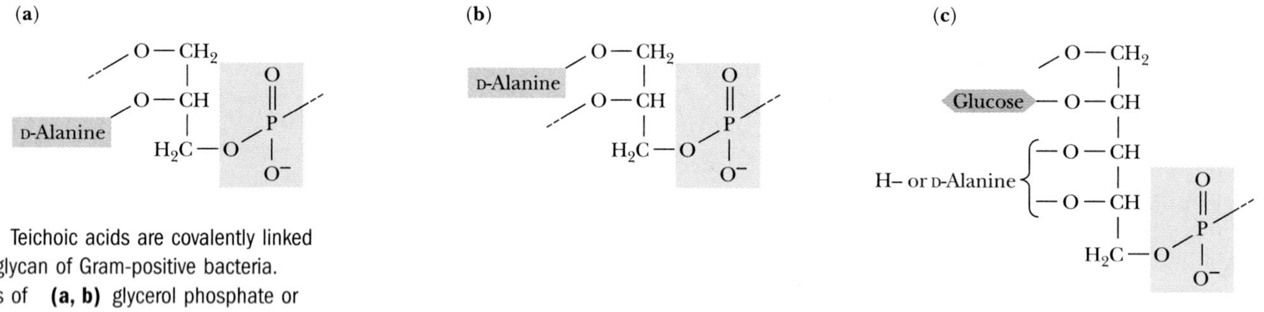

(a)

(b)

(c)

Figure 7.32 Teichoic acids are covalently linked to the peptidoglycan of Gram-positive bacteria. These polymers of **(a, b)** glycerol phosphate or **(c)** ribitol phosphate are linked by phosphodiester bonds.

information from proteins and nucleic acids, which are linear informational polymers. A few of these glycoproteins and their unique properties are described in the following sections.

7.5 Glycoproteins

Many proteins found in nature are **glycoproteins,** since they contain covalently linked oligo- and polysaccharide groups. The list of known glycoproteins includes structural proteins, enzymes, membrane receptors, transport proteins, and immunoglobulins, among others. In most cases, the precise function of the bound carbohydrate moiety is not understood.

Carbohydrate groups may be linked to polypeptide chains via the hydroxyl groups of serine, threonine, or hydroxylysine residues (in **O-linked saccharides**) (Figure 7.33a) or via the amide nitrogen of an asparagine residue (in **N-linked saccharides**) (Figure 7.33b). The carbohydrate residue linked to the protein in O-linked saccharides is usually an *N*-acetylgalactosamine, but mannose, galactose, and xylose residues linked to protein hydroxyls are also found (Figure 7.33a). Oligosaccharides O-linked to glycophorin (see Chapter 6) involve *N*-acetylgalactosamine linkages and are rich in sialic acid residues (see Figure 6.31). N-Linked saccharides always have a unique core structure composed of two *N*-acetylglucosamine residues linked to a branched mannose triad (Figure 7.33b and c). Many other sugar units may be linked to each of the mannose residues of this branched core.

O-Linked saccharides are often found in cell surface glycoproteins and in **mucins,** the large glycoproteins that coat and protect mucous membranes in the respiratory and gastrointestinal tracts in the body. Certain viral glycopro-

(a) O-linked saccharides

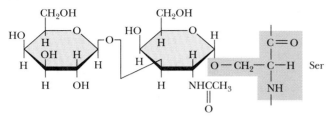

β-**Galactosyl–1,3–α-*N*-acetylgalactosyl-serine**

α-**Xylosyl-threonine**   α-**Mannosyl-serine**

(b) Core oligosaccharides in N-linked glycoproteins

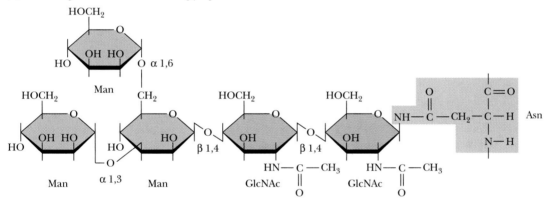

(c) N-linked glycoproteins

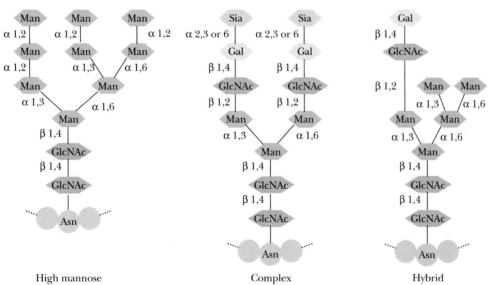

High mannose   Complex   Hybrid

Figure 7.33 The carbohydrate moieties of glycoproteins may be linked to the protein via **(a)** serine or threonine residues (in the O-linked saccharides) or **(b)** asparagine residues (in the N-linked saccharides). **(c)** N-Linked glycoproteins are of three types: high mannose, complex, and hybrid, the latter of which combines structures found in the high mannose and complex saccharides.

teins also contain O-linked sugars. O-Linked saccharides in glycoproteins are often found clustered in richly glycosylated domains of the polypeptide chain. Physical studies on mucins show that they adopt rigid, extended structures so that an individual mucin molecule ($M_r = 10^7$) may extend over a distance of 150 to 200 nm in solution. Inherent steric interactions between the sugar residues and the protein residues in these cluster regions cause the peptide core to fold into an extended and relatively rigid conformation. This interesting effect may be related to the function of O-linked saccharides in glycoproteins. It allows aggregates of mucin molecules to form extensive, intertwined networks, even at low concentrations. These viscous networks protect the mucosal surface of the respiratory and gastrointestinal tracts from harmful environmental agents.

There appear to be two structural motifs for membrane glycoproteins containing O-linked saccharides. Certain glycoproteins, such as **leukosialin,** are O-glycosylated throughout much or most of their extracellular domain (Figure 7.34). Leukosialin, like mucin, adopts a highly extended conformation, allowing it to project great distances above the membrane surface, perhaps protecting the cell from unwanted interactions with macromolecules or other cells. The second structural motif is exemplified by the **low density lipoprotein (LDL) receptor** and by the **decay accelerating factor (DAF).** These proteins contain a highly O-glycosylated stem region that separates the transmembrane domain from the globular, functional extracellular domain. The O-glycosylated stem serves to raise the functional domain of the protein far enough above the membrane surface to make it accessible to the extracellular macromolecules with which it interacts.

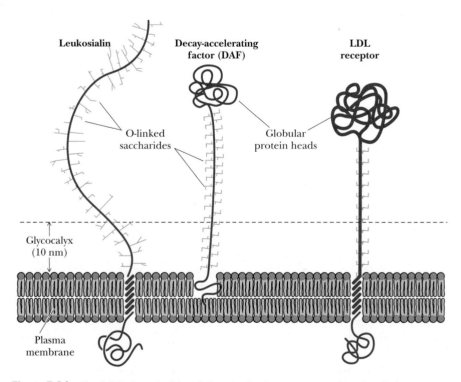

Figure 7.34 The O-linked saccharides of glycoproteins in many cases adopt extended conformations that serve to extend the functional domains of these proteins above the membrane surface. (*Adapted from Jentoft, N., 1990.* Trends in Biochemical Sciences *15:291–294.*)

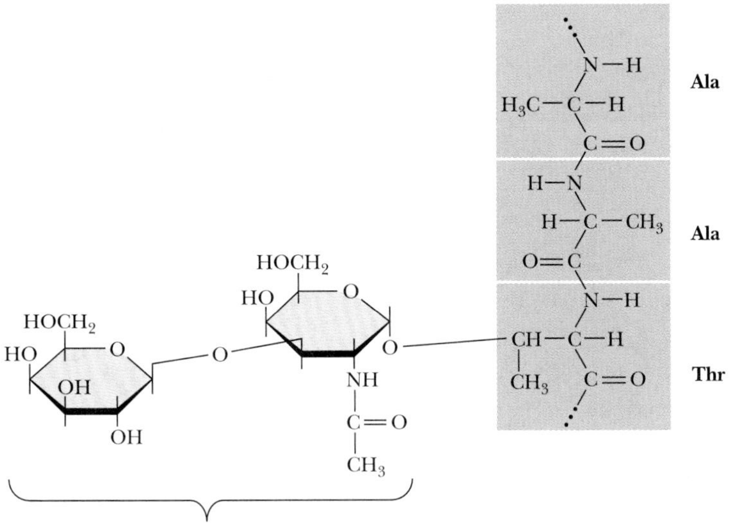

β-Galactosyl–1,3–α-*N*-acetylgalactosamine

Repeating unit of antifreeze glycoproteins

Figure 7.35 The structure of the repeating unit of antifreeze glycoproteins, a disaccharide consisting of β-galactosyl-(1 → 3)-α-*N*-acetylgalactosamine in glycosidic linkage to a threonine residue.

Antifreeze Glycoproteins

A unique family of O-linked glycoproteins permits fish to live in the icy sea water of the Arctic and Antarctic regions, where water temperature may reach as low as −1.9°C. **Antifreeze glycoproteins (AFGPs)** are found in the blood of nearly all Antarctic fish and at least five Arctic fish. These glycoproteins have the peptide structure

$$[\text{Ala-Ala-Thr}]_n\text{-Ala-Ala}$$

where n can be 4, 5, 6, 12, 17, 28, 35, 45, or 50. Each of the threonine residues is glycosylated with the disaccharide β-galactosyl-(1 → 3)-α-*N*-acetylgalactosamine (Figure 7.35). This glycoprotein adopts a **flexible rod** conformation with regions of threefold left-handed helix. The evidence suggests that antifreeze glycoproteins may inhibit the formation of ice in the fish by binding specifically to the growth sites of ice crystals, inhibiting further growth of the crystals.

N-Linked Oligosaccharides

N-Linked oligosaccharides are found in many different proteins, including immunoglobulins G and M, ribonuclease B, ovalbumin, and peptide hormones (Figure 7.36). Many different functions are known or suspected for *N*-glycosylation of proteins. Glycosylation can affect the physical and chemical properties of proteins, altering solubility, mass, and electrical charge. Carbohydrate moieties stabilize protein conformations and protect proteins against proteolysis. Eukaryotic organisms use posttranslational additions of N-linked oligosaccharides to direct selected proteins to various intracellular organelles. The step-by-step removal of monosaccharide residues from N-linked glycoproteins circulating in the blood targets these proteins for degradation by the organism.

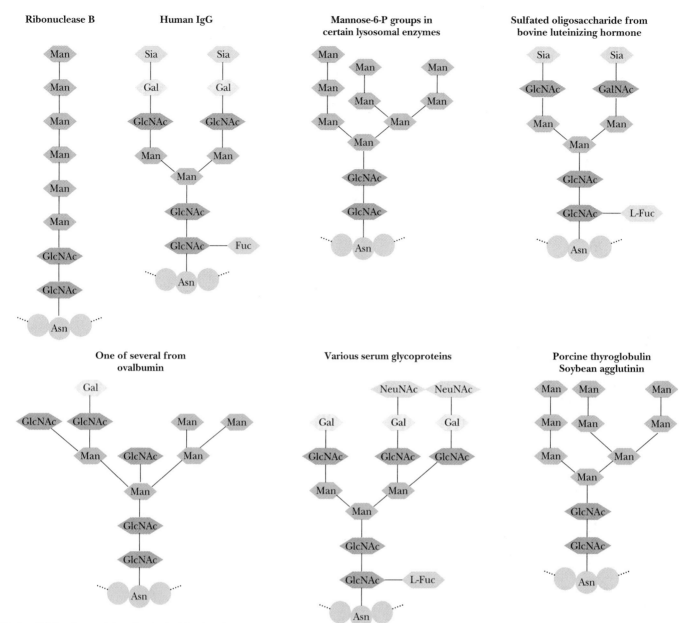

Figure 7.36 Some of the oligosaccharides found in N-linked glycoproteins.

Proteoglycans are a family of glycoproteins whose carbohydrate moieties are predominantly **glycosaminoglycans.** The structures of only a few proteoglycans are known, and even these few display considerable diversity (Figure 7.37). They range in size from **serglycin,** which has 104 amino acid residues (8.2 kD) to **versican,** which has 2409 residues (265 kD). Each of these proteoglycans contains one or two types of covalently linked glycosaminoglycans. In the known proteoglycans, the glycosaminoglycan units are O-linked to serine residues of Ser-Gly dipeptide sequences. Serglycin is named for an unusual central domain of 49 amino acids composed of alternating serine and glycine residues. The **cartilage matrix proteoglycan** contains 117 Ser-Gly pairs to which chondroitin sulfates attach. **Decorin,** a small proteoglycan secreted by fibroblasts and found in the extracellular matrix of connective tissues, contains only three Ser-Gly pairs, only one of which is normally glycosylated. In addition to glycosaminoglycan units, proteoglycans may also contain other N-linked and O-linked oligosaccharide groups.

Functions of Proteoglycans

Proteoglycans may be soluble and located in the extracellular matrix—as is the case for serglycin, versican, and the cartilage matrix proteoglycan—or they may be integral transmembrane proteins, such as **syndecan.** Both types of proteoglycan interact with a variety of other molecules through their glycosaminoglycan components and through specific receptor domains in the polypeptide itself. For example, syndecan (from the Greek *syndein,* meaning "to bind together") is a transmembrane proteoglycan that associates intracellularly with the actin cytoskeleton. Outside the cell, it interacts with **fibronectin,** an extracellular protein that binds to several cell surface proteins and to components of the extracellular matrix. The ability of syndecan to participate in multiple interactions with these target molecules allows them to act as a sort of "glue"

Figure 7.37 The known proteoglycans include a variety of structures. The carbohydrate groups of proteoglycans are predominantly glycosaminoglycans O-linked to serine residues. Proteoglycans include both soluble proteins and integral transmembrane proteins.

(e) Rat cartilage proteoglycan

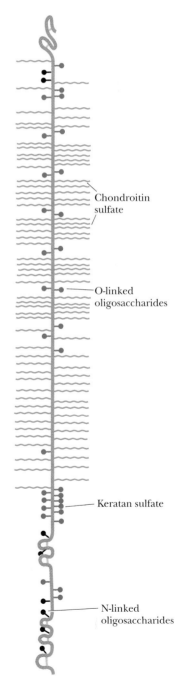

(a) Versican

NH$_3^+$

Hyaluronic acid binding domain (link-protein-like)

Chondroitin sulfate

Protein core

Epidermal growth factor–like domains

COO$^-$

(b) Serglycin

NH$_3^+$

Ser/Gly protein core

COO$^-$

Chondroitin sulfate

(c) Decorin

NH$_3^+$

Chondroitin/dermatan sulfate chain

COO$^-$

(d) Syndecan

Heparan sulfate

NH$_3^+$

Extracellular domain

Chondroitin sulfate

COO$^-$

Cytoplasmic domain

Transmembrane domain

Chondroitin sulfate

O-linked oligosaccharides

Keratan sulfate

N-linked oligosaccharides

Figure 7.38 Proteoglycans serve a variety of functions on the cytoplasmic and extracellular surfaces of the plasma membrane. Many of these functions appear to involve the binding of specific proteins to the glycosaminoglycan groups.

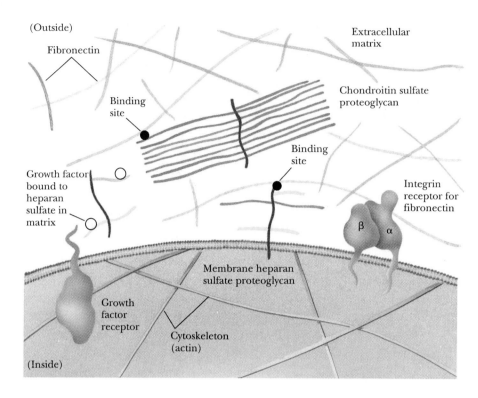

in the extracellular space, linking components of the extracellular matrix, facilitating the binding of cells to the matrix, and mediating the binding of growth factors and other soluble molecules to the matrix and to cell surfaces (Figure 7.38).

Many of the functions of proteoglycans involve the binding of specific proteins to the glycosaminoglycan groups of the proteoglycan. A particular pentasaccharide sequence in heparin, for example, binds tightly to antithrombin III (Figure 7.39), accounting for the anticoagulant properties of heparin. Other glycosaminoglycans interact much more weakly.

Proteoglycans May Modulate Cell Growth Processes

Several lines of evidence raise the possibility of modulation or regulation of cell growth processes by proteoglycans. First, heparin and heparan sulfate inhibit cell proliferation in a process involving internalization of the glycosaminoglycan moiety and its migration to the cell nucleus. Second, **fibroblast growth factor** binds tightly to heparin and other glycosaminoglycans, and the heparin–growth factor complex protects the growth factor from degradative enzymes, thus enhancing its activity. Binding of fibroblast growth factors by proteoglycans and glycosaminoglycans in the extracellular matrix creates a reservoir of growth factors for cells to use. Third, **transforming growth factor β** stimulates the synthesis and secretion of proteoglycans in certain cells. Fourth, several proteoglycan core proteins, including versican and **lymphocyte homing receptor,** have domains similar in sequence to **epidermal growth fac-**

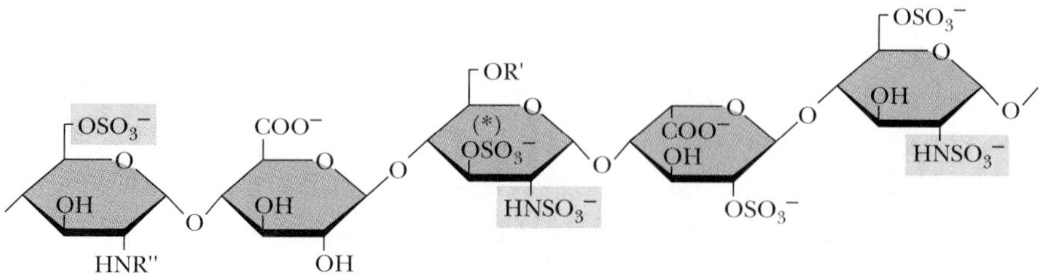

Figure 7.39 A portion of the structure of heparin, a carbohydrate having anticoagulant properties. It is used by blood banks to prevent the clotting of blood during donation and storage and also by physicians to prevent the formation of life-threatening blood clots in patients recovering from serious injury or surgery. This sulfated pentasaccharide sequence in heparin binds with high affinity to antithrombin III, accounting for its anticoagulant activity. The 3-O-sulfate marked by an asterisk is essential for high-affinity binding of heparin to antithrombin III.

tor and **complement regulatory factor.** These growth factor domains may interact specifically with growth factor receptors in the cell membrane in processes that are not yet understood.

Proteoglycans Make Cartilage Flexible and Resilient

Cartilage matrix proteoglycan is responsible for the flexibility and resilience of cartilage tissue in the body. In cartilage, long filaments of hyaluronic acid are studded or coated with proteoglycan molecules, as shown in Figure 7.40. The

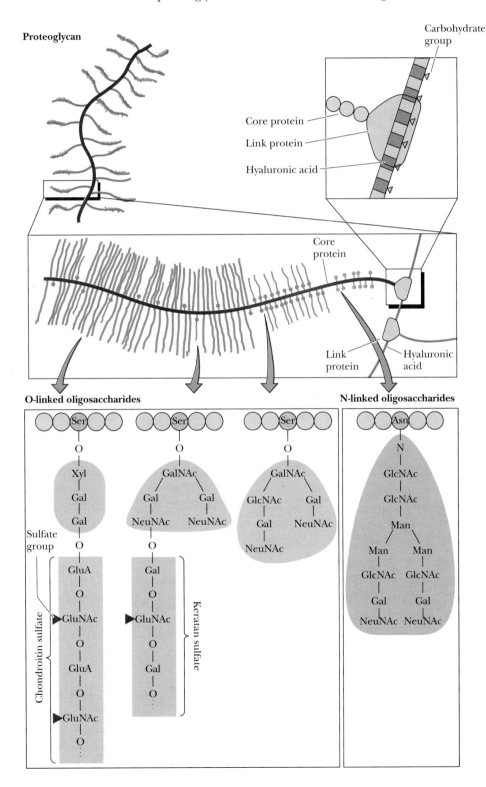

Figure 7.40 Hyaluronate (see Figure 7.27) forms the backbone of proteoglycan structures, such as those found in cartilage. The proteoglycan subunits consist of a core protein containing numerous O-linked and N-linked glycosaminoglycans. In cartilage, these highly hydrated proteoglycan structures are enmeshed in a network of collagen fibers. Release (and subsequent reabsorption) of water by these structures during compression accounts for the shock-absorbing qualities of cartilaginous tissue.

hyaluronate chains can be as long as 4 μm and can coordinate 100 or more proteoglycan units. Cartilage proteoglycan possesses a **hyaluronic acid binding domain** on the NH$_2$-terminal portion of the polypeptide, which binds to hyaluronate with the assistance of a **link protein.** The proteoglycan–hyaluronate aggregates can have molecular weights of 2 million or more.

The proteoglycan–hyaluronate aggregates are highly hydrated by virtue of strong interactions between water molecules and the polyanionic complex. When cartilage is compressed (such as when joints absorb the impact of walking or running), water is briefly squeezed out of the cartilage tissue and then reabsorbed when the stress is diminished. This reversible hydration gives cartilage its flexible, shock-absorbing qualities and cushions the joints during physical activities that might otherwise injure the involved tissues.

PROBLEMS

1. Draw Haworth structures for the two possible isomers of D-altrose (see Figure 7.2) and D-psicose (see Figure 7.3).

2. Give the systematic name for stachyose (see Figure 7.19).

3. Trehalose, a disaccharide produced in fungi, has the following structure:

   a. What is the systematic name for this disaccharide?
   b. Is trehalose a reducing sugar? Explain.

4. Draw Fischer projection structures for L-sorbose (D-sorbose is shown in Figure 7.3).

5. α-D-Glucose has a specific rotation, $[\alpha]_D^{20}$, of $+112.2°$, whereas β-D-glucose has a specific rotation of $+18.7°$. What is the composition of a mixture of α-D- and β-D-glucose, which has a specific rotation of $83.0°$?

6. A 0.2-g sample of amylopectin was analyzed to determine the fraction of the total glucose residues that are branch points in the structure. The sample was exhaustively methylated and then digested, yielding 50 μmol of 2,3-dimethylglucose and 0.4 μmol of 1, 2, 3, 6-tetramethyl glucose.
   a. What fraction of the total residues are branch points?
   b. How many reducing ends does this amylopectin have?

FURTHER READING

Aspinall, G. O., 1982. *The Polysaccharides*, vols. 1 and 2. New York: Academic Press.

Bernfield, M., et al., 1999. Functions of cell surface heparan sulfate proteoglycans. *Annual Review of Biochemistry* **68:**728–777.

Collins, P. M., 1987. *Carbohydrates*. London: Chapman and Hall.

Davison, E. A., 1967. *Carbohydrate Chemistry*. New York: Holt, Rinehart and Winston.

Feeney, R. E., Burcham, T. S., and Yeh, Y., 1986. Antifreeze glycoproteins from polar fish blood. *Annual Review of Biophysical Chemistry* **15:**59–78.

Hart, G. W., 1997. Dynamic O-linked glycosylation of nuclear and cytoskeletal proteins. *Annual Review of Biochemistry* **66:**315–335.

Jentoft, N., 1990. Why are proteins O-glycosylated? *Trends in Biochemical Sciences* **15:**291–294.

Kjellen, L., and Lindahl, U., 1991. Proteoglycans: structures and interactions. *Annual Review of Biochemistry* **60:**443–475.

Lennarz, W. J., 1980. *The Biochemistry of Glycoproteins and Proteoglycans*. New York: Plenum Press.

Lodish, H. F., 1991. Recognition of complex oligosaccharides by the multisubunit asialoglycoprotein receptor. *Trends in Biochemical Sciences* **16:**374–377.

McNeil, M., Darvill, A. G., Fry, S. C., and Albersheim, P., 1984. Structure and function of the primary cell walls of plants. *Annual Review of Biochemistry* **53:**625–664.

Pigman, W., and Horton, D. 1972. *The Carbohydrates*. New York: Academic Press.

Rademacher, T. W., Parekh, R. B., and Dwek, R. A., 1988. Glycobiology. *Annual Review of Biochemistry* **57:**785–838.

Ruoslahti, E., 1989. Proteoglycans in cell regulation. *Journal of Biological Chemistry* **264:**13369–13372.

Sharon, N., 1980. Carbohydrates. *Scientific American* **243:**90–102.

Sharon, N., 1984. Glycoproteins. *Trends in Biochemical Sciences* **9:**198–202.

Nucleotides and Nucleic Acids

Francis Crick and James Watson point out features of their model for the structure of DNA. (© Barrington Brown/Science Source/Photo Researchers, Inc.)

"We have discovered the secret of life!"

Proclamation by Francis H. C. Crick to patrons of the Eagle, a pub in Cambridge, England (1953)

Outline

Nucleotides and **nucleic acids** are biological molecules that possess heterocyclic nitrogenous bases as principal components of their structure. The biochemical roles of nucleotides are numerous; they participate as essential intermediates in virtually all aspects of cellular metabolism. Serving an even more central biological purpose are the nucleic acids, the elements of heredity and the agents of genetic information transfer. Just as proteins are linear polymers of amino acids, nucleic acids are linear polymers of nucleotides. Like the letters in this sentence, the orderly sequence of nucleotide residues in a nucleic acid can encode information. The two basic kinds of nucleic acids are **deoxyribonucleic acid (DNA)** and **ribonucleic acid (RNA).** Complete hydrolysis of nucleic acids liberates nitrogenous bases, a five-carbon sugar, and phosphoric acid in equal amounts. The five-carbon sugar in DNA is 2-deoxyribose; in RNA, it is ribose. DNA is the repository of genetic information in cells, while RNA serves in the transcription and translation of this information (Figure 8.1).

Figure 8.1 The fundamental process of information transfer in cells. Information encoded in the nucleotide sequence of DNA is transcribed through synthesis of an RNA molecule whose sequence is dictated by the DNA sequence. As the sequence of this RNA is read (as groups of three consecutive nucleotides) by the protein synthesis machinery, it is translated into the sequence of amino acids in a protein. This information transfer system is encapsulated in the dogma: DNA → RNA → protein.

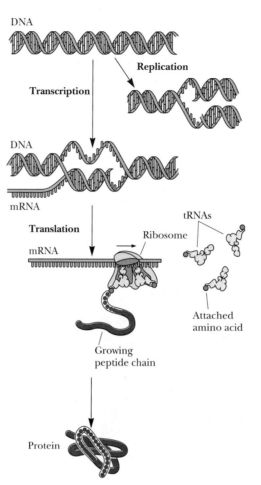

Replication
DNA replication yields two DNA molecules identical to the original one, ensuring transmission of genetic information to daughter cells with exceptional fidelity.

Transcription
The sequence of bases in DNA is recorded as a sequence of complementary bases in a single-stranded mRNA molecule.

Translation
Three-base codons on the mRNA corresponding to specific amino acids direct the sequence of building a protein. These codons are recognized by tRNAs (transfer RNAs) carrying the appropriate amino acids. Ribosomes are the "machinery" for protein synthesis.

An interesting exception to this rule is that some viruses have their genetic information stored as RNA.

This chapter describes the chemistry of nucleotides and the major classes of nucleic acids, the methods for determining nucleic acid primary structure (nucleic acid sequencing), and the major features of higher-order structure in nucleic acids. Chapter 9 introduces the **molecular biology of recombinant DNA:** the construction and uses of novel DNA molecules assembled by combining segments from other DNA molecules.

8.1 Nitrogenous Bases

The bases of nucleotides and nucleic acids are derivatives of either **pyrimidine** or **purine**. Pyrimidines are six-membered heterocyclic aromatic rings containing two nitrogen atoms (Figure 8.2a). The atoms are numbered in a clockwise fashion, as shown in the figure. The purine ring structure is represented by the combination of a pyrimidine ring with a five-membered imidazole ring to yield a fused ring system (Figure 8.2b). The nine atoms in this system are numbered according to the convention shown.

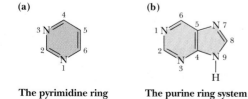

(a) The pyrimidine ring

(b) The purine ring system

◄ **Figure 8.2** **(a)** The pyrimidine ring system; by convention, atoms are numbered as indicated. **(b)** The purine ring system, atoms numbered as shown.

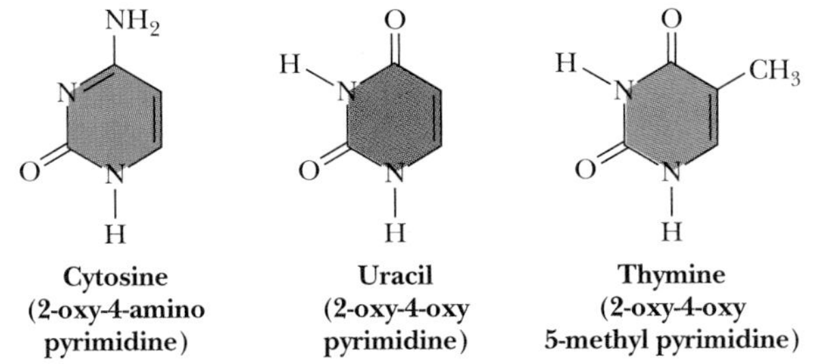

Cytosine
(2-oxy-4-amino
pyrimidine)

Uracil
(2-oxy-4-oxy
pyrimidine)

Thymine
(2-oxy-4-oxy
5-methyl pyrimidine)

Figure 8.3 The common pyrimidine bases—cytosine, uracil, and thymine—in the tautomeric forms predominant at pH 7.

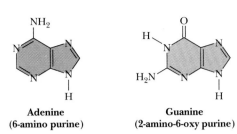

Adenine
(6-amino purine)

Guanine
(2-amino-6-oxy purine)

Figure 8.4 The common purine bases—adenine and guanine—in the tautomeric forms predominant at pH 7.

See *Interactive Biochemistry CD-ROM and Workbook,* pages 121–124

Common Pyrimidines and Purines

The common naturally occurring pyrimidines are **cytosine, uracil,** and **thymine** (5-methyluracil) (Figure 8.3). Cytosine and thymine are the pyrimidines typically found in DNA, whereas cytosine and uracil are common in RNA. To view this generality another way, the uracil component of DNA occurs as the 5-methyl variety, thymine. Various pyrimidine derivatives, such as dihydrouracil, are present as minor constituents in certain RNA molecules.

Adenine (6-amino purine) and **guanine** (2-amino-6-oxy purine), the two common purines, are found in both DNA and RNA (Figure 8.4). Other naturally occurring purines include **hypoxanthine, xanthine,** and **uric acid** (Figure 8.5). Hypoxanthine and xanthine are found only rarely as constituents of nucleic acids. Uric acid, the most oxidized state for a purine, is never found in nucleic acids.

Properties of Pyrimidines and Purines

Resonance possibilities within the pyrimidine and purine ring systems and the electron-rich nature of their —OH and —NH₂ substituents endow them with the capacity to undergo **keto–enol tautomeric shifts.** That is, pyrimidines and purines exist as tautomeric pairs, as shown in Figure 8.6 for uracil. The keto tautomer is called a **lactam,** whereas the enol form is a **lactim.** The lactam form is strongly favored at neutral pH. Similarly, tautomeric forms can be represented for purines, as given for guanine in Figure 8.7. Hydrogen bonding between purine and pyrimidine bases is fundamental to the biological functions of nucleic acids, as in the formation of the double helix structure of DNA (see Section 8.9). The important functional groups participating in H-bond formation are the amino groups of cytosine, adenine, and guanine; the ring nitrogens at position 3 of pyrimidines and position 1 of purines; and the strongly electronegative oxygen atoms attached at position 4 of uracil and thymine, position 2 of cytosine, and position 6 of guanine (see Figure 8.19).

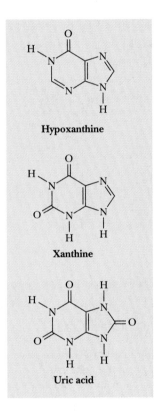

Hypoxanthine

Xanthine

Uric acid

Figure 8.5 Other naturally occurring purine derivatives—hypoxanthine, xanthine, and uric acid.

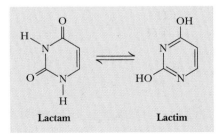

Lactam

Lactim

Figure 8.6 The keto–enol tautomerism of uracil.

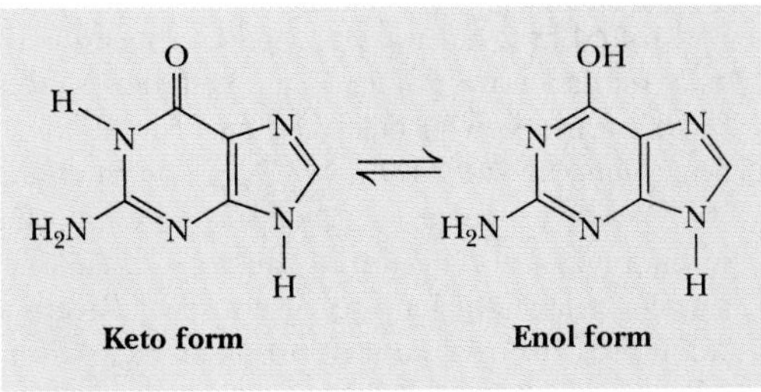

Keto form

Enol form

Figure 8.7 The tautomerism of the purine guanine.

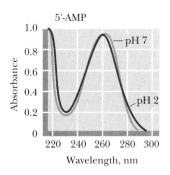

Figure 8.8 The UV absorption spectrum of adenine, a representative base in nucleic acids (in its nucleotide form, AMP).

Another property of pyrimidines and purines is their strong absorbance of ultraviolet (UV) light, a consequence of their aromatic heterocyclic ring structures. Figure 8.8 shows the characteristic absorption spectrum of a representative base, namely, adenine in its nucleotide form: AMP (see Section 8.4). This property is particularly useful in quantitative and qualitative analysis of nucleotides and nucleic acids.

8.2 The Pentoses of Nucleotides and Nucleic Acids

Five-carbon sugars are called **pentoses** (see Chapter 7). RNA contains the pentose D-ribose, while 2-deoxy-D-ribose is found in DNA. In both instances, the pentose is in the five-membered **furanose** ring form: D-ribofuranose for RNA and 2-deoxy-D-ribofuranose for DNA (Figure 8.9). When these ribofuranoses are found in nucleotides, their atoms are numbered as $1'$, $2'$, $3'$, and so on to distinguish them from the ring atoms of the nitrogenous bases. As we shall see, the seemingly minor difference of a hydroxyl group at the $2'$-position has far-reaching effects on the secondary structures available to RNA and DNA, as well as on their relative susceptibilities to chemical and enzymatic hydrolysis.

8.3 Nucleosides: Joining a Nitrogenous Base to a Sugar

Nucleosides are compounds formed when a nitrogenous base is linked to a sugar via a **glycosidic bond** (Figure 8.10). Glycosidic bonds by definition involve the carbonyl carbon atom of the sugar, which in cyclic structures is joined to

HUMAN BIOCHEMISTRY

Adenosine: A Nucleoside with Physiological Activity

For the most part, nucleosides have no biological role other than to serve as component parts of nucleotides. Adenosine is an exception. In mammals, adenosine functions as an **autocoid,** or "local hormone." This nucleoside circulates in the bloodstream, acting locally on specific cells to influence such diverse physiological phenomena as blood vessel dilation, smooth muscle contraction, neuronal discharge, neurotransmitter release, and metabolism of fat. For example, when muscles work hard, they release adenosine, causing the surrounding blood vessels to dilate, which in turn increases the flow of blood and its delivery of O_2 and nutrients to the muscles. In a different autocoid role, adenosine acts in regulating the heartbeat. The natural rhythm of the heart is controlled by a natural pacemaker, the sinoatrial node, that cyclically sends a wave of electrical excitation to the heart muscles. By blocking the flow of electrical current, adenosine slows the heart rate. *Supraventricular tachycardia* is a heart condition characterized by a rapid heartbeat. Intravenous injection of adenosine causes a momentary interruption of the rapid cycle of contraction and restores a normal heart rate. Adenosine is licensed and marketed as Adenocard to treat supraventricular tachycardia.

In addition, adenosine is implicated in sleep regulation. During periods of extended wakefulness, extracellular adenosine levels rise as a result of metabolic activity in the brain, and this increase promotes sleepiness. During sleep, adenosine levels fall. Caffeine promotes wakefulness by blocking the interaction of extracellular adenosine with its neuronal receptors.*

Caffeine

*Porrka-Heiskanan, T., et al., 1997. Adenosine: A mediator of the sleep-inducing effects of prolonged wakefulness. *Science* **276:**1265–1268.

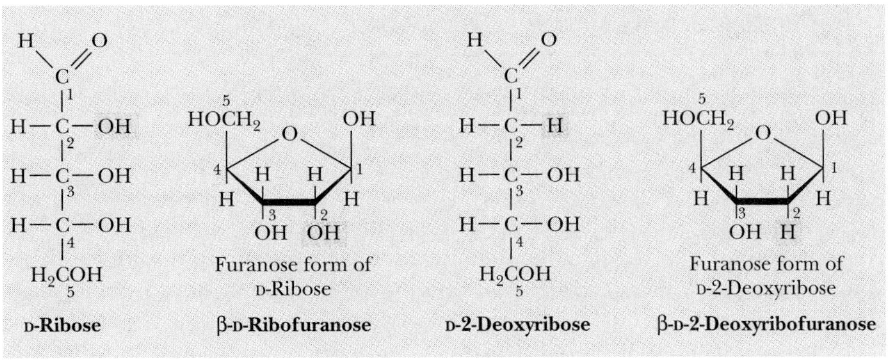

Figure 8.9 Furanose structures—ribose and deoxyribose.

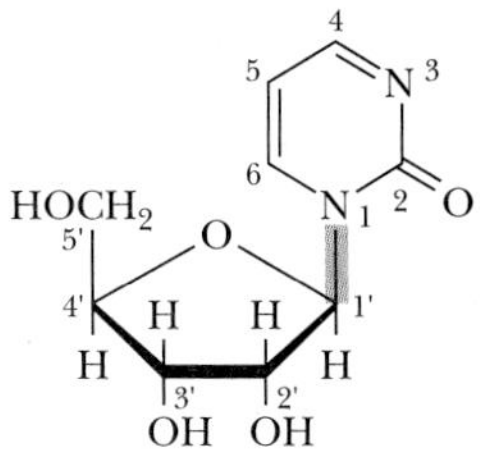

β-N$_1$-glycosidic
bond in pyrimidine
ribonucleosides

the ring O atom (see Chapter 7). Such carbon atoms are called **anomeric.** In nucleosides, the bond is an *N*-glycoside, since it connects the anomeric C-1' to N-1 of a pyrimidine or to N-9 of a purine. Glycosidic bonds can be either α or β, depending on their orientation relative to the anomeric C atom. Glycosidic bonds in nucleosides and nucleotides are always of the β-configuration, as represented in Figure 8.10. Nucleosides are named by adding the ending *-idine* to the root name of a pyrimidine or *-osine* to the root name of a purine. The common nucleosides are thus **cytidine, uridine, thymidine, adenosine,** and **guanosine.** The structures shown in Figure 8.11 are **ribonucleosides. Deoxyribonucleosides,** in contrast, lack a 2'-OH group on the pentose. The nucleoside formed by hypoxanthine and ribose is **inosine.** Nucleosides are much more water-soluble than the free bases because of the hydrophilicity of the sugar moiety.

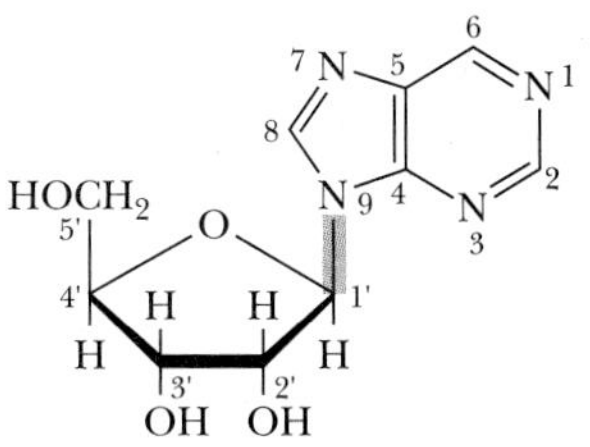

β-N$_9$-glycosidic
bond in purine
ribonucleosides

Figure 8.10 β-Glycosidic bonds link nitrogenous bases and sugars to form nucleosides.

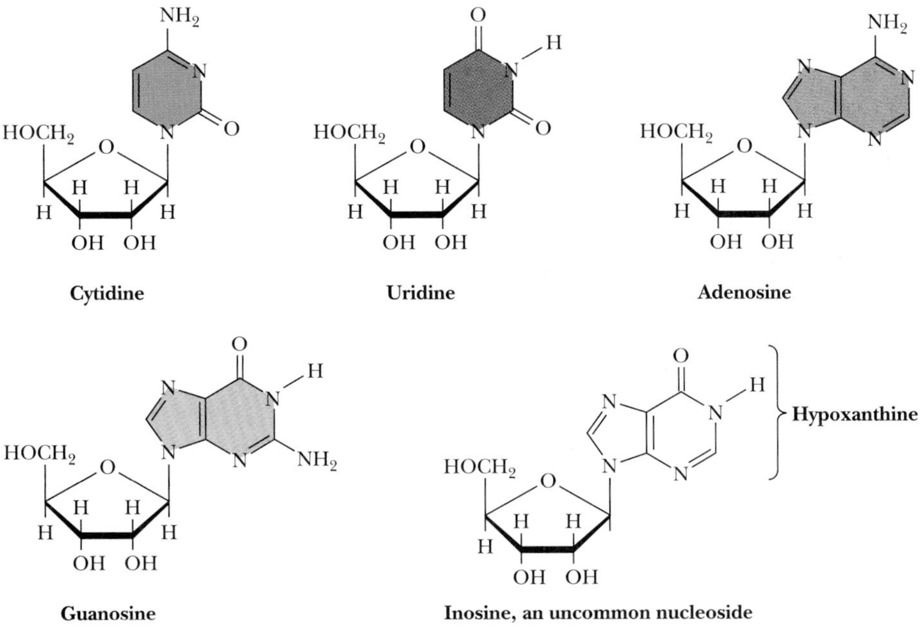

Cytidine

Uridine

Adenosine

Guanosine

Inosine, an uncommon nucleoside

Hypoxanthine

Figure 8.11 The common ribonucleosides—cytidine, uridine, adenosine, and guanosine—and inosine, an uncommon purine ribonucleotide.

 See pages 121–124

8.4 Nucleotides Are Nucleoside Phosphates

A **nucleotide** results when phosphoric acid is esterified to a sugar — OH group of a nucleoside. The nucleoside ribose ring has three — OH groups available for esterification, at C-2′, C-3′, and C-5′ (although 2′-deoxyribose has only two). The vast majority of monomeric nucleotides in the cell are **ribonucleotides** having 5′-phosphate groups. Figure 8.12 shows the structures of the common four ribonucleotides, whose formal names are **adenosine 5′-monophosphate, guanosine 5′-monophosphate, cytidine 5′-monophosphate,** and **uridine 5′-monophosphate.** These compounds are more often referred to by their abbreviations: **5′-AMP, 5′-GMP, 5′-CMP,** and **5′-UMP,** or even more simply as **AMP, GMP, CMP,** and **UMP.** Nucleoside 3′-phosphates and nucleoside 2′-phosphates (3′-NMP and 2′-NMP, where N is a generic designation for "nucleoside") do not occur naturally, but are biochemically important as products of polynucleotide or nucleic acid hydrolysis. Because the pK_a value for the first dissociation of a proton from the phosphoric acid moiety is 1.0 or less, the nucleotides have acidic properties. This acidity is implicit in the other names by which these substances are known: **adenylic acid, guanylic acid, cytidylic acid,** and **uridylic acid.** (The pK_a value for the second dissociation, pK_2, is about 6.0, so at neutral pH or above, the net charge on a nucleoside monophosphate is −2.) Nucleic acids, which are polymers of nucleoside monophosphates, derive their name from the acidity of these phosphate groups.

Cyclic Nucleotides

See pages 121–124

Nucleoside monophosphates in which the phosphoric acid is esterified to *two* of the available ribose hydroxyl groups (Figure 8.13) are found in all cells. Forming two such ester linkages with one phosphate results in a cyclic structure.

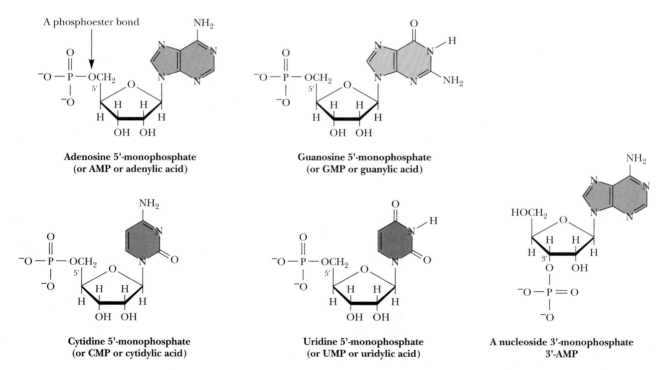

Adenosine 5'-monophosphate
(or **AMP** or adenylic acid)

Guanosine 5'-monophosphate
(or **GMP** or guanylic acid)

Cytidine 5'-monophosphate
(or **CMP** or cytidylic acid)

Uridine 5'-monophosphate
(or **UMP** or uridylic acid)

A nucleoside 3'-monophosphate
3'-**AMP**

Figure 8.12   Structures of the four common ribonucleotides—AMP, GMP, CMP, and UMP—together with their two sets of full names: for example, adenosine 5′-monophosphate and adenylic acid.

3', 5'-cyclic AMP, often abbreviated **cAMP,** and its guanine analog **3', 5'-cyclic GMP,** or **cGMP,** are important regulators of cellular metabolism (see Part 3: Metabolism and Its Regulation).

Nucleoside Diphosphates and Triphosphates

Additional phosphate groups can be linked to the phosphoryl group of a nucleotide through the formation of phosphoric anhydride linkages, as shown in Figure 8.14. Addition of a second phosphate to AMP creates **adenosine 5'-diphosphate,** or **ADP,** and adding a third yields **adenosine 5'-triphosphate,** or **ATP.** The respective phosphate groups are designated by the Greek letters α, β, and γ, starting with the α-phosphate as the one linked directly to the pentose. The abbreviations **GTP, CTP,** and **UTP** represent the other major nucleoside 5'-triphosphates. Like the nucleoside 5'-monophosphates, the nucleoside 5'-diphosphates and 5'-triphosphates all occur in the free state in the cell, as do their deoxyribonucleoside phosphate counterparts, represented as dAMP, dADP, and dATP; dGMP, dGDP, and dGTP; dCMP, dCDP, and dCTP; dUMP, dUDP, and dUTP; and dTMP, dTDP, and dTTP.

NDPs and NTPs Are Polyprotic Acids

Nucleoside 5'-diphosphates (NDPs) and **nucleoside 5'-triphosphates (NTPs)** are relatively strong *polyprotic acids,* since they dissociate three and four protons, respectively, from their phosphoric acid groups. The resulting phosphate anions on NDPs and NTPs form stable complexes with divalent cations such as Mg^{2+} and Ca^{2+}. Since Mg^{2+} is present at high concentrations (40 mM) intracellularly, NDPs and NTPs occur primarily as Mg^{2+} complexes in the cell. The phosphoric anhydride linkages in NDPs and NTPs are readily hydrolyzed by acid, liberating inorganic phosphate (often symbolized as P_i) and the corresponding NMP.

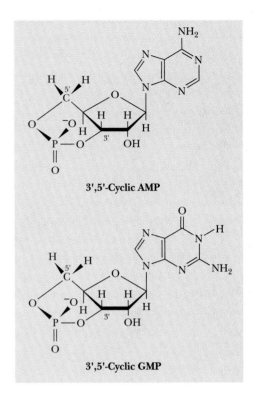

Figure 8.13 Structures of the cyclic nucleotides cAMP and cGMP.

See pages 121–124

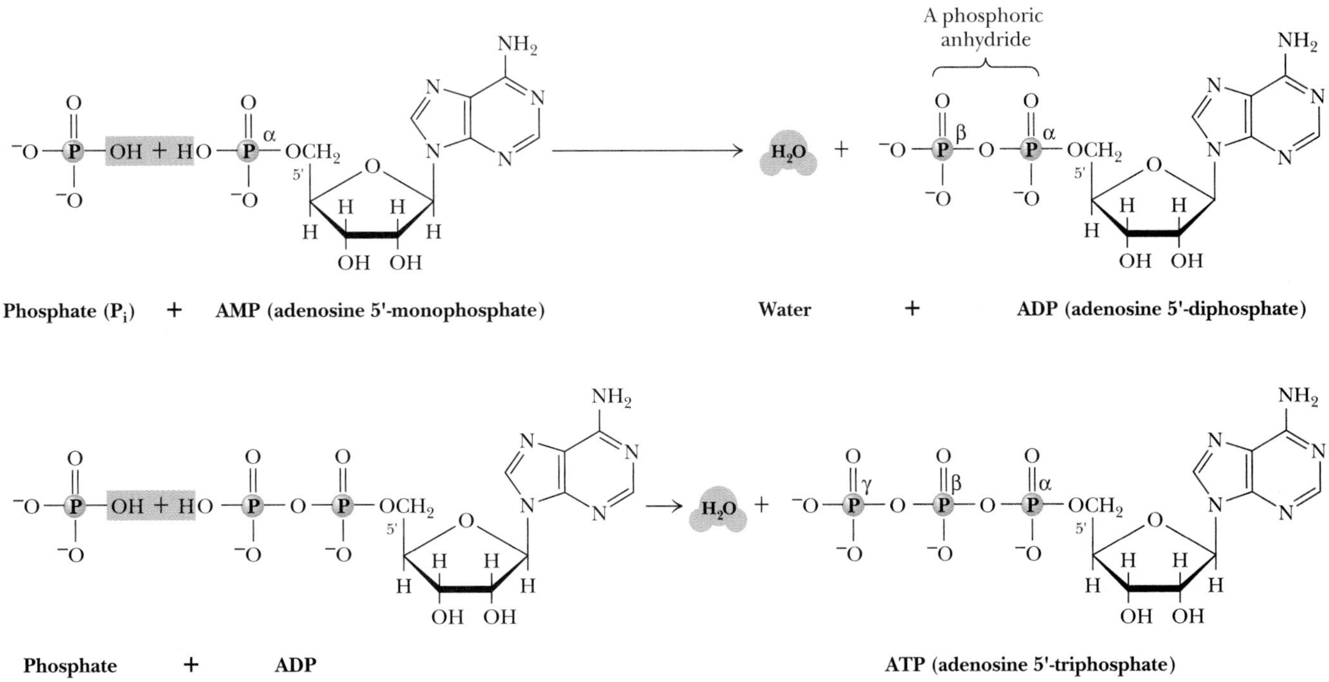

Figure 8.14 Formation of ADP and ATP by the successive addition of phosphate groups via phosphoric anhydride linkages. Note the removal of equivalents of H_2O in these dehydration synthesis reactions.

Phosphate (P_i) + AMP (adenosine 5'-monophosphate) → Water + ADP (adenosine 5'-diphosphate)

Phosphate + ADP → ATP (adenosine 5'-triphosphate)

Nucleoside 5'-Triphosphates Are Carriers of Chemical Energy

Nucleoside 5'-triphosphates are indispensable agents in metabolism because the phosphoric anhydride bonds are a prime source of chemical energy for biological work. Virtually all of the biochemical reactions of nucleotides involve either *phosphate* or *pyrophosphate group transfer:* the release of a phosphoryl group from an NTP to give an NDP, the release of a pyrophosphoryl group to give an NMP unit, or the acceptance of a phosphoryl group by an NMP or an NDP to give an NDP or an NTP (Figure 8.15). Interestingly, the pentose and the base are *not* directly involved in this chemistry. However, a "division of labor" directs ATP, the so-called "energy currency of the cell," to serve as the primary nucleotide in central pathways of energy metabolism, while GTP is used to drive protein synthesis. CTP acts in phospholipid synthesis, and UTP in complex carbohydrate biosynthesis. Thus, the various nucleotides are channeled in appropriate metabolic directions through specific recognition of the base of the nucleotide. That is, the bases of nucleotides serve solely as **information symbols,** never participating directly in the covalent bond chemistry that goes on. This role as information symbol extends to nucleotide polymers, the nucleic acids, where the bases serve as the recognition units for the code of genetic information.

8.5 Nucleic Acids Are Polynucleotides

Nucleic acids are linear polymers of nucleotides linked 3' to 5' by **phosphodiester bridges** (Figure 8.16). They are formed as 5'-nucleoside monophosphates are successively added to the 3'-OH group of the preceding nucleotide, a process that gives the polymer a directional sense. Polymers of ribonucleotides are named **ribonucleic acid,** or **RNA.** Deoxyribonucleotide polymers are called **deoxyribonucleic acid,** or **DNA.** Since C-1' and C-4' in deoxyribonucleotides

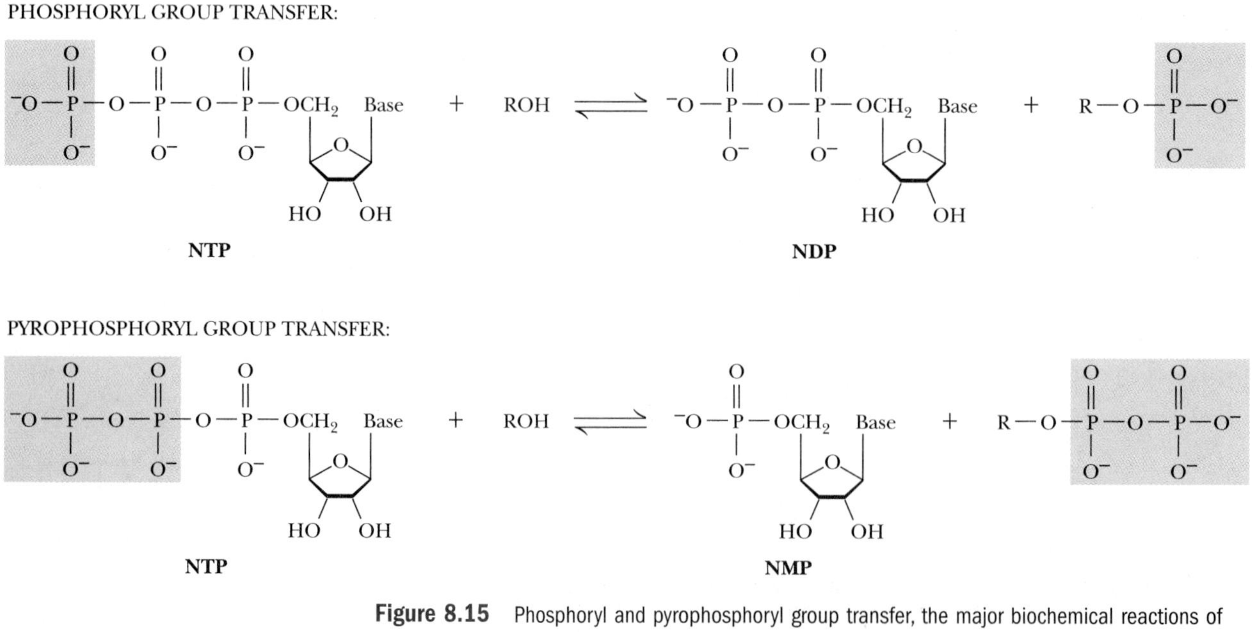

Figure 8.15 Phosphoryl and pyrophosphoryl group transfer, the major biochemical reactions of nucleotides.

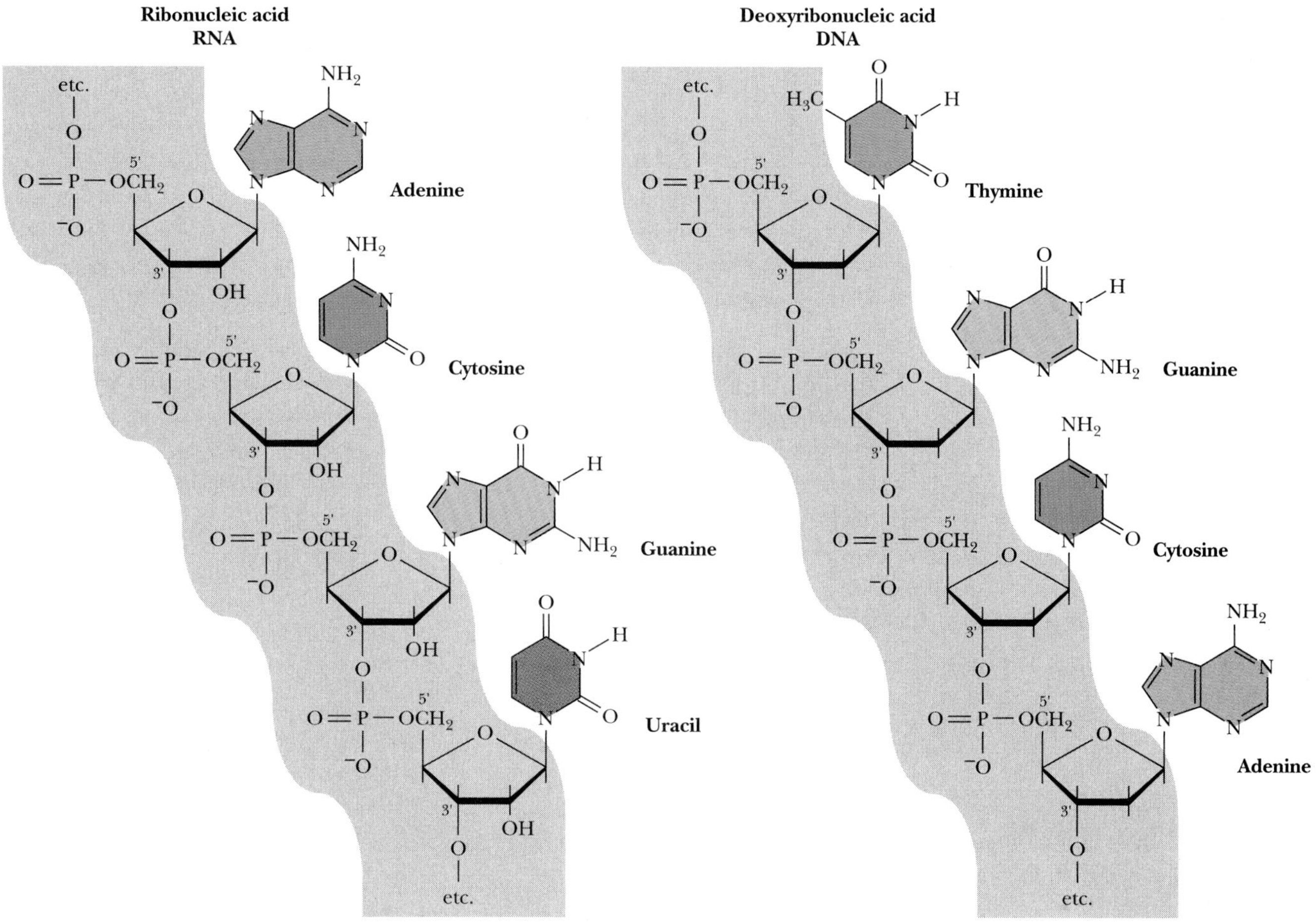

Figure 8.16 shows structures for RNA and DNA.

Ribonucleic acid
RNA

Deoxyribonucleic acid
DNA

Figure 8.16 3′-5′ phosphodiester bridges link nucleotides together to form polynucleotide chains.

are involved in furanose ring formation and since there is no 2′-OH, only the 3′- and 5′-hydroxyl groups are available for the polymerization process. In the case of DNA, a polynucleotide chain may contain hundreds of millions of nucleotide units. Any structural representation of such molecules would be cumbersome at best, even for a short oligonucleotide stretch.

Shorthand Notations for Polynucleotide Structures

Several conventions have been adopted to convey the sense of polynucleotide structures. A repetitious uniformity exists in the covalent backbone of polynucleotides, in which the chain can be visualized as running from 5′ to 3′ along the atoms of one furanose and then across the phosphodiester bridge to the furanose of the next nucleotide in line. This backbone can be symbolized using a vertical line to represent the furanose and a slash to represent the phosphodiester link, as shown in Figure 8.17. The diagonal slash runs from the middle of a furanose line to the bottom of an adjacent one to indicate the 3′-(middle) to 5′-(bottom) carbons of neighboring furanoses joined by the phosphodiester bridge. The base attached to each furanose is indicated above it by a one-letter designation: A, C, G, or U (or T). *The convention in all notations of nucleic acid structure is to read the polynucleotide chain from the 5′-end of the polymer to the 3′-end.* Note that this reading direction actually passes through each phosphodiester from 3′ to 5′.

Figure 8.17 Furanoses are represented by vertical lines; phosphodiesters are represented by diagonal slashes in this shorthand notation for nucleic acid structures.

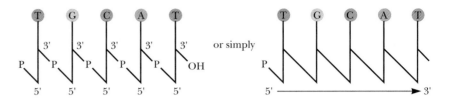

Base Sequence

The only significant variation in the chemical structure of nucleic acids is the nature of the base at each nucleotide position. These bases are not part of the sugar–phosphate backbone but instead serve as distinctive side chains, much like the R groups of amino acids along a polypeptide backbone. They give the polymer its unique identity. A simple notation system for these structures is merely to list the order of bases in the polynucleotide using single capital letters—A, G, C, and U (or T). To distinguish between RNA and DNA sequences, the latter may be preceded by a lowercase "d" to denote deoxy, as in d-GACGTA. From a simple string of letters such as this, any biochemistry student should be able to draw the unique chemical structure for a pentanucleotide, even though it may contain over 200 atoms.

8.6 Classes of Nucleic Acids

As mentioned, the two major classes of nucleic acids are DNA and RNA. DNA has only one biological role, but it is the more central one. The information to make all the functional macromolecules of the cell (even DNA itself) is preserved in DNA and accessed through transcription of the information into RNA copies. Coincident with its singular purpose, there is only a single DNA molecule (or "chromosome") in simple life forms such as viruses or bacteria. Such DNA molecules must be quite large in order to embrace enough information for making all the macromolecules necessary to maintain a living cell. The *Escherichia coli* chromosome has a molecular mass of 2.9×10^9 daltons and contains over 9 million nucleotides. Eukaryotic cells have many chromosomes, and DNA is found principally in two copies in the diploid chromosomes of the nucleus, but DNA also occurs in mitochondria and in chloroplasts, where it encodes some of the proteins and RNAs unique to these organelles. In contrast, RNA occurs in multiple copies and various forms (Table 8.1). Cells contain up to eight times as much RNA as DNA. RNA has a number of important biological functions, and on this basis, RNA molecules are categorized into several major types: **messenger RNA, ribosomal RNA,** and **transfer RNA.** Eukaryotic cells contain an additional type, **small nuclear RNA (snRNA).**

Table 8.1 Various Kinds of RNA Found in an *E. coli* Cell

Type	Sedimentation Coefficient	Molecular Weight	Number of Nucleotide Residues	Percentage of Total Cell RNA
mRNA	6–25	25,000–1,000,000	75–3,000	~2
tRNA	~4	23,000–30,000	73–94	16
rRNA	5	35,000	120	
	16	550,000	1542	82
	23	1,100,000	2904	

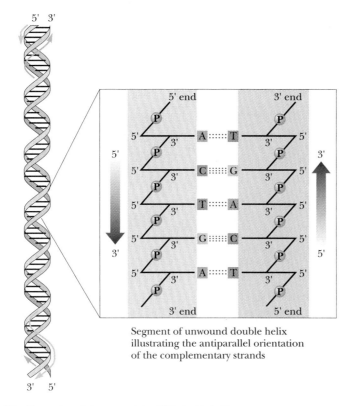

Segment of unwound double helix
illustrating the antiparallel orientation
of the complementary strands

Figure 8.18   The antiparallel nature of the DNA double helix.

With these basic definitions in mind, let us now briefly consider the chemical and structural nature of DNA and the various RNAs. We will then consider how to determine the primary structure of nucleic acids using sequencing methods and discuss the secondary and tertiary structures of DNA and RNA. Part 4, Information Transfer, is devoted to a detailed treatment of the dynamic role of nucleic acids in the molecular biology of the cell.

DNA

The DNA isolated from different cells and viruses characteristically consists of two polynucleotide strands wound together to form a long, slender, helical molecule, the **DNA double helix.** The strands run in opposite directions; that is, they are *antiparallel* and are held together in the double helical structure through *interchain hydrogen bonds* (Figure 8.18). These H bonds pair the bases of nucleotides in one chain to complementary bases in the other, a phenomenon called **base pairing.**

Chargaff's Rules

A clue to the chemical basis of base pairing in DNA came from the analysis of the base composition of various DNAs by Erwin Chargaff in the late 1940s. His data showed that the four bases commonly found in DNA (A, C, G, and T) do not occur in equimolar amounts and that the relative amounts of each vary from species to species (Table 8.2). Nevertheless, Chargaff noted that certain pairs of bases—namely, adenine and thymine, and guanine and cytosine—are always found in a 1:1 ratio and that the number of pyrimidine residues always equals the number of purine residues. These findings are known as *Chargaff's rules*: [A] = [T]; [C] = [G]; [pyrimidines] = [purines].

Table 8.2 Molar Ratios Leading to the Formulation of Chargaff's Rules

Source	Adenine to Guanine	Thymine to Cytosine	Adenine to Thymine	Guanine to Cytosine	Purines to Pyrimidines
Ox	1.29	1.43	1.04	1.00	1.1
Human	1.56	1.75	1.00	1.00	1.0
Hen	1.45	1.29	1.06	0.91	0.99
Salmon	1.43	1.43	1.02	1.02	1.02
Wheat	1.22	1.18	1.00	0.97	0.99
Yeast	1.67	1.92	1.03	1.20	1.0
Hemophilus influenzae	1.74	1.54	1.07	0.91	1.0
E. coli K-12	1.05	0.95	1.09	0.99	1.0
Avian tubercle bacillus	0.4	0.4	1.09	1.08	1.1
Serratia marcescens	0.7	0.7	0.95	0.86	0.9
Bacillus schatz	0.7	0.6	1.12	0.89	1.0

Source: After Chargaff, E., 1951. *Federation Proceedings* **10**:654–659.

Watson and Crick's Double Helix

James Watson and Francis Crick, working at the Cavendish Laboratory at Cambridge University in 1953, took advantage of Chargaff's results, and of the data obtained by Rosalind Franklin and Maurice Wilkins in X-ray diffraction studies of DNA fibers, to conclude that DNA was a *complementary double helix*. Two strands of deoxyribonucleic acid (sometimes referred to as the *Watson strand* and the *Crick strand*) are held together by hydrogen bonds formed between base pairs, always consisting of a purine in one strand and a pyrimidine in the other. Base pairing is very specific: if the purine is adenine, the pyrimidine must be thymine. Similarly, guanine pairs only with cytosine (Figure 8.19). Thus, if an A occurs in one strand of the helix, T must occupy the complementary position in the opposing strand. Likewise, a G in one dictates a C in the other. As Watson recognized from testing various combinations of bases using structurally accurate models, the A : T pair and the G : C pair form spatially equivalent units (see Figure 8.19).

The DNA molecule not only conforms to Chargaff's rules but also has a profound property relating to heredity: *the sequence of bases in one strand has a*

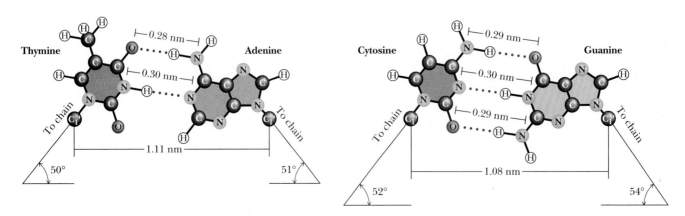

Figure 8.19 The Watson–Crick base pairs A : T and G : C. A and T are joined by two H bonds; three H bonds link G and C.

complementary relationship to the sequence of bases in the other strand. That is, the information contained in the sequence of one strand is conserved in the sequence of the other. Therefore, separation of the two strands and faithful replication of each, through a process in which base pairing specifies the sequence in the newly synthesized strand, leads to two progeny molecules identical in every respect to the parental double helix (Figure 8.20).

Elucidation of the double helical structure of DNA represented one of the most significant events in the history of science. This discovery more than any other marked the beginning of molecular biology. Indeed, upon solving the structure of DNA, Crick proclaimed in the Eagle, a pub just across from the Cavendish lab, "We have discovered the secret of life!"

Size of DNA Molecules

Because of the double helical nature of DNA molecules, their size can be represented in terms of the number of nucleotide base pairs they contain. For example, the *E. coli* chromosome consists of 4.64×10^6 base pairs (abbreviated bp) or 4.64×10^3 kilobase pairs (kbp). DNA is a threadlike molecule; the diameter of the DNA double helix is only 2 nm, but the length of the DNA molecule forming the *E. coli* chromosome is over 1.6×10^6 nm (1.6 mm) (Figure 8.21). The DNA molecule in an average mammalian chromosome is 30 times this length. Since the long dimension of an *E. coli* cell is only 2000 nm (0.002 mm), its chromosome must be highly folded. Because of their long, threadlike nature, DNA molecules are easily sheared into shorter fragments during isolation procedures, and it is difficult to obtain intact chromosomes even from the simple cells of prokaryotes.

RNA

Messenger RNA

Messenger RNA (mRNA) is synthesized during **transcription,** an enzymatic process in which an RNA copy is made of the sequence of bases along one strand of DNA. This mRNA then directs the synthesis of a polypeptide chain as the information contained within its nucleotide sequence is translated into an amino acid sequence by the protein-synthesizing machinery of the ribosomes. rRNA and tRNA molecules are also synthesized by transcription of DNA sequences, but unlike mRNA molecules, these RNAs are not subsequently

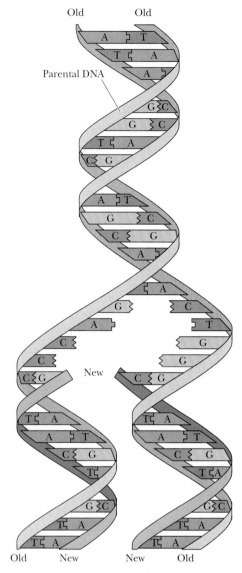

Figure 8.20 Replication of DNA gives identical progeny molecules, since base pairing is the mechanism determining the nucleotide sequence synthesized within each of the new strands during replication.

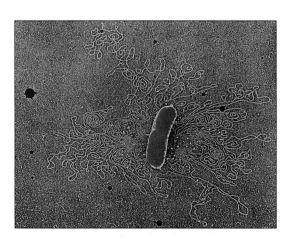

Figure 8.21 If the cell walls of bacteria such as *Escherichia coli* are partially digested and the cells are then osmotically shocked by dilution with water, the contents of the cells are extruded to the exterior. In electron micrographs, the most obvious extruded component is the bacterial chromosome, shown here surrounding the cell. (*Dr. Gopal Murti/CNRI/Phototake NYC*)

Prokaryotes:

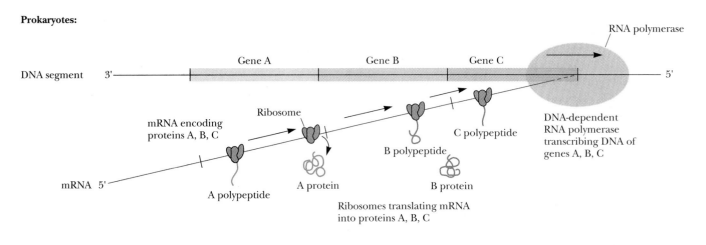

Eukaryotes:

Exons are protein-coding regions that must be joined by removing introns, the noncoding intervening sequences. The process of intron removal and exon joining is called splicing.

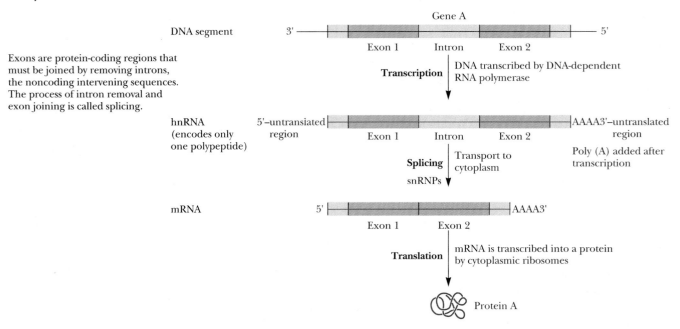

Figure 8.22 The properties of mRNA molecules in prokaryotic versus eukaryotic cells during transcription and translation.

translated to form proteins. Only the genetic units of DNA sequence that encode proteins are transcribed into mRNA molecules. In prokaryotes, a single mRNA may contain the information for the synthesis of several polypeptide chains within its nucleotide sequence (Figure 8.22). In contrast, eukaryotic mRNAs encode only one polypeptide but are more complex because they are synthesized in the nucleus in the form of much larger precursor molecules called **heterogeneous nuclear RNA,** or **hnRNA.** hnRNA molecules contain stretches of nucleotide sequence that have no protein-coding capacity. These noncoding regions are called **intervening sequences** or **introns** because they intervene between coding regions, which are called **exons.** Introns interrupt the continuity of the information specifying the amino acid sequence of a protein and must be spliced out before the message can be translated. In addition, eukaryotic hnRNA and mRNA molecules have a run of 100 to 200 adenylic acid residues attached at their 3′-ends, so-called **poly(A) tails.** This polyadenylation occurs after transcription has been completed and is believed to contribute to mRNA stability. The properties of messenger RNA molecules as they move through transcription and translation in prokaryotic versus eukaryotic cells are summarized in Figure 8.22.

Ribosomal RNA

Ribosomes, the agents of protein synthesis, are composed of two subunits of different sizes that dissociate from each other if the Mg^{2+} concentration is below 10^{-3} M. Each subunit is a supramolecular assembly of proteins and RNA and has a total mass of 10^6 daltons or more. Subcellular structures of such large size are usually described in terms of their ultracentrifugal **sedimentation coefficients,** or **S** values, which relate to how rapidly they sediment in a centrifugal field. The *E. coli* ribosomal subunits have sedimentation coefficients of 30S (the small subunit) and 50S (the large subunit). Eukaryotic ribosomes are somewhat larger than prokaryotic ribosomes, consisting of 40S and 60S subunits. Ribosomes are about 65% RNA and 35% protein. The properties of ribosomes and their rRNAs are summarized in Figure 8.23. The 30S subunit of *E. coli* contains a single RNA chain of 1542 nucleotides. This small subunit ribosomal RNA (rRNA) itself has a sedimentation coefficient of 16S. The large *E. coli* subunit has two rRNA molecules, a 23S (2904 nucleotides) and a 5S (120 nucleotides). The ribosomes of a typical eukaryote, the rat, have rRNA molecules of 18S (1874 nucleotides) and 28S (4718 bases), 5.8S (160 bases), and 5S (120 bases). The 18S rRNA is in the 40S subunit and the latter three are all part of the 60S subunit.

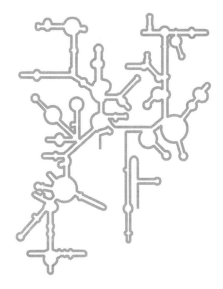

Ribosomal RNA has a complex secondary structure due to many intrastrand hydrogen bonds.

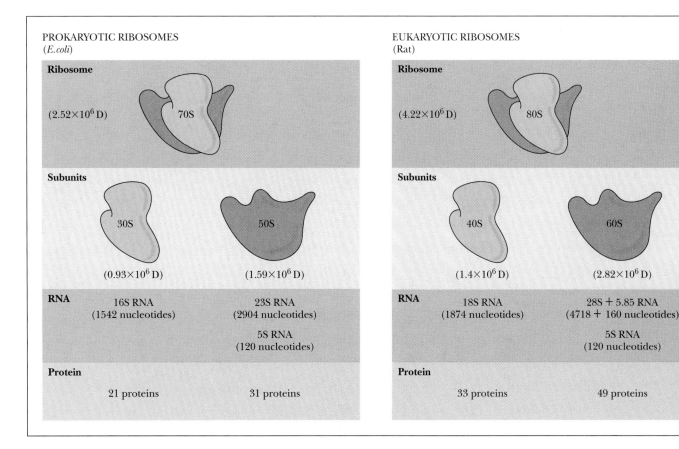

PROKARYOTIC RIBOSOMES
(*E.coli*)

Ribosome		
(2.52×10^6 D)	70S	

Subunits		
	30S	50S
	(0.93×10^6 D)	(1.59×10^6 D)

RNA		
	16S RNA (1542 nucleotides)	23S RNA (2904 nucleotides)
		5S RNA (120 nucleotides)

Protein		
	21 proteins	31 proteins

EUKARYOTIC RIBOSOMES
(Rat)

Ribosome		
(4.22×10^6 D)	80S	

Subunits		
	40S	60S
	(1.4×10^6 D)	(2.82×10^6 D)

RNA		
	18S RNA (1874 nucleotides)	28S + 5.85 RNA (4718 + 160 nucleotides)
		5S RNA (120 nucleotides)

Protein		
	33 proteins	49 proteins

Figure 8.23 The organization and composition of prokaryotic and eukaryotic ribosomes.

Ribosomal RNAs characteristically contain a number of specially modified nucleotides, including **pseudouridine** residues, **ribothymidylic acid,** and **methylated bases** (Figure 8.24). The central role of ribosomes in the biosynthesis of proteins is treated in detail in Chapter 25. Here we briefly note that genetic information in the nucleotide sequence of an mRNA is translated into the amino acid sequence of a polypeptide chain by ribosomes.

Transfer RNA

Transfer RNA (tRNA) molecules are relatively small (23–30kD) polynucleotides containing 73 to 94 residues, a substantial number of which are methylated or otherwise unusually modified. tRNA derives its name from its role as the carrier of amino acids during the process of protein synthesis. Each of the 20 amino acids of proteins has at least one unique tRNA species dedicated to chauffeuring its delivery to ribosomes for insertion into growing polypeptide chains, and some amino acids are served by several tRNAs. For example, five different tRNAs act in the transfer of leucine into proteins. In eukaryotes, there are even discrete sets of tRNA molecules for each site of protein synthesis—the cytoplasm, the mitochondrion, and, in plant cells, the chloroplast. All tRNA molecules possess a 3′-terminal nucleotide sequence that reads **-CCA,** and the amino acid is carried to the ribosome attached as an acyl ester to the free 3′-OH of the terminal A residue. These **aminoacyl-tRNAs** are the substrates of protein synthesis, the amino acid being transferred to the carboxyl end of a growing polypeptide. The peptide bond–forming reaction is a catalytic process intrinsic to ribosomes.

Small Nuclear RNAs

Small nuclear RNAs, or **snRNAs,** are a class of RNA molecule found only in eukaryotic cells, principally in the nucleus. They are neither tRNA nor small

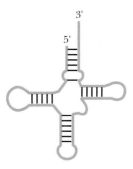

Transfer RNA also has a complex secondary stucture due to many intrastrand hydrogen bonds.

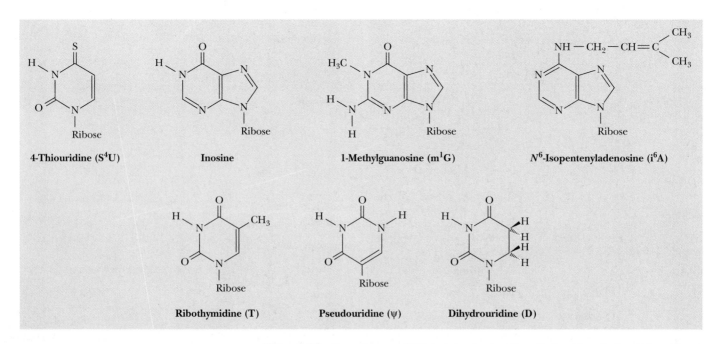

4-Thiouridine (S^4U)   **Inosine**   **1-Methylguanosine (m^1G)**   **N^6-Isopentenyladenosine (i^6A)**

Ribothymidine (T)   **Pseudouridine (ψ)**   **Dihydrouridine (D)**

Figure 8.24 Unusual bases of RNA include pseudouridine, dihydrouridine, ribothymidylic acid, and various methylated bases.

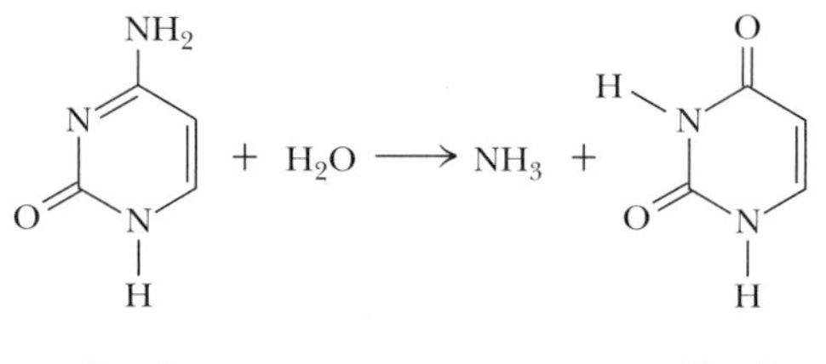

Figure 8.25 Deamination of cytosine forms uracil.

rRNA molecules, although they are similar in size to these species. They contain from 100 to about 200 nucleotides, some of which, like tRNA and rRNA, are methylated or otherwise modified. No snRNA exists as naked RNA. Instead, snRNA is found in stable complexes with specific proteins forming **small nuclear ribonucleoprotein particles,** or **snRNPs,** which are about 10S in size. Their occurrence only in eukaryotes, their location in the nucleus, and their relative abundance (1% to 10% of the number of ribosomes) are significant clues to their biological purpose: snRNPs are important in the processing of eukaryotic gene transcripts (hnRNA) into mature messenger RNA for export from the nucleus to the cytoplasm (see Figure 8.22).

Significance of Chemical Differences Between DNA and RNA

Two fundamental chemical differences distinguish DNA from RNA:

1. DNA contains 2-deoxyribose instead of ribose.
2. DNA contains thymine instead of uracil.

What are the consequences of these differences and do they hold any significance in common? An argument can be made that, because of these differences, DNA is a more stable polymeric form than RNA. The greater stability of DNA over RNA is consistent with the respective roles that these macromolecules have in heredity and information transfer.

Consider first why DNA contains thymine instead of uracil. The key observation is that *cytosine deaminates to form uracil* at a finite rate *in vivo* (Figure 8.25). Since C in one DNA strand pairs with G in the opposing strand, whereas U would pair with A, conversion of a C to a U could potentially result in a heritable change of a CG pair to a UA pair. Such a change in nucleotide sequence would constitute a *mutation* in the DNA. To prevent this reaction from causing mutations, a cellular repair mechanism "proofreads" the DNA, and when a U arising from C deamination is encountered, it is treated as inappropriate and is replaced by a C. If DNA normally contained U rather than T, this repair system could not readily distinguish U formed by C deamination from U correctly paired with A. However, the U in DNA is "5-methyl-U" or, as it is conventionally known, thymine (Figure 8.26). That is, the 5-methyl group on T labels it to say "this U belongs here; do not replace it."

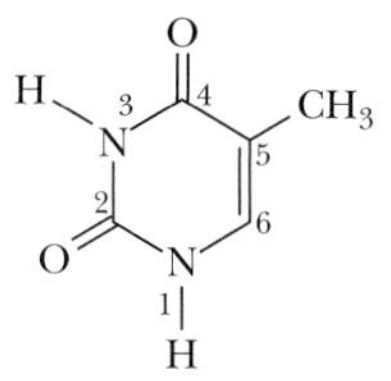

The ribose 2'-OH group of RNA is absent in DNA. Consequently, the ubiquitous 3'-O of polynucleotide backbones lacks a vicinal hydroxyl neighbor in DNA. This difference leads to a greater resistance of DNA to hydrolysis, examined in detail in the following section. To view it another way, RNA is less stable than DNA because its vicinal 2'-OH group makes the 3'-phosphodiester bond susceptible to hydrolysis (Figure 8.27). For just this reason, it is selectively advantageous for the heritable form of genetic information to be DNA rather than RNA.

Figure 8.26 The 5-methyl group on thymine labels it as a special kind of uracil.

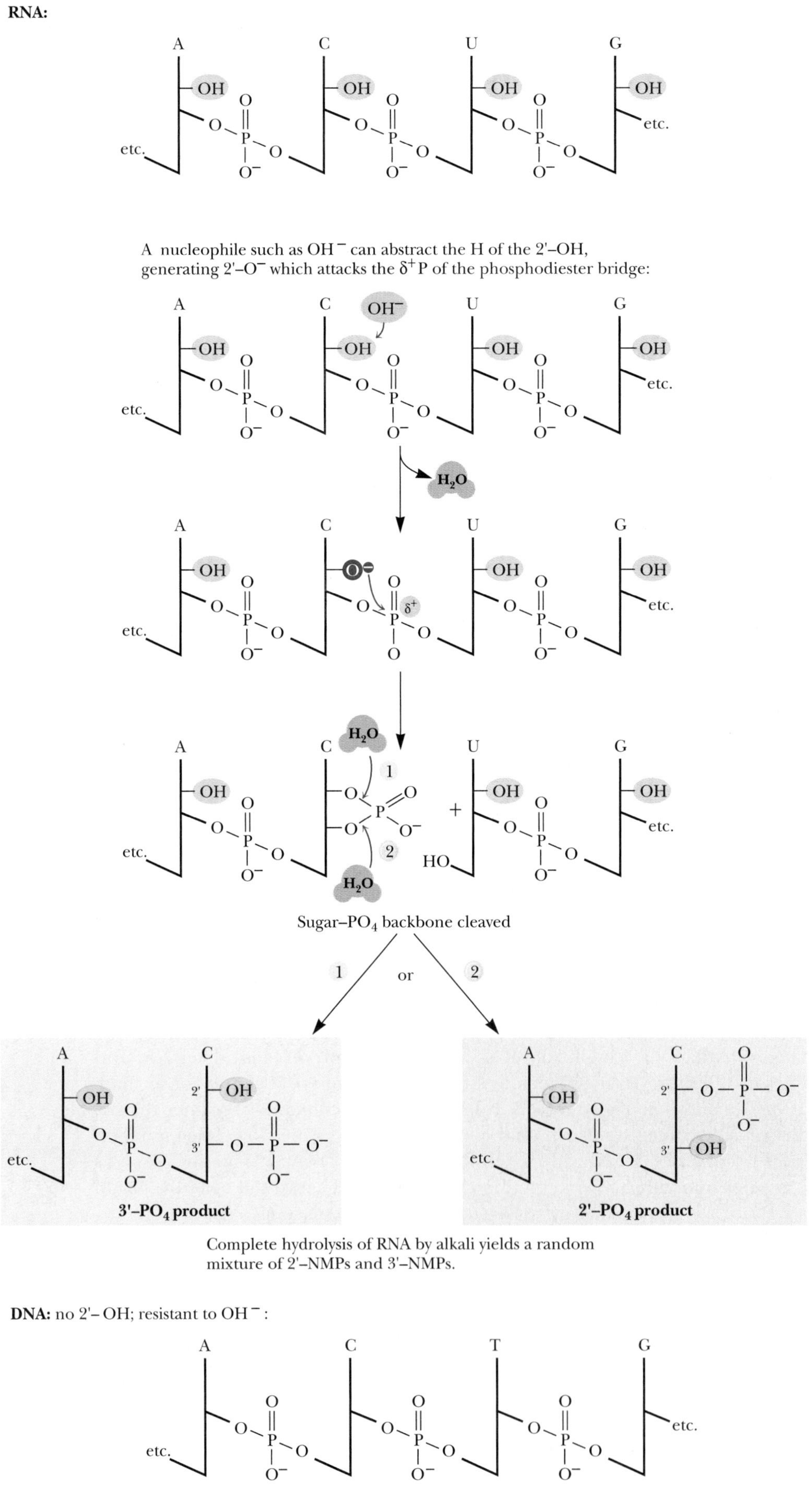

Figure 8.27 The vicinal — OH groups of RNA are susceptible to nucleophilic attack leading to hydrolysis of the phosphodiester bond and fracture of the polynucleotide chain; DNA lacks a 2'-OH vicinal to its 3'-O-phosphodiester backbone. Alkaline hydrolysis of RNA results in the formation of a mixture of 2'- and 3'-nucleoside monophosphates.

RNA:

A nucleophile such as OH⁻ can abstract the H of the 2'–OH, generating 2'–O⁻ which attacks the δ^+P of the phosphodiester bridge:

Sugar–PO₄ backbone cleaved

3'–PO₄ product

2'–PO₄ product

Complete hydrolysis of RNA by alkali yields a random mixture of 2'–NMPs and 3'–NMPs.

DNA: no 2'– OH; resistant to OH⁻ :

8.7 Hydrolysis of Nucleic Acids

The vast majority of reactions leading to nucleic acid hydrolysis break bonds in the polynucleotide backbone. Such reactions are important because they can be used to manipulate these polymeric molecules. For example, hydrolysis of polynucleotides generates smaller fragments whose nucleotide sequence can be more easily determined.

Hydrolysis by Acid or Base

RNA is relatively resistant to the effects of dilute acid, but gentle treatment of DNA with 1 mM HCl leads to selective removal of its purine bases by cleavage of the purine glycoside bonds. The glycosidic bonds between pyrimidine bases and 2′-deoxyribose are not affected, and, in this case, the polynucleotide's sugar–phosphate backbone remains intact. The purine-free polynucleotide product is called **apurinic acid.**

DNA is not susceptible to alkaline hydrolysis. On the other hand, RNA is alkali-labile and is readily hydrolyzed by dilute sodium hydroxide. Cleavage is random in RNA, and the ultimate products are a mixture of nucleoside 2′- and 3′-monophosphates. These products provide a clue to the reaction mechanism (see Figure 8.27). Abstraction of the 2′-OH proton by hydroxyl anion leaves a 2′-O⁻ that carries out a nucleophilic attack on the δ^+ phosphorus atom of the phosphate moiety, resulting in cleavage of the 5′-phosphodiester bond and formation of a cyclic 2′,3′-phosphate. This cyclic 2′,3′-phosphodiester is unstable and decomposes randomly to either a 2′- or 3′-phosphate ester. DNA, which has no 2′-OH, is alkali-stable.

A DEEPER LOOK

Peptide Nucleic Acids (PNAs) Are Synthetic Mimics of DNA and RNA

Chemists have synthesized analogs of DNA (and RNA) in which the sugar–phosphate backbone is replaced by a peptide backbone, creating a polymer appropriately termed a *peptide nucleic acid,* or *PNA.* The PNA peptide backbone was designed so that the space between successive bases is the same as in natural DNA (see figure). PNA consists of repeating units of *N*-(2-aminoethyl)-glycine residues linked by peptide bonds; the bases are attached to this backbone through methylene carbonyl linkages. This chemistry provides six bonds along the backbone between bases and three bonds between the backbone and each base, just as in natural DNA. PNA oligomers interact with DNA (and RNA) through specific base-pairing interactions, just as would be expected for a pair of complementary oligonucleotides. PNAs are resistant to nucleases and also are poor substrates for proteases. PNAs thus show great promise as specific diagnostic probes for unique DNA or RNA nucleotide sequences. PNAs also have potential application as antisense drugs (see problem 5 at the end of this chapter). Note the repeating six bonds (in blue) between the base attachments and the three-bond linker (in red) between the base (B) and backbone.

PNA

DNA

Source: Buchardt, O., et al., 1993. Peptide nucleic acids and their potential applications in biotechnology. *Trends in Biotechnology* **11**:384–386.

Enzymatic Hydrolysis

Enzymes that hydrolyze nucleic acids are called **nucleases.** Virtually all cells contain various nucleases that serve important housekeeping roles in the normal course of nucleic acid metabolism. Organs that provide digestive fluids, such as the pancreas, are rich in nucleases and secrete substantial amounts to hydrolyze dietary nucleic acids. Fungi and snake venom are often good sources of nucleases. As a class, nucleases are **phosphodiesterases** because the reaction that they catalyze is the cleavage of phosphodiester bonds by H_2O. Because each internal phosphate in a polynucleotide backbone is involved in two phosphodiester linkages, cleavage may occur on either side of the phosphorus (Figure 8.28). Convention labels the 3′-side as *a* and the 5′-side as *b*. A second convention denotes whether the nucleic acid chain was cleaved at some internal location, *endo,* or whether a terminal nucleotide residue was hydrolytically removed, *exo.*

Nuclease Specificity

Like most enzymes (see Chapter 10), nucleases exhibit selectivity or *specificity* for the nature of the substance on which they act. That is, some nucleases act only on DNA (**DNases**), while others are specific for RNA (the **RNases**). Still others act on either DNA or RNA and are referred to simply as **nucleases,** as in *nuclease S1.* Nucleases may also show specificity for only single-stranded nucleic acids or may act only on double helices. Single-stranded nucleic acids are abbreviated by an *ss* prefix, as in ssRNA; the prefix *ds* denotes double-stranded. Nucleases may also display a decided preference for acting only at certain bases in a polynucleotide, or, as we shall see for *restriction endonucleases,* some nucleases will act only at a particular nucleotide sequence four to eight

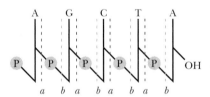

Convention: The 3'-side of each phosphodiester is termed **a**; the 5'-side is termed **b.**

Hydrolysis of the **a** bond yields 5′–PO₄ products:

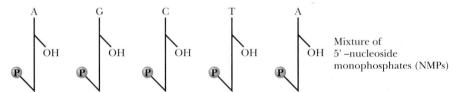

Mixture of 5′–nucleoside monophosphates (NMPs)

Hydrolysis of the **b** bond yields 3′–PO₄ products:

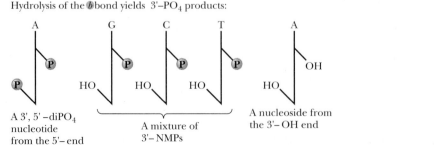

A 3′, 5′–diPO₄ nucleotide from the 5′–end

A mixture of 3′–NMPs

A nucleoside from the 3′–OH end

Figure 8.28 Cleavage in polynucleotide chains. Cleavage on the *a* side leaves the phosphate attached to the 5′-position of the adjacent nucleotide, while *b*-side hydrolysis yields 3′-phosphate products. Enzymes or reactions that hydrolyze nucleic acids are characterized as acting at either *a* or *b.*

Snake venom phosphodiesterase: an "*a*" specific exonuclease:

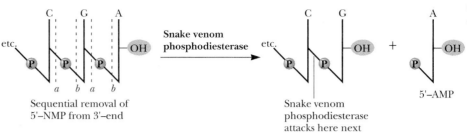

Figure 8.29 Snake venom phosphodiesterase and spleen phosphodiesterase are exonucleases that degrade polynucleotides from opposite ends.

Sequential removal of
5′–NMP from 3′–end

Snake venom phosphodiesterase attacks here next

5′–AMP

Spleen phosphodiesterase: a "*b*" specific exonuclease:

Sequential removal of
3′–NMP from 5′–end

3′–CMP

Spleen phosphodiesterase attacks here next

nucleotides in length. To the molecular biologist, nucleases are the surgical tools for the dissection and manipulation of nucleic acids in the laboratory.

Exonucleases degrade nucleic acids by sequentially removing nucleotides from their ends. Since they act on either DNA or RNA, they are referred to by the generic name *phosphodiesterase*. For example, snake venom phosphodiesterase acts by *a* cleavage and starts at the free 3′-OH end of a polynucleotide chain, liberating nucleoside 5′-monophosphates. In contrast, the bovine spleen enzyme starts at the 5′-end of a nucleic acid, cleaving *b* and releasing 3′-NMPs (Figure 8.29).

Restriction Enzymes

Restriction endonucleases are enzymes, isolated chiefly from bacteria, that have the ability to cleave double-stranded DNA. The term *restriction* comes from the capacity of prokaryotes to defend against or "restrict" the possibility of takeover by foreign DNA that might gain entry into their cells. Prokaryotes degrade foreign DNA by using their unique restriction enzymes to chop it into relatively large but noninfective fragments.

Type II Restriction Endonucleases

Type II restriction enzymes have received widespread application in the cloning and sequencing of DNA molecules. This type of restriction enzyme cuts DNA only at particular sites. These sites are specific nucleotide sequences, typically four or six nucleotides in length, with a twofold axis of symmetry. For example, *E. coli* has a restriction enzyme, *Eco*RI, that recognizes the following hexanucleotide sequence:

$$5'—G—A—A—T—T—C—3'$$
$$3'—C—T—T—A—A—G—5'$$

Note the twofold symmetry: the sequence read 5′ → 3′ is the same in both strands.

When *Eco*RI encounters this sequence in dsDNA, it causes a staggered, double-stranded break by hydrolyzing each chain between the G and A residues:

$$5'-G \quad A-A-T-T-C-3'$$
$$3'-C-T-T-A-A \quad G-5'$$

Staggered cleavage results in fragments with protruding 5'-ends:

$$5'-G \qquad A-A-T-T-C-3'$$
$$3'-C-T-T-A-A \qquad G-5'$$

Because the protruding termini of *Eco*RI fragments have complementary base sequences, they can form base pairs with one another:

$$5'-G \quad A-A-T-T-C-3'$$
$$3'-C-T-T-A-A \quad G-5'$$

Therefore, DNA restriction fragments having such "sticky" ends can be joined together to create new combinations of DNA sequence. If the fragments are derived from DNA molecules of different origin, novel recombinant forms of DNA are created.

*Eco*RI leaves staggered 5'-termini. Other restriction enzymes, such as *Pst*I, which recognizes the sequence 5'-CTGCAG-3' and cleaves between A and G, produce cohesive staggered 3'-ends. Still others, such as *Bal*I, act at the center of the twofold symmetry axis of their recognition site and generate blunt ends that are noncohesive. *Bal*I recognizes 5'-TGGCCA-3' and cuts between G and C. Table 8.3 lists many of the commonly used restriction endonucleases and their unique recognition sites. Since these sites all have twofold symmetry, only the sequence on one strand needs to be designated. Different restriction enzymes sometimes recognize and cleave within identical target sequences. Such enzymes are called **isoschizomers,** meaning that they cut at the same site; for example, *Mbo*I and *Sau*3A are isoschizomers.

schism   division, separation

Restriction Fragment Size. Assuming random distribution and equimolar proportions for the four nucleotides in DNA, a particular tetranucleotide sequence should occur once every 4^4 nucleotides, or every 256 bases. Therefore, the fragments generated by a restriction enzyme that acts at a four-nucleotide sequence should average about 250 bp in length. "Six-cutters," enzymes such as *Eco*RI or *Bam*HI, will find their unique hexanucleotide sequences on the average once in every 4096 (4^6) bp of length. Since the genetic code is a triplet code with three bases of DNA specifying one amino acid in a polypeptide sequence, and since polypeptides typically contain at most 1000 amino acid residues, the fragments generated by six-cutters are approximately the size of prokaryotic genes. This property makes these enzymes useful in the construction and cloning of genetically useful recombinant DNA molecules. For the isolation of even larger nucleotide sequences, such as those of genes encoding large polypeptides (or those of eukaryotic genes that are disrupted by large introns), partial or limited digestion of DNA by restriction enzymes can be employed. Restriction endonucleases that cut only at specific nucleotide sequences 8 or even 13 nucleotides in length are also available, such as *Not*I and *Sfi*I.

Restriction Mapping

The application of these sequence-specific nucleases to problems in molecular biology is considered in detail in Chapter 9, but one prevalent application merits discussion here. Since restriction endonucleases cut dsDNA at unique sites to generate large fragments, they provide a means for mapping DNA molecules that are many kilobase pairs in length. Restriction digestion of a DNA

Table 8.3   Restriction Endonucleases

About 1000 restriction enzymes have been characterized. They are named by italicized three-letter codes, the first a capital latter denoting the genus of the organism of origin, while the next two letters are an abbreviation of the particular species. Because prokaryotes often contain more than one restriction enzyme, the various representatives are assigned letter and number codes as they are identified. Thus, *Eco*RI is the initial restriction endonuclease isolated from *Escherichia coli,* strain R. With one exception (*Nci*I), all known type II restriction endonucleases generate fragments with $5'$-PO_4 and $3'$-OH ends.

Enzyme	Common Isoschizomers	Recognition Sequence	Compatible Cohesive Ends
*Alu*I		AG↓CT	Blunt
*Ava*I		G↓PyCGPuG	*Sal*I, *Xho*I, *Xma*I
*Avr*II		C↓CTAGG	
*Bal*I		TGG↓CCA	Blunt
*Bam*HI		G↓GATCC	*Bcl*I, *Bgl*II, *Mbo*I, *Sau*3A, *Xho*II
*Bcl*I		T↓GATCA	*Bam*HI, *Bgl*II, *Mbo*I, *Sau*3A, *Xho*II
*Bgl*II		A↓GATCT	*Bam*HI, *Bcl*I, *Mbo*I, *Sau*3A, *Xho*II
*Bst*XI		CCANNNNN↓NTGG	
*Cla*I		AT↓CGAT	*Acc*I, *Acy*I, *Asy*II, *Hpa*II, *Taq*I
*Eco*RI		G↓AATTC	
*Eco*RII	*Atu*I, *Apy*I	↓CC (^A_T)GG	
*Hae*I		(^A_T)GG↓CC(^T_A)	Blunt
*Hae*II		PuGCGC↓Py	
*Hae*III		GG↓CC	Blunt
*Hinc*II		GTPy↓PuAC	Blunt
*Hind*III		A↓AGCTT	
*Hpa*I		GTT↓AAC	Blunt
*Hpa*II		C↓CGG	*Acc*I, *Acy*I, *Asu*II, *Cla*I, *Taq*I
*Kpn*I		GGTAC↓C	
*Mbo*I	*Sau*3A	↓GATC	*Bam*HI, *Bcl*I, *Bgl*II, *Xho*II
*Not*I		GC↓GGCCGC	
*Pst*I		CTGCA↓G	
*Sac*I	*Sst*I	GAGCT↓C	
*Sal*I		G↓TCGAC	*Ava*I, *Xho*I
*Sfi*I		GGCCNNNN↓NGGCC	
*Sma*I	*Xma*I	CCC↓GGG	Blunt
*Taq*I		T↓CGA	*Acc*I, *Acy*I, *Asu*II, *Cla*I, *Hpa*II
*Xba*I		T↓CTAGA	
*Xho*I		C↓TCGAG	*Ava*I, *Sal*I
*Xho*II		(^A_G)↓GATC(^T_C)	*Bam*HI, *Bcl*I, *Bgl*II, *Mbo*I, *Sau*3A

molecule is in many ways analogous to proteolytic digestion of a protein by an enzyme such as trypsin (see Chapter 4): the restriction endonuclease acts only at its specific sites so that a discrete set of nucleic acid fragments is generated. The restriction fragments represent a unique collection of different-sized DNA pieces. Fortunately, this complex mixture can be resolved by *electrophoresis.*[1]

---

[1]Electrophoresis is a method for separating molecules based on their mobility in an electric field. Generally, electrophoresis is carried out in a porous support matrix such as polyacrylamide or agarose, which retards the movement of molecules according to their dimensions relative to the size of the pores in the matrix. For molecules whose charge per monomeric unit is constant (for DNA and RNA, the charge per monomer is −1), the relative mobility will be directly proportional to the size of the molecule (number of monomers in the polymer).

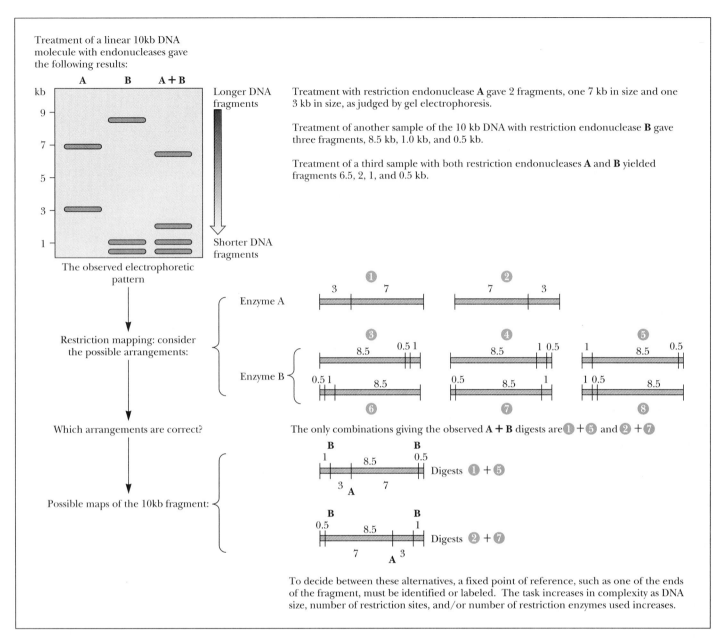

Treatment of a linear 10kb DNA molecule with endonucleases gave the following results:

The observed electrophoretic pattern

Restriction mapping: consider the possible arrangements:

Which arrangements are correct?

Possible maps of the 10kb fragment:

Treatment with restriction endonuclease **A** gave 2 fragments, one 7 kb in size and one 3 kb in size, as judged by gel electrophoresis.

Treatment of another sample of the 10 kb DNA with restriction endonuclease **B** gave three fragments, 8.5 kb, 1.0 kb, and 0.5 kb.

Treatment of a third sample with both restriction endonucleases **A** and **B** yielded fragments 6.5, 2, 1, and 0.5 kb.

The only combinations giving the observed **A + B** digests are ❶ + ❺ and ❷ + ❼

To decide between these alternatives, a fixed point of reference, such as one of the ends of the fragment, must be identified or labeled. The task increases in complexity as DNA size, number of restriction sites, and/or number of restriction enzymes used increases.

Figure 8.30 Restriction mapping of a DNA molecule as determined by an analysis of the electrophoretic pattern obtained for different restriction endonuclease digests. (Keep in mind that a dsDNA molecule will have a unique nucleotide sequence and therefore a definite polarity; thus, fragments from one end will be distinctly different from fragments derived from the other end.)

Electrophoresis of DNA molecules on gels of restricted pore size (as formed in agarose or polyacrylamide media) separates them according to size, the largest being retarded in their migration through the gel pores, while the smallest move relatively unhindered. Figure 8.30 shows a hypothetical electrophoretogram obtained for a DNA molecule treated with two different restriction nucleases, alone and in combination. Just as cleavage of a protein with different proteases to generate overlapping fragments allows an ordering of the peptides, restriction fragments can be ordered or "mapped" according to their sizes, as deduced from the patterns depicted in Figure 8.30.

<h2>8.8 The Primary Structure of Nucleic Acids</h2>

We turn now to biochemical methods that reveal the sequential order of nucleotides in a polynucleotide, the **primary structure** of nucleic acids. The sequence of nucleotides in nucleic acids is the embodiment of genetic information. Two important breakthroughs have made sequencing nucleic acids substantially easier than sequencing polypeptides. One was the discovery of *restriction endonucleases* that cleave DNA at specific oligonucleotide sites, generating unique fragments of manageable size. The second was the power of *polyacrylamide gel electrophoresis* separation methods to separate nucleic acid fragments that differ from one another in length by only a single nucleotide.

Sequencing Nucleic Acids

The most widely used protocol for nucleic acid sequencing is the **chain termination** or **dideoxy method** of Frederick Sanger. This method is carried out on nanogram amounts of DNA, since very sensitive analytical techniques can be used to detect the DNA chains following electrophoretic separation on polyacrylamide gels. Typically, the DNA molecules are labeled with radioactive ^{32}P,[2] and following electrophoresis, the pattern of their separation is visualized by **autoradiography.** A piece of X-ray film is placed over the gel and the radioactive disintegrations emanating from ^{32}P decay create a pattern on the film that is an accurate image of the resolved oligonucleotides. Recently, sensitive biochemical and chemiluminescent methods have begun to supersede the use of radioisotopes as tracers in these experiments.

Chain Termination or Dideoxy Method

To appreciate the rationale of the chain termination or dideoxy method, we first must briefly examine the biochemistry of DNA replication. DNA is a double-helical molecule. In the course of its replication, the sequence of nucleotides in one strand is copied in a complementary fashion to form a new second strand by the enzyme **DNA polymerase.** Each original strand of the double helix serves as **template** for the biosynthesis that yields two daughter DNA duplexes from the parental double helix (Figure 8.31). DNA polymerase will carry out this reaction *in vitro* in the presence of the four deoxynucleotide monomers and will copy single-stranded DNA, provided a double-stranded region of DNA is artificially generated by adding a **primer.** This primer is merely an oligonucleotide capable of forming a short stretch of dsDNA by base pairing with the ssDNA (Figure 8.32). The primer must have a free 3′-OH end from which the new polynucleotide chain can grow as the first residue is added in the initial step of the polymerization process. DNA polymerases synthesize new strands by adding successive nucleotides in the 5′ → 3′ direction.

Figure 8.31 DNA replication yields two daughter DNA duplexes identical to the parental DNA molecule. Each original strand of the double helix serves as a template, and the sequence of nucleotides in each of these strands is copied to form a new complementary strand by the enzyme DNA polymerase. By this process, biosynthesis yields two daughter DNA duplexes from the parental double helix.

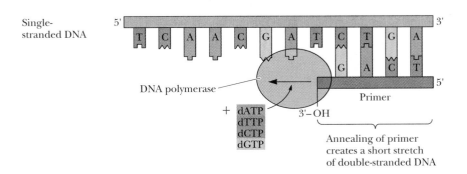

Figure 8.32 DNA polymerase will copy ssDNA *in vitro* in the presence of the four deoxynucleotide monomers, provided a double-stranded region of DNA is artificially generated by adding a primer, an oligonucleotide capable of forming a short stretch of dsDNA by base pairing with the ssDNA. The primer must have a free 3′-OH end from which the new polynucleotide chain can grow as the first residue is added in the initial step of the polymerization process.

---

[2]Because its longer half-life and lower energy make it safer to handle, ^{35}S is replacing ^{32}P as the radioactive tracer of choice in sequencing by the Sanger method. ^{35}S-α–labeled deoxynucleotide analogs provide the source for incorporating radioactivity into DNA.

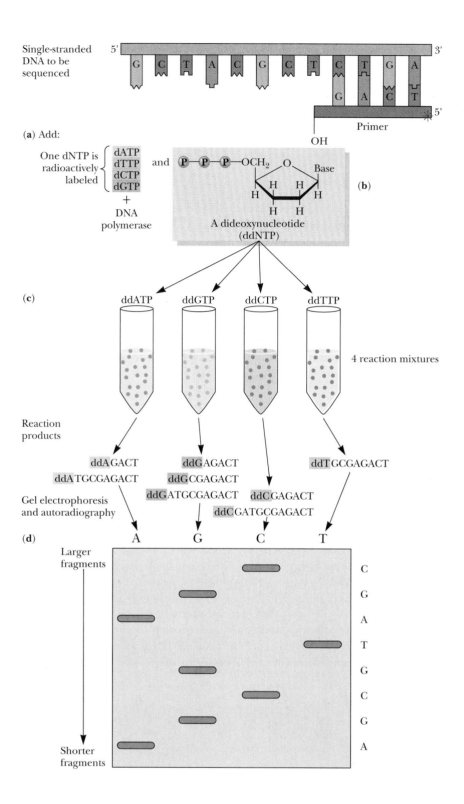

Figure 8.33 The chain termination or dideoxy method of DNA sequencing.   **(a)** DNA polymerase reaction.   **(b)** Structure of dideoxynucleotide. **(c)** Four reaction mixtures with nucleoside triphosphates plus one dideoxynucleoside triphosphate. **(d)** Electrophoretogram. Note that the nucleotide sequence as read from the bottom to the top of the gel is the order of nucleotide addition carried out by DNA polymerase.

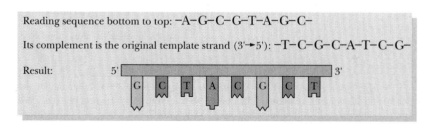

Chain Termination Protocol

In the chain termination method of DNA sequencing, a DNA fragment of unknown sequence serves as template in a polymerization reaction catalyzed by DNA polymerase. The primer requirement is met by an appropriate oligonucleotide (this method is also known as the **primed synthesis method** for this reason). Four parallel reactions are run; all four contain the four deoxynucleoside triphosphates dATP, dGTP, dCTP, and dTTP, which are the substrates for DNA polymerase (Figure 8.33). In each of the four reactions, a different 2′,3′-*di*deoxynucleotide is included, and it is these dideoxynucleotides that give the method its name.

Because dideoxynucleotides lack 3′-OH groups, these nucleotides cannot serve as acceptors for 5′-nucleotide addition in the polymerization reaction, and thus the chain is terminated where they become incorporated. The concentrations of the four deoxynucleotides and the single dideoxynucleotide in each reaction mixture are adjusted so that the dideoxynucleotide is incorporated infrequently. Therefore, base-specific premature chain termination is only a random, occasional event, and a population of new strands of varying length is synthesized. Four reactions are run, one for each dideoxynucleotide, so that terminations, although random, can occur everywhere in the sequence. In each mixture, each newly synthesized strand has a dideoxynucleotide at its 3′-end, and its presence at that position demonstrates that a base of that particular kind was specified by the template. A radioactively labeled dNTP is included in each reaction mixture to provide a tracer for the products of the polymerization process.

Reading Dideoxy Sequencing Gels

The sequencing products are visualized by autoradiography (or similar means) following their separation according to size by polyacrylamide gel electrophoresis (see Figure 8.33). Since the smallest fragments migrate fastest upon electrophoresis and since fragments differing by only single nucleotides in length are readily resolved, the autoradiogram of the gel can be read from bottom to top, noting which lane has the next largest band at each step. Thus, the gel in Figure 8.33 is read AGCGTAGC (5′ → 3′). Because of the way DNA polymerase acts, this observed sequence is complementary to the corresponding unknown template sequence. With this knowledge, the template sequence now can be written GCTACGCT (5′ → 3′).

With relatively simple technology, it is possible to read the order of as many as 400 bases from the autoradiogram of a sequencing gel (Figure 8.34). The actual chemical or enzymatic reactions, electrophoresis, and autoradiography are now routine, and a skilled technician can easily sequence several kbp per week using these manual techniques. The major effort in DNA sequencing is in the isolation and preparation of fragments of interest, such as cloned genes.

Automated DNA Sequencing

Automated DNA sequencing machines capable of identifying about 10^4 bases per day are commercially available. Automated sequencing typically employs fluorescent dyes of different colors to uniquely label the primer DNA introduced into the four sequencing reactions—for example, red for the A reaction, blue for T, green for G, and yellow for C. Then, all four reaction mixtures can be combined and run together on one electrophoretic gel slab. As the oligonucleotides are separated and pass to the bottom of the gel, each is illuminated by a low-power argon laser beam that causes the dye attached to the primer to fluoresce. The color of the fluorescence is detected

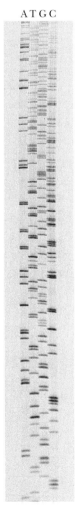

Figure 8.34 A photograph of the autoradiogram from an actual sequencing gel: A portion of the DNA sequence of *nit-6*, the *Neurospora* gene encoding the enzyme nitrite reductase. (*James D. Colandene, University of Virginia*)

Figure 8.35 Schematic diagram of the methodology used in fluorescent labeling and automated sequencing of DNA. Four reactions are set up, one for each base, and the primer in each is end-labeled with one of four different fluorescent dyes; the dyes serve to color-code the base-specific sequencing protocol (a unique dye is used in each dideoxynucleotide reaction). The four reaction mixtures are then combined and run in one lane. Thus, each lane in the gel represents a different sequencing experiment. As the differently sized fragments pass down the gel, a laser beam excites the dye in the scan area. The emitted energy passes through a rotating color filter and is detected by a fluorometer. The color of the emitted light identifies the final base in the fragment.

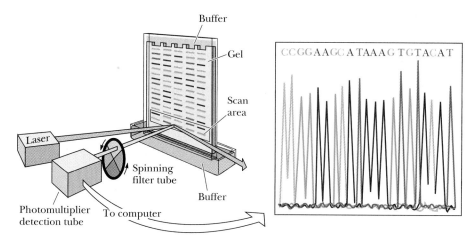

automatically, revealing the identity of the primer, and hence the base, immediately (Figure 8.35). As many as 600 to 800 bases can be sequentially read in each lane of such gels. The development of such automation made it possible to sequence the entire human genome, some 3 billion bp, in a reasonable time frame.

8.9 The ABZs of DNA Secondary Structure

Double-stranded DNA molecules assume one of three secondary structures, termed A, B, and Z. Fundamentally, double-stranded DNA is a regular two-chain structure with hydrogen bonds formed between opposing bases on the two chains (see Figures 8.18, 8.19, and 8.20). Such H bonding is possible only when the two chains are antiparallel. The polar sugar–phosphate backbones of the two chains are on the outside. The bases are stacked on the inside of the structure; these heterocyclic bases, as a consequence of their π-electron clouds, are hydrophobic on their flat sides. The two-stranded, right-handed helical "ladder" has its base-pair rungs stacked 0.34 nm apart. Since this helix repeats itself approximately every 10 bp, its **pitch** is 3.4 nm. This is the major conformation of DNA in solution and is called **B-DNA.**

The DNA Double Helix Is a Stable Structure

A number of factors account for the stability of the double-helical structure of DNA. First, both internal and external hydrogen bonds stabilize the double helix. The two strands of DNA are held together by H bonds that form between the complementary purines and pyrimidines, two in an A : T pair and three in a G : C pair, while polar atoms in the sugar–phosphate backbone form external H bonds with surrounding water molecules. Second, the negatively charged phosphate groups are all situated on the exterior surface of the helix in such a way that they have minimal effect on one another and are free to interact electrostatically with cations in solution such as Mg^{2+}. Third, the core of the helix consists of the base pairs, which, in addition to being H-bonded, stack together through hydrophobic interactions and van der Waals forces that contribute significantly to the overall stabilizing energy.

A stereochemical consequence of the way A : T and G : C base pairs form is that the sugars of the respective nucleotides have opposite orientations, and thus the sugar–phosphate backbones of the two chains run in opposite or "antiparallel" directions. Furthermore, the two glycosidic bonds holding the bases in each base pair are not directly across the helix from each other, defining a common diameter (Figure 8.36). Consequently, the sugar–phosphate backbones of the helix are not equally spaced along the helix axis and the grooves between them are not the same size. Instead, the intertwined chains create a **major groove** and a **minor groove** (see Figure 8.36). The edges of the base pairs have a specific relationship to these grooves. The "top" edges of the base pairs ("top" as depicted in Figure 8.19) are exposed along the interior surface or "floor" of the major groove; the base pair edges nearest to the glycosidic bond form the interior surface of the minor groove. Some proteins that bind to DNA can actually recognize specific nucleotide sequences by "reading" the pattern of H-bonding possibilities presented by the edges of the bases in these grooves. Such DNA–protein interactions provide one step toward understanding how cells regulate the expression of genetic information encoded in DNA (see Chapter 24).

Conformational Variation in Double-Helical Structures

In solution, DNA ordinarily assumes the structure we have been discussing: B-DNA. However, nucleic acids also occur naturally in other double-helical forms. The base-pairing arrangement remains the same, but the sugar–phosphate groupings that constitute the backbone are inherently flexible and can adopt different conformations.

Alternative Form of Right-Handed DNA

An alternative form of the right-handed double helix is **A-DNA.** A-DNA molecules differ in a number of ways from B-DNA. The pitch, or distance required to complete one helical turn, is different. In B-DNA, it is 3.4 nm, whereas in A-DNA, it is 2.46 nm. One turn in A-DNA requires 11 bp to complete. Depending on local sequence, 10 to 10.6 bp define one helical turn in B-form DNA.

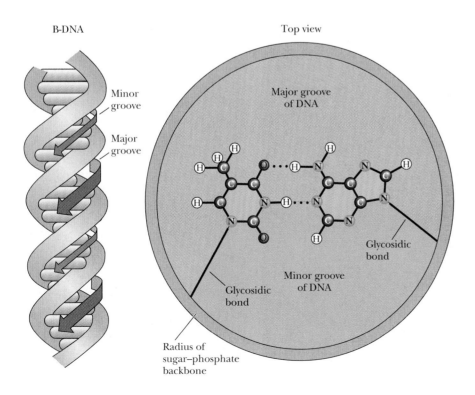

Figure 8.36 The bases in a base pair are not directly across the helix axis from one another along some diameter but rather are slightly displaced. This displacement, and the relative orientation of the glycosidic bonds linking the bases to the sugar-phosphate backbone, leads to differently sized grooves—the major groove and the minor groove—in the cylindrical column created by the double helix, each coursing along its length.

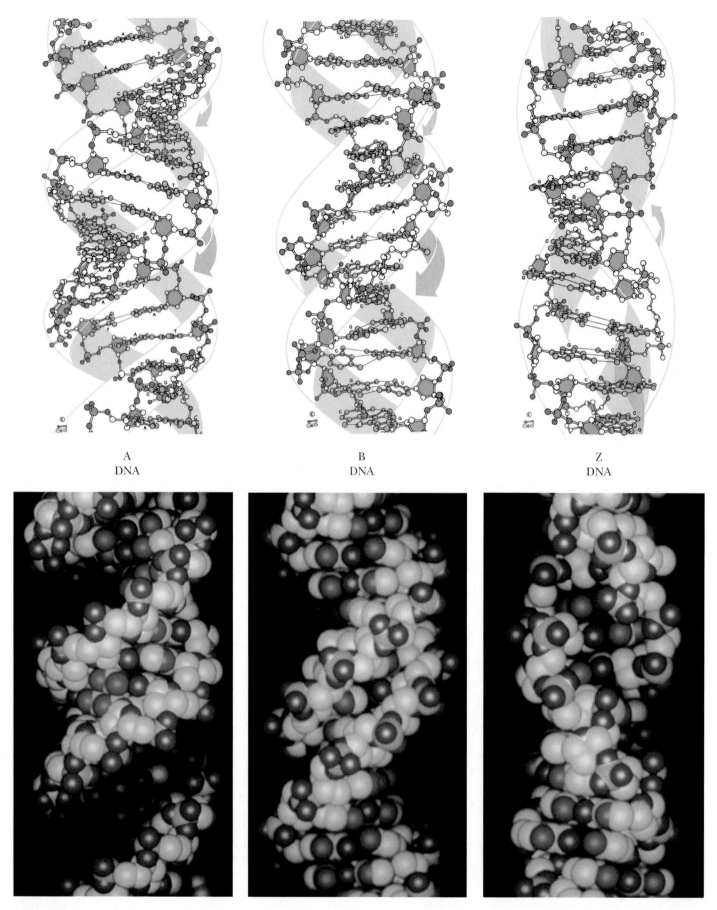

A
DNA

B
DNA

Z
DNA

Figure 8.37

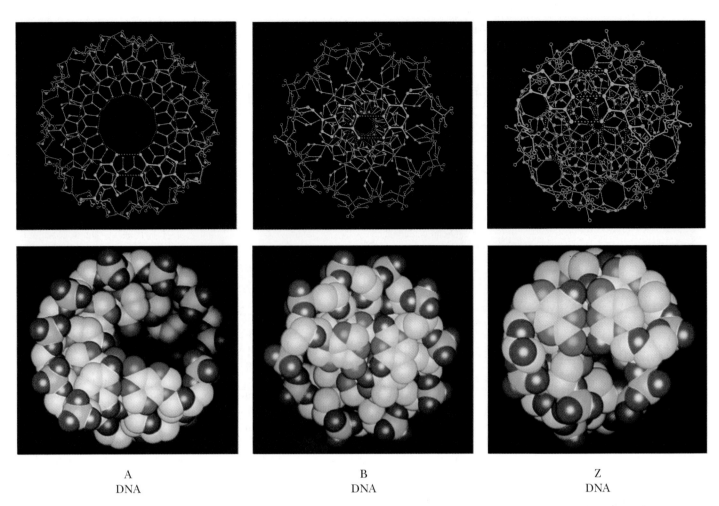

A
DNA

B
DNA

Z
DNA

Figure 8.37 continued Comparison of the A-, B-, and Z-forms of the DNA double helix. The distance required to complete one helical turn is shorter in A-DNA than it is in B-DNA. The alternating pyrimidine–purine sequence of Z-DNA is the key to the "left-handedness" of this helix. (*Robert Stodola, Fox Chase Cancer Research Center, and Irving Geis*)

In A-DNA, the base pairs are no longer nearly perpendicular to the helix axis but instead are tilted 19° with respect to this axis. Successive base pairs occur every 0.23 nm along the axis, as opposed to 0.332 nm in B-DNA. The B-form of DNA is thus longer and thinner than the short, squat A-form, which has its base pairs displaced around, rather than centered on, the helix axis. Figure 8.37 shows the relevant structural characteristics of the A- and B-forms of DNA. (Z-DNA, another form of DNA to be discussed shortly, is also depicted in Figure 8.37.) A comparison of the structural properties of A-, B-, and Z-DNA is summarized in Table 8.4.

Although relatively dehydrated DNA fibers can be shown to adopt the A-conformation under physiological conditions, it is unclear whether DNA ever assumes this form *in vivo*. However, double-helical DNA : RNA hybrids probably have an A-like conformation. The 2′-OH in RNA sterically prevents double-helical regions of RNA chains from adopting the B-form helical arrangement. Consequently, double-stranded regions in RNA chains assume an A-like conformation, with their bases strongly tilted with respect to the helix axis.

Z-DNA: A Left-Handed Double Helix

Z-DNA was first recognized during X-ray analysis of the synthetic deoxynucleotide dCpGpCpGpCpG, which crystallized into an antiparallel double helix of unexpected conformation. The alternating pyrimidine–purine (Py–Pu) sequence of this oligonucleotide is the key to its unusual properties. The *N*-glycosyl bonds of G residues in this alternating copolymer are rotated 180°

Table 8.4   Comparison of the Structural Properties of A-, B-, and Z-DNA

	Double Helix Type		
	A	**B**	**Z**
Overall proportions	Short and broad	Longer and thinner	Elongated and slim
Rise per base pair	2.3 Å	3.32 Å ± 0.19 Å	3.8 Å
Helix packing diameter	25.5 Å	23.7 Å	18.4 Å
Helix rotation sense	Right-handed	Right-handed	Left-handed
Base pairs per helix repeat	1	1	2
Base pairs per turn of helix	~11	~10	12
Mean rotation per base pair	33.6°	35.9° ± 4.2°	−60°/2
Pitch per turn of helix	24.6 Å	33.2 Å	45.6 Å
Base-pair tilt from the perpendicular	+19°	−1.2° ± 4.1°	−9°
Helix axis location	Major groove	Through base pairs	Minor groove
Glycosyl bond conformation	anti	anti	anti at C, syn at G

Adapted from Dickerson, R. L., et al., 1982. *Cold Spring Harbor Symposium on Quantitative Biology* **47**:14.

with respect to their conformation in B-DNA, so that now the purine ring is over the deoxyribose ring (syn conformation) rather than the usually favored anti conformation (Figure 8.38). Because the G ring is "flipped," the C ring must also flip to maintain normal Watson–Crick base pairing. However, pyrimidine nucleosides do not readily adopt the syn conformation because it creates steric interference between the pyrimidine C-2 oxy substituent and the pentose ring. Since the pyrimidine ring does not rotate relative to the pentose, the whole C nucleoside (base and sugar) must flip 180°. The alternating *anti*-Pyr/*syn*-Pu arrangement in each strand favors a conformational transition that realigns the sugar–phosphate backbone along a zigzag course that has a left-handed orientation—thus the designation *Z-DNA*. It is topologically possible for the G to go syn and the C nucleoside to undergo rotation by 180° without breaking and re-forming the G:C hydrogen bonds (Figure 8.39). In other words, the B to Z structural transition can take place without disruption of the bonding relationships among the atoms involved. Z-DNA is more elongated and slimmer than B-DNA. It is likely that the Z-conformation naturally occurs in specific regions of cellular DNA, which otherwise is predominantly in the B-form.

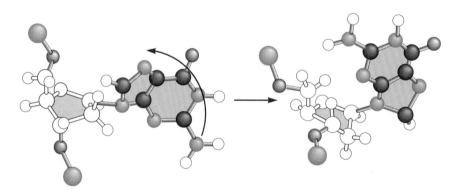

Deoxyguanosine in B-DNA (anti position)   Deoxyguanosine in Z-DNA (syn position)

Figure 8.38   Comparison of the deoxyguanosine conformation in B- and Z-DNA. The C1′–N glycosyl bond of nucleosides and nucleotides prefers the anti conformation, where the base is not over the furanose ring. All bases in B-DNA are in this anti conformation. In contrast, in the left-handed Z-DNA structure, the C1′–N-9 glycosyl bond of deoxy-guanosine rotates (as shown) to adopt the syn conformation.

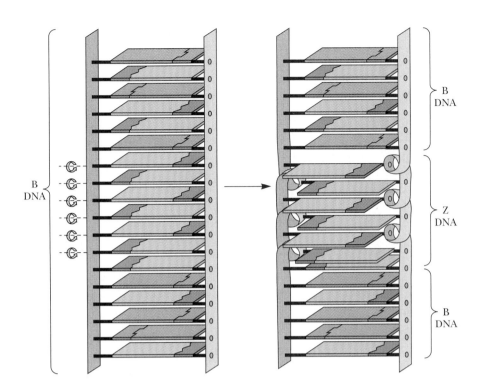

Figure 8.39 The change in topological relationships of base pairs from B- to Z-DNA. A six-base-pair segment of B-DNA is converted to Z-DNA through rotation of the base pairs, as indicated by the curved arrows. The purine rings (green) of the deoxyguanosine nucleosides rotate via an anti to syn change in the conformation of the guanine–deoxyribose glycosidic bond; the pyrimidine rings (blue) are rotated by flipping the entire deoxycytidine nucleoside (base *and* deoxyribose). As a consequence of these conformational changes, the base pairs in the Z-DNA region no longer share π, π stacking interactions with adjacent B-DNA regions.

The Double Helix in Solution

B-DNA in solution is not a rigid, linear rod but instead behaves as a dynamic, flexible molecule responding to localized thermal fluctuations that temporarily distort and deform its structure over short regions. Base and backbone ensembles of atoms undergo elastic motions on a time scale of nanoseconds. To some extent, these effects represent changes in rotational angles of the bonds comprising the polynucleotide backbone. These changes are also influenced by sequence-dependent variations in base-pair stacking. The consequence is that the helix bends gently. When these variations are summed over the great length of a DNA molecule, the net result of these bending motions is that, at any given time, the double helix assumes a roughly spherical shape, as might be expected for a long, semirigid rod undergoing apparently random coiling. It is also worth noting that, on close scrutiny, the surface of the double helix is *not* that of a totally featureless, smooth, regular "barber pole" structure. Local variations in base sequence impart their own special signature by subtle influences on such factors as the groove width, the angle between the helix axis and base planes, and the mechanical rigidity. Certain regulatory proteins bind to specific DNA sequences and participate in activating or suppressing expression of the information encoded therein. These proteins bind at unique sites by virtue of their ability to recognize novel structural characteristics imposed on the DNA by the local nucleotide sequence.

8.10 Denaturation and Renaturation of DNA

Thermal Denaturation and Hyperchromic Shift

When duplex DNA molecules are subjected to conditions of temperature, pH, or ionic strength that disrupt hydrogen bonds, the strands are no longer held together. That is, the double helix is **denatured** and the two strands separate to form single-stranded DNA. If temperature is the denaturing agent, the double helix is said to *melt.* The course of this dissociation can be followed spectrophotometrically because the relative absorbance of the DNA solution at 260 nm increases as much as 40% as the bases unstack. This absorbance increase, or **hyperchromic shift,** is due to the fact that the aromatic bases in DNA interact via their π electron clouds when stacked together in the double helix. Since the UV absorbance of the bases is a consequence of π electron transitions, and since the potential for these transitions is diminished when the bases stack, the bases in duplex DNA absorb less 260-nm radiation than expected for their numbers. Unstacking alleviates this suppression of UV absorbance. The rise in absorbance coincides with strand separation, and the midpoint of the absorbance increase is termed the **melting temperature,** T_m (Figure 8.40). DNAs differ in their T_m values because they differ in relative G + C content. The higher the G + C content of a DNA, the higher its melting temperature because G:C pairs are held by three H bonds, whereas A:T pairs have only two. pH extremes or strong H-bonding solutes also denature DNA duplexes.

DNA Renaturation

Denatured DNA will **renature** to re-form the duplex structure if the denaturing conditions are removed (that is, if the solution is cooled, the pH is returned to neutrality, or the denaturants are diluted out). Renaturation requires reassociation of the DNA strands into a double helix, a process termed **reannealing.** For this to occur, the strands must realign themselves so that their complementary bases are once again in register and the helix can be zippered up (Figure 8.41).

Nucleic Acid Hybridization

If DNA from two different species are mixed, denatured, and the solution of single-stranded DNA is allowed to cool slowly so that reannealing can occur, artificial **hybrid duplexes** may form, provided the DNA from one species is similar in nucleotide sequence to the DNA of the other. The degree of hybridization is a measure of the sequence similarity or *relatedness* between the two species. Depending on the conditions of the experiment, about 25% of the DNA from a human will form hybrids with mouse DNA, implying that some of the nucleotide sequences (genes) in humans are very similar to those in mice (Figure 8.42). Mixed RNA:DNA hybrids can be created *in vitro* if single-stranded DNA is allowed to anneal with RNA copies of itself, such as those formed when genes are transcribed into mRNA molecules.

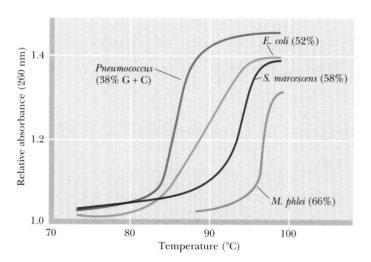

Figure 8.40 Heat denaturation of DNA from various sources, so-called melting curves. The midpoint of the melting curve is defined as the melting temperature, T_m. (*From Marmur, J., 1959. Nature* **183:***1427–1429.*)

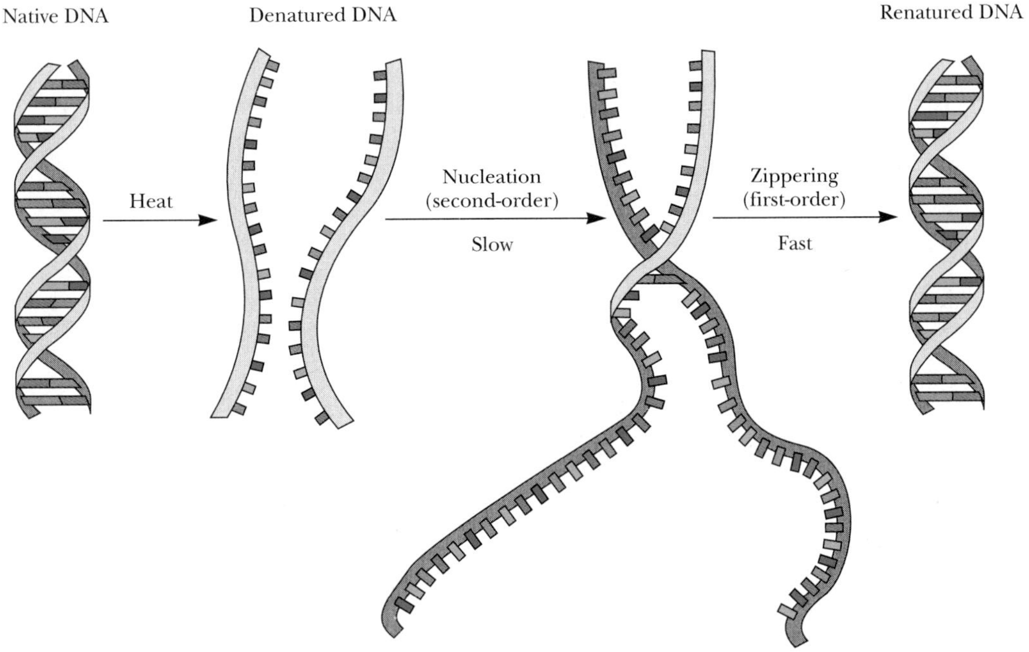

Native DNA   Denatured DNA   Renatured DNA

Heat → Nucleation (second-order) Slow → Zippering (first-order) Fast →

Figure 8.41 Steps in the thermal denaturation and renaturation of DNA. The nucleation phase of the reaction is a second-order process depending on sequence alignment of the two strands. This process takes place slowly, since it takes time for complementary sequences to encounter one another in solution and then align themselves in register. Once the sequences are aligned, the strands zipper up quickly.

Nucleic acid hybridization is a commonly employed procedure in molecular biology. First, it can reveal evolutionary relationships. Second, it gives researchers the power to identify specific genes selectively against a vast background of irrelevant genetic material. An appropriately labeled oligo- or polynucleotide, referred to as a **probe,** is constructed so that its sequence is complementary to a target gene. The probe specifically base-pairs with the target gene, allowing identification and subsequent isolation of the gene. Also, the quantitative expression of genes (in terms of the amount of mRNA synthesized) can be assayed by hybridization experiments.

8.11 Supercoils and Cruciforms: Tertiary Structure in DNA

The conformations of DNA discussed thus far are variations sharing a common secondary structural theme, the double helix, in which the DNA is assumed to be in a regular, linear form. DNA can also adopt regular structures of higher complexity in several ways. For example, many DNA molecules are circular. Most, if not all, bacterial chromosomes are covalently closed, circular DNA duplexes, as are almost all plasmid DNAs. **Plasmids** are naturally occurring, self-replicating, circular, extrachromosomal DNA molecules found in bacteria; plasmids carry genes specifying novel metabolic capacities advantageous to the host bacterium. Various animal virus DNAs are circular as well.

Supercoils

In duplex DNA, the two strands wind around each other about once every 10 bp—that is, once every turn of the helix. Double-stranded circular DNA (or

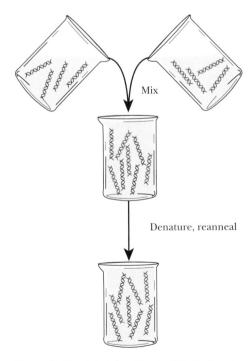

Figure 8.42 Solutions of human DNA (red) and mouse DNA (blue) are mixed, denatured, and the single strands allowed to reanneal. About 25% of the human DNA strands form hybrid duplexes (one red and one blue strand) with mouse DNA.

(a)

(b)

(c)

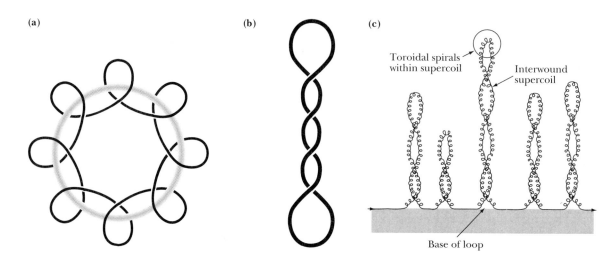

Toroidal spirals
within supercoil

Interwound
supercoil

Base of loop

Figure 8.43 Toroidal and interwound varieties of DNA supercoiling. **(a)** The DNA (black) is coiled in a spiral fashion about an imaginary toroid (yellow ring). **(b)** The DNA interwinds and wraps about itself. **(c)** Supercoils in long, linear DNA arranged into loops whose ends are restrained—a model for chromosomal DNA. *(Adapted from Figures 6.1 and 6.2 in Callandine, C. R., and Drew, H. R., 1992.* Understanding DNA: The Molecule and How It Works. *London: Academic Press.)*

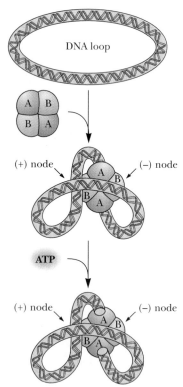

DNA loop

(+) node   (−) node

ATP

(+) node   (−) node

DNA is cut and a conformational change
allows the DNA to pass through. Gyrase
religates the DNA and then releases it.

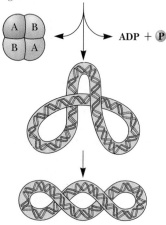

ADP + P

linear DNA duplexes whose ends are not free to rotate) can form **supercoils** if the strands are underwound (*negatively supercoiled*) or overwound (*positively supercoiled*) (Figure 8.43). DNA supercoiling is analogous to twisting or untwisting a multistranded rope so that it is torsionally stressed. Negative supercoiling introduces a torsional stress that favors unwinding of the right-handed B-DNA double helix, while positive supercoiling overwinds such a helix. Both forms of supercoiling compact the DNA into a smaller volume, in comparison to **relaxed DNA** (DNA that is not supercoiled).

Supercoiling is created by breaking one or both strands of the DNA, winding them tighter or looser, and rejoining the ends. Enzymes capable of carrying out such reactions are called **topoisomerases** because they change the topological state of DNA. Topoisomerases are important players in DNA replication (see Chapter 23).

DNA Gyrase

The bacterial enzyme **DNA gyrase** is a topoisomerase that introduces negative supercoils into DNA in the manner shown in Figure 8.44. Suppose DNA gyrase puts four negative supercoils into the 400-bp circular duplex (Figure 8.45). In actuality, the negative supercoils cause a torsional stress on the molecule so that the helix becomes a bit unwound and the base pairs are separated. The extreme would be an unwinding to the extent that the apparent supercoiling is removed. Usually the actual situation is a compromise in which these changes are distributed over the length of the circular duplex so that no localized unwinding of the helix ensues. Circular DNA isolated from natural sources is always found in the underwound, negatively supercoiled state.

◀ **Figure 8.44** A simple model for the action of bacterial DNA gyrase (topoisomerase II). The A subunits cut the DNA duplex and then hold onto the cut ends. Conformational changes occur in the enzyme that allow a continuous region of the DNA duplex to pass between the cut ends and into an internal cavity of the protein. The cut ends are then re-ligated, and the intact DNA duplex is released from the enzyme. The released intact circular DNA now contains two negative supercoils as a consequence of DNA gyrase action.

Cruciforms

Palindromes are words, phrases, or sentences that are the same when read backward or forward, such as "radar," "sex at noon taxes," "Madam, I'm Adam," and "a man, a plan, a canal, Panama." DNA sequences that are **inverted repeats,** or palindromes, have the potential to form a tertiary structure known as a **cruciform** (literally meaning "cross-shaped") if the normal interstrand base pairing is replaced by intrastrand pairing (Figure 8.46). In effect, each DNA strand folds back on itself in a hairpin structure to align the palindrome in base-pairing register. Such cruciforms are never as stable as normal DNA duplexes because an unpaired segment must exist in the loop region. However, negative supercoiling causes a localized disruption of hydrogen bonding between base pairs in DNA and may promote formation of cruciform loops. Cruciform structures have a two-fold rotational symmetry about their centers and potentially create distinctive recognition sites for specific DNA-binding proteins.

8.12 Chromosome Structure

A typical human cell is 20 μm in diameter. Its genetic material consists of 23 pairs of dsDNA molecules in the form of **chromosomes,** the average length of which is 3×10^9 bp/23 or 1.3×10^8 nucleotide pairs. At 0.34 nm/bp in B-DNA, this represents a DNA molecule 5 cm long. Together, these 46 dsDNA molecules amount to more than 2 m of DNA that must be packaged into a nucleus perhaps 5 μm in diameter! Clearly, the DNA must be condensed by a factor of more than 10^5. This remarkable task is accomplished by neatly wrapping the DNA around protein spools called **nucleosomes** and then packing the nucleosomes to form a helical filament that is arranged in loops associated with the **nuclear matrix,** the scaffold of proteins providing a structural framework within the nucleus.

Nucleosomes

The DNA in a eukaryotic cell nucleus during the interphase between cell divisions exists as a nucleoprotein complex called **chromatin.** The proteins of

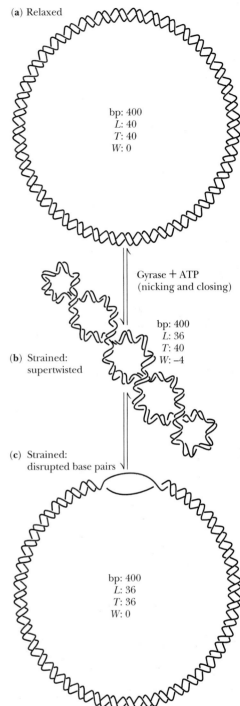

(a) Relaxed

bp: 400
L: 40
T: 40
W: 0

Gyrase + ATP
(nicking and closing)

bp: 400
L: 36
T: 40
W: −4

(b) Strained: supertwisted

(c) Strained: disrupted base pairs

bp: 400
L: 36
T: 36
W: 0

Figure 8.45 A 400-bp circular DNA molecule in different topological states: **(a)** relaxed, **(b)** negative supercoils distributed over the entire length, and **(c)** negative supercoils creating a localized single-stranded region. Negative supercoiling has the potential to cause localized unwinding of the DNA double helix so that single-stranded regions (or bubbles) are created.

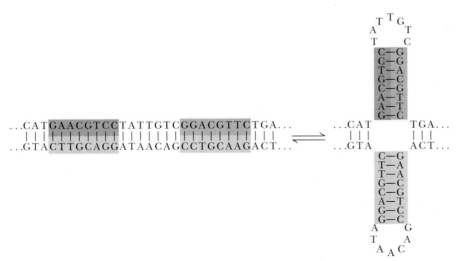

Figure 8.46 The formation of a cruciform structure from a palindromic sequence within DNA. The self-complementary inverted repeats can rearrange to form hydrogen-bonded cruciform stem-loops.

Table 8.5 Properties of Histones

Histone	Ratio of Lysine to Arginine	M_r	Copies per Nucleosome
H1	59/3	21,200	1 (not in bead)
H2A	13/13	14,100	2 (in bead)
H2B	20/8	13,900	2 (in bead)
H3	13/17	15,100	2 (in bead)
H4	11/14	11,400	2 (in bead)

chromatin fall into two classes: **histones** and **nonhistone chromosomal proteins.** Histones are abundant structural proteins, whereas the nonhistone class is represented only by a few copies each of many diverse proteins involved in genetic regulation. The histones are relatively small, positively charged arginine- or lysine-rich proteins that interact via ionic bonds with the negatively charged phosphate groups on the polynucleotide backbone. Five distinct histones are known: **H1, H2A, H2B, H3,** and **H4** (Table 8.5). Pairs of histones H2A, H2B, H3, and H4 aggregate to form an octameric structure, the core of the **nucleosome,** around which the DNA helix is wound (Figure 8.47).

If chromatin is swelled suddenly in water and prepared for viewing in the electron microscope, the nucleosomes have the appearance of "beads on a string," the dsDNA being the string (Figure 8.48). The structure of the histone octamer has been determined using X-ray crystallography both without DNA (by E. N. Moudrianakis's laboratory [see Figure 8.47]) and wrapped with DNA (by T. J. Richmond and collaborators [Figure 8.49]). The octamer (see Figure 8.47) has surface landmarks that guide the course of the DNA around the octamer. In a flat, left-handed superhelical conformation, 146 bp of B-DNA make 1.65 turns around the histone core (see Figure 8.49). The core octamer is itself a protein superhelix consisting of a spiral array of the four histone dimers. Histone 1, a three-domain protein, serves to seal the ends of the DNA turns to the nucleosome core and to organize the additional 40 to 60 bp of DNA that link two successive nucleosomes.

(a)

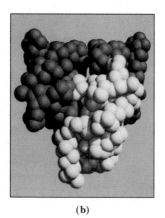
(b)

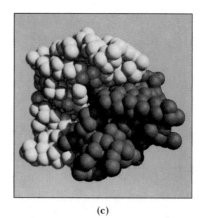
(c)

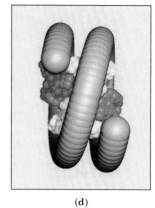
(d)

Figure 8.47 Four orthogonal views of the histone octamer as determined by X-ray crystallography: **(a)** front view, **(b)** top view, and **(c)** disk view— that is, as viewed down the long axis of the chromatin fiber. In the (c) perspective, the DNA duplex would wrap around the octamer, with the axis of the DNA supercoil perpendicular to the plane of the picture. **(d)** Suggested appearance of the nucleosome when wrapped with DNA. *(Photographs courtesy of Evangelos N. Moudrianakis of Johns Hopkins University)*

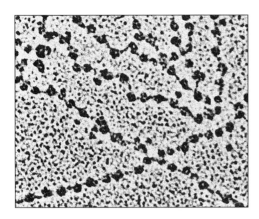

Figure 8.48 Electron micrograph of *Drosophila melanogaster* chromatin after swelling reveals the presence of nucleosomes as "beads on a string." *(Electron micrograph courtesy of Oscar L. Miller, Jr., of the University of Virginia)*

Organization of Chromatin and Chromosomes

A higher order of chromatin structure is created when the nucleosomes, in their characteristic beads-on-a-string motif, are wound in the fashion of a *solenoid* having six nucleosomes per turn (Figure 8.50). The resulting 30-nm filament contains about 1200 bp in each of its solenoid turns. Interactions between the respective H1 components of successive nucleosomes stabilize the 30-nm filament, which then forms long DNA loops of variable length, each containing on average between 60,000 and 150,000 bp. Electron microscopic analysis of human chromosome 4 suggests that 18 such loops are then arranged radially about the circumference of a single turn to form a **miniband unit** of the chromosome. According to this model, approximately 10^6 of these minibands are arranged along a central axis in each of the chromatids of human chromosome 4 that form at mitosis (see Figure 8.50). Despite intensive study, much of the higher-order structure of chromosomes remains a mystery.

solenoid   a coil wound in the form of a helix

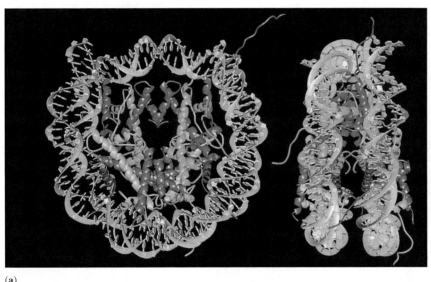

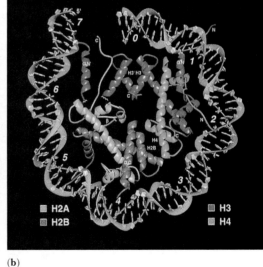

| ◼ H2A | ◼ H3 |
| ◼ H2B | ◼ H4 |

(a) (b)

Figure 8.49   **(a)** Deduced structure of the nucleosome core particle wrapped with 1.65 turns of DNA (146 bp). The DNA is shown as a ribbon. (*left*) View down the axis of the nucleosome; (*right*) view perpendicular to the axis.   **(b)** One-half of the nucleosome core particle with 73 bp of DNA, as viewed down the nucleosome axis. Note that the DNA does not wrap in a uniform circle about the histone core, but instead follows a course consisting of a series of somewhat straight segments separated by bends. (*Adapted from Luger, C., et al., 1997. Crystal structure of the nucleosome core particle at 2.8 Å resolution. Nature* **389**:251–260. *Photos courtesy of T. J. Richmond, ETH-Hönggerberg, Zurich, Switzerland.*)

Figure 8.50 A model for chromosome structure, human chromosome 4. The 2-nm DNA helix is wound twice around histone octamers to form 10-nm nucleosomes, each of which contains 160 bp (80 per turn). These nucleosomes are then wound in solenoid fashion with 6 nucleosomes per turn to form a 30-nm filament. In this model, the 30-nm filament forms long DNA loops, each containing about 60,000 bp, which are attached at their base to the nuclear matrix. Eighteen of these loops are then wound radially around the circumference of a single turn to form a miniband unit of a chromosome. Approximately 10^6 of these minibands occur in each chromatid of human chromosome 4 at mitosis.

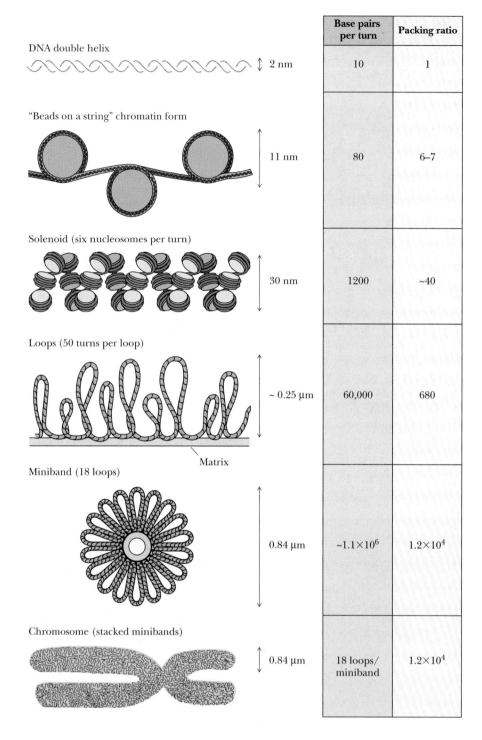

	Base pairs per turn	Packing ratio
DNA double helix — 2 nm	10	1
"Beads on a string" chromatin form — 11 nm	80	6–7
Solenoid (six nucleosomes per turn) — 30 nm	1200	~40
Loops (50 turns per loop) — ~0.25 μm — Matrix	60,000	680
Miniband (18 loops) — 0.84 μm	~1.1×10⁶	1.2×10⁴
Chromosome (stacked minibands) — 0.84 μm	18 loops/ miniband	1.2×10⁴

8.13 Chemical Synthesis of Nucleic Acids

Laboratory synthesis of oligonucleotide chains of defined sequence presents some formidable synthetic problems. First, functional groups on the monomeric units (in this case, bases) are reactive under conditions of polymerization and therefore must be protected by blocking agents. Second, to generate the desired sequence, a phosphodiester bridge must be formed between the 3′-O of one nucleotide (B) and the 5′-O of the preceding one (A) in a way that prevents the unwanted bridging of the 3′-O of A with the 5′-O of B. Finally, recoveries at each step must be high so that overall yields in the multistep process are acceptable. *Solid phase methods* are used to overcome some of these problems. Commercially available automated instruments, called **DNA synthesizers** or "gene machines," are capable of carrying out the synthesis of oligonucleotides of 150 bases or more.

Phosphoramidite Chemistry

Phosphoramidite chemistry is currently the accepted method of oligonucleotide synthesis. The general strategy involves the sequential addition of nucleotide units, as *nucleoside phosphoramidite* derivatives, to a nucleoside covalently attached to the insoluble resin. Excess reagents, starting materials, and side products are removed after each step by filtration. After the desired

---

HUMAN BIOCHEMISTRY

Telomeres and Tumors

Eukaryotic chromosomes are linear. The ends of chromosomes have specialized structures known as **telomeres.** The telomeres of virtually all eukaryotic chromosomes consist of short, tandemly repeated nucleotide sequences at the ends of the chromosomal DNA. For example, the telomeres of human germline cells contain between 1000 and 1700 copies of the hexameric repeat TTAGGG (see figure). Telomeres are believed to be responsible for maintaining chromosomal integrity by protecting against DNA degradation or rearrangement. They are added to the ends of chromosomal DNA by the enzyme **telomerase** (see Chapter 23), an unusual DNA polymerase discovered in 1985 by Elizabeth Blackburn and Carol Greider of the University of California, San Fran-

cisco. However, most normal somatic cells lack telomerase. Consequently, upon every cycle of cell division, when the cell replicates its DNA, about 50-nucleotide portions are lost from the end of each telomere. Thus, over time, the telomeres of somatic cells in animals become shorter and shorter, eventually leading to chromosome instability and cell death. This phenomenon has led some scientists to espouse a "telomere theory of aging" that implicates telomere shortening as the principal factor in cell, tissue, and even organism aging. Interestingly, cancer cells appear "immortal," since they continue to reproduce indefinitely. A survey of 20 different tumor types by Geron Corporation of Menlo Park, California, revealed that all contained telomerase activity.

(a)

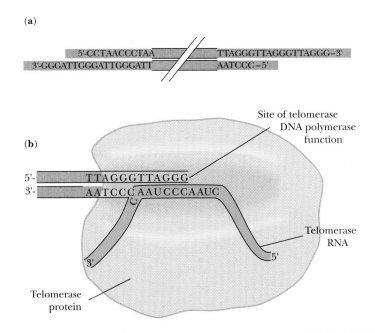

(**a**) Telomeres on human chromosomes consist of the hexanucleotide sequence TTAGGG repeated between 1000 and 1700 times. These TTAGGG tandem repeats are attached to the 3′-ends of the DNA strands and are paired with the complementary sequence 3′-AATCCC-5′ on the other DNA strand. Thus, a G-rich region is created at the 3′-end of each DNA strand and a C-rich region is created at the 5′-end of each DNA strand. Typically, at each end of the chromosome, the G-rich strand protrudes 12 to 16 nucleotides beyond its complementary C-rich strand. (**b**) Like other telomerases, human telomerase is a ribonucleoprotein. The ribonucleic acid of human telomerase is an RNA molecule 962 nucleotides long. This RNA serves as the template for the DNA polymerase activity of telomerase. Nucleotides 46 to 56 of this RNA are CUAA**CCCUAA**C and provide the template function for the telomerase-catalyzed addition of TTAGGG units to the 3′-end of a DNA strand.

Table 8.6 Some Chemically Synthesized Genes

Gene	Size (bp)
tRNA	126
α-Interferon	542
Secretin	81
γ-Interferon	453
Rhodopsin	1057
Proenkephalin	77
Connective tissue activating peptide III	280
Lysozyme	385
Tissue plasminogen activator	1610
c-Ha-ras	576
RNase T1	324
Cytochrome b_5	330
Bovine intestinal Ca-binding protein	298
Hirudin	226
RNase A	375

oligonucleotide has been formed, it is freed of all blocking groups, hydrolyzed from the resin, and purified by gel electrophoresis. The four-step cycle is shown in Figure 8.51. Chemical synthesis takes place in the $3' \rightarrow 5'$ direction (the reverse of the biological polymerization direction).

Chemically Synthesized Genes

Table 8.6 lists some of the genes that have been chemically synthesized. Since protein-coding genes are characteristically much larger than the 150-bp practical limit on oligonucleotide synthesis, total gene synthesis involves joining a series of oligonucleotides to assemble the overall sequence. A prime example is the gene for rhodopsin, which was synthesized by assembling 72 oligonucleotide fragments, 36 for each strand. (The rhodopsin gene is 1057 base pairs long and encodes the 348-amino acid photoreceptor protein of the vertebrate retina.) Theoretically, no gene is beyond the scope of these methods, opening the door to an incredibly exciting range of possibilities for investigating

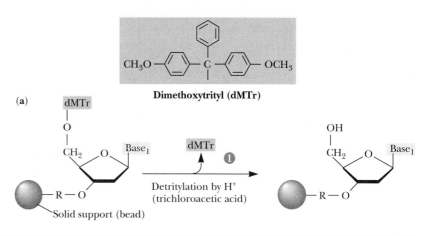

Figure 8.51 Solid phase oligonucleotide synthesis. The four-step cycle starts with the first base in nucleoside form (N-1) attached by its 3′-OH group to a bead. Its 5′-OH is blocked with a dimethoxytrityl (DMTr) group **(a)**. If the base has reactive —NH₂ functions, as in A, G, or C, then N-benzoyl or N-isobutyryl derivatives are used to prevent their reaction **(b)**. In step 1, the DMTr protecting group is removed. Step 2 is the coupling step: the second base (N-2) is added in the form of a nucleoside phosphoramidite derivative whose 5′-OH bears a DMTr blocking group so it cannot polymerize with itself **(c)**.

continued

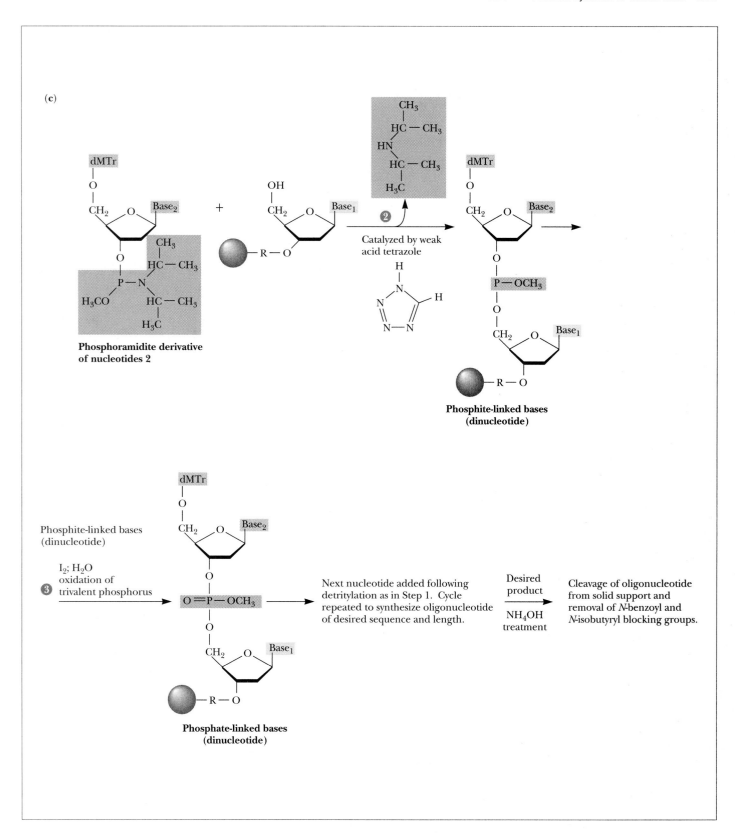

Figure 8.51 continued The presence of a weak acid activates the phosphoramidite, and it rapidly reacts with the free 5′-OH of N-1, forming a dinucleotide linked by a phosphite group. The phosphite linkage between N-1 and N-2 is highly reactive and, in step 3, it is oxidized by aqueous iodine (I_2) to form the desired more stable phosphate group. This completes the cycle. Subsequent cycles add successive residues to the resin-immobilized chain. When the chain is complete, it is cleaved from the support with NH_4OH.

structure–function relationships in the organization and expression of hereditary material.

8.14 Secondary and Tertiary Structure of RNA

RNA molecules (see Section 8.6) are typically single-stranded. Nevertheless, they are often rich in double-stranded regions that form when complementary sequences within the chain come together and join via **intrastrand hydrogen bonding.** These interactions create **stem-loop structures,** in which the base-paired regions are the stem and the unpaired regions between base pairs are the loop (see Figures 8.52 and 8.57). RNA strands cannot fold to form B-DNA–type double helices because their 2′-OH groups are a steric hindrance to this conformation. Instead, RNA double helices adopt a conformation similar to the A-form of DNA, having about 11 bp per turn, and bases strongly tilted from the plane perpendicular to the helix axis (see Figure 8.37). Both tRNA and rRNA have large amounts of A-form double-helical secondary structures. In addition, a number of defined structural motifs recur within the loops of their stem-loops. Secondary structures exist in mRNA species as well, although their nature is unique to each particular mRNA. (The functions of tRNA, rRNA, and mRNA are discussed in detail in Part 4: Information Transfer.)

Transfer RNA Structure

In tRNA molecules, which contain from 73 to 94 nucleotides in a single chain, a majority of the bases are hydrogen-bonded to one another. Figure 8.52 shows the structure that typifies tRNAs. *Hairpin turns* bring complementary stretches of bases in the chain into contact so that double-helical regions form. Because of the arrangement of these complementary stretches along the chain, the overall pattern of H bonding can be represented as a *cloverleaf.* Each cloverleaf consists of four H-bonded segments—three loops and the stem where the 3′- and 5′-ends of the molecule meet. These four segments are designated the **acceptor stem,** the **D loop,** the **anticodon loop,** and the **TψC loop.**

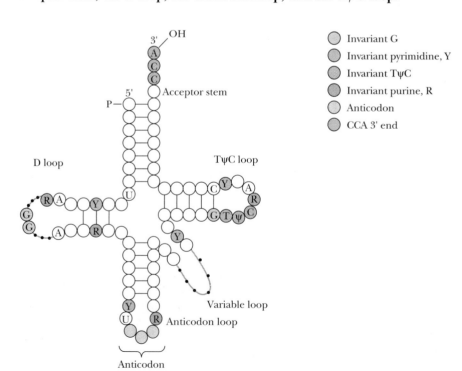

Figure 8.52 A general diagram for the structure of tRNA. The positions of invariant bases as well as bases which seldom vary are shown in color. The numbering system is based on yeast tRNA^Phe. R = purine; Y = pyrimidine. Dotted lines denote sites in the D loop and variable loop regions where varying numbers of nucleotides are found in different tRNAs.

tRNA Secondary Structure

The *acceptor stem* is where the amino acid is linked to form the aminoacyl-tRNA derivative, which serves as the amino acid–donating species in protein synthesis; this is the physiological role of tRNA. The amino acid adds to the 3′-OH of the 3′-terminal A nucleotide (see Figure 8.53). The 3′-end of tRNA is invariantly CCA-3′-OH. This CCA sequence plus a fourth nucleotide extends beyond the double-helical portion of the acceptor stem. The *D loop* is so named because this tRNA loop often contains dihydrouridine, or D, residues, in addition to other unusual bases (see Figure 8.24). The *anticodon stem-loop* consists of a double-helical segment and seven unpaired bases, three of which form the **anticodon**—a three-nucleotide unit that recognizes and base-pairs with a particular mRNA **codon,** a complementary three-base unit in mRNA which is the genetic information that specifies an amino acid. Anticodon base pairing to the codon on mRNA allows a particular tRNA species to deliver its amino acid to the protein-synthesizing apparatus. Next along the tRNA sequence in the 5′ → 3′ direction comes a loop that varies from tRNA to tRNA in the number of residues that it has—the so-called **extra** or **variable loop.** The last loop in the tRNA, reading 5′ → 3′, is the *TψC loop*, which contains seven unpaired bases among which is virtually always the sequence TψC, where ψ is the symbol for **pseudouridine.** Ribosomes bind tRNAs through recognition of this TψC loop. Almost all of the invariant residues common to tRNAs lie within the non–hydrogen-bonded regions of the cloverleaf structure (see Figure 8.52). Figure 8.54 depicts the complete nucleotide sequence and cloverleaf structure of yeast phenylalanine tRNA.

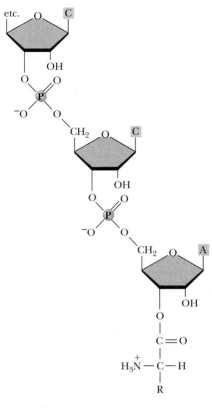

Figure 8.53 Amino acids are linked to the 3′-OH end of tRNA molecules by an ester bond formed between the carboxyl group of the amino acid and the 3′-OH of the terminal ribose of the tRNA.

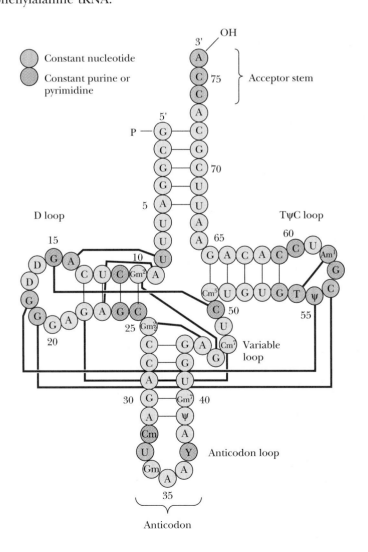

Figure 8.54 The primary structure (nucleotide sequence), secondary structure (cloverleaf), and tertiary structural interactions in yeast phenylalanine tRNA. The molecule is presented in the conventional cloverleaf secondary structure generated by intrastrand hydrogen bonding. Solid lines connect bases that are hydrogen-bonded when this cloverleaf pattern is folded into the characteristic tRNA tertiary structure.

(a)

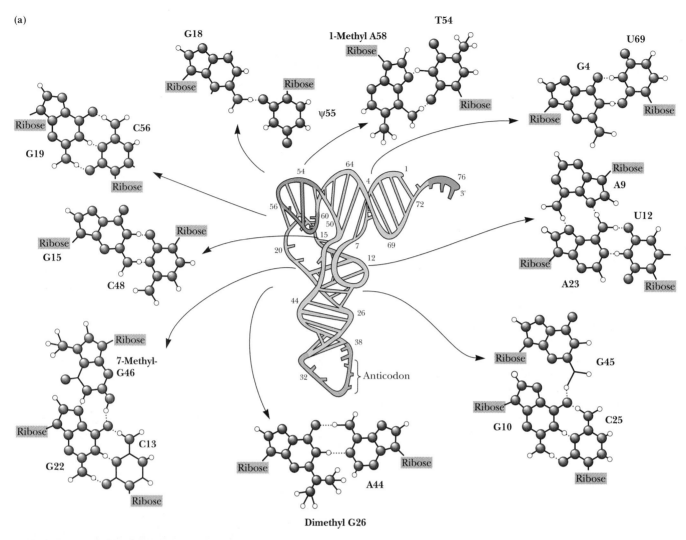

(b)

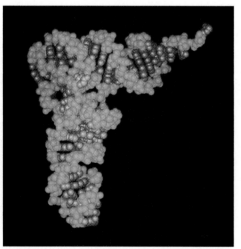

Figure 8.55 **(a)** The three-dimensional structure of yeast phenylalanine tRNA as deduced from X-ray diffraction studies of its crystals. The tertiary folding is illustrated in the center of the diagram with the ribose–phosphate backbone presented as a continuous ribbon; H bonds are indicated by crossbars. Unpaired bases are shown as short, unconnected rods. The anticodon loop is at the bottom and the -CCA 3′-OH acceptor end is at the top right. The various types of noncanonical hydrogen-bonding interactions observed between bases surround the central molecule. Three of these structures show examples of unusual H-bonded interactions involving three bases; these interactions aid in establishing tRNA tertiary structure. **(b)** A space-filling model of the molecule. *(After Kim, S. H., in Schimmel, P., Söll, D., and Abelson, J. N., eds., 1979. Transfer RNA: Structure, Properties, and Recognition. New York: Cold Spring Harbor Laboratory.)*

tRNA Tertiary Structure

Tertiary structure in tRNA arises from hydrogen-bonding interactions between bases in the D loop with bases in the variable and TψC loops, as shown for yeast phenylalanine tRNA in Figure 8.54. Note that these H bonds involve the invariant nucleotides of tRNAs, thus emphasizing the importance of the tertiary structure they create to the function of tRNAs in general. These H bonds fold the D and TψC arms together and bend the cloverleaf into the stable L-shaped tertiary form (Figure 8.55). Many of these H bonds involve base pairs that are not canonical A:T or G:C pairings. The amino acid acceptor stem is at one end of the L, separated by 7 nm or so from the anticodon at the opposite end of the L. The D and TψC loops form the corner of the L. The TψC loop contains a structural motif found in all tRNAs called a **U-turn** (named for the consensus sequence UNRN, where N is any base and R is purine). U-turns are also common in tRNA anticodon loops. In the L-conformation, the bases are oriented to maximize hydrophobic stacking interactions between their flat faces. Such stacking is a second major factor contributing to L-form stabilization.

Ribosomal RNA Structure

rRNA Secondary Structure

A large degree of *intrastrand sequence complementarity* is found in all ribosomal RNA strands, and all assume a highly folded pattern that allows base pairing between these complementary segments. Further, the loop regions of stem-loop structures adopt characteristic structural motifs, such as U-turns and **tetraloops** (another family of 4-base loop motifs). Stems of stem-loop structures may also have bulges (called **internal loops**), where the RNA strand is forced into a short single-stranded loop because a series of bases along one strand in an RNA double helix finds no base-pairing partners. Figure 8.56 shows the secondary structure assigned to the *E. coli* 16S rRNA. This structure is based on computer alignment of the nucleotide sequence into optimal H-bonding segments.

Comparison of rRNAs from Various Species

If a phylogenetic comparison is made of the 16S-like rRNAs from an archaebacterium (*Halobacterium volcanii*), a eubacterium (*E. coli*), and a eukaryote (the yeast *Saccharomyces cerevisiae*), a striking similarity in secondary structure emerges (Figure 8.57). Remarkably, these secondary structures are similar despite a low degree of similarity in the nucleotide sequences of these rRNAs. Apparently, evolution is acting at the level of rRNA secondary structure, not rRNA nucleotide sequence. Similar conserved folding patterns are seen for the 23S-like and 5S-like rRNAs that reside in the large ribosomal subunits of various species. An insightful conclusion may be drawn regarding the persistence of such strong secondary structure conservation despite the millennia that have passed since these organisms diverged: *all ribosomes are constructed to a common design and all function in a similar manner.*

rRNA Tertiary Structure

Only recently has the overall three-dimensional, or tertiary, structure of rRNAs been revealed through X-ray crystallography and cryo-electron microscopy of ribosomes (see Chapter 25). These detailed images of ribosome structure also disclose the tertiary structure of the rRNAs (Figure 8.58), as well as the quaternary interactions that must occur when ribosomal proteins combine with rRNAs and when the ensuing ribonucleoprotein complexes—the small and large subunits—come together to form the complete ribosome. An assortment of tertiary structural features characterizes the tertiary structure of rRNAs. These features arise from **coaxial stacking** (stacking of A-form double-helical regions atop one another), **pseudoknots** (short double helices formed via

(Text continued on page 290)

Figure 8.56 The proposed secondary structure for *E. coli* 16S rRNA, based on comparative sequence analysis in which the folding pattern is assumed to be conserved across different species. The molecule can be subdivided into four domains—**I, II, III,** and **IV**—on the basis of contiguous stretches of the chain that are closed by long-range base-pairing interactions. **I,** the 5′-domain, includes nucleotides 27 through 556. **II,** the central domain, runs from nucleotide 564 to 912. Two domains comprise the 3′-end of the molecule. **III,** the major one, comprises nucleotides 923 to 1391. **IV,** the 3′-terminal domain, covers residues 1392 to 1541.

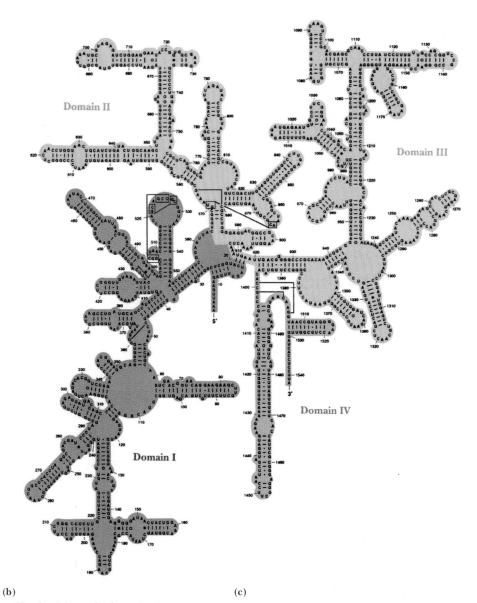

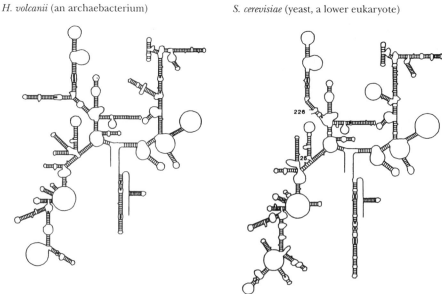

(a) *E. coli* (a eubacterium)

(b) *H. volcanii* (an archaebacterium)

(c) *S. cerevisiae* (yeast, a lower eukaryote)

Figure 8.57 Phylogenetic comparison of secondary structures of 16S-like rRNAs from **(a)** a eubacterium (*E. coli*), **(b)** an archaebacterium (*H. volcanii*), **(c)** a eukaryote (*S. cerevisiae*, a yeast).

288

(a)

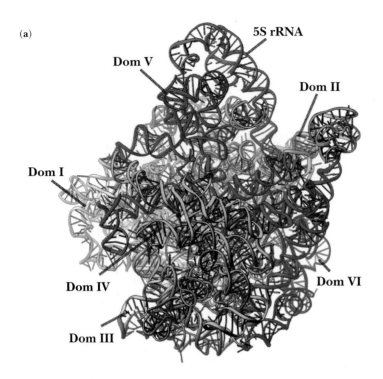

Figure 8.58 The secondary and tertiary structures of rRNAs in the 50S ribosomal subunit from the archaeon *Haloarcula marismortui.* **(a)** Tertiary structure of the rRNAs within the 50S subunit. The 5S rRNA lies atop the 23S rRNA, as indicated. Domains are color-coded according to the schematic in (b). No ribosomal proteins are shown in this illustration, only the 5S and 23S rRNAs. Note that the overall anatomy of the 50S ribosomal subunit (shown diagrammatically in Figure 8.23) is essentially the same as that of the rRNA molecules within this subunit, despite the fact that these rRNAs account for only 65% of the mass of this particle. **(b)** A schematic diagram of the secondary structure of 23S rRNA. **(c)** Secondary structure of the 5S rRNA in the 50S ribosomal subunit. *(Adapted from Figure 4 in Ban, N., et al., 2000. The complete atomic structure of the large ribosomal subunit at 2.4 Å resolution.* Science ***289****:905–920.) (Figure courtesy of Thomas A. Steitz and Peter B. Moore, Yale University.)*

(b)

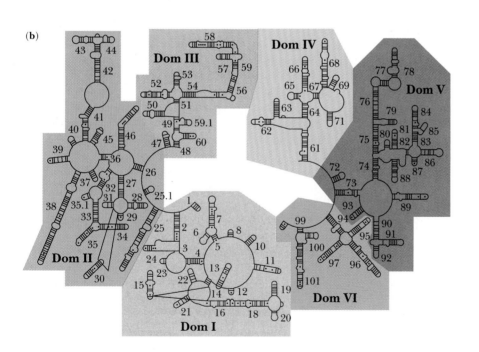

23S rRNA 5′ end　　　　　　**3′ end**

(c)

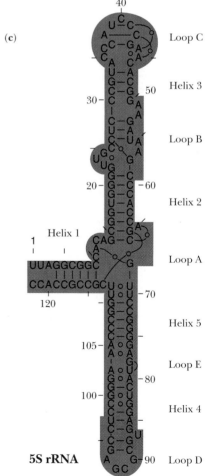

H bonding between bases in the loops of adjacent stem-loop structures), and **ribose zippers** (hydrogen-bonded networks formed between ribose units in the sugar–phosphate backbone that become juxtaposed in the highly folded RNA strand). We will consider the role of rRNA in ribosome structure and function in Chapter 25.

PROBLEMS

1. Draw the chemical structure of pACG.
2. Chargaff's results (see Table 8.2) yielded a molar ratio of 1.56 for A to G in human DNA, 1.75 for T to C, 1.00 for A to T, and 1.00 for G to C. Given these values, what are the mole fractions of A, C, G, and T in human DNA?
3. Adhering to the convention of writing nucleotide sequences in the $5' \rightarrow 3'$ direction, what is the nucleotide sequence of the DNA strand that is complementary to
   d-ATCGCAACTGTCACTA?
4. Messenger RNAs are synthesized by RNA polymerases that read along a DNA template strand in the $3' \rightarrow 5'$ direction, polymerizing ribonucleotides in the $5' \rightarrow 3'$ direction (see Figure 8.22). Give the nucleotide sequence $(5' \rightarrow 3')$ of the DNA template strand from which the following mRNA segment was transcribed: 5'-UAGUGACAGUUGCGAU-3'.
5. The DNA strand that is complementary to the template strand copied by RNA polymerase during transcription has a nucleotide sequence identical to that of the RNA being synthesized (except T residues are found in the DNA strand at sites where U residues occur in the RNA). An RNA copy of this nontemplate DNA strand would be complementary to the mRNA synthesized by RNA polymerase. Such an RNA is called antisense RNA. A promising strategy to thwart the deleterious effects of genes activated in disease states (such as cancer) is to generate antisense RNAs in affected cells. These antisense RNAs would form double-stranded hybrids with mRNAs transcribed from the activated genes and prevent their translation into protein. Suppose transcription of a cancer-activated gene yielded an mRNA whose sequence included the segment 5'-UACGGUCUAAGCUGA. What is the corresponding nucleotide sequence $(5' \rightarrow 3')$ of the template strand in a DNA duplex that might be introduced into these cells so that an antisense RNA could be transcribed from it?
6. A 10-kb DNA fragment digested with restriction endonuclease *Eco*RI yielded fragments 4 kb and 6 kb in size. When digested with *Bam*HI, fragments 1, 3.5, and 5.5 kb were generated. Concomitant digestion with both *Eco*RI and *Bam*HI yielded fragments 0.5, 1, 3, and 5.5 kb in size. Give a possible restriction map for the original fragment.
7. The oligonucleotide d-ATGCCTGACT was subjected to sequencing by Sanger's dideoxy method and the products were analyzed by electrophoresis on a polyacrylamide gel. Draw a diagram of the gel banding pattern obtained.
8. The result of sequence determination of an oligonucleotide as performed by the Sanger dideoxy chain termination method is displayed at the top of the next column.

A	C	G	T
			....
	....		
			....
		....	
....			
....			
	....		
			....
		....	
	....		
		....	
....			

What is the sequence of the original oligonucleotide?

9. X-ray diffraction studies indicate the existence of a novel double-stranded DNA helical conformation in which ΔZ (the rise per base pair) = 0.32 nm and p (the pitch) = 3.36 nm. What are the other parameters of this novel helix: (a) the number of base pairs per turn, (b) $\Delta\phi$ (the mean rotation per base pair), and (c) c (the true repeat)?
10. A 41.5-nm-long duplex DNA molecule in the B-conformation upon dehydration adopts the A-conformation. How long is it now? What is its approximate number of base pairs?
11. If 80% of the base pairs in a 12.5-kbp duplex DNA molecule are in the B-conformation and 20% are in the Z-conformation, what is the length of the molecule?
12. There is one nucleosome for every 200 bp of eukaryotic DNA. How many nucleosomes are in a diploid human cell? Nucleosomes can be approximated as disks 11 nm in diameter and 6 nm long. If all the DNA molecules in a diploid human cell are in the B-conformation, what is the sum of their lengths? If this DNA is now arrayed on nucleosomes in the "beads-on-a-string" motif, what is its approximate total length?
13. The characteristic secondary structures of tRNA and rRNA molecules are achieved through intrastrand hydrogen bonding. Even for the small tRNAs, remote regions of the primary sequence interact via H bonding when the molecule adopts the cloverleaf pattern. Using Figure 8.52 as a guide, draw the primary structure of a tRNA and label the positions of its various self-complementary regions.
14. Using the data in Table 8.2, arrange the DNAs from the following sources in order of increasing T_m: human, salmon, wheat, yeast, *E. coli*.

FURTHER READING

Adams, R. L. P., Knowler, J. T., and Leader, D. P., 1992. *The Biochemistry of the Nucleic Acids*. 11th ed. London: Chapman and Hall.

Arents, G., et al., 1991. The nucleosome core histone octamer at 3.1 Å resolution: A tripartite protein assembly and a left-hand superhelix. *Proceedings of the National Academy of Sciences, U.S.A.* **88:**10,148–10,152.

Axelrod, N., 1996. Of telomeres and tumors. *Nature Medicine* **2:**158–159.

Blackburn, E. H., 1992. Telomerases. *Annual Review of Biochemistry* **61:**113–129.

Callandine, C. R., and Drew, H. R., 1992. *Understanding DNA: The Molecule and How It Works*. London: Academic Press.

Feng, J., Funk, W. D., Wang, S-S., Weinrich, S. L., et al., 1995. The RNA component of human telomerase. *Science* **269:**1236–1241.

Ferretti, L., Karnik, S. S., Khorana, H. G., Nassal, M., and Oprian, D. D., 1986. Total synthesis of a gene for bovine rhodopsin. *Proceedings of the National Academy of Sciences, U.S.A.* **83:**599–603.

Gray, M. W., and Cedergren, R., eds., 1993. The new age of RNA. *The FASEB Journal* **7:**4–239. A collection of articles emphasizing the new appreciation for RNA in protein synthesis, in evolution, and as a catalyst.

Judson, H. F., 1979. *The Eighth Day of Creation*. New York: Simon and Schuster.

Kornberg, A., and Baker, T. A., 1991. *DNA Replication*. 2nd ed. New York: W. H. Freeman and Co.

Luger, C., et al., 1997. Crystal structure of the nucleosome core particle at 2.8 Å resolution. *Nature* **389:**251–260.

Moore, P. B., 1999. Structural motifs in RNA. *Annual Review of Biochemistry* **67:**287–300.

Noller, H. F., 1984. Structure of ribosomal RNA. *Annual Review of Biochemistry* **53:**119–162.

Pienta, K. J., and Coffey, D. S., 1984. A structural analysis of the role of the nuclear matrix and DNA loops in the organization of the nucleus and chromosomes. In Higher order structure in the nucleus. Cook, P. R., and Laskey, R. A., eds. *Journal of Cell Science Supplement* **1:**123–135.

Rhodes, D., 1997. The nucleosome core all wrapped up. *Nature* **389:**231–233.

Rich, A., Nordheim, A., and Wang, A. H.-J., 1984. The chemistry and biology of left-handed Z-DNA. *Annual Review of Biochemistry* **53:**791–846.

Sambrook, J., and Russell, D., 2000. *Molecular Cloning: A Laboratory Manual*. 3rd ed. Cold Spring Harbor, N.Y.: Cold Spring Harbor Laboratory Press.

Wang, B.-C., et al., 1994. The octameric histone core of the nucleosome. *Journal of Molecular Biology* **236:**179–188.

Watson, J. D., ed., 1983. Structures of DNA. *Cold Spring Harbor Symposia on Quantitative Biology*. Volume XLVII. Cold Spring Harbor, N.Y.: Cold Spring Harbor Laboratory.

Watson, J. D., Hopkins, N. H., Roberts, J. W., Steitz, J. A., and Weiner, A. M., 1987. *The Molecular Biology of the Gene*. Vol. I of *General Principles*. 4th ed. Menlo Park, Calif.: Benjamin Cummings.

Wu, R., 1993. Development of enzyme-based methods for DNA sequence analysis and their applications in the genome projects. *Advances in Enzymology* **67:**431–468.

Recombinant DNA: Cloning and Creation of Chimeric Genes

How many vain chimeras have you created? . . . Go and take your place with the seekers after gold.

Leonardo da Vinci, *The Notebooks* (1508–1518)

The Chimera of Arezzo, of Etruscan origin and probably from the fifth century B.C., was found near Arezzo, Italy, in 1553. Chimeric animals existed only in the imagination of the ancients. But the ability to create chimeric DNA molecules is a very real technology that has opened up a whole new field of scientific investigation. (Scala/Art Resource, Chimera, Museo Archeologico, Florence, Italy)

In the early 1970s, technologies for the laboratory manipulation of nucleic acids emerged. In turn, these technologies led to the construction of DNA molecules composed of nucleotide sequences taken from different sources. The products of these innovations, **recombinant DNA molecules,**[1] opened exciting new avenues of investigation in molecular biology and genetics, and a new field was born—**recombinant DNA technology. Genetic engineering** is the application of this technology to the manipulation of genes. These advances were made possible by methods for **amplification** of any particular DNA seg-

amplification   the production of multiple copies

---

[1]The advent of molecular biology, like that of most scientific disciplines, has generated a jargon all its own. Learning new fields often requires gaining familiarity with a new vocabulary. We will soon see that many words—*vector, amplification,* and *insert* are but a few examples—have been bent into new meanings to describe the marvels of this new biology.

ment, regardless of source, within bacterial host cells. Or, in the language of recombinant DNA technology, the **cloning** of almost any chosen DNA sequence became feasible.

9.1 ## Cloning

In classical biology, a *clone* is a population of identical organisms derived from a single parental organism. For example, the members of a colony of bacterial cells that arise from a single cell on a petri plate are a clone. Molecular biology has borrowed the term to mean a collection of molecules or cells all identical to an original molecule or cell. So, if the original cell on the petri plate harbored a recombinant DNA molecule in the form of a plasmid, the plasmids within the millions of cells in a bacterial colony represent a clone of the original DNA molecule, and these molecules can be isolated and studied. Furthermore, if the cloned DNA molecule is a gene (or part of a gene)—that is, it encodes a functional product—a new avenue to isolating and studying this product has been opened. Recombinant DNA methodology offers exciting new vistas in biochemistry.

Plasmids

Plasmids are naturally occurring, circular, extrachromosomal DNA molecules (see Chapter 8). Natural strains of the common colon bacterium *Escherichia coli* isolated from various sources harbor diverse plasmids. Often these plasmids carry genes specifying novel metabolic activities that are advantageous to the host bacterium. These activities range from catabolism of unusual organic substances to various metabolic functions that endow the host cells with resistance to antibiotics, heavy metals, or bacteriophages. Plasmids that are able to perpetuate themselves in *E. coli,* the bacterium favored by geneticists and molecular biologists, have become the darlings of recombinant DNA technology. Since restriction endonuclease digestion of plasmids can generate fragments with overlapping or "sticky" ends, artificial plasmids can be constructed by ligating different fragments together. Such artificial plasmids were among the earliest recombinant DNA molecules. These recombinant molecules can be autonomously replicated, and hence propagated, in suitable bacterial host cells, provided they possess a site signaling where DNA replication can begin (a so-called **origin of replication,** or *ori,* sequence).

Plasmids As Cloning Vectors

The idea arose that "foreign" DNA sequences could be inserted into artificial plasmids and that these foreign sequences would be carried into *E. coli* and propagated as part of the plasmid. That is, these plasmids could serve as **cloning vectors** to carry genes. (The word *vector* is used here to mean "a vehicle or carrier.") Plasmids useful as cloning vectors possess three common features: a **replicator,** a **selectable marker,** and a **cloning site** (Figure 9.1). A *replicator* is an origin of replication, or *ori.* The *selectable marker* is typically a gene conferring resistance to an antibiotic. Only those cells containing the cloning vector will grow in the presence of the antibiotic. Therefore, growth on antibiotic-containing media "selects for" plasmid-containing cells. The *cloning site* is typically a sequence of nucleotides representing one or more restriction endonuclease cleavage sites. Cloning sites are located where the insertion of foreign DNA neither disrupts the plasmid's ability to replicate nor inactivates essential markers.

Figure 9.1 One of the first widely used cloning vectors, the plasmid *pBR322*. This 4363-bp plasmid contains an origin of replication (*ori*) and genes encoding resistance to the drugs ampicillin (*amp^r*) and tetracycline (*tet^r*). The locations of restriction endonuclease cleavage sites are indicated.

See *Interactive Biochemistry CD-ROM and Workbook*, page 125

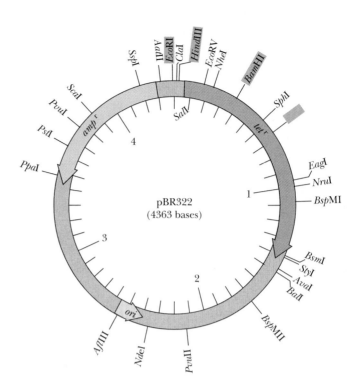

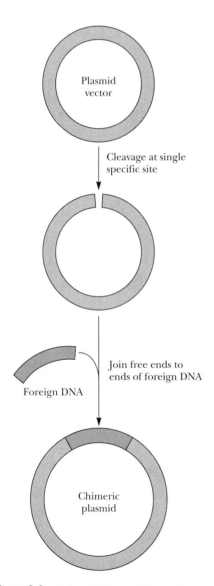

Figure 9.2 Foreign DNA sequences can be inserted into plasmid vectors by opening the circular plasmid with a restriction endonuclease. The ends of the linearized plasmid DNA are then joined with the ends of a foreign sequence, reclosing the circle to create a chimeric plasmid.

Virtually Any DNA Sequence Can Be Cloned

Nuclease cleavage at a restriction site opens, or *linearizes*, the circular plasmid so that a foreign DNA fragment can be inserted. The ends of this linearized plasmid are joined to the ends of the fragment so that the circle is closed again, creating a recombinant plasmid (Figure 9.2). **Recombinant plasmids** are hybrid DNA molecules consisting of plasmid DNA sequences plus inserted DNA elements (called *inserts*). Such hybrid molecules are also called **chimeric constructs** or **chimeric plasmids.** (The term *chimera* is borrowed from mythology and refers to a beast composed of the body and head of a lion, the heads of a goat and a snake, and the wings of a bat.) The presence of foreign DNA sequences does not adversely affect replication of the plasmid, so chimeric plasmids can be propagated in bacteria just like the original plasmid. Bacteria often harbor several hundred copies of common cloning vectors per cell. Hence, large amounts of a cloned DNA sequence can be recovered from bacterial cultures. The enormous power of recombinant DNA technology stems in part from the fact that *virtually any DNA sequence can be selectively cloned and amplified in this manner.* DNA sequences that are difficult to clone include inverted repeats, origins of replication, centromeres, and telomeres. The only practical limitation is the size of the foreign DNA segment: most plasmids with inserts larger than about 10 kbp are not replicated efficiently.

Bacterial cells may harbor one or many copies of a particular plasmid, depending on the nature of the plasmid replicator. That is, plasmids are classified as *high copy number* or *low copy number.* The copy number of most genetically engineered plasmids is high (200 or so), but some are lower.

Construction of Chimeric Plasmids

Creation of chimeric plasmids requires joining the ends of the foreign DNA insert to the ends of a linearized plasmid (see Figure 9.2). This ligation is facilitated if the ends of the plasmid and the insert have complementary, single-stranded overhangs. Then these ends can base-pair with one another, annealing the two molecules together. One way to generate such ends is to cleave the DNA with restriction enzymes that make staggered cuts; many such restriction

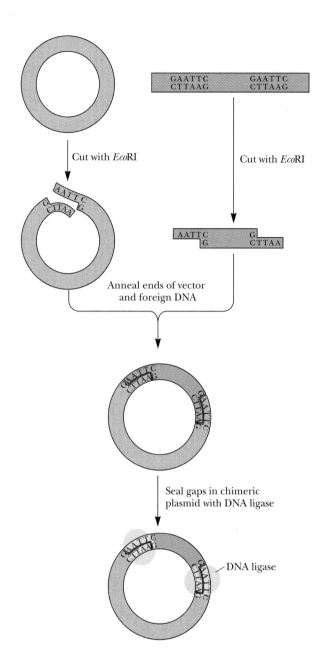

Seal gaps in chimeric
plasmid with DNA ligase

DNA ligase

Figure 9.3 Restriction endonuclease *Eco*RI cleaves double-stranded DNA. The recognition site for *Eco*RI is the hexameric sequence GAATTC:

5′ … NpNpNpNp**GpApApTpTpC**pNpNpNpNp … 3′
3′ … NpNpNpNp**CpTpTpApApG**pNpNpNpNp … 5′

Cleavage occurs at the G residue on each strand so that the DNA is cut in a staggered fashion, leaving 5′-overhanging single-stranded ends (sticky ends):

5′ … NpNpNpNp**G**　　**pApApTpTpC**pNpNpNpNp … 3′
3′ … NpNpNpNp**CpTpTpApAp**　　**G**pNpNpNpNp … 5′

An *Eco*RI restriction fragment of foreign DNA can be inserted into a plasmid having an *Eco*RI cloning site by　(a) cutting the plasmid at this site with *Eco*RI, annealing the linearized plasmid with the *Eco*RI foreign DNA fragment, and　(b) sealing the nicks with DNA ligase.

endonucleases are available (see Table 8.3). For example, if the sequence to be inserted is an *Eco*RI fragment and the plasmid is cut with *Eco*RI, the single-stranded sticky ends of the two DNAs can anneal (Figure 9.3). The interruptions in the sugar–phosphate backbone of DNA can then be sealed with DNA ligase to yield a covalently closed, circular chimeric plasmid. DNA ligase is an enzyme that covalently links adjacent 3′-OH and 5′-PO$_4$ groups. An inconvenience of this strategy is that *any* pair of *Eco*RI sticky ends can anneal with each other. So, plasmid molecules can reanneal with themselves, as can the foreign DNA restriction fragments. These DNAs can be eliminated by selection schemes designed to identify only those bacteria containing chimeric plasmids.

　　Blunt-end ligation is an alternative method for joining different DNAs. This method depends on the ability of **phage T4 DNA ligase** to covalently join the ends of any two DNA molecules (even those lacking 3′-or 5′-overhangs) (Figure 9.4). Some restriction endonucleases cut DNA so that blunt ends are formed (see Table 8.3). Since there is no control over which pair of DNAs are blunt-end–ligated by T4 DNA ligase, strategies to identify the desired products must be applied.

ligation　the act of joining

Figure 9.4 Blunt-end ligation using phage T4 DNA ligase, which catalyzes the ATP-dependent ligation of DNA molecules. AMP and PP$_i$ are by-products.

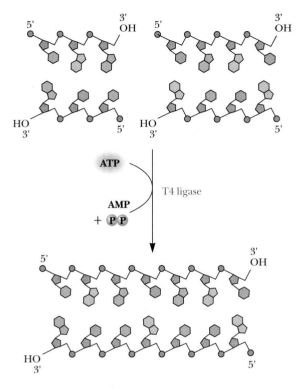

A great number of variations on these basic themes have emerged. For example, short synthetic DNA duplexes whose nucleotide sequence consists of little more than a restriction site can be blunt-end–ligated onto any DNA. These short DNAs are known as **linkers.** Cleavage of the ligated DNA with the restriction enzyme then leaves tailor-made sticky ends useful in cloning reactions (Figure 9.5). Similarly, many vectors contain a **polylinker** cloning site, a short region of DNA sequence bearing numerous restriction sites.

Promoters and Directional Cloning

Note that the strategies discussed thus far create hybrids in which the orientation of the DNA insert within the chimera is random. Sometimes, however,

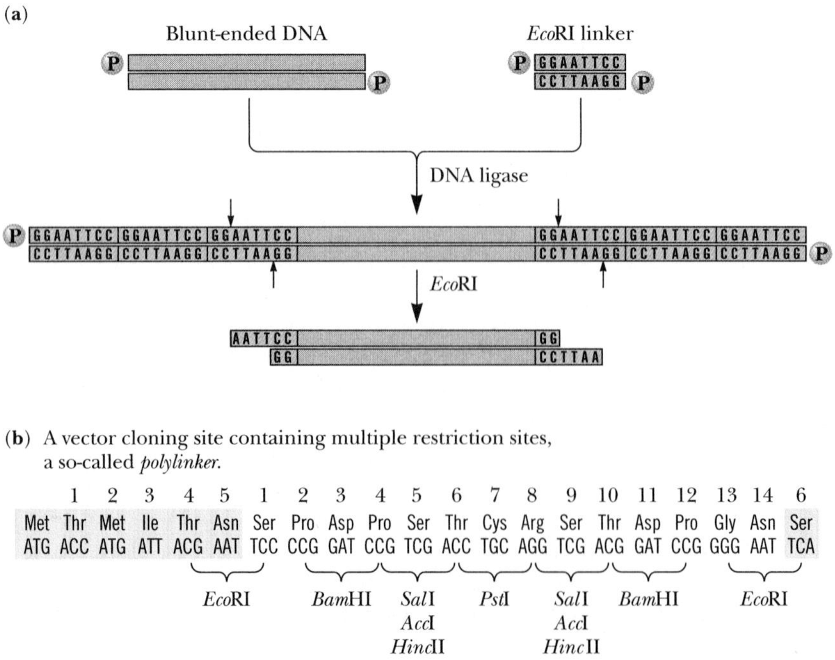

Figure 9.5 **(a)** The use of linkers to create tailor-made ends on cloning fragments. Synthetic oligonucleotide duplexes whose sequences represent *Eco*RI restriction sites are blunt-end–ligated to a DNA molecule using T4 DNA ligase. Note that the ligation reaction can add multiple linkers on each end of the blunt-ended DNA. *Eco*RI digestion removes all but the terminal one, leaving the desired 5′-overhangs. **(b)** Cloning vectors often have polylinkers consisting of a multiple array of restriction sites at their cloning sites, so restriction fragments generated by a variety of endonucleases can be incorporated into the vector. Note that the polylinker is engineered not only to have multiple restriction sites but also to have an uninterrupted sequence of codons, so this region of the vector has the potential for translation into protein. The sequence shown is the cloning site for the vectors M13mp7 and pUC7; the colored amino acid residues are contiguous with the coding sequence of the *lacZ* gene carried by this vector (see Figure 9.18). *(a, Adapted from Figure 3.16.3; b, adapted from Figure 1.14.2, in Ausubel, F. M., et al., 1987. Current Protocols in Molecular Biology. New York: John Wiley & Sons.)*

(a)

Blunt-ended DNA   *Eco*RI linker

P [============] P   P [GGAATTCC / CCTTAAGG] P

DNA ligase

P [GGAATTCC|GGAATTCC|GGAATTCC][============][GGAATTCC|GGAATTCC|GGAATTCC] P
[CCTTAAGG|CCTTAAGG|CCTTAAGG]   [CCTTAAGG|CCTTAAGG|CCTTAAGG]

*Eco*RI

[AATTCC============GG]
[GG============CCTTAA]

(b) A vector cloning site containing multiple restriction sites, a so-called *polylinker.*

1	2	3	4	5	1	2	3	4	5	6	7	8	9	10	11	12	13	14	6
Met	Thr	Met	Ile	Thr	Asn	Ser	Pro	Asp	Pro	Ser	Thr	Cys	Arg	Ser	Thr	Asp	Pro	Gly	Asn	Ser
ATG	ACC	ATG	ATT	ACG	AAT	TCC	CCG	GAT	CCG	TCG	ACC	TGC	AGG	TCG	ACG	GAT	CCG	GGG	AAT	TCA

     *Eco*RI   *Bam*HI *Sal*I *Pst*I *Sal*I *Bam*HI *Eco*RI
              *Acc*I   *Acc*I
              *Hinc*II   *Hinc*II

it is desirable to insert the DNA in a particular orientation. For example, an experimenter might wish to insert a particular DNA (a gene) in a vector so that its gene product is synthesized. To do this, the DNA must be placed downstream from a **promoter,** which is a nucleotide sequence lying upstream of a gene. The promoter controls expression of the gene. RNA polymerase molecules bind specifically at promoters and initiate transcription of adjacent genes, copying template DNA into RNA products. One way to insert DNA so that it will be properly oriented with respect to the promoter is to create DNA molecules whose ends have different overhangs. Ligation of such molecules into the plasmid vector can only take place in one orientation to give **directional cloning** (Figure 9.6).

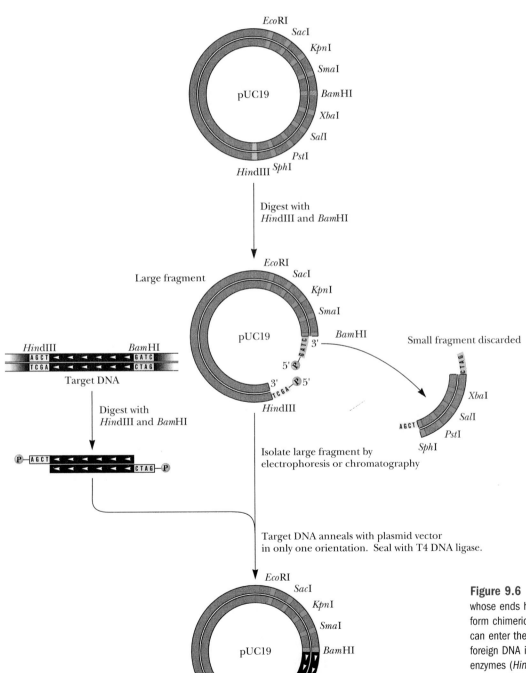

Figure 9.6 Directional cloning. DNA molecules whose ends have different overhangs can be used to form chimeric constructs in which the foreign DNA can enter the plasmid in only one orientation. The foreign DNA is digested with two different restriction enzymes (*Hind*III and *Bam*HI) and the plasmid is digested with the same two enzymes. Note that pUC19 has a polylinker or universal cloning site (see Figure 9.5b); pUC stands for universal cloning plasmid.

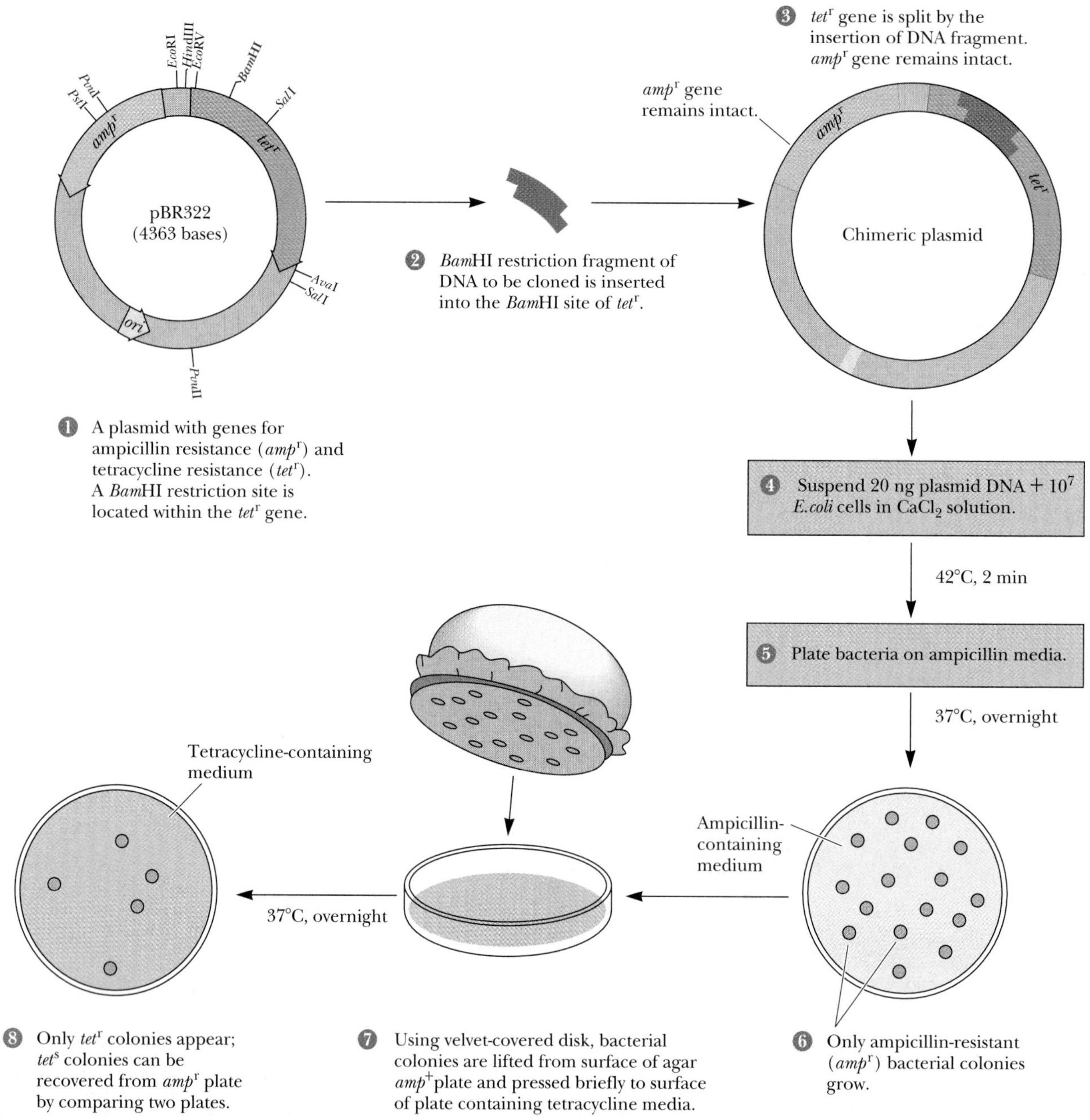

Figure 9.7 A typical bacterial transformation experiment. Here the plasmid pBR322 is the cloning vector. (1) Cleavage of pBR322 with restriction enzyme *Bam*HI, followed by (2) annealing and ligation of inserts generated by *Bam*HI cleavage of some foreign DNA, (3) creates a chimeric plasmid. (4) The chimeric plasmid is then used to transform Ca^{2+}-treated heat-shocked *E. coli* cells, and the bacterial sample is plated on a petri plate. (5) Following incubation of the petri plate overnight at 37°C, (6) colonies of amp^r bacteria will be evident. (7) Replica plating these bacteria on plates of tetracycline-containing media (8) reveals which colonies are *tet*^{r} and which are tetracycline-sensitive (*tet*^{s}). Only the *tet*^{s} colonies possess plasmids with foreign DNA inserts.

Biologically Functional Chimeric Plasmids

The first biologically functional chimeric DNA molecules constructed *in vitro* were assembled from parts of different plasmids in 1973 by Stanley Cohen, Annie Chang, Herbert Boyer, and Robert Helling. These plasmids were used to **transform** recipient *E. coli* cells (*transformation* means the uptake and replication of exogenous DNA by a recipient cell). The bacterial cells were rendered somewhat permeable to DNA by Ca^{2+} treatment and a brief 42°C heat shock. Although less than 0.1% of the Ca^{2+}-treated bacteria became competent for transformation, transformed bacteria could be selected by their resistance to certain antibiotics (Figure 9.7). Consequently, the chimeric plasmids must have been biologically functional in at least two aspects: they replicated stably within their hosts and they expressed the drug resistance markers they carried.

In general, plasmids used as cloning vectors are engineered to be small, 2.5 kbp to about 10 kbp in size, so that the size of the insert DNA can be maximized. These plasmids have only a single origin of replication, so the time necessary for complete replication depends on the size of the plasmid. Under selective pressure in a growing culture of bacteria, overly large plasmids are prone to delete any nonessential "genes," such as any foreign inserts. Such deletion would thwart the purpose of most cloning experiments. Thus, the useful upper limit on cloned inserts in plasmids is about 10 kbp. Many eukaryotic genes exceed this size.

Bacteriophage λ As a Cloning Vector

The genome of bacteriophage λ (lambda) (Figure 9.8) is a 48.5-kbp linear DNA molecule that is packaged into the head of the bacteriophage. The middle one-third of this genome is not essential to phage infection, so λ phage DNA has been engineered so that foreign DNA molecules up to 16 kbp can be inserted into this region for cloning purposes. *In vitro* packaging systems are then used to package the chimeric DNA into phage heads, which, when assembled with phage tails, form infective phage particles. Bacteria infected with these recombinant phage produce large numbers of phage progeny before they lyse, and large amounts of recombinant DNA can be easily purified from the lysate.

Cosmids

The DNA incorporated into phage heads by bacteriophage λ packaging systems must satisfy only a few criteria. It must possess a 14-bp sequence known as *cos* (which stands for *co*hesive end *s*ite) at each of its ends, and these *cos* sequences must be separated by no fewer than 36 kbp and no more than 51 kbp of DNA. Essentially any DNA satisfying these minimal requirements will be packaged and assembled into an infective phage particle. Other cloning features such as an *ori*, selectable markers, and a polylinker are joined to the *cos* sequence so that the cloned DNA can be propagated and selected in host cells. These features have been achieved by placing *cos* sequences on either side of cloning sites in plasmids to create **cosmid vectors** that are capable of carrying DNA inserts about 40 kbp in size (Figure 9.9). Since cosmids lack essential phage genes, they reproduce in host bacteria as plasmids.

Shuttle Vectors

Shuttle vectors are plasmids capable of propagating and transferring ("shuttling") genes between two different organisms, one of which is typically a prokaryote (*E. coli*) and the other a eukaryote (for example, yeast). Shuttle vectors must have unique origins of replication for each cell type as well as different markers for selection of transformed host cells harboring the vector (Figure 9.10). Shuttle vectors have the advantage that eukaryotic genes can be cloned in bacterial hosts, yet the expression of these genes can be analyzed in appropriate eukaryotic backgrounds.

Artificial Chromosomes

DNA molecules 2 megabase pairs in length have been successfully propagated in yeast by creating **yeast artificial chromosomes,** or **YACs.** Further, such YACs have been transferred into transgenic mice for the analysis of large genes or multigenic DNA sequences *in vivo* (that is, within the living animal). For these large DNAs to be replicated in the yeast cell, YAC constructs must include not

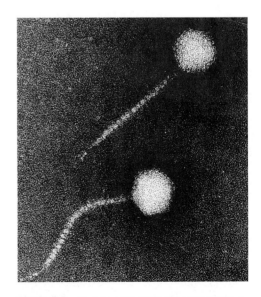

Figure 9.8 Electron micrograph of bacteriophage λ. *(Robley C. Williams, University of California/BPS)*

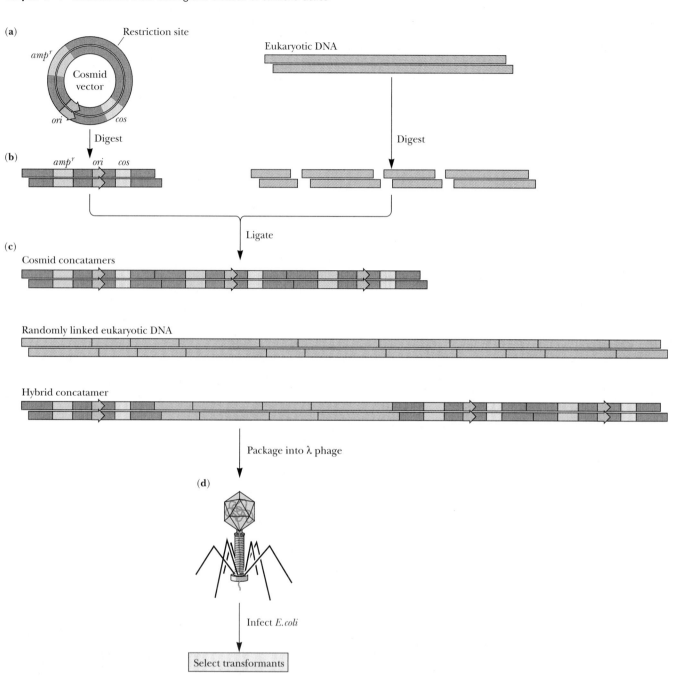

Figure 9.9 Cosmid vectors for cloning large DNA fragments. **(a)** Cosmid vectors are plasmids that carry a selectable marker such as *amp^r*, an origin of replication (*ori*), a polylinker suitable for insertion of foreign DNA, and **(b)** a *cos* sequence. Both the plasmid and the foreign DNA to be cloned are cut with a restriction enzyme and the two DNAs are then ligated together. **(c)** The ligation reaction leads to the formation of hybrid concatamers, molecules in which plasmid sequences and foreign DNAs are linked in series in no particular order. The bacteriophage λ packaging extract contains the restriction enzyme that recognizes *cos* sequences and cleaves at these sites. **(d)** DNA molecules of the proper size (36 to 51 kbp) are packaged into phage heads, forming infective phage particles. **(e)** The *cos* sequence is

$$\downarrow$$
5′-TACG**GGGCGGCGACCT**CGCG-3′

3′-ATGC**CCCGCCGCTGGA**GCGC-5′
$$\uparrow$$

Endonuclease cleavage at the sites indicated by arrows leaves 12-bp cohesive ends. (*a–d, Adapted from Figure 1.10.7, in Ausubel, et al., eds., 1987.* Current Protocols in Molecular Biology. *New York: John Wiley & Sons; e, from Figure 4, in Murialdo, 1991.* Annual Review of Biochemistry *60:136.*)

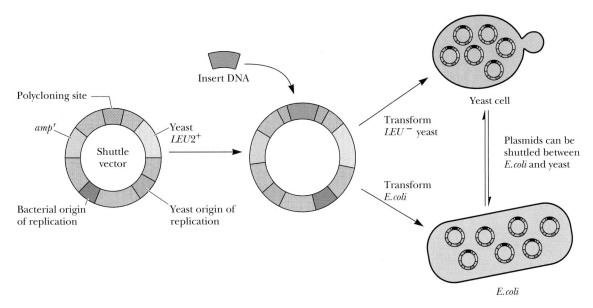

Figure 9.10 A typical shuttle vector. This vector has both yeast and bacterial origins of replication, *amp*^{r} (ampicillin resistance gene for selection in *E. coli*) and *LEU2*$^+$, a gene in the yeast pathway for leucine biosynthesis. The recipient yeast cells are *LEU2*$^-$ (defective in this gene) and thus require leucine for growth. *LEU2*$^-$ yeast cells transformed with this shuttle vector can be selected on medium lacking any leucine supplement. *(Adapted from Figure 19-5, in Watson, J. D., et al., 1987. The Molecular Biology of the Gene. Menlo Park, Calif.: Benjamin Cummings.)*

only an origin of replication (known in yeast terminology as an *autonomously replicating sequence* or *ARS*) but also a centromere and telomeres. Recall that centromeres provide the site for attachment of the chromosome to the spindle during mitosis and meiosis, and telomeres are nucleotide sequences defining the ends of chromosomes. Telomeres are essential for proper replication of the chromosome.

9.2 DNA Libraries

A DNA library is a set of cloned fragments that collectively represent the genes of a particular organism. Particular genes can be isolated from DNA libraries, much as books can be obtained from conventional libraries. The secret is knowing where and how to look.

Genomic Libraries

Any particular gene constitutes only a small part of an organism's genome. For example, if the organism is a mammal whose entire genome encompasses some 10^6 kbp and the gene is 10 kbp, then the gene represents only 0.001% of the total nuclear DNA. It is impractical to attempt to recover such rare sequences directly from isolated nuclear DNA because of the overwhelming amount of extraneous DNA sequences. Instead, a **genomic library** is prepared by isolating total DNA from the organism, digesting it into fragments of suitable size, and cloning the fragments into an appropriate vector. This approach is called *shotgun cloning*, since the strategy has no way of targeting a particular gene but instead seeks to clone all the genes of the organism at one time. The intent is that at least one recombinant clone will contain at least part of the gene of interest. Usually, the isolated DNA is only partially digested by the chosen restriction endonuclease so that not every restriction site is cleaved in every DNA molecule. Then, even if the gene of interest contains a susceptible restriction site, some intact genes might still be found in the digest. Genomic libraries have been prepared from hundreds of different species.

Many clones must be created to be confident that the genomic library contains the gene of interest. The probability, P, that some number of clones, N, contains a particular fragment representing a fraction, f, of the genome is

$$P = 1 - (1 - f)^N$$

Thus,

$$N = \frac{\ln(1 - P)}{\ln(1 - f)}$$

For example, if the library consists of 10-kbp fragments of the *E. coli* genome (4640 kbp total), over 2000 individual clones must be screened to have a 99% probability ($P = 0.99$) of finding a particular fragment. (Since $f = 10/4640 = 0.0022$ and $P = 0.99$, N is 2093). For a 99% probability of finding a particular sequence within the 3×10^6 kbp human genome, N would equal almost 1.4 million if the cloned fragments averaged 10 kbp in size. The need for cloning vectors capable of carrying very large DNA inserts becomes obvious from these numbers.

CRITICAL DEVELOPMENTS IN BIOCHEMISTRY

Combinatorial Libraries

Specific recognition and binding of other molecules is a defining characteristic of any protein or nucleic acid. Often, target ligands of a particular protein are unknown, or, in other instances, a unique ligand for a known protein may be sought in hopes of blocking the activity of the protein or otherwise perturbing its function. **Combinatorial libraries** are the products of an emerging strategy to facilitate the identification and characterization of possible ligands for a protein. The strategy is also applicable to the study of nucleic acids. Unlike genomic libraries, combinatorial libraries consist of synthetic oligomers. Arrays of such oligonucleotides printed as tiny dots on miniature solid supports are known as **DNA chips.** Specifically, combinatorial libraries contain very large numbers of chemically synthesized molecules (such as oligonucleotides or peptides) with randomized sequences or structures. Such libraries are designed and constructed with the hope that one molecule among a vast number will be recognized as a ligand by the protein (or nucleic acid) of interest. If so, perhaps that molecule will be useful in a pharmaceutical application, for instance as a drug to treat a disease involving the protein to which it binds.

An example of this strategy is the preparation of a **synthetic combinatorial library** of hexapeptides. The maximum number of sequence combinations for hexapeptides is 20^6 or 64,000,000. One approach to simplifying preparation and screening possibilities for such a library is to specify the first two amino acids in the hexapeptide while randomly choosing the next four. In this approach,

400 libraries (20^2) are synthesized, each of which is unique in the amino acids at positions 1 and 2 but random at the other four positions (as in *AAXXXX*, *ACXXXX*, *ADXXXX*, etc.), so that each of the 400 libraries contains 20^4 or 160,000 different sequence combinations. Screening these libraries with the protein of interest reveals which of the 400 libraries contains a ligand with high affinity. This library is then systematically expanded by specifying the first three amino acids (knowing from the chosen one-of-400 libraries which amino acids are best as the first two). Only 20 synthetic libraries (each containing 20^3 or 8000 hexapeptides) are made here (one for each third-position possibility, with the remaining three positions randomized). Selection for ligand binding again with the protein of interest reveals the best of these 20, and this particular library is then varied systematically at the fourth position, creating 20 more libraries (each containing 20^2 or 400 hexapeptides). This cycle of synthesis, screening, and selection is repeated until all six positions in the hexapeptide are optimized to create the best ligand for the protein. A variation on this basic strategy, using synthetic oligonucleotides rather than peptides, identified a unique 15-mer (sequence GGTTGGTGTGGTTGG) with high affinity ($K_D = 2.7$ nM) toward thrombin, a serine protease in the blood coagulation pathway. Thrombin is a major target for the pharmacological prevention of clot formation in coronary thrombosis.

From Cortese, R., 1996. *Combinatorial Libraries: Synthesis, Screening and Application Potential.* Berlin: Walter de Gruyter.

Figure 9.11 Screening a genomic library by colony hybridization (or plaque hybridization). Host ▶ bacteria transformed with a plasmid-based genomic library or infected with a bacteriophage-based genomic library are plated on a petri plate and incubated overnight to allow bacterial colonies (or phage plaques) to form. A replica of the bacterial colonies (or plaques) is then obtained by overlaying the plate with a nitrocellulose disc (1). Nitrocellulose strongly binds nucleic acids; single-stranded nucleic acids are bound more tightly than double-stranded nucleic acids. (Nylon membranes with similar nucleic acid- and protein-binding properties are also used.) Once the nitrocellulose disc has taken up an impression of the bacterial colonies (or plaques), it is removed and the petri plate is set aside and saved. The disc is treated with 2 *M* NaOH, neutralized, and dried (2). NaOH both lyses any bacteria (or phage particles) and dissociates the DNA strands. When the disc is dried, the DNA strands become immobilized on the filter. The dried disc is placed in a sealable plastic bag and a solution containing heat-denatured (single-stranded) labeled probe is added (3). The bag is incubated to allow annealing of the probe DNA to any target DNA sequences that might be present on the nitrocellulose. The filter is then washed, dried, and placed on a piece of X-ray film to obtain an autoradiogram (4). The position of any spots on the X-ray film reveals where the labeled probe has hybridized with target DNA (5). The location of these spots can be used to recover the genomic clone from the bacteria (or plaques) on the original petri plate.

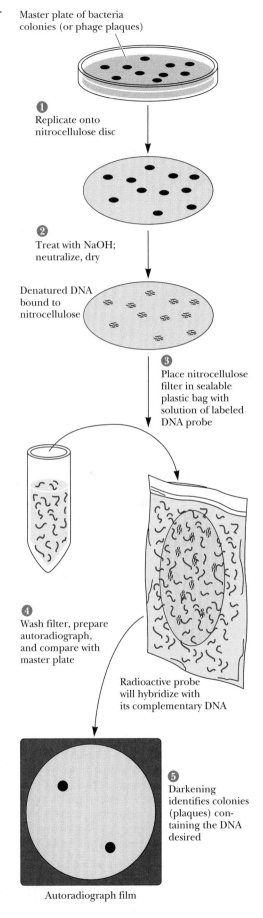

Master plate of bacteria colonies (or phage plaques)

1 Replicate onto nitrocellulose disc

2 Treat with NaOH; neutralize, dry

Denatured DNA bound to nitrocellulose

3 Place nitrocellulose filter in sealable plastic bag with solution of labeled DNA probe

4 Wash filter, prepare autoradiograph, and compare with master plate

Radioactive probe will hybridize with its complementary DNA

5 Darkening identifies colonies (plaques) containing the DNA desired

Autoradiograph film

Screening Libraries

A common method of screening plasmid-based genomic libraries is to carry out a **colony hybridization experiment.** (The protocol is similar for phage-based libraries except that bacteriophage plaques, not bacterial colonies, are screened.) In a typical experiment, host bacteria containing either a plasmid-based or bacteriophage-based library are plated out on a petri dish and allowed to grow overnight to form colonies (or plaques, in the case of phage libraries) (Figure 9.11). A replica of the bacterial colonies (or plaques) is then obtained by overlaying the plate with a nitrocellulose disc. The disc is removed, treated with alkali to dissociate bound DNA duplexes into single-stranded DNA, dried, and placed in a sealed bag with labeled probe (see *Critical Developments in Biochemistry,* page 304). If the probe DNA is duplex DNA, it must be denatured by heating at 70°C. The probe and target DNA complementary sequences must be in a single-stranded form if they are to hybridize with one another. Any DNA sequences complementary to probe DNA will be revealed by autoradiography of the nitrocellulose disc. Bacterial colonies (phage plaques) containing clones bearing target DNA are identified on the film and can be recovered from the master plate.

Probes for Southern Hybridization

Clearly, specific probes are essential reagents if the goal is to identify a particular gene against a background of innumerable DNA sequences. Usually the probes that are used to screen libraries are nucleotide sequences that are complementary to some part of the target gene. Making useful probes requires some information about the gene's nucleotide sequence. Sometimes such information is available. Alternatively, if the amino acid sequence of the protein encoded by the gene is known, it is possible to work backward through the genetic code to the DNA sequence (Figure 9.12). Since the genetic code is *degenerate* (that is, several codons may specify the same amino acid; see Chapter 25), probes designed by this approach are usually **degenerate oligonucleotides** about 17 to 50 residues long (such oligonucleotides are so-called 17- to 50-mers). The oligonucleotides are synthesized so that different bases are incorporated at sites where degeneracies occur in the codons. The final preparation thus consists of a mixture of equal-length oligonucleotides whose sequences

(Text continued on page 306)

CRITICAL DEVELOPMENTS IN BIOCHEMISTRY

Identifying Specific DNA Sequences by Southern Blotting (Southern Hybridization)

Any given DNA fragment is unique solely by virtue of its specific nucleotide sequence. The only practical way to find one particular DNA segment among a vast population of different DNA fragments (such as you might find in genomic DNA preparations) is to exploit its sequence specificity. In 1975, E. M. Southern invented a technique capable of doing just that.

Electrophoresis

Southern first fractionated a population of DNA fragments according to size by gel electrophoresis (see step 2 in figure). The electrophoretic mobility of a nucleic acid is inversely proportional to its molecular mass. Polyacrylamide gels are suitable for separation of nucleic acids of 25 to 2000 bp. Agarose gels are better if the DNA fragments range up to 10 times this size. Most preparations of genomic DNA show a broad spectrum of sizes, from less than 1 kbp to more than 20 kbp. Typically, no discrete fragments are evident following electrophoresis; just a "smear" of DNA throughout the gel is visible.

Blotting

Once the fragments have been separated by electrophoresis (step 3), the gel is soaked in a solution of NaOH. Alkali denatures duplex DNA, converting it to single-stranded DNA. After the pH of the gel is adjusted to neutrality with buffer, a sheet of nitrocellulose (or DNA-binding nylon membrane) soaked in a concentrated salt solution is then placed over the gel (d), and salt solution is drawn through the gel in a direction perpendicular to the direction of electrophoresis (step 4). The salt solution is pulled through the gel in one of three ways: capillary action (*blotting*), suction (*vacuum blotting*), or electrophoresis (*electroblotting*). The movement of salt solution through the gel carries the DNA to the nitrocellulose sheet. Nitrocellulose binds single-stranded DNA molecules very tightly, effectively immobilizing them on the sheet.* Note that the distribution pattern of the electrophoretically separated DNA is maintained when the single-stranded DNA molecules bind to the nitrocellulose sheet (step 5 in figure). Next, the nitrocellulose is dried by baking in a vacuum oven[†]; baking tightly fixes the single-stranded DNAs to the nitrocellulose. Next, in the *prehybridization step*, the nitrocellulose sheet is incubated with a solution containing protein (serum albumin, for example) or a detergent such as sodium dodecyl sulfate. The protein and detergent molecules saturate any remaining binding sites for DNA on the nitrocellulose. Thus, no more DNA can bind nonspecifically to the nitrocellulose sheet.

Hybridization

To detect a particular DNA within the electrophoretic smear of countless DNA fragments, the prehybridized nitrocellulose sheet is incubated in a sealed plastic bag with a solution of specific probe molecules (step 6 in figure). A **probe** is usually a single-stranded DNA of defined sequence that is distinctively labeled, either with a radioactive isotope (such as ^{32}P) or some other easily detectable tag. The nucleotide sequence of the probe is designed to be complementary to the sought-for or *target* DNA fragment. The single-stranded probe DNA **anneals** with the single-stranded target DNA bound to the nitrocellulose through specific base pairing to form a DNA duplex. This annealing, or **hybridization,** as it is usually called, labels the target DNA, revealing its position on the nitrocellulose. For example, if the probe is ^{32}P-labeled, its location can be detected by autoradiographic exposure of a piece of X-ray film laid over the nitrocellulose sheet (step 7 in figure).

Southern's procedure has been extended to the identification of specific RNA and protein molecules. In a play on Southern's name, the identification of particular RNAs following separation by gel electrophoresis, blotting, and probe hybridization is called **Northern blotting.** The analogous technique for identifying protein molecules is termed **Western blotting.** In Western blotting, the probe of choice is usually an antibody specific for the target protein.

The Southern blotting technique involves the transfer of electrophoretically separated DNA fragments to a nitrocellulose sheet and subsequent detection of ▶ specific DNA sequences. A preparation of DNA fragments [typically a restriction digest, (1)] is separated according to size by gel electrophoresis (2). The separation pattern can be visualized by soaking the gel in ethidium bromide to stain the DNA and then illuminating the gel with UV light (3). Ethidium bromide molecules intercalated between the hydrophobic bases of DNA are fluorescent under UV light. The gel is soaked in strong alkali to denature the DNA and then neutralized in buffer. Next, the gel is placed on a sheet of nitrocellulose (or DNA-binding nylon membrane), and concentrated salt solution is passed through the gel (4) to carry the DNA fragments out of the gel where they are bound tightly to the nitrocellulose (5). Incubation of the nitrocellulose sheet with a solution of labeled single-stranded probe DNA (6) allows the probe to hybridize with target DNA sequences complementary to it. The location of these target sequences is then revealed by an appropriate means of detection, such as autoradiography (7).

---

*The underlying cause of DNA binding to nitrocellulose is not clear, but probably involves a combination of hydrogen bonding, hydrophobic interactions, and salt bridges.

[†]Vacuum drying is essential because nitrocellulose reacts violently with O_2 if heated. For this reason, nylon-based membranes are preferred over nitrocellulose membranes.

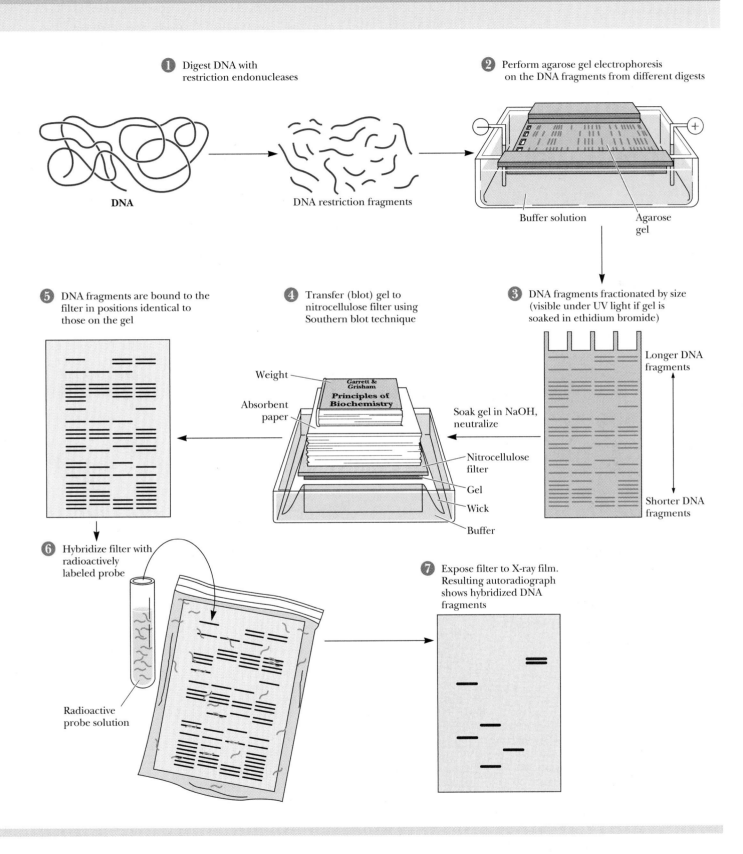

1 Digest DNA with restriction endonucleases

DNA

DNA restriction fragments

2 Perform agarose gel electrophoresis on the DNA fragments from different digests

Buffer solution

Agarose gel

3 DNA fragments fractionated by size (visible under UV light if gel is soaked in ethidium bromide)

Longer DNA fragments

Shorter DNA fragments

Soak gel in NaOH, neutralize

4 Transfer (blot) gel to nitrocellulose filter using Southern blot technique

Weight

Garrett & Grisham
Principles of Biochemistry

Absorbent paper

Nitrocellulose filter

Gel

Wick

Buffer

5 DNA fragments are bound to the filter in positions identical to those on the gel

6 Hybridize filter with radioactively labeled probe

Radioactive probe solution

7 Expose filter to X-ray film. Resulting autoradiograph shows hybridized DNA fragments

Known amino acid sequence:

| Phe | Met | Glu | Trp | His | Lys | Asn |

Possible mRNA sequence:

| UUU | AUG | GAA | UGG | CAU | AGG | AAU |
| UUC | | GAG | | CAC | AAA | AAC |

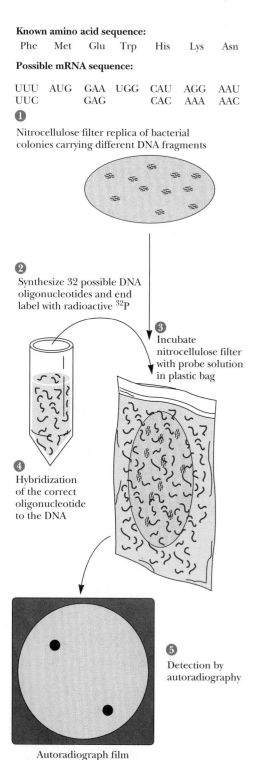

❶ Nitrocellulose filter replica of bacterial colonies carrying different DNA fragments

❷ Synthesize 32 possible DNA oligonucleotides and end label with radioactive ^{32}P

❸ Incubate nitrocellulose filter with probe solution in plastic bag

❹ Hybridization of the correct oligonucleotide to the DNA

❺ Detection by autoradiography

Autoradiograph film

Figure 9.12 Cloning genes using oligonucleotide probes designed from a known amino acid sequence. A radioactively labeled set of DNA (degenerate) oligonucleotides representing all possible mRNA coding sequences are synthesized. (In this case, there are 2^5, or 32.) The complete mixture is used to probe the genomic library by colony hybridization (see Figure 9.11). *(Adapted from Figure 19-18, in Watson, J. D., et al., 1987. Molecular Biology of the Gene. Menlo Park, Calif.: Benjamin/Cummings.)*

vary to accommodate the degeneracies. Presumably, one oligonucleotide sequence in the mixture will hybridize with the target gene. These oligonucleotide probes are at least 17-mers, because shorter degenerate oligonucleotides might hybridize with sequences unrelated to the target sequence.

A piece of DNA from the corresponding gene in a related organism can also be used as a probe in screening a library for a particular gene. Such probes are termed **heterologous probes** because they are not derived from the homologous (same) organism.

Problems arise if a complete eukaryotic gene is the cloning target; eukaryotic genes can be tens or even hundreds of kilobase pairs in size. Genes this size are fragmented in most cloning procedures. Thus, the DNA identified by the probe may represent a clone that carries only part of the desired gene. However, most cloning strategies are based on a partial digestion of the genomic DNA, a technique that generates an overlapping set of genomic fragments. This being so, DNA segments from the ends of the identified clone can now be used to probe the library for clones carrying DNA sequences that flanked the original isolate in the genome. Repeating this process ultimately yields the complete gene among a subset of overlapping clones.

cDNA Libraries

cDNAs are DNA molecules copied from mRNA templates. cDNA libraries are constructed by synthesizing cDNA from purified cellular mRNA. These libraries present an alternative strategy for gene isolation, especially eukaryotic genes. Since most eukaryotic mRNAs carry 3′-poly(A) tails, mRNA can be selectively isolated from preparations of total cellular RNA by oligo(dT)-cellulose chromatography (Figure 9.13). DNA copies of the purified mRNAs are synthesized

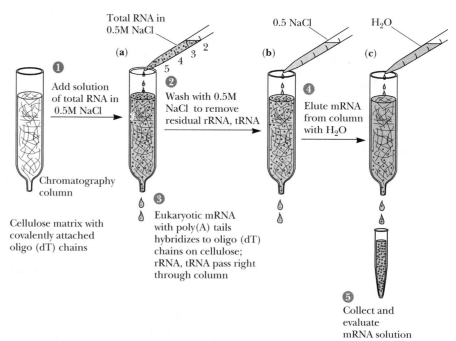

❶ Add solution of total RNA in 0.5M NaCl

Chromatography column

Cellulose matrix with covalently attached oligo (dT) chains

Total RNA in 0.5M NaCl

(a)

❷ Wash with 0.5M NaCl to remove residual rRNA, tRNA

❸ Eukaryotic mRNA with poly(A) tails hybridizes to oligo (dT) chains on cellulose; rRNA, tRNA pass right through column

0.5 NaCl

(b)

❹ Elute mRNA from column with H₂O

H₂O

(c)

❺ Collect and evaluate mRNA solution

Figure 9.13 Isolation of eukaryotic mRNA (red dots) via oligo(dT)-cellulose chromatography. **(a)** In the presence of 0.5 *M* NaCl, the poly(A) tails of eukaryotic mRNA anneal with short oligo(dT) chains covalently attached to an insoluble chromatographic matrix such as cellulose. Other RNAs, such as rRNA (green), pass right through the chromatography column. **(b)** The column is washed with more 0.5 *M* NaCl to remove residual contaminants. **(c)** Then the poly(A) mRNA is recovered by washing the column with water because the base pairs formed between the poly(A) tails of the mRNA and the oligo(dT) chains are unstable in solutions of low ionic strength.

HUMAN BIOCHEMISTRY

The Human Genome Project

The Human Genome Project is a collaborative international government- and private-sponsored effort to map and sequence the entire human genome, some 3 billion base pairs distributed among the two sex chromosomes (X and Y) and 22 **autosomes** (chromosomes that are not sex chromosomes). A primary goal was to identify and map at least 3000 genetic **markers** (genes or other recognizable loci on the DNA), which were evenly distributed throughout the chromosomes at roughly 100-kb intervals. At the same time, determination of the entire nucleotide sequence of the human genome was undertaken, and a useful version of the complete human genome was achieved in 2000. An ancillary part of the project is sequencing the genomes of other species (such as yeast, *Drosophila melanogaster* [the fruit fly], the mouse, and *Arabidopsis thaliana* [a plant]) to reveal comparative aspects of genetic and sequence organization (Table 9.1). Information about whole genome sequences of organisms has contributed to a new branch of science called **bioinformatics:** the study of the nature and organization of biological information. Bioinformatics includes such scientific approaches as **functional genomics** and **proteomics.** *Functional genomics* addresses global issues of gene expression, such as looking at all the genes that are activated during major metabolic shifts (for example, cells shifted from growth under aerobic conditions to growth under anaerobic conditions) or during embryogenesis and development of organisms. **Transcriptosome** is the word used in functional genomics to define the entire set of genes expressed (as mRNAs transcribed from DNA) in a particular cell or tissue under defined conditions. Functional genomics also provides new insights into evolutionary relationships between organisms. *Proteomics* is the study of all the proteins expressed by a certain cell or tissue under specified conditions. Typically, this set of proteins is revealed by running two-dimensional polyacrylamide gel electrophoresis on a cellular extract.

The Human Genome Project is also vital to medicine. Many human diseases have been traced to genetic defects whose positions within the human genome have been identified. Among the genes identified are

> *cystic fibrosis* gene
> the *breast cancer* genes *BRCA1* and *2*
> *Duchenne muscular dystrophy* gene* (at 2.4 megabases, the largest known gene in any organism)
> *Huntington's disease* gene
> *neurofibromatosis* gene
> *neuroblastoma* gene (a form of brain cancer)
> *amyotrophic lateral sclerosis* gene (Lou Gehrig's disease)
> *fragile X-linked mental retardation* gene*

as well as genes associated with the development of diabetes, a variety of other cancers, and affective disorders such as *schizophrenia* and *bipolar affective disorder* (manic depression).

Table 9.1 Completed Genome Nucleotide Sequences[1]

Genome	Genome Size[2] and Year Completed	
Bacteriophage ϕX174	0.0054	1977
Bacteriophage λ	0.048	1982
Marchantia[3] chloroplast genome	0.187	1986
Vaccinia virus	0.192	1990
Cytomegalovirus (CMV)	0.229	1991
Marchantia[3] mitochondrial genome	0.187	1992
Variola (smallpox) virus	0.186	1993
Hemophilus influenzae[4] (Gram-negative bacterium)	1.830	1995
Mycobacterium genitalium (mycobacterium)	0.58	1995
Escherichia coli (Gram-negative bacterium)	4.64	1996
Saccharomyces cerevisiae (yeast)	12.1	1996
Methanococcus jannaschii (archaeon)	1.66	1998
Arabidopsis thaliana (green plant)	125	2000
Caenorhabditis elegans (simple animal: nematode worm)	88	1998
Drosophila melanogaster (fruit fly)	137	2000
Homo sapiens (human)	3,120	2000

[1]Data available from the National Center for Biotechnology Information at the National Library of Medicine. Web site: http://www.ncbi.nlm.nih.gov/
[2]Genome size is given as millions of base pairs (mb).
[3]*Marchantia* is a bryophyte (a nonvascular green plant).
[4]The first complete sequence for the genome of a free-living organism.

*X-chromosome–linked gene. Over 100 disease-related genes have been mapped to the X chromosome.

by first annealing short oligo(dT) chains to the poly(A) tails. These oligo(dT) chains serve as primers for reverse transcriptase–driven synthesis of DNA (Figure 9.14). (Random oligonucleotides also can be used as primers, to avoid dependency on poly(A) tracts and to increase the likelihood of creating clones representing the 5′-ends of mRNAs.) **Reverse transcriptase** is an enzyme that synthesizes a DNA strand, copying RNA as the template. DNA polymerase is then used to copy the DNA strand and form a double-stranded (duplex DNA) molecule. Linkers are then added to the DNA duplexes rendered from the mRNA templates, and the cDNA is cloned into a suitable vector. Once a cDNA

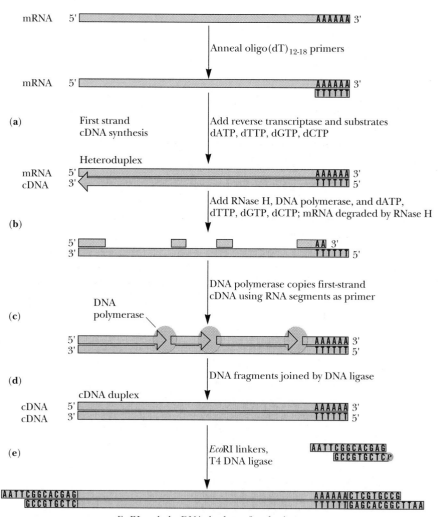

Figure 9.14 Reverse transcriptase–driven synthesis of cDNA from oligo(dT) primers annealed to the poly(A) tails of purified eukaryotic mRNA. **(a)** Oligo(dT) chains serve as primers for synthesis of a DNA copy of the mRNA by reverse transcriptase. Following completion of first-strand cDNA synthesis by reverse transcriptase, RNase H and DNA polymerase are added **(b).** RNase H specifically digests RNA strands in DNA : RNA hybrid duplexes. DNA polymerase copies the first-strand cDNA, using as primers the residual RNA segments after RNase H has created nicks and gaps **(c).** DNA polymerase has a $5′ \rightarrow 3′$ exonuclease activity that removes the residual RNA as it fills in with DNA. The nicks remaining in the second-strand DNA are sealed by DNA ligase **(d),** yielding duplex cDNA. EcoRI adapters with 5′-overhangs are then ligated onto the cDNA duplexes **(e)** using phage T4 DNA ligase to create EcoRI-ended cDNA for insertion into a cloning vector.

Figure 9.15 Expression vectors carrying the promoter recognized by the RNA polymerase of ▶ bacteriophage SP6 are useful for making RNA transcripts *in vitro*. SP6 RNA polymerase works efficiently *in vitro* and recognizes its specific promoter with high specificity. These vectors typically have a polylinker adjacent to the SP6 promoter. Successive rounds of transcription initiated by SP6 RNA polymerase at its promoter lead to the production of multiple RNA copies of any DNA inserted at the polylinker. Before transcription is initiated, the circular expression vector is linearized by a single cleavage at or near the end of the insert so that transcription terminates at a fixed point.

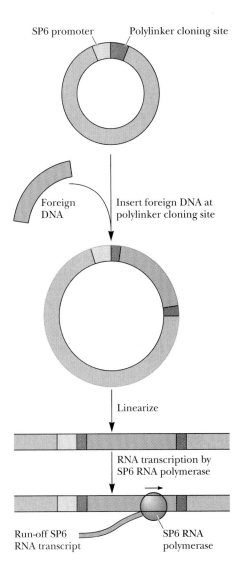

derived from a particular gene has been identified, the cDNA becomes an effective probe for screening genomic libraries for isolation of the gene itself.

Because different cell types in eukaryotic organisms express selected subsets of genes, RNA preparations from cells or tissues where genes of interest are selectively transcribed will be enriched for the desired mRNAs. cDNA libraries prepared from such mRNA are representative of the pattern and extent of gene expression that uniquely define particular kinds of differentiated cells. cDNA libraries of many normal and diseased human cell types are commercially available, including cDNA libraries of many tumor cells. Comparison of normal and abnormal cDNA libraries, in conjunction with two-dimensional gel electrophoretic analysis of the proteins produced in normal and abnormal cells, is a promising new strategy in clinical medicine to understand disease mechanisms.

Expression Vectors

Expression vectors are engineered so that any cloned insert can be transcribed into RNA and, in many instances, even translated into protein. cDNA expression libraries can be constructed in specially designed vectors derived from either plasmids or bacteriophage λ. Proteins encoded by the various cDNA clones within such expression libraries can be synthesized in the host cells, and if suitable assays are available to identify a particular protein, its corresponding cDNA clone can be identified and isolated. Expression vectors designed for RNA expression or protein expression, or both, are available.

RNA Expression

A vector for *in vitro* expression of DNA inserts as RNA transcripts can be constructed by putting a highly efficient promoter adjacent to a versatile cloning site. Figure 9.15 depicts such an expression vector. Linearized recombinant vector DNA is transcribed *in vitro* using SP6 RNA polymerase. Large amounts of RNA product can be obtained in this manner; if radioactive ribonucleotides are used as substrates, labeled RNA molecules useful as probes are made.

Protein Expression

Since cDNAs are DNA copies of mRNAs, cDNAs are uninterrupted copies of the exons of expressed genes. Because cDNAs lack introns, it is feasible to express these cDNA versions of eukaryotic genes in prokaryotic hosts that cannot process the complex primary transcripts of eukaryotic genes. To express a eukaryotic protein in *E. coli*, the eukaryotic cDNA must be cloned in an *expression vector* that contains regulatory signals for both transcription and translation. Accordingly, a *promoter*, where RNA polymerase will initiate transcription, as well as a *ribosome-binding site* to facilitate translation, are engineered into the vector just upstream from the restriction site for inserting foreign DNA. The AUG initiation codon that specifies the first amino acid in the protein (the *translation start site*) is contributed by the insert (Figure 9.16).

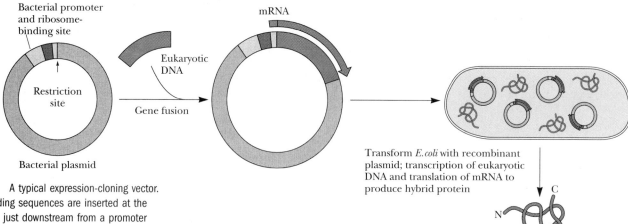

Figure 9.16 A typical expression-cloning vector. Eukaryotic coding sequences are inserted at the restriction site just downstream from a promoter region where RNA polymerase binds and initiates transcription. Transcription proceeds through a region encoding a bacterial ribosome-binding site and into the cloned insert. The presence of the bacterial ribosome-binding site in the RNA transcript ensures that the RNA can be translated into protein by the ribosomes of the host bacteria. *(Adapted from Figure 19–19, in* Molecular Biology of the Gene, *4th ed., copyright 1987 by James D. Watson. Reprinted by permission of Benjamin/Cummings Publishing Co., Inc.)*

Figure 9.17 A p_{tac} protein expression vector contains the hybrid promoter p_{tac} derived from fusion of the *lac* and *trp* promoters. Expression from p_{tac} is more than ten times greater than expression from either the *lac* or *trp* promoter alone. Isopropyl-β-D-thiogalactoside, or IPTG, induces expression from p_{tac} as well as *lac*.

Strong promoters have been constructed that drive the synthesis of foreign proteins to levels equal to 30% or more of total *E. coli* cellular protein. An example is the hybrid promoter, p_{tac}, which was created by fusing part of the promoter for the *E. coli* genes encoding the enzymes of lactose metabolism (the *lac* promoter) with part of the promoter for the genes encoding the enzymes of tryptophan biosynthesis (the *trp* promoter) (Figure 9.17). In cells carrying p_{tac} expression vectors, the p_{tac} promoter is not induced to drive transcription of the foreign insert until the cells are exposed to *inducers* that lead to its activation. Analogs of lactose (a β-galactoside) such as *isopropyl-β-thiogalactoside,* or **IPTG,** are excellent inducers of p_{tac}. Thus, expression of the foreign protein is easily controlled. (See Chapter 24 for detailed discussions of inducible gene expression.) The bacterial production of valuable eukaryotic proteins represents one of the most important uses of recombinant DNA technology. For example, human insulin for the clinical treatment of diabetes is now produced in bacteria.

Analogous systems for expression of foreign genes in eukaryotic cells include vectors carrying promoter elements derived from mammalian viruses, such as *simian virus 40 (SV40),* the *Epstein–Barr virus,* and the human *cytomegalovirus (CMV).* A system for high level expression of foreign genes uses insect cells infected with the *baculovirus* expression vector. **Baculoviruses** infect *lepidopteran* insects (butterflies and moths). In engineered baculovirus vectors, the foreign gene is cloned downstream of the promoter for **polyhedrin,** a major viral-encoded structural protein, and the recombinant vector is incorporated into insect cells grown in culture. Expression from the polyhedrin promoter can lead to accumulation of the foreign gene product to levels as high as 500 mg/L.

Screening cDNA Expression Libraries with Antibodies

Antibodies that specifically cross-react with a protein of interest are often available. If so, these antibodies can be used to screen a cDNA expression library to identify and isolate cDNA clones encoding the protein. The cDNA library is introduced into host bacteria, which are plated out and grown overnight, as in the colony hybridization scheme previously described. DNA-binding nylon membranes are placed on the plates to obtain a replica of the bacterial colonies. The nylon membrane is then incubated under conditions that induce protein expression from the cloned cDNA inserts within the bacteria, and the cells are treated to release the synthesized protein. The synthesized protein binds tightly to the nylon membrane, which can then be incubated with the specific anti-

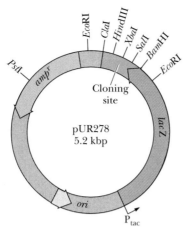

Figure 9.18 A typical expression vector for the synthesis of a hybrid protein. The cloning site is located at the end of the coding region for the protein β-galactosidase. Insertion of foreign DNAs at this site fuses the foreign sequence to the β-galactosidase coding region (the *lacZ* gene). IPTG induces the transcription of the *lacZ* gene from its promoter p_{lac}, causing expression of the fusion protein. *(Adapted from Figure 1.5.4, in Ausubel, F. M., et al., 1987.* Current Protocols in Molecular Biology. *New York: John Wiley & Sons.)*

Codon:	Cys	Gln	Lys	Gly	Asp	Pro	Ser	Thr	Leu	Glu	Ser	Leu	Ser	Met
Cloning site:	TGT	CAA	AAA	GGG	GAT	CCG	TCG	ACT	CTA	GAA	AGC	TTA	TCG	ATG

BamHI   SalI   XbaI   HindIII   ClaI

body. Binding of the antibody to its target protein reveals the position of any cells expressing the protein, and bacteria harboring the relevant cDNA clone can be recovered from the original plate. Like other libraries, expression libraries can also be screened with oligonucleotide probes.

Fusion Protein Expression

Some expression vectors carry cDNA inserts connected directly to the coding sequence of a vector-borne protein-coding gene (Figure 9.18). Translation of the recombinant sequence leads to synthesis of a *hybrid protein* or *fusion protein*. The N-terminal region of the fused protein represents amino acid sequences encoded in the vector, while the remainder of the protein is encoded by the foreign insert. Keep in mind that the triplet codon sequence within the cloned insert must be in phase with codons contributed by the vector sequences to make the right protein. The N-terminal protein sequence contributed by the vector can be chosen to suit purposes. Furthermore, adding an N-terminal signal sequence that targets the hybrid protein for secretion from the cell simplifies recovery of the fusion protein. A variety of gene fusion systems have been developed to facilitate isolation of a specific protein encoded by a cloned insert. The isolation procedures are based on affinity chromatography purification of the fusion protein through exploitation of the unique ligand-binding properties of the vector-encoded protein (Table 9.2).

Table 9.2   Gene Fusion Systems for Isolation of Cloned Fusion Proteins

Gene Product	Origin	Molecular Mass (kD)	Secreted?[1]	Affinity Ligand
β-Galactosidase	*E. coli*	116	No	*p*-Aminophenyl-β-D-thiogalactoside (APTG)
Protein A	*S. aureus*	31	Yes	Immunoglobulin G (IgG)
Chloramphenicol acetyltransferase (CAT)	*E. coli*	24	Yes	Chloramphenicol
Streptavidin	*Streptomyces*	13	Yes	Biotin
Glutathione-*S*-transferase (GST)	*E. coli*	26	No	Glutathione
Maltose-binding protein (MBP)	*E. coli*	40	Yes	Starch

[1] This indicates whether combined secretion–fusion gene systems have led to secretion of the protein product from the cells, which simplifies its isolation and purification.

Adapted from Uhlen, M., and Moks, T., 1990. Gene fusions for purpose of expression: An introduction. *Methods in Enzymology* **185**:129–143.

A DEEPER LOOK

The Two-Hybrid System to Identify Proteins Involved in Specific Protein–Protein Interactions

Specific interactions between proteins (so-called protein–protein interactions) lie at the heart of many essential biological processes. Stanley Fields, Cheng-Ting Chien, and their collaborators have invented a method to identify specific protein–protein interactions *in vivo* through expression of a reporter gene whose transcription is dependent on a functional transcriptional activator, the *GAL4* protein. The *GAL4* protein consists of two domains: A DNA-binding (or **DB**) domain and a transcriptional activation (or **TA**) domain. Even if expressed as separate proteins, these two domains will still work, provided they can be brought together. The method depends on two separate plasmids encoding two hybrid proteins, one consisting of the *GAL4* DB domain fused to protein X, and the other consisting of the *GAL4* TA domain fused to protein Y (see figure, a). If proteins X and Y interact in a specific protein–protein interaction, the *GAL4* DB and TA domains will be brought together so that transcription of a reporter gene driven by the *GAL4* promoter can take place (see figure, b). Protein X, fused to the *GAL4* DNA-binding domain (DB), serves as the "bait" to fish for the protein Y "target" and its fused *GAL4* TA domain. This method can be used to screen cells for protein "targets" that interact specifically with a particular "bait" protein. To do so, cDNAs encoding proteins from the cells of interest are inserted into the TA-containing plasmid to create fusions of the cDNA coding sequences with the *GAL4* TA domain coding sequences, so that a fusion protein library is expressed. Identification of a target of the "bait" protein by this method also directly yields a cDNA version of the gene encoding the "target" protein.

Peter Uetz, Loic Giot, and their colleagues at the University of Washington and CuraGen (a private company) have used the two-hybrid system to identify interactions between any possible pair of proteins among the approximately 6000 proteins in yeast (*Saccharomyces cerevisiae*).* This global, proteomic analysis is designed to reveal those interactions by which individual biochemical events become integrated into larger cellular processes.

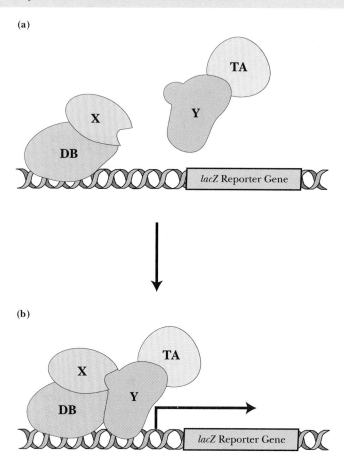

(a)

(b)

*Uetz, P, et al., 2000. A comprehensive analysis of protein–protein interactions in *Saccharomyces cerevisiae*. *Nature* **403:**623–627.

Reporter Gene Constructs

Potential regulatory regions of genes (such as promoters) can be investigated by placing these regulatory sequences into plasmids upstream of a gene, called a **reporter gene,** whose expression is easy to measure. Such chimeric plasmids are then introduced into cells of choice (including eukaryotic cells) to assess the potential function of the nucleotide sequence in regulation, since expression of the reporter gene serves as a report on the effectiveness of the regulatory element. A number of different genes have been used as reporter genes. A reporter gene with many inherent advantages is that encoding the **green fluorescent protein** (or **GFP**), which was described in Chapter 4. Unlike the protein expressed by other reporter gene systems, GFP does not require any substrate to measure its activity, nor is it dependent on any cofactor or prosthetic group. Detection of GFP only requires irradiation with near UV or blue light (400-nm light is optimal); the green fluorescence (light of 500 nm) which results is easily observed with the naked eye, although it can also be measured precisely with a fluorometer. Figure 9.19 demonstrates the use of GFP as a reporter gene.

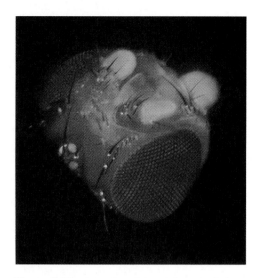

Figure 9.19 Green fluorescent protein (GFP) as a reporter gene. The promoter from the *per* gene was placed upstream of the GFP gene in a plasmid and transformed into *Drosophila* (fruit flies). The *per* gene encodes a protein involved in establishing the circadian (daily) rhythmic activity of fruit flies. The fluorescence shown here in an isolated fly head follows a 24-hour rhythmic pattern and occurs to a lesser extent throughout the entire fly, indicating that *per* gene expression can occur in cells throughout the animal. Such uniformity suggests that individual cells have their own independent clocks. *(Image courtesy of Jeffrey D. Plautz and Steve A. Kay, Scripps Research Institute, Calif. See also Plautz, J. D., et al., 1997. Independent photoreceptive circadian clocks in* Drosophila. Science **278:** 1632–1635.)

9.3 Polymerase Chain Reaction (PCR)

Polymerase chain reaction, or **PCR,** is a technique for dramatically amplifying the amount of a specific DNA segment. A preparation of denatured DNA containing the segment of interest serves as template for DNA polymerase, and two specific oligonucleotides serve as primers for DNA synthesis (as in Figure 9.20). These primers, designed to be complementary to the two 3′-ends of the specific DNA segment to be amplified, are added in excess amounts of 1000 times or greater. They prime the DNA polymerase-catalyzed synthesis of the two complementary strands of the desired segment, effectively doubling its concentration in the solution. Next, the DNA is heated to dissociate the DNA duplexes and then cooled so that primers bind to both the newly formed and the old strands. Another cycle of DNA synthesis follows. The protocol has been automated through the invention of **thermal cyclers** that alternately heat the reaction mixture to 95°C to dissociate the DNA, followed by cooling, annealing of primers, and another round of DNA synthesis. The isolation of heat-stable DNA polymerases from thermophilic bacteria (such as the *Taq* DNA polymerase from *Thermus aquaticus*) has made it unnecessary to add fresh enzyme for each round of synthesis. Since the amount of target DNA theoretically doubles each round, 25 rounds would increase its concentration about 33 million times. In practice, the increase is actually more like a million times, which is more than ample for gene isolation. Thus, starting with a tiny amount of total genomic DNA, a particular sequence can be produced in quantity in a few hours.

PCR amplification is an effective cloning strategy if sequence information for the design of appropriate primers is available. Because DNA from a single cell can be used as template, the technique has enormous potential for the clinical diagnosis of infectious diseases and genetic abnormalities. With PCR techniques, DNA from a single hair or sperm can be analyzed to identify particular individuals in criminal cases without ambiguity. **RT-PCR,** a variation on the basic PCR method, is useful when the nucleic acid to be amplified is an RNA (such as mRNA). Reverse transcriptase (RT) is used to synthesize a cDNA strand complementary to the RNA, and this cDNA serves as template for further cycles of PCR.

In Vitro Mutagenesis

The advent of recombinant DNA technology has made it possible to clone genes, manipulate them *in vitro,* and express them in a variety of cell types under various conditions. The function of any protein is ultimately dependent on its amino acid sequence, which in turn can be traced to the nucleotide sequence of its gene. The introduction of purposeful changes in the nucleotide sequence of a cloned gene represents an ideal way to make specific structural changes in a protein. The effects of these changes on the protein's function can then be studied. Such changes constitute *mutations* introduced *in vitro* into the gene. *In vitro* **mutagenesis** makes it possible to alter the nucleotide sequence of a cloned gene systematically, as opposed to relying on the chance occurrence of mutations in natural genes.

One efficient technique for *in vitro* mutagenesis is **PCR-based mutagenesis.** A mutant primer is added to a PCR reaction in which the gene (or

Figure 9.20 Polymerase chain reaction (PCR). Oligonucleotides complementary to a given DNA sequence prime the synthesis of only that sequence. Heat-stable *Taq* DNA polymerase survives many cycles of heating. Theoretically, the amount of the specific primed sequence is doubled in each cycle.

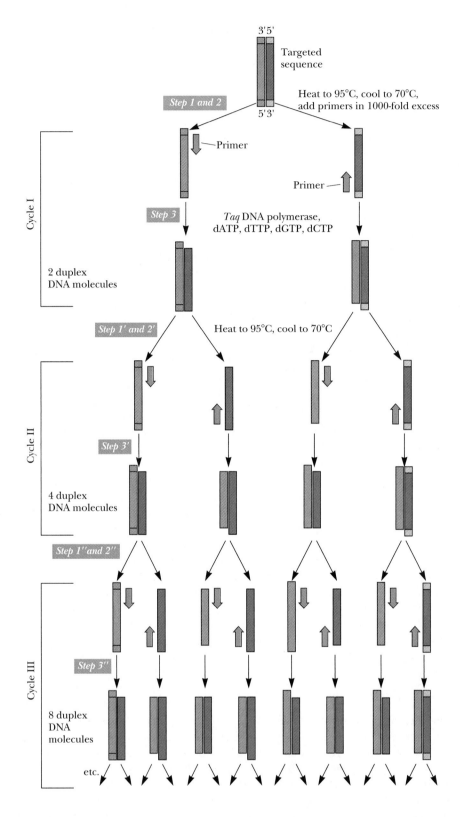

segment of a gene) is undergoing amplification. The *mutant primer* is a primer whose sequence has been specifically altered to introduce a directed change at a particular place in the nucleotide sequence of the gene being amplified (Figure 9.21). Mutant versions of the gene can then be cloned and expressed to determine any effects of the mutation on the function of the gene product.

9.4 Recombinant DNA Technology: An Exciting Scientific Frontier

The strategies and methodologies described in this chapter are but an overview of the repertoire of experimental approaches that have been devised by molecular biologists in order to manipulate DNA and the information inherent in it. The enormous success of recombinant DNA technology means that the molecular biologist's task in searching genomes for genes is now akin to that of a lexicographer compiling a dictionary, a dictionary in which the "letters"— that is, the nucleotide sequences—spell out not words but genes and what they mean. Molecular biologists have no index or alphabetical arrangement to serve as a guide through the vast volume of information in a genome; nevertheless, this information and its organization are rapidly being disclosed through the imaginative efforts and diligence of these scientists and their growing arsenal of analytical schemes.

Recombinant DNA technology now verges on the ability to engineer the genetic constitution of organisms for desired ends. The commercial production of therapeutic biomolecules in microbial cultures is already established (for example, the production of human insulin in quantity in *Escherichia coli* cells). Agricultural crops with desired attributes, such as enhanced resistance to herbicides or elevated vitamin levels, are in cultivation. The rat growth hormone gene has been cloned and transferred into mouse embryos, creating *transgenic mice* that at adulthood are twice the normal size (see Chapter 23). Already, transgenic versions of domestic animals such as pigs, sheep, and even fish have been developed for human benefit. **Gene replacement therapy** (or, more simply, *gene therapy*) to correct particular human genetic disorders has become a realistic possibility.

Human Gene Therapy

Human gene therapy seeks to repair the damage caused by a genetic deficiency through introduction of a functional version of the defective gene. To achieve this end, a cloned variant of the gene must be incorporated into the organism in such a manner that it is expressed only at the proper time *and* only in appropriate cell types. At this time, these conditions impose serious technical and clinical difficulties. Many gene therapies have received approval from the National Institutes of Health for trials in human patients, including the introduction of gene constructs into patients. Among these are constructs designed to cure SCID (*severe combined immuno deficiency*), neuroblastoma, or cystic fibrosis or to treat cancer through expression of the *E1A* and *p53* tumor suppressor genes.

A basic strategy in human gene therapy involves incorporation of a functional gene into target cells. The gene is typically in the form of an **expression cassette** consisting of a cDNA version of the gene downstream of a promoter that will drive expression of the gene. A vector carrying such an expression cassette is introduced into target cells, either *ex vivo* via gene transfer into cultured cells in the laboratory and administration of the modified cells to the patient, or *in vivo* via direct incorporation of the gene into the cells of the patient. Because retroviruses can transfer their genetic information directly into the genome of host cells, retroviruses provide one route to permanent modification of host cells *ex vivo*. A replication-deficient version of *Maloney murine leukemia virus* can serve as a vector for expression cassettes up to 9 kb in size. Figure 9.22 describes one strategy for retrovirus vector–mediated gene delivery. In this strategy, it is hoped that the expression cassette will become stably integrated into the DNA of the patient's own cells and expressed to produce the desired gene product. Scientists at the Pasteur Institute in Paris have used

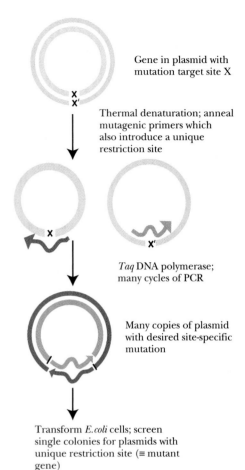

Figure 9.21 One method of PCR-based site-directed mutagenesis. Template DNA strands are separated by increased temperature and the single strands are amplified by PCR using mutagenic primers (represented as bent arrows) whose sequences will introduce a single base substitution at site *X* (and its complementary base *X'*, and thus the desired amino acid change in the protein encoded by the gene). Ideally, the mutagenic primers will also introduce a unique restriction site into the plasmid that wasn't present before. Following many cycles of PCR, the DNA product can be used to transform *E. coli* cells. Single colonies of the transformed cells can be picked. The plasmid DNA within each colony can be isolated and screened for the presence of the mutation by screening for the presence of the unique restriction site by restriction endonuclease cleavage. For example, the nucleotide sequence GGATCT within a gene codes for amino acid residues Gly-Ser. Using mutagenic primers of nucleotide sequence AGATCT (and its complement AGATCT) will change the amino acid sequence from Gly-Ser to Arg-Ser and create a *Bgl*II restriction site (Table 8.3). Gene expression of the isolated mutant plasmid in *E. coli* allows recovery and analysis of the mutant protein.

Figure 9.22 Retrovirus-mediated gene delivery *ex vivo*. Retroviruses are RNA viruses that replicate their RNA genome by first making a DNA intermediate. The Maloney murine leukemia virus (MMLV) is the retrovirus used in human gene therapy. Deletion of the essential genes *gag*, *pol*, and *env* from MMLV makes it replication-deficient (so it can't reproduce) **(a)** and creates a space for insertion of an expression cassette **(b).** The modified MMLV acts as a vector for the expression cassette; though replication-defective, it is still infectious. Infection of a packaging cell line which carries intact *gag*, *pol*, and *env* genes allows the modified MMLV to reproduce **(c)** and the packaged retroviral viruses can be collected and used to infect a patient **(d).** In the cytosol of the patient's cells, a DNA copy of the viral RNA is synthesized by viral reverse transcriptase, which accompanies the viral RNA into the cells. This DNA is then randomly integrated into the host cell genome, where its expression leads to production of the expression cassette product. (*Adapted from Figure 1, in Crystal, R. G., 1995. Transfer of genes to humans: Early lessons and obstacles to success.* Science **270:** 404–410.)

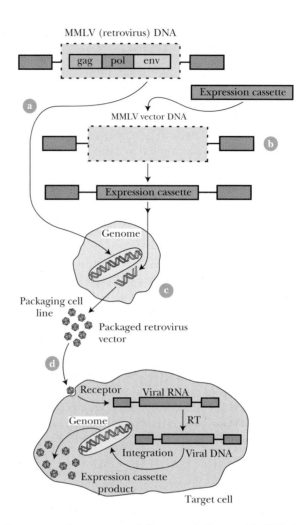

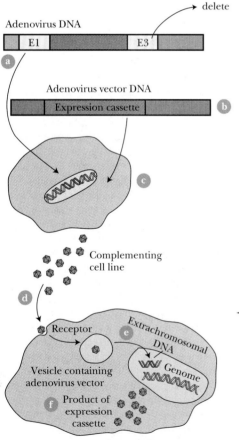

Target cell

such an *ex vivo* approach to successfully treat infants with X-linked SCID. The gene encoding the γc cytokine receptor subunit was defective in these infants, and gene therapy was used to deliver a functional γc cytokine receptor subunit gene to stem cells harvested from the infants. Transformed stem cells were reintroduced into the patients, who then became able to produce functional lymphocytes and lead normal lives. Their achievement represents the first fully successful outcome in human gene therapy.

Adenovirus vectors which can carry expression cassettes up to 7.5 kb are another possible *in vivo* approach to human gene therapy (Figure 9.23). Recombinant, replication-deficient adenoviruses enter target cells via specific receptors on the target cell surface; the transferred genetic information is expressed directly from the adenovirus recombinant DNA and is never incorporated into the host cell genome. Although many problems remain to be solved, human gene therapy as a clinical strategy is now feasible.

◀ **Figure 9.23** Adenovirus-mediated gene delivery *in vivo*. Adenoviruses are DNA viruses. The adenovirus genome (36 kb) is divided into *early genes* (E1 through E4) and *late genes* (L1 to L5) **(a).** Adenovirus vectors are generated by deleting gene E1 (and sometimes E3 if more space for an expression cassette is needed) **(b);** deletion of E1 renders the adenovirus incapable of replication unless introduced into a complementing cell line carrying the E1 gene **(c).** Adenovirus progeny from the complementing cell line can be used to infect a patient. In the patient, the adenovirus vector with its expression cassette enters the cells via specific receptors **(d).** Its linear, dsDNA ultimately gains access to the cell nucleus **(e)** where it functions extrachromosomally and expresses the product of the expression cassette **(f).** (*Adapted from Figure 2, in Crystal, R. G., 1995. Transfer of genes to humans: Early lessons and obstacles to success.* Science **270:** 404–410.)

The Biochemical Defects in Cystic Fibrosis and SCID

The gene defective in cystic fibrosis codes for CFTR (cystic fibrosis transmembrane conductance regulator), a membrane protein that pumps Cl⁻ out of cells. If this Cl⁻ pump is defective, Cl⁻ ions remain in cells, which then take up water from the surrounding mucus by osmosis. The mucus thickens and accumulates in various organs, including the lungs, where its presence favors infections such as pneumonia. Left untreated, children with cystic fibrosis seldom survive past the age of five.

SCID (*severe combined immunodeficiency*) is an inherited immunodeficiency disease in which infants, because they cannot produce antibodies, die of infection within a few weeks or months of birth. In one form of SCID (X-linked SCID, also called XSCID), the victims are deficient in both B- and T-lymphocytes. Both of these cell types originate from stem cells in the bone marrow. In X-linked SCID, these stem cells cannot respond to signals (*cytokines*) directing them to differentiate into lymphocytes because their receptors for these signals are defective. (*Cytokines* are small proteins that trigger cellular differentiation or proliferation.)

Another form of SCID, ADA⁻ SCID (adenosine deaminase–defective), is a fatal genetic disorder caused by defects in the gene that encodes adenosine deaminase (ADA). The consequence of ADA deficiency is accumulation of adenosine and 2′-deoxyadenosine, substances toxic to lymphocytes, important cells in the immune response. 2′-Deoxyadenosine is particularly toxic, since its presence leads to accumulation of its nucleotide form, dATP, an

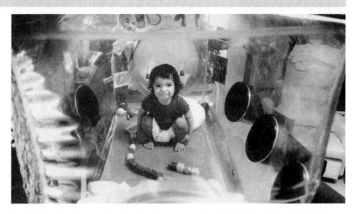

David, the Boy in the Bubble. David was born with SCID and lived all 12 years of his life inside a sterile plastic "bubble" to protect him from germs common in the environment. He died in 1984 following an unsuccessful bone marrow transplant. *(UPI/Corbis-Bettman)*

essential substrate in DNA synthesis. Elevated levels of dATP actually block DNA replication and cell division by inhibiting synthesis of the other deoxynucleoside 5′-triphosphates (see Chapter 21). Accumulation of dATP also leads to selective depletion of cellular ATP, robbing cells of energy. Children with ADA⁻ SCID fail to develop normal immune responses and are susceptible to fatal infections unless kept in protective isolation.

PROBLEMS

1. A DNA fragment isolated from an *Eco*RI digest of genomic DNA was combined with a plasmid vector linearized by *Eco*RI digestion so sticky ends could anneal. Phage T4 DNA ligase was then added to the mixture. List all possible products of the ligation reaction.

2. The nucleotide sequence of a polylinker in a particular plasmid vector is

    —GAATTCCCGGGGATCCTCTAGAGTCGACCTGCAGGCATGC—

    This polylinker contains restriction sites for *Bam*HI, *Eco*RI, *Pst*1, *Sal*I, *Sma*I, *Sph*I, and *Xba*I. Indicate the location of each restriction site in this sequence. (See Table 8.3 of restriction enzymes for their cleavage sites.)

3. A vector has a polylinker containing restriction sites in the following order: *Hind*III, *Sac*I, *Xho*I, *Bgl*II, *Xba*I, and *Cla*I.
    a. Give a possible nucleotide sequence for the polylinker.
    b. The vector is digested with *Hind*III and *Cla*I. A DNA segment contains a *Hind*III restriction site fragment 650 bases upstream from a *Cla*I site. This DNA fragment is digested with *Hind*III and *Cla*I, and the resulting *Hind*III–*Cla*I fragment is directionally cloned into the *Hind*III–*Cla*I–digested vector. Give the nucleotide sequence at each end of the vector and the insert and show that the insert can be cloned into the vector in only one orientation.

4. Yeast (*Saccharomyces cerevisiae*) has a genome size of 1.21×10^7 bp. If a genomic library of yeast DNA was con-

structed in a bacteriophage λ vector capable of carrying 16 kbp inserts, how many individual clones would have to be screened to have a 99% probability of finding a particular fragment?

5. The South American lungfish has a genome size of 1.02×10^{11} bp. If a genomic library of lungfish DNA was constructed in a cosmid vector capable of carrying inserts averaging 45 kbp in size, how many individual clones would have to be screened to have a 99% probability of finding a particular DNA fragment?

6. Given the following short DNA duplex of sequence $(5' \rightarrow 3')$

    ATGCCGTAGTCGATCATTACGATAGCATAGCACAGG
    GATCACACATGCACACACATGACATAGGACAGATAGCAT

    what oligonucleotide primers (17-mers) would be required for PCR amplification of this duplex?

7. Figure 9.5b shows a polylinker that falls within the β-galactosidase coding region of the *lacZ* gene. This polylinker serves as a cloning site in a fusion protein expression vector where the cloned insert will be expressed as a β-galactosidase fusion protein. Assume the vector polylinker was cleaved with *Bam*HI and then ligated with an insert whose sequence reads

    GATCCATTTATCCACCGGAGAGCTGGTATCCCCAAAAG
    ACGGCC . . .

    What is the amino acid sequence of the fusion protein? Where is the junction between β-galactosidase and the sequence encoded by the insert? (Consult the genetic code table on the inside front cover to decipher the amino acid sequence.)

8. The amino acid sequence across a region of interest in a protein is

Asn-Ser-Gly-Met-His-Pro-Gly-Lys-Leu-Ala-Ser-Trp-Phe-Val-Gly-Asn-Ser-

The nucleotide sequence encoding this region begins and ends with an *Eco*RI site, making it easy to clone out the sequence and amplify it by the polymerase chain reaction (PCR). Give the nucleotide sequence of this region. Suppose you wish to change the middle Ser residue to a Cys to study the effects of this change on the protein's activity. What would be the sequence of the mutant oligonucleotide you would use for PCR amplification?

FURTHER READING

Ausubel, F. M., et al., eds., 1999. *Short Protocols in Molecular Biology.* 4th ed. New York: John Wiley & Sons. A popular cloning manual.

Berger, S. L., and Kimmel, A. R., eds., 1987. Guide to molecular cloning techniques. *Methods in Enzymology* **152**:1–812.

Bolivar, F., Rodriguez, R. L., Greene, P. J., et al., 1977. Construction and characterization of new cloning vehicles. II. A multipurpose cloning system. *Gene* **2**:95–113. A paper describing one of the early plasmid cloning systems.

Cavazzana-Calvo, M., et al., 2000. Gene therapy of human severe combined immunodeficiency (SCID)-X1 disease. *Science* **288**:669–672.

Chalfie, M., et al., 1994. Green fluorescent protein as a marker for gene expression. *Science* **263**:802–805.

Chien, C.-T., et al., 1991. The two-hybrid system: A method to identify and clone genes for proteins that interact with a protein of interest. *Proceedings of the National Academy of Sciences, U.S.A.* **88**:9578–9582.

Cohen, S. N., Chang, A. C. Y., Boyer, H. W., and Helling, R. B., 1973. Construction of biologically functional bacterial plasmids in vitro. *Proceedings of the National Academy of Sciences, U.S.A.* **70**:3240–3244. The classic paper on the construction of chimeric plasmids.

Cortese, R., 1996. *Combinatorial Libraries: Synthesis, Screening and Application Potential.* Berlin: Walter de Gruyter.

Crystal, R. G., 1995. Transfer of genes to humans: Early lessons and obstacles to success. *Science* **270**:404–410.

Glorioso, J. C., and Schmidt, M. C., eds., 1999. Expression of recombinant genes in eukaryotic cells. *Methods in Enzymology* **306**:1–403.

Goeddel, D. V., ed., 1990. Gene expression technology. *Methods in Enzymology* **185**:1–681.

Grunstein, M., and Hogness, D. S., 1975. Colony hybridization: A specific method for the isolation of cloned DNAs that contain a specific gene. *Proceedings of the National Academy of Sciences, U.S.A.* **72**:3961–3965. Article describing the colony hybridization technique for specific gene isolation.

Guyer, M., 1992. A comprehensive genetic linkage map for the human genome. *Science* **258**:67–86.

Jackson, D. A., Symons, R. H., and Berg, P., 1972. Biochemical method for inserting new genetic information into DNA of simian virus 40: Circular SV40 DNA molecules containing lambda phage genes and the galactose operon of *E. coli. Proceedings of the National Academy of Sciences, U.S.A.* **69**:2904–2909.

Lander, E., Page, D., and Lifton, R., eds., 2000. *Annual Review of Genomics and Human Genetics,* Vol. 1. Palo Alto, Calif.: Annual Reviews, Inc. A new review series on genomics and human genetic diseases.

Luckow, V. A., and Summers, M. D., 1988. Trends in the development of baculovirus expression vectors. *Biotechnology* **6**:47–55.

Lyon, J., and Gorner, P., 1995. *Altered Fates. Gene Therapy and the Retooling of Human Life.* New York: Norton.

Maniatis, T., Hardison, R. C., Lacy, E., et al., 1978. The isolation of structural genes from libraries of eucaryotic DNA. *Cell* **15**:687–701.

Morgan, R. A., and Anderson, W. F., 1993. Human gene therapy. *Annual Review of Biochemistry* **62**:191–217.

Murialdo, H., 1991. Bacteriophage lambda DNA maturation and packaging. *Annual Review of Biochemistry* **60**:125–153. Review of the biochemistry of packaging DNA into nine phage heads.

Palese, P., and Roizman, B., 1996. Genetic engineering of viruses and of virus vectors: A preface. *Proceedings of the National Academy of Sciences, U.S.A.* **93**:11287–11425. A preface to a series of papers from a colloquium on genetic engineering and methods in gene therapy.

Peterson, K. R., et al., 1997. Production of transgenic mice with yeast artificial chromosomes. *Trends in Genetics* **13**:61–66.

Saiki, R. K., Gelfand, D. H., Stoeffel, B., et al., 1988. Primer-directed amplification of DNA with a thermostable DNA polymerase. *Science* **239**:487–491. Discussion of the polymerase chain reaction procedure.

Sambrook, J., and Russell, D., 2000. *Molecular Cloning.* 3rd ed. New York: Cold Spring Harbor Laboratory Press.

Scriver, C. R., Beaudet, A. L., Sly, W. S., and Valle, D., eds. 1995. *The Metabolic and Molecular Bases of Inherited Disease.* New York: McGraw-Hill. The seventh edition of this classic work contains contributions from over 300 authors. A three-volume treatise, it is the definitive source on the molecular basis of inherited diseases.

Southern, E. M., 1975. Detection of specific sequences among DNA fragments separated by gel electrophoresis. *Journal of Molecular Biology* **98**:503–517. The classic paper on the identification of specific DNA sequences through hybridization with unique probes.

Southern, E. M., 1996. DNA chips: Analysing sequence by hybridization to oligonucleotides on a large scale. *Trends in Genetics* **12**:110–115.

Thorner, J., and Emr, S., eds., 2000. Applications of chimeric genes and hybrid proteins. *Methods in Enzymology* **328**:1–690.

Timmer, W. C., and Villalobos, J. M., 1993. The polymerase chain reaction. *The Journal of Chemical Education* **70**:273–280.

Uetz, P, et al., 2000. A comprehensive analysis of protein–protein interactions in *Saccharomyces cerevisiae. Nature* **403**:623–627.

Weissman, S., ed., 1999. cDNA preparation and display. *Methods in Enzymology* **303**:1–575.

Young, R. A., and Davis, R. W., 1983. Efficient isolation of genes using antibody probes. *Proceedings of the National Academy of Sciences, U.S.A.* **80**:1194–1198. Using antibodies to screen protein expression libraries to isolate the structural gene for a specific protein.

PROTEIN DYNAMICS

Protein dynamics—the action of enzymes and molecular motors—provides the key to understanding the biochemistry of this cheetah and the grasses through which it runs. (Frank Lane/Parfitt/Tony Stone Images)

Enzymes: Their Kinetics, Specificity, and Regulation

Allostery is the key chemical process that makes possible intracellular and intercellular regulation: ". . . the molecular interactions which ensure the transmission and interpretation of [regulatory] signals rest upon [allosteric] proteins endowed with discriminatory stereospecific recognition properties."

Jacques Monod, in *Chance and Necessity*

Outline

Metabolic regulation is achieved through an exquisitely balanced interplay among enzymes and small molecules, a process symbolized by the delicate balance of forces in this mobile. (Sea Scape by Alexander Calder/Whitney Museum of American Art, NY)

Living organisms seethe with metabolic activity. Thousands of chemical reactions are proceeding very rapidly at any given instant within all living cells. Virtually all of these transformations are mediated by **enzymes,** proteins (or occasionally, RNAs) specialized to catalyze metabolic reactions. The substances transformed in these reactions are often organic compounds that show little tendency for reaction outside the cell. An excellent example is glucose, a sugar that can be stored indefinitely on the shelf with no deterioration. Most cells quickly oxidize glucose, producing carbon dioxide and water and releasing lots of energy:

$$C_6H_{12}O_6 + 6\ O_2 \rightarrow 6\ CO_2 + 6\ H_2O + 2870\ kJ\ of\ energy$$

(-2870 kJ/mol is the standard free energy change [$\Delta G°'$] for the oxidation of glucose; see Chapter 3). In chemical terms, 2870 kJ is a large amount of energy, and glucose can be viewed as an energy-rich compound, even though at

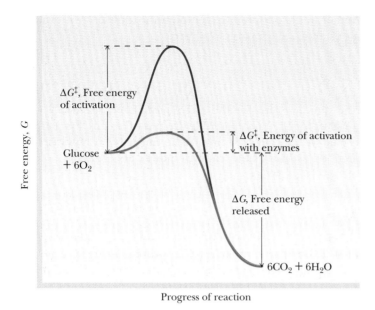

Figure 10.1 Reaction profile showing large $\Delta G^{\circ}{}'$ for glucose oxidation, free energy change of -2870 kJ/mol; catalysts lower $\Delta G^{\ddagger}$, thereby accelerating rate.

ambient temperature it does not react readily with oxygen outside of cells. Stated another way, glucose represents **thermodynamic potentiality:** its reaction with oxygen is strongly exergonic, but it just doesn't occur under normal conditions. On the other hand, enzymes can catalyze such thermodynamically favorable reactions so that they proceed at extraordinarily rapid rates (Figure 10.1). In glucose oxidation and countless other instances, enzymes provide cells with the ability to exert *kinetic control over thermodynamic potentiality*. That is, living systems use enzymes to accelerate and control the rates of vitally important biochemical reactions.

Enzymes Are the Agents of Metabolic Function

Acting in sequence, enzymes form metabolic pathways by which nutrient molecules are degraded, energy is released and converted into metabolically useful forms, and precursors are generated and transformed to create the literally thousands of distinctive biomolecules found in any living cell (Figure 10.2). Situated at key junctions of metabolic pathways are specialized **regulatory enzymes** capable of sensing the momentary metabolic needs of the cell and adjusting their catalytic rates accordingly. The responses of these enzymes ensure the harmonious integration of the diverse and often divergent metabolic activities of cells so that the living state is promoted and preserved.

10.1 Enzymes—Catalytic Power, Specificity, and Regulation

Enzymes are characterized by three distinctive features: **catalytic power, specificity,** and **regulation.**

Catalytic Power

Enzymes display enormous catalytic power, accelerating reaction rates as much as 10^{16} over uncatalyzed levels, which is far greater than any synthetic catalysts can achieve. Enzymes accomplish these astounding feats in dilute aqueous

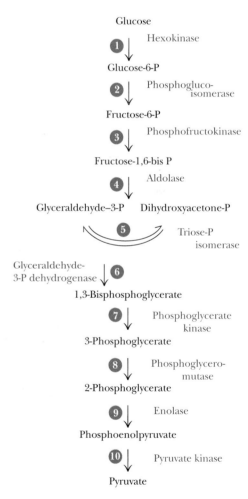

Figure 10.2 The breakdown of glucose by glycolysis provides a prime example of a metabolic pathway. Ten enzymes mediate the reactions of glycolysis. Enzyme 4, *fructose 1,6-bisphosphate aldolase,* catalyzes the C—C bond-breaking reaction in this pathway.

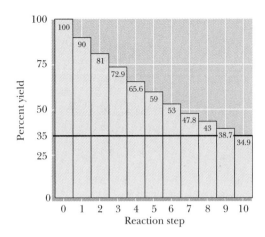

Figure 10.3 A 90% yield for each of 10 steps, for example, in a metabolic pathway gives an overall yield of 35%. Therefore, yields in biological reactions must be substantially greater; otherwise, unwanted by-products would accumulate to unacceptable levels.

solutions under mild conditions of temperature and pH. For example, the enzyme *urease* catalyzes the hydrolysis of urea:

$$N_2H-\overset{\overset{\displaystyle O}{\|}}{C}-NH_2 + 2\ H_2O + H^+ \longrightarrow 2\ NH_4^+ + HCO_3^-$$

At 20°C, the rate constant for the enzyme-catalyzed reaction is $3 \times 10^4/\text{sec}$; the rate constant for the uncatalyzed hydrolysis of urea is $3 \times 10^{-10}/\text{sec}$. Thus, 10^{14} is the ratio of the catalyzed rate to the uncatalyzed rate of reaction. Such a ratio is defined as the relative **catalytic power** of an enzyme, so the catalytic power of urease is 10^{14}.

Specificity

A given enzyme is very selective, both in the substances with which it interacts and in the reaction that it catalyzes. The substances upon which an enzyme acts are traditionally called **substrates.** In an enzyme-catalyzed reaction, none of the substrate is diverted into nonproductive side-reactions, so no wasteful by-products are produced. It follows then that the products formed by a given enzyme are also very specific. This situation can be contrasted with your own experiences in the organic chemistry laboratory, where yields of 50% or even 30% are viewed as substantial accomplishments (Figure 10.3). The selective qualities of an enzyme are collectively recognized as its **specificity.** Intimate interaction between an enzyme and its substrates occurs through molecular recognition based on structural complementarity; such mutual recognition is

---

A DEEPER LOOK

Enzyme Nomenclature

Traditionally, enzymes often were named by adding the suffix *-ase* to the name of the substrate upon which they acted, as in *urease* for the urea-hydrolyzing enzyme or *phosphatase* for enzymes hydrolyzing phosphoryl groups from organic phosphate compounds. Other enzymes acquired names bearing little resemblance to their activity, such as the peroxide-decomposing enzyme *catalase* or the proteolytic enzymes (*proteases*) of the digestive tract, *trypsin* and *pepsin*. Because of the confusion that arose from these trivial designations, an International Commission on Enzymes was established in 1956 to create a systematic basis for enzyme nomenclature. Although common names for many enzymes remain in use, all enzymes now are classified and formally named according to the reaction they catalyze. Six classes of reactions are recognized (see table). Within each class are subclasses, and under each subclass are sub-subclasses within which individual enzymes are listed. Classes, subclasses, sub-subclasses, and individual entries are each numbered, so that a series of four numbers serves to specify a particular enzyme. A systematic name, descriptive of the reaction, is also assigned to each entry. To illustrate, consider the enzyme that catalyzes this reaction:

$$\text{ATP} + \text{D-glucose} \rightarrow \text{ADP} + \text{D-glucose-6-phosphate}$$

A phosphate group is transferred from ATP to the C-6-OH group of glucose, so the enzyme is a *transferase* (see table, class 2). Sub-

class 7 of transferases is *enzymes transferring phosphorus-containing groups,* and sub-subclass 1 covers those phosphotransferases with an alcohol group as an acceptor. Entry 2 in this sub-subclass is **ATP: D-glucose-6-phosphotransferase,** and its classification number is **2.7.1.2.** In use, this number is written preceded by the letters **E.C.,** denoting the Enzyme Commission. For example, entry 1 in the same sub-subclass is **E.C.2.7.1.1,** *ATP: D-hexose-6-phosphotransferase,* an ATP-dependent enzyme that transfers a phosphate to the 6-OH of hexoses (that is, it is nonspecific regarding its hexose acceptor). These designations can be cumbersome, so in everyday usage, trivial names are employed frequently. The glucose-specific enzyme, E.C.2.7.1.2, is called *glucokinase* and the nonspecific E.C.2.7.1.1 is known as *hexokinase. Kinase* is a trivial term for enzymes that are ATP-dependent phosphotransferases.

Enzyme Commission Number	Systematic Name
1	*Oxidoreductases* (oxidation–reduction reactions)
2	*Transferases* (transfer of functional groups)
3	*Hydrolases* (hydrolysis reactions)
4	*Lyases* (addition to double bonds)
5	*Isomerases* (isomerization reactions)
6	*Ligases* (formation of bonds with ATP cleavage)

Table 10.1   Enzyme Cofactors: Some Metal Ions and Coenzymes and the Enzymes with Which They Are Associated

Metal Ions and Some Enzymes That Require Them		Coenzymes Serving As Transient Carriers of Specific Atoms or Functional Groups		Representative Enzymes Using Coenzymes
Metal Ion	**Enzyme**	**Coenzyme**	**Entity Transferred**	
Fe^{2+} or Fe^{3+}	Cytochrome Oxidase	Thiamine pyrophosphate (TPP)	Aldehydes	Pyruvate dehydrogenase
	Catalase	Flavin adenine dinucleotide (FAD)	Hydrogen atoms	Succinate dehydrogenase
	Peroxidase	Nicotinamide adenine dinucleotide (NAD)	Hydride ion (H^-)	Alcohol dehydrogenase
Cu^{2+}	Cytochrome oxidase			
Zn^{2+}	DNA polymerase	Coenzyme A (CoA)	Acyl groups	Acetyl-CoA carboxylase
	Carbonic anhydrase	Pyridoxal phosphate (PLP)	Amino groups	Aspartate aminotransferase
	Alcohol dehydrogenase	5'-Deoxyadenosylcobalamin (vitamin B_{12})	H atoms and alkyl groups	Methylmalonyl-CoA mutase
Mg^{2+}	Hexokinase			
	Glucose-6-phosphatase	Biotin (biocytin)	CO_2	Propionyl-CoA carboxylase
Mn^{2+}	Arginase	Tetrahydrofolate (THF)	Other one-carbon groups	Thymidylate synthase
K^+	Pyruvate kinase (also requires Mg^{2+})			
Ni^{2+}	Urease			
Mo	Nitrate reductase			
Se	Glutathione peroxidase			

the basis of specificity. The specific site on the enzyme where substrate binds and catalysis occurs is called the **active site.**

Regulation

Regulation of enzyme activity is achieved in a variety of ways, ranging from controls over the amount of enzyme protein produced by the cell to more rapid, reversible interactions of the enzyme with metabolic inhibitors and activators. The latter part of this chapter considers the various mechanisms of enzyme regulation in detail. Because most enzymes are proteins, we can anticipate that the remarkable properties of enzymes are due to the versatility inherent in protein structures.

Coenzymes

Many enzymes carry out their catalytic function relying solely on their protein structure. Many others require nonprotein components, called **cofactors.** Cofactors may be metal ions or organic molecules referred to as **coenzymes.** Because they are structurally less complex than proteins, cofactors tend to be stable to heat (incubation in a boiling water bath). Typically, proteins are denatured under such conditions. Many coenzymes are vitamins or contain vitamins as part of their structure. Usually coenzymes are actively involved in the catalytic reaction of the enzyme, often serving as intermediate carriers of functional groups in the conversion of substrates to products. In most cases, a coenzyme is firmly associated with its enzyme, perhaps even by covalent bonds, and it is difficult to separate the two. Such tightly bound coenzymes are referred to as **prosthetic groups** of the enzyme. The catalytically active complex of protein and prosthetic group is called the **holoenzyme.** The protein without the prosthetic group is called the **apoenzyme;** it is catalytically inactive. Table 10.1 gives examples of both cofactors and coenzymes.

10.2 Introduction to Enzyme Kinetics

Kinetics is the branch of science concerned with the rates of chemical reactions. The study of **enzyme kinetics** addresses the biological roles of enzymatic catalysts and how they accomplish their remarkable feats. In enzyme kinetics, we seek to determine the maximum reaction velocity that the enzyme can attain and its binding affinities for substrates and inhibitors. Coupled with studies on the structure and chemistry of the enzyme, analysis of the enzymatic rate under different reaction conditions yields insights regarding the enzyme's mechanism of catalytic action. Such information is essential to an overall understanding of metabolism.

Significantly, this information can be exploited to control and manipulate the course of metabolic events. The science of **pharmacology** relies on such a strategy. *Pharmaceuticals*, or *drugs*, are often special inhibitors specifically targeted at a particular enzyme in order to overcome infection or to alleviate illness. A detailed knowledge of the enzyme's kinetics is indispensable to rational drug design and successful pharmacological intervention.

A Review of Chemical Kinetics

Before beginning a quantitative treatment of enzyme kinetics, it will be fruitful to review briefly some basic principles of chemical kinetics. **Chemical kinetics** is the study of the rates of chemical reactions. Consider a reaction of overall stoichiometry

$$A \rightarrow P$$

Let us assume that $A \rightarrow P$ is an elementary reaction and that it is spontaneous and essentially irreversible. Irreversibility is easily assumed if the rate of P conversion to A is very slow *or* the concentration of P (expressed as $[P]$) is negligible under the conditions chosen. The **velocity,** v, or **rate,** of the reaction $A \rightarrow P$ is the amount of P formed or the amount of A consumed per unit time, t. That is,

$$v = \frac{d[P]}{dt} \quad \text{or} \quad v = \frac{-d[A]}{dt} \tag{10.1}$$

The mathematical relationship between reaction rate and concentration of reactant(s) is the **rate law.** For this simple case, the rate law is

$$v = \frac{-d[A]}{dt} = k[A] \tag{10.2}$$

From this expression, it is obvious that the rate is proportional to the concentration of A, and k is the proportionality constant, or **rate constant.** k has the units of $(\text{time})^{-1}$, usually sec^{-1}. v is a function of $[A]$ to the first power, or, in the terminology of kinetics, v is first-order with respect to A. For simple reactions, the **order** for any reactant is given by its exponent in the rate equation. The number of molecules that must simultaneously interact is defined as the **molecularity** of the reaction. Thus, the simple elementary reaction of $A \rightarrow P$ is a **first-order reaction.** Figure 10.4 portrays the course of a first-order reaction as a function of time. The rate of decay of a radioactive isotope, like ^{14}C or ^{32}P, is a first-order reaction. So is an intramolecular rearrangement, such as $A \rightarrow P$. Both are **unimolecular reactions** (the molecularity equals 1).

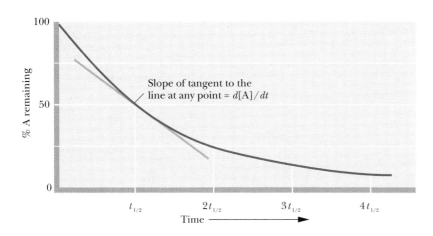

Figure 10.4 Plot of the course of a first-order reaction. The half-time, $t_{1/2}$, is the time for one-half of A to disappear.

Bimolecular Reactions

Consider the more complex reaction, in which two molecules must react to yield products:

$$A + B \rightarrow P + Q$$

Assuming this reaction is a simple one-step molecular process, its molecularity is 2; that is, it is a **bimolecular reaction.** The velocity of this reaction can be determined from the rate of disappearance of either A or B, or the rate of appearance of P or Q:

$$v = \frac{-d[A]}{dt} = \frac{-d[B]}{dt} = \frac{d[P]}{dt} = \frac{d[Q]}{dt} \tag{10.3}$$

The rate law is

$$v = k[A][B] \tag{10.4}$$

The rate is proportional to the concentrations of both A and B. Because it is proportional to the product of two concentration terms, the reaction is **second-order** overall, first-order with respect to A and first-order with respect to B. (Were this reaction $2A \rightarrow P + Q$, the rate law would be $v = k[A]^2$, second-order overall and second-order with respect to A.) Second-order rate constants have the units of (concentration)$^{-1}$ (time)$^{-1}$, as in $M^{-1}\text{sec}^{-1}$.

Molecularities greater than 2 are rarely found (and greater than 3, never). When the overall stoichiometry of a reaction is greater than 2 (for example, as in $A + B + C \rightarrow$ or $2A + B \rightarrow$), the reaction almost always proceeds via uni- or bimolecular elementary steps, and the overall rate obeys a simple first- or second-order rate law.

At this point, it may be useful to remind ourselves of an important caveat that is the first principle of kinetics: *kinetics cannot prove a hypothetical mechanism.* Kinetic experiments can only rule out various alternative hypotheses because they don't fit the data. However, through thoughtful kinetic studies, a process of elimination of alternative hypotheses leads ever closer to the reality.

Free Energy of Activation and the Action of Catalysts

In a first-order chemical reaction, the conversion of A to P occurs because, at any given instant, a fraction of the A molecules has the energy necessary to achieve a reactive condition known as the **transition state.** In this state, the

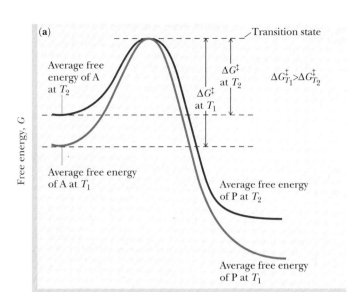

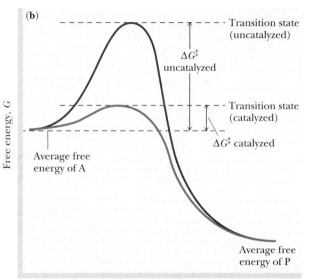

Figure 10.5 Energy diagram for a chemical reaction (A → P) and the effects of **(a)** raising the temperature from T_1 to T_2 or **(b)** adding a catalyst. Raising the temperature raises the average energy of A molecules, which increases the population of A molecules having energies equal to the activation energy for the reaction, thereby increasing the reaction rate. In contrast, the average free energy of A molecules remains the same in uncatalyzed *versus* catalyzed reactions (conducted at the same temperature). The effect of the catalyst is to lower the free energy of activation for the reaction.

probability is very high that the particular rearrangement accompanying the A → P reaction will occur. This transition state sits at the apex of the energy profile in the energy diagram describing the energetic relationship between A and P (Figure 10.5). The average free energy of A molecules defines the initial state, and the average free energy of P molecules is the final state along the reaction coordinate. The rate of any chemical reaction is proportional to the concentration of reactant molecules (A in this case) having this transition-state energy. Obviously, the higher this energy is above the average energy, the smaller the fraction of molecules that will have this energy, and the slower the reaction will proceed. The height of this energy barrier is called the **free energy of activation, $\Delta G^{\ddagger}$.** Specifically, $\Delta G^{\ddagger}$ is the energy required to raise the average energy of one mole of reactant (at a given temperature) to the transition-state energy. The relationship between activation energy and the rate constant of the reaction, k, is given by the **Arrhenius equation:**

$$k = Ae^{-\Delta G^{\ddagger}/RT} \tag{10.5}$$

where A is a constant for a particular reaction (not to be confused with the reactant species, A, that we're discussing). Another way of writing this is $1/k = (1/A)e^{-\Delta G^{\ddagger}/RT}$. That is, k is inversely proportional to $e^{-\Delta G^{\ddagger}/RT}$. As the energy of activation decreases, the reaction rate increases.

Decreasing $\Delta G^{\ddagger}$ Increases Reaction Rate

We are familiar with two general ways that rates of chemical reactions may be accelerated. First, the temperature can be raised. This will increase the average energy of reactant molecules, which in effect decreases the energy needed to reach the transition state (see Figure 10.5a). The rates of many chemical reactions are doubled by a 10°C rise in temperature. Second, the rates of chemical reactions can also be accelerated by **catalysts.** Catalysts work by lowering the energy of activation rather than by raising the average energy of the reactants (see Figure 10.5b). Catalysts accomplish this remarkable feat by combining transiently with the reactants in a way that promotes their entry into the reactive, transition-state condition. Two aspects of catalysts are worth noting: (a) they are regenerated after each reaction cycle (A → P), and so can be used over and over again; and (b) they have no effect on the overall free energy change in the reaction (the free energy difference between A and P) (see Figure 10.5b).

10.3 Kinetics of Enzyme-Catalyzed Reactions

The change in reaction velocity as the reactant concentration varies is one of the primary measurements in kinetic analysis. Returning to A → P, a plot of the reaction rate as a function of the concentration of A yields a straight line whose slope is k (Figure 10.6). The more A that is available, the greater the rate of the reaction, v. Similar analyses of enzyme-catalyzed reactions involving only a single substrate yield remarkably different results (Figure 10.7). At low concentrations of the substrate S, v is proportional to [S], as expected for a first-order reaction. However, v does not increase proportionally as [S] increases, but instead begins to level off. At high [S], v becomes virtually independent of [S] and approaches a maximal limit. The value of v at this limit is written V_{max}. Because rate is no longer dependent on [S] at these high concentrations, the enzyme-catalyzed reaction is now obeying **zero-order kinetics;** that is, the rate is independent of the reactant (substrate) concentration. This behavior is a **saturation effect:** when v shows no increase even though [S] is increased, the system is saturated with substrate. Such plots are called **substrate saturation curves.** The physical interpretation is that every enzyme molecule in the reaction mixture has its substrate-binding site (active site) occupied by S. Indeed, such curves were the initial clue that an enzyme interacts directly with its substrate by binding it.

The Michaelis–Menten Equation

In 1913 Leonor Michaelis and Maud L. Menten proposed a general theory of enzyme action consistent with observed enzyme kinetics. Their theory was based on the assumption that the enzyme, E, and its substrate, S, associate

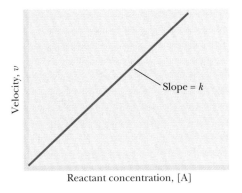

Figure 10.6 A plot of v versus [A] for the unimolecular chemical reaction A → P yields a straight line having a slope equal to k.

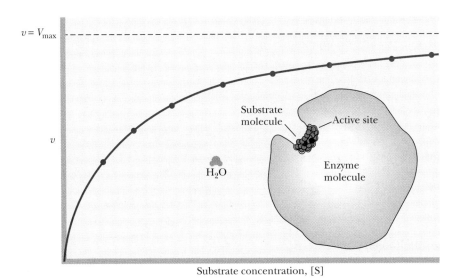

Substrate concentration, [S]

See *Interactive Biochemistry CD-ROM and Workbook*, pages 126–127

Figure 10.7 Substrate saturation curve for an enzyme-catalyzed reaction. The amount of enzyme is constant, and the velocity of the reaction is determined at various substrate concentrations. The reaction rate, v, as a function of [S] is described by a rectangular hyperbola. At very high [S], $v = V_{max}$. That is, the velocity is limited only by conditions (temperature, pH, ionic strength) and by the amount of enzyme present; v becomes independent of [S]. Such a condition is termed zero-order kinetics. Under zero-order conditions, velocity is directly dependent on [enzyme]. The H_2O molecule provides a rough guide to scale. The substrate is bound at the active site of the enzyme.

reversibly to form an enzyme–substrate complex, ES. Product, P, is formed in a second step when ES breaks down to yield E + P:

$$E + S \underset{k_{-1}}{\overset{k_1}{\rightleftharpoons}} ES \underset{k_{-2}}{\overset{k_2}{\rightleftharpoons}} E + P \tag{10.6}$$

E is then free to interact with another molecule of S.

Since enzymes increase the rate of the reverse reaction as well as that of the forward reaction, it would be helpful to ignore the back-reaction by which E + P might form ES. The best way to do this is to conduct the reaction under conditions where P is absent. For instance, if the velocity of the reaction is only measured immediately after E and S are mixed in the absence of P, the rate of any back-reaction is negligible because this rate will be proportional to [P], and [P] is essentially 0. Measurements of the velocity before appreciable concentrations of P have a chance to accumulate are called **initial velocity** measurements.

In 1925, G. E. Briggs and James B. S. Haldane refined and extended the interpretations of Michaelis and Menten by assuming the concentration of the enzyme–substrate complex ES quickly reaches a constant value in such a dynamic system. That is, ES is formed as rapidly from E + S as it disappears by its two possible fates: dissociation to regenerate E + S, and reaction to form E + P. This assumption is termed the **steady-state assumption** and is expressed as

$$\frac{d[ES]}{dt} = 0 \tag{10.7}$$

The change in concentration of ES with time, t, is 0. Figure 10.8 illustrates the time course for formation of the ES complex and establishment of the steady-state condition.

With these simplifications, we can now analyze the system described by Equation (10.6) in order to describe the initial velocity v as a function of [S] and amount of enzyme.

The total amount of enzyme is fixed and is given by the formula

$$\text{Total enzyme, } [E_T] = [E] + [ES] \tag{10.8}$$

where [E] = free enzyme and [ES] = the amount of enzyme in the enzyme–substrate complex.

The rate of [ES] formation is

$$v_f = k_1([E_T] - [ES])[S] \quad \text{where } [E_T] - [ES] = [E] \tag{10.9}$$

The rate of [ES] disappearance is

$$v_d = k_{-1}[ES] + k_2[ES] = (k_{-1} + k_2)[ES] \tag{10.10}$$

At steady state, $d[ES]/dt = 0$, and therefore, $v_f = v_d$. So,

$$k_1([E_T] - [ES])[S] = (k_{-1} + k_2)[ES] \tag{10.11}$$

Rearranging gives

$$\frac{([E_T] - [ES])[S]}{[ES]} = \frac{(k_{-1} + k_2)}{k_1} \tag{10.12}$$

The Michaelis Constant, K_m

The ratio of constants $(k_{-1} + k_2)/k_1$ is itself a constant and is defined as the **Michaelis constant, K_m:**

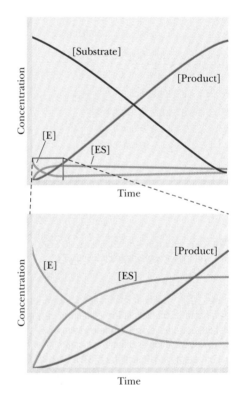

Figure 10.8 Time course for the consumption of substrate, the formation of product, and the establishment of a steady-state level of the enzyme-substrate [ES] complex for a typical enzyme obeying the Michaelis-Menten, Briggs–Haldane model for enzyme kinetics. The early stage of the time course is shown in greater magnification in the bottom graph.

$$K_m = \frac{(k_{-1} + k_2)}{k_1} \qquad (10.13)$$

Note from Equation (10.13) that K_m is given by the ratio of two concentrations—$([E_T] - [ES])$ and $[S]$—to one $([ES])$, so K_m has the units of molarity. From Equation (10.13), we can write

$$\frac{([E_T] - [ES])[S]}{[ES]} = K_m \qquad (10.14)$$

which rearranges to

$$[ES] = \frac{[E_T][S]}{K_m + S} \qquad (10.15)$$

Now, the most important parameter in the kinetics of any reaction is the **rate of product formation.** This rate is given by

$$v = \frac{d[P]}{dt} \qquad (10.16)$$

and for this reaction

$$v = k_2[ES] \qquad (10.17)$$

Substituting the expression for $[ES]$ from Equation (10.15) into Equation (10.17) gives

$$v = \frac{k_2[E_T][S]}{K_m + S} \qquad (10.18)$$

The product $k_2[E_T]$ has special meaning. When $[S]$ is high enough to saturate all of the enzyme, the velocity of the reaction, v, is maximal. At saturation, the amount of $[ES]$ complex is equal to the total enzyme concentration, E_T, its maximum possible value. From Equation (10.18), the initial velocity v then equals $k_2[E_T] = V_{max}$. Written symbolically,

when $[S] \gg [E_T]$(and K_m), $[E_T] = [ES]$ and $v = V_{max}$

Therefore,

$$V_{max} = k_2[E_T] \qquad (10.19)$$

Substituting this relationship into the expression for v gives the **Michaelis–Menten equation:**

$$v = \frac{V_{max}[S]}{K_m + [S]} \qquad (10.20)$$

This equation says that the rate of an enzyme-catalyzed reaction, v, at any moment is determined by two constants, K_m and V_{max}, and the concentration of substrate at that moment.

When $[S] = K_m$, $v = V_{max}/2$

We can find a practical definition for the constant K_m by rearranging Equation (10.20) to give

$$K_m = [S]\left(\frac{V_{max}}{v} - 1\right) \qquad (10.21)$$

Then, at $v = V_{max}/2$, $K_m = [S]$. That is, K_m is defined by the substrate concentration that gives a velocity equal to one-half the maximal velocity. Table 10.2 gives the K_m values of some enzymes for their substrates.

■ **Table 10.2** K_m **Values for Some Enzymes**

Enzyme	Substrate	K_m (mM)
Carbonic anhydrase	CO_2	12
Hexokinase	Glucose	0.15
	Fructose	1.5
β-Galactosidase	Lactose	4
Glutamate dehydrogenase	NH_4^+	57
	Glutamate	0.12
	α-Ketoglutarate	2
	NAD^+	0.025
	NADH	0.018
Aspartate aminotransferase	Aspartate	0.9
	α-Ketoglutarate	0.1
	Oxaloacetate	0.04
	Glutamate	4
Threonine deaminase	Threonine	5
Pyruvate carboxylase	HCO_3^-	1.0
	Pyruvate	0.4
	ATP	0.06
Penicillinase	Benzylpenicillin	0.05
Lysozyme	Hexa-N-acetylglucosamine	0.006

Relationships Between V_{max}, K_m, and Reaction Order

The Michaelis–Menten equation (10.20) describes a curve known from analytical geometry as a *rectangular hyperbola*. In such curves, as [S] is increased, v approaches the limiting value, V_{max}, in an asymptotic fashion. V_{max} can be approximated experimentally from a substrate saturation curve (see Figure 10.7), and K_m can be derived from $V_{max}/2$, so the two constants of the Michaelis–Menten equation can be obtained from plots of v *versus* [S]. Note, however, that actual estimation of V_{max}, and consequently K_m, is only approximate from such graphs. That is, according to Equation (10.20), in order to get $v = 0.99\ V_{max}$, [S] must equal 99 K_m, a concentration that may be difficult to achieve in practice.

From Equation (10.20), when $[S] \gg K_m$, then $v = V_{max}$. That is, v is no longer dependent on [S], so the reaction is obeying zero-order kinetics. Also, when $[S] < K_m$, then $v \approx (V_{max}/K_m)[S]$. That is, the rate, v, approximately follows a first-order rate equation, $v = k'[A]$, where $k' = V_{max}/K_m$.

K_m and V_{max}, once known explicitly, define the rate of the enzyme-catalyzed reaction, provided:

1. The reaction involves only one substrate, *or*, if the reaction is multisubstrate, the concentration of only one substrate is varied while the concentration of all other substrates is held constant.
2. The reaction $ES \rightarrow E + P$ is irreversible, *or* the experiment is limited to observing only initial velocities where $[P] = 0$.
3. The initial substrate concentration $([S]_0) > [E_T]$ and $[E_T]$ is held constant.
4. All other variables that might influence the rate of the reaction (temperature, pH, ionic strength, and so on) are constant.

Enzyme Units

In many situations, the actual molar amount of the enzyme is not known. However, its amount can be expressed in terms of the activity observed. The International Commission on Enzymes defines **One International Unit** of enzyme as *the amount that catalyzes the formation of one micromole of product in one minute.* (Because enzymes are very sensitive to factors such as pH, temperature, and ionic strength, the conditions of assay must be specified.)

Turnover Number

The **turnover number** of an enzyme, k_{cat}, is a measure of its maximal catalytic activity. k_{cat} is defined as the number of substrate molecules converted into product per enzyme molecule per unit time when the enzyme is saturated with substrate. The turnover number is also referred to as the **molecular activity** of the enzyme. For the Michaelis–Menten reaction (Equation [10.6]) under conditions of initial velocity measurements, $k_2 = k_{cat}$. Provided the concentration of enzyme, $[E_T]$, in the reaction mixture is known, k_{cat} can be determined from V_{max}. At saturating [S], $v = V_{max} = k_2[E_T]$. Thus,

$$k_2 = \frac{V_{max}}{[E_T]} = k_{cat} \qquad (10.22)$$

A DEEPER LOOK

k_{cat}/K_m: A Measure of the Catalytic Efficiency of an Enzyme

Under physiological conditions, [S] is seldom saturating, and k_{cat} itself is not particularly informative. That is, the *in vivo* ratio of $[S]/K_m$ usually falls in the range of 0.01 to 1.0, so active sites often are not filled with substrate. Nevertheless, we can derive a meaningful index of the efficiency of Michaelis–Menten–type enzymes under these conditions by employing the following equations. As presented in Equation (10.20), if

$$v = \frac{V_{max}[S]}{K_m + [S]}$$

and $V_{max} = k_{cat}[E_T]$, then

$$v = \frac{k_{cat}[E_T][S]}{[S] + K_m} \qquad (10.23)$$

When $[S] \ll K_m$, the concentration of free enzyme, [E], is approximately equal to $[E_T]$, so that

$$v = \left(\frac{k_{cat}}{K_m}\right)[E][S] \qquad (10.24)$$

That is, k_{cat}/K_m is an *apparent second-order rate constant* for the reaction of E and S to form product. This ratio provides an index of the catalytic efficiency of an enzyme operating at substrate concentrations substantially below saturation amounts.

An interesting point emerges if we restrict ourselves to the simple case where $k_{cat} = k_2$. Then

$$k_{cat}/K_m = \frac{k_1 k_2}{(k_{-1} + k_2)}$$

But k_1 must always be greater than or equal to $k_1 k_2/(k_{-1} + k_2)$. That is, the reaction can go no faster than the rate at which E and S come together. Thus, k_1 sets the upper limit for k_{cat}/K_m. In other words, *the catalytic efficiency of an enzyme cannot exceed the diffusion-controlled rate of combination of E and S to form ES*. Those enzymes that are most efficient in their catalysis have k_{cat}/K_m ratios approaching $10^9/M \cdot sec$. Their catalytic velocity is limited only by the rate at which they encounter S; enzymes this efficient have achieved so-called *catalytic perfection*. Table 10.3 lists the kinetic parameters of several enzymes in this category. Note that k_{cat} and K_m each show a substantial range of variation in this table, even though their ratio falls around $10^8/M \cdot sec$.

Table 10.3	Enzymes Whose k_{cat}/K_m Approaches the Diffusion-Controlled Rate of Association with Substrate				

Enzyme	Substrate	k_{cat} (sec^{-1})	K_m (M)	k_{cat}/K_m (sec^{-1} M^{-1})
Acetylcholinesterase	Acetylcholine	1.4×10^4	9×10^{-5}	1.6×10^8
Carbonic	CO_2	1×10^6	0.012	8.3×10^7
anhydrase	HCO_3^-	4×10^5	0.026	1.5×10^7
Catalase	H_2O_2	4×10^7	1.1	4×10^7
Crotonase	Crotonyl-CoA	5.7×10^3	2×10^{-5}	2.8×10^8
Fumarase	Fumarate	800	5×10^{-6}	1.6×10^8
	Malate	900	2.5×10^{-5}	3.6×10^7
Triosephosphate	Glyceraldehyde-	4.3×10^3	1.8×10^{-5}	2.4×10^8
isomerase	3-phosphate*			
β-Lactamase	Benzylpenicillin	2×10^3	2×10^{-5}	1×10^8

*K_m for glyceraldehyde-3-phosphate is calculated on the basis that only 3.8% of the substrate in solution is unhydrated and therefore reactive with the enzyme.

Adapted from Fersht, A. 1985. *Enzyme Structure and Mechanism*, 2nd ed. New York: W. H. Freeman & Co.

HUMAN BIOCHEMISTRY

Effects of Amino Acid Substitution on K_m and k_{cat}: Wild-Type and Mutant Forms of Human Sulfite Oxidase

Mammalian sulfite oxidase is the last enzyme in the pathway for degradation of sulfur-containing amino acids. Sulfite oxidase (SO) catalyzes the oxidation of sulfite (SO_3^{2-}) to sulfate (SO_4^{2-}), using the heme-containing protein, cytochrome c, as electron acceptor:

$$SO_3^{2-} + 2 \text{ cytochrome } c_{oxidized} \rightarrow$$

$$SO_4^{2-} + 2 \text{ cytochrome } c_{reduced} + 2 H^+$$

Isolated sulfite oxidase deficiency is a rare and often fatal genetic disorder. The disease is characterized by severe neurological abnormalities, revealed as convulsions shortly after birth. R. M. Garrett and K. V. Ragagopalan at Duke University Medical Center isolated human cDNA for sulfite oxidase from the cells of normal (*wild-type*) and SO-deficient individuals. Expression of these SO cDNAs in transformed *E.coli* cells allowed the isolation and kinetic analysis of wild-type and mutant forms of SO, including one (designated R160Q) in which the Arg at position 160 in the SO polypeptide chain is replaced by Gln. A genetically engineered version of SO (designated R160K) in which Lys replaces Arg at position 160 was

also studied.

Kinetic Constants for Wild-Type and Mutant Sulfite Oxidase

Enzyme	K_m (μM)	k_{cat} (sec^{-1})	k_{cat}/K_m(10^6M^{-1}sec^{-1})
Wild-Type	17	18	1.1
R160Q	1900	3	0.0016
R160K	360	5.5	0.015

Replacing R^{160} in sulfite oxidase by Q increases K_m, decreases k_{cat}, and markedly diminishes the catalytic efficiency (k_{cat}/K_m) of the enzyme. The R160K mutant enzyme has properties intermediate between wild-type and the R160Q mutant. The substrate, SO_3^{2-}, is strongly anionic and R^{160} is one of several Arg residues within the SO substrate-binding site. Positively charged side chains in the substrate-binding site facilitate SO_3^{2-} binding and catalysis, with Arg being optimal in this role.

The term k_{cat} represents the *kinetic efficiency* of the enzyme. Table 10.4 lists turnover numbers for some representative enzymes. Catalase has the highest turnover number known; each molecule of this enzyme can degrade 40 million molecules of H_2O_2 in 1 second! At the other end of the scale, lysozyme requires 2 seconds to cleave a glycosidic bond in its glycan substrate.

Linear Plots Can Be Derived from the Michaelis–Menten Equation

Because of the hyperbolic shape of v versus $[S]$ plots, V_{max} can only be determined from an extrapolation of the asymptotic approach of v to some limiting value as $[S]$ increases indefinitely (see Figure 10.7); and K_m is derived from that value of $[S]$ giving $v = V_{max}/2$. However, several rearrangements of the Michaelis–Menten equation transform it into a straight-line equation. The best known of these is the **Lineweaver–Burk double-reciprocal plot**.

Taking the reciprocal of both sides of the Michaelis–Menten equation, Equation (10.20), yields the equality

$$1/v = \left(\frac{K_m}{V_{max}}\right)\left(\frac{1}{[S]}\right) + \frac{1}{V_{max}} \qquad (10.26)$$

This conforms to $y = mx + b$ (the equation for a straight line), where $y = 1/v$; m, the slope, is K_m/V_{max}; $x = 1/[S]$; and $b = 1/V_{max}$. Plotting $1/v$ versus $1/[S]$ gives a straight line whose x-intercept is $-1/K_m$, whose y-intercept is $1/V_{max}$, and whose slope is K_m/V_{max} (Figure 10.9). The common advantage of such plots is that both K_m and V_{max} can be estimated accurately by extrapolation of a straight line rather than an asymptote.

Departures from Linearity—A Hint of Regulation?

If the kinetics of the reaction disobey the Michaelis–Menten equation, the violation is revealed by a departure from linearity in these straight-line graphs.

Table 10.4 Values of k_{cat} (Turnover Number) for Some Enzymes

Enzyme	k_{cat} (sec^{-1})
Catalase	40,000,000
Carbonic anhydrase	1,000,000
Acetylcholinesterase	14,000
Penicillinase	2,000
Lactate dehydrogenase	1,000
Chymotrypsin	100
DNA polymerase I	15
Lysozyme	0.5

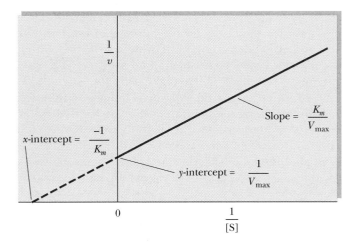

$$\frac{1}{v} = \frac{K_m}{V_{max}}\left(\frac{1}{[S]}\right) + \frac{1}{V_{max}}$$

Figure 10.9 The Lineweaver–Burk double-reciprocal plot, depicting extrapolations that allow the determination of the *x*- and *y*-intercepts and slope.

 See pages 126–130

We shall see later in this chapter that such deviations from linearity are characteristic of the kinetics of regulatory enzymes known as **allosteric enzymes.** Such regulatory enzymes are very important in the overall control of metabolic pathways.

Effect of pH on Enzymatic Activity

Enzyme–substrate recognition and the catalytic events that ensue are greatly dependent on pH. An enzyme possesses an array of ionizable side chains and prosthetic groups that not only determine its secondary and tertiary structure but may also be intimately involved in its active site. Further, the substrate itself often has ionizing groups, and one or another of the ionic forms may preferentially interact with the enzyme. Enzymes in general are active only over a limited pH range and most have a particular pH at which their catalytic activity is optimal. These effects of pH may be due to effects on K_m or V_{max} or both. Figure 10.10 illustrates the relative activity of four enzymes as a function of pH. Although the pH optimum of an enzyme often reflects the pH of its normal environment, the optimum may not be precisely the same. This difference suggests that the pH-activity response of an enzyme may be a factor in the intracellular regulation of its activity.

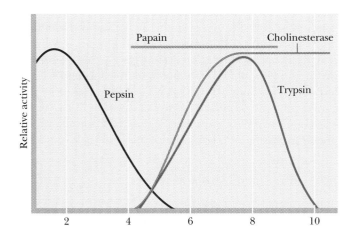

Optimum pH of Some Enzymes	
Enzyme	Optimum pH
Pepsin	1.5
Catalase	7.6
Trypsin	7.7
Fumarase	7.8
Ribonuclease	7.8
Arginase	9.7

Figure 10.10 The pH-activity profiles of four different enzymes. Trypsin, an intestinal protease, has a slightly alkaline pH optimum, whereas pepsin, a gastric protease, acts in the acidic confines of the stomach and has a pH optimum near 2. Papain, a protease found in papaya, is relatively insensitive to pHs between 4 and 8. Cholinesterase activity is pH-sensitive below pH 7 but not between pH 7 and 10. The cholinesterase pH-activity profile suggests that an ionizable group with a pK' near 6 is essential to its activity. Might it be a histidine residue within the active site?

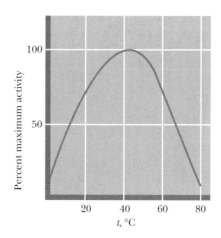

Figure 10.11 The relative activity of an enzymatic reaction as a function of temperature. The decrease in the activity above 50°C is due to thermal denaturation.

Effect of Temperature on Enzymatic Activity

Like most chemical reactions, the rates of enzyme-catalyzed reactions generally increase with increasing temperature. However, at temperatures above 50° to 60°C, enzymes typically show a decline in activity (Figure 10.11). Two effects are operating here: (a) the characteristic increase in reaction rate with temperature and (b) thermal denaturation of protein structure at higher temperatures. Most enzymatic reactions double in rate for every 10°C rise in temperature (that is, $Q_{10} = 2$, where Q_{10} is defined as *the ratio of activities at two temperatures 10° apart*) as long as the enzyme is stable and fully active. The increasing rate with increasing temperature is ultimately offset by the instability of higher orders of protein structure at elevated temperatures, where the enzyme is inactivated. Not all enzymes are quite so thermally labile. For example, the enzymes of thermophilic bacteria (*thermophilic* = "heat-loving") found in geothermal springs retain full activity at temperatures in excess of 85°C.

10.4 Enzyme Inhibition

If the velocity of an enzymatic reaction is decreased or **inhibited,** the kinetics of the reaction obviously have been perturbed. Systematic perturbations are a basic tool of experimental scientists; much can be learned about the normal workings of any system by inducing changes in it and then observing the effects of the change. The study of enzyme inhibition has contributed significantly to our understanding of enzymes.

Reversible *versus* Irreversible Inhibition

Enzyme inhibitors are classified in several ways. The inhibitor may interact either reversibly or irreversibly with the enzyme. **Reversible inhibitors** interact with the enzyme through noncovalent association/dissociation reactions. In contrast, **irreversible inhibitors** usually cause stable, covalent alterations in the enzyme. That is, the consequence of irreversible inhibition is a decrease in the concentration of active enzyme. The kinetics observed are consistent with this interpretation, as we shall see later.

Reversible Inhibition

Reversible inhibitors fall into two major categories: competitive and noncompetitive (although other more unusual and rare categories are known). **Competitive inhibitors** are characterized by the fact that the substrate and inhibitor compete for the same binding site on the enzyme, the **active site** or **S-binding site.** Thus, increasing the concentration of S favors the likelihood of S binding to the enzyme instead of the inhibitor, I. That is, high [S] can overcome the effects of I. Effects by the other major type, **noncompetitive inhibitors,** cannot be overcome by increasing [S]. The two types can be distinguished by the particular patterns obtained when the kinetic data are analyzed in linear plots, such as double-reciprocal (Lineweaver–Burk) plots. A general formulation for common inhibitor interactions in our simple enzyme kinetic model would include

$$\text{E} + \text{I} \rightleftharpoons \text{EI} \quad \text{and} \quad \text{I} + \text{ES} \rightleftharpoons \text{IES} \qquad \textbf{(10.27)}$$

That is, we consider here reversible combinations of the inhibitor with E and ES.

Competitive Inhibition

Consider the following system,

$$E + S \underset{k_{-1}}{\overset{k_1}{\rightleftharpoons}} ES \underset{k_{-2}}{\overset{k_2}{\rightleftharpoons}} E + P \qquad E + I \underset{k_{-3}}{\overset{k_3}{\rightleftharpoons}} EI \qquad (10.28)$$

where an inhibitor, I, binds **reversibly** to the enzyme at the same site as S. S-binding and I-binding are mutually exclusive, *competitive* processes. Formation of the ternary complex, EIS, where both S and I are bound, is physically impossible. This condition leads us to anticipate that S and I must share a high degree of structural similarity because they bind at the same site on the enzyme. Also notice that, in our model, EI does not react to give rise to E + P. That is, I is not changed by interaction with E. The rate of the product-forming reaction is $v = k_2[ES]$.

It is revealing to compare the equation for the uninhibited case, Equation (10.20) (the Michaelis–Menten equation) with the equation for the rate of the enzymatic reaction in the presence of a fixed concentration of the competitive inhibitor, $[I]$:

$$v = \frac{V_{max}[S]}{[S] + K_m} \qquad \text{versus} \qquad v = \frac{V_{max}[S]}{[S] + K_m\left(1 + \dfrac{[I]}{K_I}\right)} \qquad (10.29)$$

The K_m term in the denominator in the inhibited case is increased by the factor $(1 + [I]/K_I)$; thus, v is less in the presence of the inhibitor, as expected. Clearly, in the absence of I, the two equations are identical. Figure 10.12 shows a Lineweaver–Burk plot of competitive inhibition. Several features of competitive inhibition are evident. First, at a given $[I]$, v decreases ($1/v$ increases). When $[S]$ becomes infinite, $v = V_{max}$ and is unaffected by I because all of the enzyme is in the ES form. Note that the value of the $-x$-intercept decreases as $[I]$ increases. This $-x$-intercept is often termed the *apparent* K_m (or K_{mapp}) because it is the K_m apparent under these conditions. The diagnostic criterion for competitive inhibition is that V_{max} is unaffected by I; that is, all lines share a common y-intercept. This criterion is also the best experimental indication of binding at the same site by two substances. Competitive inhibitors resemble S structurally.

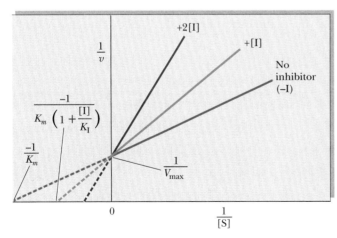

 See pages 130–135

Figure 10.12 Lineweaver-Burk plot of competitive inhibition, showing lines for no I, [I], and 2[I]. Note that when [S] is infinitely large ($1/[S] \approx 0$), V_{max} is the same whether I is present or not. In the presence of I, the x-intercept $= \dfrac{-1}{K_m\left(1 + \dfrac{[I]}{K_I}\right)}$.

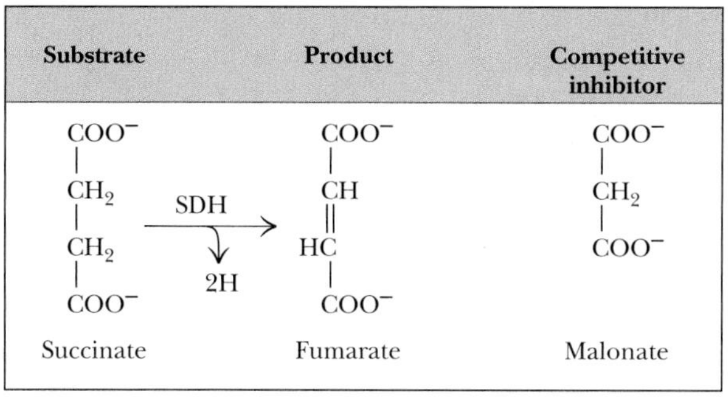

Figure 10.13 Structures of succinate, the substrate of succinate dehydrogenase (SDH), and malonate, the competitive inhibitor. Fumarate (the product of SDH action on succinate) is also shown.

Substrate	Product	Competitive inhibitor
COO⁻ | CH$_2$ | CH$_2$ | COO⁻	COO⁻ | CH || HC | COO⁻	COO⁻ | CH$_2$ | COO⁻
Succinate	Fumarate	Malonate

SDH → 2H

Succinate Dehydrogenase—A Classic Example of Competitive Inhibition. The enzyme *succinate dehydrogenase* (*SDH*) is competitively inhibited by malonate. Figure 10.13 shows the structures of succinate and malonate. The structural similarity between them is obvious and is the basis of malonate's ability to mimic succinate and bind at the active site of SDH. However, unlike succinate, which is oxidized by SDH to form fumarate, malonate cannot lose two hydrogens; consequently, it is unreactive.

Noncompetitive Inhibition

Noncompetitive inhibitors interact with both E and ES (or with S and ES, but this is a rare and specialized case). Obviously, then, the inhibitor is not binding to the same site as S, and the inhibition cannot be overcome by raising [S]. There are two types of noncompetitive inhibition: pure and mixed.

Pure Noncompetitive Inhibition. In this situation, the binding of I by E has no effect on the binding of S by E. That is, S and I bind at different sites on E, and binding of I does not affect binding of S. Consider the system

$$\text{E} + \text{I} \overset{K_I}{\rightleftharpoons} \text{EI} \quad \text{ES} + \text{I} \overset{K_I'}{\rightleftharpoons} \text{IES} \tag{10.30}$$

Pure noncompetitive inhibition occurs if $K_I = K_I'$. This situation is relatively uncommon; the Lineweaver–Burk plot for such an instance is given in Figure 10.14. Note that K_m is unchanged by I (the *x*-intercept remains the same, with or without I). Note also that V_{max} decreases. A similar pattern is seen if the amount of enzyme in the experiment is decreased. Thus, it is as if I lowered [E].

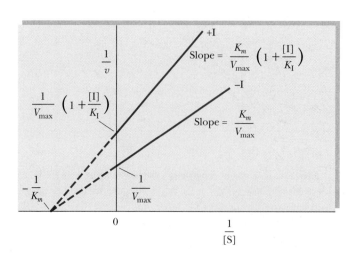

Figure 10.14 Lineweaver–Burk plot of pure noncompetitive inhibition. Note that I does not alter K_m but that it decreases V_{max}. In the presence of I, the *y*-intercept is equal to $(1/V_{max})(1 + [I]/K_I)$.

(a) $K_I < K_I'$

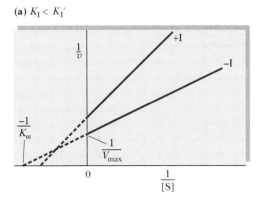

(b) $K_I' < K_I$

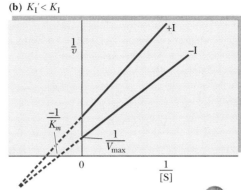

See pages 130–135

Figure 10.15 Lineweaver–Burk plot of mixed noncompetitive inhibition. Note that both intercepts and the slope change in the presence of I   **(a)** when K_I is less than K_I';   **(b)** when K_I is greater than K_I'.

Mixed Noncompetitive Inhibition.   In this situation, the binding of I by E influences the binding of S by E. Either the binding sites for I and S are near one another or conformational changes in E caused by I affect S binding. In this case, K_I and K_I', as defined previously, are not equal. Both K_m and V_{max} are altered by the presence of I, and K_m/V_{max} is not constant (Figure 10.15). This inhibitory pattern is commonly encountered. A reasonable explanation is that the inhibitor is binding at a site distinct from the active site, yet is influencing the binding of S at the active site. Presumably, these effects are transmitted via alterations in the protein's conformation.

Irreversible Inhibition

If the inhibitor combines irreversibly with the enzyme—for example, by covalent attachment—the kinetic pattern seen is like that of noncompetitive inhibition because the net effect is a loss of active enzyme. Usually, this type of inhibition can be distinguished from the noncompetitive, reversible inhibition case, since the reaction of I with E (or ES) is not instantaneous. Instead, there is a time-dependent decrease in enzymatic activity as $E + I \rightarrow EI$ proceeds, and the rate of this inactivation can be followed. Also, unlike reversible inhibitions, dilution or dialysis of the enzyme : inhibitor solution does not dissociate the EI complex and restore enzyme activity.

Suicide Substrates—Mechanism-Based Enzyme Inactivators

Suicide substrates are inhibitory substrate analogs designed so that, via normal catalytic action of the enzyme, a very reactive group is generated. This reactive group then forms a covalent bond with a functional group within the active site of the enzyme, thereby causing irreversible inhibition. Suicide substrates, also called *Trojan horse substrates*, are a type of **affinity label.** As substrate analogs, they bind with specificity and high affinity to the enzyme active site; in their reactive form, they become covalently bound to the enzyme. This covalent link effectively labels a particular functional group within the active site, identifying the group as a key player in the enzyme's catalytic cycle.

Penicillin—A Suicide Substrate.   Several drugs in current medical use are mechanism-based enzyme inactivators. For example, the antibiotic **penicillin** exerts its effects by covalently reacting with an essential serine residue in the active site of *glycoprotein peptidase,* an enzyme that acts to cross-link the

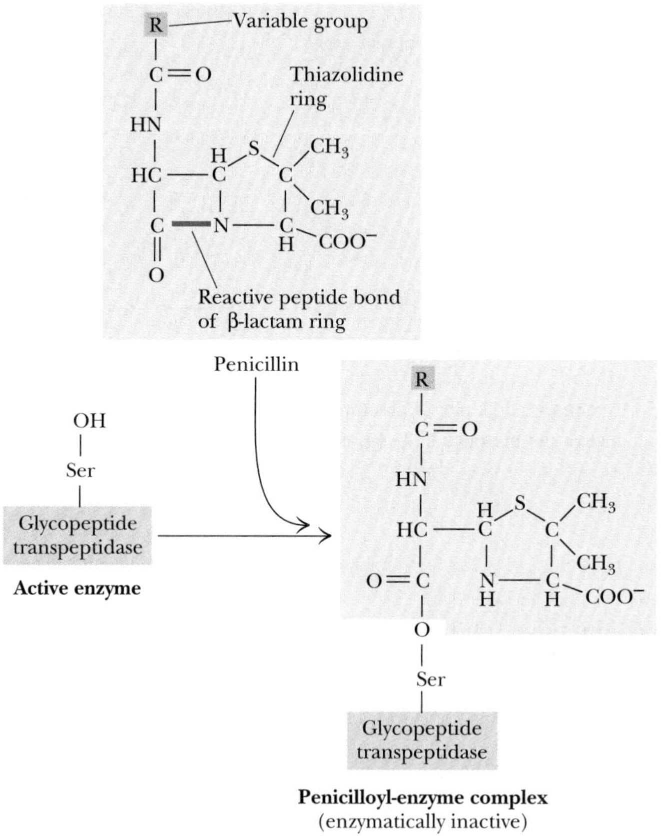

Figure 10.16 Penicillin is an irreversible inhibitor of the enzyme glycoprotein peptidase, which catalyzes an essential step in bacterial cell wall synthesis. Penicillin consists of a thiazolidine ring fused to a β-lactam ring to which a variable group R is attached. A reactive peptide bond in the β-lactam ring covalently attaches to a serine residue in the active site of the glycopeptide transpeptidase. (The conformation of penicillin around its reactive peptide bond resembles the transition state of the normal glycoprotein peptidase substrate.) The penicilloyl–enzyme complex is catalytically inactive. The bond between the enzyme and penicillin is indefinitely stable; that is, penicillin binding is irreversible.

peptidoglycan chains during synthesis of bacterial cell walls (Figure 10.16). Once cell wall synthesis is blocked, the bacterial cells are very susceptible to rupture by osmotic lysis, and bacterial growth is halted.

10.5 Kinetics of Enzyme-Catalyzed Reactions Involving Two or More Substrates

Thus far, we have considered only the simple case of enzymes that act upon a single substrate, S. This situation is not common. Usually, enzymes catalyze reactions in which two (or even more) substrates take part.

Consider the case of an enzyme catalyzing a reaction involving two substrates, A and B, and yielding the products P and Q:

$$\text{A} + \text{B} \overset{\text{enzyme}}{\rightleftharpoons} \text{P} + \text{Q} \tag{10.31}$$

HUMAN BIOCHEMISTRY

Viagra—An Unexpected Outcome in a Program of Drug Design

Prior to the accumulation of detailed biochemical information on metabolism, enzymes, and receptors, drugs were fortuitous discoveries made by observant scientists; the discovery of penicillin as a bacteria-killing substance by Alexander Fleming is an example. Today, **drug design** is the rational application of scientific knowledge and principles to the development of pharmacologically active agents. A particular target for therapeutic intervention is identified (such as an enzyme or receptor involved in illness), and chemical analogs of its substrate or ligand are synthesized, in hopes of finding an inhibitor (or activator) that can serve as a drug to treat the illness. Sometimes the outcome is unanticipated, as the story of **Viagra** (sildenafil citrate) reveals.

When the smooth muscle cells of blood vessels relax, blood flow increases and blood pressure drops. Such relaxation is the result of decreases in intracellular $[Ca^{2+}]$ triggered by increases in intracellular [cGMP] (which in turn is triggered by nitric oxide, NO; see Chapter 26). Cyclic GMP (cGMP) is hydrolyzed by *phosphodiesterases* to form 5'-GMP, and the muscles contract again. Scientists at Pfizer, the pharmaceutical company, reasoned that, if phosphodiesterase inhibitors could be found, they might be useful drugs to treat *angina* (chest pain due to inadequate blood flow to heart muscle) or *hypertension* (high blood pressure). The phosphodiesterase (PDE) prevalent in vascular muscle is PDE 5, one of at least nine different subtypes of PDE in human cells. The search was on for substances that inhibit PDE 5, but not the other prominent PDE types, and Viagra was found. Disappointingly, Viagra showed no significant benefits for angina or hypertension, but some men in clinical trials reported penile erection. Apparently, Viagra led to an increase in [cGMP] in penile vascular tissue, allowing vascular muscle relaxation, improved blood flow, and erection. A drug was born.

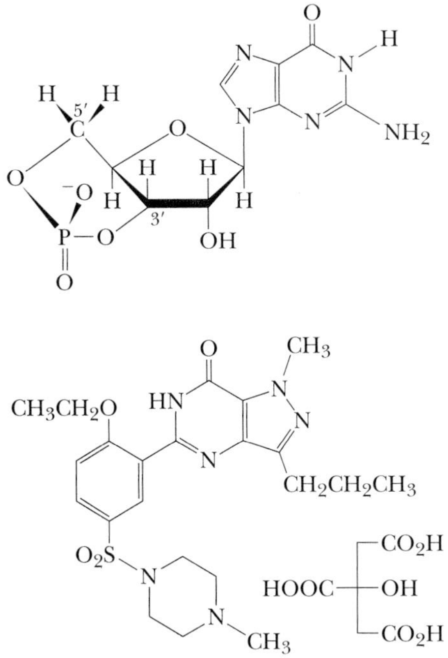

Note the structural similarity between cGMP *(top)* and Viagra *(bottom)*.

In a more focused way, detailed structural data on enzymes, receptors, and the ligands which bind to them has led to **rational drug design,** in which *computer modeling of enzyme and ligand interactions* replaces much of the initial chemical synthesis and clinical prescreening of potential therapeutic agents, saving much time and effort in drug development.

Such a reaction is termed a **bisubstrate reaction.** In general, bisubstrate reactions proceed by one of two possible routes:

1. Both A and B are bound to the enzyme and then reaction occurs to give P + Q:

$$E + A + B \rightleftharpoons AEB \rightleftharpoons PEQ \rightleftharpoons E + P + Q \qquad (10.32)$$

   Reactions of this type are defined as **sequential** or **single-displacement reactions.** They can be either of two distinct classes:

   a. random, where either A or B may bind to the enzyme first, followed by the other substrate, or
   b. ordered, where A, designated the leading substrate, must bind to E first before B can be bound.

   Both classes of single-displacement reactions are characterized by lines that intersect to the left of the $1/v$ axis in Lineweaver–Burk double-reciprocal plots (Figure 10.17).

Figure 10.17 Single-displacement bisubstrate mechanism. Double-reciprocal plots of the rates observed with different fixed concentrations of one substrate (B here) are graphed versus a series of concentrations of A. Note that, in these Lineweaver–Burk plots for single-displacement bisubstrate mechanisms, the lines intersect to the left of the 1/v axis.

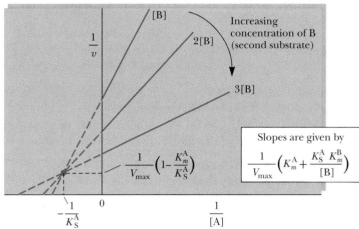

Double-reciprocal form of the rate equation:
$$\frac{1}{v} = \frac{1}{V_{max}}\left(K_m^A + \frac{K_S^A \, K_m^B}{[B]}\right)\left(\frac{1}{[A]} + \frac{1}{V_{max}}\left(1 + \frac{K_m^B}{[B]}\right)\right)$$

Slopes are given by
$$\frac{1}{V_{max}}\left(K_m^A + \frac{K_S^A \, K_m^B}{[B]}\right)$$

$$\frac{1}{V_{max}}\left(1 - \frac{K_m^A}{K_S^A}\right)$$

Increasing concentration of B (second substrate)

$-\dfrac{1}{K_S^A}$

2. The other general possibility is that one substrate, A, binds to the enzyme and reacts with it to yield a chemically modified form of the enzyme (E′) plus the product, P. The second substrate, B, then reacts with E′, regenerating E and forming the other product, Q.

$$E + A \rightleftharpoons EA \rightleftharpoons E'P \underset{P}{\rightleftharpoons} E' \underset{B}{\rightleftharpoons} E'B \rightleftharpoons EQ \rightleftharpoons E + Q \tag{10.33}$$

Reactions that fit this model are called **ping-pong** or **double-displacement reactions.** Two distinctive features of this mechanism are the obligatory formation of a modified enzyme intermediate, E′, and the pattern of parallel lines obtained in double-reciprocal plots (Figure 10.18).

Double-reciprocal form of the rate equation:
$$\frac{1}{v} = \frac{K_m^A}{V_{max}}\left(\frac{1}{[A]}\right) + \left(1 + \frac{K_m^B}{[B]}\right)\left(\frac{1}{V_{max}}\right)$$

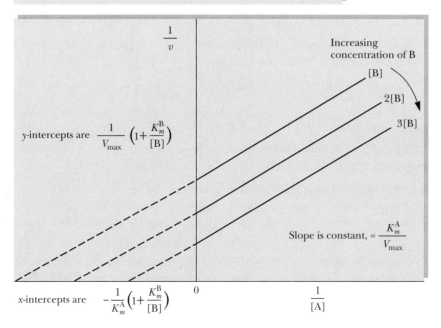

y-intercepts are $\dfrac{1}{V_{max}}\left(1 + \dfrac{K_m^B}{[B]}\right)$

Increasing concentration of B

Slope is constant, $= \dfrac{K_m^A}{V_{max}}$

x-intercepts are $-\dfrac{1}{K_m^A}\left(1 + \dfrac{K_m^B}{[B]}\right)$

Figure 10.18 Double-displacement (ping-pong) bisubstrate mechanisms are characterized by Lineweaver–Burk plots of parallel lines when double-reciprocal plots of the rates observed with different fixed concentrations of the second substrate, B, are graphed versus a series of concentrations of A.

Multisubstrate Reactions

Thus far, we have considered enzyme-catalyzed reactions involving one or two substrates. How are the kinetics described in those cases in which more than two substrates participate in the reaction? An example might be the glycolytic enzyme *glyceraldehyde-3-phosphate dehydrogenase* (see Chapter 15):

$$NAD^+ + \text{glyceraldehyde-3-P} + P_i \rightleftharpoons NADH + H^+ + \text{1,3-bisphosphoglycerate}$$

Many other multisubstrate examples abound in metabolism. In effect, these situations are managed by realizing that the interaction of the enzyme with its many substrates can be treated as a series of uni- or bisubstrate steps in a multistep reaction pathway. Thus, the complex mechanism of a multisubstrate reaction is resolved into a sequence of steps, each of which obeys the single- and double-displacement patterns just discussed.

10.6 RNA and Antibody Molecules As Enzymes: Ribozymes and Abzymes

Catalytic RNA Molecules: Ribozymes

It was long assumed that all enzymes are proteins. However, in recent years, more and more instances of biological catalysis by RNA molecules have been discovered. These catalytic RNAs, or **ribozymes,** satisfy several enzymatic criteria: they are substrate-specific, they enhance the reaction rate, and they emerge from the reaction unchanged. For example, *RNase P*, an enzyme responsible for the formation of mature tRNA molecules from tRNA precursors, requires an RNA component as well as a protein subunit for its activity in the cell. *In vitro*, the protein alone is incapable of catalyzing the maturation reaction, but the RNA component by itself can carry out the reaction under appropriate conditions.

Protein-Free 50S Ribosomal Subunits Catalyze Peptide Bond Formation *in Vitro*

Perhaps the most significant case of catalysis by RNA occurs in protein synthesis. Harry F. Noller and his colleagues have found that the **peptidyl transferase** reaction, which is the reaction of peptide bond formation during protein synthesis (Figure 10.19), can be catalyzed by 50S ribosomal subunits (see Chapter 8) from which virtually all of the protein has been removed. These experiments indicate that the 23S rRNA by itself is capable of catalyzing peptide bond formation. Detailed crystallographic studies of the ribosome structure reveal close proximity between 23S rRNA, aminoacyl-tRNA, and peptidyl-tRNA, supporting this conclusion (see Chapter 25).

The fact that RNA can catalyze certain reactions is experimental support for the idea that a primordial world dominated by RNA molecules existed before the evolution of DNA and proteins.

Catalytic Antibodies: Abzymes

Antibodies are *immunoglobulins*, which, of course, are proteins. Like other antibodies, **catalytic antibodies,** so-called **abzymes,** are elicited in an organism in response to immunological challenge by a foreign molecule called an **antigen** (see Chapter 12 for discussions on the molecular basis of immunology). In this case, however, the antigen is purposefully engineered to be *an analog of the transition-state intermediate in a reaction*. The rationale is that a protein specific for binding the transition-state intermediate of a reaction will promote entry of the normal reactant into the reactive, transition-state conformation. Thus, a catalytic antibody facilitates, or catalyzes, a reaction by forcing the conformation of its substrate in the direction of its transition state. (A prominent explanation for the remarkable catalytic power of conventional enzymes is their great affinity for the transition-state intermediates in the reactions they catalyze; see Chapter 11.)

One strategy has been to prepare ester analogs by substituting a phosphorus atom for the carbon in the ester group (Figure 10.20). The phospho-compound mimics the natural transition state of ester hydrolysis, and antibodies elicited against these analogs act like enzymes in accelerating the rate

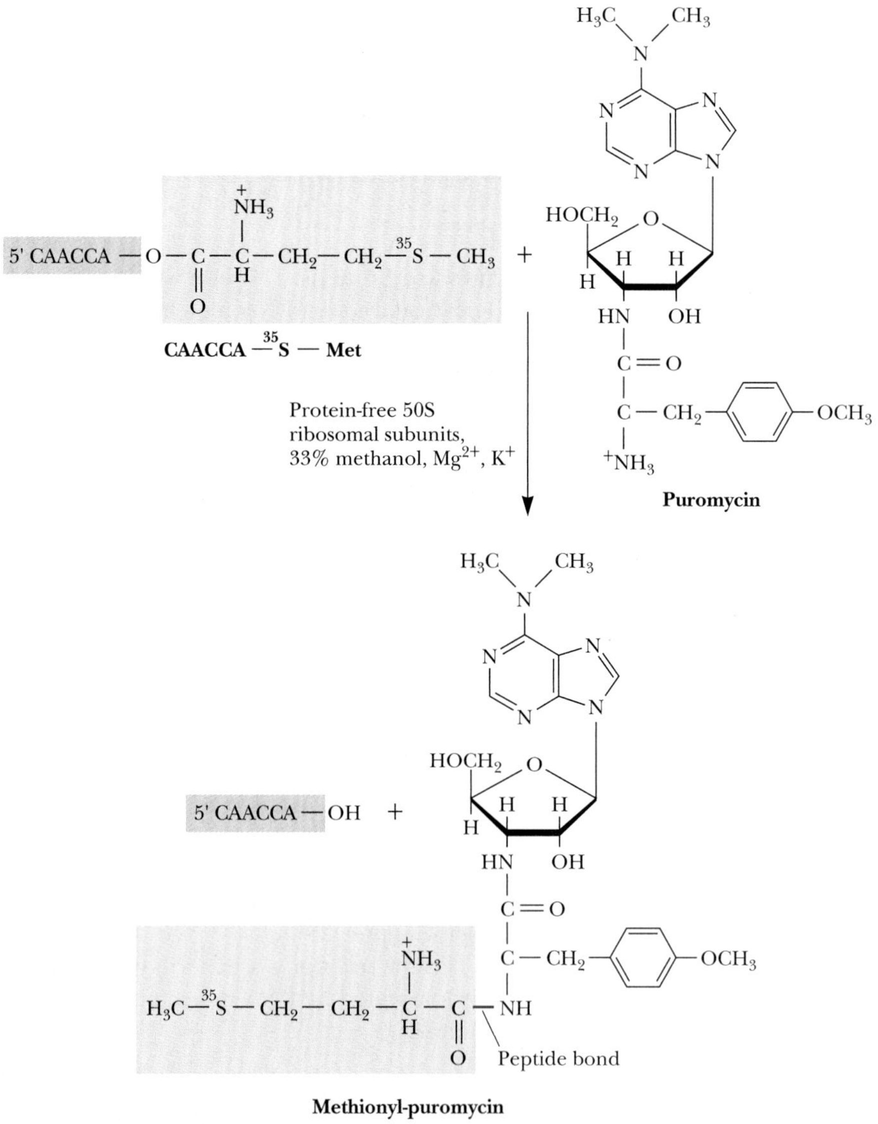

Figure 10.19 Protein-free 50S ribosomal subunits have peptidyl transferase activity. Peptidyl transferase is the name of the enzymatic function that catalyzes peptide bond formation. The presence of this activity in protein-free 50S ribosomal subunits was demonstrated using a model assay for peptide bond formation in which an aminoacyl-tRNA analog (a short RNA oligonucleotide of sequence CAACCA carrying ^{35}S-labeled methionine attached at its 3'-OH end) served as the peptidyl donor and puromycin (another aminoacyl-tRNA analog) served as the peptidyl acceptor. Activity was measured by monitoring the formation of ^{35}S-labeled methionyl-puromycin.

(a)

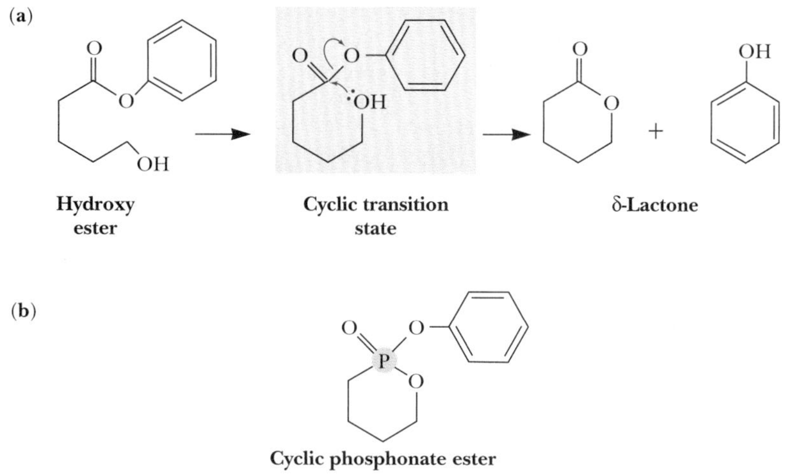

Hydroxy
ester

Cyclic transition
state

δ-Lactone

(b)

Cyclic phosphonate ester

Figure 10.20 Catalytic antibodies are designed to bind specifically the transition-state intermediate in a chemical reaction. **(a)** The intramolecular hydrolysis of a hydroxy ester to yield as products a δ-lactone and the alcohol phenol. Note the cyclic transition state. **(b)** The cyclic phosphonate ester analog of the cyclic transition state. Antibodies raised against this phosphonate ester act as enzymes: they are catalysts that markedly accelerate the rate of ester hydrolysis.

of ester hydrolysis as much as 1000-fold. This biotechnology offers the real possibility of creating **"designer enzymes,"** specially tailored enzymes designed to carry out specific catalytic processes.

10.7 Specificity Is the Result of Molecular Recognition

An enzyme molecule is typically orders of magnitude larger than its substrate. Its active site comprises only a small portion of the overall enzyme structure. The active site is part of the conformation of the enzyme molecule arranged to create a special pocket or cleft whose three-dimensional structure is complementary to the structure of the substrate. The enzyme and the substrate molecules "recognize" each other through this structural complementarity. The substrate binds to the enzyme through relatively weak forces—H bonds, ionic bonds (salt bridges), and van der Waals interactions between sterically complementary clusters of atoms. Specificity studies on enzymes entail an examination of the rates of the enzymatic reaction obtained with various **structural analogs** of the substrate. By determining which functional and structural groups within the substrate affect binding or catalysis, enzymologists can map the properties of the active site, analyzing questions such as the following: Can it accommodate sterically bulky groups? Are ionic interactions between E and S important? Are H bonds formed?

The "Lock and Key" Hypothesis

Pioneering enzyme specificity studies at the turn of the century by the great organic chemist Emil Fischer led to the notion of an enzyme resembling a **"lock"** and its particular substrate, the **"key."** This analogy captures the essence of the specificity that exists between an enzyme and its substrate, but enzymes are not rigid templates like locks (see Figure 1.15).

The "Induced Fit" Hypothesis

Enzymes are highly flexible, conformationally dynamic molecules, and many of their remarkable properties, including substrate binding and catalysis, are due to their structural pliancy. Realization of the conformational flexibility of

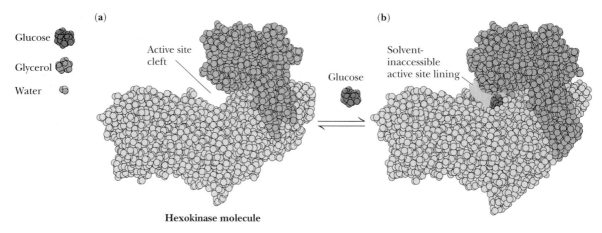

Figure 10.21 A drawing, roughly to scale, of H_2O, glycerol, glucose, and an idealized hexokinase molecule. Note the two domains of structure in hexokinase, **(a),** between which the active site is located. Binding of glucose induces a conformational change in hexokinase. The two domains close together, creating the catalytic site **(b).** The shaded area in **(b)** represents solvent-inaccessible surface area in the active site cleft that results when the enzyme binds substrate.

proteins led Daniel Koshland to hypothesize that the binding of a substrate (S) by an enzyme is an interactive process. That is, the shape of the enzyme's active site is actually modified upon binding S, in a process of dynamic recognition between enzyme and substrate aptly called **induced fit.** In essence, substrate binding alters the conformation of the protein, so that the protein and the substrate "fit" each other more precisely. The process is truly interactive in that the conformation of the substrate also changes as it adapts to the conformation of the enzyme.

This idea also helps to explain some of the mystery surrounding the enormous catalytic power of enzymes: in enzyme catalysis, precise orientation of catalytic residues within the active site is necessary for the reaction to occur; substrate binding induces this precise orientation by the changes it causes in the protein's conformation.

"Induced Fit" and the Transition-State Intermediate

The catalytically active enzyme : substrate complex is an interactive structure in which the enzyme causes the substrate to adopt a form that mimics the transition-state intermediate of the reaction. Thus, a poor substrate would be one that was less effective in directing the formation of an optimally active enzyme : transition-state intermediate conformation. This active conformation of the enzyme molecule is thought to be relatively unstable in the absence of substrate, and free enzyme thus reverts to a conformationally different state.

Specificity and Reactivity

Consider, for example, why hexokinase catalyzes the ATP-dependent phosphorylation of hexoses but not smaller phosphoryl-group acceptors such as glycerol, ethanol, or even water. Surely these smaller compounds are not sterically forbidden from approaching the active site of hexokinase (Figure 10.21). Indeed, water should penetrate the active site easily and serve as a highly effective phosphoryl-group acceptor. Accordingly, hexokinase should display high ATPase activity. It does not. Only the binding of hexoses induces hexokinase to assume its fully active conformation.

In Chapter 11 we explore in greater detail the factors that contribute to the remarkable catalytic power of enzymes and examine specific examples of enzyme reaction mechanisms. Here, we focus on another essential feature of enzymes: *the regulation of their activity.*

10.8 Controls over Enzymatic Activity: General Considerations

The activity displayed by enzymes is affected by a variety of factors, some of which are essential to the harmony of metabolism.

1. The enzymatic rate "slows down" as product accumulates and equilibrium is approached. The apparent decrease in rate is due to the conversion of P to S by the reverse reaction as [P] rises. $K_{eq} = [P]/[S]$, and enzymes exert no control over the thermodynamics of a reaction. Also, product inhibition can be a kinetically valid phenomenon: some enzymes are actually inhibited by the products of their action.

2. The availability of substrates and cofactors determines the enzymatic reaction rate. In general, enzymes have evolved such that their K_m values approximate the prevailing *in vivo* concentration of their substrates. (It is also true that the concentration of some enzymes in cells is within an order of magnitude or so of the concentrations of their substrates.)

3. There are genetic controls over the amounts of enzyme synthesized (or degraded) by cells. If the gene encoding a particular enzyme protein is turned on or off, changes in the amount of enzyme activity soon follow. By controlling the amount of an enzyme that is present at any moment, cells can either activate or terminate various metabolic routes. Genetic controls over enzyme levels have a response time ranging from minutes in rapidly dividing bacteria to hours (or longer) in higher eukaryotes.

4. Enzymes can be regulated by **covalent modification,** the reversible covalent attachment of a chemical group. For example, a fully active enzyme can be converted into an inactive form simply by the covalent attachment of a functional group, such as a phosphoryl moiety (Figure 10.22). (Alternatively, some enzymes exist in an inactive state unless specifically converted into the active form through covalent addition of a functional group.) Covalent modification reactions are catalyzed by special **converter enzymes,** which are themselves subject to metabolic regulation. Although covalent modification

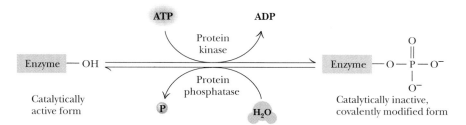

Figure 10.22 Enzymes regulated by covalent modification are called **interconvertible enzymes.** The enzymes (protein kinase and protein phosphatase, in the example shown here) catalyzing the conversion of the interconvertible enzyme between its two forms are called **converter enzymes.** In this example, the free enzyme form is catalytically active, whereas the phosphoryl-enzyme form represents an inactive state. The —OH on the interconvertible enzyme represents an —OH group on a specific amino acid side chain in the protein (for example, a particular Ser residue) capable of accepting the phosphoryl group.

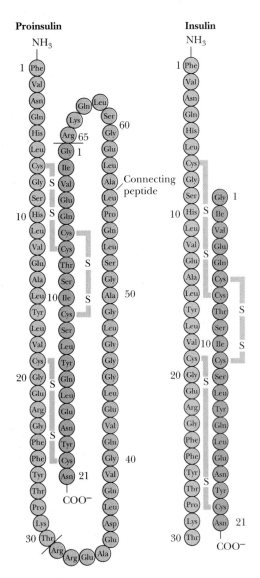

Figure 10.23 Proinsulin is an 86-residue precursor to insulin (the sequence shown here is human proinsulin). Proteolytic removal of residues 31 to 65 yields insulin. Residues 1 through 30 (the B chain) remain linked to residues 66 through 86 (the A chain) by a pair of interchain disulfide bridges.

represents a stable alteration of the enzyme, a different converter enzyme operates to remove the modification, so that when the conditions that favored modification of the enzyme are no longer present, the process can be reversed, restoring the enzyme to its unmodified state. Many examples of covalent modification at important metabolic junctions will be encountered in our discussions of metabolic pathways. Because covalent modification events are enzyme-catalyzed, they occur very quickly, with response times of seconds or even less for significant changes in metabolic activity. The 1992 Nobel Prize in physiology or medicine was awarded to Edmond Fischer and Edwin Krebs for their pioneering studies of reversible protein phosphorylation as an important means of cellular regulation.

5. Enzymatic activity can also be activated or inhibited through noncovalent interaction of the enzyme with small molecules (metabolites) other than the substrate. This form of control is termed **allosteric regulation** because the activator or inhibitor binds to the enzyme at a site *other than* (allo means "other than") the active site. Further, such allosteric regulators, or **effector molecules,** are often quite different sterically from the substrate. Because this form of regulation results simply from reversible binding of regulatory ligands to the enzyme, the cellular response time can be virtually instantaneous.

6. Enzyme regulation is an important matter to cells, and evolution has provided a variety of additional specialized controls, including zymogens, isozymes, and modulator proteins.

Zymogens

Most proteins become fully active as their synthesis is completed and they spontaneously fold into their native, three-dimensional conformations. Some proteins, however, are synthesized as inactive precursors, called **zymogens** or **proenzymes,** that only acquire full activity upon specific proteolytic cleavage of one or several of their peptide bonds. Unlike allosteric regulation or covalent modification, zymogen activation by specific proteolysis is an irreversible process. Activation of enzymes and other physiologically important proteins by specific proteolysis is a strategy frequently exploited by biological systems to switch on processes at the appropriate time and place, as the following examples illustrate.

Insulin. Some protein hormones are synthesized in the form of inactive precursor molecules, from which the active hormone is derived by proteolysis. For instance, **insulin,** an important metabolic regulator, is generated by proteolytic excision of a specific peptide from **proinsulin** (Figure 10.23).

Proteolytic Enzymes of the Digestive Tract. Enzymes of the digestive tract that serve to hydrolyze dietary proteins are synthesized in the stomach and pancreas as zymogens (Table 10.5). Only upon proteolytic activation are these enzymes able to form a catalytically active substrate-binding site. The ac-

	Table 10.5 Pancreatic and Gastric Zymogens	
Origin	**Zymogen**	**Active Protease**
Pancreas	Trypsinogen	Trypsin
Pancreas	Chymotrypsinogen	Chymotrypsin
Pancreas	Procarboxypeptidase	Carboxypeptidase
Pancreas	Proelastase	Elastase
Stomach	Pepsinogen	Pepsin

Chymotrypsinogen (inactive zymogen)

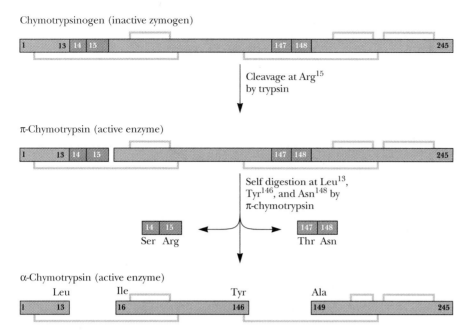

Figure 10.24   The proteolytic activation of chymotrypsinogen.

tivation of chymotrypsinogen is an interesting example (Figure 10.24). **Chymotrypsinogen** is a 245-residue polypeptide chain cross-linked by five disulfide bonds. Chymotrypsinogen is converted to an enzymatically active form called π-chymotrypsin when trypsin cleaves the peptide bond joining Arg^{15} and Ile^{16}. The enzymatically active π-chymotrypsin acts upon other π-chymotrypsin molecules, excising two dipeptides, Ser^{14}-Arg^{15} and Thr^{147}-Asn^{148}. The end product of this processing pathway is the mature protease α-**chymotrypsin,** in which the three peptide chains, A (residues 1 through 13), B (residues 16 through 146), and C (residues 149 through 245), remain together because they are linked by two disulfide bonds, one from A to B and one from B to C. It is interesting that the transformation of inactive chymotrypsinogen to active π-chymotrypsin requires the cleavage of just one particular peptide bond.

Blood Clotting.   The formation of blood clots is the result of a series of zymogen activations (Figure 10.25). The amplification achieved by this cascade of enzymatic activations allows blood clotting to occur rapidly in response to injury. Seven of the clotting factors in their active form are serine proteases: **kallikrein, XII_a, XI_a, IX_a, VII_a, X_a,** and **thrombin.** Two routes to blood clot formation exist. The **intrinsic pathway** is instigated when the blood comes into physical contact with abnormal surfaces caused by injury; the **extrinsic pathway** is initiated by factors released from injured tissues. The pathways merge at Factor X and culminate in clot formation. Thrombin excises peptides rich in negative charge from **fibrinogen,** converting it to **fibrin,** a molecule with a different surface charge distribution. Fibrin readily aggregates into ordered fibrous arrays that are subsequently stabilized by covalent cross-links. Thrombin specifically cleaves Arg-Gly peptide bonds and is homologous to trypsin, which is also a serine protease (recall that trypsin acts only at Arg and Lys residues).

Isozymes

A number of enzymes exist in more than one quaternary form, differing in their relative proportions of structurally equivalent but catalytically distinct polypeptide subunits. A classic example is mammalian **lactate dehydrogenase**

Figure 10.25 The cascade of activation steps leading to blood clotting. The intrinsic and extrinsic pathways converge at Factor X, and the final common pathway involves the activation of thrombin and its conversion of fibrinogen into fibrin, which aggregates into ordered filamentous arrays that become cross-linked to form the clot.

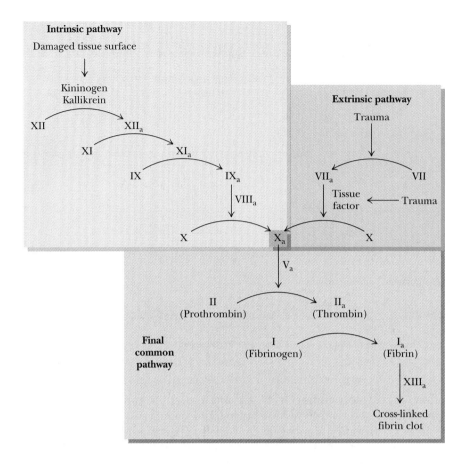

(LDH), which exists as five different isozymes, depending on the tetrameric association of two different subunits, A and B: A_4, A_3B, A_2B_2, AB_3, and B_4 (Figure 10.26). The kinetic properties of the various LDH isozymes differ in terms of their relative affinities for the various substrates and their sensitivity to inhibition by product. Different tissues express different isozyme forms, as appropriate to their particular metabolic needs. By regulating the relative amounts of A and B subunits they synthesize, the cells of various tissues control which isozymic forms are likely to assemble, and, thus, which kinetic parameters prevail.

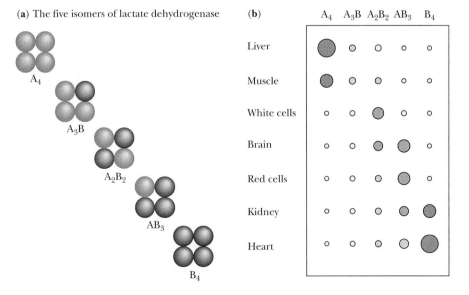

Figure 10.26 The isozymes of lactate dehydrogenase (LDH). Active muscle tissue becomes anaerobic and produces pyruvate from glucose via glycolysis (Chapter 15). It needs LDH to regenerate NAD^+ from NADH so glycolysis can continue. The lactate produced is released into the blood. The muscle LDH isozyme (A_4) works best in the NAD^+-regenerating direction. Heart tissue is aerobic and uses lactate as a fuel, converting it to pyruvate via LDH and using the pyruvate to fuel the citric acid cycle to obtain energy. The heart LDH isozyme (B_4) is inhibited by excess pyruvate so the fuel won't be wasted.

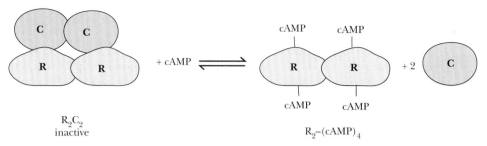

Figure 10.27 Cyclic AMP–dependent protein kinase (also known as PKA) is a 150- to 170-kD R_2C_2 tetramer in mammalian cells. The two R (regulatory) subunits bind cAMP ($K_D = 3 \times 10^{-8}$ M); cAMP binding releases the R subunits from the C (catalytic) subunits. C subunits are enzymatically active as monomers.

Modulator Proteins

Modulator proteins are yet another way that cells mediate metabolic activity. **Modulator proteins** are proteins that bind to enzymes and, by binding, influence the activity of the enzyme. For example, some converter enzymes, such as **cAMP-dependent protein kinase,** exist as dimers of catalytic subunits and regulatory subunits. These regulatory subunits are *modulator proteins* that suppress the activity of the catalytic subunits. Dissociation of the regulatory subunits (modulator proteins) activates the catalytic subunits; reassociation once again suppresses activity (Figure 10.27). We will meet other important representatives of this class as the processes of metabolism unfold in subsequent chapters. For now, let us focus our attention on the fascinating kinetics of allosteric enzymes.

 A DEEPER LOOK

Protein Kinases: Target Recognition and Intrasteric Control

Protein kinases are converter enzymes that catalyze the ATP-dependent phosphorylation of serine, threonine, or tyrosine hydroxyl groups in target proteins. Phosphorylation introduces a bulky group bearing two negative charges, causing conformational changes that alter the target protein's function. (Unlike a phosphoryl group, no amino acid side chain can provide two negative charges.) Protein kinases represent a protein superfamily whose members are widely diverse in terms of size, subunit structure, and subcellular localization. Protein kinases are classified as Ser/Thr-or Tyr-specific. They also differ in terms of the target proteins they recognize; target selection depends on the presence of an amino acid sequence within the target protein that can be recognized by the kinase. For example, cAMP-dependent protein kinase **(PKA)** phosphorylates proteins having Ser or Thr residues within an R(R/K)X(S*/T*) target consensus sequence (* denotes the residue that becomes phosphorylated). That is, PKA phosphorylates Ser or Thr residues that occur in an Arg–(Arg or Lys)–(any amino acid)–(Ser or Thr) sequence segment.

Targeting of protein kinases to particular consensus sequence elements within proteins creates a means to regulate these kinases by **intrasteric control.** Intrasteric control occurs when a regulatory subunit (or protein domain) has a **pseudosubstrate sequence** that mimics the target sequence but lacks a OH-bearing side chain at the right place. For example, the cAMP-binding regulatory subunits of PKA (R subunits in Figure 10.27) possess the pseudosubstrate sequence RRGA*I, and this sequence binds to the active site of PKA catalytic subunits, blocking their activity. This pseudosubstrate sequence has an alanine residue where serine occurs in the PKA target sequence; Ala is sterically similar to serine but lacks an OH- group to phosphorylate. When the PKA regulatory subunits bind cAMP, they undergo a conformational change and dissociate from the catalytic (C) subunits, and the active site of PKA is free to bind to and phosphorylate its targets. In other protein kinases, the pseudosubstrate regulatory sequence involved in intrasteric control and the kinase domain are part of the same polypeptide chain. In these cases, binding of an allosteric effector (like cAMP) induces a conformational change in the protein that releases the pseudosubstrate sequence from the active site of the kinase domain.

The abundance of many protein kinases in cells is an indication of the great importance of protein phosphorylation in cellular regulation. Exactly 113 protein kinase genes have been recognized in yeast, and it is estimated that the human genome encodes approximately 1000 different protein kinases. **Tyrosine kinases** (protein kinases that phosphorylate Tyr residues) occur only in multicellular organisms. Tyrosine kinases are components of signaling pathways involved in cell–cell communication (see Chapter 26).

Allosteric regulation acts to modulate enzymes situated at key steps in metabolic pathways. Consider as an illustration the following pathway, where A is the precursor for formation of an end product, F, in a sequence of five enzyme-catalyzed reactions:

$$\text{A} \xrightarrow{\text{enz 1}} \text{B} \xrightarrow{\text{enz 2}} \text{C} \xrightarrow{\text{enz 3}} \text{D} \xrightarrow{\text{enz 4}} \text{E} \xrightarrow{\text{enz 5}} \text{F}$$

In this scheme, F symbolizes an essential metabolite, such as an amino acid or a nucleotide. In such systems, F, the essential end product, inhibits enzyme 1, the *first step* in the pathway. Therefore, when sufficient F is synthesized, it blocks further synthesis of itself. This phenomenon is called **feedback inhibition** or **feedback regulation.**

General Properties of Regulatory Enzymes

Enzymes such as enzyme 1, which are subject to feedback regulation, represent a distinct class of enzymes, the **regulatory enzymes.** As a class, these enzymes have certain exceptional properties:

1. Their kinetics do not obey the Michaelis–Menten equation. Their v versus [S] plots yield **sigmoid, or S-shaped,** curves rather than rectangular hyperbolas (Figure 10.28). Such curves suggest a second-order (or higher) relationship between v and [S]; that is, v is proportional to $[S]^n$, where $n > 1$. A qualitative description of the mechanism responsible for the S-shaped curves is that binding of one S to a protein molecule makes it easier for additional substrate molecules to bind to the same protein molecule. In the jargon of allostery, substrate binding is **cooperative.**

2. Inhibition of a regulatory enzyme by a feedback inhibitor does not conform to any normal inhibition pattern, and the feedback inhibitor F bears little structural similarity to A, the substrate for the regulatory enzyme. F apparently acts at a binding site distinct from the substrate-binding site. The term allosteric is apt, because F is sterically dissimilar and, moreover, acts at a site other than the site for S. Its effect is called **allosteric inhibition.**

3. Regulatory or allosteric enzymes like enzyme 1 are, in some instances, regulated by activation. That is, whereas some effector molecules such as F exert negative effects on enzyme activity, other effectors show stimulatory, or positive, influences on activity.

4. Allosteric enzymes have an oligomeric organization. They are composed of more than one polypeptide chain (subunit) and have more than one S-binding site per enzyme molecule.

5. The working hypothesis is that, by some means, interaction of an allosteric enzyme with effectors alters the distribution of conformational possibilities or subunit interactions available to the enzyme. That is, the regulatory effects exerted on the enzyme's activity are achieved by conformational changes occurring in the protein when effector metabolites bind.

In addition to enzymes, noncatalytic proteins may exhibit many of these properties; hemoglobin is the classic example. The allosteric properties of hemoglobin are discussed in Chapter 12.

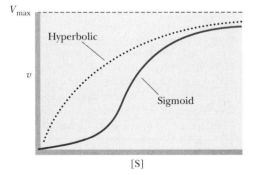

◀ **Figure 10.28** Sigmoid *v* versus [S] plot. The dotted line represents the hyperbolic plot characteristic of normal Michaelis–Menten–type enzyme kinetics.

10.10 A Model for the Allosteric Behavior of Proteins

The Symmetry Model of Monod, Wyman, and Changeux

In 1965, Jacques Monod, Jeffries Wyman, and Jean-Pierre Changeux proposed a theoretical model of allosteric transitions based on the observation that allosteric proteins are oligomers. This model is widely accepted as a reasonable approximation of the behavior of allosteric proteins. The model is based on the premise that allosteric proteins can exist in (at least) two conformational states, designated **R,** signifying "relaxed," and **T,** or "taut." Further, the model assumes that, in each protein molecule, all of the subunits have the same conformation (either R or T). That is, molecular symmetry is conserved. Molecules of mixed conformation (having subunits of both R and T states) are not allowed by this model.

In the absence of ligand, the two states of the allosteric protein are in equilibrium:

$$R_0 \rightleftharpoons T_0$$

(Note that the subscript "0" signifies "in the absence of ligand.") The equilibrium constant is termed **L:** $L = T_0/R_0$. L is assumed to be large; that is, the amount of the protein in the T conformational state is much greater than the amount in the R conformation. Let us suppose that $L = 10^4$.

The affinities of the two states for substrate, S, are characterized by the respective dissociation constants, K_R and K_T. The model supposes that $K_T \gg K_R$. That is, the affinity of R_0 for S is much greater than the affinity of T_0 for S. Let us choose the extreme where $K_R/K_T = 0$ (that is, K_T is infinitely greater than K_R). In effect, we are picking conditions where S binds only to R. (If K_T is infinite, T does not bind S.)

Given these parameters, consider what happens when S is added to a solution of the allosteric protein at conformational equilibrium (Figure 10.29). Although the relative $[R_0]$ concentration is small, S will bind "only" to R_0, forming R_1. This depletes the concentration of R_0, perturbing the T_0/R_0 equilibrium. To restore equilibrium, molecules in the T_0 conformation undergo a transition to R_0. This shift renders more R_0 available to bind S, yielding R_1, diminishing $[R_0]$, perturbing the T_0/R_0 equilibrium, and so on. Thus, these linked equilibria (see Figure 10.29) are such that S-binding by the R_0 state of the allosteric protein perturbs the T_0/R_0 equilibrium with the result that S-binding drives the conformational transition, $T_0 \rightarrow R_0$.

In just this simple system, cooperativity is achieved because each subunit has a binding site for S, and thus, each protein molecule has more than one binding site for S. Therefore, the increase in the population of R conformers gives a progressive increase in the number of sites available for S. The extent of cooperativity depends on the relative T_0/R_0 ratio and the relative affinities of R and T for S. If L is large (that is, the equilibrium lies strongly in favor of T_0) and if $K_T \gg K_R$, as in the example we have chosen, cooperativity is great (Figure 10.30). Ligands such as S here that bind in a cooperative manner, so that the binding of one equivalent enhances the binding of additional equivalents of S to the same protein molecule, are termed **positive homotropic effectors.** (The prefix "homo" indicates that the ligand influences the binding of like molecules.)

Heterotropic Effectors

This simple system also provides an explanation for the more complex substrate-binding responses to positive and negative effectors. Effectors that influence the binding of something other than themselves are termed

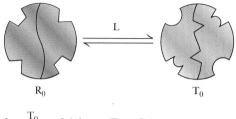

(a) A dimeric protein can exist in either of two conformational states at equilibrium.

R_0 T_0

$L = \dfrac{T_0}{R_0}$ L is large. ($T_0 \gg R_0$)

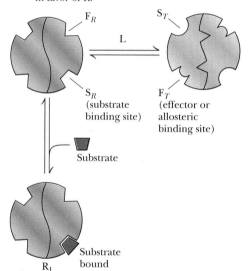

(b) Substrate binding shifts equilibrium in favor of R.

F_R S_T

L

S_R (substrate binding site) F_T (effector or allosteric binding site)

Substrate

Substrate bound

R_1

Figure 10.29 Monod–Wyman–Changeux (MWC) model for allosteric transitions. Consider a dimeric protein that can exist in either of two conformational states, R or T. Each subunit in the dimer has a binding site for substrate S and an allosteric effector site, F. The promoters are symmetrically related to one another in the protein, and symmetry is conserved regardless of the conformational state of the protein. The different states of the protein, with or without bound ligand, are linked to one another through the various equilibria. Thus, the relative population of protein molecules in the R or T state is a function of these equilibria and the concentration of the various ligands, substrate (S), and effectors (which bind at F_R or F_T). As [S] is increased, the T/R equilibrium shifts in favor of an increased proportion of R-conformers in the total population (that is, more protein molecules in the R conformational state).

Figure 10.30 The Monod-Wyman-Changeux model. Graphs of allosteric effects for a tetramer ($n = 4$) in terms of Y, the saturation function, versus [S]. Y is defined as [ligand binding sites that are occupied by ligand]/[total ligand binding sites]. **(a)** A plot of Y as a function of [S], at various L values. **(b)** Y as a function of [S], at different c, where $c = K_R/K_T$. (When $c = 0$, K_T is infinite.) (*Adapted from Monod, J., Wyman, J., and Changeux, J.-P., 1965. On the nature of allosteric transitions: A plausible model.* Journal of Molecular Biology **12**:92.)

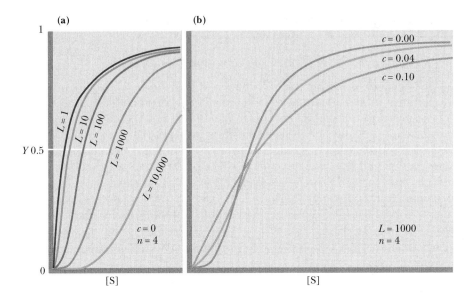

heterotropic effectors. For example, effectors that promote S-binding are termed **positive heterotropic effectors** or **allosteric activators.** Effectors that diminish S-binding are **negative heterotropic effectors** or **allosteric inhibitors.** Feedback inhibitors fit this class. Consider a protein composed of two subunits, each of which has two binding sites: one for the substrate, S, and one to which allosteric effectors bind, the allosteric site. Assume that S binds preferentially ("only") to the R conformer; further assume that the positive heterotropic effector, A, binds to the allosteric site only when the protein is in the R conformation, and the negative allosteric effector, I, binds at the allosteric site only if the protein is in the T conformation. Thus, with respect to binding at the allosteric site, A and I are competitive with each other.

Positive Effectors. If A binds to R_0, forming the new species $R_{I(A)}$, the relative concentration of R_0 is decreased and the T_0/R_0 equilibrium is perturbed (Figure 10.31). As a consequence, a relative T_0/R_0 shift occurs in order to restore equilibrium. The net effect is an increase in the number of R conformers in the presence of A, meaning that more binding sites for S are available. For this reason, A leads to a decrease in the cooperativity of the substrate saturation curve, as seen by a shift of this curve to the left (see Figure 10.31). Effectively, the presence of A lowers the apparent value of L.

Negative Effectors. The converse situation applies in the presence of I, which binds "only" to T. I-binding will lead to an increase in the population of T conformers, at the expense of R_0 (see Figure 10.31). The decline in $[R_0]$ means that it is less likely for S (or A) to bind. Consequently, the presence of I increases the cooperativity (that is, the sigmoidicity) of the substrate saturation curve, as evidenced by the shift of this curve to the right (see Figure 10.31). The presence of I raises the apparent value of L.

K Systems and *V* Systems

The allosteric model just presented is called a **K system** because the concentration of substrate giving half-maximal velocity, defined as $K_{0.5}$, changes in response to effectors (see Figure 10.31). Note that V_{max} is constant in this system.

An allosteric situation in which $K_{0.5}$ is constant but the apparent V_{max} changes in response to effectors is termed a **V system.** In a V system, all v versus [S] plots are hyperbolic rather than sigmoid (Figure 10.32). The positive heterotropic effector A activates by raising V_{max}, whereas I, the negative het-

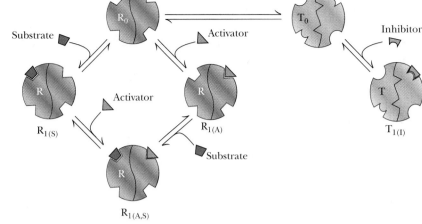

A dimeric protein which can exist in either of two states, R_0 and T_0. This protein can bind 3 ligands:

1) Substrate (S) ▰ : A positive homotropic effector that binds only to R at site S

2) Activator (A) ◣ : A positive heterotropic effector that binds only to R at site F

3) Inhibitor (I) ◸ : A negative heterotropic effector that binds only to T at site F

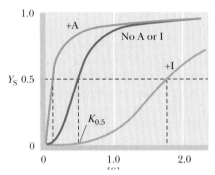

Effects of A:
$A + R_0 \longrightarrow R_{1(A)}$
Increase in number of R-conformers shifts $R_0 \rightleftharpoons T_0$ so that $T_0 \longrightarrow R_0$

1) More binding sites for S made available

2) Decrease in cooperativity of substrate saturation curve. Effector A lowers the apparent value of L.

Effects of I:
$I + T_0 \longrightarrow T_{1(I)}$
Increase in number of T-conformers (decrease in R_0 as $R_0 \longrightarrow T_0$ to restore equilibrium)

Thus, I inhibits association of S and A with R by lowering R_0 level. I increases cooperativity of substrate saturation curve. I raises the apparent value of L.

Figure 10.31 Heterotropic allosteric effects: A and I binding to R and T, respectively. The linked equilibria lead to changes in the relative amounts of R and T and, therefore, shifts in the substrate saturation curve. This behavior, depicted by the graph, defines an allosteric K system. The parameters of such a system are: (1) S and A (or I) have different affinities for R and T and (2) A (or I) modifies the apparent $K_{0.5}$ for S by shifting the relative R versus T population.

erotropic effector, decreases it. Note that neither A nor I affects $K_{0.5}$. This situation arises if R and T have the same affinity for the substrate, S, but differ in their catalytic ability and their affinities for A and I. A and I thus can shift the relative T/R distribution. Acetyl-coenzyme A carboxylase, the enzyme catalyzing the committed step in the fatty acid biosynthetic pathway, behaves as a V system in response to its allosteric activator, citrate (see Chapter 20).

K Systems and *V* Systems Fill Different Biological Roles

The *K* and *V* systems have design features that mean they work best under different physiological situations. *K*-system enzymes are adapted to conditions in which the prevailing substrate concentration is rate-limiting, as when [S] *in vivo* ≈ $K_{0.5}$. On the other hand, when the physiological conditions are such that [S] is usually saturating for the regulatory enzyme of interest, the enzyme conforms to the *V*-system mode in order to have an effective regulatory response.

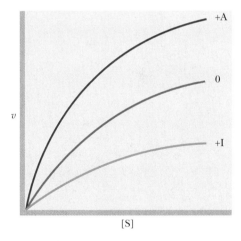

Figure 10.32 v versus [S] curves for an allosteric V system. The V system fits the model of Monod, Wyman, and Changeux, given the following conditions: (1) R and T have the same affinity for the substrate, S. (2) The effectors A and I have different affinities for R and T and thus can shift the relative T/R distribution. (That is, A and I change the apparent value of L.) Assume as before that A binds "only" to the R state and I binds "only" to the T state. (3) R and T differ in their catalytic ability. Assume that R is the enzymatically active form, whereas T is inactive. Because A perturbs the T/R equilibrium in favor of more R, A increases the apparent V_{max}. I favors transition to the inactive T state.

10.11 A Paradigm of Enzyme Regulation: Glycogen Phosphorylase

Glycogen phosphorylase is the key enzyme in liberating glucose units from stored glycogen reserves in muscle and liver in times of energy need. As such, its activity is tightly regulated. Under normal conditions, muscle glycogen phosphorylase activity is allosterically regulated by metabolites that reflect cellular

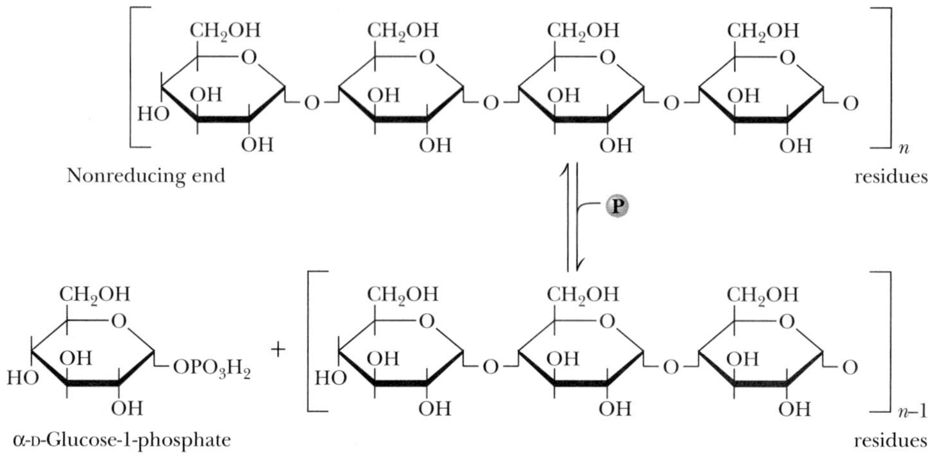

Figure 10.33 The glycogen phosphorylase reaction.

Nonreducing end

residues

α-D-Glucose-1-phosphate

residues

energy status (AMP, ATP, and glucose-6-P). These allosteric controls can be overriden during times of stress. The "fight-or-flight" hormone *adrenaline* (also known as *epinephrine*) triggers a multistep cascade of reactions that culminate in covalent modification (phosphorylation) of glycogen phosphorylase. The phosphorylated form of glycogen phosphorylase is very active; significantly, it is unresponsive to normal allosteric regulation. Thus, glucose units are rapidly released from glycogen. This glucose serves as an immediate energy source for the animal's response to a crisis situation.

The Glycogen Phosphorylase Reaction

The cleavage of glucose units from the nonreducing ends of glycogen molecules is catalyzed by **glycogen phosphorylase,** an allosteric enzyme. The enzymatic reaction involves phosphorolysis of the bond between C-1 of the departing glucose unit and the glycosidic oxygen, to yield *glucose-1-phosphate* and a glycogen molecule that is shortened by one residue (Figure 10.33). (Because the reaction involves attack by phosphate instead of H_2O, it is referred to as a **phosphorolysis** rather than a hydrolysis.) The phosphorylated sugar product, glucose-1-P, is converted to the glycolytic substrate, glucose-6-P, by *phosphoglucomutase* (Figure 10.34). In muscle, glucose-6-P proceeds into glycolysis, providing needed energy for muscle contraction. In liver, hydrolysis of glucose-6-P yields glucose, which is exported to other tissues via the circulatory system.

The Structure of Glycogen Phosphorylase

Muscle glycogen phosphorylase is a dimer of two identical subunits (842 residues, 97.44 kD). Each subunit contains a pyridoxal phosphate cofactor located within its active site (at the center of the subunit) and an allosteric effector site near the subunit interface (Figure 10.35). In addition, a regulatory phosphorylation site is located at Ser^{14} on each subunit. A glycogen-binding site on each subunit facilitates association of glycogen phosphorylase with its substrate and also exerts regulatory control on the enzymatic reaction.

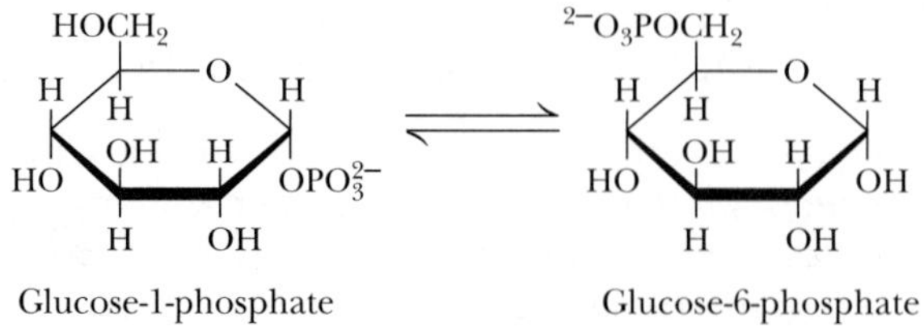

Figure 10.34 The phosphoglucomutase reaction.

Glucose-1-phosphate

Glucose-6-phosphate

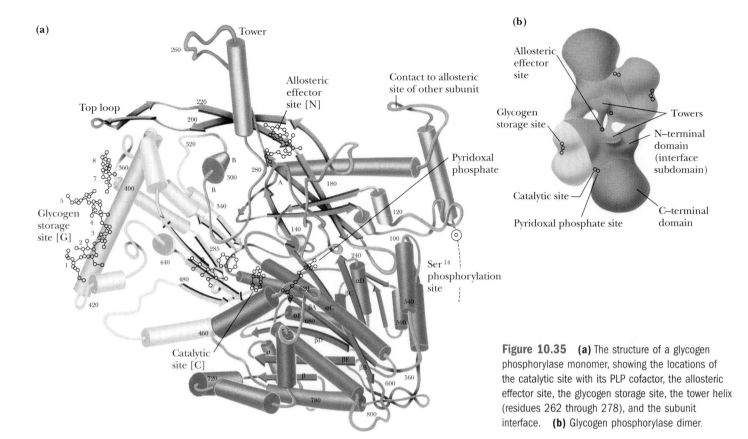

Figure 10.35 **(a)** The structure of a glycogen phosphorylase monomer, showing the locations of the catalytic site with its PLP cofactor, the allosteric effector site, the glycogen storage site, the tower helix (residues 262 through 278), and the subunit interface. **(b)** Glycogen phosphorylase dimer.

Each subunit contributes a tower helix (residues 262 to 278) to the subunit–subunit contact interface in glycogen phosphorylase. In the phosphorylase dimer, the tower helices extend from their respective subunits and pack against each other in an antiparallel manner.

Regulation of Glycogen Phosphorylase by Allosteric Effectors

Muscle Glycogen Phosphorylase Shows Cooperativity in Substrate Binding

The binding of the substrate *inorganic phosphate* (P_i) to muscle glycogen phosphorylase is highly cooperative (Figure 10.36a), which allows the enzyme activity to increase markedly over a rather narrow range of substrate concentration. P_i is a *positive homotropic effector* with regard to its interaction with glycogen phosphorylase.

ATP and Glucose-6-P Are Allosteric Inhibitors of Glycogen Phosphorylase

ATP can be viewed as the "end product" of glycogen phosphorylase action, in that the glucose-1-P liberated by glycogen phosphorylase is degraded in muscle via metabolic pathways whose purpose is energy (ATP) production. Glucose-1-P is readily converted into glucose-6-P to feed such pathways. (In the liver, glucose-1-P from glycogen is converted to glucose and released into the bloodstream to raise blood glucose levels.) Thus, feedback inhibition of glycogen phosphorylase by ATP and glucose-6-P provides a very effective way to regulate glycogen breakdown. Both ATP and glucose-6-P act by decreasing the affinity of glycogen phosphorylase for its substrate, P_i (Figure 10.36b). Because the binding of ATP or glucose-6-P has a negative effect on substrate binding, these substances act as *negative heterotropic effectors*. Note in Figure 10.36b that the

(a)

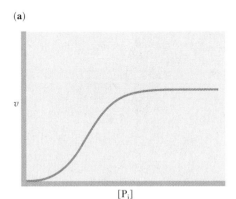

(b) (c)

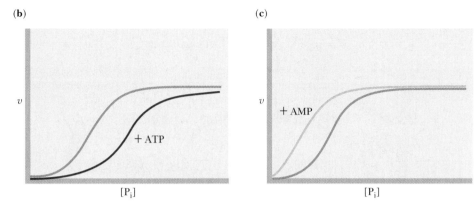

Figure 10.36 *v* versus [S] curves for glycogen phosphorylase. **(a)** The sigmoid response of glycogen phosphorylase to the concentration of the substrate phosphate (P_i) shows strong positive cooperativity. **(b)** ATP is a feedback inhibitor that affects the affinity of glycogen phosphorylase for its substrates but does not affect V_{max}. (Glucose-6-P shows similar effects on glycogen phosphorylase.) **(c)** AMP is a positive heterotropic effector for glycogen phosphorylase. It binds at the same site as ATP. AMP and ATP are competitive. Like ATP, AMP affects the affinity of glycogen phosphorylase for its substrates, but does not affect V_{max}.

substrate saturation curve is displaced to the right in the presence of ATP or glucose-6-P, and a higher substrate concentration is needed to achieve half-maximal velocity ($V_{max}/2$). When concentrations of ATP or glucose-6-P accumulate to high levels, glycogen phosphorylase is inhibited; when [ATP] and [glucose-6-P] are low, the activity of glycogen phosphorylase is regulated by availability of its substrate, P_i.

AMP Is an Allosteric Activator of Glycogen Phosphorylase

AMP also provides a regulatory signal to glycogen phosphorylase. It binds to the same site as ATP, but it stimulates glycogen phosphorylase rather than inhibiting it (see Figure 10.36c). AMP acts as a *positive heterotropic effector,* meaning that it enhances the binding of substrate to glycogen phosphorylase. Significant levels of AMP indicate that the energy status of the cell is low and that more energy (ATP) should be produced. Reciprocal changes in the cellular concentrations of ATP and AMP and their competition for binding to the same site (the *allosteric site*) on glycogen phosphorylase, with opposite effects, allow these two nucleotides to exert *rapid and reversible control* over glycogen phosphorylase activity. Such reciprocal regulation ensures that the production of energy (ATP) is commensurate with cellular needs.

To summarize, muscle glycogen phosphorylase is allosterically activated by AMP and inhibited by ATP and glucose-6-P; caffeine can also act as an allosteric inhibitor (Figure 10.37). When ATP and glucose-6-P are abundant, glycogen breakdown is inhibited. When cellular energy reserves are low (i.e., high [AMP] and low [ATP] and [G-6-P]), glycogen catabolism is stimulated.

Glycogen phosphorylase conforms to the Monod–Wyman–Changeux model of allosteric transitions, with the active form of the enzyme designated the **R state** and the inactive form denoted as the **T state** (Figure 10.37). Thus, AMP promotes the conversion to the active R state, whereas ATP, glucose-6-P, and caffeine favor conversion to the inactive T state.

X-ray diffraction studies of glycogen phosphorylase in the presence of allosteric effectors have revealed the molecular basis for the T $\rightleftharpoons$ R conversion. Although the structure of the central core of the phosphorylase subunits is identical in the T and R states, a significant change occurs at the subunit interface between the T and R states. This conformation change at the subunit interface is linked to a structural change at the active site that is important for catalysis. In the T state, the negatively charged carboxyl group of Asp[283] faces the active site, so that binding of the anionic substrate phosphate is unfavorable. In the conversion to the R state, Asp[283] is displaced from the active site and replaced by Arg[569]. The exchange of negatively charged aspartate for positively charged arginine at the active site provides a favorable binding site for phosphate.

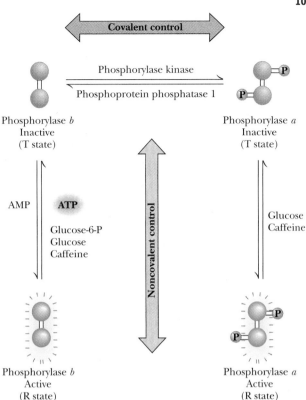

Figure 10.37 The mechanism of covalent modification and allosteric regulation of glycogen phosphorylase. The T states are blue and the R states blue-green.

These allosteric controls serve as a mechanism for adjusting the activity of glycogen phosphorylase to meet normal metabolic demands. However, in crisis situations in which abundant energy (ATP) is needed immediately, these controls can be overridden by covalent modification of glycogen phosphorylase. Covalent modification through phosphorylation of Ser^{14} in glycogen phosphorylase converts the enzyme from a less active, allosterically regulated form (the b form) to a more active, allosterically unresponsive form (the a form). Covalent modification is like a "permanent" allosteric transition that is independent of [allosteric effector], such as AMP.

Regulation of Glycogen Phosphorylase by Covalent Modification

As early as 1938, it was known that glycogen phosphorylase existed in two forms: the less active **phosphorylase b** and the more active **phosphorylase a.** In the 1950s, Edwin Krebs and Edmond Fischer demonstrated that the conversion of phosphorylase b to phosphorylase a involved covalent phosphorylation, by a "converting enzyme," as in Figure 10.37.

Phosphorylation of Ser^{14} causes a dramatic conformation change in phosphorylase. Upon phosphorylation, the amino-terminal end of the protein (including residues 10 through 22) swings through an arc of 120°, moving into the subunit interface (Figure 10.38). This conformation change moves Ser^{14} by more than 3.6 nm.

Dephosphorylation of glycogen phosphorylase is carried out by **phosphoprotein phosphatase 1.** The action of phosphoprotein phosphatase 1 inactivates glycogen phosphorylase.

Enzyme Cascades Regulate Glycogen Phosphorylase

The phosphorylation reaction that activates glycogen phosphorylase is mediated by an **enzyme cascade** (Figure 10.39). The first part of the cascade leads to hormonal stimulation (described in the next paragraph) of **adenylyl cyclase,** a membrane-bound enzyme that converts ATP to *adenosine-3′, 5′-cyclic monophos-*

Figure 10.38 In this diagram of the glycogen phosphorylase dimer, the phosphorylation site (Ser^{14}) and the allosteric (AMP) site face the viewer. Access to the catalytic site is from the opposite side of the protein. The diagram shows the major conformational change that occurs in the N-terminal residues upon phosphorylation of Ser^{14}. The solid black line shows the conformation of residues 10 to 23 in the *b*, or unphosphorylated, form of glycogen phosphorylase. The conformational change in the location of residues 10 to 23 upon phosphorylation of Ser^{14} to give the *a* (phosphorylated) form of glycogen phosphorylase is shown in yellow. Note that these residues move from intrasubunit contacts into intersubunit contacts at the subunit interface. (Sites on the two respective subunits are denoted, with those of the upper subunit designated by primes ['].) (*Adapted from Johnson, L. N., and Barford, D., 1993. The effects of phosphorylation on the structure and function of proteins.* Annual Review of Biophysics and Biomolecular Structure **22:**199–232.)

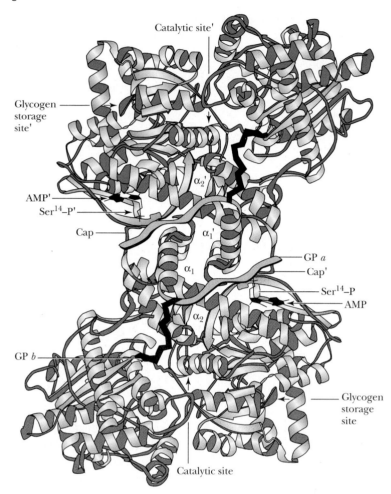

phate, denoted as *cyclic AMP* or simply *cAMP* (Figure 10.40). This regulatory molecule is found in all eukaryotic cells and acts as an intracellular messenger molecule, controlling a wide variety of processes. Cyclic AMP is known as a **second messenger** because it is the intracellular agent of a hormone (the "first messenger"). (The myriad cellular roles of cyclic AMP are described in detail in Chapter 26.)

Hormonal stimulation of adenylyl cyclase is achieved by a transmembrane signaling pathway consisting of three components, all membrane-associated. Binding of hormone to the external surface of a hormone receptor causes a conformational change in this transmembrane protein, which in turn stimu-

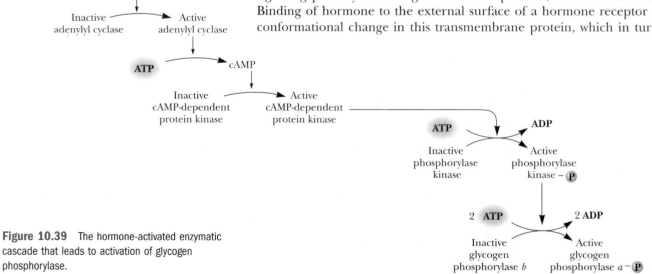

Figure 10.39 The hormone-activated enzymatic cascade that leads to activation of glycogen phosphorylase.

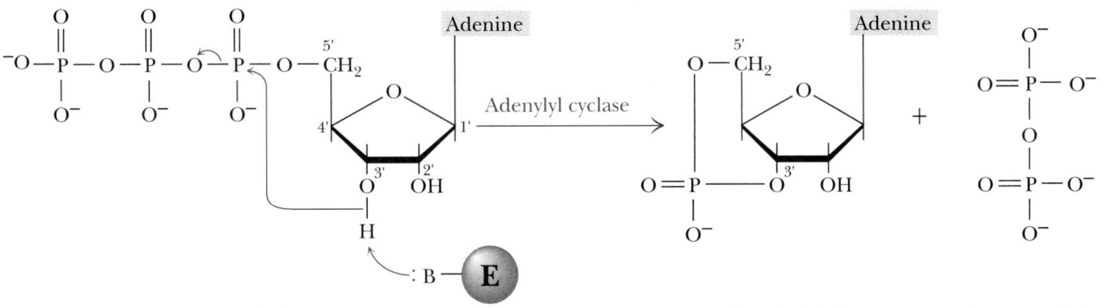

ATP

3',5'-Cyclic AMP
(cAMP)

Pyrophosphate

Figure 10.40 The adenylyl cyclase reaction yields 3',5'-cyclic AMP and pyrophosphate. The reaction is driven forward by subsequent hydrolysis of pyrophosphate by the enzyme inorganic pyrophosphatase.

lates a **GTP-binding protein** (abbreviated **G protein**). G proteins are heterotrimeric proteins consisting of α- (45–47 kD), β- (35 kD), and γ- (7–9 kD) subunits. The α-subunit binds GDP or GTP and has an intrinsic, slow GTPase activity. In the inactive state, the $G_{\alpha\beta\gamma}$ complex has GDP at the nucleotide site. When a G protein is stimulated by a hormone–receptor complex, GDP dissociates and GTP binds to G_{α}, causing it to dissociate from $G_{\beta\gamma}$ and to associate with adenylyl cyclase (Figure 10.41). *Binding of* $G_{\alpha}(GTP)$ *activates adenylyl cyclase to form cAMP from ATP.* However, the intrinsic GTPase activity of G_{α} eventually hydrolyzes GTP to GDP, leading to dissociation of $G_{\alpha}(GDP)$ from adenylyl cyclase and reassociation with $G_{\beta\gamma}$ to form the inactive $G_{\alpha\beta\gamma}$ complex. This cascade amplifies the hormonal signal because a single hormone–receptor complex can activate many G proteins before the hormone dissociates from the receptor, and because the G_{α}-activated adenylyl cyclase can synthesize many cAMP molecules before bound GTP is hydrolyzed by G_{α}. More than 100 different G protein–coupled receptors and at least 21 distinct G_{α} proteins are known (see Chapter 26).

Cyclic AMP is an essential activator of *cAMP-dependent protein kinase (PKA)*. This enzyme is normally inactive because its two catalytic subunits (C) are strongly associated with a pair of regulatory subunits (R), which serve to block activity. Binding of cyclic AMP to the regulatory subunits induces a conformation change that causes the dissociation of the C monomers from the R dimer (see Figure 10.27). The free C subunits are active and can phosphorylate other proteins. One of the many proteins phosphorylated by PKA is *phosphorylase kinase* (see Figure 10.39). Phosphorylase kinase is inactive in the unphosphorylated state and active in the phosphorylated form. As its name implies, phosphorylase kinase functions to phosphorylate (and activate) glycogen phosphorylase. Thus, stimulation of adenylyl cyclase leads to activation of glycogen breakdown.

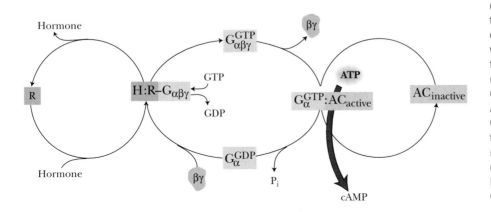

Figure 10.41 Hormone (H) binding to its receptor (R) creates a hormone : receptor complex (H : R) that catalyzes GDP–GTP exchange on the α-subunit of the heterotrimer G protein ($G_{\alpha\beta\gamma}$), replacing GDP with GTP. The G_{α}-subunit with GTP bound dissociates from the $\beta\gamma$-subunits and binds to adenylyl cyclase (AC). AC becomes active upon association with G_{α} : GTP and catalyzes the formation of cAMP from ATP. With time, the intrinsic GTPase activity of the G_{α}-subunit hydrolyzes the bound GTP, forming GDP; this leads to dissociation of G_{α} : GDP from AC, reassociation of G_{α} with the $\beta\gamma$-subunits, and cessation of AC activity. AC and the hormone receptor H are integral plasma membrane proteins; G_{α} and $G_{\beta\gamma}$ are membrane-anchored proteins.

PROBLEMS

1. According to the Michaelis–Menten equation, what is the v/V_{max} ratio when $[S] = 4K_m$?

2. If $V_{max} = 100\ \mu mol/mL \cdot sec$ and $K_m = 2\ mM$, what is the velocity of the reaction when $[S] = 20\ mM$?

3. For a Michaelis–Menten reaction, $k_1 = 7 \times 10^7/M \cdot sec$, $k_{-1} = 1 \times 10^3/sec$, and $k_2 = 2 \times 10^4/sec$. What is the value of K_m?

4. The following kinetic data were obtained for an enzyme in the absence of any inhibitor (1), and in the presence of two different inhibitors (2) and (3) at 5 mM concentration. Assume $[E_T]$ is the same in each experiment.

[S] (mM)	(1) v(μmol/mL · sec)	(2) v(μmol/mL · sec)	(3) v(μmol/mL · sec)
1	12	4.3	5.5
2	20	8	9
4	29	14	13
8	35	21	16
12	40	26	18

a. Determine V_{max} and K_m for the enzyme.

b. Determine the type of inhibition and the K_I for each inhibitor.

5. The following graphical patterns obtained from kinetic experiments have several possible interpretations depending on the nature of the experiment and the variables being plotted. Give at least two possibilities for each.

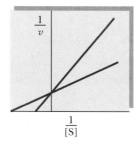
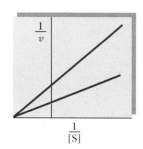

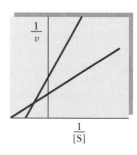

6. Liver alcohol dehydrogenase (ADH) is relatively nonspecific and will oxidize ethanol or other alcohols, including methanol. Methanol oxidation yields formaldehyde, which is quite toxic, causing, among other things, blindness. Mistaking it for the cheap wine he usually prefers, my dog Clancy ingested about 50 mL of windshield washer fluid (a solution 50% in methanol). Knowing that methanol would be excreted eventually by Clancy's kidneys if its oxidation could be blocked, and realizing that, in terms of methanol oxidation by ADH, ethanol would act as a competitive inhibitor, I decided to offer Clancy some wine. How much of Clancy's favorite vintage (12% ethanol) must he consume in order to lower the activity of his ADH on methanol to 5% of its normal value if the K_m values of canine ADH for ethanol and methanol are 1 millimolar and 10 millimolar, respectively? (The K_I for ethanol in its role as competitive inhibitor of methanol oxidation by ADH is 1 mM, the same as its K_m.) Both the methanol and ethanol will quickly distribute throughout Clancy's body fluids, which amount to about 15 L. Assume the densities of 50% methanol and the wine are both 0.9 g/mL.

7. List six general ways in which enzyme activity is controlled.

8. Why do you suppose proteolytic enzymes are often synthesized as inactive zymogens?

9. Draw a Lineweaver–Burk plot for the following: a Monod–Wyman–Changeux allosteric K enzyme system, showing separate curves for the kinetic response in (1) the absence of any effectors, (2) the presence of allosteric activator A, and (3) the presence of allosteric inhibitor I. Also draw a similar set of curves for a Monod–Wyman–Changeux allosteric V enzyme system.

10. In the Monod–Wyman–Changeux model for allosteric regulation, what values of L and relative affinities of R and T for A will lead activator A to exhibit positive homotropic effects? (That is, under what conditions will the binding of A enhance further A-binding, in the same manner that S-binding shows positive cooperativity?) Similarly, what values of L and relative affinities of R and T for I will lead inhibitor I to exhibit positive homotropic effects? (That is, under what conditions will the binding of I promote further I-binding?)

11. The cAMP formed by adenylyl cyclase (see Figure 10.40) does not persist because 5′-phosphodiesterase activity prevalent in cells hydrolyzes cAMP to give 5′-AMP. Caffeine inhibits 5′-phosphodiesterase activity. Describe the effects on glycogen phosphorylase activity that arise as a consequence of drinking lots of caffeinated coffee.

12. Enzymes have evolved such that their K_m values (or $K_{0.5}$ values) for substrate(s) are roughly equal to the *in vivo* concentration(s) of the substrate(s). Assume that glycogen phosphorylase is assayed at $[P_i] \approx K_{0.5}$ in the absence and presence of AMP or ATP. Estimate from Figure 10.36 the relative glycogen phosphorylase activity when (a) neither AMP nor ATP is present, (b) AMP is present, and (c) ATP is present.

FURTHER READING

Altman, S., 2000. The road to RNase P. *Nature Structural Biology* **7**:827–828.

Bell, J. E., and Bell, E. T., 1988. *Proteins and Enzymes.* Englewood Cliffs, N.J.: Prentice-Hall. This text describes the structural and functional characteristics of proteins and enzymes.

Caprara, M. G., and Nilsen, T. W., 2000. RNA: Versatility in form and function. *Nature Structural Biology* **7**:831–833.

Cate, J. H., et al., 1996. Crystal structure of a group I ribozyme domain: Principles of RNA packing. *Science* **273**:1678.

Cech, T. R., and Bass, B. L., 1986. Biological catalysis by RNA. *Annual Review of Biochemistry* **55**:599–629. A review of the early evidence that RNA can act like an enzyme.

Cech, T. R., et al., 1992. RNA catalysis by a group I ribozyme: Developing a model for transition-state stabilization. *Journal of Biological Chemistry* **267**:17479–17482.

Creighton, T. E., 1984. *Proteins: Structure and Molecular Properties.* New York: W. H. Freeman and Co. An advanced textbook on the structure and function of proteins.

Doherty, E. A., and Doudna, J. A., 2000. Ribozyme structures and mechanisms. *Annual Review of Biochemistry* **69**:597–615.

Fersht, A., 1985. *Enzyme Structure and Mechanism,* 2nd ed. Reading, Penn.: Freeman & Co. A monograph on the structure and action of enzymes.

Garrett, R. M., et al., 1998. Human sulfite oxidase R160Q: Identification of the mutation in a sulfite oxidase–deficient patient and expression and characterization of the mutant enzyme. *Proceedings of the National Academy of Sciences,* U.S.A. **95**:6394–6398.

Garrett, R. M., and Rajagopalan, K. V., 1996. Site-directed mutagenesis of recombinant sulfite oxidase. *Journal of Biological Chemistry* **271**:7387–7391.

Hsieh, L. C., Yonkovich, S., Kochersperger, L., and Schultz, P. G., 1993. Controlling chemical reactivity with antibodies. *Science* **260**:337–339.

International Union of Biochemistry and Molecular Biology Nomenclature Committee, 1992. *Enzyme Nomenclature.* New York: Academic Press. A reference volume and glossary on the official classification and nomenclature of enzymes. See also http://www.chem.qmw.ac.uk/iubmb/enzyme/newenz.html

Janda, K. D., 1997. Chemical selection for catalysis in combinatorial antibody libraries. *Science* **275**:945.

Janda, K. D., Shevlin, C. G., and Lerner, R. A., 1993. Antibody catalysis of a disfavored chemical transformation. *Science* **259**:490–493.

Johnson, L. N., and Barford, D., 1993. The effects of phosphorylation on the structure and function of proteins. *Annual Review of Biophysics and Biomolecular Structure* **22**:199–232. A review of protein phosphorylation and its role in regulation of enzymatic activity, with particular emphasis on glycogen phosphorylase.

Johnson, L. N., and Barford, D., 1994. Electrostatic effects in the control of glycogen phosphorylase by phosphorylation. *Protein Science* **3**:1726–1730. Discussion of the phosphate group's ability to deliver two negative charges to a protein, a property that no amino acid side chain can provide.

Kisker, C., et al., 1997. Molecular basis of sulfite oxidase deficiency from the structure of sulfite oxidase. *Cell* **91**:973–983.

Kling, J., 1998. From hypertension to angina to Viagra. *Modern Drug Discovery* **1**:31–38. The story of the serendipitous discovery of Viagra in a search for agents to treat angina and high blood pressure.

Koshland, D. E., Jr., Nemethy, G., and Filmer, D., 1966. Comparison of experimental binding data and theoretical models in proteins containing subunits. *Biochemistry* **5**:365–385. An alternative allosteric model based on ligand-induced conformational changes in proteins.

Landry, D. W., Zhao, K., Yang, G. X., et al., 1993. Antibody-catalyzed degradation of cocaine. *Science* **259**:1899–1901.

Lin, K., et al., 1996. Comparison of the activation triggers in yeast and muscle glycogen phosphorylase. *Science* **273**:1539–1541. Despite structural and regulatory differences between yeast and muscle glyogen phosphorylases, both are activated through changes in their intersubunit interface.

Lin, K., et al., 1997. Distinct phosphorylation signals converge at the catalytic center in glycogen phosphorylases. *Structure* **5**:1511–1523.

Monod, J., Wyman, J., and Changeux, J.-P., 1965. On the nature of allosteric transitions: A plausible model. *Journal of Molecular Biology* **12**:88–118. The classic paper that provided the first theoretical analysis of allosteric regulation.

Muth, G. W., et al., 2000. A single adenosine with a neutral pK_a in the ribosomal peptidyl transferase center. *Science* **289**:947–950.

Napper, A. D., Benkovic, S. J., Tramantano, A., and Lerner, R. A., 1987. A stereospecific cyclization catalyzed by an antibody. *Science* **237**:1041–1043.

Nissen, P., et al., 2000. The structural basis of ribosome activity in peptide bond synthesis. *Science* **289**:920–930. Peptide bond formation by the ribosome: The ribosome is a ribozyme. See also the paper by Muth et al.

Noller, H. F., Hoffarth, V., and Zimniak, L., 1992. Unusual resistance of peptidyl transferase to protein extraction procedures. *Science* **256**:1416–1419.

Piccirilli, J. A., McConnell, T. S., Zaug, A. J., et al., 1992. Aminoacyl esterase activity of *Tetrahymena* ribozyme. *Science* **256**:1420–1424.

Rath, V. L., et al., 1996. The evolution of an allosteric site in phosphorylase. *Structure* **4**:463–473.

Schachman, H. K., 1990. Can a simple model account for the allosteric transition of aspartate transcarbamoylase? *Journal of Biological Chemistry* **263**:18583–18586. Tests of the postulates of the allosteric models through experiments on aspartate transcarbamoylase, a key regulatory enzyme in pyrimidine biosynthesis.

Scott, W. G., and Klug, A., 1996. Ribozymes: Structure and mechanism in RNA catalysis. *Trends in Biochemical Sciences* **21**:220–224.

Scott, W. G., et al., 1996. Capturing the structure of a catalytic RNA intermediate: The hammerhead ribozyme. *Science* **274**:2065.

Segel, I. H., 1976. *Biochemical Calculations,* 2nd ed. New York: John Wiley & Sons. An excellent guide to solving problems in enzyme kinetics.

Silverman, R. B., 1988. *Mechanism-Based Enzyme Inactivation: Chemistry and Enzymology,* Vols. I and II. Boca Raton, Fla: CRC Press.

Smith, W. G., 1992. *In vivo* kinetics and the reversible Michaelis–Menten model. *Journal of Chemical Education* **12**:981–984.

Steitz, T. A., and Steitz, J. A., 1993. A general two-metal-ion mechanism for catalytic RNA. *Proceedings of the National Academy of Sciences,* U.S.A. **90**:6498–6502.

Wagner, J., Lerner, R. A., and Barbas III, C. F., 1995. Efficient adolase catalytic antibodies that use the enamine mechanism of natural enzymes. *Science* **270**:1797–1800. See also the discussion entitled "Aldolase antibody" in *Science* **270**:1737.

Watson, J. D., ed., 1987. Evolution of catalytic function. *Cold Spring Harbor Symposium on Quantitative Biology* **52:**1–955. Publications from a symposium on the nature and evolution of catalytic biomolecules (proteins and RNA) prompted by the discovery that RNA could act catalytically.

Wirsching, P., et al., 1995. Reactive immunization. *Science* **270:**1775–1783.

Description of reactive immunization, in which a highly reactive compound is used as an antigen; the antibody against it is an abzyme.

Zhang, B., and Cech, T. R., 1997. Peptide bond formation by *in vitro* selected ribozymes. *Nature* **390:**96–100.

Mechanisms of Enzyme Action

Like the workings of an ancient clock, the details of enzyme mechanisms are at once complex and simple. (David Parker/Science Photo Library/Photo Researchers, Inc.)

No single thing abides but all things flow.
Fragment to fragment clings and thus they grow
Until we know them by name.
Then by degrees they change and are no more the
  things we know.

LUCRETIUS (ca. 94 B.C.–50 B.C.)

Outline

Although the catalytic properties of enzymes may seem almost magical, it is simply chemistry—the breaking and making of bonds—that gives enzymes their prowess. This chapter will explore the unique features of this chemistry. The mechanisms of hundreds of enzymes have been studied in at least some detail. In this chapter, it will be possible to examine only a few of these. Nonetheless, the chemical principles that influence the mechanisms of these few enzymes are universal, and many other cases are understandable in light of the knowledge gained from these examples.

Although some RNA molecules act as enzymes, nearly all the enzymes in our world are proteins. It is remarkable that simple proteins, composed of only 20 different amino acids, can catalyze and accelerate many different chemical reactions. At the same time, some enzymes rely on the unique catalytic features of cofactors or coenzymes that bind to the enzyme and contribute to catalysis in some way (see Chapter 10).

11.1 The Basic Principle—Stabilization of the Transition State

In all chemical reactions, the reacting atoms or molecules pass through a state that is intermediate in structure between the reactant(s) and the product(s). Consider the transfer of a proton from a water molecule to a chloride anion:

$$H\text{—}O\text{—}H + Cl^- \rightleftharpoons HO^{\delta-} \cdots H \cdots Cl^{\delta-} \rightleftharpoons HO^- + H\text{—}Cl$$

Reactants　　　　　**Transition state**　　　　　**Products**

In the middle structure, the proton undergoing transfer is shared equally by the hydroxyl and chloride anions. This structure represents, as nearly as possible, the transition between the reactants and products, and it is known as the **transition state.** For any reaction, the transition state is a very high energy state that is very unstable. It exists only for a brief instant, typically 10^{-13} sec.[1]

Chemical reactions in which a substrate (S) is converted to a product (P) can be pictured as involving a transition state (which we henceforth denote as $X^{\ddagger}$), a species intermediate in structure between S and P (Figure 11.1). As seen in Chapter 10, the catalytic role of an enzyme is to reduce the energy barrier between substrate and transition state. This must be accomplished via formation of an **enzyme–substrate complex** (ES), which then is converted to product by passing through a transition state, $EX^{\ddagger}$ (see Figure 11.1). As shown, the energy of $EX^{\ddagger}$ is clearly lower than $X^{\ddagger}$, but there is more to the story.

The energy barrier for the uncatalyzed reaction (again see Figure 11.1) is of course the difference in energies of the S and $X^{\ddagger}$ states. Similarly, the energy barrier to be surmounted in the enzyme-catalyzed reaction, assuming that E is saturated with S, is the energy difference between ES and $EX^{\ddagger}$. Reaction rate acceleration by an enzyme means simply that the energy barrier between ES and $EX^{\ddagger}$ is less than the energy barrier between S and $X^{\ddagger}$. In terms of the free energies of activation, $\Delta G_e^{\ddagger} < \Delta G_u^{\ddagger}$.

There are important consequences for this statement. The enzyme must stabilize the transition-state complex, $EX^{\ddagger}$, more than it stabilizes the substrate complex, ES. Put another way, *enzymes are "designed" by nature to bind the transition-state structure. A corollary: Neither the substrate nor the product binds as well to the active site as does the transition state.*

11.2 Enzymes Provide Enormous Rate Accelerations

Enzymes are powerful catalysts. Enzyme-catalyzed reactions are typically 10^7 to 10^{14} times faster than their uncatalyzed counterparts (Table 11.1). (Note the remarkable rate acceleration of $>10^{16}$ for the alkaline phosphatase–catalyzed hydrolysis of methylphosphate!)

These large rate accelerations correspond to substantial changes in the free energy of activation for the reaction in question. The urease reaction, for example,

$$H_2N\text{—}\overset{\displaystyle O}{\overset{\|}{C}}\text{—}NH_2 + 2\,H_2O + H^+ \longrightarrow 2\,NH_4^+ + HCO_3^-$$

has an energy of activation some 84 kJ/mol less than the corresponding uncatalyzed reaction. To fully understand any enzyme reaction, it is important to

---

[1]It is important here to distinguish **transition states** from **intermediates.** A transition state is envisioned as an extreme distortion of a bond, and thus the lifetime of a typical transition state is viewed as being on the order of the lifetime of a bond vibration, typically 10^{-13} sec. Intermediates, on the other hand, are longer-lived, with lifetimes in the range of 10^{-13} sec to 10^{-3} sec.

(a)

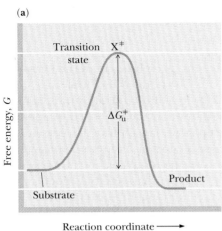

(b)

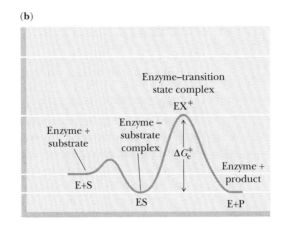

Figure 11.1 Enzymes catalyze reactions by lowering the activation energy. Here the free energy of activation for **(a)** the uncatalyzed reaction, $\Delta G_u^{\ddagger}$, is larger than that for **(b)** the enzyme-catalyzed reaction, $\Delta G_e^{\ddagger}$.

account for the rate acceleration in terms of the structure of the enzyme and its mechanism of action. There are a limited number of catalytic mechanisms or factors that contribute to the remarkable performance of enzymes. These include the following:

1. Entropy loss in the formation of ES
2. Destabilization of ES due to strain, desolvation, or electrostatic effects
3. Covalent catalysis
4. General acid or base catalysis
5. Metal ion catalysis
6. Proximity and orientation

Any or all of these mechanisms may contribute to the net rate acceleration of an enzyme-catalyzed reaction relative to the uncatalyzed reaction. A thorough understanding of any enzyme would require that the net acceleration be accounted for in terms of contributions from one or (usually) more of these mechanisms.

Table 11.1 A Comparison of Enzyme-Catalyzed Reactions and Their Uncatalyzed Counterparts

Reaction	Enzyme	Uncatalyzed Rate, v_u (sec^{-1})	Catalyzed Rate, v_e (sec^{-1})	v_e/v_u
$CH_3-O-PO_3^{2-} + H_2O \longrightarrow CH_3OH + HPO_4^{2-}$	Alkaline phosphatase	1×10^{-15}	14	1.4×10^{16}
$H_2N-\overset{\overset{O}{\|}}{C}-NH_2 + 2\,H_2O + H^+ \longrightarrow 2\,NH_4^+ + HCO_3^-$	Urease	3×10^{-10}	3×10^4	1×10^{14}
$R-\overset{\overset{O}{\|}}{C}-O-CH_2CH_3 + H_2O \longrightarrow RCOOH + HOCH_2CH_3$	Chymotrypsin	1×10^{-10}	1×10^2	1×10^{12}
Glycogen + $P_i \longrightarrow$ Glycogen + Glucose-1-P $\quad(n)\qquad\qquad\quad(n-1)$	Glycogen phosphorylase	$<5 \times 10^{-15}$	1.6×10^{-3}	$>3.2 \times 10^{11}$
Glucose + ATP $\longrightarrow$ Glucose-6-P + ADP	Hexokinase	$<1 \times 10^{-13}$	1.3×10^{-3}	$>1.3 \times 10^{10}$
$CH_3CH_2OH + NAD^+ \longrightarrow CH_3\overset{\overset{O}{\|}}{C}H + NADH + H^+$	Alcohol dehydrogenase	$<6 \times 10^{-12}$	2.7×10^{-5}	$>4.5 \times 10^6$
$CO_2 + H_2O \longrightarrow HCO_3^- + H^+$	Carbonic anhydrase	10^{-2}	10^5	1×10^7
Creatine + ATP $\longrightarrow$ Cr-P + ADP	Creatine kinase	$<3 \times 10^{-9}$	4×10^{-5}	$>1.33 \times 10^4$

Adapted from Koshland, D., 1956. *Journal of Cellular Comparative Physiology*, Supp. 1 **47**:217.

11.3 Entropy Loss in the Formation of ES

Bringing the substrate and the catalyst together results in a significant loss of entropy. On the other hand, enzyme reactions take place within the enzyme's active site. Here, all the catalytic groups are part of the same "molecule" (the enzyme–substrate complex), and there is no loss of rotational or translational entropy as the complex proceeds to the transition state. Looked at another way, the rotational and translational entropies of the substrate have been lost already during formation of the ES complex (Figure 11.2). Loss of entropy corresponds to a negative value for ΔS and a corresponding increase in ΔG, which would raise the energy of a state (ES, for example). However, this increase in the energy of ES is compensated by favorable binding interactions (thus negative values for ΔH) in the formation of ES.

In summary, as the uncatalyzed reaction proceeds from S to $X^{\ddagger}$, there is a loss of entropy which raises the energy of $X^{\ddagger}$. In the enzyme-catalyzed reaction, there is no loss of entropy as the reaction proceeds from ES to $EX^{\ddagger}$. Loss of entropy occurs in going from E + S to ES, but some of this loss is compensated by favorable binding interactions.

11.4 Destabilization of ES Due to Strain, Desolvation, and Electrostatic Effects

If the enzyme active site is a perfect mate and fit for the transition state, the substrate must bind less well to the enzyme than the transition state. The binding interactions between enzyme and transition state are perfect, or nearly so, but the binding interactions between the enzyme and the susbstrate are less than perfect. Put another way, the ES complex is destabilized, at least to some extent, by unfavorable binding interactions. Destabilization of the ES complex can involve **structural strain, desolvation,** or **electrostatic effects.** When the substrate binds, the imperfect nature of the "fit" results in distortion or strain in the substrate, the enzyme, or both. This means that the amino acid residues that make up the active site are oriented to coordinate the transition-state structure precisely but will interact with the substrate or product less effectively.

Destabilization may also involve desolvation of charged groups on the substrate upon binding in the active site. Charged groups are highly stabilized in water. For example, the transfer of Na^+ and Cl^- from the gas phase to aqueous solution is characterized by an **enthalpy of solvation, ΔH_{solv},** of -775 kJ/mol. (Energy is given off and the ions become more stable.) When

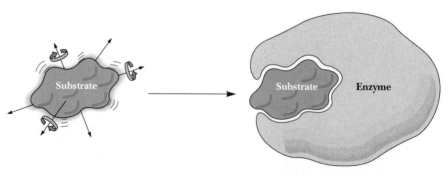

Figure 11.2 Formation of the ES complex results in a loss of entropy. Prior to binding, E and S are free to undergo translational and rotational motion. By comparison, the ES complex is a more highly ordered, low-entropy complex.

Substrate (and enzyme) are free to undergo translational motion. A disordered, high-entropy situation

The highly ordered, low-entropy complex

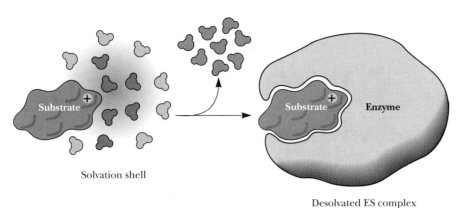

Figure 11.3 Substrates typically lose waters of hydration in the formation of the ES complex. Desolvation raises the energy of the ES complex, making it more reactive.

charged groups on a substrate move from water into an enzyme active site (Figure 11.3), they are often desolvated to some extent, becoming less stable and therefore more reactive.

When a substrate enters the active site, charged groups may be forced to interact (unfavorably) with charges of like signs, resulting in **electrostatic destabilization** (Figure 11.4). The reaction pathway acts in part to remove this stress. If the charge on the substrate is diminished or lost in the course of reaction, electrostatic destabilization can result in rate acceleration.

11.5 Transition-State Analogs Bind Very Tightly to the Active Site

Substrates typically bind to enzymes with dissociation constants that range from $10^{-6}M$ to $10^{-3}M$. How tightly do transition states bind to enzyme active sites? Since transition states are so unstable, transition-state dissociation constants cannot be measured directly, but they are estimated to be in the range of $10^{-15}M$ or even lower. Such a low value corresponds to very tight binding of the transition state by the enzyme.

It is unlikely that such tight binding in an enzyme transition state will ever be measured experimentally, however, because the transition state itself is a "moving target." It exists only for about 10^{-14} to 10^{-13} sec, less than the time required for a bond vibration. The nature of the elusive transition state can be explored, on the other hand, using **transition-state analogs,** stable molecules that are chemically and structurally similar to the transition state. Such molecules should bind more strongly than a substrate and more strongly than

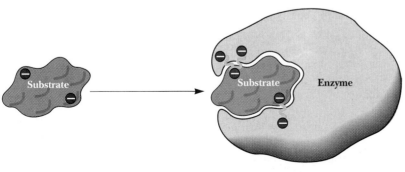

Electrostatic destabilization
in ES complex

Figure 11.4 Electrostatic destabilization of a substrate may arise from juxtaposition of like charges in the active site. If such charge repulsion is relieved in the course of the reaction, electrostatic destabilization can result in a rate increase.

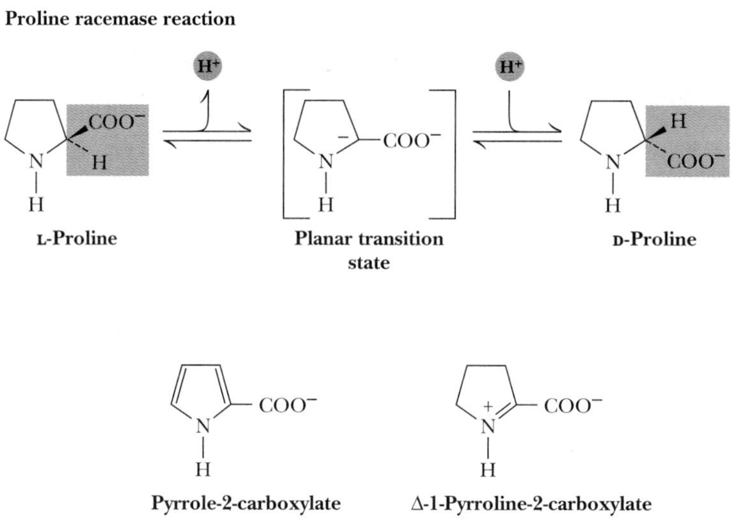

Proline racemase reaction

Figure 11.5 The proline racemase reaction. Pyrrole-2-carboxylate and Δ-1-pyrroline-2-carboxylate mimic the planar transition state of the reaction.

competitive inhibitors that bear no significant similarity to the transition state. Hundreds of examples of such behavior have been reported. For example, Robert Abeles studied a series of inhibitors of **proline racemase** (Figure 11.5) and found that pyrrole-2-carboxylate bound to the enzyme 160 times more tightly than L-proline, the normal substrate. This analog binds so tightly because it is planar and is similar in structure to the planar transition state for the racemization of proline. Two other examples of transition-state analogs are shown in Figure 11.6. Phosphoglycolohydroxamate binds 40,000 times more tightly to yeast aldolase than the substrate dihydroxyacetone phosphate. Even more remarkable, the 1,6-hydrate of purine ribonucleoside has been estimated to bind to adenosine deaminase with a K_i of $3 \times 10^{-13} M$!

It should be noted that transition-state analogs are only approximations of the transition state itself and will never bind as tightly as would be expected for the true transition state. These analogs are, after all, stable molecules and cannot be expected to resemble a true transition state too closely.

11.6 Covalent Catalysis

Some enzyme reactions derive much of their rate acceleration from the formation of **covalent bonds** between enzyme and substrate. Consider the reaction

$$BX + Y \longrightarrow BY + X$$

and an enzymatic version of this reaction involving formation of a **covalent intermediate**:

$$BX + Enz \longrightarrow E - B + X + Y \longrightarrow Enz + BY$$

If the enzyme-catalyzed reaction is to be faster than the uncatalyzed case, the acceptor group on the enzyme must be a better attacking group than Y and a better leaving group than X. Enzymes that carry out covalent catalysis have ping-pong kinetic mechanisms.

(a) **Yeast aldolase reaction**

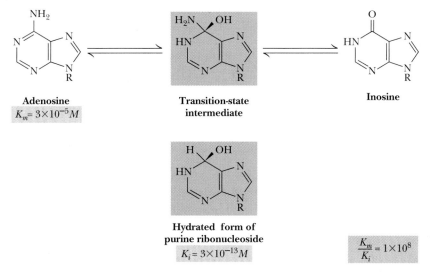

Phosphoglycolohydroxamate

$K_i = 1 \times 10^{-8} M$

$\dfrac{K_m}{K_i} = 4 \times 10^4$

(b) **Calf intestinal adenosine deaminase reaction**

Adenosine
$K_m = 3 \times 10^{-5} M$

Transition-state intermediate

Inosine

Hydrated form of purine ribonucleoside
$K_i = 3 \times 10^{-13} M$

$\dfrac{K_m}{K_i} = 1 \times 10^8$

Figure 11.6 **(a)** Phosphoglycolohydroxamate is an analog of the enediolate transition state of the yeast aldolase reaction. **(b)** Purine riboside, a potent inhibitor of the calf intestinal adenosine deaminase reaction, binds to adenosine deaminase as the 1,6-hydrate. The hydrated form of purine riboside is an analog of the proposed transition state for the reaction.

The side chains of amino acids in proteins offer a variety of **nucleophilic** centers for catalysis, including amines, carboxylates, aryl and alkyl hydroxyls, imidazoles, and thiol groups. These groups readily attack electrophilic centers of substrates, forming covalently bonded enzyme–substrate intermediates. Typical electrophilic centers in substrates include phosphoryl groups, acyl groups, and glycosyl groups (Figure 11.7). The covalent intermediates thus formed can be attacked in a subsequent step by a water molecule or a second substrate, giving the desired product. **Covalent electrophilic catalysis** is also observed but usually involves coenzyme adducts that generate electrophilic centers. Well over

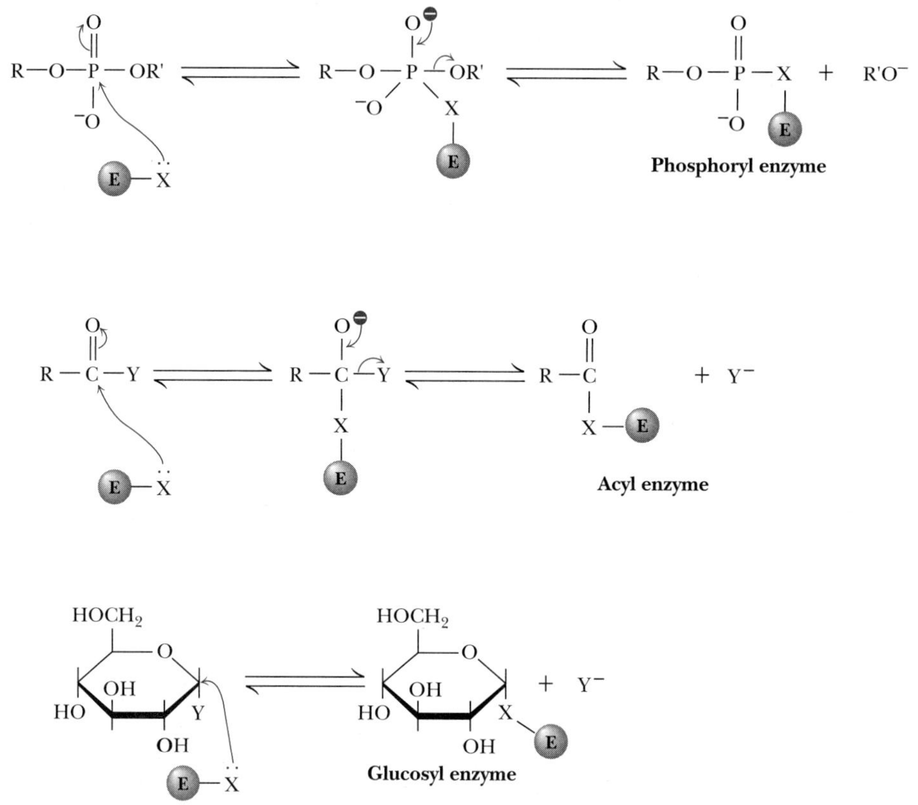

Figure 11.7 Examples of covalent bond formation between enzyme and substrate. In each case, a nucleophilic center (X:) on an enzyme attacks an electrophilic center on a substrate.

Phosphoryl enzyme

Acyl enzyme

Glucosyl enzyme

100 enzymes are now known to form covalent intermediates during catalysis. Table 11.2 lists some typical examples, including that of glyceraldehyde-3-phosphate dehydrogenase, which catalyzes the reaction:

Glyceraldehyde-3-P + NAD$^+$ + P$_I$ ⟶ 1,3-bisphosphoglycerate + NADH + H$^+$

As shown in Figure 11.8, this reaction mechanism involves nucleophilic attack by —SH on the substrate glyceraldehyde-3-P to form a covalent acylcysteine (or hemithioacetal) intermediate. Hydride transfer to NAD$^+$ generates a thioester intermediate. Nucleophilic attack by phosphate yields the desired mixed carboxylic–phosphoric anhydride product, 1,3-bisphosphoglycerate. Several examples of covalent catalysis will be discussed in detail in later chapters.

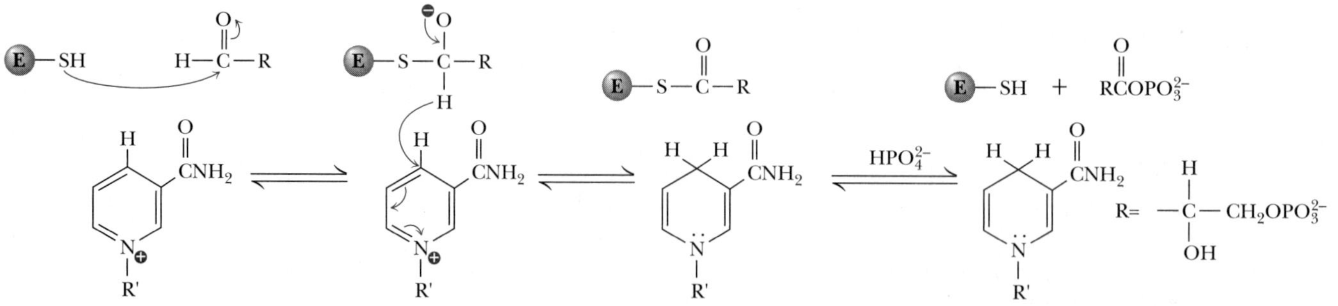

Figure 11.8 Formation of a covalent intermediate in the glyceraldehyde-3-phosphate dehydrogenase reaction. Nucleophilic attack by a cysteine—SH group forms a covalent acylcysteine intermediate. Following hydride transfer to NAD$^+$, nucleophilic attack by phosphate yields the product, 1,3-bisphosphoglycerate.

Table 11.2 **Enzymes That Form Covalent Intermediates**

Enzymes	Reacting Group	Covalent Intermediate

1. Chymotrypsin
   Elastase
   Esterases
   Subtilisin
   Thrombin
   Trypsin
	(Ser)	(Acyl-Ser)

2. Glyceraldehyde-3-phosphate
     dehydrogenase
   Papain
	(Cys)	(Acyl-Cys)

3. Alkaline phosphatase Phosphoglucomutase	(Ser)	(Phosphoserine)
4. Phosphoglycerate mutase Succinyl-CoA synthetase	(His)	(Phosphohistidine)

5. Aldolase
   Decarboxylases
   Pyridoxal phosphate–dependent
     enzymes
	$R—NH_3^+$
	(Amine)
		(Schiff base)

11.7 General Acid–Base Catalysis

Nearly all enzyme reactions involve some degree of acid or base catalysis. There are two types of acid–base catalysis: (1) **specific acid–base catalysis,** in which H^+ or OH^- accelerates the reaction, and (2) **general acid–base catalysis,** in which an acid or base other than H^+ or OH^- accelerates the reaction. For ordinary solution reactions, these two cases can be distinguished on the basis of simple experiments. In specific acid or base catalysis, the buffer concentration has no effect, as shown in Figure 11.9a. In general acid or base catalysis, however, the buffer may donate or accept a proton in the transition state and thus affect the rate (Figure 11.9b). By definition, general acid–base catalysis is catalysis in which a proton is transferred in the transition state. Consider the hydrolysis of *p*-nitrophenylacetate with imidazole acting as a general base (Figure 11.10). Proton transfer apparently stabilizes the transition state here. The water has been made more nucleophilic without generation of a high concentration of OH^- or without the formation of unstable, high-energy species. General acid or general base catalysis may increase reaction rates 10- to 100-

(a)

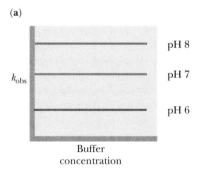

(b)

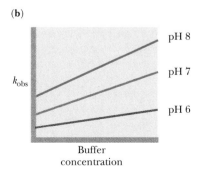

Figure 11.9 Specific and general acid–base catalysis of simple reactions in solution may be distinguished by determining the dependence of observed reaction rate constants (k_{obs}) on pH and buffer concentration. **(a)** In specific acid–base catalysis, H^+ or OH^- concentration affects the reaction rate, k_{obs} is pH-dependent, but buffers (which accept or donate H^+/OH^-) have no effect. **(b)** In general acid–base catalysis, in which an ionizable buffer may donate or accept a proton in the transition state, k_{obs} is dependent on buffer concentration.

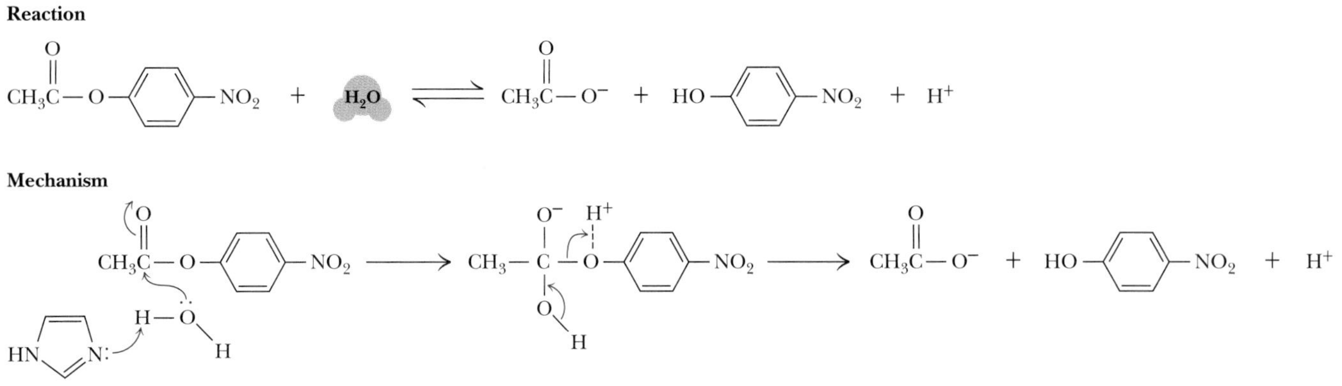

Reaction

$$CH_3\overset{\overset{O}{\|}}{C}-O-\!\!\!\langle\quad\rangle\!\!\!-NO_2 \;+\; H_2O \;\rightleftharpoons\; CH_3\overset{\overset{O}{\|}}{C}-O^- \;+\; HO-\!\!\!\langle\quad\rangle\!\!\!-NO_2 \;+\; H^+$$

Mechanism

$$CH_3\overset{\overset{O}{\|}}{C}-O-\!\!\!\langle\quad\rangle\!\!\!-NO_2 \longrightarrow CH_3-\overset{O^-}{\underset{\underset{H}{\overset{|}{O}}}{\overset{|}{C}}}-O-\!\!\!\langle\quad\rangle\!\!\!-NO_2 \longrightarrow CH_3\overset{\overset{O}{\|}}{C}-O^- \;+\; HO-\!\!\!\langle\quad\rangle\!\!\!-NO_2 \;+\; H^+$$

Figure 11.10 Catalysis of *p*-nitrophenylacetate hydrolysis by imidazole—an example of general base catalysis. Proton transfer to imidazole in the transition state facilitates hydroxyl attack on the substrate carbonyl carbon.

fold. In an enzyme, ionizable groups on the protein provide the H^+ transferred in the transition state. Clearly, an ionizable group will be most effective as an H^+ transferring agent at or near its pK_a. Because the pK_a of the histidine side chain is near 7, histidine is often the most effective general acid or base. Descriptions of several cases of general acid–base catalysis in typical enzymes follow.

11.8 Metal Ion Catalysis

Many enzymes require metal ions for maximal activity. If the enzyme binds the metal very tightly or requires the metal ion to maintain its stable, native state, it is referred to as a **metalloenzyme**. Enzymes that bind metal ions more weakly, perhaps only during the catalytic cycle, are referred to as **metal activated**. One role for metals in both metal-activated enzymes and metalloenzymes is to act as electrophilic catalysts, stabilizing the increased electron density or negative charge that can develop during reactions. Among the enzymes that function in this manner is liver alcohol dehydrogenase (Figure 11.11). Another potential function of metal ions is to provide a powerful nucleophile at neutral pH. Coordination to a metal ion can increase the acidity of a nucleophile with an ionizable proton:

$$M^{2+} + NucH \longrightarrow M^{2+}(NucH) \longrightarrow M^{2+}(Nuc^-) + H^+$$

The reactivity of the coordinated, deprotonated nucleophile is typically intermediate between that of the un-ionized and ionized forms of the nucleophile. Carboxypeptidase (Chapter 5) contains an active site Zn^{2+}, which facilitates deprotonation of a water molecule in this manner.

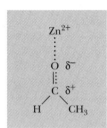

Figure 11.11 Liver alcohol dehydrogenase catalyzes the transfer of a hydride ion ($H:^-$) from NADH to acetaldehyde (CH_3CHO), forming ethanol (CH_3CH_2OH). An active-site zinc ion stabilizes negative charge development on the oxygen atom of acetaldehyde, leading to an induced partial positive charge on the carbonyl C atom. Transfer of the negatively charged hydride ion to this carbon forms ethanol.

11.9 Proximity

Chemical reactions go faster when the reactants are in proximity, that is, near each other. In solution or in the gas phase, this means that increasing the concentrations of reacting molecules, which raises the number of collisions, causes higher rates of reaction. Enzymes, which have specific binding sites for particular reacting molecules, essentially take the reactants out of dilute solution and hold them close to each other. This proximity of reactants is said to raise the "effective" concentration over that of the substrates in solution, leading to an increased reaction rate. There is more to this story, however. Enzymes not only bring substrates and catalytic groups close together but orient them in a manner suitable for catalysis as well. Comparison of the rates of reaction of the molecules shown in Figure 11.12 makes it clear that the bulky methyl groups force an orientation on the alkyl carboxylate and the aromatic hydroxyl groups that makes them approximately 250 billion times more likely to react. Enzymes function similarly by placing catalytically functional groups (from the protein side chains or from another substrate) in the proper position for reaction.

11.10 Typical Enzyme Mechanisms

The balance of this chapter will be devoted to several classic and representative enzyme mechanisms. These particular cases are well understood because the three-dimensional structures of the enzymes and the bound substrates are known at atomic resolution and because great efforts have been devoted to kinetic and mechanistic studies. They are important because they represent reaction types that appear again and again in living systems and because they demonstrate many of the catalytic principles cited previously. Enzymes are the catalytic machines that sustain life, and what follows is an intimate look at the inner workings of the machinery.

11.11 Serine Proteases

Serine proteases are a class of proteolytic enzymes whose catalytic mechanism is based on an active-site serine residue. Serine proteases are one of the best-characterized families of enzymes. This family includes trypsin, chymotrypsin, elastase, thrombin, subtilisin, plasmin, tissue plasminogen activator, and other

Reaction		Rate const. $(M^{-1}\text{sec}^{-1})$	Ratio
		$5.9\text{x}10^{-6}$	
		$1.5\text{x}10^{6}$	$2.5\text{x}10^{11}$

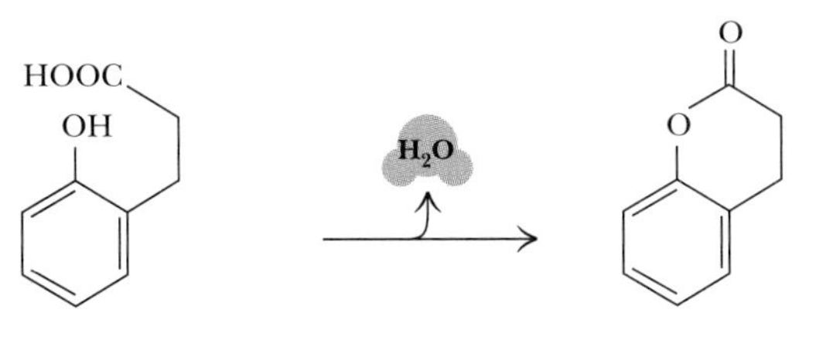

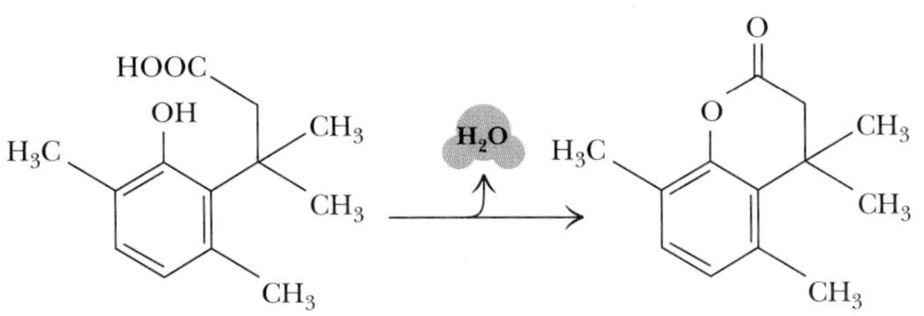

Figure 11.12 Orientation effects in intramolecular reactions can be dramatic. Steric crowding by methyl groups provides a rate acceleration of 2.5×10^{11} for the lower reaction compared to the upper reaction. (*Adapted from Milstien, S., and Cohen, L. A., 1972. Stereopopulation control I. Rate enhancements in the lactonization of o-hydroxyhydrocinnamic acid.* Journal of the American Chemical Society **94**:9158–9165.)

related enzymes. The first three of these are digestive enzymes and are synthesized in the pancreas and secreted into the digestive tract as inactive **proenzymes,** or **zymogens.** Within the digestive tract, the zymogen is converted into the active enzyme form upon removal of a portion of the peptide chain. Thrombin is a crucial enzyme in the blood-clotting cascade, subtilisin is a bacterial protease, and plasmin breaks down the fibrin polymers of blood clots. Tissue plasminogen activator (tPA) specifically cleaves the proenzyme plasminogen, yielding plasmin. Owing to its ability to stimulate breakdown of blood clots, tPA can minimize the harmful consequences of a heart attack if administered to a patient within 30 minutes of onset. Finally, although not itself a protease, acetylcholinesterase is a serine esterase related mechanistically to the serine proteases. It degrades the neurotransmitter acetylcholine in the synaptic cleft between neurons.

The Digestive Serine Proteases

Trypsin, chymotrypsin, and elastase all carry out the same reaction—the cleavage of a peptide chain—and although their structures and mechanisms are quite similar, they display very different specificities. Trypsin cleaves peptides on the carbonyl side of the basic amino acids arginine or lysine (see Table 4.4). Chymotrypsin prefers to cleave on the carbonyl side of aromatic residues, such as phenylalanine and tyrosine. Elastase is not as specific as the other two; it mainly cleaves peptides on the carbonyl side of small, neutral residues. These three enzymes all possess molecular weights in the range of 25,000, and all have similar sequences (Figure 11.13) and three-dimensional structures. The

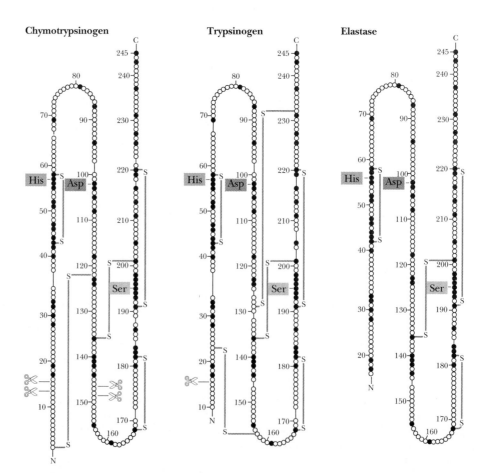

Figure 11.13 Comparison of the amino acid sequences of chymotrypsinogen, trypsinogen, and elastase. Each circle represents one amino acid. Numbering is based on the sequence of chymotrypsinogen. Filled circles indicate residues that are identical in all three proteins. Disulfide bonds are indicated in yellow. The positions of the three catalytically important active-site residues (His[57], Asp[102], and Ser[195]) are indicated.

Figure 11.14 Structure of chymotrypsin (white) in a complex with eglin C (blue ribbon structure), a target protein. The residues of the catalytic triad (His[56], Asp[102], and Ser[195]) are highlighted. His[57] (blue) is flanked above by Asp[102] (red) and on the right by Ser[195] (yellow). The catalytic site is filled by a peptide segment of eglin. Note how close Ser[195] is to the peptide that would be cleaved in a chymotrypsin reaction.

 See *Interactive Biochemistry CD-ROM and Workbook,* pages 37, 140

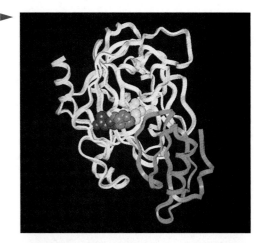

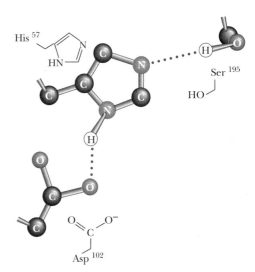

structure of chymotrypsin is typical (Figure 11.14). The molecule is ellipsoidal in shape and contains an α-helix at the C-terminal end (residues 230 to 245) and several β-sheet domains. Most of the aromatic and hydrophobic residues are buried in the interior of the protein, and most of the charged or hydrophilic residues are on the surface. Three polar residues — His[57], Asp[102], and Ser[195] — form what is known as a **catalytic triad** at the active site (Figure 11.15). These three residues are conserved in trypsin and elastase as well. The active site is actually a depression on the surface of the enzyme, with a small pocket that the enzyme uses to identify the residue for which it is specific (Figure 11.16). Chymotrypsin, for example, has a pocket surrounded by hydrophobic residues and large enough to accommodate an aromatic side chain. The pocket in trypsin has a negative charge (Asp[189]) at its bottom, facilitating the binding of positively charged arginine and lysine residues. Elastase, on the other hand, has a shallow pocket with bulky threonine and valine residues at the opening. Only small, nonbulky residues can be accommodated in its pocket. The backbone of the peptide substrate is hydrogen-bonded in antiparallel fashion to residues 215 to 219 and bent so that the peptide bond to be cleaved is bound close to His[57] and Ser[195].

The Chymotrypsin Mechanism in Detail: Kinetics

Much of what is known about the chymotrypsin mechanism is based on studies of the hydrolysis of artificial substrates—simple organic esters, such as *p*-nitrophenylacetate, and methyl esters of amino acid analogs, such as formylphenylalanine methyl ester and acetylphenylalanine methyl ester (Figure 11.17). *p*-Nitrophenylacetate is an especially useful model substrate because the nitrophenolate product is easily observed, owing to its strong absorbance at 400 nm. When large amounts of chymotrypsin are used in kinetic studies with *p*-nitrophenylacetate, a **rapid initial burst** of *p*-nitrophenolate is released (in an amount approximately equal to the enzyme concentration), followed by a much slower, linear rate of nitrophenolate release (Figure 11.18).

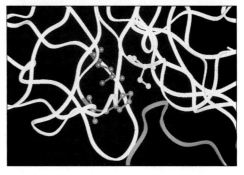

Figure 11.15 The catalytic triad of chymotrypsin.

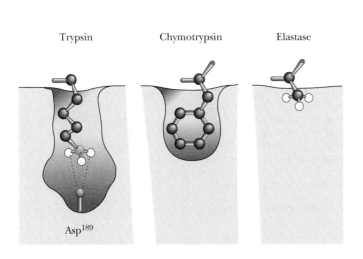

◄ Figure 11.16 The substrate-binding pockets of trypsin, chymotrypsin, and elastase. (*Irving Geis*)

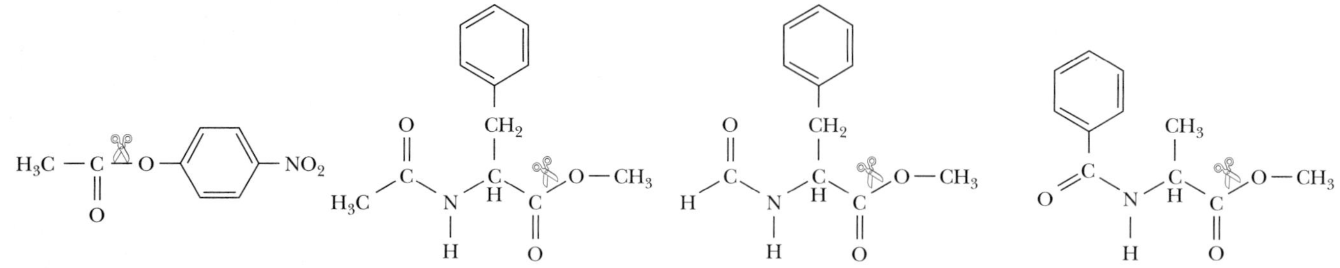

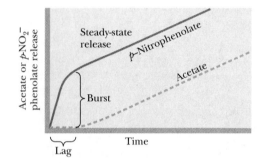

p-Nitrophenylacetate

Acetylphenylalanine methyl ester

Formylphenylalanine methyl ester

Benzoylalanine methyl ester

Figure 11.17 Artificial substrates used in studies of the mechanism of chymotrypsin.

Figure 11.18 Burst kinetics observed in the chymotrypsin reaction. A burst of nitrophenolate production is followed by a slower, steady-state release. After an initial lag period, acetate release is also observed. This kinetic pattern is consistent with rapid formation of an acyl-enzyme intermediate (and the burst of nitrophenolate). The slower, steady-state release of products corresponds to rate-limiting breakdown of the acyl-enzyme intermediate.

Observation of a burst, followed by slower, steady-state product release, is strong evidence for a multistep mechanism, with a fast first step and a slower second step.

In the chymotrypsin mechanism, the nitrophenylacetate combines with the enzyme to form an ES complex. This is followed by a rapid second step in which an **acyl-enzyme intermediate** is formed, with the acetyl group covalently bound to the very reactive Ser^{195}. The nitrophenyl moiety is released as nitrophenolate (Figure 11.19), accounting for the burst of nitrophenolate product. Attack of a water molecule on the acyl-enzyme intermediate yields acetate as the second product in a subsequent, slower step. The enzyme is now free to bind another molecule of p-nitrophenylacetate, and the p-nitrophenolate product produced at this point corresponds to the slower, steady-state formation of product in the upper right portion of Figure 11.18. In this mechanism, the release of acetate is the **rate-limiting step** and accounts for the observation of **burst kinetics**—the pattern shown in Figure 11.18.

Serine proteases like chymotrypsin are susceptible to inhibition by **organic fluorophosphates,** such as diisopropylfluorophosphate (DIFP; Figure 11.20). DIFP reacts rapidly with active-site serine residues, such as Ser^{195} of chymotrypsin and the other serine proteases (but not with any of the other serines in these proteins), to form a DIP(diisopropylphosphoryl)-enzyme. This covalent enzyme–inhibitor complex is extremely stable, and chymotrypsin is thus permanently inactivated by DIFP.

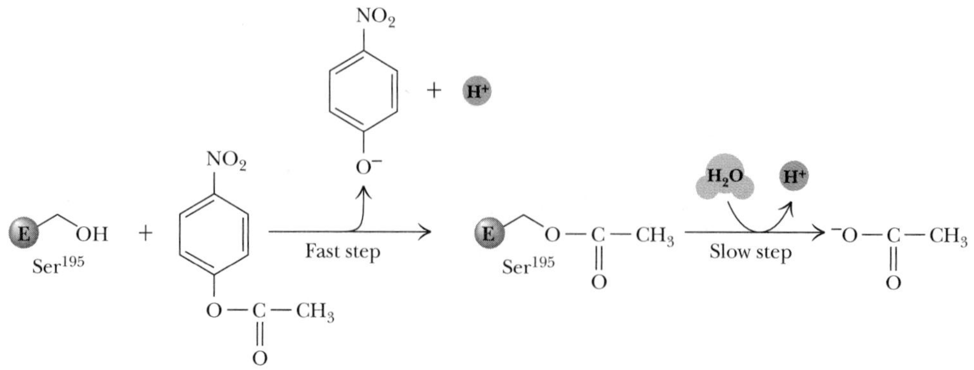

Figure 11.19 Rapid formation of the acyl-enzyme intermediate is followed by slower product release.

HUMAN BIOCHEMISTRY

Cholinesterase Inhibitors as Deadly Chemical Weapons

Neurons in the brain communicate with each other via chemical signals called neurotransmitters. Acetylcholinesterase is a serine protease that degrades one of these neurotransmitters, acetylcholine. Inhibition of acetylcholinesterase by organophosphates like DIFP (see Figure 11.20) makes these agents highly toxic, and they have been prepared and stockpiled as potential chemical weapons by many countries and terrorist groups, though they have not been widely used. An attack with the agent sarin (see figure) on a Tokyo subway in 1995 killed 12 people and injured 5000. Officials in Iraq have admitted to United Nations authorities that they have prepared and stored several tons of deadly VX gas (see figure), which is 10 times more toxic than sarin. In the popular movie *The Rock*, starring Sean Connery and Nicolas Cage, a deranged army officer steals vials of VX and holds tourists hostage on Alcatraz Island, threatening to launch missiles containing VX into the city of San Francisco. One of the most dangerous chemicals ever created, VX was developed in Pot Down, Wiltshire, England, in 1952. The British traded VX technology to the United States for information on thermonuclear weapons.

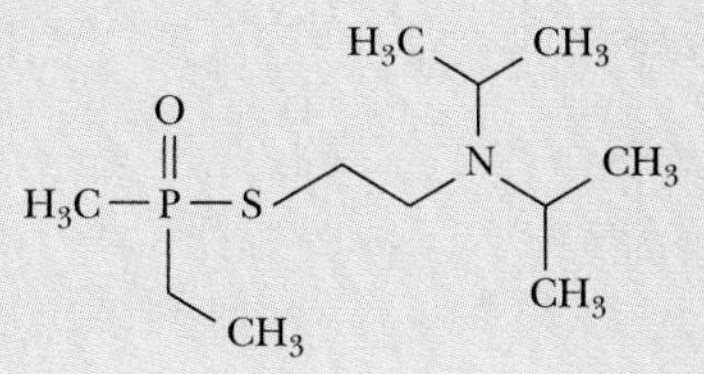

VX

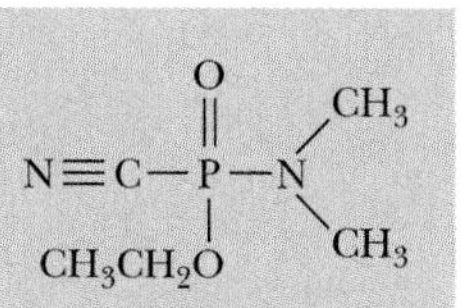

Tabun

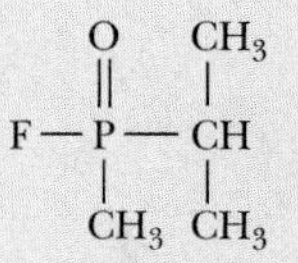

Sarin

The Serine Protease Mechanism in Detail: Events at the Active Site

A likely mechanism for peptide hydrolysis is shown in Figure 11.21. As the backbone of the substrate peptide binds adjacent to the catalytic triad, the specific side chain fits into its pocket. Asp^{102} of the catalytic triad positions His^{57} and immobilizes it through a hydrogen bond as shown. In the first step of the reaction, His^{57} acts as a general base to withdraw a proton from Ser^{195}, facilitating nucleophilic attack by Ser^{195} on the carbonyl carbon of the peptide bond to be cleaved. This is probably a concerted step, because proton transfer prior to Ser^{195} attack on the acyl carbon would leave a relatively unstable negative charge on the serine oxygen. In the next step, donation of a proton from His^{57} to the peptide's amide nitrogen creates a protonated amine on the covalent

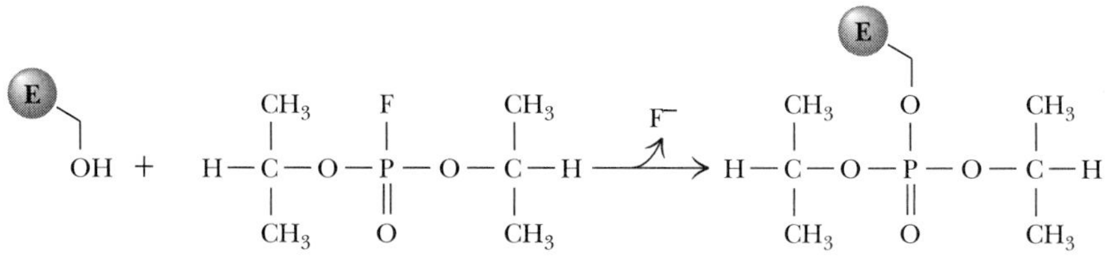

Diisopropylfluorophosphate

Diisopropylphosphoryl
derivative of chymotrypsin

Figure 11.20 Diisopropylfluorophosphate (DIFP) reacts with active-site serine residues of serine proteases and serine esterases (such as acetylcholinesterase), causing permanent inactivation.

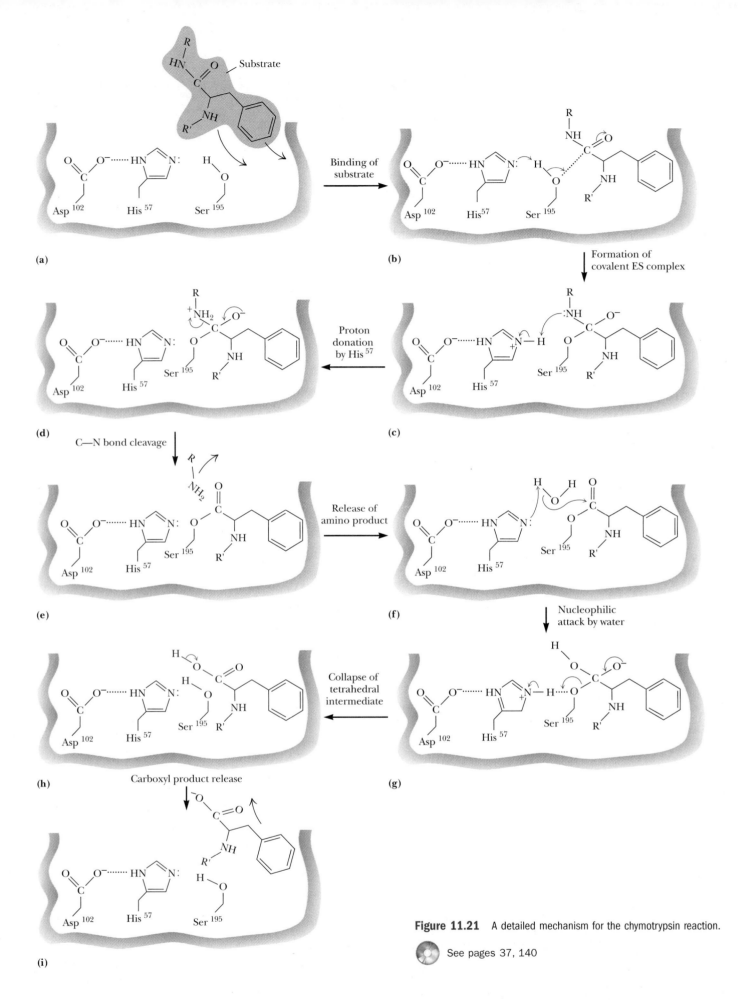

Figure 11.21 A detailed mechanism for the chymotrypsin reaction.

See pages 37, 140

tetrahedral intermediate, facilitating the subsequent bond-breaking and dissociation of the amine product. The negative charge on the peptide oxygen is unstable; the tetrahedral intermediate is short-lived and rapidly breaks down to expel the amine product. The acyl-enzyme intermediate that results is reasonably stable; it can even be isolated using substrate analogs for which further reaction cannot occur. With normal peptide substrates, however, subsequent nucleophilic attack at the carbonyl carbon by water generates another transient tetrahedral intermediate (see Figure 11.21). His[57] acts as a general base in this step, accepting a proton from the attacking water molecule. The subsequent collapse of the tetrahedral intermediate is assisted by proton donation from His[57] to the serine oxygen in a concerted manner. Deprotonation of the carboxyl group and its departure from the active site complete the reaction as shown.

Until recently, the catalytic role of Asp[102] in trypsin and the other serine proteases had been surmised on the basis of its proximity to His[57] in structures obtained from X-ray diffraction studies, but it had never been demonstrated with certainty in physical or chemical studies. As can be seen in Figure 11.14,

A DEEPER LOOK

Transition-State Stabilization in the Serine Proteases

X-ray crystallographic studies of serine protease complexes with transition-state analogs have shown how chymotrypsin stabilizes the **tetrahedral oxyanion intermediate** (structures [c] and [g] in Figure 11.21) of the protease reaction. The amide nitrogens of Ser[195] and Gly[193] form an "oxyanion hole" in which the substrate carbonyl oxygen is hydrogen-bonded to the backbone amide N—H groups.

Formation of the tetrahedral intermediate increases the interaction of the carbonyl oxygen with the amide N—H groups in two ways. Conversion of the carbonyl double bond to the longer tetrahedral single bond brings the oxygen atom closer to the amide hydrogens. Also, the hydrogen bonds between the charged oxygen and the amide hydrogens are significantly stronger than the hydrogen bonds with the uncharged carbonyl oxygen.

Transition-state stabilization in chymotrypsin also involves the side chains of the substrate. The side chain of the departing amine product forms stronger interactions with the enzyme upon formation of the tetrahedral intermediate. When the tetrahedral intermediate breaks down (see Figure 11.21d and e), steric repulsion between the product amine group and the carbonyl group of the acyl-enzyme intermediate leads to departure of the amine product. The "oxyanion hole" of chymotrypsin helps to stabilize the transition states of the mechanism in Figure 11.21.

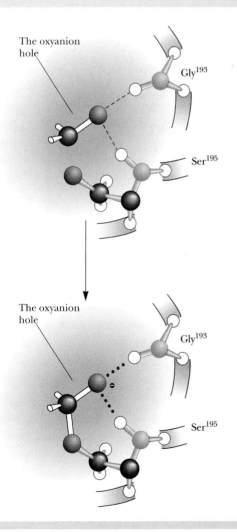

The "oxyanion hole" of chymotrypsin stabilizes the tetrahedral oxyanion ▶ intermediate of the mechanism in Figure 11.21.

Asp[102] is buried at the active site and is normally inaccessible to chemical modifying reagents. In 1987, however, Charles Craik, William Rutter, and their colleagues used site-directed mutagenesis (see Chapter 9) to prepare a mutant trypsin with an asparagine in place of Asp[102]. This mutant trypsin possessed a hydrolytic activity with ester substrates only 1/10,000 that of native trypsin, demonstrating that Asp[102] is indeed essential for catalysis and that its ability to immobilize and orient His[57] is crucial to the function of the catalytic triad.

11.12 The Aspartic Proteases

Mammals, fungi, and higher plants produce a family of proteolytic enzymes known as **aspartic proteases.** These enzymes are active at acidic (or sometimes neutral) pH, and each possesses two aspartic acid residues at the active site. Aspartic proteases carry out a variety of functions (Table 11.3), including digestion (pepsin and chymosin), lysosomal protein degradation (cathepsin D), and regulation of blood pressure (renin is an aspartic protease involved in the production of angiotensin, a hormone that stimulates smooth muscle contraction and reduces excretion of salts and fluid). Most aspartic proteases are composed of 323 to 340 amino acid residues, with molecular weights near 35,000. Aspartic protease polypeptides consist of two homologous domains that fold to produce a tertiary structure composed of two similar lobes, with approximate twofold symmetry (Figure 11.22). The two catalytic aspartate residues, residues 32 and 215 in porcine pepsin, for example, are located deep in the center of the active site cleft.

In contrast to the serine proteases, aspartic proteases do not form covalent bonds with their peptide substrates. Instead, they rely on general acid–base catalysis, with the two aspartate residues acting alternately as general acid and general base. The most widely accepted mechanism for aspartic proteases is shown in Figure 11.23. The two aspartates appear to act as a "catalytic dyad" (analogous to the catalytic triad of the serine proteases). The dyad proton may be covalently bound to either of the aspartate groups in the free enzyme or in the enzyme–substrate complex. Substrate binding is followed by a step in which two concerted proton transfers facilitate nucleophilic attack on the carbonyl carbon of the substrate by water. In the mechanism shown, Asp[32] acts as a gen-

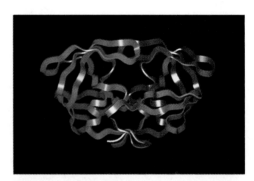

(a)

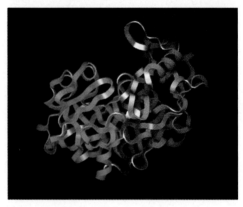

(b)

Figure 11.22 Structures of **(a)** HIV-1 protease, a dimer, and **(b)** pepsin (a monomer). Pepsin's N-terminal half is shown in red; its C-terminal half is shown in blue.

  See pages 46, 139

Table 11.3 Some Representative Aspartic Proteases

Name	Source	Function
Pepsin*	Animal stomach	Digestion of dietary protein
Chymosin[†]	Animal stomach	Digestion of dietary protein
Cathepsin D	Spleen, liver, and many other animal tissues	Lysosomal digestion of proteins
Renin[‡]	Kidney	Conversion of angiotensinogen to angiotensin I; regulation of blood pressure
HIV-protease[§]	AIDS virus	Processing of AIDS virus proteins

*The second enzyme to be crystallized (by John Northrup in 1930). Even more than urease before it, pepsin study by Northrup established that enzyme activity comes from proteins.

[†]Also known as rennin, it is the major pepsin-like enzyme in gastric juice of fetal and newborn animals.

[‡]A drop in blood pressure causes release of renin from the kidneys, which converts angiotensinogen to angiotensin.

[§]A dimer of identical monomers, homologous to pepsin.

(a) ⇌ (b) ⇌ (c) ⇌ (d)

Figure 11.23 A mechanism for the aspartic proteases. In the first step, two concerted proton transfers facilitate nucleophilic attack of water on the substrate carbonyl carbon. In the third step, one aspartate residue (Asp^{32} in pepsin) accepts a proton from one of the hydroxyl groups of the amine dihydrate, and the other aspartate (Asp^{215}) donates a proton to the nitrogen of the departing amine.

See page 139

eral base, accepting a proton from an active-site water molecule, whereas Asp^{215} acts as a general acid, donating a proton to the oxygen of the peptide carbonyl group. By virtue of these two proton transfers, nucleophilic attack occurs without explicit formation of hydroxide ion at the active site. The resulting intermediate is termed an **amide dihydrate.** Note that the protonation states of the two aspartate residues are now opposite to those in the free enzyme (see Figure 11.23).

Breakdown of the amide dihydrate occurs by a mechanism similar to its formation. The ionized aspartate carboxyl (Asp^{32} in Figure 11.23) acts as a general base to accept a proton from one of the hydroxyl groups of the amide dihydrate, while the protonated carboxyl of the other aspartate (Asp^{215} in this case) simultaneously acts as a general acid to donate a proton to the nitrogen atom of one of the departing peptide products.

The AIDS Virus HIV-1 Protease Is an Aspartic Protease

Research on acquired immune deficiency syndrome (AIDS) and its causative viral agent, the human immunodeficiency virus (HIV-1), brought a new aspartic protease to light. **HIV-1 protease** cleaves the polyprotein products of the HIV-1 genome, producing several proteins necessary for viral growth and cellular infection. HIV-1 protease cleaves several different peptide linkages in the HIV-1 polyproteins, including those shown in Figure 11.24. For example, the protease cleaves between the Tyr and Pro residues of the sequence Ser-Gln-Asn-Tyr-Pro-Ile-Val, which joins the p17 and p24 HIV-1 proteins.

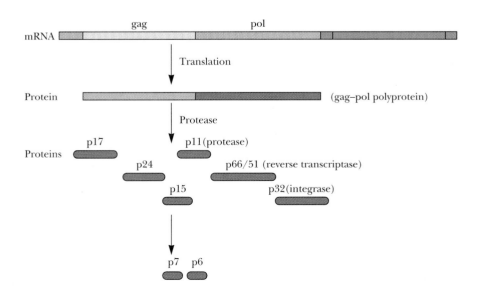

Figure 11.24 HIV mRNA provides the genetic information for synthesis of a polyprotein. Proteolytic cleavage of this polyprotein by HIV protease produces individual proteins required for viral growth and cellular infection.

Protease Inhibitors Give Life to AIDS Patients

Infection by HIV was once considered a death sentence, but the emergence of a new family of drugs called protease inhibitors has made it possible for some AIDS patients to improve their overall health and extend their lives. These drugs are all specific inhibitors of the HIV protease. By inhibiting the protease, they prevent the development of new virus particles in the cells of infected patients. Clinical testing has shown that a combination of drugs—including a protease inhibitor together with a reverse transcriptase inhibitor like AZT—(see Chapter 23) can reduce the human immunodeficiency virus (HIV) to undetectable levels in about 40% to 50% of infected individuals. Patients who respond successfully to this combination therapy have experienced dramatic improvement in their overall health and a substantially lengthened life span.

Four of the protease inhibitors approved for use in humans by the U.S. Food and Drug Administration are shown in the figure: Crixivan by Merck, Invirase by Hoffman-LaRoche, Norvir by Abbott, and Viracept by Agouron. These drugs were all developed from a "structure-based" design strategy; that is, the drug molecules were designed to bind tightly to the active site of the HIV-1 protease. The backbone HO — group in all these substances inserts between the two active-site carboxyl groups of the protease.

In the development of an effective drug, it is not sufficient merely to show that a candidate compound can cause the desired

biochemical effect. It must also be demonstrated that the drug can be effectively delivered in sufficient quantities to the desired site(s) of action in the organism, and that the drug does not cause undesirable side effects. The HIV-1 protease inhibitors shown here fulfill these criteria. Other drug candidates have been found that are even better inhibitors of HIV-1 protease in cell cultures, but many of these fail the test of bioavailability, which is the ability of a drug to be delivered to the desired site of action in the organism.

Candidate protease inhibitor drugs must be relatively specific for the HIV-1 protease. Many other aspartic proteases exist in the human body and are essential to a variety of body functions, including digestion of food and processing of hormones. An ideal drug thus must strongly inhibit the HIV-1 protease, must be delivered effectively to the lymphocytes where the protease must be blocked, and should not adversely affect the activities of the essential human aspartic proteases.

A final but important consideration is viral mutation. Certain mutant HIV strains are resistant to one or more of the protease inhibitors, and even for patients who respond initially to protease inhibitors, mutant viral forms that can eventually thrive in the infected individual may arise. The search for new and more effective protease inhibitors is ongoing.

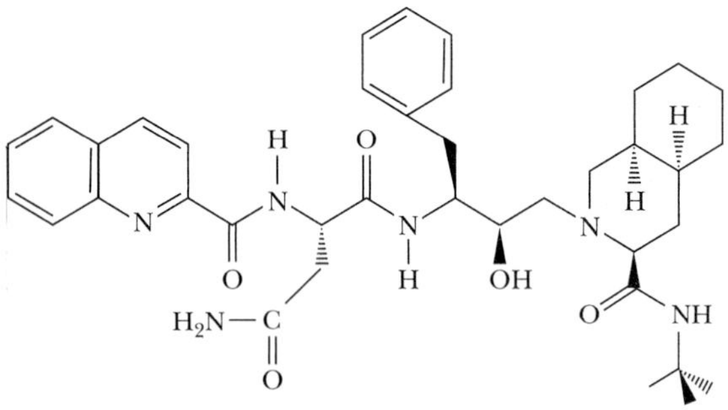

Invirase (Saquinavir)

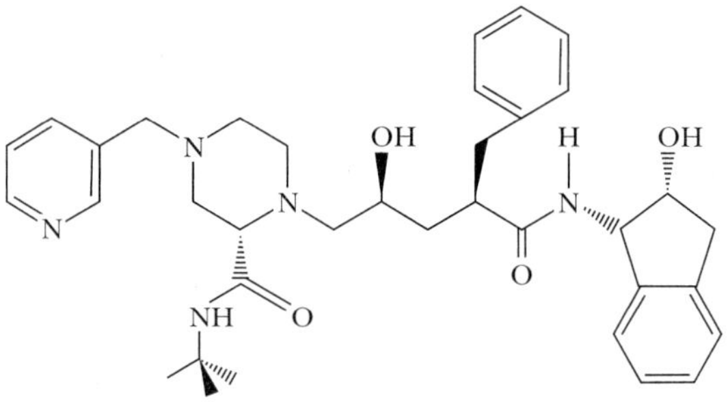

Crixivan (Indinavir)

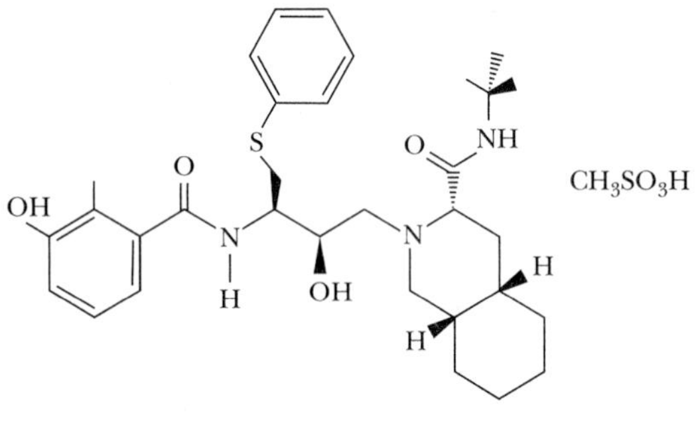

Viracept (Nelfinavir mesylate)

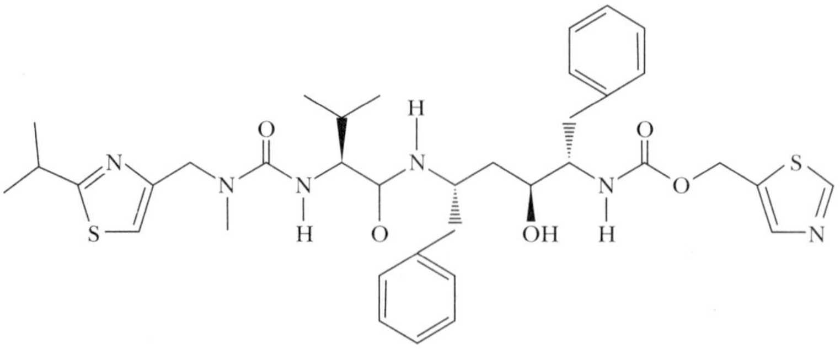

Norvir (Ritonavir)

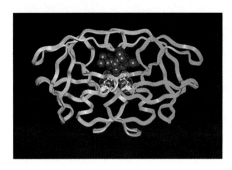 

Figure 11.25 (*left*) HIV-1 protease complexed with the inhibitor Crixivan (red) made by Merck. The flaps (residues 46–55 from each subunit) covering the active site are shown in green and the active-site aspartate residues involved in catalysis are shown in white. (*right*) The close-up of the active site shows the interaction of Crixivan with the carboxyl groups of the essential aspartate residues.

The HIV-1 protease is a remarkable viral imitation of mammalian aspartic proteases: it is a **dimer of identical subunits** that mimics the two-lobed monomeric structure of pepsin and other aspartic proteases. The HIV-1 protease subunits are 99-residue polypeptides that are homologous with the individual domains of the monomeric proteases. Structures determined by X-ray diffraction studies reveal that the active site of HIV-1 protease is formed at the interface of the homodimer and consists of two aspartate residues, designated Asp^{25} and $Asp^{25'}$, one contributed by each subunit (Figure 11.25). In the homodimer, the active site is covered by two identical "flaps," one from each subunit, in contrast to the monomeric aspartic proteases, which possess only a single active-site flap.

Enzyme kinetic measurements by Thomas Meek and his collaborators at SmithKline Beecham Pharmaceuticals have shown that the mechanism of HIV-1 protease is very similar to those of other aspartic proteases.

PROBLEMS

1. Tosyl-L-phenylalanine chloromethyl ketone (TPCK) specifically inhibits chymotrypsin by covalently labeling His^{57}.

Tosyl-L-phenylalanine chloromethyl ketone (TPCK)

a. Propose a mechanism for the inactivation reaction, indicating the structure of the product(s).

b. State why this inhibitor is specific for chymotrypsin.

c. Propose a reagent based on the structure of TPCK that might be an effective inhibitor of trypsin.

2. In this chapter, the experiment in which Craik and Rutter replaced Asp^{102} with Asn in trypsin (reducing activity 10,000-fold) was discussed.

a. On the basis of your knowledge of the catalytic triad structure in trypsin, suggest a structure for the "uncatalytic triad" of Asn-His-Ser in this mutant enzyme.

b. Explain why the structure you have proposed explains the reduced activity of the mutant trypsin.

c. Consult the original journal articles (Sprang et al., 1987. *Science* **237**:905–909; and Craik et al., 1987. *Science* **237**: 909–913) to see Craik and Rutter's answer to this question.

3. Pepstatin (see figure) is an extremely potent inhibitor of the monomeric aspartic proteases, with K_I values of less than 1 nM.

a. On the basis of the structure of pepstatin, suggest an explanation for the strongly inhibitory properties of this peptide.

b. Would pepstatin be expected to also inhibit the HIV-1 protease? Explain your answer.

Pepstatin

Iva   Val   Val   Sta   Ala   Sta

4. The k_{cat} for alkaline phosphatase–catalyzed hydrolysis of methylphosphate is approximately 14/sec at pH 8 and 25°C. The rate constant for the uncatalyzed hydrolysis of methylphosphate under the same conditions is approximately 1×10^{-15}/sec. What is the difference in the free energies of activation of these two reactions?

5. Active α-chymotrypsin is produced from chymotrypsinogen, an inactive precursor, as shown in the color figure shown below. The first intermediate, π-chymotrypsin, displays chymotrypsin activity. Suggest proteolytic enzymes that might carry out these cleavage reactions effectively. (*Hint*: See Chapter 4.)

6. Based on the reaction scheme shown, derive an expression for k_e/k_u—the ratio of the rate constants for the catalyzed and uncatalyzed reactions, respectively—in terms of the free energies of activation for the catalyzed ($\Delta G_e^{\ddagger}$) and the uncatalyzed ($\Delta G_u^{\ddagger}$) reactions.

$$S \xrightleftharpoons{K_u} X^{\ddagger} \xrightarrow{k_u'} P$$

$$E \downarrow\uparrow K_s \qquad\qquad \downarrow\uparrow E$$

$$ES \xrightleftharpoons{K_e} EX^{\ddagger} \xrightarrow{k_e'} EP$$

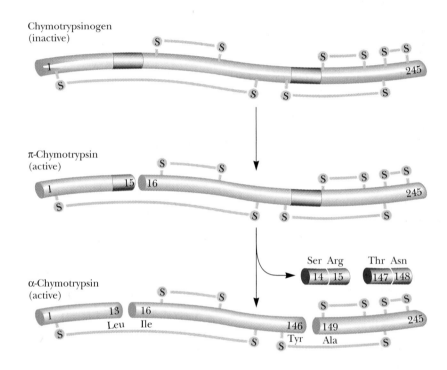

Chymotrypsinogen (inactive)

π-Chymotrypsin (active)

α-Chymotrypsin (active)

FURTHER READING

General

Cannon, W. R., Singleton, S. F., and Benkovic, S. J., 1997. A perspective on biological catalysis. *Nature Structural Biology* **3**:821–833.

Fersht, A., 1999. *Structure and Mechanism in Protein Science.* New York: W. H. Freeman and Company.

Gerlt, J. A., Kreevoy, M. M., Cleland, W. W., and Frey, P. A., 1997. Understanding enzymic catalysis: The importance of short, strong hydrogen bonds. *Chemistry and Biology* **4**:259–267.

Jencks, W. P., 1987. *Catalysis in Chemistry and Enzymology.* New York: Dover Publications.

Jencks, W. P., 1997. From chemistry to biochemistry to catalysis to movement. *Annual Review of Biochemistry* **66**:1-18.

Walsh, C., 1979. *Enzymatic Reaction Mechanisms.* San Francisco: W. H. Freeman and Company.

Transition-State Stabilization and Transition-State Analogs

Kraut, J., 1988. How do enzymes work? *Science* **242**:533–540.

Radzicka, A., and Wolfenden, R., 1995. Transition state and multisubstrate analog inhibitors. *Methods in Enzymology* **249**:284–312.

Schramm, J., 1998. Enzymatic transition states and transition state analog design. *Annual Review of Biochemistry* **67**:693–720.

Wolfenden, R., 1972. Analogue approaches to the structure of the transition state in enzyme reactions. *Accounts of Chemical Research* **5**:10–18.

Wolfenden, R., and Kati, W. M., 1991. Testing the limits of protein-ligand binding discrimination with transition-state analogue inhibitors. *Accounts of Chemical Research* **24**:209–215.

Serine Proteases

Cassidy, C. S., Lin, J., and Frey, P. A., 1997. A new concept for the mechanism of action of chymotrypsin: The role of the low-barrier hydrogen bond. *Biochemistry* **36**:4576–4584.

Craik, C. S., et al., 1987. The catalytic role of the active site aspartic acid in serine proteases. *Science* **237**:909–919.

Plotnick, M. I., Mayne, L., Schechter, N. M., and Rubin, H., 1996. Distortion of the active site of chymotrypsin complexed with a serpin. *Biochemistry* **35**:7586–7590.

Sprang, S., et al., 1987. The three-dimensional structure of Asn[102] mutant of trypsin: Role of Asp[102] in serine protease catalysis. *Science* **237**:905–909.

Aspartic Proteases

Oldziej, S., and Ciarkowski, J., 1996. Mechanism of action of aspartic proteinases: Application of transition-state analogue theory. *Journal of Computer-Aided Molecular Design* **10**:583–588.

Polgar, L., 1987. The mechanism of action of aspartic proteases involves "push-pull" catalysis. *FEBS Letters* **219**:1–4.

HIV-1 Protease

Blundell, T., et al., 1990. The 3-D structure of HIV-1 proteinase and the design of antiviral agents for the treatment of AIDS. *Trends in Biochemical Sciences* **15**:425–430.

Chen, Z., Li, Y., Chen, E., et al., 1994. Crystal structure at 1.9-Å resolution of human immunodeficiency virus (HIV) II protease complexed with L-735,524, an orally bioavailable inhibitor of the HIV proteases. *Journal of Biological Chemistry* **269**:26344–26348.

Hyland, L., et al., 1991. Human immunodeficiency virus-1 protease 1: Initial velocity studies and kinetic characterization of reaction intermediates by [18]O isotope exchange. *Biochemistry* **30**:8441–8453.

Hyland, L., Tomaszek, T., and Meek, T., 1991. Human immunodeficiency virus-1 protease 2: Use of pH rate studies and solvent isotope effects to elucidate details of chemical mechanism. *Biochemistry* **30**:8454–8463.

Wang, Y. X., Freedberg, D. I., Yamazaki, T., et al., 1997. Solution NMR evidence that the HIV-1 protease catalytic aspartyl groups have different ionization states in the complex formed with the asymmetric drug KNI-272. *Biochemistry* **35**:9945–9950.

Proteins of the Blood

A good book is . . . precious life-blood . . . ,
embalmed and treasured up on purpose to a life
beyond life.

John Milton (1608–1674)
in *Areopagitica*

Outline

Blood transfusion and administration of blood-derived products are crucial life-saving procedures so routine in today's medicine that we have a tendency to take them for granted, except in times of crisis. (American Red Cross)

The evolution of multicellular organisms removed most cells from direct interaction with the environment and with each other. Such isolation created the need for a medium of communication between the various tissues and organs and a mechanism for contact with the external environment. The circulation system carrying **blood** (and the lymph) arose to fill this need. Blood is a fluid containing a variety of cells, such as erythrocytes (red blood cells), leukocytes (white blood cells), lymphocytes, and macrophages. Its density is about 1.06 g/mL, or slightly greater than that of water, but its viscosity is five times that of water. When anticoagulants are added to prevent clotting, blood can be centrifuged to remove the cellular fraction, and the clear, pale yellow supernatant fluid remaining is the **blood plasma.**[1] Adult humans have 5 or 6 L

---

[1]Blood **serum** differs from blood plasma in that it is obtained by allowing whole blood to clot. The clear yellow fluid left after removal of the clot is blood serum. Blood serum, like plasma, lacks the cellular constituents of blood, but it differs from plasma because it lacks fibrinogen, the protein precursor to fibrin, the principal protein of clots.

of blood, accounting for 8% or so of body weight. About 7% of the blood volume is protein. Those proteins abundant in blood are easily isolated in pure form and thus they have been extensively studied by protein chemists. These proteins have a special place in biochemistry because of the many lessons learned from them.

Plasma Proteins

Table 12.1 summarizes the protein content of human blood plasma. By far the most abundant protein in plasma is **albumin,** followed in concentration by the various **globulins,** classified some time ago into groups labeled α, β, and γ on the basis of their electrophoretic mobility. The biochemistry of several prominent plasma proteins is discussed in the following sections. A discussion of hemoglobin, the O_2-transporting protein of red blood cells, concludes this chapter.

12.1 Albumin

Human serum albumin (585 amino acids, 66.4 kD) is an extremely soluble protein. Its normal level in plasma does not exceed 5 g/100 mL (0.75 mM), but solutions containing 50 g/100 mL can be prepared *in vitro*. Albumin adopts a heart-shaped tertiary structure, with two-thirds of its residues in 28 α-helical regions; it has 17 disulfide bonds, an unusually large number (Figure 12.1). Unlike many extracellular proteins, albumin is not a glycoprotein.

Albumin serves three principal functions. First, albumin stabilizes the physical environment of the blood in two ways. By providing 80% of the osmotic pressure of plasma, albumin contributes significantly to maintaining the osmotic balance between blood and cells, and as the major macromolecular anion in plasma, albumin supplies most of the acid–base buffering action of plasma proteins. Second, albumin is a major transport vehicle for hydrophobic substances, carrying fatty acids from their sites of storage in adipose tissue to their sites of oxidation in liver and muscle. Albumin also transports bilirubin and heme from the spleen to the liver, steroids from the adrenal glands to

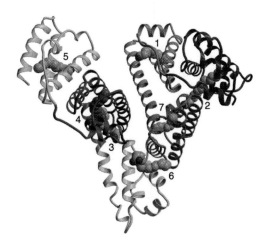

Figure 12.1 Ribbon diagram of human serum albumin, showing the seven principal fatty acid–binding sites. (*Adapted from Bhattacharya, A. A., Grüne, T., and Curry, S., 2000. Crystallographic analysis reveals common modes of binding of medium- and long-chain fatty acids to human serum albumin. Journal of Molecular Biology* ***303:***721–732. *Figure courtesy of Stephen Curry, Imperial College of Science, Technology, and Medicine, London.*)

Table 12.1 Major Human Plasma Proteins*

Protein	Amount (mg/100 mL plasma)	Function
Albumin	3500–4500	Osmoregulation; transport of fatty acids and other hydrophobic molecules
α_1- and α_2-Globulins	1500–3300	Various, including lipoprotein, protease inhibitors, hormone and metal ion transport, and response to inflammation
β-Globulins	850–1520	Various, including lipoprotein, iron and heme transport, and interactions with histocompatibility antigens
γ-Globulins	700–1500	Antibodies
Fibrinogen	300	Blood clotting

*In 1937, Arne Tiselius categorized the principal plasma proteins on the basis of a crude electrophoretic separation into albumin and several large categories which he termed α_1- and α_2-globulins, β-globulins, or γ-globulins.

Adapted from Table 1.3 in Smith, E. L., et al., 1983. *Principles of Biochemistry: Mammalian Biochemistry,* 7th ed. New York: McGraw-Hill.

other tissues, and thyroxine from the thyroid gland to its target tissues. Third, albumin serves a protective function by binding toxic metabolic by-products, wastes, and ingested substances, including bilirubin, aflatoxin, therapeutic drugs, and xenobiotics such as benzene.

The ligand-binding capacity of albumin is quite broad. Because albumin binds smaller molecules of many types and thus carries almost any "cargo," it has been called the "tramp steamer" of the blood. Hydrophobic organic anions of medium size (100–600 D)—such as long-chain fatty acids, eicosanoids, and bile acids—and steroid hormones are most strongly bound (with K_Ds in the range 10^{-4}–10^{-7} M). For many of these compounds, albumin acts as a storage depot. Such substances, when bound to albumin, become available in amounts in excess of their solubility in plasma. Potential toxins are also bound and sequestered for transport to disposal sites.

Albumin has 11 distinct fatty acid–binding sites, and circulating albumin molecules typically have one to four long-chain fatty acids bound. Long-chain fatty acids (the best albumin ligands) bind in hydrophobic pockets with their carboxylates forming salt linkages with nearby Lys or Arg. By binding to albumin, long-chain fatty acids reach a concentration of about 1 mM in plasma, despite an intrinsic solubility of only 0.1 nM. Although plasma contains steroid-specific binding proteins, the amounts of albumin are much higher, and thus most steroid hormone in plasma is bound by albumin. Binding is weak, so hormones are readily "delivered" to their targets. Further, even though binding of drugs and other exogenous substances to albumin is usually weak, albumin binds more than 90% of many drugs. Albumin has received a lot of interest from clinical biochemists because of its effects on drug pharmacokinetics.

12.2 Lipoproteins

Lipoproteins have as their primary purpose the transport of phospholipids, triacylglyerols, and cholesterol through the circulation. Lipoproteins are members of the α- and β-globulin classes of plasma proteins, a diverse set of proteins with many unrelated functions (see Table 12.1). It is now customary to classify lipoproteins according to their densities (Table 12.2). The densities are related to the relative amounts of lipid and protein in the complexes. Since most proteins have densities of about 1.3 to 1.4 g/mL, and lipid aggregates usually possess densities of about 0.8 g/mL, the more protein and the less lipid in

Table 12.2   Composition and Properties of Human Lipoprotein

Lipoprotein Class	Density (g/mL)	Diameter (nm)	Composition (% dry weight)			
			Protein	Cholesterol	Phospho-lipid	Triacyl-glycerol
HDL	1.063-1.21	5-15	33	30	29	8
LDL	1.019-1.063	18-28	25	50	21	4
IDL	1.006-1.019	25-50	18	29	22	31
VLDL	0.95-1.006	30-80	10	22	18	50
Chylomicrons	< 0.95	100-500	1-2	8	7	84

Adapted from Brown, M., and Goldstein, J., 1987. Chapter 315: The hyperlipoproteinemias and other disorders of lipid metabolism. In Braunwald, E., et al., eds., *Harrison's Principles of Internal Medicine*, 11th ed. New York: McGraw-Hill; and Vance, D., and Vance, J., eds., 1985. *Biochemistry of Lipids and Membranes*. Menlo Park, Calif.: Benjamin Cummings.

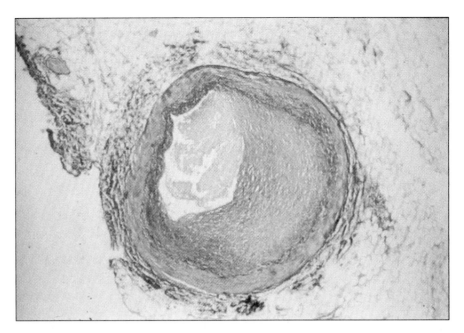

Figure 12.2 Photograph of an arterial plaque. (*Science Photo Library/Photo Researchers, Inc.*)

a complex, the denser the lipoprotein. Thus, there are **high density lipoprotein** (HDL), **low density lipoprotein** (LDL), **intermediate density lipoprotein** (IDL), **very low density lipoprotein** (VLDL), and also **chylomicrons.** Chylomicrons have the lowest protein-to-lipid ratio and thus are the lowest density lipoproteins. They are also the largest. At various sites in the body, lipoproteins interact with specific receptors and enzymes that transfer or modify their lipid cargoes.

The Structure and Synthesis of the Lipoproteins

HDL and VLDL are assembled primarily in the endoplasmic reticulum of the liver (with smaller amounts produced in the intestine), whereas chylomicrons form in the intestine. LDL is not synthesized directly but is made from VLDL. LDL appears to be the major circulatory complex for cholesterol and cholesterol esters. The primary task of chylomicrons is to transport triacylglycerols in lymph and blood. Despite all this, it is extremely important to note that each of these lipoprotein classes contains some of each type of lipid. The relative amounts of HDL and LDL are important in the disposition of cholesterol in the body and in the development of arterial plaques (Figure 12.2).

The structures of the various lipoproteins are approximately similar, consisting of a core of mobile triacylglycerols or cholesterol esters surrounded by a single layer of phospholipid, into which unesterified cholesterol is inserted. The phospholipids are oriented with their polar head groups facing outward. This lipid particle is then wrapped with protein (Figure 12.3). Together, the phospholipids, cholesterol, and protein provide an amphiphilic coating, shielding the hydrophobic lipids inside from solvent water outside. A number of different apoproteins have been identified in lipoproteins (Table 12.3), and others may exist as well. The apoproteins have high amphipathic helix content, with the nonpolar faces of the helices oriented inward toward the lipids. The proteins also function as recognition sites for the various lipoprotein receptors throughout the body.

(a)

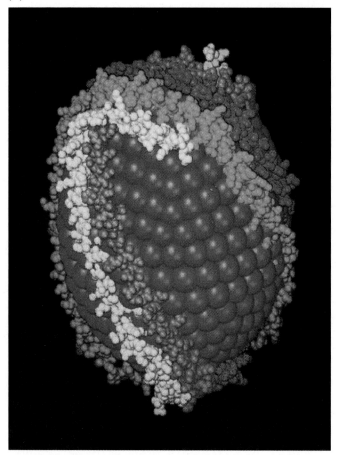

(b)

Figure 12.3 A model for the structure of a typical lipoprotein. **(a)** A core of cholesterol and cholesteryl esters is surrounded by a phospholipid (monolayer) membrane. Apolipoprotein A-1 is modeled here as a long amphipathic α-helix, with the nonpolar face of the helix embedded in the hydrophobic core of the lipid particle and the polar face of the helix exposed to solvent. **(b)** A ribbon diagram of apolipoprotein A-1. (*Adapted from Borhani, D. W., Rogers, D. P., Engler, J. A., and Brouillette, C. G., 1997. Crystal structure of truncated human apolipoprotein A-1 suggests a lipid-bound conformation.* Proceedings of the National Academy of Sciences, U.S.A. **94:**12291–12296.)

See *Interactive Biochemistry CD-ROM and Workbook,* page 36

Lipoproteins in Circulation Are Progressively Degraded

The livers and intestines of animals are the primary sources of circulating lipids. Chylomicrons carry triacylglycerol and cholesterol esters from the intestines to other tissues, and VLDLs carry lipid from the liver, as shown in Figure 12.4. At various target sites, particularly in the capillaries of muscle and adipose cells,

Table 12.3   Apoproteins of Human Lipoprotein*

Apoprotein	M_r	Concentration in plasma (mg/100 mL)	Distribution
A-1	28,300	90–120	Principal protein in HDL
A-2	8,700	30–50	Occurs as dimer mainly in HDL
B-48	240,000	< 5	Found only in chylomicrons
B-100	500,000	80–100	Principal protein in LDL
C-1	7,000	4–7	Found in chylomicrons, VLDL, HDL
C-2	8,800	3–8	Found in chylomicrons, VLDL, HDL
C-3	8,800	8–15	Found in chylomicrons, VLDL, IDL, HDL
D	32,500	8–10	Found in HDL
E	34,100	3–6	Found in chylomicrons, VLDL, IDL, HDL

*The protein components of lipoprotein are called apolipoproteins, or simply apoproteins.

Adapted from Brown, M., and Goldstein, J., 1987. Chapter 315: The hyperlipoproteinemias and other disorders of lipid metabolism. In Braunwald, E., et al., eds., *Harrison's Principles of Internal Medicine,* 11th ed. New York: McGraw-Hill; and Vance, D., and Vance, J., eds., 1985. *Biochemistry of Lipids and Membranes.* Menlo Park, Calif.: Benjamin Cummings.

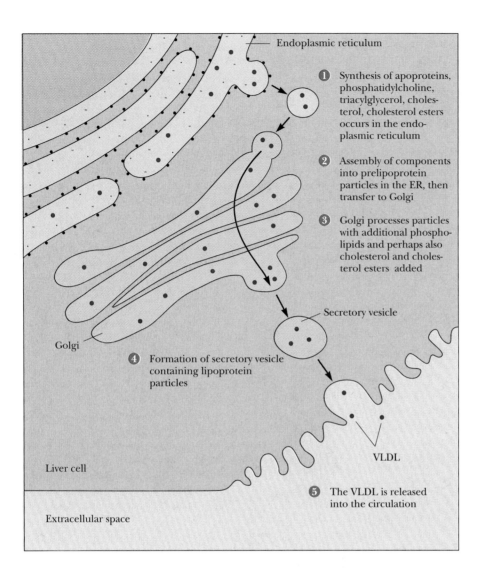

Endoplasmic reticulum

1 Synthesis of apoproteins, phosphatidylcholine, triacylglycerol, cholesterol, cholesterol esters occurs in the endoplasmic reticulum

2 Assembly of components into prelipoprotein particles in the ER, then transfer to Golgi

3 Golgi processes particles with additional phospholipids and perhaps also cholesterol and cholesterol esters added

Secretory vesicle

Golgi

4 Formation of secretory vesicle containing lipoprotein particles

Liver cell

VLDL

5 The VLDL is released into the circulation

Extracellular space

Figure 12.4 Lipoprotein components are synthesized predominantly in the ER of liver cells. Following assembly of lipoprotein particles (red dots) in the ER and processing in the Golgi, lipoproteins are packaged in secretory vesicles for export from the cell (via exocytosis) and release into the circulatory system.

these particles are degraded by **lipoprotein lipase,** which hydrolyzes triacylglycerols. Lipase action causes progressive loss of triacylglycerol (and apoprotein) and makes the lipoprotein smaller. This process gradually converts VLDL particles to IDL and then LDL particles, which are either returned to the liver for reprocessing or redirected to adipose tissues and adrenal glands. Every 24 hours, nearly half of all circulating LDL is removed from circulation in this way. LDLs bind to specific **LDL receptors,** which cluster in domains of the plasma membrane known as **coated pits.** These domains eventually invaginate to form **coated vesicles** (Figure 12.5). Within the cell, these vesicles fuse with lysosomes, and the LDLs are degraded by **lysosomal acid lipases.**

High density lipoproteins have much longer life spans in the body (five to six days) than other lipoproteins. Newly formed HDL contains virtually no cholesterol ester. However, over time, cholesterol esters are accumulated through the action of **lecithin : cholesterol acyltransferase** (LCAT), a 59-kD glycoprotein associated with HDLs. Another associated protein, **cholesterol ester transfer protein,** transfers some of these esters to VLDL and LDL. Alternatively, HDLs function to return cholesterol and cholesterol esters to the liver. This latter process apparently explains the correlation between high HDL levels and reduced risk of cardiovascular disease. (High LDL levels, on the other hand, are correlated with an *increased* risk of coronary artery and cardiovascular disease.)

Figure 12.5 Endocytosis and degradation of lipoprotein particles. (ACAT is acyl-CoA cholesterol acyltransferase.)

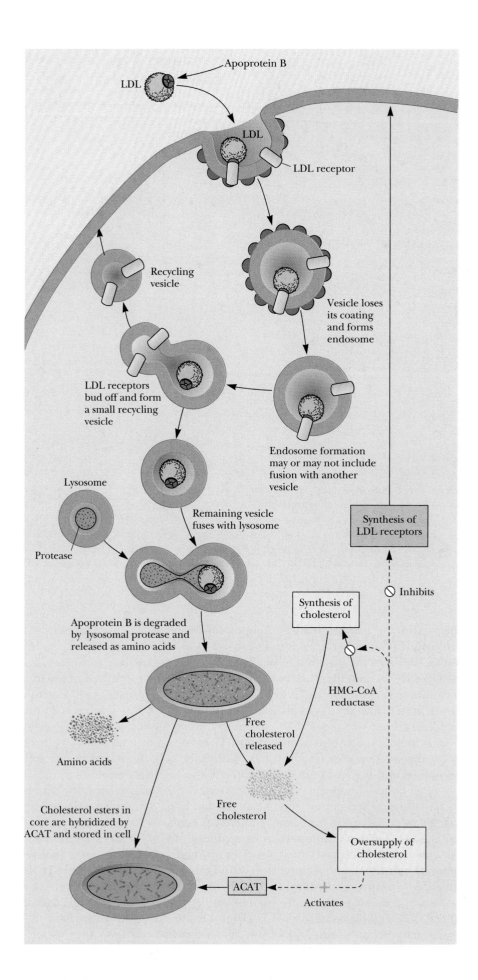

Defects in Lipoprotein Metabolism Can Lead to Elevated Serum Cholesterol

The mechanism of LDL metabolism and the various defects which can occur therein have been studied extensively by Michael Brown and Joseph Goldstein, who received the Nobel Prize in medicine or physiology in 1985. **Familial hypercholesterolemia** is the term given to a variety of inherited metabolic defects that lead to greatly elevated levels of serum cholesterol—much of it in the form of LDL particles. The general genetic defect responsible for familial hypercholesterolemia is the absence or dysfunction of LDL receptors in the body. LDL receptors in plasma membranes (see figure) consist of 839 amino acid residues organized into five domains: an LDL-binding domain of 292 residues, a segment of about 350 to 400 residues containing N-linked oligosaccharides, a 58-residue segment of O-linked oligosaccharides, a 22-residue membrane-spanning segment, and a 50-residue segment extending into the cytosol. The clustering of receptors prior to the formation of coated vesicles requires the presence of this cytosolic segment. Only about half the normal level of LDL receptors is found in individuals heterozygous for familial hypercholesterolemia. Heterozygotes are persons carrying one normal gene and one defective gene. Homozygotes (persons with two copies of the defective gene) have few if any functional LDL receptors. In such cases, LDLs (and cholesterol) cannot be absorbed, and plasma levels of LDL (and cholesterol) are very high. Typical heterozygotes display serum cholesterol levels of 300 to 400 mg/dL, but homozygotes carry serum cholesterol levels of 600 to 800 mg/dL or even higher.

There are two possible causes for an absence of LDL receptors. Either receptor synthesis does not occur at all, or the newly synthesized protein does not successfully reach the plasma membrane due to either faulty processing in the Golgi or faulty transport to the plasma membrane. Even when LDL receptors are made and reach the plasma membrane, they may fail to function for two rea-

sons. They may be unable to form clusters competent in coated pit formation because of folding or sequence anomalies in the carboxy- terminal domain, or they may be unable to bind LDL because of sequence or folding anomalies in the LDL-binding domain.

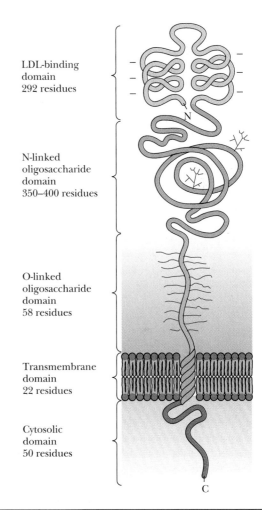

The structure of the LDL receptor. The amino-terminal binding domain is ▶ responsible for recognition and binding of LDL apoprotein. The O-linked oligosaccharide-rich domain may act as a molecular spacer, raising the binding domain above the glycocalyx. The cytosolic domain is required for aggregation of LDL receptors during endocytosis.

12.3 γ-Globulins (Immunoglobulins)

The Immune Response

The γ-globulins are plasma proteins of the immune response. Only vertebrates show an immune response. If a foreign substance, called an **antigen,** gains entry to the bloodstream of a vertebrate, the animal responds via a protective system called the *immune response.* The immune response involves production of proteins capable of recognizing and destroying the antigen. This response is mounted by certain white blood cells—the **B-** and **T-cell lymphocytes** and the **macrophages.** B-cells are so named because they mature in the bone marrow; T-cells mature in the thymus gland. Each of these cell types is capable of producing proteins essential to the immune response. **Antibodies,** which can recognize and bind antigens, are immunoglobulin proteins secreted from B-cells. Since antigens can be almost anything, the immune response must have an incredible repertoire of structural recognition. Thus, vertebrates must have the potential to produce immunoglobulins of great diversity in order to recognize virtually any antigen.

The Immunoglobulin G Molecule

Immunoglobulin G (**IgG,** or γ-**globulin**) is the major class of antibody molecules found circulating in the bloodstream. IgG is a very abundant protein, amounting to 12 mg per mL of serum (see Table 12.1). This protein is a 150-kD $\alpha_2\beta_2$-type tetramer. The α-chain or **H** (for *heavy*) chain is 50 kD; the β-chain or **L** (for *light*) chain is 25 kD. Remarkably, a preparation of IgG from serum is heterogeneous in terms of the amino acid sequences represented in its L and H chains. However, the IgG L and H chains produced from any given B-lymphocyte are homogeneous in amino acid sequence. Figure 12.6 presents a diagram of IgG organization. L chains consist of 214 amino acid residues organized into two roughly equal segments, the V_L and C_L regions. The V_L designation reflects the fact that L chains isolated from serum IgG show variations in amino acid sequence over the first 108 residues, V_L symbolizing this "variable" region of the L polypeptide. The amino acid sequence over residues 109 to 214 of the L polypeptide is constant, as represented by its designation as the "constant light" or C_L region. The heavy, or H, chains consist of 446 amino acid residues. Like L chains, the amino acid sequence over the first 108 residues of H polypeptides is variable and is designated the V_H region. Residues 109 to 446 of H polypeptides are constant in amino acid sequence. This "constant heavy" region consists of three quite equivalent domains of homology designated C_{H1}, C_{H2}, and C_{H3}. Each L chain has two intrachain disulfide bonds, one in the V_L region and the other in the C_L region. The C-terminal amino acid in L chains is cysteine and it forms an interchain disulfide bond to a neighboring H chain. Each H chain has four intrachain disulfide bonds, one in each of the four regions. Within the variable regions of the L and H chains, certain

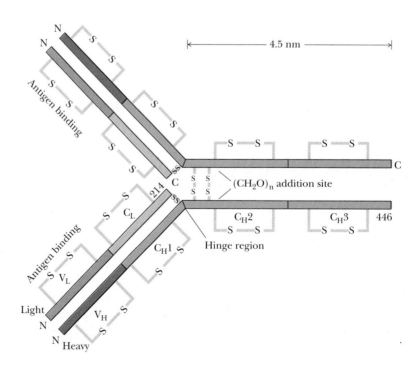

Figure 12.6 Diagram of the organization of the IgG molecule. Two identical L chains are joined with two identical H chains. Each L chain is held to an H chain via an interchain disulfide bond. The variable regions of the four polypeptides lie at the ends of the arms of the Y-shaped molecule. These regions are responsible for the antigen recognition function of antibody molecules. The actual antigen-binding site is constituted from hypervariable residues within the V_L and V_H regions.

(a)

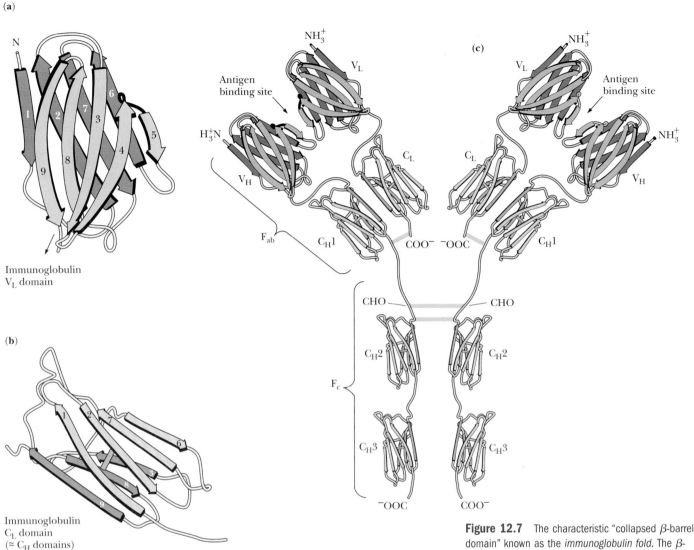

Immunoglobulin
V$_L$ domain

(b)

Immunoglobulin
C$_L$ domain
(≈ C$_H$ domains)

Figure 12.7 The characteristic "collapsed β-barrel domain" known as the *immunoglobulin fold*. The β-barrel structures for both *variable* **(a)** and *constant* **(b)** regions are shown. Part **(c)** shows a schematic diagram of the 12 collapsed β-barrel domains which make up an IgG molecule.

 See page 47

positions are **hypervariable** with regard to the amino acid found there. These hypervariable residues occur at positions 24–34, 50–55, and 89–96 in the L chains and at positions 31–35, 50–65, 81–85, and 91–102 in the H chains. The hypervariable regions are also called **complementarity-determining regions,** or CDRs, since it is these regions which form the structural site that is complementary to some part of an antigen's structure and provides the basis for antibody : antigen recognition.

In terms of its tertiary structure, the immunoglobulin G molecule is composed of 12 discrete **collapsed β-barrel domains.** The characteristic structure of this domain is referred to as the **immunoglobulin fold** (Figure 12.7). Each of the IgG heavy chains contributes four of these domains and each of its light chains contributes two.

The discovery of variability in amino acid sequence in otherwise identical polypeptide chains was surprising and almost heretical to protein chemists. For geneticists, it presented a genuine enigma. They noted that mammals can make millions of different antibodies, but they don't have millions of different antibody genes. How can the mammalian genome encode the diversity seen in L and H chains? The answer to this question is found in the rearrangements that take place within immunoglobulin genes during each lymphocyte's maturation.

12.4 Fibrinogen

Fibrinogen, a soluble, large multimeric glycoprotein of blood plasma (3–4 mg/mL), is the precursor of **fibrin,** the protein forming the basic meshwork of blood clots. Fibrinogen (340 kD) is an elongated (450 Å) molecule composed of three pairs of three different polypeptides (Aα, Bβ, and γ) linked together by a variety of disulfide bridges. The arrangement of the Aα-, Bβ-, and γ-subunits (67 kD, 55 kD, and 47 kD, respectively) is such that the fibrinogen molecule is a covalent (AαB$\beta\gamma$)$_2$ dimer. Each half of the dimer consists of a D region and an E region where the N-termini of all three chains are found; inter-region disulfide bonds link the two E regions together to form the dimer (Figure 12.8). The α, β, and γ polypeptides are homologous over the N-terminal portions of their amino acid sequences, indicating that they are evolutionarily related. Further, the β- and γ-chains are 40% homologous over their C-terminal portions, where the α-chain amino acid sequence is completely different. Fibrinogen is a member of the **kmef group** of fibrous proteins (**k** for keratin, **m** for myosin, **e** for elastin, and **f** for fibrinogen), a grouping based on the very elongated, asymmetric shapes of these proteins.

Clot Formation

In the cascade of reactions leading to blood clotting (see Figure 10.25), thrombin catalyzes the step that actually instigates clot formation by proteolytically removing a 16-residue peptide from the N-terminal end of fibrinogen α-subunits (so-called **fibrinopeptide A**). (Thrombin also removes a 14-residue **fibrinopeptide B** from the N-termini of β-chains, but at a slower rate.) The new α-chain N-termini have exposed Gly-Pro-Arg (GPR) sequences which act as "knobs" that can fit complementary "holes" in the D domain of neighboring fibrinogen molecules (Figure 12.9). The interaction between the α-chain N-terminal "knobs" in E regions with the "holes" in neighboring D regions drives a bidirectional polymerization process in which the fibrinogen molecules line up and overlap in staggered fashion (see Figure 12.9) to form a two-molecule-thick **fibrin protofibril.** A blood clot is created when these protofibrils wind about each other to form a meshwork. During the polymerization process, **Factor XIII** adds other covalent links between adjacent fibrin monomers. Factor XIII is actually a **transglutaminase** (Figure 12.10); the initial Factor XIII–catalyzed cross-links are formed between the C-terminal extremities of neighboring γ-chains.

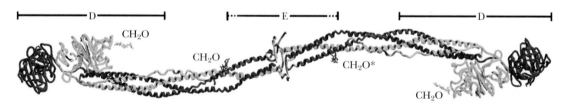

Figure 12.8 Schematic diagram of fibrinogen showing the central domain (E region) containing the N-terminal ends of all six chains, the two Aα, Bβ, γ coiled coils that tether the two terminal domains (D regions) to the central domain, and the α-chain C-terminal extensions. Aα-chains are shown in blue, Bβ-chains in green, and γ-chains in red. The two (AαB$\beta\gamma$) dimeric halves of fibrinogen are held together within the gray region of the central domain by three disulfide bridges, one between the two α-chains and two between the two γ-chains (black bars). (In addition, each α-chain is linked to its corresponding β-chain by a disulfide bridge near their respective N-termini.) Four clusters of N-linked polysaccharides (symbolized as CH$_2$O) adorn fibrinogen, two on the γ-chains near the central domain and two on the β-chains in the D regions. (*Adapted from Figure 1a in Brown, J. H., et al., 2000. The crystal structure of modified bovine fibrinogen.* Proceedings of the National Academy of Sciences, U.S.A. **97**:85–90. *Figure Courtesy of Jerry H. Brown and Carolyn Cox, Brandeis University.*)

Fibrinogen

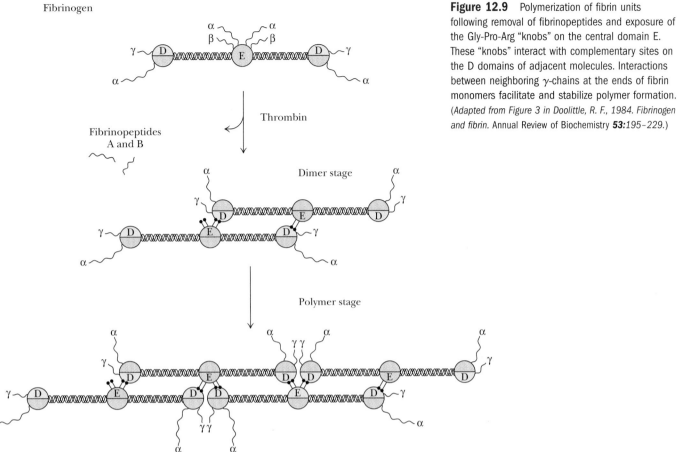

Figure 12.9 Polymerization of fibrin units following removal of fibrinopeptides and exposure of the Gly-Pro-Arg "knobs" on the central domain E. These "knobs" interact with complementary sites on the D domains of adjacent molecules. Interactions between neighboring γ-chains at the ends of fibrin monomers facilitate and stabilize polymer formation. (*Adapted from Figure 3 in Doolittle, R. F., 1984. Fibrinogen and fibrin. Annual Review of Biochemistry **53**:195–229.*)

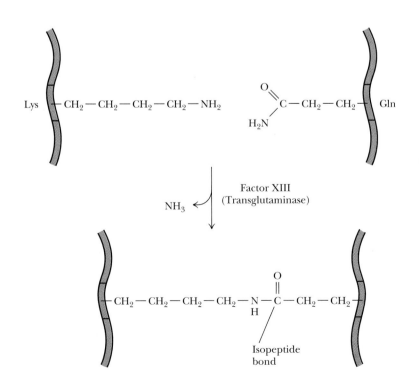

Figure 12.10 The transglutaminase activity of Factor XIII catalyzes a transamidation reaction that joins the ε-NH_2^+ of a lysine side chain in one fibrin molecule with the γ-amide group of glutamine side chains in another fibrin molecule via an isopeptide bond. Factor XIII, one of the blood-clotting factors, is a thrombin-activated protein.

Dissolving the Clot

Since the role of the blood clot is to provide a temporary scaffold upon which more permanent processes of wound healing occur, dissolving the clot (clot lysis) is an important biochemical process. Further, blood clots that arise or get loose in the blood circulation must be destroyed, lest they block blood flow upon lodging in the heart (**coronary thrombosis**) or brain (**stroke**). The fibrin matrix of clots is targeted by **plasmin,** a protease that removes exposed C-termini of α-chains and N-termini of β-chains. Succeeding proteolytic steps break down fibrin monomers into individual D and E domains which are soluble in the bloodstream. Plasmin itself is derived from its **zymogen** precursor **plasminogen** via proteolysis; **tissue plasminogen activator (tPA)** is one agent that can catalyze the conversion of plasminogen to plasmin. Emergency medical personnel often administer tPA (or **streptokinase,** another protein capable of activating the plasminogen-to-plasmin conversion) to coronary thrombosis or ischemic stroke patients as soon as possible in order to hasten clot proteolysis before Factor XIII transglutaminase has a chance to cross-link and stabilize the clot.

Myoglobin (Mb)

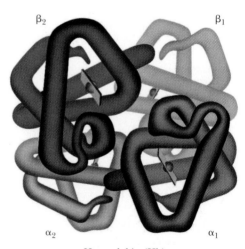

β_2 β_1

α_2 α_1

Hemoglobin (Hb)

Figure 12.11 The myoglobin and hemoglobin molecules. *Myoglobin* (sperm whale): one polypeptide chain of 153 amino acid residues (mass = 17.2 kD) has one heme and binds one O_2. *Hemoglobin* (human): four polypeptide chains, two of 141 amino acid residues (α) and two of 146 residues (β); mass = 64.45 kD. Each polypeptide has a heme; the Hb tetramer binds four O_2. *(Irving Geis)*

See pages 45, 49

12.5   Hemoglobin

The solubility of O_2 in water is low, and thus blood must contain a molecule capable of oxygen transport so that metabolic processes in the tissues are not limited by O_2 availability. This role is filled by the **hemoglobin** of erythrocytes. Hemoglobin is an $\alpha_2\beta_2$ tetramer whose α and β polypeptide chains are strikingly similar to each other and to **myoglobin,** a monomeric O_2-binding protein found in muscle. Because hemoglobin and myoglobin are two of the most studied proteins in nature, they have become paradigms of protein structure and function. Moreover, hemoglobin is a model for protein quaternary structure and allosteric function. The binding of O_2 by hemoglobin and its modulation by protons, CO_2, and 2,3-bisphosphoglycerate depend in exquisite ways on the interactions of polypeptide subunits in a tetrameric protein, revealing much about the functional significance of quaternary associations. Although myoglobin is a muscle protein, it plays an important physiological role in facilitating the transfer of O_2 from the blood to the mitochondria of actively exercising muscle. Further, the structure and function of myoglobin are usually considered along with hemoglobin because of the fundamental insights which come from such comparative biochemistry.

The Comparative Biochemistry of Myoglobin and Hemoglobin

Before examining myoglobin and hemoglobin in detail, let us first encapsulate the lesson: Myoglobin (symbolized as Mb) is a compact globular protein composed of a single polypeptide chain 153 amino acids in length; its molecular mass is 17.2 kD (Figure 12.11). It contains **heme,** a porphyrin ring system complexing an iron ion, as its prosthetic group (Figure 12.12). Oxygen binds to Mb via its heme. Hemoglobin (Hb) is also a compact globular protein, but Hb is a tetramer (Figure 12.11). It consists of four polypeptide chains, each of which is very similar structurally to the myoglobin polypeptide chain, and each bears a heme group. Thus, a hemoglobin molecule can bind four O_2 molecules. In adult human Hb, there are two identical chains of 141 amino acids, the α-chains, and two identical β-chains, each of 146 residues. The human Hb molecule is an $\alpha_2\beta_2$-type tetramer of molecular mass 64.5 kD. The tetrameric nature of Hb is crucial to its biological function: *When a molecule of O_2 binds to*

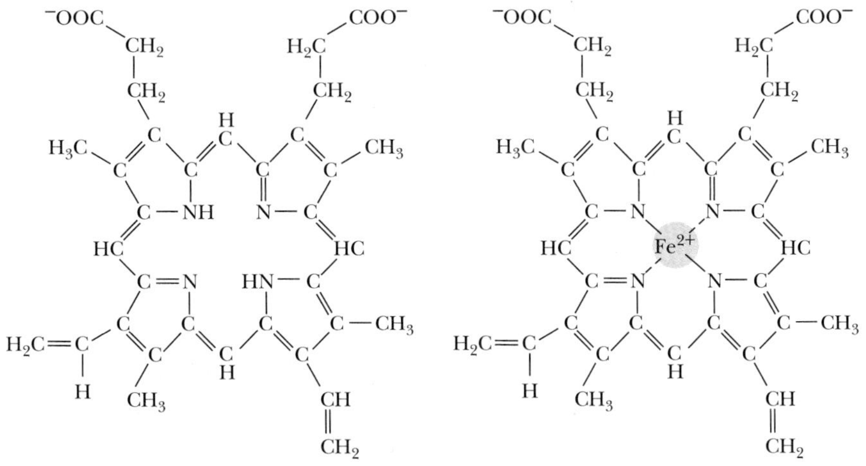

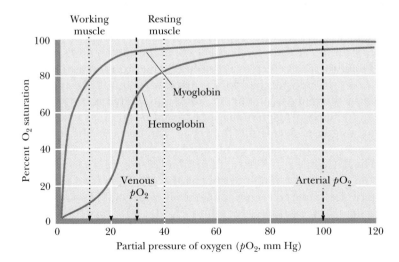

Figure 12.12 The heme prosthetic group. Porphyrin is a highly conjugated macrocycle composed of four pyrrole rings joined by methylene bridges. If the four pyrroles bear a characteristic set of eight substituents—four methyl ($-CH_3$), two vinyl ($-CH=CH_2$), and two propionic acid ($-CH_2-CH_2-COOH$) groups—the structure is known as **protoporphyrin**. Different arrangements of these eight groups around the edge of the porphyrin are possible. The one shown is **protoporphyrin IX.** Heme is the iron complex of protoporphyrin IX.

Protoporphyrin IX

Heme
(Fe-protoporphyrin IX)

a heme in Hb, the heme Fe ion is drawn into the plane of the porphyrin ring. This slight movement sets off a chain of conformational events that are transmitted to adjacent subunits, dramatically enhancing the affinity of their heme groups for O_2. That is, the binding of O_2 to one heme of Hb makes it easier for the Hb molecule to bind additional equivalents of O_2. Hemoglobin displays a sigmoid-shaped O_2-binding curve (Figure 12.13). That is, the binding of O_2 by Hb shows **cooperativity,** or a **positive cooperative effect.** Such curves closely resemble those seen in allosteric enzyme : substrate saturation graphs (see Figure 10.28). In contrast, myoglobin's interaction with oxygen obeys a classical Michaelis–Menten type of substrate saturation behavior.

Myoglobin

Myoglobin is the oxygen-storage protein of muscle. The muscles of diving mammals such as seals and whales are especially rich in this protein, which serves as a store for O_2 during the animal's prolonged periods under water. Myoglobin is abundant in skeletal and cardiac muscle of nondiving animals as well, and is the cause of the characteristic red color of muscle.

Figure 12.13 O_2-binding curves for hemoglobin and myoglobin.

Figure 12.14 Detailed structure of the myoglobin molecule. The myoglobin polypeptide chain consists of eight helical segments, designated by the letters A through H, counting from the N-terminus. These helices, ranging in length from 7 to 26 residues, are linked by short, unordered regions that are named for the helices they connect, as in the AB region or the EF region. The individual amino acids in the polypeptide are indicated according to their position within the various segments, as in His F8, the eighth residue in helix F, or Phe CD1, the first amino acid in the interhelical CD region. Occasionally, amino acids are specified in the conventional way—that is, by the relative position in the chain, as in Gly[153]. The heme group is cradled within the folded polypeptide chain. *(Irving Geis)*

See page 49

The Mb Polypeptide Cradles the Heme Group

The myoglobin polypeptide chain is folded to form a cradle ($4.4 \times 4.4 \times 2.5$ nm) that nestles the heme prosthetic group (Figure 12.14). O_2 binding depends on the heme's oxidation state. The iron ion in the heme of myoglobin is in the 2+ oxidation state, that is, the **ferrous** form. This is the form that binds O_2. Oxidation of the ferrous form to the 3+ **ferric** form yields **metmyoglobin.** Metmyoglobin will not bind O_2. It is interesting that free heme in solution will readily interact with O_2 also, but the oxygen quickly oxidizes the iron atom to the ferric state. Fe^{3+}:protoporphyrin IX is referred to as **hematin.** Thus, the polypeptide of myoglobin may be viewed as serving three critical functions: it cradles the heme group; it protects the heme iron atom from oxidation; and it provides a pocket into which the O_2 can fit.

O_2 Binding to Mb

Iron ions, whether ferrous or ferric, prefer to interact with six ligands, four of which share a common plane. The fifth and sixth ligands lie above and below this plane (Figure 12.15). In heme, four of the ligands are provided by the nitrogen atoms of the four pyrroles. A fifth ligand is donated by the imidazole side chain of amino acid residue His F8. When myoglobin binds O_2 to become **oxymyoglobin,** the O_2 molecule adds to the heme iron ion as the sixth ligand (Figure 12.15). O_2 adds end-on to the heme iron, but it is not oriented perpendicular to the plane of the heme. Rather, it is tilted about 60° to the perpendicular. In **deoxymyoglobin,** the sixth ligand position is vacant, and in metmyoglobin, a water molecule fills the O_2 site and becomes the sixth ligand for the ferric atom. On the oxygen-binding side of the heme lies another histidine residue, His E7. While its imidazole function lies too far away to interact with the Fe atom, it is close enough to contact the O_2. Therefore, the O_2–binding site is a sterically hindered region. Biologically important properties stem from this hindrance. For example, the affinity of *free heme* in solution for carbon monoxide (CO) is 25,000 times greater than its affinity for O_2. But CO binds only 250 times more tightly than O_2 to the heme of myoglobin because His E7

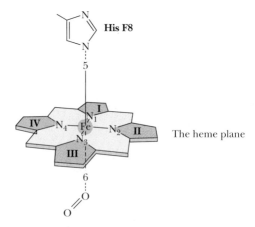

Figure 12.15 The six liganding positions of an iron ion. Four ligands lie in the same plane; the remaining two are, respectively, above and below this plane. In myoglobin, His F8 is the fifth ligand; in oxymyoglobin, O_2 becomes the sixth.

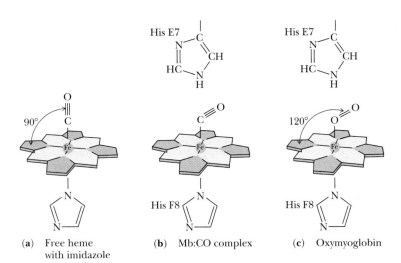

Figure 12.16 Oxygen and carbon monoxide binding to the heme group of myoglobin.

(a) Free heme with imidazole   (b) Mb:CO complex   (c) Oxymyoglobin

forces the CO molecule to tilt away from a preferred perpendicular alignment with the plane of the heme (Figure 12.16). This diminished affinity of myoglobin for CO guards against the possibility that traces of CO produced during metabolism might occupy all of the heme sites, effectively preventing O_2 from binding. Nevertheless, carbon monoxide in the environment is a potent poison, causing death by asphyxiation by precisely this mechanism.

O_2 Binding Alters Mb Conformation

What happens when the heme group of myoglobin binds oxygen? X-ray crystallography has revealed that the most profound change occurs in the position of the iron atom relative to the plane of the heme. In deoxymyoglobin, the ferrous ion has but five ligands and it is drawn out of the porphyrin plane by its interaction with His F8. Indeed, it lies 0.055 nm above the plane of the heme, in the direction of His F8. The iron : porphyrin complex is therefore dome-shaped. When O_2 binds, the iron atom is pulled back toward the porphyrin plane and is now displaced from it by only 0.026 nm (Figure 12.17). The consequences of this small motion are trivial as far as the biological role of myoglobin is concerned. However, as we shall soon see, this slight movement profoundly affects the properties of hemoglobin. Its action on His F8 is magnified through changes in polypeptide conformation that alter subunit interactions in the Hb tetramer. These changes in subunit relationships are the fundamental cause for the allosteric properties of hemoglobin.

The Physiological Significance of Cooperative Binding of Oxygen by Hemoglobin

The relative oxygen affinities of hemoglobin and myoglobin reflect their respective physiological roles (see Figure 12.13). Myoglobin, as the oxygen delivery system in muscle, has a greater affinity for O_2 than hemoglobin at all oxygen pressures. Hemoglobin, as the oxygen carrier, becomes saturated with O_2 in the lungs, where the partial pressure of O_2 (pO_2) is about 100 torr.[2] In the capillaries of tissues, pO_2 is typically 40 torr, and oxygen is released from Hb. In muscle, some of it can be bound by myoglobin, to be stored for use in times of severe oxygen deprivation, such as during strenuous exercise.

[2]The *torr* is a unit of pressure named for Evangelista Torricelli, the inventor of the barometer. 1 torr corresponds to 1 mm Hg (1/760th of an atmosphere).

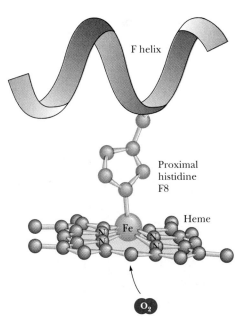

Figure 12.17 The displacement of the Fe ion of the heme of deoxymyoglobin from the plane of the porphyrin ring system by the pull of His F8. In oxymyoglobin, the bound O_2 counteracts this effect.

See page 49

A DEEPER LOOK

The Physiological Significance of the Hb:O$_2$ Interaction

We can determine quantitatively the physiological significance of the sigmoid nature of the hemoglobin oxygen-binding curve, or, in other words, the biological importance of cooperativity. The equation

$$\frac{Y}{(1-Y)} = \left(\frac{pO_2}{P_{50}}\right)^n$$

describes the relationship between pO_2, the affinity of hemoglobin for O_2 (defined as P_{50}, the partial pressure of O_2 giving half-maximal saturation of Hb with O_2), and the fraction of hemoglobin with O_2 bound, Y, versus the fraction of Hb with no O_2 bound, $(1 - Y)$. The coefficient n is the *Hill coefficient*, an index of the cooperativity (sigmoidicity) of the hemoglobin oxygen-binding curve. Taking

pO_2 in the lungs as 100 torr, P_{50} as 26 torr, and n as 2.8, Y, the fractional saturation of the hemoglobin heme groups with O_2, is 0.98. If pO_2 were to fall to 10 torr within the capillaries of an exercising muscle, Y would drop to 0.06. The oxygen delivered under these conditions would be proportional to the difference $Y_{lungs} - Y_{muscle}$, which is 0.92. That is, virtually all the oxygen carried by Hb would be released. Suppose instead that hemoglobin binding of O_2 were not cooperative; the hemoglobin oxygen-binding curve would be hyperbolic, and $n = 1.0$. Then, Y in the lungs would be 0.79 and Y in the capillaries 0.28, and now the difference in Y values is 0.51. Thus, under these conditions, the cooperativity of oxygen binding by Hb means that 0.92/0.51, or 1.8 times, as much O_2 can be delivered.

The Structure of the Hemoglobin Molecule

As noted, hemoglobin is an $\alpha_2\beta_2$ tetramer. Each of the four subunits has a conformation virtually identical to that of myoglobin. Two different types of subunits, as in α and β, are necessary to achieve cooperative O_2 binding by Hb. The β-chain at 146 amino acid residues is shorter than the myoglobin chain (153 residues), mainly because its final helical segment (the H helix) is shorter. The α-chain (141 residues) also has a shortened H helix and lacks the D helix as well (Figure 12.18). Max Perutz, who has devoted his life to elucidating the atomic structure of Hb, noted very early in his studies that the molecule was highly symmetrical. The actual arrangement of the four subunits with respect to one another is shown in Figure 12.19 for horse methemoglobin. All vertebrate hemoglobins show a three-dimensional structure essentially the same as this. The subunits pack in a tetrahedral array, creating a roughly spherical molecule, 6.4 by 5.5 by 5.0 nm. The four heme groups, nestled within the easily

Myoglobin (Mb) α-Globin (Hbα) β-Globin (Hbβ)

Figure 12.18 Conformational drawings of the α- and β-chains of Hb and the myoglobin chain, showing their conformational similarities and differences.

(a) Front view

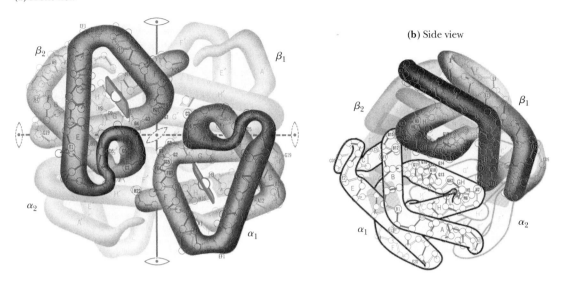

(b) Side view

Figure 12.19 The arrangement of subunits in horse methemoglobin, the first hemoglobin whose structure was determined by X-ray diffraction. The iron atoms on metHb are in the oxidized, ferric (Fe^{3+}) state. *(Irving Geis)*

See page 45

recognizable cleft formed between the E and F helices of each polypeptide, are exposed at the surface of the molecule. The heme groups are quite far apart; 2.5 nm separates the closest iron ions, those of hemes α_1 and β_2, and those of hemes α_2 and β_1. The subunit interactions are mostly between dissimilar chains: each of the α-chains is in contact with both β-chains, but there are few α–α or β–β interactions.

Oxygenation Markedly Alters the Quaternary Structure of Hemoglobin

Crystals of deoxyhemoglobin shatter when exposed to O_2. Further, X-ray crystallographic analysis reveals that oxy- and deoxyhemoglobin differ markedly in quaternary structure. In particular, specific $\alpha\beta$-subunit interactions change. The $\alpha\beta$ contacts are of two kinds. The $\alpha_1\beta_1$ and $\alpha_2\beta_2$ contacts involve helices B, G, and H and the GH corner. These contacts are extensive and important to subunit packing; they remain unchanged when hemoglobin goes from its deoxy to its oxy form. The $\alpha_1\beta_2$ and $\alpha_2\beta_1$ contacts are called *sliding contacts*. They principally involve helices C and G and the FG corner (Figure 12.20). When hemoglobin undergoes a conformational change as a result of ligand binding to the heme, these contacts are altered (Figure 12.21). Hemoglobin as a conformationally dynamic molecule consists of two dimeric halves, an $\alpha_1\beta_1$-subunit pair and an $\alpha_2\beta_2$-subunit pair. Each $\alpha\beta$ dimer moves as a rigid body, and the two halves of the molecule slide past one another upon oxygenation of the heme. The two halves rotate some 15° about an imaginary pivot passing through the $\alpha\beta$ subunits; some atoms at the interface between $\alpha\beta$-dimers are relocated by as much as 0.6 nm.

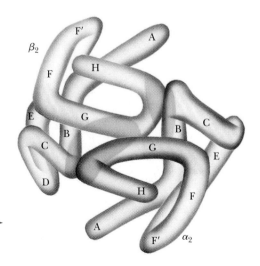

Figure 12.20 Side view of one of the two $\alpha\beta$ dimers in Hb, with packing contacts indicated in ▶ blue. The sliding contacts made with the other dimer are shown in yellow. The changes in these sliding contacts are shown in Figure 12.21.

(**a**) Deoxyhemoglobin

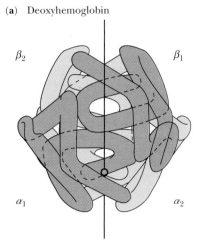

(**b**) Oxyhemoglobin

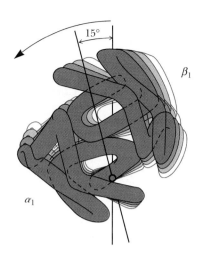

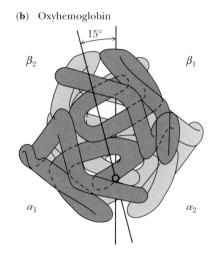

Figure 12.21 Subunit motion in hemoglobin when the molecule goes from the deoxy to the oxy form. (*Irving Geis*)

Movement of the Heme Iron Atom by 0.39 nm Triggers the Hb Conformational Change

In deoxyhemoglobin, histidine F8 is ligated to the heme iron ion, but steric constraints force the Fe^{2+} : His-N bond to be tilted about 8° to the plane of the heme. This steric repulsion between histidine F8 and the nitrogen atoms of the porphyrin ring system, combined with electrostatic repulsions between the electrons of Fe^{2+} and the porphyrin π-electrons, forces the iron atom to lie out of the porphyrin plane by about 0.06 nm. Changes in electronic and steric factors upon heme oxygenation allow the Fe^{2+} atom to move about 0.039 nm closer to the plane of the porphyrin, so now it is displaced only 0.021 nm above the plane. It is as if the O_2 were drawing the heme Fe^{2+} into the porphyrin plane (Figure 12.22). This modest displacement of 0.039 nm seems a trivial distance, but its biological consequences are far-reaching. For as the iron atom moves, it drags histidine F8 along with it, causing helix F, the EF corner, and the FG corner to follow. These shifts are transmitted to the subunit interfaces where they trigger conformational readjustments that lead to the rupture of interchain salt links.

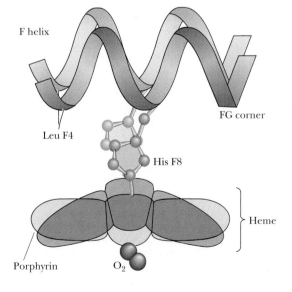

See page 45

Figure 12.22 Changes in the position of the heme iron atom upon oxygenation lead to conformational changes in the hemoglobin molecule.

The Oxy and Deoxy Forms of Hemoglobin Represent Two Different Conformational States

Hemoglobin resists oxygenation (Figure 12.13) because the deoxy form is stabilized by specific hydrogen bonds and salt bridges (ion-pair bonds). All of these interactions are broken in oxyhemoglobin, as the molecule stabilizes into a new conformation. In deoxyhemoglobin, with all of these interactions intact, the C-termini of the four subunits are restrained, and this conformational state is termed the **T,** or **taut, form.** In oxyhemoglobin, these C-termini have almost complete freedom of rotation, and the molecule is now in its **R,** or **relaxed, form.** These terms, **R** and **T,** are the same designations used in the Monod–Wyman–Changeux model (see Chapter 10) to describe alternative conformational states for allosteric proteins, the **R** form having the greater affinity for substrate.

A Model for the Allosteric Behavior of Hemoglobin

A model for the allosteric behavior of hemoglobin has been developed, based on the observation that oxygen is accessible only to the heme groups of the α-chains when hemoglobin is in the **T** conformational state. The heme environment of β-chains in the **T** state is virtually inaccessible because of steric hindrance by amino acid residues in the E helix. This hindrance disappears when the hemoglobin molecule undergoes transition to the **R** conformational state. Binding of O_2 to the β-chains is thus dependent on the **T** to **R** conformational shift, and this shift is triggered by the subtle changes that ensue when O_2 binds to the α-chain heme groups.

H$^+$ Promotes the Dissociation of Oxygen from Hemoglobin

Protons, carbon dioxide, chloride ions, and the metabolite *2,3-bisphosphoglycerate* (or *BPG*) all affect the binding of O_2 by hemoglobin. Their effects have interesting ramifications, as we shall see as we discuss them in turn. Deoxyhemoglobin has a higher affinity for protons than oxyhemoglobin. Thus, as the pH decreases, dissociation of O_2 from hemoglobin is enhanced. In simple symbolism, ignoring the stoichiometry of O_2 or H^+ involved:

$$HbO_2 + H^+ \rightleftharpoons HbH^+ + O_2$$

Expressed another way, H^+ is an antagonist of oxygen binding by Hb, and the saturation curve of Hb for O_2 is displaced to the right as acidity increases (Figure 12.23). This phenomenon is named the **Bohr effect,** after its discoverer, Danish physiologist Christian Bohr (the father of Niels Bohr, the atomic physicist). The effect has important physiological significance because actively metabolizing tissues produce acid, promoting O_2 release just where most needed. About two protons are taken up by deoxyhemoglobin. The N-termini of the two α-chains and the His β146 residues have been implicated as the major players in the Bohr effect. (The pK of a free amino terminus in a protein is about 8.0, while the pK of a protein histidine imidazole is around 6.5.) Neighboring carboxylate groups of Asp β94 residues help to stabilize the protonated state of the His β146 imidazoles that occur in deoxyhemoglobin, but changes in the conformation of β-chains upon Hb oxygenation move the negative Asp function away, and the dissociation of the imidazole protons is favored.

CO$_2$ Also Promotes the Dissociation of O$_2$ from Hemoglobin

Carbon dioxide has an effect on O_2 binding by Hb that is similar to that of H^+, partly because it produces H^+ when it dissolves in the blood:

Figure 12.23 The oxygen saturation curves for myoglobin and for hemoglobin at five different pH values: 7.6, 7.4, 7.2, 7.0, and 6.8.

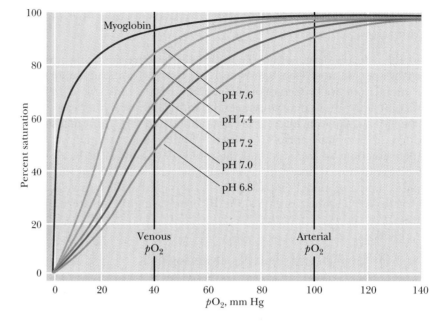

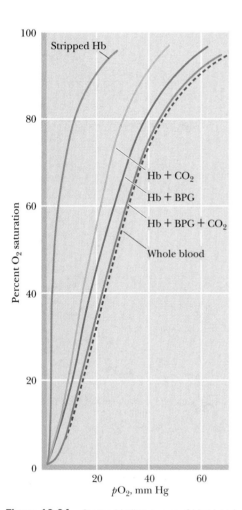

Figure 12.24 Oxygen-binding curves of blood and of hemoglobin in the absence and presence of CO_2 and BPG. *From left to right:* stripped Hb; Hb + CO_2; Hb + BPG; Hb + CO_2 + BPG, and whole blood.

$$CO_2 + H_2O \underset{\text{carbonic anhydrase}}{\overset{\downarrow}{\rightleftharpoons}} \underset{\text{carbonic acid}}{\overset{\uparrow}{H_2CO_3}} \rightleftharpoons H^+ + \underset{\text{bicarbonate}}{\overset{\uparrow}{HCO_3^-}}$$

The enzyme *carbonic anhydrase* promotes the hydration of CO_2. Many of the protons formed upon ionization of carbonic acid are picked up by Hb as O_2 dissociates. The bicarbonate ions are transported with the blood back to the lungs. When Hb becomes oxygenated again in the lungs, H^+ is released and reacts with HCO_3^- to re-form H_2CO_3, from which CO_2 is liberated and exhaled.

In addition, some CO_2 is directly transported by hemoglobin in the form of *carbamate* ($-NHCOO^-$). Free α-amino groups of Hb will react with CO_2 reversibly:

$$R-NH_2 + CO_2 \rightleftharpoons R-NH-COO^- + H^+$$

This reaction is driven to the right in tissues by the high carbon dioxide concentration; the equilibrium shifts the other way in the lungs where $[CO_2]$ is low. Carbamylation of the N-termini converts them to anionic functions which form salt links with the cationic side chains of Arg $\alpha141$ that stabilize the deoxy- or **T** state of hemoglobin.

In addition to CO_2, Cl^- and BPG also bind better to deoxyhemoglobin than to oxyhemoglobin, causing a shift in equilibrium in favor of O_2 release. These various effects are demonstrated by the shift in the oxygen saturation curves for Hb in the presence of one or more of these substances (Figure 12.24). Note that the O_2-binding curve for Hb + CO_2 + BPG fits that of whole blood very well.

2,3-Bisphosphoglycerate Is an Important Allosteric Effector for Hemoglobin

The binding of 2,3-bisphosphoglycerate (BPG) to Hb promotes the release of O_2 (Figure 12.24). Red blood cells normally contain about 4.5 mM BPG, a concentration equivalent to that of tetrameric hemoglobin molecules. Interestingly, this equivalence is maintained in the Hb : BPG binding stoichiometry,

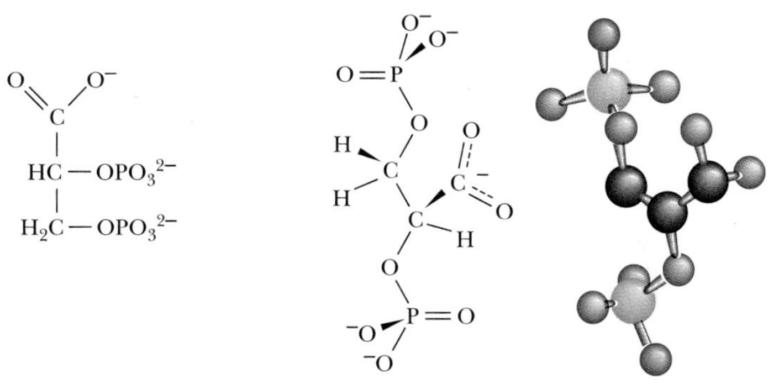

Figure 12.25 The structure, in ionic form, of BPG, or 2,3-bisphosphoglycerate, an important allosteric effector for hemoglobin.

since the tetrameric Hb molecule has but one binding site for BPG. This site is situated within the central cavity formed by the association of the four subunits. The strongly negative BPG molecule (Figure 12.25) is electrostatically bound via interactions with the positively charged functional groups of each Lys $\beta82$, His $\beta2$, His $\beta143$, and the NH_3^+- terminal group of each β-chain. These positively charged residues are arranged to form an electrostatic pocket complementary to the conformation and charge distribution of BPG (Figure 12.26).

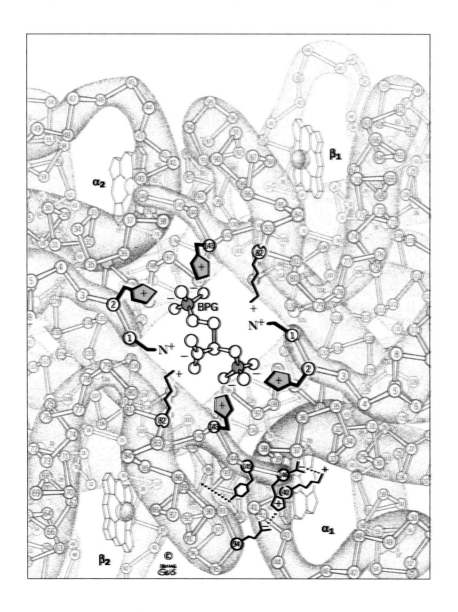

 See page 45

Figure 12.26 The ionic binding of BPG to the two β-subunits of Hb. *(Irving Geis)*

In effect, BPG cross-links the two β-subunits. The ionic bonds between BPG and the two β-chains aid in stabilizing the conformation of Hb in its deoxy form, thereby favoring the dissociation of oxygen. In oxyhemoglobin, this central cavity is too small for BPG to fit. Or, to put it another way, the conformational changes in the Hb molecule that accompany O_2 binding perturb the BPG-binding site so that BPG can no longer be accommodated. Thus, BPG and O_2 are mutually exclusive allosteric effectors for Hb, even though their binding sites are physically distinct.

The Physiological Significance of BPG Binding

The importance of the BPG effect is evident in Figure 12.24: Hemoglobin stripped of BPG is virtually saturated with O_2 at a pO_2 of only 20 torr, and could not release its oxygen within tissues, where the pO_2 is typically 40 torr. BPG shifts the oxygen saturation curve of Hb to the right, rendering it an O_2 delivery system eminently suited to the needs of the organism. BPG serves this vital function in humans, most primates, and a number of other mammals.

Fetal Hemoglobin Has a Higher Affinity for O_2 Because It Has a Lower Affinity for BPG

The fetus depends on its mother for an adequate supply of oxygen, but its circulatory system is entirely independent. Gas exchange takes place across the placenta. Ideally, then, fetal Hb should be able to absorb O_2 better than maternal Hb so that an effective transfer of oxygen can occur. Fetal Hb differs from adult Hb in that the β-chains are replaced by very similar, but not identical, 146-residue subunits called γ(*gamma*)-chains. Fetal Hb is thus $\alpha_2\gamma_2$. Recall that BPG functions through its interaction with the β-chains. BPG binds less effectively with the γ-chains of fetal Hb (also called Hb F). (Fetal γ-chains have Ser instead of His at position 143 and thus lack two of the positive charges in the central BPG-binding cavity.) Figure 12.27 compares the relative affinities of adult Hb (also known as Hb A) and Hb F for O_2 under similar conditions of pH and [BPG]. Note that Hb F will bind O_2 at pO_2 values where most of the oxygen has dissociated from Hb A. Much of the difference can be attributed to the diminished capacity of Hb F to bind BPG (compare Figures 12.24 and 12.27); Hb F thus has an intrinsically greater affinity for O_2, and oxygen transfer from mother to fetus is assured.

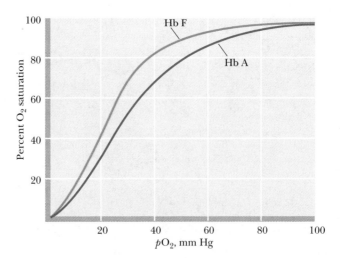

Figure 12.27 Comparison of the oxygen saturation curves of Hb A and Hb F under similar conditions of pH and [BPG].

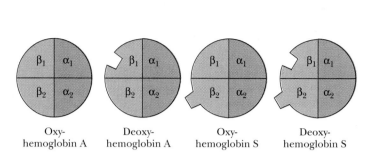

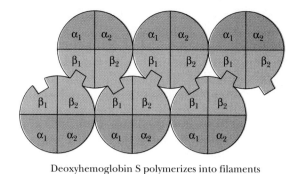

				Deoxyhemoglobin S polymerizes into filaments
Oxy-hemoglobin A	Deoxy-hemoglobin A	Oxy-hemoglobin S	Deoxy-hemoglobin S	

Figure 12.28 The polymerization of Hb S via the interactions between the hydrophobic Val side chains at position $\beta 6$ and the hydrophobic pockets in the EF corners of β-chains in neighboring Hb molecules. The protruding "block" on Oxy S represents the Val hydrophobic protrusion. The complementary hydrophobic pocket in the EF corner of the β-chains is represented by a square indentation. (This indentation is probably present in Hb A also.) Only the β_2 Val protrusions and the β_1 EF pockets are shown. (The β_1 Val protrusions and the β_2 EF pockets are not involved, though they are present.)

Sickle-Cell Anemia

In 1904, a Chicago physician treated a 20-year-old black college student complaining of headache, weakness, and dizziness. The blood of this patient revealed serious anemia; only half the normal number of red cells were present, and many of these cells were abnormally shaped. Instead of the characteristic disc shape, these erythrocytes were elongated and crescentlike in form, a feature that eventually gave name to the disease **sickle-cell anemia.** These sickle cells pass less freely through the capillaries, impairing circulation and causing tissue damage. Further, these cells are more fragile and rupture more easily than normal red cells, leading to anemia.

Sickle-Cell Anemia Is a Molecular Disease

A single amino acid substitution in the β-chains of Hb causes sickle-cell anemia. Replacement of the glutamate residue at position 6 in the β-chain by a valine residue marks the only chemical difference between Hb A and *sickle-cell hemoglobin,* Hb S. The amino acid residues at position $\beta 6$ lie at the surface of the hemoglobin molecule. In Hb A, the ionic R groups of the Glu residues fit this environment. In contrast, the aliphatic side chains of the Val residues in Hb S create hydrophobic protrusions where none existed before. To the detriment of individuals who carry this trait, a hydrophobic pocket forms in the EF corner of each β-chain of Hb when it is in the deoxy state, and this pocket nicely accommodates the Val side chain of a neighboring Hb S molecule (Figure 12.28). This interaction leads to the aggregation of Hb S molecules into long, chainlike polymeric structures. The obvious consequence is that deoxy Hb S is less soluble than deoxy Hb A. The concentration of hemoglobin in red blood cells is high (about 150 mg/mL), so that even in normal circumstances it is on the verge of crystallization. The formation of insoluble deoxy Hb S fibers distorts the red cell into the elongated sickle shape characteristic of the disease.[3]

---

[3]In certain regions of Africa, the sickle-cell trait is found in 20% of the people. Why does such a deleterious heritable condition persist in the population? For reasons as yet unknown, individuals with this trait are less susceptible to the most virulent form of malaria. The geographic distribution of malaria and the sickle-cell trait are positively correlated.

HUMAN BIOCHEMISTRY

Hemoglobin and Nitric Oxide

Nitric oxide (NO·) is a simple gaseous molecule whose many remarkable physiological functions are still being discovered. For example, NO· is known to act as a neurotransmitter and as a second messenger in signal transduction (Chapter 26). Further, **endothelial relaxing factor** (ERF, also known as endothelium-derived relaxing factor, or EDRF), an elusive hormonelike agent that acts to relax the musculature of the walls **(endothelium)** of blood vessels and lower blood pressure, has been identified as NO·. It has long been known that NO· is a high-affinity ligand for Hb, binding to its heme-Fe^{2+} atom with an affinity 10,000 times greater than O_2. An enigma thus arises: why is NO· not instantaneously bound by Hb within human erythrocytes and prevented from exerting its vasodilation properties?

The reason Hb doesn't block the action of NO· is the unique interaction between Cys 93β of Hb and NO· recently described by Li Jia, Celia and Joseph Bonaventura, and Johnathan Stamler at Duke University. Nitric oxide reacts with the sulfhydryl group of Cys 93β, forming an S-nitroso derivative (see figure). This S-nitroso group is in equilibrium with other S-nitroso compounds formed by reaction of NO· with small-molecule thiols such as free cysteine and glutathione (an isoglutamyl cysteinylglycine tripeptide). The structure of S-nitroso-glutathione is given in the second figure. These small-molecule thiols serve to transfer NO· from erythrocytes to endothelial receptors, where it acts to relax vascular tension. NO· itself is a reactive free-radical compound whose biological half-life is very short (1–5 sec). S-nitroso-glutathione has a half-life of several hours.

S-Nitroso-glutathione

The reactions between Hb and NO· are complex. NO· forms a ligand with the heme-Fe^{2+} of Hb that is quite stable in the absence of O_2. However, in the presence of O_2, NO· is oxidized to NO_3^- and the heme-Fe^{2+} of Hb is oxidized to Fe^{3+}, forming methemoglobin. Fortunately, the interaction of Hb with NO· is controlled by the allosteric transition between R-state Hb (oxy Hb) and T-state Hb (deoxy Hb). Cys 93β is more exposed and reactive in R-state Hb than in T-state Hb and binding of NO· to Cys 93β precludes reaction of NO· with heme iron. Upon release of O_2 from Hb in the tissues, Hb shifts conformation from R-state to T-state, and binding of NO· at Cys 93β is no longer favored. Consequently, NO· is released from Cys 93β and transferred to small-molecule thiols in the erythrocyte for delivery to endothelial receptors, causing capillary vasodilation. This mechanism also explains the puzzling observation that free Hb, produced by recombinant DNA methodology for use as a whole-blood substitute, causes a transient rise of 10 to 12 mm Hg in diastolic blood pressure in experimental clinical trials. (Conventional whole-blood transfusion has no such effect on blood pressure.) It is now apparent that the "synthetic" Hb, which has no bound NO·, is binding NO· in the blood and preventing its vasoregulatory function.

In the course of hemoglobin evolution, the only invariant amino acid residues in globin chains are His F8 (the obligatory heme ligand) and a Phe residue acting to wedge the heme into its pocket. However, in mammals and birds, Cys 93β is also invariant, no doubt due to its vital role in NO· delivery.

S-Nitrosocysteine residue $-CH_2-S-N=O$

PROBLEMS

1. Assume an adult human with 5 L of blood (containing 0.75 mM serum albumin) has five fatty acid–binding sites on albumin occupied by palmitates ($H_3C(CH_2)_{15}-COO^-$). How many grams of fatty acid (in the form of palmitate) is being transported in this person's blood?

2. Assume the density of the proteins in an LDL is 1.4 g/mL and the density of the lipids is 0.9 g/mL. If the density of the LDL is 1.02 g/mL, what fraction of its weight is protein and what fraction is lipid?

3. Drug X binds to serum albumin with an association constant (K_D) of 10^{-6} M. If the total concentration of drug X in the bloodstream is 10 μM, what is the concentration of free (unbound) drug X?

4. An individual has a blood cholesterol level of 200 mg/dL and 5 L of blood. Assume that half of the lipid (by weight) in LDLs is cholesterol and also that half of this cholesterol is associated with LDLs. How many grams of LDL are in this person's blood? (Use the answer to question 2 in solving this problem.)

5. The equation $Y/(1 - Y) = (pO_2/P_{50})^n$ allows the calculation of Y (the fractional saturation of hemoglobin with O_2), given P_{50} and n (see *A Deeper Look* on page 402). Let $P_{50} = 26$ torr and $n = 2.8$. Calculate Y in the lungs where $pO_2 = 100$ torr, and Y in the capillaries where $pO_2 = 40$ torr. What is the efficiency of O_2 delivery under these conditions (expressed as $Y_{lungs} - Y_{capillaries}$)? Repeat the calculations, but for $n = 1$. Compare the values for $Y_{lungs} - Y_{capillaries}$ for $n = 2.8$ versus $Y_{lungs} - Y_{capillaries}$ for $n = 1$ to determine the effect of cooperative O_2 binding on oxygen delivery by hemoglobin.

6. If no precautions are taken, blood that has been stored for some time becomes depleted in 2,3-BPG. What happens to the recipient if such blood is used in a transfusion?

FURTHER READING

Alt, F. W., et al., 1987. Development of the primary antibody repertoire. *Science* **238:**1079–1087.

Atkinson, D., 1992. Apoprotein structure: Crystals and models. *Current Opinion in Structural Biology* **2:**482–489.

Banner, D. W., 2000. Not just an active site. *Nature* **404:**449–450. A brief review of the proteins in the blood coagulation cascade and how their activities are regulated through interactions involving modular protein domains.

Bhattacharya, A. A., Grüne, T., and Curry, S., 2000. Crystallographic analysis reveals common modes of binding of medium- and long-chain fatty acids to human serum albumin. *Journal of Molecular Biology*. **303:**721–732.

Borhani, D. W., et al., 1997. Crystal structure of truncated human apolipoprotein A-1 suggests a lipid-bound conformation. *Proceedings of the National Academy of Sciences, U.S.A.* **94:**12291–12296.

Brown, J. H., et al., 2000. The crystal structure of modified bovine fibrinogen. *Proceedings of the National Academy of Sciences, U.S.A.* **97:**85–90.

Curry, S., et al., 1998. Crystal structure of human serum albumin complexed with fatty acid reveals an asymmetric distribution of binding sites. *Nature Structural Biology* **5:**827–835.

Dickerson, R. E., and Geis, I., 1983. *Hemoglobin: Structure, Function, Evolution and Pathology.* Menlo Park, Calif.: Benjamin Cummings.

Doolittle, R. F., 1984. Fibrinogen and fibrin. *Annual Review of Biochemistry* **53:**195–229.

Doolittle, R. F., et al., 1996. Human fibrinogen: Anticipating a 3-dimensional structure. *FASEB Journal* **10:**1464–1470.

Gill, S. J., et al., 1988. New twists on an old story: Hemoglobin. *Trends in Biochemical Sciences* **13:**465–467.

Jia, L., et al., 1996. *S*-Nitrosohaemoglobin: A dynamic activity of blood involved in vascular control. *Nature* **380:**221–226.

Peters, T., Jr., 1996. *All About Albumin: Biochemistry, Genetics, and Medical Applications.* New York: Academic Press.

Roberts, L. M., et al., 1997. Structural analysis of apolipoprotein A-1: Limited proteolysis of methionine-reduced and -oxidized lipid-free and lipid-bound human Apo A-1. *Biochemistry* **36:**7615–7624.

Rosseneu, M., ed., 1992. *Structure and Function of Apolipoproteins.* Boca Raton, Fla.: CRC Press.

Molecular Motors

Outline

Michelangelo's David *epitomizes the musculature of the human form. (The Firenze Academia/photo by Stephanie Colasanti/Corbis)*

Movement is an intrinsic property associated with all living things. Within cells, molecules undergo coordinated and organized movements, and cells themselves may move across a surface. At the tissue level, **muscle contraction** allows higher organisms to carry out and control crucial internal functions, such as peristalsis in the gut and the beating of the heart. Muscle contraction also enables the organism to carry out organized and sophisticated movements, such as walking, running, flying, and swimming.

13.1 Molecular Motors

Motor proteins, also known as **molecular motors,** use chemical energy (ATP) to orchestrate all these movements, transforming ATP energy into the mechanical energy of motion. In all cases, ATP hydrolysis is presumed to drive and control protein conformational changes that result in sliding, walking, or rotating movements of one molecule relative to another. To carry out directed movements, molecular motors must be able to associate and dissociate reversibly with a polymeric protein array, a surface, or a substructure in the cell. ATP hydrolysis drives the process by which the motor protein ratchets along or rotates around the protein array or surface. As fundamental and straightforward as all this sounds, elucidation of these basically simple processes has been extremely challenging for biochemists, involving the application of many sophisticated chemical and physical methods in many different laboratories. This chapter describes the structures and chemical functions of molecular motor proteins and some of the experiments by which we have come to understand them.

Molecular motors may be **linear** or **rotating.** Linear motors crawl or creep along a polymer lattice, whereas rotating motors consist of a rotating element (the "rotor") and a stationary element (the "stator"), in a fashion much like a simple electrical motor. The linear motors we will discuss include **kinesins** and **dyneins** (which crawl along microtubules), **myosin** (which slides along actin filaments in muscle), and **DNA helicases** (which move along a DNA lattice, unwinding duplex DNA to form single-stranded DNA). Rotating motors include the flagellar motor complex, described in this chapter, and the ATP synthase, which will be described in Chapter 17.

13.2 Microtubules and Their Motors

One of the simplest self-assembling structures found in biological systems is the microtubule, one of the fundamental components of the eukaryotic cytoskeleton and the primary structural element of cilia and flagella (Figure 13.1). **Microtubules** are hollow, cylindrical structures, approximately 30 nm in diameter, formed from **tubulin,** a dimeric protein composed of two similar 55-kD subunits known as α-tubulin and β-tubulin. Eva Nogales, Sharon Wolf, and Kenneth Downing have determined the structure of the bovine tubulin $\alpha\beta$ dimer to 3.7 Å resolution (Figure 13.2a). Tubulin dimers polymerize as shown in Figure 13.2b to form microtubules, which are essentially helical structures, with 13 tubulin monomer "residues" per turn. Microtubules grown *in vitro* are dynamic structures that are constantly being assembled and disassembled. Because all tubulin dimers in a microtubule are oriented similarly, microtubules are polar structures. The end of the microtubule at which growth occurs is the **plus end,** and the other is the **minus end.** Microtubules *in vitro* carry out a GTP-dependent process called **treadmilling,** in which tubulin dimers are added to the plus end at about the same rate at which dimers are removed from the minus end (Figure 13.3).

Microtubules Are Constituents of the Cytoskeleton

Although composed only of 55-kD tubulin subunits, microtubules can grow sufficiently large to span a eukaryotic cell or to form large structures such as cilia and flagella. Inside cells, networks of microtubules play many functions, including formation of the mitotic spindle that segregates chromosomes during cell division, the movement of organelles and various vesicular structures through the cell, and the variation and maintenance of cell shape. Microtubules

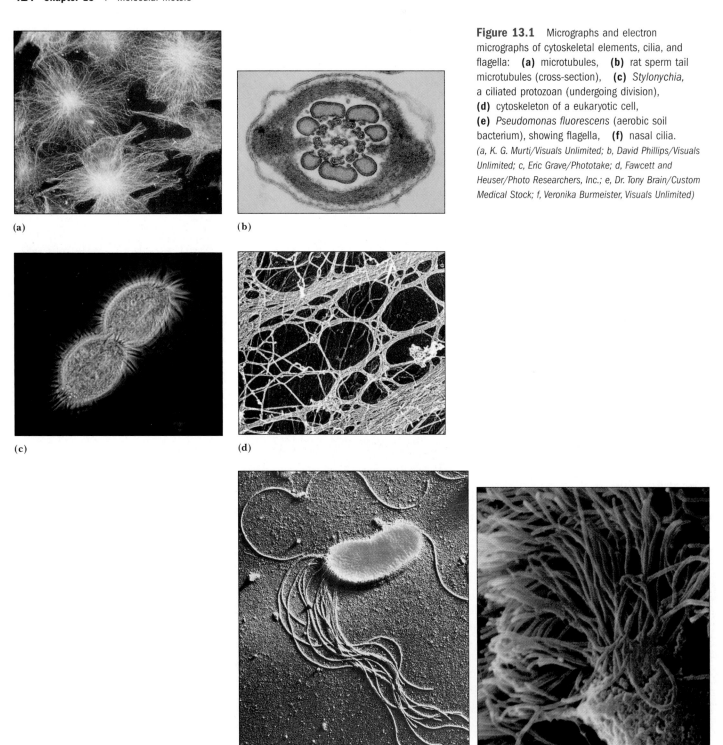

Figure 13.1 Micrographs and electron micrographs of cytoskeletal elements, cilia, and flagella: **(a)** microtubules, **(b)** rat sperm tail microtubules (cross-section), **(c)** *Stylonychia,* a ciliated protozoan (undergoing division), **(d)** cytoskeleton of a eukaryotic cell, **(e)** *Pseudomonas fluorescens* (aerobic soil bacterium), showing flagella, **(f)** nasal cilia. *(a, K. G. Murti/Visuals Unlimited; b, David Phillips/Visuals Unlimited; c, Eric Grave/Phototake; d, Fawcett and Heuser/Photo Researchers, Inc.; e, Dr. Tony Brain/Custom Medical Stock; f, Veronika Burmeister, Visuals Unlimited)*

are, in fact, a significant part of the **cytoskeleton,** a sort of intracellular scaffold formed of microtubules, *intermediate filaments,* and *microfilaments* (Figure 13.4). In most cells, microtubules are oriented with their minus ends toward the cell center and their plus ends toward the cell periphery. This consistent orientation is important for mechanisms of intracellular transport.

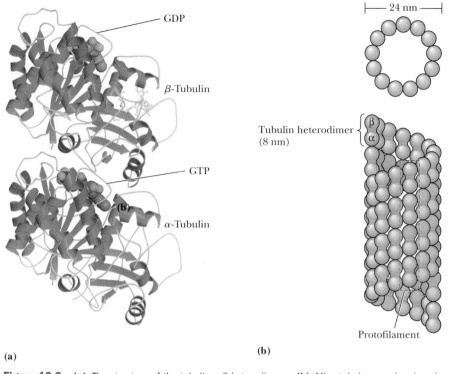

GDP

β-Tubulin

GTP

α-Tubulin

(a)

(b)

|← 24 nm →|

Tubulin heterodimer (8 nm) { β / α }

Protofilament

Dimers on

β / α

Plus end (Growing end)

Minus end

Dimers off

Figure 13.2 **(a)** The structure of the tubulin αβ heterodimer. **(b)** Microtubules may be viewed as consisting of 13 parallel, staggered protofilaments of alternating α-tubulin and β-tubulin subunits. The sequences of the α and β subunits of tubulin are homologous, and the αβ tubulin dimers are quite stable if Ca²⁺ is present. The dimer is dissociated only by strong denaturing agents.

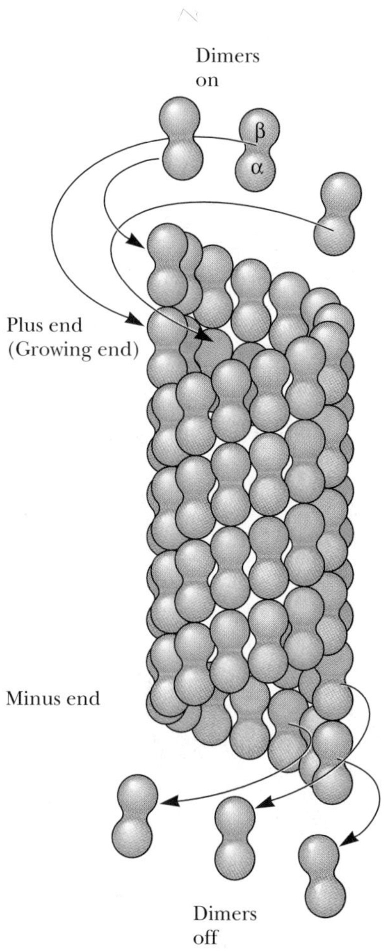

Figure 13.3 A model of the GTP-dependent treadmilling process. Both α- and β-tubulin possess ► two different binding sites for GTP. The polymerization of tubulin to form microtubules is driven by GTP hydrolysis.

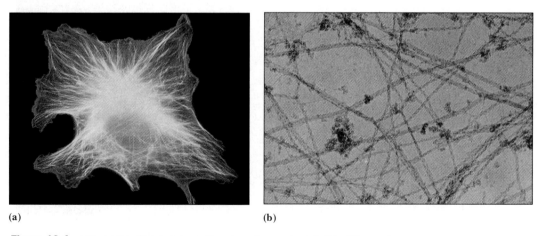

(a)

(b)

Figure 13.4 Intermediate filaments have diameters of approximately 7 to 12 nm, whereas microfilaments, which are made from actin, have diameters of approximately 7 nm. The intermediate filaments appear to play only a structural role (maintaining cell shape), but the microfilaments and microtubules play more dynamic roles. Microfilaments are involved in cell motility, whereas microtubules act as long filamentous tracks, along which cellular components may be rapidly transported by specific mechanisms. **(a)** Cytoskeleton, double-labeled with actin in red and tubulin in green. **(b)** Cytoskeletal elements in a eukaryotic cell, including microtubules (thickest strands), intermediate filaments, and actin microfilaments (smallest strands). (*a, b, M. Schliwa/ Visuals Unlimited*)

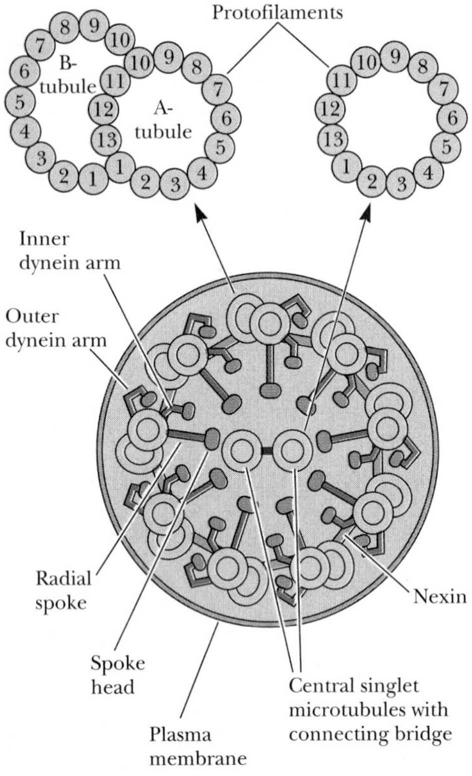

Protofilaments

B-tubule

A-tubule

Inner dynein arm

Outer dynein arm

Radial spoke

Nexin

Spoke head

Central singlet microtubules with connecting bridge

Plasma membrane

Figure 13.5 The structure of an axoneme. Note the manner in which two microtubules are joined in the nine outer pairs. The smaller-diameter tubule of each pair, which is a true cylinder, is called the A-tubule and is joined to the center sheath of the axoneme by a spoke structure. Each outer pair of tubules is joined to adjacent pairs by a nexin bridge. The A-tubule of each outer pair possesses an outer dynein arm and an inner dynein arm. The larger-diameter tubule is known as the B-tubule.

Microtubules Are the Fundamental Structural Units of Cilia and Flagella

As already noted, microtubules are also the fundamental building blocks of cilia and flagella. **Cilia** are short, cylindrical, hairlike projections on the surfaces of the cells of many animals and lower plants. The beating motion of cilia functions either to move cells from place to place or to facilitate the movement of extracellular fluid over the cell surface. Flagella are much longer structures found singly or a few at a time on certain cells (such as sperm cells). They propel cells through fluids. Cilia and flagella share a common design (Figure 13.5). The **axoneme** is a complex bundle of microtubule fibers that includes two central, separated microtubules surrounded by nine pairs of joined microtubules. The axoneme is surrounded by a plasma membrane that is continuous with the plasma membrane of the cell. Removal of the plasma membrane by detergent and subsequent treatment of the exposed axonemes with high concentrations of salt releases the **dynein** molecules (Figure 13.6), which form the dynein arms.

The Mechanism of Ciliary Motion

The motion of cilia results from the ATP-driven sliding or walking of dyneins along one microtubule while they remain firmly attached to an adjacent microtubule. The flexible stems of the dyneins remain permanently attached to A-tubules (Figure 13.6). However, the projections on the globular heads form transient attachments to adjacent B-tubules. Binding of ATP to the dynein heavy chain causes dissociation of the projections from the B-tubules. These projections then reattach to the B-tubules at a position closer to the minus end. Repetition of this process causes the sliding of A-tubules relative to B-tubules. The cross-linked structure of the axoneme dictates that this sliding motion will oc-

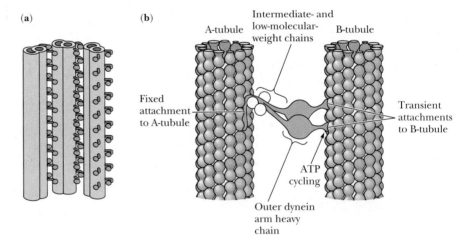

(a)

(b)

Intermediate- and low-molecular-weight chains

A-tubule

B-tubule

Fixed attachment to A-tubule

Transient attachments to B-tubule

ATP cycling

Outer dynein arm heavy chain

Figure 13.6 **(a)** Diagram showing dynein interactions between adjacent microtubule pairs. **(b)** Detailed views of dynein crosslinks between the A-tubule of one microtubule pair and the B-tubule of a neighboring pair. (The B-tubule of the first pair and the A-tubule of the neighboring pair are omitted for clarity.) Isolated axonemal dyneins, which possess ATPase activity, consist of two or three "heavy chains" with molecular masses of 400 to 500 kD, referred to as α and β (and γ when present), as well as several chains with intermediate (40 to 120 kD) and low (15 to 25 kD) molecular masses. Each outer-arm heavy chain consists of a globular domain with a flexible stem on one end and a shorter projection extending at an angle with respect to the flexible stem. In a dynein arm, the flexible stems of several heavy chains are joined in a common base, where the intermediate- and low-molecular-weight proteins are located.

cur in an asymmetric fashion, resulting in a bending motion of the axoneme, as shown in Figure 13.7.

Microtubules Also Mediate Intracellular Motion of Organelles and Vesicles

The ability of dyneins to carry out **mechano-chemical coupling**—i.e., motion coupled with a chemical reaction—is also vitally important *inside* eukaryotic cells, which, as already noted, contain microtubule networks as part of the cytoskeleton. The mechanisms of intracellular, microtubule-based transport of organelles and vesicles were first elucidated in studies of **axons,** the long projections of neurons that extend great distances away from the body of the cell. In axons, subcellular organelles and vesicles can travel at surprisingly fast rates—as great as 2 to 5 μm/sec—in either direction. Unraveling the molecular mechanism for this rapid transport turned out to be a challenging biochemical problem. The early evidence that these movements occur by association with specialized proteins on the microtubules was met with some resistance, for two reasons. First, the notion that a network of microtubules could mediate transport was novel and, like all novel ideas, difficult to accept. Second, many early attempts to isolate dyneins from neural tissue were unsuccessful, and the dyneinlike proteins that were first isolated from neurons were thought to represent contaminations from axoneme structures. However, things changed dramatically in 1985 with a report by Michael Sheetz and his coworkers of a new ATP-driven, force-generating protein, different from myosin and dynein, which they called **kinesin.** Then, in 1987, Richard McIntosh and Mary Porter described the isolation of cytosolic dynein proteins from *Caenorhabditis elegans,* a nematode worm that never makes motile axonemes at any stage of its life cycle. Kinesins have now been found in many eukaryotic cell types, and similar cytosolic dyneins have been found in fruit flies, amoebae, and slime molds; in vertebrate brain and testes; and in HeLa cells (a unique human tumor cell line).

Dyneins Move Organelles in a Plus-to-Minus Direction; Kinesins, in a Minus-to-Plus Direction

The cytosolic dyneins bear many similarities to axonemal dynein. The protein isolated from *C. elegans* includes a "heavy chain" with a molecular mass of approximately 400 kD, as well as smaller peptides with molecular masses ranging from 53 kD to 74 kD. The protein possesses a microtubule-activated ATPase activity, and, when anchored to a glass surface *in vitro,* these proteins, in the presence of ATP, can bind microtubules and move them through the solution. In the cell, cytosolic dyneins specifically move organelles and vesicles from the plus end of a microtubule to the minus end. Thus, as shown in Figure 13.8, dyneins move vesicles and organelles from the cell periphery toward the centrosome (or, in an axon, from the synaptic termini toward the cell body). The kinesins, on the other hand, assist the movement of organelles and vesicles from the minus end to the plus end of microtubules, resulting in outward movement of organelles and vesicles. Kinesin is similar to cytosolic dyneins but smaller in size (360 kD), and contains subunits of 110 kD and 65 to 70 kD. Its length is 100 nm. Like dyneins, kinesins possess ATPase activity in their globular heads, and it is the free energy of ATP hydrolysis that drives the movement of vesicles along the microtubules.

The N-terminal domain of the kinesin heavy chain (38 kD, approximately 340 residues) contains the ATP- and microtubule-binding sites and is the domain responsible for movement. Electron microscopy and image analysis of

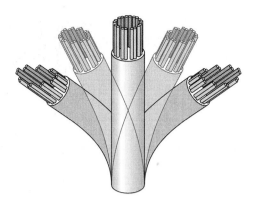

Figure 13.7 A mechanism for ciliary motion. The sliding motion of dyneins along one microtubule while attached to an adjacent microtubule results in a bending motion of the axoneme.

Effectors of Microtubule Polymerization As Therapeutic Agents

Microtubules in eukaryotic cells are important for the maintenance and modulation of cell shape and the disposition of intracellular elements during the cell cycle. It may thus come as no surprise that the inhibition of microtubule polymerization can block many normal cellular processes. The alkaloid **colchicine** (see figure), a constituent of the swollen, underground stems of the autumn crocus (*Colchicum autumnale*) and meadow saffron, inhibits the polymerization of tubulin into microtubules. This effect blocks the mitotic cycle of plants and animals. Colchicine also inhibits cell motility and intracellular transport of vesicles and organelles (which in turn blocks secretory processes of cells). Colchicine has been used for hundreds of years to alleviate some of the acute pain of gout and rheumatism. In gout, white cell lysosomes surround and engulf small crystals of uric acid. The subsequent rupture of the lysosomes and the attendant lysis of the white cells initiate an inflammatory response that causes intense pain. The mechanism of pain alleviation by colchicine is not known for certain, but appears to involve inhibition of white cell movement in tissues. Interestingly, colchicine's ability to inhibit mitosis has given it an important role in the commercial development of new varieties of agricultural and ornamental plants. When mitosis is blocked by colchicine, the treated cells may be left with an extra set of chromosomes. Plants with extra sets of chromosomes are typically larger and more vigorous than normal plants. Flowers developed in this way may grow with double the normal number of petals, and fruits may produce much larger amounts of sugar.

Another class of alkaloids, the **vinca alkaloids** from *Vinca rosea,* the Madagascar periwinkle, can also bind to tubulin and inhibit microtubule polymerization. **Vinblastine** and **vincristine** are used as potent agents for cancer chemotherapy because of their ability to inhibit the proliferation of tumor cells. For reasons that are not well understood, colchicine is not an effective chemotherapeutic agent, though it appears to act similarly to the vinca alkaloids in inhibiting tubulin polymerization.

A new antitumor drug, **taxol,** has been isolated from the bark of *Taxus brevifolia,* the Pacific yew tree. Like vinblastine and colchicine, taxol inhibits cell replication by acting on microtubules. Unlike these other antimitotic drugs, however, taxol stimulates microtubule polymerization and stabilizes microtubules. The remarkable success of taxol in treatment of breast and ovarian cancers stimulated research efforts to synthesize taxol directly and to identify new antimitotic agents that, like taxol, stimulate microtubule polymerization.

Vinblastine: R = CH₃
Vincristine: R = CHO

Colchicine

Taxol

The structures of vinblastine, vincristine, colchicine, and taxol.

tubulin–kinesin complexes reveals (Figure 13.9) that the kinesin head domain is compact and primarily contacts a single tubulin subunit on a microtubule surface, inducing a conformational change in the tubulin subunit. Optical trapping experiments demonstrate that kinesin heads move in 8-nm (80-Å) steps along the long axis of a microtubule. Kenneth Johnson and his coworkers have shown that the ability of a single kinesin tetramer to move unidirectionally for long distances on a microtubule depends upon cooperative interactions between the two mechano-chemical head domains of the protein.

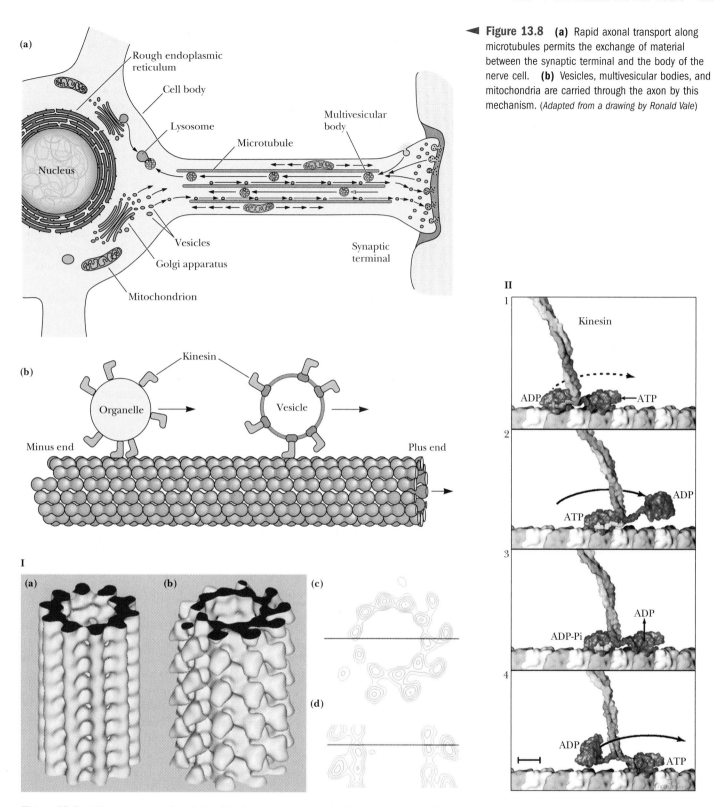

Figure 13.8 **(a)** Rapid axonal transport along microtubules permits the exchange of material between the synaptic terminal and the body of the nerve cell. **(b)** Vesicles, multivesicular bodies, and mitochondria are carried through the axon by this mechanism. (*Adapted from a drawing by Ronald Vale*)

Figure 13.9 I The structure of the tubulin–kinesin complex, as revealed by image analysis of cryoelectron microscopy data. **(a)** The computed, three-dimensional map of a microtubule, **(b)** the kinesin globular head domain-microtubule complex, **(c)** a contour plot of a horizontal section of the kinesin-microtubule complex, and **(d)** a contour plot of a vertical section of the same complex. (*Taken from Kikkawa et al., 1995.* Nature **376:**274–277. *Photo courtesy of Nobutaka Hirokawa.*) II A model for the motility cycle of kinesin. The two heads of the kinesin dimer work together to move processively along a microtubule. Frame 1: Each kinesin head is bound to the tubulin surface. The heads are connected to the coiled coil by "neck linker" segments (orange and red). Frame 2: Conformation changes in the neck linkers flip the trailing head by 160°, over and beyond the leading head and toward the next tubulin binding site. Frame 3: The new leading head binds to a new site on the tubulin surface (with ADP dissociation), completing an 80-Å movement of the coiled coil and the kinesin's cargo. During this time, the trailing head hydrolyzes ATP to ADP and P_i. Frame 4: ATP binds to the leading head, and P_i dissociates from the trailing head, completing the cycle. (*Adapted from Vale, R. D., and Milligan, R. A., 2000.* The way things move: Looking under the hood of molecular motor proteins. Science **288:**88–95.)

13.3 | *E. coli* Rep Helicase: A DNA-Unwinding Protein

DNA normally exists as a double-stranded duplex, but when DNA is to be replicated or repaired, the strands of the double helix must be unwound and separated to form single-stranded DNA intermediates. This separation is carried out by molecular motors known as **DNA helicases** that move along the length of the DNA lattice, sequentially destabilizing the hydrogen bonds between complementary base pairs. The movement along the lattice and the separation of the DNA strands are coupled to the hydrolysis of nucleoside 5′-triphosphates. An important property shared by all helicases is the ability to move along the DNA lattice for long distances without dissociating. This is termed **processive movement,** and helicases are said to have a high processivity. For example, the *E. coli* BCD helicase, which is involved in recombination processes, can unwind 33,000 base pairs before it dissociates from the DNA lattice. Processive movement is essential for helicases involved in DNA replication, where millions of base pairs must be replicated rapidly.

Helicases have evolved at least two structural and functional strategies for achieving high processivity. Certain hexameric helicases form ringlike structures that completely encircle at least one of the strands of a DNA duplex (see Chapter 23). Other helicases, notably **Rep helicase** from *E. coli,* are homodimeric and move processively along the DNA helix by means of a "hand-over-hand" movement that is remarkably similar to that of kinesin's movement along microtubules. A key feature of the hand-over-hand movement of a dimeric motor protein along a polymer is that at least one of the motor subunits must be bound to the polymer at any moment.

Negative Cooperativity Facilitates Hand-Over-Hand Movement

How does hand-over-hand movement of a motor protein along a polymer occur? Clues have come from the structures of Rep helicase and its complexes with DNA. The *E. coli* Rep helicase is a 76-kD protein that is monomeric in the absence of DNA. Binding of Rep helicase to either single-stranded or double-stranded DNA induces dimerization, and the Rep dimer is the active species in unwinding DNA. Each subunit of the Rep dimer can bind either single-stranded (ss) or double-stranded (ds) DNA. However, *the binding of Rep dimer subunits to DNA is negatively cooperative.* Once the first Rep subunit is bound, the affinity of the second subunit for DNA is at least 10,000 times weaker than for the first! This negative cooperativity provides an obvious advantage for hand-over-hand walking. When one "hand" has bound the polymer substrate, the other "hand" releases. A conformation change then moves the unbound "hand" one step farther along the polymer where it can bind again.

But what would provide the energy for such a conformation change? ATP hydrolysis is the driving force for Rep helicase movement along DNA, and the negative cooperativity of Rep binding to DNA is regulated by nucleotide binding. In the absence of nucleotide, a Rep dimer is favored, in which only one subunit is bound to ss DNA. In Figure 13.10a, this state is represented as P_2S (a Rep dimer [P_2] bound to ss DNA [S]). Timothy Lohman and his colleagues at Washington University in St. Louis have shown that binding of ATP analogs induces formation of a complex of the Rep dimer with both ss DNA and ds DNA, one to each Rep subunit (shown as P_2SD in Figure 13.10a). In their model, unwinding of the ds DNA and ATP hydrolysis occur at this point, leaving a P_2S_2 state in which both Rep subunits are bound to ss DNA. Dissociation of ADP and P_i leave the P_2S state again (see Figure 13.10a).

Work by Lohman and his colleagues has shown that coupling of ATP hydrolysis and hand-over-hand movement of Rep along the DNA involves an asym-

(a)

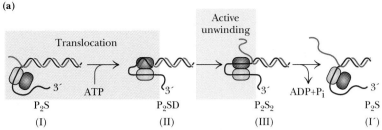

(b)

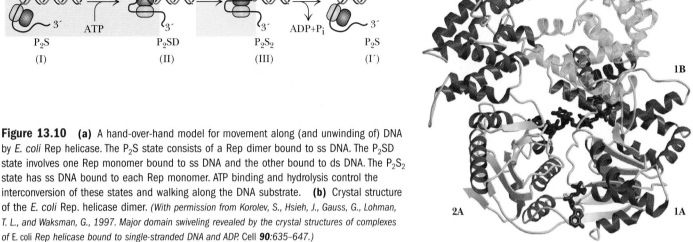

Figure 13.10 **(a)** A hand-over-hand model for movement along (and unwinding of) DNA by *E. coli* Rep helicase. The P_2S state consists of a Rep dimer bound to ss DNA. The P_2SD state involves one Rep monomer bound to ss DNA and the other bound to ds DNA. The P_2S_2 state has ss DNA bound to each Rep monomer. ATP binding and hydrolysis control the interconversion of these states and walking along the DNA substrate. **(b)** Crystal structure of the *E. coli* Rep. helicase dimer. *(With permission from Korolev, S., Hsieh, J., Gauss, G., Lohman, T. L., and Waksman, G., 1997. Major domain swiveling revealed by the crystal structures of complexes of* E. coli Rep *helicase bound to single-stranded DNA and ADP.* Cell **90**:635–647.)

metric state for the Rep dimer. A crystal structure of the Rep dimer in complex with ss DNA and ADP shows that the two Rep monomers are in different conformations (see Figure 13.10b). The two conformations differ by a 130° rotation about a hinge region between two subdomains within the monomer subunit. The hand-over-hand walking of the Rep dimer along the DNA may involve alternation of each subunit between these two conformations, with coordination of the movements by nucleotide binding and hydrolysis.

13.4 Skeletal Muscle Myosin and Muscle Contraction

Structural Features of Skeletal Muscle

The cells of skeletal muscle are long and multinucleate and are referred to as **muscle fibers.** At the microscopic level, skeletal muscle and cardiac muscle display alternating light and dark bands, and for this reason are often referred to as **striated** muscles. Skeletal muscles in higher animals consist of 100-μm-diameter **fiber bundles,** some as long as the muscle itself. Each of these muscle fibers contains hundreds of **myofibrils** (Figure 13.11), each of which spans the length of the fiber and is about 1 to 2 μm in diameter. Myofibrils are linear arrays of cylindrical **sarcomeres,** the basic structural units of muscle contraction. The sarcomeres are surrounded on each end by a membrane system that is actually an elaborate extension of the muscle fiber plasma membrane or **sarcolemma.** These extensions of the sarcolemma, which are called **transverse tubules** or **t-tubules,** enable the sarcolemmal membrane to contact the ends of each myofibril in the muscle fiber (see Figure 13.11). This topological feature is crucial to the initiation of contractions.

Between the t-tubules, the sarcomere is covered with a specialized endoplasmic reticulum called the **sarcoplasmic reticulum,** or **SR.** The SR contains high concentrations of Ca^{2+}, and the release of Ca^{2+} from the SR and its interactions within the sarcomeres trigger muscle contraction, as we will see. Each SR structure consists of two domains. **Longitudinal tubules** run the length of

Figure 13.11 The structure of a skeletal muscle cell, showing the manner in which t-tubules enable the sarcolemmal membrane to contact the ends of each myofibril in the muscle fiber. The foot structure is shown in the box.

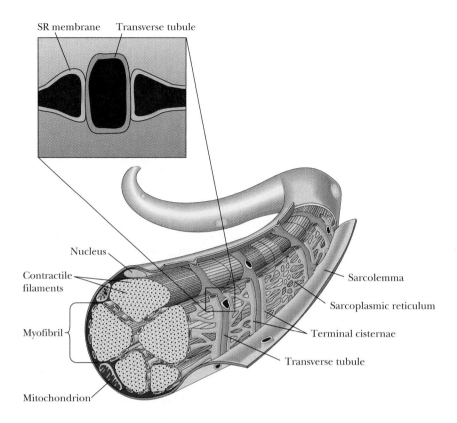

the sarcomere and are capped on either end by the **terminal cisternae** (see Figure 13.11). The structure at the end of each sarcomere, which consists of a t-tubule and two apposed terminal cisternae, is called a **triad,** and the intervening gaps of approximately 15 nm are called **triad junctions.** The junctional face of each terminal cisterna is joined to its respective t-tubule by a **foot structure.**

Skeletal muscle contractions are initiated by nerve stimuli that act directly on the muscle. Nerve impulses produce an electrochemical signal (see Chapter 26) called an action potential that spreads over the sarcolemmal membrane and into the fiber along the t-tubule network. This signal is passed across the triad junction and induces the release of Ca^{2+} ions from the SR. These Ca^{2+} ions bind to the muscle fibers and induce contraction.

The Molecular Structure of Skeletal Muscle

Examination of myofibrils in the electron microscope reveals a banded or striated structure. The bands are traditionally identified by letters (Figure 13.12). Regions of high electron density, denoted **A bands,** alternate with regions of low electron density, the **I bands.** Small, dark **Z lines** lie in the middle of the I bands, marking the ends of the sarcomere. Each A band has a central region of slightly lower electron density called the **H zone,** which contains a central **M disk** (also called an **M line**). Electron micrographs of cross-sections of each of these regions reveal molecular details. The H zone shows a regular, hexagonally arranged array of **thick filaments** (15 nm diameter), whereas the I band shows a regular, hexagonal array of **thin filaments** (7 nm diameter). In the dark regions at the ends of each A band, the thin and thick filaments interdigitate, as shown in Figure 13.12. The thin filaments are composed primarily of three proteins called **actin, troponin,** and **tropomyosin.** The thick filaments consist mainly of a protein called **myosin.** The thin and thick filaments are joined by **cross-bridges.** These cross-bridges are actually extensions of the myosin molecules, and muscle

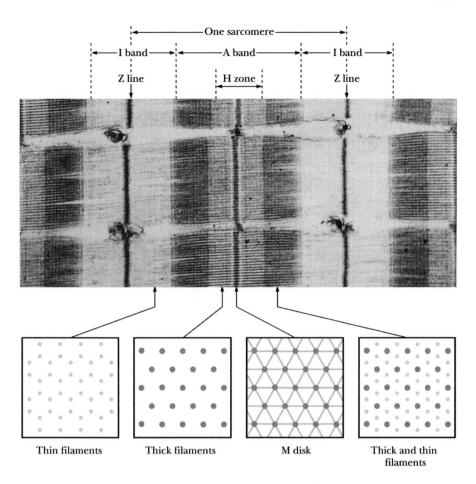

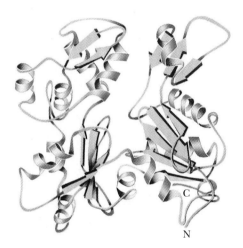

Figure 13.12 Electron micrograph of a skeletal muscle myofibril (in longitudinal section). The length of one sarcomere is indicated, as are the A and I bands, the H zone, the M disk, and the Z lines. Cross-sections from the H zone show a hexagonal array of thick filaments, whereas the I band cross-section shows a hexagonal array of thin filaments. (*Photo courtesy of Hugh Huxley, Brandeis University*)

contraction is accomplished by the sliding of the cross-bridges along the thin filaments, a mechanical movement driven by the free energy of ATP hydrolysis.

The Composition and Structure of Thin Filaments

Actin, the principal component of thin filaments, can be isolated in two forms. Under conditions of low ionic strength, actin exists as a 42-kD globular protein, denoted **G-actin.** G-actin consists of two principal lobes or domains (Figure 13.13). Under physiological conditions (higher ionic strength), G-actin polymerizes to form a fibrous form of actin, called **F-actin.** As shown in Figure 13.14, F-actin is a right-handed helical structure, with a helix pitch of about 72 nm per turn. The F-actin helix is the core of the thin filament, to which tropomyosin and the **troponin complex** also add. Tropomyosin is a dimer of homologous but nonidentical 33-kD subunits. These two subunits form long α-helices that intertwine, creating 38- to 40-nm-long coiled coils, which join in head-to-tail fashion to form long rods. These rods bind to the F-actin polymer and lie almost parallel to the long axis of the F-actin helix (Figure 13.15a–c). The troponin complex consists of three different proteins: **troponin T, or TnT** (37 kD); **troponin I, or TnI** (24 kD); and **troponin C, or TnC** (18 kD). TnT binds to tropomyosin, specifically at the head-to-tail junction. Troponin I binds both to tropomyosin and to actin. Troponin C is a Ca^{2+}-binding protein that binds to TnI. TnC shows 70% homology with the Ca^{2+} signaling protein, calmodulin (Chapter 26). The release of Ca^{2+} from the SR, which signals a contraction, raises the cytosolic Ca^{2+} concentration high enough to saturate the Ca^{2+} sites on TnC. Ca^{2+} binding induces a conformational change in the amino-terminal domain of TnC, which in turn causes a rearrangement of the troponin complex and tropomyosin with respect to the actin fiber.

Figure 13.13 The three-dimensional structure of an actin monomer from skeletal muscle. This view shows the two domains (*left and right*) of actin. (*Jane Richardson, Duke University*)

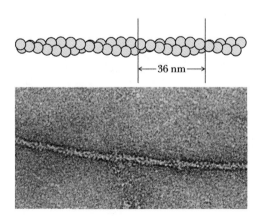

Figure 13.14 The helical arrangement of actin monomers in F-actin. The F-actin helix has a pitch of 72 nm and a repeat distance of 36 nm. (*Electron micrograph courtesy of Hugh Huxley, Brandeis University*)

Figure 13.15 **(a)** An electron micrograph of a thin filament, **(b)** a corresponding image reconstruction, and **(c)** a schematic drawing based on the images in (a) and (b). The tropomyosin coiled coil winds around the actin helix, each tropomyosin dimer interacting with seven consecutive actin monomers. Troponin T binds to tropomyosin at the head-to-tail junction. (*a and b, courtesy of Linda Rost and David DeRosier, Brandeis University; c, courtesy of George Phillips, Rice University*)

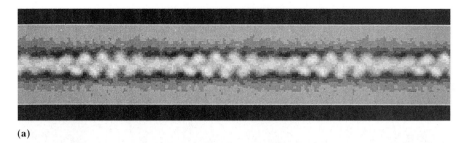

(a)

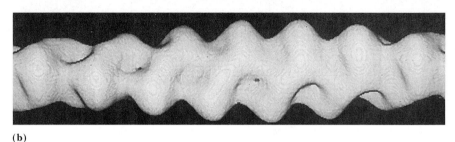

(b)

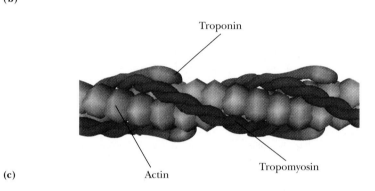

(c)

The Composition and Structure of Thick Filaments

Myosin, the principal component of muscle thick filaments, is a large protein consisting of six polypeptides, with an aggregate molecular weight of approximately 540 kD. As shown in Figure 13.16, the six peptides include two 230-kD **heavy chains,** as well as two pairs of different 20-kD **light chains,** denoted **LC1** and **LC2.** The heavy chains consist of globular amino-terminal **myosin heads,** joined to long α-helical carboxy-terminal segments, the **tails.** These tails are intertwined to form a left-handed coiled coil approximately 2 nm in diameter and 130 to 150 nm long. Each of the heads in this dimeric structure is associated with an LC1 and an LC2. The myosin heads exhibit **ATPase activity,** and hydrolysis of ATP by the myosin heads drives muscle contraction. LC1 is also known as the **essential light chain (ELC),** and LC2 is designated the **regulatory light chain (RLC).** Both light chains are homologous to calmodulin and TnC. Dissociation of LC1 from the myosin heads results in loss of the myosin ATPase activity.

Approximately 500 of the 820 amino acid residues of the myosin head are highly conserved between various species. One conserved region, located approximately at residues 170 to 214, constitutes part of the ATP-binding site. Whereas many ATP-binding proteins and enzymes employ a β-sheet–α-helix–β-sheet motif, this region of myosin forms a related α-β-α structure, beginning with an Arg at (approximately) residue 192. The β-sheet in this region of all myosins includes the amino acid sequence

Gly-Glu-Ser-Gly-Ala-Gly-Lys-Thr

The Gly-X-X-Gly-X-Gly found in this segment is found in many ATP- and nucleotide-binding enzymes. The Lys of this segment is thought to interact with the α-phosphate of bound ATP.

(a)

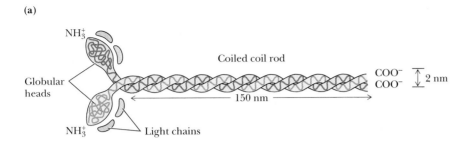

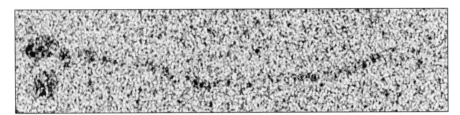

Globular heads · NH₃⁺ · Coiled coil rod · COO⁻ COO⁻ 2 nm · 150 nm · Light chains

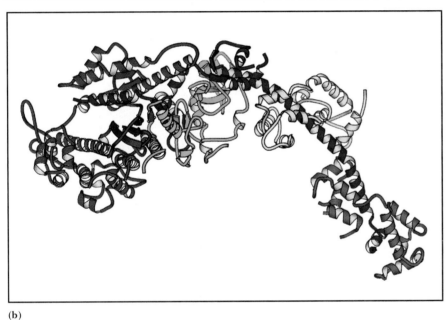

(b)

Figure 13.16 **(a)** An electron micrograph of a myosin molecule and a corresponding schematic drawing. The tail is a coiled coil of intertwined α-helices extending from the two globular heads. One of each of the myosin light chain proteins, LC1 and LC2, is bound to each of the globular heads. **(b)** A ribbon diagram shows the structure of the S1 myosin head (green, red, and purple segments) and its associated essential (yellow) and regulatory (magenta) light chains. (a, Electron micrograph courtesy of Henry Slayter, Harvard Medical School; b, courtesy of Ivan Rayment and Hazel M. Holden, University of Wisconsin, Madison)

See *Interactive Biochemistry CD-ROM and Workbook,* page 51

Repeating Structural Elements Are the Secret of Myosin's Coiled Coils

Myosin tails show less homology than the head regions, but several key features of the tail sequence are responsible for the α-helical coiled coils formed by myosin tails. *Several orders of repeating structure* are found in all myosin tails, including 7-residue, 28-residue, and 196-residue repeating units. Large stretches of the tail domain are composed of 7-residue repeating segments. The first and fourth residues of these 7-residue units are generally small, hydrophobic amino acids, whereas the second, third, and sixth are likely to be charged residues. The consequence of this arrangement is shown in Figure 13.17. Seven residues form two turns of an α-helix, and, in the coiled coil structure of the myosin tails, the first and fourth residues face the interior contact region of the coiled coil. Residues b, c, and f (2, 3, and 6) of the 7-residue repeat face the periphery, where charged residues can interact with the water solvent. Groups of four 7-residue units with distinct patterns of alternating side-chain charge form 28-residue repeats that establish alternating regions of positive and negative charge on the surface of the myosin coiled coil. These alternating charged regions interact with similar regions in the tails of adjacent myosin molecules to assist in stabilizing the thick filament.

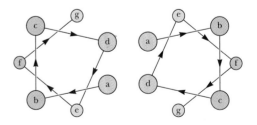

Figure 13.17 An axial view of the two-stranded, α-helical coiled coil of a myosin tail. Hydrophobic residues a and d of the 7-residue repeat sequence align to form a hydrophobic core. Residues b, c, and f face the outer surface of the coiled coil and are usually charged.

 See pages 57, 58

The Molecular Defect in Duchenne Muscular Dystrophy Involves an Actin-Anchoring Protein

Duchenne muscular dystrophy is a degenerative and fatal disorder of the muscle affecting approximately 1 in 3500 boys. Victims of Duchenne dystrophy show early abnormalities in walking and running. By the age of five, the victim cannot run and has difficulty standing, and by early adolescence, walking is difficult or impossible. The loss of muscle function progresses upward in the body, affecting next the arms and the diaphragm. Respiratory problems or infections usually result in death by the age of 30. Louis Kunkel and his coworkers identified the Duchenne muscular dystrophy gene in 1986. This gene produces a protein called **dystrophin,** a protein that appears to function by anchoring actin filaments to the sarcolemmal membrane and to the extracellar matrix. A defect in dystrophin is responsible for the muscle degeneration of Duchenne dystrophy.

(a)

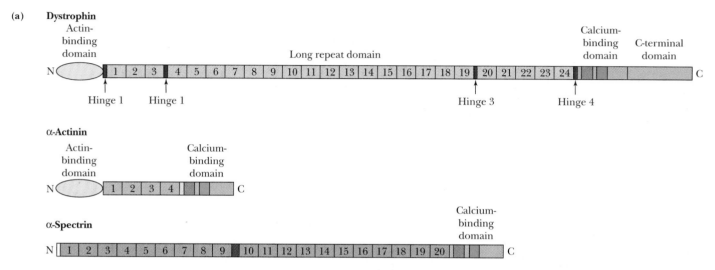

A comparison of the amino acid sequences of dystrophin, α-actinin, and spectrin. The potential hinge segments in the dystrophin structure are indicated.

(b)

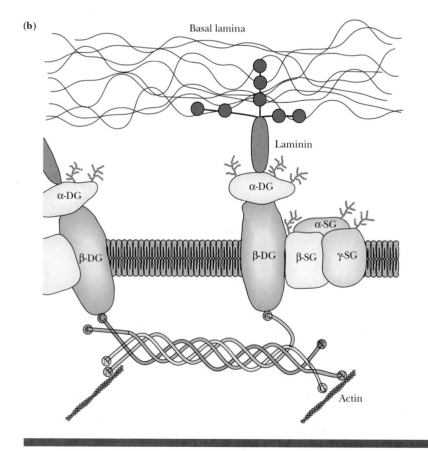

A model for the actin–dystrophin–glycoprotein complex in skeletal muscle. Dystrophin is postulated to form tetramers of antiparallel monomers that bind actin at their N-termini and a family of dystrophin-associated glycoproteins at their C-termini. This dystrophin-anchored complex may function to stabilize the sarcolemmal membrane during contraction–relaxation cycles, link the contractile force generated in the cell (fiber) with the extracellular environment, or maintain local organization of key proteins in the membrane. The dystrophin-associated membrane proteins (dystroglycans [DGs] and sarcoglycans [SGs]) range from 25 to 154 kD. (Adapted from Ahn, A. H., and Kunkel, L. M., 1993. Nature Genetics **3:**283–291, and Worton, R., 1995. Science **270:**755–756.)

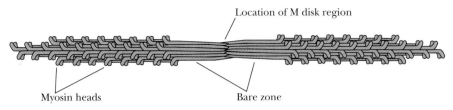

Figure 13.18 The packing of myosin molecules in a thick filament. Adjoining molecules are offset by approximately 14 nm, a distance corresponding to 98 residues of the coiled coil.

At a still higher level of organization, groups of seven of these 28-residue units—a total of 196 residues—also form a repeating pattern, and this large-scale repeating motif contributes to the packing of the myosin molecules in the thick filament. The myosin molecules in thick filaments are offset (Figure 13.18) by approximately 14 nm, a distance that corresponds to 98 residues of a coiled coil, or exactly half the length of the 196-residue repeat. Thus, several layers of repeating structure play specific roles in the formation and stabilization of the myosin coiled coil and the thick filament formed from them.

The Mechanism of Muscle Contraction

When muscle fibers contract, the thick myosin filaments slide or walk along the thin actin filaments. The basic elements of the **sliding filament model** were first described in 1954 by two different research groups, Hugh Huxley and his colleague Jean Hanson, and the physiologist Andrew Huxley and his colleague Ralph Niedergerke. Several key discoveries paved the way for this model. Electron microscopic studies of muscle revealed that sarcomeres decreased in length during contraction, and that this decrease was due to decreases in the width of both the I band and the H zone (Figure 13.19). At the same time, the width of the A band (which is the length of the thick filaments) and the distance from the Z disks to the nearby H zone (that is, the length of the thin filaments) did not change. These observations made it clear that the lengths of both the thin and thick filaments were constant during contraction. This conclusion was consistent with a sliding filament model.

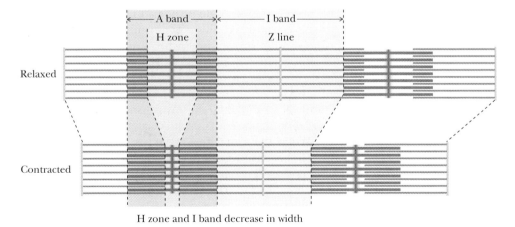

Figure 13.19 The sliding filament model of skeletal muscle contraction. The decrease in sarcomere length is due to decreases in the width of the I band and H zone, with no change in the width of the A band. These observations mean that the lengths of both the thick and thin filaments do not change during contraction. Rather, the thick and thin filaments slide along one another.

The Sliding Filament Model

The shortening of a sarcomere (see Figure 13.19) involves sliding motions in opposing directions at the two ends of a myosin thick filament. Net sliding motions in a specific direction occur because the thin and thick filaments both have **directional character.** The organization of the thin and thick filaments in the sarcomere takes particular advantage of this directional character. Actin filaments always extend outward from the Z lines in a uniform manner. Thus, between any two Z lines, the two sets of actin filaments point in opposing directions. The myosin thick filaments, on the other hand, also assemble in a directional manner. The polarity of myosin thick filaments reverses at the M disk. The nature of this reversal is not well understood, but presumably involves structural constraints provided by proteins in the M disk. The reversal of polarity at the M disk means that actin filaments on either side of the M disk are pulled toward the M disk during contraction by the sliding of the myosin heads, causing net shortening of the sarcomere.

Contraction Is Powered by Myosin ATPase Activity

The specific effect of actin on myosin ATPase becomes apparent if the product-release steps of the reaction are carefully compared. In the absence of actin, the addition of ATP to myosin produces a rapid release of H^+, one of the products of the ATPase reaction:

$$ATP^{4-} + H_2O \rightarrow ADP^{3-} + P_i^{2-} + H^+$$

However, release of ADP and P_i from myosin is much slower. Actin activates myosin ATPase activity by stimulating the release of P_i and then ADP. Product release is followed by the binding of a new ATP to the actomyosin complex, which causes actomyosin to dissociate into free actin and myosin. The cycle of ATP hydrolysis then repeats, as shown in Figure 13.20a. The crucial point of this model is that *ATP hydrolysis and the association and dissociation of actin and myosin are coupled.* It is this coupling that enables ATP hydrolysis to power muscle contraction.

The Coupling Mechanism: ATP Hydrolysis Drives Conformation Changes in the Myosin Heads

The only remaining piece of the puzzle is this: How does the close coupling of actin–myosin binding and ATP hydrolysis result in the shortening of myofibrils? Put another way, how are the models for ATP hydrolysis and the sliding filament related? The answer to this puzzle is shown in Figure 13.20b. The free energy of ATP hydrolysis is translated into a conformation change in the myosin head, so that dissociation of myosin and actin, hydrolysis of ATP, and rebinding of myosin and actin occur with stepwise movement of the myosin S1 head along the actin filament. The conformation change in the myosin head is driven by the hydrolysis of ATP.

As shown in the cycle in Figure 13.20a, the myosin heads—with the hydrolysis products ADP and P_i bound—are mainly dissociated from the actin filaments in resting muscle. When the signal to contract is presented (see following discussion), the myosin heads move out from the thick filaments to bind to actin on the thin filaments (Step 1). Binding to actin stimulates the release of phosphate, and this is followed by the crucial conformational change by the S1 myosin heads—the so-called **power stroke**—and ADP dissociation. In this step (Step 2), the thick filaments move along the thin filaments as the myosin heads relax to a lower energy conformation. In the power stroke, the myosin heads tilt by approximately 45 degrees. This moves the thick filament approximately 10 nm (100Å) along the thin filament (Step 3). Subsequent binding (Step 4) and hydrolysis (Step 5) of ATP cause dissociation of the heads from the thin filaments and also cause the myosin heads to shift back to their high-

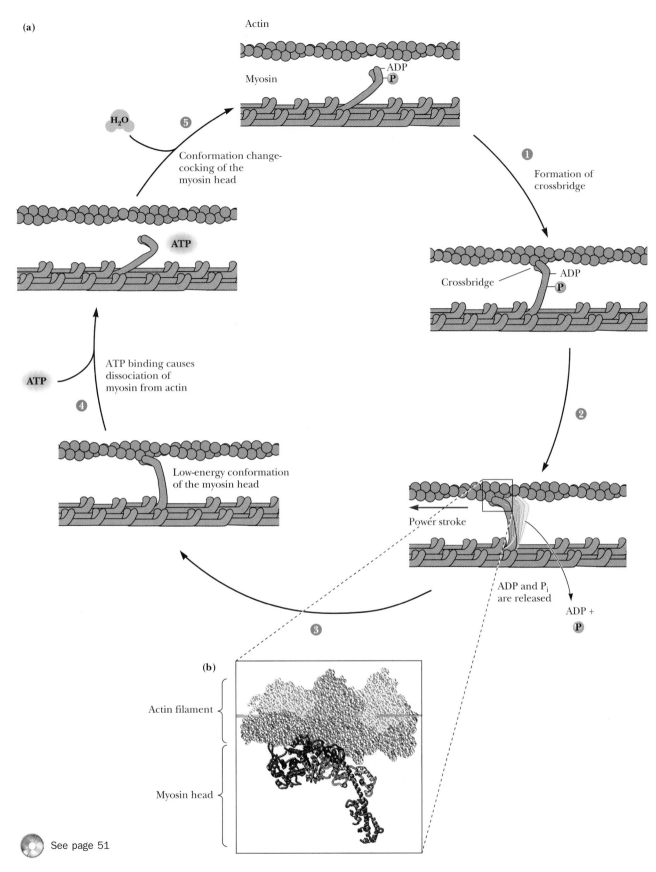

Figure 13.20 The mechanism of skeletal muscle contraction. The free energy of ATP hydrolysis drives a conformational change in the myosin head, resulting in net movement of the myosin heads along the actin filament. (*Inset*) A ribbon and space-filling representation of the actin–myosin interaction. (*S1 myosin image courtesy of Ivan Rayment and Hazel M. Holden, University of Wisconsin, Madison*)

See page 51

energy conformation with the heads' long axis nearly perpendicular to the long axis of the thick filaments. The heads may then begin another cycle by binding to actin filaments. This cycle is repeated at rates up to 5/sec in a typical skeletal muscle contraction. The conformational changes occurring in this cycle are the secret of the energy coupling that allows ATP binding and hydrolysis to drive muscle contraction.

Images of myosin, when compared with the X-ray crystal structure of myosin S1, show that the long α-helix of S1 that binds the light chains (ELC and RLC) behaves as a lever arm, and that this arm swings through an arc of 23 degrees upon release of ADP. (A glycine residue at position 770 in the S1 myosin head lies at the N-terminal end of this helix/lever arm and may act as a hinge.) *This results in a 3.5-nm (35-Å) movement of the last myosin heavy chain residue of the X-ray structure in a direction nearly parallel to the actin filament.* These two "snapshots" of the myosin S1 conformation may represent only part of the working power stroke of the contraction cycle, and the total movement of a myosin head with respect to the apposed actin filament may be as much as 5.3 nm.

The Movements of Myosin and Kinesin Are Similar

The ATP hydrolysis cycle must be linked to a conformational change cycle for motor proteins to produce directed motion. How is ATP hydrolysis coupled to the conformation change cycle? For both myosin and kinesin, a part of the protein must act as a "γ-phosphate sensor" to detect the presence or absence of the γ-P of ATP in the active site. In both myosin and kinesin, this sensor consists of two loops of the protein, termed "switch I" and "switch II," which form H bonds with the γ-P and which orient a water molecule and crucial protein residues involved in ATP hydrolysis. Small movements of the γ-P sensor are communicated to distant parts of the protein by a long "relay helix" at the amino-terminus of switch II. *The relay helix moves back and forth like a piston to link tiny movements in switch II to larger movements of the protein (Figure 13.21).*

Control of the Contraction–Relaxation Cycle by Calcium Channels and Pumps

The trigger for all muscle contraction is an increase in Ca^{2+} concentration in the vicinity of the muscle fibers of skeletal muscle or the myocytes of cardiac and smooth muscle. In all these cases, this increase in Ca^{2+} is due to the flow of Ca^{2+} through **calcium channels** (Figure 13.22). A muscle contraction ends when the Ca^{2+} concentration is reduced by specific calcium pumps (such as the SR Ca^{2+}-ATPase, Chapter 6).

Regulation of Contraction by Ca^{2+}

The importance of Ca^{2+} ion as the triggering signal for muscle contraction was described earlier. Ca^{2+} is the intermediary signal that allows striated muscle to respond to motor nerve impulses (see Figure 13.22). The Ca^{2+} signal is correctly interpreted by muscle only when tropomyosin and the troponins are present. Specifically, actomyosin prepared from pure preparations of actin and myosin (thus containing no tropomyosin and troponins) contracts when ATP is added, even in the absence of Ca^{2+}. However, actomyosin prepared directly from whole muscle contracts in the presence of ATP only when Ca^{2+} is added. Clearly muscle extracts contain a factor that confers normal Ca^{2+} sensitivity to actomyosin. The factor is the tropomyosin–troponin complex.

Actin thin filaments consist of actin, tropomyosin, and the troponins in a 7:1:1 ratio (see Figure 13.15). Each tropomyosin molecule spans seven actin

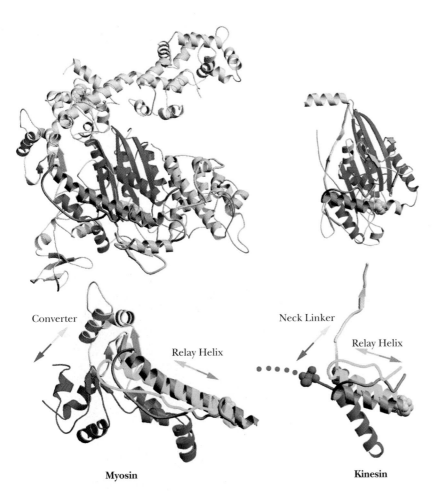

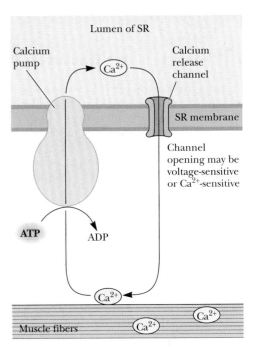

Figure 13.22 Ca^{2+} is the trigger signal for muscle contraction. Release of Ca^{2+} through voltage- or Ca^{2+}-sensitive channels activates contraction. Ca^{2+} pumps induce relaxation by reducing the concentration of Ca^{2+} available to the muscle fibers.

Figure 13.21 Ribbon structures of the myosin and kinesin motor domains and the conformational changes triggered by the γ-P sensor and the relay helix. The upper panels represent the motor domains of myosin and kinesin, respectively, in the ATP- or ADP-P_i–like state. Similar structural elements in the catalytic cores of the two domains are shown in blue, the relay helices are dark green, and the mechanical elements (neck linker for kinesin, lever arm domains for myosin) are yellow. The nucleotide is shown as a white space-filling model. The similarity of the conformation changes caused by the relay helix in going from the ATP/ADP-P_i-bound state to the ADP-bound or nucleotide-free state is shown in the lower panels. In both cases, the mechanical elements of the protein shift their positions in response to relax helix motion. Note that the direction of mechanical element motion is nearly perpendicular to the relay helix motion. (*Adapted from Vale, R. D., and Milligan, R. A., 2000. The way things move: Looking under the hood of molecular motor proteins. Science* **288**:*88–95.*)

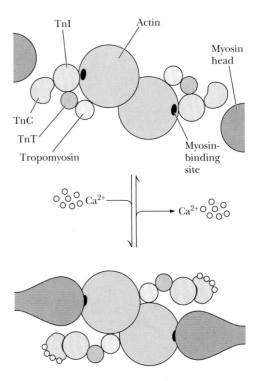

molecules, lying along the thin filament groove, between pairs of actin monomers. As shown in a cross-section view in Figure 13.23, in the absence of Ca^{2+}, troponin I is thought to interact directly with actin to prevent the interaction of actin with myosin S1 heads. Troponin I and troponin T interact with tropomyosin to keep tropomyosin away from the groove between adjacent actin monomers. However, the binding of Ca^{2+} ions to troponin C appears to increase the binding of troponin C to troponin I, simultaneously decreasing the interaction of troponin I with actin. As a result, tropomyosin slides deeper into the actin thin filament groove, exposing myosin-binding sites on actin, and initiating the muscle contraction cycle (see Figure 13.20). Because the troponin complexes can interact only with every seventh actin in the thin filament, the conformational changes that expose myosin-binding sites on actin may well be cooperative. Binding of an S1 head to an actin may displace tropomyosin and the troponin complex from myosin-binding sites on adjacent actin subunits.

Figure 13.23 A drawing of the thick and thin filaments of skeletal muscle in cross-section showing the changes that are postulated to occur when Ca^{2+} binds to troponin C.

HUMAN BIOCHEMISTRY

Smooth Muscle Effectors Are Useful Drugs

Not all vertebrate muscle is skeletal muscle. Vertebrate organisms employ smooth muscle for long, slow, and involuntary contractions in various organs, including large blood vessels, intestinal walls, the gums of the mouth, and, in the female, the uterus. In smooth muscle, contraction is initiated when Ca^{2+} binds to myosin light chain kinase (**MLCK**). MLCK is a protein kinase that phosphorylates the myosin light chains. This phosphorylation triggers smooth muscle contraction.

The action of epinephrine and related agents forms the basis of therapeutic control of smooth muscle contraction. Breathing disorders, including asthma and various allergies, can result from excessive contraction of bronchial smooth muscle tissue. Treatment with epinephrine, whether by tablets or aerosol inhalation, inhibits MLCK and relaxes bronchial muscle tissue. More specific **bronchodilators,** such as **albuterol** (see figure), act more selectively on the lungs and avoid the undesirable side effects of epinephrine on the heart. Albuterol is also used to prevent premature labor in pregnant women, owing to its relaxing effect on uterine smooth muscle. Conversely, **oxytocin,** known also as **pitocin,** stimulates contraction of uterine smooth muscle. This natural secretion of the pituitary gland is often administered to induce labor.

Albuterol

$$H_3\overset{+}{N} - Gly - Leu - Pro - Cys - Asn - Gln - Ile - Tyr - Cys - COO^-$$
$$S - S$$

Oxytocin (Pitocin)

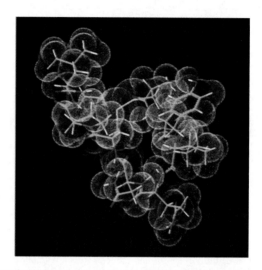

The structure of oxytocin. ▶

The Interaction of Ca^{2+} with Troponin C

There are four Ca^{2+}-binding sites on troponin C—two high-affinity sites on the carboxy-terminal end of the molecule, labeled III and IV in Figure 13.24, and two low-affinity sites on the amino-terminal end, labeled I and II. Ca^{2+} binding to sites III and IV is sufficiently strong ($K_D = 0.1\ \mu M$) that these sites are presumed to be filled under resting conditions. Sites I and II, however, where the K_D is approximately $10\ \mu M$, are empty in resting muscle. The rise of Ca^{2+} levels when contraction is signaled leads to the filling of sites I and II, causing a conformation change in the amino-terminal domain of TnC. This conformational change apparently facilitates a more intimate binding of TnI to TnC. The increased interaction between TnI and TnC results in a decreased interaction between TnI and actin.

13.5 | A Proton Gradient Drives the Rotation of Bacterial Flagella

Bacterial cells swim and move by rotating their flagella. The flagella of *Escherichia coli* are helical filaments about 10,000 nm (10 μm) in length and 15 nm in diameter. The direction of rotation of these filaments affects the

Calcium-binding domains

II

I

NH_3^+

COO^-

IV

III

(a)

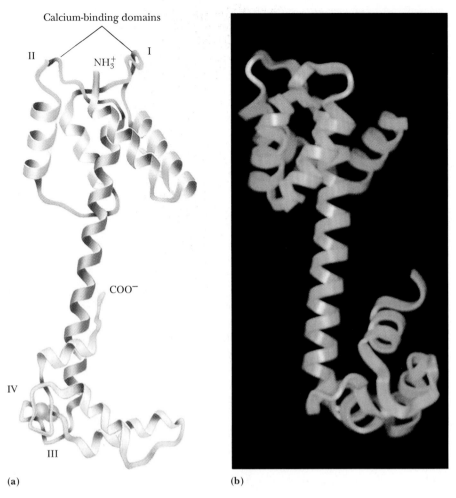

(b)

Figure 13.24 **(a)** A ribbon diagram and **(b)** a molecular graphic showing two slightly different views of the structure of troponin C. Note the long α-helical domain connecting the N-terminal and C-terminal lobes of the molecule.

 See page 105

movements of the cell. When the half-dozen filaments on the surface of the bacterial cell rotate in a counterclockwise direction, they twist and bundle together and rotate in a concerted fashion, propelling the cell through the medium. (On the other hand, clockwise-rotating flagella cannot bundle together, and under such conditions the cell merely tumbles and moves erratically.)

The rotations of bacterial flagellar filaments are the result of the rotation of motor protein complexes in the bacterial plasma membrane. The flagellar motor consists of at least two rings (including the M ring and the S ring) with diameters of about 25 nm assembled around and connected rigidly to a rod attached in turn to the helical filament (Figure 13.25). The rings are surrounded by a circular array of membrane proteins. In all, at least 40 proteins are involved in this magnificent assembly. One of these, the motB protein, lies on the edge of the M ring, where it interacts with the motA protein, located in the membrane protein array and facing the M ring.

In contrast to the many other motor proteins described in this chapter, the flagellar motor is driven by a proton gradient, not ATP hydrolysis. The concentration of protons, $[H^+]$, outside the cell is typically higher than that inside the cell. Thus, there is a thermodynamic tendency for protons to move into the cell. The motA and motB proteins together form a proton shuttling device

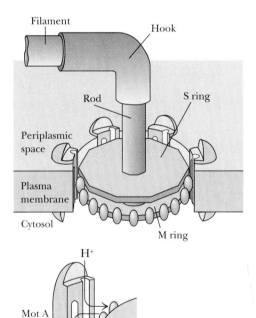

Filament

Hook

Rod

S ring

Periplasmic space

Plasma membrane

Cytosol

M ring

H^+

Mot A

H^+

Mot B

Figure 13.25 A model of the flagellar motor assembly of *Escherichia coli*. The M ring carries an array of about 100 motB proteins at its periphery. These juxtapose with motA proteins in the protein complex that surrounds the ring assembly. Movement of protons through the motA/motB complexes drives the rotation of the rings and the associated rod and helical filament.

(a)

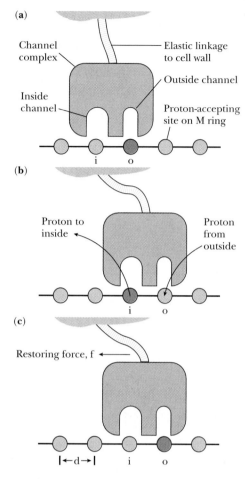

(b)

(c)

◀ **Figure 13.26** Howard Berg's model for coupling between transmembrane proton flow and rotation of the flagellar motor. A proton moves through an outside channel to bind to an exchange site on the M ring. When the channel protein slides one step around the ring, the proton is released and flows through an inside channel and into the cell, while another proton flows into the outside channel to bind to an adjacent exchange site. When the motA channel protein returns to its original position under an elastic restoring force, the associated motB protein moves with it, causing a counterclockwise rotation of the ring, rod, and helical filament. (*Adapted from Meister, M., Caplan, S. R., and Berg, H. C., 1989. Dynamics of a tightly coupled mechanism for flagellar rotation. Biophysical Journal* **55**:905–914.)

See page 167

that is coupled to motion of the motor disks. Proton movement into the cell through this protein complex or "channel" drives the rotation of the flagellar motor. A model for this coupling has been proposed by Howard Berg and his coworkers (Figure 13.26). In this model, the motB proteins possess proton-exchanging sites—for example, carboxyl groups on aspartate or glutamate residues or imidazole moieties on histidine residues. The motA proteins, on the other hand, possess a pair of "half-channels," with one half-channel facing the inside of the cell and the other facing the outside. In Berg's model, the outside edges of the motA channel protein cannot move past a proton-exchanging site on motB when that site has a proton bound, and the center of the channel protein cannot move past an exchange site when that site is empty. As shown in Figure 13.26, these constraints lead to coupling between proton translocation and rotation of the flagellar filament. For example, imagine that a proton has entered the outside channel of motA and is bound to an exchange site on motB (Figure 13.26a). An oscillation by motA, linked elastically to the cell wall, can then position the inside channel over the proton at the exchange site (Figure 13.26b), whereupon the proton can travel through the inside channel and into the cell, while another proton travels up the outside channel to bind to an adjacent exchange site. The restoring force acting on the channel protein then pulls the motA/motB complex to the left as shown (Figure 13.26c), leading to counterclockwise rotation of the disk, rod, and helical filament. The flagellar motor is driven entirely by the proton gradient. Thus, a reversal of the proton gradient (which would occur, for example, if the external medium became alkaline) would drive the flagellar filaments in a clockwise direction. Extending this picture of a single motA/motB complex to the whole motor disk array, one can imagine the torrent of protons that pass through the motor assembly to drive flagellar rotation at a typical speed of 100 rotations per second. Berg estimates that the M ring carries 100 motB proton-exchange sites, and various models predict that 800 to 1200 protons must flow through the complex during a single rotation of the flagellar filament!

PROBLEMS

1. The cheetah is generally regarded as nature's fastest mammal, but another amazing animal almost as fast as the cheetah is the pronghorn antelope, which roams the plains of Wyoming. Whereas the cheetah can maintain its top speed of 70 mph for only a few seconds, the pronghorn antelope can run at 60 mph for about an hour! (It is thought to have evolved to do so in order to elude now-extinct ancestral cheetahs that lived in North America.) What differences would you expect in the mus-cle structure and anatomy of pronghorn antelopes that could account for their remarkable speed and endurance?

2. An ATP analog, β,γ-methylene-ATP, in which a $-CH_2-$ group replaces the oxygen atom between the β- and γ-phosphorus atoms, is a potent inhibitor of muscle contraction. At which step in the contraction cycle would you expect β,γ-methylene-ATP to block contraction?

3. ATP stores in muscle are augmented or supplemented by stores of phosphocreatine. During periods of contraction, phosphocreatine is hydrolyzed to drive the synthesis of needed ATP in the creatine kinase reaction:

$$\text{phosphocreatine} + \text{ADP} \rightleftharpoons \text{creatine} + \text{ATP}$$

Muscle cells contain two different isozymes of creatine kinase, one in the mitochondria and one in the sarcoplasm. Explain.

4. Rigor is a muscle condition in which muscle fibers, depleted of ATP and phosphocreatine, develop a state of extreme rigidity and cannot be easily extended. (In death, this state is called rigor mortis, the rigor of death.) From what you have learned about muscle contraction, explain the state of rigor in molecular terms.

5. Skeletal muscle can generate approximately 3 to 4 kg of tension or force per square centimeter of cross-sectional area. This number is roughly the same for all mammals. Because many human muscles have large cross-sectional areas, the force that these muscles can (and must) generate is prodigious. The gluteus maximus (on which you are probably sitting as you read this) can generate a tension of 1200 kg! Estimate the cross-sectional area of all of the muscles in your body and the total force that your skeletal muscles could generate if they all contracted at once.

FURTHER READING

Astumian, R. D., and Bier, M., 1996. Mechanochemical coupling of the motion of molecular motors to ATP hydrolysis. *Biophysical Journal* **70**:637–653.

Berliner, E., Young, E., Anderson, K., et al., 1995. Failure of a single-headed kinesin to track parallel to microtubule protofilaments. *Nature* **373**:718–721.

Block, S. M., 1998. Kinesin: What gives? *Cell* **93**:5–8.

Boyer, P. D., 1997. The ATP synthase—a splendid molecular machine. *Annual Review of Biochemistry* **66**:717–749.

Compton, D. A., 2000. Spindle assembly in animal cells. *Annual Review of Biochemistry* **69**:95–114.

Coppin, C. M., Finer, J. T., Spudich, J. A., and Vale, R. D., 1996. Detection of sub-8-nm movements of kinesin by high-resolution optical-trap microscopy. *Proceedings of the National Academy of Sciences* **93**:1913–1917.

Corrie, J., Brandmeier, B., Ferguson, R., et al., 1999. Dynamic measurement of myosin light-chain-domain tilt and twist in muscle contraction. *Nature* **400**:425–430.

DeRosier, D. J., 1998. The turn of the screw: The bacterial flagellar motor. *Cell* **93**:17–20.

Eden, D., et al., 1995. Solution structure of two molecular motor domains: Nonclaret disjunctional and kinesin. *Biophysical Journal* **68**:59S–64S.

Engel, A., 1997. A closer look at a molecular motor by atomic force microscopy. *Biophysical Journal* **72**:988.

Finer, J. T., Simmons, R. M., and Spudich, J. A., 1994. Single myosin molecule mechanics: Piconewton forces and nanometer steps. *Nature* **368**:113–119.

Gilbert, S., Webb, M., Brune, M., and Johnson, K., 1995. Pathway of processive ATP hydrolysis by kinesin. *Nature* **373**:671–676.

Goldman, Y. E., 1998. Wag the tail: Structural dynamics of actomyosin. *Cell* **93**:1–4.

Henningsen, U., and Schliwa, M., 1997. Reversal in the direction of movement of a molecular motor. *Nature* **389**:93–96.

Hirose, K., Lockhart, A., Cross, R., and Amos, L., 1995. Nucleotide-dependent angular change in kinesin motor domain bound to tubulin. *Nature* **376**:277–279.

Hoenger, A., Sablin, E., Vale, R., et al., 1995. Three-dimensional structure of a tubulin-motor-protein complex. *Nature* **376**:271–274.

Howard, J., 1996. The movement of kinesin along microtubules. *Annual Review of Physiology* **58**:703–729.

Ishijima, A., Kojima, H., Funatsu, T., et al., 1998. Simultaneous observation of individual ATPase and mechanical events by a single myosin molecule during interaction with actin. *Cell* **92**:161–171.

Kabsch, W., and Holmes, K., 1995. The actin fold. *The FASEB Journal* **9**:167–174.

Kabsch, W., et al., 1990. Atomic structure of the actin : DNase I complex. *Nature* **347**:37–43.

Kikkawa, J., Ishikawa, T., Wakabayashi, T., and Hirokawa, N., 1995. Three-dimensional structure of the kinesin head-microtubule complex. *Nature* **376**:274–277.

Kinosita, K., Jr., Yasuda, R., Noji, H., et al., 1998. F_1-ATPase: A rotary motor made of a single molecule. *Cell* **93**:21–24.

Kitamura, K., Tokunaga, M., Iwane, A. H., and Yanagida, T., 1999. A single myosin head moves along an actin filament with regular steps of 5.3 nanometres. *Nature* **397**:129–134.

Kull, F. J., Sablin, E. P., Lau, R., et al., 1996. Crystal structure of the kinesin motor domain reveals a structural similarity to myosin. *Nature* **380**:550–555.

Lohman, T. M., Thorn, K., and Vale, R. D., 1998. Staying on track: Common features of DNA helicases and microtubule motors. *Cell* **93**:9–12.

Macnab, R. M., and Parkinson, J. S., 1991. Genetic analysis of the bacterial flagellum. *Trends in Genetics* **7**:196–200.

Meister, M., Caplan, S. R., and Berg, H., 1989. Dynamics of a tightly coupled mechanism for flagellar rotation. *Biophysical Journal* **55**:905–914.

Meyhofer, E., and Howard, J., 1995. The force generated by a single kinesin molecule against an elastic load. *Proceedings of the National Academy of Sciences* **92**:574–578.

Molloy, J., Burns, J., Kendrick-Jones, J., et al., 1995. Movement and force produced by a single myosin head. *Nature* **378**:209–213.

Nogales, E., Wolf, S., and Downing, K. H., 1998. Structure of the $\alpha\beta$ tubulin dimer by electron crystallography. *Nature* **391**:199–203.

Rayment, I., 1996. Kinesin and myosin: Molecular motors with similar engines. *Structure* **4**:501–504.

Rayment, I., and Holden, H., 1994. The three-dimensional structure of a molecular motor. *Trends in Biochemical Sciences* **19**:129–134.

Spudich, J., 1994. How molecular motors work. *Nature* **372:**515–518.

Svoboda, K., Schmidt, C., Schnapp, B., and Block, S., 1993. Direct observation of kinesin stepping by optical trapping interferometry. *Nature* **365:**721–727.

Vallee, R., and Shpetner, H., 1990. Motor proteins of cytoplasmic microtubules. *Annual Review of Biochemistry* **59:**909–932.

Walker, R., and Sheetz, M., 1993. Cytoplasmic microtubule-associated motors. *Annual Review of Biochemistry* **62:**429–451.

Whittaker, M., Wilson-Kubalek, E., Smith, J. E., et al., 1995. A 35 Å movement of smooth muscle myosin on ADP release. *Nature* **378:**748–753.

Wilson, L., and Jordan, M. A., 1995. Microtubule dynamics: Taking aim at a moving target. *Chemistry and Biology* **2:**569–573.

METABOLISM AND ITS REGULATION

Metabolism accomplishes, among other things, the conversion of food energy into the energy of motion. Regulation of metabolism allows the abrupt transition from a state of rest to the breathtaking power and grace of athletic competition. (European champs by Paul J. Sutton/Duomo; line art by J/B Woolsey Associates)

CHAPTER

14

All is flux, nothing stays still. Nothing endures but change.

HERACLITUS (C. 540–480 B.C.)

Outline

The Organization of Metabolism

Anise swallowtail butterfly (Papilio zelicans) *with its pupal case. The metamorphosis of butterflies is a dramatic example of metabolic change. (© 1986 Peter Bryant/Biological Photo Service)*

The word *metabolism* derives from the Greek word for "change." **Metabolism** represents the sum of the chemical changes that convert **nutrients,** the "raw materials" necessary to nourish living organisms, either into energy or into chemically complex cellular substances. Metabolism consists of literally hundreds of enzymatic reactions organized into discrete pathways. These pathways involve the step-by-step transformation of substrates into end products through many specific chemical **intermediates.** Metabolism is sometimes referred to as **intermediary metabolism** to reflect this aspect of the process. Metabolic maps (see Figure 1.19) portray virtually all of the principal reactions of the intermediary metabolism of carbohydrates, lipids, amino acids, nucleotides, and their derivatives. Such maps are very complex at first glance, but, despite their appearance, they become easy to follow once the major metabolic routes are known and their functions are understood. The underlying order of metabolism and the important relationships between the various pathways then appear as simple patterns against the seemingly complicated background.

The Metabolic Map As a Set of Dots and Lines

One interesting transformation of the intermediary metabolism map shows each chemical intermediate as a black dot and each enzyme that operates on an intermediate as a line (Figure 14.1). The more than 1000 different enzymes and substrates are represented by just two symbols. This chart has about 520 dots (intermediates). Table 14.1 lists the numbers of dots that have one or two or more lines (enzymes) associated with them. Thus, this table classifies intermediates by the number of enzymes that act upon them. A dot connected to just a single line must be either a nutrient, a storage form, an end product, or an excretory product of metabolism. Also, because many pathways tend to proceed in only one direction (that is, they are essentially irreversible under physiological conditions), a dot connected to just two lines is probably an intermediate in only one pathway and has only one fate in metabolism. Note that about 80% of the intermediates connect to only one or two lines and thus have only a limited purpose in the cell. However, many intermediates are subject to a variety of fates. In such instances, the pathway followed is an important regulatory choice. Indeed, whether any substrate is routed down a particular metabolic pathway is the consequence of a regulatory decision made in response to the cell's (or organism's) momentary requirements for energy or nutrition. The regulation of metabolism is an interesting and important subject to which we will return often.

Table 14.1 Number of Dots (Intermediates) in the Metabolic Map of Figure 14.1 and the Number of Lines Associated with Them

Lines	Dots
1 or 2	410
3	71
4	20
5	11
6 or more	8

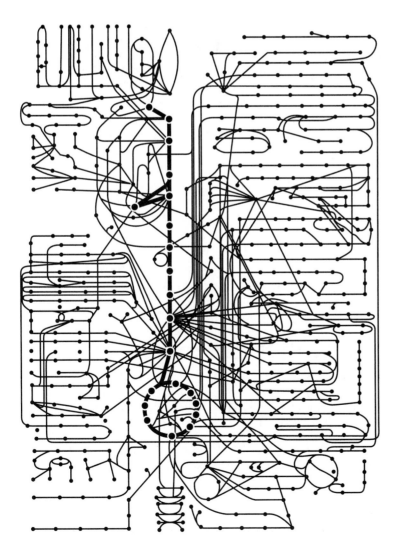

Figure 14.1 The metabolic map as a set of dots and lines. The heavy dots and lines trace the central energy-releasing pathways known as glycolysis and the citric acid cycle. *(Adapted from Alberts, B., et al., 1989. Molecular Biology of the Cell, 2nd ed. New York: Garland Publishing Co.)*

 See *Interactive Biochemistry CD-ROM and Workbook*, pages 173–178

14.1 Virtually All Organisms Have the Same Basic Set of Metabolic Pathways

One of the great unifying principles of modern biology is that organisms show marked similarity in their major pathways of metabolism. Given the almost unlimited possibilities within organic chemistry, this generality would appear most unlikely. Yet it's true, and it provides strong evidence that all life has descended from a common ancestral form. All forms of nutrition and almost all metabolic pathways evolved in early prokaryotes prior to the appearance of eukaryotes one billion years ago. For example, **glycolysis,** the metabolic pathway by which energy is released from glucose and captured in the form of ATP under anaerobic conditions, is common to almost every cell. Therefore, glycolysis is believed to be the most ancient of metabolic pathways, having arisen prior to the appearance of significant amounts of oxygen in the atmosphere. All organisms, even those that can synthesize their own glucose, are capable of glucose degradation and ATP synthesis via glycolysis. Other prominent pathways are also widely distributed among organisms.

Metabolic Diversity

Although most cells have the same basic set of central metabolic pathways, different cells (and, by extension, different organisms) are characterized by the alternative pathways they might express. These pathways offer a wide range of metabolic possibilities. For instance, organisms are often classified according to the major metabolic pathways they exploit to obtain carbon or energy. Classification based on carbon requirements defines two major groups, autotrophs and heterotrophs. **Autotrophs** are organisms that can use just carbon dioxide as their sole source of carbon. **Heterotrophs** require an organic form of carbon, such as glucose, to synthesize other essential carbon compounds.

Classification based on energy sources also gives two groups: phototrophs and chemotrophs. **Phototrophs** are photosynthetic organisms, which use light as a source of energy. **Chemotrophs** use organic compounds such as glucose, or, in some instances, oxidizable inorganic substances such as Fe^{2+}, NO_2^-, NH_4^+, or elemental sulfur, as sole sources of energy. Typically, the energy is extracted through oxidation–reduction reactions. Based on these characteristics, every organism falls into one of four categories (Table 14.2).

Metabolic Diversity Among the Five Kingdoms

Prokaryotes (the kingdom Monera—bacteria) show a greater metabolic diversity than all the four eukaryotic kingdoms (Protista [previously called Proto-

Table 14.2 Metabolic Classification of Organisms According to Their Carbon and Energy Requirements

Classification	Carbon Source	Energy Source	Electron Donors	Examples
Photoautotrophs	CO_2	Light	H_2O, H_2S, S, other inorganic compounds	Green plants, algae, cyanobacteria, photosynthetic bacteria
Photoheterotrophs	Organic compounds	Light	Organic compounds	Nonsulfur purple bacteria
Chemoautotrophs	CO_2	Oxidation–reduction reactions	Inorganic compounds: H_2, H_2S, NH_4^+, NO_2^-, Fe^{2+}, Mn^{2+}	Nitrifying bacteria; hydrogen, sulfur, and iron bacteria
Chemoheterotrophs	Organic compounds	Oxidation–reduction reactions	Organic compounds, e.g., glucose	All animals, most microorganisms, nonphotosynthetic plant tissue such as roots, photosynthetic cells in the dark

zoa], Fungi, Plants, and Animals) put together. Prokaryotes are variously chemoheterotrophic, photoautotrophic, photoheterotrophic, or chemoautotrophic. No protists are chemoautotrophs; fungi and animals are exclusively chemoheterotrophs; plants are characteristically photoautotrophs, although some are heterotrophic in their mode of carbon acquisition.

The Role of O_2 in Metabolism

A further metabolic distinction among organisms is whether or not they can use oxygen as an electron acceptor in energy-producing pathways. Those that can are called **aerobes** or aerobic organisms; others, termed **anaerobes,** can subsist without O_2. Organisms for which O_2 is obligatory for life are called **obligate aerobes;** humans are an example. Some species, the so-called **facultative anaerobes,** can adapt to anaerobic conditions by substituting other electron acceptors for O_2 in their energy-producing pathways; *Escherichia coli* is an example. Yet others cannot use oxygen at all and are even poisoned by it; these are the **obligate anaerobes.** *Clostridium botulinum,* the bacterium that produces botulin toxin, is representative of this type.

The Flow of Energy in the Biosphere and the Carbon and Oxygen Cycles Are Intimately Related

The primary source of energy for life is the sun. Photoautotrophs use light energy to drive the synthesis of organic molecules, such as carbohydrates, from atmospheric CO_2 and water (Figure 14.2). Heterotrophic cells then use these organic products of photosynthetic cells both as fuels and as building blocks for the biosynthesis of their own unique complement of biomolecules. Ultimately, CO_2 is the end product of heterotrophic carbon metabolism, and CO_2 is returned to the atmosphere for reuse by the photoautotrophs. In effect, phototrophs convert solar energy to the chemical energy of organic molecules, and heterotrophs recover this energy by metabolizing the organic substances. The flow of energy in the biosphere is thus conveyed within the carbon cycle, and the power driving the cycle is light energy.

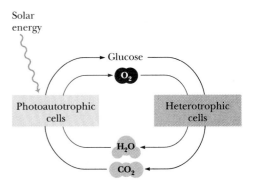

Figure 14.2   The flow of energy in the biosphere is coupled primarily to the carbon and oxygen cycles.

14.2	Metabolism Consists of Catabolism (Degradative Pathways) and Anabolism (Biosynthetic Pathways)

Metabolism serves two fundamentally different purposes: the generation of energy to drive vital processes and the synthesis of molecules necessary for the structure and function of cells. To achieve these ends, metabolism consists largely of two contrasting processes, catabolism and anabolism. Catabolic pathways are characteristically energy-yielding, whereas anabolic pathways are energy-requiring. **Catabolism** involves the oxidative degradation of complex nutrient molecules (carbohydrates, lipids, and proteins) obtained either from the environment or from cellular reserves. The breakdown of these molecules by catabolism leads to the formation of simpler molecules such as lactic acid, ethanol, carbon dioxide, urea, or ammonia. Catabolic reactions are usually exergonic, and often the chemical energy released is captured in the form of ATP (see Chapter 3). Because catabolism is oxidative, part of the chemical energy may be conserved through reduction of the coenzymes NAD^+ and $NADP^+$ to form NADH and NADPH. The reduced forms of these coenzymes have very different metabolic roles. Reduction of NAD^+ to NADH accompanies many of the oxidative reactions that occur in catabolism. Further, oxidation of NADH can be coupled to the phosphorylation of ADP in aerobic cells, and so NADH oxidation back to NAD$^+$ serves to generate more ATP. In contrast, NADPH is the source of the reducing power needed to drive reductive biosynthetic reactions, so NADPH oxidation is a feature of anabolism.

Anabolism Is Biosynthesis

Anabolism is a synthetic process in which the varied and complex biomolecules (proteins, nucleic acids, polysaccharides, and lipids) are assembled from simpler precursors. Such biosynthesis involves the formation of new covalent bonds, and an input of chemical energy is necessary to drive such endergonic processes. The ATP generated by catabolism provides this energy. Furthermore, NADPH is an excellent donor of electrons for the reductive reactions of anabolism. Despite their divergent roles, anabolism and catabolism are interrelated in that the products of one provide the substrates of the other (Figure 14.3).

Anabolism and Catabolism Are Not Mutually Exclusive

Importantly, anabolism and catabolism occur simultaneously in the cell. Despite their conflicting purposes, catabolism and anabolism can be managed concomitantly by cells. First, the cell maintains tight and separate regulation of both catabolism and anabolism, so that metabolic needs are served in an immediate and orderly fashion. Second, competing metabolic pathways are often localized within different cellular compartments. Isolating opposing activities within distinct compartments, such as separate organelles, avoids interference between them. For example, the enzymes responsible for the catabolism of fatty acids are localized within mitochondria. In contrast, fatty acid biosynthesis takes place in the cytosol.

Modes of Enzyme Organization in Metabolic Pathways

The individual metabolic pathways of anabolism and catabolism consist of sequential enzymatic steps (Figure 14.4). Several types of organization are possible. The enzymes of some multienzyme systems may exist as physically separate, soluble entities, with diffusing intermediates (Figure 14.4a). In other

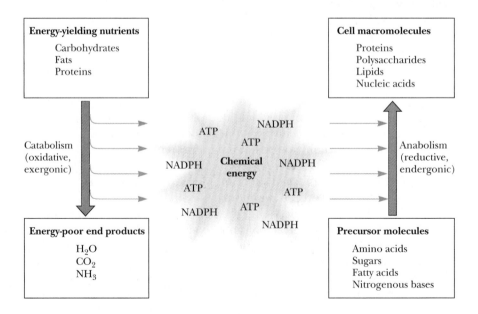

Figure 14.3 Energy relationships between the pathways of catabolism and anabolism. Oxidative, exergonic pathways of catabolism release free energy and reducing power that are captured in the form of ATP and NADPH, respectively. Anabolic processes are endergonic, consuming chemical energy in the form of ATP and using NADPH as a source of high-energy electrons for reductive purposes.

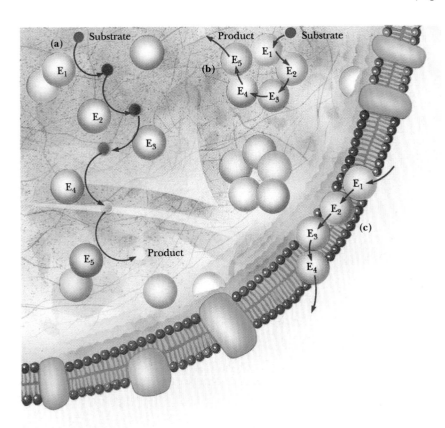

Figure 14.4 Schematic representation of types of multienzyme systems carrying out a metabolic pathway: **(a)** physically separate, soluble enzymes with diffusing intermediates; **(b)** a multienzyme complex. Substrate enters the complex, becomes covalently bound, and then is sequentially modified by enzymes E_1 to E_5 before product is released. No intermediates are free to diffuse away. **(c)** A membrane-bound multienzyme system.

instances, the enzymes of a pathway are collected to form a discrete multienzyme complex, and the substrate is sequentially modified as it is passed along from enzyme to enzyme (Figure 14.4b). This type of organization has the advantage that intermediates are not lost or diluted by diffusion. In a third pattern of organization, the enzymes common to a pathway reside together as a membrane-bound system (Figure 14.4c). In this case, the enzyme participants (and perhaps the substrates as well) must diffuse in just the two dimensions of the membrane to interact with their neighbors.

As research reveals the ultrastructural organization of the cell in ever greater detail, the individual enzymes common to a particular pathway often are found to be physically united into multifunctional complexes. Thus, many metabolic pathways are carried out by stable assemblages of consecutively acting enzymes; such assemblages are sometimes referred to as **metabolons,** a word meaning "units of metabolism."

The Pathways of Catabolism Converge to a Few End Products

If we survey the catabolism of the principal energy-yielding nutrients (carbohydrates, lipids, and proteins) in a typical heterotrophic cell, we see that the degradation of these substances involves a succession of enzymatic reactions. In the presence of oxygen (aerobic catabolism), these molecules are degraded ultimately to carbon dioxide, water, and ammonia. Aerobic catabolism consists of three distinct stages. In **stage 1,** the nutrient macromolecules are broken down into their respective building blocks. Given the great diversity of macromolecules, these building blocks represent a rather limited number of products. Proteins yield up their 20 component amino acids, polysaccharides give rise to carbohydrate units that are convertible to glucose, and lipids are broken down into glycerol and fatty acids (Figure 14.5).

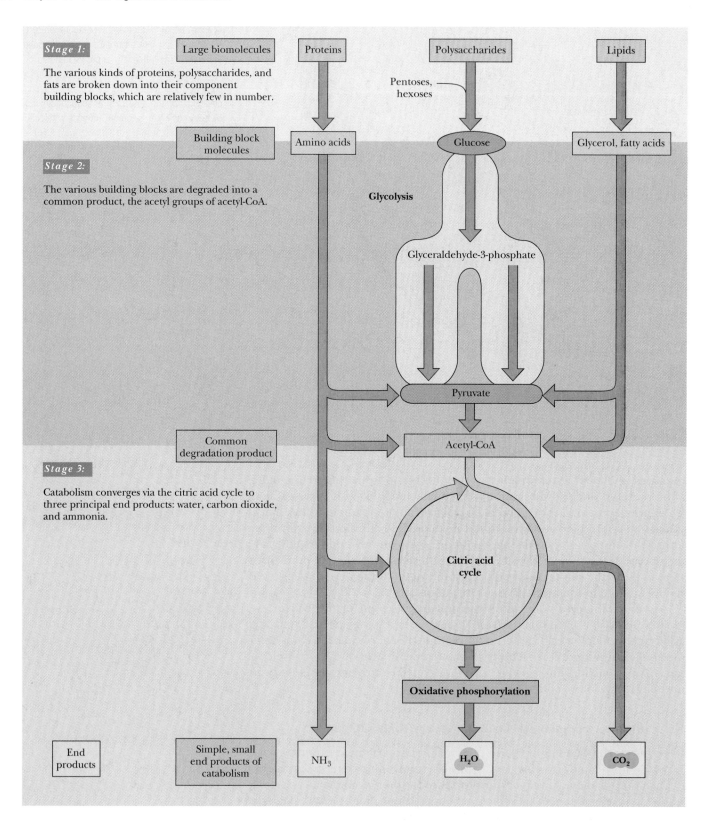

Figure 14.5 The three stages of catabolism. **Stage 1:** Proteins, polysaccharides, and lipids are broken down into their component building blocks, which are relatively few in number. **Stage 2:** The various building blocks are degraded into the common product, the acetyl groups of acetyl-CoA. **Stage 3:** Catabolism converges to three principal end products: water, carbon dioxide, and ammonia.

In **stage 2,** the collection of product building blocks generated in stage 1 is further degraded to yield an even more limited set of simpler metabolic intermediates. The deamination of amino acids gives rise to α-keto acids. Several of these α-keto acids are citric acid cycle intermediates and are fed directly into stage 3 catabolism via this cycle. Others are converted either to the three-carbon α-keto acid pyruvate or to the acetyl groups of acetyl-coenzyme A **(acetyl-CoA).** Glucose and the glycerol derived from lipids also generate pyruvate, whereas the fatty acids are broken into two-carbon units that appear as the two-carbon acetyl group of acetyl-CoA. Because pyruvate also gives rise to acetyl-CoA, we see that the degradation of macromolecular nutrients converges to a common end product, acetyl-CoA (see Figure 14.5).

The combustion of the acetyl groups of acetyl-CoA by the citric acid cycle[1] in combination with oxidative phosphorylation to produce CO_2 and H_2O represents **stage 3** of catabolism. These two simple substances are the ultimate waste products of aerobic catabolism. As we shall see in Chapter 16, the oxidation of acetyl-CoA during stage 3 metabolism generates most of the energy produced by the cell.

Anabolic Pathways Diverge, Synthesizing an Astounding Variety of Biomolecules from a Limited Set of Building Blocks

A rather limited collection of simple precursor molecules is sufficient to provide for the biosynthesis of virtually any cellular constituent—protein, nucleic acid, lipid, or polysaccharide. All of these substances are constructed from appropriate building blocks via the pathways of anabolism. In turn, the building blocks (amino acids, nucleotides, sugars, and fatty acids) can be generated from metabolites in the cell. For example, amino acids can be formed by amination of the corresponding α-keto acids, and pyruvate can be converted to hexoses for polysaccharide biosynthesis. Thus, in contrast to catabolism—which converges to the common intermediate, acetyl-CoA—the pathways of anabolism diverge from a small group of simple metabolic intermediates to yield a spectacular variety of cellular substances.

Amphibolic Intermediates

Certain of the central pathways of intermediary metabolism, such as the citric acid cycle, and many metabolites of other pathways have dual purposes: they serve in both catabolism and anabolism. This dual nature is reflected in the designation of such pathways as **amphibolic** rather than solely catabolic or anabolic.

amphi from the Greek for "on both sides"

Corresponding Pathways of Catabolism and Anabolism Differ in Important Ways

The anabolic pathway for synthesis of a given end product usually does not precisely match the pathway used for catabolism of the same substance. Some of the intermediates may be common to steps in both pathways, but different enzymatic reactions and unique metabolites characterize other steps. A good example of these differences is found in a comparison of the catabolism of glucose to pyruvate by the pathway of glycolysis and the biosynthesis of glucose from pyruvate by the pathway called gluconeogenesis. The glycolytic pathway

---

[1]The citric acid cycle is also known as the tricarboxylic acid cycle (or the TCA cycle) and the Krebs cycle, in honor of its discoverer, Sir Hans Krebs.

from glucose to pyruvate consists of 10 enzymes. Although it may seem efficient for glucose synthesis from pyruvate to proceed by a reversal of all 10 steps, gluconeogenesis uses only seven of the glycolytic enzymes in reverse, replacing the remaining three with four enzymes specific to glucose biosynthesis. In similar fashion, the pathway responsible for degrading proteins to amino acids differs from the protein synthesis system, and the oxidative degradation of fatty acids to two-carbon acetyl-CoA groups does not follow the same reaction path as the biosynthesis of fatty acids from acetyl-CoA. An obvious distinction between such oppositely directed pathways is their energetics: the catabolic direction is characteristically oxidative and energy-releasing, whereas the anabolic direction is reductive and energy-requiring.

Metabolic Regulation Favors Different Pathways for Oppositely Directed Metabolic Sequences

The maintenance of orderly metabolism in cells requires that each pathway be independently regulated and constrained in its operation, as directed by the momentary needs of the cell. If catabolism and anabolism passed along the same set of metabolic tracks, equilibrium considerations would dictate that slowing the traffic in one direction by inhibiting a particular enzymatic reaction would necessarily slow traffic in the opposite direction. Independent regulation of anabolism and catabolism can be accomplished only if these two contrasting processes move along different routes or, in the case of shared pathways, the rate-limiting steps serving as the points of regulation are catalyzed by enzymes that are unique to each opposing sequence (Figure 14.6).

The ATP Cycle

We saw in Chapter 3 that ATP is the energy currency of cells. In phototrophs, ATP is one of the two energy-rich primary products resulting from the transformation of light energy into chemical energy. (The other is NADPH; see the following discussion.) In heterotrophs, the pathways of catabolism have as their major purpose the release of free energy that can be captured in the form of

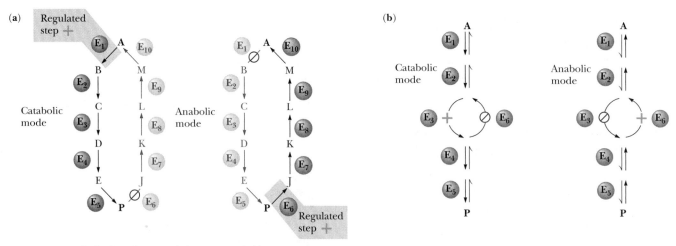

Activation of one mode is accompanied by reciprocal inhibition of the other mode.

Figure 14.6 Parallel pathways of catabolism and anabolism must differ in at least one metabolic step so they can be regulated independently. Shown here are two possible arrangements of opposing catabolic and anabolic sequences between A and P. **(a)** The parallel sequences proceed via independent routes. **(b)** Only one reaction has two different enzymes, a catabolic one (E_3) and its anabolic counterpart (E_6). These provide sites for regulation.

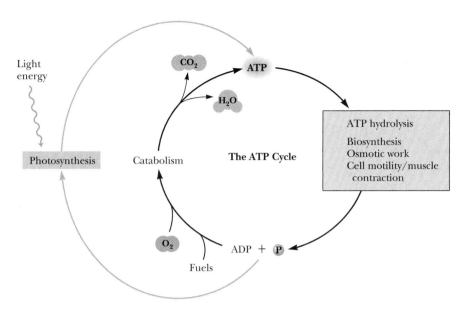

Figure 14.7   The ATP cycle in cells. ATP is formed via photosynthesis in phototrophic cells or via catabolism in heterotrophic cells. Energy-requiring cellular activities are powered by ATP hydrolysis, liberating ADP and P_i.

energy-rich phosphoric anhydride bonds in ATP. In turn, ATP provides the energy that drives the manifold activities of all living cells: the synthesis of complex biomolecules, the osmotic work involved in transporting substances into cells, the work of cell motility, the work of muscle contraction. These diverse activities are all powered by energy released in the hydrolysis of ATP to ADP and P_i. Thus, there is an energy cycle in cells where ATP serves as the vessel carrying energy from photosynthesis or catabolism to the energy-requiring processes unique to living cells (Figure 14.7).

NAD$^+$ Collects Electrons Released in Catabolism

The substrates of catabolism—proteins, carbohydrates, and lipids—are good sources of chemical energy because the carbon atoms in these molecules are relatively reduced (Figure 14.8). In the oxidative reactions of catabolism, reducing equivalents are released from these substrates, often in the form of **hydride ions** (a proton coupled with two electrons, H:$^-$). These hydride ions are transferred in enzymatic **dehydrogenase** reactions from the substrates to NAD$^+$ molecules, reducing them to NADH (Figure 14.9). A second proton accompanies these reactions, appearing in the overall equation as H$^+$. In turn, NADH is oxidized back to NAD$^+$ when it transfers its reducing equivalents to electron acceptor systems that are part of the metabolic apparatus of the mitochondria. The ultimate oxidizing agent (*e$^-$ acceptor*) is O$_2$, which is reduced to H$_2$O.

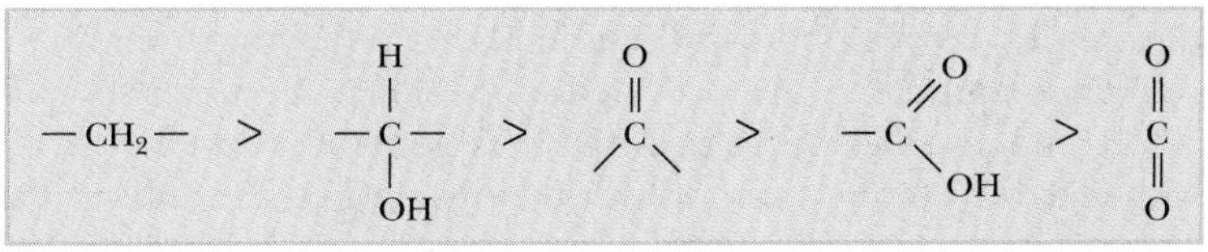

Figure 14.8   Comparison of the state of reduction of carbon atoms in biomolecules: —CH$_2$— (fats); —CHOH— (carbohydrates); $>$C$=$O (carbonyls of carbohydrates and amino acids); —COOH (carboxyls of fatty acids and amino acids); CO$_2$ (carbon dioxide, the final product of catabolism).

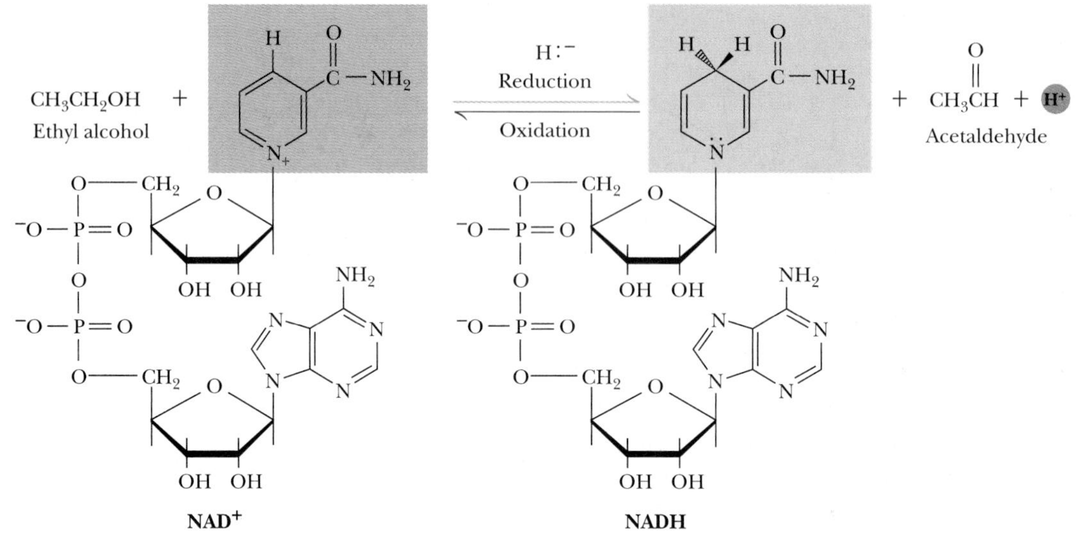

$$CH_3CH_2OH \quad + $$
Ethyl alcohol

[NAD$^+$ structure]

$$\underset{\text{Oxidation}}{\overset{\text{H}:^-}{\overset{\text{Reduction}}{\rightleftharpoons}}}$$

[NADH structure]

$$+ \quad CH_3CH \quad + \quad H^+$$
Acetaldehyde

NAD$^+$ **NADH**

Figure 14.9 Hydrogen and electrons released in the course of oxidative catabolism are transferred as hydride ions to the pyridine nucleotide, NAD$^+$, to form NADH + H$^+$ in dehydrogenase reactions of the type

$$AH_2 + NAD^+ \rightarrow A + NADH + H^+$$

The reaction shown is catalyzed by alcohol dehydrogenase.

Oxidation reactions are exergonic, and the energy released is coupled with the formation of ATP in a process called **oxidative phosphorylation.** The NAD$^+$/NADH system can be viewed as a shuttle that carries the electrons released from catabolic substrates to the mitochondria, where they are transferred to O$_2$, the ultimate electron acceptor in catabolism. In the process, the free energy released is trapped in ATP. The NADH cycle is an important player in the transformation of the chemical energy of carbon compounds into the chemical energy of phosphoric anhydride bonds. Such transformations of energy from one form to another are referred to as **energy transduction.** Oxidative phosphorylation is one cellular mechanism for energy transduction. Chapter 17 is devoted to electron transport reactions and oxidative phosphorylation.

NADPH Provides the Reducing Power for Anabolic Processes

Whereas catabolism is fundamentally an oxidative process, anabolism is, by contrast, reductive. The biosynthesis of the complex constituents of the cell begins at the level of intermediates derived from the degradative pathways of catabolism; or, less commonly, biosynthesis begins with oxidized substances available in the inanimate environment, such as carbon dioxide. When the hydrocarbon chains of fatty acids are assembled from acetyl-CoA units, activated hydrogens are needed to reduce the carbonyl (C=O) carbon of acetyl-CoA into a —CH$_2$— at every other position along the chain. When glucose is synthesized from CO$_2$ during photosynthesis in plants, reducing power is required. These reducing equivalents are provided by NADPH, the usual source of hydrogens for reductive biosynthesis. NADPH is generated when NADP$^+$ is reduced with electrons in the form of hydride ions. In heterotrophic organisms, these electrons are removed from fuel molecules by NADP$^+$-specific dehydrogenases. In these organisms, NADPH can be viewed as the carrier of electrons from catabolic reactions to anabolic reactions (Figure 14.10). In photosynthetic organisms, the energy of light is used to pull electrons from water and transfer them to NADP$^+$ (see Chapter 18); O$_2$ is a by-product of this process.

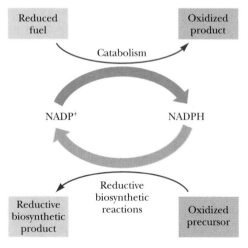

Figure 14.10 Transfer of reducing equivalents from catabolism to anabolism via the NADPH cycle.

14.3 Experimental Methods to Reveal Metabolic Pathways

Armed with the knowledge that metabolism is organized into pathways of successive reactions, we can appreciate by hindsight the techniques employed by early biochemists to reveal their sequence.

An important tool for elucidating the steps in the pathway was the use of metabolic inhibitors. Adding an enzyme inhibitor to a cell-free extract caused an accumulation of intermediates in the pathway prior to the point of inhibition (Figure 14.11). Each inhibitor was specific for a particular site in the sequence of metabolic events. As the arsenal of inhibitors was expanded, the individual steps in metabolism were revealed.

Mutations Create Specific Metabolic Blocks

Genetics provides an approach to the identification of intermediate steps in metabolism that is somewhat analogous to inhibition. Mutation in a gene encoding an enzyme often results in an inactive or absent enzyme. Such a defect leads to a block in the metabolic pathway at the point where the enzyme acts, and the enzyme's substrate accumulates. Such genetic disorders are lethal if the end product of the pathway is essential or if the accumulated intermediates have toxic effects.

Isotopic Tracers As Metabolic Probes

Another widely used approach to the elucidation of metabolic sequences is to "feed" cells a substrate or metabolic intermediate labeled with a particular isotopic form of an element that can be traced. Two sorts of isotopes are useful in this regard: **radioactive isotopes,** such as ^{14}C, and **stable ("heavy") isotopes,** such as ^{18}O or ^{15}N (Table 14.3). Because the chemical behavior of isotopically labeled compounds is rarely distinguishable from that of their unlabeled counterparts, isotopes provide reliable "tags" for observing metabolic changes. The metabolic fate of a radioactively labeled substance can be traced by determining the presence and position of the radioactive atoms in intermediates derived from the labeled compound (Figure 14.12). *Heavy isotopes* endow the

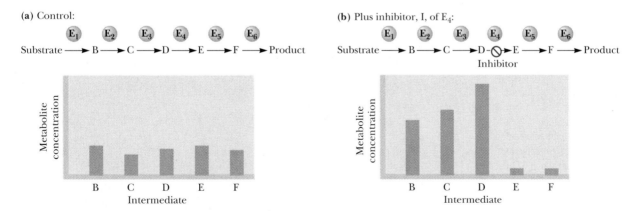

Figure 14.11 The use of inhibitors to reveal the sequence of reactions in a metabolic pathway. **(a) Control:** Under normal conditions, the steady-state concentrations of a series of intermediates will be determined by the relative activities of the enzymes in the pathway. **(b) Plus inhibitor:** In the presence of an inhibitor (in this case, an inhibitor of enzyme 4), intermediates upstream of the metabolic block (B, C, and D) accumulate, revealing themselves as intermediates in the pathway. The concentration of intermediates lying downstream (E and F) will fall.

Table 14.3 **Properties of Radioactive and Stable "Heavy" Isotopes Used As Tracers in Metabolic Studies**

Isotope	Type	Radiation Type	Half-Life	Relative Abundance*
^2H	Stable			0.0154%
^3H	Radioactive	β^-	12.1 years	
^{13}C	Stable			1.1%
^{14}C	Radioactive	β^-	5700 years	
^{15}N	Stable			0.365%
^{18}O	Stable			0.204%
^{24}Na	Radioactive	β^-, γ	15 hours	
^{32}P	Radioactive	β^-	14.3 days	
^{35}S	Radioactive	β^-	87.1 days	
^{36}Cl	Radioactive	β^-	310,000 years	
^{42}K	Radioactive	β^-	12.5 hours	
^{45}Ca	Radioactive	β^-	152 days	
^{59}Fe	Radioactive	β^-, γ	45 days	
^{131}I	Radioactive	β^-, γ	8 days	

*The relative natural abundance of a stable isotope is important because, in tracer studies, the amount of stable isotope is typically expressed in terms of atoms percent excess over the natural abundance of the isotope.

compounds in which they appear with slightly greater masses than their unlabeled counterparts. These compounds can be separated and quantitated by mass spectrometry (or density gradient centrifugation, if they are macromolecules).

NMR As a Metabolic Probe

A technology analogous to the use of isotopic tracers is provided by **nuclear magnetic resonance (NMR) spectroscopy.** The atomic nuclei of certain isotopes, such as the naturally occurring isotope of phosphorus, ^{31}P, have magnetic moments. The resonance frequency of a magnetic moment is influenced by the local chemical environment. That is, the NMR signal of the nucleus is influenced in an identifiable way by the chemical nature of its neighboring atoms in the compound. In many ways, these nuclei are ideal tracers because their signals contain a great deal of structural information about the environment around the atom and thus the nature of the compound containing the

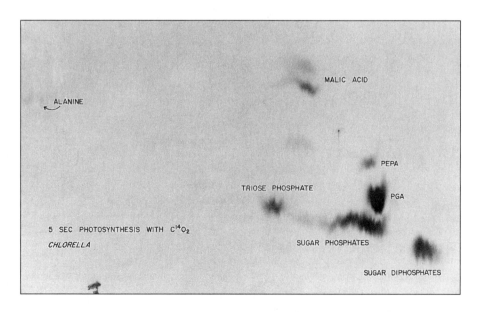

Figure 14.12 One of the earliest experiments using a radioactive isotope as a metabolic tracer. Cells of *Chlorella* (a green alga) synthesizing carbohydrate from carbon dioxide were exposed briefly (5 sec) to ^{14}C-labeled CO_2. The products of CO_2 incorporation were then quickly isolated from the cells, separated by two-dimensional paper chromatography, and observed via autoradiographic exposure of the chromatogram. Such experiments identified radioactive 3-phosphoglycerate (PGA) as the primary product of CO_2 fixation. The 3-phosphoglycerate was labeled in the 1-position (in its carboxyl group). Radioactive compounds arising from the conversion of 3-phosphoglycerate to other metabolic intermediates included phosphoenolpyruvate (PEP), malic acid, triose phosphate, alanine, and sugar phosphates and diphosphates. *(Photograph courtesy of Professor Melvin Calvin, Lawrence Berkeley Laboratory, University of California, Berkeley)*

(a)

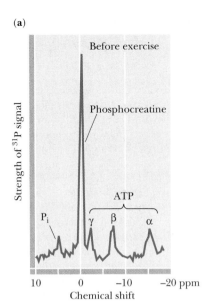

(b)

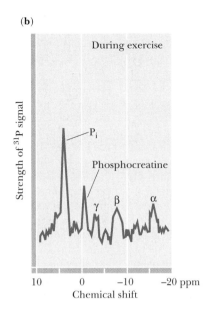

Figure 14.13 With NMR spectroscopy, one can observe the metabolism of a living subject in real time. These NMR spectra show the changes in ATP, creatine-P (phosphocreatine), and P_i levels in the forearm muscle of a human subjected to 19 minutes of exercise. Note that the three P atoms of ATP (α, β, and γ) have different chemical shifts, reflecting their different magnetic environments.

atom. Transformations of substrates and metabolic intermediates labeled with magnetic nuclei can be traced by following changes in NMR spectra. Furthermore, NMR spectroscopy is a noninvasive procedure. Whole-body NMR spectrometers are being used today in hospitals to directly observe the metabolism (and clinical condition) of living subjects (Figure 14.13).

14.4 Nutrition

The use of foods by organisms is termed **nutrition.** The ability of an organism to use a particular food material depends upon its chemical composition and upon the metabolic pathways available to the organism. In addition to essential fiber, food includes the **energy-yielding nutrients**—protein, carbohydrate, and lipid—and the **micronutrients**—vitamins and minerals. We focus here on human nutrition.

Protein

Animals must consume protein in order to make new proteins. Dietary protein is a rich source of nitrogen, and certain amino acids—the so-called **essential amino acids**—cannot be synthesized by animals and can be obtained only in the diet. The average adult in the United States consumes far more protein than is required for synthesis of necessary proteins. Any excess dietary protein becomes just another source of metabolic energy. Some of the amino acids **(glucogenic)** can be converted into glucose, whereas others, the **ketogenic** amino acids, can be converted to fatty acids or keto acids. If fat and carbohydrate are already adequate for energy needs, then both kinds of amino acids will be converted to triacylglycerol and stored in adipose tissue.

A certain percentage of an animal's own protein undergoes a constant process of degradation and resynthesis. Together with dietary protein, this recycled protein material participates in a nitrogen equilibrium, or **nitrogen balance.** A positive nitrogen balance occurs whenever dietary nitrogen intake (principally as protein) exceeds the needs for nitrogen compounds (principally amino acids for protein synthesis and nucleotides for nucleic acid

HUMAN BIOCHEMISTRY

Fad Diets Versus Recommended Intakes of Carbohydrate, Fats, and Protein

The most prevalent nutritional problem in the United States is obesity due to excessive food consumption, and many people have experimented with fad diets in the hope of losing excess weight. In truth, nutritionists recommend that caloric intake in the diet be met in the following proportions: 55–60% from carbohydrates, no more than 30% from fat, and the rest (10–15%) from protein. For the average American, these recommendations would require an *increased* consumption of carbohydrates and a *diminished* intake of fat and protein. Despite this fact, one of the most popular fad diets is the high-protein, high-fat (low-carbohydrate) diet. The premise for such diets is tantalizing: because the tricarboxylic acid (TCA) cycle (see Figure 14.5; see also Chapter 20) is the primary site of fat metabolism, and because glucose is usually needed to replenish intermediates in the TCA cycle, a restricted carbohydrate diet should result in the conversion of dietary fat into products known as ketone bodies, which would then be excreted in the urine. This so-called diet appears to work at first because a low-carbohydrate diet results in an initial water (and weight) loss. Some water loss occurs because glycogen reserves are depleted by this diet and because about 3 grams of water of hydration are lost for every gram of glycogen. Additional water loss accompanies ketone body excretion and the excretion of the urea generated by metabolism of the excess protein.

However, the long-term results of this diet are usually disappointing for several reasons. First, ketone body excretion by the human body usually does not exceed 20 grams (400 kJ) per day. Second, amino acids can function effectively to replenish TCA cycle intermediates, making the reduced carbohydrate regimen irrelevant. Third, the typical fare in a high-protein, high-fat, low-carbohydrate diet is more expensive and, ultimately, monotonous, and it is thus difficult to maintain. Finally, a high-fat diet represents a significant risk factor for atherosclerosis and coronary artery disease later in life.

synthesis). During periods of growth, a positive nitrogen balance is desirable. A negative nitrogen balance exists when dietary intake of nitrogen is insufficient.

Carbohydrate

The principal purpose of carbohydrate in the diet is production of metabolic energy. Simple sugars are metabolized in the glycolytic pathway (see Chapter 15). Complex carbohydrates are degraded into simple sugars, which then can enter the glycolytic pathway. Carbohydrates are also essential components of nucleotides, nucleic acids, glycoproteins, and glycolipids. Human metabolism can adapt to a wide range of dietary carbohydrate levels, but the brain normally depends on glucose as its fuel. When dietary carbohydrate consumption exceeds the energy needs of the organism, excess carbohydrate is converted to glycogen and triacylglycerols for long-term energy storage. On the other hand, when dietary carbohydrate intake is low, **ketone bodies** are formed from acetate units to provide metabolic fuel for the brain.

Lipid

Fatty acids and triacylglycerols can be used as fuel by many tissues in the human body, and phospholipids are essential components of all biological membranes. Even though the human body can tolerate a wide range of fat intake levels, there are disadvantages in either extreme. Excess dietary fat is stored as triacylglycerols in adipose tissue, but high levels of dietary fat can increase the risk of atherosclerosis and heart disease. Moreover, high dietary fat levels are also correlated with increased risk for cancer. When dietary fat consumption is low, there is a risk of **essential fatty acid** deficiencies. As indicated in Chapter 6, the human body cannot synthesize linoleic and linolenic acids, so these must be acquired in the diet. Additionally, arachidonic acid can be synthesized in humans only from linoleic acid, so it too is classified as essential. The essential fatty acids are key components of biological membranes, and arachidonic acid is the precursor to prostaglandins, which mediate a variety of processes in the body.

Fiber

The components of food materials that cannot be broken down by human digestive enzymes are referred to as **dietary fiber.** There are several kinds of dietary fiber, each with its own chemical and biological properties. **Cellulose** and **hemicellulose** are insoluble fiber materials that stimulate regular function of the colon. **Lignins,** another class of insoluble fibers, absorb organic molecules in the digestive system. Lignins bind cholesterol and clear it from the digestive system, reducing the risk of heart disease. Pectins and gums are water-soluble fiber materials that form viscous gel-like suspensions in the digestive system, slowing the rate of absorption of many nutrients, including carbohydrates, and lowering serum cholesterol in many cases. Insoluble fibers are prevalent in vegetable grains. Water-soluble fiber is a component of fruits, legumes, and oats.

Special Focus: Vitamins and Minerals

14.5 Vitamins

Vitamins are essential nutrients that are required in the diet, usually in trace amounts, because they cannot be synthesized by the organism itself. The requirement for any given vitamin depends on the organism. Not all vitamins are required by all organisms. Vitamins required in the human diet are listed in Table 14.4. These important substances are traditionally distinguished as being either water-soluble or fat-soluble. Except for vitamin C (ascorbic acid), the water-soluble vitamins are all components or precursors of important biological substances known as **coenzymes.** These are low molecular weight molecules that bring unique chemical functionality to certain enzyme reactions. Coenzymes may also act as carriers of specific functional groups, such as methyl groups and acyl groups. Coenzymes are needed because the side chains of the

■ Table 14.4 Vitamins and Coenzymes

Vitamin	Coenzyme Form
Water-Soluble	
Thiamine (vitamin B_1)	Thiamine pyrophosphate
Niacin (nicotinic acid)	Nicotinamide adenine dinucleotide (NAD^+)
	Nicotinamide adenine dinucleotide phosphate ($NADP^+$)
Riboflavin (vitamin B_2)	Flavin adenine dinucleotide (FAD)
	Flavin mononucleotide (FMN)
Pantothenic acid	Coenzyme A
Pyridoxal, pyridoxine, pyridoxamine (vitamin B_6)	Pyridoxal phosphate
Cobalamin (vitamin B_{12})	5′-Deoxyadenosylcobalamin
	Methylcobalamin
Biotin	Biotin–lysine conjugates (biocytin)
Lipoic acid	Lipoyl–lysine conjugates (lipoamide)
Folic acid	Tetrahydrofolate
Fat-Soluble	
Retinol (vitamin A)	
Ergocalciferol (vitamin D_2)	
Cholecalciferol (vitamin D_3)	
α-Tocopherol (vitamin E)	
Vitamin K	

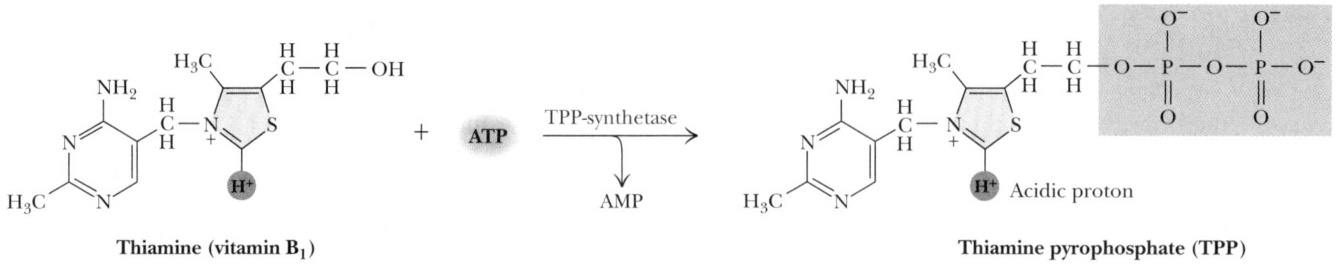

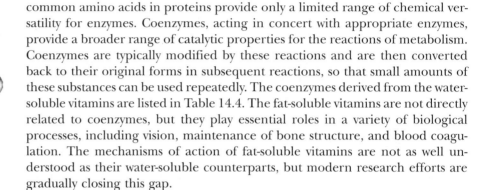

Thiamine (vitamin B$_1$)   +   **ATP** $\xrightarrow[\text{AMP}]{\text{TPP-synthetase}}$ Thiamine pyrophosphate (TPP)

H$^+$ Acidic proton

Figure 14.14 Thiamine (vitamin B$_1$) is composed of a substituted thiazole ring joined to a substituted pyrimidine by a methylene bridge. Thiamine pyrophosphate (TPP), the active form of vitamin B$_1$, is formed from thiamine by the action of TPP-synthetase on thiamine.

See pages 162–164

common amino acids in proteins provide only a limited range of chemical versatility for enzymes. Coenzymes, acting in concert with appropriate enzymes, provide a broader range of catalytic properties for the reactions of metabolism. Coenzymes are typically modified by these reactions and are then converted back to their original forms in subsequent reactions, so that small amounts of these substances can be used repeatedly. The coenzymes derived from the water-soluble vitamins are listed in Table 14.4. The fat-soluble vitamins are not directly related to coenzymes, but they play essential roles in a variety of biological processes, including vision, maintenance of bone structure, and blood coagulation. The mechanisms of action of fat-soluble vitamins are not as well understood as their water-soluble counterparts, but modern research efforts are gradually closing this gap.

Vitamin B$_1$: Thiamine and Thiamine Pyrophosphate

As shown in Figure 14.14, thiamine is the precursor of **thiamine pyrophosphate (TPP),** a coenzyme involved in reactions of carbohydrate metabolism in which bonds to carbonyl carbons (aldehydes or ketones) are synthesized or cleaved. In particular, the decarboxylations of α-keto acids and the formation and cleavage of α-hydroxyketones depend on thiamine pyrophosphate. The first of these is illustrated in Figure 14.15a by the decarboxylation of pyruvate by **yeast pyruvate decarboxylase** to yield carbon dioxide and acetaldehyde. An example of the formation and cleavage of α-hydroxyketones is illustrated in Figure 14.15b: the condensation of two molecules of pyruvate in the **acetolactate synthase** reaction.

Vitamins Containing Adenine Nucleotides

Several classes of vitamins are related to, or are precursors of, coenzymes that contain adenine nucleotide as part of their structure. These coenzymes include the **flavin dinucleotides,** the **pyridine dinucleotides,** and **coenzyme A.** The adenine nucleotide portion of these molecules does not participate actively in the reactions of these coenzymes; rather, it helps the proper enzymes to bind the

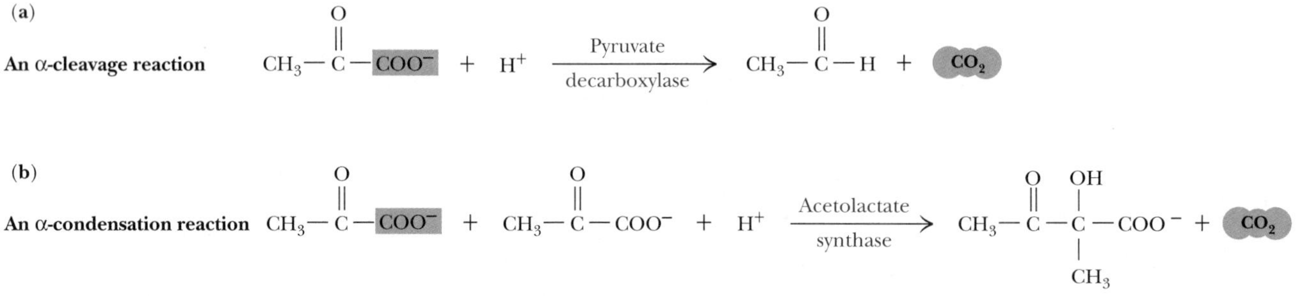

(a)

An α-cleavage reaction $CH_3 - \overset{\overset{\displaystyle O}{\|}}{C} - \boxed{COO^-} + H^+$ $\xrightarrow[\text{decarboxylase}]{\text{Pyruvate}}$ $CH_3 - \overset{\overset{\displaystyle O}{\|}}{C} - H + $ **CO$_2$**

(b)

An α-condensation reaction $CH_3 - \overset{\overset{\displaystyle O}{\|}}{C} - \boxed{COO^-} + CH_3 - \overset{\overset{\displaystyle O}{\|}}{C} - COO^- + H^+$ $\xrightarrow[\text{synthase}]{\text{Acetolactate}}$ $CH_3 - \overset{\overset{\displaystyle O}{\|}}{C} - \overset{\overset{\displaystyle OH}{|}}{\underset{\underset{\displaystyle CH_3}{|}}{C}} - COO^- + $ **CO$_2$**

Figure 14.15 Thiamine pyrophosphate participates in **(a)** the decarboxylation of α-keto acids and **(b)** the formation and cleavage of α-hydroxyketones.

HUMAN BIOCHEMISTRY

Niacin and Pellagra

Pellagra, a disease characterized by dermatitis, diarrhea, and dementia, has been known for centuries. It was once prevalent in the southern part of the United States and is still a common problem in some parts of the world. Scientists realized early in the last century that it could be cured by dietary actions, and **nicotinic acid** was identified as the relevant dietary factor. Interestingly, many animals, including humans, can synthesize nicotinic acid from tryptophan and other precursors, and nicotinic acid is thus not a true vitamin for these species. However, if dietary intake of tryptophan is low, nicotinic acid is required for optimal health. Nicotinic acid, which is beneficial, is structurally related to **nicotine,** a highly toxic tobacco alkaloid. To avoid confusion of nicotinic acid and nicotinamide with nicotine itself, **niacin** was adopted as a common name for nicotinic acid. This name comes from the letters of three words—*nic*otinic, *ac*id, and vitam*in*.

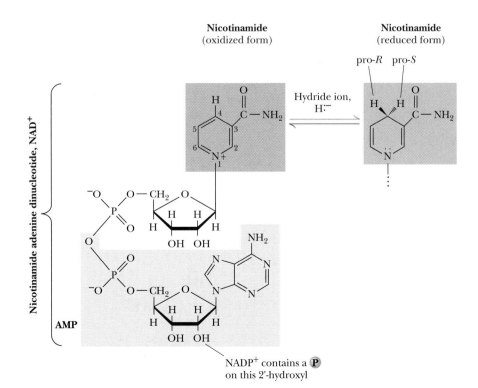

Pyridine   **Nicotinic acid**   **Nicotinamide**   **Nicotine**   The structures of pyridine, nicotinic acid, nicotinamide, and nicotine.

coenzyme. Specifically, the adenine nucleotide greatly increases both the affinity and the specificity of the coenzyme for its site on the enzyme, owing to the numerous sites for hydrogen bonding, ionic bonding, and hydrophobic interactions that the adenine nucleotide moiety brings to the coenzyme structure.

Nicotinic Acid and the Nicotinamide Coenzymes

Nicotinamide is an essential part of two important coenzymes: **nicotinamide adenine dinucleotide (NAD⁺)** and **nicotinamide adenine dinucleotide phosphate (NADP⁺)** (Figure 14.16). The reduced forms of these coenzymes are

See pages 162–164

Figure 14.16 The structures and redox states of the nicotinamide coenzymes. Hydride ion ($H:^-$, a proton with two electrons) from substrate is transferred to NAD^+ to produce NADH.

NADH and NADPH. The nicotinamide coenzymes (also known as pyridine nucleotides) are electron carriers that participate in a variety of enzyme-catalyzed oxidation–reduction reactions. These reactions involve direct transfer of hydride anion either to $NAD(P)^+$ or from $NAD(P)H$. The enzymes that facilitate such transfers are thus known as **dehydrogenases.** The hydride anion contains two electrons, and thus NAD^+ and $NADP^+$ act exclusively as **two-electron carriers.** The C-4 position of the pyridine ring, which can either accept or donate hydride ion, is the reactive center of both NAD and NADP. The quaternary nitrogen of the nicotinamide ring functions as an electron sink to facilitate hydride transfer to NAD^+ (or $NADP^+$).

Riboflavin and the Flavin Coenzymes

Riboflavin, or vitamin B$_2$, is a constituent and precursor of both **riboflavin 5′-phosphate,** also known as **flavin mononucleotide (FMN),** and **flavin adenine dinucleotide (FAD).** The name riboflavin is a synthesis of the names for the molecule's component parts, **ribitol** and **flavin.** The structures of riboflavin, FMN, and FAD are shown in Figure 14.17. The isoalloxazine ring is the core structure of the various flavins. The flavins have a characteristic bright yellow color and take their name from the Latin *flavus* for "yellow." As shown in Figure 14.18, flavin coenzymes can exist in any of three different redox states. Access to three different redox states allows flavin coenzymes to participate in one-electron transfer and two-electron transfer reactions. Partly because of this, flavoproteins catalyze many different reactions in biological systems and work together with many different electron acceptors and donors. The oxidized form of the isoalloxazine structure absorbs light around 450 nm (in the visible region) and also at 350 to 380 nm. The color is lost, however, when the ring is reduced or "bleached." Similarly, the enzymes that bind flavins, known as **flavoenzymes,** can be yellow, red, or green in their oxidized states; these enzymes also lose their color on reduction of the bound flavin group.

See pages 162–164

Figure 14.17 The structures of riboflavin, flavin mononucleotide (FMN), and flavin adenine dinucleotide (FAD). Because the ribityl group of FMN and FAD is not a true pentose sugar (it is a sugar alcohol) and is not joined to riboflavin in a glycosidic bond, FMN and FAD are not truly "nucleotides," and the terms *flavin mononucleotide* and *dinucleotide* are incorrect. Nonetheless, these designations are so deeply ingrained in common biochemical usage that the erroneous nomenclature persists. Flavin coenzymes bind tightly to the enzymes that use them, with typical dissociation constants in the range of 10^{-8} to 10^{-11} M, so that only very low levels of free flavin coenzymes occur in most cells. Even in organisms that rely on the nicotinamide coenzymes (NADH and NADPH) for many of their oxidation-reduction cycles, the flavin coenzymes fill essential roles. Flavins are stronger oxidizing agents than NAD^+ and $NADP^+$. They can be reduced by both one-electron and two-electron pathways and can be reoxidized easily by molecular oxygen. Enzymes that use flavins to carry out their reactions—flavoenzymes—are involved in many kinds of oxidation-reduction reactions.

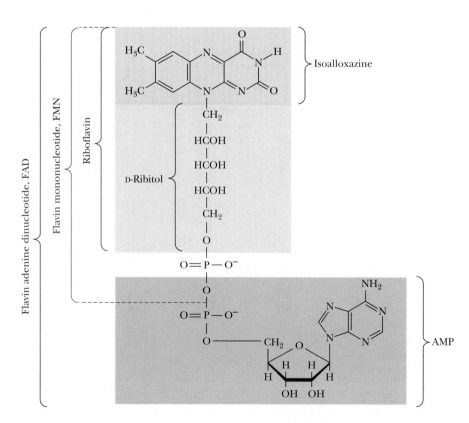

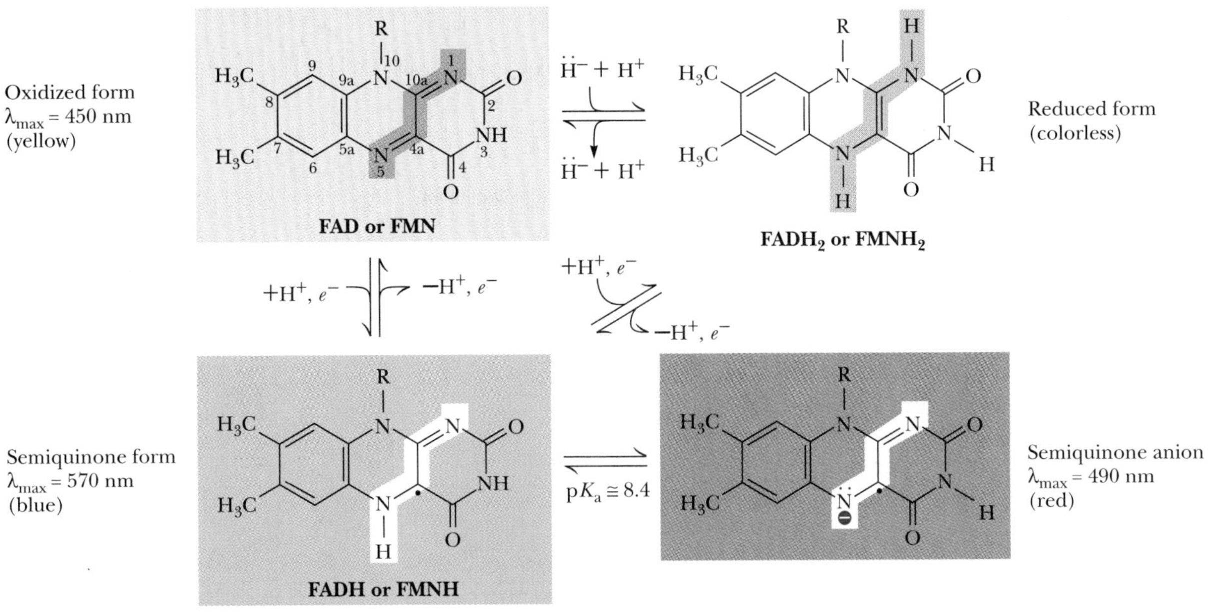

Oxidized form
λ_{max} = 450 nm
(yellow)
FAD or FMN
$\ddot{H}^- + H^+$
$\ddot{H}^- + H^+$
FADH$_2$ or FMNH$_2$
Reduced form
(colorless)
$+H^+, e^-$
$-H^+, e^-$
$+H^+, e^-$
$-H^+, e^-$
Semiquinone form
λ_{max} = 570 nm
(blue)
FADH or FMNH
$pK_a \cong 8.4$
Semiquinone anion
λ_{max} = 490 nm
(red)

Figure 14.18 The redox states of FAD and FMN. The boxes correspond to the colors of each of these forms. The atoms primarily involved in electron transfer are indicated by red shading in the oxidized form, white in the semiquinone form, and blue in the reduced form.

See pages 162–164

Pantothenic Acid and Coenzyme A

Pantothenic acid, sometimes called vitamin B_3, is a vitamin that makes up one part of a complex coenzyme called **coenzyme A (CoA)** (Figure 14.19). Pantothenic acid is also a constituent of **acyl carrier proteins.** As was the case for the nicotinamide and flavin coenzymes, the adenine nucleotide moiety of CoA acts as a recognition site, increasing the affinity and specificity of CoA binding to its enzymes.

The two main functions of coenzyme A are

a. activation of acyl groups for transfer to nucleophilic acceptors, and
b. activation of the α-hydrogen of the acyl group for removal as a proton.

Both of these functions involve the reactive sulfhydryl group on CoA through its formation of **thioester** linkages with acyl groups.

The 4-phosphopantetheine group of CoA is also used (for essentially the same purposes) in **acyl carrier proteins (ACPs)** involved in fatty acid biosynthesis (see Chapter 20). In acyl carrier proteins, the 4-phosphopantetheine is covalently linked to a serine hydroxyl group.

Vitamin B_6: Pyridoxine and Pyridoxal Phosphate

Any of several forms—pyridoxine, pyridoxal, or pyridoxol—can fulfill nutritional requirements for vitamin B_6 (Figure 14.20), but the biologically active

See pages 162–164

4-Phosphopantetheine
β-Mercaptoethylamine
Pantothenic acid
3',5'-ADP

Figure 14.19 The structure of coenzyme A. Coenzyme A consists of 3′, 5′-adenosine bisphosphate joined to 4-phosphopantetheine in a phosphoric anhydride linkage. Phosphopantetheine in turn consists of three parts: β-mercaptoethylamine linked to β-alanine, which makes an amide bond with a branched-chain dihydroxy acid. Acyl groups form thioester linkages with the –SH group of the β-mercaptoethylamine moiety.

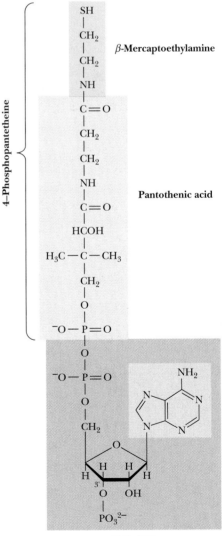

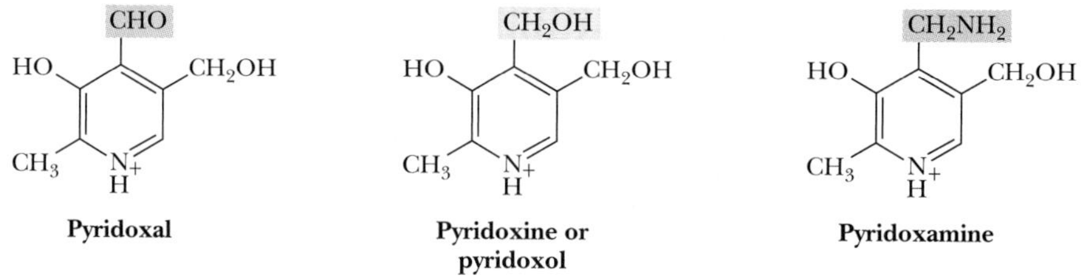

Figure 14.20 The structures of pyridoxal, pyridoxine or pyridoxol, and pyridoxamine.

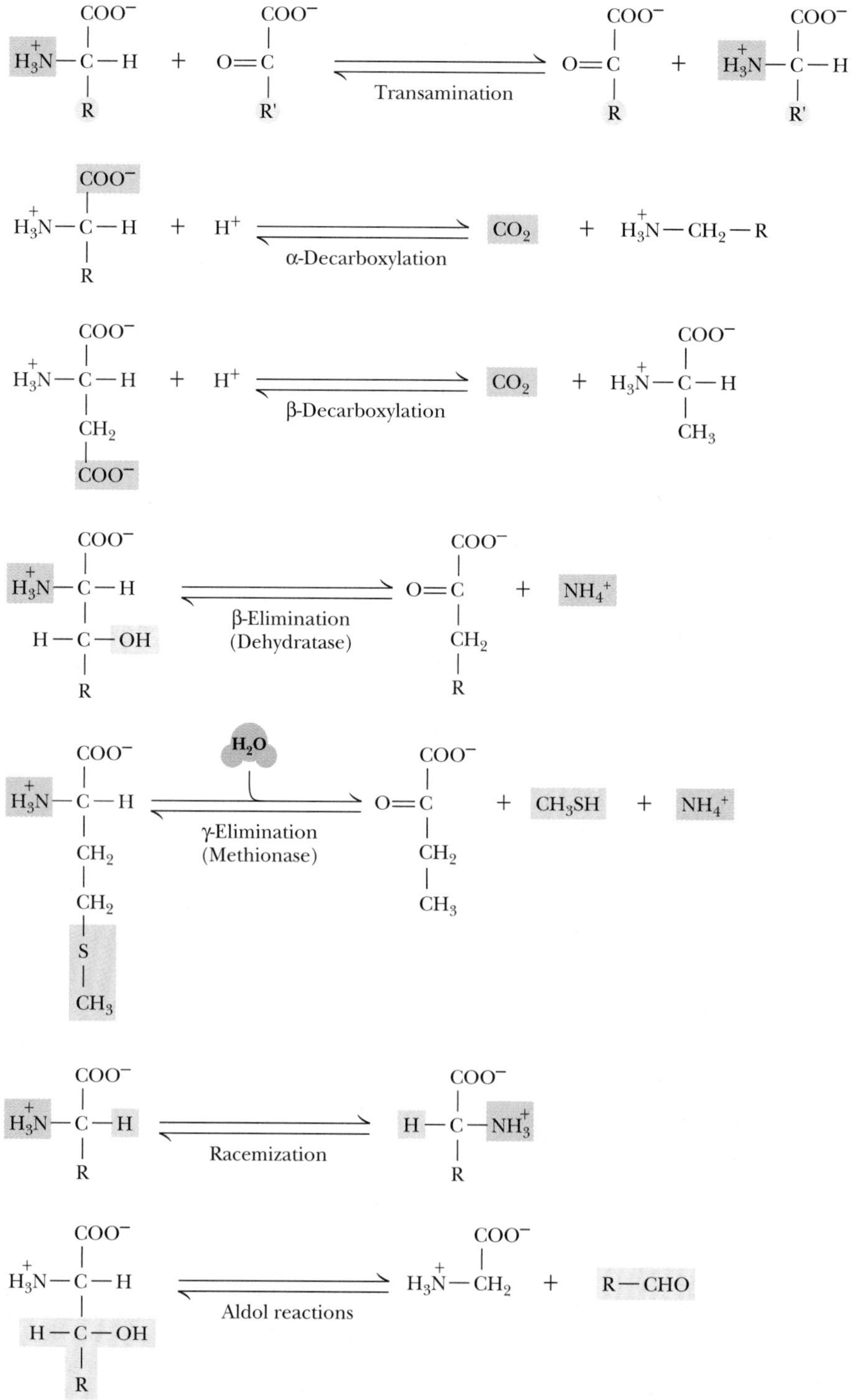

Pyridoxal

Pyridoxine or pyridoxol

Pyridoxamine

Figure 14.21 The seven classes of reactions catalyzed by pyridoxal-5-phosphate.

See pages 162–164

Transamination

α-Decarboxylation

β-Decarboxylation

β-Elimination (Dehydratase)

γ-Elimination (Methionase)

Racemization

Aldol reactions

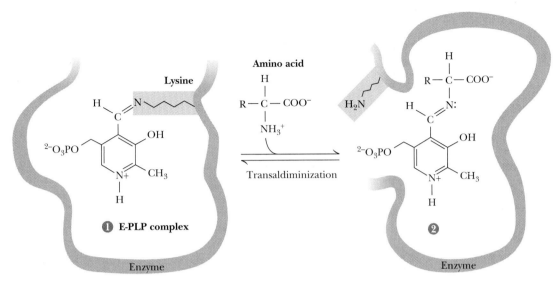

Figure 14.22 Pyridoxal-5-phosphate forms stable Schiff base adducts with amino compounds such as amino acids.

form of vitamin B_6 is **pyridoxal-5-phosphate (PLP)**. PLP participates in the catalysis of a wide variety of reactions (Figure 14.21) involving amino acids, including transaminations, α- and β-decarboxylations, β- and γ-eliminations, racemizations, and aldol reactions. Note that these reactions include cleavage of any of the bonds to the amino acid α-carbon, as well as several bonds in the side chain. The remarkably versatile chemistry of PLP is due to its ability to form stable Schiff base (aldimine) adducts with α-amino groups of amino acids.

The Schiff base formed by PLP is illustrated in Figure 14.22. In nearly all PLP-dependent enzymes, PLP in the absence of substrate is bound in a Schiff base linkage with the ε-NH_2 group of an active site lysine. Rearrangement to a Schiff base with the arriving substrate is a **transaldiminization** reaction.

Vitamin B_{12}: Cyanocobalamin

Vitamin B_{12} is not synthesized by animals or by plants. Only a few species of bacteria synthesize this complex substance. Carnivorous animals easily acquire sufficient amounts of B_{12} from meat in their diet, but herbivorous creatures typically depend on intestinal bacteria to synthesize B_{12} for them. This is sometimes not sufficient, and certain animals, including rabbits, occasionally eat their feces in order to accumulate the necessary quantities of B_{12}. **Vitamin B_{12},** or **cyanocobalamin,** is converted in the body into two coenzymes. The predominant coenzyme form is **5′-deoxyadenosylcobalamin** (Figure 14.23); small amounts of **methylcobalamin** also occur.

The B_{12} coenzymes participate in three types of reactions (Figure 14.24):

1. Intramolecular rearrangements
2. Reductions of ribonucleotides to deoxyribonucleotides for DNA synthesis in certain bacteria
3. Methyl group transfers

The first two of these involve 5′-deoxyadenosylcobalamin, whereas methyl transfers rely on methylcobalamin. The mechanism of ribonucleotide reductase is discussed in Chapter 21. Methyl group transfers that employ tetrahydrofolate as a coenzyme are described later in this chapter.

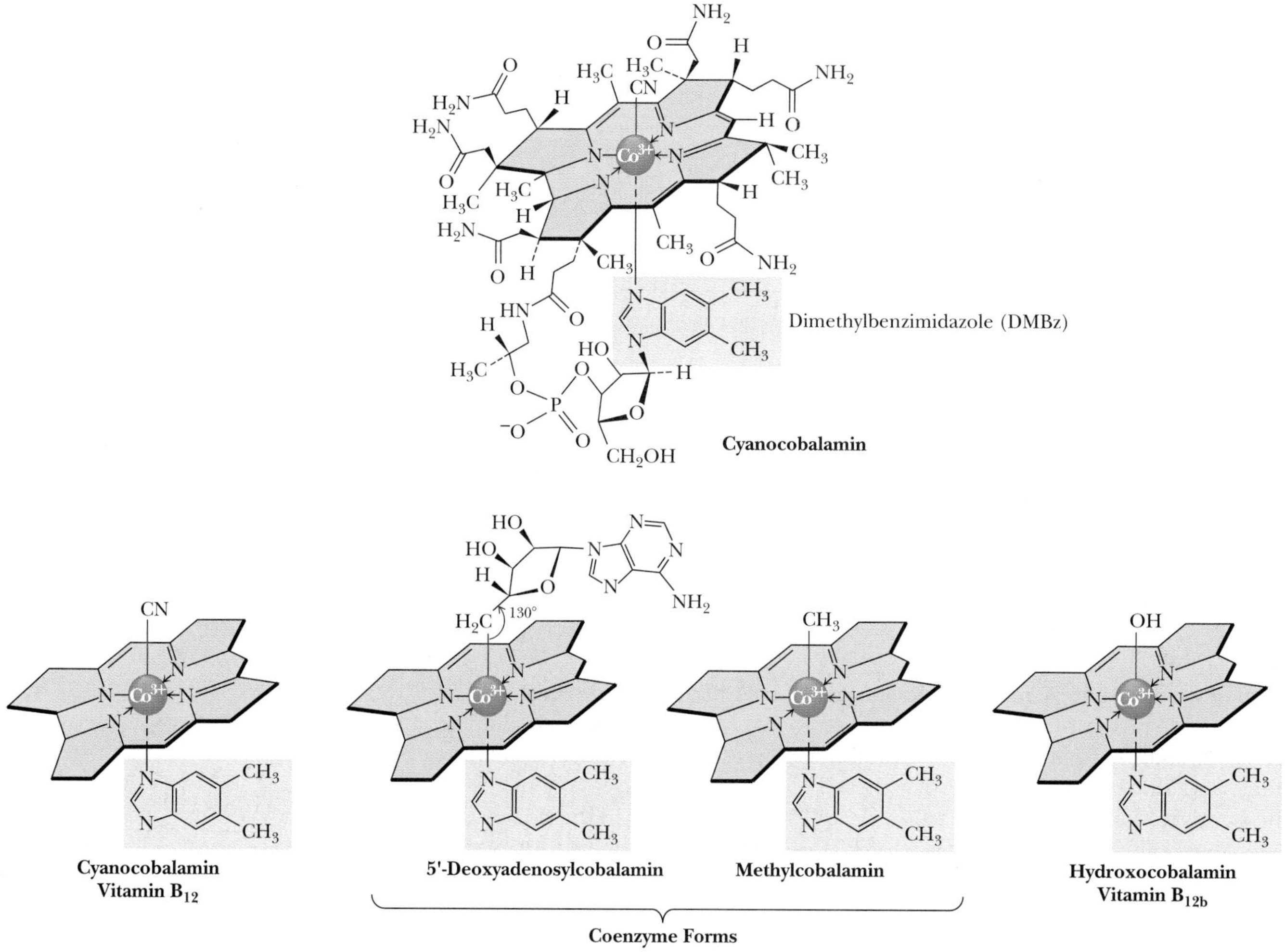

Dimethylbenzimidazole (DMBz)

Cyanocobalamin

Cyanocobalamin
Vitamin B₁₂

5′-Deoxyadenosylcobalamin

Methylcobalamin

Hydroxocobalamin
Vitamin B₁₂ᵦ

Coenzyme Forms

See pages 162–164

Figure 14.23 The structure of cyanocobalamin *(top)* and simplified structures showing several coenzyme forms of vitamin B_{12}. 5′-Deoxyadenosylcobalamin consists of a corrin ring with a cobalt ion in the center. The corrin ring, with four pyrrole groups, is similar to the heme prophyrin ring, except that two of the pyrrole rings are linked directly. Methylene bridges form the other pyrrole–pyrrole linkages, as they do for porphyrin. The cobalt is coordinated to the four (planar) pyrrole nitrogens. One of the axial cobalt ligands is a nitrogen of the dimethylbenzimidazole group. The other axial cobalt ligand may be —CN, —CH_3, —OH, or the 5′-carbon of a 5′-deoxyadenosyl group, depending on the form of the coenzyme. (The —CN form is an artifact of cobalamin purification.) The Co—C bond between Co and the 5′-deoxyadenosyl group is predominantly covalent (with a short bond length of 0.205 nm), but with some ionic character. Note that the convention of writing the cobalt atom as Co^{3+} attributes the electrons of the Co—C and Co—N bonds to carbon and nitrogen, respectively.

Vitamin C: Ascorbic Acid

L-**Ascorbic acid,** better known as **vitamin C,** has the simplest chemical structure of all the vitamins (Figure 14.25). It is widely distributed in the animal and plant kingdoms, and only a few vertebrates, including humans, are unable to synthesize it.

Ascorbic acid is a reasonably strong reducing agent, and its biochemical and physiological functions derive from its reducing properties—it functions as an electron carrier (Figure 14.25). Loss of one electron due to interactions with oxygen or metal ions leads to **semidehydro-L-ascorbate,** a reactive free radical that can be reduced back to L-ascorbic acid by various enzymes in

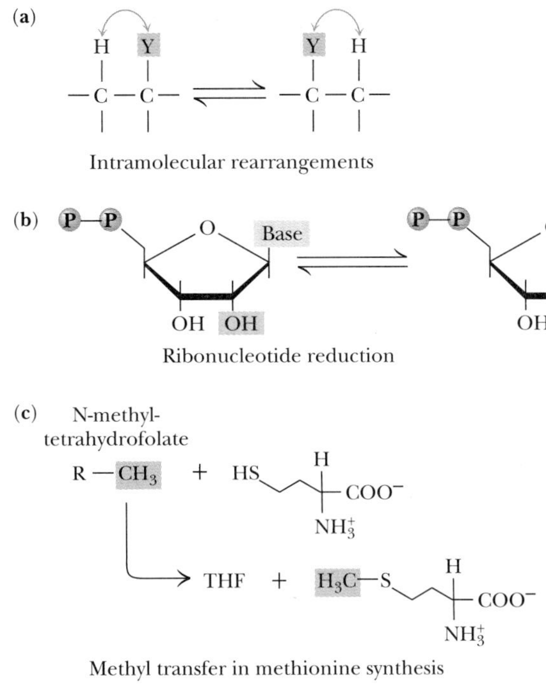

(a)

Intramolecular rearrangements

(b)

Ribonucleotide reduction

(c) N-methyl-
tetrahydrofolate

Methyl transfer in methionine synthesis

See pages 162–164

Figure 14.24 Vitamin B_{12} functions as a coenzyme in intramolecular rearrangements, in reduction of ribonucleotides, and in methyl group transfers.

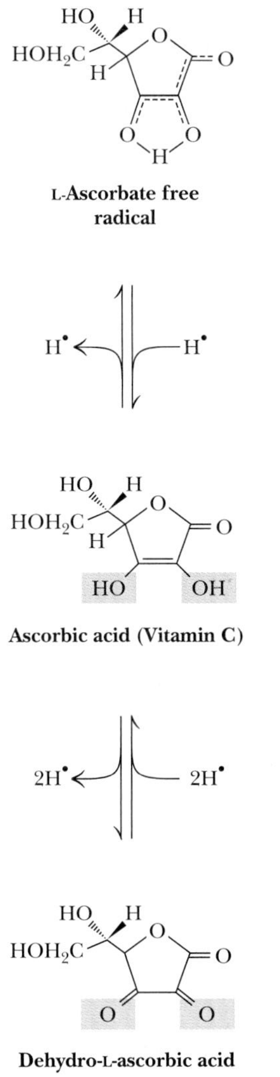

L-Ascorbate free radical

Ascorbic acid (Vitamin C)

Dehydro-L-ascorbic acid

Figure 14.25 The physiological effects of ascorbic acid (vitamin C) are the result of its action as a reducing agent. A two-electron oxidation of ascorbic acid yields dehydroascorbic acid.

animals and plants. A characteristic reaction of ascorbic acid is its oxidation to dehydro-L-ascorbic acid. Ascorbic acid and dehydroascorbic acid form an effective redox system.

In addition to its role in preventing scurvy (see *Human Biochemistry*, page 462), ascorbic acid also mobilizes iron in the body, prevents anemia, ameliorates allergic responses, and stimulates the immune system.

Biotin

Biotin (Figure 14.26) acts as a **mobile carboxyl group carrier** in a variety of enzymatic carboxylation reactions. In each of these, biotin is bound covalently to the enzyme as a prosthetic group via the ε-amino group of a lysine residue on the protein (Figure 14.27). The biotin–lysine function is referred to as a **biocytin** residue. The result is that the biotin ring system is tethered to the

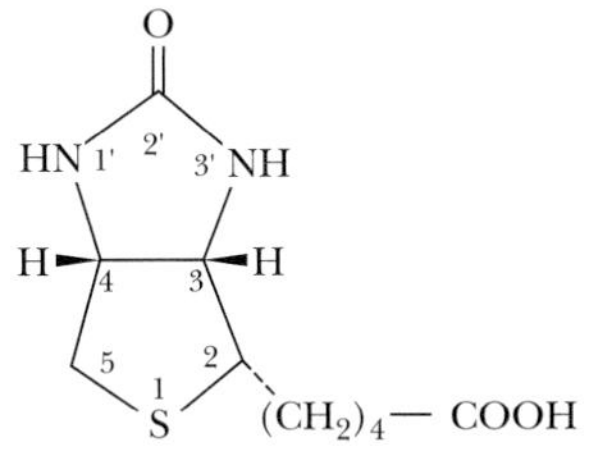

Figure 14.26 The structure of biotin.

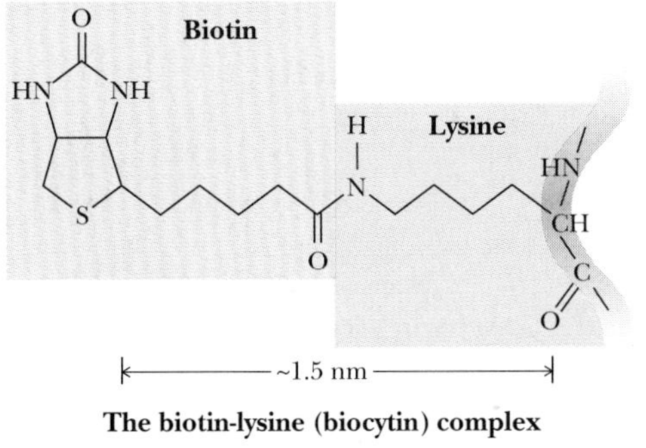

Biotin

H   Lysine

The biotin-lysine (biocytin) complex

Figure 14.27 Biotin is covalently linked to a protein via the ε-amino group of a lysine residue. The biotin ring is thus tethered to the protein by a 10-atom chain. It functions by carrying carboxyl groups between distant sites on biotin-dependent enzymes.

Ascorbic Acid and Scurvy

Ascorbic acid is effective in the treatment and prevention of **scurvy,** a potentially fatal disorder characterized by anemia, alteration of protein metabolism, and weakening of collagenous structures in bone, cartilage, teeth, and connective tissues (see Chapter 5). Western diets are now routinely so rich in vitamin C that it is easy to forget that scurvy affected many people in ancient Egypt, Greece, and Rome, and that, in the Middle Ages, it was endemic in northern Europe in winter when fresh fruits and vegetables were scarce. Ascorbic acid is a vitamin that has altered the course of history, ending ocean voyages and military campaigns when food supplies became depleted of vitamin C and fatal outbreaks of scurvy occurred.

protein by a long, flexible chain. The 10 atoms in this chain separate the biotin ring and the lysine α-carbon by approximately 1.5 nm. The mobility afforded by this chain allows biotin to acquire carboxyl groups at one subsite of the enzyme active site and deliver them to a substrate acceptor at another subsite.

Most biotin-dependent carboxylations (Table 14.5) use bicarbonate as the carboxylating agent and transfer the carboxyl group to a substrate acceptor. Bicarbonate is plentiful in biological fluids, but it is relatively unreactive and must be activated in an ATP-dependent manner in order to link it with biotin for transfer to an acceptor.

Lipoic Acid

Lipoic acid exists in two forms: a closed-ring disulfide form and an open-chain reduced form (Figure 14.28). Oxidation–reduction cycles interconvert these two species. Lipoic acid does not often occur free in nature but rather is covalently attached in amide linkage with lysine residues on enzymes.

See pages 162–164

Lipoic acid is an **acyl group carrier.** It is found in *pyruvate dehydrogenase* and *α-ketoglutarate dehydrogenase* (Figure 14.29), two multienzyme complexes involved in α-keto acid oxidation (see Chapter 16). Lipoic acid functions to couple acyl-group transfer and electron transfer during oxidation and decarboxylation of α-keto acids. No evidence exists of a dietary lipoic acid requirement in humans; strictly speaking, it is not considered a vitamin.

Table 14.5   Principal Biotin-Dependent Carboxylations

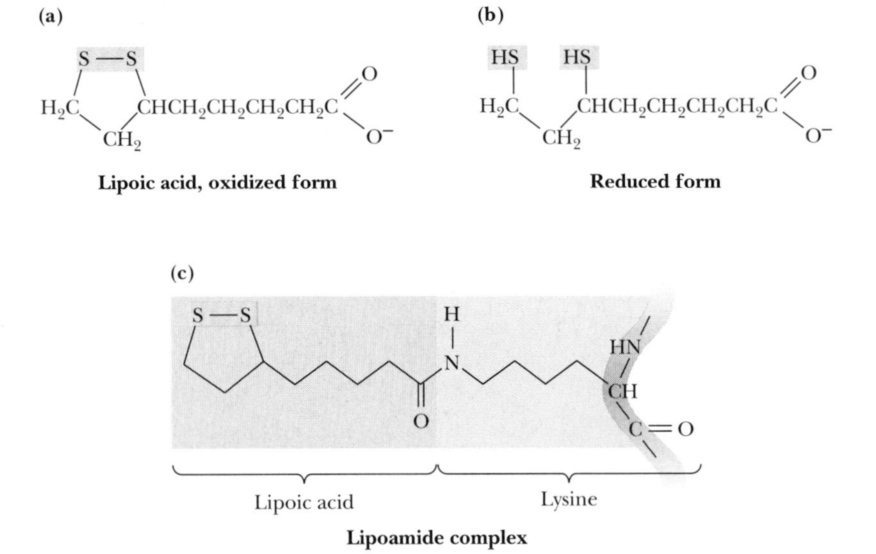

(a)

Lipoic acid, oxidized form

(b)

Reduced form

(c)

Lipoic acid   Lysine

Lipoamide complex

Figure 14.28 The oxidized and reduced forms of lipoic acid and the structure of the lipoyl-lysine conjugate.

 See pages 162–164

Folic Acid

Folic acid derivatives (folates) are acceptors and donors of one-carbon units for all oxidation levels of carbon except that of CO_2 (where biotin is the relevant carrier). The active coenzyme form of folic acid is **tetrahydrofolate (THF).** THF is formed via two successive reductions of folate by dihydrofolate reductase (Figure 14.30). One-carbon units in three different oxidation states may be bound to tetrahydrofolate at the N^5 and/or N^{10} nitrogens (Table 14.6). These one-carbon units may exist at the oxidation levels of methanol, formaldehyde, or formate (carbon atom oxidation states of -2, 0, and 2, respectively). The biosynthetic pathways for methionine and homocysteine, purines, and the pyrimidine thymine (see Chapter 21) rely on the incorporation of one-carbon units from THF derivatives.

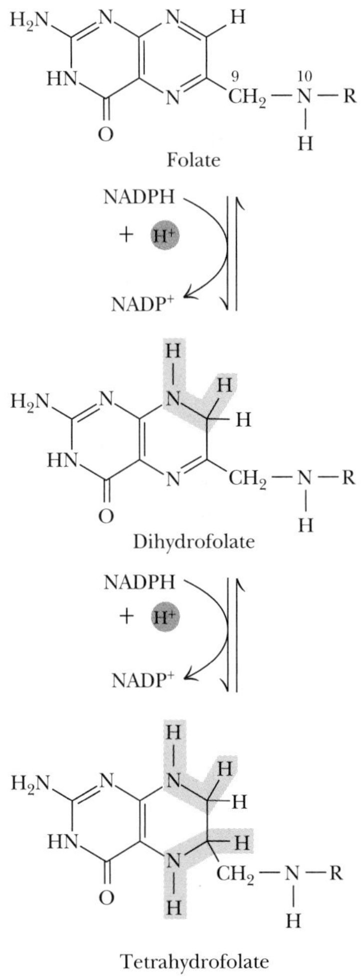

The Vitamin A Group

Vitamin A or **retinol** (Figure 14.31) often occurs in the form of esters, called **retinyl esters.** Two other forms are prominent: the aldehyde form, called **retinal** or **retinaldehyde,** and the acid form, **retinoic acid.** Like all the fat-soluble vitamins, retinol is an isoprenoid molecule and is biosynthesized from isoprene building blocks (see Chapter 6). Retinol can be obtained in the diet from animal sources or synthesized from β-carotene provided from plant sources.

Pyruvate + CoA + NAD⁺ $\xrightarrow[\substack{\text{Pyruvate} \\ \text{dehydrogenase}}]{}$ Acetyl-CoA + CO_2 + NADH + H⁺

α-Ketoglutarate + CoA + NAD⁺ $\xrightarrow[\substack{\text{α-Ketoglutarate} \\ \text{dehydrogenase}}]{}$ Succinyl-CoA + CO_2 + NADH + H⁺

Figure 14.29 Enzyme-catalyzed reactions involving lipoic acid.

 See pages 162–164

Folate

NADPH
+ H⁺

NADP⁺

Dihydrofolate

NADPH
+ H⁺

NADP⁺

Tetrahydrofolate

Figure 14.30 Formation of THF from folic acid by the dihydrofolate reductase reaction. The R group on these folate molecules symbolizes the one to seven (or more) glutamate units that folates characteristically contain. All of these glutamates are bound in γ-carboxyl amide linkages (as in the folic acid structure shown in *A Deeper Look: Folic Acid, Pterins, and Insect Wings*). The one-carbon units carried by THF are bound at N^5, or at N^{10}, or as a single carbon attached to both N^5 and N^{10}.

A DEEPER LOOK

Folic Acid, Pterins, and Insect Wings

Folic acid is a member of the vitamin B complex found in green plants, fresh fruit, yeast, and liver. Folic acid takes its name from *folium*, Latin for "leaf." Pterin compounds are named for the Greek word for "wing" because these substances were first identified in insect wings. Two pterins are familiar to any child who has seen (and chased) the common yellow sulfur butterfly and its white counterpart, the cabbage butterfly. Xanthopterin and leucopterin are the respective pigments in these butterflies' wings. Mammals cannot synthesize pterins; they obtain folates from their diet or from microorganisms active in the intestines.

Folic acid

Pterin
(2-amino-4-oxopteridine)

p-Aminobenzoic acid
(PABA)

Glutamates

Xanthopterin (yellow)   Leucopterin (white)

Pteridine

Pterin: 2-amino-4-
oxopteridine

Vitamin A is absolutely essential to vision. Retinol delivered to the retinas of the eyes is accumulated by **rod and cone cells.** In the rods (which are the better characterized of the two cell types), retinol is oxidized by a specific **retinol dehydrogenase** to become all-*trans*-retinal and then converted to 11-*cis*-retinal by **retinal isomerase** (see Figure 14.31). The aldehyde group of retinal forms a Schiff base with a lysine on **opsin,** to form **rhodopsin,** the light-sensitive pigment of vision.

Table 14.6   Oxidation States of Carbon in 1-Carbon Units Carried by Tetrahydrofolate

Oxidation Number*	Oxidation Level	One-Carbon Form[†]	Tetrahydrofolate Form
-2	Methanol (most reduced)	$-CH_3$	N^5-Methyl-THF
0	Formaldehyde	$-CH_2-$	N^5, N^{10}-Methylene-THF
2	Formate (most oxidized)	$-CH=O$	N^5-Formyl-THF
		$-CH=O$	N^{10}-Formyl-THF
		$-CH=NH$	N^5-Formimino-THF
		$-CH=$	N^5, N^{10}-Methenyl-THF

*Calculated by assigning valence bond electrons to the more electronegative atom and then counting the charge on the quasi ion. A carbon assigned four valence electrons would have an oxidation number of 0. The carbon in N^5-methyl-THF is assigned six electrons from the three C—H bonds and thus has an oxidation number of -2.
[†]Note: All vacant bonds in the structures shown are to atoms more electronegative than C.

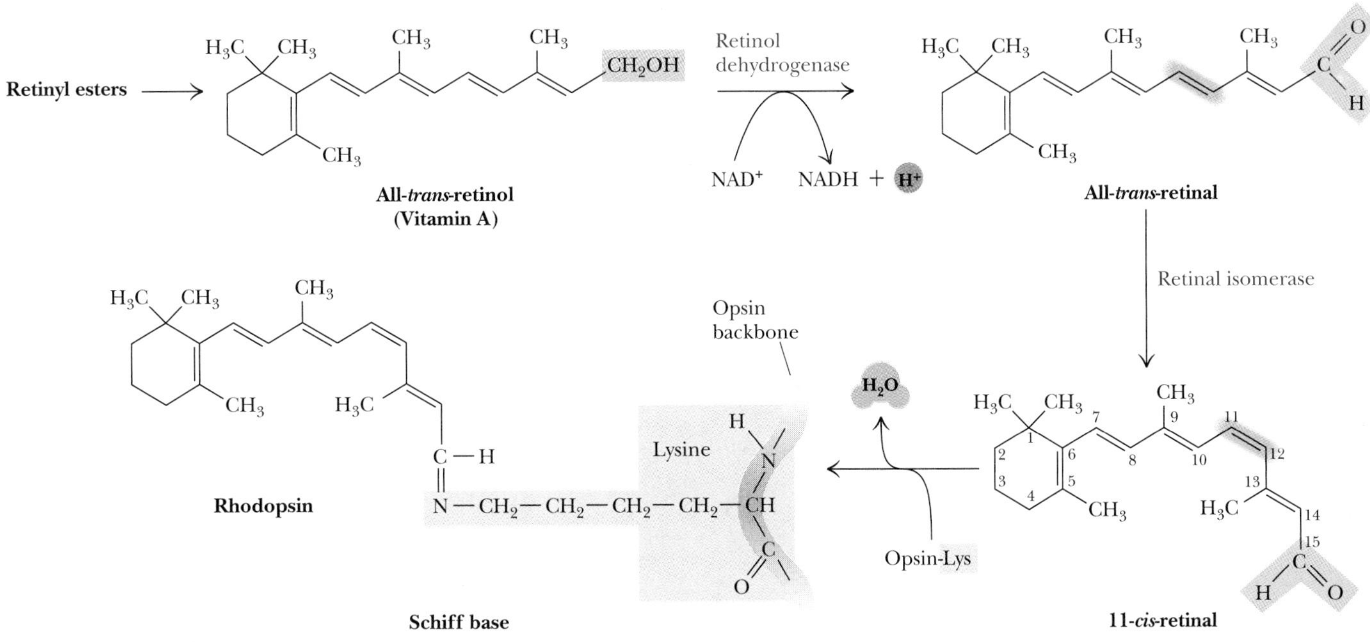

Figure 14.31 The three forms of vitamin A: retinol, retinal, and retinoic acid. The incorporation of retinal into the light-sensitive protein rhodopsin involves several steps. All-*trans*-retinol is oxidized by retinol dehydrogenase and then isomerized to 11-*cis*-retinal, which forms a Schiff base linkage with opsin to form light-sensitive rhodopsin.

Only 0.1% of the body's vitamin A is found in the retina. Much more is present in epithelial cells of the body as retinoic acid. Retinoic acid participates in processes of gene expression involved in cell growth and differentiation. Vitamin A is also essential to sperm development in males; in women, vitamin A plays an essential role in fetal development during pregnancy. Despite these benefits, vitamin A in excessive amounts is toxic.

The Vitamin D Group

The two most prominent members of the **vitamin D** family are **ergocalciferol** (known as vitamin D_2) and **cholecalciferol** (vitamin D_3). Cholecalciferol is produced in the skin of animals by the action of ultraviolet light (sunlight, for example) on its precursor molecule, 7-dehydrocholesterol (Figure 14.32). The absorption of light energy induces breakage of the 9,10 carbon bond and formation of **previtamin D_3.** The next step is a spontaneous isomerization to yield vitamin D_3, cholecalciferol. Ergocalciferol, which differs from cholecalciferol only in the side-chain structure, is similarly produced by the action of sunlight on the plant sterol **ergosterol.** Because humans can produce vitamin D_3 from 7-dehydrocholesterol by the action of sunlight on the skin, vitamin D is, strictly speaking, not a vitamin at all.

On the basis of its mechanism of action in the body, cholecalciferol should be called a **prohormone,** a hormone precursor. In the liver, cholecalciferol is hydroxylated at the C-25 position to form 25-hydroxyvitamin D_3 (that is, 25-hydroxycholecalciferol). Although this is the major circulating form of vitamin

HUMAN BIOCHEMISTRY

Vitamin A and Night Vision

Night blindness was probably the first disorder to be ascribed to a nutritional deficiency. The ancient Egyptians left records from as early as 1500 B.C. of recommendations that the juice squeezed from cooked liver could cure night blindness, and the treatment may have been known much earlier.

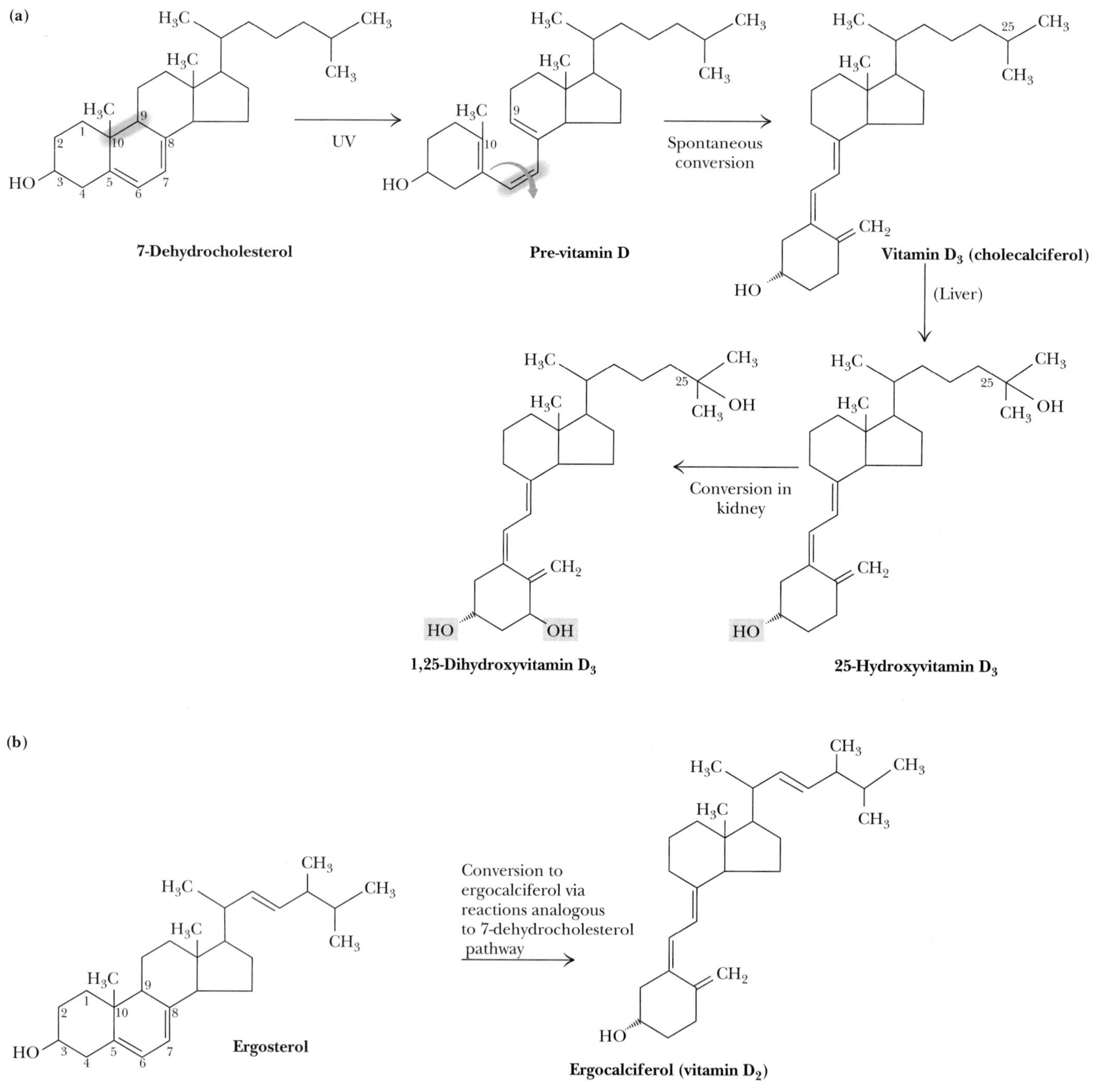

(a)

7-Dehydrocholesterol   Pre-vitamin D   Vitamin D₃ (cholecalciferol)

UV

Spontaneous conversion

(Liver)

Conversion in kidney

1,25-Dihydroxyvitamin D₃   25-Hydroxyvitamin D₃

(b)

Ergosterol

Conversion to ergocalciferol via reactions analogous to 7-dehydrocholesterol pathway

Ergocalciferol (vitamin D₂)

Figure 14.32 **(a)** Vitamin D₃ (cholecalciferol) is produced in the skin by the action of sunlight on 7-dehydrocholesterol. Subsequent enzymatic reactions in the liver and kidney produce 1,25-dihydroxyvitamin D₃, the active form of vitamin D. **(b)** Ergocalciferol is produced in analogous fashion from ergosterol.

D in the body, 25-hydroxyvitamin D₃ possesses far less biological activity than the final active form. This final form is created within kidney tissue when 25-hydroxyvitamin D₃ is hydroxylated at the C-1 position to yield 1,25-dihydroxyvitamin D₃ (that is, 1,25-dihydroxycholecalciferol), the active form of vitamin D. 1,25-Dihydroxycholecalciferol is then transported to target tissues, where it acts like a hormone to regulate calcium and phosphate metabolism.

Vitamin D and Rickets

Vitamin D is a family of closely related molecules that prevent **rickets,** a childhood disease characterized by inadequate intestinal absorption and kidney reabsorption of calcium and phosphate. These inadequacies eventually lead to the demineralization of bones. The symptoms of rickets include bowlegs, knock-knees, curvature of the spine, and pelvic and thoracic deformities—the results of normal mechanical stresses on demineralized bones. In adults, vitamin D deficiency leads to a weakening of bones and cartilage known as **osteomalacia.**

1,25-Dihydroxyvitamin D_3, together with two peptide hormones, **calcitonin** and **parathyroid hormone (PTH),** functions to regulate calcium homeostasis and plays a role in phosphorus homeostasis. Phosphorus and calcium are critically important for the formation of bones. Any disturbance of normal serum phosphorus and calcium levels will result in alterations of bone structure, as in rickets. The mechanism of calcium homeostasis involves the coordination of calcium absorption in the intestine, deposition in the bones, and excretion by the kidneys. If a decrease in serum calcium occurs, vitamin D is converted to 1,25-dihydroxyvitamin D_3, which acts in the intestine to increase calcium absorption. In conjunction with PTH, vitamin D acts on bones to enhance absorption of calcium while PTH acts on the kidney to increase calcium reabsorption. If serum calcium levels get too high, calcitonin acts to block vitamin D metabolism and PTH secretion. Calcitonin also induces calcium excretion from the kidneys and inhibits calcium mobilization from bone.

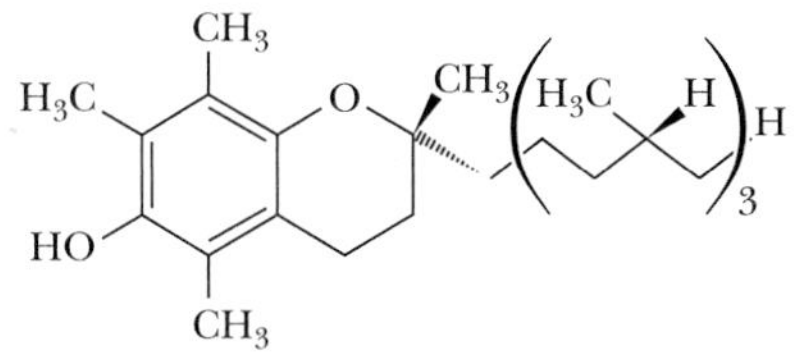

Vitamin E: Tocopherol

The structure of **vitamin E** in its most active form, **α-tocopherol,** is shown in Figure 14.33. α-Tocopherol is a potent antioxidant, and its function in animals and humans is often ascribed to this property. On the other hand, the molecular details of its function are almost entirely unknown. One possible role for vitamin E may relate to the protection of unsaturated fatty acids in membranes because these fatty acids are particularly susceptible to oxidation. When human plasma levels of α-tocopherol are low, red blood cells are increasingly prone to oxidative damage. Infants, especially premature infants, are deficient in vitamin E. When low-birth-weight infants are exposed to high oxygen levels for the purpose of alleviating respiratory distress, the risk of oxygen-induced retina damage can be reduced by administering vitamin E.

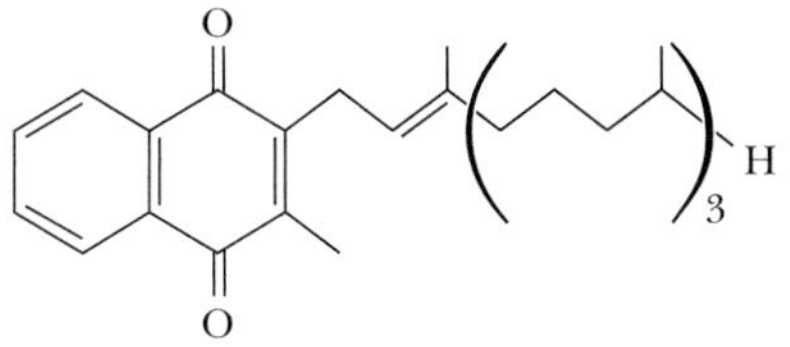

Vitamin K: A Naphthoquinone

Vitamin K (Figure 14.34) is essential to the blood-clotting process. **Prothrombin** (an essential protein in the coagulation cascade) is functional only if it is posttranslationally modified (chemically altered after it has been synthesized by ribosomes). In this modification, 10 glutamate residues on the amino terminal end of prothrombin are carboxylated to form γ-carboxyglutamyl residues. The double negative charges on these residues are effective in the coordination of calcium, which is required for the coagulation process. The enzyme responsible for this modification, a liver microsomal glutamyl carboxylase, requires vitamin K for its activity (Figure 14.35). Prothrombin—called "factor II" in the clotting pathway—and clotting factors VII, IX, and X, as well as several other plasma proteins, contain γ-carboxylglutamyl residues.

Vitamin E (α-tocopherol)

Figure 14.33 The structure of vitamin E (α-tocopherol).

Vitamin K_1 (phylloquinone)

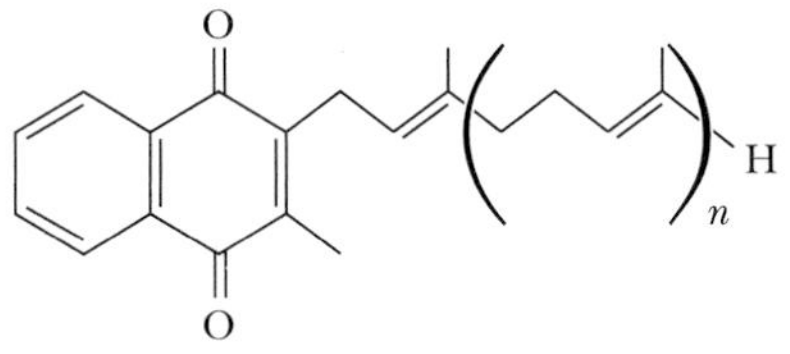

Vitamin K_2 (menaquinone series)

Figure 14.34 The structures of the K vitamins. The ring system in these compounds is 1,4-naphthoquinone.

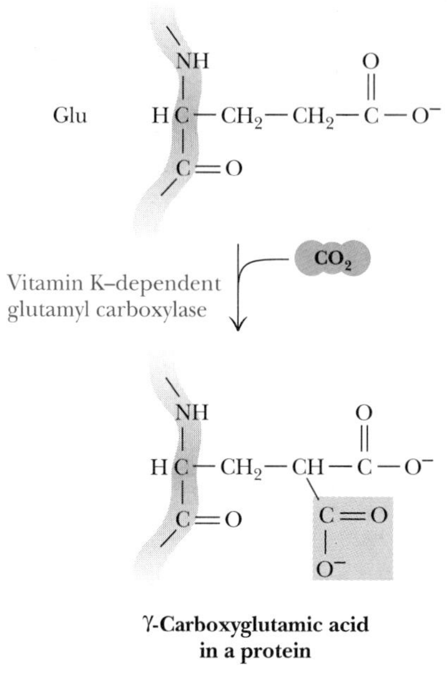

Figure 14.35 The glutamyl carboxylase reaction is vitamin K–dependent. This enzyme activity is essential for the formation of γ-carboxyglutamyl residues in several proteins of the blood-clotting cascade (see Figure 10.25), accounting for the vitamin K dependence of coagulation.

Table 14.7 **Principal Minerals in the Human Body**

Mineral	Amount*
Major	
Calcium	1150
Phosphorus	600
Potassium	210
Sulfur	150
Sodium	90
Chloride	90
Magnesium	30
Trace	
Iron	2.4
Zinc	2.0
Copper	0.09
Manganese	0.02
Iodine	0.02
Selenium	0.02

Fluoride, molybdenum, and chromium are essential trace elements present in even lower amounts in the body. Evidence for silicon, tin, vanadium, boron, aluminum, and even arsenic as essential trace elements for humans has been reported.

*Amounts given are grams present in a typical 60-kg human.

14.6 Minerals

Table 14.7 shows those minerals known to be essential to human health, classified as **major**—those present in more than 5-g quantities in the body—or **trace,** which are present at lower quantities. Calcium and phosphorus are the principal minerals in the skeleton, but both also play important roles in metabolism. Calcium ion is a prominent intracellular signaling agent. Phosphate is part of DNA and RNA, and organic phosphates are key intermediates in energy metabolism. Potassium, sodium, and chloride are crucial in maintaining fluid and electrolyte balance. K^+ is also an essential activator for some enzymes. Indeed, most of the remaining minerals in this table are necessary to human health because they serve as cofactors for enzymes.

PROBLEMS

1. Estimates indicate that 3×10^{14} kg of CO_2 are cycled through the biosphere annually. How many human equivalents (70-kg persons composed of 18% carbon by weight) could be produced from this amount of CO_2?

2. Define the differences in carbon and energy metabolism between photoautotrophs and photoheterotrophs, and between chemoautotrophs and chemoheterotrophs.

3. What are the features that generally distinguish pathways of catabolism from pathways of anabolism?

4. What are the three principal modes of enzyme organization in metabolic pathways?

5. Why do metabolic pathways have so many different steps?

6. Why is the pathway for the biosynthesis of a biomolecule at least partially different from the pathway for its catabolism? Why is the pathway for the biosynthesis of a biomolecule inherently more complex than the pathway for its degradation?

7. What are the metabolic roles of ATP, NAD^+, and NADPH?

8. What might be the advantages of compartmentalizing particular metabolic pathways (see Chapter 1) within specific organelles?

9. Maple-syrup urine disease (MSUD) is an autosomal recessive genetic disease characterized by progressive neurological dysfunction and a sweet, burnt-sugar or maple-syrup smell in the urine. Affected individuals carry high levels of branched-chain amino acids (leucine, isoleucine, and valine) and their respective branched-chain α-keto acids in cells and body fluids. The genetic defect has been traced to the mitochondrial branched-chain α-keto acid dehydrogenase (BCKD). Affected individuals exhibit mutations in their BCKD, but these mutant enzyme molecules have normal activity. Nonetheless, treatment of MSUD patients with substantial doses of thiamine can alleviate the symptoms of the disease. Suggest an explanation for the symptoms described and for the role of thiamine in ameliorating the symptoms of MSUD.

FURTHER READING

Atkinson, D. E., 1977. *Cellular Energy Metabolism and Its Regulation.* New York: Academic Press. A monograph on energy metabolism that is filled with novel insights regarding the ability of cells to generate energy in a carefully regulated fashion while contending with the thermodynamic realities of life.

Boyer, P. D., 1970. *The Enzymes,* 3rd ed. New York: Academic Press. A good source for the mechanisms of action of vitamins and coenzymes.

Boyer, P. D., 1970. *The Enzymes,* 3rd ed., Vol. 6. New York: Academic Press. See discussion of carboxylation and decarboxylation involving TPP, PLP, lipoic acid, and biotin, and of B_{12}-dependent mutases.

Boyer, P. D., 1972. *The Enzymes,* 3rd ed., Vol. 7. New York: Academic Press. See especially elimination reactions involving PLP.

Boyer, P. D., 1974. *The Enzymes,* 3rd ed., Vol. 10. New York: Academic Press. See discussion of pyridine nucleotide-dependent enzymes.

Boyer, P. D., 1976. *The Enzymes,* 3rd ed., Vol. 13. New York: Academic Press. See discussion of flavin-dependent enzymes.

Cooper, T. G., 1977. *The Tools of Biochemistry.* New York: Wiley-Interscience. Chapter 3, "Radiochemistry," discusses techniques for using radioisotopes in biochemistry.

DeLuca, H., and Schnoes, H., 1983. Vitamin D: Recent advances. *Annual Review of Biochemistry* **52**:411–439.

Jencks, W. P., 1969. *Catalysis in Chemistry and Enzymology.* New York: McGraw-Hill.

Knowles, J. R., 1989. The mechanism of biotin-dependent enzymes. *Annual Review of Biochemistry* **58**:195–221.

Page, M. I., and Williams, A., eds., 1987. *Enzyme Mechanisms.* London: Royal Society of London.

Reed, L., 1974. Multienzyme complexes. *Accounts of Chemical Research* **7**:40–46.

Srere, P. A., 1987. Complexes of sequential metabolic enzymes. *Annual Review of Biochemistry* **56**:89–124. A review of how enzymes in some metabolic pathways are organized into complexes.

Walsh, C. T., 1979. *Enzymatic Reaction Mechanisms.* San Francisco: W. H. Freeman.

Whitney, E. N., and Rolfes, S. R., 1999. *Understanding Nutrition,* 8th ed. Belmont, Calif.: International Thomson Publishing.

Glycolysis

Living organisms, like machines, conform to the law of conservation of energy, and must pay for all their activities in the currency of catabolism.

ERNEST BALDWIN, *Dynamic Aspects of Biochemistry* (1952)

Louis Pasteur in his laboratory. Pasteur's scientific investigations into the fermentation of sugar were sponsored by the French wine industry. (Albert Edelfelt, Musee d'Orsay, Paris; Giraudon/Art Resource, New York)

glycolysis from the Greek *glyk-*, "sweet," and *lysis*, "splitting"

Nearly every living cell carries out a catabolic process known as **glycolysis**—the stepwise degradation of glucose (and other simple sugars). Glycolysis is a paradigm of metabolic pathways. Carried out in the cytosol of cells, glycolysis is an anaerobic process, requiring no oxygen. Millions of years ago, living things evolved in an environment lacking O_2, and glycolysis was likely an early and important pathway for extracting energy from nutrient molecules. It played a central role in anaerobic metabolism during the first 2 billion years of biological evolution on earth. Modern organisms still employ glycolysis to provide precursor molecules for aerobic catabolic pathways (such as the tricarboxylic acid cycle) and as a short-term energy source when oxygen is limiting.

15.1 Overview of Glycolysis

An overview of the glycolytic pathway is presented in Figure 15.1. Most of the details of this pathway (the first metabolic pathway to be elucidated) were worked out in the first half of the 20th century by the German biochemists Otto Warburg, G. Embden, and O. Meyerhof. In fact, the sequence of reactions in Figure 15.1 is often referred to as the **Embden–Meyerhof pathway.**

Glycolysis consists of two phases. In the first, glucose is converted to two molecules of glyceraldehyde-3-phosphate in a series of five reactions. In the second phase, five subsequent reactions convert these two molecules of glyceraldehyde-3-phosphate into two molecules of pyruvate. Phase 1 consumes two molecules of ATP (Figure 15.2). The later stages of glycolysis result in the production of four molecules of ATP. The net is $4 - 2 = 2$ molecules of ATP produced per molecule of glucose.

Rates and Regulation of Glycolytic Reactions Vary Among Species

Microorganisms, plants, and animals (including humans) carry out the 10 reactions of glycolysis in similar fashion, although the rates of the individual reactions and the means by which they are regulated differ from species to species. The most significant difference among species is the way in which the product pyruvate is utilized. The three possible fates for pyruvate are shown in Figure 15.1. In aerobic organisms, including humans, pyruvate is oxidatively decarboxylated to give a two-carbon unit—the acetyl group of acetyl-coenzyme A. This acetyl group is fully oxidized to two molecules of CO_2 by the tricarboxylic acid cycle. The electrons removed in this process are subsequently passed through the mitochondrial electron transport system and used to generate molecules of ATP by oxidative phosphorylation, thus capturing most of the metabolic energy available in the original glucose molecule.

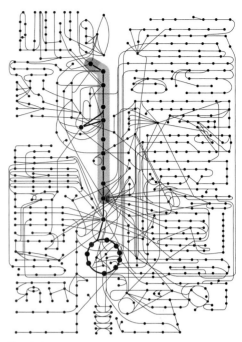

Glycolysis.

 See *Interactive Biochemistry CD-ROM and Workbook,* pages 173–178

15.2 The Importance of Coupled Reactions in Glycolysis

The process of glycolysis converts some, but not all, of the metabolic energy of the glucose molecule into ATP. The free energy change for the conversion of glucose to two molecules of lactate (the anaerobic route shown in Figure 15.1) is -183.6 kJ/mol under standard conditions:

$$C_6H_{12}O_6 \longrightarrow 2\ H_3C-\overset{\overset{\displaystyle OH}{|}}{\underset{\underset{\displaystyle H}{|}}{C}}-COO^- + 2\ H^+ \qquad \textbf{(15.1)}$$

$$\Delta G^{\circ\prime} = -183.6\ \text{kJ/mol}$$

This process occurs with no net oxidation. Although several individual steps in the pathway involve oxidation or reduction, these steps compensate each other exactly. Thus, the conversion of a molecule of glucose to two molecules of lactate involves simply a rearrangement of bonds, with no net loss or gain of electrons. The energy made available through this rearrangement into a more stable (lower energy) form is a relatively small part of the total energy obtainable from glucose.

The production of two molecules of ATP in glycolysis is an energy-requiring process:

(Text continued on page 474)

Figure 15.1   The glycolytic pathway and three possible metabolic fates of pyruvate.

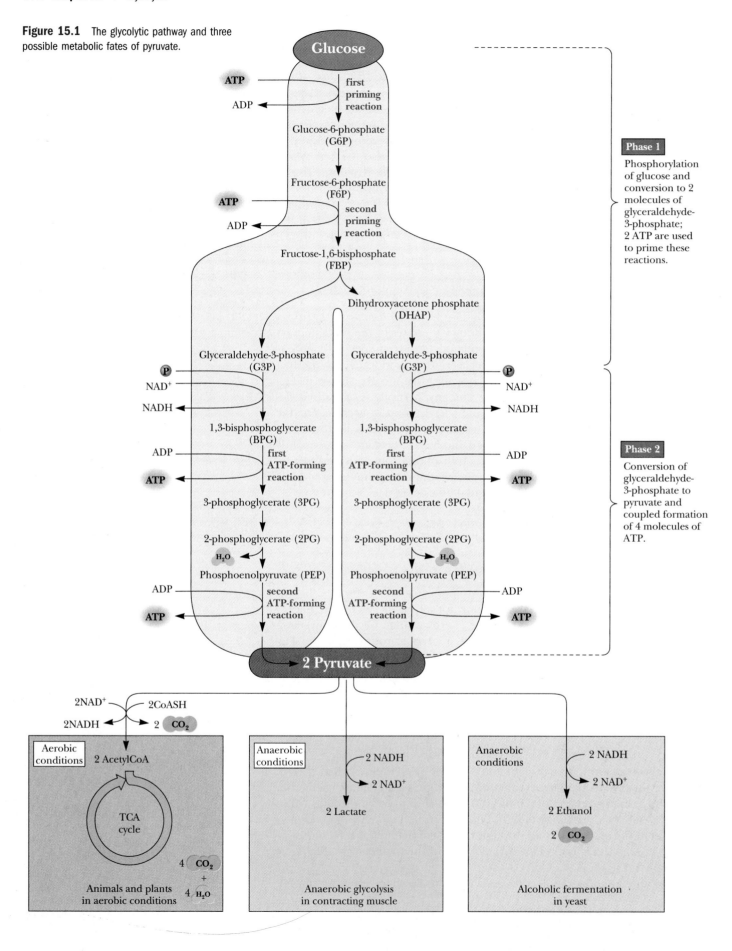

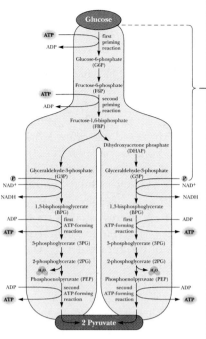

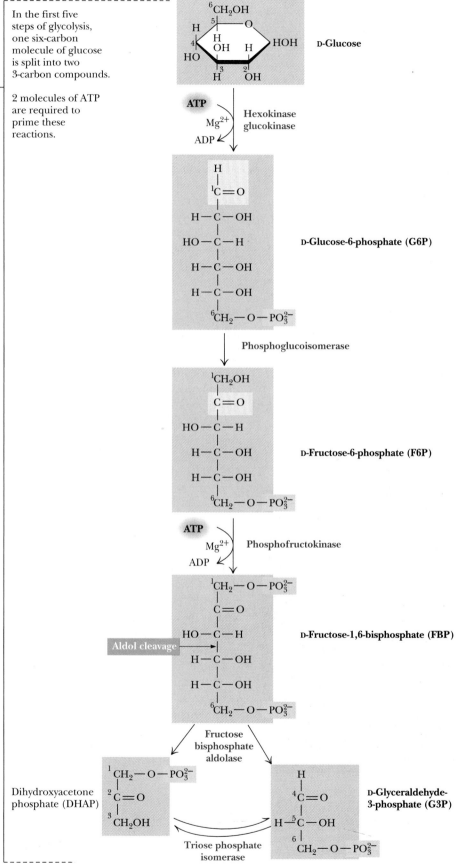

In the first five steps of glycolysis, one six-carbon molecule of glucose is split into two 3-carbon compounds.

2 molecules of ATP are required to prime these reactions.

See pages 173–178

Figure 15.2 In the first phase of glycolysis, five reactions convert a molecule of glucose to two molecules of glyceraldehyde-3-phosphate.

$$2\,\text{ADP} + 2\,\text{P}_\text{i} \longrightarrow 2\,\text{ATP} + 2\,\text{H}_2\text{O} \qquad \textbf{(15.2)}$$

$$\Delta G^{\circ\prime} = 2 \times 30.5\,\text{kJ/mol} = 61.0\,\text{kJ/mol}$$

Glycolysis couples these two reactions:

$$\text{Glucose} + 2\,\text{ADP} + 2\,\text{P}_\text{i} \longrightarrow 2\,\text{lactate} + 2\,\text{ATP} + 2\,\text{H}^+ + 2\,\text{H}_2\text{O} \qquad \textbf{(15.3)}$$

$$\Delta G^{\circ\prime} = -183.6 + 61 = -122.6\,\text{kJ/mol}$$

Thus, under standard-state conditions, $(61/183.6) \times 100\%$, or 33%, of the free energy released is preserved in the form of ATP in these reactions. However, as we discussed in Chapter 3, the various solution conditions, such as pH, concentration, ionic strength, and presence of metal ions, can substantially alter the free energy change for such reactions. Under actual cellular conditions, the free energy change for the synthesis of ATP (see Equation 15.2) is much larger, and approximately 50% of the available free energy is converted into ATP. Clearly, then, more than enough free energy is available in the conversion of glucose into lactate to drive the synthesis of two molecules of ATP.

15.3 The First Phase of Glycolysis

The synthesis of ATP using the free energy contained in the glucose molecule requires the conversion of glucose into one (or more) of the high-energy phosphates in Table 3.3 that have standard-state free energies of hydrolysis more negative than that of ATP. Among these are phosphoenolpyruvate and 1,3-bisphosphoglycerate. In the first stage of glycolysis, glucose is converted into two molecules of glyceraldehyde-3-phosphate. The energy released from this high-energy molecule in the second phase of glycolysis is then used to synthesize ATP.

Reaction 1: Phosphorylation of Glucose by Hexokinase or Glucokinase—The First Priming Reaction

Glycolysis is initiated by the phosphorylation of glucose at carbon atom 6 by either hexokinase or glucokinase. The formation of such a phosphoester is thermodynamically unfavorable and requires energy input to operate in the forward direction (Chapter 3). This energy comes from ATP, a requirement that at first seems counterproductive. Glycolysis is designed to make ATP, not consume it. However, the hexokinase/glucokinase reaction (see Figure 15.2) is one of two **priming reactions** in the pathway. Just as old-fashioned, hand-operated water pumps (Figure 15.3) have to be primed with a small amount of water to deliver more water to the thirsty pumper, glycolysis requires two priming ATP molecules to start the sequence of reactions and then delivers four molecules of ATP in the end.

The balanced equation for the first step in glycolysis is

$$\alpha\text{-D-Glucose} + \text{ATP}^{4-} \longrightarrow \text{D-glucose-6-phosphate}^{2-} + \text{ADP}^{3-} + \text{H}^+ \qquad \textbf{(15.4)}$$

$$\Delta G^{\circ\prime} = -16.7\,\text{kJ/mol}$$

Figure 15.3 Just as a water pump must be "primed" with water to get more water out, the glycolytic pathway is primed with ATP in steps 1 and 3 in order to achieve net production of ATP in the second phase of the pathway. (*Michelle Sassi/The Stock Market*)

The hydrolysis of ATP makes 30.5 kJ/mol available in this reaction, and the phosphorylation of glucose "costs" 13.8 kJ/mol (see Table 15.1). Thus, the reaction liberates 16.7 kJ/mol of free energy under standard-state conditions (1 M concentrations of reactants and products), and the equilibrium of the reaction lies far to the right ($K_{\text{eq}} = 850$ at 25°C; see Table 15.1).

Table 15.1   Reactions and Thermodynamics of Glycolysis

Reaction	Enzyme	$\Delta G^{\circ\prime}$ (kJ/mol)	ΔG (kJ/mol)
α-D-Glucose + ATP^{4-} $\rightleftharpoons$ glucose-6-phosphate^{2-} + ADP^{3-} + H$^+$	Hexokinase/Glucokinase	-16.7	-33.9^*
Glucose-6-phosphate^{2-} $\rightleftharpoons$ fructose-6-phosphate^{2-}	Phosphoglucoisomerase	$+1.67$	-2.92
Fructose-6-phosphate^{2-} + ATP^{4-} $\rightleftharpoons$ fructose-1,6-bisphosphate^{4-} + ADP^{3-} + H$^+$	Phosphofructokinase	-14.2	-18.8
Fructose-1,6-bisphosphate^{4-} $\rightleftharpoons$ dihydroxyacetone-P^{2-} + glyceraldehyde-3-P^{2-}	Fructose bisphosphate aldolase	$+23.9$	-0.23
Dihydroxyacetone-P^{2-} $\rightleftharpoons$ glyceraldehyde-3-P^{2-}	Triose phosphate isomerase	$+7.56$	$+2.41$
Glyceraldehyde-3-P^{2-} + P$_i^{2-}$ + NAD$^+$ $\rightleftharpoons$ 1,3-bisphosphoglycerate^{4-} + NADH + H$^+$	Glyceraldehyde-3-P dehydrogenase	$+6.30$	-1.29
1,3-Bisphosphoglycerate^{4-} + ADP^{3-} $\rightleftharpoons$ 3-P-glycerate^{3-} + ATP^{4-}	Phosphoglycerate kinase	-18.9	$+0.1$
3-Phosphoglycerate^{3-} $\rightleftharpoons$ 2-phosphoglycerate^{3-}	Phosphoglycerate mutase	$+4.4$	$+0.83$
2-Phosphoglycerate^{3-} $\rightleftharpoons$ phosphoenolpyruvate^{3-} + H$_2$O	Enolase	$+1.8$	$+1.1$
Phosphoenolpyruvate^{3-} + ADP^{3-} + H$^+$ $\rightleftharpoons$ pyruvate$^-$ + ATP^{4-}	Pyruvate kinase	-31.7	-23.0
Pyruvate$^-$ + NADH + H$^+$ $\rightleftharpoons$ lactate$^-$ + NAD$^+$	Lactate dehydrogenase	-25.2	-14.8

$^*\Delta G$ values calculated for 310 K (37°C) using the data in Table 15.2 for metabolite concentrations in erythrocytes. $\Delta G^{\circ\prime}$ values are assumed to be the same at 25°C and 37°C.

 See pages 1–3

Under cellular conditions, the hexokinase/glucokinase reaction of glycolysis is even more favorable than under standard-state conditions. As pointed out in Chapter 3, the free energy change for any reaction depends on the concentrations of reactants and products. Equation 3.12 in Chapter 3 and the data in Table 15.2 can be used to calculate a value for ΔG for the hexokinase/glucokinase reaction in erythrocytes:

$$\Delta G = \Delta G^{\circ\prime} + RT \ln\left(\frac{[\text{G-6-P}][\text{ADP}]}{[\text{Glu}][\text{ATP}]}\right) \qquad (15.5)$$

$$\Delta G = -16.7 \text{ kJ/mol} + (8.314 \text{ J/mol} \cdot \text{K})(310 \text{ K}) \ln\left(\frac{[0.083][0.14]}{[5.0][1.85]}\right)$$

$$\Delta G = -33.9 \text{ kJ/mol}$$

Thus, ΔG is even more favorable under cellular conditions than under standard-state conditions. As we will see later in this chapter, the hexokinase/glucokinase reaction is one of several that drive glycolysis forward.

The Importance of Phosphorylating Glucose

The incorporation of a phosphate into glucose in this energetically favorable reaction is important for several reasons. First, phosphorylation keeps the substrate in the cell. Glucose enters the cell via a specific glucose transporter protein. This transporter does not bind or transport glucose-6-phosphate, so that glucose phosphorylated by hexokinase/glucokinase must stay in the cell. Moreover, rapid conversion of glucose to glucose-6-phosphate keeps the intracellular concentration of glucose low, favoring diffusion of glucose through the transporter and into the cell. Similarly, though glucose is a neutral molecule and could diffuse across the membrane and out of the cell, glucose-6-phosphate is negatively charged, and the plasma membrane is essentially impermeable to this molecule (Figure 15.4). In addition, regulatory control can be imposed only on reactions not at equilibrium, and the large negative free energy change of this first reaction makes it an important site for regulation.

Hexokinase

In most animal, plant, and microbial cells, the enzyme that phosphorylates glucose is **hexokinase.** Magnesium ion (Mg^{2+}) is required for this reaction, as for other kinase enzymes. The true substrate for the hexokinase reaction is

Table 15.2   Steady-State Concentrations of Glycolytic Metabolites in Erythrocytes

Metabolite	m*M*
Glucose	5.0
Glucose-6-phosphate	0.083
Fructose-6-phosphate	0.014
Fructose-1,6-bisphosphate	0.031
Dihydroxyacetone phosphate	0.14
Glyceraldehyde-3-phosphate	0.019
1,3-Bisphosphoglycerate	0.001
2,3-Bisphosphoglycerate	4.0
3-Phosphoglycerate	0.12
2-Phosphoglycerate	0.030
Phosphoenolpyruvate	0.023
Pyruvate	0.051
Lactate	2.9
ATP	1.85
ADP	0.14
P$_i$	1.0

Adapted from Minakami, S., and Yoshikawa, H., 1965. Thermodynamic considerations on erythrocyte glycolysis. *Biochemical and Biophysical Research Communications* **18**:345.

Figure 15.4 Phosphorylation of glucose to glucose-6-phosphate by ATP creates a charged molecule that cannot easily cross the plasma membrane.

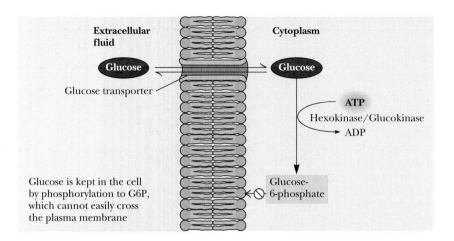

MgATP^{2-}. The apparent K_m for glucose of the animal skeletal muscle enzyme is approximately 0.1 mM, and the enzyme will operate efficiently at glucose concentrations well above this value. (In the erythrocyte, [glucose] = 5 mM; see Table 15.2.) The animal enzyme is allosterically inhibited by the product, glucose-6-phosphate. High levels of glucose-6-phosphate inhibit hexokinase activity until consumption by glycolysis lowers its concentration. The hexokinase reaction is one of three points in the glycolysis pathway that are regulated. As the generic name implies, hexokinase can phosphorylate a variety of hexoses, including glucose, mannose, and fructose.

Glucokinase

Liver contains an enzyme called **glucokinase,** which also carries out the reaction in Figure 15.4 but is highly specific for D-glucose, has a much higher K_m for glucose (approximately 10.0 mM), and is not inhibited by product. With such a high K_m for glucose, glucokinase becomes important metabolically only when liver glucose levels are high (for example, after the consumption of large amounts of carbohydrates). When glucose levels are low, hexokinase is primarily responsible for phosphorylating glucose. However, when glucose levels are high, glucose is converted to glucose-6-phosphate by glucokinase and is eventually stored in the liver as glycogen. Glucokinase is an *inducible* enzyme—the amount present in the liver is controlled by *insulin* (secreted by the pancreas). (Patients with **Type I diabetes** produce insufficient insulin. They have low levels of glucokinase, cannot tolerate high levels of blood glucose, and produce little liver glycogen.) Because glucose-6-phosphate is common to several metabolic pathways, it occupies a branch point in glucose metabolism.

Reaction 2: Phosphoglucoisomerase Catalyzes the Isomerization of Glucose-6-Phosphate

The second step in glycolysis is a common type of metabolic reaction: the isomerization of a sugar. In this particular case, the carbonyl oxygen of glucose-6-phosphate is shifted from C-1 to C-2. This amounts to isomerization of an aldose (glucose-6-phosphate) to a ketose—fructose-6-phosphate (Figure 15.5). The reaction is necessary for two reasons. First, the next step in glycolysis is phosphorylation at C-1, and the hemiacetal —OH of glucose would be more difficult to phosphorylate than a simple primary hydroxyl. Second, the isomerization to fructose (with a carbonyl group at position 2 in the linear form) activates the molecule for cleavage between C-3 and C-4 in the fourth step of

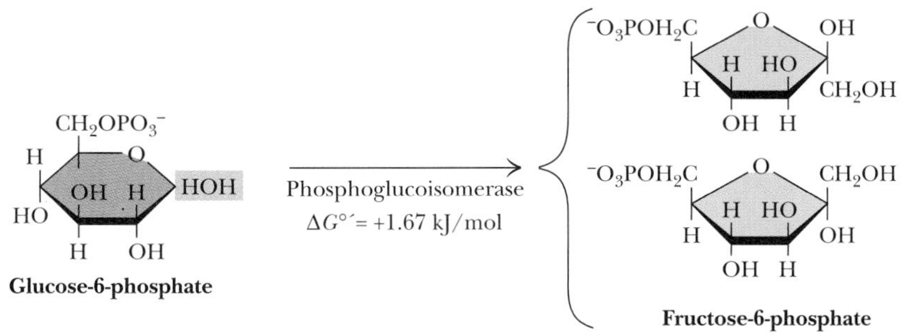

Figure 15.5 The phosphoglucoisomerase reaction.

Glucose-6-phosphate

Phosphoglucoisomerase
$\Delta G^{\circ\prime} = +1.67$ kJ/mol

Fructose-6-phosphate

glycolysis. The enzyme responsible for this isomerization is **phosphoglucoiso-merase,** also known as **glucose phosphate isomerase.** In humans, the enzyme requires Mg^{2+} for activity and is highly specific for glucose-6-phosphate. The $\Delta G^{\circ\prime}$ is 1.67 kJ/mol, and the value of ΔG under cellular conditions (see Table 15.1) is -2.92 kJ/mol. This small value means that the reaction operates near equilibrium in the cell and is readily reversible.

Reaction 3: Phosphofructokinase—The Second Priming Reaction

The action of phosphoglucoisomerase, "moving" the carbonyl group from C-1 to C-2, creates a new primary alcohol function at C-1 (see Figure 15.5). The next step in the glycolytic pathway is the phosphorylation of this group by **phosphofructokinase.** Once again, the substrate that provides the phosphoryl group is ATP. Like the hexokinase/glucokinase reaction, the phosphorylation of fructose-6-phosphate is a priming reaction and is endergonic:

$$\text{Fructose-6-P} + P_i \longrightarrow \text{fructose-1,6-bisphosphate} \qquad \textbf{(15.6)}$$

$$\Delta G^{\circ\prime} = 16.3 \text{ kJ/mol}$$

When coupled (by phosphofructokinase) with the hydrolysis of ATP, the over-all reaction (Figure 15.6) is strongly exergonic:

$$\text{Fructose-6-P} + \text{ATP} \longrightarrow \text{fructose-1,6-bisphosphate} + \text{ADP} \qquad \textbf{(15.7)}$$

$$\Delta G^{\circ\prime} = -14.2 \text{ kJ/mol}$$

$$\Delta G \text{ (in erythrocytes)} = -18.8 \text{ kJ/mol}$$

At pH 7 and 37°C, the phosphofructokinase reaction equilibrium lies far to the right. Just as the hexokinase reaction ensures that glucose remains intra-cellular, *the phosphofructokinase reaction commits glucose to glycolysis* rather than con-verting it to another sugar or storing it as glycogen. Similarly, just as the large free energy change associated with the hexokinase reaction makes it an

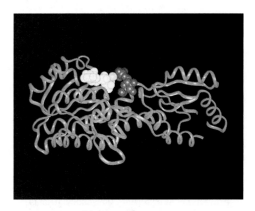

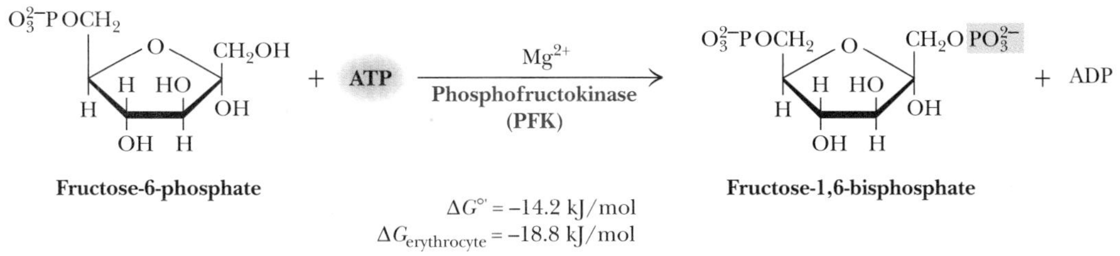

Phosphofructokinase with ADP shown in white and fructose-6-P in red.

See page 80

Fructose-6-phosphate

+

Mg^{2+}
Phosphofructokinase (PFK)

Fructose-1,6-bisphosphate

+ ADP

$\Delta G^{\circ\prime} = -14.2$ kJ/mol
$\Delta G_{\text{erythrocyte}} = -18.8$ kJ/mol

Figure 15.6 The phosphofructokinase reaction.

Phosphoglucoisomerase—A Moonlighting Protein

When someone has a day job but also works at night (i.e., under the moon) at a second job, they are said to be "moonlighting." Similarly, a number of proteins have been found to have two or more different functions, and Constance Jeffery at Brandeis University has dubbed these "moonlighting proteins." Phosphoglucoisomerase catalyzes the second step of glycolysis but also moonlights as a nerve growth factor outside animal cells. In fact, outside the cell, this protein is known as neuroleukin (NL), autocrine motility factor (AMF), and differentiation and maturation mediator (DMM). Neuroleukin is secreted by (immune system) T cells and promotes the survival of certain spinal neurons and sensory nerves.

AMF is secreted by tumor cells and stimulates cancer cell migration. DMM causes certain leukemia cells to differentiate.

How phosphoglucoisomerase is secreted by the cell for its moonlighting functions is not known, but there is evidence that the organism itself is surprised by this secretion. Diane Mathis and Christophe Benoist at the University of Strasbourg have shown that, in mice with disorders similar to rheumatoid arthritis, the immune system recognizes extracellular phosphoglucoisomerase as an antigen—i.e., a protein that is "nonself." That a protein can be vital to metabolism inside the cell and also function as a growth factor and occasionally act as an antigen outside the cell is indeed remarkable.

---

important regulatory step, so the phosphofructokinase reaction is an important site of regulation—indeed, the most important site in the glycolytic pathway.

Regulation of Phosphofructokinase

Phosphofructokinase is the "valve" controlling the rate of glycolysis. ATP is an allosteric inhibitor of this enzyme. In the presence of high ATP concentrations, phosphofructokinase behaves cooperatively, plots of enzyme activity versus fructose-6-phosphate are sigmoid, and the K_m for fructose-6-phosphate is increased (Figure 15.7). Thus, when ATP levels are sufficiently high in the cytosol, glycolysis is "turned down." AMP reverses the inhibition due to ATP. AMP levels in cells can rise dramatically when ATP levels decrease, due to the action of *adenylate kinase*, which catalyzes the reaction

$$ADP + ADP \rightleftharpoons ATP + AMP^1$$

Clearly, the activity of phosphofructokinase depends both on ATP and AMP concentrations and is a function of the cell's "energy status." Phosphofructokinase activity is increased when the energy status falls and is decreased when the energy status is high. The rate of glycolysis thus decreases when ATP is plentiful and increases when more ATP is needed.

Glycolysis and the citric acid cycle (to be discussed in Chapter 16) are coupled via phosphofructokinase, because citrate, an intermediate in the citric acid cycle, is an allosteric inhibitor of phosphofructokinase. When the citric acid cycle reaches saturation, citrate accumulates, and glycolysis (which "feeds" the citric acid cycle under aerobic conditions) slows down. The citric acid cycle directs electrons into the electron transport chain (for the purpose of ATP synthesis in oxidative phosphorylation) and also provides precursor molecules for biosynthetic pathways. Inhibition of glycolysis by citrate ensures that glucose will not be committed to these activities if the citric acid cycle is already saturated.

Phosphofructokinase is also regulated by β-D-**fructose-2,6-bisphosphate,** a potent allosteric activator that increases the affinity of phosphofructokinase for the substrate fructose-6-phosphate.

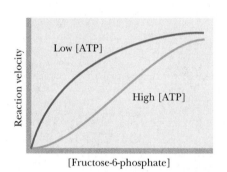

Figure 15.7 At high [ATP], phosphofructokinase (PFK) behaves cooperatively, and the plot of enzyme activity versus [fructose-6-phosphate] is sigmoid. High [ATP] thus inhibits PFK, decreasing the enzyme's affinity for fructose-6-phosphate.

Fructose-2,6-bisphosphate

---

[1]This reaction, in combination with cellular processes that convert ATP to ADP, eventually results in accumulation of AMP, even though the adenylate kinase reaction produces one ATP for each AMP.

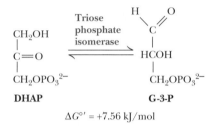

Figure 15.8 The fructose-1,6-bisphosphate aldolase reaction.

Reaction 4: Cleavage of Fructose-1,6-bisP by Fructose Bisphosphate Aldolase

Fructose bisphosphate aldolase cleaves fructose-1,6-bisphosphate between C-3 and C-4 to yield two triose phosphates. The products are dihydroxyacetone phosphate (DHAP) and glyceraldehyde-3-phosphate. The reaction (Figure 15.8) has an equilibrium constant of approximately 10^{-4} M, and a corresponding $\Delta G^{\circ\prime}$ of $+23.9$ kJ/mol. These values might imply that the reaction does not proceed effectively from left to right as written. However, the reaction makes two molecules (glyceraldehyde-3-P and dihydroxyacetone-P) from one molecule (fructose-1,6-bisphosphate), and the equilibrium is thus greatly influenced by concentration. The value of ΔG in erythrocytes is actually -0.23 kJ/mol (see Table 15.1). At physiological concentrations, the reaction is essentially at equilibrium.

Two classes of aldolase enzymes are found in nature. Animal tissues produce Class I aldolase, whose mechanism is characterized by the formation of a covalent Schiff base intermediate between an active-site lysine and the carbonyl group of the substrate. Class I aldolases do not require a divalent metal ion for activity. Class II aldolases are produced mainly in bacteria and fungi and contain an active-site metal (normally zinc, Zn^{2+}). Cyanobacteria and some other simple organisms possess both classes of aldolase.

Reaction 5: Triose Phosphate Isomerase

Of the two products of the aldolase reaction, only glyceraldehyde-3-phosphate goes directly into the second phase of glycolysis. However, the other triose phosphate, dihydroxyacetone phosphate, is converted to glyceraldehyde-3-phosphate by the enzyme **triose phosphate isomerase** (Figure 15.9). This reaction thus permits both products of the aldolase reaction to continue in the glycolytic pathway, and in essence makes the C-1, C-2, and C-3 carbons of the starting glucose molecule equivalent to the C-6, C-5, and C-4 carbons, respectively.

The triose phosphate isomerase reaction completes the first phase of glycolysis, each glucose that passes through being converted to two molecules of glyceraldehyde-3-phosphate. Although the last two steps of the pathway are energetically unfavorable, the overall five-step reaction sequence has a net $\Delta G^{\circ\prime}$ of $+2.2$ kJ/mol ($K_{eq} = 0.43$). It is the free energy of hydrolysis from the two priming molecules of ATP that brings the overall equilibrium constant close to 1 under standard-state conditions. The net ΔG under cellular conditions is quite negative (-53.4 kJ/mol in erythrocytes).

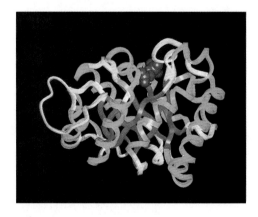

Triose phosphate isomerase with substrate analog 2-phosphoglycerate shown in red.

 See page 84

15.4 The Second Phase of Glycolysis

The second half of the glycolytic pathway involves reactions that result in the net production of ATP. Altogether, four new ATP molecules are produced. If two are considered to offset the two ATPs consumed in phase 1, a net yield of 2 ATPs per glucose is realized. Phase 2 starts with the oxidation of glyceraldehyde-3-phosphate, a reaction with a large enough energy "kick" to produce a high-energy phosphate—namely, 1,3-bisphosphoglycerate (Figure 15.10). Phosphoryl transfer from 1,3-BPG to ADP to make ATP is highly

In the second phase of glycolysis, glyceraldehyde-3-phosphate is converted to pyruvate.

These reactions yield 4 molecules of ATP, 2 for each molecule of pyruvate produced.

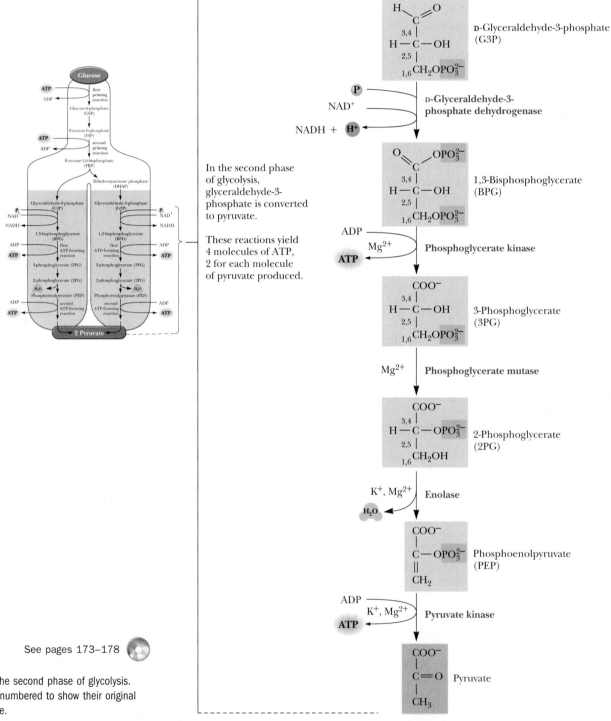

See pages 173–178

Figure 15.10 The second phase of glycolysis. Carbon atoms are numbered to show their original positions in glucose.

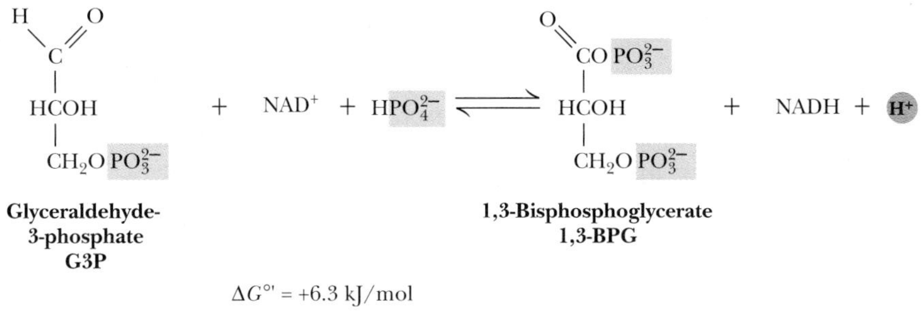

Figure 15.11 The glyceraldehyde-3-phosphate dehydrogenase reaction.

$$\Delta G^{\circ\prime} = +6.3 \text{ kJ/mol}$$

favorable. The product, 3-phosphoglycerate, is converted via several steps to phosphoenolpyruvate (PEP), another high-energy phosphate. PEP readily transfers its phosphoryl group to ADP in the pyruvate kinase reaction to make another ATP.

Reaction 6: Glyceraldehyde-3-Phosphate Dehydrogenase

In the first glycolytic reaction to involve oxidation–reduction, glyceraldehyde-3-phosphate is oxidized to 1,3-bisphosphoglycerate by **glyceraldehyde-3-phosphate dehydrogenase.** Although the oxidation of an aldehyde to a carboxylic acid is a highly exergonic reaction, the overall reaction (Figure 15.11) involves both formation of a carboxylic-phosphoric anhydride and the reduction of NAD^+ to NADH and is therefore slightly endergonic at standard state, with a $\Delta G^{\circ\prime}$ of +6.30 kJ/mol. The free energy that might otherwise be released as heat in this reaction is directed into the formation of a high-energy phosphate compound, 1,3-bisphosphoglycerate, and the reduction of NAD^+.

The glyceraldehyde-3-phosphate dehydrogenase reaction is the site of action of *arsenate (AsO_4^{3-})*, an anion analogous to phosphate. Arsenate is an effective substrate in this reaction, forming *1-arseno-3-phosphoglycerate* (Figure 15.12), but acyl arsenates are quite unstable and are rapidly hydrolyzed. 1-Arseno-3-phosphoglycerate breaks down to yield *3-phosphoglycerate,* the product of the seventh reaction of glycolysis. The result is that glycolysis continues in the presence of arsenate, but the molecule of ATP formed in reaction 7 (phosphoglycerate kinase) is not made because this step has been bypassed. The lability of 1-arseno-3-phosphoglycerate effectively uncouples the oxidation and phosphorylation events, which are normally tightly coupled in the glyceraldehyde-3-phosphate dehydrogenase reaction.

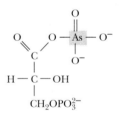

1-Arseno-3-phosphoglycerate

Figure 15.12

Reaction 7: Phosphoglycerate Kinase

The glycolytic pathway breaks even in terms of ATPs consumed and produced with this reaction. The enzyme **phosphoglycerate kinase** transfers a phosphoryl group from 1,3-bisphosphoglycerate to ADP to form an ATP (Figure 15.13). Because each glucose molecule sends two molecules of glyceraldehyde-3-

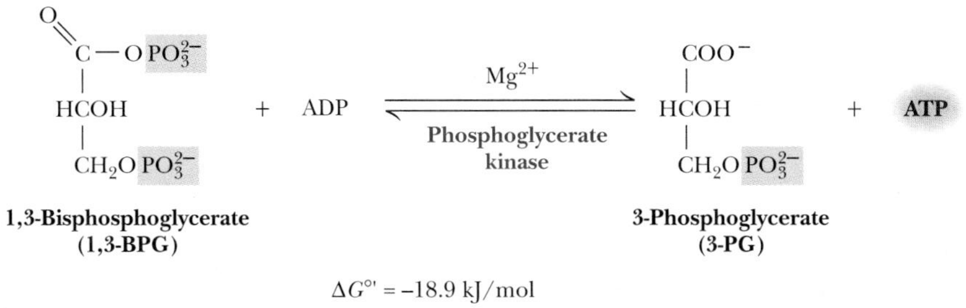

$$\Delta G^{\circ\prime} = -18.9 \text{ kJ/mol}$$

See page 81

Figure 15.13 The phosphoglycerate kinase reaction.

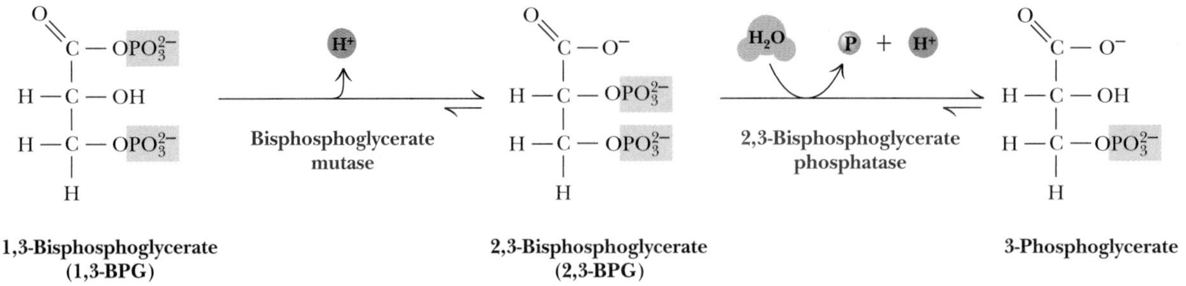

Figure 15.14 Formation and decomposition of 2,3-bisphosphoglycerate.

phosphate into the second phase of glycolysis and because two ATPs were consumed per glucose in the first-phase reactions, the phosphoglycerate kinase reaction "pays off" the ATP debt created by the priming reactions. As might be expected for a phosphoryl transfer enzyme, Mg^{2+} ion is required for activity, and the true nucleotide substrate for the reaction is $MgADP^-$. It is appropriate to view the sixth and seventh reactions of glycolysis as a coupled pair, with 1,3-bisphosphoglycerate as an intermediate. The phosphoglycerate kinase reaction is sufficiently exergonic at standard state to pull the G-3-P dehydrogenase reaction along. (In fact, the aldolase and triose phosphate isomerase reactions are also pulled forward by phosphoglycerate kinase reaction.) The net result of these coupled reactions[2] is

Glyceraldehyde-3-phosphate $+$ ADP $+$ P_i $+$ NAD^+ $\longrightarrow$

$$3\text{-phosphoglycerate} + \text{ATP} + \text{NADH} + \text{H}^+ \quad \textbf{(15.8)}$$

$$\Delta G^{\circ\prime} = -12.6 \text{ kJ/mol}$$

Another reflection of the coupling between these reactions lies in their values of ΔG under cellular conditions (see Table 15.1). In spite of its large negative $\Delta G^{\circ\prime}$, the phosphoglycerate kinase reaction operates very near equilibrium in the erythrocyte ($\Delta G = 0.1$ kJ/mol). In essence, the free energy available in the phosphoglycerate kinase reaction is used to bring the three previous reactions closer to equilibrium. Viewed in this context, it is clear that ADP has been phosphorylated to form ATP at the expense of a substrate, namely, glyceraldehyde-3-phosphate. This is an example of **substrate-level phosphorylation,** a concept that will be encountered again. (The other kind of phosphorylation, oxidative phosphorylation, is driven energetically by the transport of electrons from appropriate coenzymes and substrates to oxygen. Oxidative phosphorylation will be covered in detail in Chapter 17.) Even though the coupled reactions exhibit a very favorable $\Delta G^{\circ\prime}$, there are conditions (i.e., high ATP and 3-phosphoglycerate levels) under which Equation 15.8 can be reversed, so that 3-phosphoglycerate is phosphorylated from ATP.

An important regulatory molecule, *2,3-bisphosphoglycerate*, is synthesized and consumed by a pair of reactions that make a detour around the phosphoglycerate kinase reaction. 2,3-BPG is formed from 1,3-bisphosphoglycerate by **bisphosphoglycerate mutase** (Figure 15.14); 2,3-BPG is primarily responsible for the cooperative nature of oxygen binding by hemoglobin (see Chapter 12). Interestingly, 3-phosphoglycerate is required for this reaction, which involves phosphoryl transfer from the C-1 position of 1,3-bisphosphoglycerate to the C-2 position of 3-phosphoglycerate (Figure 15.15). Hydrolysis of 2,3-BPG is carried out by **2,3-bisphosphoglycerate phosphatase.** Although other cells contain only a trace of 2,3-BPG, erythrocytes typically contain 4 to 5 mM 2,3-BPG.

---

[2]These reactions are combined in Equation 15.8 only to illustrate their thermodynamic coupling. It is not meant to imply that glyceraldehyde-3-P is the phosphate donor which forms ATP. 1,3-Bisphosphoglycerate is the phosphate donor (in the phosphoglycerate kinase reaction itself).

Figure 15.15 The mutase that forms 2,3-BPG from 1,3-BPG requires 3-phosphoglycerate. The reaction is actually an intermolecular phosphoryl transfer from C-1 of 1,3-BPG to C-2 of 3-PG.

Reaction 8: Phosphoglycerate Mutase

The remaining steps in the glycolytic pathway prepare for synthesis of the second ATP equivalent. This begins with the **phosphoglycerate mutase** reaction (Figure 15.16), in which the phosphoryl group of 3-phosphoglycerate is moved from C-3 to C-2. The free energy change for this reaction is very small under cellular conditions ($\Delta G = 0.83$ kJ/mol in erythrocytes).

mutases enzymes that catalyze migration of a functional group within a substrate molecule

 See page 76

Reaction 9: Enolase

Recall that, prior to synthesizing ATP in the phosphoglycerate kinase reaction, it was necessary to first make a substrate having a high-energy phosphate. Reaction 9 of glycolysis similarly makes a high-energy phosphate in preparation for ATP synthesis. **Enolase** catalyzes the formation of *phosphoenolpyruvate* from 2-phosphoglycerate (Figure 15.17). The reaction in essence involves a dehydration—the removal of a water molecule—to form the enol structure of PEP. The $\Delta G^{\circ\prime}$ for this reaction is relatively small at 1.8 kJ/mol ($K_{eq} = 0.5$); and, under cellular conditions, ΔG is very close to zero. In light of this condition, it may be difficult at first to understand how the enolase reaction transforms a substrate with a relatively low free energy of hydrolysis into a product (PEP) with a very high free energy of hydrolysis. This puzzle is clarified by realizing that 2-phosphoglycerate and PEP contain about the same amount of *potential* metabolic energy with respect to decomposition to P_i, CO_2, and H_2O. What the enolase reaction does is rearrange this phosphorylated substrate into a form from which more of this potential energy can be released upon hydrolysis. Enolase is strongly inhibited by fluoride ion in the presence of phosphate. Inhibition arises from the formation of *fluorophosphate (FPO$_3^{2-}$)*, which forms a complex with Mg^{2+} at the active site of the enzyme.

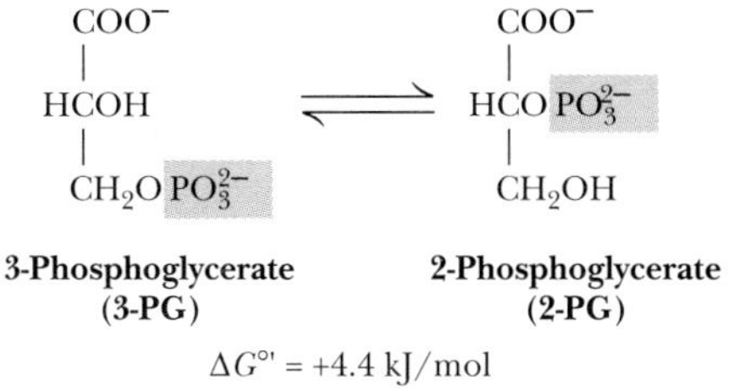

Reaction 10: Pyruvate Kinase

The second ATP-synthesizing reaction of glycolysis is catalyzed by **pyruvate kinase,** which brings the pathway to its pyruvate branch point. Pyruvate kinase mediates the transfer of a phosphoryl group from phosphoenolpyruvate to

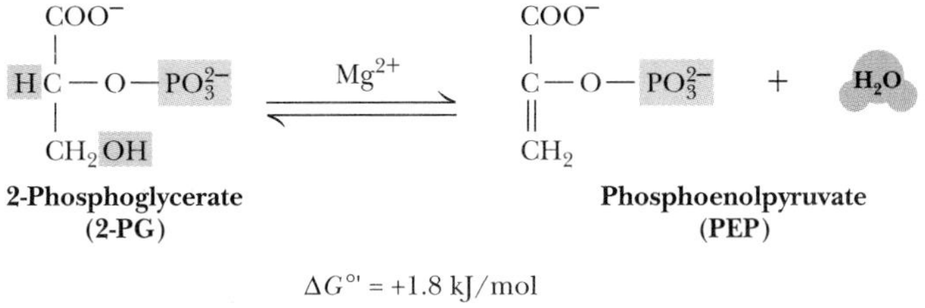

COO⁻
|
HCOH
|
CH₂O PO₃²⁻

3-Phosphoglycerate
(3-PG)

⇌

COO⁻
|
HCO PO₃²⁻
|
CH₂OH

2-Phosphoglycerate
(2-PG)

$\Delta G^{\circ\prime} = +4.4$ kJ/mol

Figure 15.16 The phosphoglycerate mutase reaction.

 See page 82

COO⁻
|
H C—O—PO₃²⁻
|
CH₂ OH

2-Phosphoglycerate
(2-PG)

Mg²⁺

COO⁻
|
C—O—PO₃²⁻
||
CH₂

+ H₂O

Phosphoenolpyruvate
(PEP)

$\Delta G^{\circ\prime} = +1.8$ kJ/mol

Figure 15.17 The enolase reaction.

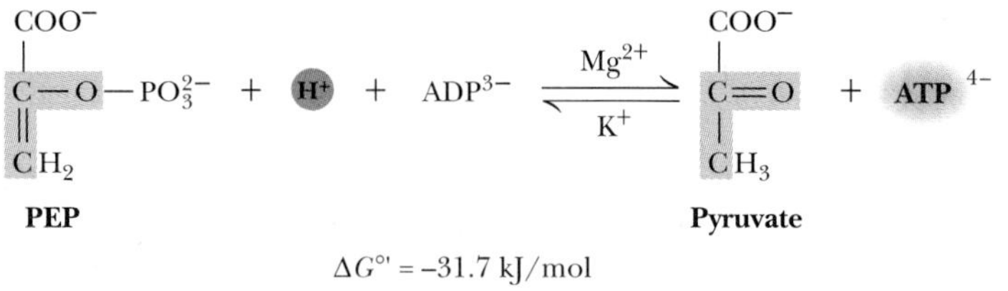

Figure 15.18 The pyruvate kinase reaction.

ADP to make ATP and pyruvate (Figure 15.18). The reaction requires Mg^{2+} ion and is stimulated by K^+ and certain other monovalent cations.

	$\Delta G^{\circ\prime}$(kJ/mol)
Phosphoenolpyruvate^{3-} + $H_2O \longrightarrow$ pyruvate$^-$ + HPO_4^{2-}	-62.2
ADP^{3-} + HPO_4^{2-} + H$^+$ $\longrightarrow$ ATP^{4-} + H_2O	$+30.5$
Phosphoenolpyruvate^{3-} + ADP^{3-} + H$^+$ $\longrightarrow$ pyruvate$^-$ + ATP^{4-}	-31.7

The equilibrium constant for the pyruvate kinase reaction lies very far to the right (K_{eq} at 25°C is 3.63×10^5). Concentration effects reduce the magnitude of the free energy change somewhat in the cellular environment, but the ΔG in erythrocytes is still quite favorable at -23.0 kJ/mol. The high free energy change for the conversion of PEP to pyruvate is due largely to the highly favorable and spontaneous conversion of the enol tautomer of pyruvate to the more stable keto form (Figure 15.19) following the phosphoryl group transfer step.

The large negative ΔG of this reaction makes pyruvate kinase a suitable target site for regulation of glycolysis. For each glucose molecule in the glycolysis pathway, two ATPs are made at the pyruvate kinase stage (because two triose molecules were produced per glucose in the aldolase reaction). Because the pathway broke even in terms of ATP at the phosphoglycerate kinase reaction (two ATPs consumed and two ATPs produced), the two ATPs produced by pyruvate kinase represent the "payoff" of glycolysis—a net yield of two ATP molecules.

Pyruvate kinase possesses allosteric sites for numerous effectors. It is activated by AMP and fructose-1,6-bisphosphate and inhibited by ATP, acetyl-CoA, and alanine. (Note that alanine is the α-amino acid counterpart of the α-keto acid, pyruvate.) Furthermore, liver pyruvate kinase is regulated by covalent modification. Hormones such as *glucagon* activate a cAMP-dependent protein kinase, which transfers a phosphoryl group from ATP to the enzyme. The phosphorylated form of pyruvate kinase is more strongly inhibited by ATP and alanine and has a higher K_m for PEP, so that, in the presence of physiological levels of PEP, the enzyme is inactive. Then PEP is used as a substrate for glucose synthesis in the *gluconeogenesis* pathway (to be described in Chapter 19), instead of going on through glycolysis and the citric acid cycle (or fermentation routes).

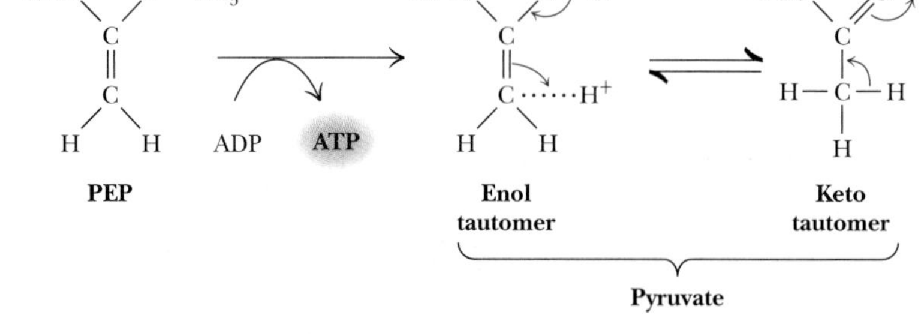

Figure 15.19 The conversion of phosphoenolpyruvate (PEP) to pyruvate catalyzed by pyruvate kinase involves two steps: phosphoryl transfer followed by an enol-keto tautomerization. The tautomerization is spontaneous ($\Delta G^{\circ\prime} \approx -35$ to -40 kJ/mol) and accounts for much of the free energy change for PEP hydrolysis.

Erythrocytes, or **red blood cells,** do not have nuclei or intracellular organelles such as mitochondria. As such, they have restricted metabolic capabilities, and their ability to adapt to changing environments and conditions is limited. At the same time, they depend upon a constant supply of energy to maintain their structural integrity. Energy is required to maintain gradients of Na^+ and K^+ across the erythrocyte membrane, and also to generate and preserve membrane lipids and proteins. If the erythocyte's energy requirements are not met, **hemolysis** (rupture of the erythrocyte membrane) can occur, and the resulting red blood cell loss is termed **hemolytic anemia.**

Glycolysis is the primary source of ATP energy for the erythrocyte, with additional energy supplied by the pentose monophosphate pathway (to be covered in Chapter 19). For this reason, deficiencies of one or more of the glycolytic enzymes are likely to result in substantial hemolysis. The most common form of hemolytic anemia results from a deficiency of pyruvate kinase. Individuals with one defective pyruvate kinase gene (*heterozygous carriers*) exhibit erythrocyte pyruvate kinase activities that are 40% to 60% of normal subjects. Those with two defective genes (and thus *homozygous* for the condition) exhibit pyruvate kinase activities that are 5% to 25% of normal.

In addition to the obvious reduction of glucose flux through glycolysis, other changes occur in cases of pyruvate kinase deficiency. Absence of pyruvate kinase activity causes glycolytic intermediates such as 3-phosphoglycerate to accumulate in affected cells. Ironically, 2,3-bisphosphoglycerate levels also rise, shifting hemoglobin's oxygen-binding curve (see Figure 12.24) to the right and releasing more oxygen to affected tissues and compensating to some extent for the attendant anemia. However, high levels of 2,3-bisphosphoglycerate also inhibit hexokinase and phosphofructokinase, further inhibiting glycolysis.

An active-site geometry for pyruvate kinase, based on NMR and EPR studies by Albert Mildvan and colleagues, is presented in Figure 15.20. The carbonyl oxygen of pyruvate and the γ-phosphorus of ATP lie within 0.3 nm of each other at the active site, consistent with direct transfer of the phosphoryl group without formation of a phosphoenzyme intermediate.

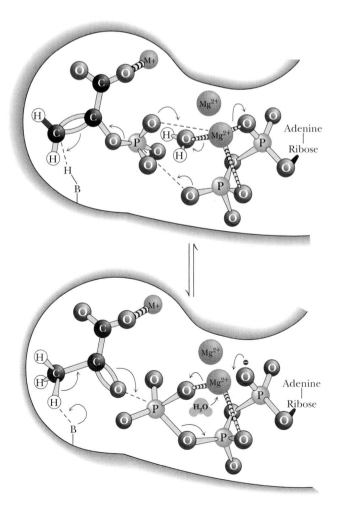

Figure 15.20 A mechanism for the pyruvate kinase reaction, based on NMR and EPR studies by Albert Mildvan and colleagues. Phosphoryl transfer from phosphoenolpyruvate (PEP) to ADP occurs in four steps: (a) a water on the Mg^{2+} ion coordinated to ADP is replaced by the phosphoryl group of PEP; (b) Mg^{2+} dissociates from the α-P of ADP; (c) the phosphoryl group is transferred; and (d) the enolate of pyruvate is protonated. (*Adapted from Mildvan, A., 1979.* Advances in Enzymology *49:103–126.*)

 See page 83

15.5 The Metabolic Fates of NADH and Pyruvate: The Products of Glycolysis

In addition to ATP, the products of glycolysis are NADH and pyruvate. Their processing depends upon other cellular pathways. NADH must be recycled to NAD^+, lest NAD^+ become limiting in glycolysis. NADH can be recycled by both aerobic and anaerobic paths, either of which results in further metabolism of pyruvate. What a given cell does with the pyruvate produced in glycolysis depends in part on the availability of oxygen. Under aerobic conditions, pyruvate can be sent into the citric acid cycle (also known as the tricarboxylic acid cycle; see Chapter 16), where it is oxidized to CO_2 with the production of additional NADH, as well as $FADH_2$ and ATP/GTP. Under aerobic conditions, the NADH produced in glycolysis and the citric acid cycle is reoxidized to NAD^+ in the mitochondrial electron transport chain (Chapter 17).

15.6 Anaerobic Pathways for Pyruvate

Under anaerobic conditions, the pyruvate produced in glycolysis is processed differently. In yeast, it is reduced to ethanol; in other microorganisms and in animals, it is reduced to lactate. These processes are examples of **fermentation**—the production of ATP energy by reaction pathways in which organic molecules function as donors and acceptors of electrons. In either case, reduction of pyruvate provides a means of reoxidizing the NADH produced in the glyceraldehyde-3-phosphate dehydrogenase reaction of glycolysis (Figure 15.21). **Alcoholic fermentation,** exemplified by yeast, is a two-step process. Pyruvate is decarboxylated to acetaldehyde by **pyruvate decarboxylase** in an essentially irreversible reaction. Thiamine pyrophosphate is a required cofactor for this enzyme. The second step, the reduction of acetaldehyde to ethanol by NADH, is catalyzed by **alcohol dehydrogenase** (see Figure 15.21). At pH 7, the

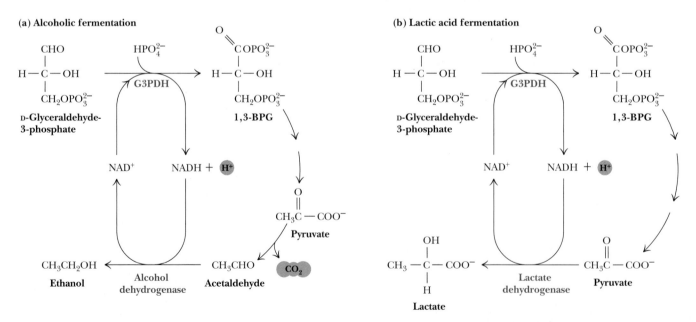

Figure 15.21 **(a)** Pyruvate reduction to ethanol in yeast provides a means for regenerating NAD^+ consumed in the glyceraldehyde-3-P dehydrogenase reaction. **(b)** In oxygen-depleted muscle, NAD^+ is regenerated in the lactate dehydrogenase reaction.

reaction equilibrium strongly favors ethanol. The end products of alcoholic fermentation are thus ethanol and carbon dioxide. Alcoholic fermentations are the basis for the brewing of beers and the making of wine. Lactate produced by anaerobic microorganisms during **lactic acid fermentation** is responsible for the taste of sour milk and yogurt, and for the characteristic taste and fragrance of sauerkraut, which in reality is fermented cabbage.

Lactate Accumulates Under Anaerobic Conditions in Animal Tissues

In animal tissues experiencing anaerobic conditions, pyruvate is reduced to lactate. Pyruvate reduction occurs in O_2-limited tissues, such as rapidly contracting skeletal muscle and tissues with minimal access to blood flow (e.g., the cornea of the eye). When skeletal muscles are exercised strenuously, the available tissue oxygen is consumed, and the pyruvate generated by glycolysis can no longer be oxidized in the TCA cycle. Instead, excess pyruvate is reduced to lactate by **lactate dehydrogenase** (see Figure 15.21). In anaerobic muscle tissue, lactate represents the end of glycolysis. Anyone who exercises to the point of consuming all available muscle oxygen stores knows the cramps and muscle fatigue associated with the buildup of lactic acid in the muscle. Most of this lactate must be carried out of the muscle by the blood and transported to the liver, where it can be resynthesized into glucose by gluconeogenesis. Moreover, because glycolysis generates only a fraction of the total energy available from the breakdown of glucose (the rest is generated by the TCA cycle and oxidative phosphorylation), the onset of anaerobic conditions in skeletal muscle also means a reduction in the energy available from the breakdown of glucose.

15.7 The Energetic Elegance of Glycolysis

The elegance of nature's design for the glycolytic pathway may be appreciated through an examination of Figure 15.22. The standard-state free energy changes for the 10 reactions of glycolysis and the lactate dehydrogenase reaction (see Figure 15.22a) are variously positive and negative and, taken together, offer little insight into the coupling that occurs in the cellular milieu. On the other hand, the values of ΔG under cellular conditions (see Figure 15.22b) fall into two distinct classes. For reactions 2 and 4 through 9, ΔG is very close to zero, so that these reactions operate essentially at equilibrium. Small changes in the concentrations of reactants and products could "push" any of these reactions either forward or backward. By contrast, the hexokinase, phosphofructokinase, and pyruvate kinase reactions all exhibit large negative ΔG values under cellular conditions. These reactions are thus the sites of glycolytic regulation. When these three enzymes are active, glycolysis proceeds and glucose is readily metabolized to pyruvate or lactate. Inhibition of the three key enzymes by allosteric effectors brings glycolysis to a halt. When we consider **gluconeogenesis**—the biosynthesis of glucose—in Chapter 19, we will see that different enzymes are used to carry out reactions 1, 3, and 10 in reverse, achieving the net synthesis of glucose. The maintenance of reactions 2 and 4 through 9 at or near equilibrium permits these reactions to operate effectively in either the forward or reverse direction.

15.8 Utilization of Other Substrates in Glycolysis

The glycolytic pathway described in this chapter begins with the breakdown of glucose, but other sugars, both simple and complex, can enter the cycle if they

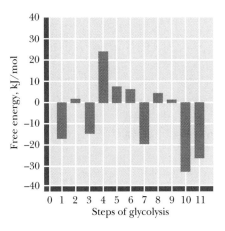

(a) ΔG at standard state ($\Delta G^{\circ \prime}$)

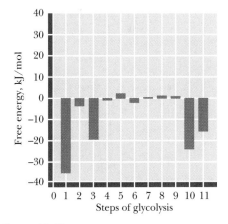

(b) ΔG in erythrocytes (ΔG)

Figure 15.22 A comparison of free energy changes for the reactions of glycolysis (step 1 = hexokinase) under **(a)** standard-state conditions and **(b)** actual intracellular conditions in erythrocytes. The values of $\Delta G^{\circ \prime}$ provide little insight into the actual free energy changes that occur in glycolysis. On the other hand, under intracellular conditions, seven of the glycolytic reactions operate near equilibrium (with ΔG near zero). The driving force for glycolysis lies in the hexokinase (1), phosphofructokinase (3), and pyruvate kinase (10) reactions. The lactate dehydrogenase (step 11) reaction also exhibits a large negative ΔG under cellular conditions.

can be converted by appropriate enzymes to one of the intermediates of glycolysis. Figure 15.23 shows the mechanisms by which several simple metabolites can enter the glycolytic pathway. **Fructose,** for example, which is produced by breakdown of sucrose, may participate in glycolysis by at least two different routes. In the liver, fructose is phosphorylated at C-1 by the enzyme **fructokinase:**

$$\text{D-Fructose} + \text{ATP}^{4-} \longrightarrow \text{D-fructose-1-phosphate}^{2-} + \text{ADP}^{3-} + \text{H}^+ \quad \textbf{(15.9)}$$

Subsequent action by **fructose-1-phosphate aldolase** cleaves fructose-1-P in a manner like the fructose bisphosphate aldolase reaction to produce dihydroxyacetone phosphate and D-glyceraldehyde:

$$\text{D-Fructose-1-P}^{2-} \longrightarrow \text{D-glyceraldehyde} + \text{dihydroxyacetone phosphate}^{2-}$$
$$\textbf{(15.10)}$$

Dihydroxyacetone phosphate is of course an intermediate in glycolysis. D-Glyceraldehyde can be phosphorylated by **triose kinase** in the presence of ATP to form D-glyceraldehyde-3-phosphate, another glycolytic intermediate.

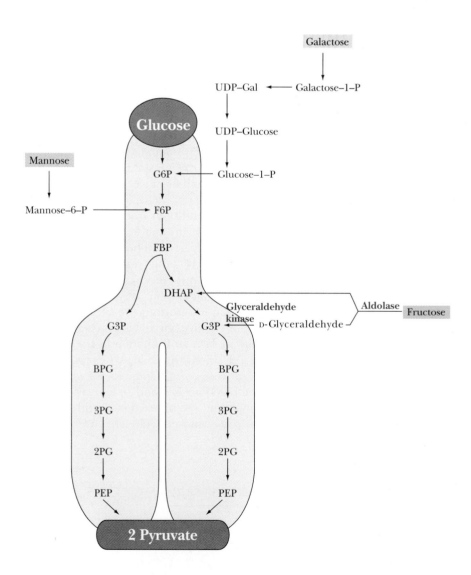

Figure 15.23 Mannose, galactose, fructose, and other simple metabolites can enter the glycolytic pathway.

In the kidney and in muscle tissues, fructose is readily phosphorylated by hexokinase, which, as pointed out earlier, can utilize several different hexose substrates. The free energy of hydrolysis of ATP drives the reaction forward:

$$\text{D-Fructose} + \text{ATP}^{4-} \longrightarrow \text{D-fructose-6-phosphate}^{2-} + \text{ADP}^{3-} + \text{H}^+ \quad \textbf{(15.11)}$$

Fructose-6-phosphate generated in this way enters the glycolytic pathway directly in step 3, the second priming reaction. This is the principal means for channeling fructose into glycolysis in adipose tissue, which contains high levels of fructose.

The Entry of Mannose into Glycolysis

Another simple sugar that enters glycolysis at the same point as fructose is **mannose,** which occurs in many glycoproteins, glycolipids, and polysaccharides (Chapter 7). Mannose is also phosphorylated from ATP by hexokinase, and the mannose-6-phosphate thus produced is converted to fructose-6-phosphate by **phosphomannoisomerase.**

$$\text{D-Mannose} + \text{ATP}^{4-} \longrightarrow \text{D-mannose-6-phosphate}^{2-} + \text{ADP}^{3-} + \text{H}^+ \quad \textbf{(15.12)}$$

$$\text{D-Mannose-6-phosphate}^{2-} \longrightarrow \text{D-fructose-6-phosphate}^{2-} \quad \textbf{(15.13)}$$

The Special Case of Galactose

A somewhat more complicated route into glycolysis is followed by **galactose,** another simple hexose sugar. The process, called the **Leloir pathway** after Luis Leloir, its discoverer, begins with phosphorylation from ATP at the C-1 position by **galactokinase:**

$$\text{D-Galactose} + \text{ATP}^{4-} \longrightarrow \text{D-galactose-1-phosphate}^{2-} + \text{ADP}^{3-} + \text{H}^+ \quad \textbf{(15.14)}$$

Galactose-1-phosphate is then converted into *UDP-galactose* (a sugar nucleotide) by **galactose-1-phosphate uridylyltransferase** (Figure 15.24), with concurrent production of glucose-1-phosphate and consumption of a molecule of UDP-glucose. The glucose-1-phosphate produced by the transferase reaction is a substrate for the **phosphoglucomutase** reaction (see Figure 15.24), which produces glucose-6-phosphate, a glycolytic substrate. The other transferase product, UDP-galactose, is converted to UDP-glucose by **UDP-glucose-4-epimerase.** The combined action of the uridylyltransferase and epimerase thus produces glucose-1-P from galactose-1-P, with regeneration of UDP-glucose.

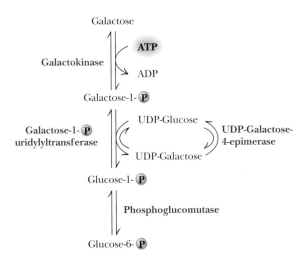

Figure 15.24 Galactose metabolism via the Leloir pathway.

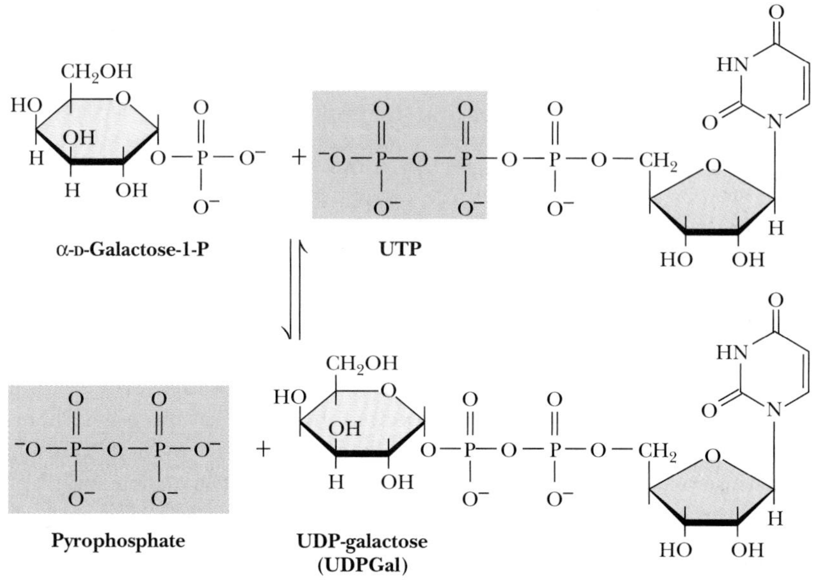

Figure 15.25 The UDP–glucose pyrophosphorylase reaction.

α-D-**Galactose-1-P**   **UTP**

Pyrophosphate   **UDP-galactose**
(UDPGal)

A rare hereditary condition known as **galactosemia** involves defects in galactose-1-P uridylyltransferase that render the enzyme inactive. Toxic levels of galactose accumulate in afflicted individuals, causing cataracts and permanent neurological disorders. These problems can be prevented by removing galactose and lactose from the diet. In adults, the toxicity of galactose appears to be less severe, due in part to the metabolism of galactose-1-P by **UDP-glucose pyrophosphorylase,** which apparently can accept galactose-1-P in place of glucose-1-P (Figure 15.25). The levels of this enzyme may increase in galactosemic individuals, in order to accommodate the metabolism of galactose.

Glycerol Can Also Enter Glycolysis

Glycerol is another simple substance whose ability to enter the glycolytic pathway merits consideration. This metabolite, which is produced in substantial amounts by the hydrolysis of *triacylglycerols* (see Chapter 20), can be converted to glycerol-3-phosphate by the action of **glycerol kinase** and then oxidized to dihydroxyacetone phosphate by the action of **glycerol phosphate dehydrogenase,** with NAD^+ as the required coenzyme (Figure 15.26). The dihydroxyacetone phosphate thereby produced enters the glycolytic pathway via triose phosphate isomerase.

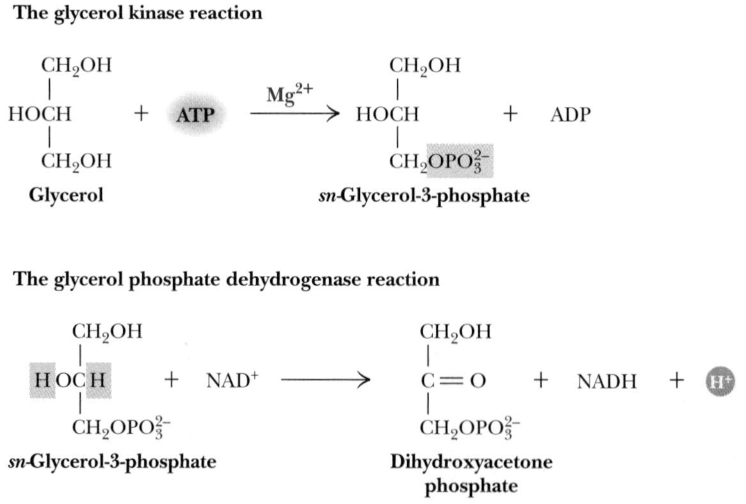

The glycerol kinase reaction

Glycerol   *sn*-**Glycerol-3-phosphate**

The glycerol phosphate dehydrogenase reaction

sn-**Glycerol-3-phosphate**   **Dihydroxyacetone phosphate**

Figure 15.26 How glycerol enters the glycolytic pathway.

Lactose—from Mother's Milk to Yogurt—and Lactose Intolerance

Lactose is an interesting sugar in many ways. In placental mammals, it is synthesized only in the mammary gland, and then only during late pregnancy and lactation. The synthesis is carried out by **lactose synthase,** a dimeric complex of two proteins, galactosyl transferase and α-lactalbumin. Galactosyl transferase is present in all human cells and it is normally involved in incorporation of galactose into glycoproteins. In late pregnancy, the pituitary gland in the brain releases a protein hormone, prolactin, which triggers production of α-lactalbumin by certain cells in the breast. α-Lactalbumin, a 123-residue protein, associates with galactosyl transferase to form lactose synthase, which catalyzes the reaction:

$$\text{UDP-galactose} + \text{glucose} \longrightarrow \text{lactose} + \text{UDP}$$

Lactose breakdown by **lactase** in the small intestine provides newborn mammals with essential galactose for many purposes, including the synthesis of gangliosides in the developing brain. Lactase is a **β-galactosidase** that cleaves lactose to yield galactose and glucose; in fact, it is the only human enzyme that can cleave a β-glycosidic linkage:

Breakdown of lactose to galactose and glucose by lactase

Lactase is an inducible enzyme in mammals, and it appears in the fetus only during the late stages of gestation. Lactase activity peaks shortly after birth, but by the age of three to five years, declines to a low level in nearly all human children. Low levels of lactase make many adults **lactose-intolerant.** Lactose intolerance occurs commonly in most parts of the world (with the notable exception of some parts of Africa and northern Europe; see table). The symptoms of lactose intolerance, including diarrhea and general discomfort, can be relieved by eliminating milk from the diet. Alternatively, products containing β-galactosidase are available commercially.

Certain bacteria, including several species of *Lactobacillus,* thrive on the lactose in milk and carry out lactic acid fermentation, converting lactose to lactate via glycolysis. This is the basis of production of yogurt, which is now popular in the Western world but is of Turkish origin. Other cultures also produce yogurtlike foods. Nomadic Tatars in Siberia and Mongolia use camel milk to make *koumiss,* which is consumed for medicinal purposes. In the Caucasus, *kefir* is made much like yogurt, except that the starter culture contains (in addition to *Lactobacillus*) *Streptococcus lactis* and yeast, which convert some of the glucose to ethanol and CO_2, producing an effervescent and slightly intoxicating brew.

Percentage of Population with Lactase Persistence

Country	Lactase persistence (%)
Sweden	99
Denmark	97
United Kingdom (Scotland)	95
Germany	88
Switzerland	84
Australia	82
United States (Iowa)	81
Bedouin tribes (North Africa)	75
Spain	72
France	58
Italy	49
India	36
Japan	10
China (Shanghai)	8
China (Singapore)	0

Adapted from Bloch, K., 1994. *Blondes in Venetian Paintings, the Nine-Banded Armadillo, and Other Essays in Biochemistry.* New Haven: Yale University Press.

Portions adapted from Hill, R., and Brew, K., 1975. Lactose synthetase. *Advances in Enzymology* **43:**411–485; and Bloch, K. 1994. *Blondes in Venetian Paintings, the Nine-Banded Armadillo, and Other Essays in Biochemistry.* New Haven: Yale University Press.

HUMAN BIOCHEMISTRY

Tumor Diagnosis Using Positron Emission Tomography (PET)

More than 70 years ago, Otto Warburg at the Kaiser Wilhelm Institute of Biology in Germany demonstrated that most animal and human tumors display a very high rate of glycolysis compared to normal tissue. This observation from long ago is the basis of a very modern diagnostic method for tumor detection called **positron emission tomography** or **PET.** PET uses molecular probes that contain a neutron-deficient, radioactive element such as carbon-11 or

fluorine-18. An example is 2-[^{18}F]fluoro-2-deoxy-glucose (FDG), a molecular mimic of glucose. The ^{18}F nucleus is unstable and spontaneously decays by emission of a positron (an antimatter particle) from a proton, thus converting a proton to a neutron, and transforming the ^{18}F to ^{18}O. The emitted positron typically travels a short distance (less than a millimeter) and collides with an electron, annihilating both particles and creating a pair of high-energy photons—gamma rays. Detection of the gamma rays with special cameras can be used to construct three-dimensional models of the location of the radio-labeled molecular probe in the tissue of interest.

FDG is taken up by human cells and converted by hexokinase to 2-[^{18}F]fluoro-2-deoxy-glucose-6-phosphate in the first step of glycolysis. Cells of a human brain, for example, accumulate FDG in direct proportion to the amount of glycolysis occurring in those cells. Tumors can be identified in PET scans as sites of unusually high FDG accumulation.

(a)

CH$_2$OH
O
OH · · · · HOH
HO
^{18}F

2-[^{18}F]Fluoro-2-deoxy-glucose

(b)

511 kev Photon

^{18}F

Emitted positron

^{18}O

Electron in tissue

511 kev Photon

(c) PET image of a human brain following administration of ^{18}FDG. Red area indicates a large malignant tumor. (*NIH/Science Source/Photo Researchers, Inc.*)

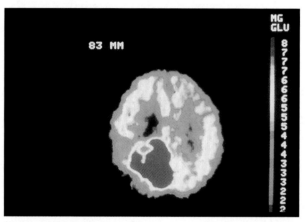

83 MM

MG GLU

8 7 7 7 6 6 6 6 5 5 5 5 4 4 4 3 3 3 2 2 2

PROBLEMS

1. List the reactions of glycolysis that
   a. are energy-consuming (under standard-state conditions).
   b. are energy-yielding (under standard-state conditions).
   c. consume ATP.
   d. yield ATP.
   e. are strongly influenced by changes in concentration of substrate and product because of their molecularity.
   f. are at or near equilibrium in the erythrocyte (see Table 15.2).

2. Determine the anticipated location in pyruvate of labeled carbons if glucose molecules labeled (in separate experiments) with ^{14}C at each position of the carbon skeleton proceed through the glycolytic pathway.

3. In an erythrocyte undergoing glycolysis, what would be the effect of a sudden increase in the concentration of
   a. ATP?
   b. AMP?
   c. fructose-1,6-bisphosphate?

  d. fructose-2,6-bisphosphate?

  e. citrate?

  f. glucose-6-phosphate?

4. Discuss the cycling of NADH and NAD$^+$ in glycolysis and the related fermentation reactions.

5. Write the reactions that permit galactose to be utilized in glycolysis. Write a suitable mechanism for one of these reactions.

6. If ^{32}P-labeled inorganic phosphate were introduced to erythrocytes undergoing glycolysis, would you expect to detect ^{32}P in glycolytic intermediates? If so, describe the relevant reactions and the ^{32}P incorporation you would observe.

7. Sucrose can enter glycolysis by either of two routes:

Sucrose phosphorylase:

$$\text{Sucrose} + \text{P}_i \rightleftharpoons \text{fructose} + \text{glucose-1-phosphate}$$

Invertase:

$$\text{Sucrose} + \text{H}_2\text{O} \rightleftharpoons \text{fructose} + \text{glucose}$$

Would either of these reactions offer an advantage over the other in the preparation of hexoses for entry into glycolysis?

8. What would be the consequences of a Mg^{2+} ion deficiency for the reactions of glycolysis?

9. Triose phosphate isomerase catalyzes the conversion of dihydroxyacetone-P to glyceraldehyde-3-P. The standard free energy change, $\Delta G°'$, for this reaction is $+7.6$ kJ/mol. However, the observed free energy change (ΔG) for this reaction in erythrocytes is $+2.4$ kJ/mol.

  a. Calculate the ratio of [dihydroxyacetone-P]/[glyceraldehyde-3-P] in erythrocytes from ΔG.

  b. If [dihydroxyacetone-P] = 0.2 mM, what is [glyceraldehyde-3-P]?

10. Enolase catalyzes the conversion of 2-phosphoglycerate to phosphoenolpyruvate + H$_2$O. The standard free energy change, $\Delta G°'$, for this reaction is $+1.8$ kJ/mol. If the concentration of 2-phosphoglycerate is 0.045 mM and the concentration of phosphoenolpyruvate is 0.034 mM, what is ΔG, the free energy change for the enolase reaction, under these conditions?

11. The standard free energy change ($\Delta G°'$) for hydrolysis of phosphoenolpyruvate (PEP) is -61.9 kJ/mol. The standard free energy change ($\Delta G°'$) for ATP hydrolysis is -30.5 kJ/mol.

  a. What is the standard free energy change for the pyruvate kinase reaction

$$\text{ADP} + \text{phosphoenolpyruvate} \longrightarrow \text{ATP} + \text{pyruvate}$$

  b. What is the equilibrium constant for this reaction?

  c. Assuming the intracellular concentrations of [ATP] and [ADP] remain fixed at 8 mM and 1 mM, respectively, what will be the ratio of [pyruvate]/[phosphoenolpyruvate] when the pyruvate kinase reaction reaches equilibrium?

12. The standard free energy change ($\Delta G°'$) for hydrolysis of fructose-1,6-bisphosphate (FBP) to fructose-6-phosphate (F-6-P) and P$_i$ is -16.7 kJ/mol:

$$\text{FBP} + \text{H}_2\text{O} \rightarrow \text{fructose-6-P} + \text{P}_i$$

The standard free energy change ($\Delta G°'$) for ATP hydrolysis is -30.5 kJ/mol:

$$\text{ATP} + \text{H}_2\text{O} \longrightarrow \text{ADP} + \text{P}_i$$

  a. What is the standard free energy change for the phosphofructokinase reaction

$$\text{ATP} + \text{fructose-6-P} \longrightarrow \text{ADP} + \text{FBP}$$

  b. What is the equilibrium constant for this reaction?

  c. Assuming the intracellular concentrations of [ATP] and [ADP] are maintained constant at 4 mM and 1.6 mM, respectively, in a rat liver cell, what will be the ratio of [FBP]/[fructose-6-P] when the phosphofructokinase reaction reaches equilibrium?

FURTHER READING

Arkin, A., Shen, P., and Ross, J., 1997. A test case of correlation metric construction of a reaction pathway from measurements. *Science* **277**:1275–1279.

Beitner, R., 1985. *Regulation of Carbohydrate Metabolism.* Boca Raton, Fla.: CRC Press.

Bioteux, A., and Hess, A., 1981. Design of glycolysis. *Philosophical Transactions,* Royal Society of London B **293**:5–22.

Bodner, G. M., 1986. Metabolism: Part I, Glycolysis. *Journal of Chemical Education* **63**:566–570.

Bosca, L., and Corredor, C., 1984. Is phosphofructokinase the rate-limiting step of glycolysis? *Trends in Biochemical Sciences* **9**:372–373.

Boyer, P. D., 1972. *The Enzymes,* 3rd ed., Vols. 5–9. New York: Academic Press.

Braun, L., Puskas, F., Csala, M., et al., 1997. Ascorbate as a substrate for glycolysis or gluconeogenesis: Evidence for an interorgan ascorbate cycle. *Free Radical Biology and Medicine* **23**:804–808.

Brosnan, J. T., 1999. Comments on metabolic needs for glucose and the role of gluconeogenesis. *European Journal of Clinical Nutrition* **53**:S107–S111.

Conley, K. E., Blei, M. L., Richards, T. L., et al., 1997. Activation of glycolysis in human muscle *in vivo. American Journal of Physiology* **273**:C306–C315.

Fothergill-Gilmore, L., 1986. The evolution of the glycolytic pathway. *Trends in Biochemical Sciences* **11**:47–51.

Goncalves, P. M., Giffioen, G., Bebelman, J. P., and Planta, R. J., 1997. Signalling pathways leading to transcriptional regulation of genes involved in the activation of glycolysis in yeast. *Molecular Microbiology* **25**:483–493.

Green, H. J., 1997. Mechanisms of muscle fatigue in intense exercise. *Journal of Sports Sciences* **15**:247–256.

Jeffrey, C. J., et al., 2000. Crystal structure of rabbit phosphoglucose isomerase, a glycolytic enzyme that moonlights as neuroleukin, autocrine motility factor, and differentiation mediator. *Biochemistry* **39**:955–964.

Jucker, B. M., Rennings, A. J., Cline, G. W., et al., 1997. *In vivo* NMR investigation of intramuscular glucose metabolism in conscious rats. *American Journal of Physiology* **273**:E139–E148.

Knowles, J., and Albery, W., 1977. Perfection in enzyme catalysis: The energetics of triose phosphate isomerase. *Accounts of Chemical Research* **10**:105–111.

Luczak-Szczurek, A., and Flisinska-Bojanowska, A., 1977. Effect of high-protein diet on glycolytic processes in skeletal muscles of exercising rats. *Journal of Physiology and Pharmacology* **48**:119–126.

Matsumoto, I., et al., 1999. Arthritis provoked by linked T and B cell recognition of a glycolytic enzyme. *Science* **286**:1732–1735.

Newsholme, E., Challiss, R., and Crabtree, B., 1984. Substrate cycles: Their role in improving sensitivity in metabolic control. *Trends in Biochemical Sciences* **9**:277–280.

Pilkus, S., and El-Maghrabi, M., 1988. Hormonal regulation of hepatic gluconeogenesis and glycolysis. *Annual Review of Biochemistry* **57**:755–783.

Saier, M., Jr., 1987. *Enzymes in Metabolic Pathways.* New York: Harper and Row.

Sparks, S., 1997. The purpose of glycolysis. *Science* **277**:459–460.

Vertessy, B. G., Orosz, F., Kovacs, J., and Ovadi, J., 1997. Alternative binding of two sequential glycolytic enzymes to microtubules: Molecular studies in the phosphofructokinase/aldolase/microtubule system. *Journal of Biological Chemistry* **272**:25542–25546.

Wackerhage, H., Mueller, K., Hoffmann, U., et al., 1996. Glycolytic ATP production estimated from ^{31}P magnetic resonance spectroscopy measurements during ischemic exercise in vivo. *Magma* **4**:151–155.

Waddell, T. G., et al., 1997. Optimization of glycolysis: A new look at the efficiency of energy coupling. *Biochemical Education* **25**:204–205.

Walsh, C. T., 1979. *Enzymatic Reaction Mechanisms.* San Francisco: W. H. Freeman.

The Tricarboxylic Acid Cycle

A time-lapse photograph of a Ferris wheel at night. Aerobic cells use a metabolic wheel—the tricarboxylic acid cycle—to generate energy by acetyl-CoA oxidation. (Ferris Wheel, Del Mar Fair © Corbis/Richard Cummins)

Thus times do shift, each thing his turn does hold;
New things succeed, as former things grow old.

ROBERT HERRICK, *Hesperides* (1648), "Ceremonies for Christmas Eve"

The glycolytic pathway converts glucose to pyruvate and produces two molecules of ATP per glucose—only a small fraction of the potential energy available from glucose. Under anaerobic conditions, pyruvate is reduced to lactate in animals and to ethanol in yeast, and much of the potential energy of the glucose molecule remains untapped. *In the presence of oxygen,* however, a much more interesting and thermodynamically complete story unfolds. Under aerobic conditions, NADH is oxidized in the electron transport chain, rather than becoming oxidized through reduction of pyruvate to lactate or acetaldehyde to ethanol, for example. Further, pyruvate is converted to acetyl-coenzyme A and oxidized to CO_2 in the **tricarboxylic acid (TCA) cycle** (also called the **citric acid cycle**). The electrons liberated by this oxidative process are then passed through an elaborate, membrane-associated **electron transport pathway** to O_2, the final electron acceptor. Electron transfer is coupled to creation of a proton gradient across the membrane. Such a gradient represents an energized

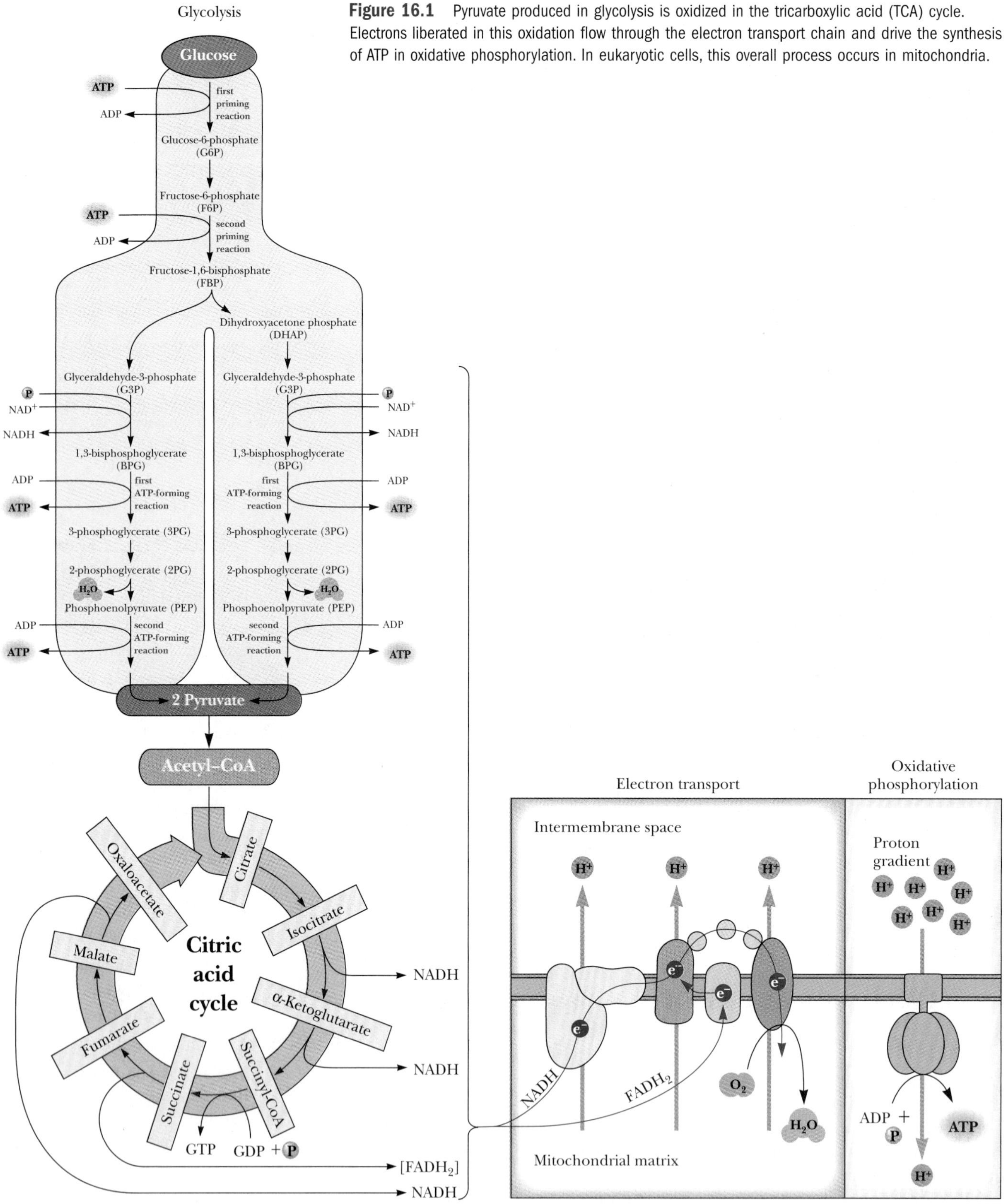

Figure 16.1 Pyruvate produced in glycolysis is oxidized in the tricarboxylic acid (TCA) cycle. Electrons liberated in this oxidation flow through the electron transport chain and drive the synthesis of ATP in oxidative phosphorylation. In eukaryotic cells, this overall process occurs in mitochondria.

state, and the energy stored in this gradient is used to drive the synthesis of many equivalents of ATP.

ATP synthesis as a consequence of electron transport is termed **oxidative phosphorylation;** the complete process is diagrammed in Figure 16.1. *Aerobic pathways* permit the production of 30 to 38 molecules of ATP per glucose oxidized. Athough two molecules of ATP come from glycolysis and two more directly out of the TCA cycle, most of the ATP arises from oxidative phosphorylation. Specifically, reducing equivalents released in the oxidative reactions of glycolysis, pyruvate decarboxylation, and the TCA cycle are captured in the form of NADH and enzyme-bound $FADH_2$, and these reduced coenzymes fuel the electron transport pathway and oxidative phosphorylation. The path to oxidative phosphorylation winds through the TCA cycle, and we will examine this cycle in detail in this chapter.

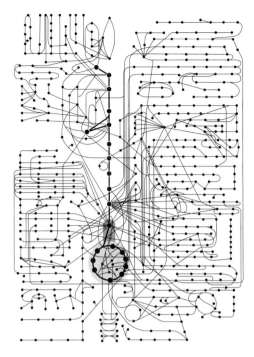

Tricarboxylic acid cycle.

See *Interactive Biochemistry CD-ROM and Workbook,* pages 173–178

16.1 The TCA Cycle: A Brief Summary

The entry of new carbon units into the TCA cycle is through acetyl-CoA (Figure 16.2). This entry metabolite can be formed either from pyruvate (from glycolysis) or from oxidation of fatty acids (discussed in Chapter 20). *Citrate synthase* catalyzes the transfer of the two-carbon acetyl group from acetyl-CoA to the four-carbon oxaloacetate to yield six-carbon citrate. Aconitase catalyzes the isomerization of citrate to isocitrate by a dehydration–rehydration mechanism. Two successive decarboxylations produce α-ketoglutarate and then succinyl-CoA, a CoA-conjugate of a four-carbon unit. Several steps later, oxaloacetate is regenerated and can combine with another acetyl-CoA to repeat the cycle. Thus, carbon enters the cycle as acetyl-CoA and exits as CO_2. In the process, metabolic energy is captured in the form of ATP, NADH, and enzyme-bound $FADH_2$ (symbolized as $[FADH_2]$).

The Chemical Logic of the TCA Cycle

The cycle shown in Figure 16.2 at first appears to be a complicated way to oxidize acetate units to CO_2, but there is a chemical basis for the apparent complexity. Oxidation of an acetyl group to a pair of CO_2 molecules requires C — C cleavage:

$$CH_3COO^- + 2\ H_2O + H^+ \longrightarrow CO_2 + CO_2 + 8\ H^+$$

In many instances, C — C cleavage reactions in biological systems occur between carbon atoms α and β to a carbonyl group:

$$-\overset{\overset{\displaystyle O}{\|}}{C}-C_\alpha-C_\beta-$$

$\uparrow$
Cleavage

A good example of such a cleavage is the fructose bisphosphate aldolase reaction (see Chapter 15, Figure 15.8).

Another common type of C — C cleavage is α-cleavage of an α-hydroxyketone:

$$-\overset{\overset{\displaystyle O}{\|}}{C}-\overset{\overset{\displaystyle OH}{|}}{C_\alpha}-$$

$\uparrow$
Cleavage

(We see this type of cleavage in the *transketolase* reaction described in Chapter 19.)

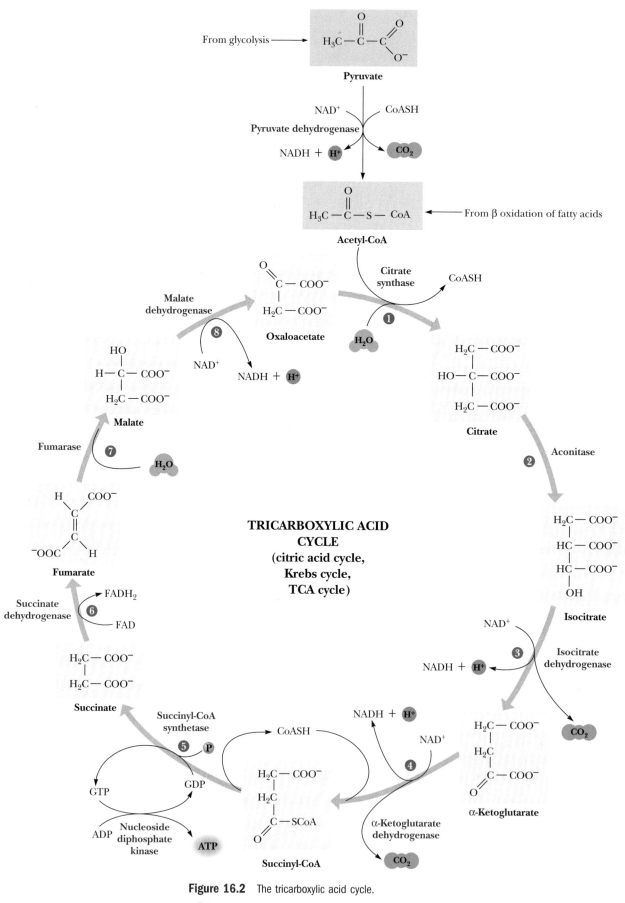

Figure 16.2 The tricarboxylic acid cycle.

See pages 173–178

Neither of these cleavage strategies is suitable for acetate. It has no β-carbon, and the second method would require hydroxylation—not a favorable reaction for acetate. Instead, living things have evolved the clever chemistry of condensing acetate with oxaloacetate and then carrying out a β-cleavage. *The TCA cycle combines this β-cleavage reaction with oxidation to form CO_2, regenerate oxaloacetate, and capture the liberated metabolic energy in NADH and ATP.*

16.2 The Bridging Step: Oxidative Decarboxylation of Pyruvate

Pyruvate produced by glycolysis is a significant source of acetyl-CoA for the TCA cycle. Because, in eukaryotic cells, glycolysis occurs in the cytoplasm, whereas the TCA cycle reactions and all subsequent steps of aerobic metabolism take place in the mitochondria, pyruvate must first enter the mitochondria to enter the TCA cycle. The oxidative decarboxylation of pyruvate to acetyl-CoA,

$$\text{Pyruvate} + \text{CoA} + \text{NAD}^+ \longrightarrow \text{acetyl-CoA} + CO_2 + \text{NADH} + \text{H}^+$$

is the connecting link between glycolysis and the TCA cycle. The reaction is catalyzed by pyruvate dehydrogenase, a multienzyme complex.

The **pyruvate dehydrogenase complex (PDC)** is a noncovalent assembly of multiple copies of three different enzymes operating in concert to catalyze successive steps in the conversion of pyruvate to acetyl-CoA. The active sites of all three enzymes are not far removed from one another, and the product of the first enzyme is passed directly to the second enzyme and so on, without diffusion of substrates and products through the solution. The overall reaction (see *A Deeper Look*, page 500) involves a total of five coenzymes: thiamine pyrophosphate, coenzyme A, lipoic acid, NAD$^+$, and FAD.

16.3 Entry into the Cycle: The Citrate Synthase Reaction

The first reaction within the TCA cycle, the one by which carbon atoms are introduced, is the **citrate synthase reaction** (Figure 16.3). Two proton transfers promote this reaction: proton donation by a general acid activates the carbonyl carbon of oxaloacetate for nucleophilic attack, and the C_α carbon of acetyl-CoA is deprotonated by a general base to form a nucleophilic carbanion. The citrate synthase reaction depends upon both modes of activation. As shown in

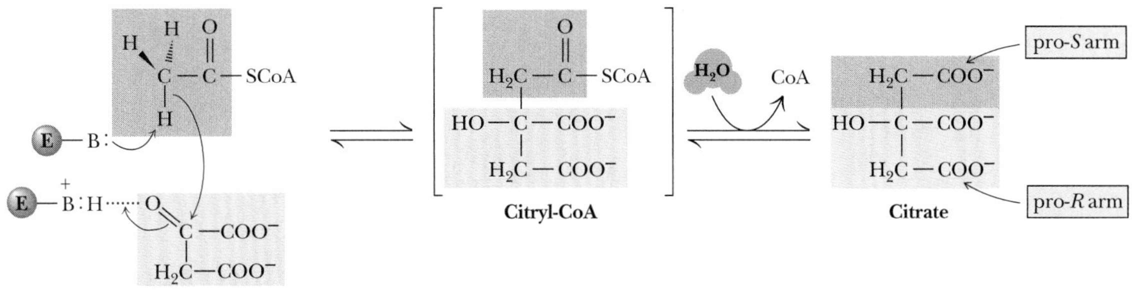

Figure 16.3 Citrate is formed in the citrate synthase reaction from oxaloacetate and acetyl-CoA. The mechanism involves nucleophilic attack by the carbanion of acetyl-CoA on the carbonyl carbon of oxaloacetate, followed by thioester hydrolysis.

A DEEPER LOOK

Reaction Mechanism of Pyruvate Dehydrogenase Complex

The mechanism of the pyruvate dehydrogenase reaction is a tour de force of mechanistic chemistry, involving a total of three enzymes (figure, a) and five different coenzymes—thiamine pyrophosphate, lipoic acid, coenzyme A, FAD, and NAD^+ (figure, b).

The first step of this reaction, decarboxylation of pyruvate and transfer of the acetyl group to lipoic acid, begins with nucleophilic attack by a TPP carbanion on the carbonyl carbon of pyruvate. Decarboxylation yields a resonance-stabilized intermediate—**hydroxyethyl-TPP.**

The reaction of hydroxyethyl-TPP with the oxidized form of lipoic acid yields the energy-rich thiol ester of reduced lipoic acid and results in oxidation of the hydroxyl-carbon of the two-carbon substrate unit to a carbonyl (step 2). Then, nucleophilic attack by coenzyme A on the carbonyl-carbon results in transfer of the acetyl group from lipoic acid to CoA (step 3). The subsequent oxidation of lipoic acid is catalyzed by the FAD-dependent dihydrolipoyl dehydrogenase and NAD^+ is reduced (step 4).

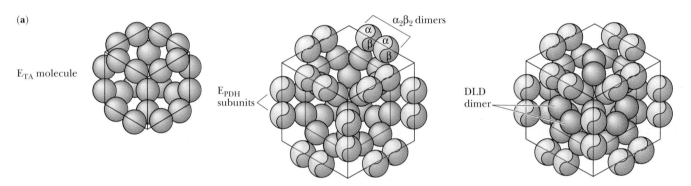

(a) The structure of the pyruvate dehydrogenase complex. This complex consists of three enzymes: pyruvate dehydrogenase (PDH), dihydrolipoyl transacetylase (TA), and dihydrolipoyl dehydrogenase (DLD). (i) 24 dihydrolipoyl transacetylase subunits form a cubic core structure. (ii) 24 $\alpha\beta$ dimers of pyruvate dehydrogenase are added to the cube (two per edge). (iii) Addition of 12 dihydrolipoyl dehydrogenase subunits (two per face) completes the complex.

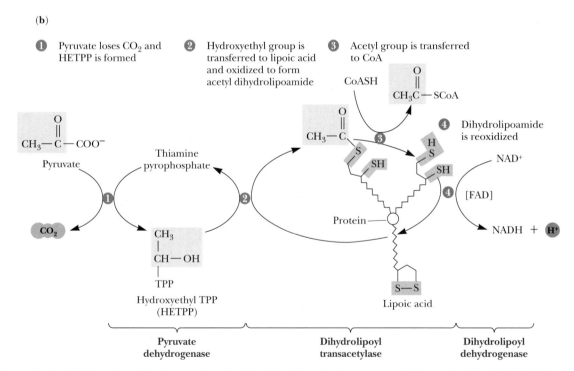

(b) The reaction mechanism of the pyruvate dehydrogenase complex. Decarboxylation of pyruvate occurs with formation of hydroxyethyl-TPP (step 1). Transfer of the two-carbon unit to lipoic acid in step 2 is followed by formation of acetyl-CoA in step 3. Lipoic acid is reoxidized in step 4 of the reaction.

Figure 16.3, a general base on the enzyme accepts a proton from the methyl group of acetyl-CoA, producing a stabilized α-carbanion of acetyl-CoA. This strong nucleophile attacks the α-carbonyl of oxaloacetate, yielding citryl-CoA. This part of the reaction has an equilibrium constant near 1, but the overall reaction is driven to completion by the subsequent hydrolysis of the high-energy thioester to citrate and free CoA. The overall $\Delta G^{\circ\prime}$ is -31.4 kJ/mol, and under standard conditions the reaction is essentially irreversible. Although the mitochondrial concentration of oxaloacetate is very low (much less than 1 μM; see the example in Section 16.10), the strong, negative $\Delta G^{\circ\prime}$ drives the reaction forward.

The Structure and Regulation of Citrate Synthase

Citrate synthase in mammals is a dimer of 49-kD subunits. On each subunit, oxaloacetate and acetyl-CoA bind to the active site, which lies in a cleft between two domains and is surrounded mainly by α-helical segments (Figure 16.4). Binding of oxaloacetate induces a conformational change that facilitates the binding of acetyl-CoA and closes the active site, so that the reactive carbanion of acetyl-CoA is protected from protonation by water.

Citrate synthase catalyzes the first step in the TCA cycle, and, given the large negative $\Delta G^{\circ\prime}$ of this reaction, the activity of the synthase is highly regulated. NADH, a product of the TCA cycle, is an allosteric inhibitor of citrate synthase, as is succinyl-CoA, the product of the fifth step in the cycle (and an acetyl-CoA analog).

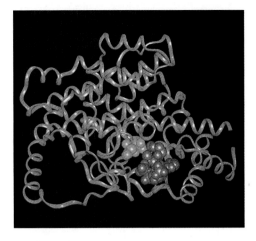

Figure 16.4 Citrate synthase monomer. Citrate is shown in green, and CoA is pink.

 See page 74

16.4 | The Isomerization of Citrate by Aconitase

Citrate itself poses a problem: it is a poor candidate for further oxidation because it is a tertiary alcohol, which can be oxidized only by breaking a carbon–carbon bond. This problem is solved by isomerization of the tertiary alcohol to a secondary alcohol, a process catalyzed by **aconitase.**

The isomerization reaction is a two-step process involving aconitate as an intermediate (Figure 16.5). The elements of water are first abstracted from citrate to yield aconitate, which is then rehydrated by reversal of the positions of the H— and HO— groups to produce isocitrate. The net effect is the conversion of a tertiary alcohol (citrate) to a secondary alcohol (isocitrate). Oxidation of the secondary alcohol of isocitrate involves breakage of a C—H bond, an easier process than the C—C cleavage required for the direct oxidation of citrate.

(b)

(a)

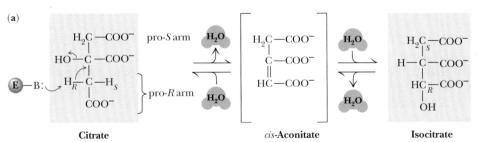

Citrate   *cis*-**Aconitate**   **Isocitrate**

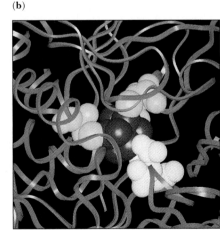

Figure 16.5 **(a)** The aconitase reaction converts citrate to *cis*-aconitate and then to isocitrate. Aconitase is stereospecific and removes the pro-*R* hydrogen from the pro-*R* arm of citrate. **(b)** The active site of aconitase. The iron–sulfur cluster (red) is coordinated by cysteines (yellow) and isocitrate (white).

 See page 73

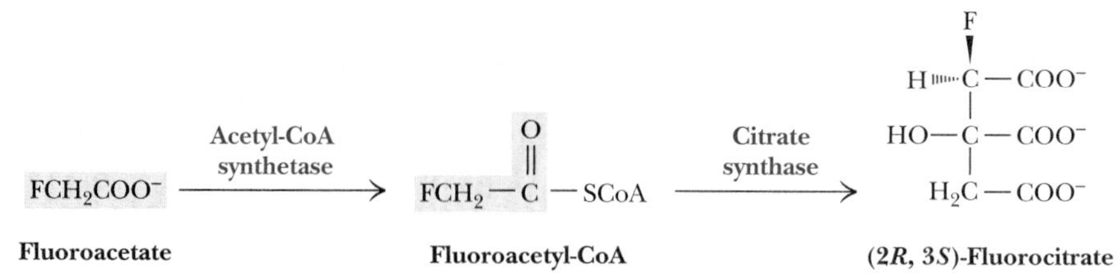

Figure 16.6 The conversion of fluoroacetate to fluorocitrate.

Inspection of the citrate structure shows a total of four chemically equivalent hydrogens, but only one of these—the pro-R H atom of the pro-R arm of citrate—is abstracted by aconitase. Thus the enzyme is stereospecific with respect to proton abstraction, an indication of the precise alignment of groups within the active site. Formation of the double bond of aconitate following proton abstraction requires departure of hydroxide ion from the C-3 position. Hydroxide is a relatively poor leaving group, and its departure is facilitated in the aconitase reaction by coordination with an iron atom in an iron–sulfur cluster.

Fluoroacetate Blocks the TCA Cycle

Fluoroacetate is an extremely poisonous agent that blocks the TCA cycle *in vivo, although it has no apparent effect on any of the isolated enzymes.* Its LD_{50}, the lethal dose for 50% of animals consuming it, is 0.2 mg per kilogram of body weight; it has been used as a rodent poison. Fluoroacetate readily crosses both the cellular and mitochondrial membranes, and in mitochondria it is converted to fluoroacetyl-CoA by *acetyl-CoA synthetase.* Fluoroacetyl-CoA is a substrate for citrate synthase, which condenses it with oxaloacetate to form fluorocitrate (Figure 16.6). Fluorocitrate is a potent inhibitor of aconitase. Fluoroacetate may thus be viewed as a **trojan horse inhibitor.** Analogous to the giant Trojan Horse of legend—which the soldiers of Troy took into their city, not knowing that Greek soldiers were hidden inside it and waiting to attack—fluoroacetate enters the TCA cycle innocently enough in the citrate synthase reaction. Citrate synthase converts fluoroacetate to inhibitory fluorocitrate for its TCA cycle partner, aconitase, blocking the cycle.

16.5 Isocitrate Dehydrogenase: The First Oxidation in the Cycle

In the next step of the TCA cycle, isocitrate is oxidatively decarboxylated to yield α-ketoglutarate, with concomitant reduction of NAD^+ to NADH in the isocitrate dehydrogenase reaction (Figure 16.7). The reaction has a net $\Delta G^{\circ\prime}$ of -8.4 kJ/mol, and it is sufficiently exergonic to pull the aconitase reaction forward. This two-step reaction involves (1) oxidation of the C-2 alcohol of isocitrate to form oxalosuccinate, followed by (2) a β-decarboxylation reaction that expels the central carboxyl group as CO_2, leaving the product α-ketoglutarate. Oxalosuccinate, the β-keto acid produced by the initial dehydrogenation reaction, is unstable and thus is readily decarboxylated.

Isocitrate Dehydrogenase Links the TCA Cycle and Electron Transport

Isocitrate dehydrogenase provides the first connection between the TCA cycle and the electron transport pathway and oxidative phosphorylation, via its production of NADH. As a connecting point between two metabolic pathways, isocitrate dehydrogenase is a regulated reaction. NADH and ATP are allosteric

(a)

$$H_2C - COO^-$$
$$|$$
$$H - C - COO^-$$
$$|$$
$$H - C - COO^-$$
$$|$$
$$OH$$

NAD$^+$

Isocitrate
dehydrogenase

NADH
+ **H$^+$**

$$H_2C - COO^-$$
$$|$$
$$H - C - C \overset{O}{\underset{O^-}{\diagdown}}$$
$$\quad\quad\quad H^+$$
$$\quad\quad |$$
$$\quad\quad C$$
$$\quad O \quad COO^-$$
Oxalosuccinate

 CO$_2$

$$H_2C - COO^-$$
$$|$$
$$H_2C$$
$$|$$
$$C$$
$$O \quad COO^-$$
α-Ketoglutarate

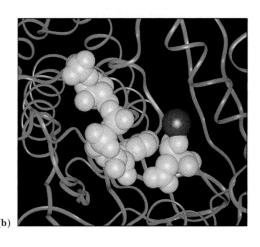

(b)

Figure 16.7 **(a)** The isocitrate dehydrogenase reaction. **(b)** The active site of isocitrate dehydrogenase. Isocitrate is shown in green, NADP$^+$ is shown in gold, with Ca^{2+} in red.

See page 78

inhibitors, whereas ADP acts as an allosteric activator, lowering the K_m for isocitrate by a factor of 10. The enzyme is virtually inactive in the absence of ADP. Also, the product, α-ketoglutarate, is a crucial α-keto acid for aminotransferase reactions (see Chapter 21), connecting the TCA cycle (that is, carbon metabolism) with nitrogen metabolism.

16.6 α-Ketoglutarate Dehydrogenase: A Second Decarboxylation

A second oxidative decarboxylation occurs in the **α-ketoglutarate dehydrogenase reaction** (Figure 16.8). Like the pyruvate dehydrogenase complex, α-ketoglutarate dehydrogenase is a multienzyme complex—consisting of *α-ketoglutarate dehydrogenase, dihydrolipoyl transsuccinylase,* and *dihydrolipoyl dehydrogenase*—that employs five different coenzymes. The dihydrolipoyl dehydrogenase in this reaction is identical to that in the pyruvate dehydrogenase reaction. The mechanism is directly analogous to that of pyruvate dehydrogenase, and the free energy change for this reaction is −30 kJ/mol. As with the pyruvate dehydrogenase reaction, this reaction produces NADH and a thioester product—in this case, succinyl-CoA. Succinyl-CoA and NADH are energy-rich species that are important sources of metabolic energy in subsequent cellular processes.

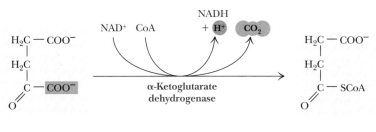

α-Ketoglutarate

Succinyl-CoA

Figure 16.8 The α-ketoglutarate dehydrogenase reaction.

16.7 Succinyl-CoA Synthetase: A Substrate-Level Phosphorylation

The NADH produced in the foregoing steps can be routed through the electron transport pathway to make high-energy phosphates via oxidative phosphorylation. However, succinyl-CoA is itself a high-energy intermediate and is utilized in the next step of the TCA cycle to drive the phosphorylation of GDP to GTP (in mammals) or ADP to ATP (in plants and bacteria). The reaction (Figure 16.9) is catalyzed by **succinyl-CoA synthetase,** sometimes called **succinate thiokinase.** The free energies of hydrolysis of succinyl-CoA and GTP or ATP are similar, and the net reaction has a $\Delta G°'$ of -3.3 kJ/mol. Succinyl-CoA synthetase provides another example of a **substrate-level phosphorylation** (Chapter 15), in which a substrate, rather than an electron transport chain or proton gradient, provides the energy for phosphorylation. It is the only such reaction in the TCA cycle. The GTP produced by mammals in this reaction can exchange its terminal phosphoryl group with ADP via the **nucleoside diphosphate kinase reaction:**

$$\text{GTP} + \text{ADP} \xrightleftharpoons[]{\substack{\text{Nucleoside diphosphate}\\\text{kinase}}} \text{ATP} + \text{GDP}$$

The First Five Steps of the TCA Cycle Produce NADH, CO$_2$, GTP (ATP), and Succinate

This is a good point to pause in our trip through the TCA cycle and see what has happened. A two-carbon acetyl group has been introduced as acetyl-CoA and linked to oxaloacetate, and two CO$_2$ molecules have been liberated. The cycle has produced two molecules of NADH and one of GTP or ATP, and has left a molecule of succinate.

The TCA cycle can now be completed by converting succinate to oxaloacetate. This latter process represents a net oxidation. The TCA cycle breaks it down into (consecutively) an oxidation step, a hydration reaction, and a second oxidation step. The oxidation steps are accompanied by the reduction of an [FAD] and an NAD$^+$. The reduced coenzymes, [FADH$_2$] and NADH, subsequently provide reducing power in the electron transport chain. (We see in Chapter 20 that virtually the same chemical strategy is used in β-oxidation of fatty acids.)

16.8 Succinate Dehydrogenase: An Oxidation Involving FAD

The oxidation of succinate to fumarate (Figure 16.10) is carried out by **succinate dehydrogenase,** a membrane-bound enzyme. As will be seen in Chapter 17, succinate dehydrogenase (also known as succinate–coenzyme Q reductase) is

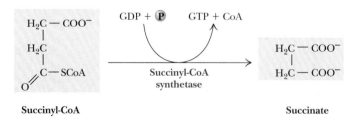

Succinyl-CoA

Succinate

Figure 16.9 The succinyl-CoA synthetase reaction.

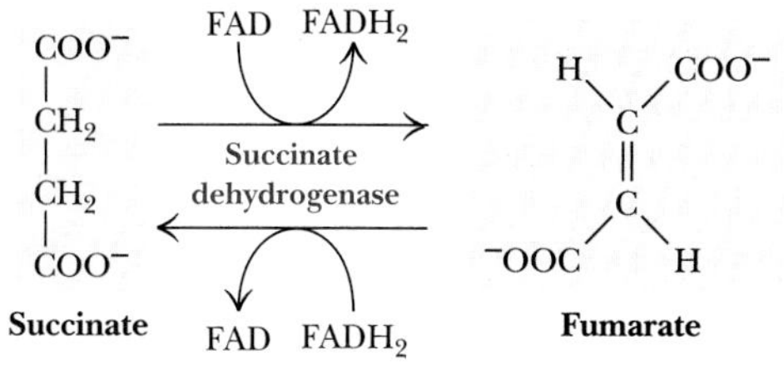

Figure 16.10 The succinate dehydrogenase reaction. Oxidation of succinate occurs with ▶ reduction of [FAD]. Reoxidation of [FADH$_2$] transfers electrons to coenzyme Q.

part of the electron transport chain. Succinate dehydrogenase is an integral membrane protein tightly associated with the inner mitochondrial membrane. In contrast, the other enzymes of the TCA cycle are water-soluble proteins found in the mitochondrial matrix. Succinate oxidation involves removal of H atoms across a C—C bond, rather than a C—O or C—N bond, and produces *fumarate* (or *trans*-butenedioic acid). This reaction (the oxidation of an alkane to an alkene) is not sufficiently exergonic to reduce NAD$^+$, but it does yield enough energy to reduce [FAD].

The electrons captured by [FAD] in this reaction are passed to *coenzyme Q (UQ)*. The covalently bound FAD is first reduced to [FADH$_2$] and then reoxidized to form [FAD] and the reduced form of coenzyme Q, *UQH$_2$*. Electrons captured by UQH$_2$ then flow through the rest of the electron transport chain in a series of events that is discussed in detail in Chapter 17.

Note that flavin coenzymes can carry out either one-electron or two-electron transfers. The succinate dehydrogenase reaction represents a net two-electron reduction of FAD.

16.9 Fumarase Catalyzes *Trans*-Hydration of Fumarate

Fumarase catalyzes the stereospecific hydration of fumarate to give L-malate (Figure 16.11). This reaction involves the *trans*-addition of the elements of water across the double bond, so that the configuration at C-2 is *L*. Recall that aconitase carries out a similar reaction, and that *trans*-addition of —H and —OH occurs across the double bond of *cis*-aconitate.

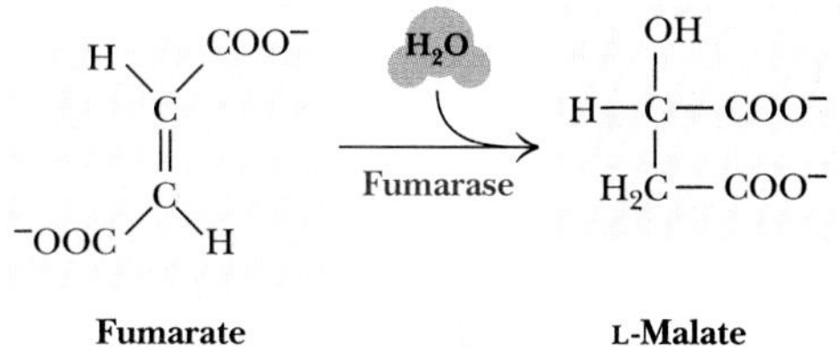

Figure 16.11 The fumarase reaction.

16.10 Malate Dehydrogenase: Completing the Cycle

In the last step of the TCA cycle, L-malate is oxidized to oxaloacetate by **malate dehydrogenase** (Figure 16.12). This reaction is very endergonic, with a $\Delta G°'$ of +30 kJ/mol. Consequently, the concentration of oxaloacetate in the mitochondrial matrix is usually quite low (see the following example). The reaction, however, is pulled forward by the favorable citrate synthase reaction. Oxidation of malate is coupled to reduction of yet another molecule of NAD$^+$, the third one of the cycle. Counting the [FAD] reduced by succinate dehydrogenase, this makes the fourth coenzyme reduced through oxidation of a single acetate unit.

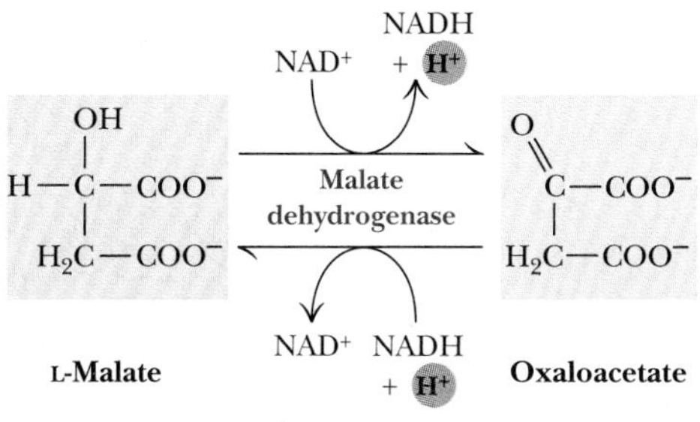

Figure 16.12 The malate dehydrogenase reaction.

▼ **EXAMPLE**

The intramitochondrial concentration of malate is typically 0.22 mM. If the [NAD$^+$]/[NADH] ratio in mitochondria is 20 and if the malate dehydrogenase reaction is at equilibrium, calculate the intramitochondrial concentration of oxaloacetate at 25°C.

SOLUTION

For the malate dehydrogenase reaction,

$$\text{Malate} + \text{NAD}^+ \rightarrow \text{oxaloacetate} + \text{NADH} + \text{H}^+$$

the value of $\Delta G^{\circ\prime}$ is +30 kJ/mol. Then

$$\Delta G^{\circ\prime} = -\text{RT} \ln K_{eq}$$

$$= -(8.314 \, \text{J/mol} \cdot \text{K})(298 \, \text{K}) \ln \left(\frac{[1]x}{[20][2.2 \times 10^{-4}]} \right)$$

$$\frac{-30,000 \, \text{J/mol}}{2478 \, \text{J/mol}} = \ln \left(\frac{x}{4.4 \times 10^{-3}} \right)$$

$$-12.1 = \ln \left(\frac{x}{4.4 \times 10^{-3}} \right)$$

$$x = (5.6 \times 10^{-6})(4.4 \times 10^{-3})$$

$$x = [\text{oxaloacetate}] = 0.024 \, \mu M$$

▲

Malate dehydrogenase is structurally and functionally similar to other dehydrogenases, notably lactate dehydrogenase (Figure 16.13).

Figure 16.13 **(a)** The structure of malate dehydrogenase. Like lactate dehydrogenase, MDH consists of alternating β-sheet and α-helical segments. **(b)** The active site of malate dehydrogenase. Malate is shown in red; NAD$^+$ is blue.

See page 79

16.11 A Summary of the Cycle

The TCA cycle results in the production of two molecules of CO_2, one ATP, and four reduced coenzymes per acetate group oxidized. The cycle is exergonic, with a net $\Delta G^{\circ\prime}$ for one pass around the cycle of approximately

(a)

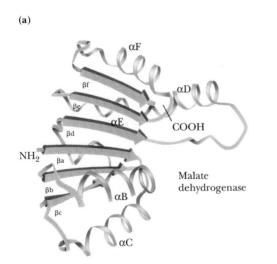

αF

βf

αD

βe

αE

COOH

βd

NH₂ βa

Malate
dehydrogenase

βb

αB

βc

αC

(b)

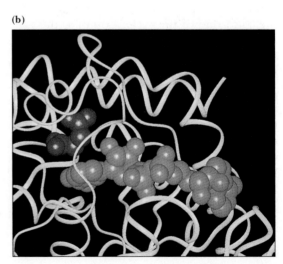

Table 16.1 The Enzymes and Reactions of the TCA Cycle

Reaction	Enzyme	$\Delta G^{\circ\prime}$ (kJ/mol)	ΔG (kJ/mol)
1. Acetyl-CoA + oxaloacetate + H_2O $\rightleftharpoons$ CoASH + citrate	Citrate synthase	-31.4	-53.9
2. Citrate $\rightleftharpoons$ isocitrate	Aconitase	$+6.7$	$+0.8$
3. Isocitrate + NAD^+ $\rightleftharpoons$ α-ketoglutarate + NADH + CO_2 + H^+	Isocitrate dehydrogenase	-8.4	-17.5
4. α-Ketoglutarate + CoASH + NAD^+ $\rightleftharpoons$ succinyl-CoA + NADH + CO_2 + H^+	α-Ketoglutarate dehydrogenase complex	-30	-43.9
5. Succinyl-CoA + GDP + P_i $\rightleftharpoons$ succinate + GTP + CoASH	Succinyl-CoA synthetase	-3.3	≈ 0
6. Succinate + $[FAD]$ $\rightleftharpoons$ fumarate + $[FADH_2]$	Succinate dehydrogenase	$+0.4$	≈ 0
7. Fumarate + H_2O $\rightleftharpoons$ L-malate	Fumarase	-3.8	≈ 0
8. L-Malate + NAD^+ $\rightleftharpoons$ oxaloacetate + NADH + H^+	Malate dehydrogenase	$+29.7$	≈ 0

Net for reactions 1–8:

Actelyl-CoA + 3 NAD^+ + [FAD] + GDP + P_i + 2 H_2O $\rightleftharpoons$ CoASH + 3 NADH + $[FADH_2]$ + GTP + 2 CO_2 + 3 H^+		-40	$\approx(-115)$
Simple combustion of acetate: Acetate + 2 O_2 + H^+ $\rightleftharpoons$ 2 CO_2 + 2 H_2O		-849	

ΔG values from Newsholme, E. A., and Leech, A. R., 1983. *Biochemistry for the Medical Sciences.* New York: Wiley & Sons.

 See pages 1–3

-40 kJ/mol. Table 16.1 compares the $\Delta G^{\circ\prime}$ values for the individual reactions with the overall $\Delta G^{\circ\prime}$ for the net reaction.

$$\text{Acetyl-CoA} + 3\ NAD^+ + [FAD] + ADP + P_i + 2\ H_2O \longrightarrow$$
$$2\ CO_2 + 3\ NADH + 3\ H^+ + [FADH_2] + ATP + CoASH$$

$$\Delta G^{\circ\prime} = -40\ \text{kJ/mol}$$

Glucose metabolized via glycolysis produces two molecules of pyruvate and thus two molecules of acetyl-CoA, which can enter the TCA cycle. Combining glycolysis and the TCA cycle gives the net reaction shown:

$$\text{Glucose} + 2\ H_2O + 10\ NAD^+ + 2\ [FAD] + 4\ ADP + 4\ P_i \longrightarrow$$
$$6\ CO_2 + 10\ NADH + 10\ H^+ + 2\ [FADH_2] + 4\ ATP$$

All six carbons of glucose are liberated as CO_2, and a total of four molecules of ATP are formed in substrate-level phosphorylations. Further, the 12 reduced coenzymes produced up to this point can eventually produce a maximum of 34 molecules of ATP in the electron transport and oxidative phosphorylation pathways. A stoichiometric relationship for these subsequent processes is

HUMAN BIOCHEMISTRY

Mitochondrial Diseases Are Rare

Diseases arising from defects in mitochondrial enzymes are quite rare because major defects in the TCA cycle (and the respiratory chain) are incompatible with life, and affected embryos rarely survive to birth. Even so, about 150 different hereditary mitochondrial diseases have been reported. Even though mitochondria carry their own DNA, many of these inherited diseases map to the nuclear genome because most of the mitochondrial proteins are imported from the cytosol.

An interesting disease linked to mitochondrial DNA mutations is Leber's hereditary optic neuropathy (LHON), in which the genetic defects are located primarily in the mitochondrial DNA coding for the subunits of NADH-CoQ dehydrogenase, also known as Complex I of the electron transport chain (Chapter 17). Leber's disease is the most common form of blindness in otherwise healthy young men; it occurs less often in women.

$$NADH + H^+ + \tfrac{1}{2}O_2 + 3\,ADP + 3\,P_i \rightleftharpoons NAD^+ + 3\,ATP + 4\,H_2O$$

$$[FADH_2] + \tfrac{1}{2}O_2 + 2\,ADP + 2\,P_i \rightleftharpoons [FAD] + 2\,ATP + 3\,H_2O$$

Thus, a total of 3 ATP per NADH and 2 ATP per $FADH_2$ may be produced through the processes of electron transport and oxidative phosphorylation.

16.12 The TCA Cycle Provides Intermediates for Biosynthetic Pathways

Until now we have viewed the TCA cycle as a catabolic process because it oxidizes acetate units to CO_2 and converts the liberated energy to ATP and reduced coenzymes. The TCA cycle is, after all, the end point for breakdown of food materials, at least in terms of carbon turnover. However, as shown in Figure 16.14, four-, five-, and six-carbon species produced in the TCA cycle also fuel a variety of **biosynthetic processes.** α-Ketoglutarate, succinyl-CoA, fumarate,

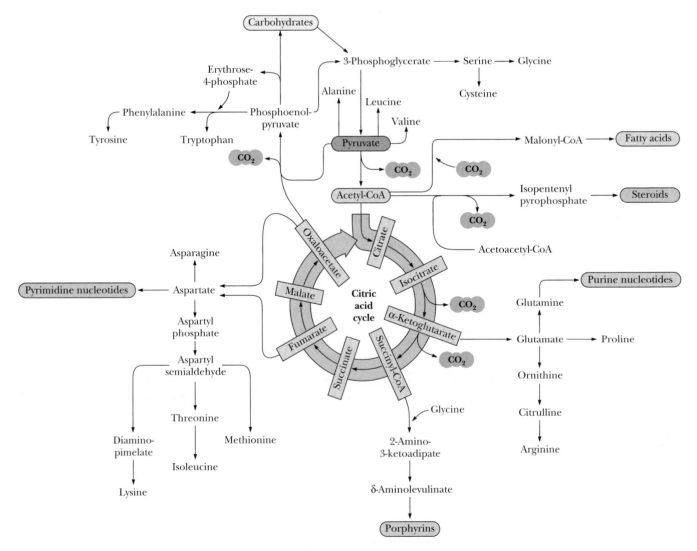

Figure 16.14 The TCA cycle provides intermediates for numerous biosynthetic processes in the cell.

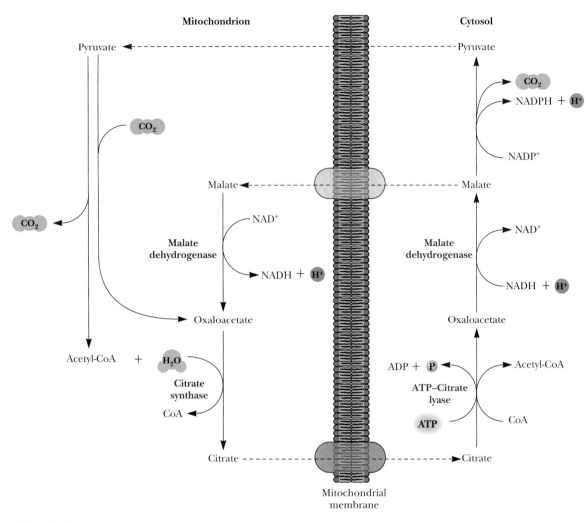

Figure 16.15 Export of citrate from mitochondria and cytosolic breakdown produces oxaloacetate and acetyl-CoA. Oxaloacetate is recycled to malate or pyruvate, which re-enters the mitochondria. This cycle provides acetyl-CoA for fatty acid synthesis in the cytosol.

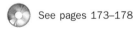 See pages 173–178

and oxaloacetate are all precursors of important cellular species. (To participate in eukaryotic biosynthetic processes, however, they must first be transported out of the mitochondria.) A transamination reaction converts α-ketoglutarate directly to glutamate, which can then serve as a versatile precursor for proline, arginine, and glutamine (as described in Chapter 21). Succinyl-CoA provides most of the carbon atoms of the porphyrins. Oxaloacetate can be transaminated to produce aspartate. Aspartic acid itself is a precursor of the pyrimidine nucleotides and, in addition, is a key precursor for the synthesis of asparagine, methionine, lysine, threonine, and isoleucine. Oxaloacetate can undergo GTP-dependent decarboxylation to yield PEP, which is a key element of several pathways, namely (1) synthesis (in plants and microorganisms) of the aromatic amino acids phenylalanine, tyrosine, and tryptophan; (2) formation of 3-phosphoglycerate and conversion to the amino acids serine, glycine, and cysteine; and (3) gluconeogenesis, which, as we will see in Chapter 19, is the pathway that synthesizes new glucose and many other carbohydrates.

Finally, citrate can be exported from the mitochondria and then broken down by **ATP-citrate lyase** to yield oxaloacetate and acetyl-CoA, a precursor of fatty acids (Figure 16.15). Oxaloacetate produced in this reaction is rapidly reduced to malate, which can then be processed in either of two ways: it may be transported into mitochondria, where it is reoxidized to oxaloacetate, or it may be oxidatively decarboxylated to pyruvate by **malic enzyme,** with subsequent mitochondrial uptake of pyruvate. This cycle permits citrate to provide acetyl-CoA for biosynthetic processes in the cytosol, with return of the malate and pyruvate by-products to the mitochondria.

16.13 The Anaplerotic, or "Filling Up," Reactions

In a sort of reciprocal arrangement, the cell also feeds many intermediates back into the TCA cycle from other reactions. Since such reactions replenish the TCA cycle intermediates, Hans Kornberg proposed that they be called **anaplerotic reactions** (literally, the "filling up" reactions). Thus, **PEP carboxylase** and **pyruvate carboxylase** synthesize oxaloacetate from pyruvate (Figure 16.16).

Pyruvate carboxylase is the most important of the anaplerotic enzymes. It exists in the mitochondria of animal cells but not in plants, and it provides a direct link between glycolysis and the TCA cycle. The enzyme contains covalently bound biotin. (It is examined in greater detail in our discussion of gluconeogenesis in Chapter 19.) Pyruvate carboxylase has an absolute allosteric requirement for acetyl-CoA. Thus, when acetyl-CoA levels exceed the oxaloacetate supply, allosteric activation of pyruvate carboxylase by acetyl-CoA raises oxaloacetate levels, so that the excess acetyl-CoA can enter the TCA cycle.

PEP carboxylase occurs in yeast, bacteria, and higher plants, but not in animals. The enzyme is specifically inhibited by aspartate, which is produced by transamination of oxaloacetate. Thus, organisms utilizing this enzyme control aspartate production by regulation of PEP carboxylase. Malic enzyme is found in the cytosol or mitochondria of many animal and plant cells and is an NADPH-

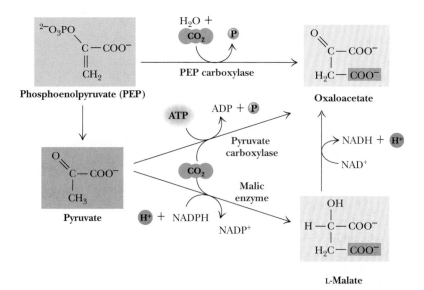

Figure 16.16 Phosphoenolpyruvate (PEP) carboxylase, pyruvate carboxylase, and malic enzyme catalyze anaplerotic reactions, replenishing TCA cycle intermediates.

A DEEPER LOOK

Fool's Gold and the Reductive Citric Acid Cycle: The First Metabolic Pathway?

How did life arise on the planet Earth? It was once supposed that a reducing atmosphere, together with random synthesis of organic compounds, gave rise to a prebiotic "soup," in which the first living things appeared. However, certain key compounds, such as arginine, lysine, and histidine, the straight-chain fatty acids, porphyrins, and essential coenzymes, have not been convincingly synthesized under simulated prebiotic conditions. This and other problems have led researchers to consider other models for the evolution of life.

One of these alternate models, postulated by Günter Wächtershäuser, involves an archaic version of the TCA cycle running in the reverse (reductive) direction. Reversal of the TCA cycle results in assimilation of CO_2 and fixation of carbon as shown. For each turn of the reversed cycle, two carbons are fixed in the formation of isocitrate and two more are fixed in the reductive transformation of acetyl-CoA to oxaloacetate. Thus, for every succinate that enters the reversed cycle, two succinates are returned, making the cycle highly autocatalytic. Because TCA cycle intermediates are involved in many biosynthetic pathways (see Section 16.12), a reversed TCA cycle would be a bountiful and broad source of metabolic substrates.

A reversed, reductive TCA cycle would require energy input to drive it. What might have been the thermodynamic driving force for such a cycle? Wächtershäuser hypothesizes that the anaerobic reaction of FeS and H_2S to form insoluble FeS_2 (iron pyrite, also known as fool's gold) in the prebiotic milieu could have been the driving reaction:

$$FeS + H_2S \rightarrow FeS_2 \downarrow + H_2$$

This reaction is highly exergonic, with a standard-state free energy change ($\Delta G°'$) of -38 kJ/mol. Under conditions that might have existed in a prebiotic world, this reaction would have been sufficiently exergonic to drive the reductive steps of a reversed TCA cycle. In addition, in an H_2S-rich prebiotic environment, organic compounds would have been in equilibrium with their thio-organic counterparts. High-energy thioesters formed in this way may have played key roles in the energetics of early metabolic pathways.

Wächtershäuser has also suggested that early metabolic processes first occurred on the surface of pyrite and other related mineral materials. The iron–sulfur chemistry that prevailed on these mineral surfaces may have influenced the evolution of the iron–sulfur proteins that control and catalyze many reactions in modern pathways (including the succinate dehydrogenase and aconitase reactions of the TCA cycle).

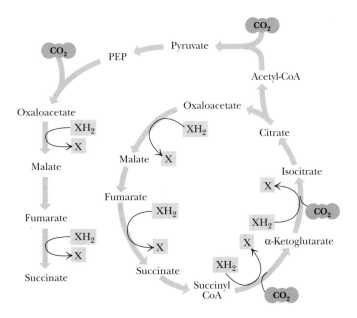

A reductive, reversed TCA cycle.

dependent enzyme. Malate produced by the anaplerotic action of malic enzyme is converted to oxaloacetate by malate dehydrogenase.

It is worth noting that the reaction catalyzed by **PEP carboxykinase** (Figure 16.17) could also function as an anaplerotic reaction, were it not for the particular properties of the enzyme. CO_2 binds weakly to PEP carboxykinase,

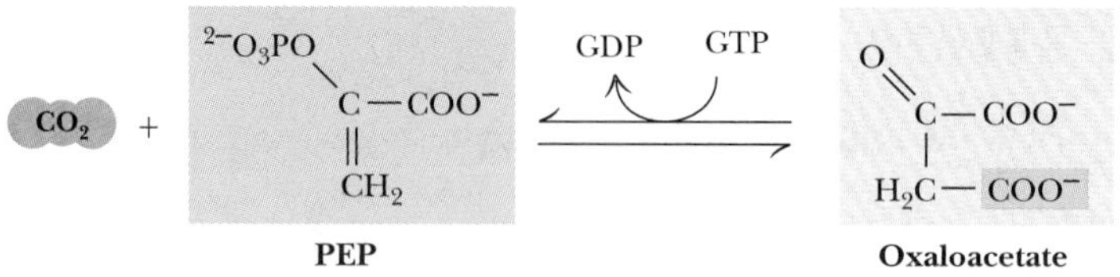

Figure 16.17 The phosphoenolpyruvate carboxykinase reaction.

whereas oxaloacetate binds very tightly ($K_D = 2 \times 10^{-6} M$), and, as a result, the enzyme favors formation of PEP from oxaloacetate.

The catabolism of amino acids provides pyruvate, acetyl-CoA, oxaloacetate, fumarate, α-ketoglutarate, and succinate, all of which may be oxidized by the TCA cycle. In this way, proteins may serve as excellent sources of nutrient energy, as seen in Chapter 21.

16.14 Regulation of the TCA Cycle

Situated as it is between glycolysis and the electron transport chain, the TCA cycle must be carefully controlled by the cell. If the cycle were permitted to run unchecked, large amounts of metabolic energy could be wasted in overproduction of reduced coenzymes and ATP; conversely, if it ran too slowly, ATP would not be produced rapidly enough to satisfy the needs of the cell. Also, as just seen, the TCA cycle is an important source of precursors for biosynthetic processes and must be able to provide them as needed.

What are the sites of regulation in the TCA cycle? Based upon our experience with glycolysis (Figure 15.22), we might anticipate that some of the reactions of the TCA cycle would operate near equilibrium under cellular conditions (with $\Delta G \approx 0$), whereas others—the sites of regulation—would be characterized by large, negative ΔG values. Estimates for the values of ΔG in mitochondria, based on mitochondrial concentrations of metabolites, are summarized in Table 16.1. Three reactions of the cycle—citrate synthase, isocitrate dehydrogenase, and α-ketoglutarate dehydrogenase—operate with large, negative ΔG values under mitochondrial conditions and are thus the primary sites of regulation in the cycle.

The regulatory actions that control the TCA cycle are shown in Figure 16.18. As one might expect, the principal regulatory "signals" are the concentrations of acetyl-CoA, ATP, NAD^+, and NADH, with additional effects provided by several other metabolites. The main sites of regulation are pyruvate dehydrogenase, citrate synthase, isocitrate dehydrogenase, and α-ketoglutarate dehydrogenase. All of these enzymes are inhibited by NADH, so that when the cell has produced all the NADH that can be oxidized to drive ATP synthesis, the cycle shuts down. For similar reasons, ATP is an inhibitor of pyruvate dehydrogenase and isocitrate dehydrogenase. The TCA cycle is turned on, however, when either the ADP/ATP or NAD^+/NADH ratio is high, an indication that the cell has run low on ATP or NADH. Regulation of the TCA cycle by NADH, NAD^+, ATP, and ADP thus reflects the energy status of the cell. On the other hand, succinyl-CoA is an *intracycle regulator,* inhibiting citrate synthase and α-ketoglutarate dehydrogenase. Acetyl-CoA acts as a signal to the TCA cycle that glycolysis or fatty acid breakdown is producing two-carbon units. Acetyl-CoA activates pyruvate carboxylase, the anaplerotic reaction that provides

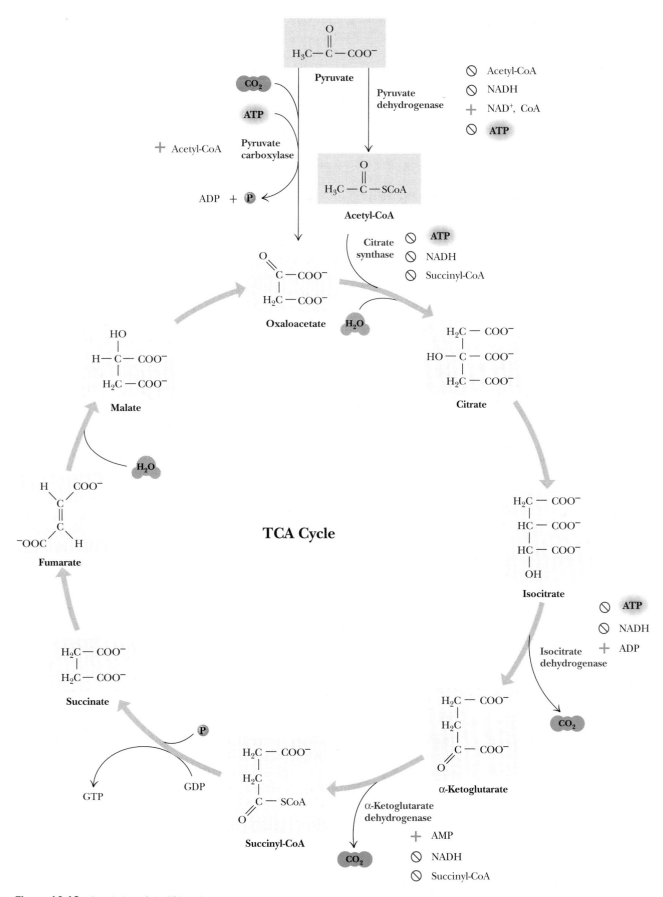

Figure 16.18 Regulation of the TCA cycle.

oxaloacetate, the four-carbon acceptor necessary for increased flux of acetyl-CoA into the TCA cycle.

Regulation of Pyruvate Dehydrogenase

As we shall see in Chapter 19, most organisms can synthesize sugars such as glucose from pyruvate. In animals, however, both carbons of acetyl-CoA are irreversibly lost as CO_2 and thus net synthesis of sugars from acetyl-CoA cannot occur. For this reason, the pyruvate dehydrogenase complex, which converts pyruvate to acetyl-CoA, plays a pivotal role in metabolism. Conversion to acetyl-CoA commits nutrient carbon atoms either to oxidation in the TCA cycle or to fatty acid synthesis (see Chapter 20). Because this choice is so crucial to the organism, pyruvate dehydrogenase is a carefully regulated enzyme. It is subject to product inhibition and is further regulated by nucleotides. Finally, activity of pyruvate dehydrogenase is regulated by phosphorylation and dephosphorylation of the enzyme complex itself.

High levels of either product, acetyl-CoA or NADH, allosterically inhibit the pyruvate dehydrogenase complex. Acetyl-CoA specifically blocks dihydrolipoyl transacetylase, and NADH acts on dihydrolipoyl dehydrogenase. The mammalian pyruvate dehydrogenase is also regulated by covalent modifications. As shown in Figure 16.19, a Mg^{2+}-dependent **pyruvate dehydrogenase kinase** is associated with the enzyme in mammals. This kinase is allosterically activated by NADH and acetyl-CoA, and when levels of these metabolites rise in the mitochondrion, they stimulate phosphorylation of a serine residue on the PDH α-subunit, blocking the first step of the pyruvate dehydrogenase reaction, the decarboxylation of pyruvate. Inhibition of the dehydrogenase in this manner eventually lowers the levels of NADH and acetyl-CoA in the matrix of the mitochondrion. Reactivation of the enzyme is carried out by **pyruvate dehydrogenase phosphatase,** a Ca^{2+}-activated enzyme that binds to the dehydrogenase complex and hydrolyzes the phosphoserine moiety on the α-subunit of PDH. At low ratios of NADH to NAD^+ and low acetyl-CoA levels, the phosphatase maintains the PDH in an activated state, but high level of acetyl-CoA or NADH once again activates the kinase, leading to the inhibition of the PDH. Insulin and Ca^{2+} ions activate dephosphorylation, and pyruvate inhibits the phosphorylation reaction.

Pyruvate dehydrogenase is also sensitive to the energy status of the cell. AMP activates pyruvate dehydrogenase, whereas GTP inhibits it. High levels of AMP are a sign that the cell may become energy-poor. Activation of pyruvate dehydrogenase under such conditions commits pyruvate to energy production.

Regulation of Isocitrate Dehydrogenase

The mechanism of regulation of isocitrate dehydrogenase is in some respects like that of pyruvate dehydrogenase. The mammalian isocitrate dehydrogenase is subject only to allosteric activation by ADP and NAD^+ and to inhibition by ATP and NADH. Thus, high $NAD^+/NADH$ and ADP/ATP ratios stimulate isocitrate dehydrogenase and TCA cycle activity. The *Escherichia coli* enzyme, on the other hand, is regulated by covalent modification. Serine residues on each subunit of the dimeric enzyme are phosphorylated by a protein kinase, causing inhibition of the isocitrate dehydrogenase activity. Activity is restored by the action of a specific phosphatase. When TCA cycle and glycolytic intermediates—such as isocitrate, 3-phosphoglycerate, pyruvate, PEP, and oxaloac-

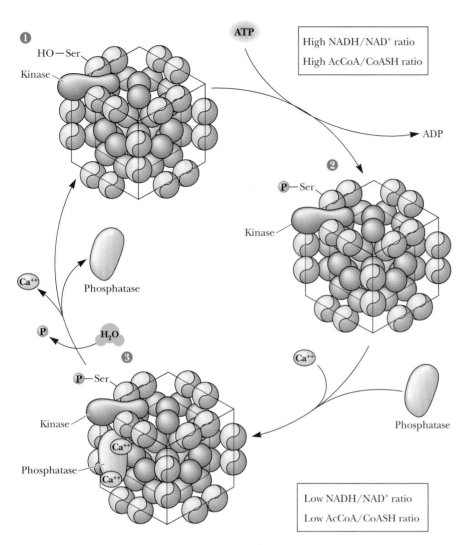

High NADH/NAD⁺ ratio
High AcCoA/CoASH ratio

Low NADH/NAD⁺ ratio
Low AcCoA/CoASH ratio

Figure 16.19   Regulation of the pyruvate dehydrogenase reaction.

HUMAN BIOCHEMISTRY

Therapy for Heart Attacks by Alterations of Heart Muscle Metabolism?

Ischemia, the state of reduced blood flow to a tissue, occurs in heart muscle during a heart attack. One strategy for minimizing tissue damage during and immediately following a heart attack involves interventions to alter heart muscle metabolism. One drug being studied for this purpose is dichloroacetate, which specifically inhibits pyruvate dehydrogenase kinase, thus activating pyruvate dehydrogenase. The primary energy source for heart muscle is normally the oxidation of long-chain fatty acids. By contrast, carbohydrate metabolism is less important, except during periods of

high workload. However, treatment with dichloroacetate activates pyruvate dehydrogenase in the heart muscle and increases the flow of carbon from pyruvate (and thus glucose) into the TCA cycle. This is advantageous for the ischemic heart (for which oxygen is limited), because the ATP yield per oxygen consumed is higher for glucose oxidation than for fatty acid oxidation (see Chapter 20). Research to date indicates that heart function is improved with intravenous dichloroacetate in animal models.

A DEEPER LOOK

The Glyoxylate Cycle of Plants and Bacteria

Plants (particularly seedlings, which cannot yet accomplish efficient photosynthesis), as well as some bacteria and algae, can use acetate as the *only* source of carbon for all the carbon compounds they produce. Although the TCA cycle can supply intermediates for some biosynthetic processes, the cycle gives off 2 CO_2 for every two-carbon acetate group that enters and cannot accomplish the *net synthesis* of TCA cycle intermediates. Thus, it would not be possible for the cycle to produce the massive amounts of biosynthetic intermediates needed for acetate-based growth unless alternative reactions were available. In essence, the TCA cycle is geared primarily to energy production, and it "wastes" carbon units by giving off CO_2. Modification of the cycle to support acetate-based growth would require eliminating the CO_2-producing reactions and enhancing the net production of four-carbon units (i.e., oxaloacetate).

Plants and bacteria employ a modification of the TCA cycle called the **glyoxylate cycle** to produce four-carbon dicarboxylic acids (and eventually even sugars) from two-carbon acetate units. The glyoxylate cycle bypasses the two oxidative decarboxylations of the TCA cycle, and instead routes isocitrate through the **isocitrate lyase** and **malate synthase** reactions (see figure). Glyoxylate produced by isocitrate lyase reacts with a second molecule of acetyl-CoA to form L-malate. The net effect is to conserve carbon units, using two acetyl-CoA molecules per cycle to generate oxaloacetate. Some of this is converted to PEP and then to glucose by pathways discussed in Chapter 19.

The enzymes of the glyoxylate cycle in plants are contained in **glyoxysomes,** organelles devoted to this cycle. Yeast and algae carry out the glyoxylate cycle in the cytoplasm. The enzymes common to both the TCA and glyoxylate pathways exist as isozymes, with spatially and functionally distinct enzymes operating independently in the two cycles.

The isocitrate lyase reaction produces succinate, a four-carbon product of the cycle, as well as glyoxylate, which can then combine with a second molecule of acetyl-CoA. Isocitrate lyase catalyzes an aldol cleavage similar to the reaction mediated by aldolase in glycolysis. The **malate synthase** reaction, a Claisen condensation of acetyl-CoA with the aldehyde of glyoxylate to yield malate, is quite similar to the citrate synthase reaction. Compared with the TCA cycle, the glyoxylate cycle (a) contains only five steps (as opposed to eight), (b) lacks the CO_2-liberating reactions, (c) consumes two molecules of acetyl-CoA per cycle, and (d) produces four-carbon units (oxaloacetate) as opposed to one-carbon units.

The existence of the glyoxylate cycle explains how certain seeds grow underground (or in the dark), where photosynthesis is impossible. Many seeds (peanuts, soybeans, and castor beans, for example) are rich in lipids; and, as we will see in Chapter 20, most organisms degrade the fatty acids of lipids to acetyl-CoA. Glyoxysomes form in seeds as germination begins, and the glyoxylate cycle uses the acetyl-CoA produced in fatty acid oxidation to provide large amounts of oxaloacetate and other intermediates for carbohydrate synthesis. Once the growing plant begins photosynthesis and can fix CO_2 to produce carbohydrates (see Chapter 18), the glyoxysomes disappear.

The glyoxylate cycle. The first two steps are identical to TCA cycle reactions. The third step bypasses the CO_2-evolving steps of the TCA cycle to produce ▶ succinate and glyoxylate. The malate synthase reaction forms malate from glyoxylate and another acetyl-CoA. The result is that one turn of the cycle consumes one oxaloacetate and two acetyl-CoA molecules but produces two molecules of oxaloacetate. The net for this cycle is one oxaloacetate from two acetyl-CoA molecules.

 See pages 173–178

etate—are high, the kinase is inhibited, the phosphatase is activated, and the TCA cycle operates normally. When levels of these intermediates fall, the kinase is activated, isocitrate dehydrogenase is inhibited, and isocitrate is diverted to the glyoxylate pathway (see *A Deeper Look*, above).

It may seem surprising that isocitrate dehydrogenase is strongly regulated, because it is not an apparent branch point within the TCA cycle. However, the citrate/isocitrate ratio controls the rate of production of cytosolic acetyl-CoA because acetyl-CoA in the cytosol is derived from citrate exported from the mitochondrion. (Breakdown of cytosolic citrate produces oxaloacetate and acetyl-CoA, which can be used in a variety of biosynthetic processes.) Thus, isocitrate dehydrogenase activity in the mitochondrion favors catabolic TCA cycle activity over anabolic utilization of acetyl-CoA in the cytosol.

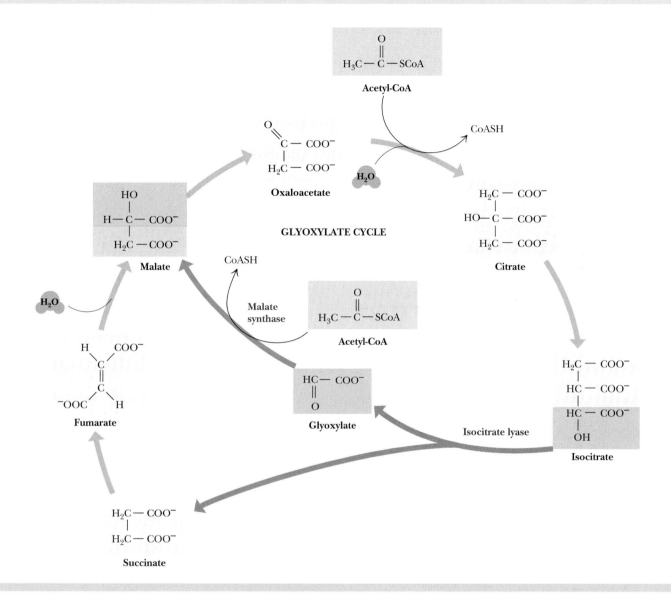

PROBLEMS

1. Describe the labeling pattern that would result from the introduction into the TCA cycle of glutamate labeled at C_γ with ^{14}C.

2. Describe the effect on the TCA cycle of (a) increasing the concentration of NAD^+, (b) reducing the concentration of ATP, and (c) increasing the concentration of isocitrate.

3. The serine residue of isocitrate dehydrogenase that is phosphorylated by protein kinase lies within the active site of the enzyme. This situation contrasts with most other examples of covalent modification by protein phosphorylation, where the phosphorylation occurs at a site remote from the active site. What direct effect do you think such active-site phosphorylation might have on the catalytic activity of isocitrate dehydrogenase? (See Barford, D., 1991. Molecular mechanisms for the control of enzymic activity by protein phosphorylation. *Biochimica et Biophysica Acta* **1133**:55–62.)

4. The first step of the α-ketoglutarate dehydrogenase reaction involves decarboxylation of the substrate and leaves a covalent TPP intermediate. Write a reasonable mechanism for this reaction.

5. In a tissue where the TCA cycle has been inhibited by fluoroacetate, what difference in the concentration of each TCA cycle metabolite would you expect, compared with a normal, uninhibited tissue?

6. On the basis of the description in Chapter 14 of the physical properties of FAD and $FADH_2$, suggest a method for the measurement of the enzyme activity of succinate dehydrogenase.

7. Starting with citrate, isocitrate, α-ketoglutarate, and succinate, state which of the individual carbons of the molecule undergo oxidation in the next step of the TCA cycle. Which molecules undergo a net oxidation?

8. In addition to fluoroacetate, consider whether other analogs of TCA cycle metabolites or intermediates might be introduced to inhibit other, specific reactions of the cycle. Explain your reasoning.

9. Based on the action of thiamine pyrophosphate in catalysis of the pyruvate dehydrogenase reaction, suggest a suitable chemical mechanism for the pyruvate decarboxylase reaction in yeast:

$$pyruvate \rightarrow acetaldehyde + CO_2$$

10. Aconitase catalyzes the citric acid cycle reaction:

$$citrate \rightleftharpoons isocitrate$$

The standard free energy change, $\Delta G°'$, for this reaction is $+6.7$ kJ/mol. However, the observed free energy change (ΔG) for this reaction in pig heart mitochondria is $+0.8$ kJ/mol. What is the ratio of [isocitrate]/[citrate] in these mitochondria? If [isocitrate] $= 0.03$ mM, what is [citrate]?

FURTHER READING

Akiyama, S. K., and Hammes, G. G., 1980. Elementary steps in the reaction mechanism of the pyruvate dehydrogenase multienzyme complex from *Escherichia coli:* Kinetics of acetylation and deacetylation. *Biochemistry* **19:**4208–4213.

Akiyama, S. K., and Hammes, G. G., 1981. Elementary steps in the reaction mechanism of the pyruvate dehydrogenase multienzyme complex from *Escherichia coli:* Kinetics of flavin reduction. *Biochemistry* **20:**1491–1497.

Atkinson, D. E., 1977. *Cellular Energy Metabolism and Its Regulation.* New York: Academic Press.

Bodner, G. M., 1986. The tricarboxylic acid (TCA), citric acid or Krebs cycle. *Journal of Chemical Education* **63:**673–677.

Frey, P. A., 1982. Mechanism of coupled electron and group transfer in *Escherichia coli* pyruvate dehydrogenase. *Annals of the New York Academy of Sciences* **378:**250–264.

Gibble, G. W., 1973. Fluoroacetate toxicity. *Journal of Chemical Education* **50:**460–462.

Hansford, R. G., 1980. "Control of Mitochondrial Substrate Oxidation." *In Current Topics in Bioenergetics,* vol. 10, pp. 217–278. New York: Academic Press.

Hawkins, R. A., and Mans, A. M., 1983. "Intermediary Metabolism of Carbohydrates and Other Fuels." In *Handbook of Neurochemistry,* 2nd ed., A. Lajtha, ed., pp. 259–294. New York: Plenum Press.

Kelly, R. M., and Adams, M. W., 1994. Metabolism in hyperthermophilic microorganisms. *Antonie van Leeuwenhoek* **66:**247–270.

Krebs, H. A., 1970. The history of the tricarboxylic acid cycle. *Perspectives in Biology and Medicine* **14:**154–170.

Krebs, H. A., 1981. *Reminiscences and Reflections.* Oxford, England: Oxford University Press.

Lowenstein, J. M., ed., 1969. *Citric Acid Cycle: Control and Compartmentation.* New York: Marcel Dekker.

Maden, B. E., 1995. No soup for starters? Autotrophy and the origins of metabolism. *Trends in Biochemical Sciences* **20:**337–341.

Newsholme, E. A., and Leech, A. R., 1983. *Biochemistry for the Medical Sciences.* New York: John Wiley & Sons.

Perham, R. N., 2000. Swinging arms and swinging domains in multifunctional enzymes: Catalytic machines for multistep reactions. *Annual Review of Biochemistry* **69:**961–1004.

Srere, P. A., 1975. The enzymology of the formation and breakdown of citrate. *Advances in Enzymology* **43:**57–101.

Srere, P. A., 1987. Complexes of sequential metabolic enzymes. *Annual Review of Biochemistry* **56:**89–124.

Walsh, C., 1979. *Enzymatic Reaction Mechanisms.* San Francisco: W. H. Freeman.

Wiegand, G., and Remington, S. J., 1986. Citrate synthase: Structure, control and mechanism. *Annual Review of Biophysics and Biophysical Chemistry* **15:**97–117.

Williamson, J. R., 1980. Mitochondrial metabolism and cell regulation. In *Mitochondria: Bioenergetics, Biogenesis and Membrane Structure,* L. Packer and A. Gomez-Puyou, eds. New York: Academic Press.

Electron Transport and Oxidative Phosphorylation

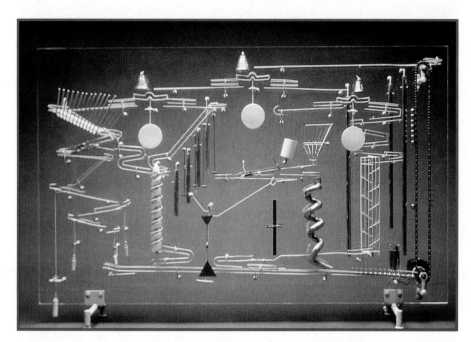

Wall Piece #IV (1985), a kinetic sculpture by George Rhoads. This complex mechanical art form can be viewed as a metaphor for the molecular apparatus underlying electron transport and ATP synthesis by oxidative phosphorylation. (© 1985 by George Rhoads)

In all things of nature there is something of the marvelous.

ARISTOTLE (384–322 B.C.)

Living cells save up metabolic energy predominantly in the form of fats and carbohydrates, and they "spend" this energy for biosynthesis, membrane transport, and movement. In both directions, energy is exchanged and transferred in the form of ATP. In Chapters 15 and 16 we saw that glycolysis and the TCA cycle convert some of the energy available from stored and dietary sugars directly to ATP. However, most of the metabolic energy that is obtainable from substrates entering glycolysis and the TCA cycle is funneled via oxidation–reduction reactions into NADH and reduced flavoproteins, the latter symbolized by [FADH$_2$]. We now embark on the discovery of how cells convert the stored metabolic energy of NADH and [FADH$_2$] into ATP.

Whereas ATP made in glycolysis and the TCA cycle is the result of substrate-level phosphorylation, NADH-dependent ATP synthesis is the result of **oxidative phosphorylation.** Electrons stored in the form of the reduced coenzymes, NADH or [FADH$_2$], are passed through an elaborate and highly organized

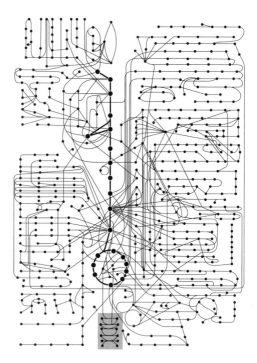

Electron transport and oxidative phosphorylation.

See *Interactive Biochemistry CD-ROM and Workbook,* pages 173–178

chain of proteins and coenzymes, the so-called **electron transport chain,** finally reaching O_2 (molecular oxygen), the terminal electron acceptor. Each component of the chain can exist in (at least) two oxidation states, and each component is successively reduced and reoxidized as electrons move through the chain from NADH (or [$FADH_2$]) to O_2. In the course of electron transport, a proton gradient is established across the inner mitochondrial membrane. It is the energy of this proton gradient that drives ATP synthesis.

17.1 Electron Transport and Oxidative Phosphorylation Are Membrane-Associated Processes

The processes of electron transport and oxidative phosphorylation are **membrane-associated.** Bacteria are the simplest life form, and bacterial cells typically consist of a single cellular compartment surrounded by a plasma membrane and a more rigid cell wall. In such a system, the conversion of energy from NADH and [$FADH_2$] to the energy of ATP via electron transport and oxidative phosphorylation is carried out at (and across) the plasma membrane. In eukaryotic cells, electron transport and oxidative phosphorylation are localized in mitochondria, which are also the sites of TCA cycle activity and (as we shall see in Chapter 20) fatty acid oxidation. Mammalian cells contain from 800 to 2500 mitochondria; other types of cells may have as few as one or two or as many as half a million mitochondria. Human erythrocytes, whose purpose is simply to transport oxygen to tissues, contain *no* mitochondria at all. The typical mitochondrion is about 0.5 ± 0.3 micron in diameter and from 0.5 micron to several microns long; its overall shape is sensitive to metabolic conditions in the cell.

Mitochondria are surrounded by a simple **outer membrane** and a more complex **inner membrane** (Figure 17.1). The space between the inner and outer membranes is referred to as the **intermembrane space.** Several enzymes that utilize ATP (such as creatine kinase and adenylate kinase) are found in the intermembrane space. The smooth outer membrane is about 30% to 40% lipid and 60% to 70% protein, and has a relatively high concentration of phosphatidylinositol. The outer membrane contains significant amounts of **porin**— a transmembrane protein, rich in β-sheets, that forms large channels across

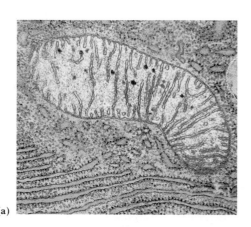

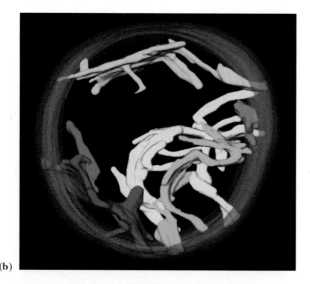

Figure 17.1 **(a)** Electron micrograph of a mitochondrion. **(b)** Tomography of a rat liver mitochondrion. The tubular structures in red, yellow, green, magenta and aqua represent individual cristae formed from the inner mitochondrial membrane. *(a, B. King/BPS; b, Frey, T. G., and Mannella, C.A., 2000. The internal structure of mitochondria.* Trends in Biochemical Sciences **25***:319–324.)*

the membrane, permitting free diffusion of molecules with molecular weights of about 10,000 or less. Apparently, the outer membrane functions mainly to maintain the shape of the mitochondrion. The inner membrane is richly packed with proteins, which account for nearly 80% of its weight; thus, its density is higher than that of the outer membrane. The fatty acids of inner membrane lipids are highly unsaturated. Cardiolipin and diphosphatidylglycerol (see Chapter 6) are abundant. The inner membrane lacks cholesterol and is quite impermeable to molecules and ions. Species that must cross the mitochondrial inner membrane—ions, substrates, fatty acids for oxidation, and so on—are carried by specific transport proteins in the membrane. Notably, the inner membrane is elaborated in a variety of folds and tubular structures (see Figure 17.1). The structures, known as **cristae,** provide the inner membrane with a large surface area in a small volume. Cristae are often connected to each other by a network of tubes and channels. During periods of active respiration, the inner membrane appears to shrink significantly, leaving a comparatively large intermembrane space.

The Mitochondrial Matrix Contains the Enzymes of the TCA Cycle

The soluble phase inside the inner mitochondrial membrane is called the **matrix,** and it contains most of the enzymes of the TCA cycle and fatty acid oxidation. (An important exception, succinate dehydrogenase of the TCA cycle, is located in the inner membrane itself.) In addition, mitochondria contain circular DNA molecules, along with ribosomes and the enzymes required to synthesize proteins coded within the mitochondrial genome. Although some of the mitochondrial proteins are made this way, most are encoded by nuclear DNA and synthesized by cytosolic ribosomes.

17.2	Reduction Potentials: An Accounting Device for Free Energy Changes in Redox Reactions

On numerous occasions in earlier chapters, we have stressed that NADH and reduced flavoproteins ($[FADH_2]$) are forms of metabolic energy. These reduced coenzymes have a strong tendency to be oxidized—that is, to transfer electrons to other species. The electron transport chain converts the energy of electron transfer into the energy of phosphoryl transfer stored in the phosphoric anhydride bonds of ATP. Just as the *group transfer potential* was used in Chapter 3 to quantitate the energy of phosphoryl transfer, the **standard reduction potential,** denoted by $\mathscr{E}_\circ$, quantitates the tendency of chemical species to be reduced or oxidized. The standard reduction potential describing electron transfer between two species,

$$
\begin{array}{cc}
\text{Reduced donor} & \text{Oxidized acceptor} \\[4pt]
\Big\downarrow & ne^- \quad \Big\downarrow \\[4pt]
\text{Oxidized donor} & \text{Reduced acceptor}
\end{array}
\qquad \textbf{(17.1)}
$$

is related to the free energy change for the process by

$$\Delta G^\circ = -n\mathscr{F}\Delta\mathscr{E}_\circ \qquad \textbf{(17.2)}$$

where n represents the number of electrons transferred; $\mathscr{F}$ is Faraday's constant (96,485 J/V · mol); and $\Delta\mathscr{E}_\circ$ is the difference in reduction potentials between the donor and acceptor. This relationship is straightforward, but it depends on a *standard* of reference by which reduction potentials are defined.

Measurement of Standard Reduction Potentials

Standard reduction potentials are determined by measuring the voltages generated in **reaction half-cells** (Figure 17.2). A half-cell consists of a solution containing 1 M concentrations of both the oxidized and reduced forms of the substance whose reduction potential is being measured, and a simple electrode. (Together, the oxidized and reduced forms of the substance are referred to as a **redox couple**.) Such a **sample half-cell** is connected to a **reference half-cell** and electrode via a conductive bridge (usually a salt-containing agar gel). A sensitive potentiometer (voltmeter) connects the two electrodes so that the electrical potential (voltage) between them can be measured. The reference half-cell normally contains 1 M H^+ in equilibrium with H_2 gas at a pressure of 1 atm. The H^+/H_2 reference half-cell is arbitrarily assigned a standard reduction potential of 0.0 V. The standard reduction potentials of all other redox couples are defined relative to the H^+/H_2 reference half-cell on the basis of the sign and magnitude of the voltage (electromotive force, emf) registered on the potentiometer (see Figure 17.2).

If electron flow between the electrodes is toward the sample half-cell, reduction (electron gain) occurs spontaneously in the sample half-cell, and the reduction potential is said to be positive. If electron flow between the electrodes is away from the sample half-cell and toward the reference cell, the reduction potential is said to be negative because electron loss (oxidation) is occurring in the sample half-cell. Note that the reduction potential of the hydrogen half-cell is pH-dependent. The standard reduction potential, $\mathcal{E}_\circ$, 0.0 V, assumes 1 M H^+. The standard reduction potential, $\mathcal{E}_\circ'$, is the electromotive force generated at 25°C and pH 7.0 by a sample half-cell (containing 1 M concentrations of the oxidized and reduced species) with respect to a reference half-cell. The hydrogen half-cell measured at pH 7.0 has an $\mathcal{E}_\circ'$ of -0.421 V.

The Significance of $\mathcal{E}_\circ'$

Some typical half-cell reactions and their respective standard reduction potentials are listed in Table 17.1. Whenever reactions of this type are tabulated, they are uniformly written as *reduction* reactions, regardless of what occurs in

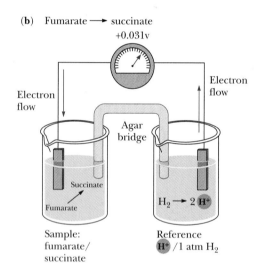

(a) Ethanol ⟶ acetaldehyde
−0.197v

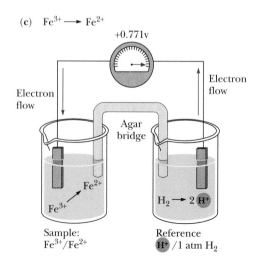

(b) Fumarate ⟶ succinate
+0.031v

(c) Fe^{3+} ⟶ Fe^{2+}
+0.771v

◀ **Figure 17.2** Experimental apparatus used to measure the standard reduction potential of the indicated redox couples: **(a)** the acetaldehyde/ethanol couple, **(b)** the fumarate/succinate couple, **(c)** the Fe^{3+}/Fe^{2+} couple. Figure 17.2a shows a sample/reference half-cell pair for measurement of the standard reduction potential of the acetaldehyde/ethanol couple. Because electrons flow toward the reference half-cell and away from the sample half-cell, the standard reduction potential is negative, specifically −0.197 V. In contrast, the fumarate/succinate couple (b) and the Fe^{3+}/Fe^{2+} couple (c) both accept electrons from the reference half-cell; that is, reduction occurs spontaneously in each system, and the reduction potentials of both are thus positive. For each half-cell, a **half-cell reaction** describes the reaction taking place. For the fumarate/succinate half-cell coupled to a H^+/H_2 reference half-cell (b), the reaction occurring is indeed a reduction of fumarate.

$$\text{Fumarate} + 2\,H^+ + 2\,e^- \rightarrow \text{succinate}$$
$$\mathcal{E}_\circ' = +0.031 \text{ V} \tag{17.3}$$

Similarly, for the Fe^{3+}/Fe^{2+} half-cell (c),

$$Fe^{3+} + e^- \rightarrow Fe^{2+}$$
$$\mathcal{E}_\circ' = +0.771 \text{ V} \tag{17.4}$$

However, the reaction occurring in the acetaldehyde/ethanol half-cell (a) is the oxidation of ethanol:

$$\text{Ethanol} \rightarrow \text{acetaldehyde} + 2\,H^+ + 2\,e^-$$
$$\mathcal{E}_\circ' = -0.197 \text{ V} \tag{17.5}$$

Table 17.1   Standard Reduction Potentials for Several Biological Reduction Half-Reactions

Reduction Half-Reaction	$\mathcal{E}_0'$ (V)
$\frac{1}{2}O_2 + 2\ H^+ + 2\ e^- \longrightarrow H_2O$	0.816
$Fe^{3+} + e^- \longrightarrow Fe^{2+}$	0.771
Photosystem P700	0.430
Cytochrome $f(Fe^{3+}) + e^- \longrightarrow$ cytochrome $f(Fe^{2+})$	0.365
Cytochrome $a_3(Fe^{3+}) + e^- \longrightarrow$ cytochrome $a_3(Fe^{2+})$	0.350
Cytochrome $a(Fe^{3+}) + e^- \longrightarrow$ cytochrome $a(Fe^{2+})$	0.290
Rieske Fe-S$(Fe^{3+}) + e^- \longrightarrow$ Rieske Fe-S(Fe^{2+})	0.280
Cytochrome $c(Fe^{3+}) + e^- \longrightarrow$ cytochrome $c(Fe^{2+})$	0.254
Cytochrome $c_1(Fe^{3+}) + e^- \longrightarrow$ cytochrome $c_1(Fe^{2+})$	0.220
$UQH \cdot + H^+ + e^- \longrightarrow UQH_2(UQ = $ coenzyme Q$)$	0.190
$UQ + 2\ H^+ + 2\ e^- \longrightarrow UQH_2$	0.060
Cytochrome $b_H(Fe^{3+}) + e^- \longrightarrow$ cytochrome $b_H(Fe^{2+})$	0.050
Fumarate $+ 2\ H^+ + 2\ e^- \longrightarrow$ succinate	0.031
$UQ + H^+ + e^- \longrightarrow UQH \cdot$	0.030
Cytochrome $b_5(Fe^{3+}) + e^- \longrightarrow$ cytochrome $b_5(Fe^{2+})$	0.020
$[FAD] + 2\ H^+ + 2\ e^- \longrightarrow [FADH_2]$	0.003–0.091*
Cytochrome $b_L(Fe^{3+}) + e^- \longrightarrow$ cytochrome $b_L(Fe^{2+})$	−0.100
Oxaloacetate $+ 2\ H^+ + 2\ e^- \longrightarrow$ malate	−0.166
Pyruvate $+ 2\ H^+ + 2\ e^- \longrightarrow$ lactate	−0.185
Acetaldehyde $+ 2\ H^+ + 2\ e^- \longrightarrow$ ethanol	−0.197
$FMN + 2\ H^+ + 2\ e^- \longrightarrow FMNH_2$	−0.219
$FAD + 2\ H^+ + 2\ e^- \longrightarrow FADH_2$	−0.219
1,3-Bisphosphoglycerate $+ 2\ H^+ + 2\ e^- \longrightarrow$ glyceraldehyde-3-phosphate $+ P_i$	−0.290
$NAD^+ + 2\ H^+ + 2\ e^- \longrightarrow NADH + H^+$	−0.320
$NADP^+ + 2\ H^+ + 2\ e^- \longrightarrow NADPH + H^+$	−0.320
α-Ketoglutarate $+ CO_2 + 2\ H^+ + 2\ e^- \longrightarrow$ isocitrate	−0.380
$2\ H^+ + 2\ e^- \longrightarrow H_2\ (pH\ 7)$	−0.421
Ferredoxin (spinach) $(Fe^{3+}) + e^- \longrightarrow$ ferredoxin (spinach) (Fe^{2+})	−0.430
Succinate $+ CO_2 + 2\ H^+ + 2\ e^- \longrightarrow \alpha$-ketoglutarate $+ H_2O$	−0.670

*Typical values for reduction of bound FAD in flavoproteins such as succinate dehydrogenase (see Bonomi, F., Pagani, S., Cerletti, P., and Giori, C., 1983. *European Journal of Biochemistry* **134**:439–445).

the given half-cell. The sign of the standard reduction potential indicates which reaction really occurs when the given half-cell is combined with the reference hydrogen half-cell. Redox couples that have large positive reduction potentials have a strong tendency to accept electrons, and the oxidized form of such a couple (O_2, for example) is a strong oxidizing agent. Redox couples with large negative reduction potentials have a strong tendency to undergo oxidation (that is, donate electrons), and the reduced form of such a couple (NADPH, for example) is a strong reducing agent.

Coupled Redox Reactions

The half-reactions and reduction potentials in Table 17.1 can be used to analyze energy changes in redox reactions at physiological pH. The reduction of NAD^+ to NADH can be coupled with the oxidation of isocitrate to α-ketoglutarate.

$$NAD^+ + \text{isocitrate} \longrightarrow NADH + H^+ + \alpha\text{-ketoglutarate} + CO_2 \quad (17.6)$$

This is the isocitrate dehydrogenase reaction of the TCA cycle. Writing the two half-cell reactions, we have

$$NAD^+ + 2\ H^+ + 2\ e^- \longrightarrow NADH + H^+ \qquad \mathcal{E}_0' = -0.32\ V \quad (17.7)$$

$$\alpha\text{-ketoglutarate} + CO_2 + 2\ H^+ + 2\ e^- \longrightarrow \text{isocitrate} \qquad \mathcal{E}_0' = -0.38V \quad (17.8)$$

In a spontaneous reaction, electrons are donated by (flow away from) the half-reaction with the more negative reduction potential and are accepted by (flow toward) the half-reaction with the more positive reduction potential. Thus, in the present case, isocitrate donates electrons and NAD^+ accepts electrons. The convention defines $\Delta\mathscr{E}_\circ'$ as

$$\Delta\mathscr{E}_\circ' = \mathscr{E}_\circ'(\text{acceptor}) - \mathscr{E}_\circ'(\text{donor}) \tag{17.9}$$

In the present case, isocitrate is the donor and NAD^+ the acceptor, so we write

$$\Delta\mathscr{E}_\circ' = -0.32\text{V} - (-0.38\text{V}) = +0.06\text{V} \tag{17.10}$$

From Equation 17.2, we can now calculate $\Delta G^{\circ\prime}$ as

$$\begin{aligned} \Delta G^{\circ\prime} &= -(2)(96.485 \text{ kJ/V} \cdot \text{mol})(0.06\text{V}) \\ \Delta G^{\circ\prime} &= -11.58 \text{ kJ/mol} \end{aligned} \tag{17.11}$$

Note that a reaction with a net positive $\Delta\mathscr{E}_\circ'$ yields a negative $\Delta G^{\circ\prime}$, indicating a spontaneous reaction.

The Dependence of the Reduction Potential on Concentration

We have already noted that the standard free energy change for a reaction, $\Delta G^{\circ\prime}$, does not reflect the actual conditions in a cell, where reactants and products are not at standard-state concentrations ($1M$). Equation 3.12 was introduced to permit calculations of actual free energy changes under non–standard-state conditions. Similarly, standard reduction potentials for redox couples must be modified to account for the actual concentrations of the oxidized and reduced species. For any redox couple,

$$\text{ox} + ne^- \rightleftharpoons \text{red} \tag{17.12}$$

the actual reduction potential is given by

$$\mathscr{E} = \mathscr{E}_\circ' + \frac{RT}{n\mathscr{F}} \ln \frac{[\text{ox}]}{[\text{red}]} \tag{17.13}$$

Reduction potentials can also be quite sensitive to molecular environment. The influence of environment is especially important for flavins, such as $FAD/FADH_2$ and $FMN/FMNH_2$. These species are normally bound to their respective flavoproteins; the reduction potential of bound FAD, for example, can be very different from the value shown in Table 17.1 for the free $FAD–FADH_2$ couple of -0.219 V. A problem at the end of the chapter addresses this case.

17.3 | The Electron Transport Chain: An Overview

The metabolic energy from oxidation of food materials—sugars, fats, and amino acids—is funneled into formation of reduced coenzymes (NADH) and reduced flavoproteins ($[FADH_2]$). The electron transport chain reoxidizes the coenzymes, and channels the free energy obtained from these reactions into the synthesis of ATP. This reoxidation process involves the removal of both protons and electrons from the coenzymes. Electrons move from NADH and $[FADH_2]$ to molecular oxygen, O_2, which is the terminal acceptor of electrons in the chain. The reoxidation of NADH,

$$\text{NADH(reductant)} + H^+ + \tfrac{1}{2}O_2(\text{oxidant}) \longrightarrow NAD^+ + H_2O \tag{17.14}$$

involves the following half-reactions:

$$NAD^+ + 2\,H^+ + 2\,e^- \longrightarrow NADH + H^+ \qquad \mathscr{E}_o' = -0.32\;V \qquad \textbf{(17.15)}$$

$$\tfrac{1}{2}\,O_2 + 2\,H^+ + 2\,e^- \longrightarrow H_2O \qquad \mathscr{E}_o' = +0.816\;V \qquad \textbf{(17.16)}$$

Here, half-reaction (17.16) is the electron acceptor and half-reaction (17.15) is the electron donor. Then

$$\Delta\mathscr{E}_o' = 0.816 - (-0.32) = 1.136\;V$$

and, according to Equation (17.2), the standard-state free energy change, $\Delta G^{\circ\prime}$, is -219 kJ/mol. Molecules along the electron transport chain have reduction potentials between the values for the $NAD^+/NADH$ couple and the oxygen/H_2O couple, so that electrons move down the energy scale toward progressively more positive reduction potentials (Figure 17.3).

The Electron Transport Chain Can Be Isolated in Four Complexes

The electron transport chain involves several different molecular species, including:

(a) **Flavoproteins,** which contain tightly bound FMN or FAD as prosthetic groups, and which (as noted in Chapter 16) may participate in one- or two-electron transfer events.

(b) **Coenzyme Q,** also called **ubiquinone** (and abbreviated **CoQ** or **UQ**) (Figure 6.17), which can function in either one- or two-electron transfer reactions.

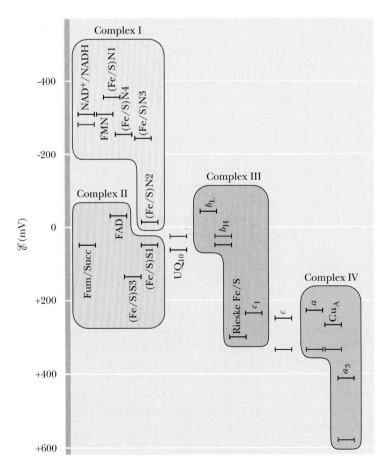

Figure 17.3 $\mathscr{E}_o'$ and $\mathscr{E}$ values for the components of the mitochondrial electron transport chain. Values indicated are consensus values for animal mitochondria. Black bars represent $\mathscr{E}_o'$; red bars, $\mathscr{E}$.

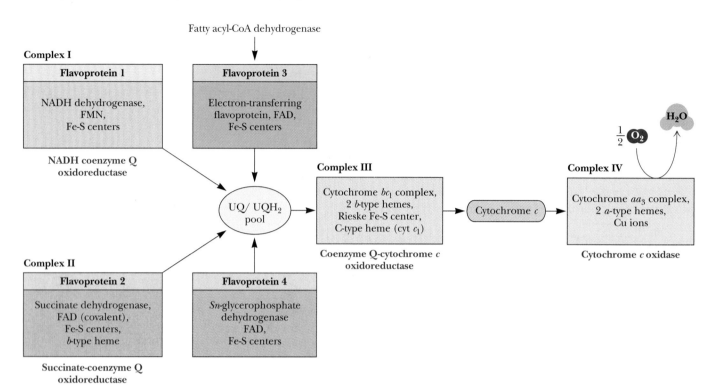

Figure 17.4 An overview of the complexes and pathways in the mitochondrial electron transport chain. *(Adapted from Nicholls, D. G., and Ferguson, S. J., 1992. Bioenergetics 2. London: Academic Press.)*

(c) Several **cytochromes** (proteins containing heme prosthetic groups [see Chapter 4], which function by carrying or transferring electrons), including cytochromes b, c, c_1, a, and a_3. Cytochromes are one-electron transfer agents, in which the heme iron is converted from Fe^{2+} to Fe^{3+} and back.

(d) A number of **iron–sulfur proteins,** which participate in one-electron transfers involving the Fe^{2+} and Fe^{3+} states.

(e) Protein-bound **copper,** a one-electron transfer site, which converts between Cu^+ and Cu^{2+}.

All these intermediates except for cytochrome c are membrane-associated (either in the mitochondrial inner membrane of eukaryotes or in the plasma membrane of prokaryotes). All three types of proteins involved in this chain—flavoproteins, cytochromes, and iron–sulfur proteins—possess electron-transferring **prosthetic groups.**

The components of the electron transport chain can be purified from the mitochondrial inner membrane. Solubilization of the membranes containing the electron transport chain results in the isolation of four distinct protein complexes, and the complete chain can thus be considered to be composed of four parts*: (I) **NADH–coenzyme Q reductase,** (II) **succinate–coenzyme Q reductase,** (III) **coenzyme Q–cytochrome c reductase,** and (IV) **cytochrome c oxidase** (Figure 17.4). Complex I accepts electrons from NADH, serving as a link between glycolysis, the TCA cycle, fatty acid oxidation, and the rest of the electron transport chain. Complex II includes succinate dehydrogenase and thus forms a direct link between the TCA cycle and electron transport. Complexes I and II produce a common product, reduced coenzyme Q (UQH$_2$), which is the substrate for coenzyme Q–cytochrome c reductase (Complex III). As shown in Figure 17.4, there are two other ways to feed electrons to UQ: the **electron-transferring flavoprotein,** which transfers electrons from the flavoprotein-linked step of fatty acyl-CoA dehydrogenase, and ***sn*-glycerophosphate**

*These four enzyme complexes, three "reductases" and an "oxidase," are sometimes also called "oxidoreductases."

dehydrogenase. Complex III oxidizes UQH_2 while reducing cytochrome c, which in turn is the substrate for Complex IV, cytochrome c oxidase. Complex IV is responsible for reducing molecular oxygen. Each of the complexes shown in Figure 17.4 is a large multisubunit complex embedded within the inner mitochondrial membrane.

17.4 Complex I: NADH–Coenzyme Q Reductase

As its name implies, this complex transfers a pair of electrons from NADH to coenzyme Q, a small, hydrophobic, yellow compound. Another common name for this enzyme complex is *NADH dehydrogenase*. The complex (with an estimated mass of 850 kD) involves more than 30 polypeptide chains, one molecule of flavin mononucleotide (FMN), and as many as seven Fe-S clusters, together containing a total of 20 to 26 iron atoms (Table 17.2). By virtue of its dependence on FMN, NADH–UQ reductase is a *flavoprotein.*

Although the precise mechanism of the NADH–UQ reductase is not known, the first step involves binding of NADH to the enzyme on the matrix side of the inner mitochondrial membrane, and transfer of electrons from NADH to tightly bound FMN:

$$NADH + [FMN] + H^+ \longrightarrow [FMNH_2] + NAD^+ \qquad (17.17)$$

The second step involves the transfer of electrons from the reduced $[FMNH_2]$ to a series of Fe-S proteins, including both 2Fe-2S and 4Fe-4S clusters. The unique redox properties of the flavin group of FMN are probably important here. NADH is a two-electron donor, whereas the Fe-S proteins are one-electron transfer agents. The flavin of FMN has three redox states—the oxidized, semiquinone, and reduced states. It can act as either a one-electron or a two-electron transfer agent and may serve as a critical link between NADH and the Fe-S proteins.

The final step of the reaction involves the transfer of two electrons from iron–sulfur clusters to coenzyme Q. Coenzyme Q is a **mobile electron carrier.**

Table 17.2 Protein Complexes of the Mitochondrial Electron Transport Chain

Complex	Mass (kD)	Subunits	Prosthetic Group	Binding Site for:
NADH–UQ reductase	850	>30	FMN Fe-S	NADH (matrix side) UQ (lipid core)
Succinate–UQ reductase	140	4	FAD Fe-S	Succinate (matrix side) UQ (lipid core)
UQ–Cyt c reductase monomer	248	11	Heme b_L Heme b_H Heme c_1 Fe-S	Cyt c (intermembrane space side)
Cytochrome c	13	1	Heme c	Cyt c_1 Cyt a
Cytochrome c oxidase	204	13	Heme a Heme a_3 Cu_A Cu_B	Cyt c (intermembrane space side)

Adapted from: Hatefi, Y., 1985. The mitochondrial electron transport chain and oxidative phosphorylation system. *Annual Review of Biochemistry* **54**:1015–1069; and DePierre, J., and Ernster, L.,1977. Enzyme topology of intracellular membranes. *Annual Review of Biochemistry* **46**:201–262.

(a)

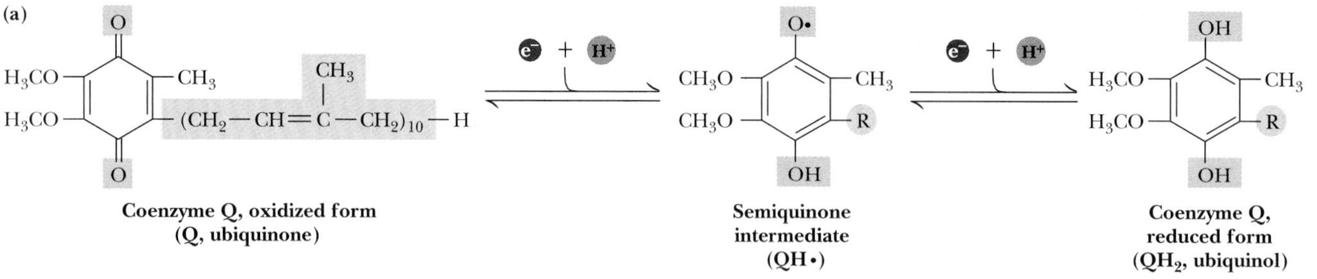

Coenzyme Q, oxidized form
(Q, ubiquinone)

Semiquinone
intermediate
(QH·)

Coenzyme Q,
reduced form
(QH$_2$, ubiquinol)

(b)

Figure 17.5 (a) The three oxidation states of coenzyme Q. **(b)** A space-filling model of coenzyme Q.

Its isoprenoid tail makes it highly hydrophobic, and it diffuses freely in the hydrophobic core of the inner mitochondrial membrane. As a result, it shuttles electrons from Complexes I and II to Complex III. The redox reactions of UQ are shown in Figure 17.5, and its reduction by Complex I is shown schematically in Figure 17.6.

Complex I Transports Protons from the Matrix to the Cytosol

The oxidation of one NADH and the reduction of one UQ by NADH–UQ reductase results in the net transport of protons from the matrix side to the cytosolic side of the inner membrane. The cytosolic side, where H$^+$ accumulates, is referred to as the **P** (for *positive*) face; similarly, the matrix side is the **N** (for *negative*) face. Some of the energy liberated by the flow of electrons through this complex is used in a *coupled process* to drive the transport of protons across the membrane. Experimental evidence suggests a stoichiometry of four H$^+$ transported per two electrons passed from NADH to UQ.

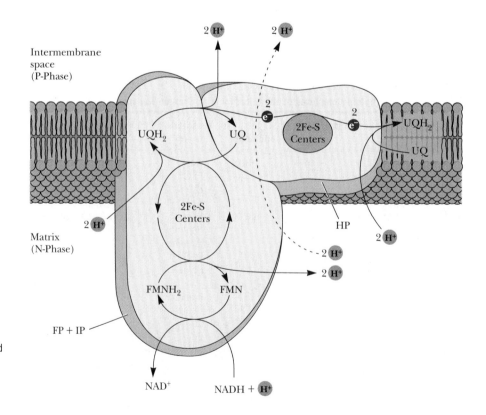

Figure 17.6 Structural model and electron transport pathway for Complex I. Three protein complexes have been isolated, including the **flavoprotein (FP), iron–sulfur protein (IP), and hydrophobic protein (HP).** FP contains three peptides (of mass 51, 24, and 10 kD) and bound FMN and has two Fe-S centers (a 2Fe-2S center and a 4Fe-4S center). IP contains six peptides and at least three Fe-S centers. HP contains at least seven peptides and one Fe-S center.

HUMAN BIOCHEMISTRY

Solving a Medical Mystery Revolutionized Our Treatment of Parkinson's Disease

A tragedy among illegal drug users was the impetus for a revolutionary treatment of Parkinson's disease. In 1982, several mysterious cases of paralysis came to light in Southern California. The victims, some of them teenagers, were frozen like living statues, unable to talk or move. The case was baffling at first, but it was soon traced to a batch of synthetic heroin that contained MPTP (1-methyl-4-phenyl-1,2,3,6-tetrahydropyridine) as a contaminant. MPTP is rapidly converted in the brain to MPP$^+$ (1-methyl-4-phenylpyridine) by the enzyme monoamine oxidase B. MPP$^+$ is a potent inhibitor of mitochondrial Complex I (NADH–UQ reductase), and it acts preferentially in the *substantia nigra*, an area of the brain that is essential to movement and also the region of the brain that deteriorates slowly in Parkinson's disease.

Parkinson's disease results from the inability of the brain to produce sufficient quantities of dopamine, a neurotransmitter. Neurologist J. William Langston, asked to consult on the treatment of some of these patients, recognized that the symptoms of this drug-induced disorder were in fact similar to those of Parkinsonism. He began treatment of the patients with L-dopa, which is decarboxylated in the brain to produce dopamine. The treated patients immediately regained movement. Langston then took a bold step. He implanted fetal brain tissue into the brains of several of the affected patients, prompting substantial recovery from the Parkinsonlike symptoms. Langston's innovation sparked a revolution in the use

Cell death
in substantia nigra

of tissue implantation for the treatment of neurodegenerative diseases.

Other toxins may cause similar effects in neural tissue. Timothy Greenamyre at Emory University has shown that rats exposed to the pesticide rotenone (see Figure 17.24) over a period of weeks experience a gradual loss of function in dopaminergic neurons and then develop symptoms of Parkinsonism, including limb tremors and rigidity. This finding supports earlier research that links long-term pesticide exposure to Parkinson's disease.

17.5 | Complex II: Succinate–Coenzyme Q Reductase

Complex II is perhaps better known by its other name—**succinate dehydrogenase,** the only TCA cycle enzyme that is an integral membrane protein in the inner mitochondrial membrane. This enzyme has a mass of approximately 100 to 140 kD and is composed of four subunits: two Fe-S proteins of masses 70 kD and 27 kD, and two other peptides of masses 15 kD and 13 kD. Also known as *flavoprotein 2 (FP$_2$)*, it contains an FAD covalently bound to a histidine residue, and three Fe-S centers: a 4Fe-4S cluster, a 3Fe-4S cluster, and a 2Fe-2S cluster. When succinate is converted to fumarate in the TCA cycle, concomitant reduction of bound FAD to FADH$_2$ occurs in succinate dehydrogenase. This FADH$_2$ transfers its electrons immediately to Fe-S centers, which pass them on to UQ. Electron flow from succinate to UQ,

$$\text{Succinate} \longrightarrow \text{fumarate} + 2\ \text{H}^+ + 2\ e^- \qquad \text{(17.18)}$$

$$\text{UQ} + 2\ \text{H}^+ + 2\ e^- \longrightarrow \text{UQH}_2 \qquad \text{(17.19)}$$

$$\text{Net rxn:} \quad \text{Succinate} + \text{UQ} \longrightarrow \text{fumarate} + \text{UQH}_2 \qquad \text{(17.20)}$$

yields a net reduction potential of 0.029 V. (Note that the first half-reaction is written in the direction of the e^- flow. As always, $\Delta\mathscr{E}_o'$ is calculated according to Equation 17.9.) The small free energy change of this reaction is not sufficient to drive the transport of protons across the inner mitochondrial membrane.

This is a crucial point because (as we will see) proton transport is coupled with ATP synthesis. Oxidation of one FADH$_2$ in the electron transport chain

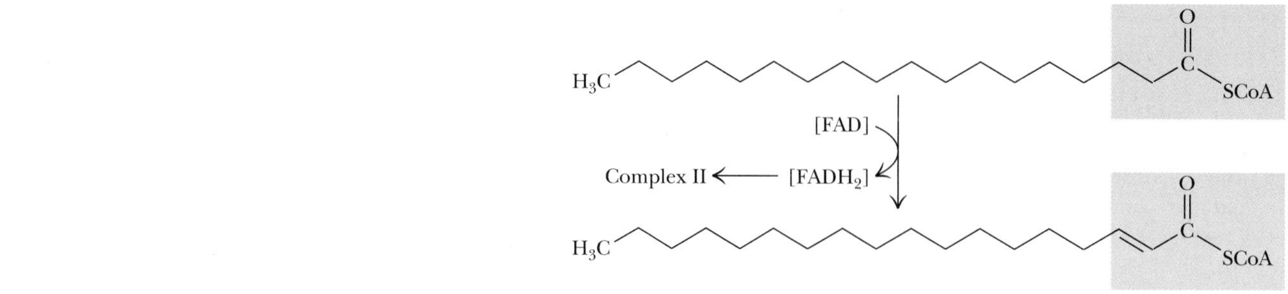

Figure 17.7 The fatty acyl-CoA dehydrogenase reaction, emphasizing that the reaction involves reduction of enzyme-bound FAD (indicated by brackets).

results in synthesis of approximately two molecules of ATP, compared with the approximately three ATPs produced by the oxidation of one NADH.

Other enzymes can also supply electrons to UQ, including mitochondrial *sn*-glycerophosphate dehydrogenase, an inner membrane-bound shuttle enzyme, and the fatty acyl-CoA dehydrogenases, three soluble matrix enzymes involved in fatty acid oxidation (Figure 17.7; see also Chapter 20). The path of electrons from succinate to UQ is shown in Figure 17.8.

17.6 Complex III: Coenzyme Q–Cytochrome c Reductase

In the third complex of the electron transport chain, reduced coenzyme Q (UQH$_2$) passes its electrons to cytochrome *c* via a unique redox pathway known as the **Q cycle.** *UQ–cytochrome c reductase (UQ–cyt c reductase),* as this complex is known, involves three different cytochromes and an Fe-S protein. In the cytochromes of these and similar complexes, the iron atom at the center of the porphyrin ring cycles between the reduced Fe^{2+} (ferrous) and oxidized Fe^{3+} (ferric) states.

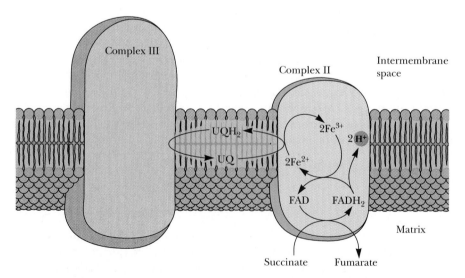

Figure 17.8 A scheme for electron flow in Complex II. Oxidation of succinate occurs with reduction of [FAD]. Electrons are then passed to Fe-S centers and then to coenzyme Q (UQ). Proton transport does not occur in this complex.

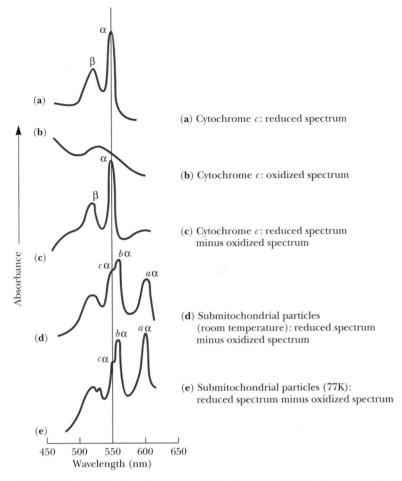

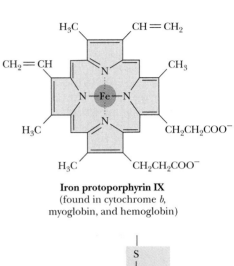

Iron protoporphyrin IX
(found in cytochrome *b*,
myoglobin, and hemoglobin)

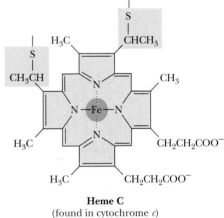

Heme C
(found in cytochrome *c*)

Figure 17.9 Typical visible absorption spectra of cytochromes. **(a)** Cytochrome c, reduced spectrum; **(b)** cytochrome c, oxidized spectrum; **(c)** the difference spectrum: (a) minus (b); **(d)** beef heart mitochondrial particles: room temperature difference (reduced minus oxidized) spectrum; **(e)** same as (d) but at liquid nitrogen temperature.

(a) Cytochrome *c*: reduced spectrum

(b) Cytochrome *c*: oxidized spectrum

(c) Cytochrome *c*: reduced spectrum minus oxidized spectrum

(d) Submitochondrial particles (room temperature): reduced spectrum minus oxidized spectrum

(e) Submitochondrial particles (77K): reduced spectrum minus oxidized spectrum

Cytochromes were first named and classified on the basis of their absorption spectra (Figure 17.9), which depend upon the structure and environment of their heme groups. The **b cytochromes** contain *iron–protoporphyrin IX* (Figure 17.10), the same heme found in hemoglobin and myoglobin. The **c cytochromes** contain *heme c*, derived from iron–protoporphyrin IX by the covalent attachment of cysteine residues from the associated protein. UQ–cyt *c* reductase contains a *b*-type cytochrome with two different heme sites (Figure 17.11) and one *c*-type cytochrome. (One other variation, heme *a*, contains a 15-carbon isoprenoid chain on a modified vinyl group, and a formyl group in place of one of the methyls [see Figure 17.10]. **Cytochrome a** is found in two forms in Complex IV of the electron transport chain, as we shall see.) The two hemes on the *b* cytochrome polypeptide in UQ–cyt *c* reductase are distinguished by their reduction potentials and the wavelength (λ_{max}) of the so-called **α band** (see Figure 17.9). One of these hemes, known as b_L or b_{566}, has a standard reduction potential, $\mathscr{E}_o'$, of -0.100 V and a wavelength of maximal absorbance (λ_{max}) of 566 nm. The other, known as b_H or b_{562}, has a standard reduction potential of $+0.050$ V and a λ_{max} of 562 nm. (*H* and *L* here refer to high and low reduction potential.)

The structure of the UQ–cyt *c* reductase, also known as the cytochrome bc_1 complex, has been determined by Johann Deisenhofer and his colleagues.

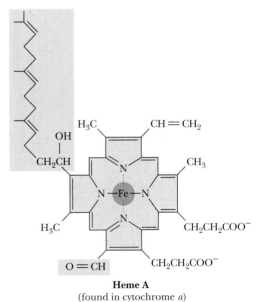

Heme A
(found in cytochrome *a*)

Figure 17.10 The structures of iron protoporphyrin IX, heme c, and heme a.

Figure 17.11 The structure of UQ–cyt *c* reductase, also known as the cytochrome bc_1 complex. The alpha helices of cytochrome *b* (blue-green) define the transmembrane domain of the protein. The bottom of the structure as shown extends approximately 75 Å into the mitochondrial matrix, and the top of the structure as shown extends about 38 Å into the intermembrane space. The functional core of the protein consists of cytochrome *b*, the Rieske iron–sulfur protein (yellow), and cytochrome c_1 (dark blue). The Q_P and Q_N sites where ubiquinone binds in the Q cycle are located at the interface between the two subunits of the dimer. Each monomer contributes one Q_P and one Q_N. *(Photograph kindly provided by Di Xia and Johann Deisenhofer [From Xia, D., et al., 1997. The crystal structure of the cytochrome bc_1 complex from bovine heart mitochondria. Science **277:**60–66.])*

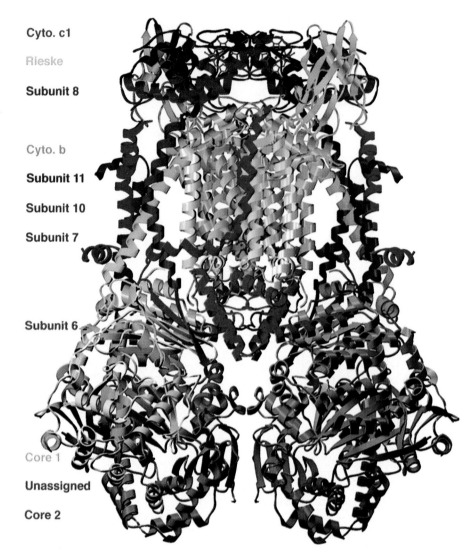

Cyto. c1
Rieske
Subunit 8

Cyto. b
Subunit 11
Subunit 10
Subunit 7

Subunit 6

Core 1
Unassigned
Core 2

The complex is a dimer, with each monomer consisting of 11 protein subunits and 2165 amino acid residues (monomer mass, 248 kD). The dimeric structure is pear-shaped and consists of a large domain that extends 75 Å into the mitochondrial matrix, a transmembrane domain consisting of 13 transmembrane α-helices in each monomer and a small domain that extends 38 Å into the intermembrane space (see Figure 17.11). Most of the **Rieske protein** (an Fe-S protein named for its discoverer) is mobile in the crystal (only 62 of 196 residues are shown in the structure in Figure 17.11), and Deisenhofer has postulated that mobility of this subunit could be required for electron transfer by this complex.

Complex III Drives Proton Transport

As with Complex I, passage of electrons through the Q cycle of Complex III is accompanied by proton transport across the inner mitochondrial membrane. The pathway for electrons in this system is shown in Figure 17.12. A large pool of UQ and UQH_2 exists in the inner mitochondrial membrane. The Q cycle is initiated when a molecule of UQH_2 from this pool diffuses to a site (called Q_P) on Complex III near the cytosolic face of the membrane.

Oxidation of this UQH_2 occurs in two steps. First, an electron from UQH_2 is transferred to the Rieske protein and then to cytochrome c_1. This releases two H^+ to the cytosol and leaves $UQ \cdot^-$, a semiquinone anion form of UQ, at the Q_P site. The second electron is then transferred to the b_L heme, converting $UQ \cdot^-$ to UQ. The Rieske protein and cytochrome c_1 are similar in structure; each has a globular domain and each is anchored to the inner membrane by a hydrophobic segment.

The electron on the b_L heme facing the cytosolic side of the membrane is now passed to the b_H heme on the matrix side of the membrane. The electron is then passed from b_H to a molecule of UQ at a second quinone-binding site, $\mathbf{Q_N}$, converting this UQ to $UQ \cdot^-$. The resulting $UQ \cdot^-$ remains firmly bound to the Q_N site. This completes the first half of the Q cycle (Figure 17.12a).

The second half of the cycle (Figure 17.12b) is similar to the first half. A second molecule of UQH_2 is oxidized at the Q_P site, and one electron is passed to cytochrome c_1, while the other is transferred to heme b_L and then to heme b_H. In this latter half of the Q cycle, however, the b_H electron is transferred to the semiquinone anion, $UQ \cdot^-$, at the Q_N site. With the addition of two H^+ from the mitochondrial matrix, this produces a molecule of UQH_2, which is released from the Q_N site and returns to the coenzyme Q pool, completing the Q cycle.

The Q Cycle Is an Unbalanced Proton Pump

Why has nature chosen this rather convoluted path for electrons in Complex III? First of all, Complex III takes up two protons on the matrix side of the

(a) First half of Q cycle

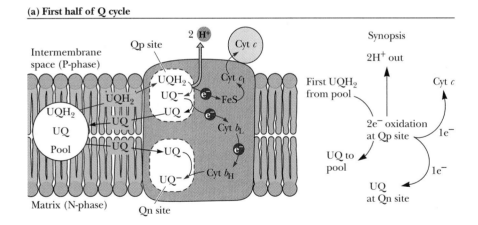

(b) Second half of Q cycle

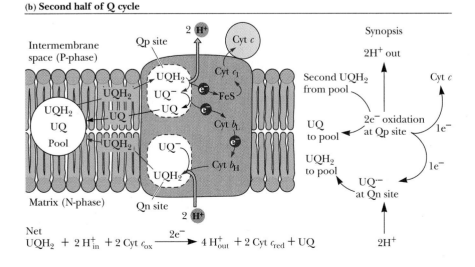

Figure 17.12 The Q cycle in mitochondria. **(a)** The electron transfer pathway following oxidation of the first UQH_2 at the Q_P site near the cytosolic face of the membrane. **(b)** The pathway following oxidation of a second UQH_2.

Cytochrome c Is a Mediator of Apoptosis (Programmed Cell Death)

Nearly all cells are programmed to die eventually. The process is termed **apoptosis** (where the second "p" is silent). Many cellular entities are involved in this process, including mitochondria. Acting on signals from molecules in the cytosol, transport channels open in the mitochondrial outer membrane, releasing cytochrome c, which then activates caspases—a family of proteases that contain a cysteine residue at their active site. Caspase activation triggers a series of proteolytic reactions that cause the eventual death of the cell.

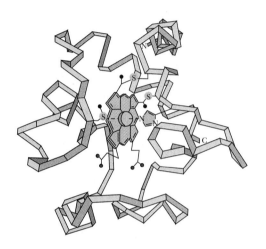

Figure 17.13 The structure of mitochondrial cytochrome c. The heme at the center of the structure is covalently linked to the protein via its two sulfur atoms (yellow). The iron in the porphyrin ring is coordinated both to a histidine nitrogen and to the sulfur atom of a methionine residue. Coordination with ligands in this manner on both sides of the porphyrin plane precludes the binding of oxygen and other ligands, a feature that distinguishes most cytochromes from hemoglobin (see Chapter 12).

inner membrane and releases four protons on the cytoplasmic side for each pair of electrons that passes through the Q cycle. The other significant feature of this mechanism is that it offers a convenient way for a two-electron carrier, UQH_2, to interact with the b_L and b_H hemes, the Rieske protein Fe-S cluster, and cytochrome c_1, all of which are one-electron carriers.

Cytochrome c Is a Mobile Electron Carrier

Electrons traversing Complex III are passed through cytochrome c_1 to cytochrome c. Cytochrome c is the only one of the cytochromes that is water-soluble. Its structure, determined by X-ray crystallography (Figure 17.13), is globular; the planar heme group lies near the center of the protein, surrounded predominantly by hydrophobic protein residues.

Cytochrome c, like UQ, is a mobile electron carrier. It associates loosely with the inner mitochondrial membrane (in the *intermembrane space* on the cytosolic side of the inner membrane) to acquire electrons from the Fe-S-cyt c_1 aggregate of Complex III, and then it migrates along the membrane surface in the reduced state, carrying electrons to *cytochrome c oxidase*, the fourth complex of the electron transport chain.

17.7 Complex IV: Cytochrome c Oxidase

Complex IV is called cytochrome c oxidase because it accepts electrons from cytochrome c and directs them to the four-electron reduction of O_2 to form H_2O:

$$4 \text{ cyt } c\,(Fe^{2+}) + 4\,H^+ + O_2 \longrightarrow 4 \text{ cyt } c\,(Fe^{3+}) + 2\,H_2O \qquad (17.21)$$

Thus, O_2 and cytochrome c oxidase are the final destination for the electrons derived from the oxidation of food materials. In concert with this process, cytochrome c oxidase also drives transport of protons across the inner mitochondrial membrane. These important functions are carried out by a transmembrane protein complex consisting of 13 subunits (see Table 17.2). The total mass of the protein in the complex is 204 kD. Subunits I through III, the largest ones, are encoded by mitochondrial DNA, synthesized in the mitochondrion, and inserted into the inner membrane from the matrix side. The smaller subunits are coded by nuclear DNA and synthesized in the cytosol. The essential Fe and Cu sites are contained entirely within the structures of subunits I, II, and III (Figure 17.14). None of the 10 nuclear DNA–derived subunits directly impinges on the essential metal sites. The implication is that subunits I to III actively participate in the events of electron transfer, but that

Figure 17.14 Molecular graphic image of subunits I, II, and III, the core of cytochrome c oxidase. Subunit I (green) coordinates hemes a and a_3 and Cu_B, with the iron atom of a_3 and Cu_B forming a binuclear center. Subunit II (gold) contains the binuclear Cu_A center—two Cu atoms coordinated by the—SH groups of two cysteines. Subunit I is cylindrical and consists of 12 transmembrane helices, without any significant extramembrane parts. Hemes a and a_3, which lie perpendicular to the membrane plane, are cradled by the helices of subunit I. Subunits II and III lie on opposite sides of subunit I and do not contact each other. Subunit II has an extramembrane domain on the cytosolic face of the mitochondrial membrane. This domain consists of a 10-strand β-barrel that holds Cu_A 7 Å from the nearest surface atom of the subunit. Subunit III consists of seven transmembrane helices with no significant extramembrane domains.

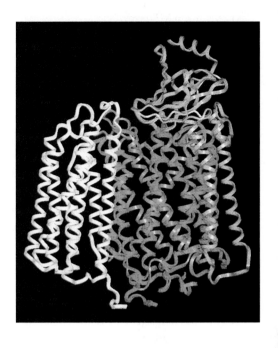

the other 10 subunits play regulatory roles in this process. Figure 17.15 presents a molecular graphic image of the entire cytochrome c oxidase.

Electron Transfer in Complex IV Involves Two Hemes and Two Copper Sites

Cytochrome c oxidase contains two heme centers (cytochromes a and a_3) as well as two copper atoms (Figure 17.16). The copper sites, Cu_A and Cu_B, are associated with cytochromes a and a_3, respectively. The copper sites participate in electron transfer by cycling between the reduced (*cuprous*) Cu^+ state and the oxidized (*cupric*) Cu^{2+} state. (Remember, the cytochromes and copper sites are one-electron transfer agents.) Reduction of one oxygen molecule requires passage of four electrons through these carriers—one at a time (see Figure 17.16).

Complex IV Also Transports Protons Across the Inner Mitochondrial Membrane

The reduction of oxygen in Complex IV is accompanied by transport of protons across the inner mitochondrial membrane. Transfer of four electrons through this complex drives the uptake of four protons on the matrix side and the delivery of two protons on the cytosolic side of the membrane (see Figure 17.16).

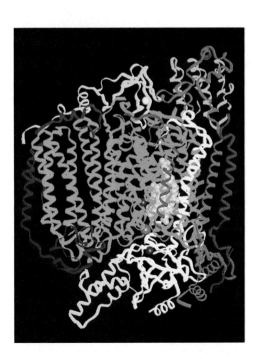

Figure 17.15 Molecular graphic image of cytochrome c oxidase. Seven of the 10 nuclear DNA-derived subunits (IV, VIa, VIc, VIIa, VIIb, VIIc, and VIII) possess transmembrane segments. Three (Va, Vb, and VIb) do not.

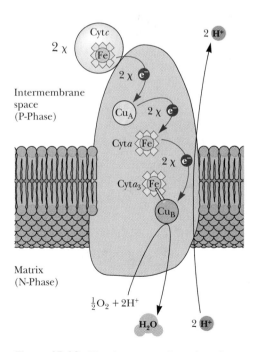

Figure 17.16 The electron transfer pathway for cytochrome oxidase. Cytochrome c binds on the cytosolic side, transferring electrons through the copper and heme centers to reduce O_2 on the matrix side of the membrane.

Independence of the Four Complexes

It should be emphasized here that the four major complexes of the electron transport chain operate quite independently in the inner mitochondrial membrane. Each is a multiprotein aggregate maintained by numerous strong associations between peptides of the complex, but there is no evidence that the complexes associate with one another in the membrane.

A Dynamic Model of Electron Transport

The model that emerges for electron transport is shown in Figure 17.17. The four complexes are independently mobile in the membrane. Coenzyme Q collects electrons from NADH–UQ reductase and succinate–UQ reductase and delivers them (by diffusion through the membrane core) to UQ–cyt c reductase. Cytochrome c is water-soluble and moves freely, carrying electrons from UQ–cyt c reductase to cytochrome c oxidase. In the process of these electron transfers, protons are driven across the inner membrane (from the matrix side to the intermembrane space). The proton gradient generated by electron transport represents an enormous source of potential energy. As seen in the next section, this potential energy is used to synthesize ATP as protons flow back into the matrix.

Peter Mitchell's Chemiosmotic Hypothesis

In 1961, Peter Mitchell, a British biochemist, proposed that the energy derived from electron transport is stored in a proton gradient across the inner mito-

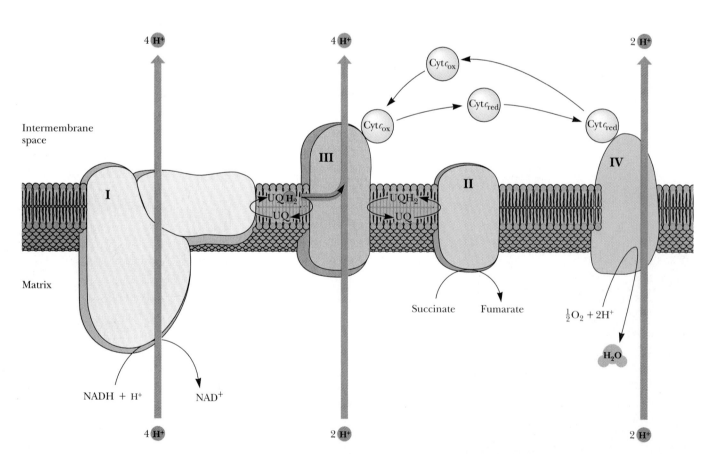

Figure 17.17 A model for electron transport and proton translocation in the mitochondrial inner membrane. UQ/UQH$_2$ and cytochrome c are mobile electron carriers and function by transferring electrons between the complexes. Proton transport is driven by Complexes I, III, and IV as indicated.

chondrial membrane, and that this energy—termed an **electrochemical potential**—drives the synthesis of ATP in cells. The proposal became known as **Mitchell's chemiosmotic hypothesis.** In this model, as protons are driven out of the matrix, the matrix pH rises, and the matrix becomes negatively charged with respect to the cytosol (Figure 17.18). Proton pumping thus creates a pH gradient as well as an electrical potential across the inner membrane, both of which tend to attract protons back into the matrix from the cytosol. Flow of protons down this electrochemical gradient, an energetically favorable process, then drives the synthesis of ATP.

The ratio of protons transported per pair of electrons passed through the chain—the so-called **$H^+/2\ e^-$ ratio**—has been an object of great interest for many years. Nevertheless, the ratio has remained extremely difficult to determine. The consensus estimate for the electron transport pathway from succinate to O_2 is $6\ H^+/2\ e^-$. The ratio for Complex I by itself remains uncertain, but estimates place it as high as $4\ H^+/2\ e^-$. On the basis of this value, the stoichiometry of transport for the pathway from NADH to O_2 is $10\ H^+/2\ e^-$. Although this value is assumed in Figure 17.17, it is important to realize that this consensus, drawn from many experiments, is not yet proven.

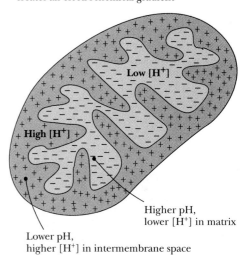

Electron transport drives H^+ out and creates an electrochemical gradient

Low [H⁺]

High [H⁺]

Higher pH, lower [H⁺] in matrix

Lower pH, higher [H⁺] in intermembrane space

Figure 17.18 The proton and electrochemical gradients existing across the inner mitochondrial membrane. The electrochemical gradient is generated by the transport of protons across the membrane.

17.8 The Thermodynamic View of Chemiosmotic Coupling

Peter Mitchell's chemiosmotic hypothesis revolutionized our thinking about the energy coupling that drives ATP synthesis by means of an electrochemical gradient. How much energy is stored in this electrochemical gradient? For the transfer of protons across the inner membrane (from inside [matrix] to outside), we could write

$$H^+_{in} \longrightarrow H^+_{out} \qquad (17.22)$$

The free energy difference for protons across the inner mitochondrial membrane includes a term for the concentration difference and a term for the electrical potential. This is expressed as

$$\Delta G = RT \ln\left(\frac{[c_2]}{[c_1]}\right) + Z\mathscr{F}\Delta\psi \qquad (17.23)$$

where c_1 and c_2 are the proton concentrations on the two sides of the membrane, Z is the charge on a proton, $\mathscr{F}$ is Faraday's constant, and $\Delta\psi$ is the potential difference across the membrane. For the case at hand, this equation becomes

$$\Delta G = RT \ln \frac{[H^+_{out}]}{[H^+_{in}]} + \mathscr{F}\Delta\psi \qquad (17.24)$$

In terms of the matrix and cytoplasm pH values, the free energy difference is

$$\Delta G = -2.303RT(pH_{out} - pH_{in}) + \mathscr{F}\Delta\psi \qquad (17.25)$$

Reported values for $\Delta\psi$ and ΔpH vary, but the membrane potential is always found to be positive outside and negative inside, and the pH is always more acidic outside (more basic inside). Taking typical values of $\Delta\psi = 0.18$ V and ΔpH = 1 unit, the free energy change associated with the movement of one mole of protons from inside to outside is

$$\Delta G = 2.3RT + \mathscr{F}(0.18\ \text{V}) \qquad (17.26)$$

With $\mathscr{F} = 96.485$ kJ/V · mol, the value of ΔG at 37°C is

$$\Delta G = 5.9\ \text{kJ} + 17.4\ \text{kJ} = +23.3\ \text{kJ} \qquad (17.27)$$

which is the free energy change for movement of a mole of protons across a typical inner membrane. Note that the free energy terms for *both* the pH difference and the potential difference are unfavorable for the outward transport of protons, with the latter term making the greater contribution. On the other hand, the ΔG for inward flow of protons is -23.3 kJ/mol. It is this energy that drives the synthesis of ATP, in accord with Mitchell's model. Peter Mitchell was awarded the Nobel Prize in chemistry in 1978.

17.9 ATP Synthase: A Remarkable Molecular Motor

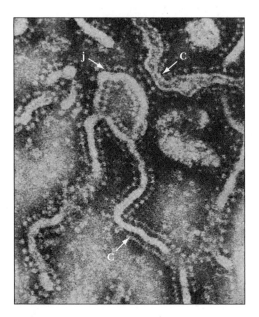

Figure 17.19 Electron micrograph of submitochondrial particles showing the 8.5-nm projections or particles on the inner membrane, eventually shown to be the F_1 component of the ATP synthase. *(Parsons, D. F., 1963. Science **140**:985.)*

The mitochondrial complex that carries out ATP synthesis is called $\mathbf{F_1F_0}$**–ATP synthase** or sometimes $\mathbf{F_1F_0}$**–ATPase** (for the reverse reaction it catalyzes). ATP synthase was observed in early electron micrographs of submitochondrial particles (prepared by sonication of inner membrane preparations) as round, 8.5-nm-diameter projections or particles on the inner membrane (Figure 17.19). In micrographs of native mitochondria, the projections appear on the matrix-facing surface of the inner membrane. Mild agitation removes the particles from isolated membrane preparations, and the isolated spherical particles (the F_1 part of the complex) catalyze ATP hydrolysis, the reverse reaction of the ATP synthase.

ATP Synthase Consists of Two Complexes—F_1 and F_0

As noted, the spheres observed in electron micrographs make up the $\mathbf{F_1}$ **unit,** which catalyzes ATP synthesis. These F_1 spheres are attached to an integral membrane protein aggregate called the $\mathbf{F_0}$ **unit.** F_1 consists of five polypeptide chains named $\alpha, \beta, \gamma, \delta,$ and $\varepsilon,$ with a subunit stoichiometry $\alpha_3\beta_3\gamma\delta\varepsilon$ (Table 17.3). F_0 consists of three hydrophobic subunits denoted by a, b, and c, with an apparent stoichiometry of $a_1b_2c_{10}$. F_0 forms the transmembrane pore or channel through which protons move to drive ATP synthesis. The $\alpha, \beta, \gamma, \delta,$ and ε subunits of F_1 contain 510, 482, 272, 146, and 50 amino acids, respectively, with a total molecular mass for F_1 of 371 kD. The α and β subunits are homologous, and each binds a single adenine nucleotide. The β subunits have the catalytic sites for ATP synthesis. The ATP-binding sites in the α subunits are not essential to ATP synthesis.

John Walker and his colleagues have determined the structures of the bovine F_1 complex and the F_1F_0 complex from yeast (Figure 17.20). The F_1-ATPase is an inherently asymmetrical structure, with the three β subunits having three different conformations. In the bovine structure solved by Walker, one of the β-subunit ATP sites contains AMP–PNP (a nonhydrolyzable analog

Table 17.3 *Escherichia coli* F_1F_0 ATP Synthase Subunit Organization

Complex	Protein Subunit	Mass (kD)	Stoichiometry
F_1	α	55	3
	β	52	3
	γ	30	1
	δ	15	1
	ε	5.6	1
F_0	a	30	1
	b	17	2
	c	8	10

Figure 17.20 Structure of bovine F_1-ATP synthase (*a* and *b*) and yeast F_1F_0-ATP synthase (*c* and *d*). **(a)** Side view and **(b)** top view of the bovine F_1-ATP synthase showing the individual component peptides. The α-subunits are yellow; the β-subunits are blue. One β-subunit has AMP-PNP bound (red), the second β-subunit has ADP (purple), and the third has P_i (white) bound. (The adenine nucleotides bound to the α-subunits are not shown.)

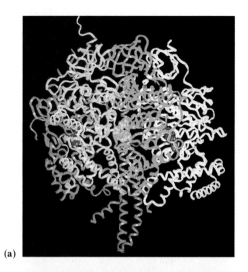

(a)

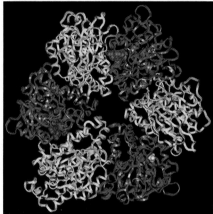

(b)

of ATP), and another contains ADP, with the third site empty. This structure confirms the predictions made in the **binding change mechanism** for ATP synthesis proposed by Paul Boyer, in which each of the three reaction sites cycle in turn through the three intermediate states of ATP synthesis (take a look at Figure 17.22 on page 540).

How might such cycling occur? Important clues have emerged from several experiments that show that the γ subunit rotates with respect to the $\alpha\beta$ complex. How such rotation might be linked to transmembrane proton flow and ATP synthesis is shown in Figure 17.21. In this model, based on the yeast F_1F_0 structure, the $10c$ subunits of F_0 are arranged in a ring. Each c subunit consists of a pair of antiparallel transmembrane helices with a short hairpin loop on the cytosolic side of the membrane. The ring of c subunits forms a **rotor** that turns with respect to the a subunit, a **stator** consisting of five transmembrane α-helices with proton access channels on either side of the membrane. The γ subunit serves as the link between F_1 and F_0. Several experiments have shown

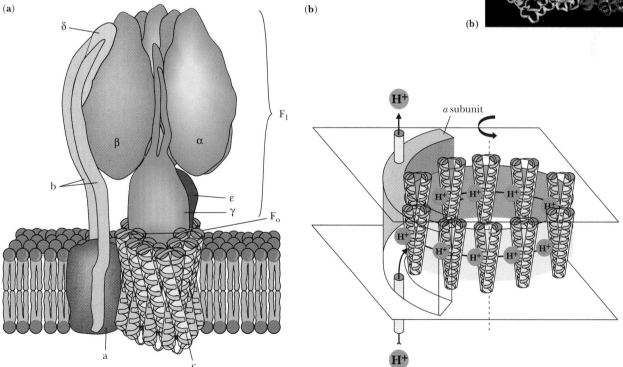

Figure 17.21 A model of the F_1 and F_0 components of the ATP synthase, a rotating molecular motor. **(a)** The a, b, α, β, and δ subunits constitute the stator of the motor, and the c, γ, and ε subunits form the rotor. Flow of protons through the structure turns the rotor **(b)** and drives the cycle of conformational changes in α and β that synthesize ATP.

 See page 166

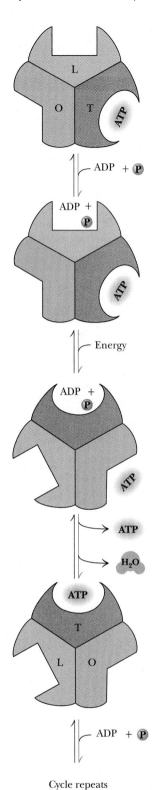

ADP + Ⓟ

Energy

ATP

H₂O

ADP + Ⓟ

Cycle repeats

◄ **Figure 17.22**   The binding change mechanism for ATP synthesis by ATP synthase. This model assumes that F_1 has three interacting and conformationally distinct active sites. The open (O) conformation is inactive and has a low affinity for ligands; the L conformation (with "loose" affinity for ligands) is also inactive; the tight (T) conformation is active and has a high affinity for ligands. Synthesis of ATP is initiated (step 1) by binding of ADP and P_i to an L site. In the second step, an energy-driven conformational change converts the L site to a T conformation and also converts T to O and O to L. In the third step, ATP is synthesized at the T site and released from the O site. Two more such steps complete a cycle, produce two more ATPs, and return the enzyme to its original state. Thus, one full cycle yields three ATP, one per site.

that γ rotates within the $(\alpha\beta)_3$ complex during ATP synthesis. Because γ is anchored to the c subunit rotor, the c rotor-γ complex rotates together, relative to the $(\alpha\beta)_3$ complex. Subunit b possesses a single transmembrane segment and a long hydrophilic head domain, and the complete stator consists of the b subunits anchored at one end to the a subunit and linked at the other end to the $(\alpha\beta)_3$ complex via the δ subunit, as shown in Figure 17.21.

What, then, is the mechanism for rotation? The c rotor subunits each carry an essential residue, Asp^{61}. (Changing this residue to Asn abolishes ATP synthase activity.) Rotation of the c rotor relative to the stator depends upon neutralization of the negative charge on each c subunit Asp^{61} as the rotor turns. Protons taken up from the cytosol by one of the proton access channels in subunit a protonate Asp^{61} and then ride the rotor until they reach the other proton access channel on a, from which they would be released into the matrix. Such rotation causes the γ subunit to turn relative to the three β-subunit nucleotide sites of F_1, changing the conformation of each in sequence, so that ADP is first bound, then phosphorylated, then released, according to Boyer's binding change mechanism. Thus, conformational changes in the β subunits are the direct source of energy for ATP synthesis.

In summary, the energy provided by electron transport creates a proton gradient that drives enzyme conformational changes resulting in the binding of substrates on ATP synthase, ATP synthesis, and the release of products. The mechanism involves catalytic cooperativity between three interacting sites (Figure 17.22). Paul Boyer and John Walker shared the 1997 Nobel Prize in chemistry for their work on the structure and mechanism of ATP synthase.

Racker and Stoeckenius Confirmed the Mitchell Model in a Reconstitution Experiment

When Mitchell first described his chemiosmotic hypothesis in 1961, little evidence existed to support it, and it was met with considerable skepticism by the scientific community. Eventually, however, considerable evidence accumulated to support this model. It is now clear that the electron transport chain generates a proton gradient, and careful measurements have shown that ATP is synthesized when a pH gradient is applied to mitochondria that cannot carry out electron transport. Even more relevant is a simple but crucial experiment reported in 1974 by Efraim Racker and Walther Stoeckenius, which provided specific confirmation of the Mitchell hypothesis. In this experiment, the bovine mitochondrial ATP synthase was reconstituted in simple lipid vesicles with **bacteriorhodopsin,** a light-driven proton pump from *Halobacterium halobium.* As shown in Figure 17.23, upon illumination, bacteriorhodopsin pumped protons into these vesicles, and the resulting proton gradient was sufficient to drive ATP synthesis by the ATP synthase. Because the only two kinds of proteins present were one that produced a proton gradient and one that used such a gradient to make ATP, this experiment essentially verified Mitchell's chemiosmotic hypothesis.

17.10 Inhibitors of Oxidative Phosphorylation

The unique properties and actions of an inhibitory substance can often help to identify aspects of an enzyme mechanism. Many details of electron transport and oxidative phosphorylation mechanisms have been gained from studying the effects of particular inhibitors. Figure 17.24 presents the structures of some electron transport and oxidative phosphorylation inhibitors. The sites of inhibition by these agents are indicated in Figure 17.25.

Inhibitors of Complexes I, II, and III Block Electron Transport

Rotenone is a common insecticide that strongly inhibits Complex I (NADH–UQ reductase). Rotenone is obtained from the roots of several species of plants. Tribes in certain parts of the world have made a practice of beating the roots of trees along riverbanks to release rotenone into the water, where it paralyzes fish and makes them easy prey. (See *Human Biochemistry*, page 529.) Ptericidin, Amytal, and other barbiturates, mercurial agents, and the widely prescribed painkiller Demerol also exert inhibitory actions on this enzyme complex. All these substances appear to inhibit reduction of coenzyme Q and the oxidation of the Fe-S clusters of NADH–UQ reductase.

2-Thenoyltrifluoroacetone and carboxin and its derivatives specifically block Complex II, the succinate–UQ reductase. Antimycin, an antibiotic produced by *Streptomyces griseus*, inhibits the UQ–cytochrome *c* reductase by

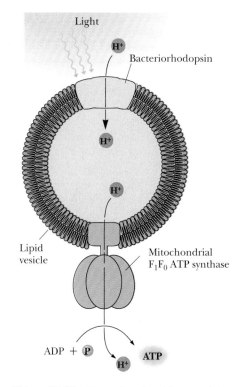

Figure 17.23 Reconstituted vesicles containing ATP synthase and bacteriorhodopsin were used by Stoeckenius and Racker to confirm the Mitchell chemiosmotic hypothesis.

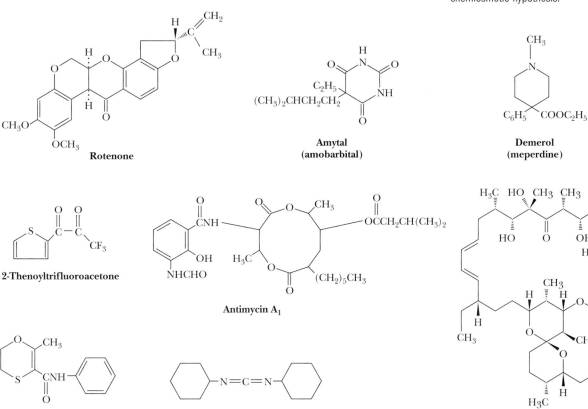

Figure 17.24 The structures of several inhibitors of electron transport and oxidative phosphorylation.

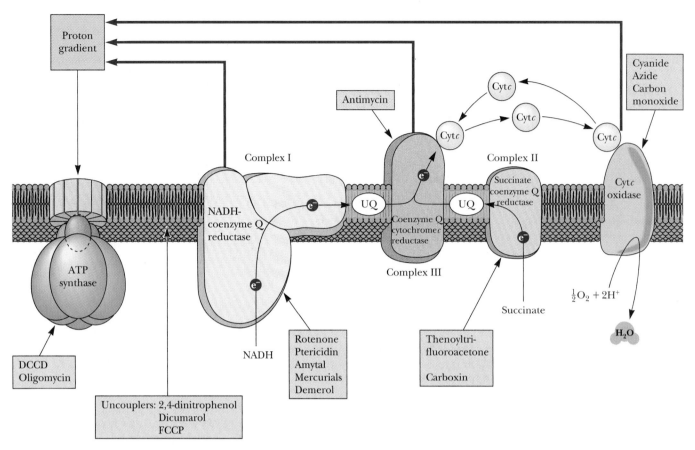

Figure 17.25 The sites of action of several inhibitors of electron transport and oxidative phosphorylation.

blocking electron transfer between b_H and coenzyme Q in the Q_n site. Myxothiazol inhibits the same complex by acting at the Q_p site.

Cyanide, Azide, and Carbon Monoxide Inhibit Complex IV

Complex IV, the cytochrome c oxidase, is specifically inhibited by cyanide (CN^-), azide (N_3^-), and carbon monoxide (CO). Cyanide and azide bind tightly to the ferric form of cytochrome a_3, whereas carbon monoxide binds only to the ferrous form. The inhibitory actions of cyanide and azide at this site are very potent, whereas the principal toxicity of carbon monoxide arises from its affinity for the iron of hemoglobin. Herein lies an important distinction between the poisonous effects of cyanide and carbon monoxide. Because animals (including humans) carry many, many hemoglobin molecules, they must inhale a large quantity of carbon monoxide to die from it. These same organisms, however, possess comparatively few molecules of cytochrome a_3. Consequently, a limited exposure to cyanide can be lethal. The sudden action of cyanide attests to the organism's constant and immediate need for the energy supplied by electron transport.

Oligomycin and DCCD Are ATP Synthase Inhibitors

Inhibitors of ATP synthase include dicyclohexylcarbodiimide (DCCD) and oligomycin (see Figure 17.24). DCCD bonds covalently to carboxyl groups in hydrophobic domains of proteins in general, and to a glutamic acid residue in the c subunit of F_0, the proteolipid forming the proton channel of the ATP synthase, in particular. If the c subunit is labeled with DCCD, proton flow through F_0 is blocked and ATP synthase activity is inhibited. Likewise, oligomycin acts directly on the ATP synthase. By binding to a subunit of F_0, oligomycin also blocks the movement of protons through F_0.

17.11 Uncouplers Disrupt the Coupling of Electron Transport and ATP Synthase

Another important class of reagents affects ATP synthesis, but in a manner that does not involve direct binding to any of the proteins of the electron transport chain or the F_1F_0-ATP synthase. These agents are known as **uncouplers** because they disrupt the tight coupling between electron transport and ATP synthesis. Uncouplers act by dissipating the proton gradient across the inner mitochon-

HUMAN BIOCHEMISTRY

Endogenous Uncouplers Enable Organisms to Generate Heat

Certain cold-adapted animals, hibernating animals, and newborn animals generate large amounts of heat by uncoupling oxidative phosphorylation. These organisms have a type of fat known as brown adipose tissue, so called for the color imparted by the many mitochondria this adipose tissue contains. The inner membrane of brown adipose tissue mitochondria contains large amounts of an endogenous protein called **thermogenin** (literally, "heat maker"), or **uncoupling protein 1 (UCP1).** UCP1 creates a passive proton channel through which protons flow from the cytosol to the matrix. Mice that lack UCP1 cannot maintain their body temperature in cold conditions, whereas normal animals produce larger amounts of UCP1 when they are cold-adapted. Two other mitochondrial proteins, designated UCP2 and UCP3, have sequences similar to UCP1.

Since the function of UCP1 is so closely linked to energy utilization, there has been great interest in the possible roles of UCP1, UCP2, and UCP3 as metabolic regulators and as factors in obesity.

Under fasting conditions, expression of UCP1 mRNA is decreased, but expression of UCP2 and UCP3 is increased. There is no indication, however, that UCP2 and UCP3 actually function as uncouplers. There has also been interest in the possible roles of UCP2 and UCP3 in the development of obesity, expecially since the genes for these proteins lie on chromosome 7 of the mouse, close to other genes linked to obesity.

Certain plants use the heat of uncoupled proton transport for a special purpose. Skunk cabbage and philodendron contain floral spikes that are maintained as much as 20 degrees above ambient temperature in this way. The warmth of the spikes serves to vaporize odiferous molecules, which attract insects that fertilize the flowers. Red tomatoes have very small mitochondrial membrane proton gradients compared with green tomatoes—evidence that uncouplers are more active in red tomatoes, whose energy needs are less.

Grizzly bear
(© Ralph A. Clevenger/CORBIS)

Skunk cabbage
(© Deni Brown OSF/
Animals Animals/
Earth Scenes)

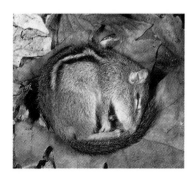

Chipmunk
(© Breck P. Kent/Animals Animals)

Philodendron
(© Deni Brown OSF/Animals Animals/Earth
Scenes)

Dinitrophenol

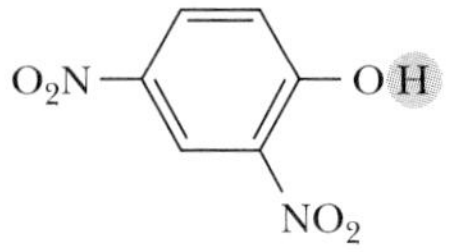

Dicumarol

Carbonyl cyanide-p-trifluoro-methoxyphenyl hydrazone
—best known as **FCCP**; for **F**luoro **C**arbonyl **C**yanide **P**henylhydrazone

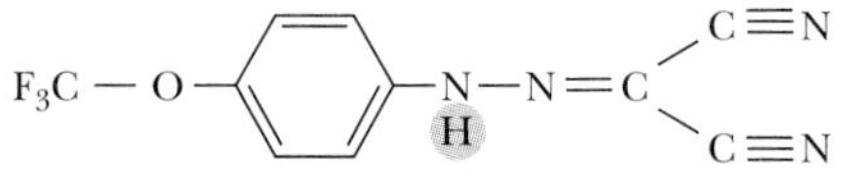

Figure 17.26 Structures of several uncouplers, molecules that dissipate the proton gradient across the inner mitochondrial membrane and thereby destroy the tight coupling between electron transport and ATP synthesis.

drial membrane created by the electron transport system. Typical examples include 2,4-dinitrophenol, dicumarol, and carbonyl cyanide-*p*-trifluoro-methoxyphenyl hydrazone (perhaps better known as fluorocarbonyl-cyanide phenylhydrazone, or FCCP) (Figure 17.26). These compounds share two common features: hydrophobic character and a dissociable proton. As uncouplers, they function by carrying protons across the inner membrane. Their tendency is to acquire protons on the cytosolic surface of the membrane (where the proton concentration is high) and carry them to the matrix side, thereby destroying the proton gradient that couples electron transport and the ATP synthase. In mitochondria treated with uncouplers, electron transport continues and protons are driven out through the inner membrane. However, they leak back in so rapidly via the uncouplers that ATP synthesis does not occur. Instead, the energy released in electron transport is dissipated as heat.

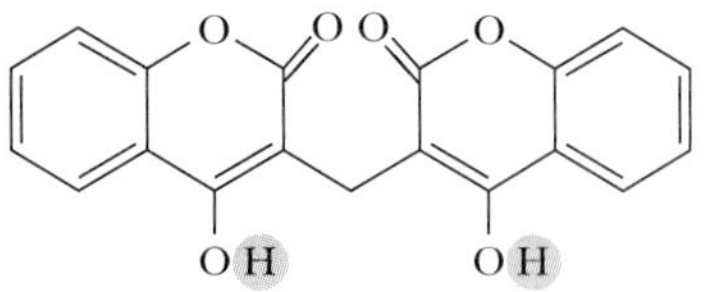

17.12 ATP Exits the Mitochondria via an ATP-ADP Translocase

ATP, the cellular energy currency, must exit the mitochondria to carry energy throughout the cell, and ADP must be brought into the mitochondria for reprocessing. Neither of these processes occurs readily because the highly charged ATP and ADP molecules do not readily cross biological membranes. Instead, these processes are mediated by a transport system, the **ATP–ADP translocase.** This protein tightly couples the exit of ATP with the entry of ADP so that the mitochondrial nucleotide levels remain approximately constant. For each ATP transported out, one ADP is transported into the matrix. The translocase, which accounts for approximately 14% of the total mitochondrial membrane protein, is a homodimer of 30-kD subunits. Transport occurs via a single nucleotide-binding site, which alternately faces the matrix and the cytosol (Figure 17.27). ATP binds on the matrix side, the site reorients to face the cytosol, and ATP is exchanged for ADP, with subsequent reorientation back to the matrix face of the inner membrane.

Outward Movement of ATP Is Favored over Outward ADP Movement

The charge on ATP at pH 7.2 or so is about -4, and the charge on ADP at the same pH is about -3. Thus, net exchange of an ATP (out) for an ADP (in) results in the net movement of one negative charge from the matrix to the cy-

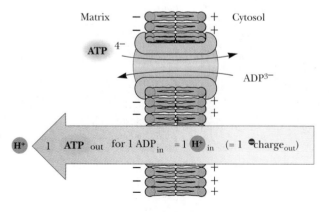

Figure 17.27 Outward transport of ATP (via the ATP/ADP translocase) is favored by the membrane electrochemical potential.

tosol. (This process is equivalent to the movement of a proton from the cytosol to the matrix.) Recall that the inner membrane is positive outside (see Figure 17.18), and it becomes clear that outward movement of ATP is favored over outward ADP transport, ensuring that ATP will be transported out (see Figure 17.27). Inward movement of ADP is favored over inward movement of ATP for the same reason. Thus, the membrane electrochemical potential itself controls the specificity of the ATP–ADP translocase. However, the electrochemical potential is diminished by the ATP–ADP translocase cycle and therefore operates with an energy cost to the cell.

What is the cost of ATP–ADP exchange relative to the energy cost of ATP synthesis itself? We already noted that moving one ATP out and one ADP in is the equivalent of one proton moving from the cytosol to the matrix. *Since synthesis of an ATP by F_1F_0–ATP synthase results from the movement of approximately three protons from the cytosol into the matrix through F_0, approximately four protons are transported into the matrix per ATP synthesized.* Thus, approximately one-fourth of the energy derived from the respiratory chain (electron transport and oxidative phosphorylation) is expended as the electrochemical energy devoted to mitochondrial ATP–ADP transport.

17.13 What Is the P/O Ratio for Mitochondrial Electron Transport and Oxidative Phosphorylation?

The **P/O ratio** is the number of molecules of ATP formed in oxidative phosphorylation per two electrons flowing through a defined segment of the electron transport chain. In spite of intense study of this ratio, its actual value remains a matter of contention. If we accept the value of $10 \, H^+$ transported out of the matrix per $2 \, e^-$ passed from NADH to O_2 through the electron transport chain, and also agree that $4 \, H^+$ are transported into the matrix per ATP synthesized and translocated, then the mitochondrial P/O ratio is 10/4, or 2.5, for the case of electrons entering the electron transport chain as NADH. This is somewhat lower than earlier estimates, which placed the P/O ratio at 3 for mitochondrial oxidation of NADH. For the portion of the chain from succinate to O_2, the $H^+/2 \, e^-$ ratio is 6 (as noted earlier), and the P/O ratio in this case would be 6/4, or 1.5; earlier estimates placed this number at 2. Experimental measurements of P/O ratios for these two cases tend to validate the values of 2.5 and 1.5. At some point, as we learn more about these complex coupled processes, it may be necessary to reassess the numbers.

$$\left(\frac{1 \, \text{ATP}}{4 \, H^+}\right)\left(\frac{10 \, H^+}{2 \, e^- \, [\text{NADH} \rightarrow \frac{1}{2}O_2]}\right) = \frac{10}{4} = \frac{P}{O}$$

17.14 Shuttle Systems Feed the Electrons of Cytosolic NADH into Electron Transport

Most of the NADH used in electron transport is produced in the mitochondrial matrix, an appropriate site because NADH is oxidized by Complex I on the matrix side of the inner membrane. Furthermore, the inner mitochondrial membrane is impermeable to NADH. Recall, however, that NADH is produced in glycolysis by glyceraldehyde-3-P dehydrogenase in the cytosol. If this NADH were not oxidized to regenerate NAD^+, the glycolytic pathway would cease to function due to NAD^+ limitation. Eukaryotic cells have a number of shuttle systems that harvest the electrons of cytosolic NADH for delivery to mitochondria without actually transporting NADH across the inner membrane (Figures 17.28 and 17.29).

Figure 17.28 The glycerophosphate shuttle (also known as the glycerol phosphate shuttle) couples the cytosolic oxidation of NADH with mitochondrial reduction of [FAD].

See pages 173–178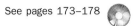

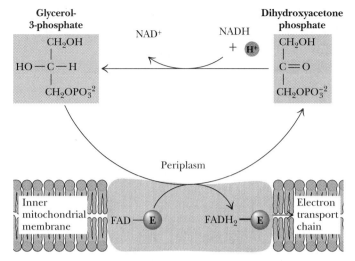

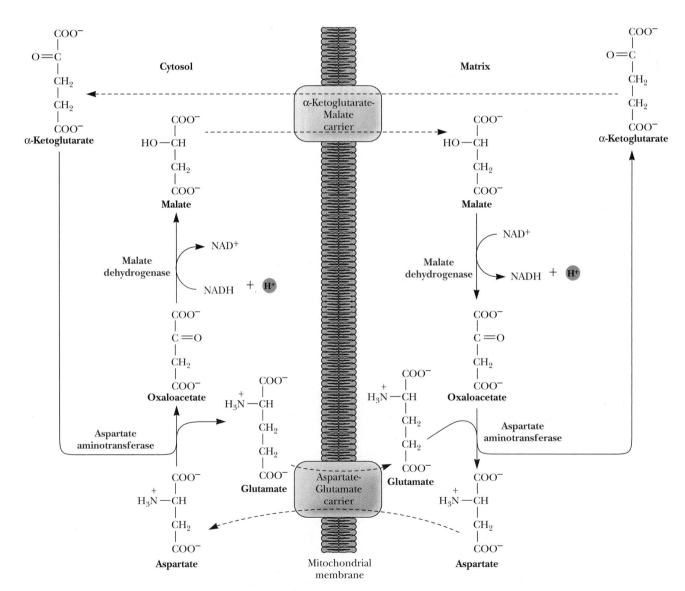

Figure 17.29 The malate (oxaloacetate)–aspartate shuttle, which operates across the inner mitochondrial membrane.

The Glycerophosphate Shuttle Ensures Efficient Use of Cytosolic NADH

In the **glycerophosphate shuttle,** two different **glycerophosphate dehydrogenases**—one in the cytoplasm and one on the outer face of the mitochondrial inner membrane—work together to carry electrons into the mitochondrial matrix (see Figure 17.28). NADH produced in the cytosol transfers its electrons to dihydroxyacetone phosphate, thus reducing it to *glycerol-3-phosphate.* This metabolite is reoxidized by the FAD-dependent mitochondrial membrane enzyme to re-form dihydroxyacetone phosphate and enzyme-bound $FADH_2$. The two electrons of $[FADH_2]$ are passed directly to UQ, forming UQH_2. Thus, via this shuttle, cytosolic NADH can be used to produce mitochondrial $[FADH_2]$ and, subsequently, UQH_2. Cytosolic NADH oxidized via this shuttle route yields only 1.5 molecules of ATP. The cell "pays" with a potential ATP molecule for the convenience of getting cytosolic NADH into the mitochondria. Although this may seem wasteful, there is an important payoff. The glycerophosphate shuttle is essentially irreversible, and even when NADH levels are very low relative to NAD^+, the cycle operates effectively.

The Malate–Aspartate Shuttle Is Reversible

The second electron shuttle system, called the **malate–aspartate shuttle,** is shown in Figure 17.29. Oxaloacetate is reduced in the cytosol, acquiring the electrons of NADH (which is oxidized to NAD^+). Malate is transported across the inner membrane, where it is reoxidized by malate dehydrogenase, converting NAD^+ to NADH in the matrix. This mitochondrial NADH readily enters the electron transport chain. The oxaloacetate produced in this reaction cannot cross the inner membrane and must be transaminated to form aspartate, which can be transported across the membrane to the cytosolic side. Transamination in the cytosol recycles aspartate back to oxaloacetate. In contrast to the glycerol phosphate shuttle, the malate–aspartate cycle is reversible, and it operates as shown in Figure 17.29 only if the $NADH/NAD^+$ ratio in the cytosol is higher than the ratio in the matrix. Because this shuttle produces NADH in the matrix, 2.5 ATPs are recovered per NADH.

The Net Yield of ATP from Glucose Oxidation Depends on the Shuttle Used

The complete route for the conversion of the metabolic energy of glucose to ATP has now been described in Chapters 15 through 17. Assuming appropriate P/O ratios, the number of ATP molecules produced by the complete oxidation of a molecule of glucose can be estimated. Keeping in mind that P/O ratios must be viewed as approximate, for all the reasons previously cited, we will assume the values of 2.5 and 1.5 for the mitochondrial oxidation of NADH and succinate, respectively. In eukaryotic cells, the combined pathways of glycolysis, the TCA cycle, electron transport, and oxidative phosphorylation then yield a net of approximately 30 to 32 molecules of ATP per molecule of glucose oxidized, depending on the shuttle route employed (Table 17.4).

The net stoichiometric equation for the oxidation of glucose, using the glycerophosphate shuttle, is

$$\text{Glucose} + 6\,O_2 + {\sim}30\,\text{ADP} + {\sim}30\,P_i \longrightarrow \\ 6\,CO_2 + {\sim}30\,\text{ATP} + {\sim}36\,H_2O \qquad \textbf{(17.28)}$$

Because the 2 NADH formed in glycolysis are "transported" by the glycerophosphate shuttle in this case, they each yield only 1.5 ATP, as already described. On the other hand, if these 2 NADH take part in the malate–aspartate shuttle, each yields 2.5 ATP, giving a total (in this case) of 32 ATP formed per glucose oxidized. Most of the ATP—26 out of 30 or 28 out of 32—

◼ Table 17.4 Yield of ATP from Glucose Oxidation		
	ATP Yield per Glucose	
Pathway	**Glycerol— Phosphate Shuttle**	**Malate— Aspartate Shuttle**
Glycolysis: glucose to pyruvate (cytosol)		
Phosphorylation of glucose	−1	−1
Phosphorylation of fructose-6-phosphate	−1	−1
Dephosphorylation of 2 molecules of 1,3-BPG	+2	+2
Dephosphorylation of 2 molecules of PEP	+2	+2
Oxidation of 2 molecules of glyceraldehyde-3-phosphate yields 2 NADH		
Pyruvate conversion to acetyl-CoA (mitochondria)		
2 NADH		
Citric acid cycle (mitochondria)		
2 molecules of GTP from 2 molecules of succinyl-CoA	+2	+2
Oxidation of 2 molecules each of isocitrate, α-ketoglutarate, and malate yields 6 NADH		
Oxidation of 2 molecules of succinate yields 2 $[FADH_2]$		
Oxidative phosphorylation (mitochondria)		
2 NADH from glycolysis yield 1.5 ATP each if NADH is oxidized by glycerol-phosphate shuttle; 2.5 ATP by malate-aspartate shuttle	+3	+5
Oxidative decarboxylation of 2 pyruvate to 2 acetyl-CoA: 2 NADH produce 2.5 ATP each	+5	+5
2 $[FADH_2]$ from citric acid cycle produce 1.5 ATP each	+3	+3
6 NADH from citric acid cycle produce 2.5 ATP each	+15	+15
Net Yield	30	32

Note: These P/O ratios of 2.5 and 1.5 for mitochondrial oxidation of NADH and $[FADH_2]$ are "consensus values." Because they may not reflect actual values and because these ratios may change depending on metabolic conditions, these estimates of ATP yield from glucose oxidation are approximate.

is produced by oxidative phosphorylation; only 4 ATP molecules result from direct synthesis during glycolysis and the TCA cycle.

The situation in bacteria is somewhat different. Prokaryotic cells need not carry out ATP/ADP exchange. Thus, bacteria have the potential to produce approximately 38 ATP per glucose.

3.5 Billion Years of Evolution Have Resulted in a System That Is 54% Efficient

Hypothetically speaking, how much energy does a eukaryotic cell extract from the glucose molecule? With a value of 50 kJ/mol for the hydrolysis of ATP under cellular conditions (see Chapter 3), the production of 32 ATP per glucose oxidized consumes 1600 kJ/mol of glucose. The cellular oxidation (combustion) of glucose yields $\Delta G = -2937$ kJ/mol. We can calculate an efficiency for the pathways of glycolysis, the TCA cycle, electron transport, and oxidative phosphorylation of

$$\left(\frac{1600}{2937}\right) \times 100\% = 54\%$$

This efficiency is the result of approximately 3.5 billion years of evolution, and it represents a compromise between energy yield and the need to drive metabolic reactions to completion.

PROBLEMS

1. For the following reaction,

$$[FAD] + 2 \text{ cyt } c(Fe^{2+}) + 2 H^+ \rightarrow [FADH_2] + 2 \text{ cyt } c(Fe^{3+})$$

   determine which of the redox couples is the electron acceptor and which is the electron donor under standard-state conditions, calculate the value of $\Delta\mathcal{E}_\circ'$, and determine the free energy change for the reaction.

2. Calculate the value of $\Delta\mathcal{E}_\circ'$ for the glyceraldehyde-3-phosphate dehydrogenase reaction, and calculate the free energy change for the reaction under standard-state conditions.

3. For the following redox reaction,

$$NAD^+ + 2 H^+ + 2 e^- \rightarrow NADH + H^+$$

   suggest an equation (analogous to Equation 17.13) that predicts the pH dependence of this reaction, and calculate the reduction potential for this reaction at pH 8.

4. Sodium nitrite ($NaNO_2$) is used by emergency medical personnel as an antidote for cyanide poisoning (for this purpose, it must be administered immediately). Based on the discussion of cyanide poisoning in Section 17.10, suggest a mechanism for the life-saving effect of sodium nitrite.

5. A wealthy investor has come to you for advice. She has been approached by a biochemist who seeks financial backing for a company that would market dinitrophenol and dicumarol as weight-loss medications. The biochemist has explained to her that these agents are uncouplers and that they would dissipate metabolic energy as heat. The investor wants to know if you think she should invest in the biochemist's company. How do you respond?

6. Assuming that $3 H^+$ are transported per ATP synthesized in the mitochondrial matrix, and that the membrane potential difference is 0.18 V (negative inside) and the pH difference is 1 unit (acid outside, basic inside), calculate the largest ratio of $[ATP]/[ADP][P_i]$ under which synthesis of ATP can occur.

7. Of the dehydrogenase reactions in glycolysis and the TCA cycle, all but one use NAD^+ as the electron acceptor. The lone exception is the succinate dehydrogenase reaction, which uses covalently bound FAD of a flavoprotein as the electron acceptor. The standard reduction potential for this bound FAD is in the range of 0.003 to 0.091 V (see Table 17.1). Compared with the other dehydrogenase reactions of glycolysis and the TCA cycle, what is unique about succinate dehydrogenase? Why is bound FAD a more suitable electron acceptor in this case?

8.
   a. What is the standard free energy change ($\Delta G^{\circ\prime}$) for the reduction of coenzyme Q by NADH as carried out by Complex I (NADH–coenzyme Q reductase) of the electron transport pathway if $\mathcal{E}_\circ'(NAD^+/NADH + H^+) = -0.320$ volts and $\mathcal{E}_\circ'(CoQ/CoQH_2) = +0.060$ volts?
   b. What is the equilibrium constant (K_{eq}) for this reaction?
   c. Assume that (1) the actual free energy release accompanying the NADH–coenzyme Q reductase reaction is equal to the amount released under standard conditions (as cal-

culated in [a]), (2) this energy can be converted into the synthesis of ATP with an efficiency = 0.75 (that is, 75% of the energy released upon NADH oxidation is captured in ATP synthesis), and (3) the oxidation of 1 equivalent of NADH by coenzyme Q leads to the phosphorylation of 1 equivalent of ATP.

   Under these conditions, what is the maximum ratio of $[ATP]/[ADP]$ attainable by oxidative phosphorylation when $[P_i] = 1 \text{ m}M$? (Assume $\Delta G^{\circ\prime}$ for ATP synthesis = +30.5 kJ/mol.)

9. Consider the oxidation of succinate by molecular oxygen as carried out via the electron transport pathway

$$\text{succinate} + \tfrac{1}{2}O_2 \rightarrow \text{fumarate} + H_2O$$

   a. What is the standard free energy change ($\Delta G^{\circ\prime}$) for this reaction if $\mathcal{E}_\circ'(\text{fum/succ}) = +0.031$ volts and $\mathcal{E}_\circ'(\tfrac{1}{2}O_2/H_2O) = +0.816$ volts?
   b. What is the equilibrium constant (K_{eq}) for this reaction?
   c. Assume that (1) the actual free energy release accompanying succinate oxidation by the electron transport pathway is equal to the amount released under standard conditions (as calculated in [a]), (2) this energy can be converted into the synthesis of ATP with an efficiency = 0.7 (that is, 70% of the energy released upon succinate oxidation is captured in ATP synthesis), and (3) the oxidation of one succinate leads to the phosphorylation of two equivalents of ATP.

   Under these conditions, what is the maximum ratio of $[ATP]/[ADP]$ attainable by oxidative phosphorylation when $[P_i] = 1 \text{ m}M$? (Assume $\Delta G^{\circ\prime}$ for ATP synthesis = +30.5 kJ/mol.)

10. Consider the oxidation of NADH by molecular oxygen as carried out via the electron transport pathway

$$NADH + H^+ + \tfrac{1}{2}O_2 \rightarrow NAD^+ + H_2O$$

   a. What is the standard free energy change ($\Delta G^{\circ\prime}$) for this reaction if $\mathcal{E}_\circ'(NAD^+/NADH) = -0.320$ volts and $\mathcal{E}_\circ'(\tfrac{1}{2}O_2/H_2O) = +0.816$ volts?
   b. What is the equilibrium constant (K_{eq}) for this reaction?
   c. Assume that (1) the actual free energy release accompanying NADH oxidation by the electron transport pathway is equal to the amount released under standard conditions (as calculated in [a]), (2) this energy can be converted into the synthesis of ATP with an efficiency = 0.75 (that is, 75% of the energy released upon NADH oxidation is captured in ATP synthesis), and (3) the oxidation of 1 NADH leads to the phosphorylation of 3 equivalents of ATP.

   Under these conditions, what is the maximum ratio of $[ATP]/[ADP]$ attainable by oxidative phosphorylation when $[P_i] = 2 \text{ m}M$? (Assume $\Delta G^{\circ\prime}$ for ATP synthesis = +30.5 kJ/mol.)

11. Write a balanced equation for the reduction of molecular oxygen by reduced cytochrome c as carried out by Complex IV (cytochrome oxidase) of the electron transport pathway.

a. What is the standard free energy change $(\Delta G^{\circ\prime})$ for this reaction if

$$\mathscr{E}_{\circ}{}'(\text{cyt } c(Fe^{3+})/\text{cyt } c(Fe^{2+})) = +0.254 \text{ volts} \quad \text{and}$$

$$\mathscr{E}_{\circ}{}'(\tfrac{1}{2}O_2/H_2O) = 0.816 \text{ volts}$$

b. What is the equilibrium constant (K_{eq}) for this reaction?

c. Assume that (1) the actual free energy release accompanying cytochrome c oxidation by the electron transport pathway is equal to the amount released under standard conditions (as calculated in [a]), (2) this energy can be converted into the synthesis of ATP with an efficiency = 0.6 (that is, 60% of the energy released upon cytochrome c oxidation is captured in ATP synthesis), and (3) the reduction of one molecule of O_2 by reduced cytochrome c leads to the phosphorylation of 2 equivalents of ATP.

   Under these conditions, what is the maximum ratio of [ATP]/[ADP] attainable by oxidative phosphorylation when $[P_i]$ = 3 mM? (Assume $\Delta G^{\circ\prime}$ for ATP synthesis = +30.5 kJ/mol.)

12. The standard reduction potential for $(NAD^+/NADH)$ is -0.320 volts, and the standard reduction potential for (pyruvate/lactate) is -0.185 volts.

a. What is the standard free energy change, $\Delta G^{\circ\prime}$, for the lactate dehydrogenase reaction:

$$NADH + H^+ + \text{pyruvate} \rightarrow \text{lactate} + NAD^+$$

b. What is the equilibrium constant, K_{eq}, for this reaction?

c. If [pyruvate] = 0.05 mM and [lactate] = 2.9 mM and ΔG for the lactate dehydrogenase reaction = -15 kJ/mol in erythrocytes, what is the $[NAD^+]/[NADH]$ ratio under these conditions?

13. Assume that the free energy change, ΔG, associated with the movement of one mole of protons from the outside to the inside of a bacterial cell is -23 kJ/mol and 3 H$^+$ must cross the bacterial plasma membrane per ATP formed by the bacterial F_1F_0–ATP synthase. ATP synthesis thus takes place by the coupled process:

$$3 H_{out}{}^+ + ADP + P_i \rightleftharpoons 3 H_{in}{}^+ + ATP + H_2O$$

a. If the overall free energy change $(\Delta G_{overall})$ associated with ATP synthesis in these cells by the coupled process is -21 kJ/mol, what is the equilibrium constant, K_{eq}, for the process?

b. What is $\Delta G_{synthesis}$, the free energy change for ATP synthesis, in these bacteria under these conditions?

c. The standard free energy change for ATP hydrolysis, $\Delta G^{\circ\prime}_{hydrolysis}$, is -30.5 kJ/mol. If $[P_i]$ = 2 mM in these bacterial cells, what is the [ATP]/[ADP] ratio in these cells?

FURTHER READING

Abraham, J. P., Leslie, A. G. W., Lutter, R., and Walker, J. E., 1994. Structure at 2.8 Å resolution of F_1-ATPase from bovine heart mitochondria. *Nature* **370**:621–628.

Babcock, G. T., and Wikström, M., 1992. Oxygen activation and the conservation of energy in cell respiration. *Nature* **356**:301–309.

Bonomi, F., Pagani, S., Cerletti, P., and Giori, C., 1983. Modification of the thermodynamic properties of the electron-transferring groups in mitochondrial succinate dehydrogenase upon binding of succinate. *European Journal of Biochemistry* **134**:439–445.

Boyer, P. D., 1989. A perspective of the binding change mechanism for ATP synthesis. *The FASEB Journal* **3**:2164–2178.

Boyer, P. D., 1997. The ATP synthase—a splendid molecular machine. *Annual Review of Biochemistry* **66**:717–750.

Cross, R. L., 1994. Our primary source of ATP. *Nature* **370**:594–595.

Dickerson, R. E., 1980. Cytochrome c and the evolution of energy metabolism. *Scientific American* **242**(3):137–153.

Ernster, L., ed., 1980. *Bioenergetics.* New York: Elsevier Press.

Ferguson-Miller, S., 1996. Mammalian cytochrome c oxidase, a molecular monster subdued. *Science* **272**:1125.

Fillingame, R. H., 1980. The proton-translocating pump of oxidative phosphorylation. *Annual Review of Biochemistry* **49**:1079–1113.

Frey, T. G., and Mannella, C. A., 2000. The internal structure of mitochondria. *Trends in Biochemical Sciences* **25**:319–324.

Futai, M., Noumi, T., and Maeda, M., 1989. ATP synthase: Results by combined biochemical and molecular biological approaches. *Annual Review of Biochemistry* **58**:111–136.

Harold, F. M., 1986. *The Vital Force: A Study of Bioenergetics.* New York: W. H. Freeman and Company.

Iwata, S., Ostermeier, C., Ludwig, B., and Michel, H., 1995. Structure at 2.8 Å resolution of cytochrome c oxidase from *Paracoccus denitrificans. Nature* **376**:660–669.

Junge, W., Lill, H., and Engelbrecht, S., 1997. ATP synthase: An electrochemical transducer with rotatory mechanics. *Trends in Biochemical Sciences* **22**:420–423.

Kinosita, K., Yasuda, R., Noji, H., Ishiwata, S., and Yoshida, M., 1998. F_1-ATPase: A rotatory motor made of a single molecule. *Cell* **93**: 21–24.

Mitchell, P., 1979. Keilin's respiratory chain concept and its chemiosmotic consequences. *Science* **206**:1148–1159.

Mitchell, P., and Moyle, J., 1965. Stoichiometry of proton translocation through the respiratory chain and adenosine triphosphatase systems of rat mitochondria. *Nature* **208**:147–151.

Moser, C. C., et al. 1992. Nature of biological electron transfer. *Nature* **355**:796–802.

Naqui, A., Chance, B., and Cadenas, E., 1986. Reactive oxygen intermediates in biochemistry. *Annual Review of Biochemistry* **55**:137.

Nicholls, D. G., and Ferguson, S. J., 1992. *Bioenergetics 2.* London: Academic Press.

Nicholls, D. G., and Rial, E., 1984. Brown fat mitochondria. *Trends in Biochemical Sciences* **9**:489–491.

Noji, H., Yasuda, R., Yoshida, M., and Kinosita, K., 1997. Direct observation of the rotation of F_1-ATPase. *Nature* **386**:299–302.

Pedersen, P., and Carafoli, E., 1987. Ion-motive ATPases. I. Ubiquity, properties and significance to cell function. *Trends in Biochemical Sciences* **12**:146–150.

Sabbert, D., Engelbrecht, S., and Junge, W., 1996. Intersubunit rotation in active F_1-ATPase. *Nature* **381**:623–625.

Sambongi, Y., Iko, Y., Tanabe, M., et al., 1999. Mechanical rotation of the c subunit oligomer in ATP synthase (F_0F_1): Direct observation. *Science* **286:** 1722–1724.

Slater, E. C., 1983. The Q cycle: An ubiquitous mechanism of electron transfer. *Trends in Biochemical Sciences* **8**:239–242.

Stock, D., Leslie, A., and Walker, J., 1999. Molecular architecture of the rotary motor in ATP synthase. *Science* **286:** 1700–1705.

Trumpower, B. L., 1990. Cytochrome bc_1 complexes of microorganisms. *Microbiological Reviews* **54**:101–129.

Trumpower, B. L., 1990. The protonmotive Q cycle—energy transduction by coupling of proton translocation to electron transfer by the cytochrome bc_1 complex. *Journal of Biological Chemistry* **265**:11409–11412.

Tsukihara, T., Aoyama, H., Yamashita, E., et al., 1996. The whole structure of the 13-subunit oxidized cytochrome c oxidase at 2.8 Å. *Science* **272**:1136–1144.

Vignais, P. V., and Lunardi, J., 1985. Chemical probes of mitochondrial ATP synthesis and translocation. *Annual Review of Biochemistry* **54**:977–1014.

von Jagow, G., 1980. b-Type cytochromes. *Annual Review of Biochemistry* **49**:281–314.

Walker, J. E., 1992. The NADH:ubiquinone oxidoreductase (Complex I) of respiratory chains. *Quarterly Reviews of Biophysics* **25**:253–324.

Weiss, H., Friedrich, T., Hofhaus, G., and Preis, D., 1991. The respiratory-chain NADH dehydrogenase (Complex I) of mitochondria. *European Journal of Biochemistry* **197**:563–576.

Wilkens, S., Dunn, S. D., Chandler, J., et al., 1997. Solution structure of the N-terminal domain of the δ subunit of the *E. coli* ATP synthase. *Nature Structural Biology* **4**:198–201.

Yasuda, R., Noji, H., Kinosita, K., and Yoshida, M., 1998. F_1-ATPase is a highly efficient molecular motor that rotates with discrete 120° steps. *Cell* **93:** 1117–1124.

Xia, D., Yu, C-A., Kim, H., et al., 1997. The crystal structure of the cytochrome bc_1 complex from bovine heart mitochondria. *Science* **277:** 60–66.

Photosynthesis

In a sun-flecked sylvan lane,
Beside a path where cattle trod,
Blown by wind, beaten by rain,
Drawing substance from air and sod;

In ruggedness, it stands aloof,
The ragged grass and puerile leaves,
Lending a thread to fill the woof
In the pattern that beauty makes.

What mystery, this, hath been wrought:
Beauty from sunshine, air and sod!
Could we thus gain the ends we sought—
Tell us thy secret, Goldenrod.

Rosa Staubus, Oklahoma pioneer (1866-1966)

Field of goldenrod. (©Richard Hamilton Smith/CORBIS)

The vast majority of energy consumed by living organisms stems from solar energy captured by the process of photosynthesis. Only chemoautotrophic bacteria (Chapter 14) are independent of this energy source. Of the 1.5×10^{22} kJ of energy reaching the earth each day from the sun, 1% is absorbed by photosynthetic organisms and transduced into chemical energy.[1] This energy, in the form of biomolecules, becomes available to other members of the biosphere through food chains. The transduction of solar, or light, energy into chemical energy is often expressed in terms of **carbon dioxide fixation,** in which hexose is formed from carbon dioxide and oxygen is evolved:

$$6\,CO_2 + 6\,H_2O \xrightarrow{\text{Light}} C_6H_{12}O_6 + 6\,O_2 \qquad (18.1)$$

Estimates indicate that 10^{11} tons of carbon dioxide are fixed globally per year.

---

[1] Of the remaining 99%, two-thirds is absorbed by the earth and oceans, thereby heating the planet; the remaining one-third is lost as light reflected back into space.

Although photosynthesis is traditionally equated with CO_2 fixation, light energy—or rather, the chemical energy derived from it—is used to drive virtually all plant cell processes. The assimilation of inorganic forms of nitrogen and sulfur into organic molecules represents two other metabolic conversions driven by light energy in green plants. Our previous considerations of aerobic metabolism (Chapters 15 through 17) treated cellular respiration (precisely the reverse of Equation [18.1]) as the central energy-releasing process in life. It necessarily follows that the formation of hexose from carbon dioxide and water, the products of cellular respiration, requires energy. The necessary energy comes from light. Note that in the carbon dioxide fixation reaction described, light is used to drive a chemical reaction against its thermodynamic potential.

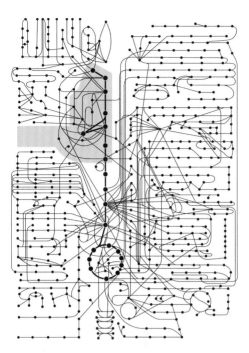

Photosynthesis.

See *Interactive Biochemistry CD-ROM and Workbook,* pages 173–178

18.1 General Aspects of Photosynthesis

Photosynthesis Occurs in Membranes

Organisms capable of photosynthesis are very diverse, ranging from simple prokaryotic forms to the largest organisms of all, *Sequoia gigantea,* the giant redwood trees of California. Despite this diversity, we find certain generalities regarding photosynthesis. An important one is that *photosynthesis occurs in membranes.* In photosynthetic prokaryotes, the photosynthetic membranes fill up the cell interior; in photosynthetic eukaryotes, the photosynthetic membranes are localized in large organelles known as **chloroplasts** (Figures 18.1 and 18.2). Chloroplasts are one member in a family of related plant-specific organelles known as **plastids.** Chloroplasts themselves show a range of diversity, from the single, spiral chloroplast that gives *Spirogyra* its name to the multitude of ellipsoidal plastids typical of higher plant cells (Figure 18.3).

Characteristic of all chloroplasts, however, is the organization of the inner membrane system, the so-called **thylakoid membrane.** The thylakoid membrane is organized into paired folds that extend throughout the organelle, as in Figure 18.1. These paired folds, or **lamellae,** give rise to flattened sacs or disks, **thylakoid vesicles** (from the Greek *thylakos,* meaning "sack"), which occur in stacks called **grana.** A single stack, or **granum,** appears by electron microscopy to contain many thylakoid vesicles joined by lamellae that run through the soluble portion, or **stroma,** of the organelle. However, this view is due to the cross-section nature of electron microscopic images; the grana actually arise from extensive folding and packing of the thylakoid membrane. Indeed, it is likely that chloroplasts contain only a few highly folded thylakoid vesicles, or even just one. Chloroplasts thus possess three membrane-bound aqueous compartments: the intermembrane space, the stroma, and the interior of the thylakoid vesicles, the so-called **thylakoid space** (also known as the **thylakoid lumen**). The thylakoid membrane has a highly characteristic lipid composition and, like the inner membrane of the mitochondrion, is impermeable to most ions and molecules. Chloroplasts, like their mitochondrial counterparts, possess DNA, RNA, and ribosomes and consequently display a considerable amount of autonomy. However, as is also the case with mitochondria, 90% or more of the proteins in chloroplasts are encoded by nuclear genes, so autonomy is far from absolute.

Photosynthesis Consists of Both Light Reactions and Dark Reactions

If a chloroplast suspension is illuminated in the absence of carbon dioxide, oxygen is evolved. Furthermore, if the illuminated chloroplasts are then placed

Figure 18.1 Electron micrograph of a representative chloroplast. *(James Dennis/CNRI/Phototake NYC)*

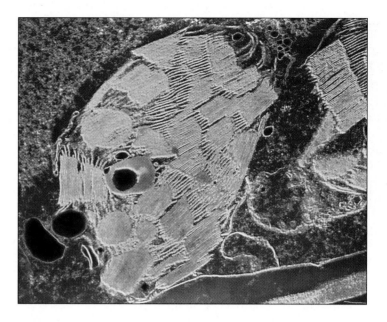

temporally with regard to time

in the dark and supplied with CO_2, net hexose synthesis can be observed (Figure 18.4). Thus, the evolution of oxygen can be temporally separated from CO_2 fixation. O_2 evolution also has a light dependency that CO_2 fixation lacks. The **light reactions** of photosynthesis, of which O_2 evolution is only one part, are associated with the thylakoid membranes. In contrast, the light-independent reactions, or so-called **dark reactions,** notably CO_2 fixation, are located in the stroma. A concise summary of the photosynthetic process is that radiant electromagnetic energy (light) is transformed by a specific photochemical system located in the thylakoids to yield chemical energy in the form of reducing potential (NADPH) and high-energy phosphate (ATP). NADPH and ATP can then be used to drive the endergonic process of hexose formation from CO_2 by a series of enzymatic reactions found in the stroma (see Equation 18.3, which follows).

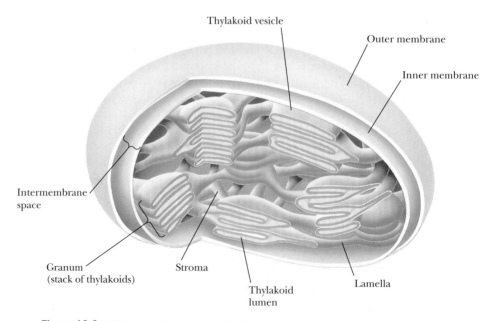

Figure 18.2 Schematic diagram of an idealized chloroplast.

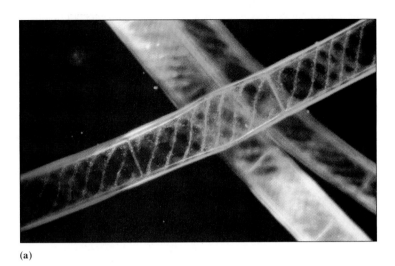

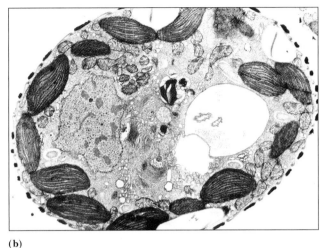

(a)

(b)

Figure 18.3 **(a)** *Spirogyra* freshwater green alga. **(b)** A higher plant cell. *(a, Michael Siegel/Phototake NYC; b, Biophoto Associates/Science Source)*

Water Is the Ultimate e^- Donor for Photosynthetic NADP$^+$ Reduction

In green plants, water serves as the ultimate electron donor for the photosynthetic generation of reducing equivalents. The reaction sequence

$$2\,H_2O + 2\,NADP^+ + x\,ADP + x\,P_i \xrightarrow{nh\nu} O_2 + 2\,NADPH + 2\,H^+ + x\,ATP + x\,H_2O \tag{18.2}$$

describes the process, where $nh\nu$ symbolizes *light energy* (n is some number of photons of energy $h\nu$, where h is Planck's constant and ν is the frequency of the light). Light energy is necessary to make the unfavorable reduction of NADP$^+$ by H_2O ($\Delta\mathscr{E}_o' = -1.136\ V$; $\Delta G^{\circ\prime} = +219\ kJ/mol\ NADP^+$) thermodynamically favorable. Thus, the light energy input, $nh\nu$, must exceed $219\ kJ/mol\ NADP^+$. The stoichiometry of ATP formation depends on the pattern of photophosphorylation operating in the cell at the time and on the ATP yield in terms of the chemiosmotic ratio, ATP/H$^+$, as we will see later. On the other hand, the stoichiometry for the metabolic pathway of CO$_2$ fixation is certain:

$$12\,NADPH + 12\,H^+ + 18\,ATP + 6\,CO_2 + 12\,H_2O \longrightarrow C_6H_{12}O_6 + 12\,NADP^+ + 18\,ADP + 18\ P_i \tag{18.3}$$

Figure 18.4 The light-dependent and light-independent reactions of photosynthesis. Light reactions are associated with the thylakoid membranes, and light-independent reactions are associated with the stroma.

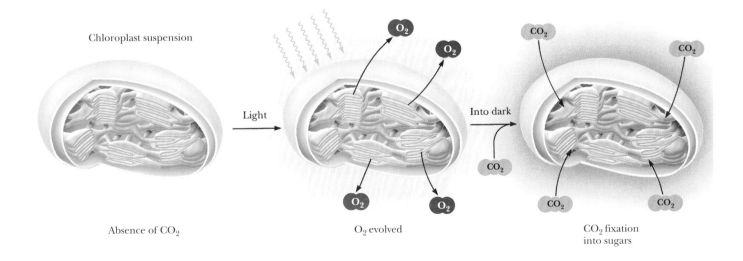

Chloroplast suspension

Light

Into dark

O$_2$

O$_2$

CO$_2$

CO$_2$

CO$_2$

O$_2$

O$_2$

CO$_2$

CO$_2$

CO$_2$

Absence of CO$_2$

O$_2$ evolved

CO$_2$ fixation into sugars

Figure 18.5 Structures of chlorophylls *a* and *b*. Chlorophylls are structurally related to hemes, except Mg^{2+} replaces Fe^{2+} and ring II is more reduced than the corresponding ring of the porphyrins. The chlorophyll tetrapyrrole ring system is known as a chlorin. $R = CH_3$ in chlorophyll *a*; $R = CHO$ in chlorophyll *b*. Note that the aldehyde $C=O$ bond of chlorophyll *b* introduces an additional double bond into conjugation with the double bonds of the tetrapyrrole ring system. Ring V is the additional ring created by interaction of the substituent of the methine bridge between pyrroles III and IV with the side chain of ring III. The phytyl side chain of ring IV provides a hydrophobic tail to anchor the chlorophyll in membrane protein complexes.

R=
Chlorophyll *a* —CH_3
Chlorophyll *b* —CHO

Hydrophobic phytyl side chain

18.2 Photosynthesis Depends on the Photoreactivity of Chlorophyll

Chlorophylls are magnesium-containing substituted tetrapyrroles whose basic structure is reminiscent of heme, the iron-containing porphyrin (Chapters 4, 12, and 17). The structures of chlorophyll *a* and *b* are shown in Figure 18.5. Chlorophylls are excellent light absorbers. When light energy is absorbed, an electron is promoted to a higher orbital, enhancing the potential for transfer of this electron to a suitable acceptor. Loss of such a photo-excited electron to an acceptor is an oxidation–reduction reaction. The net result is the transduction of light energy into the chemical energy of a redox reaction.

Chlorophylls and Accessory Light-Harvesting Pigments

The absorption spectra of chlorophylls *a* and *b* (Figure 18.6) differ somewhat. Plants that possess both chlorophylls can harvest a wider spectrum of incident energy. Other pigments in photosynthetic organisms, so-called **accessory light-harvesting pigments** (Figure 18.7), increase the possibility for absorption of light of wavelengths not absorbed by the chlorophylls. These accessory pigments include carotenoids and phycobilins.

The Fate of Light Energy Absorbed by Photosynthetic Pigments

Each photon represents a quantum of light energy. A quantum of light energy absorbed by a photosynthetic pigment has four possible fates (Figure 18.8):

A. **Loss as heat.** The energy can be dissipated as heat through redistribution into atomic vibrations within the pigment molecule.

B. **Loss of light.** Energy of excitation reappears as **fluorescence** (light emission); a photon of fluorescence is emitted as the e^- returns to a lower orbital.

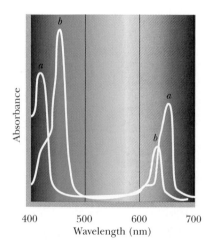

Figure 18.6 Absorption spectra of chlorophylls *a* and *b*.

(a)

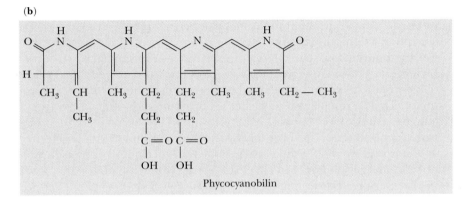

β-Carotene

(b)

Phycocyanobilin

Figure 18.7 Structures of representative accessory light-harvesting pigments in photosynthetic cells. **(a)** β-Carotene, a pigment in leaves. Note the many conjugated double bonds. Carotenoids function primarily as protectants against photo-oxidative damage. **(b)** Phycocyanobilin, a blue pigment found in cyanobacteria. It is a linear or open pyrrole.

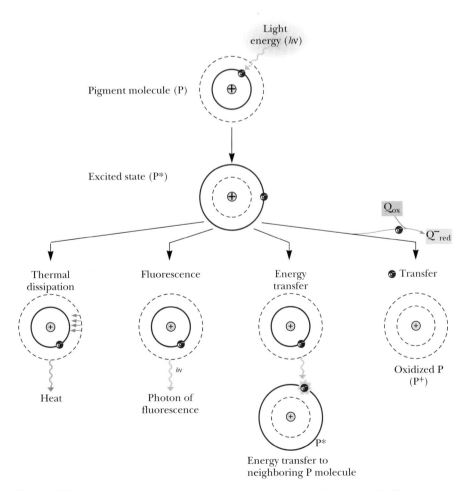

Figure 18.8 Possible fates of the quantum of light energy absorbed by photosynthetic pigments.

This fate is common only in saturating light intensities. For thermodynamic reasons, the photon of fluorescence is of longer wavelength and hence lower energy than the quantum of excitation.

C. Resonance energy transfer. The excitation energy can be transferred by resonance energy transfer to a neighboring molecule if the energy level difference between the two corresponds to the quantum of excitation energy. In this process, the energy transferred raises an electron in the receptor molecule to a higher energy state as the photo-excited e^- in the original absorbing molecule returns to ground state. This so-called *Förster resonance energy transfer* is the mechanism whereby quanta of light falling anywhere within an array of pigment molecules can be transferred ultimately to specific photochemically reactive sites.

D. Energy transduction. The energy of excitation, in raising an electron to a higher energy orbital, dramatically changes the standard reduction potential, $\mathscr{E}_\circ'$, of the pigment such that it becomes a much more effective electron donor. That is, the excited-state species, by virtue of having an electron at a higher energy level through light absorption, has become a potent electron donor. Reaction of this excited-state electron donor with an electron acceptor situated in its vicinity leads to the transformation, or **transduction,** of light energy (photons) to chemical energy (reducing power, the potential for electron-transfer reactions). *Transduction of light energy into chemical energy, the photochemical event, is the essence of photosynthesis.*

A Simple Model of Photosynthetic Energy Transduction

The diagram presented in Figure 18.9 illustrates the fundamental transduction of light energy into chemical energy (an oxidation–reduction reaction) that is the basis of photosynthesis. Chlorophyll (Chl) resides in a membrane in close association with molecules competent in e^- transfer, symbolized here as A and B. Chl absorbs a photon of light, becoming activated to Chl* in the process. Electron transfer from Chl* to A leads to oxidized Chl (Chl·$^+$, a **cationic free radical**) and reduced A (A$^-$ in the diagram). Subsequent oxidation of A$^-$ culminates eventually in reduction of NADP$^+$ to NADPH. The electron "hole" in oxidized Chl (Chl·$^+$) is filled by transfer of an electron from B to Chl·$^+$, restoring Chl and creating B$^+$. B$^+$ is restored to B by an e^- donated by water. O$_2$ is the product of water oxidation. Note that the system is restored to its original state once NADPH is formed and H$_2$O is oxidized.

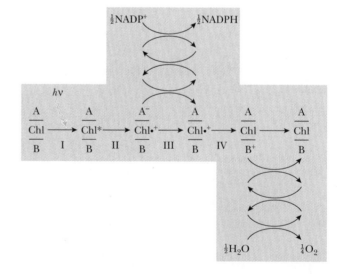

Figure 18.9   Model for light absorption by chlorophyll and transduction of light energy into an oxidation–reduction reaction. Not shown here are the H$^+$ translocations that accompany these light-driven electron transport reactions; such proton translocations establish a chemiosmotic gradient across the photosynthetic membrane that can drive ATP synthesis.

Photosynthetic Units Consist of Many Chlorophyll Molecules but Only a Single Reaction Center

Chlorophyll serves two roles in photosynthesis. It is involved in light harvesting and the transfer of light energy to photoreactive sites by resonance energy transfer, and it participates directly in the photochemical events whereby light energy becomes chemical energy. A **photosynthetic unit** consists of an antenna of several hundred light-harvesting chlorophyll molecules plus a special pair of photochemically reactive chlorophyll *a* molecules called the **reaction center.** The vast majority of chlorophyll in a photosynthetic unit harvests light incident within the unit and funnels it, via resonance energy transfer, to special reaction-center chlorophyll molecules that are photochemically active. Most chlorophyll thus acts as a large light-collecting antenna, and it is at the reaction centers that the photochemical event occurs (Figure 18.10). The chlorophyll Mg^{2+} ion does not change in valence during these redox reactions.

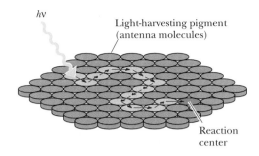

Figure 18.10 Schematic diagram of a photosynthetic unit. The light-harvesting pigments, or antenna molecules (green), absorb and transfer light energy to the specialized chlorophyll dimer that constitutes the reaction center (orange).

18.3   Plants Possess Two Distinct Photosystems

All photosynthetic cells contain some form of photosystem. Photosynthetic bacteria have only one photosystem. Cyanobacteria and eukaryotic phototrophs (green algae and higher plants) have two distinct photosystems, **Photosystem I** (PSI) and **Photosystem II** (PSII). PSI is defined by reaction center chlorophylls with maximal red light absorption at 700 nm; PSII uses reaction centers that exhibit maximal red light absorption at 680 nm. The reaction center Chl of PSI is referred to as P700 because it absorbs light of 700-nm wavelength; the reaction center Chl of PSII is called P680 for analogous reasons. Both P700 and P680 are chlorophyll *a* dimers situated within specialized protein complexes. A distinct property of PSII is its role in light-driven O_2 evolution. Interestingly, bacterial photosystems resemble eukaryotic PSII more than PSI, even though photosynthetic bacteria lack O_2-evolving capacity.

Chlorophyll Exists in Plant Membranes in Association with Proteins

Detergent treatment of a suspension of thylakoids dissolves the membranes, releasing complexes containing both chlorophyll and protein. These chlorophyll–protein complexes represent integral components of the thylakoid membrane, and their organization reflects their roles as either **light-harvesting complexes** (LHC), **PSI complexes,** or **PSII complexes.** All chlorophyll is apparently localized within these three macromolecular assemblies.

The Roles of PSI and PSII

What are the roles of the two photosystems, and what is their relationship to each other? Photosystem I provides reducing power in the form of NADPH. Photosystem II splits water, producing O_2, and feeds the electrons released into an electron transport chain that couples PSII to PSI. Electron transfer between PSII and PSI pumps protons for chemiosmotic ATP synthesis. As summarized by Equation (18.2), photosynthesis involves the reduction of $NADP^+$, using electrons derived from water and activated by light, $h\nu$. ATP is generated in the process. The standard reduction potential for the $NADP^+/NADPH$ couple is -0.32 V. Thus, a strong reductant with $\mathcal{E}_o'$ more negative than -0.32 V is required to reduce $NADP^+$ under standard conditions. By similar reasoning, a very strong oxidant will be required to oxidize water to oxygen because $\mathcal{E}_o'(\frac{1}{2} O_2/H_2O)$ is $+0.82$ V. Separation of the oxidizing and reducing aspects of

Figure 18.11   Roles of the two photosystems, PSI and PSII.

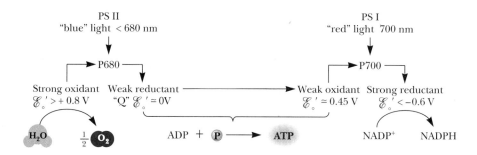

Equation (18.2) is accomplished in nature by devoting PSI to $NADP^+$ reduction and PSII to water oxidation. PSI and PSII are linked via an electron transport chain so that the weak reductant generated by PSII can provide an electron to reduce the weak oxidant side of P700 (Figure 18.11). Thus, electrons flow from H_2O to $NADP^+$, driven by light energy absorbed at the reaction centers. Oxygen is a by-product of the **photolysis**—literally, "light-splitting"—of water. Light-driven electron flow through these systems produces a transmembrane proton gradient that in turn can drive chemiosmotic ATP synthesis (see Section 18.6). This light-driven phosphorylation is termed **photophosphorylation.**

18.4   The Z Scheme of Photosynthetic Electron Transfer

Photosystems I and II contain unique complements of electron carriers, and these carriers mediate the stepwise transfer of electrons from water to $NADP^+$. When the individual redox components of PSI and PSII are arranged as an e^- transport chain according to their standard reduction potentials, the zig-zag result resembles the letter *Z* laid sideways (Figure 18.12). The various electron carriers in this chain are summarized in Table 18.1.

Overall photosynthetic electron transfer through these carriers is accomplished by three **membrane-spanning supramolecular complexes,** composed of intrinsic and extrinsic polypeptides (represented by shaded boxes bounded by solid black lines in Figure 18.12). These complexes are the PSII complex, the cytochrome b_6/cytochrome f complex, and the PSI complex. The PSII

Figure 18.12   The *Z* scheme of photosynthesis.   **(a)** The *Z* scheme is a diagrammatic ▶ representation of photosynthetic electron flow from H_2O to $NADP^+$. The energy relationships can be derived from the $\mathscr{E}_o'$ scale beside the *Z* diagram, with lower standard potentials and hence greater energy as you go from bottom to top. Energy input as light is indicated by two broad arrows, one photon appearing in P680 and the other in P700. P680* and P700* represent photoexcited states. Electron loss from P680* and P700* creates $P680^+$ and $P700^+$. The representative components of the three supramolecular complexes (PSI, PSII, and the cytochrome b_6/cytochrome f complex) are in shaded boxes enclosed by solid black lines. Proton translocations that establish the proton-motive force driving ATP synthesis are illustrated as well.   **(b)** Figure showing the functional relationships among PSII, the cytochrome b_6/cytochrome f complex, PSI, and the photosynthetic CF_1CF_0-ATP synthase within the thylakoid membrane. Note that e^- acceptors Q_A (for PSII) and A_1 (for PSI) are at the stromal side of the thylakoid membrane, whereas the e^- donors to $P680^+$ and $P700^+$ are situated at the lumenal side of the membrane. The consequence is charge separation ($-_{stroma}$, $+_{lumen}$) across the membrane. Also note that protons are translocated into the thylakoid lumen, giving rise to a chemiosmotic gradient that is the driving force for ATP synthesis by CF_1CF_0-ATP synthase.

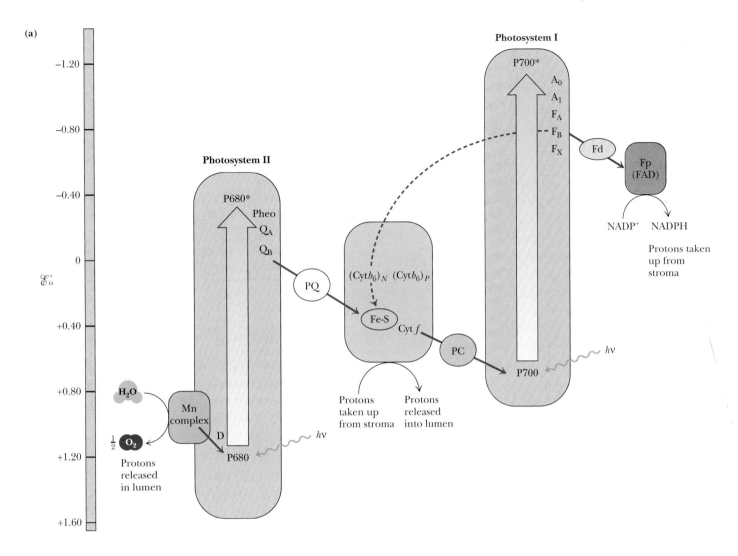

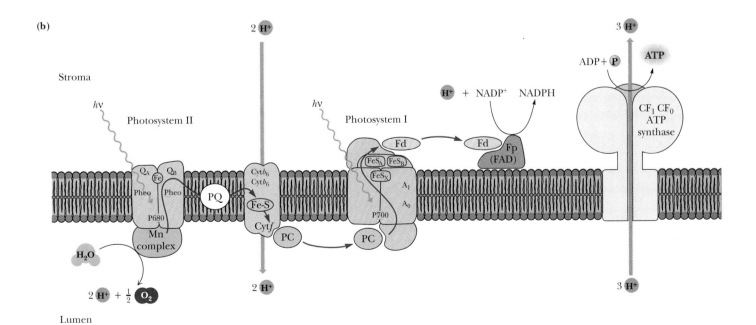

Table 18.1

Electron Carrier	Description of Its Properties
Mn complex	The manganese-containing O_2- evolving complex
D	The immediate e^- acceptor for the Mn complex and e^- donor to $P680^+$
Q_A and Q_B	Special plastoquinone molecules (see Figure 18.13)
PQ	The plastoquinone pool in the thylakoid membrane
Fe-S	The Reiske iron–sulfur center of the cytochrome b/f complex
cyt f	Cytochrome f
PC	Plastocyanin, a Cu-containing protein of the thylakoid lumen that is the immediate e^- donor to $P700^+$
A_0	A special chlorophyll a
A_1	A special PSI quinone
F_A, F_B, F_x	Membrane-associated ferredoxins downstream from A_0
Fd	The soluble ferredoxin pool that serves as e^- donor to Fp
Fp	Ferredoxin-$NADP^+$ reductase, the enzyme that reduces $NADP^+$ to NADPH
$(Cytb_6)_n/(Cytb_6)_p$	The cytochrome b moieties that transfer e^- back to $P700^+$ during cyclic photophosphorylation (shown by dashed arrow)

complex is aptly described as a light-driven **water : plastoquinone oxidoreductase;** it is the enzyme system responsible for photolysis of water, and as such, it is also referred to as the **oxygen-evolving complex,** or **OEC.** As strong evidence of evolutionary relationship, the cytochrome b_6/cytochrome f complex is strikingly similar in structure and function to the mitochondrial cytochrome bc_1 complex described in Chapter 17.

Light-Driven Electron Flow from H₂O Through PSII

The events intervening between H_2O and P680 involve D, a specific protein **tyrosine residue** that mediates e^- transfer from H_2O via the Mn complex to $P680^+$ (see Figure 18.12). To begin the cycle, light energy excites P680 to $P680^*$, changing its $\mathscr{E}_o'$ from $+1.2$ V to -0.4 V. $P680^*$ donates an electron to a special molecule of **pheophytin,** symbolized by "Pheo" in Figure 18.12. Pheophytin is like chlorophyll a, except 2 H^+ replace the centrally coordinated Mg^{2+} ion. This special pheophytin is the direct electron acceptor from $P680^*$. Loss of an electron from $P680^*$ creates $P680^+$, an electron acceptor with an $\mathscr{E}_o' = +1.2$ V. D is the electron donor to $P680^+$. Electrons flow from reduced Pheo via specialized molecules of **plastoquinone,** represented by Q in Figure 18.12, to a pool of plastoquinone within the membrane. Because of its lipid nature, plastoquinone is mobile within the membrane and hence serves to shuttle electrons from the PSII complex to the cytochrome b_6/cytochrome f complex. Alternate oxidation–reduction of plastoquinone to its hydroquinone form involves the uptake of protons (Figure 18.13). The asymmetry of the thylakoid membrane is designed to exploit this proton uptake and release so that protons (H^+) accumulate within the thylakoid vesicle, establishing an electrochemical gradient. Note that plastoquinone is an analog of coenzyme Q, the mitochondrial electron carrier (Chapter 17).

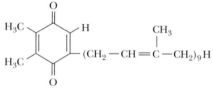

Plastoquinone A

$+2\ \text{H}^+, 2\ e^-$ ⇌ $-2\ \text{H}^+, 2\ e^-$

Plastohydroquinone A

Figure 18.13 The structures of plastoquinone and its reduced form, plastohydroquinone (or plastoquinol). The oxidation of the hydroquinone releases 2 H^+ as well as 2 e^-. The form shown (plastoquinone A) has nine isoprene units and is the most abundant plastoquinone in plants and algae. Other plastoquinones have different numbers of isoprene units and may vary in the substitutions on the quinone ring.

Electron Transfer Within the Cytochrome b_6/Cytochrome f Complex

The cytochrome b_6/cytochrome f complex, or **plastoquinol : plastocyanin oxidoreductase,** is a large (210 kD) multimeric protein possessing 22 to 24 transmembrane α-helices. It includes the two heme-containing electron transfer proteins for which it is named, as well as iron–sulfur clusters (Chapter 17) which also participate in electron transport. The purpose of this complex is to transfer electrons from PSII to PSI and pump protons across the thylakoid membrane via a plastoquinone-mediated Q cycle, analogous to that found in mitochondrial e^- transport (Chapter 17). Cytochrome f (f from the Latin *folium*, meaning "foliage") is a c-type cytochrome. Under certain conditions, electrons derived from P700* are not passed on to $NADP^+$ but instead cycle down an alternative e^- transfer pathway via ferredoxins in the PSI complex to cytochrome b_6, plastoquinone, and ultimately back to P700$^+$. This cyclic flow yields no O_2 evolution or $NADP^+$ reduction but can lead to ATP synthesis via so-called cyclic photophosphorylation, discussed later.

Electron Transfer from the Cytochrome b_6/Cytochrome f Complex to PSI

Plastocyanin (PC in Figure 18.12) is a small (10.4 kD) protein that functions as a single-electron carrier as its copper atom undergoes alternate oxidation–reduction between the cuprous (Cu^+) and cupric (Cu^{2+}) states. As an electron carrier capable of diffusion along the inside of the thylakoid and migration in and out of the membrane, plastocyanin is aptly suited to its role in shuttling electrons between the cytochrome b_6/cytochrome f complex and PSI. PSI is a light-driven **plastocyanin : ferredoxin oxidoreductase.** When P700, the specialized chlorophyll a dimer of PSI, is excited by light, P700* ($\mathscr{E}_o' = -0.6$ V) results. P700* is oxidized by transferring its e^- to an adjacent chlorophyll a molecule that serves as its immediate e^- acceptor; P700$^+$ is formed. P700$^+$ ($\mathscr{E}_o' = +0.4$ V) readily gains an electron from plastocyanin.

The immediate electron acceptor for P700* is a special molecule of chlorophyll. This unique Chl a (A_0) rapidly passes the electron to a specialized quinone (A_1), which in turn passes the e^- to the first in a series of membrane-bound ferredoxins (Fd; see Chapter 17). This Fd series ends with a soluble form of ferredoxin, Fd$_s$, which serves as the immediate electron donor to the flavoprotein (Fp) that catalyzes $NADP^+$ reduction, namely, **ferredoxin : NADP$^+$ reductase.**

18.5 The Molecular Architecture of Photosynthetic Reaction Centers

What molecular architecture couples the absorption of light energy to rapid electron-transfer events, in turn coupling these e^- transfers to proton translocations so that ATP synthesis is possible? Part of the answer to this question lies in the membrane-associated nature of the photosystems. Membrane proteins are difficult to study due to their insolubility in the usual aqueous solvents employed in protein biochemistry. A major breakthrough occurred in 1984 when Johann Deisenhofer, Hartmut Michel, and Robert Huber reported the first X-ray crystallographic analysis of a membrane protein. To the great benefit of photosynthesis research, this protein was the reaction center from the photosynthetic purple bacterium *Rhodopseudomonas viridis*. This research earned these three scientists the 1984 Nobel Prize in chemistry.

The *R. viridis* Photosynthetic Reaction Center Is a Paradigm for the Structure and Function of Photosystems

Rhodopseudomonas viridis is a photosynthetic bacterium with a single type of photosystem. The reaction center (145 kD) of the *R. viridis* photosystem is localized in the plasma membrane and is composed of four different polypeptides, designated L (273 amino acid residues), M (323 residues), H (258 residues), and *cytochrome* (333 amino acid residues). L and M each consist of five membrane-

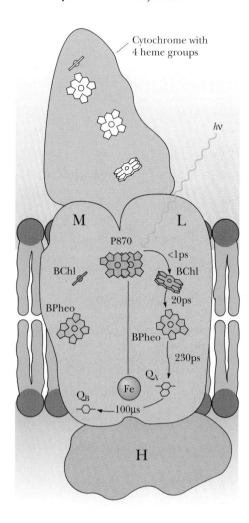

Note: The cytochrome subunit is membrane-associated via a diacylglycerol moiety on its N-terminal Cys residue:

Membrane anchor

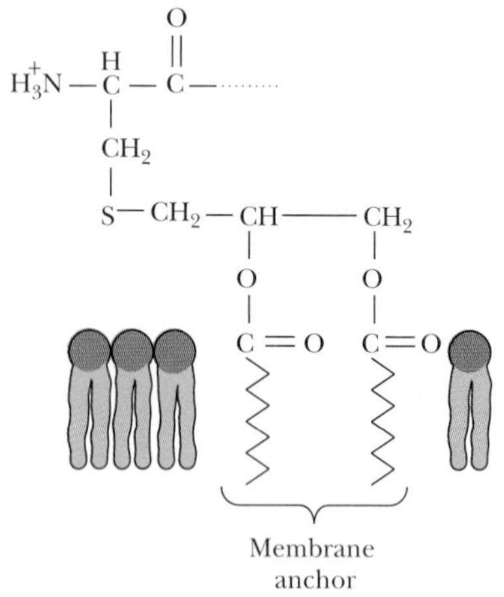

◀ **Figure 18.14** Model of the structure and activity of the *R. viridis* reaction center. Four polypeptides (designated *cytochrome, M, L,* and *H*) make up the reaction center, an integral membrane complex. The cytochrome maintains its association with the membrane via a diacylglyceryl group linked to its N-terminal Cys residue by a thioether bond. *M* and *L* both consist of five membrane-spanning α-helices; *H* has a single membrane-spanning α-helix. The prosthetic groups are spatially situated so that rapid e^- transfer from P870* to Q_B is facilitated. Photoexcitation of P870 leads in less than 1 picosecond (psec) to reduction of the *L*-branch BChl only. $P870^+$ is re-reduced via an e^- provided through the heme groups of the cytochrome.

spanning α-helical segments; *H* has one such helix, the majority of the protein forming a globular domain in the cytoplasm (Figure 18.14). The cytochrome subunit contains four heme groups; the N-terminal amino acid of this protein is cysteine. This cytochrome is anchored to the periplasmic face of the membrane via the hydrophobic chains of a diacylglycerol covalently attached to the Cys (see Figure 18.14). *L* and *M* each bear two bacteriochlorophyll molecules (the bacterial version of Chl) and one bacteriopheophytin. *L* also has a bound quinone molecule, Q_A. Together, *L* and *M* coordinate a Fe atom. The photochemically active species of the *R. viridis* reaction center, **P870,** is composed of two bacteriochlorophylls, one contributed by *L* and the other by *M*.

Photosynthetic Electron Transfer in the *R. viridis* Reaction Center

The prosthetic groups of the *R. viridis* reaction center (P870, BChl, BPheo, and the bound quinones) are fixed in a spatial relationship to one another that favors photosynthetic e^- transfer (see Figure 18.14). Photoexcitation of P870 (creation of P870*) leads to e^- loss ($P870^+$) via electron transfer to the nearby bacteriochlorophyll (BChl). The e^- is then transferred via the *L* bacteriopheophytin (BPheo) to Q_A, which is also an *L* prosthetic group. The corresponding site on *M* is occupied by a loosely bound quinone, Q_B, and electron transfer from Q_A to Q_B takes place. An interesting aspect of the system is that *no* electron transfer occurs through *M*, even though it has components apparently symmetrical to and identical with the *L* e^- transfer pathway.

The reduced quinone formed at the Q_B site is free to diffuse to a neighboring cytochrome *b*/cytochrome c_1 membrane complex, where its oxidation is coupled to H^+ translocation (and, hence, ultimately to ATP synthesis) (Figure 18.15). Cytochrome c_2, a periplasmic protein, serves to cycle electrons back to $P870^+$ via the four hemes of the reaction-center cytochrome subunit. A specific tyrosine residue of *L* (Tyr^{162}) is situated between P870 and the closest cytochrome heme. This Tyr is the immediate e^- donor to $P870^+$ and completes the light-driven electron transfer cycle. The structure of the *R. viridis* reaction center (derived from X-ray crystallographic data) is modeled in Figure 18.16.

Eukaryotic Photosystems

Photosystems I and II of eukaryotic phototrophs (and cyanobacteria) are considerably more complex than the *R. viridis* reaction center. For example, PSII contains more than 20 subunits and PSI more than 11. Nevertheless, the

Figure 18.16 Model of the *R. viridis* reaction center. Two views **(a, b)** of the ribbon diagram ▶ of the reaction center. *M* and *L* subunits appear in purple and blue, respectively. Cytochrome subunit is brown; *H* subunit is green. These proteins provide a scaffold upon which the prosthetic groups of the reaction center are situated for effective photosynthetic electron transfer. Panel **(c)** shows the spatial relationship between the various prosthetic groups (4 hemes, P870, 2 BChl, 2 BPheo, 2 quinones, and the Fe atom) in the same view as in (b), but with protein chains deleted.

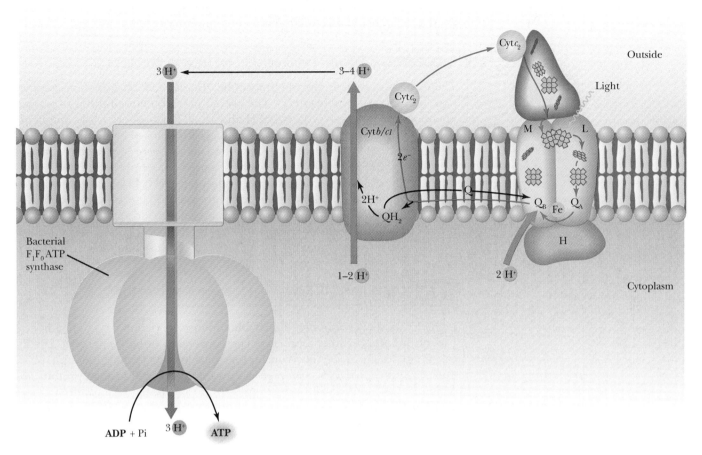

Figure 18.15 The *R. viridis* reaction center is coupled to the cytochrome b/c_1 complex through the quinone pool (Q). Quinone molecules are photoreduced at the reaction center Q_B site ($2\ e^-$ [$2\ h\nu$] per Q reduced) and then diffuse to the cytochrome b/c_1 complex, where they are reoxidized. Note that e^- flow from cytochrome b/c_1 back to the reaction center occurs via the periplasmic protein cytochrome c_2. Note also that 3 to 4 H^+ are translocated into the periplasmic space for each Q molecule oxidized at cytochrome b/c_1. The resultant proton-motive force drives ATP synthesis by the bacterial F_1F_0-ATP synthase. *(Adapted from Deisenhofer, J., and Michel, H., 1989. The photosynthetic reaction center from the purple bacterium Rhodopseudomonas viridis. Science **245**:1463.)*

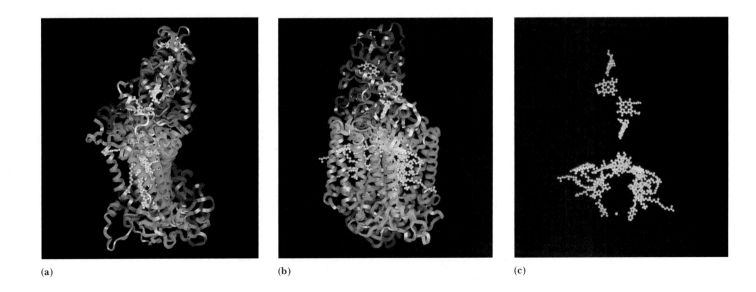

(a) (b) (c)

R. viridis reaction center is a fairly good model for the core structure and function of both PSI and PSII. These photosystems are surrounded by and connected to large Chl-based light-harvesting antenna systems.

Photosynthetic Energy Requirements for Hexose Synthesis

The fixation of carbon dioxide to form hexose, *the dark reactions of photosynthesis,* requires considerable energy. The overall stoichiometry of this process (see Equation [18.3]) involves 12 NADPH and 18 ATP. To generate 12 equivalents of NADPH necessitates the consumption of 48 Einsteins of light, minimally 170 kJ each. If $1\frac{1}{3}$ ATP are formed per NADPH (see *A Deeper Look,* below), only 16 ATP for CO_2 fixation would be produced from 48 Einsteins. Six additional Einsteins would provide the necessary two additional ATP. From 54 Einsteins, or 9180 kJ, one mole of hexose could be synthesized. The standard free energy change, $\Delta G°'$, for hexose formation from carbon dioxide and water (the exact reverse of cellular respiration) is +2870 kJ/mol.

18.6   Light-Driven ATP Synthesis: Photophosphorylation

Light-driven ATP synthesis, termed **photophosphorylation,** is a fundamental part of the photosynthetic process. The conversion of light energy to chemical energy results in electron-transfer reactions leading to the generation of reducing power (NADPH). Coupled with these electron transfers, protons are driven across the thylakoid membranes from the stromal side to the lumenal side. These proton translocations occur in a manner analogous to the proton translocations accompanying mitochondrial electron transport that provide the driving force for oxidative phosphorylation (see Chapter 17). Figure 18.12 indicates that proton translocations can occur at a number of sites. For example, protons can be translocated by reactions between H_2O and PSII as a consequence of the photolysis of water. The oxidation–reduction events as electrons pass through the plastoquinone pool and the Q cycle are another source of proton translocations. The proton transfer accompanying $NADP^+$ reduction also can be envisioned as protons being taken from the stromal side of the thylakoid vesicle. The current view is that two protons are translocated for each

---

A DEEPER LOOK

The Quantum Yield of Photosynthesis

The **quantum yield** of photosynthesis is defined as *the amount of product formed per equivalent of light input.* The quantum yield of photosynthesis traditionally has been expressed as the ratio of CO_2 fixed or O_2 evolved per quantum absorbed. At each reaction center, one photon or quantum yields one electron. Interestingly, an overall stoichiometry of one H^+ translocated into the thylakoid vesicle for each photon has also been observed. Two photons per center would allow a pair of electrons to flow from H_2O to $NADP^+$ (see Figure 18.12), resulting in the formation of 1 NADPH and $\frac{1}{2}O_2$. If 1 ATP were formed for every 3 H^+ translocated during photosynthetic electron transport, $1\frac{1}{3}$ ATP would be synthesized. More appropriately, 4 $h\nu$ per center (8 quanta total)

would drive the evolution of 1 O_2, the reduction of 2 $NADP^+$, and the phosphorylation of ATP.

The energy of a photon depends on its wavelength, according to the equation $E = h\nu = hc/\lambda$, where E is energy, c is the speed of light, and λ is its wavelength. Expressed in molar terms, the amount of energy in Avogadro's number (N) of photons is given by $E = Nhc/\lambda$; the energy in a mole of photons is called an *Einstein.* Light of 700-nm wavelength is the longest-wavelength and the lowest-energy light acting in the eukaryotic photosystems discussed here. An Einstein of 700-nm light is equivalent in energy to approximately 170 kJ. Eight Einsteins of this light, 1360 kJ, theoretically generate 2 moles of NADPH, $2\frac{2}{3}$ moles of ATP, and 1 mole of O_2.

electron that flows from H_2O to $NADP^+$. Because this electron transfer requires two photons—one falling at PSII and one at PSI—the overall yield is one proton per quantum of light.

The Mechanism of Photophosphorylation Is Chemiosmotic

The thylakoid membrane is asymmetrically organized, or "sided," like the mitochondrial membrane. It also shares the property of being a barrier to the passive diffusion of H^+ ions. Photosynthetic electron transport thus establishes an electrochemical gradient, or proton-motive force, across the thylakoid membrane with the interior, or lumen, side accumulating H^+ ions relative to the stroma of the chloroplast. Like oxidative phosphorylation, the mechanism of photophosphorylation is chemiosmotic.

A proton-motive force of approximately -180 mV is needed to achieve ATP synthesis. This proton-motive force, Δp, is composed of a membrane potential, $\Delta \Psi$ (negative outside), and a pH gradient, ΔpH ($pH_{out} - pH_{in}$) (see Chapter 17). The proton-motive force is defined as the free energy difference, ΔG, divided by $\mathcal{F}$, Faraday's constant:

$$\Delta p = \frac{\Delta G}{\mathcal{F}} = \Delta \Psi - \left(\frac{2.3\,RT}{\mathcal{F}}\right)\Delta pH \qquad (18.4)$$

At room temperature, $2.3RT/\mathcal{F} = 0.059V$, and

$$\Delta p = \Delta \Psi - 0.059\Delta pH \qquad (18.5)$$

In chloroplasts, the steady-state value of $\Delta \Psi$ is close to 0 mV, and the pH gradient is typically greater than 3 pH units, so that $\Delta p \approx -180$ mV. This situation contrasts with the mitochondrial proton-motive force, where the membrane potential contributes relatively more to Δp than does the pH gradient.

CF_1CF_0–ATP Synthase Is the Chloroplast Equivalent of the Mitochondrial F_1F_0–ATP Synthase

The transduction of the electrochemical gradient into the chemical energy represented by ATP is carried out by the chloroplast ATP synthase, which is highly analogous to the mitochondrial F_1F_0–ATP synthase. The chloroplast enzyme complex is called **CF_1CF_0–ATP synthase**, "C" symbolizing chloroplast. Like the mitochondrial complex, CF_1CF_0–ATP synthase is a heteromultimer of α, β, γ, δ, and ε subunits, as well as a, b, and c subunits (see Chapter 17), consisting of a knoblike structure some 9 nm in diameter (CF_1) attached to a stalked base (CF_0) embedded in the thylakoid membrane. The mechanism of action of CF_1CF_0–ATP synthase in coupling ATP synthesis to the potential energy of the pH gradient is similar to that of the mitochondrial ATP synthase described in Chapter 17. The mechanism of photophosphorylation is summarized schematically in Figure 18.17.

Cyclic and Noncyclic Photophosphorylation

Photosynthetic electron transport, which pumps H^+ into the thylakoid lumen, can occur in two modes: **cyclic photophosphorylation** and **noncyclic photophosphorylation.** Both modes lead to the establishment of a transmembrane proton-motive force. Thus, both modes are coupled to ATP synthesis. The two modes are distinguished by differences in their electron transfer pathways. Noncyclic photophosphorylation has been the focus of our discussion thus far and is represented by the scheme in Figure 18.17, where electrons activated by quanta at PSII and PSI flow from H_2O to $NADP^+$, with concomitant

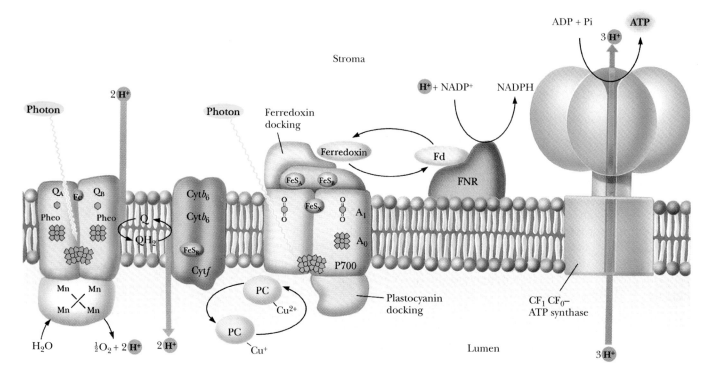

Figure 18.17 The mechanism of photophosphorylation. Photosynthetic electron transport establishes a proton gradient that is tapped by the CF_1CF_0–ATP synthase to drive ATP synthesis. Critical to this mechanism is the fact that the membrane-bound components of light-induced electron transport and ATP synthesis are asymmetrical with respect to the thylakoid membrane so that vectorial discharge and uptake of H^+ ensue, generating the proton-motive force.

establishment of the proton-motive force driving ATP synthesis. Note that in noncyclic photophosphorylation, O_2 is evolved and $NADP^+$ is reduced.

Cyclic Photophosphorylation

In cyclic photophosphorylation, the "electron hole" in $P700^+$ created by electron loss from $P700^*$ is filled not by an electron derived from H_2O via PSII but by a cyclic pathway in which the photo-excited electron returns ultimately to $P700^+$. This pathway is schematically represented in Figure 18.12 by the dashed line connecting F_B and cytochrome b_6. Thus, one function of cytochrome b_6 (b_{563}) is to couple the bound ferredoxin carriers of the PSI complex with the cytochrome b_6/cytochrome f complex via the plastoquinone pool. This pathway diverts the activated e^- from $NADP^+$ reduction back through plastocyanin to re-reduce $P700^+$ (Figure 18.18).

Proton translocations accompany these cyclic electron transfer events, so ATP synthesis can be achieved. In cyclic photophosphorylation, ATP is the sole product of energy conversion. No NADPH is generated, and, because PSII is not involved, no oxygen is evolved. Cyclic photophosphorylation depends only on PSI. The maximal rate of cyclic photophosphorylation is less than 5% of the rate of noncyclic photophosphorylation.

18.7 Carbon Dioxide Fixation

As we began this chapter, we saw that photosynthesis traditionally is equated with the process of CO_2 fixation—that is, the net synthesis of carbohydrate from CO_2. Indeed, the capacity to perform net accumulation of carbohydrate from CO_2 distinguishes the phototrophic (and autotrophic) organisms from

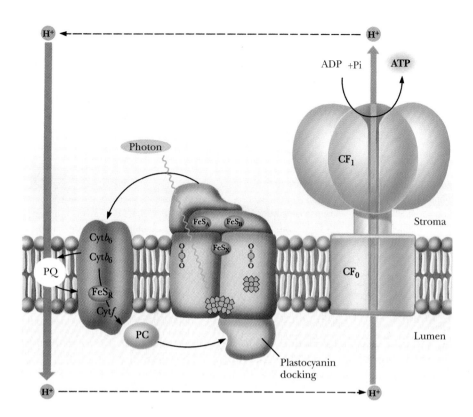

Figure 18.18 The pathway of cyclic photophosphorylation by PSI. *(Adapted from Arnon, D. I., 1984.* Trends in Biochemical Sciences *9:258.)*

heterotrophs. Although animals possess enzymes capable of linking CO_2 to organic acceptors, they cannot achieve a *net* accumulation of organic material by these reactions. For example, fatty acid biosynthesis is primed by covalent attachment of CO_2 to acetyl-CoA to form malonyl-CoA (see Chapter 20). Nevertheless, this "fixed CO_2" is liberated in the very next reaction, so no net CO_2 incorporation occurs.

Elucidation of the pathway of CO_2 fixation represents one of the earliest applications of radioisotope tracers to the study of biology. In 1945, Melvin Calvin and his colleagues at the University of California at Berkeley were investigating photosynthetic CO_2 fixation in *Chlorella*. Using $^{14}CO_2$, they traced the incorporation of radioactive ^{14}C into organic products and found that the earliest labeled product was **3-phosphoglycerate** (see Figure 14.12). Although this result suggested that the CO_2 acceptor was a two-carbon compound, further investigation revealed that, in reality, two equivalents of 3-phosphoglycerate were formed following addition of CO_2 to a five-carbon (pentose) sugar:

$$CO_2 + 5\text{-carbon acceptor} \rightarrow [6\text{-carbon intermediate}] \rightarrow \text{two 3-phosphoglycerates}$$

Ribulose-1,5-Bisphosphate Is the CO_2 Acceptor in CO_2 Fixation

The five-carbon CO_2 acceptor was identified as **ribulose-1,5-bisphosphate** (RuBP), and the enzyme catalyzing this key reaction of CO_2 fixation is **ribulose bisphosphate carboxylase/oxygenase,** or, in the jargon used by workers in this field, **rubisco.** The name ribulose bisphosphate carboxylase/oxygenase reflects the fact that rubisco catalyzes the reaction of either CO_2 or, alternatively, O_2 with RuBP. Rubisco is found in the chloroplast stroma. It is a very abundant enzyme, constituting more than 15% of the total chloroplast protein (Figure 18.19). Given the preponderance of plant material in the biosphere, rubisco is probably the world's most abundant protein.

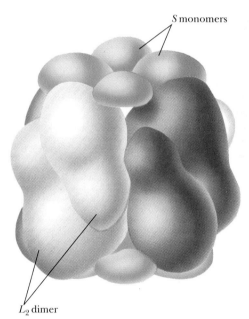

S monomers

L_2 dimer

Figure 18.19 Schematic diagram of the subunit organization of ribulose bisphosphate carboxylase as revealed by X-ray crystallography. Higher plant rubisco is a large heteromultimeric enzyme consisting of eight equivalents each of two types of subunits, large L (55 kD) and small S (15 kD). Clusters of four small subunits are located at each end of the symmetrical octamer formed by four L_2 dimers. The large subunit is the catalytic unit of the enzyme. It binds both substrates (CO_2 and RuBP) and Mg^{2+} (a divalent cation essential for enzymatic activity). The small subunit modulates the activity of the enzyme, increasing its k_{cat} more than 100-fold. *(From Knight, S., Andersson, I., and Branden, C. I., 1990.* Journal of Molecular Biology *215:113–160.)*

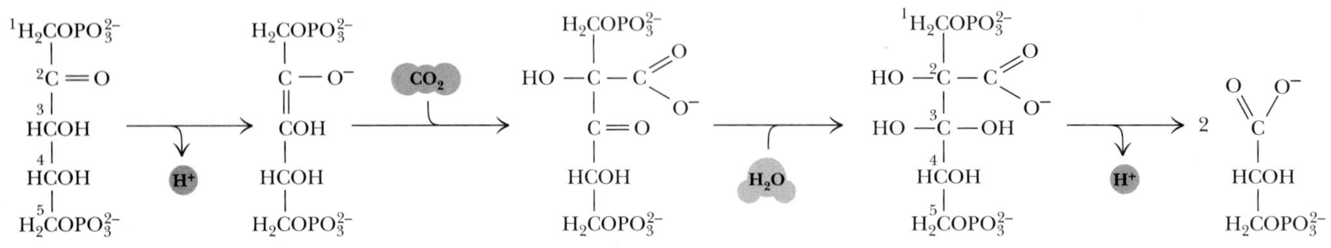

Figure 18.20 The ribulose bisphosphate carboxylase reaction. Enzymatic abstraction of the C-3 proton of RuBP yields a 2,3-enediol intermediate (I). CO_2 is added at C-2 to create the six-carbon β-keto acid intermediate (II) known as 2-carboxy, 3-keto-arabinitol. Intermediate II is rapidly hydrated to give III. Deprotonation of the C-3 hydroxyl and cleavage yield two 3-phosphoglycerates. Mg^{2+} at the active site aids in stabilizing the transition state (I) for CO_2 addition and in facilitating the carbon-carbon bond cleavage that leads to product formation.

The Ribulose-1,5-Bisphosphate Carboxylase Reaction

The addition of CO_2 to ribulose-1,5-bisphosphate results in the formation of an enzyme-bound intermediate, **2-carboxy, 3-keto-arabinitol** (Figure 18.20). This intermediate arises when CO_2 adds to the enediol intermediate generated from ribulose-1,5-bisphosphate. Hydrolysis of the C_2—C_3 bond of the intermediate generates two molecules of 3-phosphoglycerate. The CO_2 ends up as the carboxyl group of one of the two product molecules.

18.8 The Calvin–Benson Cycle

The immediate product of CO_2 fixation, 3-phosphoglycerate, undergoes a series of transformations leading to accumulation of carbohydrate products. Among carbohydrates, hexoses (particularly glucose) occupy center stage. Glucose is the building block for both cellulose and starch synthesis. These plant polymers constitute the most abundant organic material in the living world; thus, the central focus on glucose as the ultimate end product of CO_2 fixation is amply justified. Also, sucrose (α-D-glucopyranosyl-$(1 \rightarrow 2)$-β-D-fructofuranoside) is the major carbon form translocated out of leaves to other plant tissues. In nonphotosynthetic tissues, sucrose is metabolized via glycolysis and the TCA cycle to produce ATP.

The set of reactions that transforms 3-phosphoglycerate into hexose is named the **Calvin–Benson cycle** (often referred to simply as the Calvin cycle) for its discoverers, Melvin Calvin and Andrew Benson. The reaction series is indeed cyclic because not only must carbohydrate appear as an end product, but the 5-carbon acceptor, RuBP, must be regenerated to provide for continual CO_2 fixation. Balanced equations that schematically represent this situation are

$$6(1) + 6(5) \longrightarrow 12(3)$$

$$12(3) \longrightarrow 1(6) + 6(5)$$

$$Net: \quad 6(1) \longrightarrow 1(6)$$

Each number in parentheses represents the number of carbon atoms in a compound, and the number preceding the parentheses indicates the stoichiometry of the reaction. Thus, 6(1), or 6 CO_2, condense with 6(5) or 6 RuBP to give 12 3-phosphoglycerates. These 12(3)s are then rearranged in the Calvin cycle to form one hexose, 1(6), and regenerate the six 5-carbon (RuBP) acceptors. The overall net reaction is the fixation of 6 CO_2 molecules to form one glucose molecule.

The Enzymes of the Calvin Cycle

The Calvin cycle enzymes serve three important ends:

1. They constitute the predominant CO_2 fixation pathway in nature.
2. They accomplish the reduction of 3-phosphoglycerate, the primary product of CO_2 fixation, to glyceraldehyde-3-phosphate so that carbohydrate synthesis becomes feasible.
3. They catalyze reactions that transform 3-carbon compounds into 4-, 5-, 6-, and 7-carbon compounds.

Most of the enzymes mediating the reactions of the Calvin cycle also participate in either glycolysis (see Chapter 15) or the pentose phosphate pathway (see Chapter 19). The aim of the Calvin scheme is to account for hexose formation from 3-phosphoglycerate. In the course of this metabolic sequence, the NADPH and ATP produced in the light reactions are consumed, as indicated earlier in Equation (18.3).

The Calvin cycle of reactions starts with ribulose bisphosphate carboxylase catalyzing formation of 3-phosphoglycerate from CO_2 and RuBP and concludes with **ribulose-5-phosphate kinase** (also called phosphoribulose kinase), which forms RuBP (Figure 18.21 and Table 18.2). The carbon balance is given at the right side of the table. Several features of the reactions shown merit discussion. Note that the 18 equivalents of ATP consumed in hexose formation are expended in reactions 2 and 15: 12 to form 12 equivalents of 1,3-bisphosphoglycerate from 3-phosphoglycerate by a reversal of the normal glycolytic reaction catalyzed by **3-phosphoglycerate kinase,** and 6 to phosphorylate Ru-5-P to regenerate 6 RuBP. All 12 NADPH equivalents are used in reaction 3. Plants possess an **NADPH-specific glyceraldehyde-3-phosphate dehydrogenase,** which contrasts with its glycolytic counterpart in its specificity for NADP over NAD and in the direction in which the reaction normally proceeds.

Balancing the Calvin Cycle Reactions to Account for Net Hexose Synthesis

When carbon rearrangements are balanced to account for net hexose synthesis, five of the glyceraldehyde-3-phosphate molecules are converted to dihydroxyacetone phosphate (DHAP). Three of these DHAPs then condense with three glyceraldehyde-3-P via the aldolase reaction to yield three hexoses in the form of fructose bisphosphate (see Figure 18.21). (Recall that the $\Delta G^{\circ\prime}$ for the aldolase reaction in the glycolytic direction is +23.9 kJ/mol. Thus, the aldolase reaction running "in reverse" in the Calvin cycle would be thermodynamically favored under standard-state conditions.) Taking one FBP to glucose, the desired product of this scheme, leaves 30 carbons, distributed as two fructose-6-phosphates, four glyceraldehyde-3-phosphates, and two DHAP. These 30 Cs are reorganized into six RuBP by reactions 9 through 15. Step 9 and steps 12 through 14 involve carbohydrate rearrangements like those in the pentose phosphate pathway (see Chapter 19). Reaction 11 is mediated by **sedoheptulose-1,7-bisphosphatase.** This phosphatase is unique to plants; it generates sedoheptulose-7-P, the seven-carbon sugar serving as the transketolase substrate. Likewise, **phosphoribulose kinase** carries out the unique plant function of providing RuBP from Ru-5-P (reaction 15). The net conversion accounts for the fixation of six equivalents of carbon dioxide into one hexose at the expense of 18 ATP and 12 NADPH.

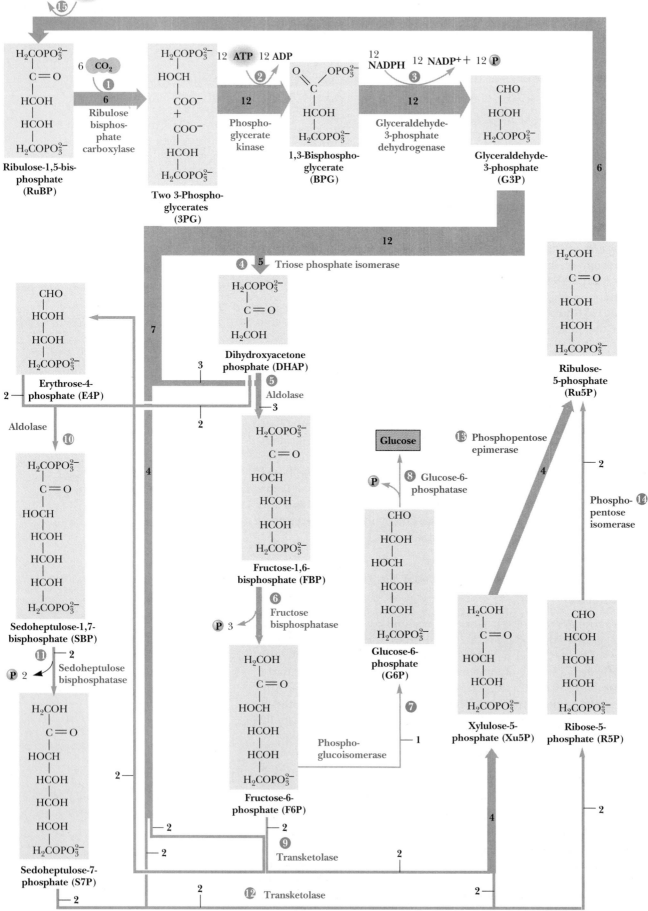

6 ADP   6 ATP   Phosphoribulose kinase
⑮

$H_2COPO_3^{2-}$ $H_2COPO_3^{2-}$ 12 ATP   12 ADP   12   12 NADP$^+$ + 12 P
| | ②   NADPH   ③
C=O   6 CO$_2$ HOCH   O   OPO$_3^{2-}$ CHO
| ① | 12   \ //   12 |
HCOH   COO$^-$ C   HCOH
| 6   +   12 | 12 |
HCOH   COO$^-$ HCOH $H_2COPO_3^{2-}$
| | Phospho- | Glyceraldehyde-
$H_2COPO_3^{2-}$ HCOH   glycerate $H_2COPO_3^{2-}$ 3-phosphate
                          | kinase   dehydrogenase
Ribulose-1,5-bis- $H_2COPO_3^{2-}$ 1,3-Bisphospho-   Glyceraldehyde-
phosphate   glycerate   3-phosphate
(RuBP)   Two 3-Phospho-   (BPG)   (G3P)
                          glycerates
                          (3PG)   6

                                                    12

                                            ④   5   Triose phosphate isomerase
CHO $H_2COPO_3^{2-}$ H_2COH
| | |
HCOH   7   C=O   C=O
| | |
HCOH H_2COH HCOH
| |
$H_2COPO_3^{2-}$ Dihydroxyacetone   HCOH
                                    3   phosphate (DHAP) |
Erythrose-4-   ⑤ $H_2COPO_3^{2-}$
2 — phosphate (E4P)   3   Aldolase
                                    2   Ribulose-
                                                                                                         5-phosphate
Aldolase $H_2COPO_3^{2-}$ (Ru5P)
       ⑩ |
                                              C=O   Glucose   ⑬ Phosphopentose
$H_2COPO_3^{2-}$ | epimerase
| HOCH   P   2
C=O | ⑧ Glucose-6-
| HCOH   phosphatase   4
HOCH | 2
| 4   HCOH   CHO
HCOH | | Phospho- ⑭
| $H_2COPO_3^{2-}$ HCOH   pentose
HCOH | isomerase
| Fructose-1,6-   HOCH
$H_2COPO_3^{2-}$ bisphosphate (FBP) |
                                               ⑥   HCOH
Sedoheptulose-1,7-   Fructose |
bisphosphate (SBP)   P 3   bisphosphatase   HCOH
     ⑪   2 | H_2COH
                                                                         $H_2COPO_3^{2-}$ |
P 2   Sedoheptulose   C=O   CHO
            bisphosphatase H_2COH Glucose-6- | |
                                                | phosphate   HOCH   HCOH
H_2COH C=O   (G6P) | |
| | HCOH   HOCH
C=O   HOCH | |
| | ⑦ $H_2COPO_3^{2-}$ HCOH
HOCH   HCOH |
| 2 | Xylulose-5- $H_2COPO_3^{2-}$
HCOH   HCOH   1   phosphate
| | (Xu5P)   Ribose-5-
HCOH $H_2COPO_3^{2-}$ Phospho-   phosphate (R5P)
| glucoisomerase   4
HCOH   Fructose-6-
| phosphate (F6P)
$H_2COPO_3^{2-}$ ⑨   4
                                     2   Transketolase   2
Sedoheptulose-7-   2   2
phosphate (S7P)   2   ⑫   Transketolase   2
         2   2

◀ **Figure 18.21** The Calvin–Benson cycle of reactions. The number associated with the arrow at each step indicates the number of molecules reacting in a turn of the cycle that produces one molecule of glucose. Reactions are numbered as in Table 18.2.

 See pages 173–178

Regulation of Carbon Dioxide Fixation

Plant cells contain mitochondria and can carry out cellular respiration (glycolysis, the citric acid cycle, and oxidative phosphorylation) to provide energy in the dark. Futile cycling of carbohydrate to CO_2 by glycolysis and the citric acid cycle in one direction, and CO_2 to carbohydrate by the CO_2 fixation pathway in the opposite direction, is thwarted through regulation of the Calvin cycle (Figure 18.22). In this regulation, the activities of key Calvin cycle enzymes are coordinated with the output of photosynthesis. In effect, these enzymes respond indirectly to light activation. Thus, when light energy is available to generate ATP and NADPH for CO_2 fixation, the Calvin cycle proceeds. In the dark, when ATP and NADPH cannot be produced by photosynthesis, fixation of CO_2 ceases.

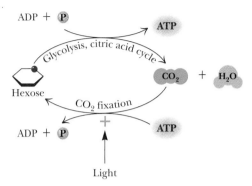

Figure 18.22 Light regulation of CO_2 fixation prevents a substrate cycle between cellular respiration and hexose synthesis by CO_2 fixation. Because plants possess mitochondria and are capable of deriving energy from hexose catabolism (glycolysis and the citric acid cycle), regulation of photosynthetic CO_2 fixation by light activation controls the net flux of carbon between these opposing routes.

18.9 The Ribulose Bisphosphate Oxygenase Reaction: Photorespiration

As indicated, ribulose bisphosphate carboxylase/oxygenase catalyzes an alternative reaction in which O_2 replaces CO_2 as the substrate added to RuBP (Figure 18.23). The *ribulose-1,5-bisphosphate oxygenase* reaction diminishes plant productivity because it leads to loss of RuBP, the essential CO_2 acceptor. The K_m for O_2 in this oxygenase reaction is about 200 μM. Given the relative abundance of CO_2 and O_2 in the atmosphere and their relative K_m values in these rubisco-mediated reactions, the ratio of carboxylase to oxygenase activity *in vivo* is about 3 or 4 to 1.

▌ **Table 18.2 The Calvin Cycle Series of Reactions**

Reactions 1 through 15 constitute the cycle that leads to the formation of one equivalent of glucose. The enzyme catalyzing each step, a concise reaction, and the overall carbon balance is given. Numbers in parentheses show the numbers of carbon atoms in the substrate and product molecules. Prefix numbers indicate in a stoichiometric fashion how many times each step is carried out in order to provide a balanced net reaction.

1. Ribulose bisphosphate carboxylase: 6 CO_2 + 6 H_2O + 6 RuBP ⟶ 12 3-PG   6(1) + 6(5) ⟶ 12(3)
2. 3-Phosphoglycerate kinase: 12 3-PG + 12 ATP ⟶ 12 1,3-BPG + 12 ADP   12(3) ⟶ 12(3)
3. NADP⁺-glyceraldehyde-3-P dehydrogenase:
    12 1,3-BPG + 12 NADPH ⟶ 12 NADP⁺ + 12 G3P + 12 P_i 12(3) ⟶ 12(3)
4. Triose-P isomerase: 5 G3P ⟶ 5 DHAP   5(3) ⟶ 5(3)
5. Aldolase: 3 G3P + 3 DHAP ⟶ 3 FBP   3(3) + 3(3) ⟶ 3(6)
6. Fructose bisphosphatase: 3 FBP + 3 H_2O ⟶ 3 F6P + 3 P_1 3(6) ⟶ 3(6)
7. Phosphoglucoisomerase: 1 F6P ⟶ 1 G6P   1(6) ⟶ 1(6)
8. Glucose phosphatase: 1 G6P + 1 H_2O ⟶ 1 GLUCOSE + 1 P_i 1(6) ⟶ 1(6)
    The remainder of the pathway involves regenerating six RuBP acceptors (= 30 C) from the leftover two F6P (12 C), four G3P (12 C), and two DHAP (6 C).
9. Transketolase: 2 F6P + 2 G3P ⟶ 2 Xu5P + 2 E4P   2(6) + 2(3) ⟶ 2(5) + 2(4)
10. Aldolase: 2 E4P + 2 DHAP ⟶ 2 sedoheptulose-1,7-bisphosphate (SBP)   2(4) + 2(3) ⟶ 2(7)
11. Sedoheptulose bisphosphatase: 2 SBP + 2 H_2O ⟶ 2 S7P + 2 P_i 2(7) ⟶ 2(7)
12. Transketolase: 2 S7P + 2 G3P ⟶ 2 Xu5P + 2 R5P   2(7) + 2(3) ⟶ 4(5)
13. Phosphopentose epimerase: 4 Xu5P ⟶ 4 Ru5P   4(5) ⟶ 4(5)
14. Phosphopentose isomerase: 2 R5P ⟶ 2 Ru5P   2(5) ⟶ 2(5)
15. Phosphoribulose kinase: 6 Ru5P + 6 ATP ⟶ 6 RuBP + 6 ADP   6(5) ⟶ 6(5)
Net: 6 CO_2 + 18 ATP + 12 NADPH + 12 H⁺ + 12 H_2O ⟶ glucose + 18 ADP + 18 P_i + 12 NADP⁺   6(1) ⟶ 1(6)

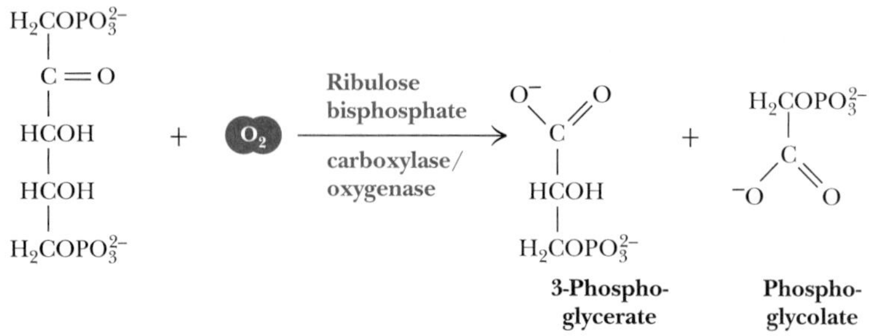

Figure 18.23 The oxygenase reaction of rubisco. The reaction of ribulose bisphosphate carboxylase with O_2 in the presence of ribulose bisphosphate leads to wasteful cleavage of RuBP to yield 3-phosphoglycerate and phosphoglycolate. Breakdown of phosphoglycolate in mitochondria leads to CO_2 release, the source of the CO_2 evolved in photorespiration.

The products of ribulose bisphosphate oxygenase activity are 3-phosphoglycerate and phosphoglycolate. Phosphoglycolate breakdown can lead to CO_2 release. Obviously, agricultural productivity is dramatically lowered by this CO_2 loss, which, because it is a light-related uptake of O_2 and release of CO_2, is termed **photorespiration.** Certain plants, particularly tropical grasses, have evolved the means to circumvent photorespiration. These plants are more efficient users of light for carbohydrate synthesis.

18.10 The C-4 Pathway of CO_2 Fixation

Tropical grasses are less susceptible to the effects of photorespiration, as noted earlier. Studies employing $^{14}CO_2$ as a tracer indicated that the first organic intermediate labeled in these plants was not a three-carbon compound but a four-carbon compound. Marshall Hatch and Rodger Slack, two Australian biochemists, first discovered this C-4 product of CO_2 fixation, and the C-4 pathway of CO_2 incorporation is named the *Hatch–Slack pathway* after them. The C-4 pathway is not an alternative to the Calvin cycle series of reactions or even a net CO_2 fixation scheme. Instead, it functions as a CO_2 delivery system, carrying carbon dioxide from the relatively oxygen-rich surface of the leaf to interior cells where oxygen is lower in concentration and hence less effective in competing with CO_2 in the rubisco reaction. Thus, the C-4 pathway is a means of avoiding photorespiration by sheltering the rubisco reaction in a cellular compartment away from high $[O_2]$. The C-4 compounds serving as CO_2 transporters are *malate* or *aspartate.*

Compartmentation of these reactions to prevent photorespiration involves the interaction of two cell types, *mesophyll cells* and *bundle sheath cells.* The mesophyll cells take up CO_2 at the leaf surface, where O_2 is abundant, and use it to carboxylate phosphoenolpyruvate (PEP) to yield oxaloacetate (OAA) in a reaction catalyzed by **PEP carboxylase** (Figure 18.24). This four-carbon dicarboxylic acid (following conversion into either malate or aspartate) is then transported to the bundle sheath cells, where it is decarboxylated to yield CO_2 and a 3-C product. The CO_2 is then fixed into organic carbon by the Calvin cycle localized within the bundle sheath cells, and the 3-C product is returned to the mesophyll cells, where it is reconverted to PEP in preparation to accept another CO_2 (see Figure 18.24). Plants that use the C-4 pathway are termed **C4 plants,** in contrast to those plants with the conventional pathway of CO_2 uptake **(C3 plants).** The transport of each CO_2 requires the expenditure of two high-energy phosphate bonds. The energy of these bonds is expended in the formation of PEP from pyruvate, ATP, and P_i by the plant enzyme **pyruvate-P_i dikinase;** the products are PEP, AMP, and pyrophosphate (PP_i). The overall reaction is: ATP + pyruvate + P_i → AMP + PEP + PP_i.

Crassulacean Acid Metabolism

In contrast to C4 plants, which have separated CO_2 uptake and fixation into distinct cells in order to minimize photorespiration, succulent plants native to semiarid and tropical environments separate CO_2 uptake and fixation in time. Carbon dioxide (as well as O_2) enters the leaf through microscopic pores known as **stomata,** and water vapor escapes from plants via these same openings. In nonsucculent plants, the stomata are open during the day, when light can drive photosynthetic CO_2 fixation, and closed at night. Succulent plants, such as the *Cactaceae* (cacti) and *Crassulaceae,* cannot open their stomata during the heat of day because any loss of precious H_2O in their arid habitats would doom them. Instead, these plants open their stomata to take up CO_2 only at night, when temperatures are lower and water loss is less likely. This carbon dioxide is immediately incorporated into PEP to form four-carbon dicarboxylic acids, which are stored within vacuoles until morning. During the day, these acids are released from the vacuoles and decarboxylated to yield CO_2 and a 3-C product. The CO_2 is then fixed into organic carbon by rubisco and the reactions of the Calvin cycle. Because this process involves the accumulation of organic acids (OAA, malate) and is common to succulents of the *Crassulaceae* family, it is referred to as *crassulacean acid metabolism,* and plants capable of it are called *CAM plants.*

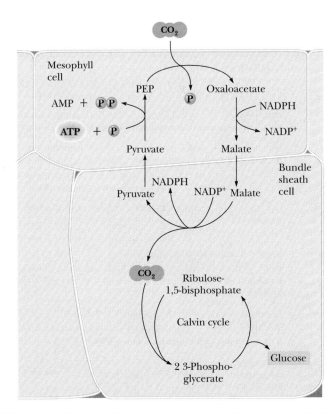

Figure 18.24 Essential features of the compartmentation and biochemistry in the Hatch–Slack pathway of carbon dioxide uptake in C4 plants. Carbon dioxide is fixed into organic linkage by PEP carboxylase of mesophyll cells, forming OAA. Either malate (the reduced form of OAA) or aspartate (the aminated form) serves as the carrier transporting CO_2 to the bundle sheath cells. Within the bundle sheath cells, CO_2 is liberated by decarboxylation of malate or aspartate; the 3-C product is returned to the mesophyll cell. Formation of PEP by pyruvate-P_i dikinase completes the cycle. The CO_2 liberated in the bundle sheath cell is used to synthesize hexose by the conventional rubisco–Calvin cycle series of reactions.

PROBLEMS

1. In photosystem I, P700 in its ground state has an $\mathscr{E}_\circ' = 0.4$ V. Excitation of P700 by a photon of 700-nm light alters the $\mathscr{E}_\circ'$ of P700* to -0.6 V. What is the efficiency of energy capture in this light reaction of P700?

2. What is the $\mathscr{E}_\circ'$ for the light-generated primary oxidant of photosystem II if the light-induced oxidation of water (which leads to O_2 evolution) proceeds with a $\Delta G^{\circ\prime}$ of -25 kJ/mol?

3. Assuming that the concentrations of ATP, ADP, and P_i in chloroplasts are 3 mM, 0.1 mM, and 10 mM, respectively, what is the ΔG for ATP synthesis under these conditions? Photosynthetic electron transport establishes the proton-motive force driving photophosphorylation. What redox potential difference is necessary to achieve ATP synthesis under the foregoing conditions, assuming an electron pair is transferred per molecule of ATP generated?

4. Write a balanced equation for the synthesis of a glucose molecule from ribulose-1,5-bisphosphate and CO_2 that involves the first three reactions of the Calvin cycle and subsequent conversion of the two glyceraldehyde-3-P molecules into glucose.

5. If noncyclic photosynthetic electron transport leads to the translocation of 3 H^+/e^- and cyclic photosynthetic electron transport leads to the translocation of 2 H^+/e^-, what is the relative photosynthetic efficiency of ATP synthesis (expressed as the number of photons absorbed per ATP synthesized) for noncyclic versus cyclic photophosphorylation? (Assume that the CF_1CF_0–ATP synthase yields 1 ATP/3 H^+.)

6. The overall equation for photosynthetic CO_2 fixation is

$$6\,CO_2 + 6\,H_2O \rightarrow C_6H_{12}O_6 + 6\,O_2$$

All the O atoms evolved as O_2 come from water; none comes from carbon dioxide. But 12 O atoms are evolved as 6 O_2, and only 6 O atoms appear as 6 H_2O in the equation. Also, 6 CO_2 have 12 O atoms, yet there are only 6 O atoms in $C_6H_{12}O_6$. How can you account for these discrepancies? (*Hint*: Consider the partial reactions of photosynthesis: ATP synthesis, $NADP^+$ reduction, photolysis of water, and the overall reaction for hexose synthesis in the Calvin–Benson cycle.)

FURTHER READING

Bendall, D. S., and Manasse, R. S., 1995. Cyclic photophosphorylation and electron transport. *Biochimica et Biophysica Acta* **1229:**23–38.

Berry, S., and Rumberg, B., 1996. H^+/ATP coupling ratio at the unmodulated CF_1CF_0–ATP synthase determined by proton flux measurements. *Biochimica et Biophysica Acta* **1276:**51–56.

Blankenship, R. E., and Hartman, H., 1998. The origin and evolution of oxygenic photosynthesis. *Trends in Biochemical Sciences* **23:**94–97.

Cramer, W. A., and Knaff, D. B., 1990. *Energy Transduction in Biological Membranes: A Textbook of Bioenergetics.* New York: Springer-Verlag. 545 pages. A textbook on bioenergetics by two prominent workers in photosynthesis.

Cushman, J. C., and Bohnert, H. J., 1999. Crassulacean acid metabolism: Molecular genetics. *Annual Review of Plant Physiology and Plant Molecular Biology* **50:**305–332.

Deisenhofer, J., and Michel, H., 1989. The photosynthetic reaction center from the purple bacterium *Rhodopseudomonas viridis. Science* **245:**1463–1473. Published version of the Nobel laureate address by the researchers who first elucidated the molecular structure of a photosynthetic reaction center.

Deisenhofer, J., Michel, H., and Huber, R., 1985. The structural basis of light reactions in bacteria. *Trends in Biochemical Sciences* **10:**243–248.

Green, B. R., and Durnford, D. G., 1996. The chlorophyll-carotenoid proteins of oxygenic photosynthesis. *Annual Review of Plant Physiology and Plant Molecular Biology* **47:**685–714.

Hankamer, B., Barber, J., and Boekema, E. J., 1997. Structure and membrane organization of photosystem II in green plants. *Annual Review of Plant Physiology and Plant Molecular Biology* **48:**641–671.

Harold, F. M., 1986. *The Vital Force: A Study of Bioenergetics.* Chapter 8: Harvesting the Light. San Francisco: Freeman & Company.

Hatch, M. D., 1987. C_4 photosynthesis: A unique blend of modified bio-

chemistry, anatomy and ultrastructure. *Biochimica et Biophysica Acta* **895:**81–106. A review of C4 biochemistry by its discoverer.

Hoganson, C. W., and Babcock, G. T., 1997. A metalloradical mechanism for the generation of oxygen in photosynthesis. *Science* **277:**1953–1956.

Jagendorf, A. T., and Uribe, E., 1966. ATP formation caused by acid–base transition of spinach chloroplasts. *Proceedings of the National Academy of Sciences, U.S.A.* **55:**170–177. The classic paper providing the first experimental verification of Mitchell's chemiosmotic hypothesis.

Kaplan, A., and Reinhold, L., 1999. CO_2-concentrating mechanisms in photosynthetic organisms. *Annual Review of Plant Physiology and Plant Molecular Biology* **50:**539–570.

Knight, S., Andersson, I., and Branden, C. I., 1990. Crystallographic analysis of ribulose 1,5-bisphosphate carboxylase from spinach at 2.4 Å resolution. Subunit interactions and active site. *Journal of Molecular Biology* **215:**113–160.

Krauss, N., et al., 1996. Photosystem I at 4 Å resolution represents the first structural model of a joint photosynthetic reaction centre and core antenna system. *Nature Structural Biology* **3:**965–973.

Miziorko, H. M., and Lorimer, G. H., 1983. Ribulose-1,5-bisphosphate carboxylase/oxygenase. *Annual Review of Biochemistry* **52:**507–535. An early review of the enzymological properties of rubisco.

Portis, A. R., Jr., 1992. Regulation of ribulose 1,5-bisphosphate carboxylase/oxygenase activity. *Annual Review of Plant Physiology and Plant Molecular Biology* **43:**415–437.

Rogner, M., Boekema, E. J., and Barber, J., 1996. How does photosystem 2 split water? The structural basis of energy conversion. *Trends in Biochemical Sciences* **21:**44–49.

Ting, I. P., 1985. Crassulacean acid metabolism. *Annual Review of Plant Physiology* **36:**595–622.

Gluconeogenesis, Glycogen Metabolism, and the Pentose Phosphate Pathway

Con pan y vino se anda el camino. (With bread and wine you can walk your road.)

Spanish proverb

Bread and pastries on a rack at a French bakery in Paris. Carbohydrates such as these provide a significant portion of human caloric intake. (Steven Rothfeld/Tony Stone Images)

As shown in Chapters 14 and 15, the metabolism of sugars is an important source of energy for cells. Animals, including humans, typically obtain substantial amounts of glucose and other sugars from the breakdown of starch and glycogen in their diets. Alternatively, glucose is supplied via breakdown of cellular reserves of glycogen (in animals) or starch (in plants). Significantly, glucose also can be synthesized from noncarbohydrate precursors by a process known as gluconeogenesis. Each of these important pathways, as well as the synthesis of glycogen from glucose, will be examined in this chapter.

Another pathway of glucose catabolism, the pentose phosphate pathway, is the primary source of NADPH, the reduced coenzyme essential to most reductive biosynthetic processes. For example, NADPH is crucial to the biosynthesis of fatty acids (see Chapter 20) and amino acids (see Chapter 21). The pentose phosphate pathway also results in the production of ribose-5-

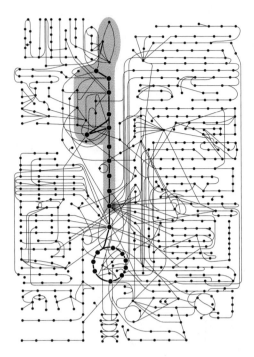

Gluconeogenesis, glycogen metabolism, and the pentose phosphate pathway.

See *Interactive Biochemistry CD-ROM and Workbook,* pages 173–178

phosphate, an essential component of ATP, NAD$^+$, FAD, coenzyme A, and particularly DNA and RNA. This important pathway will also be considered in this chapter.

19.1 Gluconeogenesis

The ability to synthesize glucose from common metabolites is very important to most organisms. Human metabolism, for example, utilizes about 160 ± 20 grams of glucose per day, about 75% of this in the brain. Body fluids carry only about 20 grams of free glucose, and glycogen stores normally can provide only about 180 to 200 grams of free glucose. Thus, the body carries only a little more than a one-day supply of glucose. If glucose is not obtained in the diet, the body must produce new glucose from noncarbohydrate precursors. The term for this activity is **gluconeogenesis,** which means the generation (genesis) of new (neo) glucose.

Further, muscles consume large amounts of glucose via glycolysis, producing large amounts of pyruvate. In vigorous exercise, muscle cells become anaerobic and pyruvate is converted to lactate. Gluconeogenesis salvages this pyruvate and lactate and reconverts it to glucose.

The Substrates of Gluconeogenesis

In addition to pyruvate and lactate, other noncarbohydrate precursors can be used as substrates for gluconeogenesis in animals. These include most of the amino acids, as well as glycerol and all the TCA cycle intermediates. On the other hand, fatty acids are not substrates for gluconeogenesis in animals, because most fatty acids yield only acetyl-CoA upon degradation, and animals cannot carry out net synthesis of sugars from acetyl-CoA. Lysine and leucine are the only amino acids that are not substrates for gluconeogenesis. These amino acids produce only acetyl-CoA upon degradation. Acetyl-CoA becomes a substrate for gluconeogenesis only in tissues where the glyoxylate cycle is operating (see Chapter 16).

Nearly All Gluconeogenesis Occurs in the Liver and Kidneys in Animals

Interestingly, the mammalian organs that consume the most glucose—namely, brain and muscle—carry out very little glucose synthesis. The major sites of gluconeogenesis are the liver and kidneys, which account for about 90% and 10% of the body's gluconeogenic activity, respectively. Glucose produced by gluconeogenesis in the liver and kidney is released into the blood and is subsequently absorbed by brain, heart, muscle, and red blood cells to meet their metabolic needs. In turn, pyruvate and lactate produced in these tissues are returned to the liver and kidney to be used as gluconeogenic substrates.

Gluconeogenesis Is Not Merely the Reverse of Glycolysis

In some ways, gluconeogenesis is the reverse, or antithesis, of glycolysis. Glucose is synthesized, not catabolized; ATP is consumed, not produced; and NADH is oxidized to NAD$^+$ rather than the other way around. However, gluconeogenesis cannot be merely the reversal of glycolysis for two reasons. First, glycolysis is exergonic, with a $\Delta G^{\circ\prime}$ of approximately -74 kJ/mol. If gluconeogenesis were merely the reverse, it would be a strongly endergonic process

HUMAN BIOCHEMISTRY

The Chemistry of Glucose-Monitoring Devices

Individuals with diabetes must measure their serum glucose concentration frequently, often several times a day. The advent of computerized, automated devices for glucose monitoring has made this necessary chore easier, and far more accurate and convenient than it once was. These devices all use a simple chemical scheme for glucose measurement that involves oxidation of glucose to gluconic acid by glucose oxidase. This reaction produces two molecules of hydrogen peroxide per molecule of glucose oxidized. The H_2O_2 is then used to oxidize a dye, such as o-dianisidine, to create a colored product that can be measured:

$$\text{Glucose} + 2\,H_2O + O_2 \longrightarrow \text{gluconic acid} + 2\,H_2O_2$$
$$o\text{-dianisidine (colorless)} + H_2O_2 \longrightarrow$$
$$\text{oxidized } o\text{-anisidine (colored)} + H_2O$$

The amount of colored dye produced is directly proportional to the amount of glucose in the sample.

The patient typically applies a drop of blood (from a finger prick) to a plastic test strip, which is then inserted into the glucose monitor.* Within half a minute, a digital readout indicates the blood glucose value. Modern glucose monitors store several days of glucose measurements and the data can be easily transferred to a computer for analysis and graphing.

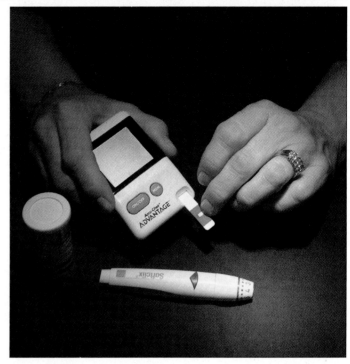

(Charles Grisham)

*How does the monitor "know" it has gotten the right amount of blood? The blood flows up an absorbent "wick" by capillary action. It is not possible to overfill this device, but the monitor will give an error signal if not enough blood flows up the strip.

and could not occur spontaneously. Somehow the energetics of the process must be augmented so that gluconeogenesis can proceed spontaneously. Second, the processes of glycolysis and gluconeogenesis must be regulated in a reciprocal fashion so that when glycolysis is active, gluconeogenesis is inhibited, and when gluconeogenesis is proceeding, glycolysis is turned off. Both these limitations are overcome by having unique reactions within the routes of glycolysis and gluconeogenesis, rather than a completely shared pathway.

Gluconeogenesis: Something Borrowed, Something New

The complete route of gluconeogenesis is shown in Figure 19.1, side by side with the glycolytic pathway. Gluconeogenesis employs three different reactions, catalyzed by three different enzymes, for the three steps of glycolysis that are highly exergonic (and highly regulated). In essence, seven of the ten steps of glycolysis are merely reversed in gluconeogenesis. The six reactions between PEP and fructose-1,6-bisphosphate are shared by the two pathways, as is the isomerization between glucose-6-P and fructose-6-P. The three exergonic regulated reactions—the hexokinase (glucokinase), phosphofructokinase, and pyruvate kinase reactions—are replaced by alternative reactions in the gluconeogenic pathway.

The conversion of pyruvate to PEP that initiates gluconeogenesis is accomplished by two unique reactions. **Pyruvate carboxylase** catalyzes the first, converting pyruvate to oxaloacetate (OAA). Then, **PEP carboxykinase** catalyzes

Figure 19.1 The pathways of gluconeogenesis and glycolysis. Species in blue, green, and peach-colored shaded boxes indicate other entry points for gluconeogenesis (in addition to pyruvate).

See pages 173–178

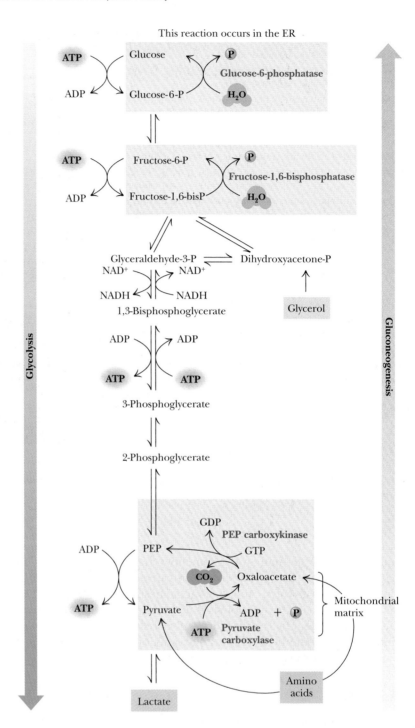

the conversion of oxaloacetate to PEP. Conversion of fructose-1,6-bisphosphate to fructose-6-phosphate is catalyzed by a specific phosphatase, **fructose-1,6-bisphosphatase.** The final step to produce glucose—hydrolysis of glucose-6-phosphate—is mediated by **glucose-6-phosphatase.** Each of these steps is considered in detail in the following paragraphs. The overall conversion of pyruvate to PEP by pyruvate carboxylase and PEP carboxykinase has a $\Delta G°'$ close to zero but is pulled along by subsequent reactions. The conversion of fructose-1,6-bisphosphate to glucose in the last three steps of gluconeogenesis is strongly exergonic with a $\Delta G°'$ of about -30.5 kJ/mol. This sequence of two phosphatase reactions separated by an isomerization accounts for most of the free energy release that makes the gluconeogenesis pathway spontaneous.

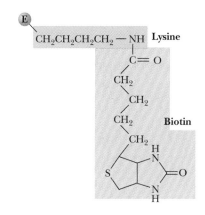

Figure 19.2 The pyruvate carboxylase reaction.

The Unique Reactions of Gluconeogenesis

(1) Pyruvate Carboxylase: A Biotin-Dependent Enzyme

Initiation of gluconeogenesis occurs in the **pyruvate carboxylase reaction**—the conversion of pyruvate to oxaloacetate (Figure 19.2). The reaction takes place in two discrete steps, involves ATP and bicarbonate as substrates, and utilizes biotin as a coenzyme and acetyl-coenzyme A as an allosteric activator. Pyruvate carboxylase is a tetrameric enzyme (with a molecular mass of about 500 kD). Each monomer possesses a biotin covalently linked to the ε-amino group of a lysine residue at the active site (Figure 19.3).

Pyruvate Carboxylase Is Allosterically Activated By Acyl-Coenzyme A. Two particularly interesting aspects of the pyruvate carboxylase reaction are (a) allosteric activation of the enzyme by acyl-coenzyme A derivatives, and (b) compartmentation of the reaction in the mitochondrial matrix. The carboxylation of biotin requires the presence (at an allosteric site) of acetyl-coenzyme A or another acylated coenzyme A derivative. The second half of the carboxylase reaction—the attack by pyruvate to form oxaloacetate—is not affected by CoA derivatives.

Activation of pyruvate carboxylase by acetyl-CoA provides an important physiological regulation. Acetyl-CoA is the primary substrate for the TCA cycle, and oxaloacetate (formed by pyruvate carboxylase) is an important intermediate in both the TCA cycle and the gluconeogenesis pathway. If levels of ATP or acetyl-CoA (or other acyl-CoAs) are low, pyruvate is directed primarily into the TCA cycle, which eventually promotes the synthesis of ATP. If ATP or acetyl-CoA levels are high, pyruvate is converted to oxaloacetate and consumed in gluconeogenesis. Clearly, high levels of ATP and CoA derivatives are signs that energy is abundant and that metabolites will be converted to glucose (and perhaps even glycogen). If the energy status of the cell is low (in terms of ATP and CoA derivatives), pyruvate is consumed in the TCA cycle. Also, as noted in Chapter 16, pyruvate carboxylase is an important anaplerotic enzyme. Its activation by acetyl-CoA leads to oxaloacetate formation, replenishing the level of TCA cycle intermediates.

Compartmentalized Pyruvate Carboxylase Depends on Metabolite Conversion and Transport. The second interesting feature of pyruvate carboxylase is that it is found only in the matrix of the mitochondria (Figure 19.4). By contrast, the next enzyme in the gluconeogenic pathway, PEP carboxykinase, may be localized in the cytosol or in the mitochondria or both. For example, rabbit liver PEP carboxykinase is predominantly mitochondrial, whereas the rat liver enzyme is strictly cytosolic. In human liver, PEP carboxykinase is found both in the cytosol and in the mitochondria. Pyruvate is transported into the mitochondrial matrix, where it can be converted to acetyl-CoA (for use in the TCA cycle) and then to citrate (for fatty acid synthesis; see Figure 20.21). Alternatively, it may be converted directly to OAA by pyruvate carboxylase and used in gluconeogenesis. In tissues where PEP carboxykinase is found only in the mitochondria, oxaloacetate is converted to PEP, which is then transported to the cytosol for gluconeogenesis. However, in tissues that must convert some

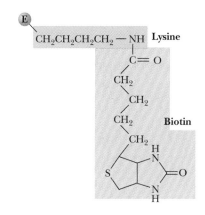

Figure 19.3 Covalent linkage of biotin to an active-site lysine in pyruvate carboxylase.

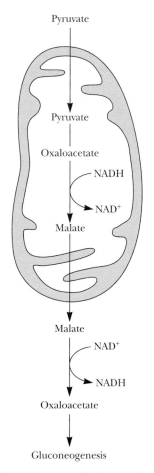

Figure 19.4 Pyruvate carboxylase catalyzes a compartmentalized reaction. Pyruvate is converted to oxaloacetate in the mitochondria. Because oxaloacetate cannot be transported across the mitochondrial membrane, it must be reduced to malate, transported to the cytosol, and then oxidized back to oxaloacetate before gluconeogenesis can continue.

Figure 19.5 The PEP carboxykinase reaction.

Oxaloacetate   PEP

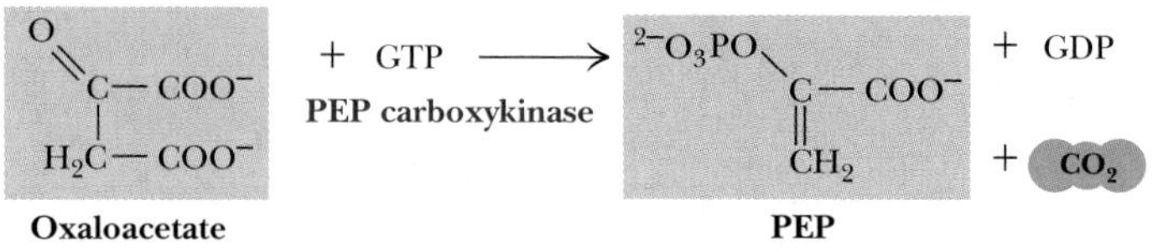

oxaloacetate to PEP in the cytosol, a problem arises. Oxaloacetate cannot be transported directly across the mitochondrial membrane. Instead, it must first be transformed into malate or aspartate (see Figure 19.4). Cytosolic malate and aspartate are reconverted to oxaloacetate before continuing along the gluco-neogenic route.

(2) PEP Carboxykinase

The second reaction in the gluconeogenic pyruvate–PEP bypass is the conversion of oxaloacetate to PEP. Production of a high-energy metabolite such as PEP requires energy. The energetic requirements are handled in two ways here. First, the CO_2 added to pyruvate in the pyruvate carboxylase step is removed in the PEP carboxykinase reaction. Decarboxylation is a favorable process and helps to drive the formation of the very high-energy enol phosphate in PEP. This decarboxylation drives a reaction that would otherwise be highly endergonic. Note the inherent metabolic logic in this pair of reactions: pyruvate carboxylase consumed an ATP to drive a carboxylation, so that the PEP carboxykinase could use the decarboxylation to facilitate formation of PEP. Second, as shown in Figure 19.5, another high-energy phosphate is consumed by the carboxykinase. Mammals and several other species use GTP in this reaction, rather than ATP. The use of GTP here is equivalent to the consumption of an ATP, due to the activity of the **nucleoside diphosphate kinase** (see Figure 16.2). The substantial free energy of the hydrolysis of GTP is crucial to the synthesis of PEP in this step. The overall ΔG for the pyruvate carboxylase and PEP carboxykinase reactions under physiological conditions in the liver is -22.6 kJ/mol. Once PEP is formed in this way, the phosphoglycerate mutase, phosphoglycerate kinase, glyceraldehyde-3-P dehydrogenase, aldolase, and triose phosphate isomerase reactions act to eventually form fructose-1,6-bisphosphate, as in Figure 19.1.

(3) Fructose-1,6-Bisphosphatase

The hydrolysis of fructose-1,6-bisphosphate to fructose-6-phosphate (Figure 19.6), like all phosphate ester hydrolyses, is a thermodynamically favorable (exergonic) reaction under standard-state conditions ($\Delta G^{\circ\prime} = -16.7 \text{ kJ/mol}$). Under physiological conditions in the liver, the reaction is also exergonic ($\Delta G = -8.6 \text{ kJ/mol}$). **Fructose-1,6-bisphosphatase** is an allosterically regulated enzyme. Citrate stimulates bisphosphatase activity, but fructose-2,6-bisphosphate is a potent allosteric inhibitor. AMP also inhibits the bisphosphatase, and the inhibition by AMP is enhanced by fructose-2,6-bisphosphate.

Fructose-1,6-bisphosphate   Fructose-6-phosphate

$$\Delta G^{\circ\prime} = -16.7 \text{ kJ/mol}$$

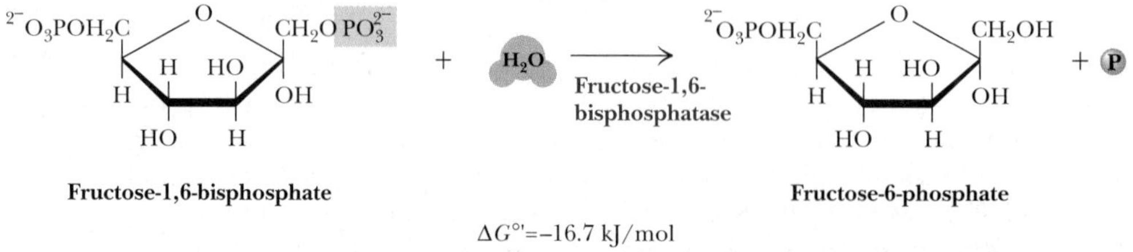

Figure 19.6 The fructose-1,6-bisphosphatase reaction.

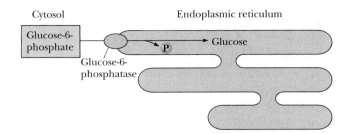

Cytosol · Endoplasmic reticulum

Glucose-6-phosphate → Glucose

Glucose-6-phosphatase

Figure 19.7 Glucose-6-phosphatase is localized in the endoplasmic reticulum (ER) membrane. Conversion of glucose-6-phosphate to glucose occurs during transport into the ER.

(4) Glucose-6-Phosphatase

The final step in the gluconeogenesis pathway is the conversion of glucose-6-phosphate to glucose by the action of **glucose-6-phosphatase.** Its membrane association is important to its function because the substrate is hydrolyzed as it passes into the endoplasmic reticulum itself (Figure 19.7). Vesicles form from the endoplasmic reticulum membrane and move to the plasma membrane, where they fuse with it, releasing their glucose contents into the bloodstream. The glucose-6-phosphatase reaction involves a phosphohistidine intermediate (Figure 19.8). The ΔG for the glucose-6-phosphatase reaction in liver is -5.1 kJ/mol.

Coupling with Hydrolysis of ATP and GTP Drives Gluconeogenesis. The net reaction for the conversion of pyruvate to glucose in gluconeogenesis is

$$2 \text{ Pyruvate} + 4 \text{ ATP} + 2 \text{ GTP} + 2 \text{ NADH} + 2 \text{ H}^+ + 6 \text{ H}_2\text{O} \longrightarrow$$
$$\text{glucose} + 4 \text{ ADP} + 2 \text{ GDP} + 6 \text{ P}_i + 2 \text{ NAD}^+$$

The net free energy change, $\Delta G^{\circ\prime}$, for this conversion is -37.7 kJ/mol. The consumption of a total of six nucleoside triphosphates drives this process forward. If glycolysis were merely reversed to achieve the net synthesis of glucose from pyruvate, the net reaction would be

$$2 \text{ Pyruvate} + 2 \text{ ATP} + 2 \text{ NADH} + 2 \text{ H}^+ + 2 \text{ H}_2\text{O} \longrightarrow$$
$$\text{glucose} + 2 \text{ ADP} + 2 \text{ P}_i + 2 \text{ NAD}^+$$

and the overall $\Delta G^{\circ\prime}$ would be about $+74$ kJ/mol. Such a process would be highly endergonic, and therefore thermodynamically unfeasible. Hydrolysis of four additional high-energy phosphate bonds makes gluconeogenesis thermodynamically favorable. Under physiological conditions, however, gluconeogenesis is somewhat less favorable than at standard state, with an overall ΔG of -15.6 kJ/mol for the conversion of pyruvate to glucose.

Lactate Formed in Muscles Is Recycled to Glucose in the Liver. A final point on the redistribution of lactate and glucose in the body serves to emphasize the metabolic interactions between organs. Vigorous exercise can lead to oxygen shortage (anaerobic conditions), so that energy requirements must be met

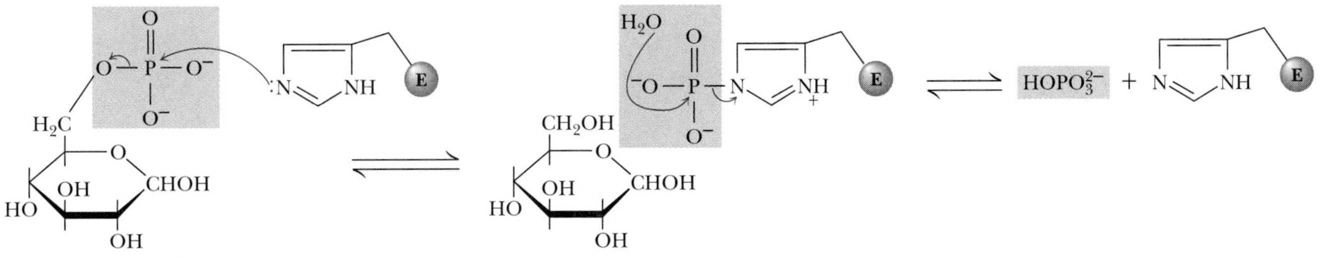

Figure 19.8 The glucose-6-phosphatase reaction involves formation of a phosphohistidine intermediate.

HUMAN BIOCHEMISTRY

Gluconeogenesis Inhibitors and Other Diabetes Therapy Strategies

Diabetes, which is the inability to assimilate and metabolize serum glucose, afflicts millions of people. Type I diabetics are unable to produce insulin. On the other hand, Type II diabetics make sufficient insulin, but the cellular response to insulin is defective. Many Type II diabetics exhibit a condition called **insulin resistance** even before the onset of diabetes. **Metformin** (see figure) is a drug that improves sensitivity to insulin, primarily by stimulating glucose uptake by glucose transporters in peripheral tissues. It also increases binding of insulin to insulin receptors, stimulates tyrosine kinase activity of the insulin receptor (see Chapter 26), and inhibits gluconeogenesis in the liver.

Gluconeogenesis inhibitors may be the next wave in diabetes therapy. Drugs that block gluconeogenesis without affecting glycolysis would target one of the enzymes unique to gluconeogenesis. 3-Mercaptopicolinate and hydrazine specifically inhibit PEP carboxykinase, and **chlorogenic acid,** a natural product found in the skin of peaches, inhibits the transport activity of the glucose-6-phosphatase system (but not the glucose-6-phosphatase enzyme activity). The drug S-3483, a derivative of chlorogenic acid, also inhibits the glucose-6-phosphatase transport activity and binds a thousand times more tightly to the transporter than chlorogenic acid. Drugs of this type may be useful in treating Type II diabetes.

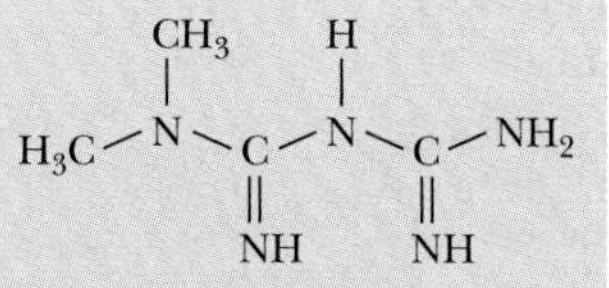

Metformin

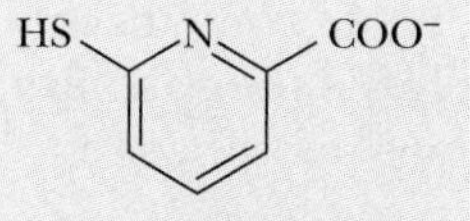

3-Mercaptopicolinate

Hydrazine

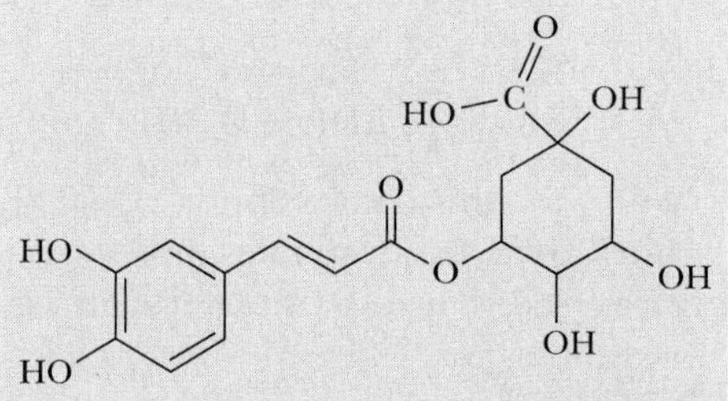

Chlorogenic acid

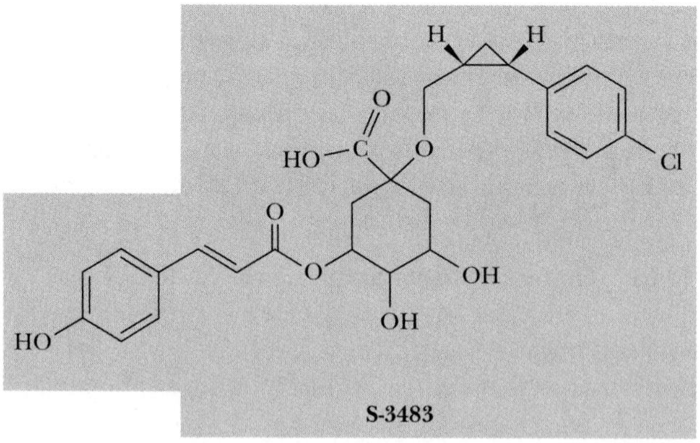

S-3483

by increased levels of glycolysis. Under such conditions, glycolysis converts NAD^+ to NADH, yet O_2 is unavailable for regeneration of NAD^+ via cellular respiration. Instead, large amounts of NADH are reoxidized by the reduction of pyruvate to lactate. The lactate thus produced can be transported from muscle to the liver, where it is reoxidized by lactate dehydrogenase to yield pyruvate, which is converted eventually to glucose. In this way, the liver shares in the metabolic stress created by vigorous exercise. It exports glucose to muscle, which produces lactate, which can be processed by the liver into new glucose.

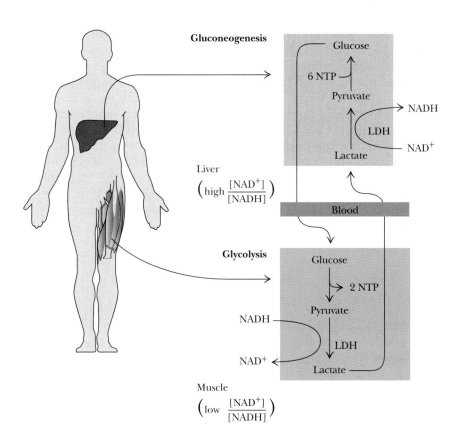

Figure 19.9 The Cori cycle.

This is referred to as the **Cori cycle** (Figure 19.9). Liver, with a typically high $NAD^+/NADH$ ratio (about 700), readily produces more glucose than it can use. Muscle that is vigorously exercising will enter anaerobiosis and show a decreasing $NAD^+/NADH$ ratio, which favors reduction of pyruvate to lactate.

19.2 Regulation of Gluconeogenesis

Nearly all of the reactions of glycolysis and gluconeogenesis take place in the cytosol. If metabolic control were not exerted over these reactions, glycolytic degradation of glucose and gluconeogenic synthesis of glucose could operate

CRITICAL DEVELOPMENTS IN BIOCHEMISTRY

The Pioneering Studies of Carl and Gerty Cori

The Cori cycle is named for Carl and Gerty Cori, who received the Nobel Prize in physiology or medicine in 1947 for their studies of glycogen metabolism and blood glucose regulation. Carl Ferdinand Cori and Gerty Theresa Radnitz were both born in Prague (then part of Austria). They earned medical degrees from the German University of Prague in 1920 and were married later that year. They joined the faculty of the Washington University School of Medicine in St. Louis, Missouri, in 1931. Their remarkable collaboration resulted in many fundamental advances in carbohydrate and glycogen metabolism. They were credited with the discovery of glucose-1-phosphate, also known at the time as the "Cori ester."

They also showed that glucose-6-phosphate was produced from glucose-1-P by the action of phosphoglucomutase. They isolated and crystallized glycogen phosphorylase and elucidated the pathway of glycogen breakdown. In 1952, they showed that absence of glucose-6-phosphatase in the liver was the enzymatic defect in von Gierke's disease, an inherited glycogen-storage disease. Six eventual Nobel laureates received training in their laboratory. Gerty Cori was the first American woman to receive a Nobel Prize. Carl Cori said of their remarkable collaboration: "Our efforts have been largely complementary and one without the other would not have gone so far."

simultaneously, with no net benefit to the cell and with considerable consumption of ATP. This is prevented by a sophisticated system of **reciprocal control,** so that glycolysis is inhibited when gluconeogenesis is active, and vice versa. Reciprocal regulation of these two pathways depends largely on the energy status of the cell. When the energy status of the cell is low, glucose is rapidly degraded to produce needed energy. When the energy status is high, pyruvate and other metabolites are utilized for synthesis (and storage) of glucose.

In glycolysis, the three regulated enzymes are those catalyzing the strongly exergonic reactions: hexokinase (glucokinase), phosphofructokinase, and pyruvate kinase. As noted, the gluconeogenic pathway replaces these three reactions with corresponding reactions that are exergonic in the direction of glucose synthesis: glucose-6-phosphatase, fructose-1,6-bisphosphatase, and the pyruvate carboxylase–PEP carboxykinase pair, respectively. These are the three most appropriate sites of regulation in gluconeogenesis.

Gluconeogenesis Is Regulated by Allosteric and Substrate-Level Control Mechanisms

The mechanisms of regulation of gluconeogenesis are shown in Figure 19.10. Control is exerted at all of the predicted sites, but in different ways. Glucose-6-phosphatase is not under allosteric control. However, the K_m for the sub-

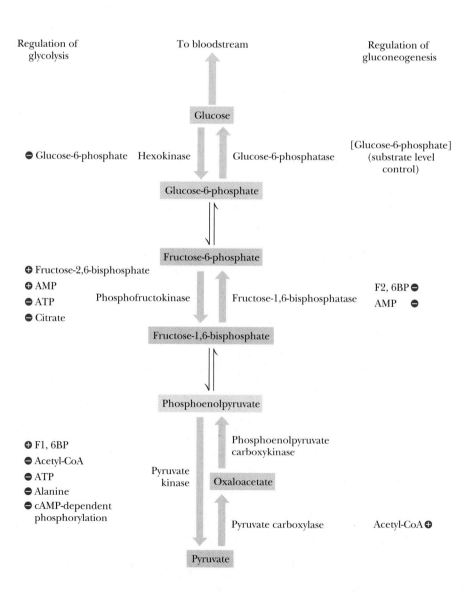

Figure 19.10 The principal regulatory mechanisms in glycolysis and gluconeogenesis. Activators are indicated by plus signs and inhibitors by minus signs.

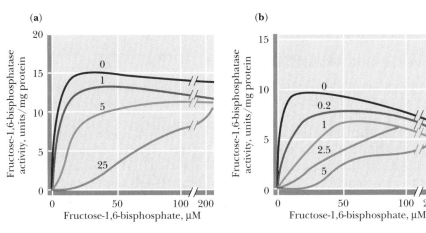

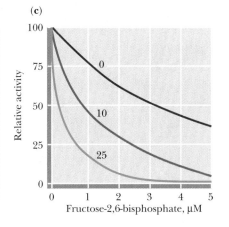

Figure 19.11 Inhibition of fructose-1,6-bisphosphatase by fructose-2,6-bisphosphate in the **(a)** absence and **(b)** presence of 25 μM AMP. In (a) and (b), enzyme activity is plotted against substrate (fructose-1,6-bisphosphate) concentration. Concentrations of fructose-2,6-bisphosphate (in μM) are indicated above each curve. **(c)** The effect of AMP (0, 10, and 25 μM) on the inhibition of fructose-1,6-bisphosphatase by fructose-2,6-bisphosphate. Activity was measured in the presence of 10 μM fructose-1,6-bisphosphate. *(Adapted from Van Schaftingen, E., and Hers, H.-G., 1981. Inhibition of fructose-1,6-bisphosphatase by fructose-2,6-bisphosphate. Proceedings of the National Academy of Sciences, U.S.A.* **78:***2861–2863.)*

strate, glucose-6-phosphate, is considerably higher than the normal range of substrate concentrations. As a result, glucose-6-phosphatase displays a near-linear dependence of activity on substrate concentrations and is thus said to be under **substrate-level control** by glucose-6-phosphate.

Acetyl-CoA is a potent allosteric effector of glycolysis and gluconeogenesis. It allosterically inhibits pyruvate kinase (as noted in Chapter 15) and activates pyruvate carboxylase. Because it also allosterically inhibits pyruvate dehydrogenase (the enzymatic link between glycolysis and the TCA cycle), the cellular fate of pyruvate is strongly dependent on acetyl-CoA levels. A rise in [acetyl-CoA] indicates that cellular energy levels are high and that carbon metabolites can be directed to glucose synthesis and storage. When acetyl-CoA levels drop, the activities of pyruvate kinase and pyruvate dehydrogenase increase and flux through the TCA cycle increases, providing needed energy for the cell.

Fructose-1,6-bisphosphatase is another important site of gluconeogenic regulation. This enzyme is inhibited by AMP and activated by citrate. These effects by AMP and citrate are the opposite of those exerted on phosphofructokinase in glycolysis, providing another example of reciprocal regulatory effects. When AMP levels increase, gluconeogenic activity is diminished and glycolysis is stimulated. An increase in citrate concentration signals that TCA cycle activity can be curtailed and that pyruvate should be directed to sugar synthesis instead.

Fructose-2,6-Bisphosphate: Allosteric Regulator of Gluconeogenesis

As described in Chapter 15, Emile Van Schaftingen and Henri-Géry Hers demonstrated in 1980 that fructose-2,6-bisphosphate is a potent stimulator of phosphofructokinase. Cognizant of the reciprocal nature of regulation in glycolysis and gluconeogenesis, Van Schaftingen and Hers also considered the possibility of an opposite effect—inhibition—for fructose-1,6-bisphosphatase. In 1981 they reported that fructose-2,6-bisphosphate was indeed a powerful inhibitor of fructose-1,6-bisphosphatase (Figure 19.11). Inhibition occurs in either the presence or absence of AMP, and the effects of AMP and fructose-2,6-bisphosphate are synergistic.

Fructose-2,6-bisphosphate

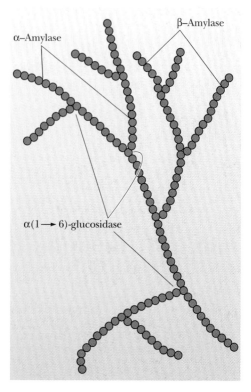

Figure 19.12 Hydrolysis of glycogen and starch by α-amylase and β-amylase. In contrast to α-amylase (an endoglucosidase), β-amylase is an exoglucosidase.

19.3 Glycogen Catabolism

Dietary Glycogen and Starch Breakdown

As noted earlier, well-fed adult human beings normally metabolize about 160 g of carbohydrates each day. A balanced diet easily provides this amount, mostly in the form of starch, with smaller amounts of glycogen. If too little carbohydrate is supplied by the diet, glycogen reserves in liver and muscle tissue can also be mobilized. The reactions by which ingested starch and glycogen are digested are shown in Figure 19.12. The enzyme known as **α-amylase** is an important component of saliva and pancreatic juice. (**β-Amylase** is found in plants. The α- and β-designations for these enzymes serve only to distinguish the two, and do not refer to glycosidic linkage nomenclature.) α-Amylase is an endoglycosidase that hydrolyzes $\alpha(1 \rightarrow 4)$ linkages of amylopectin and glycogen at random positions, eventually producing a mixture of maltose, maltotriose [with three $\alpha(1 \rightarrow 4)$-linked glucose residues], and other small oligosaccharides. α-Amylase can cleave on either side of a glycogen or amylopectin branch point, but activity is reduced in highly branched regions of the polysaccharide and stops four residues from any branch point.

Metabolism of Tissue Glycogen

Digestion itself is a highly efficient process in which almost 100% of ingested food is absorbed and metabolized. Digestive breakdown of starch and glycogen is an unregulated process. On the other hand, tissue glycogen represents an important reservoir of potential energy, and it should be no surprise that the reactions involved in its degradation and synthesis are carefully controlled and regulated. Glycogen reserves in liver and muscle tissue are stored in the cytosol as granules exhibiting a molecular weight range from 6×10^6 to 1600×10^6 (1.6×10^9). These granular aggregates contain the enzymes required to synthesize and catabolize the glycogen as well as all the enzymes of glycolysis. The principal enzyme of glycogen catabolism is **glycogen phosphorylase,** a highly regulated enzyme that was discussed extensively in Chapter 10.

19.4 Glycogen Synthesis

Animals synthesize and store glycogen when glucose levels are high, but the synthetic pathway is not merely a reversal of the glycogen phosphorylase reaction. High levels of phosphate in the cell favor glycogen breakdown and prevent the phosphorylase reaction from synthesizing glycogen *in vivo*, in spite of the fact that $\Delta G°'$ for the phosphorylase reaction actually favors glycogen synthesis. Hence, another reaction pathway must be employed in the cell for the net synthesis of glycogen. In essence, this pathway must activate glucose units for transfer to glycogen chains.

Glucose Units Are Activated for Transfer by Formation of Sugar Nucleotides

We are familiar with several examples of chemical activation as a strategy for group transfer reactions. Acetyl-CoA is an activated form of acetate; biotin and tetrahydrofolate activate one-carbon groups for transfer; and ATP is an activated form of phosphate. Luis Leloir, a biochemist in Argentina, showed in the 1950s that glycogen synthesis depended upon **sugar nucleotides,** which may be

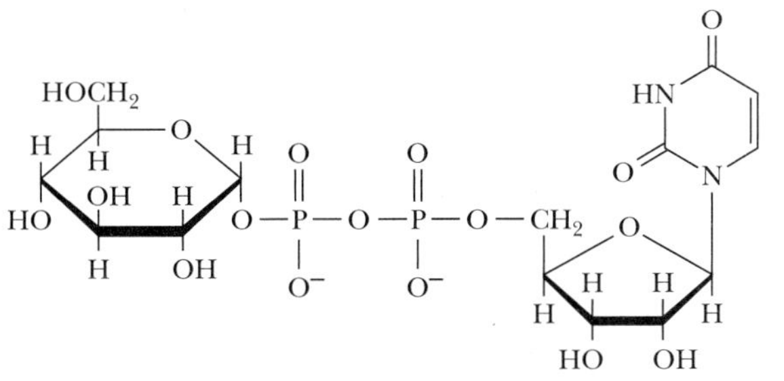

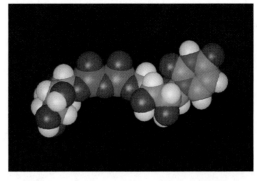

**Uridine diphosphate glucose
(UDPG)**

Figure 19.13 The structure of UDP-glucose, a sugar nucleotide.

thought of as activated forms of sugar units (Figure 19.13). For example, formation of an ester linkage between the C-1 hydroxyl group and the β-phosphate of UDP activates the glucose moiety of **UDP-glucose.**

UDP-Glucose Synthesis Is Driven by Pyrophosphate Hydrolysis

Sugar nucleotides are formed from sugar-1-phosphates and nucleoside triphosphates by specific **pyrophosphorylase** enzymes (Figure 19.14). For example, **UDP-glucose pyrophosphorylase** catalyzes the formation of UDP-glucose from glucose-1-phosphate and uridine 5′-triphosphate:

$$\text{Glucose-1-P} + \text{UTP} \longrightarrow \text{UDP-glucose} + \text{pyrophosphate}$$

The reaction proceeds via attack by a phosphate oxygen of glucose-1-phosphate on the α-phosphorus of UTP, with departure of the pyrophosphate anion. The

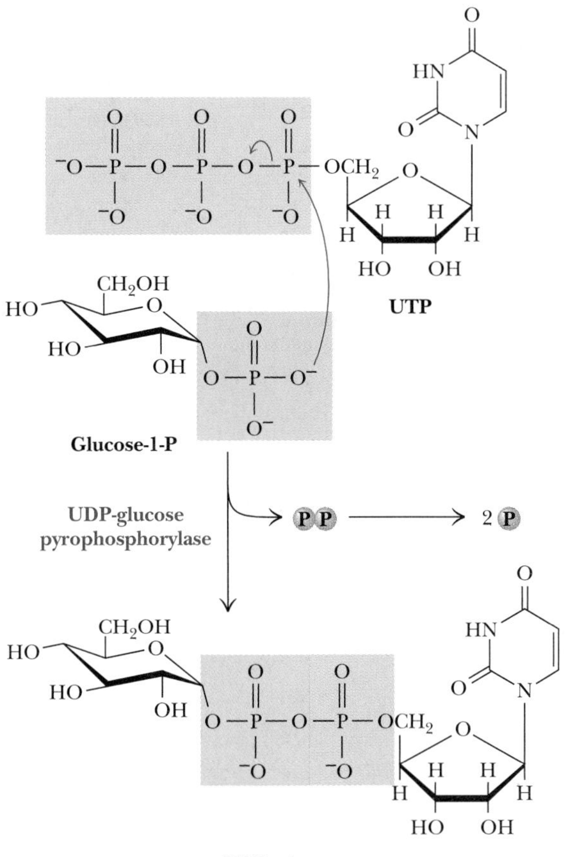

UTP

Glucose-1-P

UDP-glucose
pyrophosphorylase

UDP–glucose

Figure 19.14 The UDP-glucose pyrophosphorylase reaction.

reaction is a reversible one, but—as is the case for many biosynthetic reactions—it is driven forward by subsequent hydrolysis of pyrophosphate:

$$\text{Pyrophosphate} + H_2O \longrightarrow 2\,P_i$$

The net reaction for sugar nucleotide formation (combining the preceding two equations) is thus

$$\text{Glucose-1-P} + \text{UTP} + H_2O \longrightarrow \text{UDP-glucose} + 2\,P_i$$

HUMAN BIOCHEMISTRY

Advanced Glycation End-Products: A Serious Complication of Diabetes

Covalent linkage of sugars to proteins to form glycoproteins normally occurs through the action of enzymes that use sugar nucleotides as substrates. However, sugars may also react nonenzymatically with proteins. The C-1 carbonyl group of glucose forms Schiff base linkages with lysine side chains of proteins. These Schiff base adducts undergo rearrangement and subsequent oxidation to form irreversible "glycation" products, including carboxymethyllysine and pentosidine derivatives (see figure). These **advanced glycation end-products (AGEs)** can alter the function of the protein. Such AGE-dependent changes are thought to contribute to circulation, joint, and vision problems in diabetes.

Nonenzymatic glycation of hemoglobin is a useful diagnostic yardstick of long-term serum glucose levels. Red blood cells have an average life expectancy of about four months. By measuring the concentration of "glycated hemoglobin" in a patient, it is possible to determine the average glucose concentration in the blood over the past several months.

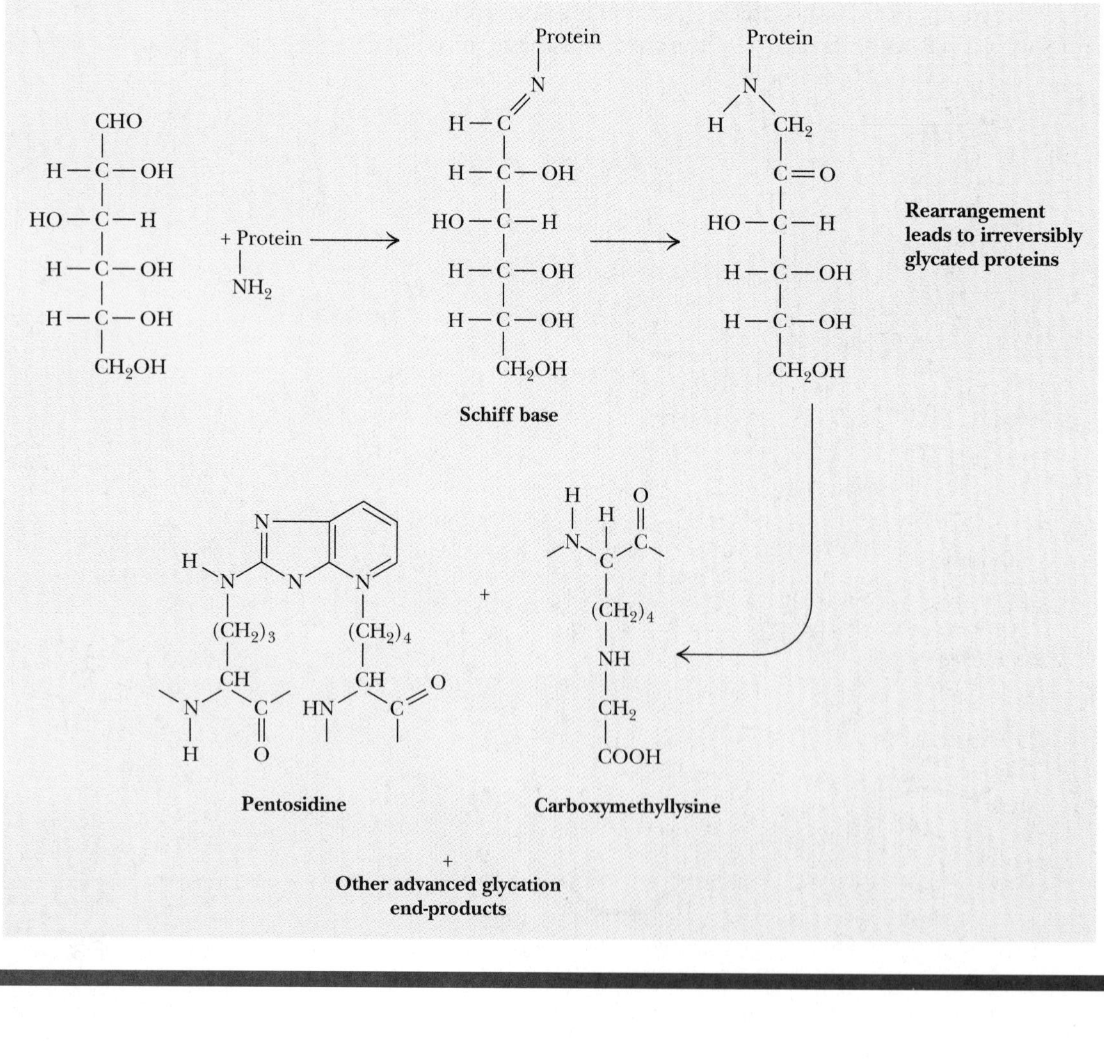

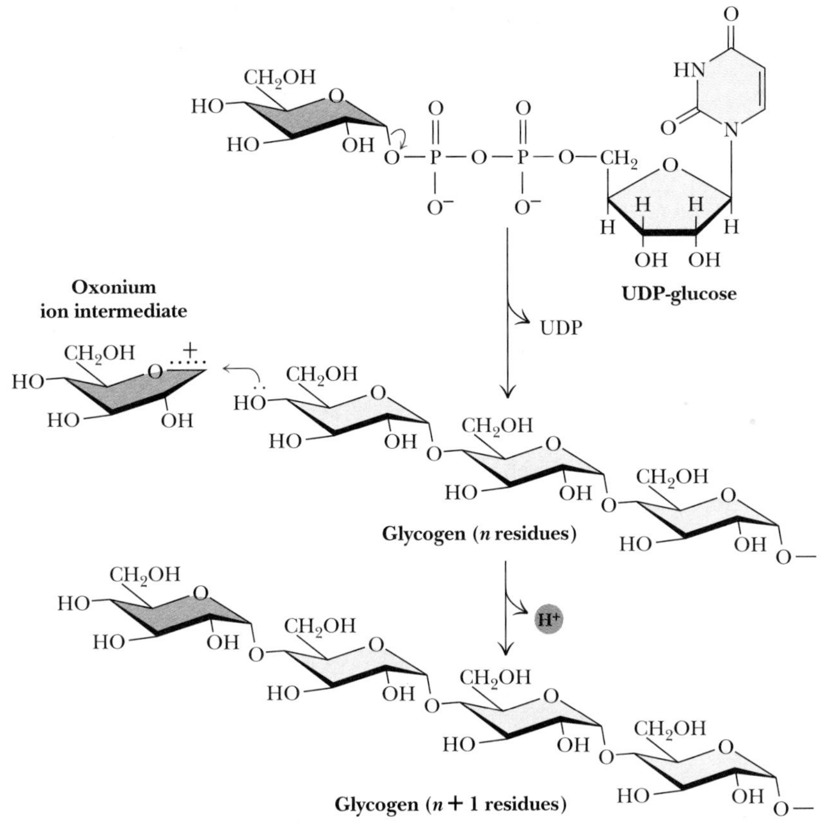

Figure 19.15 The glycogen synthase reaction.

Sugar nucleotides of this type act as donors of sugar units in the biosynthesis of oligo- and polysaccharides. In animals, UDP-glucose is the donor of glucose units for glycogen synthesis, but ADP-glucose is the glucose source for starch synthesis in plants.

Glycogen Synthase Catalyzes Formation of $\alpha(1 \rightarrow 4)$ Glycosidic Bonds in Glycogen

The very large glycogen polymer is built around a tiny protein core. The first glucose residue is covalently joined to the protein **glycogenin** via an acetal linkage to a tyrosine–OH group on the protein. Sugar units are added to the glycogen polymer by the action of **glycogen synthase.** The reaction involves transfer of a glucosyl unit from UDP-glucose to the C-4 hydroxyl group at a nonreducing end of a glycogen strand (Figure 19.15).

19.5 Control of Glycogen Metabolism

Synthesis and degradation of glycogen must be carefully controlled so that this important energy reservoir can properly serve the metabolic needs of the organism. Glucose is the principal metabolic fuel for the brain, and the concentration of glucose in circulating blood must be maintained at about 5 mM for this purpose. Glucose derived from glycogen breakdown is also a primary energy source for muscle contraction. Control of glycogen metabolism is

HUMAN BIOCHEMISTRY

Carbohydrate Utilization in Exercise

Animals have a remarkable ability to "shift gears" metabolically during periods of strenuous exercise or activity. Metabolic adaptations allow the body to draw on different sources of energy (all of which produce ATP) for different types of activity. During periods of short-term, high-intensity exercise (e.g., a 100-m dash), most of the required energy is supplied directly by existing stores of ATP and creatine phosphate (see figure, part a). Long-term, low-intensity exercise (a 10-km run or a 42.2-km marathon) is fueled almost entirely by aerobic metabolism. Between these extremes is a variety of activities (an 800-m run, for example) that rely heavily on anaerobic glycolysis—conversion of glucose to lactate in the muscles and utilization of the Cori cycle.

For all these activities, the breakdown of muscle glycogen provides much of the needed glucose. The rate of glycogen con-

sumption depends upon the intensity of the exercise (see figure, part b), By contrast, glucose derived from gluconeogenesis makes only small contributions to total glucose consumed during exercise. During prolonged mild exercise, gluconeogenesis accounts for only about 8% of the total glucose consumed. During heavy exercise, this percentage becomes even lower.

Choice of diet has a dramatic effect on glycogen recovery following exhaustive exercise. A diet consisting mainly of protein and fat results in very little recovery of muscle glycogen, even after five days (see figure, part c). On the other hand, a high-carbohydrate diet provides faster restoration of muscle glycogen. Even in this case, however, complete recovery of glycogen stores takes about two days.

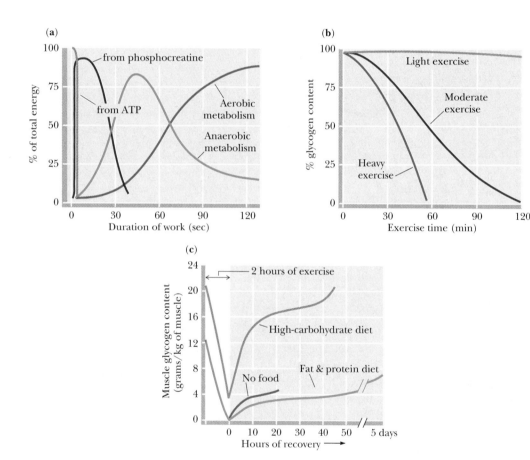

(a) Contributions of the various energy sources to muscle activity during mild exercise. (b) Consumption of glycogen stores in fast-twitch muscles during light, moderate, and heavy exercise. (c) Rate of glycogen replenishment following exhaustive exercise. (a and c, adapted from Rhoades and Pflanzer, 1992. Human Physiology. Philadelphia: Saunders College Publishing; b, adapted from Horton and Terjung, 1988. Exercise, Nutrition and Energy Metabolism. New York: Macmillan.)

achieved via reciprocal regulation of glycogen phosphorylase and glycogen synthase. Thus, activation of glycogen phosphorylase is tightly linked to inhibition of glycogen synthase, and vice versa. Regulation involves both allosteric control and covalent modification, with the latter being under hormonal control. The regulation of glycogen phosphorylase is discussed in detail in Chapter 10.

Regulation of Glycogen Synthase by Covalent Modification

Glycogen synthase exists in two distinct forms which can be interconverted by the action of specific enzymes: active, dephosphorylated **glycogen synthase I** (glucose-6-P-independent) and less active, phosphorylated **glycogen synthase D** (glucose-6-P-dependent). Phosphorylation of glycogen synthase is complex. As many as nine serine residues on the enzyme appear to be subject to phosphorylation, with each site's phosphorylation having some effect on enzyme activity.

Dephosphorylation of both glycogen phosphorylase and glycogen synthase is carried out by **phosphoprotein phosphatase 1.** The action of phosphoprotein phosphatase 1 inactivates glycogen phosphorylase and activates glycogen synthase.

Hormones Regulate Glycogen Synthesis and Degradation

Storage and utilization of tissue glycogen, maintenance of blood glucose concentration, and other aspects of carbohydrate metabolism are meticulously regulated by hormones, including insulin, glucagon, epinephrine, and the glucocorticoids.

Insulin Is a Response to Increased Blood Glucose

The primary hormone responsible for conversion of glucose to glycogen is **insulin** (see Figure 4.21). Insulin is secreted by β cells in a part of the pancreas called the **islets of Langerhans.** Secretion of insulin is a response to increased glucose in the blood. When blood glucose levels rise (after a meal, for example), insulin is secreted from the pancreas into the pancreatic vein, which empties into the **portal vein system** (Figure 19.16), so that insulin traverses the liver before it enters the systemic blood supply. Insulin acts to rapidly lower blood glucose concentration in several ways. Insulin stimulates glycogen synthesis and inhibits glycogen breakdown in liver and muscle.

Glucagon and Epinephrine Stimulate Glycogen Breakdown

Catabolism of tissue glycogen is triggered by the actions of the hormones **epinephrine** and **glucagon.** In response to decreased blood glucose, glucagon (Figure 19.17) is released from the α cells in the pancreatic islets of Langerhans. This peptide hormone travels through the blood to specific receptors on

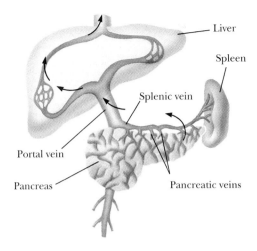

Figure 19.16 The portal vein system carries pancreatic secretions such as insulin and glucagon to the liver and then into the rest of the circulatory system.

HUMAN BIOCHEMISTRY

Von Gierke's Disease: A Glycogen Storage Disease

In 1929 the physician Edgar von Gierke treated a patient with a very distended abdomen. The patient's liver and kidneys were severely enlarged due to massive accumulations of glycogen, and von Gierke appropriately called the condition "hepato-nephromegalia glycogenica." Now termed **von Gierke's disease,** this condition results from the absence of glucose-6-phosphatase activity in the affected organs. This simple genetic defect causes a host of difficult complications, including a striking elevation of serum triglycerides, excess adipose tissue in the cheeks, thin extremities, short stature, excessive curvature of the lumbar spine, and delay of puberty.

The absence of glucose-6-phosphatase activity in the liver blocks the last steps of glycogenolysis and gluconeogenesis, interrupting the recycling of glucose and causing affected individuals to be hypoglycemic. The accumulation of glucose-6-phosphate in the liver leads to greatly increased glycolytic activity, with consequent elevation of lactic acid, a condition known more commonly as **lactic acidosis.** Large amounts of uric acid and lipids are produced, and the high rates of glycolysis produce excess NADH.

The treatment of von Gierke's disease consists of trying to maintain normal levels of glucose in the patient's serum. This often requires administration of large amounts of glucose in its various forms, including, for example, uncooked cornstarch.

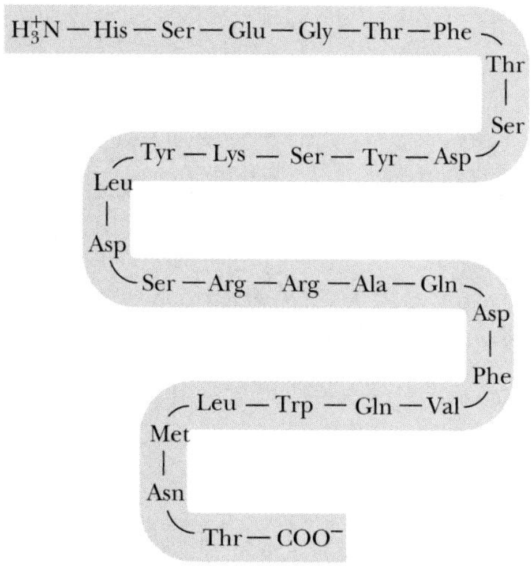

Figure 19.17 The amino acid sequence of glucagon.

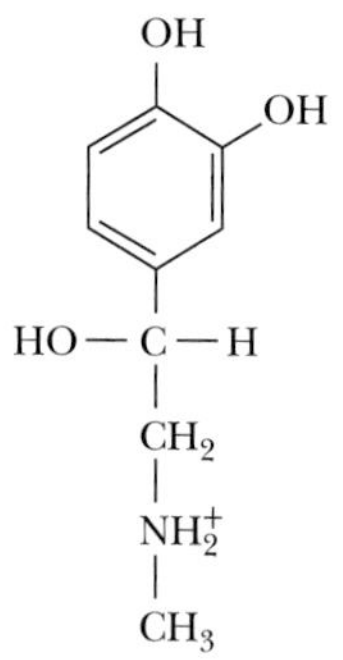

Figure 19.18 The structure of epinephrine.

liver cell membranes. (Glucagon is active in liver and adipose tissue, but not in other tissues.) Similarly, signals from the central nervous system cause release of epinephrine (Figure 19.18)—also known as adrenaline—from the adrenal glands into the bloodstream. Epinephrine acts on liver and muscles. When either hormone binds to its receptor on the outside surface of the cell membrane, a cascade is initiated that activates glycogen phosphorylase and inhibits glycogen synthase. The result of these actions is tightly coordinated stimulation of glycogen breakdown and inhibition of glycogen synthesis.

The Difference Between Epinephrine and Glucagon

Although both epinephrine and glucagon exert glycogenolytic effects, they do so for quite different reasons. Epinephrine is secreted as a response to anger or fear and may be viewed as an alarm or danger signal for the organism. Called the "fight or flight" hormone, it prepares the organism for mobilization of large amounts of energy. Among the many physiological changes elicited by epinephrine, one is the initiation of the enzyme cascade, as in Figure 10.39, which leads to rapid breakdown of glycogen, inhibition of glycogen synthesis, stimulation of glycolysis, and production of energy. The burst of energy produced is the result of a 2000-fold amplification of the rate of glycolysis. Because a fear or anger response must include generation of energy (in the form of glucose)—both immediately in localized sites (the muscles) and eventually throughout the organism (as supplied by the liver)—epinephrine must be able to activate glycogenolysis in both the liver and muscles.

Glucagon is involved in the long-term maintenance of steady-state levels of glucose in the blood and other tissues. It performs this function by stimulating the liver to release glucose from glycogen stores into the bloodstream. To further elevate glucose levels, glucagon also activates liver gluconeogenesis. It is important to note, however, that stabilization of blood glucose levels is managed almost entirely by the liver. Glucagon does not activate the phosphorylase cascade in muscle (muscle membranes do not contain glucagon receptors). Muscle glycogen breakdown occurs only in response to epinephrine release, and muscle tissue does not participate in maintenance of steady-state glucose levels in the blood.

19.6 **The Pentose Phosphate Pathway**

Cells require a constant supply of NADPH for reductive reactions vital to biosynthetic purposes. Much of this requirement is met by a glucose-based metabolic sequence variously called the **pentose phosphate pathway,** the **hexose monophosphate shunt,** or the **phosphogluconate pathway.** In addition to providing NADPH for biosynthetic processes, this pathway produces ribose-5-phosphate, which is essential for nucleotide and nucleic acid synthesis. Several metabolites of the pentose phosphate pathway can also be shuttled into glycolysis.

An Overview of the Pathway

The pentose phosphate pathway begins with glucose-6-phosphate, a six-carbon sugar, and produces three-, four-, five-, six-, and seven-carbon sugars (Figure 19.19). As we will see, two successive oxidations lead to the reduction of NADP$^+$ to NADPH and the release of CO_2. Five subsequent nonoxidative steps produce a variety of carbohydrates, some of which may enter the glycolytic pathway. The enzymes of the pentose phosphate pathway are particularly

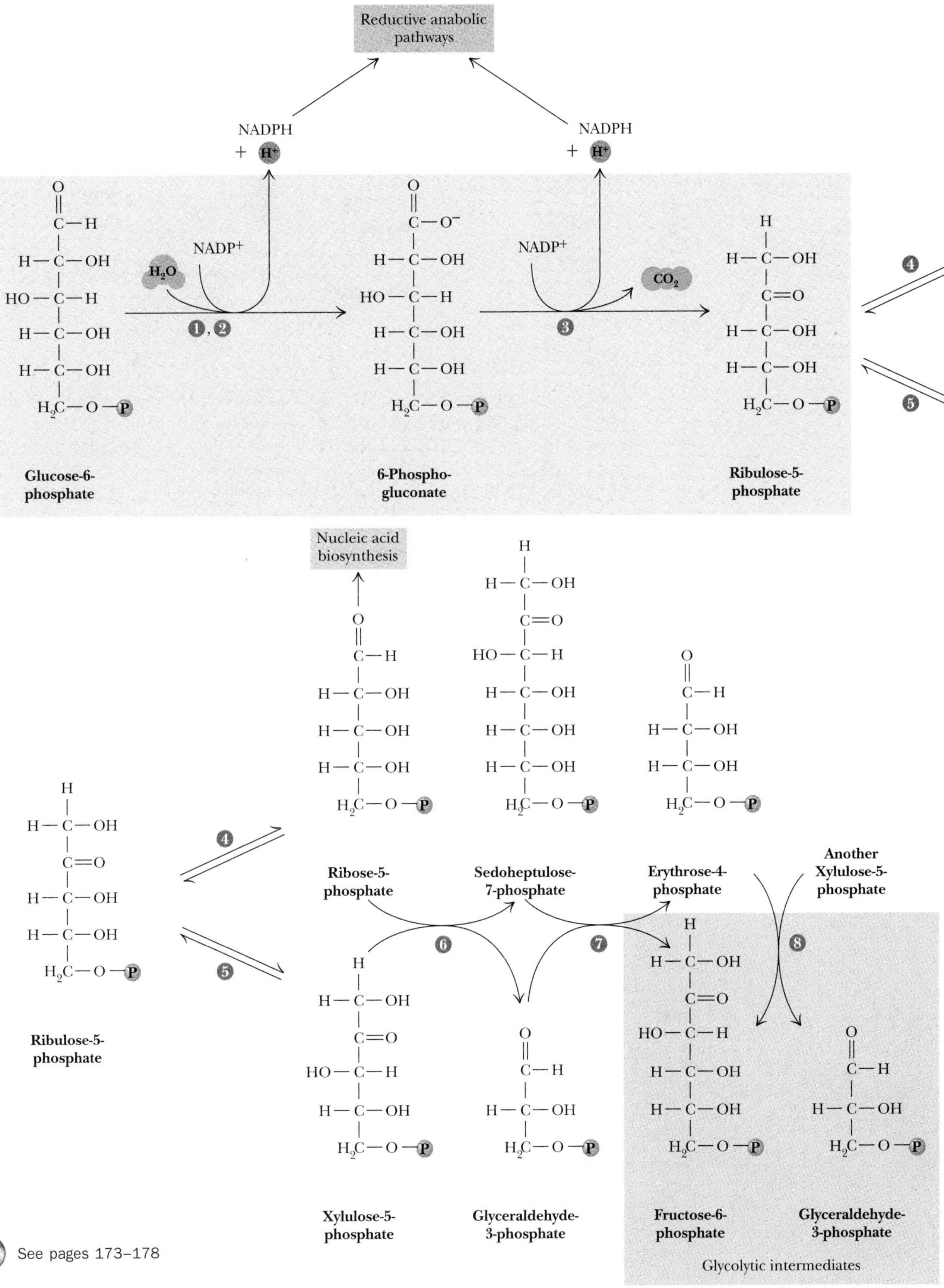

Reductive anabolic
pathways

NADPH
+ H⁺

NADPH
+ H⁺

NADP⁺

H₂O

❶, ❷

NADP⁺

CO₂

❸

❹

❺

Glucose-6-phosphate

6-Phospho-gluconate

Ribulose-5-phosphate

Nucleic acid
biosynthesis

❹

❺

Ribulose-5-phosphate

Ribose-5-phosphate

Sedoheptulose-7-phosphate

Erythrose-4-phosphate

Another
Xylulose-5-phosphate

❻

❼

❽

Xylulose-5-phosphate

Glyceraldehyde-3-phosphate

Fructose-6-phosphate

Glyceraldehyde-3-phosphate

Glycolytic intermediates

See pages 173–178

Figure 19.19 The pentose phosphate pathway. The numerals in the blue circles indicate the steps discussed in the text.

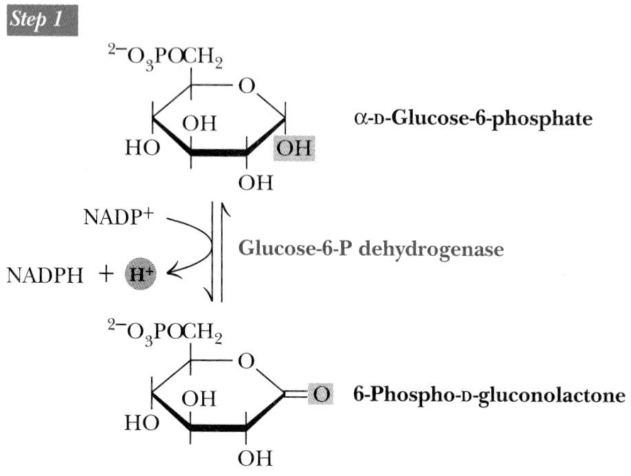

Figure 19.20 The glucose-6-phosphate dehydrogenase reaction is the committed step in the pentose phosphate pathway.

abundant in the cytoplasm of liver and adipose cells. These enzymes are largely absent in muscle, where glucose-6-phosphate is utilized primarily for energy production via glycolysis and the TCA cycle. These pentose phosphate pathway enzymes are located in the cytosol, which is the site of fatty acid synthesis, a pathway heavily dependent on NADPH for reductive reactions.

The Oxidative Steps of the Pentose Phosphate Pathway

(1) Glucose-6-Phosphate Dehydrogenase

The pentose phosphate pathway begins with the oxidation of glucose-6-phosphate. The products of the reaction are a cyclic ester (the lactone of phosphogluconic acid) and NADPH (Figure 19.20). **Glucose-6-phosphate dehydrogenase,** which catalyzes this reaction, is highly specific for $NADP^+$. As the first step of a major pathway, the reaction is irreversible and highly regulated. Glucose-6-phosphate dehydrogenase is strongly inhibited by the product coenzyme, NADPH, and also by fatty acid esters of coenzyme A (which are intermediates of fatty acid biosynthesis). Inhibition due to NADPH depends upon the cytosolic $NADP^+/NADPH$ ratio, which in the liver is about 0.015 (compared to about 725 for the $NAD^+/NADH$ ratio in the cytosol).

(2) Gluconolactonase

The gluconolactone produced in step 1 is hydrolytically unstable and readily undergoes a spontaneous ring-opening hydrolysis, although an enzyme, **gluconolactonase,** accelerates this reaction (Figure 19.21). The linear product, the sugar acid 6-phospho-D-gluconate, is further oxidized in step 3.

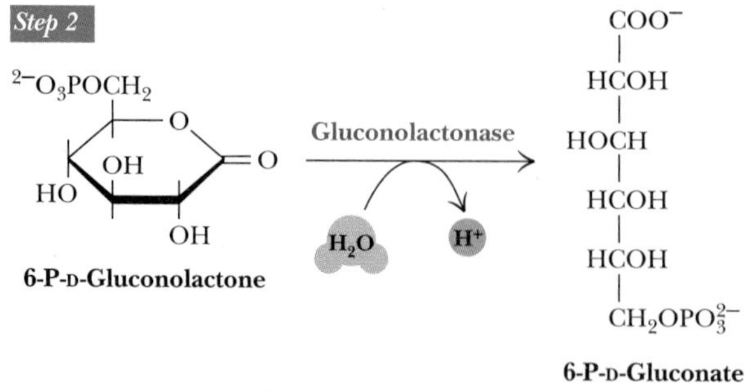

Figure 19.21 The gluconolactonase reaction.

Step 3

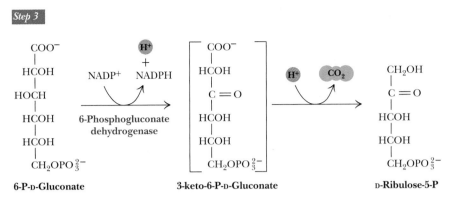

Figure 19.22 The 6-phosphogluconate dehydrogenase reaction.

(3) 6-Phosphogluconate Dehydrogenase

The oxidative decarboxylation of 6-phosphogluconate by **6-phosphogluconate dehydrogenase** yields D-ribulose-5-phosphate and another equivalent of NADPH. There are two distinct steps in this reaction (Figure 19.22): the initial NADP$^+$-dependent dehydrogenation yields a β-keto acid, 3-keto-6-phosphogluconate, which is very susceptible to decarboxylation (the second step). The resulting product, D-ribulose-5-P, is the substrate for the nonoxidative reactions composing the rest of this pathway.

The Nonoxidative Steps of the Pentose Phosphate Pathway

This portion of the pathway begins with an isomerization and an epimerization, and it leads to the formation of either D-ribose-5-phosphate or D-xylulose-5-phosphate. These intermediates can then be converted into glycolytic intermediates or directed to biosynthetic processes.

(4) Phosphopentose Isomerase

This enzyme interconverts ribulose-5-P and ribose-5-P via an enediol intermediate (Figure 19.23). The reaction (and mechanism) is quite similar to the phosphoglucoisomerase reaction of glycolysis, which interconverts glucose-6-P and fructose-6-P. The ribose-5-P produced in this reaction is utilized in the biosynthesis of coenzymes (including NADH, NADPH, FAD, and B$_{12}$), nucleotides, and nucleic acids (DNA and RNA). The net reaction for the first four steps of the pentose phosphate pathway is

$$\text{Glucose-6-P} + 2\ \text{NADP}^+ \longrightarrow \text{ribose-5-P} + 2\ \text{NADPH} + 2\ \text{H}^+ + \text{CO}_2$$

Step 4

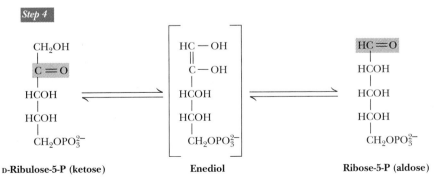

Figure 19.23 The phosphopentose isomerase reaction involves an enediol intermediate.

HUMAN BIOCHEMISTRY

Aldose Reductase and Diabetic Cataract Formation

The complications of diabetes include a high propensity for cataract formation in later life, in both Type I and Type II diabetics. Hyperglycemia is the suspected cause, but by what mechanism? Several lines of evidence point to the **polyol pathway**, in which glucose and other simple sugars are reduced in NADPH-dependent reactions. Glucose, for example, is reduced by *aldose reductase* to sorbitol, which accumulates in lens fiber cells, increasing the intracellular osmotic pressure and eventually rupturing the cells (see

figure, a). The involvement of aldose reductase in this process is supported by the fact that animals (such as rats and dogs) that have high levels of this enzyme in their lenses are prone to developing diabetic cataracts, whereas mice that have low levels of lens aldose reductase activity are not. Moreover, aldose reductase inhibitors such as **tolrestat** and **epalrestat** suppress cataract formation (see figure, b). These drugs or derivatives of them may provide an effective therapy in preventing cataract formation in diabetics.

(a)

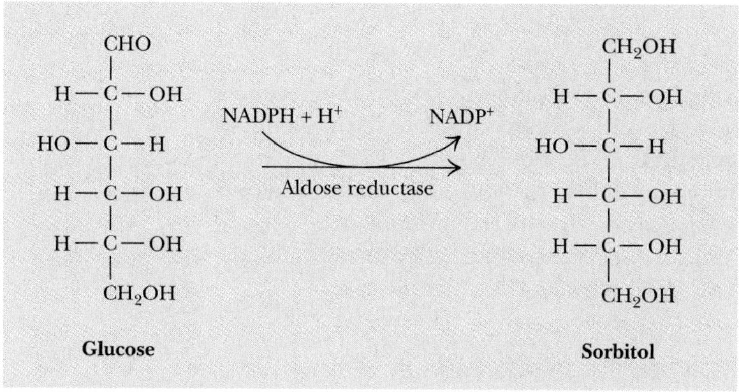

Glucose   Sorbitol

(b)

Tolrestat   Epalrestat

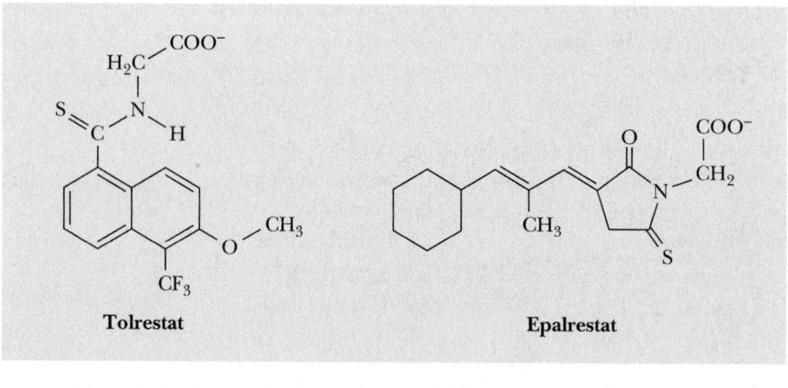

(5) Phosphopentose Epimerase

This reaction converts ribulose-5-P to another ketose, namely, xylulose-5-P. This reaction also proceeds by an enediol intermediate, but involves an inversion at C-3 (Figure 19.24). In the reaction, an acidic proton located α- to a carbonyl carbon is removed to generate the enediolate, but the proton is added back to the same carbon from the opposite side. Note the distinction in nomenclature here. Interchange of groups on a single carbon is an epimerization, and interchange of groups between carbons is referred to as an isomerization.

To this point, the pathway has generated a pool of pentose phosphates. The $\Delta G^{\circ\prime}$ for each of the last two reactions is small, and the three pentose-5-

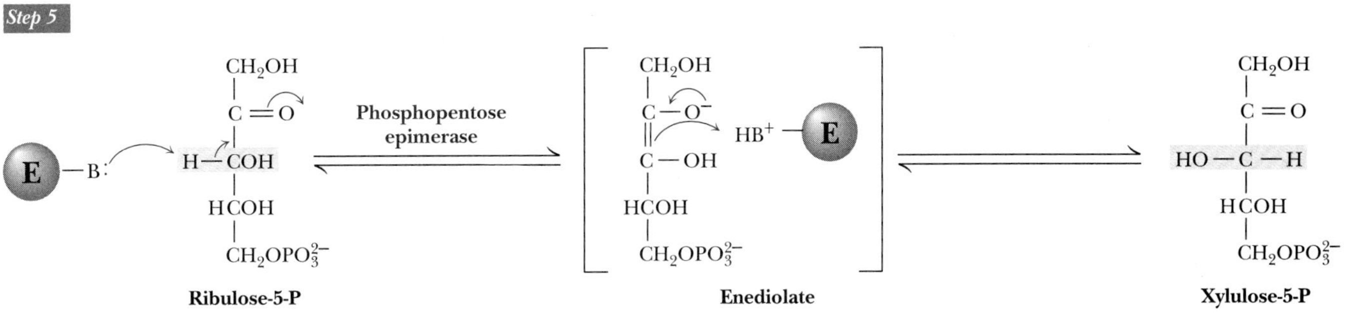

Step 5

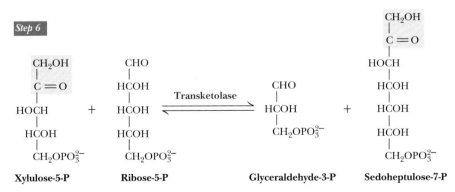

Ribulose-5-P   Enediolate   Xylulose-5-P

Figure 19.24 The phosphopentose epimerase reaction interconverts ribulose-5-P and xylulose-5-P.

phosphates coexist at equilibrium. The pathway has also produced two molecules of NADPH for each glucose-6-P converted to pentose-5-phosphate. The next three steps rearrange the five-carbon skeletons of the pentoses to produce three-, four-, six-, and seven-carbon units, which can be used for various metabolic purposes. Why should the cell do this? Very often, the cellular need for NADPH is considerably greater than the need for ribose-5-phosphate. The next three steps thus return some of the five-carbon units to glyceraldehyde-3-phosphate and fructose-6-phosphate, which can enter the glycolytic pathway. The advantage of this is that the cell has met its needs for NADPH and ribose-5-phosphate in a single pathway, yet at the same time it can return the excess carbon metabolites to glycolysis.

(6) and (8) Transketolase

The transketolase enzyme acts at both steps 6 and 8 of the pentose phosphate pathway. In both cases, the enzyme catalyzes the transfer of two-carbon units. In these reactions (and also in step 7, the transaldolase reaction, which transfers three-carbon units), the donor molecule is a ketose and the recipient is an aldose. In step 6, xylulose-5-phosphate transfers a two-carbon unit to ribose-5-phosphate to form glyceraldehyde-3-phosphate and sedoheptulose-7-phosphate (Figure 19.25). Step 8 involves a two-carbon transfer from xylulose-5-phosphate to erythrose-4-phosphate to produce another glyceraldehyde-3-phosphate and a fructose-6-phosphate (Figure 19.26). Three of these products enter directly into the glycolytic pathway. (The sedoheptulose-7-phosphate is taken care of in step 7, as we shall see.) Transketolase is a thiamine pyrophosphate–dependent enzyme, which can process a variety of 2-keto sugar phosphates in a similar manner.

Step 6

Figure 19.25 The transketolase reaction of step 6 in the pentose phosphate pathway.

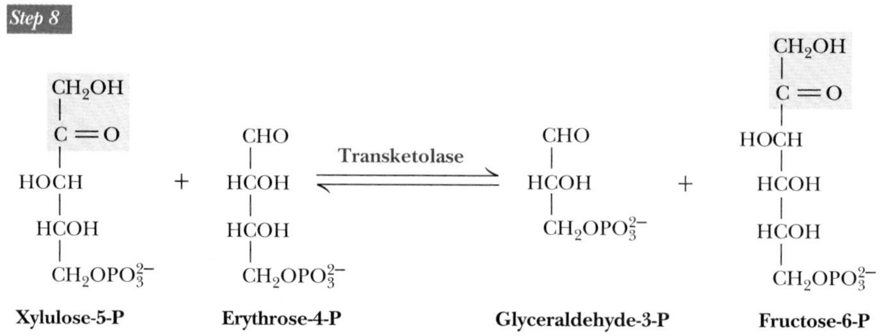

Figure 19.26 The transketolase reaction of step 8 in the pentose phosphate pathway.

(7) Transaldolase

The transaldolase functions primarily to make a useful glycolytic substrate from the sedoheptulose-7-phosphate produced by the first transketolase reaction. This reaction (Figure 19.27) is quite similar to the aldolase reaction of glycolysis.

Utilization of Glucose-6-P Depends on the Cell's Need for ATP, NADPH, and Ribose-5-P

It is clear that glucose-6-phosphate can be used as a substrate either for glycolysis or for the pentose phosphate pathway. The cell makes this choice on the basis of its relative needs for biosynthesis and for energy from metabolism. ATP can be produced in abundance if glucose-6-phosphate is channeled into glycolysis. On the other hand, if NADPH or ribose-5-phosphate is needed, glucose-6-phosphate can be directed to the pentose phosphate pathway. The molecular basis for this regulatory decision depends on the enzymes that metabolize glucose-6-phosphate in glycolysis and the pentose phosphate pathway. In glycolysis, phosphoglucoisomerase converts glucose-6-phosphate to fructose-6-phosphate, which is utilized by phosphofructokinase (a highly regulated enzyme) to produce fructose-1,6-bisphosphate. In the pentose phosphate pathway, glucose-6-phosphate dehydrogenase (also highly regulated) produces gluconolactone from glucose-6-phosphate. Thus, the fate of glucose-6-phosphate is determined to a large extent by the relative activities of phosphofructokinase (PFK) and glucose-6-P dehydrogenase. Recall (see Chapter 15) that PFK is inhibited when the ATP/AMP ratio increases, and that it is inhibited by citrate but activated by fructose-2,6-bisphosphate. Thus, when the energy charge is high, glycolytic flux decreases. Glucose-6-P dehydrogenase, on the other

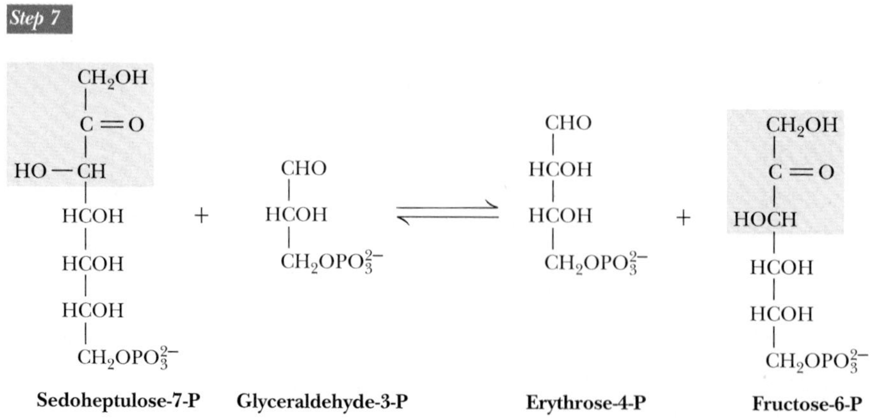

Figure 19.27 The transaldolase reaction.

hand, is inhibited by high levels of NADPH and also by the intermediates of fatty acid biosynthesis. Both of these are indicators that biosynthetic demands have been satisfied. If that is the case, glucose-6-phosphate dehydrogenase and the pentose phosphate pathway are inhibited. If NADPH levels drop, the pentose phosphate pathway turns on, and NADPH and ribose-5-phosphate are made for biosynthetic purposes.

Even when the latter choice has been made, however, the cell must still be able to meet the relative needs for ribose-5-phosphate and NADPH (as well as ATP). Depending on these relative needs, the reactions of glycolysis and the pentose phosphate pathway can be combined in novel ways to emphasize the synthesis of needed metabolites.

PROBLEMS

1. Consider the balanced equation for gluconeogenesis in Section 19.1. Account for each of the components of this equation and the indicated stoichiometry.

2. Calculate $\Delta G^{\circ\prime}$ and ΔG for gluconeogenesis in the erythrocyte, using data in Table 15.2 (assume $NAD^+/NADH = 20$, $[GTP] = [ATP]$, and $[GDP] = [ADP]$). See how closely your values match those in Section 19.1.

3. Use the data of Figure 19.11 to calculate the percent inhibition of fructose-1,6-bisphosphatase by 25 mM fructose-2,6-bisphosphate when fructose-1,6-bisphosphate is (a) 25 mM and (b) 100 mM.

4. Suggest an explanation for the exergonic nature of the glycogen synthase reaction ($\Delta G^{\circ\prime} = -13.3$ kJ/mol). Consult Chapter 3 to review the energetics of high-energy phosphate compounds if necessary.

5. Using the values in Table 20.1 for body glycogen content and the data in part (b) of the figure in *A Deeper Look* (page 592), calculate the rate of energy consumption by muscles in heavy exercise (in J/sec). Use the data for fast-twitch muscle.

6. What would be the distribution of carbon from positions 1, 3, and 6 of glucose after one pass through the pentose phosphate pathway if the primary need of the organism is for ribose-5-P

and the oxidative steps are bypassed via use of glycolytic reactions?

7. What is the fate of carbon from positions 2 and 4 of glucose-6-P after one pass through the scheme shown in Figure 19.19?

8. Imagine a glycogen molecule with 8000 glucose residues. If branches occur every eight residues, how many reducing ends does the molecule have? If branches occur every 12 residues, how many reducing ends does it have? How many nonreducing ends does it have in each of these cases?

9. Explain the effects of each of the following on the rates of gluconeogenesis and glycogen metabolism:
   a. Increasing the concentration of tissue fructose-1,6-bisphosphate
   b. Increasing the concentration of blood glucose
   c. Increasing the concentration of blood insulin
   d. Increasing the amount of blood glucagon
   e. Decreasing levels of tissue ATP
   f. Increasing the concentration of tissue AMP
   g. Decreasing the concentration of fructose-6-phosphate

10. The free-energy change of the glycogen phosphorylase reaction is $\Delta G^{\circ\prime} = +3.1$ kJ/mol. If $[P_i] = 1$ mM, what is the concentration of glucose-1-P when this reaction is at equilibrium?

FURTHER READING

Akermark, C., Jacobs, I., Rasmusson, M., and Karlsson, J., 1996. Diet and muscle glycogen concentration in relation to physical performance in Swedish elite ice hockey players. *International Journal of Sport Nutrition* **6**:272–284.

Browner, M. F., and Fletterick, R. J., 1992. Phosphorylase: A biological transducer. *Trends in Biochemical Sciences* **17**:66–71.

Fox, E. L., 1984. *Sports Physiology,* 2nd ed. Philadelphia: Saunders College Publishing.

Hanson, R. W., and Reshef, L., 1997. Regulation of phosphoenolpyruvate carboxykinase (GTP) gene expression. *Annual Review of Biochemistry* **66**:581–611.

Hargreaves, M., 1997. Interactions between muscle glycogen and blood glucose during exercise. *Exercise and Sport Sciences Reviews* **25**:21–39.

Hers, H.-G., and Hue, L., 1983. Gluconeogenesis and related aspects of glycolysis. *Annual Review of Biochemistry* **52**:617–653.

Horton, E. S., and Terjung, R. L., eds., 1988. *Exercise, Nutrition and Energy Metabolism.* New York: Macmillan.

Huang, D., Wilson, W. A., and Roach, P. J., 1997. Glucose-6-P control of glycogen synthase phosphorylation in yeast. *Journal of Biological Chemistry* **272**:22495–22501.

Johnson, L. N., 1992. Glycogen phosphorylase: Control by phosphorylation and allosteric effectors. *FASEB Journal* **6**:2274–2282.

Larner, J., 1990. Insulin and the stimulation of glycogen synthesis: The road from glycogen structure to glycogen synthase to cyclic AMP-dependent protein kinase to insulin mediators. *Advances in Enzymology* **63**:173–231.

Newsholme, E. A., Chaliss, R. A. J., and Crabtree, B., 1984. Substrate cycles: Their role in improving sensitivity in metabolic control. *Trends in Biochemical Sciences* **9**:277–280.

Newsholme, E. A., and Leech, A. R., 1983. *Biochemistry for the Medical Sciences.* New York: John Wiley & Sons.

Pilkis, S. J., El-Maghrabi, M. R., and Claus, T. H., 1988. Hormonal regulation of hepatic gluconeogenesis and glycolysis. *Annual Review of Biochemistry* **57**:755–783.

Rhoades, R., and Pflanzer, R., 1992. *Human Physiology.* Philadelphia: Saunders College Publishing.

Rolfe, D. F., and Brown, G. C., 1997. Cellular energy utilization and molecular origin of standard metabolic rate in mammals. *Physiological Reviews* **77**:731–758.

Rybicka, K. K., 1996. Glycosomes: The organelles of glycogen metabolism. *Tissue and Cell* **28**:253–265.

Shulman, R. G., and Rothman, D. L., 1996. Nuclear magnetic resonance studies of muscle and applications to exercise and diabetes. *Diabetes* **45**:S93–S98.

Sies, H., ed., 1982. *Metabolic Compartmentation.* London: Academic Press.

Stalmans, W., Cadefau, J., Wera, S., and Bollen, M., 1997. New insight into the regulation of liver glycogen metabolism by glucose. *Biochemical Society Transactions* **25**:19–25.

Sukalski, K. A., and Nordlie, R. C., 1989. Glucose-6-phosphatase: Two concepts of membrane-function relationship. *Advances in Enzymology* **62**:93–117.

Taylor, S. S., et al., 1993. A template for the protein kinase family. *Trends in Biochemical Sciences* **18**:84–89.

Van Schaftingen, E., and Hers, H.-G., 1981. Inhibition of fructose-1,6-bisphosphatase by fructose-2,6-bisphosphate. *Proceedings of the National Academy of Sciences, U.S.A.* **78**:2861–2863.

Williamson, D. H., Lund, P., and Krebs, H. A., 1967. The redox state of free nicotinamide–adenine dinucleotide in the cytoplasm and mitochondria of rat liver. *Biochemical Journal* **103**:514–527.

Woodget, J. R., 1991. A common denominator linking glycogen metabolism, nuclear oncogenes, and development. *Trends in Biochemical Sciences* **16**:177–181.

Lipid Metabolism

The fat is in the fire.

Proverbs, John Heywood (1497–1580)

*The hummingbird's tremendous capacity to store and use fatty acids enables it to make migratory journeys of remarkable distances. (*Two Hummingbirds, *lithograph; Academy of Natural Sciences of Philadelphia/Corbis Images)*

Fatty acids represent the principal form of stored energy for many organisms. There are two important advantages to storing energy in the form of fatty acids. (1) The carbon in fatty acids (mostly $-CH_2-$ groups) is almost completely reduced compared to the carbon in other simple biomolecules (sugars, amino acids); therefore, oxidation of fatty acids will yield more energy (in the form of ATP) than any other form of carbon. (2) Fatty acids are not generally hydrated as mono- and polysaccharides are, and thus can pack more closely in storage tissues. This chapter will be devoted to several important aspects of fatty acid catabolism and lipid biosynthesis.

Fatty acid catabolism.

See *Interactive Biochemistry CD-ROM and Workbook,* pages 173–178

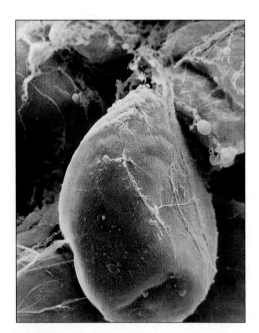

Figure 20.1 Scanning electron micrograph of an adipose cell (fat cell). Globules of triacylglycerols occupy most of the volume of such cells. *(Prof. P. Motta, Dept. of Anatomy, University "La Sapienza," Rome/Science Photo Library/Photo Researchers, Inc.)*

20.1 Mobilization of Fats from Dietary Intake and Adipose Tissue

Modern Diets Are Often High in Fat

Fatty acids are acquired readily in the diet and can also be made from carbohydrates and the carbon skeletons of amino acids. Fatty acids provide 30% to 60% of the calories in the diets of most Americans. For our caveman and cavewoman ancestors, the figure was probably closer to 20%. Dairy products were apparently not part of their diet, and the meat they consumed (from fast-moving animals) was low in fat. In contrast, modern domesticated cows and pigs are actually bred for high fat content (and better taste). However, woolly-mammoth burgers and saber-toothed tiger steaks are hard to find these days—even in the gourmet sections of grocery stores—and so, by default, we consume (and metabolize) large quantities of fatty acids.

Triacylglycerols Are a Major Form of Stored Energy in Animals

Although some of the fat in our diets is in the form of phospholipids, **triacylglycerols** are a major source of fatty acids. Triacylglycerols are also our principal stored energy reserve. As shown in Table 20.1, the energy available in stores of fat in the average person far exceeds the energy available from protein, glycogen, and glucose. Overall, fat accounts for approximately 83% of available energy, partly because more fat is stored than protein and carbohydrate and partly because of the substantially higher energy yield per gram for fat compared with protein and carbohydrate. In animals, fat is stored mainly as triacylglycerols in specialized cells called **adipocytes** or **adipose cells.** As shown in Figure 20.1, triacylglycerols, aggregated to form large globules, occupy most of the volume of adipose cells. Much smaller amounts of triacylglycerols are stored as small, aggregated globules in muscle tissue.

Hormones Signal the Release of Fatty Acids from Adipose Tissue

The pathways for liberation of fatty acids from triacylglycerols, either from adipose cells or from the diet, are shown in Figures 20.2 and 20.3. Fatty acids are mobilized from adipocytes in response to hormone messengers such as adrenaline, glucagon, and ACTH (adrenocorticotropic hormone). These signal molecules bind to receptors on the plasma membrane of adipose cells and lead to the activation of adenylyl cyclase, which forms cyclic AMP from ATP. (Second messengers and hormonal signaling are discussed in Chapters 10 and 26.) In adipose cells, cAMP activates protein kinase A, which phosphorylates and activates a **triacylglycerol lipase** (also termed **hormone-sensitive lipase**) that hy-

Table 20.1 **Stored Metabolic Fuel in a 70-kg Person**

Energy Constituent	Dry Weight (kJ/g dry weight)	Available (g)	Energy (kJ)
Fat (adipose tissue)	37	15,000	555,000
Protein (muscle)	17	6,000	102,000
Glycogen (muscle)	16	120	1,920
Glycogen (liver)	16	70	1,120
Glucose (extracellular fluid)	16	20	320
Total			660,360

Sources: Owen, O. E., and Reichard, G. A., Jr., 1971. *Progress in Biochemistry and Pharmacology* **6:**177; Newsholme, E. A., and Leech, A. R., 1983. *Biochemistry for the Medical Sciences.* New York: John Wiley & Sons.

drolyzes a fatty acid from C-1 or C-3 of triacylglycerols. Subsequent actions of **diacylglycerol lipase** and **monoacylglycerol lipase** yield fatty acids and glycerol. The cell then releases the fatty acids into the blood, where they are carried (in complexes with serum albumin) to sites of utilization.

Degradation of Dietary Fatty Acids Occurs Primarily in the Duodenum

Dietary triacylglycerols are degraded to a small extent (via fatty acid release) by lipases in the low-pH environment of the stomach, but mostly pass untouched into the duodenum. Alkaline pancreatic juice secreted into the duodenum (see Figure 20.3a) raises the pH of the digestive mixture, allowing hydrolysis of the triacylglycerols by pancreatic lipase and nonspecific esterases. Pancreatic lipase cleaves fatty acids from the C-1 and C-3 positions of triacylglycerols, and other lipases and esterases attack the C-2 position (see Figure 20.3b). **Bile salts,** a family of carboxylic acid salts with steroid backbones, act

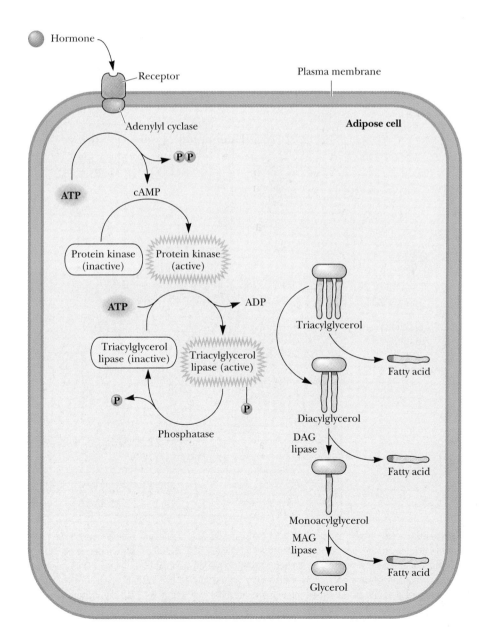

Figure 20.2 Liberation of fatty acids from triacylglycerols in adipose tissue is hormone-dependent.

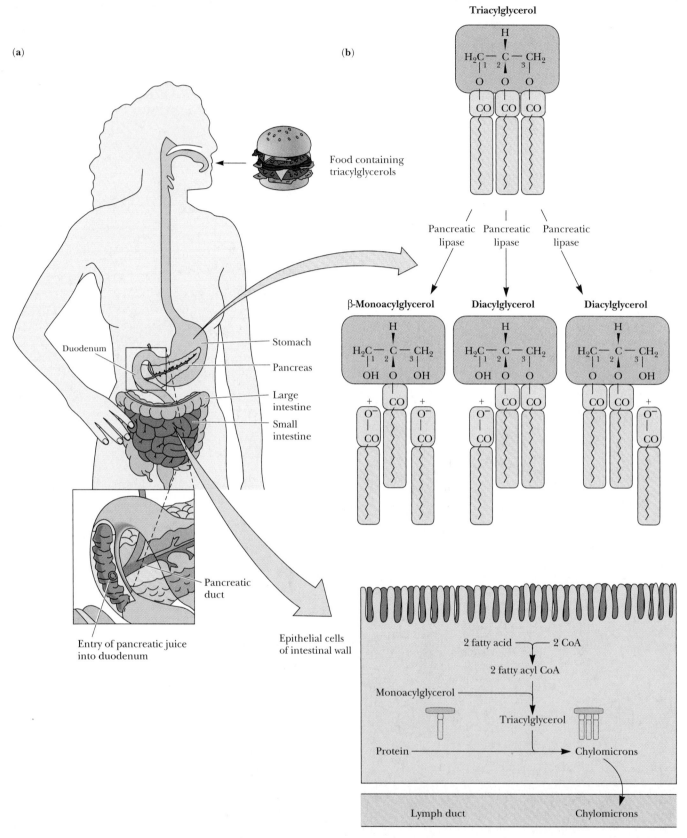

Figure 20.3 **(a)** A duct at the junction of the pancreas and duodenum secretes pancreatic juice into the duodenum, the first portion of the small intestine. **(b)** Hydrolysis of triacylglycerols by pancreatic and intestinal lipases. Pancreatic lipases cleave fatty acids at the C-1 and C-3 positions. Resulting monoacylglycerols with fatty acids at C-2 are hydrolyzed by intestinal lipases. Fatty acids and monoacylglycerols are absorbed through the intestinal wall and assembled into lipoprotein aggregates termed chylomicrons.

as detergents to emulsify the triacylglycerols and facilitate the hydrolytic activity of the lipases and esterases. The fatty acids pass into the epithelial cells, where they are condensed with glycerol to form new triacylglycerols. These triacylglycerols aggregate with lipoproteins to form particles called **chylomicrons,** which are then transported into the lymphatic system and on to the bloodstream, where they circulate to the liver, lungs, heart, muscles, and other organs. At these sites, the triacylglycerols are hydrolyzed to release fatty acids, which can then be oxidized in a highly exergonic metabolic pathway known as β-oxidation.

20.2 β-Oxidation of Fatty Acids

The Discovery of β-Oxidation

In the early 1900s, Franz Knoop carried out experiments with dogs and rabbits which showed that fatty acids must be degraded by oxidation at the β-carbon (Figure 20.4), followed by cleavage of the C_α—C_β bond. Repetition of this process yielded two-carbon units, which Knoop assumed must be acetate. Much later, Albert Lehninger showed that this degradative process took place in the mitochondria, and F. Lynen and E. Reichart showed that the two-carbon unit released is acetyl-CoA, not free acetate. Because the entire process begins with oxidation of the carbon that is "β" to the carboxyl carbon, it is called **β-oxidation.**

Coenzyme A Activates Fatty Acids for Degradation

β-Oxidation begins with the formation of a thiol ester bond between the fatty acid and the thiol group of coenzyme A. This reaction, shown in Figure 20.5, is catalyzed by **acyl-CoA synthetase,** which is also called **acyl-CoA ligase** or **fatty acid thiokinase.** This condensation with CoA activates the fatty acid for reaction in the β-oxidation pathway. For long-chain fatty acids, this reaction normally occurs at the outer mitochondrial membrane, prior to entry of the fatty acid into the mitochondrion, but it may also occur at the surface of the endoplasmic reticulum. Short- and medium-length fatty acids undergo this activating reaction in the mitochondria. In all cases, the reaction is accompanied

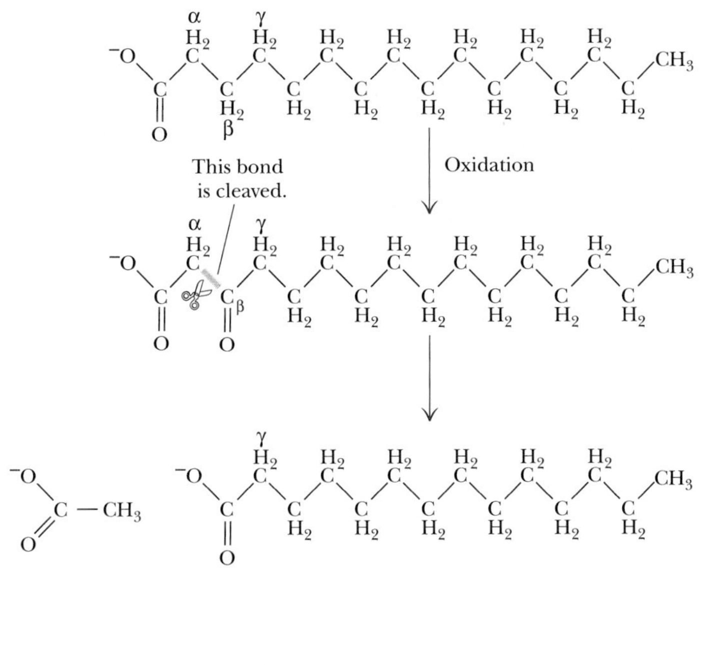

Figure 20.4 Fatty acids are degraded by repeated cycles of oxidation at the β-carbon and cleavage of the C_α—C_β bond to yield acetate units.

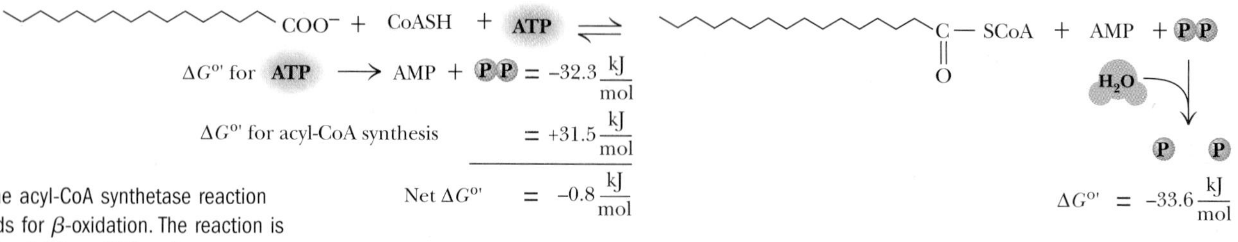

$$\text{\large\textasciitilde\textasciitilde\textasciitilde\textasciitilde COO}^- + \text{CoASH} + \boxed{\text{ATP}} \rightleftharpoons \qquad \text{\large\textasciitilde\textasciitilde\textasciitilde C} - \text{SCoA} + \text{AMP} + \boxed{\text{P}}\boxed{\text{P}}$$

$$\Delta G^{\circ\prime} \text{ for } \boxed{\text{ATP}} \longrightarrow \text{AMP} + \boxed{\text{P}}\boxed{\text{P}} = -32.3 \frac{\text{kJ}}{\text{mol}}$$

$$\Delta G^{\circ\prime} \text{ for acyl-CoA synthesis} = +31.5 \frac{\text{kJ}}{\text{mol}}$$

$$\text{Net } \Delta G^{\circ\prime} = -0.8 \frac{\text{kJ}}{\text{mol}}$$

$$\Delta G^{\circ\prime} = -33.6 \frac{\text{kJ}}{\text{mol}}$$

Figure 20.5 The acyl-CoA synthetase reaction activates fatty acids for β-oxidation. The reaction is driven by hydrolysis of ATP to AMP and pyrophosphate and by the subsequent hydrolysis of pyrophosphate.

by the hydrolysis of ATP to form AMP and pyrophosphate. As shown in Figure 20.5, the two combined reactions have a net $\Delta G^{\circ\prime}$ of about -0.8 kJ/mol, so that the reaction is favorable but easily reversible. However, there is more to the story. As we have seen in several similar cases, the pyrophosphate produced in this reaction is rapidly hydrolyzed by inorganic pyrophosphatase to two molecules of phosphate, with a net $\Delta G^{\circ\prime}$ of about -33.6 kJ/mol. Thus, pyrophosphate is maintained at a low concentration in the cell (usually less than 1 mM) and the synthetase reaction is strongly promoted.

Carnitine Carries Fatty Acyl Groups Across the Inner Mitochondrial Membrane

The remaining enzymes of the β-oxidation pathway are located in the mitochondrial matrix. Short-chain fatty acids are transported into the matrix as free acids and form the acyl-CoA derivatives there. However, long-chain fatty acyl-CoA derivatives cannot be transported into the matrix directly. These long-chain derivatives must first be converted to acylcarnitine derivatives, as shown

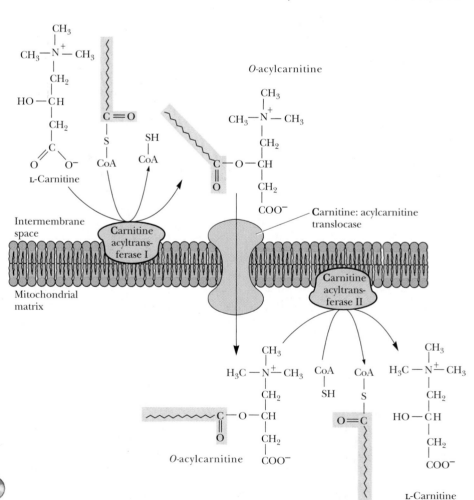

Figure 20.6 The formation of acylcarnitines and their transport across the inner mitochondrial membrane. The process involves the coordinated actions of carnitine acyltransferases on both sides of the membrane and of a translocase that shuttles *O*-acylcarnitines across the membrane.

See pages 173–178

in Figure 20.6. **Carnitine acyltransferase I,** located on the outer side of the inner mitochondrial membrane, catalyzes the formation of the O-acylcarnitine, which is then transported across the inner membrane by a **translocase.** At this point, the acylcarnitine is passed to **carnitine acyltransferase II** on the matrix side of the inner membrane, which transfers the fatty acyl group back to CoA to re-form the fatty acyl-CoA. This frees carnitine, which can return across the membrane via the translocase.

Several additional points should be made. The O—acyl bonds in acylcarnitines have high group-transfer potentials, and the transesterification reactions mediated by the acyltransferases have equilibrium constants close to 1. Note that eukaryotic cells maintain separate pools of CoA in the mitochondria and in the cytosol. The cytosolic pool is utilized principally in fatty acid biosynthesis, and the mitochondrial pool is important in the oxidation of fatty acids and pyruvate, as well as some amino acids.

β-Oxidation Involves a Repeated Sequence of Four Reactions

For saturated fatty acids, β-oxidation involves a recurring cycle of four steps, as shown in Figure 20.7. The overall strategy in the first three steps is to create a carbonyl group on the β-carbon by oxidizing the C_α—C_β bond to form an olefin, with subsequent hydration and oxidation. In essence, this cycle is directly analogous to the sequence of reactions converting succinate to oxaloacetate in the TCA cycle. The fourth reaction of the cycle cleaves the β-keto ester, producing an acetate unit and leaving a fatty acid chain two carbons shorter.

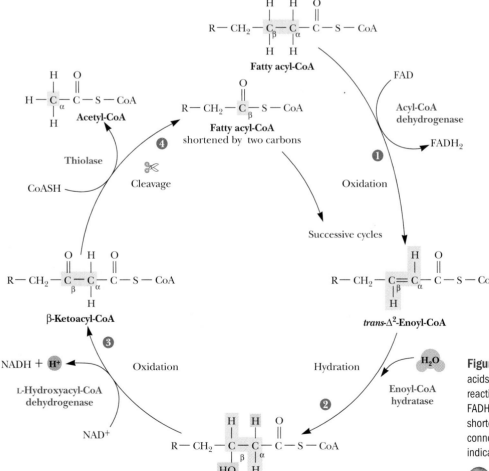

Figure 20.7 The β-oxidation of saturated fatty acids involves a cycle of four enzyme-catalyzed reactions. Each cycle produces single molecules of FADH₂, NADH, and acetyl-CoA and yields a fatty acid shortened by two carbons. (The delta [Δ] symbol connotes a double bond, and its superscript indicates the lower-numbered carbon involved.)

 See pages 173–178

Figure 20.8 The subunit structure of medium-chain acyl-CoA dehydrogenase from pig-liver mitochondria. Note the location of the bound FAD (red). *(Adapted from Kim, J.-T., and Wu, J., 1988. Structure of the medium-chain acyl-CoA dehydrogenase from pig liver mitochondria at 3-Å resolution.* Proceedings of the National Academy of Sciences, U.S.A. **85:**6671–6681.*)*

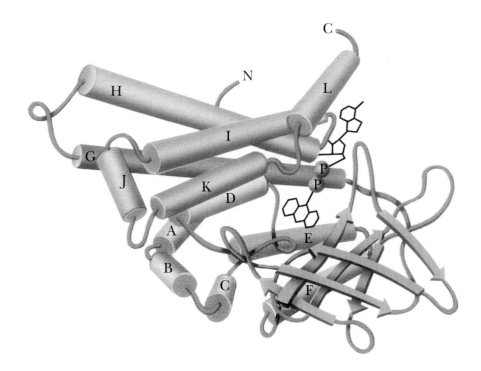

Acyl-CoA Dehydrogenase: The First Reaction of β-Oxidation

The first reaction, the oxidation of the C_α—C_β bond, is catalyzed by **acyl-CoA dehydrogenases** (Figure 20.8), a family of three soluble matrix enzymes, which differ in their specificity for either long-, medium-, or short-chain acyl-CoAs. They carry noncovalently (but tightly) bound FAD, which is reduced during the oxidation of the fatty acid. As shown in Figure 17.4, electrons from the acyl-CoA dehydrogenase reaction are passed directly to *flavoprotein 3*, an electron-transferring flavoprotein associated with the mitochondrial electron transfer pathway.

Enoyl-CoA Hydratase Adds Water Across the Double Bond

The next step in β-oxidation is the addition of the elements of H_2O across the new double bond in a stereospecific manner, yielding the corresponding hydroxyacyl-CoA (Figure 20.9). The reaction is catalyzed by **enoyl-CoA hydratase.** At least three different enoyl-CoA hydratase activities have been detected in various tissues. Also called **crotonases,** these enzymes specifically convert *trans*-enoyl-CoA derivatives to L-β-hydroxyacyl-CoA.

L-Hydroxyacyl-CoA Dehydrogenase Oxidizes the β-Hydroxyl Group

The third reaction of this cycle is the oxidation of the hydroxyl group at the β-position to produce a β-ketoacyl-CoA derivative. This second oxidation reaction is catalyzed by **L-hydroxyacyl-CoA dehydrogenase,** an enzyme that requires NAD^+ as a coenzyme. Each NADH produced in mitochondria by this reaction drives the synthesis of 2.5 molecules of ATP in the electron transport

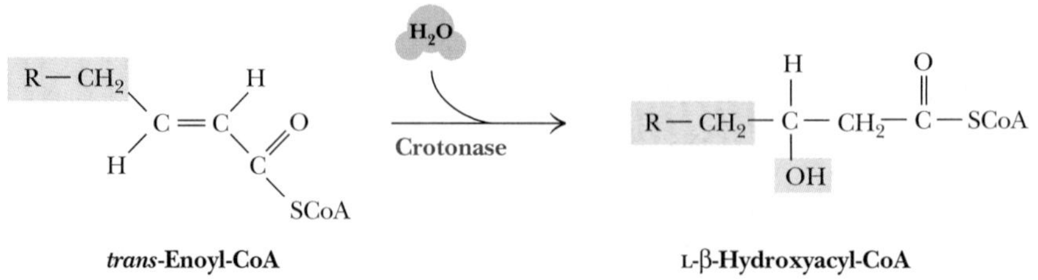

Figure 20.9 The conversion of *trans*-enoyl CoA to L-β-hydroxyacyl CoA. This reaction is catalyzed by enoyl-CoA hydratases (also called crotonases), enzymes that vary in their acyl-chain length specificity.

trans-Enoyl-CoA $\xrightarrow[\text{Crotonase}]{H_2O}$ L-β-Hydroxyacyl-CoA

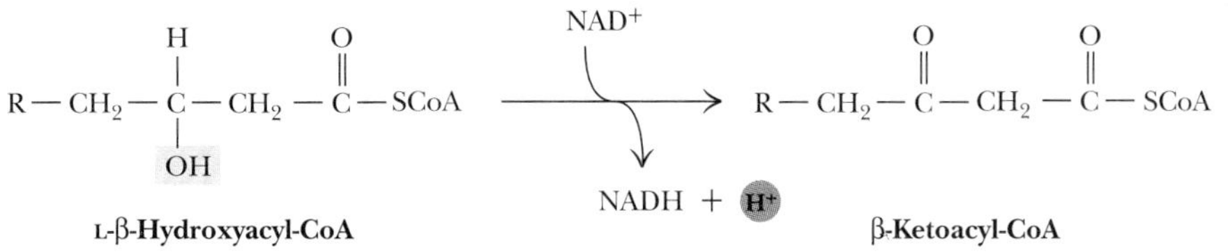

Figure 20.10 The L-β-hydroxyacyl-CoA dehydrogenase reaction.

pathway. L-Hydroxyacyl-CoA dehydrogenase shows absolute specificity for the L-hydroxyacyl isomer of the substrate (Figure 20.10). (D-Hydroxyacyl isomers, which arise mainly from oxidation of unsaturated fatty acids, are handled differently.)

β-Ketoacyl-CoA Intermediates Are Cleaved in the Thiolase Reaction

The final step in the β-oxidation cycle is the cleavage of the β-ketoacyl-CoA. This reaction, catalyzed by **thiolase** (also known as **β-ketothiolase**), involves the attack of a cysteine thiolate from the enzyme on the β-carbonyl carbon. Despite the formation of a second thioester, this reaction has a very favorable K_{eq}, and it drives the three previous reactions of β-oxidation.

Repetition of the β-Oxidation Cycle Yields a Succession of Acetate Units

In essence, this series of four reactions has yielded a fatty acid (as a CoA ester) that has been shortened by two carbons, and one molecule of acetyl-CoA. The shortened fatty acyl-CoA can now go through another β-oxidation cycle, as shown in Figure 20.7. Repetition of this cycle with a fatty acid with an even number of carbons eventually yields two molecules of acetyl-CoA in the final step. As noted in the first reaction in Table 20.2, complete β-oxidation of palmitic acid yields eight molecules of acetyl-CoA as well as seven molecules of $FADH_2$ and seven molecules of NADH. The acetyl-CoA can be further metabolized in the TCA cycle (as we have already seen). Alternatively, acetyl-CoA can also be used as a substrate in amino acid biosynthesis (see Chapter 21). As noted in Chapter 19, however, acetyl-CoA cannot be used as a substrate for gluconeogenesis.

Complete β-Oxidation of One Palmitic Acid Yields 106 Molecules of ATP

If an acetyl-CoA is directed entirely to the TCA cycle in mitochondria, it can eventually generate approximately 10 high-energy phosphate bonds—that is, 10 molecules of ATP synthesized from ADP (see Table 20.2). Including the ATP formed from $FADH_2$ and NADH, complete β-oxidation of a molecule of

Table 20.2 Equations for the Complete Oxidation of Palmitoyl-CoA to CO_2 and H_2O

Equation	ATP Yield	Free Energy Yield(kJ/mol)
$CH_3(CH_2)_{14}CO\text{-}CoA + 7[FAD] + 7 H_2O + 7 NAD^+ + 7 CoA \longrightarrow 8 CH_3CO\text{-}CoA + 7[FADH_2] + 7 NADH + 7H^+$		
$7[FADH_2] + 10.5 P_i + 10.5 ADP + 3.5 O_2 \longrightarrow 7[FAD] + 17.5 H_2O + 10.5 ATP$	10.5	320
$7 NADH + 7 H^+ + 17.5 P_i + 17.5 ADP + 3.5 O_2 \longrightarrow 7 NAD^+ + 24.5 H_2O + 17.5 ATP$	17.5	534
8-Acetyl-CoA $+ 16 O_2 + 80 ADP + 80 P_i \longrightarrow 8 CoA + 88 H_2O + 16 CO_2 + 80 ATP$	80	2440
$CH_3-(CH_2)_{14}CO\text{-}CoA + 108 P_i + 108 ADP + 23 O_2 \longrightarrow 108 ATP + 16 CO_2 + 130 H_2O + CoA$	108	3294
Energetic "cost" of forming palmitoyl-CoA from palmitate and CoA	−2	−61
Total	106	3233

(a) Gerbil

(b) Ruby-throated hummingbird

palmitoyl-CoA in mitochondria yields 108 molecules of ATP. Subtracting the two high-energy bonds needed to form palmitoyl-CoA, the substrate for β-oxidation, one concludes that β-oxidation of a molecule of palmitic acid yields 106 molecules of ATP. The $\Delta G^{\circ\prime}$ for complete combustion of palmitate to CO_2 is -9790 kJ/mol. The hydrolytic energy embodied in 106 ATPs is 106×30.5 kJ/mol $= 3233$ kJ/mol, so the overall efficiency of β-oxidation under standard-state conditions is approximately 33%. The large energy yield from fatty acid oxidation is a reflection of the highly reduced state of the carbon in fatty acids. Sugars, in which the carbon is already partially oxidized, produce much less energy, carbon for carbon, than do fatty acids. The breakdown of fatty acids is regulated by a variety of metabolites and hormones.

Migratory Birds Travel Long Distances on Energy from Fatty Acid Oxidation

Because they represent the most highly concentrated form of stored biological energy, fatty acids are the metabolic fuel of choice for sustaining the incredibly long flights of many migratory birds. Although some birds migrate over land masses and eat frequently, other species fly long distances without stopping to eat. The American golden plover flies directly from Alaska to Hawaii, a 3300-kilometer flight requiring 35 hours (at an average speed of nearly 60 miles/hr) and more than 250,000 wing beats! The ruby-throated hummingbird, which winters in Central America and nests in southern Canada, often flies nonstop across the Gulf of Mexico. These and similar birds accomplish these prodigious feats by storing large amounts of fatty acids (as triacylglycerols) in the days before their migratory flights. The percentage of dry-weight body fat in these birds may be as high as 70% when migration begins (compared with values of 30% and less for nonmigratory birds).

Fatty Acid Oxidation Is an Important Source of Metabolic Water for Some Animals

Large amounts of metabolic water are generated by β-oxidation (130 H_2O per palmitoyl-CoA). For certain animals—including desert animals, such as gerbils,

(c) Golden plover

(d) Orca

(e) Camels

Figure 20.11 Animals whose existence is strongly dependent on fatty acid oxidation: **(a)** gerbil, **(b)** ruby-throated hummingbird, **(c)** golden plover, **(d)** orca (killer whale), and **(e)** camels.
(a, Photo Researchers, Inc.; b, Tom J. Ulrich/Visuals Unlimited; c, S. J. Krasemann/Photo Researchers, Inc.; d, © Francois Gohier/Photo Researchers, Inc.; e, © George Holton/Photo Researchers, Inc.)

and killer whales (which do not drink seawater)—the oxidation of fatty acids can be a significant source of dietary water. A striking example is the camel (Figure 20.11), whose hump is essentially a large deposit of fat. Metabolism of fatty acids from this store provides needed water (as well as metabolic energy) during periods when drinking water is not available. It might well be said that "the ship of the desert" sails on its own metabolic water!

20.3 β-Oxidation of Odd-Carbon Fatty Acids

β-Oxidation of Odd-Carbon Fatty Acids Yields Propionyl-CoA

Fatty acids with odd numbers of carbon atoms are rare in mammals but fairly common in plants and marine organisms. Humans and animals whose diets include these food sources metabolize odd-carbon fatty acids via the β-oxidation pathway. The final product of β-oxidation in this case is the three-carbon propionyl-CoA instead of acetyl-CoA. Three specialized enzymes then carry out the reactions that convert propionyl-CoA to succinyl-CoA, a TCA cycle intermediate. (Because propionyl-CoA is a degradation product of methionine, valine, and isoleucine, this sequence of reactions is also important in amino acid catabolism.) The pathway involves an initial carboxylation at the α-carbon of propionyl-CoA to produce D-methylmalonyl-CoA (Figure 20.12). The reaction is catalyzed by a biotin-dependent enzyme, **propionyl-CoA carboxylase.** The mechanism involves ATP-driven carboxylation of biotin at N_1, followed by nucleophilic attack by the α-carbanion of propionyl-CoA in a stereospecific manner.

D-Methylmalonyl-CoA, the product of this reaction, is converted to the L-isomer by **methylmalonyl-CoA epimerase** (see Figure 20.12). (This enzyme has often and incorrectly been called "methylmalonyl-CoA racemase." It is not a racemase because the CoA moiety contains five other asymmetric centers.) The epimerase reaction is readily reversible and involves a reversible dissociation of the acidic α-proton. The L-isomer is the substrate for methylmalonyl-CoA mutase. Methylmalonyl-CoA epimerase is an impressive catalyst. The pK_a for the proton that must dissociate to initiate this reaction is approximately 21! If binding of a proton to the α-anion is diffusion-limited, with $k_{on} = 10^9\ M^{-1}\ sec^{-1}$, then the initial proton dissociation must be rate-limiting, and the rate constant must be

$$k_{off} = K_a \cdot k_{on} = (10^{-21}\ M) \cdot (10^9\ M^{-1}\ sec^{-1}) = 10^{-12}\ sec^{-1}$$

The turnover number of methylmalonyl-CoA epimerase is 100 sec^{-1}, and thus the enzyme enhances the reaction rate by a factor of 10^{14}.

A B₁₂-Catalyzed Rearrangement Yields Succinyl-CoA from L-Methylmalonyl-CoA

The third reaction, catalyzed by **methylmalonyl-CoA mutase,** is quite unusual because it involves a migration of the carbonyl-CoA group from one carbon to its neighbor (see Figure 20.12). The mutase reaction is vitamin B₁₂–dependent.

Net Oxidation of Succinyl-CoA Requires Conversion to Acetyl-CoA

Succinyl-CoA derived from propionyl-CoA can enter the TCA cycle. Oxidation of succinate to oxaloacetate provides a substrate for glucose synthesis. Thus, although the acetate units produced in β-oxidation cannot be utilized in gluconeogenesis by animals, the occasional propionate produced from oxidation

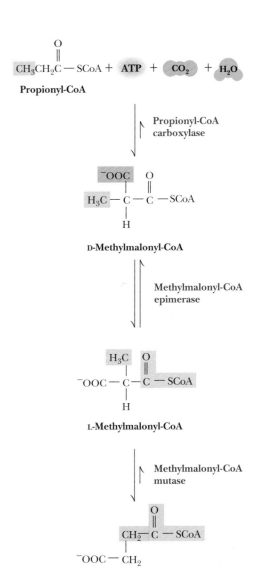

Figure 20.12 The conversion of propionyl-CoA (formed from β-oxidation of odd-carbon fatty acids) to succinyl-CoA is carried out by a trio of enzymes as shown. Succinyl-CoA can enter the TCA cycle or be converted to acetyl-CoA.

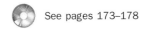 See pages 173–178

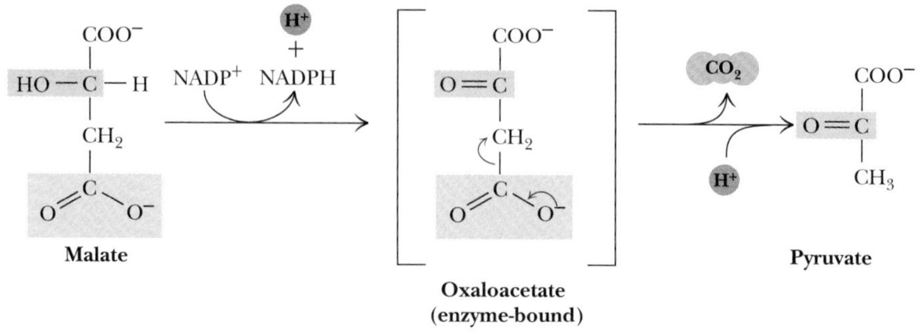

Figure 20.13 The malic enzyme reaction proceeds by oxidation of malate to oxaloacetate, followed by decarboxylation to yield pyruvate.

$$\text{Malate} \xrightarrow[\text{NADP}^+ \quad \text{NADPH}]{\text{H}^+} [\text{Oxaloacetate (enzyme-bound)}] \xrightarrow[\text{H}^+]{\text{CO}_2} \text{Pyruvate}$$

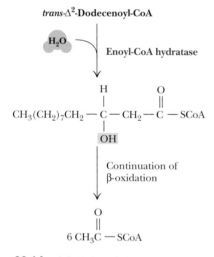

Oleoyl-CoA

β-oxidation (three cycles) → $3\ \text{CH}_3\text{—C—SCoA}$

cis-Δ³-Dodecenoyl-CoA

Enoyl-CoA isomerase

trans-Δ²-Dodecenoyl-CoA

H_2O — **Enoyl-CoA hydratase**

Continuation of β-oxidation

$6\ \text{CH}_3\text{C—SCoA}$

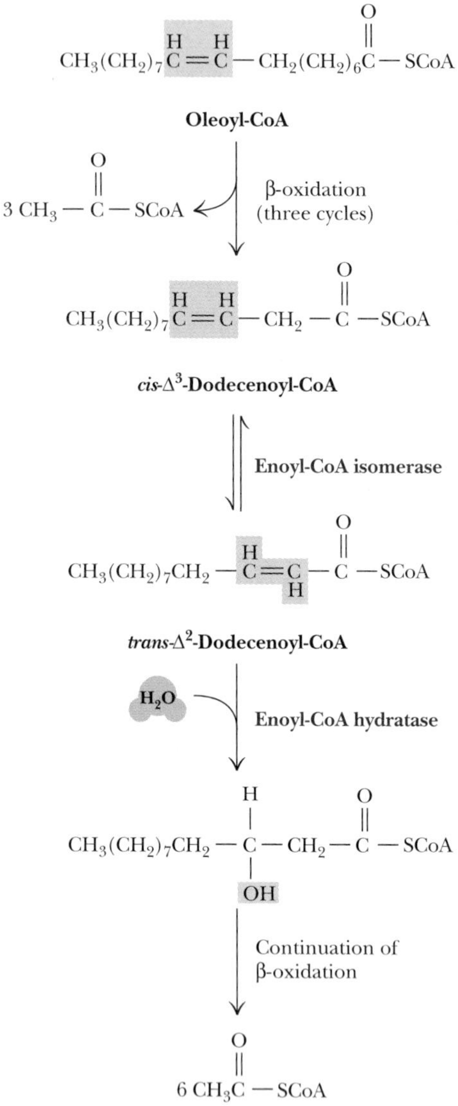

Figure 20.14 β-Oxidation of unsaturated fatty acids. In the case of oleoyl-CoA, three β-oxidation cycles produce three molecules of acetyl-CoA and leave cis-Δ³-dodecenoyl-CoA. Rearrangement of enoyl-CoA isomerase gives the trans-Δ² species, which then proceeds normally through the β-oxidation pathway.

See pages 173–178

of odd-carbon fatty acids can be used for sugar synthesis. Alternatively, succinate introduced to the TCA cycle from odd-carbon fatty acid oxidation may be oxidized to CO_2. However, all of the four-carbon intermediates in the TCA cycle are regenerated in the cycle and thus should be viewed as catalytic species. Net consumption of succinyl-CoA therefore does not occur directly in the TCA cycle. Rather, to account for combustion of the succinyl-CoA generated from β-oxidation of odd-carbon fatty acids, malate derived from succinyl-CoA must be converted to pyruvate and then to acetyl-CoA (which is completely oxidized in the TCA cycle). To follow this latter route, malate from succinyl-CoA is transported from the mitochondrial matrix to the cytosol, where it is oxidatively decarboxylated to pyruvate and CO_2 by **malic enzyme,** as shown in Figure 20.13. Pyruvate can then be transported back to the mitochondrial matrix, where it enters the TCA cycle via pyruvate dehydrogenase.

20.4 β-Oxidation of Unsaturated Fatty Acids

An Isomerase and a Reductase Facilitate the β-Oxidation of Unsaturated Fatty Acids

Unsaturated fatty acids are also catabolized by β-oxidation, but two additional mitochondrial enzymes—an isomerase and a novel reductase—are required to handle the *cis*-double bonds of naturally occurring fatty acids. As an example, consider the breakdown of oleic acid, an 18-carbon chain with a double bond at the 9,10-position. The reactions of β-oxidation proceed normally through three cycles, producing three molecules of acetyl-CoA and leaving the degradation product *cis*-Δ³-dodecenoyl-CoA, shown in Figure 20.14. This intermediate is not a substrate for acyl-CoA dehydrogenase. With a double bond at the 3,4-position, it is not possible to form another double bond at the 2,3-(or β) position. As shown in Figure 20.14, this problem is solved by **enoyl-CoA isomerase,** an enzyme that rearranges this *cis*-Δ³ double bond to a *trans*-Δ² double bond. This latter species can proceed through the normal route of β-oxidation.

Degradation of Polyunsaturated Fatty Acids Requires 2,4-Dienoyl-CoA Reductase

Polyunsaturated fatty acids pose a slightly more complicated situation for the cell. Consider, for example, the case of linoleic acid shown in Figure 20.15. As with oleic acid, β-oxidation proceeds through three cycles, and enoyl-CoA isomerase converts the *cis*-Δ³ double bond to a *trans*-Δ² double bond to permit one more round of β-oxidation. What results this time, however, is a *cis*-Δ⁴

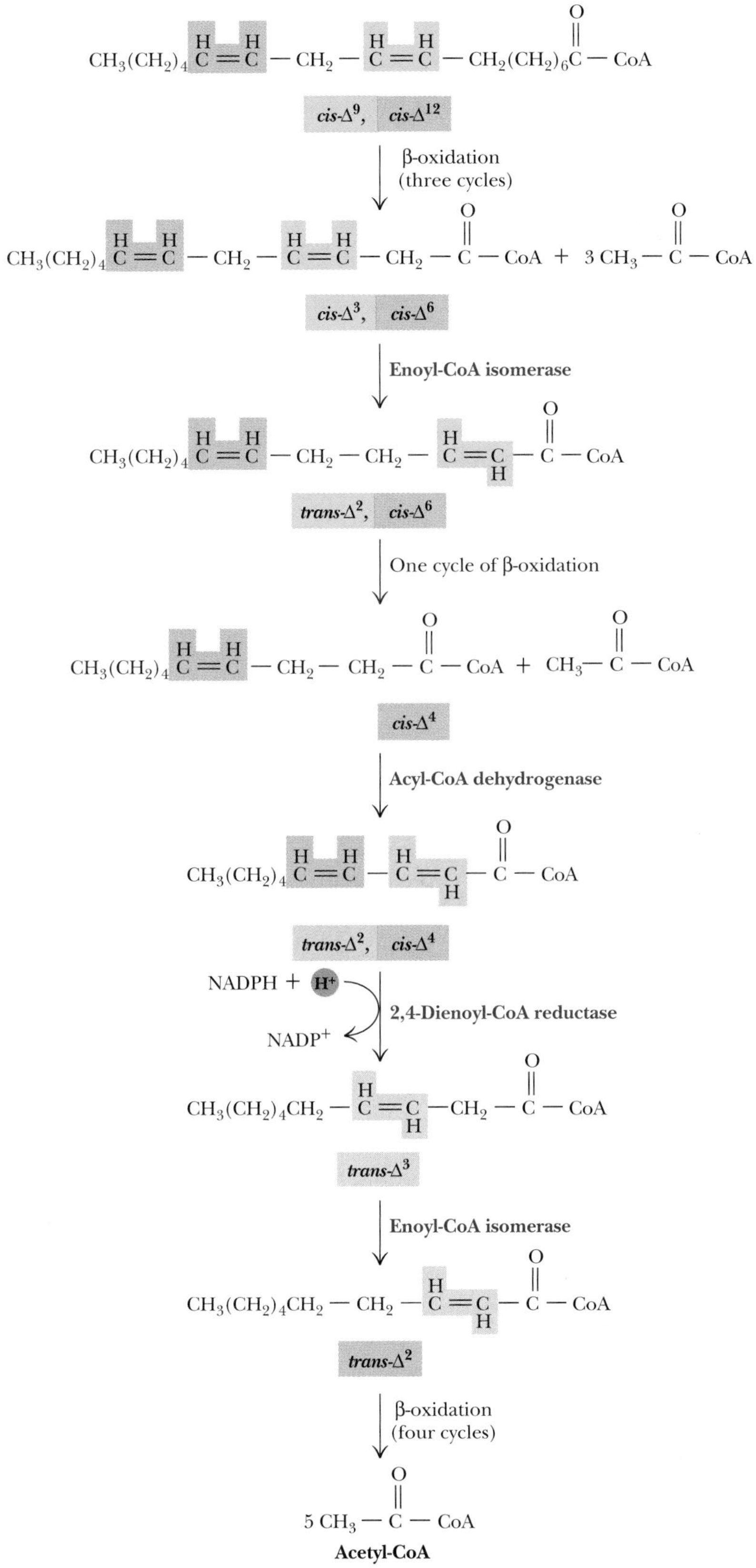

$$CH_3(CH_2)_4\;C{=}C - CH_2 - C{=}C - CH_2(CH_2)_6\overset{\displaystyle O}{\overset{\|}{C}} - CoA$$

cis-Δ^9,   *cis*-Δ^{12}

β-oxidation
(three cycles)

$$CH_3(CH_2)_4\;C{=}C - CH_2 - C{=}C - CH_2 - \overset{\displaystyle O}{\overset{\|}{C}} - CoA\; +\; 3\;CH_3 - \overset{\displaystyle O}{\overset{\|}{C}} - CoA$$

cis-Δ^3,   *cis*-Δ^6

Enoyl-CoA isomerase

$$CH_3(CH_2)_4\;C{=}C - CH_2 - CH_2 - C{=}C - \overset{\displaystyle O}{\overset{\|}{C}} - CoA$$

trans-Δ^2,   *cis*-Δ^6

One cycle of β-oxidation

$$CH_3(CH_2)_4\;C{=}C - CH_2 - CH_2 - \overset{\displaystyle O}{\overset{\|}{C}} - CoA\; +\; CH_3 - \overset{\displaystyle O}{\overset{\|}{C}} - CoA$$

cis-Δ^4

Acyl-CoA dehydrogenase

$$CH_3(CH_2)_4\;C{=}C - C{=}C - \overset{\displaystyle O}{\overset{\|}{C}} - CoA$$

trans-Δ^2,   *cis*-Δ^4

NADPH +

2,4-Dienoyl-CoA reductase

NADP$^+$

$$CH_3(CH_2)_4CH_2 - C{=}C - CH_2 - \overset{\displaystyle O}{\overset{\|}{C}} - CoA$$

trans-Δ^3

Enoyl-CoA isomerase

$$CH_3(CH_2)_4CH_2 - CH_2 - C{=}C - \overset{\displaystyle O}{\overset{\|}{C}} - CoA$$

trans-Δ^2

β-oxidation
(four cycles)

$$5\;CH_3 - \overset{\displaystyle O}{\overset{\|}{C}} - CoA$$

Acetyl-CoA

Figure 20.15 The oxidation pathway for polyunsaturated fatty acids, illustrated for linoleic acid. Three cycles of β-oxidation on linoleoyl-CoA yield the *cis*-Δ^3, *cis*-Δ^6 intermediate, which is converted to a *trans*-Δ^2, *cis*-Δ^6 intermediate. An additional round of β-oxidation gives *cis*-Δ^4 enoyl-CoA, which is oxidized to the *trans*-Δ^2, *cis*-Δ^4 species by acyl-CoA dehydrogenase. The subsequent action of 2,4-dienoyl-CoA reductase yields the *trans*-Δ^3 product, which is converted by enoyl-CoA isomerase to the *trans*-Δ^2 form. Normal β-oxidation then produces five molecules of acetyl-CoA.

See pages 173–178

enoyl-CoA, which is converted normally by acyl-CoA dehydrogenase to a *trans*-Δ^2, *cis*-Δ^4 species. This, however, is a poor substrate for the enoyl-CoA hydratase. This problem is solved by **2,4-dienoyl-CoA reductase,** the product of which depends on the organism. The mammalian form of this enzyme produces a *trans*-Δ^3 enoyl product, as shown in Figure 20.15, which can be converted by an

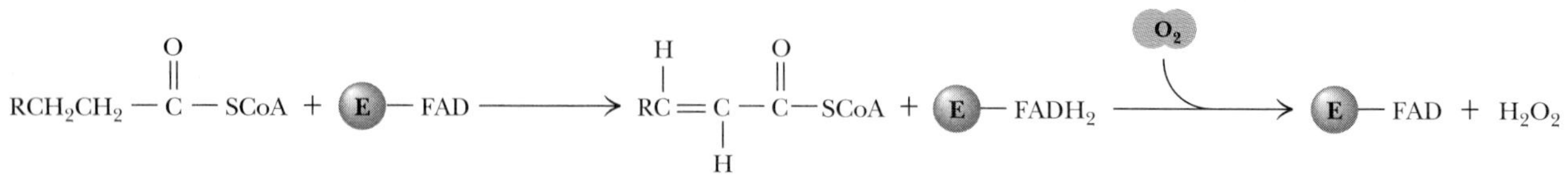

$$RCH_2CH_2 - \overset{\overset{\displaystyle O}{\|}}{C} - SCoA + \text{(E)} - FAD \longrightarrow RC = \overset{\overset{\displaystyle H}{|}}{\underset{\underset{\displaystyle H}{|}}{C}} - \overset{\overset{\displaystyle O}{\|}}{C} - SCoA + \text{(E)} - FADH_2 \xrightarrow{\;O_2\;} \text{(E)} - FAD + H_2O_2$$

Figure 20.16 The acyl-CoA oxidase reaction in peroxisomes.

enoyl-CoA isomerase to the *trans*-Δ^2 enoyl-CoA, which can then proceed normally through the β-oxidation pathway. *Escherichia coli* possesses a 2,4-dienoyl-CoA reductase that reduces the double bond at the 4,5-position to yield the *trans*-Δ^2 enoyl-CoA product in a single step.

20.5 Other Aspects of Fatty Acid Oxidation

Peroxisomal β-Oxidation Requires FAD-Dependent Acyl-CoA Oxidase

Although β-oxidation in mitochondria[1] is the principal pathway of fatty acid catabolism, several other minor pathways play important roles in fat catabolism. For example, organelles other than mitochondria carry out β-oxidation processes, including peroxisomes and glyoxysomes. **Peroxisomes** are so named because they carry out a variety of flavin-dependent oxidation reactions, regenerating oxidized flavins by reaction with oxygen to produce hydrogen peroxide, H_2O_2. Peroxisomal β-oxidation is similar to mitochondrial β-oxidation, except that the initial double bond formation is catalyzed by an FAD-dependent **acyl-CoA oxidase** (Figure 20.16). The action of this enzyme in the peroxisomes transfers the liberated electrons directly to oxygen instead of the electron transport chain. As a result, each two-carbon unit oxidized in peroxisomes produces fewer ATPs. The enzymes responsible for fatty acid oxidation in peroxisomes are inactive with chains of eight or fewer carbons. Such short-chain products must be transferred to the mitochondria for further breakdown. Similar β-oxidation enzymes are also found in **glyoxysomes** in plants (see Chapter 16).

Branched-Chain Fatty Acids and α-Oxidation

Although β-oxidation is universally important, there are some instances in which it cannot operate effectively. For example, branched-chain fatty acids with alkyl branches at odd-numbered carbons are not effective substrates for β-oxidation. For such species, α-**oxidation** is a useful alternative. Consider **phytol**, a breakdown product of chlorophyll that occurs in the fat of ruminant animals such as sheep and cows and also in dairy products. Ruminants oxidize phytol to phytanic acid, and digestion of phytanic acid in dairy products is thus an important dietary consideration for humans. The methyl group at C-3 will block β-oxidation, but as shown in Figure 20.17, **phytanic acid α-hydroxylase** places an —OH group at the α-carbon, and **phytanic acid α-oxidase** decarboxylates it to yield pristanic acid. The CoA ester of this metabolite can undergo β-oxidation in the normal manner. The terminal product, isobutyryl-CoA, can be sent into the TCA cycle by conversion to succinyl-CoA.

---

[1]β-Oxidation does not occur significantly in plant mitochondria.

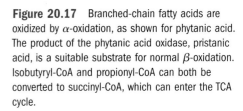

Figure 20.17 Branched-chain fatty acids are oxidized by α-oxidation, as shown for phytanic acid. The product of the phytanic acid oxidase, pristanic acid, is a suitable substrate for normal β-oxidation. Isobutyryl-CoA and propionyl-CoA can both be converted to succinyl-CoA, which can enter the TCA cycle.

See pages 173–178

Phytol

Phytanic acid

Phytanic acid α-hydroxylase

Phytanic acid α-oxidase

Pristanic acid

Acyl-CoA synthetase

Six cycles of β-oxidation

Isobutyryl-CoA   **3 Acetyl-CoA**   **3 Propionyl-CoA**

ω-Oxidation of Fatty Acids Yields Small Amounts of Dicarboxylic Acids

In the endoplasmic reticulum of eukaryotic cells, the oxidation of the terminal carbon of a normal fatty acid—a process termed ω-oxidation—can lead to the synthesis of small amounts of dicarboxylic acids (Figure 20.18). A **cytochrome P-450**–containing[2] monooxygenase enzyme that requires NADPH as a coenzyme

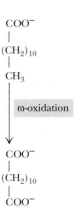

Figure 20.18 Dicarboxylic acids can be formed by oxidation of the methyl group of fatty acids in a cytochrome P-450–dependent reaction.

---
[2]The name cytochrome P-450 refers to a family of heme proteins that resemble mitochondrial cytochrome oxidase in being able to bind both O_2 and CO. The name refers to the fact that the reduced form of the heme complexed with CO absorbs light strongly at 450 nm. Cytochromes P-450 are located in the endoplasmic reticulum of eukaryotic cells and participate in a wide variety of reactions, including hydroxylation, epoxidation, dealkylation, and dehalogenation, particularly in liver. Pesticides and drugs (including the barbiturates) induce the synthesis of cytochrome P-450, which in turn metabolizes these substances, in a process known as **detoxification.**

Refsum's Disease Is a Result of Defects in α-Oxidation

The α-oxidation pathway is defective in **Refsum's disease,** an inherited metabolic disorder that results in impaired night vision, tremors, and other neurologic abnormalities. These symptoms are caused by accumulation of phytanic acid in the body. Treatment of

Refsum's disease requires a diet free of chlorophyll, the precursor of phytanic acid. This regimen is difficult to implement because all green vegetables and even meat from plant-eating animals, such as cows, pigs, and poultry, must be excluded from the diet.

and uses O_2 as a substrate places a hydroxyl group at the terminal carbon. Subsequent oxidation to a carboxyl group produces a dicarboxylic acid. Either end can form an ester linkage to CoA and can undergo β-oxidation, producing a variety of smaller dicarboxylic acids.

20.6 | Ketone Bodies

Ketone Bodies Are a Significant Source of Fuel and Energy for Certain Tissues

Most of the acetyl-CoA produced by the oxidation of fatty acids in liver mitochondria undergoes further oxidation in the TCA cycle, as stated earlier. However, some of this acetyl-CoA is converted to three important metabolites: acetone, acetoacetate, and β-hydroxybutyrate. The process is known as **ketogenesis,** and these three metabolites are traditionally known as **ketone bodies,** in spite of the fact that β-hydroxybutyrate does not contain a ketone function. These three metabolites are synthesized primarily in the liver but are important sources of fuel and energy for many peripheral tissues, including brain, heart, and skeletal muscle. The brain, for example, normally uses glucose as its source of metabolic energy. However, during periods of starvation, ketone bodies may be the major energy source for the brain. Acetoacetate and 3-hydroxybutyrate are the preferred and normal substrates for kidney cortex and can also be utilized by heart muscle.

Ketone body synthesis occurs only in the mitochondrial matrix. The reactions responsible for the formation of ketone bodies are shown in Figure 20.19. The first reaction—the condensation of two molecules of acetyl-CoA to form acetoacetyl-CoA—is catalyzed by **thiolase,** which is also known as **acetoacetyl-CoA thiolase** or **acetyl-CoA acetyltransferase.** This is the same enzyme that

Ketone Bodies and Diabetes Mellitus

Diabetes mellitus is the most common endocrine disease and the third leading cause of death in the United States, with approximately 6 million diagnosed cases and an estimated 4 million more borderline but undiagnosed cases. In both Type I (insulin-dependent) and Type II (insulin-independent) diabetes mellitus, transport of glucose into muscle, liver, and adipose tissue is significantly reduced, and, despite abundant glucose in the blood, the cells are metabolically starved. They respond by turning to increased gluconeogenesis and catabolism of fat and protein. In Type

I diabetes, increased gluconeogenesis consumes most of the available oxaloacetate, but breakdown of fat (and, to a lesser extent, protein) produces large amounts of acetyl-CoA. This increased acetyl-CoA would normally be directed into the TCA cycle, but with oxaloacetate in short supply, it is used instead for production of unusually large amounts of ketone bodies. Acetone can often be detected on the breath of Type I diabetics, an indication of high plasma levels of ketone bodies.

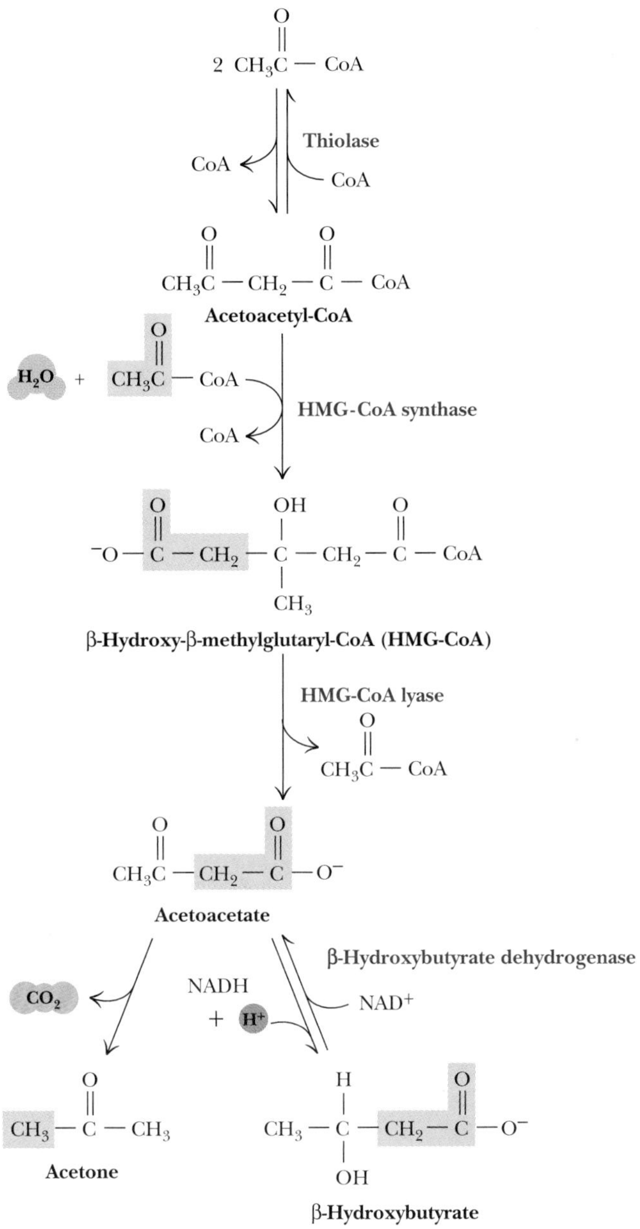

Figure 20.19 The formation of ketone bodies, synthesized primarily in liver mitochondria.

 See pages 173–178

carries out the thiolase reaction in β-oxidation, but here it runs in reverse. The second reaction adds another molecule of acetyl-CoA to give β-hydroxy-β-methylglutaryl-CoA, commonly abbreviated HMG-CoA. These two mitochondrial matrix reactions are analogous to the first two steps in cholesterol biosynthesis occurring in the cytosol (see Section 20.10). HMG-CoA is converted to acetoacetate and acetyl-CoA by the action of **HMG-CoA lyase.** This reaction is mechanistically similar to the reverse of the citrate synthase reaction in the TCA cycle. A membrane-bound enzyme, **β-hydroxybutyrate dehydrogenase,** then can reduce acetoacetate to β-hydroxybutyrate.

Acetoacetate and β-hydroxybutyrate are transported through the blood from liver to target organs and tissues, where they are converted to acetyl-CoA (Figure 20.20). Ketone bodies are easily transportable forms of fatty acids that move through the circulatory system without the need for complexation with serum albumin and other fatty acid–binding proteins. Ketone bodies also do not require insulin to enter target organs and tissues.

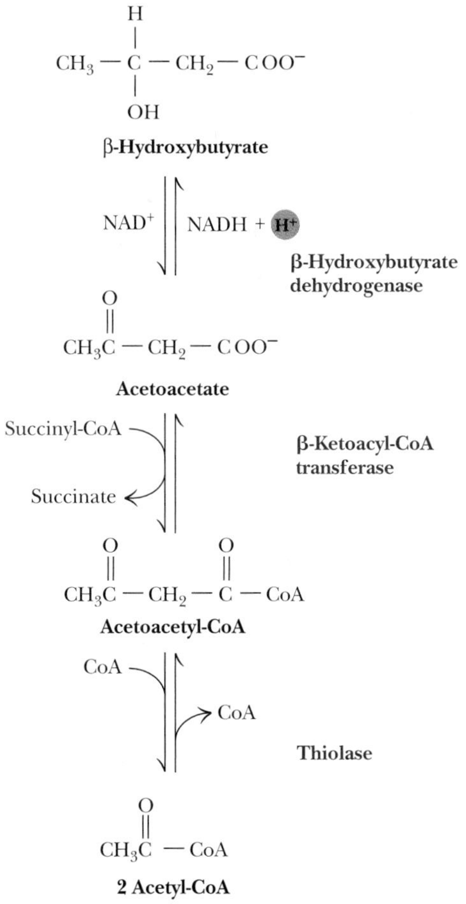

Figure 20.20 Reconversion of ketone bodies to acetyl-CoA in the mitochondria of many tissues (other than liver) provides significant metabolic energy.

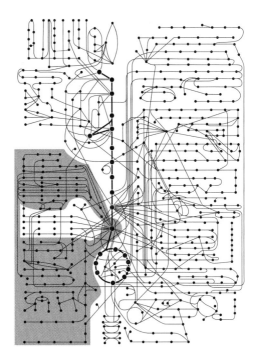

Lipid biosynthesis.

See pages 173–178

We have already seen several cases in which the synthesis of a class of biomolecules is conducted differently from degradation (glycolysis versus gluconeogenesis and glycogen or starch breakdown versus polysaccharide synthesis, for example). Likewise, the synthesis of fatty acids and other lipid components is different from their degradation. Fatty acid synthesis involves a set of reactions that follow a strategy different in several ways from the corresponding degradative process:

1. Intermediates in fatty acid synthesis are linked covalently to the sulfhydryl groups of special proteins, the **acyl carrier proteins.** In contrast, fatty acid breakdown intermediates are bound to the —SH group of coenzyme A.
2. Fatty acid synthesis occurs in the cytosol, whereas fatty acid degradation takes place in mitochondria.
3. In animals, the enzymes of fatty acid synthesis are components of one long polypeptide chain, the **fatty acid synthase,** whereas no similar association exists for the degradative enzymes. (Plants and bacteria employ separate enzymes to carry out the biosynthetic reactions.)
4. The coenzyme for the oxidation–reduction reactions of fatty acid synthesis is $NADP^+/NADPH$, whereas degradation involves the $NAD^+/NADH$ couple.

Formation of Malonyl-CoA Activates Acetate Units for Fatty Acid Synthesis

Fatty acid synthesis has the following features:

a. Fatty acid chains are constructed by the addition of two-carbon units derived from acetyl-CoA.
b. The acetate units are activated by formation of malonyl-CoA (at the expense of ATP).
c. The addition of two-carbon units to the growing chain is driven by decarboxylation of malonyl-CoA.
d. The elongation reactions are repeated until the growing chain reaches 16 carbons in length (palmitic acid).
e. Other enzymes then may add double bonds or additional carbon units to the chain.

Fatty Acid Biosynthesis Depends on the Reductive Power of NADPH

The net reaction for the formation of palmitate from acetyl-CoA is

$$\text{Acetyl-CoA} + 7\text{ malonyl-CoA}^- + 14\text{ NADPH} + 14\text{ H}^+ \longrightarrow$$
$$\text{palmitoyl-CoA} + 7\text{ HCO}_3^- + 7\text{ CoASH} + 14\text{ NADP}^+ \tag{20.1}$$

(Levels of free fatty acids are very low in the typical cell. The palmitate made in this process is rapidly converted to CoA esters in preparation for the formation of triacylglycerols and phospholipids.)

Providing Cytosolic Acetyl-CoA and Reducing Power for Fatty Acid Synthesis

Eukaryotic cells face a dilemma in providing suitable amounts of substrate for fatty acid synthesis. Sufficient quantities of acetyl-CoA, malonyl-CoA, and NADPH must be generated in the cytosol for fatty acid synthesis. Malonyl-CoA is made by carboxylation of acetyl-CoA, so the problem reduces to generating sufficient acetyl-CoA and NADPH.

There are three principal sources of acetyl-CoA (Figure 20.21):

1. Amino acid degradation produces cytosolic acetyl-CoA.
2. Fatty acid oxidation produces mitochondrial acetyl-CoA.
3. Glycolysis yields cytosolic pyruvate, which (after transport into the mitochondria) is converted to acetyl-CoA by pyruvate dehydrogenase.

The acetyl-CoA derived from amino acid degradation is normally insufficient for fatty acid biosynthesis, and the acetyl-CoA produced by pyruvate dehydrogenase and by fatty acid oxidation cannot cross the mitochondrial membrane to participate directly in fatty acid synthesis. Instead, acetyl-CoA is linked with oxaloacetate to form citrate, which is transported from the mitochondrial matrix to the cytosol (see Figure 20.21). Here it can be converted back into acetyl-CoA and oxaloacetate by **ATP–citrate lyase.** In this manner, mitochondrial acetyl-CoA becomes the substrate for cytosolic fatty acid synthesis. (Oxaloacetate returns to the mitochondria in the form of either pyruvate or malate, which is then reconverted to acetyl-CoA or oxaloacetate, respectively.)

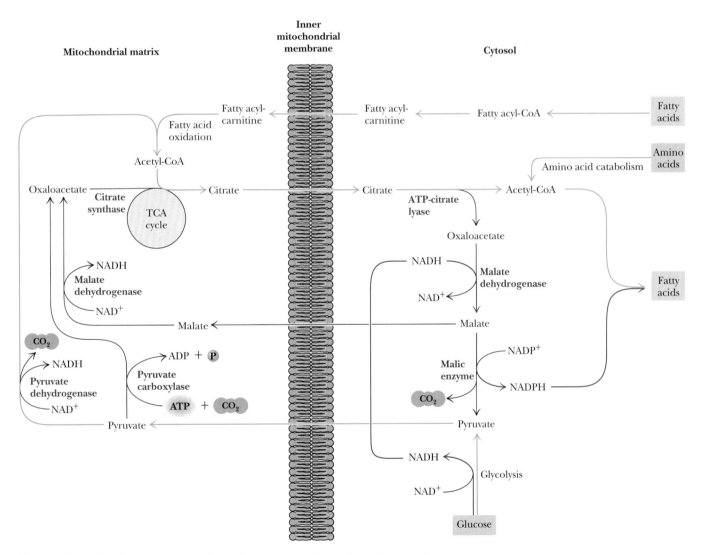

Figure 20.21 The citrate–malate–pyruvate shuttle provides cytosolic acetate units and reducing equivalents (electrons) for fatty acid synthesis. The shuttle collects carbon substrates, primarily from glycolysis but also from fatty acid oxidation and amino acid catabolism. Most of the reducing equivalents are glycolytic in origin. Pathways that provide carbon for fatty acid synthesis are shown in blue; pathways that supply electrons for fatty acid synthesis are shown in red.

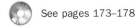 See pages 173–178

NADPH can be produced in the pentose phosphate pathway as well as by malic enzyme (see Figure 20.21). Reducing equivalents (electrons) derived from glycolysis in the form of NADH can be transformed into NADPH by the combined action of malate dehydrogenase and malic enzyme:

$$\text{Oxaloacetate} + \text{NADH} + \text{H}^+ \longrightarrow \text{malate} + \text{NAD}^+$$
$$\text{Malate} + \text{NADP}^+ \longrightarrow \text{pyruvate} + \text{CO}_2 + \text{NADPH} + \text{H}^+$$

Additional NADPH is provided by the pentose phosphate pathway.

Acetate Units Are Committed to Fatty Acid Synthesis by Formation of Malonyl-CoA

David Rittenberg and Konrad Bloch showed in the late 1940s that acetate units are the building blocks of fatty acids. Their work, together with the discovery by Salih Wakil that bicarbonate is required for fatty acid biosynthesis, eventually made clear that this pathway involves synthesis of malonyl-CoA. The carboxylation of acetyl-CoA to form malonyl-CoA is essentially irreversible and is the **committed step** in the synthesis of fatty acids (Figure 20.22). The reaction is catalyzed by **acetyl-CoA carboxylase,** which contains a biotin prosthetic group. This carboxylase is the only enzyme of fatty acid synthesis in animals that is not part of the multienzyme complex called fatty acid synthase.

Acetyl-CoA Carboxylase in Animals Is a Multifunctional Protein

Acetyl-CoA carboxylase from *Escherichia coli* has three subunits: (1) a **biotin carboxyl carrier protein** (a dimer of 22.5-kD subunits); (2) **biotin carboxylase** (a dimer of 51-kD subunits), which adds CO_2 to the prosthetic group; and (3) **transcarboxylase** (an $\alpha_2\beta_2$ tetramer with 30-kD and 35-kD subunits), which transfers the activated CO_2 unit to acetyl-CoA. In animals, acetyl-CoA carboxylase (ACC) is a filamentous polymer (4 to 8×10^6 D) composed of 230-kD protomers. Each of these subunits contains the biotin carboxyl carrier moiety, biotin carboxylase, and transcarboxylase activities, as well as allosteric regulatory sites. Animal ACC is thus a multifunctional protein. The polymeric form is active, but the 230-kD protomers are inactive. The activity of ACC is thus dependent upon the position of the equilibrium between these two forms:

$$\text{Inactive protomers} \rightleftharpoons \text{active polymer}$$

Because this enzyme catalyzes the committed step in fatty acid biosynthesis, it is carefully regulated. Palmitoyl-CoA, the final product of fatty acid biosynthesis, shifts the equilibrium toward the inactive protomers, whereas citrate, an important allosteric activator of this enzyme, shifts the equilibrium toward the active polymeric form of the enzyme. Acetyl-CoA carboxylase shows the kinetic behavior of a Monod–Wyman–Changeux V-system allosteric enzyme (see Chapter 10).

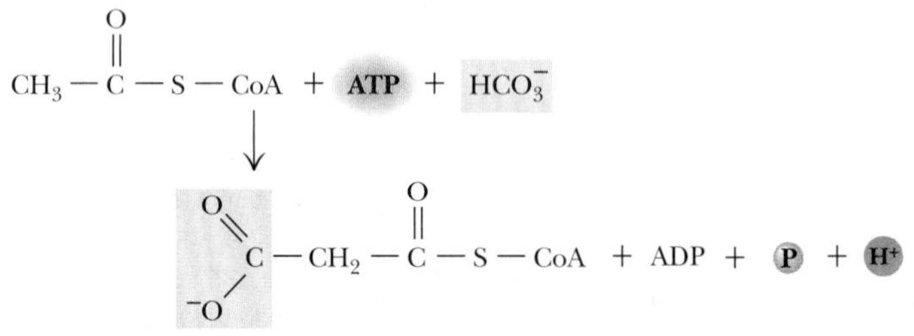

Figure 20.22 The acetyl-CoA carboxylase reaction produces malonyl-CoA for fatty acid synthesis.

Phosphorylation of ACC Modulates Activation by Citrate and Inhibition by Palmitoyl-CoA

The regulatory effects of citrate and palmitoyl-CoA are dependent on the phosphorylation state of acetyl-CoA carboxylase. Unphosphorylated acetyl-CoA carboxylase binds citrate with high affinity and thus is active at very low citrate concentrations (Figure 20.23). Phosphorylation of the regulatory sites decreases the affinity of the enzyme for citrate, and in this case high levels of citrate are required to activate the carboxylase. The inhibition by fatty acyl-CoAs operates in a similar but opposite manner. Thus, low levels of fatty acyl-CoA inhibit the phosphorylated carboxylase, but the dephosphoenzyme is inhibited only by high levels of fatty acyl-CoA. Specific phosphatases act to dephosphorylate ACC, thereby increasing the sensitivity to citrate.

Acyl Carrier Proteins Carry the Intermediates in Fatty Acid Synthesis

The basic building blocks of fatty acid synthesis are acetyl and malonyl groups, but they are not transferred directly from CoA to the growing fatty acid chain. Rather, they are first passed to **acyl carrier protein** (or simply ACP), discovered by P. Roy Vagelos. This protein consists (in *E. coli*) of a single polypeptide chain of 77 residues to which is attached (on a serine residue) a **phosphopantetheine group,** the same group that forms the "business end" of coenzyme A. Thus, acyl carrier protein is a somewhat larger version of coenzyme A, specialized for use in fatty acid biosynthesis (Figure 20.24).

Fatty Acid Synthesis in Eukaryotes Occurs on a Multienzyme Complex

The enzymes that catalyze formation of acetyl-ACP and malonyl-ACP and the subsequent reactions of fatty acid synthesis are organized quite differently in different organisms. In bacteria and plants, the various reactions are catalyzed by separate, independent proteins. Fatty acid biosynthesis in animals, on the other hand, involves a single multienzyme complex called **fatty acid synthase.** The individual steps in the elongation of the fatty acid chain are quite similar

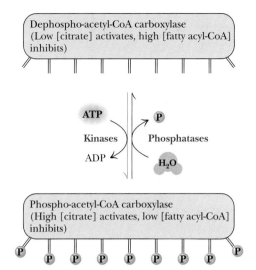

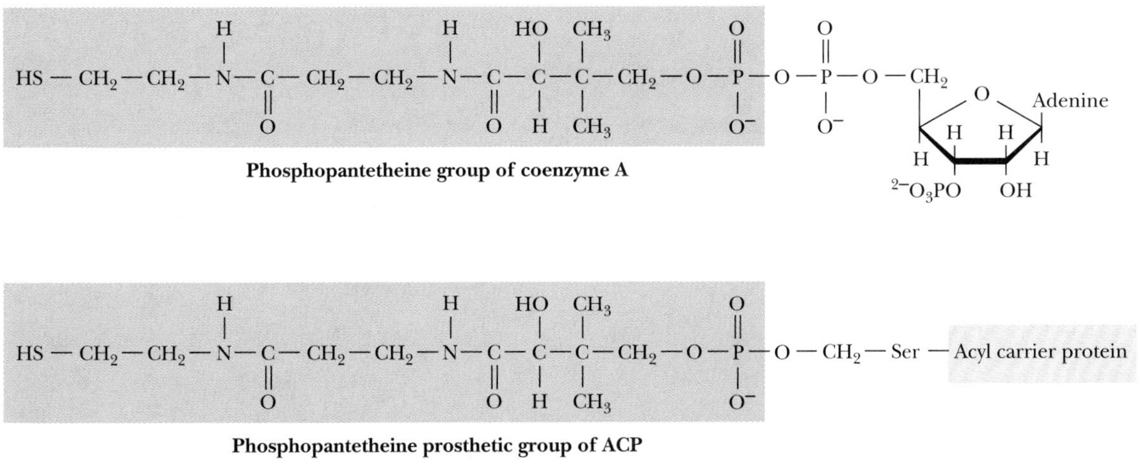

Figure 20.23 The activity of acetyl-CoA carboxylase is modulated by phosphorylation and dephosphorylation. The dephospho form of the enzyme is activated by low [citrate] and inhibited only by high levels of fatty acyl-CoA. In contrast, the phosphorylated form of the enzyme is activated only by high levels of citrate but is very sensitive to inhibition by fatty acyl-CoA.

Phosphopantetheine group of coenzyme A

Phosphopantetheine prosthetic group of ACP

Figure 20.24 Fatty acids are conjugated both to coenzyme A and to acyl carrier protein through the sulfhydryl of phosphopantetheine prosthetic groups.

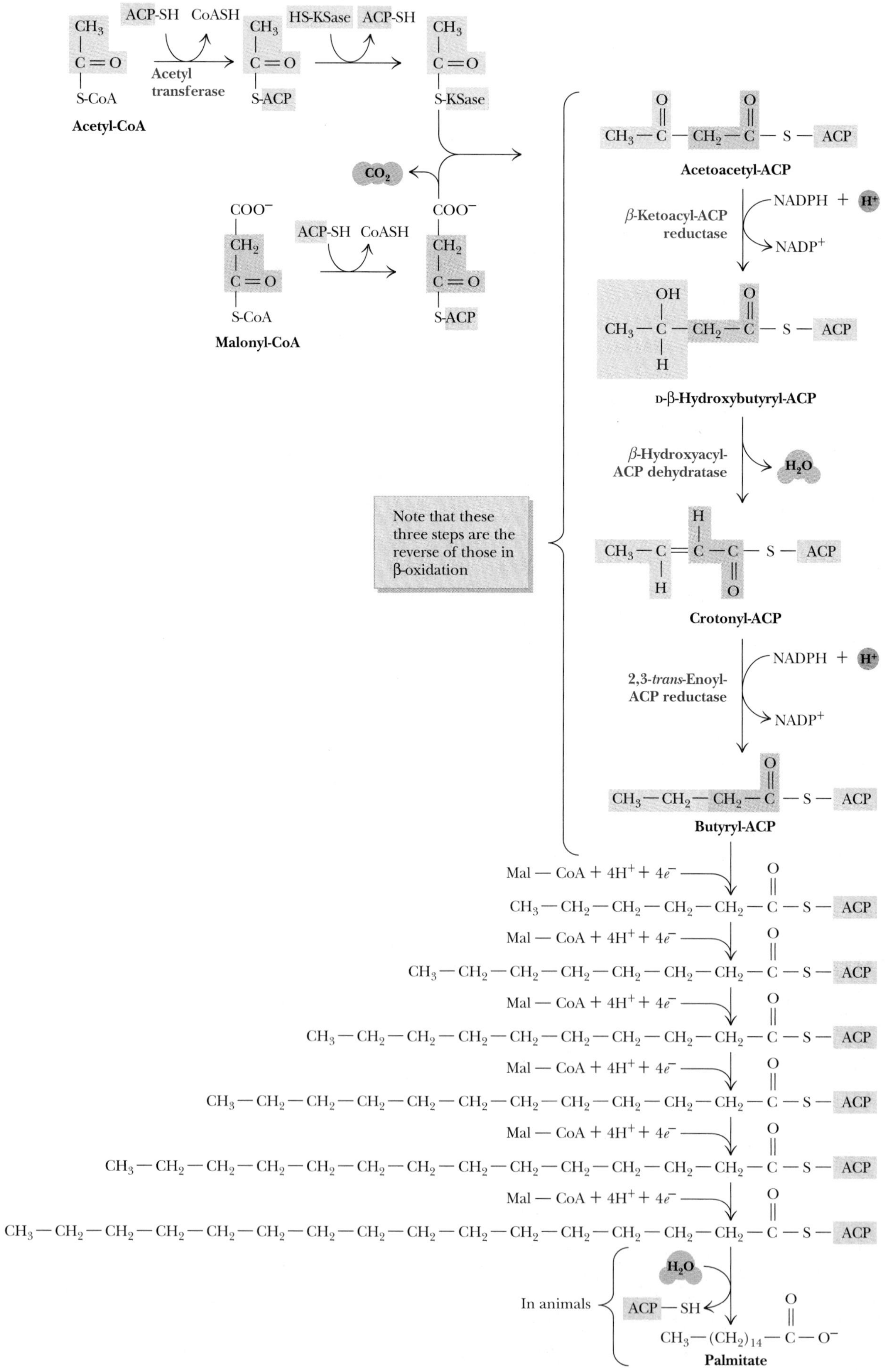

Acetyl-CoA

ACP-SH CoASH

Acetyl transferase

S-ACP

HS-KSase ACP-SH

S-KSase

CH_3
|
$C = O$
|
S-CoA

Acetyl-CoA

CH_3
|
$C = O$
|
S-ACP

CH_3
|
$C = O$
|
S-KSase

CO_2

COO^-
|
CH_2
|
$C = O$
|
S-CoA

Malonyl-CoA

ACP-SH CoASH

COO^-
|
CH_2
|
$C = O$
|
S-ACP

Note that these three steps are the reverse of those in β-oxidation

$$CH_3 - \overset{\displaystyle O}{\overset{\|}{C}} - CH_2 - \overset{\displaystyle O}{\overset{\|}{C}} - S - ACP$$

Acetoacetyl-ACP

NADPH + H⁺

β-Ketoacyl-ACP reductase

NADP⁺

$$CH_3 - \overset{\displaystyle OH}{\underset{\displaystyle H}{\overset{\displaystyle |}{\underset{\displaystyle |}{C}}}} - CH_2 - \overset{\displaystyle O}{\overset{\|}{C}} - S - ACP$$

D-β-Hydroxybutyryl-ACP

β-Hydroxyacyl-ACP dehydratase

H_2O

$$CH_3 - \overset{\displaystyle H}{\underset{\displaystyle H}{\overset{\displaystyle |}{\underset{\displaystyle |}{C}}}} = \overset{}{C} - \overset{\displaystyle }{\underset{\displaystyle O}{\overset{\displaystyle }{\underset{\displaystyle \|}{C}}}} - S - ACP$$

Crotonyl-ACP

NADPH + H⁺

2,3-*trans*-Enoyl-ACP reductase

NADP⁺

$$CH_3 - CH_2 - CH_2 - \overset{\displaystyle O}{\overset{\|}{C}} - S - ACP$$

Butyryl-ACP

Mal — CoA + 4H⁺ + 4e⁻

$$CH_3 - CH_2 - CH_2 - CH_2 - CH_2 - \overset{\displaystyle O}{\overset{\|}{C}} - S - ACP$$

Mal — CoA + 4H⁺ + 4e⁻

$$CH_3 - CH_2 - CH_2 - CH_2 - CH_2 - CH_2 - CH_2 - \overset{\displaystyle O}{\overset{\|}{C}} - S - ACP$$

Mal — CoA + 4H⁺ + 4e⁻

$$CH_3 - CH_2 - CH_2 - CH_2 - CH_2 - CH_2 - CH_2 - CH_2 - CH_2 - \overset{\displaystyle O}{\overset{\|}{C}} - S - ACP$$

Mal — CoA + 4H⁺ + 4e⁻

$$CH_3 - CH_2 - CH_2 - CH_2 - CH_2 - CH_2 - CH_2 - CH_2 - CH_2 - CH_2 - CH_2 - \overset{\displaystyle O}{\overset{\|}{C}} - S - ACP$$

Mal — CoA + 4H⁺ + 4e⁻

$$CH_3 - CH_2 - CH_2 - CH_2 - CH_2 - CH_2 - CH_2 - CH_2 - CH_2 - CH_2 - CH_2 - CH_2 - CH_2 - \overset{\displaystyle O}{\overset{\|}{C}} - S - ACP$$

Mal — CoA + 4H⁺ + 4e⁻

$$CH_3 - CH_2 - CH_2 - CH_2 - CH_2 - CH_2 - CH_2 - CH_2 - CH_2 - CH_2 - CH_2 - CH_2 - CH_2 - CH_2 - CH_2 - \overset{\displaystyle O}{\overset{\|}{C}} - S - ACP$$

In animals

H_2O

ACP — SH

$$CH_3 - (CH_2)_{14} - \overset{\displaystyle O}{\overset{\|}{C}} - O^-$$

Palmitate

◀ **Figure 20.25** The pathway of palmitate synthesis from acetyl-CoA and malonyl-CoA. Acetyl and malonyl building blocks are introduced as acyl carrier protein conjugates. Decarboxylation drives the β-ketoacyl-ACP synthase and results in the addition of two-carbon units to the growing chain. Concentrations of free fatty acids are extremely low in most cells, and newly synthesized fatty acids exist primarily as acyl-CoA esters.

 See pages 173–178

in all organisms (Figure 20.25). The process begins with the formation of acetyl-ACP and malonyl-ACP, which are formed by **acetyl transacylase (acetyl transferase)** and **malonyl transacylase (malonyl transferase),** respectively.

Decarboxylation Drives the Condensation of Malonyl-CoA with Acetyl-CoA

Another transacylase reaction transfers the acetyl group from ACP to **β-ketoacyl-ACP synthase (KSase),** also known as **acyl-malonyl-ACP condensing enzyme.** The first actual elongation reaction involves the condensation of acetyl-ACP and malonyl-ACP by the β-ketoacyl-ACP synthase to form acetoacetyl-ACP (see Figure 20.25). One might ask at this point: Why is the three-carbon malonyl group used here as a two-carbon donor? The answer is that this is yet another example of a decarboxylation driving a desired but otherwise thermodynamically unfavorable reaction. The decarboxylation that accompanies this reaction drives the synthesis of acetoacetyl-ACP. Note that hydrolysis of ATP drove the carboxylation of acetyl-CoA to form malonyl-ACP, so, indirectly, ATP is responsible for the condensation reaction to form acetoacetyl-ACP. Malonyl-CoA can be viewed as a form of stored energy for driving fatty acid synthesis.

It is also worth noting that the carbon of the carboxyl group that was added to drive this reaction is the one removed by the condensing enzyme. Thus, all the carbons of acetoacetyl-ACP (and of the fatty acids to be made) are derived from acetate units of acetyl-CoA.

Reduction of the β-Carbonyl Group Follows a Now-Familiar Route

The next three steps—reduction of the β-carbonyl group to form a β-alcohol, followed by dehydration and reduction to saturate the chain (see Figure 20.25)—look very similar to the fatty acid degradation pathway in reverse. However, there are two crucial differences between fatty acid biosynthesis and fatty acid oxidation (besides the fact that different enzymes are involved): first, the alcohol formed in the first step has the D configuration rather than the L form

 **A DEEPER LOOK**

Choosing the Best Organism for the Experiment

The selection of a suitable and relevant organism is an important part of any biochemical investigation. The studies that revealed the secrets of fatty acid synthesis are a good case in point.

The paradigm for fatty acid synthesis in plants has been the avocado, which has one of the highest fatty acid contents in the plant kingdom. Early animal studies centered primarily on pigeons, which are easily bred and handled and which possess high levels of fats in their tissues. Other animals, rich in fatty tissues, might be even more attractive but more challenging to maintain. Grizzly bears, for example, carry very large fat reserves but are difficult to work with in the lab!

Figure 20.26 Fatty acid synthase in animals contains all the functional groups and enzyme activities on a single multifunctional subunit. The active enzyme is a head-to-tail dimer of identical subunits. *(Adapted from Wakil, S. J., Stoops, J. K., and Joshi, V. C., 1983. Annual Review of Biochemistry **52**:556.)*

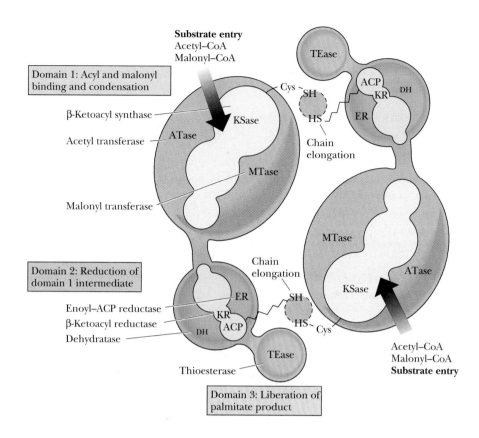

seen in catabolism, and, second, the reducing coenzyme is NADPH, although NAD^+ and FAD are the oxidants in the catabolic pathway.

The net result of this biosynthetic cycle is the synthesis of a four-carbon unit, a butyryl group, from two smaller building blocks. In the next cycle of the process, this butyryl-ACP condenses with another malonyl-ACP to make a six-carbon β-ketoacyl-ACP and CO_2. Subsequent reduction to a β-alcohol, dehydration, and another reduction yield a six-carbon saturated acyl-ACP. This cycle continues with the net addition of a two-carbon unit in each turn until the chain is 16 carbons long (Figure 20.25). The β-ketoacyl-ACP synthase cannot accommodate larger substrates, so the reaction cycle ends with a 16-carbon chain. Hydrolysis of the C_{16}-acyl-ACP yields a palmitic acid and the free ACP.

In the end, seven malonyl-CoA molecules and one acetyl-CoA yield a palmitate (shown here as palmitoyl-CoA):

$$\text{Acetyl-CoA} + 7 \text{ malonyl-CoA}^- + 14 \text{ NADPH} + 14 \text{ H}^+ \longrightarrow$$
$$\text{palmitoyl-CoA} + 7 \text{ HCO}_3^- + 14 \text{ NADP}^+ + 7 \text{ CoASH}$$

The formation of seven malonyl-CoA molecules requires

$$7 \text{ Acetyl-CoA} + 7 \text{ HCO}_3^- + 7 \text{ ATP}^{4-} \longrightarrow$$
$$7 \text{ malonyl-CoA}^- + 7 \text{ ADP}^{3-} + 7 \text{ P}_i^{2-} + 7 \text{ H}^+$$

Thus, the overall reaction of acetyl-CoA to yield palmitic acid is

$$8 \text{ Acetyl-CoA} + 7 \text{ ATP}^{4-} + 14 \text{ NADPH} + 7 \text{ H}^+ \longrightarrow$$
$$\text{palmitoyl-CoA} + 14 \text{ NADP}^+ + 7 \text{ CoASH} + 7 \text{ ADP}^{3-} + 7 \text{ P}_i^{2-}$$

Note: These equations are stoichiometric and are charge balanced.

As noted earlier, the reactions of fatty acid synthesis beyond the acetyl-CoA carboxylase in animal systems are carried out by a special multienzyme complex called **fatty acid synthase (FAS).** Animal FAS is a dimer of identical 250-kD multifunctional polypeptides. Studies of the action of proteolytic enzymes on this polypeptide have led to a model involving three separate domains joined by flexible connecting sequences (see Figure 20.26). The first domain of one subunit of FAS interacts with the second and third domains of the other subunit; that is, the subunits are arranged in head-to-tail fashion. The first domain is responsible for the binding of acetyl and malonyl building blocks and for the condensation of these units. This domain includes the acetyl transferase, the malonyl transferase, and the acyl-malonyl-ACP condensing enzyme (the β-ketoacyl synthase). The second domain is primarily responsible for the reduction of the intermediate synthesized in domain 1, and contains the acyl carrier protein, the β-ketoacyl reductase, the dehydratase, and the enoyl-ACP reductase. The third domain contains the thioesterase that liberates the product palmitate when the growing acyl chain reaches its limit length of 16 carbons. The close association of activities in this complex permits efficient exposure of intermediates to one active site and then the next. The presence of all these activities on a single polypeptide ensures that the cell will simultaneously synthesize all the enzymes needed for fatty acid synthesis.

Further Processing of C$_{16}$ Fatty Acids

Additional Elongation

As seen already, palmitate is the primary product of the fatty acid synthase. Cells synthesize many other fatty acids. Shorter chains are easily made if the chain is released before reaching 16 carbons in length. Longer chains (Figure 20.27) are made through special elongation reactions, which occur both

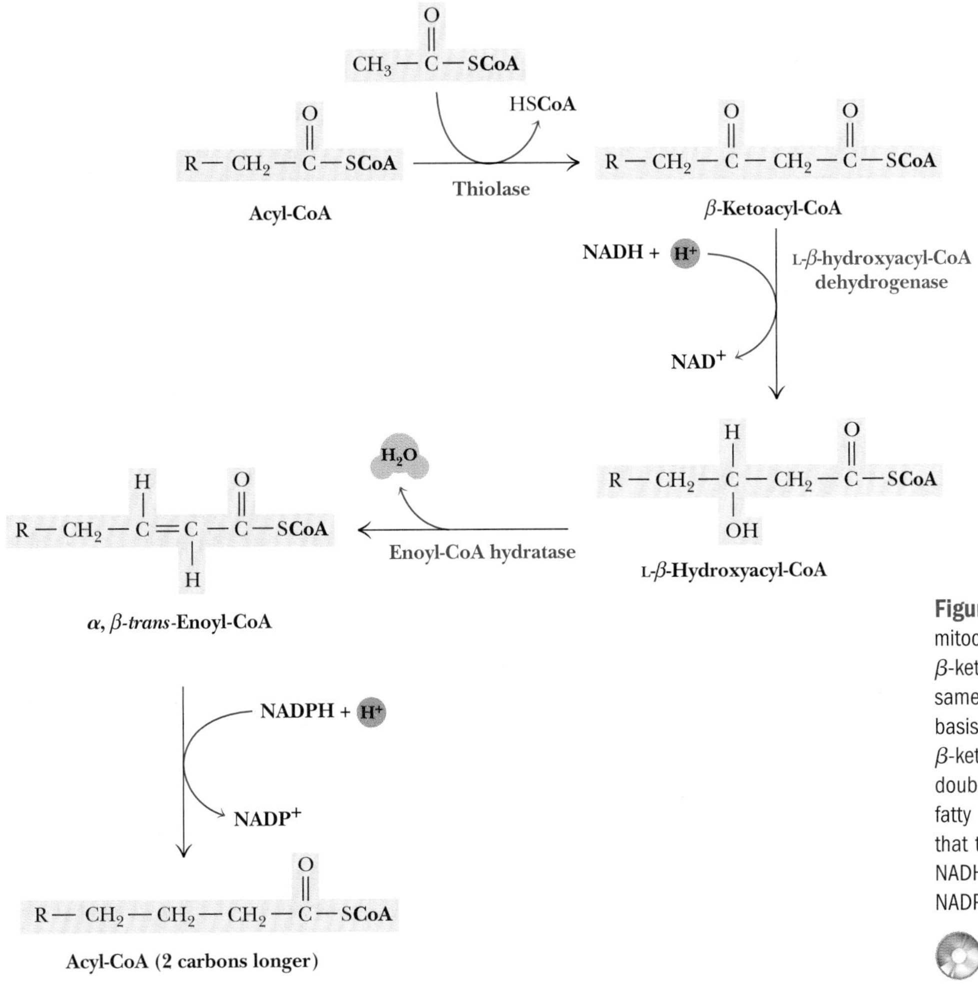

Figure 20.27 Elongation of fatty acids in mitochondria is initiated by the thiolase reaction. The β-ketoacyl intermediate thus formed undergoes the same three reactions (in reverse order) that are the basis of β-oxidation of fatty acids. Reduction of the β-keto group is followed by dehydration to form a double bond. Reduction of the double bond yields a fatty acyl-CoA that is elongated by two carbons. Note that the reducing coenzyme for the second step is NADH, whereas the reductant for the fourth step is NADPH.

 See pages 173–178

in the mitochondria and at the surface of the endoplasmic reticulum. These reactions are similar to those of palmitate synthesis, except that coenzyme A (rather than ACP) is the acyl carrier and different enzyme systems are involved.

Introduction of a Single *cis* Double Bond

Both prokaryotes and eukaryotes are capable of introducing a single *cis* double bond in a newly synthesized fatty acid. Bacteria such as *E. coli* carry out this process in an O_2-independent pathway, whereas eukaryotes have adopted an O_2-dependent pathway. There is a fundamental chemical difference between the two. The O_2-dependent reaction can occur anywhere in the fatty acid chain, with no (additional) need to activate the desired bond toward dehydrogenation. However, in the absence of O_2, some other means must be found to activate the bond in question. Thus, in the bacterial reaction, dehydrogenation occurs while the bond of interest is still near the β-carbonyl or β-hydroxy group and the thioester group at the end of the chain.

In *E. coli*, the biosynthesis of a monounsaturated fatty acid begins with four normal cycles of elongation to form a 10-carbon intermediate, β-hydroxydecanoyl-ACP (Figure 20.28). At this point, **β-hydroxydecanoyl thioester dehydrase** forms a double bond β, γ to the thioester and in the *cis* configuration. This is followed by three rounds of the normal elongation reactions to form palmitoleoyl-ACP. Elongation may terminate at this point or may be followed by additional biosynthetic events. The principal unsaturated fatty acid in *E. coli*, *cis*-vaccenic acid, is formed by an additional elongation step, using palmitoleoyl-ACP as a substrate.

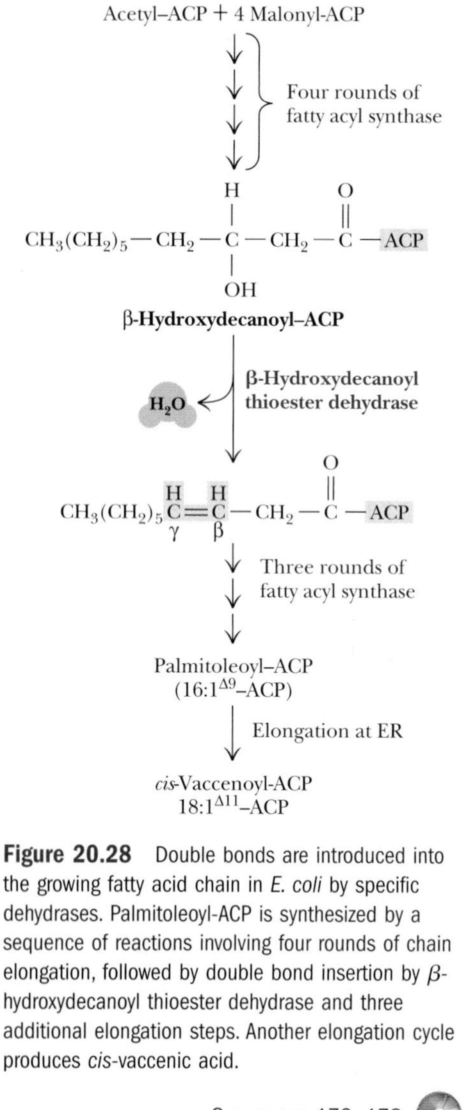

Unsaturation Reactions Occur in Eukaryotes in the Middle of an Aliphatic Chain

The addition of double bonds to fatty acids in eukaryotes does not occur until the fatty acyl chain has reached its full length (usually 16 to 18 carbons). Dehydrogenation of stearoyl-CoA occurs in the middle of the chain despite the absence of any useful functional group on the chain to facilitate activation:

$$CH_3-(CH_2)_{16} CO-SCoA \longrightarrow CH_3-(CH_2)_7 CH=CH(CH_2)_7 CO-SCoA$$

This impressive reaction is catalyzed by **stearoyl-CoA desaturase**, a 53-kD enzyme containing a nonheme iron center. NADH and oxygen (O_2) are required, as are two other proteins: **cytochrome b_5 reductase** (a 43-kD flavoprotein) and **cytochrome b_5** (16.7 kD). All three proteins are associated with the endoplasmic reticulum membrane. Cytochrome b_5 reductase transfers a pair of electrons from NADH through FAD to cytochrome b_5 (Figure 20.29). Oxidation of reduced cytochrome b_5 is coupled to reduction of nonheme Fe^{3+} to Fe^{2+} in the desaturase. The Fe^{3+} accepts a pair of electrons (one at a time in a cycle) from cytochrome b_5 and creates a *cis* double bond at the 9,10-position of the stearoyl-CoA substrate. O_2 is the terminal electron acceptor in this fatty acyl desaturation cycle. Note that two water molecules are made, which means that four electrons are transferred overall. Two of these come through the reaction sequence from NADH, and two come from the fatty acyl substrate that is being dehydrogenated.

The Unsaturation Reaction May Be Followed by Chain Elongation

Additional chain elongation can occur following this single desaturation reaction. The oleoyl-CoA produced can be elongated by two carbons to form a 20 : 1 *cis*-Δ^{11} fatty acyl-CoA. If the starting fatty acid is palmitate, reactions sim-

Acetyl–ACP + 4 Malonyl-ACP

Four rounds of fatty acyl synthase

$$CH_3(CH_2)_5-CH_2-\overset{\overset{\text{H}}{|}}{C}-CH_2-\overset{\overset{\text{O}}{\|}}{C}-\boxed{ACP}$$
$$\underset{\text{OH}}{|}$$

β-Hydroxydecanoyl–ACP

β-Hydroxydecanoyl thioester dehydrase

H_2O

$$CH_3(CH_2)_5\underset{\gamma}{\overset{\overset{\text{H}}{|}}{C}}=\underset{\beta}{\overset{\overset{\text{H}}{|}}{C}}-CH_2-\overset{\overset{\text{O}}{\|}}{C}-\boxed{ACP}$$

Three rounds of fatty acyl synthase

Palmitoleoyl–ACP
(16:1$^{\Delta 9}$–ACP)

Elongation at ER

cis-Vaccenoyl-ACP
18:1$^{\Delta 11}$–ACP

Figure 20.28 Double bonds are introduced into the growing fatty acid chain in *E. coli* by specific dehydrases. Palmitoleoyl-ACP is synthesized by a sequence of reactions involving four rounds of chain elongation, followed by double bond insertion by β-hydroxydecanoyl thioester dehydrase and three additional elongation steps. Another elongation cycle produces *cis*-vaccenic acid.

See pages 173–178

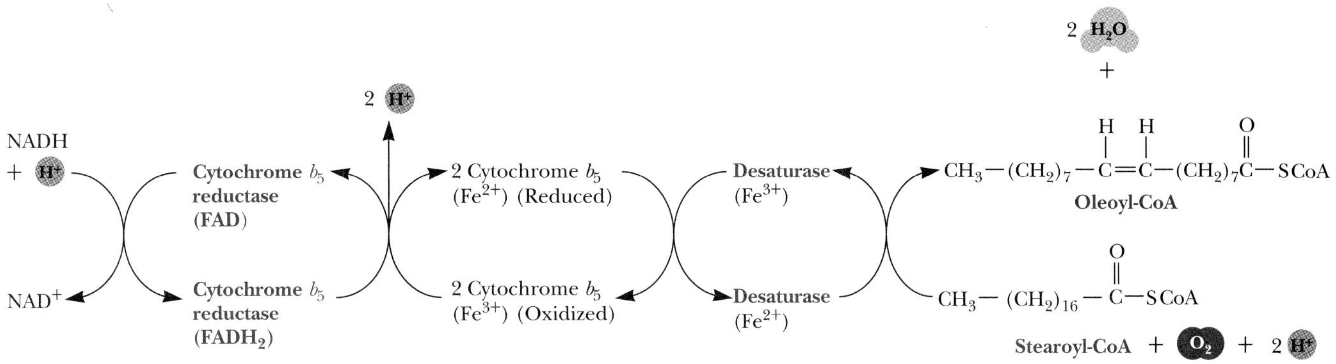

NADH
+ H⁺

Cytochrome b_5
reductase
(FAD)

NAD⁺

Cytochrome b_5
reductase
(FADH₂)

2 H⁺

2 Cytochrome b_5
(Fe^{2+}) (Reduced)

2 Cytochrome b_5
(Fe^{3+}) (Oxidized)

Desaturase
(Fe^{3+})

Desaturase
(Fe^{2+})

2 H₂O
+

$$CH_3-(CH_2)_7-\overset{\overset{H}{|}}{C}=\overset{\overset{H}{|}}{C}-(CH_2)_7\overset{\overset{O}{||}}{C}-SCoA$$
Oleoyl-CoA

$$CH_3-(CH_2)_{16}-\overset{\overset{O}{||}}{C}-SCoA$$

Stearoyl-CoA + O₂ + 2 H⁺

Figure 20.29 The conversion of stearoyl-CoA to oleoyl-CoA in eukaryotes is catalyzed by stearoyl-CoA desaturase in a reaction sequence that also involves cytochrome b_5 and cytochrome b_5 reductase. Two electrons are passed from NADH through the chain of reactions as shown, and two electrons are also derived from the fatty acyl substrate.

 See pages 173–178

ilar to the preceding scheme yield palmitoleoyl-CoA (16 : 1 cis-Δ^9), which subsequently can be elongated to yield cis-vaccenic acid (18 : 1 cis-Δ^{11}). Similarly, C_{16} and C_{18} fatty acids can be elongated to yield C_{22} and C_{24} fatty acids, such as are often found in sphingolipids.

Biosynthesis of Polyunsaturated Fatty Acids

Organisms differ with respect to formation, processing, and utilization of polyunsaturated fatty acids. *E. coli*, for example, does not have any polyunsaturated fatty acids. Eukaryotes do synthesize a variety of polyunsaturated fatty acids, certain organisms more than others. For example, plants manufacture double bonds between the Δ^9 and the methyl end of the chain, but mammals cannot. Plants readily desaturate oleic acid at the 12-position (to give linoleic acid) or at both the 12- and 15-positions (producing linolenic acid). Mammals require polyunsaturated fatty acids but must acquire them in their diet. As such, they are referred to as **essential fatty acids.** On the other hand, mammals can introduce double bonds between the double bond at the 8- or 9-position and the carboxyl group. Enzyme complexes in the endoplasmic reticulum desaturate the 5-position, provided a double bond exists at the 8-position, and form a double bond at the 6-position if one already exists at the 9-position. Thus, oleate can be unsaturated at the 6,7-position to give an 18 : 2 cis-Δ^6,Δ^9 fatty acid.

Arachidonic Acid Is Synthesized from Linoleic Acid by Mammals

Mammals can add additional double bonds to unsaturated fatty acids in their diets. Their ability to make arachidonic acid from linoleic acid is one example (Figure 20.30). This fatty acid is the precursor for prostaglandins and other biologically active derivatives such as leukotrienes. Synthesis involves formation of a linoleoyl ester of CoA from dietary linoleic acid, followed by introduction of a double bond at the 6-position. The triply unsaturated product is then elongated (by malonyl-CoA with a decarboxylation step) to yield a 20-carbon fatty acid with double bonds at the 8-, 11-, and 14-positions. A second desaturation reaction at the 5-position followed by an **acyl-CoA synthetase** reaction liberates the product, a 20-carbon fatty acid with double bonds at the 5-, 8-, 11-, and 14-positions.

HUMAN BIOCHEMISTRY

Docosahexaenoic Acid: A Major Polyunsaturated Fatty Acid in the Retina and Brain

Along with α-linolenic acid, two long-chain polyunsaturated fatty acids, docosahexaenoic acid (DHA) and eicosapentaenoic acid (EPA), are known as "omega-3 fatty acids" because, counting from the end (omega) of the chain, the first double bond is at the third position (see figure). These fatty acids have beneficial effects in a variety of organs and biological processes, including growth regulation, modulation of inflammation, platelet activation, and lipoprotein metabolism. Interestingly, especially high levels of DHA have been found in rod cell membranes in animal retina and in neural tissue. DHA comprises approximately 22% of total fatty acids in the animal retina and 35% to 40% of the fatty acids in retinal phosphatidylethanolamine. DHA supports neural and visual development, in part because it is a precursor for eicosanoids that regulate numerous cell and organ functions. Infants can synthesize DHA and other polyunsaturated fatty acids, but the rates of synthesis are low. Strong evidence exists for the importance of these fatty acids in infant nutrition.

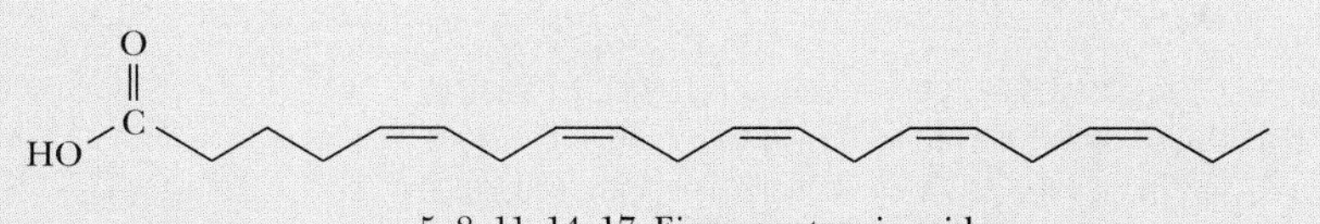
O
C
HO

5, 8, 11, 14, 17- Eicosapentanoic acid

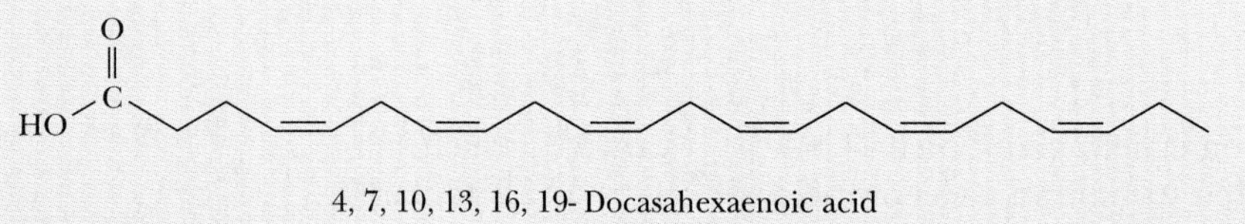
O
C
HO

4, 7, 10, 13, 16, 19- Docasahexaenoic acid

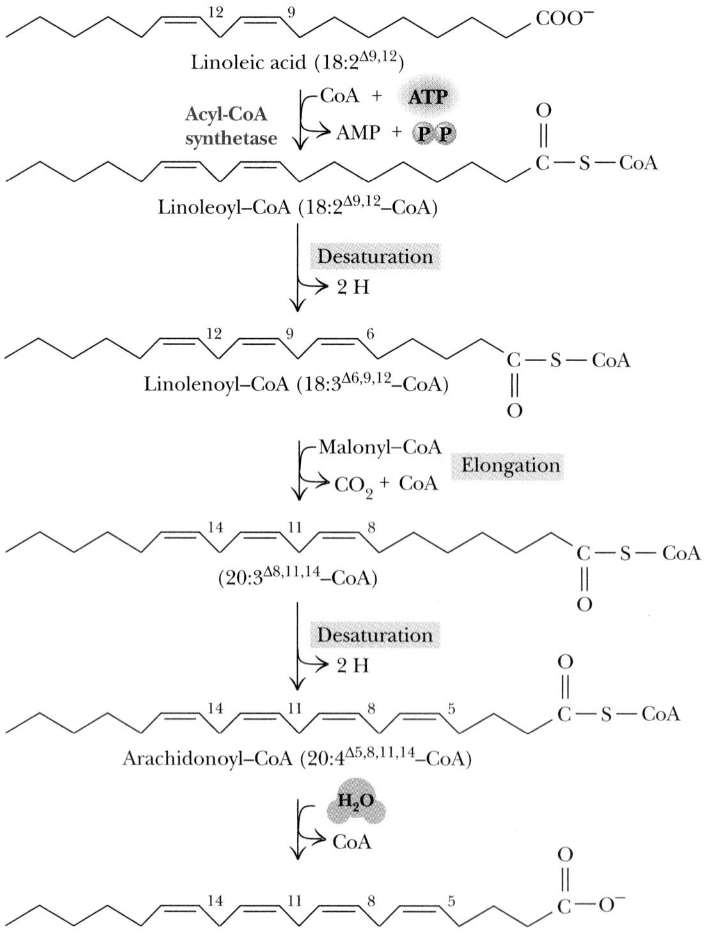
12
9
COO⁻
Linoleic acid (18:2Δ9,12)
CoA +
ATP
Acyl-CoA synthetase
AMP +
P P
O
C—S—CoA
Linoleoyl–CoA (18:2Δ9,12–CoA)
Desaturation
2 H
12
9
6
C—S—CoA
O
Linolenoyl–CoA (18:3Δ6,9,12–CoA)
Malonyl–CoA
Elongation
CO2 + CoA
14
11
8
C—S—CoA
O
(20:3Δ8,11,14–CoA)
Desaturation
2 H
O
14
11
8
5
C—S—CoA
Arachidonoyl–CoA (20:4Δ5,8,11,14–CoA)
H2O
CoA
O
14
11
8
5
C—O⁻
Arachidonic acid

Figure 20.30 Arachidonic acid is synthesized from linoleic acid in eukaryotes. This is the only means by which animals can synthesize fatty acids with double bonds at positions beyond C-9.

See pages 173–178

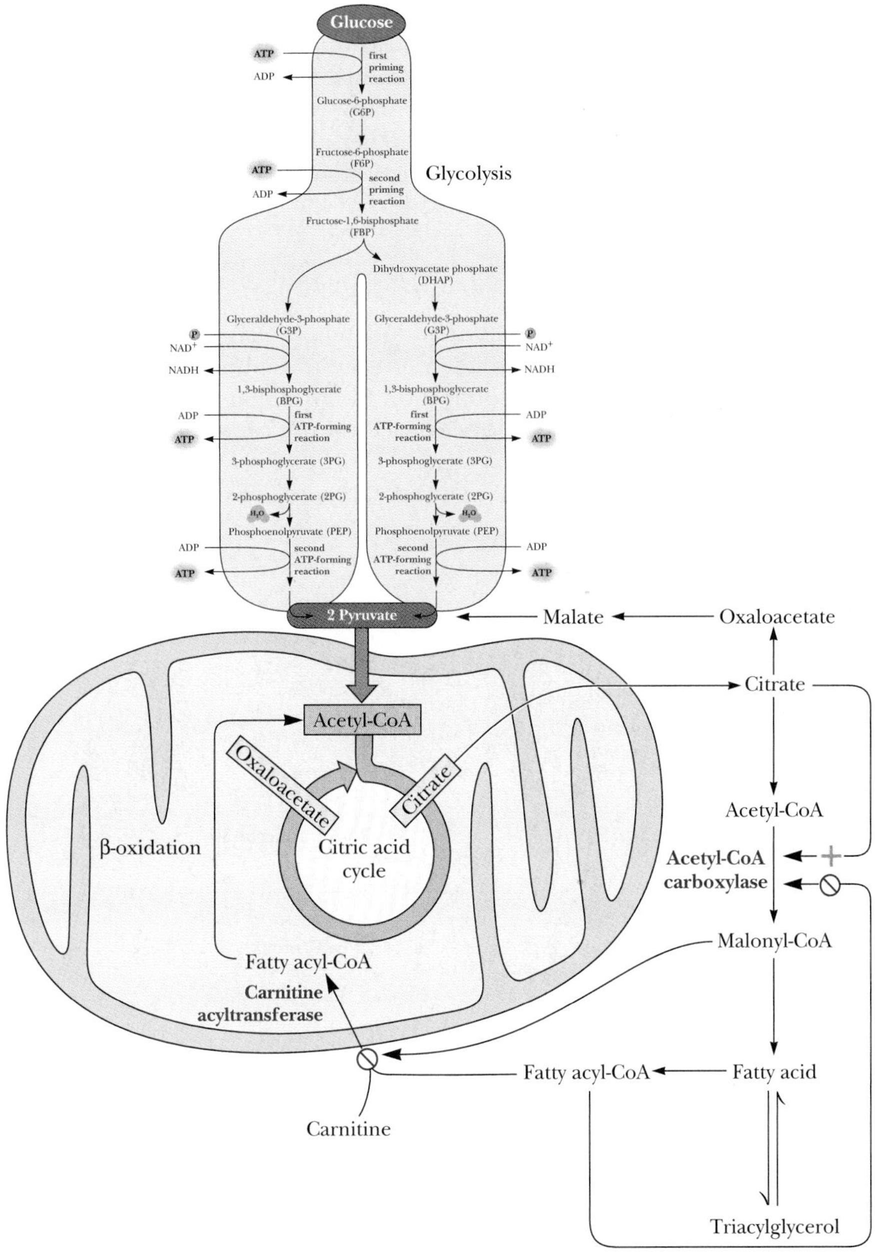

Figure 20.31 Regulation of fatty acid synthesis and fatty acid oxidation are coupled as shown. Malonyl-CoA, produced during fatty acid synthesis, inhibits the uptake of fatty acylcarnitine (and thus fatty acid oxidation) by mitochondria. When fatty acyl-CoA levels rise, fatty acid synthesis is inhibited and fatty acid oxidation activity increases. Rising citrate levels (which reflect an abundance of acetyl-CoA) similarly signal the initiation of fatty acid synthesis.

See pages 173–178

Regulatory Control of Fatty Acid Metabolism: An Interplay of Allosteric Modifiers and Phosphorylation–Dephosphorylation Cycles

The control and regulation of fatty acid synthesis is intimately related to regulation of fatty acid breakdown, glycolysis, and the TCA cycle. Acetyl-CoA is an important metabolic intermediate in all these processes. In these terms, it is easy to appreciate the interlocking relationships in Figure 20.31. Malonyl-CoA can act to prevent fatty acyl-CoA derivatives from entering the mitochondria by inhibiting the carnitine acyltransferase that is responsible for this transport.

In this way, when fatty acid synthesis is turned on (as signaled by higher levels of malonyl-CoA), β-oxidation is inhibited. As we pointed out earlier, citrate is an important allosteric activator of acetyl-CoA carboxylase, and fatty acyl-CoAs are inhibitors. The degree of inhibition is proportional to the chain length of the fatty acyl-CoA; longer chains show a higher affinity for the allosteric inhibition site on acetyl-CoA carboxylase. Palmitoyl-CoA, stearoyl-CoA, and arachidyl-CoA are the most potent inhibitors of the carboxylase.

Hormonal Signals Regulate ACC and Fatty Acid Biosynthesis

As described earlier, citrate activation and palmitoyl-CoA inhibition of acetyl-CoA carboxylase are strongly dependent on the phosphorylation state of the enzyme. This provides a crucial connection to hormonal regulation. Many of the enzymes that act to phosphorylate acetyl-CoA carboxylase (see Figure 20.23) are controlled by hormonal signals. Glucagon is a good example (Figure 20.32). As noted in Chapter 19, glucagon binding to membrane receptors

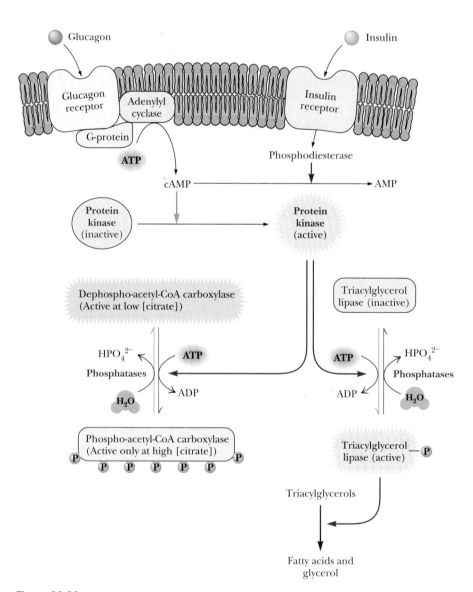

Figure 20.32 Hormonal signals regulate fatty acid synthesis, primarily through actions on acetyl-CoA carboxylase. Availability of fatty acids also depends upon hormonal activation of triacylglycerol lipase.

activates an intracellular cascade involving activation of adenylyl cyclase. Cyclic AMP produced by the cyclase activates a protein kinase, which then phosphorylates acetyl-CoA carboxylase. Unless citrate levels are high, phosphorylation causes inhibition of fatty acid biosynthesis. The carboxylase (and fatty acid synthesis) can be reactivated by a specific phosphatase, which dephosphorylates the carboxylase. Also indicated in Figure 20.32 is the simultaneous activation by glucagon of triacylglycerol lipases, which hydrolyze triacylglycerols, releasing fatty acids for β-oxidation. Both the inactivation of acetyl-CoA carboxylase and the activation of triacylglycerol lipase are counteracted by insulin, whose receptor acts to stimulate a phosphodiesterase that converts cAMP to AMP.

20.8 | Biosynthesis of Complex Lipids

Complex lipids consist of backbone structures to which fatty acids are covalently bound. Principal classes include the **glycerolipids,** for which glycerol is the backbone, and **sphingolipids,** which are built on a sphingosine backbone. The two major classes of glycerolipids are **glycerophospholipids** and **triacylglycerols.** The **phospholipids,** which include both glycerophospholipids and sphingomyelins, are crucial components of membrane structure. They are also precursors of hormones such as the eicosanoids (e.g., prostaglandins) and signal molecules, such as the breakdown products of phosphatidylinositol.

Different organisms possess greatly different complements of lipids and therefore invoke somewhat different lipid biosynthetic pathways. For example, sphingolipids and triacylglycerols are produced only in eukaryotes. In contrast, bacteria usually have rather simple lipid compositions.

Glycerolipid Biosynthesis

A common pathway operates in nearly all organisms for the synthesis of **phosphatidic acid,** the precursor to other glycerolipids. **Glycerokinase** catalyzes the phosphorylation of glycerol to form glycerol-3-phosphate, which is then acylated at both the 1- and 2-positions to yield phosphatidic acid (Figure 20.33). The first acylation, at position 1, is catalyzed by **glycerol-3-phosphate acyltransferase,** an enzyme that in most organisms is specific for saturated fatty acyl groups. Eukaryotic systems can also utilize **dihydroxyacetone phosphate** as a starting point for synthesis of phosphatidic acid (see Figure 20.33). Again, a specific acyltransferase adds the first acyl chain, followed by reduction of the backbone keto group by **acyldihydroxyacetone phosphate reductase,** using NADPH as the reductant. Alternatively, dihydroxyacetone phosphate can be reduced to glycerol-3-phosphate by **glycerol-3-phosphate dehydrogenase.**

Eukaryotes Synthesize Glycerolipids from CDP-Diacylglycerol or Diacylglycerol

In eukaryotes, phosphatidic acid is converted directly either to diacylglycerol or to cytidine diphosphodiacylglycerol (or simply CDP-diacylglycerol; Figure 20.34). From these two precursors, all other glycerophospholipids in eukaryotes are derived. Diacylglycerol is a precursor for synthesis of triacylglycerol, phosphatidylethanolamine, and phosphatidylcholine. Triacylglycerol biosynthesis in liver and adipose tissue occurs via **diacylglycerol acyltransferase,** an enzyme bound to the cytoplasmic face of the endoplasmic reticulum. A different route is used, however, in intestines. Recall (see Figure 20.3) that triacylglycerols from the diet are broken down to 2-monoacylglycerols by specific

Figure 20.33 Synthesis of glycerolipids in eukaryotes begins with the formation of phosphatidic acid, which may be formed from dihydroxyacetone phosphate or glycerol as shown.

See pages 173–178

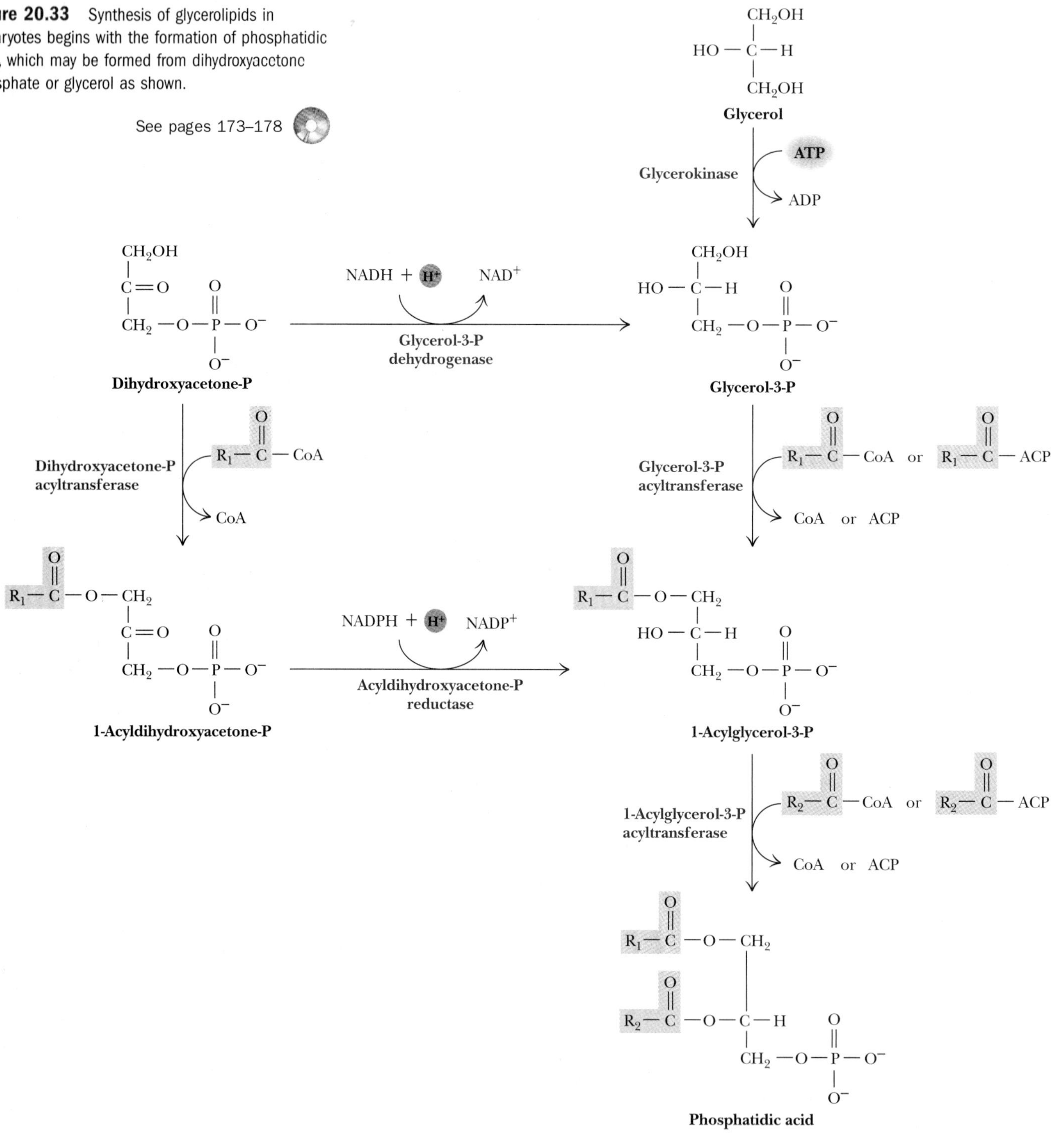

lipases. Acyltransferases then acylate 2-monoacylglycerol to produce new triacylglycerols (Figure 20.35).

Phosphatidylethanolamine Is Synthesized from Diacylglycerol and CDP-Ethanolamine

Phosphatidylethanolamine synthesis begins with phosphorylation of ethanolamine to form phosphoethanolamine (see Figure 20.34). The next reaction involves transfer of a cytidylyl group from CTP to form CDP-ethanolamine and pyrophosphate. As always, PP$_i$ hydrolysis drives this reaction forward. A specific **phosphoethanolamine transferase** then links phosphoethanolamine to the diacylglycerol backbone. Biosynthesis of phosphatidylcholine is entirely analogous. The choline used for this reaction in animals must be acquired from the diet. Phosphatidylethanolamine can also be converted to phosphatidylcholine

Figure 20.34 Diacylglycerol and CDP-diacylglycerol are the principal precursors of glycerolipids in eukaryotes. Phosphatidylethanolamine and phosphatidylcholine are formed by reaction of diacylglycerol with CDP-ethanolamine or CDP-choline, respectively.

See pages 173–178

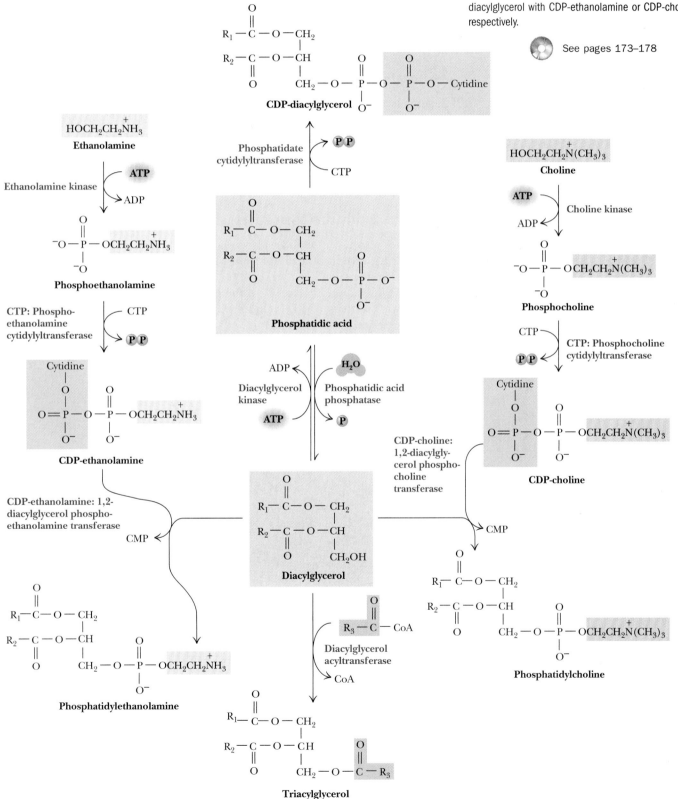

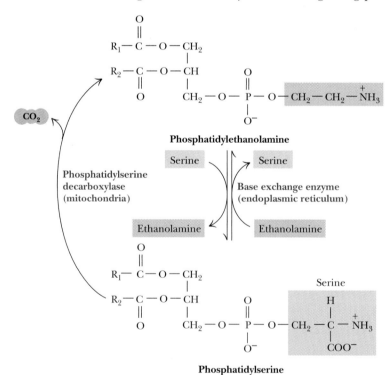

Figure 20.35 Triacylglycerols are formed primarily by the action of acyltransferases on mono- and diacylglycerol. Acyltransferase in *E. coli* is an integral membrane protein (83 kD) that can utilize either fatty acyl-CoAs or acylated acyl carrier proteins as substrates. It shows a particular preference for palmitoyl groups. Eukaryotic acyltransferases use only fatty acyl-CoA molecules as substrates.

See pages 173–178

by methylation reactions involving *S*-adenosylmethionine in the liver (see Chapter 21).

Exchange of Ethanolamine for Serine Converts Phosphatidylethanolamine to Phosphatidylserine

Mammals synthesize phosphatidylserine (PS) in a calcium ion–dependent reaction involving amino alcohol exchange (Figure 20.36). The enzyme catalyzing this reaction is associated with the endoplasmic reticulum and will accept phosphatidylethanolamine (PE) and other phospholipid substrates. A mitochondrial **PS decarboxylase** can subsequently convert PS to PE. No other pathway converting serine to ethanolamine has been found.

Eukaryotes Synthesize Other Phospholipids via CDP-Diacylglycerol

Eukaryotes also use CDP-diacylglycerol, derived from phosphatidic acid, as a precursor for several other important phospholipids, including phosphatidylinositol (PI), phosphatidylglycerol (PG), and cardiolipin (Figure 20.37). PI accounts for only about 2% to 8% of the lipids in most animal membranes, but breakdown products of PI, including inositol-1,4,5-trisphosphate and diacylglycerol, are second messengers in a vast array of cellular signaling processes.

Figure 20.36 The interconversion of phosphatidylethanolamine and phosphatidylserine in mammals.

 See pages 173–178

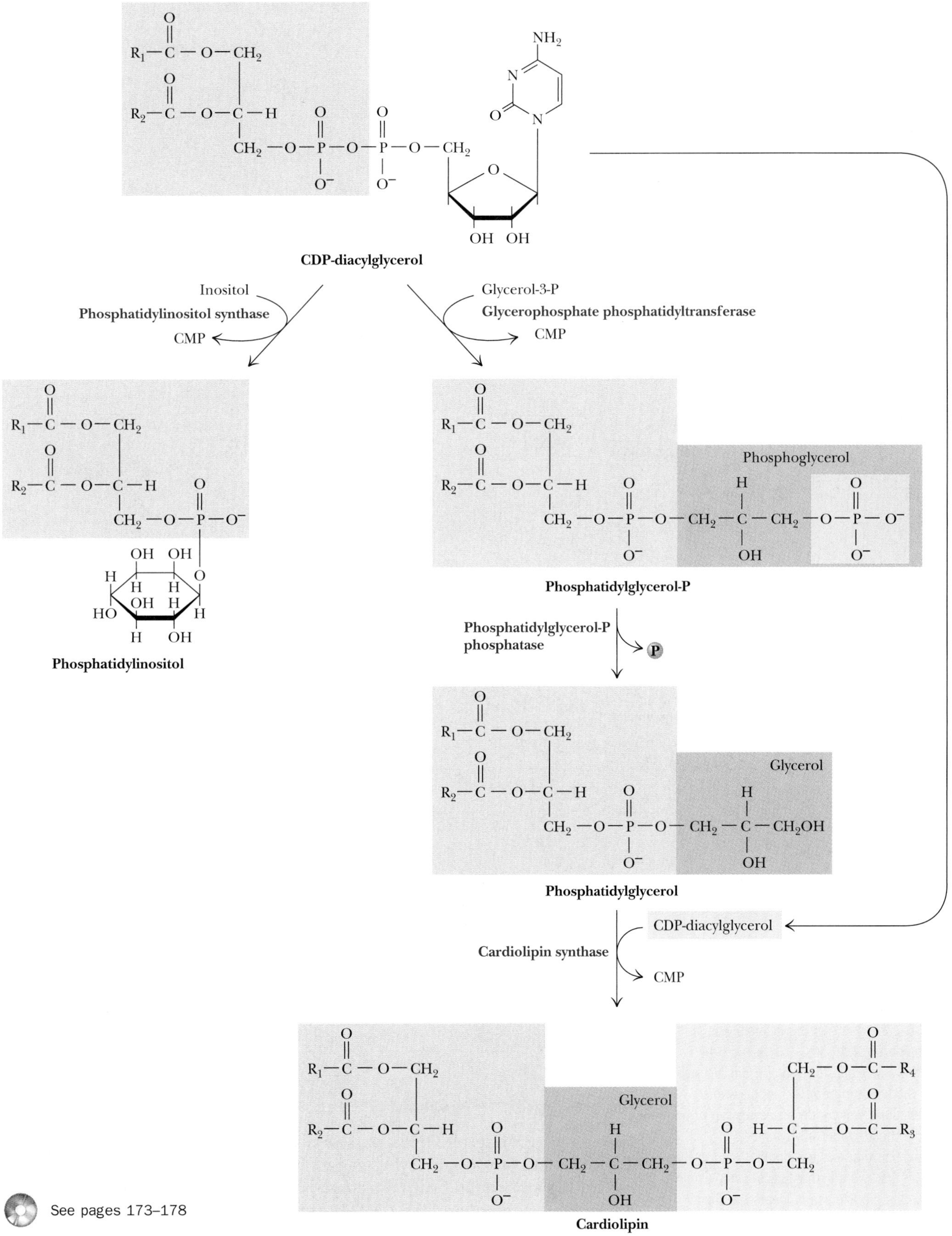

See pages 173–178

Figure 20.37 CDP-diacylglycerol is a precursor of phosphatidylinositol, phosphatidylglycerol, and cardiolipin in eukaryotes.

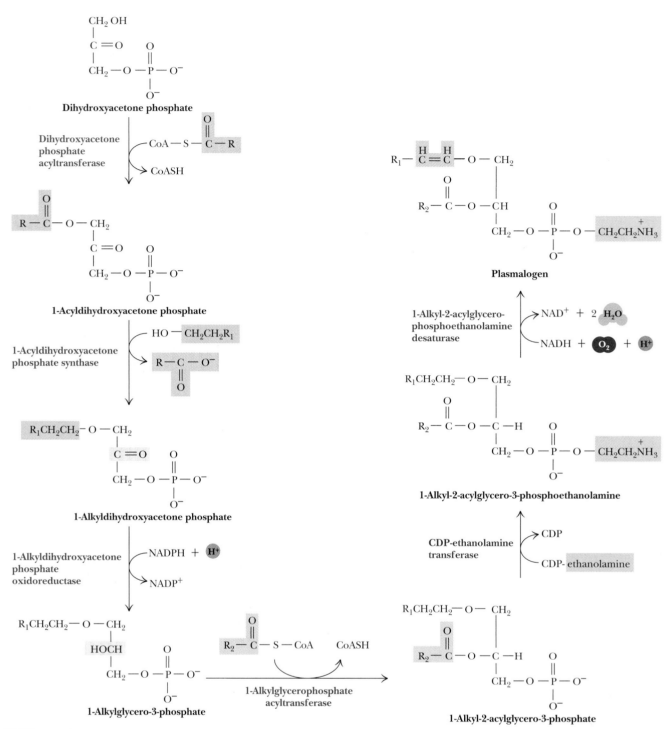

Figure 20.38 Biosynthesis of plasmalogens in animals. Acylation of DHAP at C-1 is followed by exchange of the acyl group for a long-chain alcohol. Reduction of the keto group at C-2 is followed by transferase reactions, which add an acyl group at C-2 and a polar head-group moiety, and a desaturase reaction that forms a double bond in the alkyl chain. The first two enzymes are of cytoplasmic origin, and the last transferase is located at the endoplasmic reticulum.

See pages 173–178

Dihydroxyacetone Phosphate Is a Precursor to the Plasmalogens

Certain glycerophospholipids possess alkyl or alkenyl ether groups at the 1-position in place of an acyl ester group. These glyceroether phospholipids are synthesized from dihydroxyacetone phosphate (Figure 20.38). Acylation of dihydroxyacetone phosphate (DHAP) is followed by an exchange reaction, in which the acyl group is removed as a carboxylic acid and a long-chain alcohol is added to the 1-position. This long-chain alcohol is derived from the corresponding acyl-CoA by means of an **acyl-CoA reductase** reaction involving oxidation of two molecules of NADH. The 2-keto group of the DHAP backbone is then reduced to an alcohol, followed by acylation. The resulting 1-alkyl-2-acylglycero-3-phosphate can react in a manner similar to phosphatidic acid to produce ether analogs of phosphatidylcholine, phosphatidylethanolamine, and so forth (see Figure 20.38). In addition, specific **desaturase** enzymes associated with the endoplasmic reticulum can desaturate the alkyl ether chains of these lipids as shown. The products, which contain α,β-unsaturated ether-linked chains at the C-1 position, are **plasmalogens;** they are abundant in cardiac tissue and in the central nervous system. The desaturases catalyzing these reactions are distinct from but similar to those that introduce unsaturations in fatty acyl-CoAs. These enzymes use cytochrome b_5 as a cofactor, NADH as a reductant, and O_2 as a terminal electron acceptor.

Sphingolipid Biosynthesis

Sphingolipids, ubiquitous components of eukaryotic cell membranes, are present at high levels in neural tissues. The myelin sheath that insulates nerve axons is particularly rich in sphingomyelin and other related lipids. Prokaryotic organisms normally do not contain sphingolipids. Sphingolipids are built upon sphingosine backbones rather than glycerol. The initial reaction, which involves condensation of serine and palmitoyl-CoA with release of bicarbonate, is catalyzed by **3-ketosphinganine synthase,** a PLP-dependent enzyme (Figure 20.39). Reduction of the ketone product to form sphinganine is catalyzed by **3-ketosphinganine reductase,** with NADPH as a reactant. In the next step, sphinganine is acylated to form N-acyl sphinganine, which is then desaturated to form ceramide. Sphingosine itself does not appear to be an intermediate in this pathway in mammals. Ceramide is the building block for all other sphingolipids, including sphingomyelin, cerebrosides, and gangliosides (see Figure 20.39).

20.9 Eicosanoid Biosynthesis and Function

Eicosanoids, so named because they are all derived from 20-carbon fatty acids, are ubiquitous breakdown products of phospholipids. In response to appropriate stimuli, cells activate the breakdown of selected phospholipids (Figure 20.40). Phospholipase A_2 (see Chapter 6) selectively cleaves fatty acids from the C-2 position of phospholipids. Often these are unsaturated fatty acids, among which is arachidonic acid. Arachidonic acid may also be released from phospholipids by the combined actions of phospholipase C (which yields diacylglycerols) and diacylglycerol lipase (which releases fatty acids).

Eicosanoids Are Local Hormones

Animal cells can modify arachidonic acid and other polyunsaturated fatty acids, in processes often involving cyclization and oxygenation, to produce so-called local hormones that (1) exert their effects at very low concentrations and (2) usually act near their sites of synthesis. These substances include the **prostaglandins** (PG) (see Figure 20.40) as well as **thromboxanes** (Tx), **leukotrienes,** and other **hydroxyeicosanoic acids.** Thromboxanes, discovered in blood platelets (thrombocytes), are cyclic ethers (TxB_2 is actually a hemiacetal; see Figure 20.40) with a hydroxyl group at C-15.

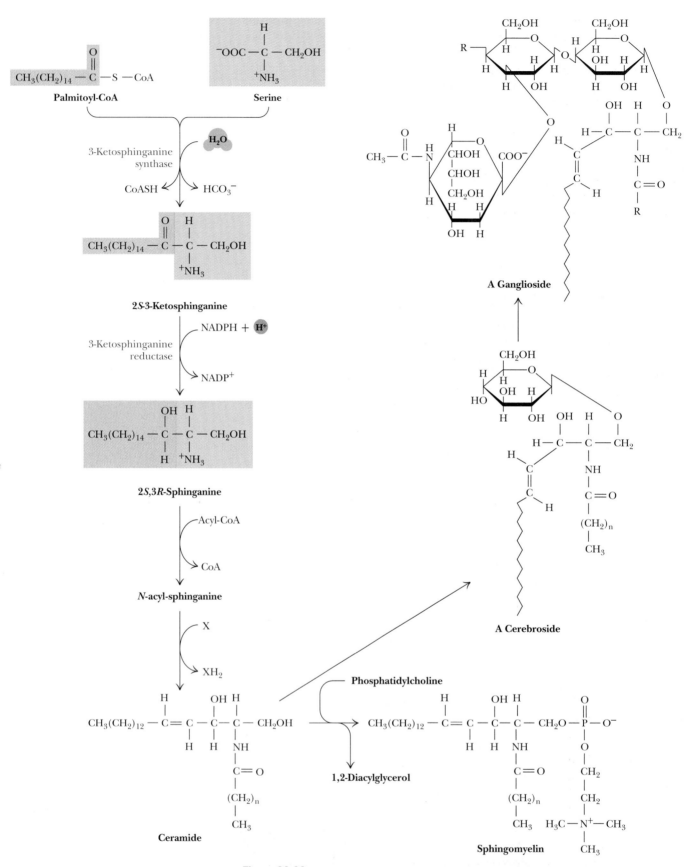

See pages 173–178

Figure 20.39 Biosynthesis of sphingolipids in animals begins with the 3-ketosphinganine synthase reaction, a PLP-dependent condensation of palmitoyl-CoA and serine. Subsequent reduction of the keto group, acylation, and desaturation (via reduction of an electron acceptor, X) form ceramide, the precursor of other sphingolipids.

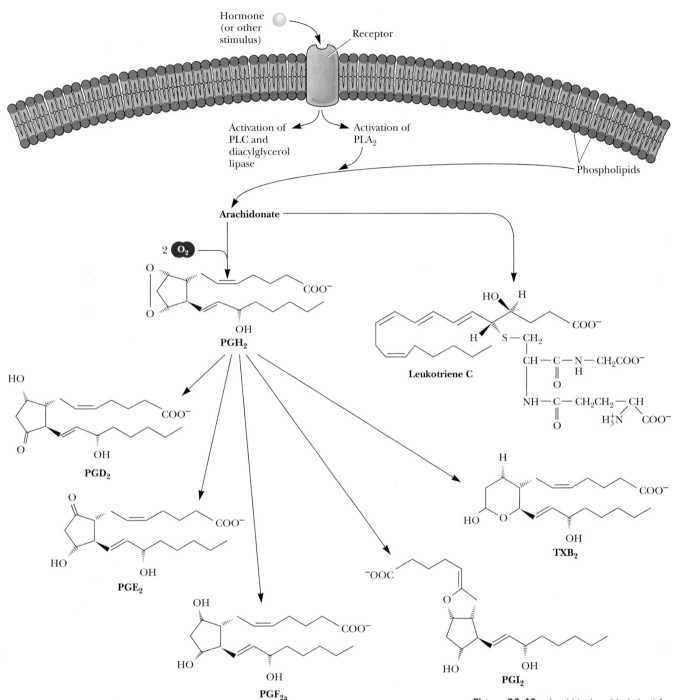

Figure 20.40 Arachidonic acid, derived from breakdown of phospholipids (PL), is the precursor of prostaglandins, thromboxanes, and leukotrienes. The letters used to name the prostaglandins are assigned on the basis of similarities in structure and physical properties. The class denoted PGE, for example, consists of β-hydroxyketones that are soluble in ether, whereas PGF denotes 1,3-diols that are soluble in phosphate buffer. PGA denotes prostaglandins possessing α,β-unsaturated ketones. The number following the letters refers to the number of carbon-carbon double bonds. Thus, PGE_2 contains two double bonds.

 See pages 173–178

Prostaglandins Are Formed from Arachidonate by Oxidation and Cyclization

All prostaglandins are cyclopentanoic acids derived from arachidonic acid. The biosynthesis of prostaglandins is initiated by an enzyme associated with the endoplasmic reticulum, called **prostaglandin endoperoxide synthase,** also known as **cyclooxygenase.** The enzyme catalyzes simultaneous oxidation and cyclization of arachidonic acid. The enzyme is viewed as having two distinct activities, cyclooxygenase and peroxidase, as shown in Figure 20.41.

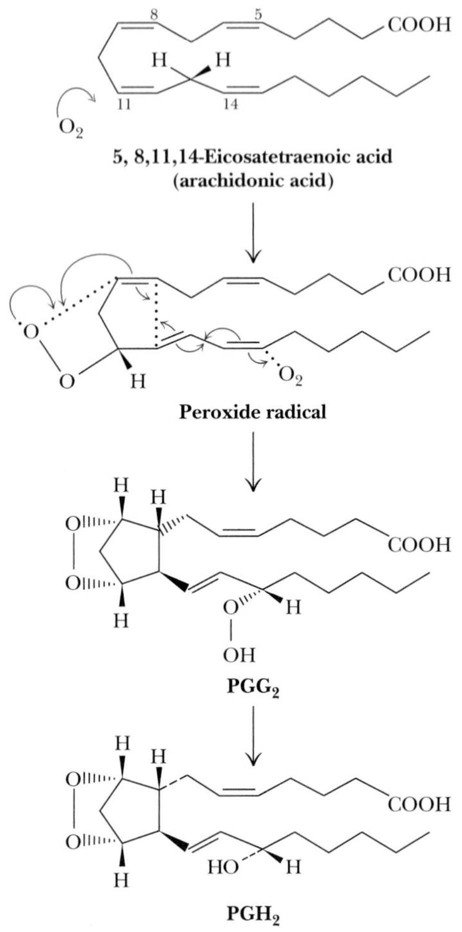

◀ **Figure 20.41** Prostaglandin endoperoxide synthase, the enzyme that converts arachidonic acid to prostaglandin PGH_2, possesses two distinct activities: cyclooxygenase (steps 1 and 2) and glutathione (GSSG)–dependent hydroperoxidase (step 3). Cyclooxygenase is the site of action of aspirin and many other analgesic agents.

5, 8,11,14-Eicosatetraenoic acid (arachidonic acid)

Peroxide radical

PGG$_2$

PGH$_2$

A Variety of Stimuli Trigger Arachidonate Release and Eicosanoid Synthesis

The release of arachidonate and the synthesis or interconversion of eicosanoids can be initiated by a variety of stimuli, including histamine, hormones such as epinephrine and bradykinin, proteases such as thrombin, and even serum albumin. An important mechanism of arachidonate release and eicosanoid synthesis involves tissue injury and inflammation. When tissue damage or injury occurs, special inflammatory cells, **monocytes** and **neutrophils,** invade the injured tissue and interact with the resident cells (e.g., smooth muscle cells and fibroblasts). This interaction typically leads to arachidonate release and eicosanoid synthesis. Examples of tissue injury in which eicosanoid synthesis has been characterized include heart attack (myocardial infarction), rheumatoid arthritis, and ulcerative colitis.

Take Two Aspirin and ... Inhibit Your Prostaglandin Synthesis

In 1971, biochemist Sir John Vane was the first to show that **aspirin** (acetylsalicylate; see Figure 20.42) exerts most of its effects by inhibiting the biosynthesis of prostaglandins. Its site of action is prostaglandin endoperoxide synthase. Cyclooxygenase activity is destroyed when aspirin O-acetylates Ser[530] on the enzyme. From this you may begin to infer something about how prostaglandins (and aspirin) function. Prostaglandins are known to enhance inflammation in animal tissues. Aspirin exerts its powerful antiinflammatory effect by inhibiting this first step in their synthesis. Aspirin does not have any measurable effect on the peroxidase activity of the synthase. Other nonsteroidal antiinflammatory agents, such as **ibuprofen** (Figure 20.42) and phenylbutazone, inhibit the cyclooxygenase by competing at the active site with arachidonate or with the peroxyacid intermediate (PGG$_2$; see Figure 20.41). (See *Human Biochemistry,* page 643.)

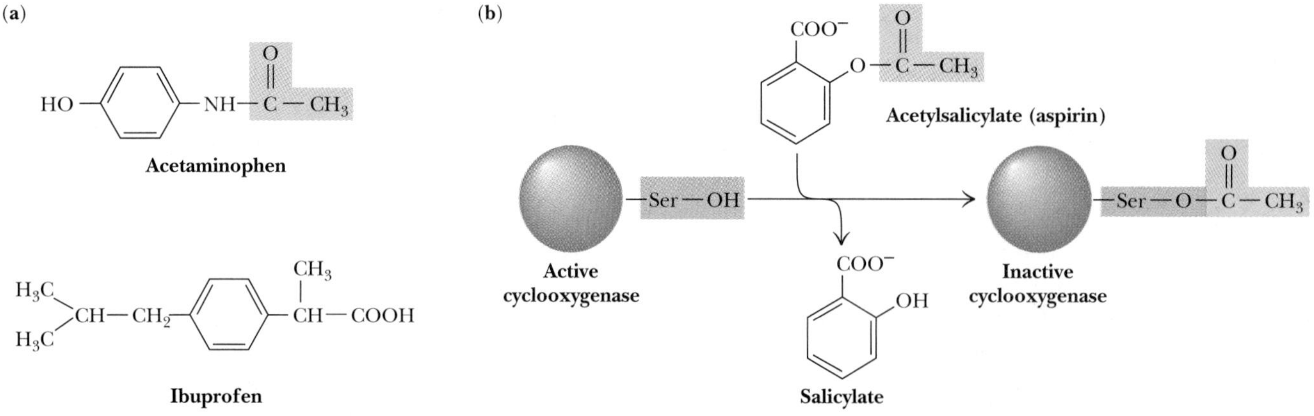

(a)

Acetaminophen

Ibuprofen

See page 75

(b)

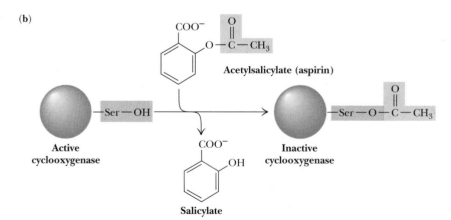

Acetylsalicylate (aspirin)

Active cyclooxygenase

Inactive cyclooxygenase

Salicylate

Figure 20.42 **(a)** The structures of several common analgesic agents. Acetaminophen is marketed under the tradename Tylenol. Ibuprofen is sold as Motrin, Nuprin, and Advil. **(b)** Acetylsalicylate (aspirin) inhibits the cyclooxygenase activity of endoperoxide synthase via acetylation (covalent modification) of Ser[530].

HUMAN BIOCHEMISTRY

The Discovery of Prostaglandins

The name prostaglandin was given to this class of compounds by Ulf von Euler, their discoverer, in Sweden in the 1930s. He extracted fluids containing these components from human semen. Because he thought they originated in the prostate gland, he named them prostaglandins. Actually, they were synthesized in the seminal vesicles, and it is now known that similar substances are synthesized in most animal tissues (both male and female). Von Euler observed that injection of these substances into animals caused smooth muscle contraction and dramatic lowering of blood pressure.

Von Euler (and others) soon found that it is difficult to analyze and characterize these obviously interesting compounds because they are present at extremely low levels. Prostaglandin E_{2a}, or PGE_{2a},

is present in human serum at a level of less than 10^{-14} M! In addition, they often have half-lives of only 30 seconds to a few minutes, not lasting long enough to be easily identified. Moreover, most animal tissues upon dissection and homogenization rapidly synthesize and degrade a variety of these substances, so the amounts obtained in isolation procedures are extremely sensitive to the methods used and highly variable even when procedures are carefully controlled.

Sune Bergstrom and his colleagues described the first structural determinations of prostaglandins in the late 1950s. In the early 1960s, dramatic advances in laboratory techniques such as NMR spectroscopy and mass spectrometry made further characterization possible.

20.10 Cholesterol Biosynthesis

The most prevalent steroid in animal cells is **cholesterol** (Figure 20.43). Plants contain no cholesterol, but they do contain other steroids very similar to cholesterol in structure (see page 169). Liver is the primary site of cholesterol biosynthesis.

The cholesterol biosynthetic pathway begins in the cytosol with the synthesis of **mevalonate** from acetyl-CoA (Figure 20.44). The first step is the **β-ketothiolase**–catalyzed condensation of two molecules of acetyl-CoA to form acetoacetyl-CoA. In the next reaction, acetyl-CoA and acetoacetyl-CoA join to form 3-hydroxy-3-methylglutaryl-CoA, which is abbreviated HMG-CoA. The reaction is catalyzed by **HMG-CoA synthase.** The third step in the pathway is the rate-limiting step in cholesterol biosynthesis. Here, HMG-CoA undergoes two NADPH-dependent reductions to produce 3R-mevalonate. The reaction is catalyzed by **HMG-CoA reductase,** a 97-kD glycoprotein that traverses the

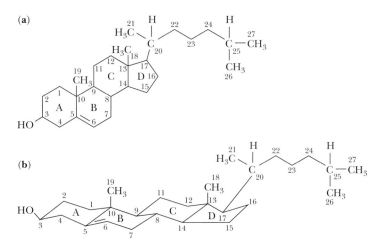

Figure 20.43 The structure of cholesterol, drawn **(a)** in the traditional planar motif and **(b)** in a form that more accurately describes the conformation of the ring system.

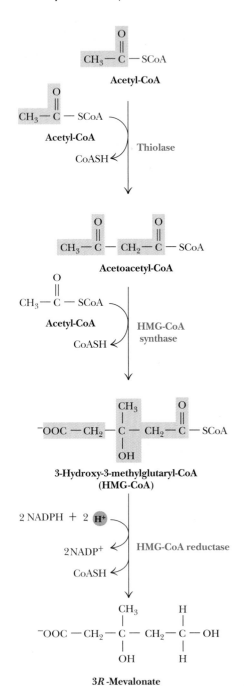

Figure 20.44 The biosynthesis of 3R-mevalonate from acetyl-CoA.

See pages 173–178

endoplasmic reticulum membrane with its active site facing the cytosol. As the rate-limiting step, HMG-CoA reductase is the principal site of regulation in cholesterol synthesis.

Three different regulatory mechanisms are involved:

1. Covalent modification. Phosphorylation by cAMP-dependent protein kinases inactivates the reductase. This inactivation can be reversed by two specific phosphatases (Figure 20.45).
2. Degradation of HMG-CoA reductase. This enzyme has a half-life of only three hours, and the half-life itself depends on cholesterol levels: high [cholesterol] means a short half-life for HMG-CoA reductase.
3. Gene expression. Cholesterol levels control the amount of mRNA. If [cholesterol] is high, levels of mRNA coding for the reductase are reduced. If [cholesterol] is low, more mRNA is made. (Regulation of gene expression is discussed in Chapter 24.)

Squalene Is Synthesized from Mevalonate

In 1952, Konrad Bloch and Robert Langdon showed that **squalene** is an intermediate in the cholesterol biosynthetic pathway. The formation of squalene involves conversion of mevalonate to two key five-carbon intermediates, **isopentenyl pyrophosphate** and **dimethylallyl pyrophosphate,** which join to yield farnesyl pyrophosphate and then squalene. A series of four reactions converts mevalonate to isopentenyl pyrophosphate and then to dimethylallyl pyrophosphate (Figure 20.46). The first three steps each consume an ATP, two for the

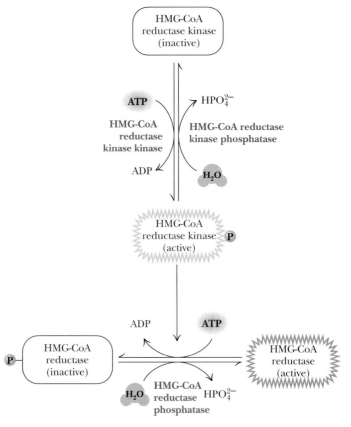

Figure 20.45 HMG-CoA reductase activity is modulated by a cycle of phosphorylation and dephosphorylation.

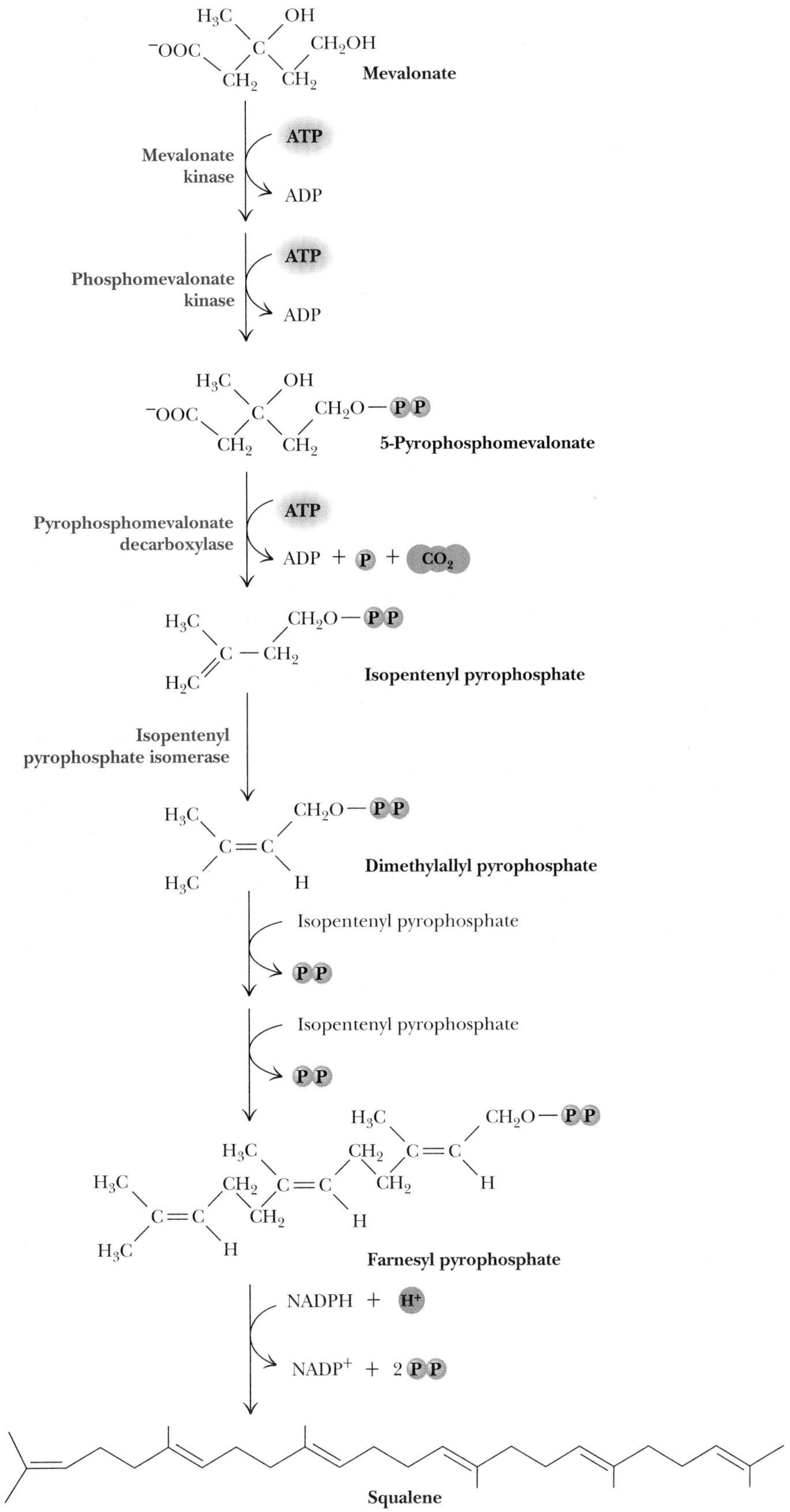

Figure 20.46 The conversion of mevalonate to squalene.

See pages 173–178

HUMAN BIOCHEMISTRY

The Molecular Basis for the Action of Nonsteroidal Antiinflammatory Drugs

Prostaglandins are potent mediators of inflammation. The first and committed step in the production of prostaglandins from arachidonic acid is the bis-oxygenation of arachidonate to prostaglandin PGG_2. This is followed by reduction to PGH_2 in a peroxidase reaction. Both these reactions are catalyzed by prostaglandin endoperoxide synthase, also known as PGH_2 synthase or cyclooxygenase, thus abbreviated COX. This enzyme is inhibited by the family of drugs known as nonsteroidal anti-inflammatory drugs, or NSAIDs. Aspirin, ibuprofen, flurbiprofen, and acetaminophen (trade name: Tylenol) are all NSAIDs.

There are two isoforms of COX in animals: COX-1 (figure a), which carries out normal, physiological production of prostaglandins, and COX-2 (figure b), which is induced by cytokines, mitogens, and endotoxins in inflammatory cells and is responsible for the production of prostaglandins in inflammation.

The enzyme structure shown in (a) is that of residues 33 to 583 of COX-1 from sheep, inactivated by bromoaspirin. Residues 73 to 116 (yellow) form a membrane-binding motif. The helical segments are amphipathic, with most of the hydrophobic residues facing away from the protein, where they can interact with a lipid bilayer. The third domain of the COX enzyme, the catalytic domain (in blue), is a globular structure that contains both the cyclooxygenase and peroxidase active sites.

The cyclooxygenase active site lies at the end of a long, narrow, hydrophobic tunnel or channel. In this bromoaspirin-inactivated structure, Ser^{530}, which lies along the wall of the tunnel, is bromoacetylated, and a molecule of salicylate is also bound in the tunnel. Deep in the tunnel, at the far end, lies Tyr^{385}, a catalytically important residue. Heme-dependent peroxidase activity apparently creates a Tyr^{385} radical, which is required for cyclooxygenase ac-

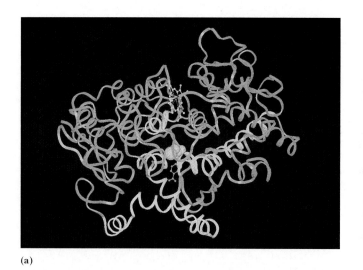

(a)

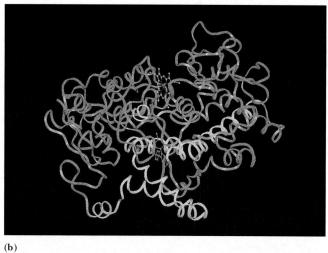

(b)

See page 75

purpose of forming a pyrophosphate at the 5-position and the third to drive the decarboxylation and double bond formation in the third step. **Pyrophosphomevalonate decarboxylase** phosphorylates the 3-hydroxyl group, and this is followed by *trans* elimination of the phosphate and carboxyl groups to form the double bond in isopentenyl pyrophosphate. Isomerization of the double bond yields the dimethylallyl pyrophosphate. Condensation of these two five-carbon intermediates produces **geranyl pyrophosphate;** addition of another five-carbon isopentenyl group gives **farnesyl pyrophosphate.** Both steps in the production of farnesyl pyrophosphate occur with release of pyrophosphate, hydrolysis of which drives these reactions forward. Note too that the linkage of isoprene units to form farnesyl pyrophosphate occurs in a head-to-tail fashion. This is the general rule in biosynthesis of molecules involving isoprene linkages. The next step—the joining of two farnesyl pyrophosphates to produce squalene—is a head-to-head condensation and represents an important exception to the general rule.

tivity. Aspirin and other NSAIDs block the synthesis of prostaglandins by filling and blocking the tunnel, preventing the migration of arachidonic acid to Tyr[385] in the active site at the back of the tunnel.

There are thought to be at least four different mechanisms of action for NSAIDs. Aspirin (and also bromoaspirin) also irreversibly inactivate both COX-1 and COX-2 by covalently modifying a Ser residue in the tunnel. Ibuprofen acts instead by competing in a reversible fashion for the substrate-binding site in the tunnel.

Flurbiprofen and indomethacin, which comprise the third class of inhibitors, cause a slow, time-dependent inhibition of COX-1 and COX-2, apparently via formation of a salt bridge between a carboxylate on the drug and Arg[120], which lies in the tunnel.

The new drugs Vioxx (from Merck) and Celebrex (from Pharmacia) act by a fourth mechanism, specifically inhibiting COX-2. Selective COX-2 inhibitors will likely be the drugs of the future, because they selectively block the inflammation mediated by COX-2, without the potential for stomach lesions and renal toxicity that arise from COX-1 inhibition.

Aspirin   Bromoaspirin   Celebrex

Flurbiprofen   Indomethacin   Vioxx   Ibuprofen

Squalene monooxygenase, an enzyme bound to the endoplasmic reticulum, converts squalene to squalene-2,3-epoxide (Figure 20.47). This reaction employs FAD and NADPH as coenzymes and requires O_2 as well as a cytosolic protein called **soluble protein activator.** A second ER membrane enzyme, **2,3-oxidosqualene lanosterol cyclase,** catalyzes the second reaction, which involves shifts of hydride ions and methyl groups.

Conversion of Lanosterol to Cholesterol Requires 20 Additional Steps

Although lanosterol may appear similar to cholesterol in structure, another 20 steps are required to convert lanosterol to cholesterol (see Figure 20.47). The enzymes responsible for this are all associated with the ER. The primary pathway involves 7-dehydrocholesterol as the penultimate intermediate. An alternative pathway, also composed of many steps, produces the intermediate

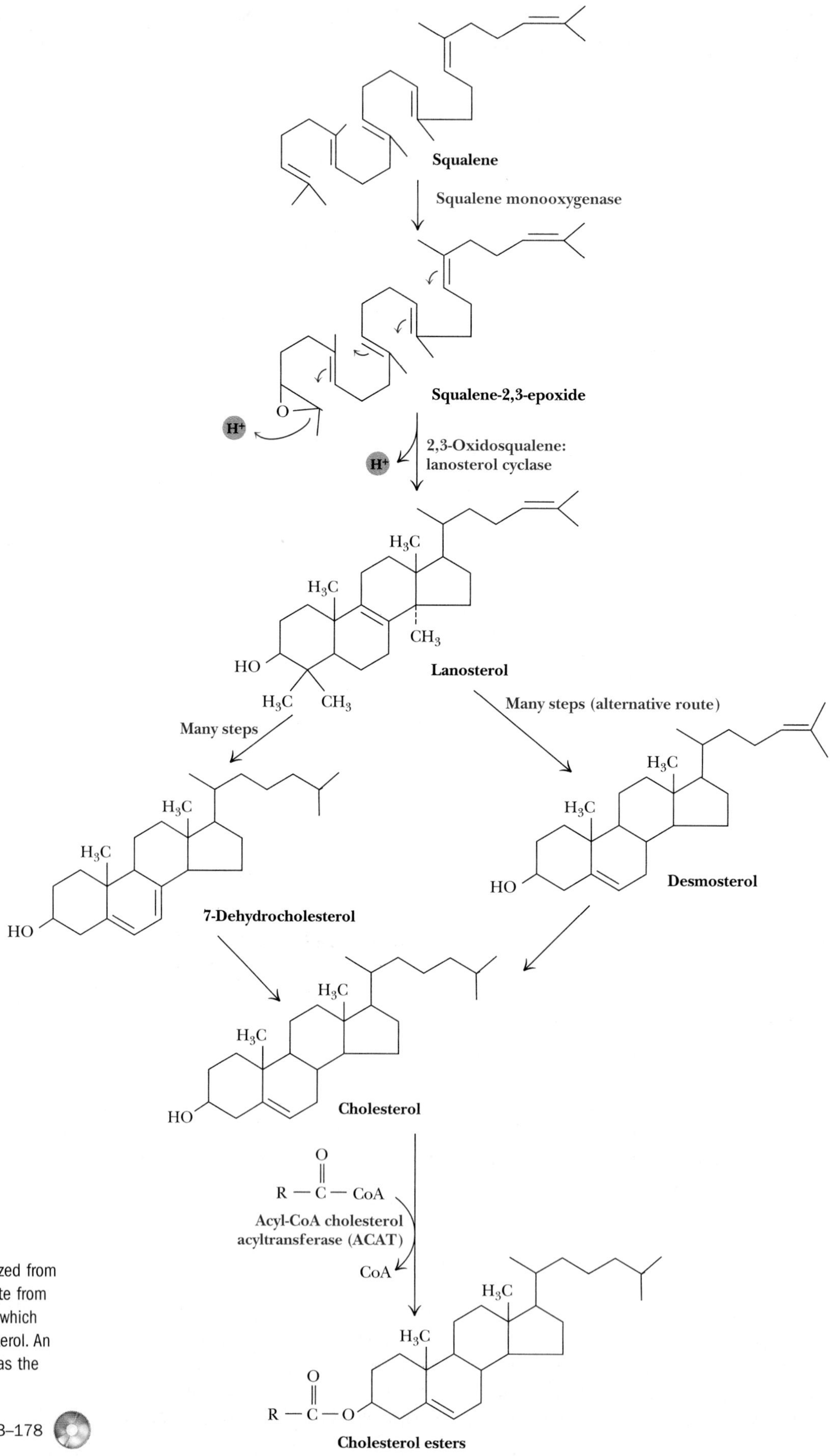

Squalene

Squalene monooxygenase

Squalene-2,3-epoxide

H⁺

2,3-Oxidosqualene:
lanosterol cyclase

H⁺

H₃C

H₃C

CH₃

HO

H₃C   CH₃

Lanosterol

Many steps

Many steps (alternative route)

H₃C

H₃C

H₃C

HO

Desmosterol

H₃C

H₃C

HO

7-Dehydrocholesterol

H₃C

H₃C

HO

Cholesterol

O
||
R — C — CoA

**Acyl-CoA cholesterol
acyltransferase (ACAT)**

CoA

H₃C

H₃C

O
||
R — C — O

Cholesterol esters

Figure 20.47 Cholesterol is synthesized from squalene via lanosterol. The primary route from lanosterol involves 20 steps, the last of which converts 7-dehydrocholesterol to cholesterol. An alternative route produces desmosterol as the penultimate intermediate.

See pages 173–178

desmosterol. Reduction of the double bond at C-24 yields cholesterol. Cholesterol esters—a principal form of circulating cholesterol—are synthesized by **acyl-CoA : cholesterol acyltransferases (ACAT)** on the cytoplasmic face of the endoplasmic reticulum.

20.11 Biosynthesis of Bile Acids

Bile acids, which exist mainly as **bile salts,** are polar carboxylic acid derivatives of cholesterol that are important in the digestion of food, especially the solubilization of ingested fats. The Na^+ and K^+ salts of glycocholic acid and taurocholic acid are the principal bile salts (Figure 20.48). Glycocholate and taurocholate are conjugates of cholic acid with glycine and taurine, respectively. Because they contain both nonpolar and polar domains, these bile salt conjugates are highly effective as detergents. These substances are made in the liver, stored in the gallbladder, and secreted as needed into the intestines (see Figure 20.3).

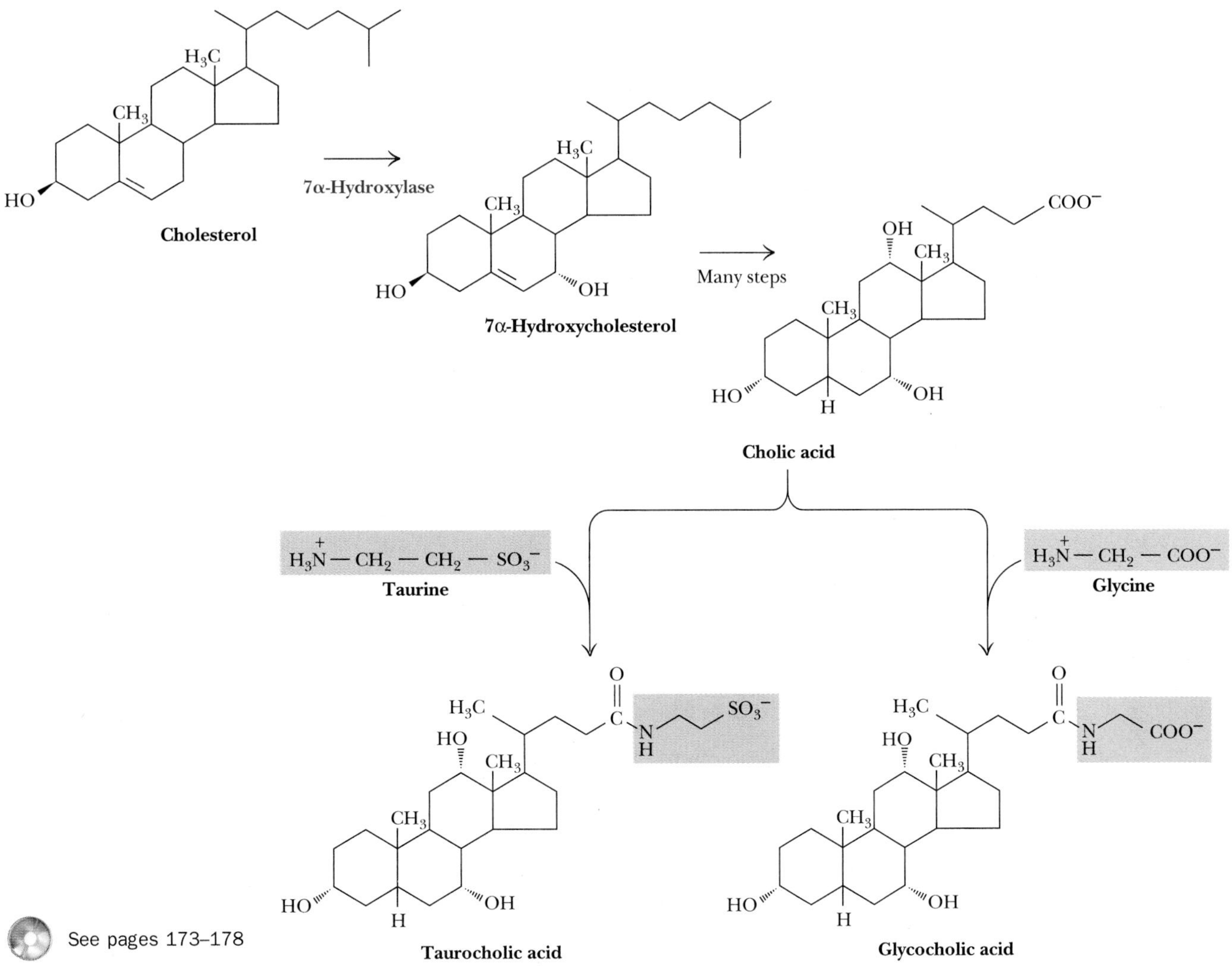

See pages 173–178

Figure 20.48 Cholic acid, a bile salt, is synthesized from cholesterol via 7α-hydroxycholesterol. Conjugation with taurine or glycine produces taurocholic acid and glycocholic acid, respectively. Taurocholate and glycocholate are freely water-soluble and are highly effective detergents.

Statins and the Lowering of Human Serum Cholesterol Levels

Chemists and biochemists have long sought a means of reducing serum cholesterol levels to reduce the risk of heart attack and cardiovascular disease. Because HMG-CoA reductase is the rate-limiting step in cholesterol biosynthesis, this enzyme is a likely drug target. **Mevinolin,** also known as **lovastatin** (see figure), was isolated from the fungus *Aspergillus terreus* and developed at Merck, Sharpe and Dohme for this purpose. It is now a widely prescribed cholesterol-lowering drug. Dramatic reductions of serum cholesterol are observed at doses of 20 to 80 mg per day.

Lovastatin is administered as an inactive lactone. After oral ingestion, it is hydrolyzed to the active **mevinolinic acid,** a competitive inhibitor of the reductase with a K_I of 0.6 nM. Mevinolinic acid is thought to behave as a transition-state analog (see Chapter 11) of the tetrahedral intermediate formed in the HMG-CoA reductase reaction (see figure).

Derivatives of lovastatin have been found to be even more potent in cholesterol-lowering trials. **Synvinolin** lowers serum cholesterol levels at much lower doses than lovastatin.

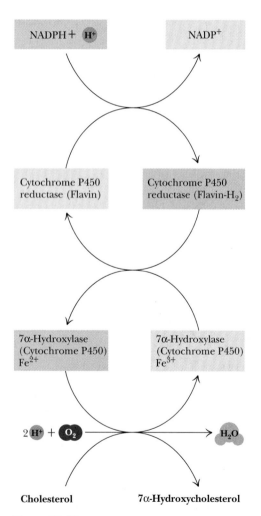

Figure 20.49 The mixed-function oxidase activity of 7α-hydroxylase.

The formation of bile salts represents the major pathway for cholesterol degradation. The first step involves hydroxylation at C-7 (Figure 20.49). **7α-Hydroxylase,** which catalyzes the reaction, is a mixed-function oxidase involving cytochrome P450. **Mixed-function oxidases** use O_2 as substrate. One oxygen atom goes to hydroxylate the substrate, while the other is reduced to water (see Figure 20.49). The function of cytochrome P450 is to activate O_2 for the hydroxylation reaction. Such hydroxylations are quite common in the synthetic routes for cholesterol, bile acids, and steroid hormones and also in detoxification pathways for aromatic compounds. Several of these are considered in the next section. 7α-Hydroxycholesterol is the precursor for cholic acid.

20.12 Synthesis and Metabolism of Steroid Hormones

Steroid hormones are crucial signal molecules in mammals. (The details of their physiological effects are described in Chapter 26.) Their biosynthesis begins with the **desmolase** reaction, which converts cholesterol to pregnenolone (Figure 20.50). Desmolase is found in the mitochondria of tissues that synthesize steroids (mainly the adrenal glands and gonads). Desmolase activity includes two hydroxylases and utilizes cytochrome P450.

Pregnenolone and Progesterone Are the Precursors of All Other Steroid Hormones

Pregnenolone is transported from the mitochondria to the ER, where a hydroxyl oxidation and migration of the double bond yield progesterone. Pregnenolone synthesis in the adrenal cortex is activated by **adrenocorticotropic hormone** (ACTH), a peptide of 39 amino acid residues secreted by the anterior pituitary gland.

Progesterone is secreted from the corpus luteum during the latter half of the menstrual cycle and prepares the lining of the uterus for attachment of a fertilized ovum. If an ovum attaches, progesterone secretion continues to ensure the successful maintenance of a pregnancy. Progesterone is also the precursor for synthesis of the **sex hormone steroids** and the **corticosteroids.** Male sex hormone steroids are called **androgens,** and female hormones, **estrogens.**

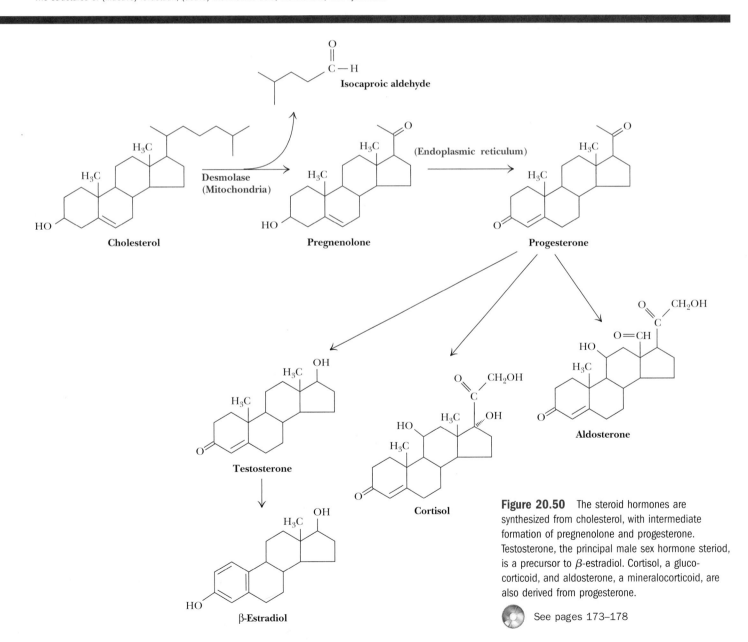

1 R=H Mevinolin (Lovastatin, MEVACOR®)
2 R=CH₃ Synvinolin (Simnastatin, ZOCOR®)

Mevinolinic acid

Mevalonate

Tetrahedral intermediate in HMG-CoA reductase mechanism

The structures of (inactive) lovastatin, (active) mevinolinic acid, mevalonate, and synvinolin.

Isocaproic aldehyde

Cholesterol

Desmolase (Mitochondria)

Pregnenolone

(Endoplasmic reticulum)

Progesterone

Testosterone

Cortisol

Aldosterone

β-Estradiol

Figure 20.50 The steroid hormones are synthesized from cholesterol, with intermediate formation of pregnenolone and progesterone. Testosterone, the principal male sex hormone steriod, is a precursor to β-estradiol. Cortisol, a glucocorticoid, and aldosterone, a mineralocorticoid, are also derived from progesterone.

See pages 173–178

Steroid 5α-Reductase: A Factor in Male Baldness, Prostatic Hyperplasia, and Prostate Cancer

An enzyme that metabolizes testosterone may be involved in the benign conditions of male pattern baldness (also known as androgenic alopecia) and benign prostatic hyperplasia (prostate gland enlargement), as well as potentially fatal prostate cancers. Steroid 5α-reductases are membrane-bound enzymes that catalyze the NADPH-dependent reduction of testosterone to dihydrotestosterone (DHT) (see figure). Two isoforms of 5α-reductase have been identified. In humans, the type I enzyme predominates in the sebaceous glands of skin and liver, whereas type II is most abundant in the prostate, seminal vesicles, liver, and epididymis. Dihydrotestosterone is a contributory factor in male baldness and prostatic hyperplasia, and it has also been shown to act as a mitogen (a stimulator of cell division). For these reasons, 5α-reductase inhibitors are potential candidates for treatment of these human conditions.

Finasteride (see figure) is a specific inhibitor of type II 5α-reductase. It has been used clinically for treatment of benign prostatic hyperplasia, and it is also marketed under the trade name Propecia by Merck as a treatment for male baldness. Type II 5α-reductase inhibitors may also be potential therapeutic agents for treatment of prostate cancer. Somatic mutations occur in the gene for type II 5α-reductase during prostate cancer progression.

Dihydrotestosterone   Finasteride

Because type I 5α-reductase is the predominant form of the enzyme in human scalp, the mechanism of finasteride's promotion of hair growth in men with androgenic alopecia has been uncertain. However, scientists at Merck have shown that whereas type I 5α-reductase predominates in sebaceous ducts of the skin, type II 5α-reductase is the only form of the enzyme present in hair follicles. Thus finasteride's therapeutic effects may arise from inhibition of the type II enzyme in the hair follicle itself.

Testosterone is an androgen synthesized in males primarily in the testes (and in much smaller amounts in the adrenal cortex). Androgens are necessary for sperm maturation. Even nonreproductive tissue (liver, brain, and skeletal muscle) is susceptible to the effects of androgens.

Testosterone is also produced primarily in the ovaries (and in much smaller amounts in the adrenal glands) of females as a precursor for the estrogens. β-Estradiol is the most important estrogen (see Figure 20.50).

Steroid Hormones Modulate Transcription in the Nucleus

Steroid hormones act in a different manner from most hormones we have considered. In many cases, they do not bind to plasma membrane receptors but rather pass easily across the plasma membrane. Steroids may bind directly to receptors in the nucleus or may bind to cytosolic steroid hormone receptors, which then enter the nucleus. In the nucleus, the hormone–receptor complex binds directly to specific nucleotide sequences in DNA, increasing transcription of DNA to RNA (see Chapters 24 and 26).

Cortisol and Other Corticosteroids Regulate a Variety of Body Processes

Corticosteroids, including the glucocorticoids and mineralocorticoids, are made by the cortex of the adrenal glands on top of the kidneys. **Cortisol** (see Figure 20.50) is representative of the **glucocorticoids,** a class of compounds that (1) stimulate gluconeogenesis and glycogen synthesis in the liver (by signaling the synthesis of PEP carboxykinase, fructose-1,6-bisphosphatase, glucose-6-phosphatase, and glycogen synthase); (2) inhibit protein synthesis and stim-

HUMAN BIOCHEMISTRY

Salt and Water Balances and Deaths in Marathoners

Marathon running puts unusual demands on the human body. Dehydration is an obvious problem, and runners are often encouraged to drink substantial amounts of water during a race. However, as Allen Arleff of the University of California at San Francisco has shown, drinking too much water during a race can lead to hyponatremia—a shortage of salt in the blood. Ordinarily, aldosterone carefully regulates salt balances in the body. However, exercise leads the body to release antidiuretic hormone (arginine vasopressin), which blocks normal elimination of excess fluid as urine. At the same time, during a race, so much blood is diverted to the muscles that the intestines are unable to absorb any water that the runner drinks. Once the race is over, however, the water floods into the bloodstream, disrupting the body's salt balance. This can cause the brain to swell and also results in release of water into the lungs. On the occasions when a runner collapses during or after a run, the observation of fluid in the lungs and the circumstances of the collapse often lead emergency room doctors to suspect heart trouble and to order treatments that make the problem worse.

Allen Arleff has suggested that the amount of water a runner should drink during a race will vary from person to person, and that the amount of water consumed should probably not exceed the amount of water actually lost during the race.

Runners at Nice, France, Marathon, 1986. (©Jean Marc Barey, Vandystadt, Agence de Presse/Photo Researchers, Inc.)

---

ulate protein degradation in peripheral tissues such as muscle; (3) inhibit allergic and inflammatory responses; (4) exert an immunosuppressive effect, inhibiting DNA replication and mitosis and repressing the formation of antibodies and lymphocytes; and (5) inhibit formation of fibroblasts involved in healing wounds and slow the healing of broken bones.

Aldosterone, the most potent of the **mineralocorticoids** (see Figure 20.50), is involved in the regulation of sodium and potassium balances in tissues. Aldosterone increases the kidney's capacity to absorb Na^+, Cl^-, and H_2O from the glomerular filtrate in the kidney tubules.

Anabolic Steroids Have Been Used Illegally to Enhance Athletic Performance

The dramatic effects of androgens on protein biosynthesis have led many athletes to the use of synthetic androgens, which go by the blanket term **anabolic steroids.** Despite numerous warnings from the medical community about side effects, which include kidney and liver disorders, sterility, and heart disease, abuse of such substances is epidemic. **Stanozolol** (Figure 20.51) was one of the agents found in the blood and urine of Ben Johnson following his record-setting performance in the 100-meter dash in the 1988 Olympic Games. Because use of such substances is disallowed, Johnson lost his gold medal and Carl Lewis was declared the official winner.

Stanozolol

Figure 20.51 The structure of stanozolol, an anabolic steroid.

HUMAN BIOCHEMISTRY

Androstenedione: A Steroid of Uncertain Effects

In 1998, Mark McGwire and Sammy Sosa electrified the baseball world by hitting unprecedented numbers of home runs for the St. Louis Cardinals and Chicago Cubs, respectively. McGwire hit an astounding and unprecedented 70 home runs and Sosa followed close behind with 68. In the course of this historic performance, it was reported that McGwire used androstenedione as a dietary supplement related to his weight-lifting regimen. "Andro," as it is called by athletes, is a precursor to testosterone, the male sex steroid hormone. Its use is permitted by Major League Baseball but is banned by the National Football League, the National Collegiate Athletic Association, and the International Olympic Committee. In spite of its precursor relationship to testosterone, little actual research on the conversion of orally administered andro to testosterone has been reported, and its efficacy for body building and the enhancement of performance in sports is uncertain at best. In

Mark McGwire's case, the issue is certainly moot. McGwire's use of andro was confined to the 1998 season, but his home-run production both before and after this time have been prodigious. McGwire hit an unprecedented total of 245 home runs over the four seasons of 1996 to 1999.

(©AFP/CORBIS)

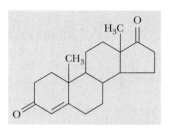

4-Androstene-3,17-dione

PROBLEMS

1. Calculate the volume of metabolic water available to a camel through fatty acid oxidation if it carries 30 lb of triacylglycerol in its hump.

2. Calculate the approximate number of ATP molecules that can be obtained from the oxidation of *cis*-11-heptadecenoic acid to CO_2 and water.

3. Phytanic acid, the product of chlorophyll that causes problems for individuals with Refsum's disease, is 3,7,11,15-tetramethyl hexadecanoic acid. Suggest a route for its oxidation that is consistent with what you have learned in this chapter. (*Hint:* The methyl group at C-3 effectively blocks hydroxylation and normal β-oxidation. You may wish to initiate breakdown in some other way.)

4. Even though acetate units, such as those obtained from fatty acid oxidation, cannot be used for net synthesis of carbohydrate in animals, labeled carbon from ^{14}C-labeled acetate can be found in newly synthesized glucose (for example, in liver glycogen) in animal tracer studies. Explain how this can be. Which carbons of glucose would you expect to be the first to be labeled by ^{14}C-labeled acetate?

5. Overweight individuals who diet to lose weight often view fat in negative ways because adipose tissue is the repository of excess caloric intake. However, the "weighty" consequences might be even worse if excess calories were stored in other forms. Consider a person who is 10 lb "overweight," and estimate how much more he or she would weigh if excess energy were stored in the form of carbohydrate instead of fat.

6. What would be the consequences of a deficiency in vitamin B_{12} for fatty acid oxidation? What metabolic intermediates might accumulate?

7. Write properly balanced chemical equations for the oxidation to CO_2 and water of (a) myristic acid, (b) stearic acid, (c) α-linolenic acid, and (d) arachidonic acid.

8. How many tritium atoms are incorporated into acetate if a molecule of palmitic acid is oxidized in 100% tritiated water?

9. What would be the consequences of a carnitine deficiency for fatty acid oxidation?

10. Use the relationships shown in Figure 20.21 to determine which carbons of glucose will be incorporated into palmitic acid. Consider the cases of both citrate that is immediately ex-

ported to the cytosol following its synthesis and citrate that enters the TCA cycle.

11. Based on the information presented in the text and in Figure 20.23, suggest a model for the regulation of acetyl-CoA carboxylase. Consider the possible roles of subunit interactions, phosphorylation, and conformation changes in your model.

12. Consider the role of the pantothenic acid groups in animal fatty acyl synthase and the size of the pantothenic acid group itself, and estimate a maximal separation between the malonyl transferase and the ketoacyl-ACP synthase active sites.

13. Write a balanced, stoichiometric reaction for the synthesis of phosphatidylethanolamine from glycerol, fatty acyl-CoA, and ethanolamine. Make an estimate of the $\Delta G^{\circ\prime}$ for the overall process.

14. Write a balanced, stoichiometric reaction for the synthesis of cholesterol from acetyl-CoA.

15. Trace each of the carbon atoms of mevalonate through the synthesis of cholesterol, and determine the source (the position in the mevalonate structure) of each carbon in the final structure.

16. Identify the lipid synthetic pathways that would be affected by abnormally low levels of CTP.

FURTHER READING

Athappilly, F. K., and Hendrickson, W. A., 1995. Structure of the biotinyl domain of acetyl-CoA carboxylase determined by MAD phasing. *Structure* **3**:1407.

Bennett, M. J., 1994. The enzymes of mitochondrial fatty acid oxidation. *Clinica Chimica Acta* **226**:211–224.

Bieber, L. L., 1988. Carnitine. *Annual Review of Biochemistry* **88**:261–283.

Bloch, K., 1965. The biological synthesis of cholesterol. *Science* **150**:19–28.

Bloch, K., 1987. Summing up. *Annual Review of Biochemistry* **56**:1–19.

Boyer, P. D., ed., 1983. *The Enzymes,* 3rd ed., vol. 16. New York: Academic Press.

Carman, G. M., and Henry, S. A., 1989. Phospholipid biosynthesis in yeast. *Annual Review of Biochemistry* **58**:635–669.

Carman, G. M., and Zeimetz, G. M., 1996. Regulation of phospholipid biosynthesis in the yeast *Saccharomyces cerevisiae. Journal of Biological Chemistry* **271**:13292–13296.

Chang, S. I., and Hammes, G. G., 1990. Structure and mechanism of action of a multifunctional enzyme: Fatty acid synthase. *Accounts of Chemical Research* **23**:363–369.

Dunne, S. J., Cornell, R. B., Johnson, J. E., et al., 1996. Structure of the membrane-binding domain of CTP phosphocholine cytidylyltransferase. *Biochemistry* **35**:1975.

Eder, M., Krautle, F., Dong, Y., et al., 1997. Characterization of human and pig kidney long-chain-acyl-CoA dehydrogenases and their role in beta-oxidation. *European Journal of Biochemistry* **245**:600–607.

Friedman, J. M., 1997. The alphabet of weight control. *Nature* **385**:119–120.

Grynberg, A., and Demaison, L., 1996. Fatty acid oxidation in the heart. *Journal of Cardiovascular Pharmacology* **28**:S11–S17.

Halpern, J., 1985. Mechanisms of coenzyme B₁₂–dependent rearrangements. *Science* **227**:869–875.

Hansen, H. S., 1985. The essential nature of linoleic acid in mammals. *Trends in Biochemical Sciences* **11**:263–265.

Hiltunen, J. K., Palosaari, P., and Kunau, W.-H., 1989. Epimerization of 3-hydroxyacyl-CoA esters in rat liver. *Journal of Biological Chemistry* **264**:13535–13540.

Jackowski, S., 1996. Cell cycle regulation of membrane phospholipid metabolism. *Journal of Biological Chemistry* **271**:20219–20222.

Jeffcoat, R., 1979. The biosynthesis of unsaturated fatty acids and its control in mammalian liver. *Essays in Biochemistry* **15**:1–36.

Kent, C., 1995. Eukaryotic phospholipid biosynthesis. *Annual Review of Biochemistry* **64**:315–343.

Kim, K-H., et al., 1989. Role of reversible phosphorylation of acetyl-CoA carboxylase in long-chain fatty acid synthesis. *The FASEB Journal* **3**:2250–2256.

Lardy, H., and Shrago, E., 1990. Biochemical aspects of obesity. *Annual Review of Biochemistry* **59**:689–710.

Liscum, L., and Underwood, K. W., 1995. Intracellular cholesterol transport and compartmentation. *Journal of Biological Chemistry* **270**:15443–15446.

Lopaschuk, G. D., and Gamble, J., 1994. The 1993 Merck Frosst Award. Acetyl-CoA carboxylase: an important regulator of fatty acid oxidation in the heart. *Canadian Journal of Physiology and Pharmacology* **72**:1101–1109.

McCarthy, A. D., and Hardie, D. G., 1984. Fatty acid synthase: An example of protein evolution by gene fusion. *Trends in Biochemical Sciences* **9**:60–63.

McGarry, J. D., and Foster, D. W., 1980. Regulation of hepatic fatty acid oxidation and ketone body production. *Annual Review of Biochemistry* **49**:395–420.

Needleman, P., et al., 1986. Arachidonic acid metabolism. *Annual Review of Biochemistry* **55**:69–102.

Newsholme, E. A., and Leech, A. R., 1983. *Biochemistry for the Medical Sciences.* New York: John Wiley & Sons.

Pollitt, R. J., 1995. Disorders of mitochondrial long-chain fatty acid oxidation. *Journal of Inherited Metabolic Disease* **18**:473–490.

Romijn, J. A., Coyle, E. F., Sidossis, L. S., et al., 1996. Relationship between fatty acid delivery and fatty acid oxidation during strenuous exercise. *Journal of Applied Physiology* **79**:1939–1945.

Samuelsson, B., et al., 1987. Leukotrienes and lipoxins: Structures, biosynthesis and biological effects. *Science* **237**:1171–1176.

Schewe, T., and Kuhn, H., 1991. Do 15-lipoxygenases have a common biological role? *Trends in Biochemical Sciences* **16**:369–373.

Schulz, H., 1987. Inhibitors of fatty acid oxidation. *Life Sciences* **40**:1443–1449.

Schulz, H., and Kunau, W.-H., 1987. β-Oxidation of unsaturated fatty acids: A revised pathway. *Trends in Biochemical Sciences* **12**:403–406.

Scriver, C. R., et al., 1995. *The Metabolic and Molecular Bases of Inherited Disease,* 7th ed. New York: McGraw-Hill.

Sherratt, H. S., 1994. Introduction: The regulation of fatty acid oxidation in cells. *Biochemical Society Transactions* **22:**421–422.

Sherratt, H. S., and Spurway, T. D., 1994. Regulation of fatty acid oxidation in cells. *Biochemical Society Transactions* **22:**423–427.

Smith, W. L., Garavito, R. M., and DeWitt, D. L., 1996. Prostaglandin endoperoxide H synthases (cyclooxygenases)-1 and -2. *Journal of Biological Chemistry* **271:**33157–33160.

Srere, P. A., and Sumegi, B., 1994. Processivity and fatty acid oxidation. *Biochemical Society Transactions* **22:**446–450.

Tataranni, P. A., and Ravussin, E., 1997. Effect of fat intake on energy balance. *Annals of the New York Academy of Sciences* **819:**37–43.

Tolbert, N. E., 1981. Metabolic pathways in peroxisomes and glyoxysomes. *Annual Review of Biochemistry* **50:**133–157.

Vance, D. E., and Vance, J. E., eds., 1985. *Biochemistry of Lipids and Membranes.* Menlo Park, Calif.: Benjamin Cummings.

Wakil, S., 1989. Fatty acid synthase, a proficient multifunctional enzyme. *Biochemistry* **28:**4523–4530.

Wakil, S., Stoops, J. K., and Joshi, V. C., 1983. Fatty acid synthesis and its regulation. *Annual Review of Biochemistry* **52:**537–579.

Yagoob, P., Newsholme, E. A., and Calder, P. C., 1994. Fatty acid oxidation by lymphocytes. *Biochemical Society Transactions* **22:**116S.

Amino Acid and Nucleotide Metabolism

Pigeon drinking at Gaia Fountain, Siena, Italy. The basic features of purine biosynthesis were elucidated initially from metabolic studies of nitrogen metabolism in pigeons. Pigeons excrete excess N as uric acid, a purine analog. (Arte & Imamagini srl/Corbis Images)

Guano, a substance found on some coasts frequented by sea birds, is composed chiefly of the birds' partially decomposed excrement. . . . The name for the purine guanine derives from the abundance of this base in guano.

The Chincha Islands . . . are . . . off the coast of Peru. . . . Myriads of birds, after satisfying their voracious appetites upon the finny tenants of the deep, have for ages made the islands their nightly abode, and the receptacle of their fecal offerings.

J. C. Nesbit, *On Agricultural Chemistry and the Nature and Properties of Peruvian Guano* (1850)

Nitrogen is a vital macronutrient for all life. Nitrogen atoms are the defining characteristic of amino acids and nucleotides, and, in turn, amino acids and nucleotides are the building blocks for the informational macromolecules that are the definitive substances of cells. In this chapter we consider the reactions that incorporate ammonium ions (or ammonia) into metabolic intermediates and the pathways that carry out the synthesis and degradation of the primary nitrogenous building blocks of cells: amino acids and nucleotides.

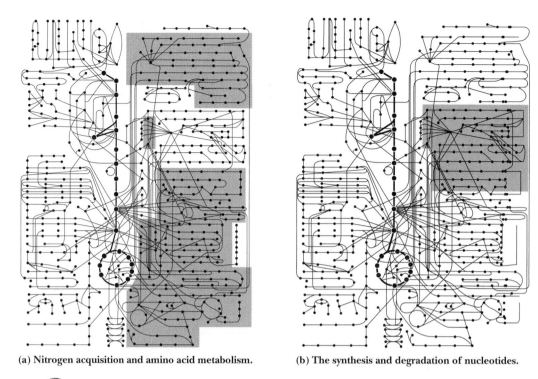

(a) Nitrogen acquisition and amino acid metabolism.

(b) The synthesis and degradation of nucleotides.

See *Interactive Biochemistry CD-ROM and Workbook*, pages 173–178

AMMONIUM FORMATION AND AMINO ACID METABOLISM

The Nitrogen Cycle

In the inanimate environment, nitrogen exists predominantly in an oxidized state, occurring principally as N_2 in the atmosphere or as nitrate ion (NO_3^-) in the soils and oceans. Its acquisition by biological systems is accompanied by its reduction to ammonium ion (NH_4^+) and the incorporation of NH_4^+ into organic linkage as amino or amido groups (Figure 21.1). The reduction of NO_3^- to NH_4^+ occurs in green plants, various fungi, and certain bacteria in a two-step metabolic pathway known as **nitrate assimilation.** The formation of NH_4^+ from N_2 gas is termed **nitrogen fixation.** N_2 fixation is an exclusively prokaryotic process, although bacteria in symbiotic association with certain green plants also carry out nitrogen fixation. Since no animals are capable of either

Figure 21.1 The nitrogen cycle. Organic nitrogenous compounds are formed by the incorporation of NH_4^+ into carbon skeletons. Ammonium can be formed from oxidized inorganic percursors by reductive reactions: **nitrogen fixation** reduces N_2 to NH_4^+; **nitrate assimilation** reduces NO_3^- to NH_4^+. Nitrifying bacteria can oxidize NH_4^+ back to NO_3^- and obtain energy for growth in the process of **nitrification.** **Denitrification** is a form of bacterial respiration whereby nitrogen oxides serve as electron acceptors in the place of O_2 under anaerobic conditions.

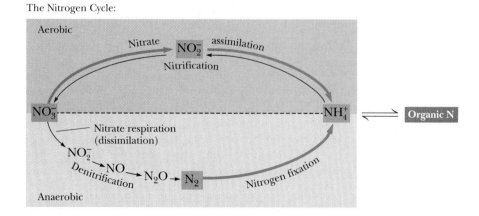

The Nitrogen Cycle:

nitrogen fixation or nitrate assimilation, they are totally dependent on plants and microorganisms for the synthesis of organic nitrogenous compounds, such as amino acids and proteins, to satisfy their requirements for this essential element.

Animals release excess nitrogen in a reduced form, either as NH_4^+ or as organic nitrogenous compounds such as urea. The release of N occurs both during life and as a consequence of microbial decomposition following death. Various bacteria return the reduced forms of nitrogen back to the environment by oxidizing them. The oxidation of NH_4^+ to NO_3^- by **nitrifying bacteria,** a group of chemoautotrophs, provides the sole source of chemical energy for the life of these microbes. Nitrate nitrogen also returns to the atmosphere as N_2 as a result of the metabolic activity of **denitrifying bacteria.** These bacteria are capable of using NO_3^- and similar oxidized inorganic forms of nitrogen as electron acceptors in place of O_2 in energy-producing pathways. The NO_3^- is reduced ultimately to dinitrogen (N_2). Such bacterial activity is being exploited in water treatment plants to reduce the load of combined nitrogen that might otherwise enter lakes, streams, and bays.

21.1 Ammonium Incorporation into Metabolic Intermediates

Ammonium enters organic linkage via three major reactions that are widely distributed throughout nature. The enzymes mediating these reactions are (1) carbamoyl-phosphate synthetase I, (2) glutamate dehydrogenase, and (3) glutamine synthetase.

Carbamoyl-phosphate synthetase I catalyzes one of the steps in the urea cycle (see also Figure 21.11). Two ATPs are consumed, one in the activation of HCO_3^- for reaction with ammonium, and the other in the phosphorylation of the carbamate formed:

$$NH_4^+ + HCO_3^- + 2\,ATP \longrightarrow H_2NC \overset{\displaystyle O}{\overset{\|}{}} - O - PO_3^{2-} + 2\,ADP + P_i + 2\,H^+$$

N-acetylglutamate is an essential allosteric activator for this enzyme.

Glutamate dehydrogenase (GDH) catalyzes the reductive amination of α-ketoglutarate to yield glutamate. Reduced pyridine nucleotides (NADH or NADPH) provide the reducing power:

$$NH_4^+ + \alpha\text{-ketoglutarate} + NADPH + H^+ \longrightarrow glutamate + NADP^+ + H_2O$$

This reaction (Figure 21.2) provides an important interface between nitrogen metabolism and cellular pathways of carbon and energy metabolism because α-ketoglutarate is a citric acid cycle intermediate.

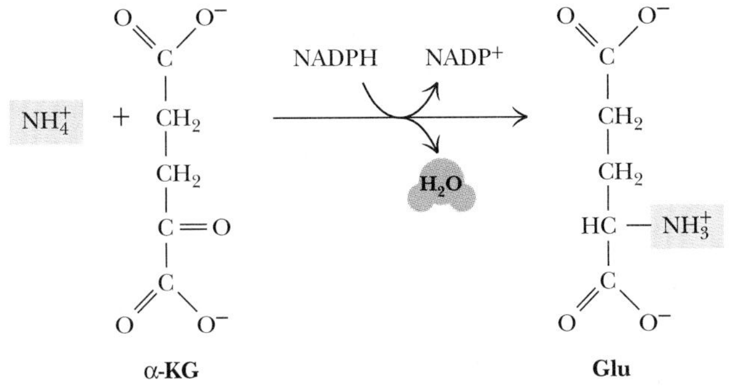

Figure 21.2 The glutamate dehydrogenase reaction.

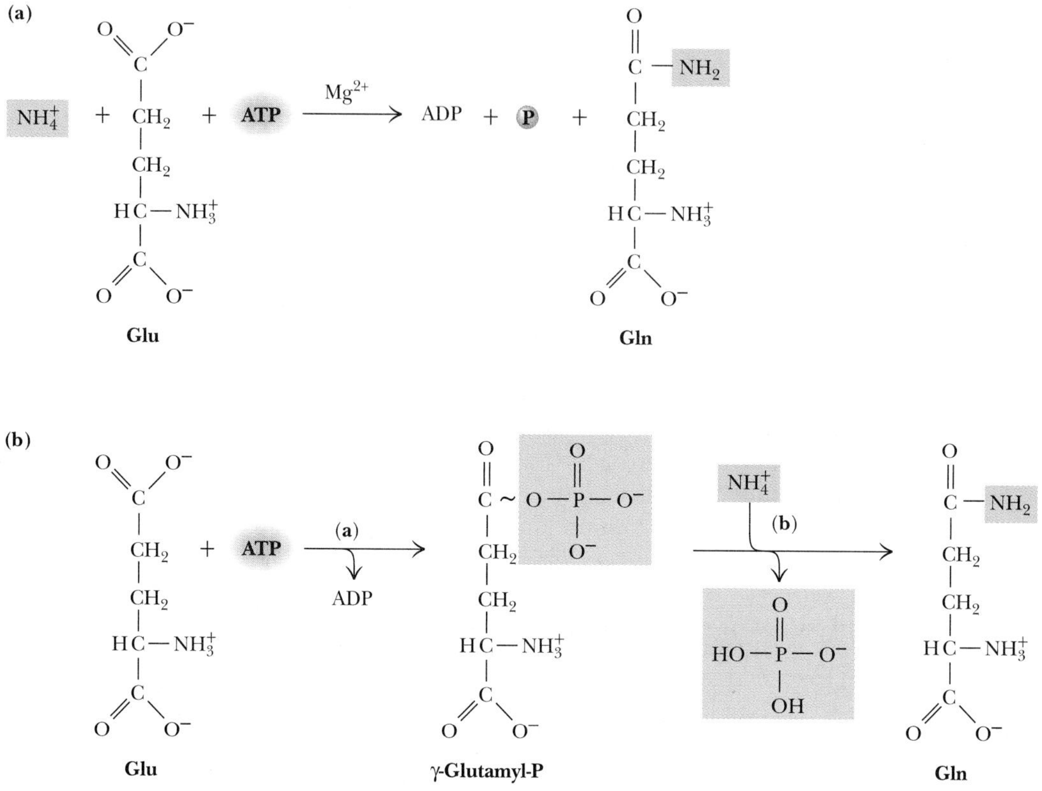

Figure 21.3 **(a)** The overall reaction catalyzed by glutamine synthetase. **(b)** (a) Activation of the γ-carboxyl group of Glu by ATP precedes (b) amidation by NH_4^+.

Glutamine synthetase (GS) catalyzes the ATP-dependent amidation of the γ-carboxyl group of glutamate to form glutamine. The reaction proceeds via a γ-glutamyl-phosphate intermediate, and GS activity depends on the presence of divalent cations such as Mg^{2+} (Figure 21.3). **Glutamine** is a major N donor in the biosynthesis of many organic N compounds, such as purines, pyrimidines, and other amino acids; in many cells, GS activity is tightly regulated. The amide-N of glutamine provides the nitrogen atom in these biosyntheses. In quantitative terms, GDH and GS are responsible for most of the ammonium assimilated into organic compounds.

21.2 ***Escherichia coli* Glutamine Synthetase: A Case Study in Enzyme Regulation**

As indicated, glutamine plays a pivotal role in nitrogen metabolism by donating its amide nitrogen to the biosynthesis of many important organic N compounds. Consistent with its metabolic importance, GS (Figure 21.4) in enteric bacteria such as *E. coli* and *S. typhimurium* is regulated allosterically by feedback inhibition, and GS is interconverted between active and inactive forms by covalent modification. Eukaryotic versions of glutamine synthetase are not so intricately regulated.

Allosteric Regulation of GS

Nine distinct feedback inhibitors (Gly, Ala, Ser, His, Trp, CTP, AMP, carbamoyl-P, and glucosamine-6-P) act on GS. Gly, Ala, and Ser are key indicators of amino acid metabolism in the cell; each of the other six compounds represents an end product of a biosynthetic pathway dependent on Gln (Figure 21.5). AMP competes with ATP for binding at the ATP substrate site. Gly, Ala, and Ser compete with Glu for binding at the active site. Carbamoyl-P binds at a site that overlaps both the Glu site and the site occupied by the γ-PO_4 of ATP.

Covalent Modification of GS

Each GS subunit can be adenylylated at a specific tyrosine residue (Tyr^{397}) in an ATP-dependent reaction (Figure 21.6). Adenylylation inactivates GS. If n is the average number of adenylyl groups per GS molecule, GS activity is inversely proportional to n. The number n varies from 0 (no adenylyl groups) to 12 (every subunit in each GS molecule is adenylylated). Adenylylation of GS is catalyzed by the converter enzyme **ATP : GS : adenylyl transferase,** or simply *adenylyl transferase* **(AT).** However, whether or not this covalent modification occurs is determined by a highly regulated cycle (Figure 21.7). AT not only catalyzes adenylylation of GS, it also catalyzes **deadenylylation**—the P_i-coupled removal of the Tyr-linked adenylyl groups as ADP. The direction in which AT operates depends on the nature of a regulatory protein associated with it called P_{II}. P_{II} is a 44-kD protein (tetramer of 11-kD subunits). If P_{II} is in its so-called P_{IIA} form, the AT : P_{IIA} complex acts to adenylylate GS. When P_{II} is in its so-called P_{IID} form, the AT : P_{IID} complex catalyzes the deadenylylation of GS. The active sites of AT : P_{IIA} and AT : P_{IID} are different, consistent with the difference in their catalytic roles. In addition, the AT : P_{IIA} and AT : P_{IID} complexes

(a)   (b)

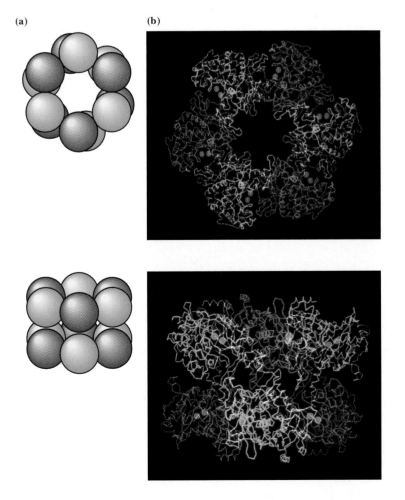

Figure 21.4 **(a)** The subunit organization of bacterial glutamine synthetase. Bacterial GS is a 600-kD dodecamer (α_{12}-type subunit organization) of identical 52-kD monomers (each monomer contains 468 amino acid residues). These monomers are arranged as a stack of two hexagons. The active sites are located at subunit interfaces within the hexagons; these active sites are recognizable in the X-ray crystallographic structure **(b)** by the pair of divalent cations that occupy them. Adjacent subunits contribute to each active site, thus accounting for the fact that GS monomers are catalytically inactive. The crystal structure of glutamine synthetase shown is that for GS from *Salmonella typhimurium* (a close relative of *E. coli*). (*From Almassy, R. J., Janson, C. A., Hamlin, R., Xuong, N.-H., and Eisenberg, D., 1986. Nature* **323:**304. *Photos courtesy of S.-H. Liaw and D. Eisenberg.*)

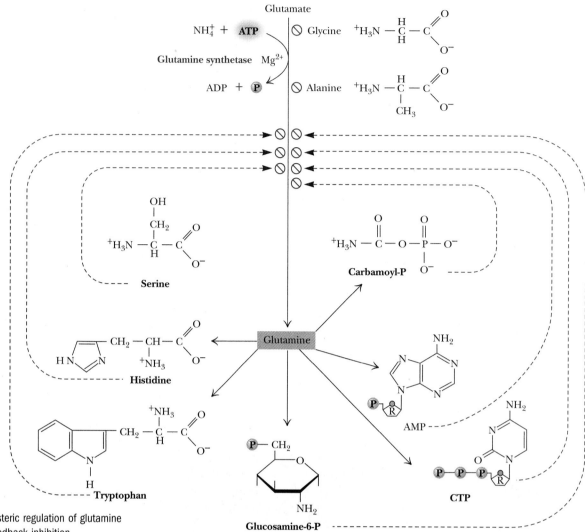

Figure 21.5 The allosteric regulation of glutamine synthetase activity by feedback inhibition.

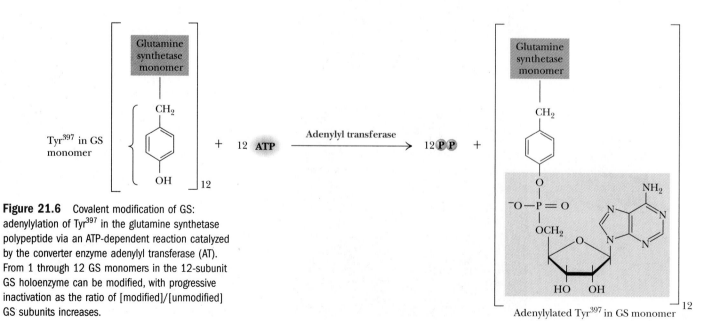

Figure 21.6 Covalent modification of GS: adenylylation of Tyr^{397} in the glutamine synthetase polypeptide via an ATP-dependent reaction catalyzed by the converter enzyme adenylyl transferase (AT). From 1 through 12 GS monomers in the 12-subunit GS holoenzyme can be modified, with progressive inactivation as the ratio of [modified]/[unmodified] GS subunits increases.

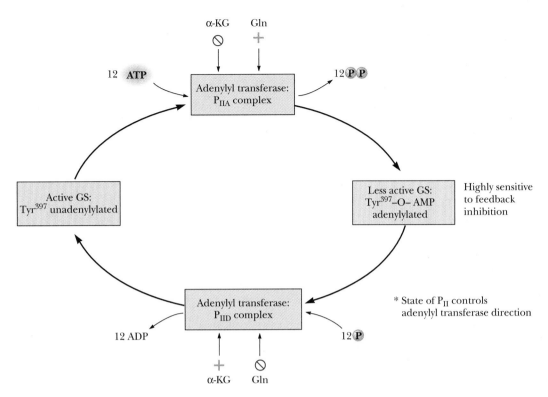

Figure 21.7 The cyclic cascade system regulating the covalent modification of GS.

are allosterically regulated in a reciprocal fashion by the effectors α-KG and Gln. Gln stimulates AT : P_{IIA} activity and inhibits AT : P_{IID} activity; the effect of α-KG on the activities of these two complexes is diametrically opposite (see Figure 21.7).

Clearly, the determining factor regarding the degree of adenylylation, n, and hence the relative activity of GS, is the $[\text{Gln}]/[\alpha\text{-KG}]$ ratio. A high $[\text{Gln}]$ level signals cellular nitrogen sufficiency, and GS becomes adenylylated and inactivated. In contrast, a high $[\alpha\text{-KG}]$ level is an indication of nitrogen limitation and a need for ammonium fixation by GS.

21.3 Amino Acid Biosynthesis

Organisms show substantial differences in their capacity to synthesize the 20 amino acids common to proteins. Typically, plants and microorganisms can form all of their nitrogenous metabolites, including all of the amino acids, from inorganic forms of N such as NH_4^+ and NO_3^-. In these organisms, the α-amino group for all amino acids is derived from glutamate, usually via transamination of the corresponding α-keto acid analog of the amino acid (Figure 21.8). In many cases, amino acid biosynthesis is thus a matter of synthesizing the appropriate α-keto acid carbon skeleton, followed by transamination with Glu. The amino acids can be classified according to the source of intermediates for the α-keto acid biosynthesis (Table 21.1). For example, the amino acids Glu, Gln, Pro, and Arg (and, in some instances, Lys) are all members of the α-ketoglutarate family because they are all derived from the citric acid cycle intermediate, α-ketoglutarate.

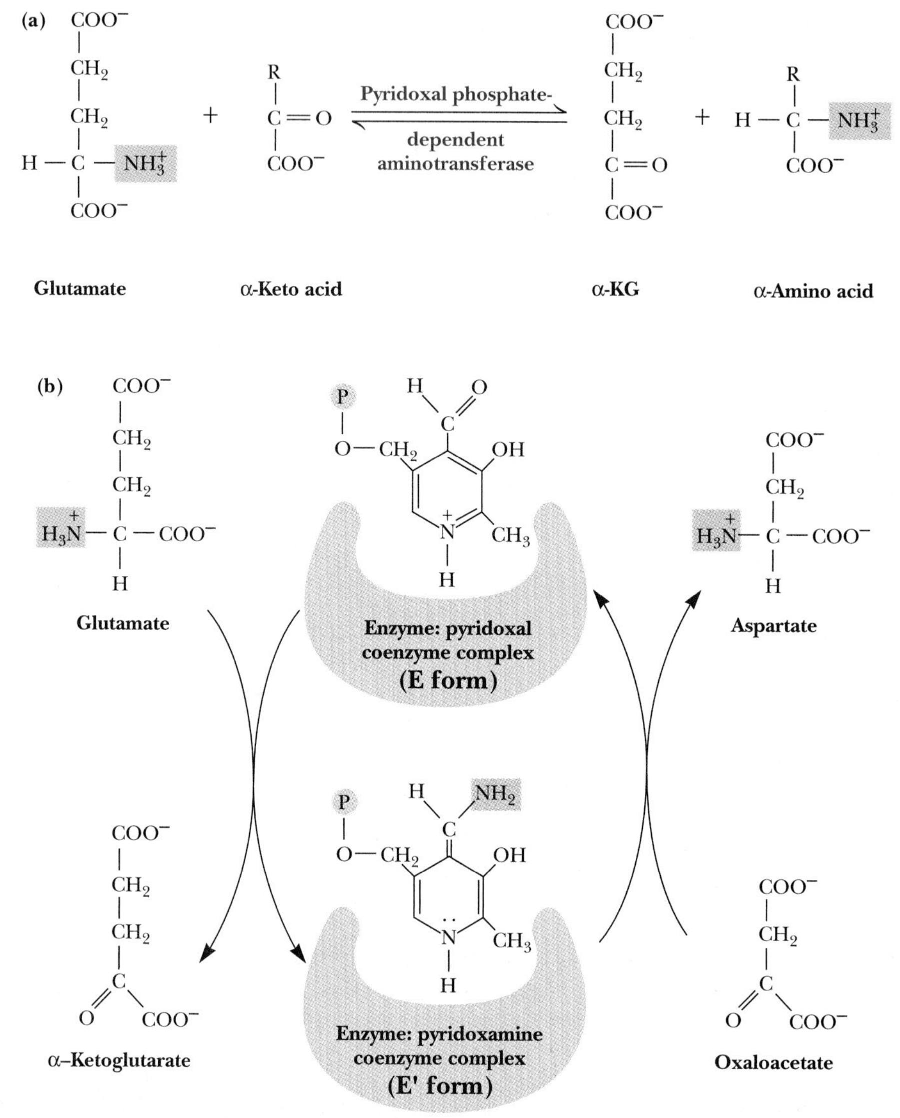

Figure 21.8 Glutamate-dependent transamination of α-keto acid carbon skeletons is a primary mechanism for amino acid synthesis. **(a)** The generic transamination–aminotransferase reaction involves transfer of the α-amino group of glutamate to an α-keto acid acceptor. **(b)** The transamination of oxaloacetate by glutamate to yield aspartate and α-ketoglutarate is a prime example. *Glutamate : aspartate aminotransferase* is a pyridoxal phosphate–dependent enzyme. Enzyme-bound pyridoxal serves as the —NH_2 acceptor from glutamate to form pyridoxamine. Pyridoxamine is then the amino donor to oxaloacetate to form aspartate and regenerate the pyridoxal coenzyme form. This aminotransferase mechanism is a classic example of a double-displacement (ping-pong) bisubstrate kinetic mechanism (see Chapter 10). In this mechanism, the pyridoxamine : enzyme form is the covalently modified (E') form of the enzyme.

Transamination

Transamination involves transfer of an α-amino group from an amino acid to the α-keto position of an α-keto acid (see Figure 21.8). In the process, the amino donor becomes an α-keto acid while the α-keto acid acceptor becomes an α-amino acid:

$$\text{Amino acid}_1 + \alpha\text{-keto acid}_2 \longrightarrow \alpha\text{-keto acid}_1 + \text{amino acid}_2$$

Table 21.1 The Grouping of Amino Acids into Families According to the Metabolic Intermediates That Serve As Their Progenitors

α-Ketoglutarate Family	Aspartate Family
Glutamate	Aspartate
Glutamine	Asparagine
Proline	Methionine
Arginine	Threonine
Lysine*	Isoleucine
	Lysine*

Pyruvate Family	3-Phosphoglycerate Family
Alanine	Serine
Valine	Glycine
Leucine	Cysteine

Phosphoenolpyruvate and Erythrose-4-P Family

The aromatic amino acids:
  Phenylalanine
  Tyrosine
  Tryptophan
The remaining amino acid, *histidine*, is derived from PRPP (5-phosphoribosyl-1-pyrophosphate) and ATP.

*Different organisms use different precursors to synthesize lysine.

The predominant amino acid/α-keto acid pair in these reactions is glutamate/α-ketoglutarate, with the net effect that glutamate is the primary amino donor for the synthesis of amino acids. Transamination reactions are catalyzed by **aminotransferases** (the preferred name for enzymes formerly termed transaminases). Aminotransferases are named according to their amino acid substrates, as in **glutamate–aspartate aminotransferase.**

HUMAN BIOCHEMISTRY

Human Dietary Requirements for Amino Acids

Humans can synthesize only 10 of the 20 common amino acids (see table); the others must be obtained in the diet. Those that can be synthesized are classified as **nonessential,** meaning it is not essential that these amino acids be part of the diet. In effect, humans can synthesize the α-keto acid analogs of nonessential amino acids and form the amino acids by transamination. In contrast, humans are incapable of constructing the carbon skeletons of **essential** amino acids, and so must rely on dietary sources for these essential metabolites. Excess dietary amino acids cannot be stored for future use, nor are they excreted unused. Instead, they are converted to common metabolic intermediates that can be either oxidized by the citric acid cycle or used to form glucose or fat (see Section 21.4).

Essential and Nonessential Amino Acids in Humans

Essential	Nonessential
Arginine*	Alanine
Histidine*	Asparagine
Isoleucine	Aspartate
Leucine	Cysteine
Lysine	Glutamate
Methionine	Glutamine
Phenylalanine	Glycine
Threonine	Proline
Tryptophan	Serine
Valine	Tyrosine[†]

*Arginine and histidine are essential in the diets of juveniles, not adults.

[†]Tyrosine is classified as nonessential only because it is readily formed from essential phenylalanine.

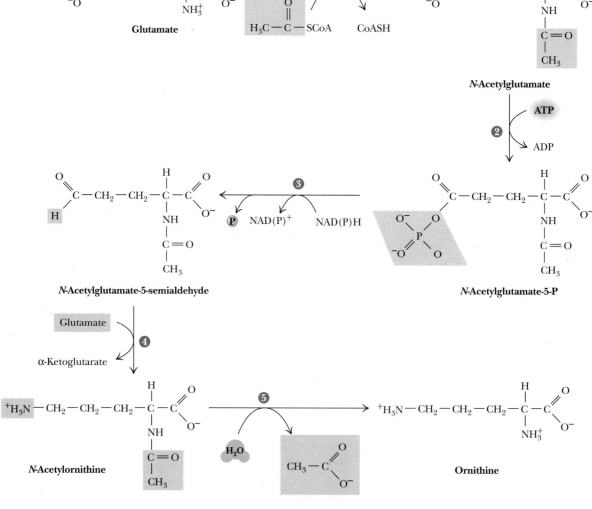

Figure 21.9 The pathway of proline biosynthesis from glutamate. The enzymes are (1) γ-glutamyl kinase, (2) glutamate-5-semialdehyde dehydrogenase, and (4) Δ¹-pyrroline-5-carboxylate reductase; reaction (3) occurs nonenzymatically.

See pages 173–178

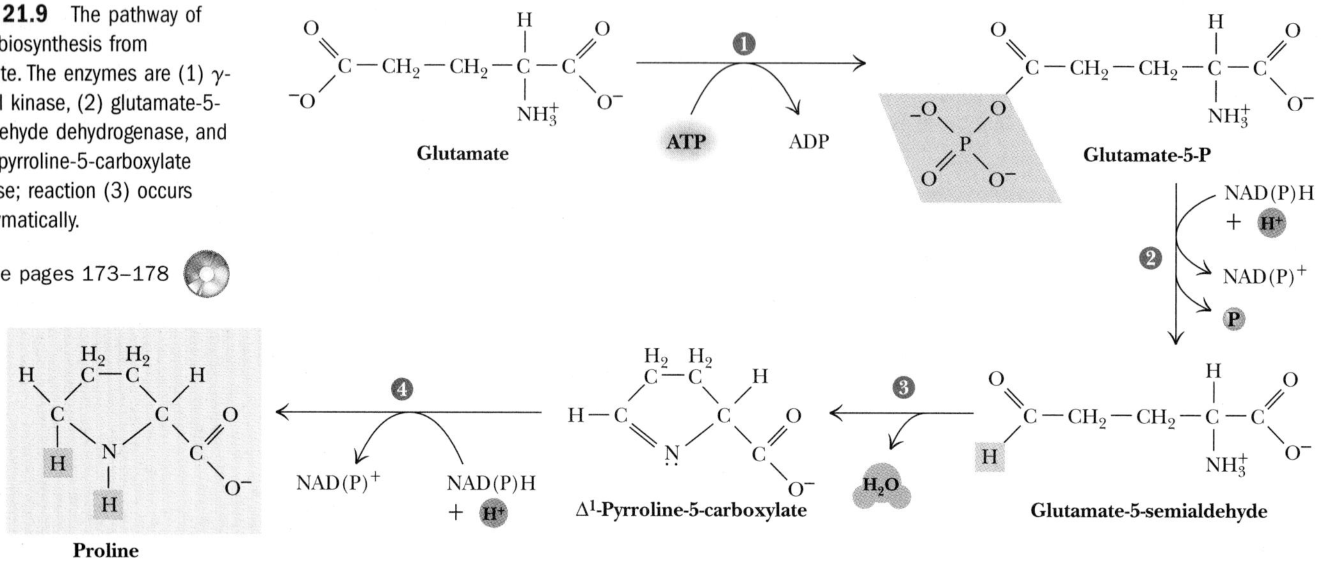

Figure 21.10 The bacterial pathway of ornithine biosynthesis from glutamate. The enzymes are (1) *N*-acetylglutamate synthase, (2) *N*-acetylglutamate kinase, (3) *N*-acetylglutamate-5-semialdehyde dehydrogenase, (4) *N*-acetylornithine δ-aminotransferase, and (5) *N*-acetylornithine deacetylase. In mammals, ornithine is synthesized directly from glutamate-5-semialdehyde by a pathway that does not involve an *N*-acetyl-blocked form of the substrate.

The Pathways for Biosynthesis of the Nonessential Amino Acids

The α-Ketoglutarate Family

Nonessential amino acids derived from α-ketoglutarate include glutamate (Glu), glutamine (Gln), proline (Pro), and, for adults, arginine (Arg). Glu and Gln synthesis are achieved by GDH and GS, the enzymes discussed earlier under pathways of ammonium assimilation (see Section 21.1).

Proline is derived from glutamate via a series of four reactions involving activation, then reduction, of the γ-carboxyl group to an aldehyde (*glutamate-5-semialdehyde*), which spontaneously cyclizes to yield the internal Schiff base, Δ^1-*pyrroline-5-carboxylate* (Figure 21.9). NADPH-dependent reduction of the pyrroline double bond gives proline.

Arginine biosynthesis involves enzymatic steps that are also part of the **urea cycle,** a metabolic pathway that functions in N excretion in certain animals. Net synthesis of arginine depends on the formation of **ornithine.** Interestingly, ornithine is synthesized from glutamate via a reaction pathway reminiscent of the proline biosynthetic pathway (Figure 21.10).

Ornithine has three metabolic roles: (1) to serve as a precursor to arginine, (2) to function as an intermediate in the urea cycle, and (3) to act as an intermediate in Arg degradation. In any case, the δ-NH_3^+ of ornithine is carbamoylated in a reaction catalyzed by **ornithine transcarbamoylase.** The carbamoyl group is derived from carbamoyl-P synthesized by **carbamoyl phosphate synthetase** I (*CPS-I*). CPS-I is the mitochondrial CPS isozyme; it uses two ATPs in catalyzing the formation of carbamoyl-P from NH_3 and HCO_3^- (Figure 21.11). CPS-I represents the committed step in the urea cycle, and CPS-I is allosterically activated by *N*-acetylglutamate. Because *N*-acetylglutamate is

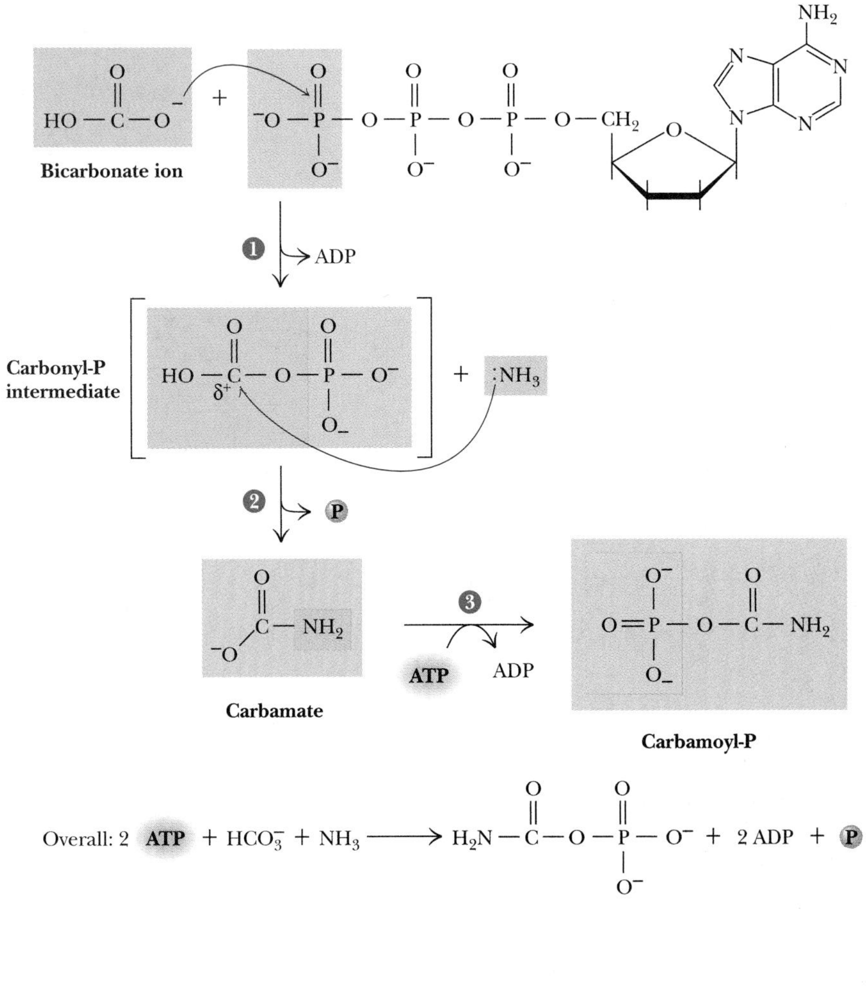

Figure 21.11 The mechanism of action of CPS-I, the NH_3-dependent mitochondrial CPS isozyme. (1) HCO_3^- is activated via an ATP-dependent phosphorylation. (2) Ammonia attacks the carbonyl carbon of carbonyl-P, displacing P_i to form carbamate. (3) Carbamate is phosphorylated via a second ATP to give carbamoyl-P.

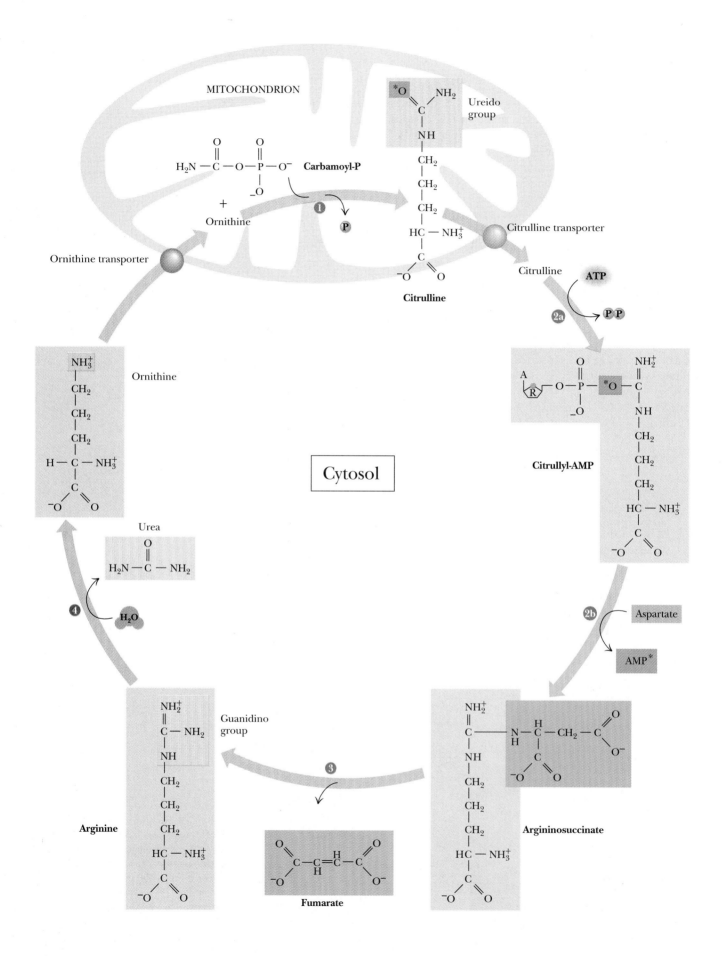

◀ **Figure 21.12** The urea cycle series of reactions: Transfer of the carbamoyl group of carbamoyl-P to ornithine by *ornithine transcarbamoylase* (*OTCase*, reaction 1) yields citrulline. The citrulline ureido group is then activated by reaction with ATP to give a citrullyl-AMP intermediate (reaction 2a); AMP is then displaced by aspartate, which is linked to the carbon framework of citrulline via its α-amino group (reaction 2b). The course of reaction 2 was verified using ^{18}O-labeled citrulline. The ^{18}O label (indicated by the asterisk,*) was recovered in AMP. Citrulline and AMP are joined via the ureido *O atom. The product of this reaction is argininosuccinate; the enzyme catalyzing the two steps of reaction 2 is *argininosuccinate synthetase.* The next step (reaction 3) is carried out by *argininosuccinase,* which catalyzes the nonhydrolytic removal of fumarate from argininosuccinate to give arginine. Hydrolysis of Arg by *arginase* (reaction 4) yields urea and ornithine, completing the urea cycle.

 See pages 173–178

both a precursor to ornithine synthesis and essential to the operation of the urea cycle, it serves to coordinate these related pathways.

The product of the ornithine transcarbamoylase reaction is *citrulline* (Figure 21.12). Ornithine and citrulline are two α-amino acids of metabolic importance that nevertheless are *not* among the 20 α-amino acids commonly found in proteins. Like CPS-I, ornithine transcarbamoylase is a mitochondrial enzyme. The reactions of ornithine synthesis and the rest of the urea cycle enzymes occur in the cytosol.

The pertinent feature of the citrulline side chain is the **ureido group.** In a complex reaction catalyzed by **argininosuccinate synthetase,** this ureido group is first activated by ATP to yield a citrullyl-AMP derivative, followed by displacement of AMP by aspartate to give argininosuccinate (see Figure 21.12). The formation of arginine is then accomplished by **argininosuccinase,** which catalyzes the nonhydrolytic elimination of fumarate from argininosuccinate. This reaction completes the biosynthesis of Arg.

The Urea Cycle—Excretion of Excess N Through Arg Breakdown

The carbon skeleton of arginine is derived principally from α-ketoglutarate, but the N and C atoms composing the **guanidino group** (see Figure 21.12) of the Arg side chain come from NH_4^+, HCO_3^- (as carbamoyl-P), and the α-NH_3^+ groups of glutamate and aspartate. The circle of the urea cycle is closed when ornithine is regenerated from Arg by the **arginase**-catalyzed hydrolysis of arginine. Urea is the other product of this reaction and lends its name to the cycle. In humans, urea synthesis is required to excrete excess nitrogen generated by increased amino acid catabolism, such as occurs following dietary consumption of more than adequate amounts of protein. Urea formation is basically confined to the liver. Increases in amino acid catabolism lead to elevated glutamate levels and a rise in *N*-acetylglutamate, the allosteric activator of CPS-I. Stimulation of CPS-I raises overall urea cycle activity because activities of the remaining enzymes of the cycle simply respond to increased substrate availability. Removal of NH_3, which is toxic, by CPS-I is an important aspect of this regulation. The urea cycle is linked to the citric acid cycle through fumarate, a by-product of the action of argininosuccinase (see Figure 21.12, reaction 3).

The Aspartate Family

The only members of the aspartate family of amino acids that are not essential are aspartate (Asp) and asparagine (Asn).

Aspartate is formed from the citric acid cycle intermediate, oxaloacetate, by transfer of an amino group from glutamate via an aminotransferase reaction (Figure 21.13). Like glutamate synthesis from α-ketoglutarate, aspartate synthesis is a drain on the citric acid cycle. As we already saw, the Asp amino group serves as the N donor in the conversion of citrulline to arginine. Later

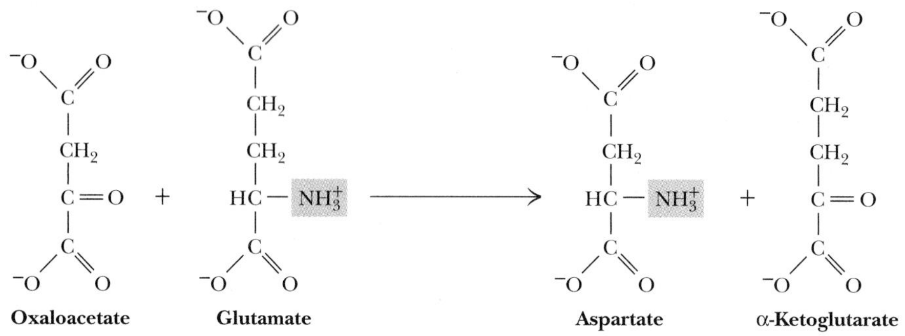

Figure 21.13 Aspartate biosynthesis via transamination of oxaloacetate by glutamate. The enzyme responsible is PLP-dependent glutamate : aspartate aminotransferase.

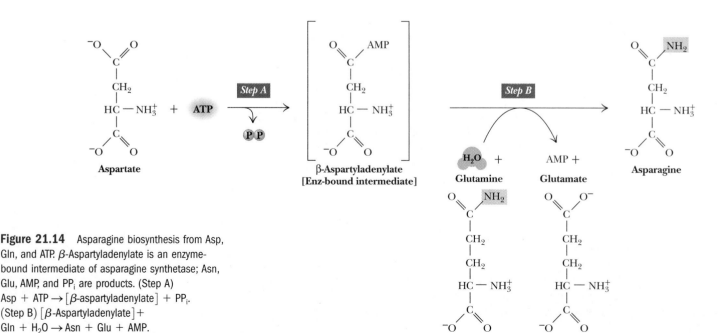

in this chapter we shall see that this $-NH_3^+$ is also the source not only of one of the N atoms of the purine ring system during nucleotide biosynthesis, but also of the 6-amino group of the major purine, adenine. In addition, the entire aspartate molecule is used in the biosynthesis of pyrimidine nucleotides.

Asparagine is formed by amidation of the β-carboxyl group of aspartate. Analogous to glutamine synthesis, the nitrogen added in this amidation in bacteria comes directly from NH_4^+. In higher organisms, **asparagine synthetase** catalyzes the ATP-dependent transfer of the amido-N of glutamine to aspartate to yield glutamate and asparagine (Figure 21.14).

The remaining members of the aspartate family, **threonine, methionine,** and **lysine,** are formed from aspartate but are essential amino acids because humans do not have the ability to synthesize the enzymes of these pathways. Humans do form **homocysteine** by demethylation of methionine, and from homocysteine, they form cysteine (see Figure 21.17).

The Pyruvate Family

The pyruvate family of amino acids includes alanine (Ala), valine (Val), and leucine (Leu). Of these three, only Ala is nonessential. Transamination of pyruvate, with glutamate as amino donor, gives **alanine.** Because these transamination reactions are readily reversible, alanine degradation occurs via the reverse route, with α-ketoglutarate serving as amino acceptor.

Figure 21.14 Asparagine biosynthesis from Asp, Gln, and ATP. β-Aspartyladenylate is an enzyme-bound intermediate of asparagine synthetase; Asn, Glu, AMP, and PP_i are products. (Step A) Asp + ATP → [β-aspartyladenylate] + PP_i. (Step B) [β-Aspartyladenylate] + Gln + H_2O → Asn + Glu + AMP.

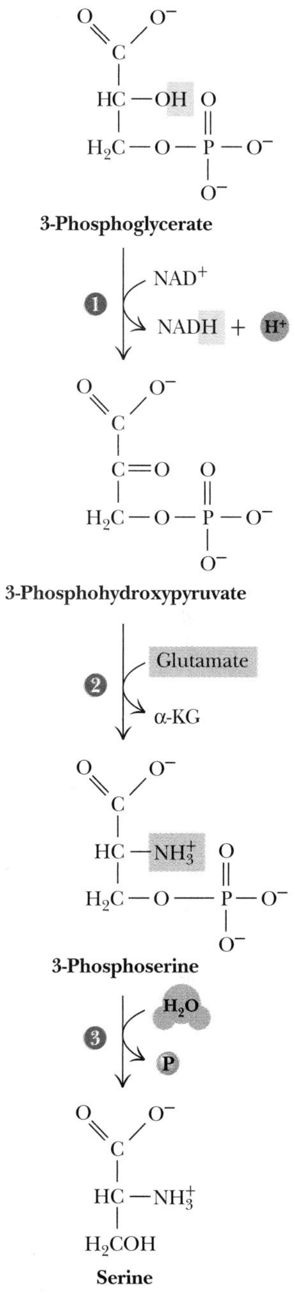

Figure 21.15 Biosynthesis of serine from 3-phosphoglycerate. The enzymes are (1) 3-phosphoglycerate dehydrogenase, (2) 3-phosphoserine aminotransferase, and (3) phosphoserine phosphatase.

 See pages 173–178

The 3-Phosphoglycerate Family

Serine, glycine, and cysteine are derived from the glycolytic intermediate 3-phosphoglycerate (3-PG). The diversion of 3-PG from glycolysis is achieved via **3-phosphoglycerate dehydrogenase** (Figure 21.15). This NAD^+-dependent oxidation of 3-PG yields *3-phosphohydroxypyruvate*—which, as an α-keto acid, is a substrate for transamination by glutamate to give *3-phosphoserine*. **Serine phosphatase** then generates **serine**. Serine inhibits the first enzyme, 3-PG dehydrogenase, and thereby feedback-regulates its own synthesis.

Glycine is made from serine via two related enzymatic processes. In the first, **serine hydroxymethyltransferase,** a PLP-dependent enzyme, catalyzes the transfer of the serine β-carbon to tetrahydrofolate (THF), the principal agent of one-carbon metabolism (Figure 21.16a; see also Table 14.6 and *A Deeper Look,* page 681). Glycine and N^5, N^{10}-methylene-THF are the products. In addition, glycine can be synthesized by a reversal of the **glycine oxidase** reaction (see Figure 21.16b). Here, glycine is formed when N^5, N^{10}-methylene-THF condenses with NH_4^+ and CO_2. Via this route, the β-carbon of serine becomes part of glycine. The conversion of serine to glycine is a prominent means of generating one-carbon derivatives of THF, which are so important for the biosynthesis of purines (see Section 21.6) and the C-5 methyl group of thymine (a pyrimidine; see Section 21.12), as well as the amino acid methionine. Glycine itself contributes to both purine and heme synthesis.

Cysteine Synthesis. In humans, cysteine is a nonessential amino acid only because it can be made from essential methionine. Cysteine formation involves demethylation of Met to yield homocysteine (see Figure 21.18). The —SH group of homocysteine is then used to replace the —OH group on serine, giving cysteine (Figure 21.17).

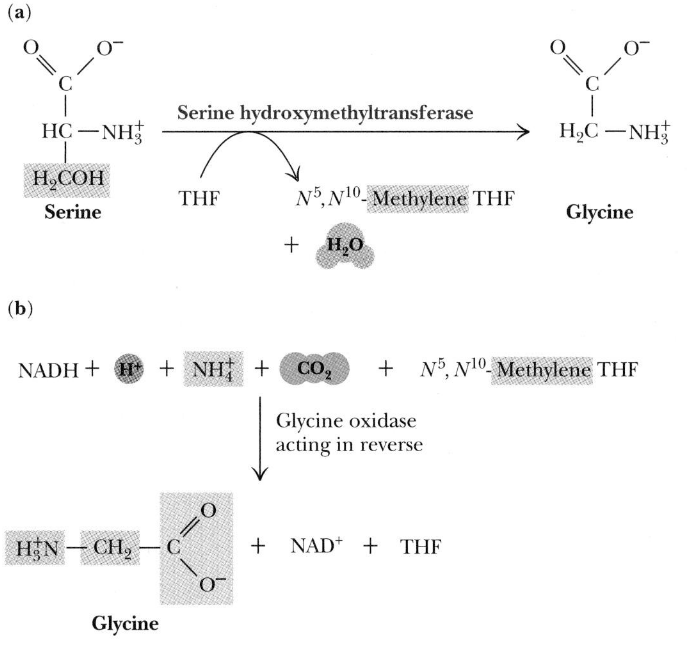

Figure 21.16 Biosynthesis of glycine from serine via **(a)** serine hydroxymethyltransferase and **(b)** glycine oxidase.

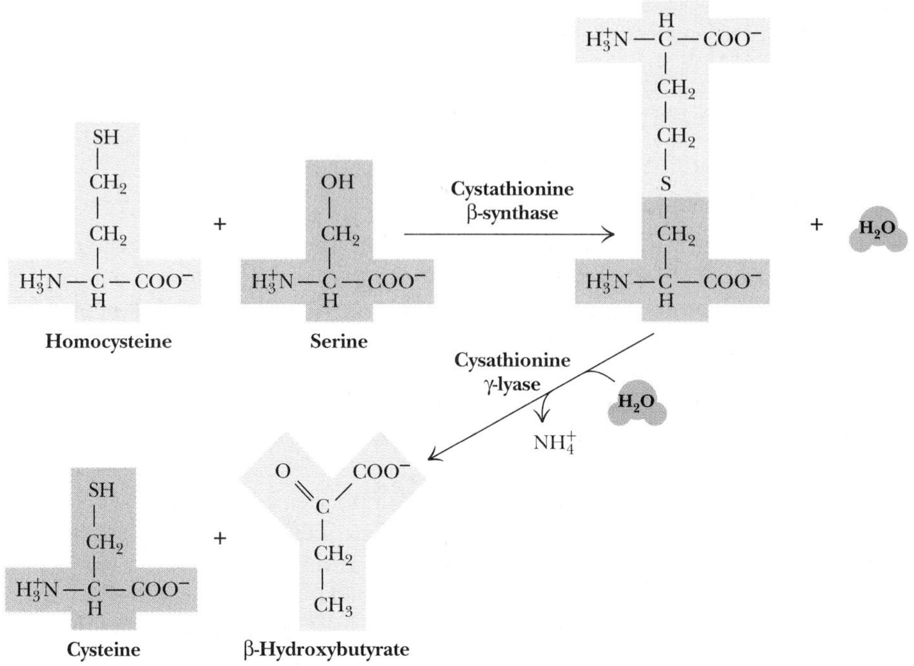

Figure 21.17 Homocysteine (formed by demethylation of methionine [see Figure 21.18]) is combined with serine to form cystathionine by *cystathionine β-synthase*. Hydrolysis of cystathionine by *cystathionine γ-lyase* gives cysteine and β-hydroxbutyrate. Both of these enzymes are pyridoxal phosphate–dependent.

See pages 173–178

Methionine, the "Other" Sulfur-Containing Amino Acid, Is an Important Methyl-Group Donor

Although it is an essential amino acid, methionine deserves mention here because of its role as a methyl-group donor. To serve as a methyl donor, methionine must first be converted to **S-adenosylmethionine (SAM)** by *S-adenosylmethionine synthase,* the adenosyl group coming from ATP (Figure 21.18). SAM then acts as the methyl-group donor in many methylation reactions, such as formation of phosphatidylcholine from phosphatidylethanolamine (see Figure 6.5). Methyl transfer results in formation of S-adenosylhomocysteine, which

HUMAN BIOCHEMISTRY

Homocysteine and Heart Attacks

A rare inherited disease known as homocysteinuria results in very high levels of homocysteine in the bloodstream. Children born with this disease seldom survive to be teenagers and die of such cardiovascular problems as stroke and arteriosclerosis (hardening of the arteries), diseases usually associated with old age. Further, studies indicate that adults with elevated levels of homocysteine in their blood are at higher risk for heart attack and stroke.

Damage to blood vessels by homocysteine is the basis of the disease. As early as 1969, Dr. Kilmer McCully, a Harvard-educated

physician interested in homocysteinuria, suggested that homocysteine might cause heart disease, and that many people may have high plasma levels of homocysteine because their diets are deficient in folic acid. His work went unheeded for 25 years. Fortunately, supplementing the amounts of folic acid (a B vitamin) in the diet reduces blood concentrations of homocysteine to a safe level, presumably by enhancing the conversion of homocysteine to methionine by homocysteine methyltransferase (methionine synthase) (see Figure 21.18).

is then hydrolyzed to yield adenosine and L-**homocysteine.** Alternatively, SAM can be decarboxylated by S-adenosylmethionine decarboxylase, forming "decarboxylated SAM," a propylamine donor in the biosynthesis of certain **polyamines** (Figure 21.18), cationic compounds found in association with nucleic acids in cells.

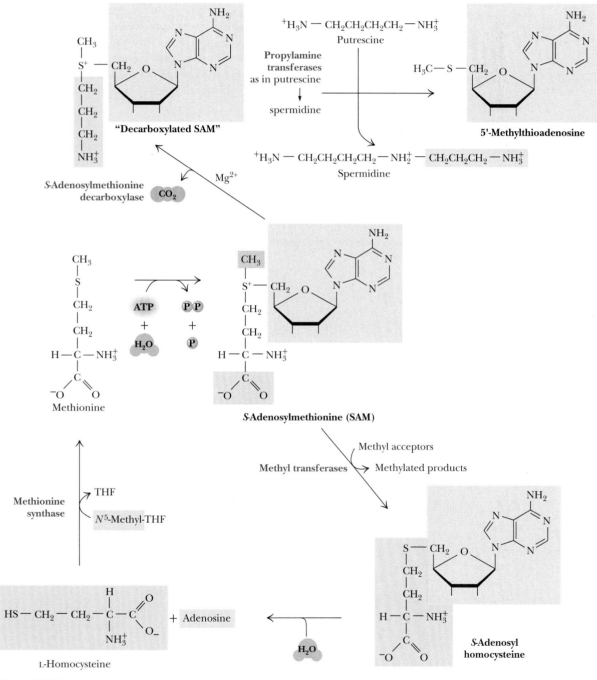

Figure 21.18 The synthesis of S-adenosylmethionine (SAM) from methionine plus ATP, and the role of SAM as a substrate in methyl donor reactions and in propylamine transfer reactions, as in spermidine synthesis. Methionine can be recovered from homocysteine in a remethylation reaction catalyzed by homocysteine methyltransferase (also known as methionine synthase) that uses N^5-methyl-THF as the —CH_3 donor. The human form of methionine synthase is a vitamin B_{12}-dependent enzyme.

 See pages 173–178

Biosynthesis of Phenylalanine and Tyrosine

Phenylalanine is an essential amino acid for humans, and tyrosine is classified as nonessential only because it can be made from Phe obtained in the diet. In organisms that can synthesize them, these two amino acids (as well as tryptophan, the other aromatic amino acid) are derived from a shared pathway that has **chorismic acid** (Figure 21.19) as a key intermediate. Indeed, chorismate is common to the synthesis of many cellular compounds having benzene rings, including these amino acids, the fat-soluble vitamins E and K, folic acid, and coenzyme Q and plastoquinone (the two quinones necessary to electron transport during respiration and photosynthesis, respectively). Thus, the synthesis of chorismate, phenylalanine, and tyrosine are worthy of mention here. **Lignin,** a polymer of nine-carbon aromatic units, is also a derivative of chorismate. Lignin and related compounds can account for as much as 35% of the dry weight of higher plants; clearly, enormous amounts of carbon pass through the chorismate biosynthetic pathway.

The Shikimate Pathway. Chorismate biosynthesis occurs via the **shikimate pathway** (Figure 21.20). Chorismate is made from two equivalents of *phosphoenolpyruvate* (PEP) and one of *erythrose-4-phosphate* (E-4-P). To begin, one PEP and E-4-P are linked to form *3-deoxy-D-arabino-heptulosonate-7-phosphate* (DAHP) by **DAHP synthase.** Although this reaction is remote from the ultimate aromatic amino acid end products, it is an important point for regulation of aromatic amino acid biosynthesis. In the next step on the way to chorismate, DAHP is cyclized to form a six-membered saturated ring compound, *3-dehydroquinate.*

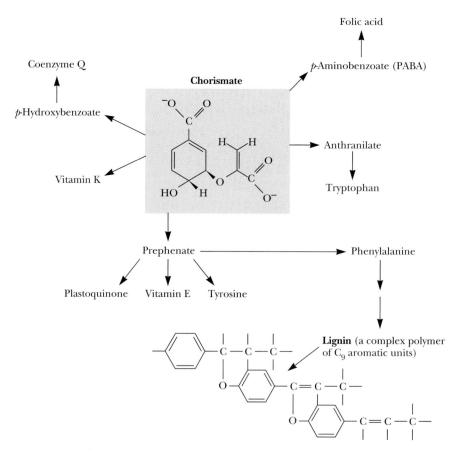

Figure 21.19 Some of the aromatic compounds derived from chorismate.

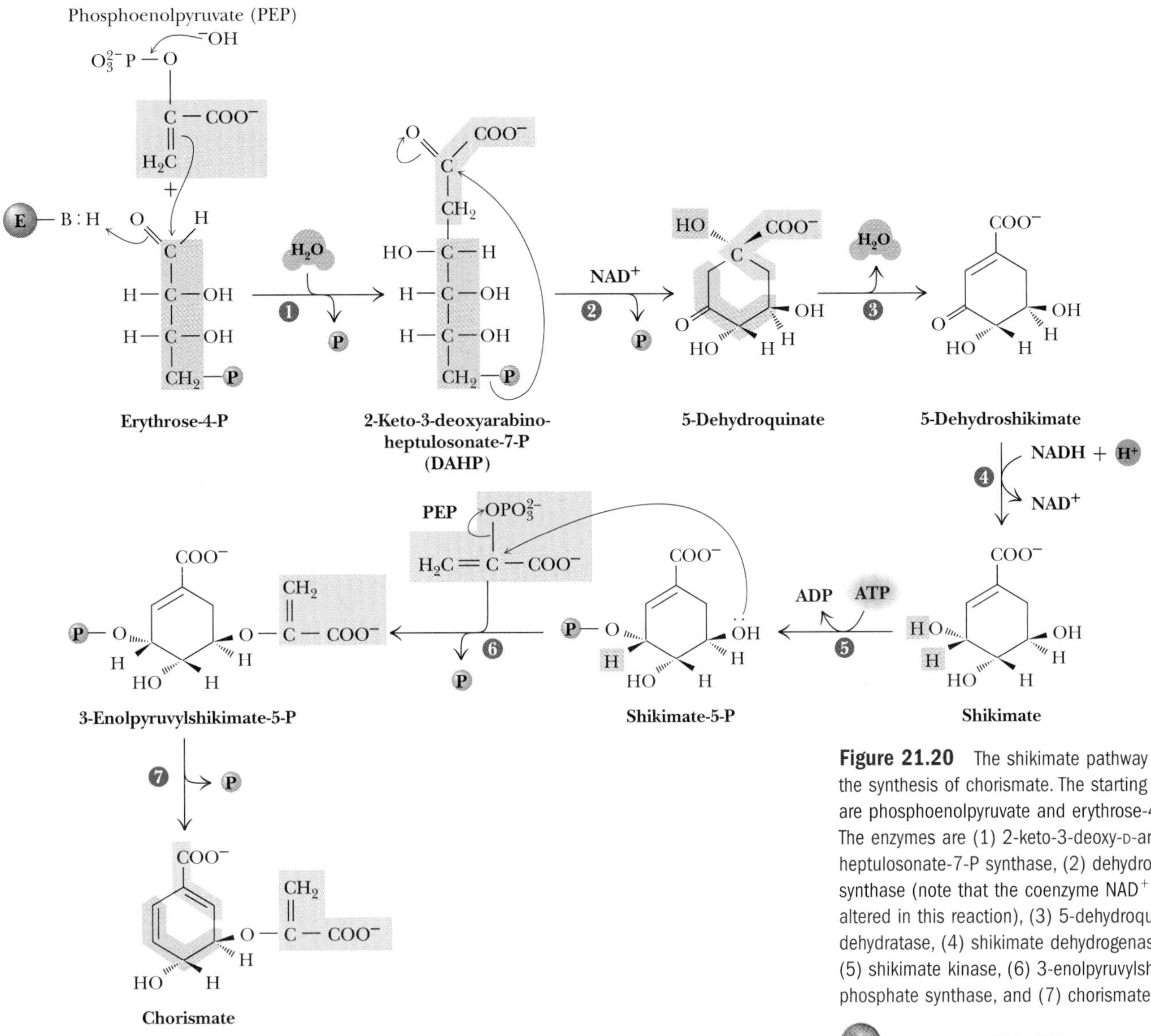

Figure 21.20 The shikimate pathway leading to the synthesis of chorismate. The starting substrates are phosphoenolpyruvate and erythrose-4-phosphate. The enzymes are (1) 2-keto-3-deoxy-D-arabino-heptulosonate-7-P synthase, (2) dehydroquinate synthase (note that the coenzyme NAD$^+$ is not altered in this reaction), (3) 5-dehydroquinate dehydratase, (4) shikimate dehydrogenase, (5) shikimate kinase, (6) 3-enolpyruvylshikimate-5-phosphate synthase, and (7) chorismate synthase.

See pages 173–178

A sequence of reactions ensues that introduces unsaturations into the ring, yielding *shikimate,* then *chorismate.* Note that the side chain of chorismate is derived from the second equivalent of phosphoenolpyruvate.

Phenylalanine and Tyrosine. At chorismate, the pathway separates into three branches, each leading specifically to one of the aromatic amino acids. The branches leading to phenylalanine and tyrosine both pass through *prephenate* (Figure 21.21). **Chorismate mutase** is the first reaction leading to Phe or Tyr. In the Phe branch, the —OH group *para* to the prephenate carboxyl is removed by a **dehydratase;** in the Tyr branch, this —OH is retained and becomes the phenolic —OH of Tyr. Glutamate-dependent aminotransferases introduce the amino groups into the two α-keto acids *phenylpyruvate* and *4-hydroxy-phenylpyruvate* to give Phe and Tyr, respectively. Humans can synthesize Tyr from Phe obtained in the diet via **phenylalanine-4-monooxygenase,** using O_2 and tetrahydrobiopterin, an analog of tetrahydrofolic acid, as cosubstrates (Figure 21.22).

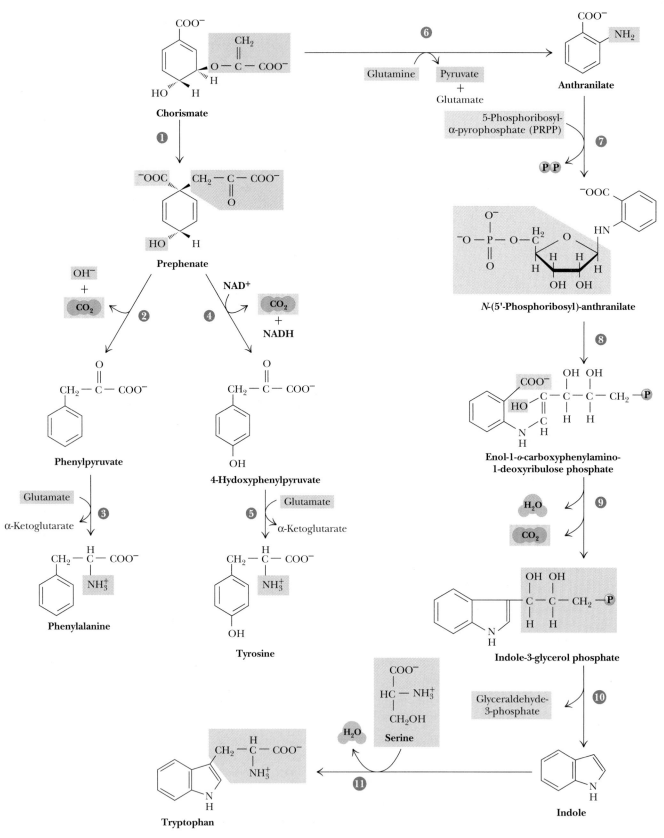

Figure 21.21 The biosynthesis of phenylalanine, tyrosine, and tryptophan from chorismate. The enzymes are (1) chorismate mutase, (2) prephenate dehydratase, (3) phenylalanine aminotransferase, (4) prephenate dehydrogenase, (5) tyrosine aminotransferase, (6) anthranilate synthase, (7) anthranilate-phosphoribosyl transferase, (8) N-(5′-phosphoribosyl)-anthranilate isomerase, (9) indole-3-glycerol phosphate synthase, (10) tryptophan synthase (α-subunit), and (11) tryptophan synthase (β-subunit).

See pages 173–178

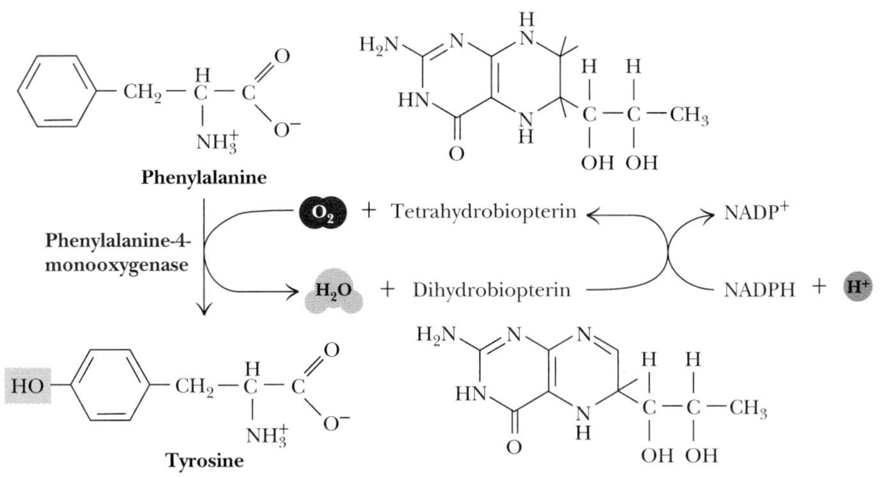

Figure 21.22 The formation of tyrosine from phenylalanine. This reaction is normally the first step in phenylalanine degradation in most organisms; in mammals, however, it provides a route for the biosynthesis of Tyr from Phe. (Phenylalanine-4-monooxygenase is also known as phenylalanine hydroxylase.)

See pages 173–178

21.4 Metabolic Degradation of Amino Acids

In normal human adults, close to 90% of the caloric energy requirement is met by oxidation of carbohydrates and fats; the remainder comes from oxidation of the carbon skeletons of amino acids. The primary physiological purpose of amino acids is to serve as the building blocks for protein biosynthesis. The dietary amount of free amino acids is trivial under most circumstances.

A DEEPER LOOK

Amino Acid Biosynthesis Inhibitors As Herbicides

Unlike animals, plants can synthesize all 20 of the common amino acids. Inhibitors acting specifically on the plant enzymes that are capable of carrying out the biosynthesis of the "essential" amino acids (i.e., enzymes that animals lack) have been developed. These substances appear to be ideal for use as herbicides because they should show no effect on animals. **Glyphosate,** sold commercially as Roundup, is a PEP analog that acts as a specific inhibitor of 3-enolpyruvylshikimate-5-P synthase (see Figure 21.20). **Sulfmeturon**

methyl, a sulfonylurea herbicide that inhibits acetohydroxy acid synthase, an enzyme common to Val, Leu, and Ile biosynthesis, is the active ingredient in the product Oust. **Aminotriazole,** sold as Amitrole, blocks His biosynthesis. **PPT (phosphinothricin)** is a potent inhibitor of glutamine synthetase. Although Gln is a nonessential amino acid and glutamine synthetase is a ubiquitous enzyme, PPT is relatively safe for animals because it does not cross the blood–brain barrier and is rapidly cleared by the kidneys.

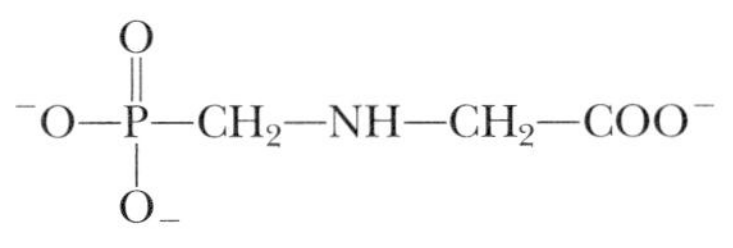

Glyphosate

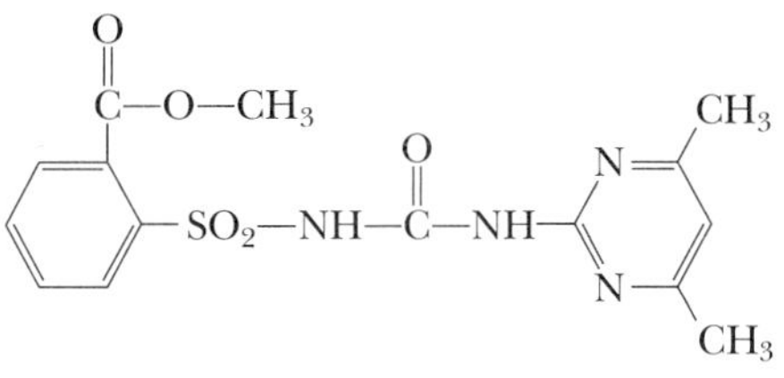

Sulfmeturon methyl

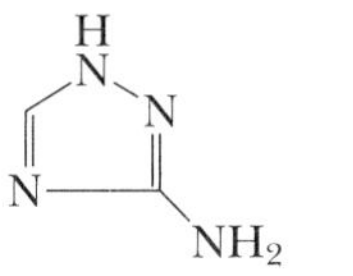

Aminotriazole

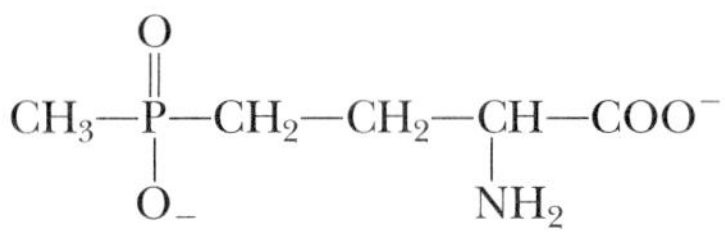

DL-Phosphinothricin (PPT)

However, if excess protein is consumed in the diet or if the amount of amino acids released during normal turnover of cellular proteins exceeds the requirements for new protein synthesis, the amino acid surplus must be catabolized. Also, if carbohydrate intake is insufficient (as during fasting or starvation) or if carbohydrates cannot be appropriately metabolized due to disease (as in diabetes mellitus), body protein becomes an important fuel for metabolic energy.

The 20 Common Amino Acids Are Degraded by 20 Different Pathways That Converge to Just 7 Metabolic Intermediates

Because the 20 common amino acids of proteins have distinctive carbon skeletons, each amino acid requires its own unique degradative pathway. And because amino acid degradation normally supplies only 10% of the body's energy, degradation of any given amino acid will satisfy, on average, less than 1% of energy needs. Therefore, we will not discuss these pathways in detail. It so happens, however, that degradation of the carbon skeletons of the 20 common α-amino acids converges to just 7 metabolic intermediates: acetyl-CoA, succinyl-CoA, pyruvate, α-ketoglutarate, fumarate, oxaloacetate, and acetoacetate (Figure 21.23). Because succinyl-CoA, pyruvate, α-ketoglutarate, fumarate, and oxaloacetate can serve as precursors for glucose synthesis, amino acids giving rise to these intermediates are termed **glucogenic.** Those degraded to yield acetyl-CoA or acetoacetate are termed **ketogenic** because these substances can be used to synthesize fatty acids or ketone bodies. Some amino acids are both glucogenic and ketogenic (see Figure 21.23).

Figure 21.13 Metabolic degradation of the common amino acids. The 20 common amino acids can be classified according to their degradation products. Those that give rise to precursors for glucose synthesis, such as α-ketoglutarate, succinyl-CoA, fumarate, oxaloacetate, and pyruvate, are termed *glucogenic* (shown in pink). Those degraded to acetyl-CoA or acetoacetate are called *ketogenic* (shown in blue) because they can be converted to fatty acids or ketone bodies. Some amino acids are both glucogenic and ketogenic.

See pages 173–178

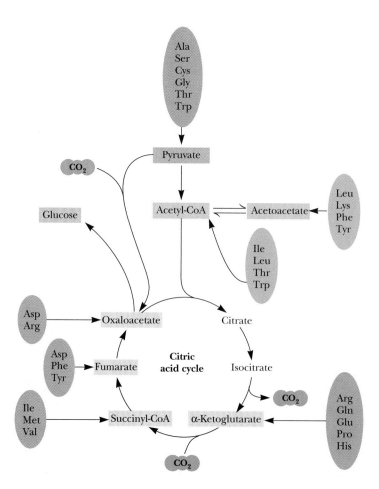

HUMAN BIOCHEMISTRY

Hereditary Defects in Phe Catabolism: Alkaptonuria and Phenylketonuria

The reactions for phenylalanine and tyrosine degradation are shown in the figure, a. **Alkaptonuria** and **phenylketonuria** are two human genetic diseases arising from specific enzyme defects in phenylalanine degradation. Alkaptonuria is characterized by urinary excretion of large amounts of *homogentisate* and results from a deficiency in **homogentisate dioxygenase.** Air oxidation of homogentisate causes urine to turn dark on standing, but the only malady suffered by carriers of this disease is a tendency toward arthritis later in life.

In contrast, **phenylketonurics,** whose urine contains excessive phenylpyruvate (see figure, b), suffer severe mental retardation if the defect is not recognized immediately after birth and treated by putting the victim on a diet low in phenylalanine. These individu-

als are deficient in phenylalanine hydroxylase, and the excess Phe that accumulates loses its $-NH_3^+$ group through an aminotransferase reaction, forming phenylpyruvate, which is then excreted.

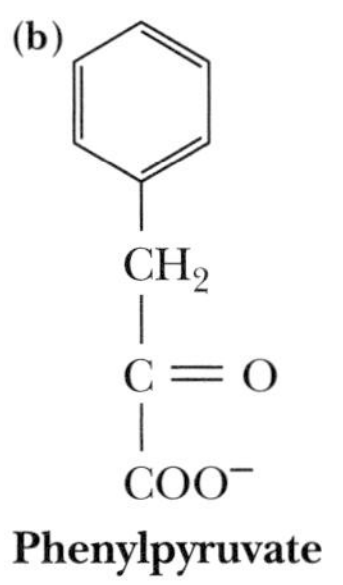

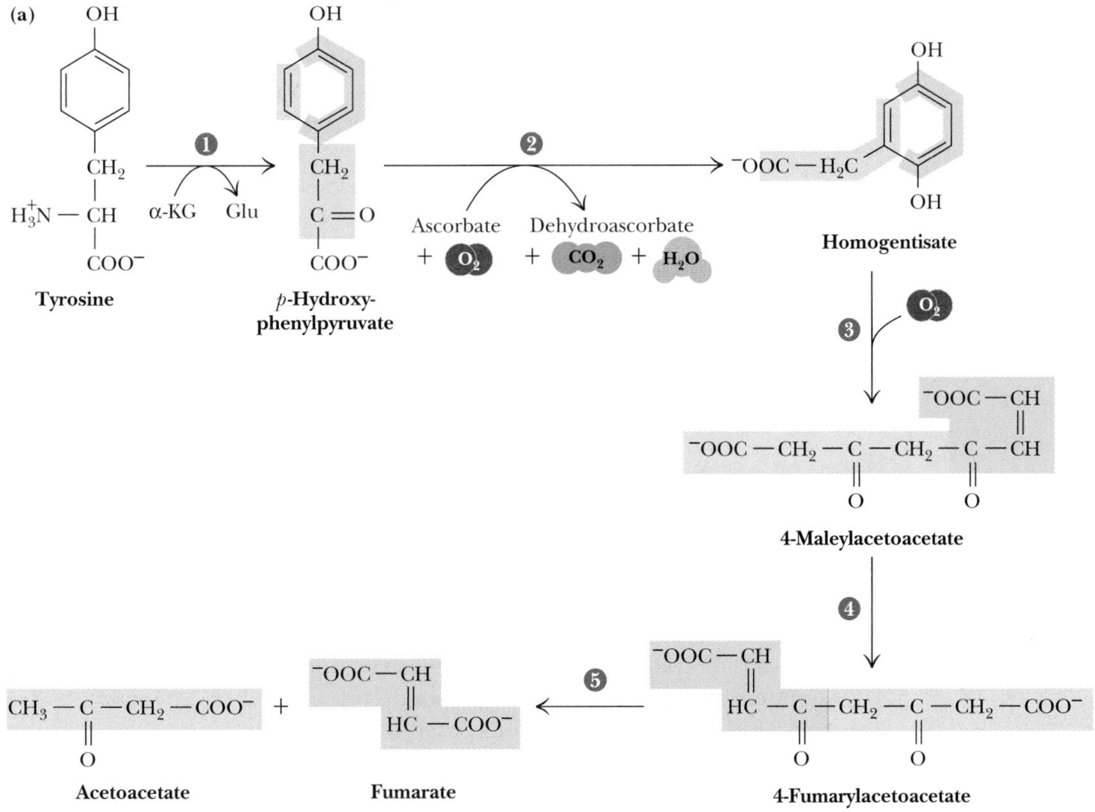

(b)

Phenylpyruvate

(a) Phenylalanine and tyrosine degradation. The first step in Phe degradation is actually the phenylalanine hydroxylase reaction (Figure 21.22) of *tyrosine biosynthesis.* (1) Transamination of Tyr gives *p*-hydroxyphenylpyruvate, which (2) is oxidized to homogentisate by *p*-hydroxyphenylpyruvate dioxygenase in an ascorbic acid (vitamin C)–dependent reaction. (3) The ring opening of homogentisate by homogentisate dioxygenase gives 4-maleylacetoacetate. (4) 4-Maleylacetoacetate isomerase gives 4-fumarylacetoacetate, which (5) is hydrolyzed by fumarylacetoacetase. **(b)** The structure of phenylpyruvate.

See pages 173–178

Nitrogen Excretion

Animals often enjoy a dietary surplus of nitrogen. Excess nitrogen liberated upon metabolic degradation of amino acids is excreted by animals in three different ways, in accord with the availability of water. Aquatic animals simply release free ammonia to the surrounding water; such animals are termed **ammonotelic** (from the Greek *telos,* "end"). On the other hand, terrestrial and aerial species employ mechanisms that convert ammonium to less toxic waste compounds that require little H_2O for excretion. Humans and other terrestrial vertebrates are **ureotelic,** meaning that they excrete excess N as **urea,** a highly

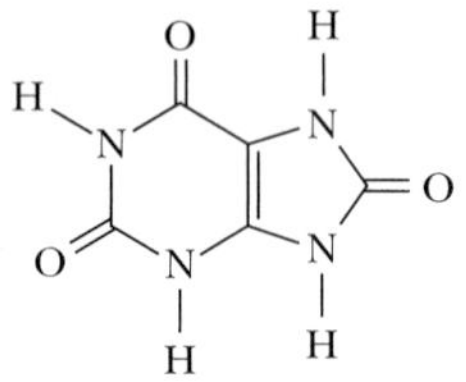

Figure 21.24 Nitrogen waste is excreted by birds principally as the purine analog uric acid.

water-soluble nonionic substance. Urea is formed by ureoteles via the urea cycle. The **uricotelic** organisms are those animals using the third means of N excretion—conversion to **uric acid** (Figure 21.24), a rather insoluble purine analog. Birds, terrestrial reptiles, and many insects are uricoteles. Uric acid metabolism is discussed in Section 21.8. Some animals can switch from ammonotelic to ureotelic to uricotelic metabolism, depending on water availability.

NUCLEOTIDE METABOLISM

Nucleotides are ubiquitous constituents of life, actively participating in the majority of biochemical reactions. Recall that ATP is the "energy currency" of the cell, that uridine nucleotide derivatives of carbohydrates are common intermediates in cellular transformations of carbohydrates (Chapter 19), and that biosynthesis of phospholipids proceeds via cytosine nucleotide derivatives (Chapter 20). In Chapter 25, we will see that GTP serves as the immediate energy source driving the endergonic reactions of protein synthesis. Many of the coenzymes (such as coenzyme A, NAD, NADP, and FAD) are derivatives of nucleotides. Nucleotides also act in metabolic regulation, as in the response of key enzymes of intermediary metabolism to the relative concentrations of AMP, ADP, and ATP (PFK [phosphofructokinase] is a prime example here; see also Chapter 15). Further, cyclic derivatives of purine nucleotides, cAMP and cGMP, have no role in metabolism other than regulation. Last and not least, nucleotides are the monomeric units of nucleic acids. Deoxynucleoside triphosphates (dNTPs) and nucleoside triphosphates (NTPs) serve as the immediate substrates for the biosynthesis of DNA and RNA, respectively (see Part 4, Information Transfer).

21.5 Nucleotide Biosynthesis

Nearly all organisms can make the purine and pyrimidine nucleotides via so-called *de novo* biosynthetic pathways. (*De novo* means "anew"; a less literal but more apt translation might be "from scratch," because *de novo* pathways are metabolic sequences that form complex end products from rather simple precursors.) Many organisms also have salvage pathways to recover purine and pyrimidine compounds obtained in the diet or released during nucleic acid turnover and degradation. While the ribose of nucleotides can be catabolized to generate energy, the nitrogenous bases do not serve as energy sources; their catabolism does not lead to products used by pathways of energy conservation. Compared to slowly dividing cells, rapidly proliferating cells synthesize larger amounts of DNA and RNA per unit time. To meet the increased demand for nucleic acid synthesis, substantially greater quantities of nucleotides must be produced. The pathways of nucleotide biosynthesis thus become attractive targets for the clinical control of rapidly dividing cells such as cancers or infectious bacteria. Many antibiotics and anticancer drugs are inhibitors of purine or pyrimidine nucleotide biosynthesis.

21.6 The Biosynthesis of Purines

Substantial insight into the *de novo* pathway for purine biosynthesis was provided in 1948 by John Buchanan, who cleverly exploited the fact that birds excrete excess nitrogen principally in the form of *uric acid,* a water-insoluble purine

A DEEPER LOOK

Tetrahydrofolate (THF) and One-Carbon Units

An elaborate ensemble of enzymatic reactions serves to introduce one-carbon units into THF and to interconvert the various oxidation states (see figure and Table 14.6). N^5-methyltetrahydrofolate can be oxidized directly to N^5, N^{10}-methylenetetrahydrofolate, which can be further oxidized to N^5, N^{10}-methenyltetrahydrofolate. The N^5-formimino-, N^5-formyl-, and N^{10}-formyltetrahydrofolates can be formed from N^5, N^{10}-methenyltetrahydrofolate (all of these being at the same oxidation level), or they can be formed by one-carbon addition reactions from tetrahydrofolate itself. The principal pathway for incorporation of one-carbon units into tetrahydrofolate is the **serine hydroxymethyltransferase** reaction (see

Figure 21.16), which converts serine to glycine and forms N^5, N^{10}-methylenetetrahydrofolate.

The biosynthetic pathways for methionine, purines (see Figure 21.26), and the pyrimidine thymine (see Figure 21.48) all rely on the incorporation of one-carbon units from tetrahydrofolate derivatives. N^5, N^{10}-Methylene-THF is the source of carbons 2 and 8 of the purine ring and the 5-CH_3 group of the pyrimidine thymine.

The reactions that introduce one-carbon units into tetrahydrofolate (THF) link seven different folate intermediates that carry one-carbon units in three different oxidation states (-2, 0, and $+2$).

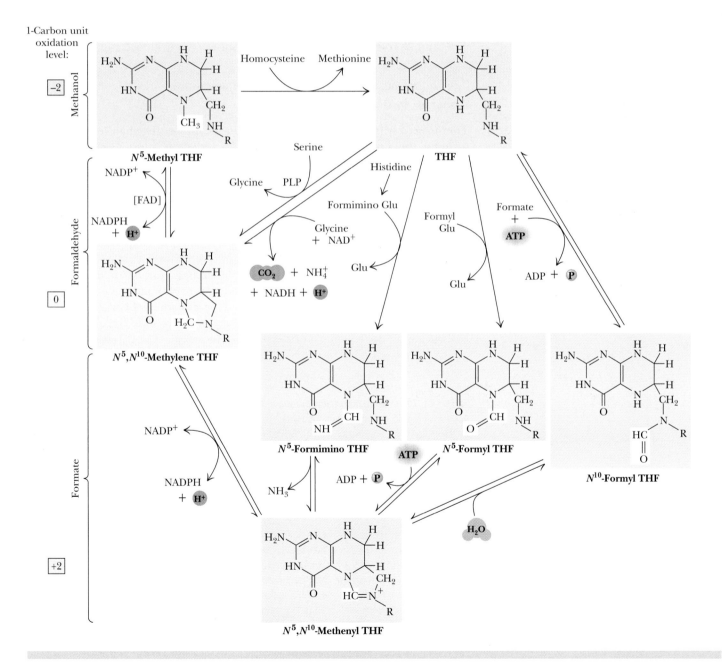

Adapted from Brody, T., et al., 1984. In Handbook of Vitamins. *By Machlin, L. J. New York: Marcel Dekker.*

 See pages 173–178

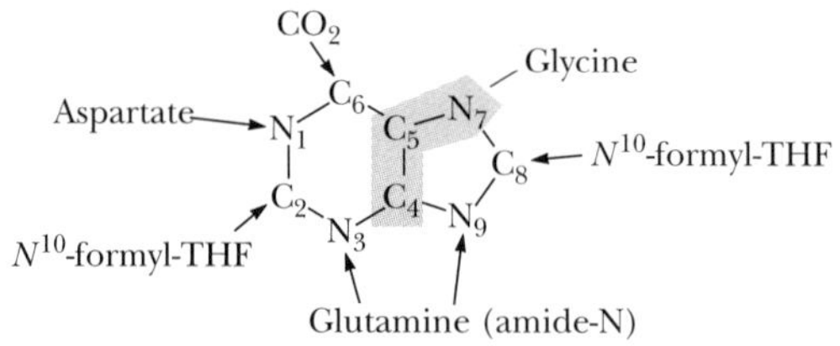

Figure 21.25 The metabolic origin of the nine atoms in the purine ring system.

analog. Buchanan fed isotopically labeled compounds to pigeons and then examined the distribution of the labeled atoms in uric acid (see Figure 21.24). By tracing the metabolic source of the various atoms in this end product, he showed that the nine atoms of the purine ring system (Figure 21.25) are contributed by aspartic acid (N-1), glutamine (N-3 and N-9), glycine (C-4, C-5, and N-7), CO_2 (C-6), and THF one-carbon derivatives (C-2 and C-8). The coenzyme THF and its role in one-carbon metabolism were introduced in Chapter 14.

IMP Biosynthesis: Inosinic Acid Is the Immediate Precursor to GMP and AMP

The *de novo* synthesis of purines occurs in an interesting manner: the atoms forming the purine ring are successively added to *ribose-5-phosphate;* thus, purines are directly synthesized as nucleotide derivatives by assembling the atoms that comprise the purine ring system directly on the ribose. In step 1, ribose-5-phosphate is activated via the direct transfer of a pyrophosphoryl group from ATP to C-1 of the ribose, yielding *5-phosphoribosyl-α-pyrophosphate (PRPP)* (Figure 21.26). The enzyme is **ribose-5-phosphate pyrophosphokinase.** PRPP is the limiting substance in purine biosynthesis. The two major purine nucleoside diphosphates, ADP and GDP, are negative effectors of ribose-5-phosphate pyrophosphokinase. However, because PRPP serves additional metabolic needs, the next reaction is actually the committed step in the pathway.

Step 2 (see Figure 21.26) is catalyzed by **glutamine phosphoribosyl pyrophosphate amidotransferase.** The product is a β-glycoside (recall that all the biologically important nucleotides are β-glycosides). The N atom of this *N*-glycoside becomes N-9 of the nine-membered purine ring; it is the first atom

Figure 21.26 The *de novo* pathway for purine synthesis. The first purine product of this pathway, ▶ IMP (inosinic acid or inosine monophosphate), serves as a precursor to AMP and GMP. *Step 1*: PRPP synthesis from ribose-5-phosphate and ATP by ribose-5-phosphate pyrophosphokinase. *Step 2*: 5-Phosphoribosyl-β-1-amine synthesis from α-PRPP, glutamine, and H_2O by glutamine phosphoribosyl pyrophosphate amidotransferase. *Step 3*: Glycinamide ribonucleotide (GAR) synthesis from glycine, ATP, and 5-phosphoribosyl-β-amine by glycinamide ribonucleotide synthetase. *Step 4*: Formylglycinamide ribonucleotide synthesis from N^{10}-formyl-THF and GAR by GAR transformylase. *Step 5*: Formylglycinamidine ribonucleotide (FGAM) synthesis from FGAR, ATP, glutamine, and H_2O by FGAM synthetase (FGAR amidotransferase). The other products are ADP, P_i, and glutamate. *Step 6*: 5-Aminoimidazole ribonucleotide (AIR) synthesis is achieved via the ATP-dependent closure of the imidazole ring, as catalyzed by FGAM cyclase (AIR synthetase). (Note that the ring closure changes the numbering system.) *Step 7*: Carboxyaminoimidazole ribonucleotide (CAIR) synthesis from CO_2, ATP, and AIR by AIR carboxylase. *Step 8*: N-succinylo-5-aminoimidazole-4-carboxamide ribonucleotide (SAICAR) synthesis from aspartate, CAIR, and ATP by SAICAR synthetase. *Step 9*: 5-Aminoimidazole carboxamide ribonucleotide (AICAR) formation by the nonhydrolytic removal of fumarate from SAICAR. The enzyme is adenylosuccinase. *Step 10*: 5-Formylaminoimidazole carboxamide ribonucleotide (FAICAR) formation from AICAR and N^{10}-formyl-THF by AICAR transformylase. *Step 11*: Dehydration/ring closure yields the authentic purine ribonucleotide IMP. The enzyme is IMP synthase.

See pages 173–178

Inosine monophosphate (IMP)

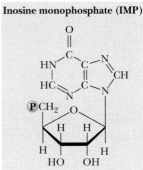

1. Ribose-5-phosphate pyrophosphokinase — ATP, AMP

α-D-Ribose-5-phosphate

5-Phosphoribosyl-α-pyrophosphate (PRPP)

2. Gln: PRPP amido-transferase — Glutamine + H_2O → Glutamate + P P

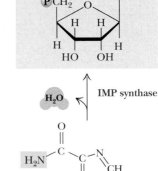

Phosphoribosyl-β-amine

3. GAR synthetase — Glycine + ATP → ADP + P

Glycinamide ribonucleotide (GAR)

4. GAR transformylase — N^{10}-Formyl-THF → THF

Formylglycinamide ribonucleotide (FGAR)

5. FGAM synthetase — ATP + Glutamine + H_2O → ADP + Glutamate + P

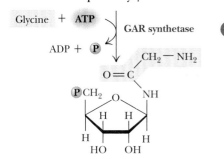

Formylglycinamidine ribonucleotide (FGAM)

6. AIR synthetase — ATP → ADP + P

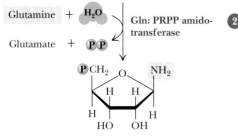

5-Aminoimidazole ribonucleotide (AIR)

7. AIR carboxylase — CO_2 + ATP → ADP + P

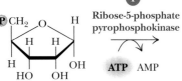

Carboxyaminoimidazole ribonucleotide (CAIR)

8. SAICAR synthetase — Aspartate + ATP → ADP + P

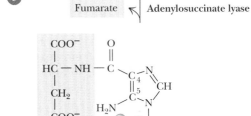

N-succinylo-5-aminoimidazole-4-carboxamide ribonucleotide (SAICAR)

9. Adenylosuccinate lyase — Fumarate

5-Aminoimidazole-4-carboxamide ribonucleotide (AICAR)

10. AICAR transformylase — N^{10}-Formyl-THF → THF

N-formylaminoimidazole-4-carboxamide ribonucleotide (FAICAR)

11. IMP synthase — H_2O

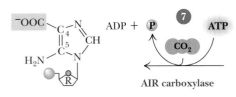
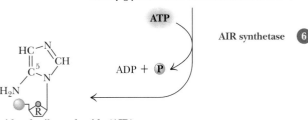

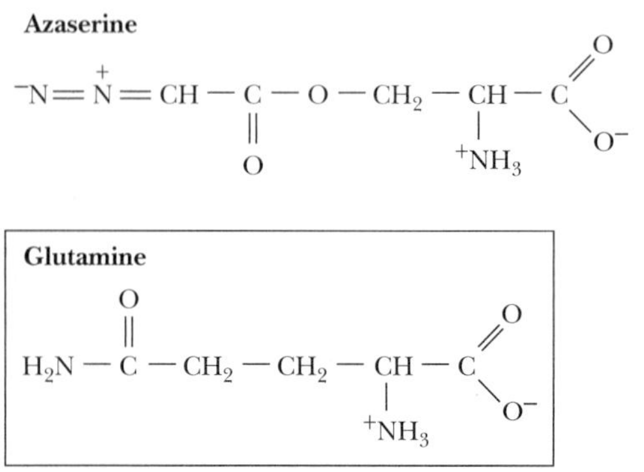

Azaserine

$$^-N \equiv \overset{+}{N} = CH - \underset{\underset{O}{\parallel}}{C} - O - CH_2 - \underset{\underset{+NH_3}{|}}{CH} - C\overset{\displaystyle O}{\underset{\displaystyle O^-}{\diagup}}$$

Glutamine

$$H_2N - \underset{\underset{O}{\parallel}}{C} - CH_2 - CH_2 - \underset{\underset{+NH_3}{|}}{CH} - C\overset{\displaystyle O}{\underset{\displaystyle O^-}{\diagup}}$$

Figure 21.27 The structure of azaserine. Azaserine acts as an irreversible inhibitor of glutamine-dependent enzymes by covalently attaching to nucleophilic groups in the glutamine-binding site.

added in the construction of this ring. Glutamine phosphoribosyl pyrophosphate amidotransferase is subject to feedback inhibition by GMP, GDP, and GTP, as well as AMP, ADP, and ATP. The G series of nucleotides interacts at a guanine-specific allosteric site on the enzyme, whereas the adenine nucleotides act at an A-specific site. The pattern of inhibition by these nucleotides is competitive, thus ensuring that residual enzyme activity is expressed until sufficient amounts of both adenine and guanine nucleotides are synthesized. Glutamine phosphoribosyl pyrophosphate amidotransferase is also sensitive to inhibition by the glutamine analog **azaserine** (Figure 21.27). Azaserine has been employed as an antitumor agent because it causes inactivation of glutamine-dependent enzymes in the purine biosynthetic pathway.

Step 3 is carried out by **glycinamide ribonucleotide synthetase** (GAR synthetase) via its ATP-dependent condensation of the glycine carboxyl group with the amine of *5-phosphoribosyl-β amine* (see Figure 21.26). Glycine contributes C-4, C-5, and N-7 of the purine.

Step 4 is the first of two THF-dependent reactions in the purine pathway. **GAR transformylase** transfers the N^{10}-formyl group of N^{10}-formyl-THF to the free amino group of GAR to yield *α-N-formylglycinamide ribonucleotide (FGAR)*. Although all of the atoms of the imidazole portion of the purine ring are now present, the ring is not closed until reaction 6.

Step 5 is catalyzed by **FGAR amidotransferase** (also known as *FGAM synthetase*). ATP-dependent transfer of the glutamine amido group to the C-4-carbonyl of FGAR yields *formylglycinamidine ribonucleotide (FGAM)*. As a glutamine-dependent enzyme, FGAR amidotransferase is also irreversibly inactivated by azaserine. The imino-N becomes N-3 of the purine.

Step 6 is an ATP-dependent dehydration that leads to formation of the imidazole ring. Because the product is *5-aminoimidazole ribonucleotide*, or *AIR*, this enzyme is called **AIR synthetase.** In avian liver, the enzymatic activities for steps 3, 4, and 6 (GAR synthetase, GAR transformylase, and AIR synthetase) reside on a single, 110-kD multifunctional polypeptide.

In step 7, carbon dioxide is added at the C-4 position of the imidazole ring by **AIR carboxylase** in an ATP-dependent reaction; the carbon of CO_2 will become C-6 of the purine ring. The product is *carboxyaminoimidazole ribonucleotide (CAIR)*.

In step 8, the amino-N of aspartate provides N-1 through linkage to the C-6 carboxyl function of CAIR. ATP hydrolysis drives the condensation of Asp with CAIR. The product is *N-succinylo-5-aminoimidazole-4-carboxamide ribonucleotide (SAICAR)*. **SAICAR synthetase** catalyzes the reaction. The enzymatic activities for steps 7 and 8 reside on a single, bifunctional polypeptide in avian liver.

Step 9 removes the four carbons of Asp as fumarate in a nonhydrolytic cleavage. The product is *5-aminoimidazole-4-carboxamide ribonucleotide (AICAR)*; the enzyme is **adenylosuccinase** (*adenylosuccinate lyase*). Adenylosuccinase acts again in that part of the purine pathway leading from IMP to AMP and takes its name from this latter reaction (see Figure 21.28). AICAR is also an intermediate in the pathway for histidine biosynthesis.

Step 10 adds the formyl carbon of N^{10}-formyl-THF as the ninth and last atom necessary for forming the purine nucleus. The enzyme is called **AICAR transformylase;** the products are THF and *N-formylaminoimidazole-4-carboxamide ribonucleotide*, or *FAICAR*.

Step 11 involves dehydration and ring closure and completes the initial phase of purine biosynthesis. The enzyme is **IMP cyclohydrolase** (also known as *IMP synthase* and *inosinicase*). Unlike step 6, this ring closure does not require ATP. In avian liver, the enzymatic activities catalyzing steps 10 and 11 (AICAR transformylase and inosinicase) reside on 67-kD bifunctional polypeptides organized into 135-kD dimers.

Folate Analogs As Anticancer and Antimicrobial Agents

The requirement for folic acid compounds at steps 4 and 10 of the purine biosynthetic pathway (see Figure 21.26) means that antagonists of folic acid metabolism (for example, methotrexate; see Figure 21.49) indirectly inhibit purine formation and, in turn, nucleic acid synthesis, cell growth, and cell division. Rapidly dividing cells such as malignancies or infective bacteria are more susceptible to

these antagonists than slower-growing normal cells. Also among the folic acid antagonists are *sulfonamides* (see figure). Folic acid is a vitamin for animals and is obtained in the diet. In contrast, bacteria synthesize folic acid from precursors, including *p-aminobenzoic acid (PABA)*, and thus are more susceptible to sulfonamides than are animal cells.

Sulfonamides have the generic structure:

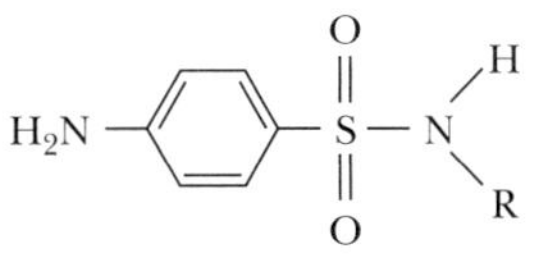

PABA (*p*-aminobenzoic acid)

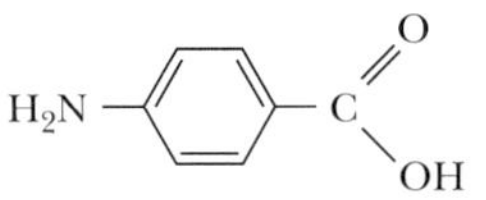

THF (tetrahydrofolate)

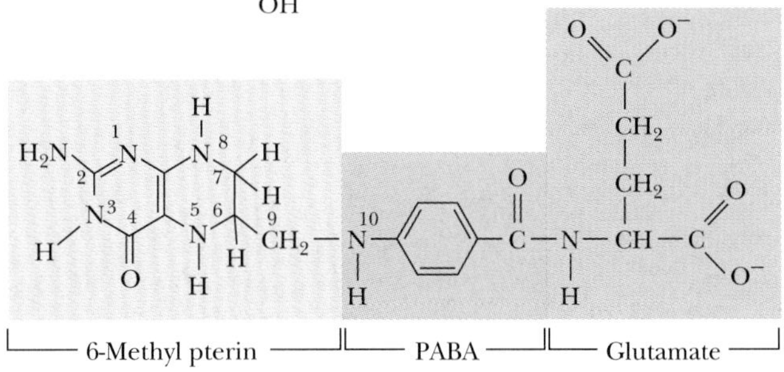

Sulfa drugs, or sulfonamides, owe their antibiotic properties to their similarity to *p*-aminobenzoate (PABA), an important precursor in folic acid synthesis. Sulfonamides block folic acid formation by competing with PABA.

Note that six ATPs are required in the purine biosynthetic pathway from ribose-5-phosphate to IMP: one each at steps 1, 3, 5, 6, 7, and 8. However, seven high-energy phosphate bonds (equal to seven ATP equivalents) are consumed because α-PRPP formation in reaction 1 followed by PP_i release in reaction 2 represents the loss of two ATP equivalents.

AMP and GMP Are Synthesized from IMP

IMP is the precursor to both AMP and GMP. These major purine nucleotides are formed via distinct two-step metabolic pathways that diverge from IMP. The branch leading to AMP (adenosine 5′-monophosphate) involves the displacement of the 6-O group of inosine with aspartate (Figure 21.28) in a GTP-dependent reaction, followed by the removal of the 4-carbon skeleton of Asp as fumarate; the Asp amino group remains as the 6-amino group of AMP. **Adenylosuccinate synthetase** and **adenylosuccinase** are the two enzymes. Recall that adenylosuccinase also acted at step 9 in the pathway from ribose-5-phosphate to IMP. Fumarate production provides a connection between purine synthesis and the citric acid cycle.

The formation of GMP from IMP requires oxidation at C-2 of the purine ring, followed by a glutamine-dependent amidotransferase reaction that replaces the oxygen on C-2 with an amino group to yield *2-amino,6-oxy purine*

Figure 21.28 The synthesis of AMP and GMP from IMP. **(a)** AMP synthesis: The two reactions of AMP synthesis mimic steps 8 and 9 in the purine pathway leading to IMP. In *step 1*, the 6-O of inosine is displaced by aspartate to yield adenylosuccinate. The energy required to drive this reaction is derived from GTP hydrolysis. The enzyme is adenylosuccinate synthetase. AMP is a competitive inhibitor (with respect to the substrate IMP) of adenylosuccinate synthetase. In *step 2*, adenylosuccinase (also known as adenylosuccinate lyase, the same enzyme catalyzing step 9 in the purine pathway) carries out the nonhydrolytic removal of fumarate from adenylosuccinate, leaving AMP. **(b)** GMP synthesis: The two reactions of GMP synthesis are an NAD^+-dependent oxidation followed by an amidotransferase reaction. In *step 1*, IMP dehydrogenase employs the substrates NAD^+ and H_2O in catalyzing oxidation of IMP at C-2. The products are xanthylic acid (XMP or xanthosine monophosphate), NADH, and H^+. GMP is a competitive inhibitor (with respect to IMP) of IMP dehydrogenase. In *step 2*, transfer of the amido-N of glutamine to the C-2 position of XMP yields GMP. This ATP-dependent reaction is catalyzed by GMP synthetase. Besides GMP, the products are glutamate, AMP, and PP_i. Hydrolysis of PP_i to two P_i by ubiquitous pyrophosphatases pulls this reaction to completion.

See pages 173–178

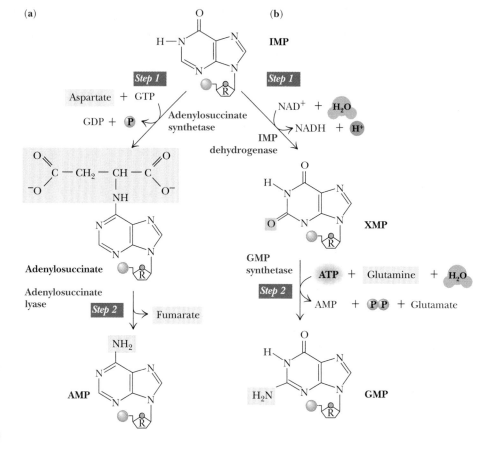

nucleoside monophosphate, or, as this compound is commonly known, *guanosine monophosphate*. The enzymes in the GMP branch are **IMP dehydrogenase** and **GMP synthetase.** Note that, starting from ribose-5-phosphate, eight ATP equivalents are consumed in the synthesis of AMP and nine in the synthesis of GMP.

Regulation of the Purine Biosynthetic Pathway

The regulatory network that controls purine synthesis is schematically represented in Figure 21.29. To recapitulate, the purine biosynthetic pathway from ribose-5-phosphate to IMP is allosterically regulated at the first two steps. Ribose-5-phosphate pyrophosphokinase, although not the committed step in purine synthesis, is subject to feedback inhibition by ADP and GDP. The enzyme catalyzing the next step, glutamine phosphoribosyl pyrophosphate amidotransferase, has two allosteric sites: one where the "A" series of nucleoside phosphates (AMP, ADP, and ATP) binds and feedback-inhibits, and another where the corresponding "G" series binds and inhibits. Further, PRPP is a "feed-forward" activator of this enzyme. Thus, the rate of IMP formation by this pathway is governed by the levels of the final end products, the adenine and guanine nucleotides.

The purine pathway splits at IMP. The first enzyme in the AMP branch, adenylosuccinate synthetase, is competitively inhibited by AMP. Its counterpart in the GMP branch, IMP dehydrogenase, is inhibited in like fashion by GMP. Thus, the fate of IMP is determined by the relative levels of AMP and GMP, so that any deficiency in the amount of either of the principal purine nucleotides is self-correcting. This reciprocity of regulation is an effective mechanism for balancing the formation of AMP and GMP to satisfy cellular needs. Note also that reciprocity is even manifested at the level of energy input: GTP provides the energy to drive AMP synthesis, while ATP serves this role in GMP synthesis (see Figure 21.29).

ATP-Dependent Kinases Form Nucleoside Diphosphates and Triphosphates from the Nucleoside Monophosphates

The products of *de novo* purine biosynthesis are the nucleoside monophosphates AMP and GMP. These nucleotides are converted by successive phosphorylation reactions into their metabolically prominent triphosphate forms, ATP and GTP. The first phosphorylation, to give the nucleoside diphosphate forms, is carried out by two base-specific, ATP-dependent kinases, **adenylate kinase** and **guanylate kinase:**

$$\text{Adenylate kinase: AMP + ATP} \longrightarrow \text{2 ADP}$$

$$\text{Guanylate kinase: GMP + ATP} \longrightarrow \text{GDP + ADP}$$

These nucleoside monophosphate kinases also act on deoxynucleoside monophosphates to give dADP or dGDP.

Oxidative phosphorylation (see Chapter 17) is primarily responsible for the conversion of ADP into ATP. ATP then serves as the phosphoryl donor for synthesis of the other nucleoside triphosphates from their corresponding NDPs

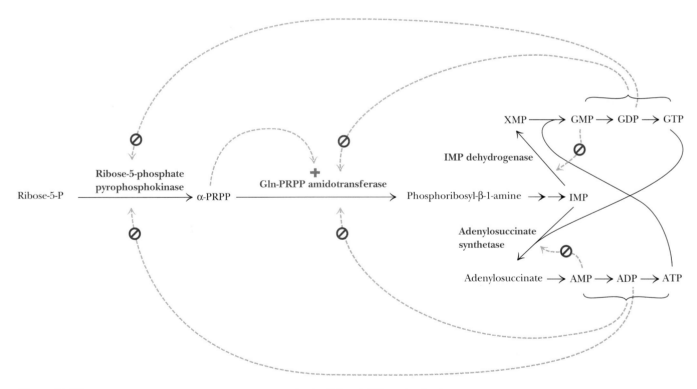

Figure 21.29 The regulatory circuit controlling purine biosynthesis. ADP and GDP are feedback inhibitors of ribose-5-phosphate pyrophosphokinase, the first reaction in the pathway. The second enzyme, glutamine phosphoribosyl pyrophosphate amidotransferase, has two distinct feedback inhibition sites, one for A nucleotides and one for G nucleotides. Also, this enzyme is allosterically activated by PRPP. In the branch leading from IMP to AMP, the first enzyme is feedback-inhibited by AMP, while the corresponding enzyme in the branch from IMP to GMP is feedback-inhibited by GMP. Last, ATP is the energy source for GMP synthesis, whereas GTP is the energy source for AMP synthesis.

in a reaction catalyzed by **nucleoside diphosphate kinase,** a nonspecific enzyme. For example,

$$GDP + ATP \rightleftharpoons GTP + ADP$$

Because this enzymatic reaction is readily reversible and nonspecific with respect to both phosphoryl acceptor and donor, in effect, any NDP can be phosphorylated by any NTP and vice versa. The preponderance of ATP over all other nucleoside triphosphates means that, in quantitative terms, it is the principal nucleoside diphosphate kinase substrate. The enzyme does not discriminate between the ribose moieties of nucleotides and thus functions in phosphoryl transfers involving deoxy-NDPs and deoxy-NTPs as well.

21.7 Purine Salvage

Nucleic acid turnover (synthesis and degradation) is an ongoing metabolic process in most cells. Messenger RNA in particular is actively synthesized and degraded. These degradative processes can lead to the release of free purines in the form of adenine, guanine, and hypoxanthine (the base in IMP). These substances represent a metabolic investment by cells. So-called salvage pathways exist to recover them in useful form. Salvage reactions involve resynthesis of nucleotides from bases via **phosphoribosyltransferases:**

$$\text{Base} + \text{PRPP} \longrightarrow \text{nucleoside-5'-phosphate} + \text{PP}_i$$

The subsequent hydrolysis of PP_i to inorganic phosphate by pyrophosphatases renders the phosphoribosyltransferase reaction effectively irreversible.

The purine phosphoribosyltransferases are **adenine phosphoribosyltransferase (APRT),** which mediates AMP formation, and **hypoxanthine-guanine phosphoribosyltransferase (HGPRT),** which can act on either hypoxanthine to form IMP or guanine to form GMP (Figure 21.30).

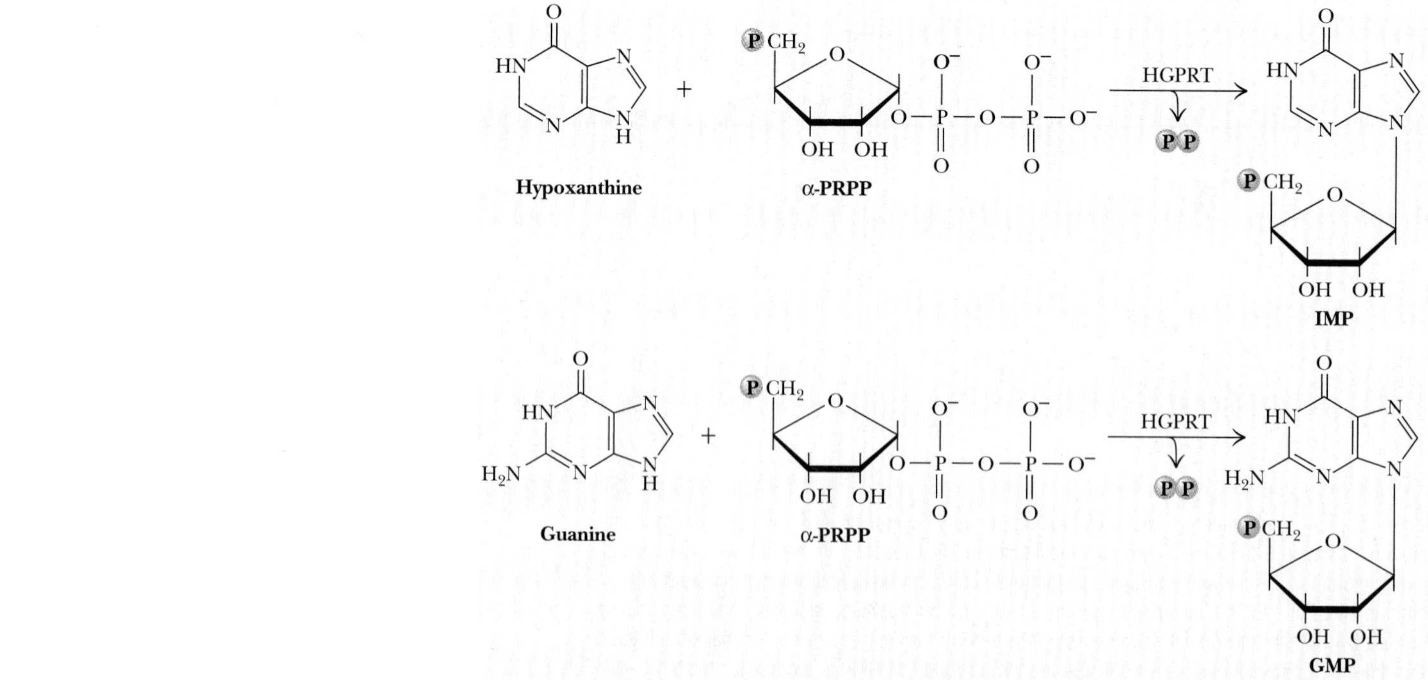

See pages 173–178

Figure 21.30 Purine salvage by the HGPRT reaction.

Because nucleic acids are ubiquitous in cellular material, significant amounts are ingested in the diet. Nucleic acids are degraded in the digestive tract to nucleotides by various nucleases and phosphodiesterases. Nucleotides are then converted to nucleosides by base-specific nucleotidases and nonspecific phosphatases:

$$\text{NMP} + \text{H}_2\text{O} \longrightarrow \text{nucleoside} + \text{P}_i$$

Nucleosides are hydrolyzed by nucleosidases or nucleoside phosphorylases to release the purine base:

Nucleosidase: Nucleoside + $\text{H}_2\text{O} \longrightarrow$ base + ribose

Nucleosidase phosphorylase: Nucleoside + $\text{P}_i \longrightarrow$ base + ribose-1-P

The pentoses liberated in these reactions provide the only source of metabolic energy available from purine nucleotide degradation.

Feeding experiments using radioactively labeled nucleic acids as metabolic tracers have demonstrated that little of the nucleotide ingested in the diet is incorporated into cellular nucleic acids. These findings confirm the *de novo* pathways of nucleotide biosynthesis as the primary source of nucleic acid precursors. Ingested bases are, for the most part, excreted. Nevertheless, cellular nucleic acids do undergo degradation in the course of the continuous recycling of cellular constituents.

The Major Pathways of Purine Catabolism Lead to Uric Acid

The major pathways of purine catabolism in animals are outlined in Figure 21.31. The various nucleotides are first converted to nucleosides by **intracellular nucleotidases.** These nucleotidases are under strict metabolic regulation so that their substrates, which act as intermediates in many vital processes, are not depleted below critical levels. Nucleosides are then degraded by the enzyme **purine nucleoside phosphorylase (PNP)** to release the purine base and ribose-l-P. Note that neither adenosine nor deoxyadenosine is a substrate for PNP. Instead, these nucleosides are first converted to inosine by **adenosine deaminase.** The PNP products are merged into *xanthine* by **guanine deaminase** and **xanthine oxidase,** and xanthine is then oxidized to uric acid by this latter enzyme.

The Purine Nucleoside Cycle: An Anaplerotic Pathway in Skeletal Muscle

Deamination of AMP to IMP by **AMP deaminase** (see Figure 21.31), followed by resynthesis of AMP from IMP by the *de novo* purine pathway enzymes *adenylosuccinate synthetase* and *adenylosuccinate lyase,* constitutes a purine nucleoside cycle (Figure 21.32). This cycle has the net effect of converting aspartate to fumarate plus NH_4^+. Although this cycle might seem like senseless energy consumption, it plays an important role in energy metabolism in skeletal muscle: the fumarate that it generates replenishes the levels of citric acid cycle intermediates lost in amphibolic side reactions (see Chapter 16). Skeletal muscle lacks the usual complement of anaplerotic enzymes and relies on enhanced levels of AMP deaminase, adenylosuccinate synthetase, and adenylosuccinate lyase (adenylosuccinase) to compensate.

Xanthine Oxidase, Uric Acid, and Gout

Xanthine oxidase (see Figure 21.31) is present in large amounts in liver, intestinal mucosa, and milk. It oxidizes hypoxanthine to xanthine and xanthine to uric acid. Xanthine oxidase is a rather indiscriminate enzyme, using molecular oxygen to oxidize a wide variety of purines, pteridines, and aldehydes, producing H_2O_2 as a product. Xanthine oxidase possesses FAD, nonheme Fe-S centers, and *molybdenum cofactor* (a molybdenum-containing pterin complex) as electron-transferring prosthetic groups.

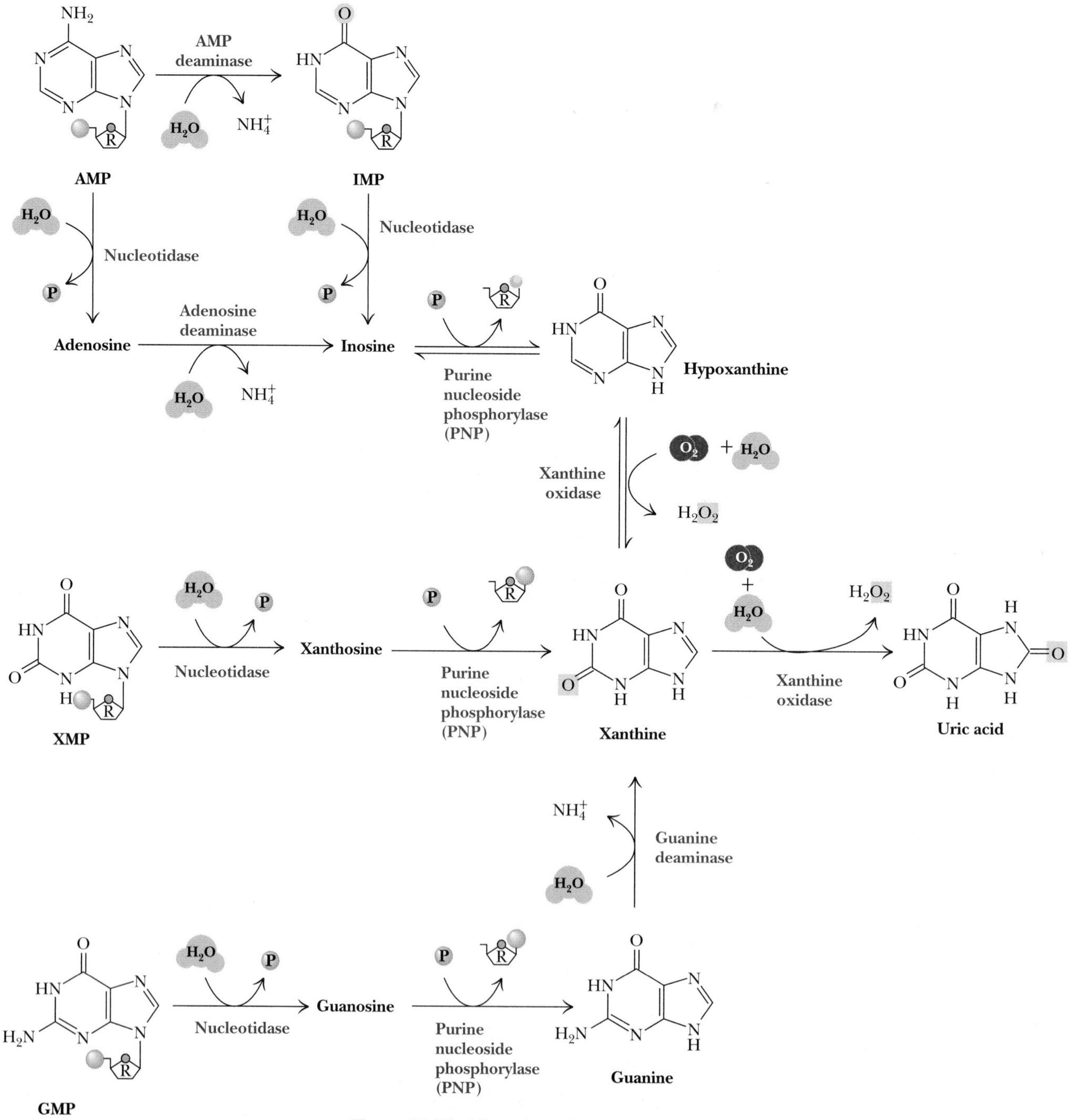

Figure 21.31 The major pathways for purine catabolism. Catabolism of the different purine nucleotides converges in the formation of uric acid.

See pages 173–178

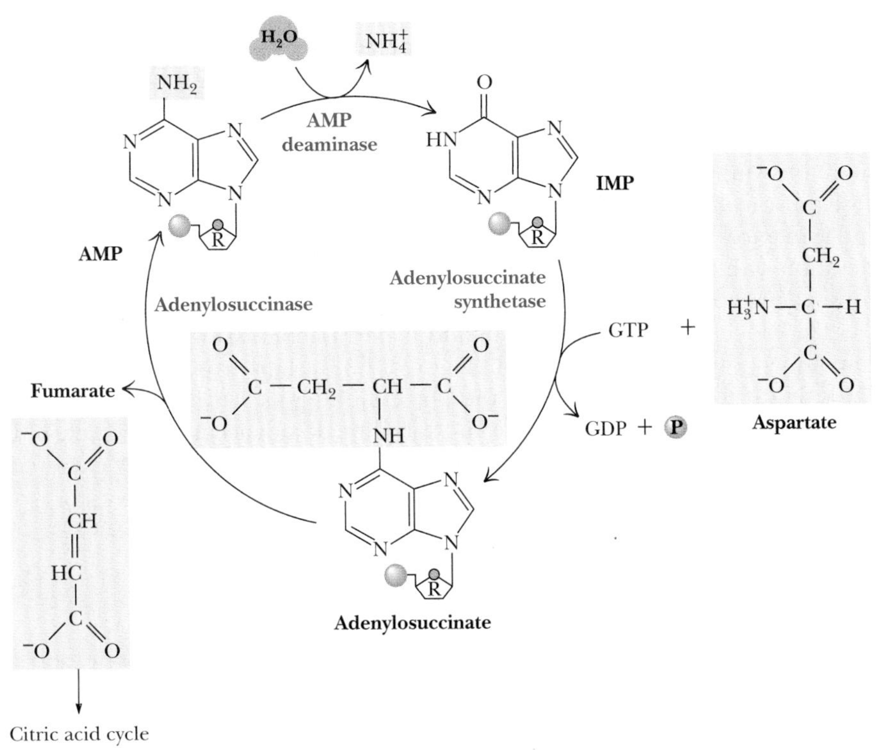

Figure 21.32 The purine nucleoside cycle for anaplerotic replenishment of citric acid cycle intermediates in skeletal muscle.

See pages 173–178

HUMAN BIOCHEMISTRY

Severe Combined Immunodeficiency Syndrome: A Lack of Adenosine Deaminase Is One Cause of This Inherited Disease

Severe combined immunodeficiency syndrome, or **SCID,** is a group of related inherited disorders characterized by the lack of an immune response to infectious disease. This immunological insufficiency is attributable to the inability of B and T lymphocytes to proliferate and produce antibodies in response to an antigenic challenge. About 30% of SCID patients suffer from a deficiency in the enzyme *adenosine deaminase (ADA)*. ADA deficiency is also implicated in a variety of other diseases, including AIDS, anemia, and various lymphomas and leukemias. **Gene therapy,** the repair of a genetic deficiency by introduction of a functional recombinant ver-

sion of the gene, has been attempted on individuals with SCID due to a defective ADA gene. Since ADA is a Zn^{2+}-dependent enzyme, Zn^{2+} deficiency can also lead to reduced immune function.

In the absence of ADA, deoxyadenosine is not degraded but instead is converted into dAMP and then into dATP. dATP is a potent feedback inhibitor of deoxynucleotide biosynthesis (discussed later in this chapter). Without deoxyribonucleotides, DNA cannot be replicated and cells cannot divide (see figure). Rapidly proliferating cell types such as lymphocytes are particularly susceptible if DNA synthesis is impaired.

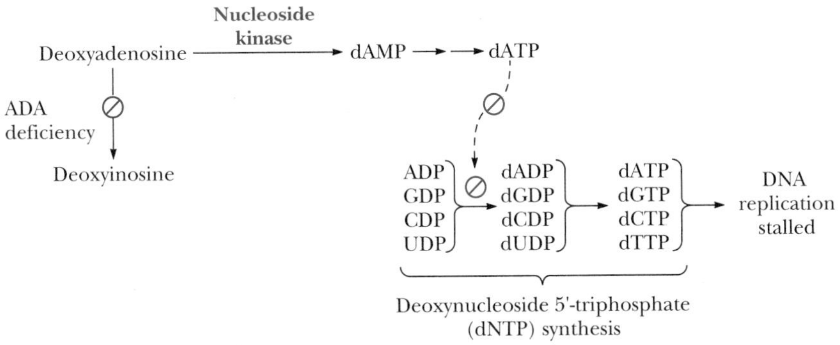

The effect of elevated levels of deoxyadenosine on purine metabolism. If ADA is deficient or absent, deoxyadenosine is not converted into deoxyinosine as normal (see Figure 21.31). Instead, it is salvaged by a nucleoside kinase, which converts it to dAMP, leading to accumulation of dATP and inhibition of deoxynucleotide synthesis (see Figure 21.45). Thus, DNA replication is stalled.

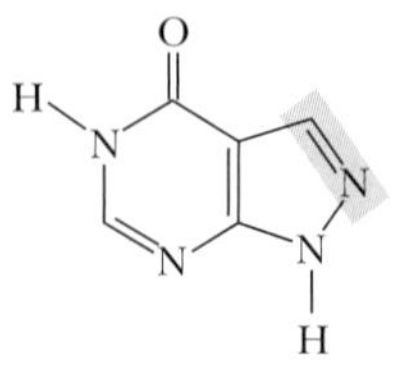

Allopurinol

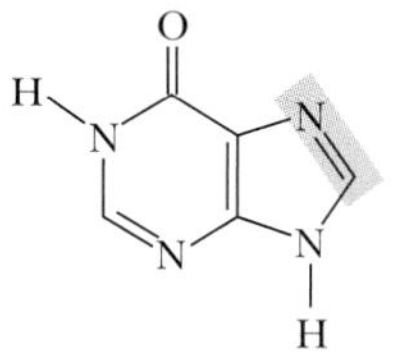

Hypoxanthine

Figure 21.33 Allopurinol, an analog of hypoxanthine, is a potent inhibitor of xanthine oxidase.

Uric acid is the end product of purine catabolism in humans and other primates and is excreted in the urine. Uric acid and urate salts are rather insoluble in water and tend to precipitate from solution if produced in excess. **Gout,** a human disease, is the clinical term describing the physiological consequences accompanying *excessive uric acid accumulation in body fluids.* The most common symptom of gout is arthritic pain in the joints as a result of urate deposits in cartilaginous tissue. Urate crystals may also appear as kidney stones and lead to painful obstruction of the urinary tract. **Hyperuricemia,** the chronic elevation of blood uric acid levels, occurs in about 3% of the population as a consequence of impaired excretion of uric acid or overproduction of purines.

The biochemical causes of gout and hyperuricemia are varied. However, a common treatment is **allopurinol** (Figure 21.33). This hypoxanthine analog binds tightly to xanthine oxidase, thereby inhibiting its activity and preventing uric acid formation. Hypoxanthine and xanthine do not accumulate to harmful concentrations because they are more soluble and thus more easily excreted.

21.9 The Biosynthesis of Pyrimidines

In contrast to purines, pyrimidines are not synthesized as nucleotide derivatives. Instead, the pyrimidine ring system is completed before a ribose-5-P moiety is attached. Also, only two precursors, carbamoyl-P and aspartate, contribute atoms to the six-membered pyrimidine ring (Figure 21.34), compared to seven precursors for the nine purine atoms.

Mammals have two enzymes for carbamoyl phosphate synthesis. Carbamoyl phosphate for pyrimidine biosynthesis is formed by **carbamoyl phosphate synthetase II (CPS II),** a cytosolic enzyme. Recall that carbamoyl phosphate synthetase I is a mitochondrial enzyme dedicated to the urea cycle and arginine biosynthesis (see Figure 21.11). The substrates of carbamoyl phosphate synthetase II are HCO_3^-, H_2O, glutamine, and two ATPs (Figure 21.35). Because carbamoyl phosphate made by CPS II in mammals has no fate other than incorporation into pyrimidines, mammalian CPS II can be viewed as the committed step in the pyrimidine *de novo* pathway. Bacteria have but one CPS, and its carbamoyl phosphate product is incorporated into arginine as well as pyrimidines. Thus, the committed step in bacterial pyrimidine synthesis is the next reaction, which is mediated by **aspartate transcarbamoylase (ATCase).**

ATCase catalyzes the condensation of carbamoyl phosphate with aspartate to form carbamoyl-aspartate (step 2, Figure 21.36). No ATP input is required at this step because carbamoyl phosphate represents an "activated" carbamoyl group.

Step 3 of pyrimidine synthesis involves ring closure and dehydration via linkage of the —NH_2 group introduced by carbamoyl-P with the former β-COO^- of aspartate; this reaction is mediated by the enzyme **dihydroorotase.** The product of the reaction is *dihydroorotate,* a six-membered ring compound. Dihydroorotate (DHO) is not a true pyrimidine, but its oxidation yields *orotate,* which is. This oxidation (step 4) is catalyzed by **dihydroorotate dehydrogenase.** Eukaryotic dihydroorotate dehydrogenase is a protein component of the inner mitochondrial membrane; its immediate e^- acceptor is a quinone, and oxidation of the reduced quinone by the mitochondrial e^- transport chain can drive ATP synthesis via oxidative phosphorylation. At this stage, ribose-5-phosphate is joined to N-1 of orotate, giving the pyrimidine nucleotide *orotidine-5'-monophosphate,* or *OMP* (step 5, Figure 21.36). The ribose phosphate donor is PRPP; the enzyme is **orotate phosphoribosyltransferase.** The next reaction is catalyzed by **OMP decarboxylase.** Decarboxylation of OMP gives *UMP* (*uridine-5'-monophosphate,* or *uridylic acid*), one of the two common pyrimidine ribonucleotides.

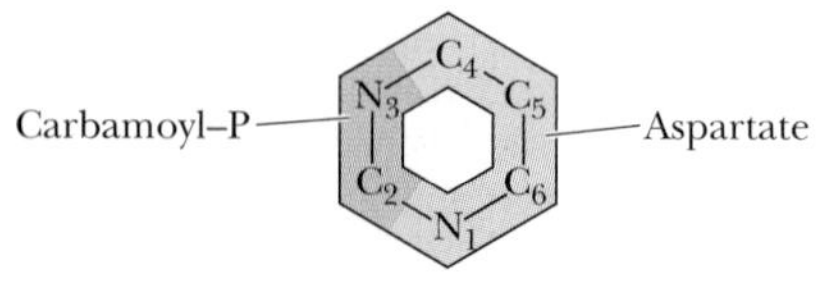

Carbamoyl–P —— Aspartate

Figure 21.34 The metabolic origin of the six atoms of the pyrimidine ring.

Pyrimidine Biosynthesis in Mammals Is Another Example of "Metabolic Channeling"

In bacteria, the six enzymes of *de novo* pyrimidine biosynthesis exist as distinct proteins, each independently catalyzing its specific step in the overall pathway. In contrast, in mammals, the six enzymatic activities are distributed among only three proteins, two of which are **multifunctional polypeptides:** single polypeptide chains having two or more enzymic centers. The first three steps of pyrimidine synthesis, CPS-II, aspartate transcarbamoylase, and dihydroorotase, are all localized on a single 210-kD cytosolic polypeptide. This multifunctional enzyme is the product of a solitary gene, yet it is equipped with the active sites for all three enzymatic activities. Step 4 (see Figure 21.36) is catalyzed by DHO dehydrogenase, a separate enzyme associated with the outer surface of the inner mitochondrial membrane, but the enzymatic activities mediating steps 5 and 6—namely, orotate phosphoribosyltransferase and OMP decarboxylase in mammals—are also found on a single cytosolic polypeptide known as **UMP synthase.**

The purine biosynthetic pathway of avian liver also provides examples of metabolic channeling. Recall that steps 3, 4, and 6 of *de novo* purine synthesis

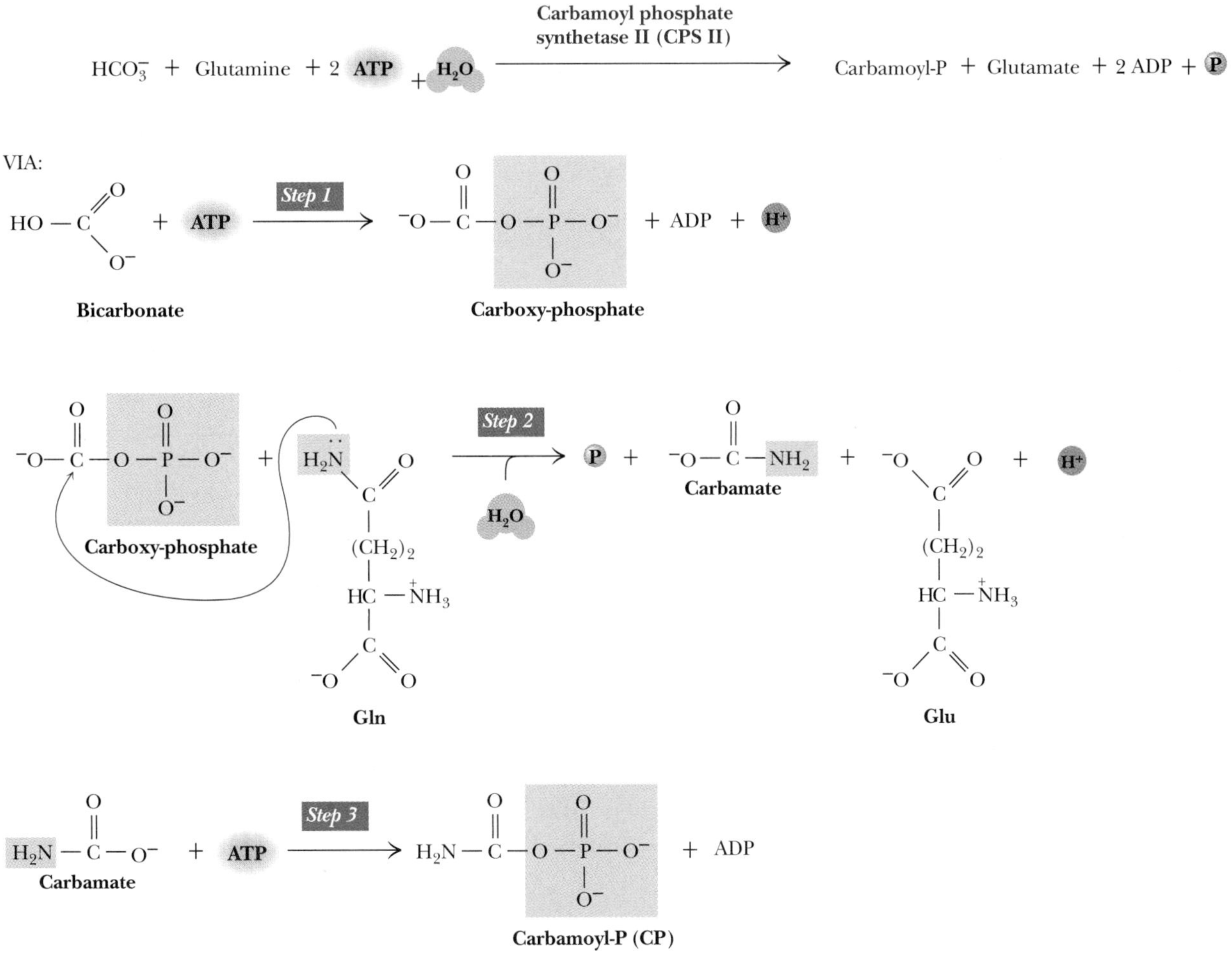

Figure 21.35 The reaction catalyzed by carbamoyl phosphate synthetase II (CPS II). Note that, in contrast to carbamoyl phosphate synthetase I, CPS II uses the amide of glutamine, not NH_4^+, to form carbamoyl-P. *Step 1*: The first ATP consumed in carbamoyl phosphate synthesis is used in forming carboxy-phosphate, an activated form of CO_2. *Step 2*: Carboxy-phosphate (also called carbonyl-phosphate) then reacts with the glutamine amide to yield carbamate and glutamate. *Step 3*: Carbamate is phosphorylated by the second ATP to give ADP and carbamoyl phosphate.

See pages 173–178

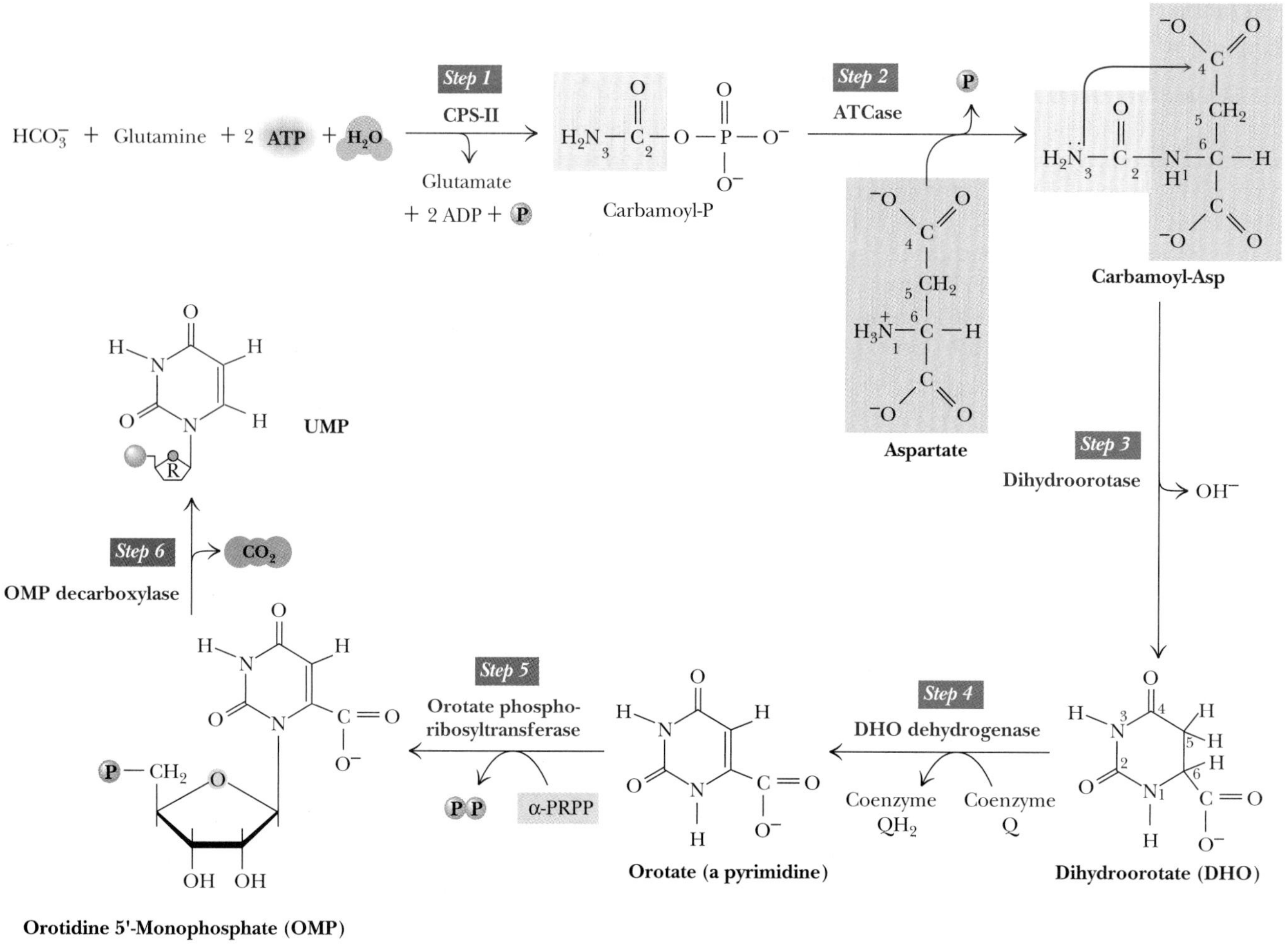

Figure 21.36 The *de novo* pyrimidine biosynthetic pathway. *Step 1*: Carbamoyl-P synthesis. *Step 2*: Condensation of carbamoyl phosphate and aspartate to yield carbamoyl-aspartate is catalyzed by aspartate transcarbamoylase (ATCase). *Step 3*: An intramolecular condensation catalyzed by dihydroorotase gives the six-membered heterocyclic ring characteristic of pyrimidines. The product is dihydroorotate (DHO). *Step 4*: The oxidation of DHO by dihydroorotate dehydrogenase gives orotate. (In bacteria, NAD$^+$ is the electron acceptor from DHO.) *Step 5*: PRPP provides the ribose-5-P moiety that transforms orotate into orotidine-5′-monophosphate, a pyrimidine nucleotide. Note that orotate phosphoribosyltransferase joins N-1 of the pyrimidine to the ribosyl group in appropriate β-configuration. PP$_i$ hydrolysis renders this reaction thermodynamically favorable. *Step 6*: Decarboxylation of OMP by OMP decarboxylase yields UMP.

See pages 173–178

are catalyzed by three enzymatic activities localized on a single multifunctional polypeptide and steps 7 and 8 and steps 10 and 11 by respective bifunctional polypeptides (see Figure 21.26).

Such multifunctional enzymes confer an advantage: The product of one reaction in a pathway is the substrate for the next. In multifunctional enzymes, such products remain bound and are channeled directly to the next active site rather than dissociated into the surrounding medium for diffusion to the next enzyme. This **metabolic channeling** is more efficient because substrates are not diluted into the milieu and no pools of intermediates accumulate.

Synthesis of the Prominent Ribonucleotides UTP and CTP

The two prominent pyrimidine ribonucleotide products are derived from UMP via the same unbranched pathway. First, UDP is formed from UMP via an ATP-dependent *nucleoside monophosphate kinase*:

$$UMP + ATP \rightleftharpoons UDP + ADP$$

Then, UTP is formed by *nucleoside diphosphate kinase*:

$$UDP + ATP \rightleftharpoons UTP + ADP$$

Amination of UTP at the 6-position gives CTP. The enzyme, **CTP synthetase,** is a glutamine amidotransferase (Figure 21.37). ATP hydrolysis provides the energy to drive the reaction.

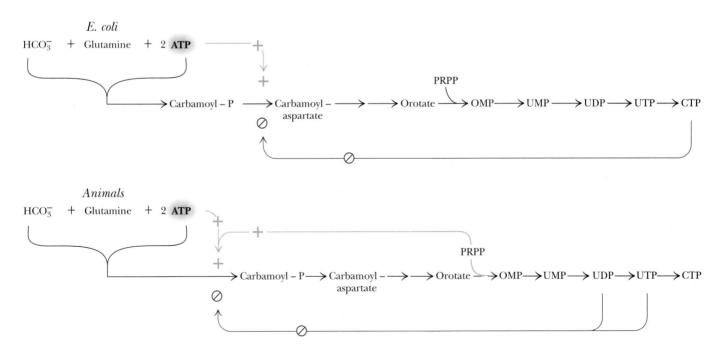

Figure 21.37 CTP synthesis from UTP. CTP synthetase catalyzes amination of the 4-position of the UTP pyrimidine ring, yielding CTP. In eukaryotes, this NH_2 comes from the amide-N of glutamine; in bacteria, NH_4^+ serves this role.

Regulation of Pyrimidine Biosynthesis

Pyrimidine biosynthesis in bacteria is allosterically regulated at aspartate trans-carbamoylase (ATCase). *Escherichia coli* ATCase is feedback-inhibited by the end product, CTP. ATP, which can be viewed as a signal of both energy availability and purine sufficiency, is an allosteric activator of ATCase. CTP and ATP compete for a common allosteric site on the enzyme. In many bacteria, UTP, not CTP, acts as the ATCase feedback inhibitor.

In animals, CPS-II catalyzes the committed step in pyrimidine synthesis and serves as the focal point for allosteric regulation. UDP and UTP are feedback inhibitors of CPS-II, while PRPP and ATP are allosteric activators. With the exception of ATP, none of these compounds are substrates of CPS-II or of either of the two other enzymic activities residing with it on the trifunctional polypeptide. Figure 21.38 compares the regulatory circuits governing pyrimidine synthesis in bacteria and animals.

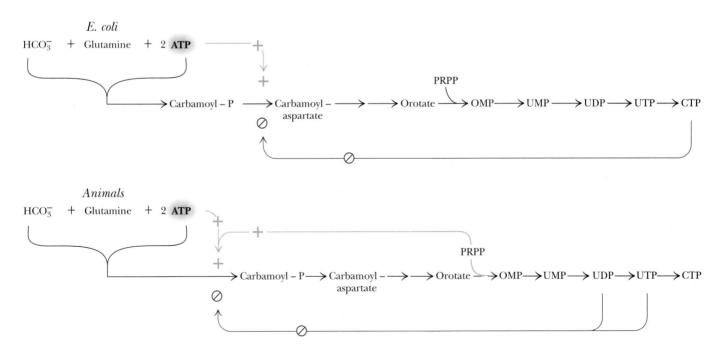

Figure 21.38 A comparison of the regulatory circuits that control pyrimidine synthesis in *E. coli* and animals.

Mammalian CPS-II Is Activated *In Vitro* by MAP Kinase and *In Vivo* by Epidermal Growth Factor

The rate-limiting step in mammalian *de novo* pyrimidine synthesis is catalyzed by CPS-II, and proliferating cells require lots of pyrimidine nucleotides for growth and cell division. Normally, CPS-II is feedback-inhibited by UTP, but *in vitro* phosphorylation of CPS-II by **MAP kinase** (*m*itogen-*a*ctivated *p*rotein kinase) creates a covalently modified (phosphorylated) CPS-II that is no longer sensitive to UTP inhibition. Further, phosphorylated CPS-II is more responsive to PRPP activation. Both of these responses favor enhanced pyrimidine biosynthesis. This regulation occurs *in vivo* when **epidermal growth factor (EGF),** a **mitogen** (mitogen = a hormone that stimulates mitosis [cell division]), activates an intracellular cascade of reactions that culminates in MAP kinase activation. Thus, pyrimidine biosynthesis in mammals is activated in concert with stimulation of cell proliferation by EGF.

21.10 Pyrimidine Degradation

Like purines, free pyrimidines can be salvaged and recycled to form nucleotides via phosphoribosyltransferase reactions similar to those discussed earlier. Pyrimidine catabolism results in degradation of the pyrimidine ring to products reminiscent of the original substrates, aspartate, CO_2, and ammonia (Figure 21.39). β-Alanine can be recycled into the synthesis of coenzyme A. Catabolism of the pyrimidine base, thymine (5-methyluracil) yields β-amino isobutyric acid instead of β-alanine.

Pathways presented thus far in this chapter account for the synthesis of the four principal ribonucleotides: ATP, GTP, UTP, and CTP. These compounds serve important coenzymic functions in metabolism and are the immediate precursors for ribonucleic acid (RNA) synthesis. Roughly 90% of the total nucleic acid in cells is RNA; the remainder is deoxyribonucleic acid (DNA). DNA differs from RNA in that it is a polymer of deoxyribonucleotides, one of which is deoxythymidylic acid. We now turn to the synthesis of these compounds.

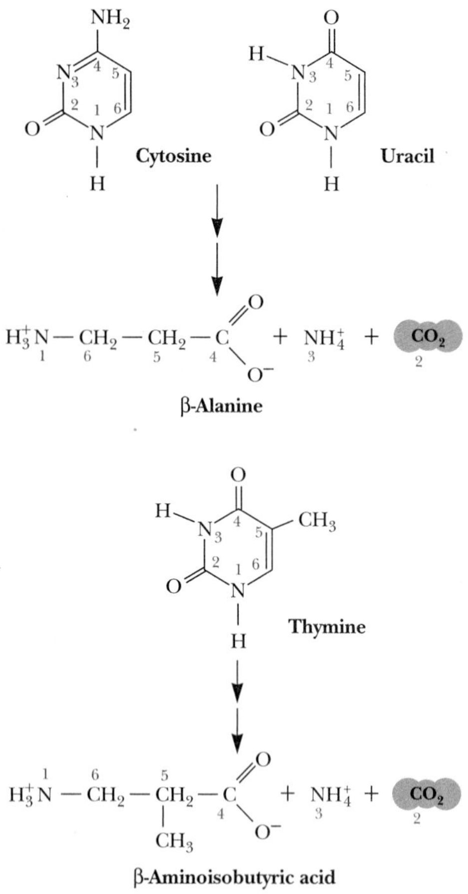

21.11 Deoxyribonucleotide Biosynthesis

The deoxyribonucleotides have only one metabolic purpose: to serve as precursors for DNA synthesis. In most organisms, ribonucleoside diphosphates (NDPs) are the substrates for deoxyribonucleotide formation. Reduction at the 2′-position of the ribose ring in NDPs produces 2′-deoxy forms of these nucleotides (Figure 21.40). This reaction involves replacement of the 2′-OH by a hydride ion ($H:^-$) and is catalyzed by an enzyme known as **ribonucleotide reductase.** Enzymatic ribonucleotide reduction involves a free radical mechanism, and three classes of ribonucleotide reductases are known, differing from each other in their mechanisms of free radical generation. Class I enzymes, found in *E. coli* and virtually all eukaryotes, are Fe-dependent and generate the required free radical on a specific tyrosyl side chain.

E. coli Ribonucleotide Reductase

The enzyme system for dNDP formation consists of four proteins, two of which constitute the ribonucleotide reductase proper, an enzyme of the $\alpha_2\beta_2$ type. The other two proteins, **thioredoxin** and **thioredoxin reductase,** function in the delivery of reducing equivalents, as we shall see shortly. The two proteins of ribonucleotide reductase are designated **R1** (86 kD) and **R2** (43.5 kD), and

Figure 21.39 Pyrimidine degradation. Carbons 4, 5, and 6 plus N-1 are released as β-alanine, N-3 as NH_4^+, and C-2 as CO_2. (The pyrimidine thymine yields β-aminoisobutyric acid.) Recall that aspartate was the source of N-1 and C-4, -5, and -6, while C-2 came from CO_2 and N-3 from NH_4^+ via glutamine.

See pages 173–178

each is a homodimer in the holoenzyme (Figure 21.41). The R1 homodimer carries two types of regulatory sites in addition to the catalytic site. Substrates (ADP, CDP, GDP, UDP) bind at the catalytic site. One regulatory site—the **substrate specificity site**—binds ATP, dATP, dGTP, or dTTP, and the nucleotide that is bound there determines which nucleoside diphosphate is bound at the catalytic site. The other regulatory site, the **overall activity site,** binds either the activator ATP or the negative effector dATP; the nucleotide bound here determines whether the enzyme is active or inactive. Activity depends also on residues Cys^{439}, Cys^{225}, and Cys^{462} in R1. The two Fe atoms within the single active site formed by the R2 homodimer generate the free radical required for ribonucleotide reduction on a specific R2 residue, Tyr^{122}, which in turn generates a thiyl free radical ($Cys-S\cdot$) on Cys^{439}. $Cys^{439}-S\cdot$ initiates ribonucleotide reduction by abstracting the $3'-H$ from the ribose ring of the nucleoside diphosphate substrate (Figure 21.42) and forming a free radical on C-3'. Subsequent dehydration forms the deoxyribonucleotide product.

The Reducing Power for Ribonucleotide Reductase

NADPH is the ultimate source of reducing equivalents for ribonucleotide reduction, but the immediate source is reduced **thioredoxin,** a small (12 kD) protein with reactive Cys-sulfhydryl groups situated next to one another in the sequence Cys-Gly-Pro-Cys. These Cys residues are able to undergo reversible oxidation–reduction between ($-S-S-$) and ($-SH \; HS-$) and, in their reduced form, serve as primary electron donors to regenerate the reactive $-SH$ pair of the ribonucleotide reductase active site (Figure 21.43). In turn, the sulfhydryls of thioredoxin must be restored to the ($-SH \; HS-$) state for another catalytic cycle. **Thioredoxin reductase,** an α_2-type enzyme composed of 58-kD flavoprotein subunits, mediates the NADPH-dependent reduction of thioredoxin (see Figure 21.43). Thioredoxin functions in a number of metabolic roles besides deoxyribonucleotide synthesis, the common denominator of which is reversible sulfide : sulfhydryl transitions. Another sulfhydryl protein

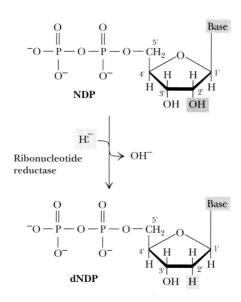

Figure 21.40 Deoxyribonucleotide synthesis involves reduction at the 2'-position of the ribose ring of nucleoside diphosphates.

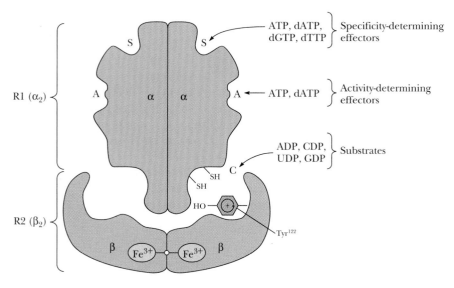

Figure 21.41 *E. coli* ribonucleotide reductase: its binding sites and subunit organization. Two proteins, R1 and R2 (each a dimer of identical subunits), combine to form the holoenzyme. The holoenzyme has three classes of nucleotide binding sites: S, the specificity-determining sites; A, the activity-determining sites; and C, the catalytic or active site. These various sites bind different nucleotide ligands. Note that the holoenzyme apparently possesses only one active site formed by interaction between Fe^{3+} atoms in each R2 subunit.

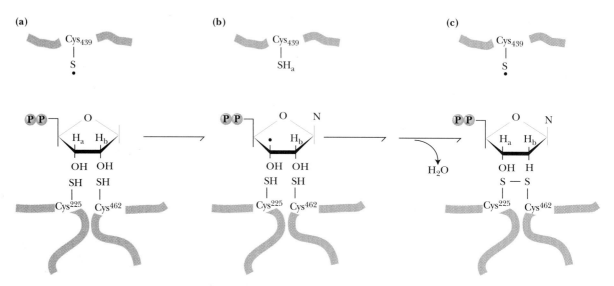

Figure 21.42 The free radical mechanism of ribonucleotide reduction. H_a designates the C-3′ hydrogen and H_b the C-2′ hydrogen atom. Formation of a thiyl radical on Cys^{439} **(a)** of the *E. coli* ribonucleotide reductase R1 homodimer through reaction with a Tyr^{122} free radical on R2 leads to removal of the H_a hydrogen and creation of a C-3′· radical **(b)**. Dehydration via removal of the C-2′ — OH group, accompanied by oxidation of R1 Cys^{225} and Cys^{462} — SH groups to form a disulfide, and restoration of H_a to C-3′ forms the dNDP product **(c)**. (*Adapted from Reichard, P., 1997. The evolution of ribonucleotide reduction.* Trends in Biochemical Sciences **22**:81–85. *This free radical mechanism of ribonucleotide reduction was originally proposed by JoAnn Stubbe of MIT.*)

similar to thioredoxin, called **glutaredoxin,** can also function in ribonucleotide reduction. Oxidized glutaredoxin is re-reduced by two equivalents of **glutathione** (γ-glutamylcysteinylglycine; Figure 21.44), which in turn is re-reduced by glutathione reductase, another NADPH-dependent flavoenzyme.

The substrates for ribonucleotide reductase are CDP, UDP, GDP, and ADP, and the corresponding products are dCDP, dUDP, dGDP, and dADP. Because CDP is not an intermediate in pyrimidine nucleotide synthesis, it must arise by dephosphorylation of CTP, for instance, via nucleoside diphosphate kinase action. Although uridine nucleotides do not occur in DNA, UDP is a substrate. The formation of dUDP is justified because it is a precursor to dTTP, a necessary substrate for DNA synthesis (see following discussion).

Regulation of Ribonucleotide Reductase Specificity and Activity

Ribonucleotide reductase activity must be modulated in two ways in order to maintain an appropriate balance of the four deoxynucleotides essential to DNA synthesis—namely, dATP, dGTP, dCTP, and dTTP. First, the overall activity of the enzyme must be turned on and off in response to the need for dNTPs. Second, the relative amounts of each NDP substrate transformed into dNDP must be controlled in order that the right balance of dATP : dGTP : dCTP : dTTP is produced. The two different sets of effector binding sites on ribonucleotide reductase, *discrete from the substrate-binding active site,* are designed to serve these purposes. These two regulatory sites are the *overall activity site* and the *substrate specificity site.* Only ATP and dATP are able to bind at the overall activity site. ATP is an allosteric activator and dATP is an allosteric inhibitor, and they compete for the same site. If ATP is bound, the enzyme is active; if its deoxy counterpart, dATP, occupies this site, the enzyme is inactive.

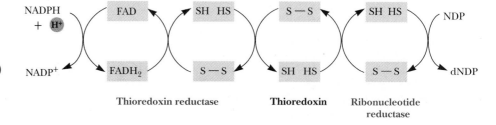

Figure 21.43 The (— S — S —)/(— SH HS —) oxidation–reduction cycle involving ribonucleotide reductase, thioredoxin, thioredoxin reductase, and NADPH.

Figure 21.44 The structure of glutathione.

Reduced glutathione (γ-glutamylcysteinylglycine, GSH)

Oxidized glutathione (GSSG) or glutathione disulfide

NADPH + H^+ → NADP$^+$

Glutathione reductase

GSSG → 2GSH

The second effector site, the *substrate specificity site,* can bind either ATP, dTTP, dGTP, or dATP, and the substrate specificity of the enzyme is determined by which of these nucleotides occupies this site. If ATP is in the substrate specificity site, ribonucleotide reductase preferentially binds pyrimidine nucleotides (UDP or CDP) at its active site and reduces them to dUDP and dCDP. With dTTP in the specificity-determining site, GDP is the preferred substrate. When dGTP binds to this specificity site, ADP becomes the favored substrate for reduction. The rationale for these varying affinities (Figure 21.45) is as follows: High [ATP] is consistent with cell growth and division and, consequently, the need for DNA synthesis. Thus, ATP binds in the activity-determining site of ribonucleotide reductase, turning it on and promoting production of dNTPs for DNA synthesis. Under these conditions, ATP is also likely to occupy the substrate specificity site, so that UDP and CDP are reduced to dUDP and dCDP. Both of these pyrimidine deoxynucleoside diphosphates are precursors to dTTP. Thus, elevation of dUDP and dCDP levels leads to an increase in [dTTP]. High dTTP levels increase the likelihood that it will occupy the substrate specificity site, in which case GDP becomes the preferred substrate, and dGTP levels rise. Upon dGTP association with the substrate specificity site, ADP is the

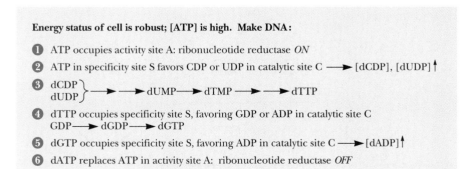

Figure 21.45 Regulation of deoxynucleotide biosynthesis—the rationale for the various affinities displayed by the two nucleotide-binding regulatory sites on ribonucleotide reductase.

Figure 21.46 Pathways of dTMP synthesis. dTMP production is dependent on dUMP formation from dCDP and dUDP synthesis. If the dCDP pathway is traced from the common pyrimidine precursor, UMP, it will proceed as follows:

UMP ⟶ UDP ⟶ UTP ⟶ CTP ⟶ CDP
⟶ dCDP ⟶ dCMP ⟶ dUMP ⟶ dTMP

From dUDP: dUDP ⟶ dUTP ⟶ dUMP ⟶ dTMP
From dCDP: dCDP ⟶ dCMP ⟶ dUMP ⟶ dTMP

favored substrate, leading to ADP reduction and the eventual accumulation of dATP. Binding of dATP to the overall activity site then shuts the enzyme down. In summary, the relative affinities of the three classes of nucleotide binding sites in ribonucleotide reductase for the various substrates, activators, and inhibitors are such that the formation of dNDPs proceeds in an orderly and balanced fashion. As these dNDPs are formed, they are phosphorylated by nucleoside diphosphate kinases to produce dNTPs, the actual substrates of DNA synthesis.

21.12 Synthesis of Thymine Nucleotides

The synthesis of thymine nucleotides proceeds from other pyrimidine deoxyribonucleotides. Cells have no requirement for free thymine ribonucleotides and do not synthesize them. Small amounts of thymine ribonucleotides do occur in tRNA (tRNA is notable for having unusual nucleotides), but these Ts arise via methylation of U residues already incorporated into the tRNA. Both dUDP and dCDP can lead to formation of dUMP, the immediate precursor for dTMP synthesis (Figure 21.46). Interestingly, formation of dUMP from dUDP passes through dUTP, which is then cleaved by **dUTPase,** a pyrophosphatase that removes PP$_i$ from dUTP. The action of dUTPase prevents dUTP from serving as a substrate in DNA synthesis. An alternative route to dUMP formation starts with dCDP, which is dephosphorylated to dCMP and then deaminated by **dCMP deaminase** (Figure 21.47), leaving dUMP. dCMP deaminase provides a second point for allosteric regulation of dNTP synthesis; it is allosterically activated by dCTP and feedback-inhibited by dTTP. Of the four dNTPs, only dCTP does not interact with either of the regulatory sites on ribonucleotide reductase (see Figure 21.45). Instead, it acts upon dCMP deaminase.

Synthesis of dTMP from dUMP is catalyzed by **thymidylate synthase** (Figure 21.48). This enzyme methylates dUMP at the 5-position to create dTMP; the methyl donor is the one-carbon folic acid derivative N^5, N^{10}-methylene-THF. The reaction is actually a reductive methylation in which the one-carbon unit is transferred at the methylene level of reduction and then reduced to the methyl level. The THF cofactor is oxidized at the expense of methylene reduction to yield dihydrofolate, or DHF. Dihydrofolate reductase then reduces DHF back to THF for service again as a one-carbon vehicle (see Figure 14.30).

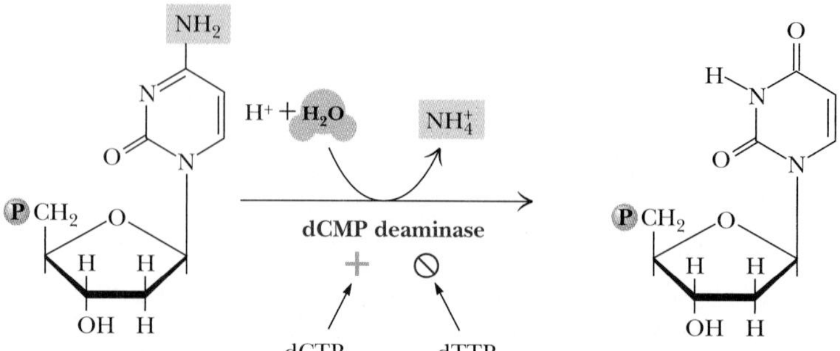

Figure 21.47 The dCMP deaminase reaction.

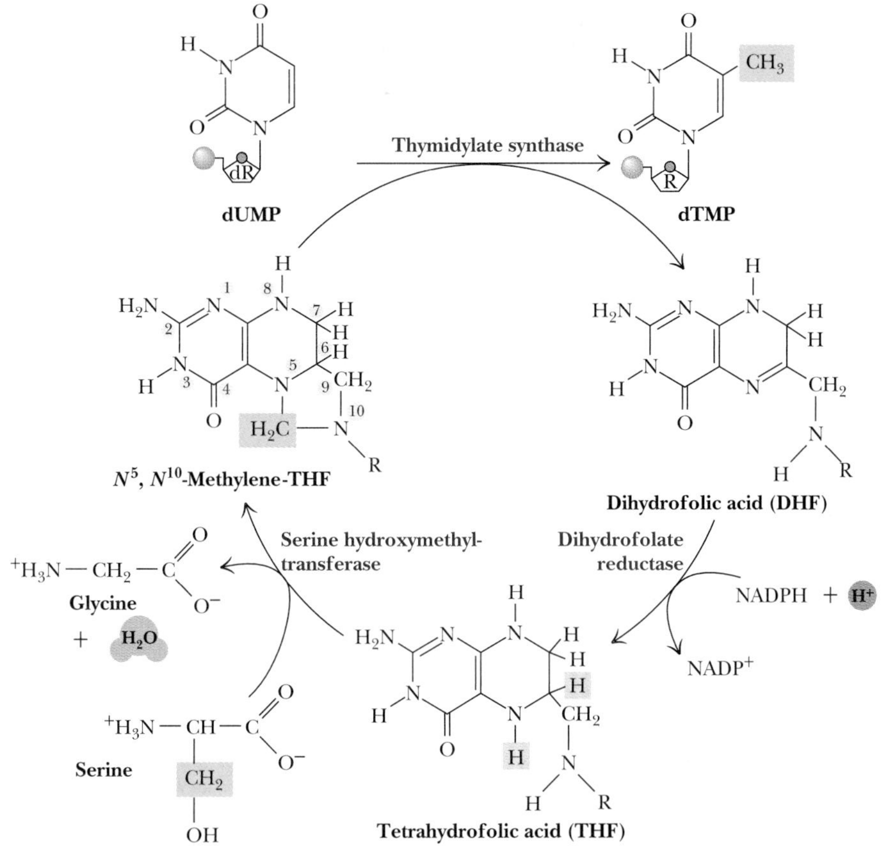

Figure 21.48 The thymidylate synthase reaction. The 5-CH$_3$ group is ultimately derived from the β-carbon of serine.

 See pages 173–178

dUMP

Thymidylate synthase

dTMP

N^5, N^{10}-**Methylene-THF**

Dihydrofolic acid (DHF)

Serine hydroxymethyl-transferase

Dihydrofolate reductase

NADPH + H$^+$

NADP$^+$

Glycine

+ **H$_2$O**

Serine

Tetrahydrofolic acid (THF)

Thymidylate synthase sits at a junction connecting dNTP synthesis with folate metabolism. It has become a preferred target for inhibitors designed to disrupt DNA synthesis. An indirect approach is to employ folic acid precursors or analogs as antimetabolites of dTMP synthesis (Figure 21.49). Purine synthesis is affected as well because it is also dependent on THF (see Figure 21.26).

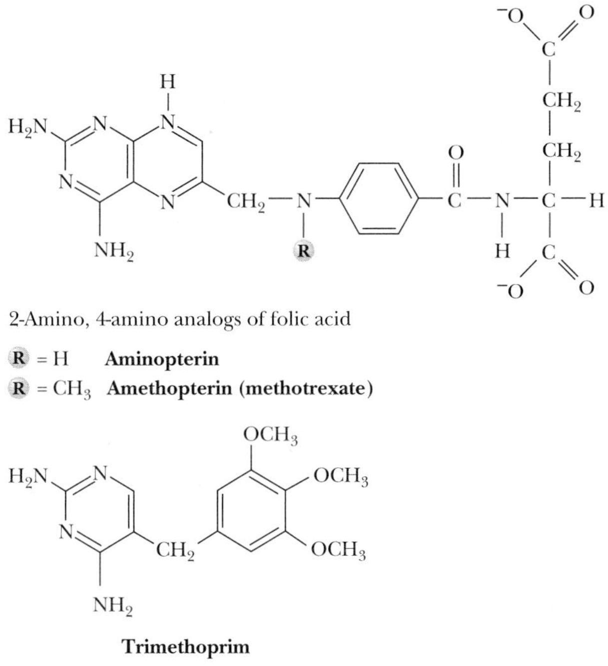

2-Amino, 4-amino analogs of folic acid

R = H   **Aminopterin**

R = CH$_3$ **Amethopterin (methotrexate)**

Trimethoprim

Figure 21.49 Precursors and analogs of folic acid employed as antimetabolites : sulfonamides (see *Human Biochemistry*, page 685), methotrexate, aminopterin, and trimethoprim. The latter three compounds bind to dihydrofolate reductase with about one thousand–fold greater affinity than DHF and thus act as virtually irreversible inhibitors.

HUMAN BIOCHEMISTRY

Fluoro-Substituted Pyrimidine Analogs As Therapeutic Agents

Fluoro-substituted compounds are potent inhibitors in reactions that involve proton abstraction from a substrate. If F is substituted for H at the position of H^+ abstraction, the fluoro-substituted analog strongly inhibits the reaction because F^+ is such a poor leaving group. *5-Fluorouracil* (*5-FU*; see figure, a) is a thymine analog. It is converted *in vivo* to *5′-fluorouridylate* by a PRPP-dependent phosphoribosyltransferase, and passes through the reactions of dNTP synthesis, culminating ultimately as *2′-deoxy-5-fluorouridylic acid,* a potent inhibitor of dTMP synthase. 5-FU is used as a chemotherapeutic agent in the treatment of human cancers. Similarly, *5-fluorocytosine* (see figure, b) is used as an antifungal drug because fungi, unlike mammals, can convert it to 2′-deoxy-5-fluorouridylate. Further, malarial parasites can use exogenous orotate to make pyrimidines for nucleic acid synthesis, whereas mammals cannot. Thus, *5-fluoroorotate* (see figure, c) is an effective antimalarial drug because it is selectively toxic to these parasites.

(a) **5-Fluorouracil**

(b) **5-Fluorocytosine**

(c) **5-Fluoroorotate**

The structures of
(a) 5-fluorouracil (5-FU),
(b) 5-fluorocytosine, and
(c) 5-fluoroorotate.

PROBLEMS

1. Suppose at certain specific metabolite concentrations *in vivo* the cyclic cascade regulating *E. coli* glutamine synthetase has reached a dynamic equilibrium where the average state of GS adenylylation is poised at $n = 6$. Predict what change in n will occur if
   a. [ATP] increases.
   b. P_{IID}/P_{IIA} increases.
   c. $[\alpha\text{-KG}]/[\text{Gln}]$ increases.
   d. $[P_i]$ decreases.

2. How many ATP equivalents are consumed in the production of one equivalent of urea by the urea cycle?

3. Why are persons on a high-protein diet advised to drink lots of water?

4. If PEP labeled with ^{14}C in the 2-position serves as the precursor to chorismate synthesis, which C atom in chorismate is radioactive?

5. Write a balanced equation for the synthesis of glucose (by gluconeogenesis) from aspartate.

6. Draw the purine and pyrimidine ring structures, indicating the metabolic source of each atom in the rings.

7. Starting from glutamine, aspartate, glycine, CO_2, and N^{10}-formyl-THF, how many ATP equivalents are expended in the synthesis of (a) ATP, (b) GTP, (c) UTP, and (d) CTP?

8. Illustrate the key points of regulation in (a) the biosynthesis of IMP, AMP, and GMP; (b) *E. coli* pyrimidine biosynthesis; and (c) mammalian pyrimidine biosynthesis.

9. Indicate which reactions of purine or pyrimidine metabolism are affected by the inhibitors (a) azaserine, (b) methotrexate, (c) sulfonamides, (d) allopurinol, and (e) 5-fluorouracil.

10. Since dUTP is not a normal component of DNA, why do you suppose ribonucleotide reductase has the capacity to convert UDP to dUDP?

11. Describe the underlying rationale for the regulatory effects exerted on ribonucleotide reductase by ATP, dATP, dTTP, and dGTP.

12. By what pathway(s) does the ribose released upon nucleotide degradation enter intermediary metabolism and become converted to cellular energy? How many ATP equivalents can be recovered from one equivalent of ribose?

FURTHER READING

Abeles, R. H., and Alston, T. A., 1990. Enzyme inhibition by fluoro compounds. *Journal of Biological Chemistry* **265**:16705–16708. A brief review of the usefulness of fluoro derivatives for probing reaction mechanisms.

Atkinson, D. E., and Camien, M. N., 1982. The role of urea synthesis in the removal of metabolic bicarbonate and the regulation of blood pH.

Current Topics in Cellular Regulation **21**:261–302. Describes the reasoning behind the proposal that the urea cycle eliminates bicarbonate as well as ammonium when urea is formed.

Bender, D. A., 1985. *Amino Acid Metabolism.* New York: John Wiley & Sons. A general review of amino acid metabolism.

Benkovic, S. J., 1984. The transformylase enzymes in *de novo* purine biosynthesis. *Trends in Biochemical Sciences* **9:**320–322. These enzymes provide an instance of metabolic channeling in one-carbon metabolism.

Boushey, C. J., et al., 1995. A quantitative assessment of plasma homocysteine as a risk factor for vascular disease. *Journal of the American Medical Association* **274:**1049–1057.

Fernandez-Canon, J. M., et al., 1996. The molecular basis of alkaptonuria. *Nature Genetics* **14:**19–24.

Graves, L. M., et al., 2000. Regulation of carbamoyl phosphate synthetase by MAP kinase. *Nature* **403:**328–331.

Henikoff, S., 1987. Multifunctional polypeptides for purine *de novo* synthesis. *BioEssays* **6:**8–13.

Holmgren, A., 1989. Thioredoxin and glutaredoxin systems. *Journal of Biological Chemistry* **264:**13963–13966.

Hudson, R. C., and Daniel, R. M., 1993. L-Glutamate dehydrogenases: Distribution, properties, and mechanism. *Comparative Biochemistry* **106B:**767–792.

Kishore, G. M., and Shah, D. M., 1988. Amino acid biosynthesis inhibitors as herbicides. *Annual Review of Biochemistry* **57:**627–663.

Liaw, S.-H., Pan, C., and Eisenberg, D. S., 1993. Feedback inhibition of fully unadenylylated glutamine synthetase from *Salmonella typhimurium* by glycine, alanine, and serine. *Proceedings of the National Academy of Sciences, U.S.A* **90:**4996–5000.

Licht, S., Gerfen, G. J., and Stubbe, J., 1996. Thiyl radicals in ribonucleotide reductases. *Science* **271:**477–481.

Marsh, E. N. G., 1995. A radical approach to enzyme catalysis. *BioEssays* **17:**431–441. A discussion of enzymatic mechanisms proceeding via carbon-based free radicals, the most prominent examples of which are ribonucleotide reductases.

Nordlund, P., and Eklund, H., 1993. Structure and function of the *Escherichia coli* ribonucleotide reductase protein R2. *Journal of Molecular Biology* **232:**123–164.

Reichard, P., 1988. Interactions between deoxyribonucleotide and DNA synthesis. *Annual Review of Biochemistry* **57:**349–374. A review of the regulation of ribonucleotide reductase by the scientist who discovered these phenomena.

Reichard, P., 1997. The evolution of ribonucleotide reduction. *Trends in Biochemical Sciences* **22:**81–85.

Rhee, C., and Stadtman, E. R., 1989. Regulation of *E. coli* glutamine synthetase. *Advances in Enzymology* **62:**37–92.

Scriver, C. R., et al., 1995. *The Metabolic and Molecular Bases of Inherited Disease*, 7th ed. New York: McGraw-Hill. A three-volume treatise on the biochemistry and genetics of inherited metabolic disorders, including disorders of amino acid, purine, and pyrimidine metabolism.

Srere, P. A., 1987. Complexes of sequential metabolic enzymes. *Annual Review of Biochemistry* **56:**89–124. A review of the evidence that enzymes in a pathway sequence are often physically associated with one another *in vivo*, particularly in eukaryotic cells.

Stubbe, J., 1990. Ribonucleotide reductases: Amazing and confusing. *Journal of Biological Chemistry* **265:**5329–5332. An analysis of the complex catalytic mechanism that underlies the overtly simple reaction mediated by ribonucleotide reductase.

Uhlin, U., and Eklund, H., 1996. The ten-stranded β/α barrel in ribonucleotide reductase protein R1. *Journal of Molecular Biology* **262:**358–369.

Wilson, D. K., Rudolph, F. B., and Quiocho, F. A., 1991. Atomic structure of adenosine deaminase complexed with a transition-state analog: Understanding catalysis and immunodeficient mutations. *Science* **252:**1279–1284.

CHAPTER
22

Study of an enzyme, a reaction, or a sequence can be biologically relevant only if its position in the hierarchy of function is kept in mind.

DANIEL E. ATKINSON, *Cellular Energy Metabolism and Its Regulation*

Outline

22.1 A Systems Analysis of Metabolism

22.2 Metabolic Stoichiometry and ATP Coupling

22.3 Energy Storage in the Adenylate System

22.4 Metabolic Specialization in the Organ Systems of Animals

Metabolic Integration and Organ Specialization

Ballet class, University of San Francisco. Metabolic integration is achieved through the highly regulated choreography of thousands of enzymatic reactions. (Phil Schermeister/Tony Stone Images)

I n the preceding chapters, we have explored the major metabolic pathways—glycolysis, the citric acid cycle, electron transport and oxidative phosphorylation, photosynthesis, gluconeogenesis, fatty acid oxidation, lipid biosynthesis, amino acid metabolism, and nucleotide metabolism. Several of these pathways are catabolic and serve to generate chemical energy useful to the cell; others are anabolic and use this energy to drive the synthesis of essential biomolecules. Despite their opposing purposes, these reactions typically occur at the same time, so that food molecules are broken down to provide the building blocks and energy for ongoing biosynthesis. Cells maintain a dynamic steady state through processes that involve considerable metabolic flux. We can gain a broader understanding of metabolism and biological processes in general if we step back and consider intermediary metabolism at a systems level of organization. At this level, we see that metabolic pathways merge into an integrated, orderly, responsive whole that operates in accord with the needs of the cell.[1]

---

[1]Many of the ideas presented in this chapter are derived from an insightful book by Daniel E. Atkinson of the University of California, Los Angeles, entitled *Cellular Energy Metabolism and Its Regulation* (New York: Academic Press, 1977).

The metabolism of a typical heterotrophic cell can be portrayed by a schematic diagram consisting of just three interconnected functional blocks: (1) catabolism, (2) anabolism, and (3) macromolecular synthesis and growth (Figure 22.1).

1. Catabolism. Foods are oxidized to CO_2 and H_2O in catabolism, and most of the electrons liberated are passed to oxygen via an electron transport pathway coupled to oxidative phosphorylation, so that ATP is formed. Some electrons go to reduce $NADP^+$ to NADPH, the source of reducing power for anabolism. Glycolysis, the citric acid cycle, electron transport and oxidative phosphorylation, and the pentose phosphate pathway are the principal pathways within this block. The metabolic intermediates in these pathways also serve as substrates for processes within the anabolic block.

2. Anabolism. The biosynthetic reactions that form the many cellular molecules collectively comprise anabolism. For thermodynamic reasons, the chemistry of anabolism is more complex than that of catabolism (i.e., it takes more

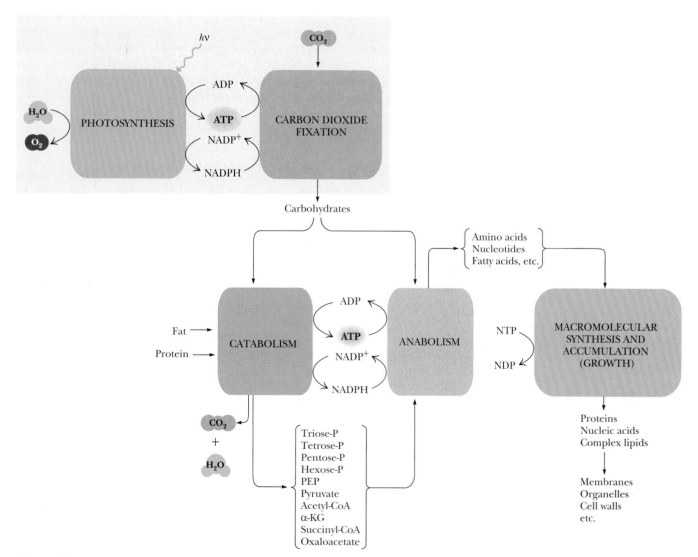

Figure 22.1 Block diagram of intermediary metabolism.

energy [and often more steps] to synthesize a molecule than can be produced from its degradation). Metabolic intermediates derived from glycolysis and the citric acid cycle are the precursors for this synthesis, with NADPH supplying the reducing power and ATP the coupling energy.

3. Macromolecular Synthesis and Growth. The organic molecules produced in anabolism are the building blocks for the creation of macromolecules. Like anabolism, macromolecular synthesis is driven by energy from ATP, although indirectly in some cases: GTP is the principal energy source for protein synthesis, CTP for phospholipid synthesis, and UTP for polysaccharide synthesis. However, keep in mind that ATP is the ultimate phosphorylating agent for formation of GTP, CTP, and UTP from GDP, CDP, and UDP, respectively. Macromolecules are the agents of biological function and information—proteins, nucleic acids, lipids that self-assemble into membranes, and so on. Growth can be represented as cellular accumulation of macromolecules and the partitioning of these materials of function and information into daughter cells in the process of cell division.

Only a Few Intermediates Interconnect the Major Metabolic Systems

Despite the complexity of processes taking place within each block, the connections between blocks involve only a limited number of substances. Just 10 or so kinds of catabolic intermediates from glycolysis, the pentose phosphate pathway, and the citric acid cycle serve as the raw material for most of anabolism: four kinds of sugar phosphates (triose-P, tetrose-P, pentose-P, hexose-P), three α-keto acids (pyruvate, oxaloacetate, and α-ketoglutarate), two coenzyme A derivatives (acetyl-CoA and succinyl-CoA), and PEP (phosphoenolpyruvate).

ATP and NADPH Couple Anabolism and Catabolism

Metabolic intermediates are consumed by anabolic reactions and must be continuously replaced by catabolic processes. In contrast, the energy-rich compounds ATP and NADPH are recycled rather than replaced. When these substances are used in biosynthesis, the products are ADP and $NADP^+$, and ATP and NADPH are regenerated from them by the energy released in oxidative reactions that occur in catabolism. ATP and NADPH are unique in that they are the only compounds whose purpose is to couple the energy-yielding processes of catabolism to the energy-consuming reactions of anabolism. Certainly, other coupling agents serve essential roles in metabolism. For example, NADH and [FADH$_2$] participate in the transfer of electrons from substrates to O_2 during oxidative phosphorylation. However, these reactions are solely catabolic, and the functions of NADH and [FADH$_2$] are fulfilled within the block called catabolism.

Phototrophs Have an Additional Metabolic System— The Photochemical Apparatus

The systems in Figure 22.1 reviewed thus far are representative only of metabolism as it exists in aerobic heterotrophs. The photosynthetic production of ATP and NADPH in photoautotrophic organisms entails a fourth block, the photochemical system (also shown in Figure 22.1). This block consumes H_2O and releases O_2. When this fourth block operates, energy production within the catabolic block can be largely eliminated. Yet another block, one to account for the fixation of carbon dioxide into carbohydrates, is also required for photoautotrophs. The inputs to this fifth block are the products of the photochemical system (ATP and NADPH) and CO_2 derived from the environment. The carbohydrate products of this block may enter catabolism, but not primarily for energy production. In photoautotrophs, carbohydrates are fed into catabolism to generate the metabolic intermediates needed to supply the block of anabolism. Although these diagrams are oversimplifications of the total metabolic processes in heterotrophic or phototrophic cells, they are useful illustrations of functional relationships between the major metabolic subdivisions. This general pattern provides an overall perspective on metabolism, making its purpose easier to understand.

22.2 Metabolic Stoichiometry and ATP Coupling

Virtually every metabolic pathway either consumes or produces ATP. The amount of ATP involved—that is, the stoichiometry of ATP synthesis or hydrolysis—lies at the heart of metabolic relationships. It is the ATP stoichiometry that determines the overall thermodynamics of metabolic sequences. By this we mean that the overall reaction mediated by any metabolic pathway is energetically favorable because of its particular ATP stoichiometry. A significant part of the energy released in the highly exergonic reactions of catabolism is captured in ATP synthesis. In turn, energy released upon ATP hydrolysis drives the thermodynamically unfavorable reactions of anabolism. Ultimately, then, the overall thermodynamic efficiency of metabolism is determined through coupling with ATP.

To illustrate this principle, we must first consider the three types of stoichiometries. The first two are fixed by the laws of chemistry, but the third is unique to living systems and reveals a fundamental difference between the inanimate world of chemistry and physics and the world of functional design—that is, the world of living organisms. The fundamental difference is the stoichiometry of ATP coupling.

1. Reaction Stoichiometry

This is simple chemical stoichiometry—the number of each kind of atom in any chemical reaction remains the same; thus, equal numbers must be present on both sides of the equation. This requirement holds even for a process as complex as cellular respiration:

$$C_6H_{12}O_6 + 6\,O_2 \longrightarrow 6\,CO_2 + 6\,H_2O$$

The six carbons in glucose appear as $6\,CO_2$, the 12 H of glucose appear as the 12 H in six molecules of water, and the 18 oxygens are distributed between CO_2 and H_2O.

2. Obligate Coupling Stoichiometry

Cellular respiration is an oxidation–reduction process, and the oxidation of glucose is coupled to the reduction of NAD^+ and [FAD]. (Brackets here denote that the relevant FAD is covalently linked to succinate dehydrogenase; see Chapter 17.) The NADH and $[FADH_2]$ thus formed are oxidized in the electron transport pathway:

(a) $C_6H_{12}O_6 + 10\,NAD^+ + 2\,[FAD] + 6\,H_2O \longrightarrow$
$$6\,CO_2 + 10\,NADH + 10\,H^+ + 2\,[FADH_2]$$

(b) $10\,NADH + 10\,H^+ + 2\,[FADH_2] + 6\,O_2 \longrightarrow$
$$12\,H_2O + 10\,NAD^+ + 2\,[FAD]$$

Sequence (a) accounts for the oxidation of glucose via glycolysis and the citric acid cycle. Sequence (b) is the overall equation for electron transport per

stoichiometry measurement of the amounts of chemical elements and molecules involved in chemical reactions (from the Greek *stoicheion*, "element," and *metria*, "measure")

glucose. The stoichiometry of coupling by the biological e^- carriers NAD^+ and FAD is fixed by the chemistry of electron transfer; each of the coenzymes serves as an e^- pair acceptor. Reduction of each O atom takes an e^- pair. Metabolism must obey these facts of chemistry: biological oxidation of glucose releases 12 e^- pairs, creating a requirement for 12 equivalents of e^- pair acceptors, which transfer the electrons to 12 O atoms.

3. Evolved Coupling Stoichiometries

The participation of ATP is fundamentally different from the role played by pyridine nucleotides and flavins. The stoichiometry of adenine nucleotides in metabolic sequences is not fixed by chemical necessity. Instead, the "stoichiometries" we observe are the consequences of evolutionary design. The overall equation for cellular respiration,[2] including the coupled formation of ATP by oxidative phosphorylation, is

$$C_6H_{12}O_6 + 6\,O_2 + 38\,ADP + 38\,P_i \longrightarrow 6\,CO_2 + 38\,ATP + 44\,H_2O$$

The "stoichiometry" of ATP formation, $38\,ADP + 38\,P_i \rightarrow 38\,ATP + 38\,H_2O$, cannot be predicted from any chemical considerations. The value of 38 ATP is an end result of biological adaptation. It is a phenotypic character of organisms—that is, a trait acquired through interaction of heredity and environment over the course of evolution. Like any evolved phenotypic character, this ATP stoichiometry is the result of compromise. The final trait is one particularly suited to the fitness of the organism.

The number 38 is not magical. Recall that in eukaryotes, the net yield of ATP per glucose is 30–32, not 38 (see Table 17.4). Also, the value of 38 was established a long time ago in evolution, when the prevailing atmospheric O_2 and CO_2 levels and the competitive situation were undoubtedly very different from those today. The significance of this number is that it provides a high yield of ATP for every glucose.

The Thermodynamic Role of ATP in Metabolism

The fundamental biological purpose of ATP as an energy coupling agent is to drive thermodynamically unfavorable reactions. (As a corollary, metabolic sequences composed of thermodynamically favorable reactions are exploited to drive the phosphorylation of ADP to make ATP.) Nature has devised enzymatic mechanisms that couple unfavorable reactions with ATP hydrolysis. In effect, the energy release accompanying ATP hydrolysis is transmitted to the unfavorable reaction so that the overall free energy change for the coupled process is negative, and thus the overall reaction is favorable. The involvement of ATP serves to alter the free energy change for a reaction; or, to put it another way, the role of ATP is to change the equilibrium ratio of [reactants] to [products] for a reaction.

Another way of viewing these relationships is to note that, at equilibrium, the concentrations of ADP and P_i will be vastly greater than that of ATP because $\Delta G^{\circ\prime}$ for ATP hydrolysis is a large negative number.[3] However, the cell

---

[2]This overall equation for cellular respiration is for the reaction within an uncompartmentalized (bacterial) cell. In eukaryotes, where much of the cellular respiration is compartmentalized within mitochondria, mitochondrial ADP/ATP exchange imposes a metabolic cost on the proton gradient of 1 H^+ per ATP, so that the overall yield of ATP per glucose is 32, not 38.

[3]Since $\Delta G^{\circ\prime} = -30.5$ kJ/mol, ln $K_{eq} = 12.3$. So $K_{eq} = 2.2 \times 10^5$. Choosing starting conditions of [ATP] = 8 mM, [ADP] = 8 mM, and [P_i] = 1 mM, we can assume that, at equilibrium, [ATP] has fallen to some insignificant value x, [ADP] = approximately 16 mM, and [P_i] = approximately 9 mM. The concentration of ATP at equilibrium, x, then is calculated to be about 1 nM.

where this reaction is at equilibrium is a dead cell. The living cell metabolizes food molecules to generate ATP. These catabolic reactions proceed with a very large overall decrease in free energy. Kinetic controls over the rates of the catabolic pathways are designed to ensure that the $[ATP]/([ADP][P_i])$ ratio is maintained very high. *The cell, by employing kinetic controls over the rates of metabolic pathways, maintains a very high $[ATP]/([ADP][P_i])$ ratio so that ATP hydrolysis can serve as the driving force for virtually all biochemical events.*

ATP Has Two Metabolic Roles

The role of ATP in metabolism is twofold:

1. ATP serves in a stoichiometric role to establish large equilibrium constants for metabolic conversions and to render metabolic sequences thermodynamically favorable. This is the role referred to when we call ATP the *energy currency* of the cell.
2. ATP also serves as an important allosteric effector in the kinetic regulation of metabolism. Its concentration (relative to those of ADP and AMP) is an index of the energy status of the cell and determines the rates of regulatory enzymes situated at key points in metabolism, such as PFK in glycolysis and FBPase in gluconeogenesis.

22.3 Energy Storage in the Adenylate System

Energy transduction and energy storage in the *adenylate system*—ATP, ADP, and AMP—lie at the very heart of metabolism. The amount of ATP used per minute by a cell is roughly equivalent to the steady-state amount of ATP it contains. Thus, the metabolic lifetime of an ATP molecule is brief. ATP, ADP, and AMP are all important effectors in exerting kinetic control on regulatory enzymes situated at key points in metabolism, so uncontrolled changes in their concentrations could have drastic consequences. The regulation of metabolism by adenylates in turn requires close control of the relative concentrations of ATP, ADP, and AMP. Some ATP-consuming reactions produce ADP; PFK and hexokinase are examples. Others lead to the formation of AMP, as in fatty acid activation by acetyl-CoA synthetases:

$$\text{Fatty acid} + \text{ATP} + \text{coenzyme A} \longrightarrow \text{AMP} + \text{PP}_i + \text{fatty acyl-CoA}$$

Adenylate Kinase Interconverts ATP, ADP, and AMP

Adenylate kinase (see Chapter 15), by catalyzing the reversible phosphorylation of AMP by ATP, provides a direct connection among all three members of the adenylate pool:

$$\text{ATP} + \text{AMP} \rightleftharpoons 2\,\text{ADP}$$

The free energy of hydrolysis of a phosphoanhydride bond is essentially the same in ADP and ATP (see Chapter 3), and the standard free energy change for this reaction is close to zero ($K_{eq} = 2.27$).

Energy Charge

The role of the adenylate system is to provide phosphoryl groups at **high group-transfer potential** in order to drive thermodynamically unfavorable reactions. The capacity of the adenylate system to fulfill this role depends on how fully

group-transfer potential free energy change for transfer of a functional group. High group-transfer potential means the transfer is thermodynamically favored.

A DEEPER LOOK

ATP Changes the K_{eq} for a Process by a Factor of 10^8

Consider a process, A $\rightleftharpoons$ B. It could be a biochemical reaction, or the transport of an ion against a concentration gradient, or even a mechanical process (such as muscle contraction). Assume that it is a thermodynamically unfavorable reaction. Let's say, for purposes of illustration, that $\Delta G^{\circ\prime} = +13.8$ kJ/mol. From the equation,

$$\Delta G^{\circ\prime} = -RT \ln K_{eq}$$

we have

$$+13,800 = -(8.31 \text{ J/K} \cdot \text{mol})(298 \text{ K}) \ln K_{eq}$$

which yields

$$\ln K_{eq} = -5.57$$

Therefore,

$$K_{eq} = 0.0038 = \frac{[B_{eq}]}{[A_{eq}]}$$

This reaction is clearly unfavorable (as we could have foreseen from its positive $\Delta G^{\circ\prime}$). At equilibrium, there is one molecule of product B for every 263 molecules of reactant A. Not much A was transformed to B.

Now suppose the reaction A $\rightleftharpoons$ B is coupled to ATP hydrolysis, as is often the case in metabolism:

$$A + ATP \rightleftharpoons B + ADP + P_i$$

The thermodynamic properties of this coupled reaction are the same as the sum of the thermodynamic properties of the partial reactions:

$$A \rightleftharpoons B \qquad \Delta G^{\circ\prime} = +13.8 \text{ kJ/mol}$$

$$ATP + H_2O \rightleftharpoons ADP + P_i \qquad \Delta G^{\circ\prime} = -30.5 \text{ kJ/mol}$$

$$A + ATP + H_2O \rightleftharpoons B + ADP + P \qquad \Delta G^{\circ\prime} = -16.7 \text{ kJ/mol}$$

That is,

$$\Delta G^{\circ\prime}_{overall} = -16.7 \text{ kJ/mol}$$

So

$$-16,700 = -RT \ln K_{eq} = -(8.31)(298) \ln K_{eq}$$

$$\ln K_{eq} = \frac{-16,700}{-2476} = 6.75$$

$$K_{eq} = 850$$

Using this equilibrium constant, let us now consider the cellular situation in which the concentrations of A and B are brought to equilibrium in the presence of typical prevailing concentrations of ATP, ADP, and P_i.*

$$K_{eq} = \frac{[B_{eq}][ADP][P_i]}{[A_{eq}][ATP]}$$

$$850 = \frac{[B_{eq}](8 \times 10^{-3})(10^{-3})}{[A_{eq}](8 \times 10^{-3})}$$

$$\frac{[B_{eq}]}{[A_{eq}]} = 850,000$$

Comparison of the $[B_{eq}]/[A_{eq}]$ ratio for the simple A $\rightleftharpoons$ B reaction with the coupling of this reaction to ATP hydrolysis gives

$$\frac{850,000}{0.0038} = 2.2 \times 10^8$$

The equilibrium ratio of B to A is more than 10^8 greater when the reaction is coupled to ATP hydrolysis. A reaction that was clearly unfavorable ($K_{eq} = 0.0038$) has become emphatically spontaneous!

The involvement of ATP has raised the equilibrium ratio of B/A more than 200 million–fold. It is informative to realize that this multiplication factor does not depend on the nature of the reaction. (Recall that we defined A $\rightleftharpoons$ B in the most general terms.) Also, the value of this equilibrium constant ratio (2.2×10^8) is not at all dependent on the particular reaction chosen or its standard free energy change, $\Delta G^{\circ\prime}$. You can satisfy yourself on this point by choosing some value for $\Delta G^{\circ\prime}$ other than $+13.8$ kJ/mol and repeating these calculations (keeping the concentrations of ATP, ADP, and P_i at 8, 8, and 1 mM, as before).

*The concentrations of ATP, ADP, and P_i in a normal, healthy bacterial cell growing at 25°C are maintained at roughly 8 mM, 8 mM, and 1 mM, respectively. Therefore, the ratio $[ADP][P_i]/[ATP]$ is about 10^{-3}. Under these conditions, ΔG for ATP hydrolysis is approximately -47.6 kJ/mol.

charged it is with phosphoric anhydrides. **Energy charge** is an index of this capacity:

$$\text{Energy charge} = \frac{1}{2} \frac{(2[ATP] + [ADP])}{([ATP] + [ADP] + [AMP])}$$

The denominator represents the total adenylate pool ([ATP] + [ADP] + [AMP]); the numerator is the number of phosphoric anhydride bonds in the pool, two for each ATP and one for each ADP. The factor $\frac{1}{2}$ normalizes the equation so that energy charge, or **E.C.,** has the range 0 to 1.0. If all the adenylate is in the form of ATP, then E.C. = 1.0, and the potential for phosphoryl transfer is maximal. At the other extreme, if AMP is the only adenylate form present, then E.C. = 0. It is reasonable to assume that the adenylate kinase reaction is never far from equilibrium in the cell. Then the relative amounts of the three adenine nucleotides are fixed by the energy charge. Figure 22.2 shows the relative changes in the concentrations of the adenylates as energy charge varies from 0 to 1.0.

The Response of Enzymes to Energy Charge

Regulatory enzymes typically respond in reciprocal fashion to adenine nucleotides. For example, PFK is stimulated by AMP and inhibited by ATP. If the activities of various regulatory enzymes are examined *in vitro* as a function of energy charge, an interesting relationship appears. Regulatory enzymes in energy-producing catabolic pathways show greater activity at low energy charge, but the activity falls off abruptly as E.C. approaches 1.0. In contrast, regulatory enzymes of anabolic sequences are not very active at low energy charge, but their activities increase exponentially as E.C. nears 1.0 (Figure 22.3). These contrasting responses are termed **R,** for ATP-regenerating, and **U,** for ATP-utilizing. Regulatory enzymes such as PFK and pyruvate kinase in glycolysis follow the **R** response curve as E.C. is varied. Note that PFK itself is an ATP-utilizing enzyme, using ATP to phosphorylate fructose-6-phosphate to yield fructose-1,6-bisphosphate. Nevertheless, because PFK acts physiologically as the valve controlling the flux of carbohydrate down the catabolic pathways of cellular respiration that lead to ATP regeneration, it responds as an **"R"** enzyme to energy charge. Regulatory enzymes in anabolic pathways, such as acetyl-CoA carboxylase, which initiates fatty acid biosynthesis, respond as **"U"** enzymes.

The overall purposes of the **R** and **U** pathways are diametrically opposite in terms of ATP involvement. Note in Figure 22.3 that the **R** and **U** curves intersect at a rather high E.C. value. As E.C. increases past this point, **R** activities decline precipitously and **U** activities rise. That is, when E.C. is very high, biosynthesis is accelerated while catabolism diminishes. The consequence of these effects is that ATP is used up faster than it is regenerated, and so E.C. begins to fall. As E.C. drops below the point of intersection, **R** processes are favored over **U.** Then, ATP is generated faster than it is consumed, and E.C. rises again. The net result is that the value of energy charge oscillates about a point of **steady state** (Figure 22.4). The experimental results obtained from careful measurement of the relative amounts of AMP, ADP, and ATP in living cells reveals that normal cells have an energy charge in the neighborhood of 0.85 to 0.88. Maintenance of this steady-state value is one criterion of cell health and normalcy.

22.4 Metabolic Specialization in the Organ Systems of Animals

In animals, organ systems carry out specific physiological functions. Each organ expresses a repertoire of metabolic pathways that is consistent with its physiological purpose. Such specialization depends on coordination of metabolic

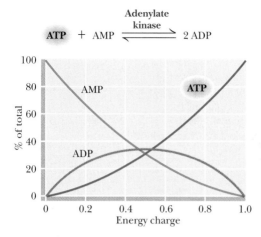

Figure 22.2 Relative concentrations of AMP, ADP, and ATP as a function of energy charge. (This graph was constructed assuming that the adenylate kinase reaction is at equilibrium and that $\Delta G°'$ for the reaction is −473 J/mol; K_{eq} = 1.2.)

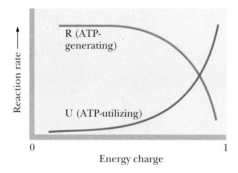

Figure 22.3 Responses of regulatory enzymes to variation in energy charge. Enzymes in catabolic pathways have as their ultimate metabolic purpose the regeneration of ATP from ADP. Such enzymes show an **R** pattern of response to energy charge. Enzymes in biosynthetic pathways utilize ATP to drive anabolic reactions; these enzymes follow the **U** curve in response to energy charge.

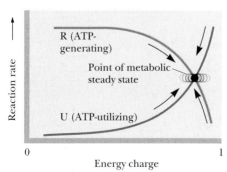

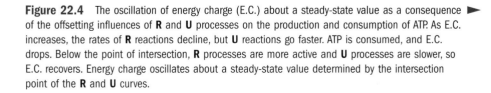

Figure 22.4 The oscillation of energy charge (E.C.) about a steady-state value as a consequence ▶ of the offsetting influences of **R** and **U** processes on the production and consumption of ATP. As E.C. increases, the rates of **R** reactions decline, but **U** reactions go faster. ATP is consumed, and E.C. drops. Below the point of intersection, **R** processes are more active and **U** processes are slower, so E.C. recovers. Energy charge oscillates about a steady-state value determined by the intersection point of the **R** and **U** curves.

responsibilities among organs so that the organism as a whole may thrive. Essentially all cells in animals have the set of enzymes common to the central pathways of intermediary metabolism, especially the enzymes involved in the formation of ATP and the synthesis of glycogen and lipid reserves. Nevertheless, organs differ in the metabolic fuels they prefer as substrates for energy production. Important differences also occur in the ways ATP is used to fulfill the organs' specialized metabolic functions. To illustrate these relationships, we will consider the metabolic interactions among the major organ systems found in humans: brain, skeletal muscle, heart, adipose tissue, and liver. In particular, the focus will be on energy metabolism in these organs (Figure 22.5). The major fuel depots in animals are glycogen in liver and muscle, triacylglycerols (fats) stored in adipose tissue, and protein, most of which is in skeletal muscle. In general, the order of preference for the use of these fuels is the order given: glycogen > triacylglycerol > protein. Nevertheless, the tissues of the body work together to maintain **caloric homeostasis,** defined as a constant availability of fuels in the blood.

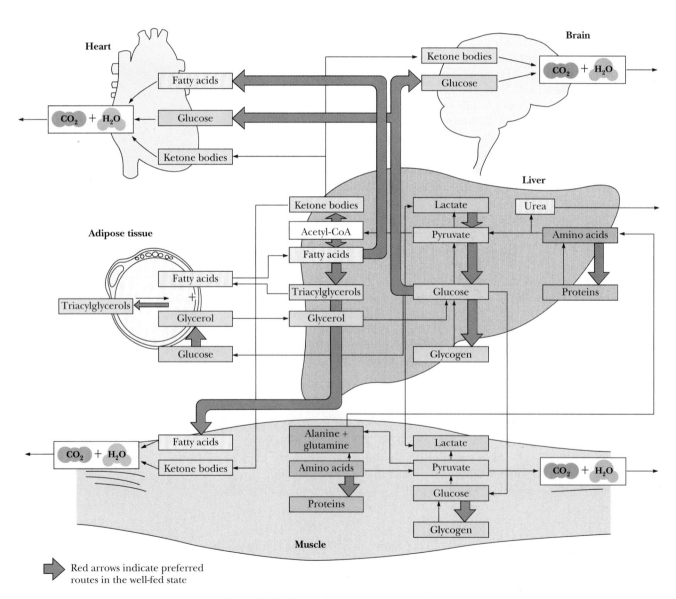

Figure 22.5 Metabolic relationships among the major human organs: brain, muscle, heart, adipose tissue, and liver.

Table 22.1   Energy Metabolism in Major Vertebrate Organs

Organ	Energy Reservoir	Preferred Substrate	Energy Sources Exported
Brain	None	Glucose (ketone bodies during starvation)	None
Skeletal muscle (resting)	Glycogen	Fatty acids	None
Skeletal muscle (prolonged exercise)	None	Glucose	Lactate
Heart muscle	Glycogen	Fatty acids	None
Adipose tissue	Triacylglycerol	Fatty acids	Fatty acids, glycerol
Liver	Glycogen, triacylglycerol	Amino acids, glucose, fatty acids	Fatty acids, glucose, ketone bodies

Organ Specializations

Table 22.1 summarizes the energy metabolism of the major human organs.

Brain.   The brain has two remarkable metabolic features. First, it has a very high respiratory metabolism. In resting adult humans, 20% of the oxygen consumed is used by the brain, even though it constitutes only 2% or so of body mass. Interestingly, this level of oxygen consumption is independent of mental activity, continuing even during sleep. Second, the brain is an organ with no significant fuel reserves—no glycogen, usable protein, or fat (even in "fatheads"!). Normally, the brain uses only glucose as a fuel and is totally dependent on the blood for a continuous incoming supply. Interruption of glucose supply for even brief periods of time (as in a stroke) can lead to irreversible losses in brain function. The brain uses glucose to carry out ATP synthesis via cellular respiration. High rates of ATP production are necessary to power the plasma membrane Na^+,K^+-ATPase so that the membrane potential essential for transmission of nerve impulses is maintained.

During prolonged fasting or starvation, the body's glycogen reserves are depleted. Under such conditions, the brain adapts to use β-hydroxybutyrate (Figure 22.6) as a source of fuel, converting it to acetyl-CoA for energy production via the citric acid cycle. β-Hydroxybutyrate (see Chapter 20) is formed from fatty acids in the liver. Although the brain cannot use free fatty acids or lipids directly from the blood as fuel, the conversion of these substances to β-hydroxybutyrate in the liver allows the brain to use body fat as a source of energy. The brain's other potential source of fuel during starvation is glucose obtained from gluconeogenesis in the liver (see Chapter 19), using the carbon skeletons of amino acids derived from muscle protein breakdown. The ability of the brain to use β-hydroxybutyrate from fat spares protein from degradation until lipid reserves are exhausted.

Muscle.   Skeletal muscle is responsible for about 30% of the O_2 consumed by the human body at rest. During periods of maximal exertion, skeletal muscle can account for over 90% of the total metabolism. Muscle metabolism is primarily dedicated to the production of ATP as the source of energy for contraction and relaxation. Muscle contraction occurs when a motor nerve impulse causes Ca^{2+} release from specialized endomembrane compartments (the transverse tubules and sarcoplasmic reticulum). Ca^{2+} floods the *sarcoplasm* (the cytosolic compartment of muscle cells), where it binds to **troponin C,** a regulatory protein, initiating a series of events that culminate in the sliding of *myosin*

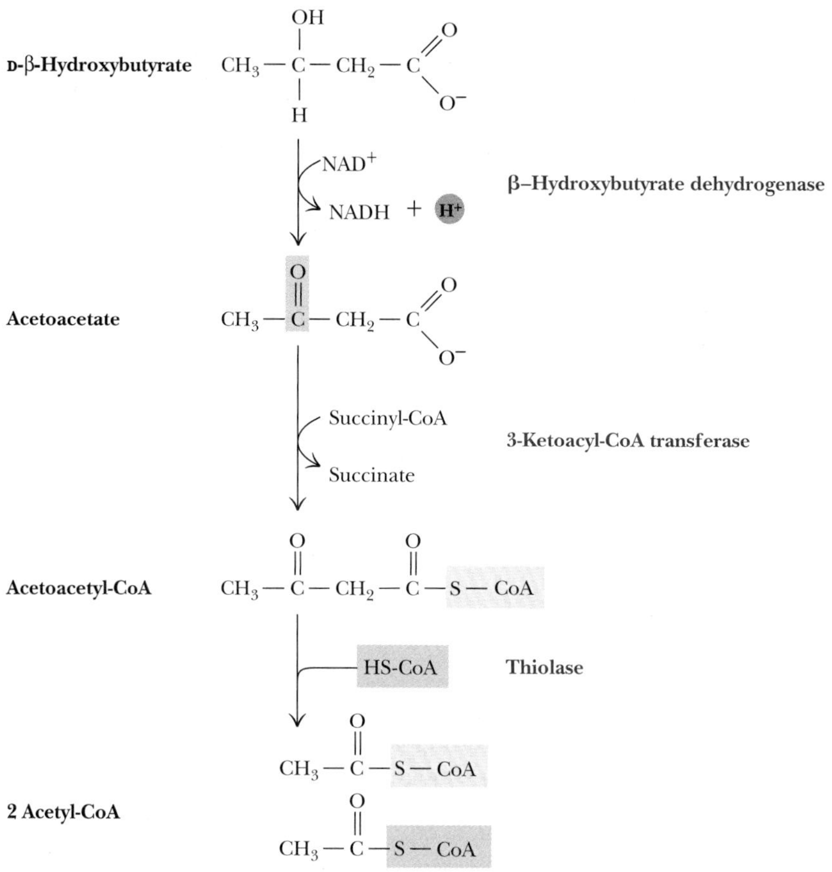

Figure 22.6 The structure of β-hydroxybutyrate and its conversion to acetyl-CoA for combustion in the citric acid cycle.

thick filaments along *actin* thin filaments. This mechanical movement is driven by energy released upon hydrolysis of ATP (see Chapter 13). The net result is that the muscle shortens. Relaxation occurs when the Ca^{2+} ions are pumped back into the sarcoplasmic reticulum by the action of a Ca^{2+}-transporting membrane ATPase. Two Ca^{2+} are translocated per ATP hydrolyzed. The amount of ATP used during relaxation is almost as much as that consumed during contraction.

Because muscle contraction is an intermittent process that occurs upon demand, muscle metabolism is designed for a demand response. Muscle at rest uses free fatty acids, glucose, or ketone bodies as fuel and produces ATP via oxidative phosphorylation. Resting muscle also contains about 2% glycogen by weight and an amount of phosphocreatine (Figure 22.7) capable of providing

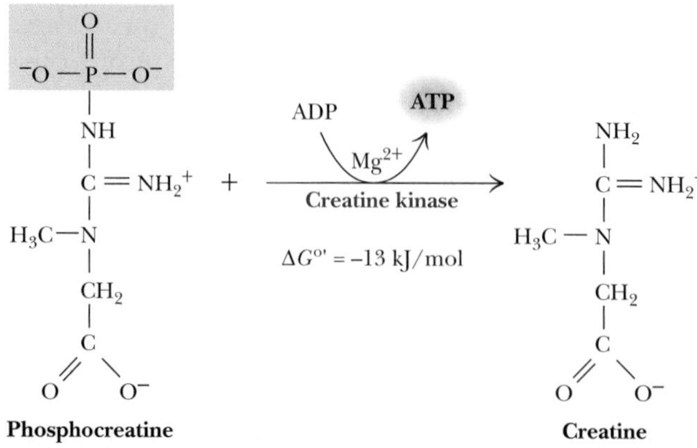

Figure 22.7 Phosphocreatine serves as a reservoir of ATP-synthesizing potential. When ADP accumulates as a consequence of ATP hydrolysis, creatine kinase catalyzes the formation of ATP at the expense of phosphocreatine. During periods of rest, when ATP levels are restored by oxidative phosphorylation, creatine kinase acts in reverse to restore the phosphocreatine supply.

HUMAN BIOCHEMISTRY

Athletic Performance Enhancement with Creatine Supplements?

The creatine pool in a 70-kg (154-lb) human body is about 120 grams. This pool includes dietary creatine (from meat) and creatine synthesized by the human body from its precursors (arginine, glycine, and methionine). Ninety-five percent of this creatine is stored in the skeletal and smooth muscles, about 70% of it in the form of phosphocreatine. Supplementing the diet with 20 to 30 grams of creatine per day for 4 to 21 days can increase the muscle creatine pool by as much as 50% in someone with a previously low creatine level. Thereafter, supplements of 2 g/day will maintain elevated creatine stores. Studies indicate that creatine supplementa-

tion gives some improvement in athletic performance during high-intensity, short-duration events (such as weight lifting), but no benefit in endurance events (such as distance running). The distinction makes sense in light of phosphocreatine's role as the substrate that creatine kinase uses to regenerate ATP from ADP. Intense muscular activity quickly (in less than 2 sec) exhausts ATP supplies; [phosphocreatine]$_{muscle}$ is sufficient to restore ATP levels for a few extra seconds, but no more. The Food and Drug Administration advises consumers to consult their doctors before using creatine as a dietary supplement.

---

enough ATP to power about 4 seconds of exertion. During strenuous exertion, such as a 100-meter sprint, once the phosphocreatine is depleted, muscle relies solely on its glycogen reserves, making ATP for contraction via glycolysis. In contrast to the citric acid cycle and oxidative phosphorylation pathways, glycolysis is capable of explosive bursts of activity, and the flux of glucose-6-phosphate through this pathway can increase 2000-fold almost instantaneously. The triggers for this activation are Ca^{2+} and the "fight or flight" hormone *epinephrine* (see Chapters 19 and 26). Little inter-organ cooperation occurs during strenuous (anaerobic) exercise.

Muscle fatigue is the inability of a muscle to maintain power output. During maximum exertion, the onset of fatigue takes only 20 seconds or so. Fatigue is not the result of exhaustion of the glycogen reserves, nor is it a consequence of lactate accumulation in the muscle. Instead, it is caused by a decline in intramuscular pH as protons are generated during glycolysis. (The overall conversion of glucose to two lactate in glycolysis is accompanied by the release of two H^+.) The pH may fall as low as 6.4. It is likely that the decline in PFK activity at low pH leads to a lowered flux of hexose through glycolysis and inadequate ATP levels, causing a feeling of fatigue. One benefit of PFK inhibition is that the ATP remaining is not consumed in the PFK reaction, and the cell is spared the more serious consequences of losing all of its ATP.

During fasting or excessive activity, skeletal muscle protein is degraded to amino acids so that their carbon skeletons can be used as fuel. Many of the skeletons are converted to pyruvate, which can be transaminated back into alanine for export via the circulation (Figure 22.8). Alanine is carried to the liver,

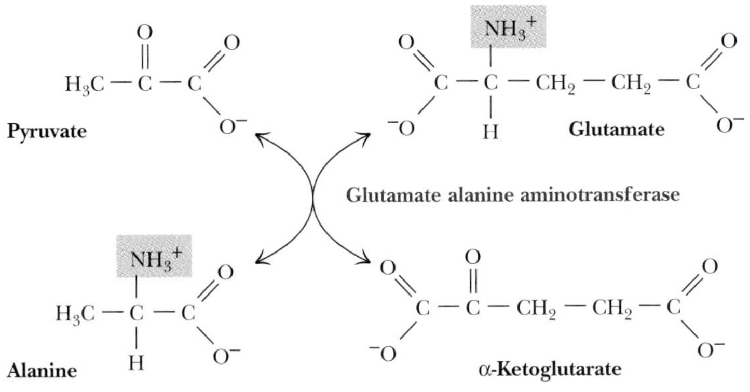

Figure 22.8 The transamination of pyruvate to alanine by glutamate : alanine aminotransferase.

which in turn deaminates it back into pyruvate so that it can serve as a substrate for gluconeogenesis. Although muscle protein can be mobilized as an energy source, it is not efficient for an organism to consume its muscle and lower its overall fitness for survival. Muscle protein represents a fuel of last resort.

Heart. In contrast to the intermittent work of skeletal muscle, the activity of heart muscle is constant and rhythmic. The range of activity in heart is also much less than that in muscle. Consequently, the heart functions as a completely aerobic organ and, as such, is very rich in mitochondria. Roughly half the cytoplasmic volume of heart muscle cells is occupied by mitochondria. Under normal working conditions, the heart prefers fatty acids as fuel, oxidizing acetyl-CoA units via the citric acid cycle and producing ATP for contraction via oxidative phosphorylation. Heart tissue has minimal energy reserves: a small amount of phosphocreatine and limited quantities of glycogen. As a result, the heart must be continually nourished with oxygen and free fatty acids, glucose, or ketone bodies as fuel.

Adipose Tissue. Adipose tissue is an amorphous tissue that is widely distributed about the body—around blood vessels, in the abdominal cavity and mammary glands, and, most prevalently, as deposits under the skin. It consists principally of cells known as **adipocytes** that no longer replicate. However, adipocytes can increase in number as adipocyte precursor cells divide, and obese individuals tend to have more of them. As much as 65% of the weight of adipose tissue is triacylglycerol that is stored in adipocytes, essentially as oil droplets. A normal 70-kg man has enough caloric reserve stored as fat to sustain a 6000-kJ/day rate of energy production for three months, which is adequate for survival, assuming no serious metabolic aberrations (such as nitrogen, mineral, or vitamin deficiencies). Despite their role as energy storage depots, adipocytes have a high rate of metabolic activity, synthesizing and breaking down triacylglycerol so that the average turnover time for a triacylglycerol molecule is just a few days. Adipocytes actively carry out cellular respiration, transforming glucose to energy via glycolysis, the citric acid cycle, and oxidative phosphorylation. If glucose levels in the diet are high, glucose is converted to acetyl-CoA for fatty acid synthesis. However, under most conditions, free fatty acids for triacylglycerol synthesis are obtained from the liver. Because adipocytes lack glycerol kinase, they cannot recycle the glycerol of triacylglycerol but depend instead on glycolytic conversion of glucose to dihydroxyacetone-3-phosphate (DHAP) and the reduction of DHAP to glycerol-3-phosphate for triacylglycerol biosynthesis. Adipocytes also require glucose to feed the pentose phosphate pathway for NADPH production.

Glucose plays a pivotal role for adipocytes. If glucose levels are adequate, glycerol-3-phosphate is formed in glycolysis, and the free fatty acids liberated in triacylglycerol breakdown are reesterified to glycerol to re-form triacylglycerols. However, if glucose levels are low, [glycerol-3-phosphate] falls, and free fatty acids are released to the bloodstream (see Chapter 20).

"Brown Fat." A specialized type of adipose tissue called **brown fat** is found in newborns and hibernating animals. The abundance of mitochondria, which are rich in cytochromes, is responsible for the brown color of this fat. As usual, these mitochondria are very active in electron transport–driven proton translocation, but these particular mitochondria contain in their inner membranes a protein, **thermogenin,** also known as *uncoupler protein-1* (see Chapter 17), that creates a passive proton channel, permitting the H^+ ions to re-enter the mitochondrial matrix without generating ATP. Instead, the energy of oxidation is dissipated as heat. Indeed, brown fat is specialized to oxidize fatty acids for heat production rather than ATP synthesis.

HUMAN BIOCHEMISTRY

Fat-Free Mice: A Model for One Form of Diabetes

Scientists at the National Institutes of Health have created transgenic mice that lack white adipose tissue throughout their lifetimes. These mice were created by blocking the normal differentiation of stem cells into adipocytes, so that essentially no white adipose tissue can be formed in these animals. These "fat-free" mice have double the food intake and five times the water intake of normal mice. "Fat-free" mice also show decreased physical activity and must be kept warm on little heating pads to survive because they lack insulating fat. They are also diabetic, with three times the normal blood glucose and triacylglycerol levels and only 5% of normal leptin levels; they die prematurely. Like Type II diabetics, "fat-free"

mice have markedly elevated insulin levels (50–400 times normal) but are unresponsive to insulin. These mice serve as an excellent model for the disease *lipoatropic diabetes,* an inherited disease characterized by the absence of adipose tissue and by severe diabetic symptoms. Indeed, transplantation of adipose tissue into these "fat-free" mice cured their diabetes. As the major organ for triacylglycerol storage, white adipose tissue helps control energy homeostasis (food intake and energy expenditure) via leptin (see *Human Biochemistry,* below). Clearly, the absence of adipose tissue has widespread, harmful consequences for metabolism.

Liver.   The liver serves as the major metabolic processing center in vertebrates. Except for dietary triacylglycerols, which are metabolized principally by adipose tissue, most of the incoming nutrients that pass through the intestinal tract are routed via the portal vein to the liver for processing and distribution. Much of the liver's activity centers around conversions involving glucose-6-phosphate (Figure 22.9). Glucose-6-phosphate can be converted to glycogen, released as blood glucose, used to generate NADPH and pentoses via the

HUMAN BIOCHEMISTRY

Controlling Obesity—Leptin Is a Protein That Stimulates "Fat Burning"

Leptin (from the Greek *lepto,* meaning "thin") is a 16-kD, 146-amino-acid residue protein produced principally in adipocytes (fat cells). Leptin has a four-helix bundle tertiary structure similar to cytokines (protein hormones involved in cell–cell communication). Leptin acts as a signal molecule that limits food intake and increases energy expenditure. When high levels of leptin are injected daily into obese mice, the animals have diminished food intake and increased rates of fat oxidation, losing 40% of their body weight in a month. Accordingly, leptin arouses great interest in terms of its potential to cause weight loss in humans. Half of the American population is overweight and one-third is clinically obese (overweight by 20% or more). The desire to lose weight is a national obsession. Normally, as fat deposits accumulate in adipocytes, more and more leptin is produced in these cells and spewed into the bloodstream. Leptin levels in the blood communicate the status of triacylglycerol levels in the adipocytes to the central nervous system so that appropriate changes in appetite take place. If leptin levels are low ("starvation"), appetite increases; if leptin levels are high ("overfeeding"), appetite is suppressed. Leptin also regulates fat metabolism in adipocytes, inhibiting fatty acid biosynthesis and stimulating fat metabolism. In the latter case, leptin induces synthesis of the enzymes in the fatty acid oxidation pathway and in-

creases expression of *uncoupling protein-2 (UCP-2),* a mitochondrial protein that uncouples oxidation from phosphorylation so that the energy of oxidation is lost as heat (thermogenesis).

Interestingly, obese humans typically have high blood leptin levels in proportion to their mass of body fat. Nevertheless, leptin injection into obese human subjects has been shown to induce weight loss. Leptin receptors are located in the hypothalamus, a region of the lower brain involved in integrating the body's hormonal and nervous systems. Leptin binding to its receptors inhibits hypothalamic release of *neuropeptide Y,* a potent *orexic* (appetite-stimulating) peptide hormone; thus, leptin is an *anorexic* (appetite-suppressing) agent. Functional leptin receptors are also essential for pituitary function, growth hormone secretion, and normal puberty.

The mechanisms underlying appetite regulation by the hypothalamus remain elusive. Thus far, genetic studies have implicated, among other things, leptin, the leptin receptor, α-melanocyte–stimulating hormone (α-MSH), and the melanocortin-4 receptor (MC4R). MC4R is a member of the 7-TMS G protein–coupled receptor (GPCR) family of membrane receptors; MC4R triggers cellular responses through adenylyl cyclase activation (see Chapters 11 and 26).

pentose phosphate cycle, or catabolized to acetyl-CoA for fatty acid synthesis or for energy production via oxidative phosphorylation. Most of the liver glucose-6-phosphate arises from dietary carbohydrate, from degradation of glycogen reserves, or from muscle lactate that enters the gluconeogenic pathway.

The liver plays an important regulatory role in metabolism by buffering the level of blood glucose. Liver has two enzymes for glucose phosphorylation, hexokinase and glucokinase. Unlike hexokinase, glucokinase has a low affinity for glucose. Its K_m for glucose is high, on the order of 10 mM. When blood glucose levels are high, glucokinase activity augments hexokinase in phosphorylating glucose as an initial step leading to its storage in glycogen. The major metabolic hormones—epinephrine, glucagon, and insulin—all influence glucose metabolism in the liver to keep blood glucose levels relatively constant (see Chapter 26).

The liver is a major center for fatty acid turnover. When the demand for metabolic energy is high, triacylglycerols are broken down and fatty acids are degraded in the liver to acetyl-CoA to form ketone bodies, which are exported to the heart, brain, and other tissues. If energy demands are low, fatty acids are incorporated into triacylglycerols that are carried to adipose tissue for deposition as fat. Cholesterol is also synthesized in the liver from two-carbon units derived from acetyl-CoA.

In addition to these central functions in carbohydrate and fat-based energy metabolism, the liver serves other purposes. For example, the liver can use amino acids as metabolic fuels. Amino acids are first converted to their corresponding α-keto acids by aminotransferases. The amino group is excreted after incorporation into urea in the urea cycle. The carbon skeletons of gluconeogenic amino acids can be used for glucose synthesis, whereas those of ketogenic amino acids appear in ketone bodies (see Figure 21.23). The liver is also the principal detoxification organ in the body. The endoplasmic retic-

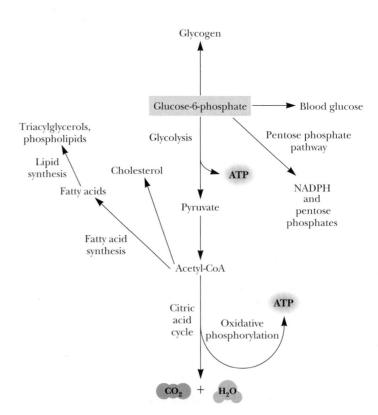

Figure 22.9 Metabolic conversions of glucose-6-phosphate in the liver.

HUMAN BIOCHEMISTRY

The Metabolic Effects of Alcohol Consumption

Ethanol metabolism alters the $NAD^+/NADH$ ratio. Ethanol is metabolized to acetate in the liver by alcohol dehydrogenase and aldehyde dehydrogenase:

$$CH_3CH_2OH + NAD^+ \longrightarrow CH_3CHO + NADH + H^+$$

$$CH_3CHO + NAD^+ \longrightarrow CH_3COO^- + NADH + H^+$$

The excess NADH produced inhibits NAD^+-requiring reactions, such as gluconeogenesis and fatty acid oxidation. Inhibition of fatty acid oxidation causes elevated triacylglycerol levels in the liver. Over

time, these triacylglycerols accumulate as fatty deposits, which ultimately contribute to cirrhosis of the liver. Inhibition of gluconeogenesis leads to buildup of this pathway's substrate, lactate. Lactic acid accumulation in the blood causes acidosis. A further consequence is that acetaldehyde can form adducts with protein $-NH_2$ groups, which may impair protein function. Because gluconeogenesis is impaired, alcohol consumption can cause *hypoglycemia* (low blood-sugar levels) in someone who is undernourished. In turn, hypoglycemia can cause irreversible damage to the central nervous system.

ulum of liver cells is rich in enzymes that convert biologically active substances such as hormones, poisons, and drugs into less harmful by-products.

Liver disease leads to serious metabolic derangements, particularly in amino acid metabolism. In cirrhosis, the liver becomes defective in converting NH_4^+ to urea for excretion, and blood levels of NH_4^+ rise. Ammonia is toxic to the central nervous system, and coma ensues.

PROBLEMS

1. Assume the following intracellular concentrations in muscle tissue: ATP = 8 mM, ADP = 0.9 mM, AMP = 0.04 mM, P_i = 8 mM. What is the *energy charge* in muscle?

2. Strenuous muscle exertion (as in the 100-meter dash) rapidly depletes ATP levels. How long will 8 mM ATP last if 1 gram of muscle consumes 300 μmol of ATP per minute? (Assume muscle is 70% water.) Muscle contains phosphocreatine as a reserve of phosphorylation potential. Assuming [phosphocreatine] = 40 mM, [creatine] = 4 mM, and $\Delta G^{\circ\prime}$(phosphocreatine + H_2O → creatine + P_i) = −43.3 kJ/mol, how low must [ATP] become before it can be replenished by the reaction: phosphocreatine + ADP → ATP + creatine? [Remember, $\Delta G^{\circ\prime}$ (ATP hydrolysis) = −30.5 kJ/mol.]

3. The standard reduction potentials for the $(NAD^+/NADH)$ and $(NADP^+/NADPH)$ couples are identical, namely, −320 mV. Assuming the *in vivo* concentration ratios $NAD^+/NADH$ = 20 and $NADP^+/NADPH$ = 0.1, what is ΔG for the following reaction?

$$NADPH + NAD^+ \longrightarrow NADP^+ + NADH$$

4. Assume the total intracellular pool of adenylates (ATP + ADP + AMP) = 8 mM, 90% of which is ATP. What are [ADP] and [AMP] if the adenylate kinase reaction is at equilibrium? Suppose [ATP] drops suddenly by 10%. What are the concentrations now for ADP and AMP, assuming that the adenylate kinase reaction is at equilibrium? By what factor has the AMP concentration changed?

5. Leptin not only induces synthesis of fatty acid oxidation enzymes and uncoupler protein-2 in adipocytes, but it also causes inhibition of acetyl-CoA carboxylase, resulting in a decline in fatty acid biosynthesis. This effect on acetyl-CoA carboxylase, as an additional consequence, enhances fatty acid oxidation. Explain how leptin-induced inhibition of acetyl-CoA carboxylase might promote fatty acid oxidation.

FURTHER READING

Atkinson, D. E., 1977. *Cellular Energy Metabolism and Its Regulation.* New York: Academic Press. A very readable book on the design and purpose of cellular energy metabolism. Its emphasis is the evolutionary design of metabolism within the constraints of chemical thermodynamics. The book is filled with novel insights regarding why metabolism is organized as it is and why ATP occupies a central position in biological energy transformations.

Barinaga, M., 1995. "Obese" protein slims mice. *Science* **269**:475–476. See also the references therein.

Clement, K., et al., 1998. A mutation in the human leptin receptor gene causes obesity and pituitary dysfunction. *Nature* **392**:398–401.

Ekblom, B., 1999. Effects of creatine supplementation on performance. *American Journal of Sports Medicine* **24**:S-38.

Erickson, J. C., et al., 1996. Attenuation of the obesity syndrome of *ob/ob* mice by the loss of neuropeptide Y. *Science* **274:**1704–1707.

Flier, J. S., 1997. Leptin expression and action: New experimental paradigms. *Proceedings of the National Academy of Sciences, U.S.A.* **94:**4242–4245.

Gavrilova, O., et al., 2000. Surgical implantation of adipose tissue reverses diabetes in lipoatrophic mice. *Journal of Clinical Investigation* **105:**271–278.

Gura, T., 1997. Obesity sheds its secrets. *Science* **275:**751–753.

Harris, R., and Crabb, D. W., 1997. Metabolic interrelationships. In *Textbook of Biochemistry with Clinical Correlations*, 4th ed., edited by T. M. Devlin. New York: Wiley-Liss. A synopsis of the interdependence of metabolic processes in the major tissues of the human body—brain, liver, muscle, kidney, gut, and adipose tissue. Metabolic aberrations that occur in certain disease states are also discussed.

Heymsfield, S. B., et al., 1999. Recombinant leptin for weight loss in obese and lean adults: A randomized, controlled, dose-escalation trial. *Journal of the American Medical Association* **282:**1568–1575.

Kreider, R., 1998. Creatine supplementation: Analysis of ergogenic value, medical safety, and concerns. *Journal of Exercise Physiology* **1:** an international on-line journal at http://www.css.edu/users/tboone2/asep/jan3.htm.

Lonnqvist, F., et al., 1999. Leptin and its potential role in human obesity. *Journal of Internal Medicine* **245:**643–652.

Moitra, J., et al., 1998. Life without white fat: A transgenic mouse. *Genes and Development* **12:**3168–3181.

Sugden, M. C., Holness, M. J., and Palmer, T. N., 1989. Fuel selection and carbon flux during the starved-to-fed transition. *Biochemical Journal* **263:**313–323. Changes in lipid and carbohydrate metabolism of rats upon carbohydrate feeding after prolonged starvation.

Tartaglia, L. A., 1997. The leptin receptor. *Journal of Biological Chemistry* **272:**6093–6096.

Vaisse, C., et al., 2000. Melanocortin-4 receptor mutations are a frequent and heterogeneous cause of morbid obesity. *Journal of Clinical Investigation* **106:**253–262.

Zhou, Y.-T., et al., 1997. Induction by leptin of uncoupling protein-2 and enzymes of fatty acid oxidation. *Proceedings of the National Academy of Sciences, U.S.A.* **94:**6386–6390.

INFORMATION TRANSFER

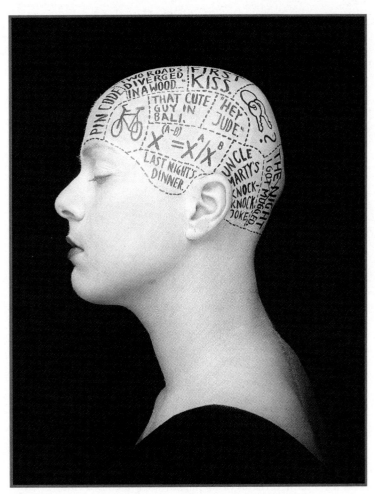

The biochemistry of information transfer not only includes the central dogma of molecular biology (DNA → RNA → protein) but also encompasses those molecular mechanisms that allow cell–cell communication in multicellular organisms and those that serve to detect and integrate information about the environment. The biochemistry of learning and memory in humans depends on all these forms of information transfer.

("Manipulating Memory" © 1998 Karen Kuehn/Matrix)

DNA: Replication, Recombination, and Repair

Our revels now are ended. These our actors
As I foretold you, were all spirits, and
Are melted into air, into thin air,
And like the baseless fabric of this vision,
The cloud-capp'd towers, the gorgeous palaces,
The solemn temples, the great globe itself,
Yea, all which it inherit, shall dissolve,
And like this insubstantial pageant faded
Leave not a rack behind. We are such stuff
As dreams are made on; and our little life
Is rounded with a sleep.

WILLIAM SHAKESPEARE (1564–1616), *The Tempest* 4.1.
148-58 (1611–1612)

Shakespeare's quote speaks to the ephemeral nature of life. Inheritance, throughout life's long history a matter of chance, now can be manipulated by scientists. Dolly, the first mammal to be cloned from an adult cell, is shown here with her naturally conceived first lamb, Bonnie. Hello, Dolly! (AP/Wide World Photos)

Heredity, which we can define generally as the tendency of an organism to possess the characteristics of its parent(s), is clearly evident throughout nature, and since the dawn of history has served to justify the classification of organisms according to shared similarities. The basis of heredity, however, was a mystery. Early in the twentieth century, geneticists demonstrated that **genes,** the elements or units carrying and transferring inherited characteristics from parent to offspring, are contained within the nuclei of cells in association with the chromosomes. Yet the chemical identity of genes remained unknown, and genetics was an abstract science. Even the realization that chromosomes are composed of proteins and nucleic acids did little to define the molecular nature of the gene because, at the time, no one understood either of these substances.

The material of heredity must have certain properties. It must be very stable so that genetic information can be stored in it and transmitted countless times to subsequent generations. It must be capable of precise copying or replication so that its information is not lost or altered. And, although stable, it must also be subject to change in order to account, in the short term, for the appearance of mutant forms and, in the long term, for evolution. DNA is the material of heredity.

23.1 DNA Replication

Transfer of genetic information from generation to generation requires the faithful reproduction of the parental DNA. DNA reproduction produces two identical copies of the original DNA in a process termed **DNA replication.**

The 1953 publication of Watson and Crick's famous paper titled *Molecular Structure of Nucleic Acids: A Structure for Deoxyribose Nucleic Acid* (Figure 23.1) marked the dawn of a new scientific epoch: the age of molecular biology. As these authors drew to a close their brief but far-reaching description of the DNA double helix, they pointedly commented, "It has not escaped our notice that the specific [base] pairing we have postulated immediately suggests a possible copying mechanism for the genetic material." The mechanism for DNA replication that Watson and Crick viewed as intuitively obvious is strand separation followed by the copying of each strand. In the process, each separated

No. 4356 April 25, 1953 NATURE 737

MOLECULAR STRUCTURE OF NUCLEIC ACIDS

A Structure for Deoxyribose Nucleic Acid

WE wish to suggest a structure for the salt of deoxyribose nucleic acid (D.N.A.). This structure has novel features which are of considerable biological interest.

A structure for nucleic acid has already been proposed by Pauling and Corey[1]. They kindly made their manuscript available to us in advance of publication. Their model consists of three intertwined chains, with the phosphates near the fibre axis, and the bases on the outside. In our opinion, this structure is unsatisfactory for two reasons: (1) We believe that the material which gives the X-ray diagrams is the salt, not the free acid. Without the acidic hydrogen atoms it is not clear what forces would hold the structure together, especially as the negatively charged phosphates near the axis will repel each other. (2) Some of the van der Waals distances appear to be too small.

Another three-chain structure has also been suggested by Fraser (in the press). In his model the phosphates are on the outside and the bases on the inside, linked together by hydrogen bonds. This structure as described is rather ill-defined, and for his reason we shall not comment on it.

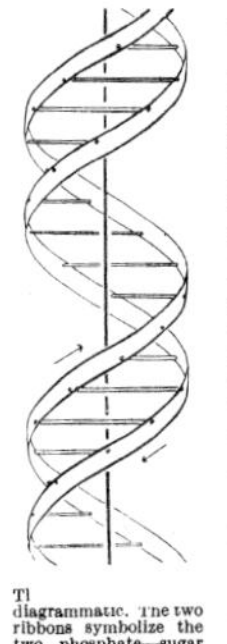

Tl diagrammatic. The two ribbons symbolize the two phosphate—sugar chains, and the horizontal rods the pairs of bases holding the chains together. The vertical line marks the fibre axis

We wish to put forward a radically different structure for the salt of deoxyribose nucleic acid. This structure has two helical chains each coiled round the same axis (see diagram). We have made the usual chemical assumptions, namely, that each chain consists of phosphate diester groups joining β-D-deoxyribofuranose residues with 3′,5′ linkages. The two chains (but not their bases) are related by a dyad perpendicular to the fibre axis. Both chains follow right-handed helices, but owing to the dyad the sequences of the atoms in the two chains run in opposite directions. Each chain loosely resembles Furberg's[2] model No. 1; that is, the bases are on the inside of the helix and the phosphates on the outside. The configuration of the sugar and the atoms near it is close to Furberg's 'standard configuration', the sugar being roughly perpendicular to the attached base. There is a residue on each chain every 3·4 A. in the z-direction. We have assumed an angle of 36° between adjacent residues in the same chain, so that the structure repeats after 10 residues on each chain, that is, after 34 A. The distance of a phosphorus atom from the fibre axis is 10 A. As the phosphates are on the outside, cations have easy access to them.

The structure is an open one, and its water content is rather high. At lower water contents we would expect the bases to tilt so that the structure could become more compact.

The novel feature of the structure is the manner in which the two chains are held together by the purine and pyrimidine bases. The planes of the bases are perpendicular to the fibre axis. They are joined together in pairs, a single base from one chain being hydrogen-bonded to a single base from the other chain, so that the two lie side by side with identical z-co-ordinates. One of the pair must be a purine and the other a pyrimidine for bonding to occur. The hydrogen bonds are made as follows: purine position 1 to pyrimidine position 1; purine position 6 to pyrimidine position 6.

If it is assumed that the bases only occur in the structure in the most plausible tautomeric forms (that is, with the keto rather than the enol configurations) it is found that only specific pairs of bases can bond together. These pairs are: adenine (purine) with thymine (pyrimidine), and guanine (purine) with cytosine (pyrimidine).

In other words, if an adenine forms one member of a pair, on either chain, then on these assumptions the other member must be thymine; similarly for guanine and cytosine. The sequence of bases on a single chain does not appear to be restricted in any way. However, if only specific pairs of bases can be formed, it follows that if the sequence of bases on one chain is given, then the sequence on the other chain is automatically determined.

It has been found experimentally[3,4] that the ratio of the amounts of adenine to thymine, and the ratio of guanine to cytosine, are always very close to unity for deoxyribose nucleic acid.

It is probably impossible to build this structure with a ribose sugar in place of the deoxyribose, as the extra oxygen atom would make too close a van der Waals contact.

The previously published X-ray data[5,6] on deoxyribose nucleic acid are insufficient for a rigorous test of our structure. So far as we can tell, it is roughly compatible with the experimental data, but it must be regarded as unproved until it has been checked against more exact results. Some of these are given in the following communications. We were not aware of the details of the results presented there when we devised our structure, which rests mainly though not entirely on published experimental data and stereochemical arguments.

It has not escaped our notice that the specific pairing we have postulated immediately suggests a possible copying mechanism for the genetic material.

Full details of the structure, including the conditions assumed in building it, together with a set of co-ordinates for the atoms, will be published elsewhere.

We are much indebted to Dr. Jerry Donohue for constant advice and criticism, especially on interatomic distances. We have also been stimulated by a knowledge of the general nature of the unpublished experimental results and ideas of Dr. M. H. F. Wilkins, Dr. R. E. Franklin and their co-workers at King's College, London. One of us (J. D. W.) has been aided by a fellowship from the National Foundation for Infantile Paralysis.

J. D. WATSON
F. H. C. CRICK

Medical Research Council Unit for the Study of the Molecular Structure of Biological Systems,
Cavendish Laboratory, Cambridge.
April 2.

1 Pauling, L., and Corey, R. B., *Nature*, **171**, 346 (1953); *Proc. U.S. Nat. Acad. Sci.*, **39**, 84 (1953).
2 Furberg, S., *Acta Chem. Scand.*, **6**, 634 (1952).
3 Chargaff, E., for references see Zamenhof, S., Brawerman, G., and Chargaff, E., *Biochim. et Biophys. Acta*, **9**, 402 (1952).
4 Wyatt, G. R., *J. Gen. Physiol.*, **36**, 201 (1952).
5 Astbury, W. T., Symp. Soc. Exp. Biol. 1, Nucleic Acid, 66 (Camb. Univ. Press, 1947).
6 Wilkins, M. H. F., and Randall, J. T., *Biochim. et Biophys. Acta*, **10**, 192 (1953).

Figure 23.1 Watson and Crick's famous paper, in its entirety. *(Watson, J. D., and Crick, F. H. C., 1953.* Nature ***171****:737–738.)*

template something whose edge is shaped in a particular way so that it can serve as a guide in making a similar object with a corresponding contour

strand acts as a **template** for the synthesis of a new complementary strand whose nucleotide sequence is fixed by the base-pairing rules Watson and Crick proposed. Strand separation is achieved by untwisting the double helix (Figure 23.2). Base pairing then dictates an accurate replication of the original DNA double helix.

DNA Replication Is Semiconservative

Actually, three basic models for DNA replication are consistent with the requirement that the nucleotide sequence in one strand dictate, through Watson and Crick's base-pairing rules, the sequence of nucleotides in the other. These three models—conservative, semiconservative, and dispersive—are diagrammed in Figure 23.3. In 1958, Matthew Meselson and Franklin Stahl provided the experimental proof for the **semiconservative model** of DNA replication. *Escherichia coli* cells were grown for many generations in a medium containing $^{15}NH_4Cl$ as the sole nitrogen source. Thus, the nitrogen atoms in the purine and pyrimidine bases of the DNA in these cells were mostly ^{15}N, the stable heavy isotope of nitrogen. Then, a tenfold excess of ordinary $^{14}NH_4Cl$

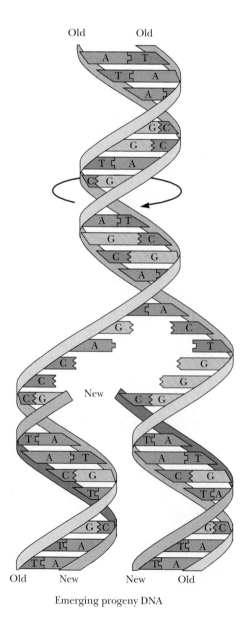

Old   Old

New

Old   New   New   Old

Emerging progeny DNA

Figure 23.2 Untwisting DNA strands exposes their bases for hydrogen bonding. Base pairing ensures that appropriate nucleotides are inserted in the correct positions as the new complementary strands are synthesized. By this mechanism, the nucleotide sequence of one strand dictates a complementary sequence in its daughter strand. The original strands untwist by rotating about the axis of the unreplicated DNA double helix.

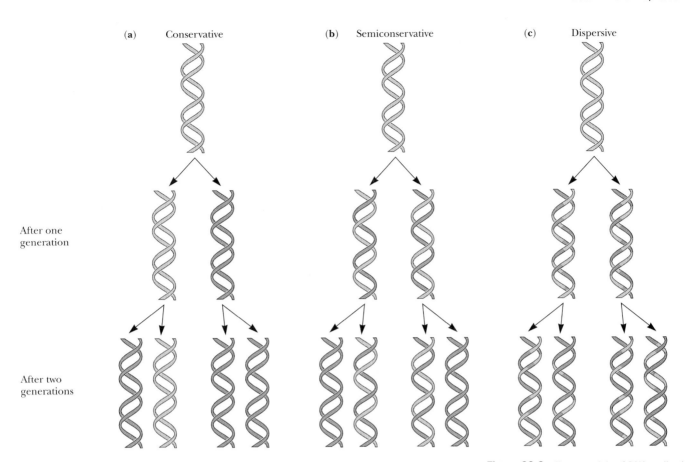

| (a) Conservative | (b) Semiconservative | (c) Dispersive |

After one generation

After two generations

Figure 23.3 Three models of DNA replication prompted by Watson and Crick's double helix structure for DNA. **(a)** *Conservative:* Each strand of the DNA duplex is replicated and the two newly synthesized strands join to form one DNA double helix while the two parental strands remain associated with each other. The products are one completely new DNA duplex and the original DNA duplex. **(b)** *Semiconservative:* The two strands separate, and each strand is copied to generate a complementary strand. Each parental strand remains associated with its newly synthesized complement, so that each DNA duplex contains one parental strand and one new strand. **(c)** *Dispersive:* This model predicts that each of the four strands in the two daughter DNA duplexes contains both newly synthesized segments and segments derived from the parental strands.

was added to the growing culture, and, at appropriate intervals, cells were collected from the culture and lysed. The DNA they contained was analyzed by CsCl density-gradient ultracentrifugation. This technique can resolve macromolecules differing in density by less than 0.01 g/mL.

DNA isolated from cells grown on $^{15}NH_4{}^+$ (the "0" generation cells) banded in the ultracentrifuge at a density corresponding to 1.724 g/mL, whereas DNA from cells grown for 4.1 generations on ^{14}N had a density of 1.710 g/mL (Figure 23.4). These bands represent "heavy" and "light" DNA, respectively. Significantly, DNA isolated from cells grown for just one generation on ^{14}N yielded just a single band corresponding to a density of 1.717 g/mL, halfway between heavy and light DNA, indicating that each DNA duplex molecule contained equal amounts of ^{15}N and ^{14}N. This result is consistent with the semiconservative model for DNA replication, at the same time ruling out the conservative model. After approximately two generations on ^{14}N, cells yielded DNA that gave two essentially equal bands upon ultracentrifugation: one at the intermediate density of 1.717 g/mL and the other at the light position, 1.710 g/mL, also in accord with semiconservative replication.

It remained conceivable that DNA replication might follow a dispersive model, so that both strands of DNA in the chromosomes of cells after one generation on ^{14}N might be intermediate in density. Meselson and Stahl eliminated this possibility by the following experiment: DNA isolated from ^{15}N-labeled cells kept one generation on ^{14}N was heated at 100°C so that the DNA duplexes were denatured into their component single strands. When this heat-denatured DNA was analyzed by CsCl density-gradient ultracentrifugation, two distinct bands were observed, one ^{15}N-labeled and the other ^{14}N-labeled. A dispersive mode of replication would have yielded a single band of DNA of intermediate density.

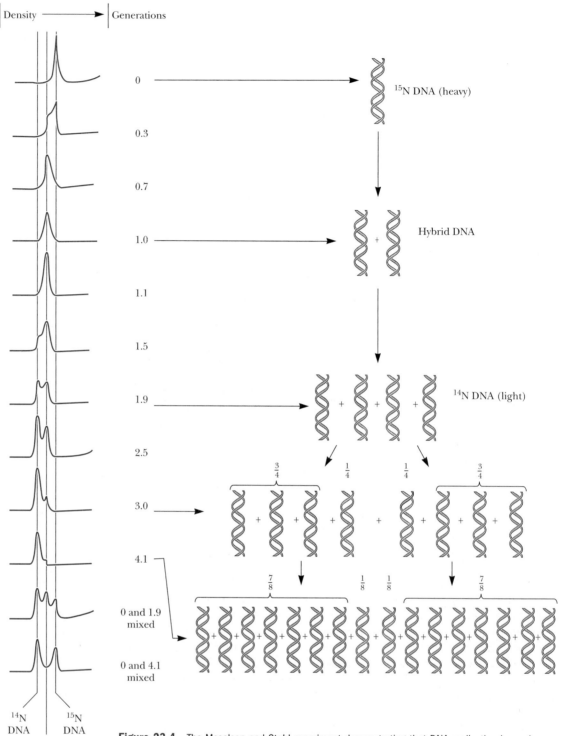

Figure 23.4 The Meselson and Stahl experiment demonstrating that DNA replication is semi-conservative. On the left are densitometric traces made of UV absorption photographs taken of the ultracentrifugation cells containing DNA isolated from *E. coli* grown for various generation times after ^{15}N-labeling. The photographs were taken once the migration of the DNA in the density gradient had reached equilibrium. Density increases from left to right. The peaks reveal the positions of the banded DNA with respect to the density of the solution. The number of generations that the *E. coli* cells were grown (following 14 generations of ^{15}N density-labeling) is shown down the middle of the figure. A schematic representation interpreting the pattern expected of semiconservative replication is shown on the right side of this figure. *(Adapted from Meselson, M., and Stahl, F. W., 1958.* Proceedings of the National Academy of Sciences, U.S.A. **44:**671–682.)

23.2 General Features of DNA Replication

Replication Is Bidirectional

Replication of DNA molecules begins at one or more specific regions called **origin(s) of replication** (discussed in Section 23.3), and, excepting certain bacteriophage chromosomes and plasmids, proceeds in both directions from this origin (Figure 23.5). For example, replication of *E. coli* DNA begins at *oriC*, a unique 245-bp chromosomal site. From this site, replication advances in both directions around the circular chromosome. That is, bidirectional replication involves two **replication forks,** which move in opposite directions.

Unwinding the DNA Helix

Semiconservative replication depends on unwinding the DNA double helix to expose single-stranded templates to polymerase action. For a double helix to unwind, either it must rotate about its axis (while the end of its strands are held fixed), or positive supercoils must form, one for each turn of the helix unwound (see Chapter 8). If the chromosome is circular, as in *E. coli*, only the latter alternative is available. Because DNA replication in *E. coli* proceeds at a rate approaching 1000 nucleotides per second, and since there are about 10 bp per helical turn, the chromosome would accumulate 100 positive supercoils per second! In effect, the DNA would become too tightly supercoiled to allow the strands to unwind.

   DNA gyrase, a Type II topoisomerase, acts to overcome the torsional stress imposed upon unwinding by introducing negative supercoils at the expense of ATP hydrolysis. The unwinding reaction is driven by **helicases** (see also Chapter 13), a class of proteins that catalyze the ATP-dependent unwinding of DNA double helices. Unlike topoisomerases that alter the linking number of dsDNA through phosphodiester bond breakage and reunion (see Chapter 8), helicases simply disrupt the hydrogen bonds that hold the two strands of duplex DNA

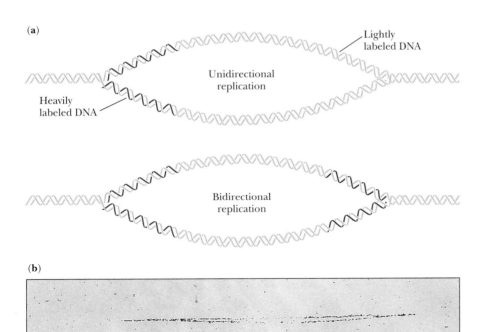

(a)

Heavily labeled DNA

Unidirectional replication

Lightly labeled DNA

Bidirectional replication

(b)

Figure 23.5 Bidirectional replication of the *E. coli* chromosome. **(a)** If replication is bidirectional, autoradiograms of radioactively labeled replicating chromosomes should show two replication forks heavily labeled with radioactive thymidine. **(b)** An autoradiogram of the chromosome from a dividing *E. coli* cell shows bidirectional replication. *(Photo courtesy of David M. Prescott, University of Colorado)*

together. A helicase molecule requires a single-stranded region for binding. It then moves along the single strand, its translocation coupled to ATP hydrolysis and to unwinding the double-stranded DNA it encounters. SSB (ssDNA-binding protein) binds to the unwound strands, preventing their re-annealing. At least 10 distinct DNA helicases involved in different aspects of DNA and RNA metabolism have been found in *E. coli* alone. DnaB is the DNA helicase acting in *E. coli* DNA replication. DnaB helicase assembles as a hexameric (α_6) "doughnut"-shaped protein ring, with DNA passing through its hole.

Replication Is Semidiscontinuous

Incorporation of radioactively labeled thymine nucleotides into DNA during replication, followed by autoradiography of the replicating DNA (see Figure 23.5), reveals that the two strands of duplex DNA are both replicated at each advancing replication fork by **DNA polymerase.** DNA polymerase uses ssDNA as a template and makes a complementary strand by polymerizing deoxynucleotides in the order specified by their base-pairing with bases in the template. DNA polymerases synthesize DNA only in a $5' \rightarrow 3'$ direction, reading the antiparallel template strand in a $3' \rightarrow 5'$ sense. A dilemma arises: how does DNA polymerase copy the parent strand that runs in the $5' \rightarrow 3'$ direction at the replication fork? It turns out that *replication is semidiscontinuous* (Figure 23.6): as the DNA helix is unwound during its replication, the $3' \rightarrow 5'$ strand (as defined by the direction that the replication fork is moving) can be copied continuously by DNA polymerase proceeding in the $5' \rightarrow 3'$ direction behind the replication fork. The other parental strand is copied only when a sufficient stretch of its sequence has been exposed for DNA polymerase to move along it in the $5' \rightarrow 3'$ mode. Thus, one parental strand is copied continuously to give a newly synthesized copy, called the **leading strand,** at each replication fork. The other parental strand is copied in an intermittent, or discontinuous, mode to yield a set of fragments. These fragments are then joined to form an intact **lagging strand.** Overall, each of the two DNA duplexes produced in DNA replication contains one "old" and one "new" DNA strand, and half of the new strand was formed by leading-strand synthesis and the other half by lagging-strand synthesis.

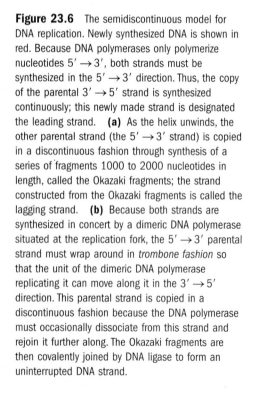

Figure 23.6 The semidiscontinuous model for DNA replication. Newly synthesized DNA is shown in red. Because DNA polymerases only polymerize nucleotides $5' \rightarrow 3'$, both strands must be synthesized in the $5' \rightarrow 3'$ direction. Thus, the copy of the parental $3' \rightarrow 5'$ strand is synthesized continuously; this newly made strand is designated the leading strand. **(a)** As the helix unwinds, the other parental strand (the $5' \rightarrow 3'$ strand) is copied in a discontinuous fashion through synthesis of a series of fragments 1000 to 2000 nucleotides in length, called the Okazaki fragments; the strand constructed from the Okazaki fragments is called the lagging strand. **(b)** Because both strands are synthesized in concert by a dimeric DNA polymerase situated at the replication fork, the $5' \rightarrow 3'$ parental strand must wrap around in *trombone fashion* so that the unit of the dimeric DNA polymerase replicating it can move along it in the $3' \rightarrow 5'$ direction. This parental strand is copied in a discontinuous fashion because the DNA polymerase must occasionally dissociate from this strand and rejoin it further along. The Okazaki fragments are then covalently joined by DNA ligase to form an uninterrupted DNA strand.

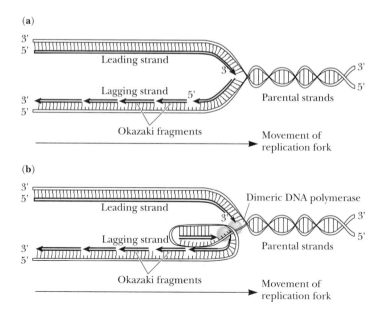

The Lagging Strand Is Formed from Okazaki Fragments

In 1968, Tuneko and Reiji Okazaki provided biochemical verification of the semidiscontinuous pattern of DNA replication just described. The Okazakis exposed a rapidly dividing *E. coli* culture to ^{3}H-labeled thymidine for 30 seconds, quickly collected the cells, and found that half of the label incorporated into nucleic acid appeared in short ssDNA chains just 1000 to 2000 nucleotides in length. (The other half of the radioactivity was recovered in very large DNA molecules.) Subsequent experiments demonstrated that with time the newly synthesized short ssDNA **Okazaki fragments** became covalently joined to form longer polynucleotide chains, in accord with a semidiscontinuous mode of replication. The generality of this mode of replication has been corroborated with electron micrographs of DNA undergoing replication in eukaryotic cells.

23.3   DNA Polymerases: The Enzymes That Replicate DNA

All DNA polymerases, whether from prokaryotic or eukaryotic sources, share the following properties: (a) The incoming base is selected within the DNA polymerase active site, as determined by Watson–Crick geometric interactions with its corresponding base in the template strand, (b) chain growth is in the $5' \rightarrow 3'$ direction and is antiparallel to the template strand, and (c) DNA polymerases cannot initiate DNA synthesis *de novo;* all require a primer oligonucleotide with a free 3'-OH to build upon.

E. coli DNA Polymerases

Table 23.1 compares the properties of the various DNA polymerases in *E. coli*. These enzymes are numbered **I** through **V** in order of their discovery. DNA polymerases I, II, and V function principally in DNA repair; DNA polymerase III is the chief DNA-replicating enzyme of *E. coli*. Only 10 to 20 copies of this enzyme are present per cell.

E. coli DNA Polymerase I

In 1957, Arthur Kornberg and his colleagues discovered the first DNA polymerase, **DNA polymerase I.** DNA polymerase I catalyzes the synthesis of DNA *in vitro* if provided with all four deoxynucleoside-5'-triphosphates (dATP, dTTP, dCTP, dGTP), a template DNA strand to copy, and a **primer.** A primer is essential because DNA polymerases can elongate only pre-existing chains; they

Table 23.1   Properties of the DNA Polymerases of *E. coli*

Property	Pol I	Pol II	Pol III (core)*
Mass (kD)	103	90	130(α), 27.5(ε), 8.6(θ)
Molecules/cell	400	?	40
Turnover number[†]	600	30	1200
Polymerization $5' \rightarrow 3'$	Yes	Yes	Yes
Exonuclease $3' \rightarrow 5'$	Yes	Yes	Yes
Exonuclease $5' \rightarrow 3'$	Yes	No	No

*α, ε, and θ subunits.
[†]Nucleotides polymerized at 37°/minute/molecule of enzyme.
Source: Adapted from Kornberg, A., and Baker, T. A., 1991, *DNA Replication,* 2nd ed., New York: W. H. Freeman and Co.; and Kelman, Z., and O'Donnell, M., 1995. DNA polymerase III holoenzyme: Structure and function of a chromosomal replicating machine. *Annual Review of Biochemistry* **64**:171–200.

Figure 23.7 The chain elongation reaction catalyzed by DNA polymerase. DNA polymerase I joins deoxynucleoside monophosphate units to the 3'-OH end of the primer, employing dNTPs as substrates. The 3'-OH carries out a nucleophilic attack on the α-phosphoryl group of the incoming dNTP to form a phosphoester bond, and PP$_i$ is released. The subsequent hydrolysis of PP$_i$ by inorganic pyrophosphatase renders the reaction effectively irreversible.

See *Interactive Biochemistry CD-ROM and Workbook,* page 91

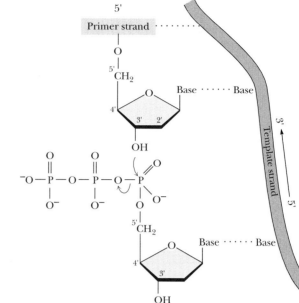

cannot join two deoxyribonucleoside-5'-phosphates together to make the initial phosphodiester bond. The primer base-pairs with the template DNA, forming a short, double-stranded region. This primer must possess a free 3'-OH end to which an incoming deoxynucleoside monophosphate is added. All four dNTPs are substrates; pyrophosphate (PP$_i$) is released and the dNMP is linked to the 3'-OH of the primer chain through formation of a phosphoester bond (Figure 23.7). The deoxynucleoside monophosphate to be incorporated is chosen through its geometric fit with the template base to form a Watson–Crick base pair. As DNA polymerase I catalyzes the successive addition of deoxynucleotide units to the 3'-end of the primer, the chain is elongated in the $5' \rightarrow 3'$ direction, forming a polynucleotide sequence that runs antiparallel to the template but complementary to it. DNA polymerase I can proceed along the template strand, synthesizing a complementary strand of about 20 bases before it "falls off" (dissociates from) the template. The degree to which the enzyme remains associated with the template through successive cycles of nucleotide addition is referred to as its **processivity.** As DNA polymerases go, DNA polymerase I is a modestly processive enzyme. Arthur Kornberg was awarded the Noble Prize in physiology or medicine in 1959 for his discovery of this DNA polymerase. DNA polymerase I is the best-characterized of these enzymes.

E. coli DNA Polymerase I Has Three Active Sites on Its Single Polypeptide Chain

In addition to its $5' \rightarrow 3'$ polymerase activity, *E. coli* DNA polymerase I has two other catalytic functions: a $3' \rightarrow 5'$ *exonuclease (3'-exonuclease)* activity and a $5' \rightarrow 3'$ *exonuclease (5'-exonuclease)* activity. The three distinct catalytic activities of DNA polymerase I reside in separate active sites in the enzyme.

E. coli DNA Polymerase I Is Its Own Proofreader and Editor

The exonuclease activities of *E. coli* DNA polymerase I are functions that enhance the accuracy of DNA replication. The 3'-exonuclease activity removes nucleotides from the 3'-end of the growing chain (Figure 23.8), an action that negates the action of the polymerase activity. Its purpose, however, is to remove incorrect (mismatched) bases. Although the 3'-exonuclease works slowly when compared to the polymerase, the polymerase cannot elongate an improperly base-paired primer terminus. Thus, the relatively slow 3'-exonuclease has time

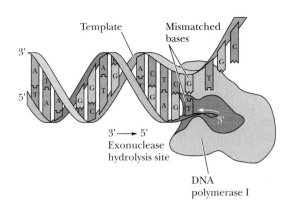

Template   Mismatched bases

$3' \longrightarrow 5'$
Exonuclease
hydrolysis site

DNA
polymerase I

Figure 23.8 The $3' \rightarrow 5'$ exonuclease activity of DNA polymerase I removes nucleotides from the $3'$-end of the growing DNA chain.

to act and remove the mispaired nucleotide. Therefore, the polymerase active site is a proofreader and the $3'$-exonuclease activity is an editor. This check on the accuracy of base pairing enhances the overall precision of the process.

The $5'$-exonuclease of DNA polymerase I acts upon duplex DNA, degrading it from the $5'$-end by releasing mono- and oligonucleotides. It can remove distorted (mispaired) segments lying in the path of the advancing polymerase. Its biological roles depend on the ability of DNA polymerase I to bind at nicks (single-stranded breaks) in double-stranded DNA and move in the $5' \rightarrow 3'$ direction, removing successive nucleotides with its $5'$-exonucleolytic activity. This $5'$-exonuclease activity plays an important role in primer removal during DNA replication, as we shall soon see. DNA polymerase I is also involved in DNA repair processes (see Section 23.8).

E. coli DNA Polymerase III

In its holoenzyme form, DNA polymerase III is the enzyme responsible for replication of the *E. coli* chromosome. "Core" DNA polymerase III, the simplest form of DNA polymerase III showing any DNA-synthesizing activity *in vitro*, is 165 kD in size and consists of three polypeptides: α (130 kD), ε (27.5 kD), and θ (10 kD). *In vivo*, "core" DNA polymerase III functions as part of a multisubunit complex, the **DNA polymerase III holoenzyme,** which is composed of 10 different kinds of subunits (Table 23.2). The various auxiliary subunits increase both the polymerase activity of the "core" enzyme and its processivity. DNA polymerase III holoenzyme synthesizes DNA strands at a speed of nearly 1 kb/sec. DNA polymerase III holoenzyme is organized in the following way: two "core" ($\alpha\varepsilon\theta$) DNA polymerase III units and one γ complex are held together by a dimer of τ subunits in a structure known as **DNA polymerase III*.**

Table 23.2 Subunits of *E. coli* DNA Polymerase III Holoenzyme

Subunit	Mass (kD)	Function
α	130	Polymerase
ε	27.5	$3'$-exonuclease
θ	8.6	α, ε assembly?
τ	71	Assembly of holoenzyme on DNA
β	41	Sliding clamp, processivity
γ	47.5	Part of the γ complex*
δ	39	Part of the γ complex*
δ'	37	Part of the γ complex*
χ	17	Part of the γ complex*
ψ	15	Part of the γ complex*

*Subunits γ, δ, δ', χ, and ψ form the so-called γ complex responsible for adding β-subunits (the sliding clamp) to DNA. The γ complex is referred to as the clamp loader.

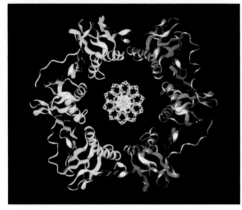

(a)

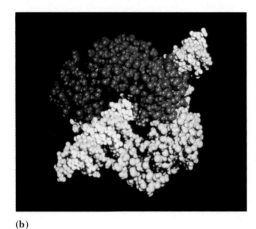

(b)

◄ **Figure 23.9** **(a)** Ribbon diagram of the β-subunit dimer of the DNA polymerase III holoenzyme on B-DNA, viewed down the axis of the DNA. One monomer of the β-subunit dimer is colored red and the other yellow. The centrally located DNA is mostly blue. **(b)** Space-filling model of the β-subunit dimer of the DNA polymerase III holoenzyme on B-DNA. One monomer is shown in red, the other in yellow. The B-DNA has one strand colored white and the other blue. The hole formed by the β-subunits (diameter $\approx$ 3.5 nm) is large enough to easily accommodate DNA (diameter $\approx$ 2.5 nm) with no steric repulsion. The rest of pol III holoenzyme ("core" polymerase + γ complex) associates with this sliding clamp to form the replicative polymerase (not shown). *(Adapted from Kong, X.-P., et al., 1992. Cell **69**:425–437; photos courtesy of John Kuriyan of the Rockefeller University.)*

In turn, each "core" polymerase within DNA polymerase III* binds to a β-subunit dimer to create DNA polymerase III holoenzyme. The γ complex is responsible for assembly of the DNA polymerase III holoenzyme complex onto DNA. The γ complex of the holoenzyme acts as a **clamp loader** by catalyzing the ATP-dependent transfer of a pair of β subunits to each strand of the DNA template. Each β-subunit dimer forms a closed ring around a DNA strand and acts as a tight clamp that can slide along the DNA (Figure 23.9). Each β_2 **sliding clamp** tethers a "core" polymerase to the template, accounting for the great processivity of the DNA polymerase III holoenzyme. This complex can replicate an entire strand of the *E. coli* genome (more than 4.6 megabases) without dissociating. Compare this to the processivity of DNA polymerase I, which is only 20!

General Features of a Replication Fork

We now can present a snapshot of the enzymatic apparatus assembled at a replication fork (Figure 23.10 and Table 23.3). DNA gyrase (topoisomerase) and DnaB helicase unwind the DNA double helix, and the unwound, single-stranded regions of DNA are maintained through interaction with SSB. Primase (DnaG) synthesizes an RNA primer on the lagging strand; the leading strand, which needs priming only once, was primed when replication was ini-

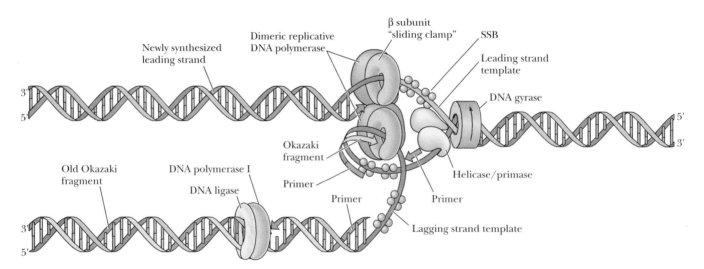

Figure 23.10 General features of a replication fork. The DNA duplex is unwound by the action of DNA gyrase and helicase, and the single strands are coated with SSB (ssDNA-binding protein). Primase periodically primes synthesis on the lagging strand. Each half of the dimeric replicative polymerase is a "core" polymerase bound to its template strand by a β-subunit sliding clamp. DNA polymerase I and DNA ligase act downstream on the lagging strand to remove RNA primers, replace them with DNA, and ligate the Okazaki fragments.

Table 23.3   Proteins Involved in DNA Replication in *E. coli*

Protein	Function
DNA gyrase	Unwinding DNA
SSB	Single-stranded DNA binding
DnaA	Initiation factor; origin-binding protein
DnaB	$5' \rightarrow 3'$ helicase (DNA unwinding)
DnaC	DnaB chaperone; loading DnaB on DNA
DnaT	Assists DnaC in delivery of DnaB
Primase (DnaG)	Synthesis of RNA primer
DNA polymerase III holoenzyme	Elongation (DNA synthesis)
DNA polymerase I	Excises RNA primer, fills in with DNA
DNA ligase	Covalently links Okazaki fragments
Tus	Termination

tiated. The lagging strand template is looped around, and each replicative DNA polymerase moves $5' \rightarrow 3'$ relative to its strand, copying the template and synthesizing a new DNA strand. Each replicative polymerase is tethered to the DNA by its β-subunit sliding clamp. The DNA polymerase III γ complex periodically unclamps and then reclamps β subunits on the lagging strand as the primer for each new Okazaki fragment is encountered. Downstream on the lagging strand, DNA polymerase I excises the RNA primer and replaces it with DNA, and DNA ligase seals the remaining nick.

DNA Ligase

DNA ligase (see Chapter 8) seals nicks in double-stranded DNA where a 3'-OH and a 5'-phosphate are juxtaposed. This enzyme is responsible for joining Okazaki fragments together to make the lagging strand a covalently contiguous polynucleotide chain.

Termination

Located diametrically opposite from *oriC* on the *E. coli* circular map is a terminus region, the ***Ter,*** or ***t,*** locus. The oppositely moving replication forks meet here and replication is terminated. The *Ter* region contains a number of short DNA sequences with a consensus core element 5'-GTGTGTTGT. These *Ter* sequences act as terminators; clusters of three or four *Ter* sequences are organized into two sets inversely oriented with respect to one another. One set blocks the clockwise-moving replication fork, and its inverted counterpart blocks the counterclockwise-moving replication fork. During termination, a specific replication termination protein, **Tus protein,** is bound to *Ter.* Tus protein is a **contrahelicase;** that is, it prevents the DNA duplex from unwinding by blocking progression of the replication fork and inhibiting the ATP-dependent DnaB helicase activity. Final synthesis of both duplexes is completed.

DNA Polymerases Are Immobilized in Replication Factories

Most illustrations of DNA replication (such as Figure 23.10) suggest that the DNA polymerases are tracking along the DNA like locomotives along train tracks, synthesizing DNA as they go. Recent evidence, however, favors the view that the DNA polymerases are immobilized, via attachment either to the cell membrane in prokaryotic cells or to the nuclear matrix in eukaryotic cells. All the associated proteins of DNA replication, as well as the proteins necessary to hold DNA polymerase at its fixed location, constitute so-called **replication**

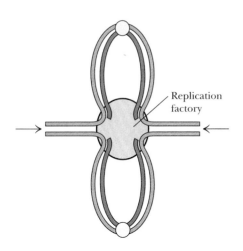

Figure 23.11 A replication factory "fixed" to a cellular substructure extrudes loops of newly synthesized DNA as parental DNA duplex is fed in from the sides. Parental DNA strands are green; newly synthesized strands are blue. Small circles indicate origins of replication. *(Adapted from Cook, P. R., 1999. The organization of replication and transcription. Science 284:1790–1795.)*

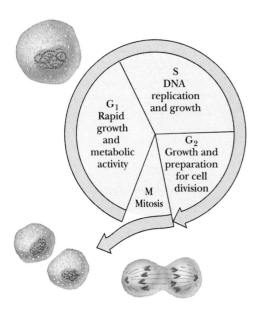

Figure 23.12 The eukaryotic cell cycle. The stages of mitosis and cell division define the M phase ("M" for mitosis). G_1 ("G" for gap, not growth) is typically the longest part of the cell cycle; G_1 is characterized by rapid growth and metabolic activity. Cells that are quiescent—that is, not growing and dividing (such as neurons)—are said to be in G_0. The S phase is the time of DNA synthesis. S is followed by G_2, a relatively short period of growth when the cell prepares for cell division. Cell cycle times vary from less than 24 hours (for rapidly dividing cells such as the epithelial cells lining the mouth and gut) to hundreds of days.

factories. The DNA is then fed through the DNA polymerases within the replication factory, much like tape is fed past the heads of a tape player, with all four strands of newly replicated DNA looping out from this fixed structure (Figure 23.11).

23.4 **Eukaryotic DNA Replication**

DNA replication in eukaryotic cells shows strong parallels with prokaryotic DNA replication, but it is vastly more complex. First of all, eukaryotic DNA is organized into chromosomes, which are compartmentalized within the nucleus. Further, these chromosomes must be duplicated with high fidelity once (and only once!) each cell cycle. For example, in a dividing human cell, a carefully choreographed replication of 6 billion bp of DNA distributed among 46 chromosomes occurs. The events associated with cell growth and division in eukaryotic cells fall into a general sequence having four distinct phases: M, G_1, S, and G_2 (Figure 23.12). Eukaryotic cells have solved the problem of replicating their enormous genomes in the few hours allotted to the S phase by initiating DNA replication at multiple origins distributed along each chromosome. Depending on the organism and cell type, replication origins, also called **replicators,** are DNA regions 0.5 to 2 kbp in size that occur every 3 to 300 kbp (for example, an average human chromosome has several hundred replication origins). So that eukaryotic DNA replication can proceed concomitantly throughout the genome, each eukaryotic chromosome contains many units of replication, called **replicons.**

Cell Cycle Control of DNA Replication

Checkpoints, Cyclins, and CDKs. Progression through the cell cycle is regulated through a series of **checkpoints** that control whether the cell continues into the next phase. These checkpoints are situated to ensure that *all* the necessary steps in each phase of the cycle have been satisfactorily completed before the next phase is entered. If conditions for advancement to the next phase are not met, the cycle is arrested until they are. Checkpoints depend on **cyclins** and **cyclin-dependent protein kinases (CDKs).** *Cyclin* is the name given to a class of proteins synthesized at one phase of the cell cycle and degraded at another. Thus, cyclins appear and then disappear at specific times during the cell cycle. Cyclins bind to CDKs and are essential for the protein kinase activity displayed by these CDKs. In turn, these CDKs control events at each phase of the cycle by targeting specific proteins for phosphorylation. Destruction of the phase-specific cyclin at the end of the phase inactivates the CDK.

Initiation of Replication. Initiation of replication involves replicators and the **origin recognition complex,** or **ORC,** a protein complex that binds to replicators. Yeast, a simple eukaryote, provides an informative model for initiation of eukaryotic DNA replication. Early in G_1 (just after M), ORC serves as a "landing pad" for proteins essential to replication control. Proteins binding to ORC establish a **pre-replication complex (pre-RC),** but the pre-RC can be formed only during a window of opportunity during G_1. One of the principal proteins in assembly of the pre-RC in yeast is **Cdc6p,** the **replication activator protein** (Figure 23.13). Once Cdc6p binds to ORC, **MCM proteins** then bind to the chromosomes. MCM proteins are essential replication initiation factors; they are also known as **replication licensing factors (RLFs)** because they "license" or permit DNA replication to occur. The MCM proteins render the chromo-

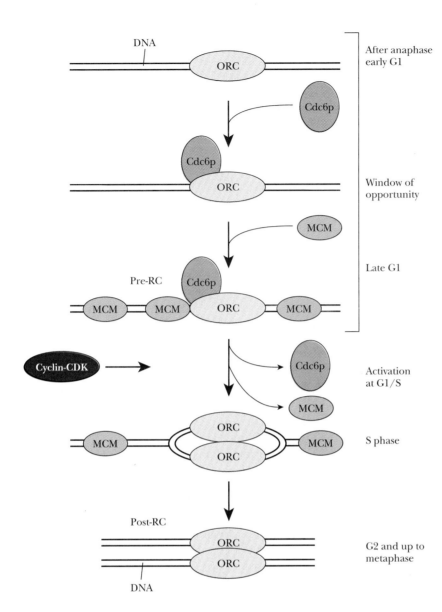

DNA

ORC

After anaphase
early G1

Cdc6p

Cdc6p

ORC

Window of
opportunity

MCM

Pre-RC   Cdc6p

MCM   MCM   ORC   MCM

Late G1

Cyclin-CDK

Cdc6p

MCM

Activation
at G1/S

ORC

MCM   MCM

ORC

S phase

Post-RC

ORC

ORC

G2 and up to
metaphase

DNA

Figure 23.13 Model for initiation of the DNA
replication cycle in eukaryotes. ORC is present at the
replicators throughout the cell cycle. The pre-
replication complex (pre-RC) is assembled through
the sequential addition of Cdc6p and MCM proteins
during a window of opportunity defined by the state
of the cyclin-CDKs. After initiation, a post-RC state is
established. *(Adapted from Figure 2 in Stillman, B., 1996.
Cell cycle control of DNA replication.* Science ***274**:1659–
1663.)*

somes competent for DNA replication. The pre-RC therefore consists of ORC,
Cdc6p, the MCM complex, and other proteins.

At this point, two protein kinases act upon the pre-RC to directly trigger
DNA replication. One of these protein kinases is the **cyclin B-CDK** complex.
B-Cyclin is a cyclin that accumulates to high levels just before S phase. Cyclin
B-CDK can phosphorylate sites in ORC, Cdc6p, and several MCM subunits.
Phosphorylation of Cdc6p causes it to dissociate from ORC, whereupon it is
degraded. Cyclin B-CDK also phosphorylates other protein kinases essential to
activation of DNA replication (see Figure 23.13). The MCM complex is one
phosphorylation target of these CDKs, and some of the MCM proteins dissociate from the chromosomes when they are phosphorylated. The consequence
of these actions brings the cell into S phase.

These phosphorylation events serve as a **replication switch** because once
proteins in the pre-RC are phosphorylated (and perhaps destroyed, as Cdc6p
is), the **post-RC** state is achieved. The post-RC state is incapable of re-
initiating DNA replication. This transformation ensures that eukaryotic DNA
replication occurs once, and only once, per cell cycle.

Table 23.4 **Biochemical Properties of the Principal Eukaryotic DNA Polymerases**

	α	δ	ε	β	γ
Mass (kD)					
Native	>250	170	256	36–38	160–300
Catalytic core	165–180	125	215	36–38	125
Other subunits	70, 50, 60	48	55	None	35, 47
Location	Nucleus	Nucleus	Nucleus	Nucleus	Mitochondria
Associated functions					
$3' \rightarrow 5'$ exonuclease	No	Yes	Yes	No	Yes
Primase	Yes	No	No	No	No
Properties					
Processivity	Low	High	High	Low	High
Fidelity	High	High	High	Low	High
Replication	Yes	Yes	Yes	No	Yes
Repair	No	?	Yes	Yes	No

Source: Adapted from Kornberg, A., and Baker, T. A., 1992. *DNA Replication,* 2nd ed. New York: W. H. Freeman and Co.

(a)

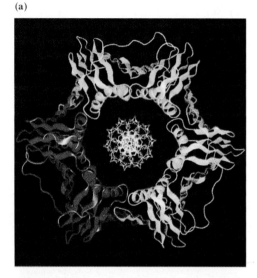

(b)

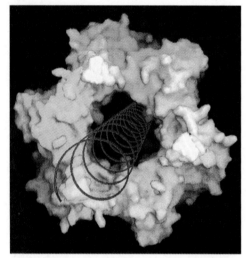

Figure 23.14 Structure of the PCNA homotrimer. Note that the trimeric PCNA ring of eukaryotes is remarkably similar to its prokaryotic counterpart, the dimeric β_2 sliding clamp (see Figure 23.9). **(a)** Ribbon representation of the PCNA trimer with an axial view of a B-form DNA duplex in its center. **(b)** Molecular surface of the PCNA trimer with each monomer colored differently. The red spiral represents the sugar–phosphate backbone of a strand of B-form DNA. *(Adapted from Figure 3 in Krishna, T. S., et al., 1994. Crystal structure of the eukaryotic DNA polymerase processivity factor PCNA.* Cell **79**:1233–1243. *Photos courtesy of John Kuriyan, Rockefeller University.)*

Eukaryotic DNA Polymerases

A number of different DNA polymerases have been described in animal cells and assigned Greek letters in the order of their discovery (Table 23.4). Three of these DNA polymerases—α, δ, and ε—share the task of genome replication. **DNA polymerase α** has an associated primase subunit, and *DNA polymerase α-primase* functions in the initiation of nuclear DNA replication. Given a template, it not only synthesizes an RNA primer, but it then adds deoxynucleotides to extend the chain in the $5' \rightarrow 3'$ direction. **DNA polymerase δ** is the principal DNA polymerase in eukaryotic DNA replication. It interacts with **PCNA** (*proliferating cell nuclear antigen*) protein. In association with PCNA, DNA polymerase δ carries out highly processive DNA synthesis. PCNA is the eukaryotic counterpart of the *E. coli* β_2 sliding clamp; it clamps DNA polymerase δ to the DNA. Like β_2, PCNA encircles the double helix, but in contrast to the prokaryotic β_2 sliding clamp, PCNA is a homotrimer of 37-kD subunits (Figure 23.14). **DNA polymerase ε** also plays a major role in DNA replication; its precise role is unclear, but it cooperates with DNA polymerase δ in lagging-strand synthesis. **DNA polymerase β** functions in DNA repair; **DNA polymerase γ** is the DNA-replicating enzyme of mitochondria.

Other proteins involved in eukaryotic DNA replication include **replication protein A (RPA),** a ssDNA-binding host cell protein that is the eukaryotic counterpart of SSB, and **replication factor C (RFC),** the eukaryotic counterpart of the prokaryotic γ complex.

23.5 Replicating the Ends of Chromosomes: Telomeres and Telomerase

Telomeres are short (5–8 bp), tandemly repeated, G-rich nucleotide sequences that form protective caps 1–12 kbp long on the ends of chromosomes (see Chapter 8). Vertebrate telomeres have a TTAGGG consensus sequence. Telomeres are necessary for chromosome maintenance and stability. DNA polymerases cannot replicate the extreme 5'-ends of chromosomes because these enzymes require a template and a primer and replicate only in the $5' \rightarrow 3'$ direction. Thus, lagging-strand synthesis at the 3'-ends of chromosomes is primed by RNA primase to form Okazaki fragments, but these RNA primers are subsequently removed, resulting in gaps in the progeny 5'-terminal strands

A DEEPER LOOK

Protein Rings in DNA Metabolism

Many of the proteins involved in DNA metabolism (DNA replication, recombination, and repair) adopt a ring-shaped (toroidal) quaternary structure (see table). An obvious advantage of this quaternary structure is that the DNA is enclosed within the "hole" of the ring and therefore remains stably associated with and contained within the protein that is operating on it.

Protein	Example	Action on DNA
Sliding clamps	E. coli β; human PCNA	Protein ring encircles DNA, tethers DNA polymerase, and increases its processivity
Helicases	E. coli DnaB; SV40 T antigen; MCM proteins	Hexameric protein ring unwinds dsDNA to form ssDNA substrates for replication, recombination, or repair
Topoisomerases	E. coli and human topoisomerase I	Changes the superhelicity of DNA
DNA recombination proteins	Human RAD51, 52	Facilitates homologous DNA recombination

Adapted from Hingorani, M. M., and O'Donnell, M., 2000. A tale of toroids in DNA metabolism. *Nature Reviews Molecular Cell Biology* **1**:22–30.

at each end of the chromosome after each round of replication ("primer gap"; Figure 23.15).

Telomerase is an RNA-dependent DNA polymerase. Telomerase maintains telomere length by restoring telomeres at the 3′-ends of chromosomes. The RNA upon which telomerase activity depends is actually part of the enzyme's structure. That is, telomerase is a ribonucleoprotein, and its RNA component contains a 9- to 30-nucleotide–long region that serves as a template for the synthesis of telomeric repeats at DNA ends. The human telomerase RNA component is 450 nucleotides long; its template sequence is CUAACCCUAAC. Telomerase uses the 3′-end of the DNA as a primer and adds successive TTAGGG repeats to it, employing its RNA as a template over and over again (see Figure 23.15; see also figure in Chapter 8, *Human Biochemistry*, page 281).

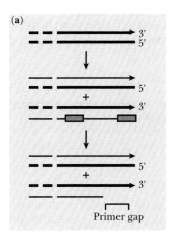

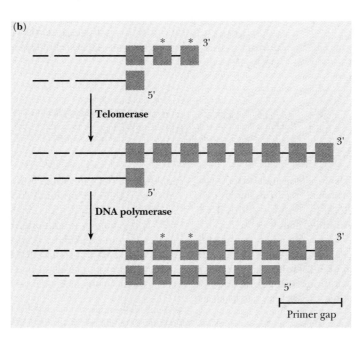

Figure 23.15 Telomere replication. **(a)** In replication of the lagging strand, short RNA primers are added (pink) and extended by DNA polymerase. When the RNA primer at the 5′-end of each strand is removed, there is no nucleotide sequence to read in the next round of DNA replication. The result is a gap (primer gap) at the 5′-end of each strand (only one end of a chromosome is shown in this figure). **(b)** Asterisks indicate sequences at the 3′-end that cannot be copied by conventional DNA replication. Synthesis of telomeric DNA by telomerase extends the 5′-ends of DNA strands, allowing the strands to be copied by normal DNA replication.

Telomeres—A Timely End to Chromosomes?

Mammalian cells in culture undergo only 50 or so cell divisions before they die. Somatic cells are known to lack telomerase activity and thus they inevitably lose bits of their telomeres with each cell division. Telomerase activity is missing because the telomerase-reverse transcriptase gene (the *TRT* gene) is switched off. This fact has led to a telomere theory of cell aging, which suggests that cells senesce and die when their telomeres are gone. In support of this

notion, a team of biologists headed by Calvin B. Harley at Geron Corporation used recombinant DNA techniques to express the catalytic subunit of human telomerase in skin cells in culture and observed that such cells divide 40 times more after cells lacking this treatment have become senescent. These results, though controversial, may have relevance to the aging process.

23.6 Reverse Transcriptase: An RNA-Directed DNA Polymerase

Many viruses have genomes composed of RNA, not DNA. How then is the information in these RNA genomes replicated? In 1964, Howard Temin noted that inhibitors of DNA synthesis prevented infection of cells in culture by RNA tumor viruses such as avian sarcoma virus. On the basis of this observation, Temin made the bold proposal that DNA is an intermediate in the replication of such viruses; that is, an RNA tumor virus can use viral RNA as the template for DNA synthesis:

RNA viral chromosome → DNA intermediate → RNA viral chromosome

In 1970, Temin and David Baltimore independently discovered a viral enzyme capable of mediating such a process, namely, an **RNA-directed DNA polymerase,** or, as it is usually called, **reverse transcriptase.** All RNA tumor viruses contain such an enzyme within their virions (viral particles), so they are now classified as **retroviruses.**

Like other DNA and RNA polymerases, reverse transcriptase synthesizes polynucleotides in the $5' \rightarrow 3'$ direction, and, like all DNA polymerases, reverse transcriptase requires a primer. Interestingly, the primer is a specific tRNA molecule captured by the virion from the host cell in which it was produced. The 3'-end of the tRNA is base-paired with the viral RNA template at the site where DNA synthesis initiates and its free 3'-OH accepts the initial deoxynucleotide once transcription commences. Reverse transcriptase then transcribes the RNA template into a complementary DNA (cDNA) strand to form a double-stranded DNA : RNA hybrid.

RNA As Genetic Material

Whereas the genetic material of cells is double-stranded DNA, virtually all plant viruses, several bacteriophages, and many animal viruses have genomes consisting of RNA. In most cases, this RNA is single-stranded. Viruses with single-stranded genomes use the single strand as a template for synthesis of a complementary strand, which can then serve as template in replicating the original strand. **Retroviruses** are an interesting group of eukaryotic viruses having single-stranded RNA genomes that replicate through a double-stranded DNA intermediate. Further, the life cycle of retroviruses

includes an obligatory step in which the dsDNA is inserted into the host cell genome in a transposition event. Retroviruses are responsible for many diseases, including tumors and other disorders. **HIV-1,** the **human immunodeficiency virus** that causes **AIDS** (acquired immunodeficiency syndrome), is a retrovirus. **Tobacco mosaic virus (TMV),** a rodlike RNA virus infecting plants, consists of an RNA genome packaged in a protein coat made of 2130 identical protein chains of 18 kD each (see Figure 1.22).

The Enzymatic Activities of Reverse Transcriptases

Reverse transcriptases possess three enzymatic activities, all of which are essential to viral replication:

1. *RNA-directed DNA polymerase activity,* for which the enzyme is named (see Figure 9.14).
2. *RNase H activity.* Recall that RNase H is an exonuclease activity that specifically degrades RNA chains in DNA:RNA hybrids (see Figure 9.14). The RNase H function of reverse transcriptase degrades the template genomic RNA and also removes the priming tRNA after DNA synthesis is completed.
3. *DNA-directed DNA polymerase activity.* This activity replicates the ssDNA remaining after RNase H degradation of the viral genome, yielding a DNA duplex. This DNA duplex directs the remainder of the viral infection process or it becomes integrated into the host chromosome, where it can lie dormant for many years as a **provirus.** Activation of the provirus restores the infectious state.

HIV reverse transcriptase is of great clinical interest because it is the enzyme for replication of the AIDS virus. DNA synthesis by HIV reverse transcriptase is blocked by the drug AZT (Figure 23.16). HIV reverse transcriptase is error-prone: it incorporates the wrong base at a frequency of 1 per 2000 to 4000 nucleotides polymerized. This high error rate during replication of the HIV genome means that the virus is ever changing, a feature that makes it difficult to devise an effective vaccine.

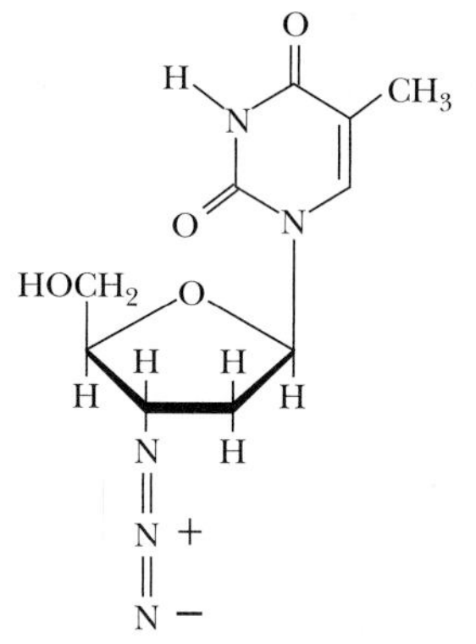

Figure 23.16 The structure of AZT (3'-azido-2', 3'-dideoxythymidine). This nucleoside is a major drug in the treatment of AIDS. AZT is phosphorylated *in vivo* to give AZTTP (AZT 5'-triphosphate), a substrate analog that binds to HIV reverse transcriptase. HIV reverse transcriptase incorporates AZTTP into growing DNA chains in place of dTTP. Incorporated AZTMP blocks further chain elongation because its 3'-azido group cannot form a phosphodiester bond with an incoming nucleotide. Host cell DNA polymerases have little affinity for AZTTP.

23.7 Genetic Recombination

Genetic recombination is the natural process by which genetic information is rearranged to form new associations. For example, compared to their parents, progeny may have new combinations of traits because of genetic recombination. At the molecular level, genetic recombination is the exchange (or incorporation) of one DNA sequence with (or into) another. If recombination involves reaction between very similar sequences (homologous sequences) of DNA, the process is called **homologous recombination.** If very different nucleotide sequences recombine, it is **nonhomologous recombination. Transposition**—the enzymatic insertion of a *transposon* (a mobile segment of DNA; see page 747) into a new location in the genome—and nonhomologous recombination (incorporation of a DNA segment whose sequence differs greatly from the DNA at the point of insertion) are two types of recombination that play a significant evolutionary role.

Nonhomologous recombination occurs at a low frequency in all cells and serves as a powerful genetic force that reshapes the genomes of all organisms. Homologous recombination involves an exchange of DNA sequences between homologous chromosomes, resulting in the arrangement of genes into new combinations. Homologous recombination is generally used to fix the DNA so that information is not lost. For example, large lesions in DNA are repaired via recombination of the damaged chromosome with a homologous chromosome.

The process underlying homologous recombination is termed **general recombination** because the enzymatic machinery that mediates the exchange can use essentially any pair of homologous DNA sequences as substrates. Homologous recombination occurs in all organisms and is particularly prevalent during the production of gametes (meiosis) in diploid organisms. In higher animals—that is, those with immune systems—recombination also occurs in the DNA of somatic cells responsible for expressing proteins of the immune

Prions: Proteins As Genetic Agents?

DNA is the genetic material in organisms, though some viruses have RNA genomes. The idea that proteins could carry genetic information was considered early in the history of molecular biology but was dismissed for lack of evidence. *Prions* may be an exception to this rule.

Prion is an acronym derived from the words "protein infectious particle." The term was coined to distinguish such particles, which are pathogenic and thus capable of causing disease, from nucleic acid–containing infectious particles such as viruses and virions. Prions are transmissible agents (genetic material?) that are apparently composed of only a protein that has adopted an abnormal conformation. They produce fatal degenerative diseases of the central nervous system in mammals and are believed to be the agents responsible for the human diseases kuru, Creutzfeldt–Jakob disease, Gerstmann–Strussler–Sheinker syndrome, and fatal familial insomnia. Prions also cause diseases in animals, including scrapie (in sheep), "mad cow disease" (bovine spongiform encephalopathy), and chronic wasting disease (in elk and mule deer). All attempts to show that the infectivity of these diseases is due to a nucleic acid–

carrying agent have been unsuccessful. Prion diseases are novel in that they are both genetic and infectious; their occurrence may be sporadic, dominantly inherited, or acquired by infection. Their inheritability questions the principle that nucleic acids are the sole genetic agents.

PrP, the prion protein, comes in various forms, such as **Prp^c,** the normal cellular prion protein, and **Prp^{sc}** (the scrapie form of PrP), a conformational variant of Prp^c that is protease-resistant. These two forms are thought to differ only in their secondary and tertiary structure. One model suggests that PrP^c is dominated by α-helical elements (see figure, a), whereas PrP^{sc} has both α-helices and β-strands (see figure, b). It has been hypothesized that the presence of PrP^{sc} can cause PrP^c to adopt the PrP^{sc} conformation. The various diseases are a consequence of the accumulation of the abnormal PrP^{sc} form, which accumulates as amyloid plaques (amyloid = starchlike), causing vacuolarization of tissues in the central nervous system. The 1997 Nobel Prize in physiology or medicine was awarded to Stanley B. Prusiner for his discovery of prions.

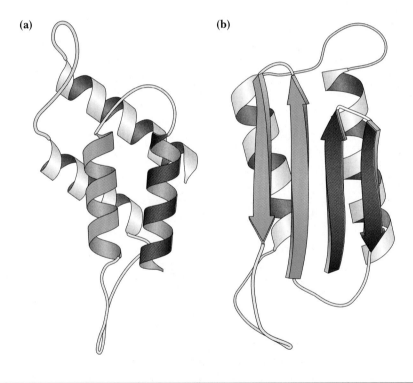

(a)   **(b)**

 See page 53

Adapted from Figure 1 in Prusiner, S. B., 1996. Molecular biology and the pathogenesis of prion diseases. *Trends in Biochemical Sciences* **21:**482–487.

response, such as the immunoglobulins. This **somatic recombination** rearranges the immunoglobulin genes, dramatically increasing the potential diversity of immunoglobulins available from a fixed amount of genetic information. Homologous recombination can also occur in bacteria. Indeed, even viral chromosomes undergo recombination. For example, if two mutant viral particles simultaneously infect a host cell, a recombination event between the two viral genomes can lead to the formation of a virus chromosome that is wild-type.

Figure 23.17 Meselson and Weigle's experiment demonstrated that a physical exchange of chromosome parts actually occurs during recombination. Density-labeled "heavy" phage, symbolized as ABC phage in the diagram, was used to coinfect bacteria along with "light" phage, the abc phage. The progeny from the infection were collected and subjected to CsCl density gradient centrifugation. Parental-type ABC and abc phage were well-separated in the gradient, but recombinant phage (ABc, Abc, aBc, aBC, and so on) were distributed diffusely between the two parental bands because they contained chromosomes constituted from fragments of both "heavy" and "light" DNA. These recombinant chromosomes formed by breakage and reunion of parental "heavy" and "light" chromosomes.

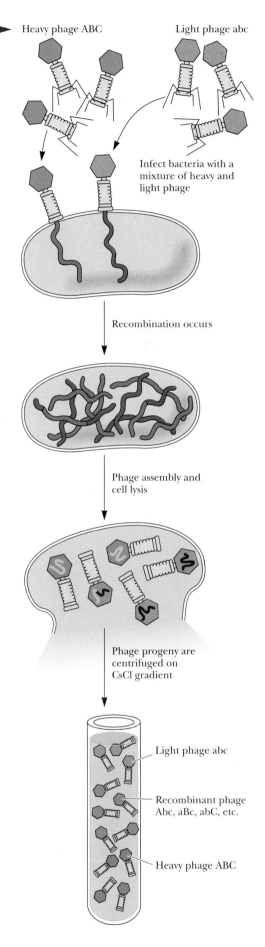

General Recombination

Recombination occurs by the breakage and reunion of DNA strands so that a physical exchange of parts takes place. Matthew Meselson and J. J. Weigle demonstrated this in 1961 by coinfecting *E. coli* with two genetically distinct bacteriophage λ strains, one of which had been density-labeled by growth in ^{13}C- and ^{15}N-containing media (Figure 23.17). The phage progeny were recovered and separated by CsCl density gradient centrifugation. Phage particles that displayed recombinant genotypes were distributed throughout the gradient while parental (nonrecombinant) genotypes were found within discrete "heavy" and "light" bands in the density gradient. The results showed that the recombinant phage contained DNA derived in varying proportions from both parents. The obvious explanation is that these recombinant DNAs arose via the breakage and rejoining of DNA molecules.

A second important observation made during this type of experiment was that some of the plaques formed by the phage progeny contained phage of two different genotypes, even though each plaque was caused by a single phage infecting one bacterium. Therefore, some infecting phage chromosomes must have contained a region of **heteroduplex DNA,** duplex DNA in which a part of each strand is contributed by a different parent (Figure 23.18).

The Holliday Model

In 1964, Robin Holliday proposed a model for homologous recombination that has proven to be correct in its essential features (Figure 23.19). The two homologous DNA duplexes are first juxtaposed so that their sequences are aligned. This process of **chromosome pairing** is called **synapsis** (Figure 23.19a).

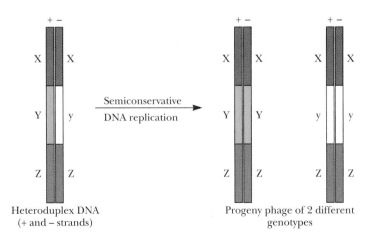

Figure 23.18 The generation of progeny bacteriophage of two different genotypes from a single phage particle carrying a heteroduplex DNA region within its chromosome. The heteroduplex DNA is composed of one strand that is genotypically XYZ (the + strand), and the other strand that is genotypically XyZ (the − strand). That is, the genotype of the two parental strands for gene Y is different (one is Y, the other y).

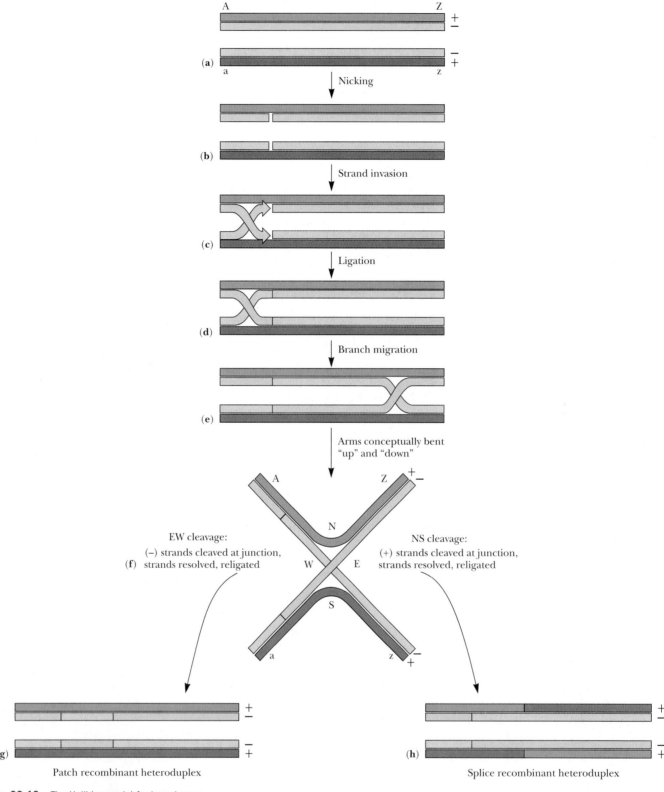

Figure 23.19 The Holliday model for homologous recombination. The + signs and − signs label strands of like polarity. For example, assume that the two strands running 5′ → 3′ as read left to right are labeled +, and the two strands running 3′ → 5′ as read left to right are labeled −. Only strands of like polarity exchange DNA during recombination (see text for detailed description).

Holliday suggested that recombination begins by introduction of single-stranded nicks in the DNA at homologous sites on the two paired chromosomes (Figure 23.19b). The two duplexes partially unwind, and the free, single-stranded end of one duplex begins to base-pair with its nearly complementary, single-stranded region along the intact strand in the other duplex, and vice versa (Figure 23.19c). This **strand invasion** is followed by ligation of the free ends from different duplexes to create a cross-stranded intermediate known as a **Holliday junction;** (Figure 23.19d). The cross-stranded junction can now migrate in either direction (**branch migration**) by unwinding and rewinding the two duplexes (Figure 23.19e). Branch migration results in **strand exchange;** heteroduplex regions of varying length are possible. In order for the joint molecule formed by strand exchange to be resolved into two DNA duplex molecules, another pair of nicks must be introduced. Resolution can be represented best if the duplexes are drawn with the chromosome arms bent "up" or "down" to give a planar representation (Figure 23.19f). Nicks then take place, either at E and W—that is, in the − strands that were originally nicked (see Figure 23.19b)—or at N and S, in the + strands (the strands not previously nicked). Duplex resolution is most easily kept straight by remembering that + strands are complementary to − strands and any resulting duplex must have one of each. Nicks made in the strands originally nicked lead to DNA duplexes in which one strand of each remains intact. Although these duplexes contain heteroduplex regions, they are not recombinant for the markers (AZ, az) that flank the heteroduplex region; such heteroduplexes are called **patch recombinants** (Figure 23.19g). Nicks introduced into the two strands not previously nicked yield DNA molecules that are both heteroduplex and recombinant for the markers A/a and Z/z; these heteroduplexes are termed **splice recombinants** (Figure 23.19h). Although this Holliday model explains the outcome of recombination, it provides no mechanistic explanation for the strand exchange reactions and other molecular details of the process.

The Enzymology of General Recombination

To illustrate recombination mechanisms, we focus on general recombination as it occurs in *E. coli*. The principal players in the process are the **RecBCD** enzyme complex, which initiates recombination; the **RecA** protein, which binds single-stranded DNA, forming a nucleoprotein filament capable of strand invasion and homologous pairing; and the **RuvA, RuvB,** and **RuvC** proteins, which drive branch migration and process the Holliday junction into recombinant products. Eukaryotic homologs of these prokaryotic recombination proteins have been identified, indicating that the fundamental process of general recombination is conserved across all organisms.

The RecBCD Enzyme Complex

The RecBCD complex is comprised of the proteins **RecB** (140 kD; 1180 amino acids), **RecC** (130 kD; 1122 amino acids), and **RecD** (67 kD; 608 amino acids). This multifunctional enzyme complex has both helicase and nuclease activity and initiates recombination by attaching to the end of a DNA duplex (or at a double-stranded break in the DNA) and using its ATP-dependent helicase function to unwind the dsDNA (Figure 23.20a). As RecBCD progresses along unwinding the duplex, its nuclease activity cleaves both of the newly formed single strands (although the strand that provided the 3′-end at the RecBCD entry site is cut more frequently than the 5′-terminal strand [Figure 23.20b]).

SSB (and some RecA protein) readily binds to the emerging single strands. Sooner or later, RecBCD encounters a particular nucleotide sequence, a so-called Chi (or χ) site, characterized by the sequence **5′-GCTGGTGG-3′.** These χ sites are recombinational "hot spots"; 1009 χ sites have been identified in the *E. coli* genome (on average, about one every 4.5 kb of DNA). When a χ sequence is encountered by a RecBCD complex approaching its 3′-side (the ..G-3′-side), RecBCD cleaves the χ-bearing DNA strand four to six bases to the 3′-side of χ (Figure 23.20c). The D subunit of RecBCD becomes irreversibly altered such that the RecBCD complex no longer expresses nuclease activity

Figure 23.20 Model of RecBCD-dependent initiation of recombination. **(a)** RecBCD binds to a duplex DNA end and its helicase activity begins to unwind the DNA double helix. "Rabbit ears" of ssDNA loop out from RecBCD because the rate of DNA unwinding exceeds the rate of ssDNA release by RecBCD. **(b)** As it unwinds the DNA, SSB (and some RecA) binds to the single-stranded regions; the RecBCD endonuclease activity randomly cleaves the ssDNA, showing a greater tendency to cut the 3'-terminal strand rather than the 5'-terminal strand. **(c)** When RecBCD encounters a properly oriented χ site, the 3'-terminal strand is cleaved just below the 3'-end of χ. **(d)** RecBCD now directs the binding of RecA to the 3'-terminal strand, as RecBCD endonuclease activity now acts more often on the 5'-terminal strand. **(e)** A nucleoprotein filament consisting of RecA-coated 3'-strand ssDNA is formed. This nucleoprotein filament is capable of homologous pairing with a dsDNA and strand invasion. *(Adapted from Figure 2 in Eggleston, A. K., and West, S. C., 1996. Exchanging partners: recombination in E. coli. Trends in Genetics* **12**:*20–25; and Figure 3 in Eggleston, A. K., and West, S. C., 1997. Recombination initiation: Easy as A, B, C, D . . . χ? Current Biology* **7**:*R745–R749.)*

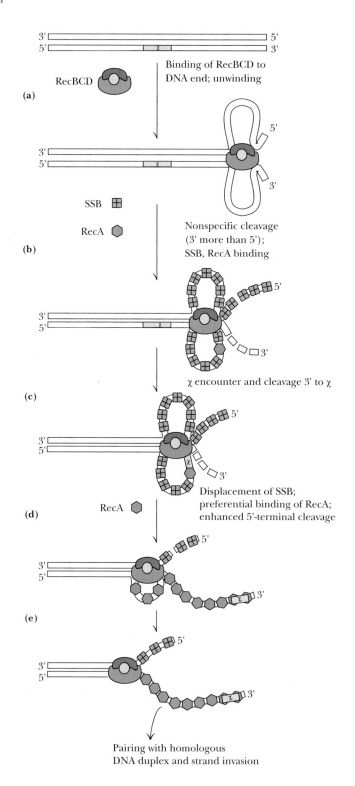

against the 3'-terminal strand, but nuclease activity against the 5'-terminal strand increases (Figure 23.20d).

Resuming its helicase function, RecBCD unwinds the dsDNA, and collectively these processes generate a ssDNA tail bearing a χ site at its 3'-terminal end. This ssDNA may reach several kilobases in length. RecA protein now binds to the 3'-terminal strand to form a **nucleoprotein filament** (Figure 23.20e). This filament is active in pairing and strand invasion with a homologous region in another dsDNA molecule.

The RecA Protein

The **RecA** protein, or **recombinase,** is a multifunctional protein that acts in general recombination to catalyze the ATP-dependent **DNA strand exchange reaction,** leading to formation of a Holliday junction (see Figure 23.19b–f). RecA protein (Figure 23.21a) crystallizes in the absence of DNA to form a helical filament having six monomers per turn (Figure 23.21b). This filament has a deep spiral groove large enough to accommodate three strands of DNA. The nucleoprotein filament formed by binding of RecA protein to the 3′-terminal ssDNA has affinity for other DNA molecules. In fact, binding of multiple DNA strands is the hallmark of RecA function. In recombination, RecA uses its high-

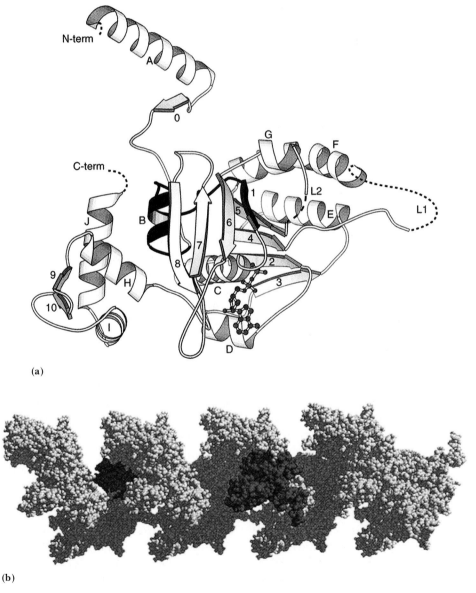

(a)

(b)

Figure 23.21   The structure of RecA, a 352-residue, 38-kD protein.   **(a)** Ribbon diagram of the RecA monomer. Note the ADP bound at the site near helices C and D.   **(b)** RecA filament. Four turns of a helical filament that has six RecA monomers per turn. A RecA monomer is highlighted in red. *(Adapted from Figures 2 and 3 in Roca, A. I., and Cox, M. M. 1997. RecA protein: Structure, function, and role in recombinational DNA repair. Progress in Nucleic Acid Research and Molecular Biology* **56:***127–223. Photos courtesy of Michael M. Cox, University of Wisconsin.)*

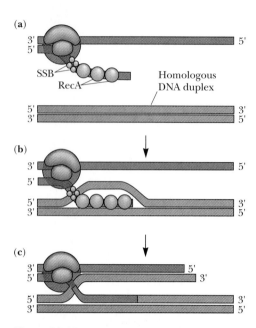

Figure 23.22 Model for homologous recombination as promoted by RecA enzyme. **(a)** RecA protein (and SSB) aids strand invasion of the 3′-ssDNA into a homologous DNA duplex, **(b)** forming a D-loop. **(c)** The D-loop strand that has been displaced by strand invasion pairs with its complementary strand in the original duplex to form a Holliday junction as strand invasion continues.

affinity primary DNA-binding site to bind ssDNA. This complex then interacts with other DNA molecules through a secondary DNA-binding site within RecA.

Procession of strand separation of dsDNA and the re-pairing into hybrid strands along the DNA duplex initiates branch migration (Figure 23.22b). Branch migration drives the displacement of the homologous DNA strand from the DNA duplex and its replacement with the ssDNA strand, a process known as **single-strand assimilation** (or **single-strand uptake**). Strand assimilation does not occur if there is no sequence homology between the ssDNA and the invaded DNA duplex. The DNA strand displaced by the invading 3′-terminal ssDNA is free to anneal with the 5′-terminal strand in the original DNA, a step that is also mediated by RecA protein and SSB (Figure 23.22c). The result is a Holliday junction, the classic intermediate in genetic recombination.

Resolution of the Holliday Junction of the RuvA, RuvB, and RuvC Proteins

The Holliday junction is processed into recombination products by **RuvA** (203 amino acids), **RuvB** (336 amino acids), and **RuvC** (173 amino acids). Specifically, RuvA and RuvB work together as a Holliday junction–specific helicase complex which dissociates the RecA filament and catalyzes branch migration. An RuvA tetramer (Figure 23.23a) fits precisely within the junction point (Figure 23.23b), which has a square-planar geometry, and this RuvA tetramer targets the assembly of RuvB around opposite arms of the DNA junction. The RuvB protein binds to form two oppositely oriented hexameric [(RuvB)$_6$] ring structures encircling the dsDNAs, one on each side of the Holliday junction. Rotation of the dsDNAs by the RuvB hexameric rings pulls the dsDNAs through (RuvB)$_6$ and unwinds the DNA strands across the "spool" of RuvA, which threads the separated single strands into newly forming hybrid (recombinant) duplexes (Figure 23.23b). The RuvA tetramer is a disklike structure, one face of which has an overall positive charge (Figure 23.23c), with the exception of four negatively charged central pins, each contributed by an RuvA monomer. These four pins fit neatly into the hole at the center of the Holliday junction. The negatively charged sugar–phosphate backbones of the four DNA duplexes of the Holliday junction are threaded along grooves in the positively charged RuvA face, with the negatively charged central pins appropriately situated to transiently separate the dsDNA molecules into their component single strands through repulsive electrostatic interactions with the phosphate backbones of the DNA. The separated strands of each parental duplex are then channeled into grooves in the RuvA face, where they are led into hydrogen-bonding interactions with bases contributed by strands of the other parental DNA to form the two daughter hybrid duplexes flowing out from the RuvAB complex (Figure 23.23b). Figure 23.23d illustrates a model for the RuvA tetramer with the square-planar Holliday junction.

Depending on how the strands in the Holliday junction are cleaved and resolved, patch or splice recombinant duplexes result (Figure 23.19g and h). RuvC is an endonuclease that resolves Holliday junctions into heteroduplex recombinant products (**RuvC resolvase**). An RuvC dimer binds at the Holliday junction and cuts pairs of DNA strands of similar polarity (Figure 23.23b); whether a patch or a splice recombinant results depends on which DNA pair is cleaved.

Hexameric ring helicases such as RuvB are DNA-driving molecular motors; similar motors act during DNA replication to propel strand separation and initiate DNA synthesis. Thus, the RuvABC system for processing Holliday junctions may represent a general paradigm for DNA manipulation in all cells.

(a)

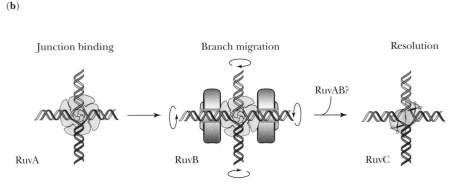

(b)

Junction binding   Branch migration   Resolution

RuvAB?

RuvA   RuvB   RuvC

Figure 23.23 Model for the resolution of a Holliday junction in *E. coli* by the RuvA, RuvB, and RuvC proteins. **(a)** Ribbon diagram of the RuvA tetramer. RuvA monomers have an overall L shape (one of them is outlined by the dashed white line); four of them form a tetramer with fourfold rotational symmetry, a structure reminiscent of a four-petaled flower. **(b)** Model for RuvA/RuvB action (first suggested by Parsons, C. A., et al., 1995. Structure of a multisubunit complex that promotes DNA branch migration. *Nature* **374:**375–378). *(left):* The RuvA tetramer fits snugly within the Holliday junction point. *(center):* Oppositely facing RuvB hexameric rings assemble on the heteroduplexes, with the DNA passing through their centers. These RuvB hexamers act as motors to promote branch migration by driving the passage of the DNA duplexes through themselves. *(right):* Binding of RuvC at the Holliday junction and strand scission by its nuclease activity. The locations of the RuvC active sites are indicated by the scissors. **(c)** Charge distribution on the concave surface of an RuvA tetramer. Blue indicates positive charge and red, negative charge. Note the overall positive charge on this surface of $(RuvA)_4$, with the exception of the four red (negatively charged) pins at its center. **(d)** Structural model for the interaction of $(RuvA)_4$ with the hypothesized square-planar Holliday junction center. *(Adapted from Figures 1, 2, and 3 in Rafferty, J. B., et al., 1996. Crystal structure of DNA recombination protein RuvA and a model for its binding to the Holliday junction. Science **274:**415–421. Figures courtesy of J. B. Rafferty, University of Sheffield, England.)*

(c)

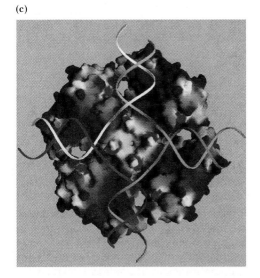

(d)

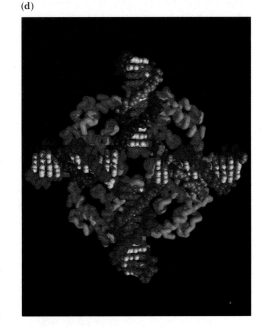

Recombination-Dependent Replication

It is likely that most replication forks that begin at the *E. coli oriC* initiation sites (or analogous initiation sites in eukaryotes) are derailed by nicks or more extensive DNA damage lying downstream. However, DNA replication can be re-initiated (and genome replication can be completed) following **replication fork restart.** Such restart depends on the recombination proteins RecA and RecBCD and the formation of a D loop (see Figure 23.22). The *E. coli* protein **PriA** recognizes and binds with high affinity to D loops. Once bound, PriA recruits DnaB helicase to this D loop, and a replication fork complete with two copies of the replicative DNA polymerase is re-established so that DNA replication resumes.

Transposons

In 1950, Barbara McClintock reported the results of her studies on an **activator gene** in maize (*Zea mays* or, as it's usually called, corn) that was recognizable principally by its ability to cause mutations in a second gene. Activator genes were thus an internal source of mutation. A most puzzling property was their ability to move relatively freely about the genome. Scientists had labored to

A DEEPER LOOK

The Three Rs of Genomic Manipulation: Replication, Recombination, and Repair

DNA replication, recombination, and repair have traditionally been treated as separate aspects of DNA metabolism. In recent years, however, scientists have come to realize that DNA replication is an essential component of both DNA recombination and DNA repair processes. Further, recombination mechanisms play an absolutely vital role in restarting replication forks that become halted at breaks or other lesions in the DNA strands. If a double-stranded break (DSB) or a nick in just one of the DNA strands (called a single-stranded gap, or SSG) is present in the DNA undergoing replication, the replication fork stalls and the replication complex dissociates (*replication fork "collapse"*). Significantly, the whole process of

homologous recombination can only initiate at SSGs or DSBs, and establishment of homologous recombination at such sites can rescue DNA replication. This **recombination-dependent replication (RDR)** has the interesting property of initiating DNA replication at sites other than the *oriC* site, and thus RDR is an important mechanism for restarting DNA replication if the replication fork is disrupted for any reason. As might be expected from the close relationships between replication, recombination, and repair, many of the same proteins are involved in all three, and all three must be viewed as essential processes in the perpetuation of the genome.

establish that chromosomes consisted of genes arrayed in a fixed order, so most geneticists viewed as incredible this idea of genes moving around. The recognition that McClintock so richly deserved for her explanation of this novel phenomenon had to await verification by molecular biologists. In 1983, Barbara McClintock was finally awarded the Nobel Prize in physiology or medicine. By

A DEEPER LOOK

"Knockout" Mice: A Method to Investigate the Essentiality of a Gene

Homologous recombination can be used to replace a gene with an inactivated equivalent of itself. Inactivation is accomplished by inserting a foreign gene, such as *neo*, which encodes resistance to the drug G418, within one of the exons of a copy of the gene of interest. Homologous recombination between the *neo*-bearing transgene and the gene of interest in wild-type mouse embryonic stem (ES) cells replaces the target gene with the inactive transgene (see figure). ES cells in which homologous recombination has occurred are resistant to G418, and such cells can be selected. These re-

combinant ES cells can be injected into early-stage mouse embryos, where they have a chance of becoming the germline cells of the newborn mouse. If they do, an inactivated target gene is then present within the gametes of this mouse. Mating between male and female mice with inactive target genes yields a generation of homozygous "knockout" mice—mice lacking a functional copy of the targeted gene. Characterization of these "knockout" mice reveals which physiological functions the gene directs.

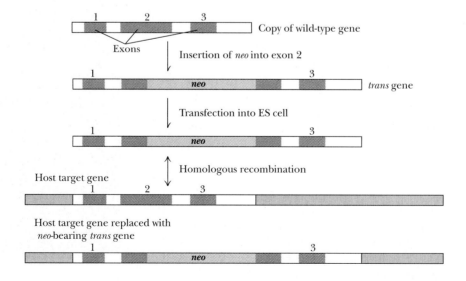

this time, it was appreciated that many organisms, from bacteria to humans, possessed similar "jumping genes" that were able to move from one site to another in the genome. This mobility led to their designation as **mobile elements, transposable elements,** or, simply, **transposons.**

Transposons are segments of DNA that are moved enzymatically from place to place in the genome (Figure 23.24). That is, their location within the DNA is unstable. Transposons range in size from several hundred bp to more than 8 kbp. Transposons contain a gene encoding an enzyme necessary for insertion into a chromosome and for the remobilization of the transposon to different locations. These movements are termed **transposition events.** The smallest transposons are called **insertion sequences,** or **ISs,** signifying their ability to insert apparently at random in the genome. Insertion into a new site can cause

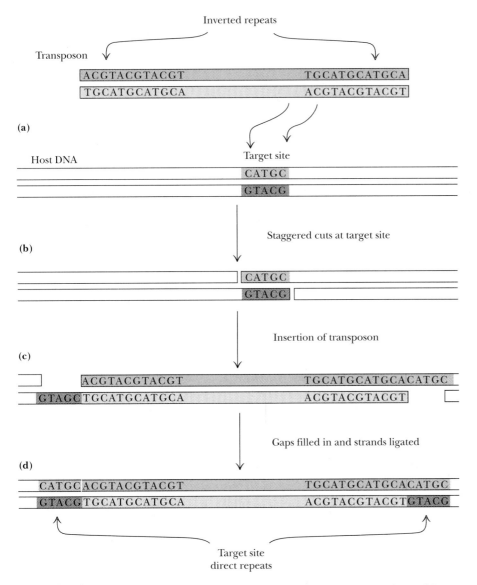

Figure 23.24 The typical transposon has inverted nucleotide-sequence repeats at its termini, represented here as the 12-bp sequence ACGTACGTACGT **(a)**. It acts at a target sequence (shown here as the sequence CATGC) within host DNA by creating a staggered cut **(b)** whose protruding single-stranded ends are then ligated to the transposon **(c)**. The gaps at the target site are then filled in, and the filled-in strands are ligated **(d)**. Transposon insertion thus generates direct repeats of the target site in the host DNA, and these direct repeats flank the inserted transposon.

A DEEPER LOOK

Transgenic Animals

Experimental advances in gene transfer techniques have made it possible to introduce genes into animals by **transfection.** Transfection is defined as the uptake or injection of plasmid DNA into recipient cells. Animals that have acquired new genetic information as a consequence of the introduction of foreign genes are termed **transgenic.** Plasmids carrying the gene of interest are in-jected into the nucleus of an oocyte or fertilized egg, and then the egg is implanted in a receptive female. The technique has been perfected for mice (see figure, a). In a small number of cases—10% or so—the mice that develop from the injected eggs carry the transfected gene integrated into a single chromosomal site. The gene is subsequently inherited by the progeny of the transfected

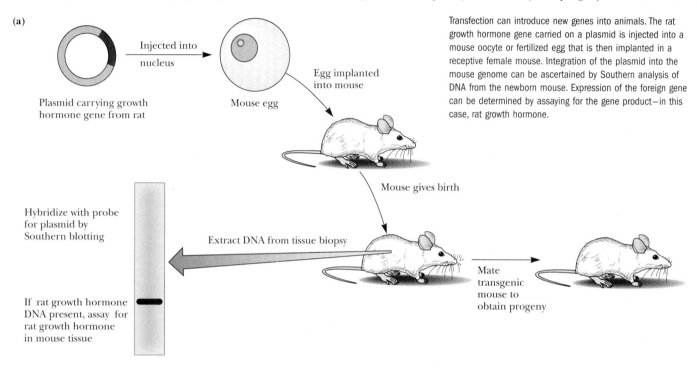

(a)

Plasmid carrying growth
hormone gene from rat

Injected into
nucleus

Mouse egg

Egg implanted
into mouse

Mouse gives birth

Hybridize with probe
for plasmid by
Southern blotting

Extract DNA from tissue biopsy

Mate
transgenic
mouse to
obtain progeny

If rat growth hormone
DNA present, assay for
rat growth hormone
in mouse tissue

Transfection can introduce new genes into animals. The rat growth hormone gene carried on a plasmid is injected into a mouse oocyte or fertilized egg that is then implanted in a receptive female mouse. Integration of the plasmid into the mouse genome can be ascertained by Southern analysis of DNA from the newborn mouse. Expression of the foreign gene can be determined by assaying for the gene product—in this case, rat growth hormone.

a mutation if a gene or regulatory region at the site is disrupted. Because transposition events can move genes to new places or lead to the duplication of existing genes, transposition is a major force in evolution.

23.8 | DNA Repair

Biological macromolecules are susceptible to chemical alterations that arise from environmental damage or errors during synthesis. For RNAs, proteins, or other cellular molecules, most consequences of such damage are avoided by replacement of these molecules through normal turnover (synthesis and degradation). However, the integrity of DNA is vital to cell survival and reproduction. Its information content must be protected over the life span of the cell and preserved from generation of generation. Safeguards include (a) high-fidelity replication systems and (b) repair systems that correct DNA damage that might alter its information content. DNA is the only molecule that, if damaged, is repaired by the cell. Such repair is possible because the information content of duplex DNA is inherently redundant. The most common forms of

animal as if it were a normal gene. Expression of the donor gene in the transgenic animals is variable because the gene is randomly integrated into the host genome and gene expression is often influenced by chromosomal location. Nevertheless, transfection of animals has produced some startling results, as in the case of the transfection of mice with the gene encoding the **rat growth hormone (rGH).** The transgenic mice grew to nearly twice the normal size (see figure, b). Growth hormone levels in these animals were several hundred times greater than normal. Similar results were ob-

tained in transgenic mice transfected with the **human growth hormone (hGH)** gene. The biotechnology of transfection has been extended to farm animals, and transgenic chickens, cows, pigs, rabbits, sheep, and even fish have been produced.

The first animal cloned from an adult cell, a sheep named Dolly, represented a milestone in cloning technology. Subsequent accomplishments include incorporation of the human gene encoding blood coagulation Factor IX into fetal sheep fibroblast cells, and successful transfer of nuclei from these cells into sheep oocytes lacking nuclei. These transgenic oocytes were placed in the uterus of receptive female sheep, which subsequently gave birth to transgenic lambs. The introduced Factor IX transgene was specifically designed so that Factor IX protein, a medically useful product for the treatment of hemophiliacs, would be expressed in the milk of the transgenic sheep. Similar successes in cows, which produce much more milk, has brought the potential for commercial production of virtually any protein into the realm of reality.

Transfection technology also holds promise as a mechanism for "gene therapy" by replacing defective genes with functional genes in animals (see Chapter 9). Problems concerning delivery, integration, and regulation of the transfected gene, including its appropriate expression in the right cells at the proper time during development and growth of the organism, must be brought under control before gene therapy becomes commonplace in humans.

(b)

◄ Photograph showing a transgenic mouse with an active rat growth hormone gene *(left)*. This transgenic mouse is twice the size of a normal mouse *(right)*. *(Photo courtesy of Ralph L. Brinster; School of Veterinary Medicine, University of Pennsylvania)*

damage are (a) replication errors resulting in a missing or incorrect base; (b) bulges due to deletions or insertions; (c) UV-induced base alterations, such as pyrimidine dimers (Figure 23.25); (d) strand breaks at phosphodiester bonds or within deoxyribose rings; and (e) covalent cross-linking of strands. Cells have extraordinarily diverse and effective DNA repair systems to deal with these problems. When repair fails, the genome may still be preserved if an "error-prone" mode of replication allows the lesion to be bypassed.

Human DNA replication has an error rate of about three base-pair mistakes during copying of the 6 billion base pairs in the diploid human genome. The low error rate is due to DNA repair systems that review and edit the newly replicated DNA. Further, about 10^4 bases (mostly purines) are lost per cell per day from spontaneous breakdown in human DNA; the repair systems must replace these bases to maintain the fidelity of the encoded information.

Usually, the complementary structure of duplex DNA ensures that information lost through damage to one strand can be recovered from the other. However, even errors involving both strands can be corrected. For example, deletions or insertions can be repaired by replacing the region through recombination (see Section 23.7). Double-stranded breaks, potentially the most serious lesions, can be repaired by DNA ligases or recombination events.

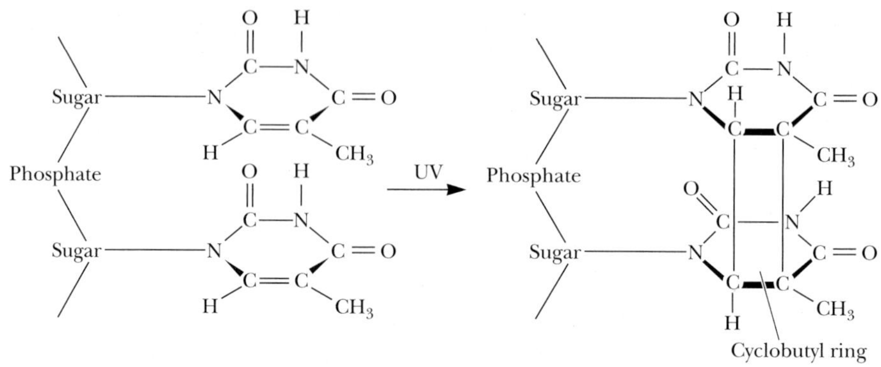

Figure 23.25 UV irradiation causes dimerization of adjacent thymine bases. A cyclobutyl ring is formed between carbons 5 and 6 of the pyrimidine rings. Normal base pairing is disrupted by the presence of such dimers.

Molecular Mechanisms of DNA Repair

Several fundamental types of molecular mechanisms for DNA repair can be distinguished: mechanisms that excise and replace damaged regions by recombination-dependent replication (discussed earlier), **mismatch repair,** and mechanisms that reverse damaging chemical changes in DNA, such as **excision repair** systems.

Mismatch Repair

The *mismatch repair system* corrects errors introduced when DNA is replicated. It scans newly synthesized DNA for mispaired bases, excises the mismatched region, and then replaces it by DNA polymerase–mediated local replication. The key to such replacement is to know which base of the mismatched pair is correct.

The *E. coli* **methyl-directed pathway** of mismatch repair relies on methylation patterns in the DNA to determine which strand is the newly synthesized one and which one was the parental (template) strand. DNA methylation, often an identifying and characteristic feature of a prokaryote's DNA, occurs just after DNA replication. During methylation, methyl groups are added to certain bases along the new DNA strand. However, a window of opportunity exists between the start of methylation and the end of replication, when only the parental strand of a dsDNA is methylated. This window in time provides an opportunity for the mismatch repair system to review the dsDNA for mismatched bases that arose as a consequence of replication errors. By definition, the newly synthesized strand is the one containing the error, and the methylated strand is the one having the correct nucleotide sequence. When the methyl-directed mismatch repair system encounters a mismatched base pair, it searches along the DNA—through thousands of base pairs, if necessary—until it finds a methylated base.

The system identifies the strand bearing the methylated base as parental, assumes its sequence is the correct one, and replaces the entire stretch of nucleotides within the new strand from this recognition point to and including the mismatched base. Mismatch repair does this by using an endonuclease to cut the new, unmethylated strand and an exonuclease to remove the mismatched bases, creating a gap in the newly synthesized strand. DNA polymerase III holoenzyme then fills in the gap, using the methylated strand as template. Finally, DNA ligase re-seals the strand.

Reversing Chemical Damage

Photoreactivation of Pyrimidine Dimers. UV irradiation promotes the formation of covalent bonds between adjacent thymine residues in a DNA strand, creating a cyclobutyl ring (see Figure 23.25). Because the C — C bonds in this ring are shorter than the normal 0.34-nm base stacking in B-DNA, the DNA is distorted at this spot and is no longer a proper template for either replication or transcription. **Photolyase** (also called **photoreactivating enzyme**), a flavin- and pterin-dependent enzyme, binds at the dimer and uses the energy of visible light to break the cyclobutyl ring, restoring the pyrimidines to their original form.

Excision Repair.   Replacement of many damaged or modified bases occurs via **excision repair systems.** There are two fundamental excision repair systems: **base excision** and **nucleotide excision.** *Base excision repair* acts on single bases that have been damaged through oxidation or other chemical modification during normal cellular processes. The damaged base is removed by **DNA glycosylase,** which cleaves the glycosidic bond, creating an AP site where the sugar–phosphate backbone is intact but a purine (*apurinic site*) or a pyrimidine (*apyrimidinic site*) is missing. An **AP endonuclease** then cleaves the backbone, an exonuclease removes the deoxyribose-P and a number of additional residues, and the gap is repaired by DNA polymerase and DNA ligase (Figure 23.26). The information on the complementary strand is used to dictate which bases are to be added in re-filling the gap. The DNA polymerase I binds at the gap and moves in the $5' \rightarrow 3'$ direction, removing nucleotides with its $5'$-exonuclease activity. The $5' \rightarrow 3'$ DNA polymerase activity of DNA polymerase I fills in the sequence behind the $5'$-exonuclease action. No net synthesis of DNA results, but this action of DNA polymerase I "edits out" sections of damaged DNA. Excision is coordinated with $5' \rightarrow 3'$ polymerase-catalyzed replacement of the damaged nucleotides so that DNA of the right sequence is restored.

Nucleotide excision repair recognizes and repairs larger regions of damaged DNA than does base excision repair. The nucleotide excision repair system cuts the sugar–phosphate backbone of a DNA strand in two places, one on each side of the lesion, and removes the region. The region removed in prokaryotic nucleotide excision repair spans 12 or 13 nucleotides; in eukaryotic excision repair, an oligonucleotide stretch 27 to 29 units long is removed. The resulting gap is then filled in using DNA polymerase (DNA polymerase I in prokaryotes or DNA polymerase δ or ε and PCNA plus RFC in eukaryotes), and the sugar–phosphate backbone is covalently closed by DNA ligase.

In mammalian cells, nucleotide excision repair is the main pathway for removal of carcinogenic (cancer-causing) lesions caused by sunlight or other mutagenic agents. Such lesions are recognized by **XPA** protein, named for *xeroderma pigmentosum,* an inherited human syndrome whose victims suffer serious skin lesions if exposed to sunlight. At sites recognized by XPA, a multiprotein endonuclease is assembled and the damaged strand is cleaved and repaired.

23.9   The Molecular Nature of Mutation

Genes are normally transmitted unchanged from generation to generation, owing to the great precision and fidelity with which genes are copied during chromosome duplication. However, on rare occasions, genetically heritable changes (**mutations**) occur that result in altered forms. Most mutated genes function less effectively than the unaltered, wild-type allele, but occasionally mutations arise that give the organism a selective advantage. When this occurs, they are propagated to many offspring. Together with recombination, mutation provides for genetic variability within species and, ultimately, the evolution of new species.

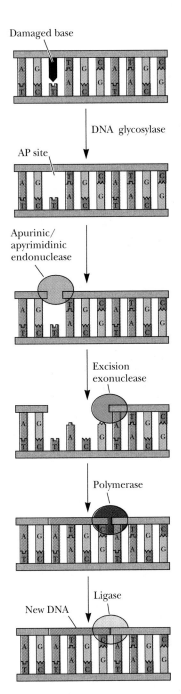

Figure 23.26 Base excision repair. A damaged base (■) is excised from the sugar–phosphate backbone by DNA glycosylase, creating an AP site. Then, an apurinic/apyrimidinic endonuclease severs the DNA strand, and an excision nuclease removes the AP site and several nucleotides. DNA polymerase I and DNA ligase then repair the gap.

Mutations change the sequence of bases in DNA either by the substitution of one base pair for another (so-called **point mutations**) or by the insertion or deletion of one or more base pairs (**insertions** and **deletions**).

Point Mutations

Point mutations arise when a base pairs with an inappropriate partner. The two possible kinds of point mutations are **transitions,** in which one purine (or pyrimidine) is replaced by another, as in A → G (or T → C), and **transversions,** in which a purine is substituted for a pyrimidine or *vice versa.*

Point mutations arise by the pairing of bases with inappropriate partners during DNA replication, by the introduction of base analogs into DNA, or by chemical mutagens. Bases may rarely mispair (Figure 23.27), either because of their tautomeric properties (see Chapter 8) or because of other influences. Even in mispairing, the $C_{1'}$–$C_{1'}$ distances between bases must still be close to that of a Watson–Crick base pair (11 nm or so; see Figure 8.19) to maintain the mismatched base pair in the double helix. In tautomerization, for example, an amino group ($-NH_2$), usually an H-bond donor, can tautomerize to an imino form ($=NH$), and become an H-bond acceptor. Or a keto group ($C=O$), normally an H-bond acceptor, can tautomerize to an enol $C-OH$, an H-bond donor. Proofreading mechanisms operating during DNA replication catch most mispairings. The frequency of spontaneous mutation in both *E. coli* and fruit flies (*Drosophila melanogaster*) is about 10^{-10} per base pair per replication.

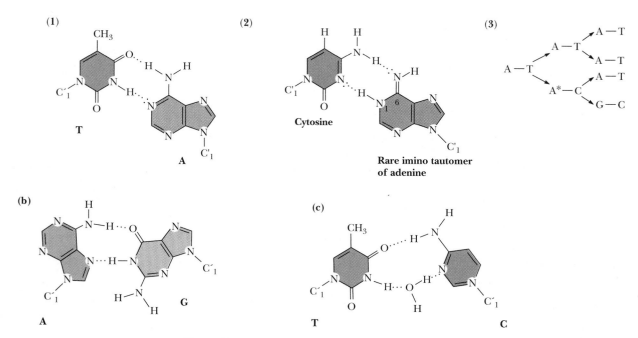

Figure 23.27 Point mutations due to base mispairings. **(a)** An example based on tautomeric properties. The rare imino tautomer of adenine base-pairs with cytosine rather than thymine. (1) The normal A-T base pair. (2) The A*-C base pair is possible for the adenine tautomer in which a proton has been transferred from the 6-NH$_2$ of adenine to N-1. (3) Pairing of C with the imino tautomer of A (A*) leads to a transition mutation (A–T to G–C) appearing in the next generation. **(b)** A in the syn conformation pairing with G (G is in the usual anti conformation). **(c)** T and C form a base pair by H-bonding interactions mediated by a water molecule.

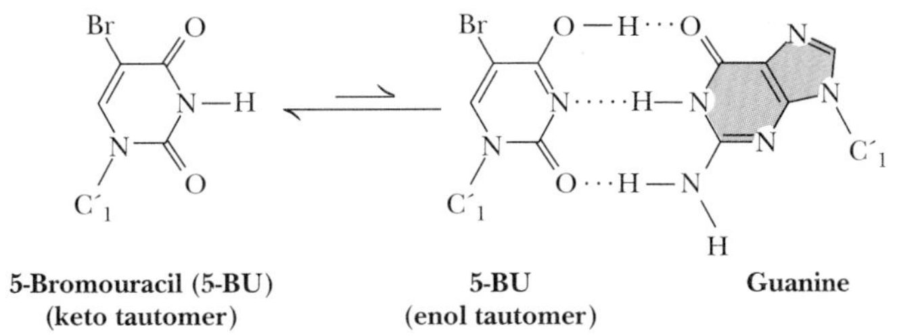

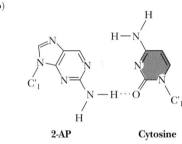

Figure 23.28 5-Bromouracil usually favors the keto tautomer that mimics the base-pairing properties of thymine, but it frequently shifts to the enol form, whereupon it can base-pair with guanine, causing a T–A to C–G transition.

5-Bromouracil (5-BU)
(keto tautomer)

5-BU
(enol tautomer)

Guanine

Mutations Induced by Base Analogs

Base analogs that become incorporated into DNA can induce mutations through changes in base-pairing possibilities. Two examples are **5-bromouracil (5-BU)** and **2-aminopurine (2-AP)**. 5-Bromouracil is a thymine analog and becomes inserted into DNA at sites normally occupied by T; its 5-Br group sterically resembles thymine's 5-methyl group. However, because 5-BU frequently assumes the enol tautomeric form and pairs with G instead of A, a point mutation of the transition type may be induced (Figure 23.28). Less often, 5-BU is inserted into DNA at cytosine sites, not T sites. Then, if it base-pairs in its keto form, mimicking T, a C-G to T-A transition ensues. The adenine analog, 2-aminopurine (recall that adenine is 6-aminopurine) normally behaves like A and base-pairs with T. However, 2-AP can form a single H bond of sufficient stability with cytosine (Figure 23.29) that occasionally C replaces T in DNA replicating in the presence of 2-AP. Hypoxanthine (Figure 23.30) is an adenine analog that arises *in situ* in DNA through oxidative deamination of A. Hypoxanthine base-pairs with cytosine, creating an A-T to G-C transition.

Chemical Mutagens

Chemical mutagens are agents that chemically modify bases so that their base-pairing characteristics are altered. For instance, nitrous acid (HNO_2) causes the oxidative deamination of primary amine groups, found in adenine and cytosine. Oxidative deamination of cytosine yields uracil, which base-pairs the way

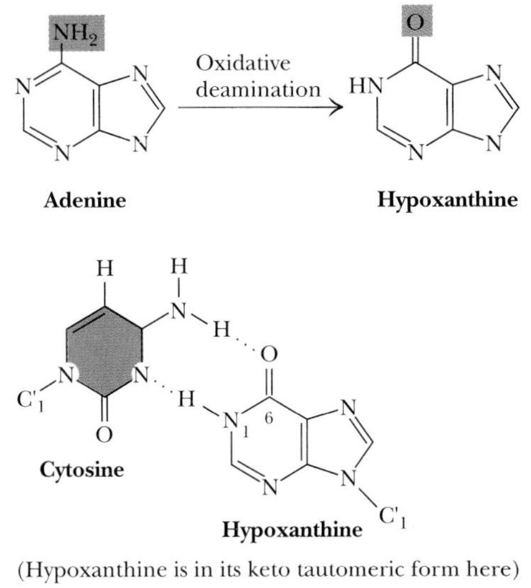

(a)

2-Aminopurine (2-AP)   **Thymine**

(b)

2-AP   **Cytosine**

Figure 23.29 **(a)** 2-Aminopurine normally base-pairs with T, but **(b)** may also pair with cytosine through a single hydrogen bond.

Adenine

Oxidative deamination →

Hypoxanthine

Cytosine

Hypoxanthine

(Hypoxanthine is in its keto tautomeric form here)

Figure 23.30 Oxidative deamination of adenine in DNA yields hypoxanthine, which base-pairs with cytosine, resulting in an A–T to G–C transition.

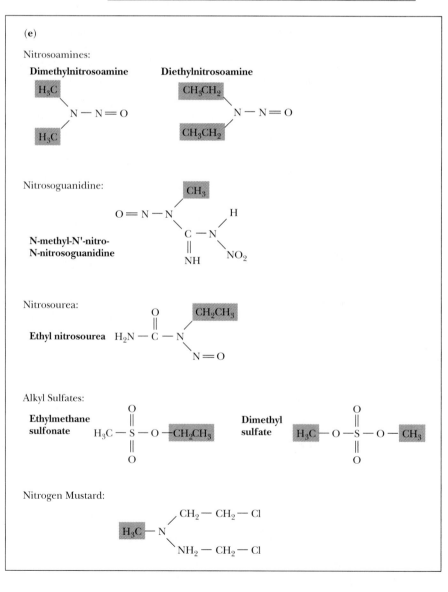

(a)

Cytosine → (HNO$_2$) → Uracil ··· Adenine

Adenine → (HNO$_2$) → Hypoxanthine ··· Cytosine

(b) Generic structure of nitrosoamines

(c)

Cytosine → (NH$_2$OH) → Adenine

(d)

Pairs normally with cytosine

Guanine

Alkylating agent ↓

Sometimes pairs with thymine

O^6-methylguanine

(e)

Nitrosoamines:

Dimethylnitrosoamine

Diethylnitrosoamine

Nitrosoguanidine:

N-methyl-N'-nitro-N-nitrosoguanidine

Nitrosourea:

Ethyl nitrosourea

Alkyl Sulfates:

Ethylmethane sulfonate

Dimethyl sulfate

Nitrogen Mustard:

◄ Figure 23.31 Chemical mutagens. **(a)** HNO_2 (nitrous acid) converts cytosine to uracil and adenine to hypoxanthine. **(b)** Nitrosoamines, organic compounds that react to form nitrous acid, also lead to the oxidative deamination of A and C. **(c)** Hydroxylamine (NH_2OH) reacts with cytosine, converting it to a derivative that base-pairs with adenine instead of guanine. The result is a C–G to T–A transition. **(d)** Alkylation of G residues to give O_6-methylguanine, which base-pairs with T. **(e)** Alkylating agents include nitrosoamines, nitrosoguanidines, nitrosoureas, alkyl sulfates, and nitrogen mustards. Note that nitrosoamines are mutagenic in two ways: they can react to yield HNO_2 or they can act as alkylating agents. The nitrosoguanidine, *N*-methyl-*N'*-nitro-*N*-nitrosoguanidine, is a very potent mutagen used in laboratories to induce mutations in experimental organisms such as *Drosophila melanogaster*. Ethylmethane sulfate (EMS) and dimethyl sulfate are also favorite mutagens among geneticists.

T does and gives a C-G to T-A transition (Figure 23.31a). Hydroxylamine specifically causes C-G to T-A transitions because it reacts specifically with cytosine, converting it to a derivative that base-pairs with adenine instead of guanine. **Alkylating agents** are also chemical mutagens. Alkylation of reactive sites on the bases to add methyl or ethyl groups alters their H-bonding and hence base pairing. For example, methylation of O^6 on guanine (giving O^6-methylguanine) causes this G to mispair with thymine, resulting in a G-C to A-T transition (Figure 23.31d). Alkylating agents can also induce point mutations of the transversion type. Alkylation of N^7 of guanine labilizes its N-glycosidic bond, which leads to elimination of the purine ring, creating a gap in the base sequence. The enzyme **AP endonuclease** then cleaves the sugar–phosphate backbone of the DNA on the 5′-side, and the gap can be repaired by enzymatic removal of the 5′-sugar phosphate and insertion of a new nucleotide. A transversion results if a pyrimidine nucleotide is inserted in place of the purine during enzymatic repair of this gap.

Insertions and Deletions

The addition or removal of one or more base pairs leads to insertion or deletion mutations, respectively. Either possibility shifts the triplet reading frame of codons, causing **frameshift mutations** (misincorporation of all subsequent amino acids) in the protein encoded by the gene. Such mutations can arise if flat aromatic molecules such as acridine orange insert themselves between successive bases in one or both strands of the double helix. This insertion, or, more aptly, **intercalation,** doubles the distance between the bases as measured along the helix axis. This distortion of the DNA results in inappropriate insertion or deletion of bases when the DNA is replicated. Disruptions that arise from the insertion of a transposon within a gene also fall into this category of mutation.

PROBLEMS

1. If ^{15}N-labeled *E. coli* DNA has a density of 1.724 g/mL, ^{14}N-labeled DNA has a density of 1.710 g/mL, and *E. coli* cells grown for many generations on $^{14}NH_4^+$ as a nitrogen source are transferred to media containing $^{15}NH_4^+$ as sole N-source, what will be the density of the DNA after one generation, because replication is semiconservative? Assume the mode of replication is dispersive: what would be the density of DNA after one generation? Design an experiment to distinguish between semiconservative and dispersive modes of replication.

2. What are the respective roles of the 5′-exonuclease and 3′-exonuclease activities of DNA polymerase I? What would be a feature of an *E. coli* strain that lacked DNA polymerase I 3′-exonuclease activity?

3. Assuming DNA replication proceeds at a rate of 750 base pairs per second, how long it will take to replicate the entire *E. coli* genome? Under optimal conditions, *E. coli* cells divide every 20 minutes. What is the minimal number of replication forks required per *E. coli* chromosome in order to sustain such a rate of cell division?

4. It is estimated that there are 10 molecules of DNA polymerase III per *E. coli* cell. Is it likely that *E. coli* growth rate is limited by DNA polymerase III levels?

5. Approximately how many Okazaki fragments are synthesized in the course of replicating an *E. coli* chromosome? How many are synthesized in replicating an "average" human chromosome?

6. How do DNA gyrases and helicases differ in their respective functions and modes of action?

7. If DNA replication proceeds at a rate of 100 base pairs per second in human cells, and origins of replication occur every 300 kbp, how long would it take to replicate the entire diploid human genome? How many molecules of DNA polymerase does each cell need to carry out this task?

8. From the information in Figure 23.19, diagram the recombinational event leading to the formation of a heteroduplex DNA region within a bacteriophage chromosome.

9. Homologous recombination in *E. coli* leads to the formation of regions of heteroduplex DNA. By definition, such regions contain mismatched bases. Why doesn't the mismatch repair system of *E. coli* eliminate these mismatches?

10. If RecA protein unwinds duplex DNA so that there are about 18.6 bp per turn, what is the change in $\Delta\phi$, the helical twist of DNA, compared to its value in B-DNA?

11. Diagram a Holliday junction between two duplex DNA molecules and show how the action of resolvase might give rise to either patch or splice recombinant DNA molecules.

12. Show the nucleotide sequence changes that might arise in a dsDNA (coding strand segment GCTA) upon mutagenesis with (a) HNO_2, (b) bromouracil, and (c) 2-aminopurine.

13. Transposons are mutagenic agents. Why?

14. Give a plausible explanation for the genetic and infectious properties of PrP[sc].

FURTHER READING

Anderson, D. G., and Kowalczykowski, S. C., 1997. The translocating RecBCD enzyme stimulates recombination by directing RecA protein onto ssDNA in a χ-regulated manner. *Cell* **90**:77–86.

Baker, T. A., and Bell, S. P., 1998. Polymerases and the replisome: Machines within machines. *Cell* **92**:295–305.

Bambara, R. A., and Huang, L., 1995. Reconstitution of mammalian DNA replication. *Progress in Nucleic Acid Research and Molecular Biology* **51**:93–122.

Bambara, R. A., Murante, R. S., and Henricksen, L. A., 1997. Enzymes and reactions at the eukaryotic replication fork. *Journal of Biological Chemistry* **272**:4647–4650.

Baumann, P., and West, S. C., 1998. Role of the human RAD51 protein in homologous recombination and double-stranded-break repair. *Trends in Biochemical Sciences* **23**:247–252.

Beernink, H. T. H., and Morrical, S. W., 1999. RMPs: Recombination/replication proteins. *Trends in Biochemical Sciences* **24**:385–389.

Behringer, R. R., et al., 1989. Synthesis of functional human hemoglobin in transgenic mice. *Science* **245**:971–979.

Bik, T., 1999. MCM proteins in DNA replication. *Annual Review of Biochemistry* **68**:649–686.

Blackburn, E. H., 1992. Telomerases. *Annual Review of Biochemistry* **61**:113–129.

Boehmer, P. E., and Lehman, I. R., 1997. Herpes simplex virus DNA replication. *Annual Review of Biochemistry* **66**:347–384.

Botchan, M., 1996. Coordinating DNA replication with cell division: Current status of the licensing concept. *Proceedings of the National Academy of Sciences, U.S.A.* **93**:9997–10000.

Chong, J. P. J., et al., 1996. The role of MCM/P1 proteins in the licensing of DNA replication. *Trends in Biochemical Sciences* **21**:102–106.

Cohen, F. E., and Prusiner, S. B., 1998. Pathological conformations of prion proteins. *Annual Review of Biochemistry* **67**:793–819.

Collins, K., 1999. Ciliate telomerase biochemistry. *Annual Review of Biochemistry* **68**:187–218.

Cook, P. R., 1999. The organization of replication and transcription. *Science* **284**:1790–1795.

Cox, M. M., 1999. Recombinational DNA repair in bacteria and the RecA protein. *Progress in Nucleic Acid Research and Molecular Biology* **63**:311–366.

Critchlow, S. E., and Jackson, S. P., 1998. DNA end-joining: From yeast to man. *Trends in Biochemical Sciences* **23**:394–398.

DePamphilis, M. E., 1998. Initiation of DNA replication in eukaryotic chromosomes. *Journal of Cellular Biochemistry* **S30/31**:8–17.

Eggleston, A. K., and West, S. C., 1996. Exchanging partners: Recombination in *E. coli*. *Trends in Genetics* **12**:20–25.

Eggleston, A. K., and West, S. C., 1997. Recombination initiation: Easy as A, B, C, D . . . χ? *Current Biology* **7**:R745–R749.

Eggleston, A. K., Mitchell, A. H., and West, S. C., 1997. *In vitro* reconstitution of the late steps of recombination in *E. coli*. *Cell* **89**:607–617.

Friedberg, E. C., 1995. Out of the shadows and into the light: The emergence of DNA repair. *Trends in Biochemical Sciences* **20**:381. (October 1995 [**20**:10] is a special issue on DNA repair.)

Friedberg, E. C., Walker, G. C., and Siede, W., 1995. *DNA Repair and Mutagenesis*. Washington, D.C.: American Society for Microbiology Press.

Goodman, M., 2000. Coping with replication "train wrecks" in *Escherichia coli*. *Trends in Biochemical Sciences* **25**:185–189.

Haber, J. E., 1999. DNA recombination: The replication connection. *Trends in Biochemical Sciences* **24**:271–275.

Hingorani, M. M., and O'Donnell, M., 2000. A tale of toroids in DNA metabolism. *Nature Reviews Molecular Cell Biology* **1**:22–30.

Hiom, K., and Gellert, M., 1997. A stable RAG1-RAG2-DNA complex that is active in V(D)J cleavage. *Cell* **88**:65–72.

Holliday, R., 1964. A mechanism for gene conversion in fungi. *Genetic Research* **5**:282–304. The classic model for the mechanism of DNA strand exchange during homologous recombination.

Hübscher, U., et al., 2000. Eukaryotic DNA polymerases—a growing family. *Trends in Biochemical Sciences* **25**:143–147.

Jallepalli, P. V., and Kelly, T. J., 1997. Cyclin-dependent kinase and initiation at eukaryotic origins: a replication switch? *Current Opinion in Cell Biology* **9**:358–363.

Kamada, K., et al., 1996. Structure of a replication-terminator protein complexed with DNA. *Nature* **383**:598–603.

Kearsey, S. E., et al., 1996. The role of MCM proteins in the cell cycle control of genome duplication. *BioEssays* **18**:183–190.

Keck, J. L., 2000. Structure of the RNA polymerase domain of the *E. coli* primase. *Science* **287**:2482–2486.

Kelly, T. J., and Brown, G. W., 2000. Regulation of chromosome replication. *Annual Review of Biochemistry* **69**:829–880.

Kelman, Z., and O'Donnell, M., 1995. DNA polymerase III holoenzyme: structure and function of a chromosomal replicating machine. *Annual Review of Biochemistry* **64**:171–200.

Kim, N. W., 1994. Specific association of human telomerase activity with immortal cells and cancer. *Science* **266**:2011–2015.

Kim, S., et al., 1996. Coupling of a replicative polymerase and helicase: a τ-DnaB interaction mediates rapid replication fork movement. *Cell* **84**:643–650.

Kong, X.-P., et al., 1992. Three-dimensional structure of the β-subunit of *E. coli* DNA polymerase III holoenzyme: a sliding clamp model. *Cell* **69**:425–437. The crystal structure of the β-subunit of DNA pol III holoenzyme reveals that this processivity factor forms a closed ring around DNA, tightly clamping the pol III enzyme to its substrate.

Kornberg, A., and Baker, T. A., 1992. *DNA Replication*, 2nd ed. New York: W. H. Freeman and Co. A comprehensive detailed account of the enzymology of DNA metabolism, including replication, recombination, repair, and more.

Kowalczykowski, S. C., 2000. Initiation of genetic recombination and recombination-dependent replication. *Trends in Biochemical Sciences* **25**:156–165.

Krishna, T. S., et al., 1994. Crystal structure of the eukaryotic DNA polymerase processivity factor PCNA. *Cell* **79**:1233–1243.

Krude, T., et al., 1997. Cyclin/Cdk–dependent initiation of DNA replication in a human cell-free system. *Cell* **88**:109–119.

Kunkel, T. A., and Bebenek, K., 2000. DNA replication fidelity. *Annual Review of Biochemistry* **69**:497–529.

Lewis, S. M., and Wu, G. E., 1997. The origins of V(D)J recombination. *Cell* **88**:159–162.

Lieber, M. R., 1996. The FEN-1 family of structure-specific nucleases in eukaryotic DNA replication, recombination, and repair. *BioEssays* **19**:233–240.

Lilley, D. M. J., ed., 1995. *DNA-Protein: Structural Interactions*. Oxford: Oxford University Press.

Malkas, L. H., 1998. DNA replication machinery of the mammalian cell. *Journal of Cellular Biochemistry* **S30/31**:18–29.

Marians, K. J., 2000. PriA-directed replication fork restart in *Escherichia coli*. *Trends in Biochemical Sciences* **25**:185–189.

Mazin, A. V., and Kowalczykowski, S. C., 1996. The specificity of the secondary DNA binding site of RecA protein defines its role in DNA strand exchange. *Proceedings of the National Academy of Sciences, U.S.A.* **93**:10673–10678.

McCollough, A. K., et al., 1999. Initiation of base excision repair: Glycosylase mechanisms and structures. *Annual Review of Biochemistry* **68**:255–285.

Meselson, M., and Stahl, F. W., 1958. The replication of DNA in *Escherichia coli*. *Proceedings of the National Academy of Sciences, U.S.A.* **44**:671–682. The classic paper showing that DNA replication is semiconservative.

Meselson, M., and Weigle, J. J., 1961. Chromosome breakage accompanying genetic recombination in bacteriophage. *Proceedings of the National Academy of Sciences, U.S.A.* **47**:857–869. The experiments demonstrating that physical exchange of DNA occurs during recombination.

Michel, B., 2000. Replication fork arrest and DNA recombination. *Trends in Biochemical Sciences* **25**:173–178.

Modrich, P., and Lahue, R. 1996. Mismatch repair in replication fidelity, genetic recombination, and cancer biology. *Annual Review of Biochemistry* **65**:101–133.

Mol, C. D., et al., 1999. DNA repair mechanisms for the recognition and removal of damaged DNA bases. *Annual Review of Biophysics and Biomolecular Structure* **28**:101–128.

Morgan, A. R., 1993. Base mismatches and mutagenesis: How important is tautomerism? *Trends in Biochemical Sciences* **18**:160–163.

Morgan, R. A., and Anderson, W. F., 1993. Human gene therapy. *Annual Review of Biochemistry* **62**:192–217.

Nakamura, T. M., et al., 1997. Telomerase catalytic subunit homologs from fission yeast and human. *Science* **277**:955–959.

Newport, J., and Yan, H., 1996. Organization of DNA into foci during replication. *Current Opinion in Cell Biology* **8**:365–368.

Ogawa, T., and Okazaki T., 1980. Discontinuous DNA replication. *Annual Review of Biochemistry* **49**:421–457. Okazaki fragments and their implications for the mechanism of DNA replication.

Page, A. M., and Hieter, P., 1999. The anaphase-promoting complex: New subunits and regulators. *Annual Review of Biochemistry* **68**:583–609.

Palmiter, R. D., et al., 1982. Dramatic growth of mice that develop from eggs microinjected with metallothionein–growth hormone fusion genes. *Nature* **300**:611–615.

Parikh, S. S., et al., 1999. Envisioning the molecular choreography of DNA base excision repair. *Current Opinion in Structural Biology* **9**:37–47.

Park, H.-W., 1995. Crystal structure of DNA photolyase from *Escherichia coli*. *Science* **268**:1866–1872.

Patel, S. S., and Picha, K. M., 2000. Structure and function of hexameric helicases. *Annual Review of Biochemistry* **69**:651–697.

Prusiner, S. B., 1996. Molecular biology and pathogenesis of prion diseases. *Trends in Biochemical Sciences* **21**:482–487.

Prusiner, S. B., 1997. Prion diseases and the BSE crisis. *Science* **278**:245–251.

Rafferty, J. B., et al., 1996. Crystal structure of DNA recombination protein RuvA and a model for its binding to the Holliday junction. *Science* **274**:415–421.

Rao, B. J., et al., 1995. How specific is the first step in homologous recombination? *Trends in Biochemical Sciences* **20**:109–113.

Roca, A. I., and Cox, M. M., 1997. RecA protein: Structure, function, and role in recombinational DNA repair. *Progress in Nucleic Acid Research and Molecular Biology* **56**:127–223.

Russell, P., 1998. Checkpoints on the road to mitosis. *Trends in Biochemical Sciences* **23**:399–402.

Sancar, A., 1994. Mechanisms of DNA excision repair. *Science* **266**:1954–1956. (*Science* named the extended family of DNA repair enzymes its "Molecules of the Year" in 1994. See the 23 December 1994 issue for additional readings.)

Schnieke, A. E., et al., 1997. Human Factor IX transgenic sheep produced by transfer of nuclei from transfected fetal fibroblasts. *Science* **278**:2130–2133.

Shinagawa, H., and Iwasaki, H., 1996. Processing the Holliday junction in homologous recombination. *Trends in Biochemical Sciences* **21**:107–111.

Steitz, T. A., 1998. A mechanism for all polymerases. *Nature* **391**:231–232.

Stellwagen, A. E., and Craig, N. L., 1998. Mobile DNA elements: Controlling transposition with ATP-dependent molecular switches. *Trends in Biochemical Sciences* **23:**486–490.

Stillman, B., 1996. Cell cycle control of DNA replication. *Science* **274:**1659–1663.

Story, R. M., Weber, I. T., and Steitz, T. A., 1992. The structure of the *E. coli* RecA protein monomer and polymer. *Nature* **355:**318–325. Also, Story, R. M., and Steitz, T. A., 1992. Structure of the RecA protein-ADP complex. *Nature* **355:**374–376.

Tonegawa, S., 1983. Somatic generation of antibody diversity. *Nature* **302:**575–581.

Tye, B. K., 1999. MCM proteins in DNA replication. *Annual Review of Biochemistry* **68:**649–686.

Waga, S., and Stillman, B., 1998. The DNA replication fork in eukaryotic cells. *Annual Review of Biochemistry* **67:**721–751.

Wang, T. A., and Li, J. J., 1995. Eukaryotic DNA replication. *Current Opinion in Cell Biology* **7:**414–420.

Wilmut, I., et al., 1997. Viable offspring derived from fetal and adult mammalian cells. *Nature* **385:**810–818. See also Campbell, K. H. S., et al., 1996. Sheep cloned by nuclear transfer from a cultured cell line. *Nature* **380:**64–66.

Wold, M. S., 1997. Replication protein A: a heterotrimeric, single-stranded DNA-binding protein required for eukaryotic DNA metabolism. *Annual Review of Biochemistry* **66:**61–92.

Wyman, C., and Botchan, M., 1995. DNA replication: a familiar ring to DNA polymerase processivity. *Current Biology* **5:**334–337.

Transcription and the Regulation of Gene Expression

Monk Transcribing Manuscript, *ca. 1470. Jean Mielot (author of* Miracles de Notre Dame) *at his desk using a quill and scraping knife. (Bibliotheque National de Paris/Mary Evans Picture Library/London)*

A day will come when some laborious monk
Will bring to light my zealous, nameless toil,
Kindle . . . his lamp, and from the parchment
Shaking the dust of ages, will transcribe
My chronicles.

ALEXANDER PUSHKIN, *Boris Godunov* (1825)

Outline

In 1958, Francis Crick enunciated the "central dogma of molecular biology" (Figure 24.1). This scheme outlined the residue-by-residue transfer of biological information as encoded in the primary structure of the informational biopolymers, nucleic acids and proteins. The predominant path of information transfer, DNA → RNA → protein, postulated that RNA was an information carrier between DNA and proteins, the agents of biological function. In 1961, François Jacob and Jacques Monod extended this hypothesis to predict that the RNA intermediate, which they dubbed **messenger RNA,** or **mRNA,** would have the following properties:

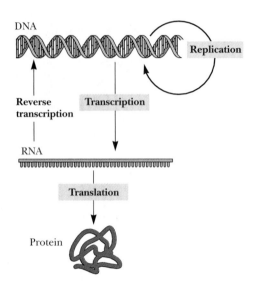

Figure 24.1 Crick's 1958 view of the "central dogma of molecular biology": directional flow of detailed sequence information includes DNA → DNA (replication), DNA → RNA (transcription), RNA → protein (translation), RNA → DNA (reverse transcription). Note that no pathway exists for the flow of information from proteins to nucleic acids— that is, protein → RNA or DNA. A possible path from DNA to protein has since been discounted. Interestingly, in 1958, mRNA had not yet been discovered.

1. Its base composition would reflect the base composition of DNA (a property consistent with genes as protein-encoding units).
2. It would be very heterogeneous with respect to molecular mass, yet the average molecular mass would be several hundred kD. (A 200-kD RNA contains roughly 750 nucleotides, which could encode a protein of about 250 amino acids, approximately 30 kD, a reasonable estimate for the average size of polypeptides.)
3. It would be able to associate with ribosomes because ribosomes are the site of protein synthesis.
4. It would have a high rate of turnover. (That is, mRNA would be rapidly degraded. Turnover of mRNA would allow the rate of mRNA synthesis to control the rate of protein synthesis.)

Since Jacob and Monod's 1961 hypothesis, cells have been found to contain three major classes of RNA—mRNA, ribosomal RNA (rRNA), and transfer RNA (tRNA)—all of which participate in protein synthesis (see Chapters 8 and 25). All of these RNAs are synthesized from DNA templates by **DNA-dependent RNA polymerases** in the process known as **transcription.** However, only mRNAs direct the synthesis of proteins. Thus, not all genes encode proteins; some encode rRNAs or tRNAs. Protein synthesis occurs via the process of **translation,** wherein the instructions encoded in the sequence of bases in mRNA are translated into a specific amino acid sequence by ribosomes, the "workbenches" of polypeptide synthesis (see Chapter 25).

Transcription is tightly regulated in all cells. In prokaryotes, only about 3% of the genes are undergoing transcription at any given time. The metabolic conditions and the growth status of the cell dictate which gene products are needed at any moment. In a differentiated eukaryotic cell, the figure is around 0.01%. Such differentiated cells express only the information needed for their biological functions, not the full genetic potential encoded in their chromosomes.

24.1 Transcription in Prokaryotes

In prokaryotes, virtually all RNA is synthesized by a single species of DNA-dependent RNA polymerase. (The only exception is the short RNA primers formed by primase during DNA replication.) Like DNA polymerases, RNA polymerase links ribonucleoside 5′-triphosphates (ATP, GTP, CTP, and UTP, represented generically as NTPs) in an order specified by base-pairing with a DNA template:

$$n\,\text{NTP} \longrightarrow (\text{NMP})_n + n\,\text{PP}_i$$

The enzyme moves along a DNA strand in the 3′ → 5′ direction, joining the 5′-phosphate of an incoming ribonucleotide to the 3′-OH of the previous residue. Thus, the RNA chain grows 5′ → 3′ during transcription, just as DNA chains do during replication. The reaction is driven by subsequent hydrolysis of PP_i to inorganic phosphate by the pyrophosphatases present in all cells.

The Structure and Function of *Escherichia coli* RNA Polymerase

The RNA polymerase of *E. coli*, so-called **RNA polymerase holoenzyme,** is a complex multimeric protein (450 kD) large enough to be visible in the electron microscope. Its subunit composition is $\alpha_2\beta\beta'\sigma$. The largest subunit, β' (155 kD), functions in DNA binding; β (151 kD) binds the nucleoside triphosphate

A DEEPER LOOK

Conventions Employed in Expressing the Sequences of Nucleic Acids and Proteins

Certain conventions are useful in tracing the course of information transfer from DNA to protein. The strand of duplex DNA that is read by RNA polymerase is termed the **template strand.** Thus, the strand that is not read is the **nontemplate strand.** Because the template strand is read by the RNA polymerase moving $3' \rightarrow 5'$ along it, the RNA product, the so-called **transcript,** grows in the $5' \rightarrow 3'$ direction (see figure). Note that the nontemplate strand has a nucleotide sequence and direction identical to the RNA transcript, except that the transcript has U residues in place of T. The RNA transcript will eventually be translated into the amino acid sequence of a protein (see Chapter 25) by a process in which suc-

cessive triplets of bases (termed **codons**), read $5' \rightarrow 3'$, specify a particular amino acid. Polypeptide chains are synthesized in the $N \rightarrow C$ direction, and the 5'-end of mRNA encodes the N-terminus of the protein.

By convention, when the order of nucleotides in DNA is specified, it is the $5' \rightarrow 3'$ sequence of nucleotides in the nontemplate strand that is presented. Consequently, if convention is followed, DNA sequences are written in terms that correspond directly to mRNA sequences, which correspond in turn to the amino acid sequences of proteins as read beginning with the N-terminus.

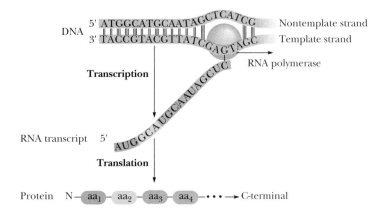

substrates and interacts with the σ (sigma)–subunit. A number of related proteins, **the sigma factors,** can serve as the σ-subunit. The different σ factors allow the RNA polymerase holoenzyme to recognize different DNA sequences that act as **promoters.** Promoters are nucleotide sequences that identify the location of *transcription start sites,* where transcription begins. Both β and β' contribute to formation of the catalytic site for RNA synthesis. The two α-subunits (36.5 kD each) are essential for assembly of the enzyme and activation by some regulatory proteins. Dissociation of the σ-subunit from the holoenzyme leaves the so-called **core polymerase** ($\alpha_2\beta\beta'$), which can transcribe DNA into RNA but is unable to recognize promoters and initiate transcription.

The Steps of Transcription in Prokaryotes

Transcription can be divided into four stages: (a) binding of RNA polymerase holoenzyme at promoter sites, (b) initiation of polymerization, (c) chain elongation, and (d) chain termination.

Binding of RNA Polymerase to Template DNA

The process of transcription begins when the σ-subunit of RNA polymerase recognizes a promoter sequence (Figure 24.2), and RNA polymerase holoenzyme and the promoter form a so-called **closed promoter complex** (Figure 24.2, Step 2). This stage in RNA polymerase : DNA interaction is referred to as the *closed* promoter complex because the dsDNA has not yet been "opened"

Figure 24.2 Sequence of events in the initiation and elongation phases of transcription as it occurs in prokaryotes. Nucleotides in this region are numbered with reference to the base at the transcription start site, which is designated +1.

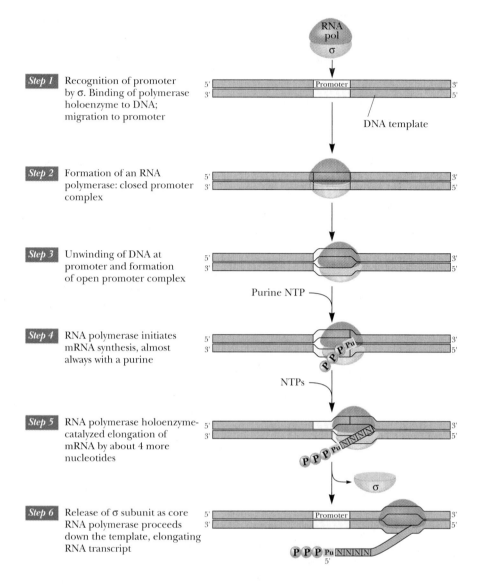

Step 1 Recognition of promoter by σ. Binding of polymerase holoenzyme to DNA; migration to promoter

Step 2 Formation of an RNA polymerase: closed promoter complex

Step 3 Unwinding of DNA at promoter and formation of open promoter complex

Step 4 RNA polymerase initiates mRNA synthesis, almost always with a purine

Step 5 RNA polymerase holoenzyme-catalyzed elongation of mRNA by about 4 more nucleotides

Step 6 Release of σ subunit as core RNA polymerase proceeds down the template, elongating RNA transcript

(unwound) so that the RNA polymerase can read the base sequence of the DNA template strand and transcribe it into a complementary RNA sequence.

Once the closed promoter complex is established, the RNA polymerase holoenzyme unwinds about 14 base pairs of DNA (base pairs located at positions −10 to +2, relative to the transcription start site; see later), forming the very stable **open promoter complex** (Figure 24.2, Step 3).

Promoter sequences can be identified *in vitro* by **DNA footprinting:** RNA polymerase holoenzyme is bound to a putative promoter sequence in a DNA duplex, and the DNA:protein complex is treated with DNase I. DNase I cleaves the DNA at sites not protected by bound protein, and the set of DNA fragments left after DNase I digestion reveals the promoter (by definition, the promoter is the RNA polymerase holoenzyme binding site).[1]

RNA polymerase binding typically protects a nucleotide sequence spanning the region from −70 to +20, where the +1 position is defined as the **transcription start site:** that base in DNA that specifies the first base in the RNA transcript. The next base, +2, specifies the second base in the transcript. Bases

---

[1]Promoters can also be defined genetically in terms of mutations (nucleotide changes) in this region that block gene expression because they are no longer recognizable by the σ-subunit.

in the 5′ or "minus" direction from the transcription start site are numbered −1, −2, and so on. (Note that there is no zero.) Nucleotides in the "minus" direction are said to lie **upstream** of the transcription start site, whereas nucleotides in the 3′ or "plus" direction are **downstream** of the transcription start site. The transcription start site on the template strand is almost always a pyrimidine, so almost all transcripts begin with a purine. RNA polymerase binding

A DEEPER LOOK

DNA Footprinting: Identifying the Nucleotide Sequence in DNA Where a Protein Binds

DNA footprinting is a widely used technique to identify the nucleotide sequence within DNA where a specific protein binds, such as the promoter sequence(s) bound by RNA polymerase holoenzyme. In this technique, the protein is incubated with a labeled (*) DNA fragment containing the nucleotide sequence where the protein is believed to bind. (The DNA fragment is labeled at only one end.) Then, a DNA cleaving agent, such as DNase I, is added to the solution containing the DNA:protein complex. DNase I cleaves the DNA backbone in exposed regions—that is, wherever the pres-

ence of the DNA-binding protein does not prevent DNase I from binding. A control solution containing naked DNA (a sample of the same labeled DNA fragment with no DNA-binding protein added) is also treated with DNase I. When these DNase I digests are analyzed by gel electrophoresis, a difference is found between the set of labeled fragments from the DNA:protein complex and the set from naked DNA. The absence of certain fragments in the digest of the DNA:protein complex reveals the location of the protein-binding site on the DNA (see figure).

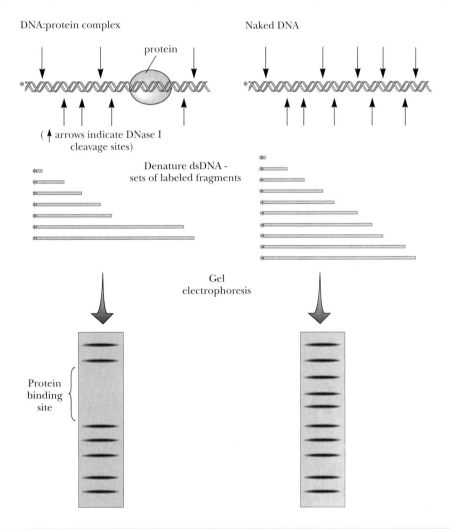

Adapted from Rhodes, D., and Fairall, L., 1997. Analysis of sequence-specific DNA-binding proteins, in *Protein Function: A Practical Approach*, T. E. Creighton, ed. Oxford: IRL Press at Oxford University Press.

Gene	−35 region		Pribnow box (−10 region)	Initiation site (+1)

Gene	Sequence
araBAD	G G A T C C T A C C T G A C G C T T T T T A T C G C A A C T C T C T A C T G T T T C T C C A T A C C C G T T T T T
araC	G C C G T G A T T A T A G A C A C T T T T G T T A C G C G T T T T T G T C A T G G C T T T G G T C C C G C T T T G
bioA	T T C C A A A A C G T G T T T T T T G T T G T T A A T T C G G T G T A G A C T T G T A A A C C T A A A T C T T T T
bioB	C A T A A T C G A C T T G T A A A C C A A A T T G A A A A G A T T T A G G T T T A C A A G T C T A C A C C G A A T
galP2	A T T T A T T C C A T G T C A C A C T T T C G C A T C T T T G T T A T G C T A T G G T T A T T T C A T A C C A T
lac	A C C C C A G G C T T T A C A C T T T A T G C T T C C G G C T C G T A T G T T G T G T G G A A T T G T G A G C G G
lacI	C C A T C G A A T G G C G C A A A A C C T T T C G C G G T A T G G C A T G A T A G C G C C C G G A A G A G A G T C
rrnA1	A A A A T A A A T G C T T G A C T C T G T A G C G G G A A G G C G T A T T A T C A C A C A C C C G C G C C G C T G
rrnD1	C A A A A A A A T A C T T G T G C A A A A A A T T G G G A T C C C T A T A A T G C G C C T C C G T T G A G A C G A
rrnE1	C A A T T T T T C T A T T G C G G C C T G C G G A G A A C T C C C T A T A A T G C G C C T C C A T C G A C A C G G
*t*RNA^Tyr	C A A C G T A A C A C T T T A C A G C G G C G C G T C A T T T G A T A T G A T G C G C C C C G C T T C C C G A T A
trp	A A A T G A G C T G T T G A C A A T T A A T C A T C G A A C T A G T T A A C T A G T A C G C A A G T T C A C G T A

	−35 region	Pribnow box	Initiation site

Consensus sequence:

−35 region									[11–15 bp]	Pribnow box						[5–8 bp]	Initiation site
T	C	T	T	G	A	C	A	T		T	A	T	A	A	T		A 51 / C 55 / G 42 / T 48
42	38	82	84	79	64	53	45	41		79	95	44	59	51	96		

Figure 24.3 The nucleotide sequences of representative *E. coli* promoters. (In accordance with convention, these sequences are those of the nontemplate strand where RNA polymerase binds.) Consensus sequences for the −35 region, the Pribnow box, and the initiation site are shown at the bottom. The numbers represent the percent occurrence of the indicated base. (*Note*: The −35 region is only roughly 35 nucleotides from the transcription start site; the Pribnow box [the −10 region] likewise is located at approximately position −10.) In this figure, sequences are aligned relative to the Pribnow box.

protects 90 bp of DNA, equivalent to a distance of 30 nm along B-DNA. Since RNA polymerase is only 16 nm in its longest dimension, the DNA must be wrapped around the enzyme.

Properties of Prokaryotic Promoters. Promoters recognized by the principal σ factor, σ^{70}, serve as the paradigm for prokaryotic promoters. These promoters vary in size from 20 to 200 bp but typically consist of a 40-bp region located on the 5′-side of the transcription start site. Within the promoter are two **consensus sequence elements.** (A consensus sequence can be defined as *the bases that appear with highest frequency at each position when several sequences believed to have common function are compared.*) These two elements are the **Pribnow box**[2] near −10, whose consensus sequence is the hexameric TATAAT, and a sequence in the **−35 region** containing the hexameric consensus TTGACA (Figure 24.3). The Pribnow box and the −35 region are separated by about 17 bp of nonconserved sequence. RNA polymerase holoenzyme uses its σ-subunit to bind to the conserved sequences, and the more closely the −35 region sequence corresponds to its consensus sequence, the greater is the efficiency of transcription of the gene. The highly expressed *rrn* genes in *E. coli* which encode ribosomal RNA (rRNA) have a third sequence element in their promoters, the **upstream element** (**UP** element), located about 20 bp immediately upstream of the −35 region. (Transcription from the *rrn* genes accounts for more than 60% of total RNA synthesis in rapidly growing *E. coli* cells.) Whereas the σ-subunit recognizes the −10 and −35 elements, the C-terminal domains (CTD) of the α-subunits of RNA polymerase recognize and bind the UP element.

In order for transcription to begin, the DNA duplex must be "opened" so that RNA polymerase has access to single-stranded template. The efficiency of initiation is inversely proportional to the melting temperature, T_m, in the Pribnow box, suggesting that the A : T-rich nature of this region is aptly suited for easy "melting" of the DNA duplex and creation of the open promoter complex (see Figure 24.2). Negative supercoiling facilitates transcription initiation by favoring DNA unwinding.

---

[2]Named for David Pribnow, who, along with David Hogness, first recognized the importance of this sequence element in transcription.

The RNA polymerase σ-subunit is directly involved in melting the dsDNA. Interaction of the σ-subunit with the nontemplate strand maintains the open complex formed between RNA polymerase and promoter DNA, with the σ-subunit acting as a sequence-specific single-stranded DNA-binding protein. Association of the σ-subunit with the nontemplate strand stabilizes the open promoter complex and leaves the bases along the template strand available to the catalytic site of the RNA polymerase.

Initiation of Polymerization

RNA polymerase has two binding sites for NTPs—the initiation site and the elongation site. The **initiation site** binds the purine nucleotides ATP and GTP preferentially; most RNAs begin with a purine at the 5′-end. The first nucleotide binds at the initiation site, H-bonding with the +1 base exposed within the *open promoter complex* (Figure 24.2, Step 4). The second incoming nucleotide binds at the elongation site, H-bonding with the +2 base. The ribonucleotides are then united when the 3′-O of the first nucleotide makes a phosphoester bond with the α-phosphorus atom of the second nucleotide, and PP_i is eliminated. Note that the 5′-end of the transcript starts out with a triphosphate attached to it. Movement of RNA polymerase along the template strand (*translocation*) to the next base prepares the RNA polymerase to add the next nucleotide (Figure 24.2, Step 5). Once an oligonucleotide 6 to 10 residues long has been formed, the σ-subunit dissociates from RNA polymerase, signaling the completion of initiation (Figure 24.2, Step 6). The core RNA polymerase goes on to synthesize the remainder of the mRNA. As the core RNA polymerase progresses, advancing the 3′-end of the RNA chain, the DNA duplex is unwound just ahead of it. About 12 base pairs of the growing RNA remain base-paired to the DNA template at any time, with the RNA strand becoming displaced as the DNA duplex rewinds behind the advancing RNA polymerase.

Chain Elongation

Elongation of the RNA transcript is catalyzed by the *core polymerase*, because once a short oligonucleotide chain has been synthesized, the σ-subunit dissociates. The accuracy of transcription is such that, about once every 10^4 nucleotides, an error is made and the wrong base is inserted. Because many transcripts are made per gene and most transcripts are smaller than 10 kb, this error rate is acceptable.

Two possibilities can be envisioned for the course of the new RNA chain. In one, the RNA chain is wrapped around the DNA as the RNA polymerase follows the template strand around the axis of the DNA duplex, but this possibility seems unlikely due to its potential for tangling the nucleic acid strands (Figure 24.4a). The more likely possibility involves supercoiling of the DNA, so that positive supercoils are created ahead of the transcription bubble and negative supercoils are created behind it (Figure 24.4b). To prevent torsional stress from inhibiting transcription, topoisomerases act to remove these supercoils from the DNA segment undergoing transcription (see Figure 24.4b).

Chain Termination

Two types of transcription termination mechanisms operate in bacteria: one that is dependent on a specific protein called **rho termination factor** (for the Greek letter, ρ) and another that is not dependent on this protein. In the latter, termination of transcription is determined by specific sequences in the DNA called **termination sites.** These sites are not characterized by a unique base where transcription halts. Instead, these sites consist of three structural features whose base-pairing possibilities lead to termination:

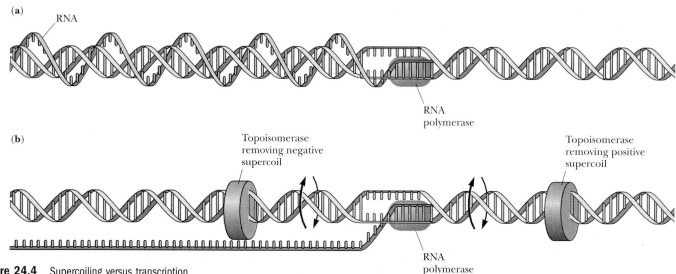

(a)

RNA

RNA
polymerase

(b)

Topoisomerase
removing negative
supercoil

Topoisomerase
removing positive
supercoil

RNA
polymerase

Figure 24.4 Supercoiling versus transcription.
(a) If the RNA polymerase followed the template strand around the axis of the DNA duplex, no supercoiling of the DNA would occur, but the RNA chain would be wrapped around the double helix once every 10 bp. This possibility seems unlikely because it would be difficult to disentangle the transcript from the DNA duplex. **(b)** Alternatively, topoisomerases could remove the supercoils. A topoisomerase capable of relaxing positive supercoils situated ahead of the advancing transcription bubble would "relax" the DNA. A second topoisomerase behind the bubble would remove the negative supercoils. *(Adapted from Futcher, B., 1988. Supercoiling and transcription, or vice versa? Trends in Genetics **4**:271–272.)*

1. Inverted repeats, which are typically G:C-rich, so a stable **stem-loop structure** can form in the transcript via intrachain hydrogen bonding (Figure 24.5).
2. A nonrepeating segment that punctuates the inverted repeats.
3. A run of 6 to 8 As in the DNA template, coding for Us in the transcript.

Termination then occurs as follows: a G:C-rich, stem-loop structure, or "hairpin," forms in the transcript. The hairpin apparently causes the RNA polymerase to pause, whereupon the A:U base pairs between the transcript and the DNA template strand are displaced through formation of somewhat more stable A:T base pairs between the template and nontemplate strands of the DNA. The result is spontaneous dissociation of the nascent transcript from DNA.

The alternative mechanism of termination—factor-dependent termination—is less common and mechanistically more complex. Rho factor is an ATP-dependent helicase (hexamer of 50-kD subunits) that catalyzes the unwinding of RNA:DNA hybrid duplexes (or RNA:RNA duplexes). The rho factor recognizes and binds to C-rich regions in the RNA transcript. These regions must be unoccupied by translating ribosomes for rho factor to bind. Once bound,

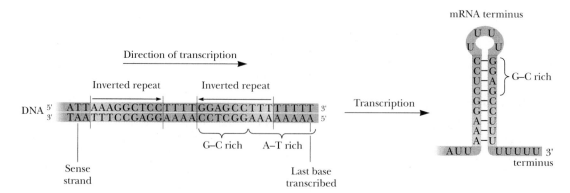

Figure 24.5 The termination site for the *E. coli trp* operon (the *trp* operon encodes the enzymes of tryptophan biosynthesis). The inverted repeats give rise to a stem-loop, or "hairpin," structure ending in a series of U residues.

(a)

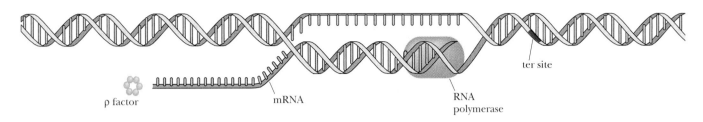

ρ factor   mRNA   RNA polymerase   ter site

(b)

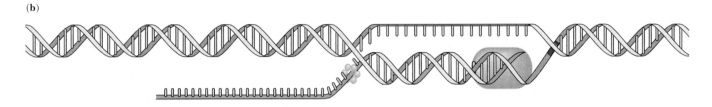

(c)

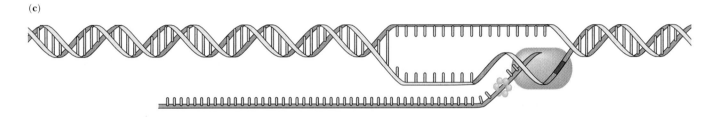

(d)

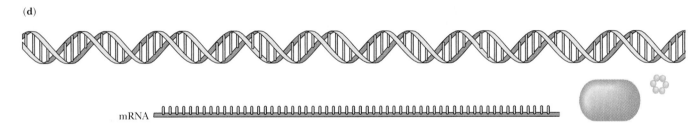

mRNA

rho factor advances in the $5' \rightarrow 3'$ direction until it reaches the transcription bubble (Figure 24.6). There it catalyzes the unwinding of the transcript and template, releasing the nascent RNA chain. It is likely that the RNA polymerase stalls in a G:C-rich termination region, allowing rho factor to overtake it.

Figure 24.6 The rho factor mechanism of transcription termination. Rho factor **(a)** attaches to a recognition site on mRNA and **(b)** moves along it behind RNA polymerase. **(c)** When RNA polymerase pauses at the termination site, rho factor unwinds the DNA:RNA hybrid in the transcription bubble, **(d)** releasing the nascent mRNA.

24.2 Transcription Regulation in Prokaryotes

In bacteria, genes encoding the enzymes of a particular metabolic pathway are often grouped adjacent to one another in a cluster on the chromosome. This pattern of organization allows all of the genes in the group to be expressed in a coordinated fashion through transcription into a **single polycistronic mRNA** encoding all the enzymes of the metabolic pathway.[3] A regulatory sequence

---

[3]A polycistronic mRNA is a single RNA transcript that encodes more than one polypeptide. "Cistron" is a genetic term for a DNA region encoding a protein: "cistron" and "gene" are essentially equivalent terms.

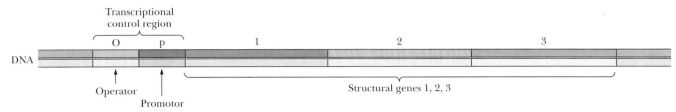

Figure 24.7 The general organization of operons. Operons consist of transcriptional control regions and a set of related structural genes, all organized in a contiguous linear array along the chromosome. The transcriptional control regions are the promoter and the operator, which lie next to, or overlap, each other, upstream from the structural genes they control. Operators may lie at various positions relative to the promoter, either upstream or downstream. Expression of the operon is determined by access of RNA polymerase to the promoter, and occupancy of the operator by regulatory proteins influences this access. Induction activates transcription from the promoter; repression prevents it.

lying adjacent to the DNA being transcribed determines whether transcription takes place. This sequence is termed the **operator** (Figure 24.7). The operator is located next to a promoter. Interaction of a **regulatory protein** with the operator controls transcription of the gene cluster by controlling access of RNA polymerase to the promoter.[4] Such co-expressed gene clusters, together with the operator and promoter sequences that control their transcription, are called **operons.**

Transcription of Operons Is Controlled by Induction and Repression

In prokaryotes, regulation is ultimately responsive to small molecules serving as signals of the nutritional or environmental conditions confronting the cell. Increased synthesis of enzymes in response to the presence of a particular substrate is termed **induction.** For example, lactose (Figure 24.8) can serve as both a carbon and an energy source for *E. coli.* Metabolism of lactose depends on hydrolysis into its component sugars, glucose and galactose, by the enzyme **β-galactosidase.** In the absence of lactose, *E. coli* cells contain very little β-galactosidase (fewer than 5 molecules per cell). However, lactose availability *induces* the synthesis of β-galactosidase by activating transcription of the *lac* **operon.** One of the genes in the *lac* operon, *lacZ*, is the structural gene for β-galactosidase. When its synthesis is fully induced, β-galactosidase can amount to almost 10% of the total soluble protein in *E. coli.* When lactose is removed from the culture, synthesis of β-galactosidase halts.

The alternative to induction—namely, *decreased* synthesis of enzymes in response to a specific metabolite—is termed **repression.** For example, the enzymes of tryptophan biosynthesis in *E. coli* are encoded in the ***trp* operon.** If sufficient Trp is available to the growing bacterial culture, the *trp* operon is not transcribed, so the Trp biosynthetic enzymes are not made; that is, their synthesis is *repressed.* Repression of the *trp* operon in the presence of Trp is an eminently logical control mechanism: if the end product of the pathway is present, why waste cellular resources making unneeded enzymes?

Induction and repression are two faces of the same phenomenon. In induction, a substrate activates enzyme synthesis. Substrates capable of activating synthesis of the enzymes that metabolize them are called **co-inducers,** or, often, simply **inducers.** Some substrate analogs can induce enzyme synthesis even though the enzymes are incapable of metabolizing them. These analogs are called **gratuitous inducers.** A number of thiogalactosides, such as **IPTG** (isopropyl β-thiogalactoside; Figure 24.9), are excellent gratuitous inducers of β-galactosidase activity in *E. coli.* In repression, a metabolite, typically an end product, depresses synthesis of its own biosynthetic enzymes. Such metabolites are called **co-repressors.**

Lactose
(O-β-D-galactopyranosyl (1 → 4) β-D-glucopyranose)

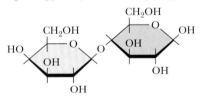

Figure 24.8 The structure of lactose, a β-galactoside.

Isopropyl β-thiogalactoside (IPTG)

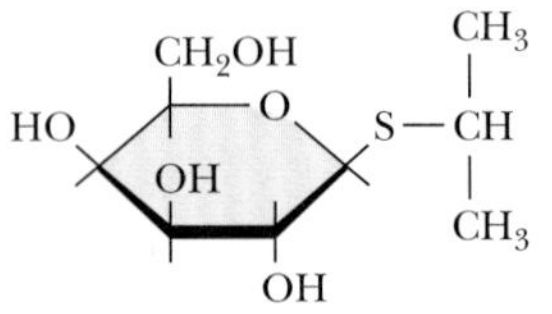

Figure 24.9 The structure of IPTG (isopropyl β-thiogalactoside).

---

[4]Although this is the paradigm for prokaryotic gene regulation, it must be emphasized that many regulated prokaryotic genes do not contain operators and are regulated in ways that do not involve protein : operator interactions.

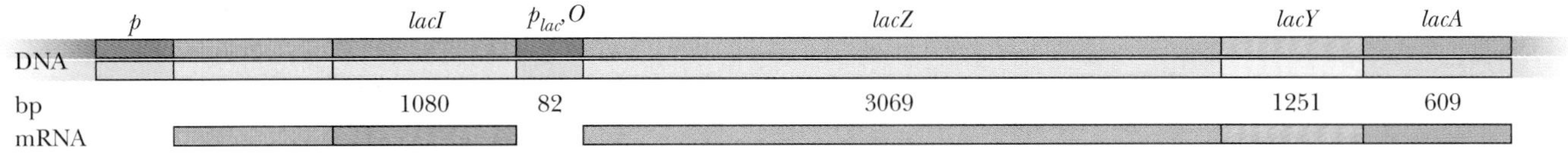

		lacI		lacZ		lacY	lacA
	p		p_{lac} O				
DNA							
bp		1080	82	3069		1251	609
mRNA							
Polypeptide Amino acids		360		1023		417	203
kD		38.6		116.4		46.5	22.7
Protein Structure		Tetramer		Tetramer		Membrane protein	Dimer
kD		154.4		465		46.5	45.4
Function		Repressor		β-Galactosidase		Permease	Trans-acetylase

Figure 24.10 The *lac* operon. The operon consists of two transcription units. In one unit, there are three structural genes, *lacZ*, *lacY*, and *lacA*, under control of the promoter, p_{lac}, and the operator *O*. In the other unit, there is a regulator gene, *lacI*, with its own promoter, p_{lacI}. *lacI* encodes a 360-residue, 38.6-kD polypeptide that forms a tetrameric *lac* repressor protein. *lacZ* encodes β-galactosidase, a tetrameric enzyme of 116-kD subunits. *lacY* is the β-galactoside permease structural gene, a 46.5-kD integral membrane protein active in β-galactoside transport into the cell. The remaining structural gene encodes a 22.7-kD polypeptide that forms a dimer displaying thiogalactoside transacetylase activity *in vitro*, transferring an acetyl group from acetyl-CoA to the C-6 OH of thiogalactosides, but the metabolic role of this protein *in vivo* remains uncertain. *lacA* mutants show no identifiable metabolic deficiency. Perhaps the *lacA* protein acts to detoxify toxic analogs of lactose through acetylation.

lac: The Paradigm of Operons

In 1961, François Jacob and Jacques Monod proposed the **operon hypothesis** to account for the coordinate regulation of related metabolic enzymes. The operon was considered to be the unit of gene expression, consisting of two classes of genes: the **structural genes** for the enzymes and **regulatory genes** that control expression of the structural genes. The two kinds of genes could be distinguished by mutation. Mutations in a structural gene would abolish one particular enzymatic activity, but mutations in a regulatory gene would affect all of the different enzymes under its control. Mutations of both kinds were known in *E. coli* for lactose metabolism. Bacteria with mutations in either the *lacZ* gene or the *lacY* gene (Figure 24.10) could no longer metabolize lactose—the *lacZ* mutants (*lacZ⁻* strains) because β-galactosidase activity was absent, the *lacY* mutants because lactose was no longer transported into the cell. Other mutations defined another gene, the *lacI* gene. *lacI* mutants were different because they both expressed β-galactosidase activity and immediately transported lactose *without prior exposure to an inducer.* That is, a single mutation led to the expression of lactose metabolic functions independently of inducer. Expression of genes independently of regulation is termed **constitutive expression.** Thus, *lacI* had the properties of a regulatory gene. The *lac* operon includes the regulatory gene *lacI*, its promoter *p*, and three structural genes, *lacZ*, *lacY*, and *lacA*, with their own promoter p_{lac} and operator *O* (see Figure 24.10).

Negative Regulation of the *lac* Operon

The structural genes of the *lac* operon are controlled by **negative regulation.** That is, they are transcribed to give an mRNA unless turned off by the *lacI* gene product. This gene product is the **lac repressor,** a tetrameric protein (Figure 24.11). The *lac* repressor has two kinds of binding sites—one for inducer and another for DNA. In the absence of inducer, *lac* repressor blocks *lac* gene expression by binding to the operator DNA site upstream from the *lac* structural genes. The *lac* operator is a palindromic DNA sequence (Figure 24.12). **Palindromes,** or "inverted repeats" (see Chapter 8), provide a twofold, or dyad, symmetry, a structural feature common at sites in DNA where proteins specifically bind. Despite the presence of *lac* repressor, RNA polymerase can still initiate transcription at the promoter (p_{lac}), but *lac* repressor blocks elongation of transcription, so initiation is aborted. In *lacI* mutants, the *lac* repressor is absent or defective in binding to operator DNA, *lac* gene transcription is not blocked, and the *lac* operon is constitutively expressed in these mutants. Note that *lacI* is normally expressed constitutively from its promoter, so that *lac* repressor protein is always available to fill its regulatory role. About 10 molecules of *lac* repressor are present in an *E. coli* cell.

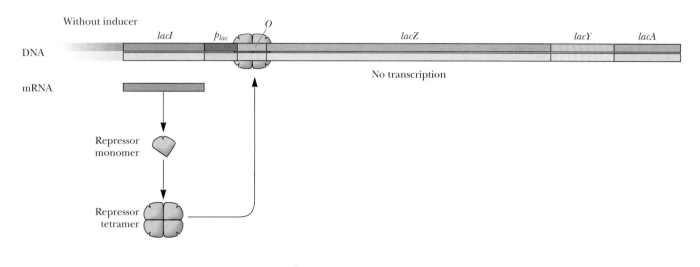

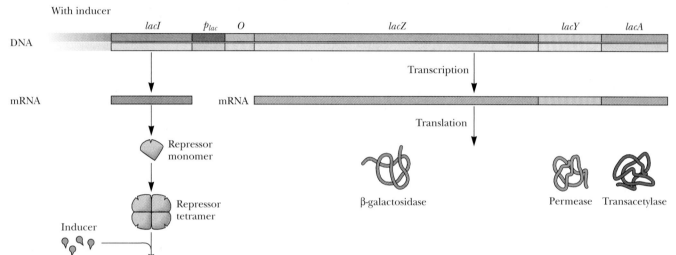

Figure 24.11 The mode of action of *lac* repressor.

Derepression of the *lac* operon occurs when appropriate β-galactosides occupy the inducer site on *lac* repressor, causing a conformational change in the protein that lowers the repressor's affinity for operator DNA. As a tetramer, *lac* repressor has four inducer binding sites, and its response to inducer shows cooperative allosteric effects. Thus, as a consequence of the "inducer"-induced conformational change, the inducer : *lac* repressor complex dissociates from the DNA, and RNA polymerase transcribes the structural genes (see Figure 24.11). Induction reverses rapidly: *lac* mRNA has a half-life of only 3 minutes, and once the inducer is used up through metabolism by the enzymes, free *lac* repressor re-associates with the operator DNA, transcription of the operon is halted, and the residual *lac* mRNA decays.

Figure 24.12 The nucleotide sequence of the *lac* operator. This sequence comprises 36 bp showing nearly palindromic symmetry. The inverted repeats that constitute this approximate twofold symmetry are shaded in rose. The bases are numbered relative to the +1 start site for transcription. The G : C base pair at position +11 represents the axis of symmetry. *In vitro* studies show that bound *lac* repressor protects a 26-bp region from −5 to +21 against nuclease digestion. Bases that interact with bound *lac* repressor are indicated below the operator. Note the symmetry of protection at +1 through +4 TTAA to +18 through +21 AATT.

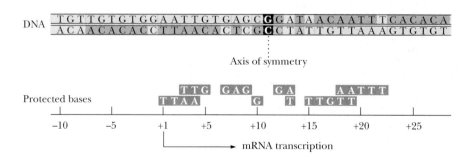

Positive Control of the *lac* Operon by CAP

Transcription by RNA polymerase from some promoters proceeds with low efficiency unless assisted by an accessory protein that acts as a *positive regulator.* One such protein is **CAP, or catabolite activator protein.** Its name derives from the phenomenon of catabolite repression in *E. coli.* Catabolite repression is a global control that coordinates gene expression with the total physiological state of the cell: as long as glucose is available, *E. coli* catabolizes it in preference to any other energy source, such as lactose or galactose. Catabolite repression ensures that the operons necessary for metabolism of these alternative energy sources—that is, the *lac* and *gal* (galactose) operons—remain repressed until the supply of glucose is exhausted. Catabolite repression overrides the influence of any inducers that might be present.

Catabolite repression is maintained until the *E. coli* cells become glucose-starved. Glucose starvation leads to activation of adenylyl cyclase and the cells begin to make cAMP. (In contrast, glucose uptake is accompanied by deactivation of adenylyl cyclase.) The action of CAP as a positive regulator is cAMP-dependent. cAMP is a small-molecule inducer for CAP, and cAMP binding enhances CAP's affinity for DNA. CAP, also referred to as **CRP** (for **cAMP receptor protein**), is a dimer of identical 22.5-kD polypeptides. The N-terminal domains bind cAMP; the C-terminal domains constitute the DNA-binding site. Two molecules of cAMP are bound per dimer. The CAP–(cAMP)$_2$ complex binds to specific target sites near the promoters of operons (Figure 24.13). Binding of CAP–(cAMP)$_2$ to DNA causes the DNA to bend more than 80° (Figure 24.14). This CAP-induced DNA bending near the promoter assists RNA polymerase holoenzyme binding and closed promoter complex formation. Contacts made between the CAP–(cAMP)$_2$ complex and the α-subunit of RNA polymerase holoenzyme activate transcription.

Positive Versus Negative Control

Negative- and positive-control systems are fundamentally different (although in some instances both govern the expression of the same gene). Genes un-

 See *Interactive Biochemistry CD-ROM and Workbook,* page 90

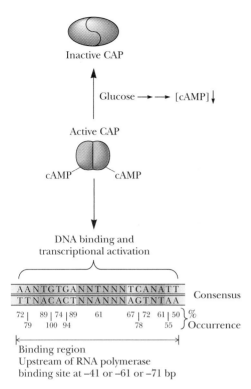

Figure 24.13 The mechanism of catabolite repression and CAP action. Glucose instigates catabolite repression by lowering cAMP levels. cAMP is necessary for CAP binding near promoters of operons whose gene products are involved in the metabolism of alternative energy sources such as lactose, galactose, and arabinose. The binding sites for the CAP-(cAMP)$_2$ complex are consensus DNA sequences containing the conserved pentamer TGTGA and a less well conserved inverted repeat, TCANA (where N is any nucleotide).

Figure 24.14 Binding of CAP-(cAMP)$_2$ induces a severe bend in DNA about the center of dyad symmetry at the CAP-binding site. The CAP dimer with two molecules of cAMP bound interacts with 27 to 30 base pairs of duplex DNA. Two α-helices of the CAP dimer insert into the major groove of the DNA at the dyad-symmetric CAP-binding site. The cAMP-binding domain of CAP protein is shown in blue and the DNA-binding domain in purple. The two cAMP molecules bound by the CAP dimer are indicated in red. For DNA, the bases are shown in white and the sugar–phosphate backbone in yellow. DNA phosphates that interact with CAP are highlighted in red. Binding of CAP-(cAMP)$_2$ to its specific DNA site involves H bonding and ionic interactions between protein functional groups and DNA phosphates, as well as H-bonding interactions in the DNA major groove between amino acid side chains of CAP and DNA base pairs. *(Adapted from Schultz, S. C., Shields, G. C., and Steitz, T. A., 1991. Crystal structure of a CAP-DNA complex: The DNA is bent by 90°. Science **253**:1001–1007. Photograph courtesy of Professor Thomas A. Steitz of Yale University.)*

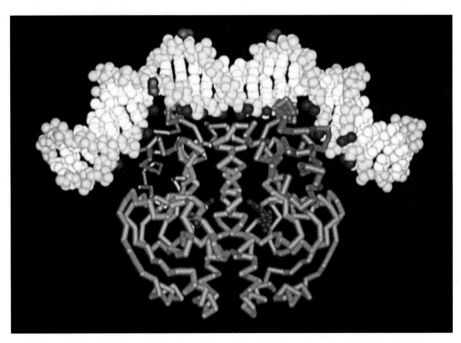

der negative control are transcribed unless they are turned off by the presence of a repressor protein. Often, transcription activation is merely the release from negative control. In contrast, genes under positive control are expressed only if an active regulator protein is present. The *lac* operon illustrates these differences. The action of *lac* repressor is negative. It binds to operator DNA and blocks transcription; expression of the operon only occurs when this negative control is lifted through release of the repressor. In contrast, regulation of the *lac* operon by CAP is positive: transcription of the operon by RNA polymerase is stimulated by CAP's action as a positive regulator.

Operons can also be classified as **inducible** or **repressible,** or both, depending on how they respond to the small molecules that mediate their expression. Repressible operons are expressed only in the absence of their co-repressors. Inducible operons are transcribed only in the presence of small-molecule co-inducers (Figure 24.15).

The *trp* Operon: Regulation Through a Co-Repressor–Mediated Negative Control Circuit

The *trp* operon of *E. coli* (and *S. typhimurium*) encodes the five polypeptides (*trpE* through *trpA* [Figure 24.16]) that assemble into the three enzymes catalyzing tryptophan synthesis from chorismate (see Chapter 21). Expression of

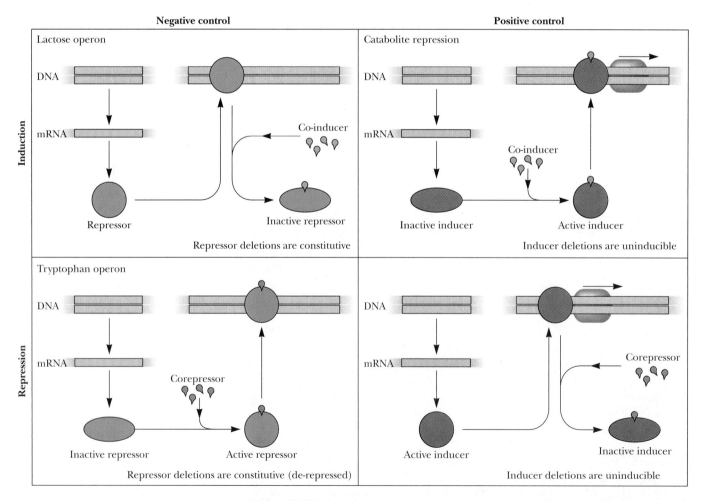

Figure 24.15 Control circuits governing the expression of genes. These circuits can be either negative or positive, inducible or repressible.

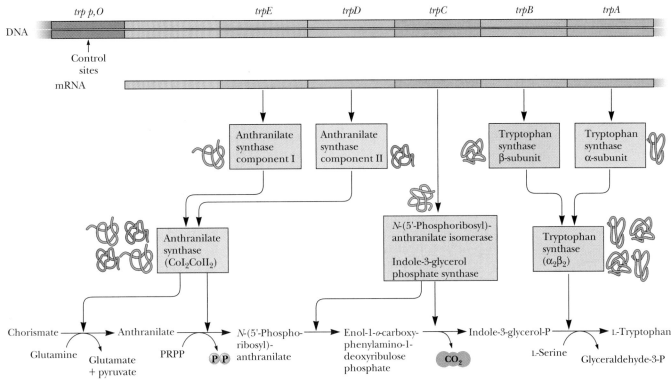

Figure 24.16 The *trp* operon of *E. coli*.

the *trp* operon is under the control of **Trp repressor,** a dimer of 108-residue polypeptide chains. When tryptophan is plentiful, Trp repressor binds two molecules of tryptophan and associates with the *trp* operator that is located within the *trp* promoter. Trp repressor binding excludes RNA polymerase from the promoter, preventing transcription of the *trp* operon. When Trp becomes limiting, repression is lifted because Trp repressor lacking bound Trp (Trp apo-repressor) has a lowered affinity for the *trp* promoter. Thus, the behavior of Trp repressor corresponds to a co-repressor–mediated, negative control circuit (see Figure 24.16). Trp repressor not only is encoded by the *trpR* operon, but it also regulates expression of the *trpR* operon. This is an example of **autogenous regulation (autoregulation)**: regulation of gene expression by the product of the gene.

DNA: Protein Interactions and Protein: Protein Interactions in Transcription Regulation

Quite a variety of control mechanisms regulate transcription in prokaryotes. Several organizing principles are apparent. First, **DNA:protein interactions** are a central feature in transcriptional control, and the DNA sites where regulatory proteins bind commonly display at least partial dyad symmetry or inverted repeats. Further, DNA-binding proteins themselves are generally even-numbered oligomers (for example, dimers, tetramers) that have an innate twofold rotational symmetry. Second, **protein: protein interactions** are an essential component of transcriptional activation. We see this latter feature in the activation of RNA polymerase by CAP–(cAMP)$_2$, for example. Third, the regulator proteins receive cues that signal the status of the environment (for example, Trp, lactose, cAMP) and act to communicate this information to the genome, typically via the medium of conformational changes and DNA: protein interactions.

Proteins That Activate Transcription Work Through Protein : Protein Contacts with RNA Polymerase

Although transcriptional control is governed by a variety of mechanisms, an underlying principle of transcriptional activation has emerged. Transcriptional activation can take place when a **transcriptional activator** protein [such as CAP–(cAMP)$_2$] bound to DNA makes protein:protein contacts with RNA polymerase, and the degree of transcriptional activation is proportional to the strength of the protein : protein interaction. Generally speaking, a nucleotide sequence that provides a binding site for a DNA-binding protein can serve as an **activator site** if the DNA-binding protein bound there can interact with promoter-bound RNA polymerase. These interactions can involve either the α-, β-, β'-, or σ-subunits of RNA polymerase. Further, if the DNA-bound transcriptional activator makes contacts with two different components of RNA polymerase, a synergistic effect takes place such that transcription is markedly elevated. Thus, transcriptional activation at specific genes relies on the presence of one or more activator sites where one or more transcriptional activator proteins can bind and make contacts with RNA polymerase bound at the promoter of the gene. Indeed, transcriptional activators may facilitate the recruitment and binding of RNA polymerase to the promoter. This general principle applies to transcriptional activation in both prokaryotic and eukaryotic cells. In eukaryotes, transcriptional activators typically have discrete domains of protein structure dedicated to DNA binding (DB domains) and transcriptional activation (TA domains).

DNA Looping

Because transcription must respond to a variety of regulatory signals, multiple proteins are essential for appropriate regulation of gene expression. These regulatory proteins are the **sensors** of cellular circumstances, and they communicate this information to the genome by binding at specific nucleotide sequences. However, DNA is virtually a one-dimensional polymer, and there is little space for a lot of proteins to bind at (or even near) a transcription initiation site. DNA looping permits additional proteins to convene at the initiation site and to exert their influence on creating and activating an RNA polymerase initiation complex (Figure 24.17). The repertoire of transcriptional regulation is greatly expanded by DNA looping.

24.3 Transcription in Eukaryotes

Although the mechanism of transcription in prokaryotes and eukaryotes is fundamentally similar, transcription is substantially more complicated in eukaryotes. The significant difference is that the DNA of eukaryotes is wrapped around histones to form nucleosomes, and the nucleosomes are further organized into chromatin (see Chapter 8). *Nucleosomes repress all genes; only those genes activated by specific positive regulatory mechanisms are transcribed.* General repression occurs because nucleosomes severely restrict access of the transcription apparatus to promoters. In considering transcription in eukaryotes, the following features merit emphasis:

- The three classes of RNA polymerase in eukaryotes: RNA pol I, II, and III
- The structure and function of RNA polymerase II, the mRNA-synthesizing RNA polymerase

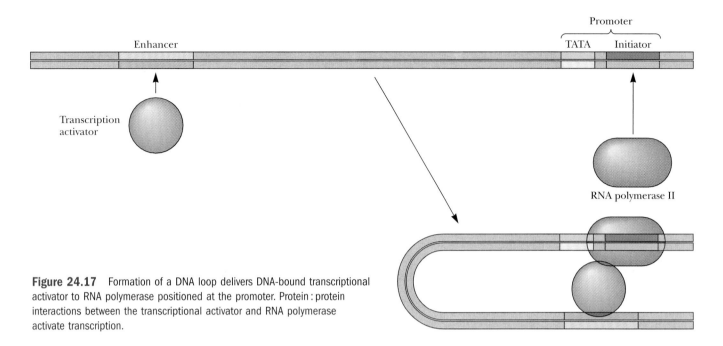

Figure 24.17 Formation of a DNA loop delivers DNA-bound transcriptional activator to RNA polymerase positioned at the promoter. Protein : protein interactions between the transcriptional activator and RNA polymerase activate transcription.

- Transcription regulation in eukaryotes, including:
    General features of gene regulatory sequences: promoters, enhancers, and response elements
    Transcription initiation by RNA polymerase II
        The general transcription factors (GTFs)
        Alleviating the repression due to nucleosomes
            Histone acetyl transferases (HATs)
            Chromatin remodeling complexes
- A general model for eukaryotic gene activation, based on the above

We turn now to a review of these various features.

The Three Classes of Eukaryotic RNA Polymerases

Eukaryotic cells have three classes of RNA polymerase, each of which synthesizes a different class of RNA. All three enzymes are found in the nucleus. **RNA polymerase I** is localized to the nucleolus and transcribes the major ribosomal RNA genes. **RNA polymerase II** transcribes protein-encoding genes, and thus it is responsible for the synthesis of mRNA. **RNA polymerase III** transcribes tRNA genes, the ribosomal RNA genes encoding 5S rRNA, and a variety of other small RNAs, including several involved in mRNA processing and protein transport.

All three RNA polymerase types are large, complex multimeric proteins (500 to 700 kD), consisting of 10 or more types of subunits. Although the three differ in overall subunit composition, they have several smaller subunits in common. Further, all possess two large subunits (each 140 kD or greater) having sequence similarity to the large β- and β'-subunits of *E. coli* RNA polymerase, indicating that the fundamental catalytic site of RNA polymerase is conserved among its various forms.

In addition to their different functions, the three classes of RNA polymerase can be distinguished by their sensitivity to **α-amanitin** (Figure 24.18), a bicyclic octapeptide produced by the poisonous mushroom *Amanita phalloides* (the "destroying angel" mushroom). α-Amanitin blocks RNA chain elongation.

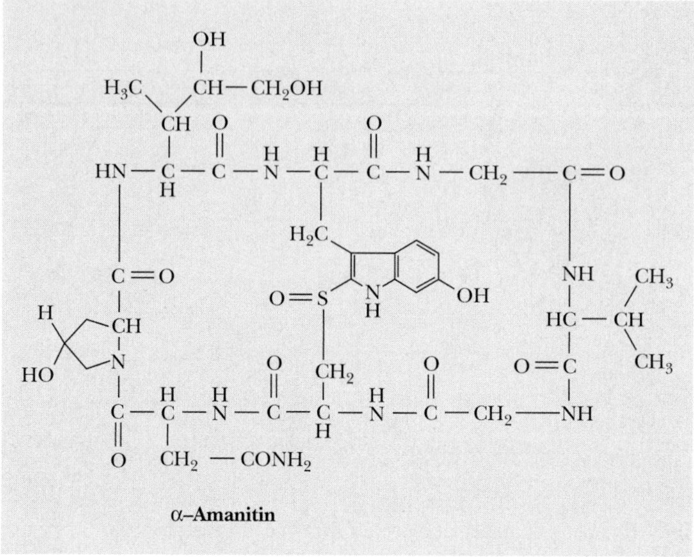

Figure 24.18 The structure of α-amanitin, one of a series of toxic compounds known as amatoxins that are found in the mushroom *Amanita phalloides.*

α–Amanitin

Although RNA polymerase I is resistant to this compound, RNA polymerase II is very sensitive and RNA polymerase III is less sensitive.

The existence of three classes of RNA polymerases acting on three distinct sets of genes implies that at least three categories of promoters exist to maintain this specificity. Eukaryotic promoters are very different from prokaryotic promoters. All three eukaryotic RNA polymerases interact with their promoters via so-called **transcription factors,** DNA-binding proteins that recognize and accurately initiate transcription at specific promoter sequences. For RNA polymerase I, its templates are the rRNA genes. Ribosomal RNA genes are present in multiple copies. Optimal expression of these genes requires the first 150 nucleotides in the immediate 5′-upstream region.

RNA polymerase III interacts with transcription factors **TFIIIA, TFIIIB,** and **TFIIIC.** Interestingly, TFIIIA and TFIIIC bind to specific recognition sequences that in some instances are located *within* the coding regions of the genes, not in the 5′-untranscribed region upstream from the transcription start site. TFIIIB associates with TFIIIA or TFIIIC already bound to the DNA. RNA pol III then binds to TFIIIB to establish an initiation complex.

The Structure and Function of RNA Polymerase II

As the enzyme responsible for the regulated synthesis of mRNA, RNA polymerase II has aroused greater interest than RNA pol I and pol III. RNA pol II must be capable of transcribing a great diversity of genes, yet it must carry out its function at any moment only on those genes whose products are appropriate to the needs of the cell in its everchanging metabolism and growth. The RNA pol II from yeast (*Saccharomyces cerevisiae*) has been extensively characterized and its structure has been solved by X-ray crystallography (Figure 24.19). Strong homology between yeast and human RNA pol II subunits suggests that the yeast RNA pol II is an excellent model for human RNA pol II. The yeast RNA pol II consists of 12 different polypeptides, designated RPB1 through RPB12, ranging in size from 192 to 8 kD (Table 24.1). RPB 3, 4, and 7 are unique to RNA pol II, whereas RPB 5, 6, 8, and 10 are common to all three eukaryotic RNA polymerases.

The RPB1 subunit has an unusual structural feature not found in prokaryotes: its **C-terminal domain (CTD)** contains 27 repeats of the amino acid se-

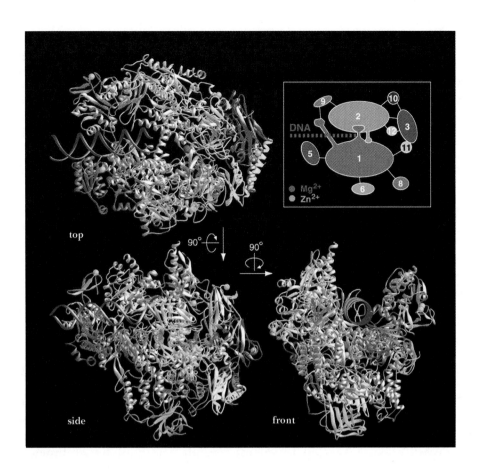

Figure 24.19 Crystal structure of a 10-subunit version of the yeast RNA pol II. (Subunits RPB4 and RPB7 are absent from this structure.) Protein is shown in the form of ribbon diagrams; a length of DNA corresponding to 20 base pairs is shown as a blue double-helical coil. The inset shows the color code and location of principal RPB subunits. The active site Mg^{2+} atom (pink), as well as eight Zn^{2+} atoms (green) associated with this enzyme preparation, is also highlighted. *(Adapted from Figure 3 in Cramer, P., et al., 2000. Architecture of RNA polymerase II and implications for the transcription mechanism. Science **288**:640–649. Image courtesy of Patrick Cramer and Roger D. Kornberg, Stanford University.)*

quence PTSPSYS. (The analogous subunit in RNA pol II enzymes of other eukaryotes has this heptapeptide tandemly repeated as many as 52 times.) Note that the side chains of five of the seven residues in this repeat have —OH groups, endowing the CTD with considerable hydrophilicity *and* multiple sites for phosphorylation. This domain may project more than 50 nm from the surface of RNA pol II.

Table 24.1   Yeast* RNA Polymerase II Subunits

Subunit	Size (kD)	Features	Prokaryotic Homolog
RPB1[†]	192	PTSPSYS CTD	β'
RPB2	139	NTP binding	β
RPB3	35	Core assembly	α
RPB4	25	Promoter recognition	σ
RPB5	25	In pol I, II, and III	
RPB6	18	In pol I, II, and III	
RPB7	19	Unique to pol II	
RPB8	17	In pol I, II, and III	
RPB9	14	Nonessential	
RPB10	8	In pol I, II, and III	
RPB11	14		
RPB12	8	In pol I, II, and III	

*A very similar RNA polymerase II can be isolated from human cells.
[†]RPB stands for RNA polymerase B; RNA pol I, II, and III are sometimes called RPA, RPB, and RPC.

Source: Adapted from Myer, V. E., and Young, R. A., 1998. RNA polymerase II holoenzymes and subcomplexes. *The Journal of Biological Chemistry* **273**:27757–27760.

The CTD is essential to RNA pol II function, since only RNA pol II molecules whose CTD is not phosphorylated can initiate transcription. However, transcription elongation proceeds only after protein phosphorylation within the CTD, suggesting that phosphorylation triggers the conversion of an initiation complex into an elongation complex. Such a mechanism would allow protein phosphorylation to regulate gene expression. Following termination of transcription, a phosphatase recycles RNA pol II to its unphosphorylated form.

Transcription Regulation in Eukaryotes

In addition to metabolic activity and cell division, complex patterns of embryonic development and cell differentiation must be coordinated through transcriptional regulation. All this coordinated regulation takes place in cells where the relative quantity (and diversity) of DNA is very great: a typical mammalian cell has 1500 times as much DNA as an *E. coli* cell. The structural genes of eukaryotes are rarely organized in clusters akin to operons. Each eukaryotic gene typically possesses a discrete set of regulatory sequences appropriate to the requirements for its expression. Certain of these sequences provide sites of interaction for general transcription factors, whereas others endow the gene with great specificity in expression by providing targets for specific transcription factors.

General Features of Gene Regulatory Sequences in Eukaryotes

RNA polymerase II promoters commonly consist of two separate sequence features: the **core** element, near the transcription start site, where **general transcription factors (GTFs)** bind; and more distantly located **regulatory elements,** known variously as **enhancers** (or **silencers**). These latter elements are recognized by specific DNA-binding proteins that activate transcription above basal levels (*enhancers* bind *transcriptional activators*) or repress transcription (*silencers* bind *repressors*). The core region often consists of a **TATA box** (a TATAAA consensus element) indicating the transcription start site; the TATA motif is usually located at position −25 (Figure 24.20). Those genes that lack a TATA often have an **initiator element,** or *Inr,* where transcription is initiated. The *Inr* sequence encompasses the transcription start site, but this sequence is not highly conserved among eukaryotic genes. Other regulatory elements include short nucleotide sequences (sometimes called **response elements**) found near the promoter (the *promoter-proximal region*) that can bind certain specific transcription factors, such as proteins that trigger expression of a related set of genes in response to some physiological signal (hormone) or challenge (temperature shock).

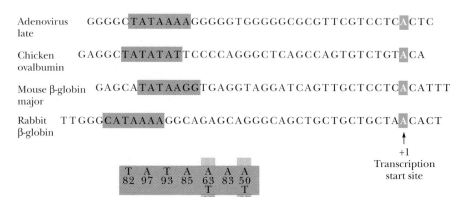

Figure 24.20 The TATA box in selected eukaryotic genes. The consensus sequence of a number of such promoters is presented in the lower part of the figure, the numbers giving the percent occurrence of various bases at the positions indicated.

(a)

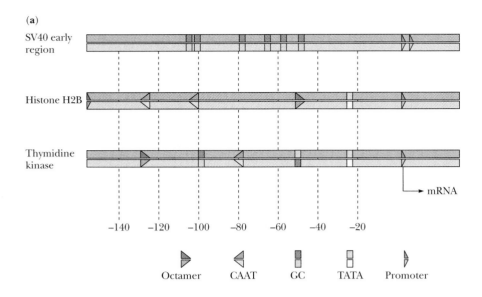

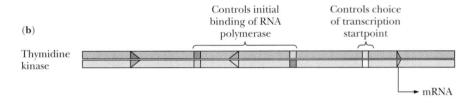

Figure 24.21 Promoter regions of several representative eukaryotic genes. **(a)** The SV40 early genes, the histone H2B gene, and the thymidine kinase gene. Note that these promoters contain different combinations of the various modules. In **(b)**, the function of the modules within the thymidine kinase gene is shown.

Promoters

The promoters of eukaryotic genes encoding proteins can be quite complex and variable, but they typically contain **modules** of short conserved sequences, such as the **TATA box,** the **CAAT box,** and the **GC box.** Sets of such modules embedded in the upstream region collectively define the promoter. The presence of a CAAT box, usually located around −80 relative to the transcription start site, signifies a strong promoter. One or more copies of the sequence GGGCGG or its complement (referred to as the GC box) have been found upstream from the transcription start sites of so-called "housekeeping genes." Housekeeping genes encode proteins commonly present in all cells and essential to normal function; such genes are typically transcribed at more or less steady levels. Figure 24.21 depicts the promoter regions of several representative eukaryotic genes. Table 24.2 lists transcription factors that bind to

Table 24.2 **A Selection of Consensus Sequences That Define Various RNA Polymerase II Promoter Modules and the Transcription Factors That Bind to Them**

Sequence Module	Consensus Sequence	DNA Bound	Factor	Size (kD)	Abundance (molecules/cell)
TATA box	TATAAAA	~10 bp	TBP	27	?
CAAT box	GGCCAATCT	~22 bp	CTF/NF1	60	300,000
GC box	GGGCGG	~20 bp	SP1	105	60,000
Octamer	ATTTGCAT	~20 bp	Oct-1	76	?
"	"	23 bp	Oct-2	52	?
κB	GGGACTTTCC	~10 bp	NFκB	44	?
"	"	~10 bp	H2-TF1	?	?
ATF	GTGACGT	~20 bp	ATF	?	?

Source: Adapted from Lewin, B., 1994. *Genes V.* Cambridge, Mass.: Cell Press.

respective modules. These transcription factors typically behave as positive regulatory proteins essential to transcriptional activation by RNA polymerase II at these promoters.

Enhancers

Eukaryotic genes have, in addition to promoters, regulatory sequences known as enhancers (also called **upstream activation sequences,** or **UASs**) which assist initiation. Enhancers differ from promoters in two fundamental ways. First, the location of enhancers relative to the transcription start site is not fixed. Enhancers may be several thousand nucleotides away from the promoter, and they act to enhance transcription initiation even if positioned *downstream* from the gene. Second, enhancer sequences are *bidirectional* in that they function in either orientation. That is, enhancers can be removed and then reinserted in the reverse sequence orientation without impairing their function. Like promoters, enhancers represent modules of consensus sequence. They are "promiscuous" because they stimulate transcription from any promoter that happens to be in their vicinity. Nevertheless, *enhancer function is dependent on recognition by a specific transcription factor.* A specific transcription factor bound at an enhancer element stimulates transcription by interacting with RNA pol II at a nearby promoter.

Response Elements

Promoter modules in genes responsive to common regulation are termed response elements. Examples include the **heat shock element (HSE),** the **glucocorticoid response element (GRE),** and the **metal response element (MRE).** These various elements are found in the promoter regions of genes whose transcription is activated in response to a sudden increase in temperature (heat shock), glucocorticoid hormones, or toxic heavy metals, respectively (Table 24.3). HSE sequences are recognized by a specific transcription factor, **HSTF** (for **heat shock transcription factor**). HSEs are located about 15 bp upstream from the transcription start site of a variety of genes whose expression is dramatically enhanced in response to elevated temperature. Similarly, the response to steroid hormones depends on the presence of a GRE positioned 250 bp upstream of the transcription start point. Activation of the **steroid receptor** (a specific transcription factor) at a GRE occurs when certain steroids bind to the steroid receptor.

Many genes are subject to multiple regulatory influences. Regulation of such genes is achieved through the presence of an array of different regulatory elements. The **metallothionein** gene is a good example (Figure 24.22). Metallothionein is a metal-binding protein that protects against metal toxicity by binding excess amounts of heavy metals and removing them from the cell. This protein is always present at low levels, but its concentration increases in

---

Table 24.3 Response Elements That Identify Genes Coordinately Regulated in Response to Particular Physiological Challenges

Physiological Challenge	Response Element	Consensus Sequence	DNA Bound	Factor	Size (kD)
Heat shock	HSE	CNNGAANNTCCNNG	27 bp	HSTF	93
Glucocorticoid	GRE	TGGTACAAATGTTCT	20 bp	Receptor	94
Cadmium	MRE	CGNCCCGGNCNC	?	?	?
Phorbol ester	TRE	TGACTCA	22 bp	AP1	39
Serum	SRE	CCATATTAGG	20 bp	SRF	52

Source: Adapted from Lewin, B., 1994. *Genes V.* Cambridge, Mass.: Cell Press.

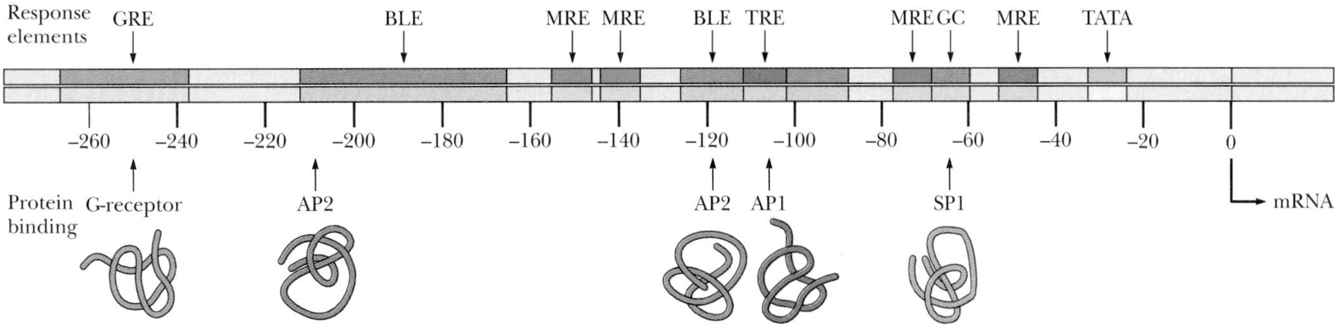

Figure 24.22 The metallothionein gene possesses several constitutive elements in its promoter (the TATA and GC boxes) as well as specific response elements such as MREs and a GRE. The BLEs are elements involved in *basal level* expression (constitutive expression). TRE is a *tumor response* element activated in the presence of tumor-promoting phorbol esters such as TPA (tetradecanoyl phorbol acetate).

response to heavy metal ions such as cadmium or in response to glucocorticoid hormones. The metallothionein gene promoter consists of two general promoter elements (namely, a TATA box and a GC box), two basal-level enhancers, four MREs, and one GRE. These elements function independently of one another; any one is able to activate transcription of the gene.

Transcription Initiation by RNA Polymerase II: TBP and the GTFs

A eukaryotic transcription initiation complex consists of RNA pol II, five general transcription factors, and an 18-subunit complex called **Mediator** (or **Srb/Med**). The CTD of RNA pol II anchors Mediator to the polymerase. There are six GTFs (Table 24.4), five of which are required for transcription: **TFIIB, TFIID, TFIIE, TFIIF,** and **TFIIH.** The sixth, **TFIIA,** stimulates transcription by stabilizing the interaction of TFIID with the TATA box. TFIID consists of **TBP** (*TATA-b*inding *p*rotein), which directly recognizes the TATA box within the core promoter, and a set of *T*BP-*a*ssociated *f*actors (**TAFs** or **TAF$_{II}$s**)[5]. The TBP–TAF$_{II}$ complexes serve as a bridge between the promoter and RNA pol II. Some are capable of recognizing core promoters lacking a TATA box. TBP binds to the core promoter through contacts made with the minor groove of the DNA, distorting and bending the DNA so that DNA sequences upstream and downstream of the TATA box come into closer proximity (Figure 24.23a). Once TBP/TFIID is bound at the core promoter, a complex containing RNA polymerase II and the remaining GTFs convenes at this site, establishing a competent transcription *preinitiation complex* (Figure 24.23b). An *open complex* then forms, and transcription begins.

Alleviating the Repression Due to Nucleosomes

The basic structural unit of nucleosomes, the histone "core octamer" (see Figure 8.47), is constructed from the eight *histone-fold protein domains* of the eight various histone monomers comprising the octamer. Successive histone octamers are linked via histone H1, which is not part of the octamer. Each histone monomer in the core octamer has an unstructured N-terminal tail that extends outside the core octamer. Interactions between histone tails contributed by core histones in adjacent nucleosomes are an important influence

---

[5]For many genes, another transcription factor complex called SAGA (which also contains TAF$_{II}$s) can serve instead of TFIID to initiate transcription.

Table 24.4   General Transcription Initiation Factors from Human Cells

Factor	Number of Subunits	Function
TFIID		
TBP	1	Core promoter recognition (TATA); TFIIB recruitment
TAFs	12	Core promoter recognition (non-TATA elements); positive and negative regulatory functions; HAT (histone acetyltransferase) activity
TFIIA	3	Stabilization of TBP binding; stabilization of TAF-DNA interactions
TFIIB	1	RNA pol II–TFIIF recruitment; start-site selection by RNA pol II
TFIIF	2	Promoter targeting of pol II; destabilization of nonspecific RNA pol II–DNA interactions
RNA pol II	12	Enzymatic synthesis of RNA; TFIIE recruitment
TFIIE	2	TFIIH recruitment; modulation of TFIIH helicase, ATPase, and kinase activities; promoter melting
TFIIH	9	Promoter melting using helicase activity; promoter clearance via CTD phosphorylation (2 subunits of TFHII are a cyclin : Cdk pair)

Adapted from Table 1 in Roeder, R. G., 1996. The role of general initiation factors in transcription by RNA polymerase II. *Trends in Biochemical Sciences* **21:**327–335.

in establishing the higher orders of chromatin organization. Activation of eukaryotic transcription is dependent on *two* sets of circumstances: (1) relief from the repression imposed by chromatin structure, and (2) interaction of RNA pol II with the promoter and transcription regulatory proteins. Relief from repression requires factors that can reorganize the chromatin and then alter the nucleosomes so that promoters become accessible to the transcriptional

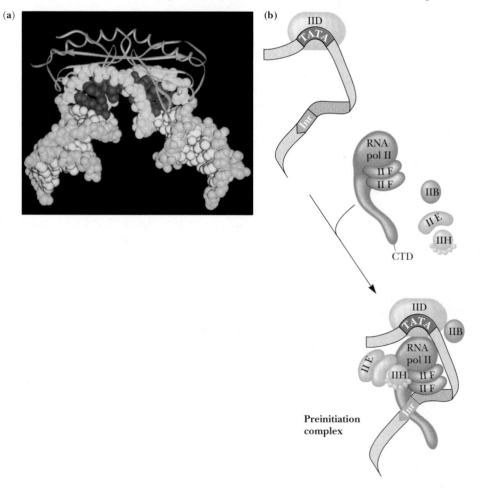

Figure 24.23 Transcription initiation. **(a)** Model of the yeast TATA-binding protein (TBP) in complex with a yeast DNA TATA sequence. The sugar–phosphate backbone of the TATA box is shown in yellow; the TATA base pairs are in red; adjacent DNA segments are in blue. The saddle-shaped TBP (green) is unusual in that it binds in the minor groove of DNA, sitting on the DNA like a saddle on a horse. TBP-binding pries open the minor groove, creating a 100° bend in the DNA axis and unwinding the DNA within the TATA sequence. The other subunits of the TFIID complex (see Table 24.4) sit on TBP, like a "cowboy on a saddle." All known eukaryotic genes (those lacking a TATA box as well as those transcribed by RNA polymerase I or III) rely on TBP. *(Photo courtesy of Paul B. Sigler of Yale University)* **(b)** Formation of a preinitiation complex at a TATA-containing promoter. TFIID bound to the TATA motif recruits RNA pol IIA (the nonphosphorylated form of RNA pol II) and the other GTFs to form the preinitiation complex. Melting of the DNA duplex around *Inr* generates the open complex and transcription ensues.

See page 96

machinery. Two such factors are important: **chromatin-remodeling complexes** that mediate ATP-dependent conformational (noncovalent) changes in nucleosome structure and **HATs** that covalently modifiy histones. HATs is the acronym for **histone acetyltransferases** (enzymes that acetylate the $\varepsilon - NH_3^+$ groups of lysine residues in the histone tails).

HATs

Initial events in transcriptional activation include acetyl-CoA-dependent acetylation of histone tails by HATs. The histone transacetylases responsible are essential components of several megadalton-size complexes known to be required for transcription co-activation (*co-activation* in the sense that they are required along with RNA pol II and other components of the transcriptional apparatus). Examples of such complexes include the TFIID (some of whose TAF$_{II}$s have HAT activity), the **SAGA complex** (which also contains TAF$_{II}$s), and the **ADA complex.** *N*-Acetylation suppresses the positive charge in histone tails, disrupting their ability to interact with DNA and with neighboring nucleosomes. It is believed that phosphorylation of Ser residues in histone tails may also contribute to transcription regulation. In contrast to HATs, **histone deacetylase complexes,** or **HDACs,** catalyze the deacetylation of histone tails, restoring the chromatin to a repressed state.

Chromatin Remodeling Complexes

Although acetylation of histone tails disrupts chromatin structure, it does not expose the DNA within nucleosome "core" particles for transcription. Such exposure requires *chromatin remodeling complexes,* huge (1–2 megadalton) multisubunit entities that mediate noncovalent changes within the "core" particle of the nucleosome in an ATP-dependent manner. These changes disrupt DNA : histone interactions, allowing DNA-binding proteins such as RNA pol II, GTFs, and other transcription factors to gain access to the DNA. Prominent among the various chromatin remodeling complexes are **Swi/Snf, RSC,** and **ISWI.**

A General Model for Eukaryotic Gene Activation

Gene activation (the initiation of transcription) can thus be viewed as a process requiring two principal steps: (1) alterations in chromatin (and nucleosome) structure that relieve the general repressed state imposed by chromatin structure, followed by (2) the interaction of RNA pol II and the GTFs with the

cognition act or process of knowing; acquisition of knowledge

---

HUMAN BIOCHEMISTRY

Storage of Long-Term Memory Depends on Gene Expression Activated by CREB-Type Transcription Factors

Learning can be defined as the process whereby new information is acquired, and memory as the process by which this information is retained. Short-term memory (which lasts minutes or hours) requires only the covalent modification of pre-existing proteins, but long-term memory (which lasts days, weeks, or a lifetime) depends on gene expression, protein synthesis, and the establishment of new neuronal connections.

The macromolecular synthesis underlying long-term memory storage requires *cAMP response element-binding* (**CREB**) protein-related transcription factors and the activation of cAMP-dependent gene expression. Serotonin (5-hydroxytryptamine, or 5-HT, a hormone implicated in learning and memory) acting on neurons

promotes cAMP synthesis, which in turn stimulates protein kinase A to phosphorylate CREB-related transcription factors that activate transcription of cAMP-inducible genes. These genes are characterized by the presence of **CRE** (*cAMP response element*) consensus sequences containing the 8-bp TGACGTCA palindrome. CREB transcription factors are *bZIP*-type proteins (see Section 24.4).

These exciting findings represent the opening of a new arena in molecular biology: the molecular biology of **cognition.** Eric Kandel shared in the 2000 Nobel Prize in physiology or medicine for, among other things, his discovery of the role of CREB-type transcription factors in long-term memory storage.

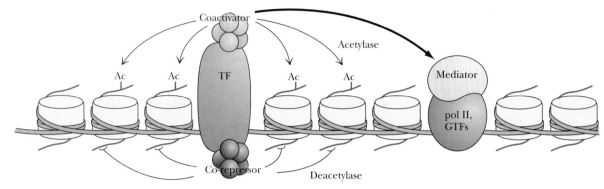

Figure 24.24 A model for the transcriptional regulation of eukaryotic genes. The DNA is a green ribbon wrapped around disklike nucleosomes. A specific transcription factor (TF, in pink) is bound to a regulatory element (either an enhancer or silencer). RNA pol II and its associated GTFs (blue) is bound at the promoter. The N-terminal tails of histones are shown as wavy lines (blue) emanating from the nucleosome disks. A specific transcription factor that is a transcription activator stimulates transcription both through interaction with a coactivator whose HAT activity renders the DNA more accessible *and* through interactions with the Mediator complex associated with RNA pol II. A specific transcription factor that is a repressor interacts with a co-repressor that has HDAC activity that deacetylates histones, restructuring the nucleosomes into a repressed state. *(From Figure 1 in Kornberg, R. D., 1999. Eukaryotic transcription control.* Trends in Biochemical Sciences *24:M46–M49.)*

promoter. Transcription activators (proteins that bind to enhancers and response elements) initiate the process by recruiting chromatin-altering proteins (the chromatin remodeling complexes and histone acetyltransferases described above). Once these alterations have occurred, promoter DNA is accessible to TBP:TFIID, the other GTFs, and RNA pol II. Transcription activation, however, requires communication between RNA pol II and the transcription activator for transcription to take place. Mediator (or Srb/med) fulfills this function. Mediator interacts with *both* the transcription activator *and* the CTD of RNA pol II. This Mediator bridge provides an essential interface for communication between enhancers and promoters, triggering RNA pol II to begin transcription. A general model for transcription initiation is shown in Figure 24.24. Once transcription begins, Mediator is replaced by another complex called **Elongator.** Elongator has HAT subunits whose activity remodels downstream nucleosomes as RNA pol II progresses along the chromatin-associated DNA.

24.4 Structural Motifs in DNA-Binding Regulatory Proteins

Proteins that recognize nucleic acids do so by the basic rule of macromolecular recognition. That is, such proteins present a three-dimensional shape or contour that is structurally and chemically complementary to the surface of a DNA sequence. When the two molecules come into close contact, the numerous atomic interactions that underlie recognition and binding can take place between the two. Nucleotide sequence–specific recognition by the protein involves a set of atomic contacts with the bases and the sugar–phosphate backbone. Hydrogen bonding is critical for recognition, with amino acid side chains providing most of the critical contacts with DNA. Protein contacts with the bases of DNA usually occur within the major groove (but not always). Protein contacts with the DNA backbone involve both H bonds and salt bridges with electronegative oxygen atoms of the phosphodiester linkages. Structural studies on regulatory proteins that bind to specific DNA sequences have revealed that roughly 80% of such proteins can be assigned to one of three principal classes based on their possession of one of three kinds of small, distinctive structural motifs: the **helix-turn-helix** (or **HTH**), the **zinc finger** (or **Zn-finger**), and the **basic region–leucine zipper** (or **bZIP**). The latter two motifs are found only in DNA-binding proteins from eukaryotic organisms.

In addition to their DNA-binding domains, these proteins commonly possess other structural domains that function in protein:protein recognitions essential to oligomerization (for example, dimer formation), DNA looping, transcriptional activation, and signal reception (for example, effector binding).

α-Helices Fit Snugly into the Major Groove of B-DNA

A recurring structural feature in DNA-binding proteins is the presence of α-helical segments that fit directly into the major groove of B-form DNA. The diameter of an α-helix (including its side chains) is about 1.2 nm. The dimensions of the major groove in B-DNA are 1.2 nm wide by 0.6 to 0.8 nm deep. Thus, one side of an α-helix can fit snugly into the major groove. Although examples of β-sheet DNA recognition elements in proteins are known, the α-helix and B-form DNA are the predominant structures involved in protein : DNA interactions. Significantly, proteins can recognize specific sites in "normal" B-DNA; the DNA need not assume any unusual, alternative conformation (such as Z-DNA).

Proteins with the Helix-Turn-Helix Motif

The helix-turn-helix (HTH) motif is a protein structural domain consisting of two successive α-helices separated by a sharp β-turn (Figure 24.25). Within this domain, the α-helix situated more toward the C-terminal end of the protein, the so-called **helix 3,** is the DNA recognition helix; it fits nicely into the major groove, with several of its side chains touching DNA base pairs. **Helix 2,** the helix at the beginning of the HTH motif, through hydrophobic interactions with helix 3, creates a stable structural domain that locks helix 3 into its DNA interface. Proteins with HTH motifs bind to DNA as dimers. In the dimer, the two helix 3 cylinders are antiparallel to each other, such that their N → C orientations match the inverted relationship of nucleotide sequence in the dyad-symmetric DNA-binding site. An example is *Antp*. *Antp* is a member of a family of eukaryotic proteins involved in the regulation of early embryonic development that have in common an amino acid sequence element known as the **homeobox**[6] **domain.** The homeobox is a DNA motif that encodes a related 60–amino acid sequence (the homeobox domain) found among proteins of virtually every eukaryote, from yeast to human. Embedded within the homeobox domain is an HTH motif (Figure 24.26). Homeobox domain proteins act as **sequence-specific transcription factors.** Typically, the homeobox portion comprises only 10% or so of the protein's mass, with the remainder of the protein serving in protein : protein interactions essential to transcription regulation.

How Does the Recognition Helix Recognize Its Specific DNA-Binding Site?

The edges of base pairs in dsDNA present a pattern of hydrogen-bond donor and acceptor groups within the major and minor grooves, but only the pattern displayed on the major-groove side is distinctive for each of the four base pairs A : T, T : A, C : G, and G : C. (Satisfy yourself on this point by inspecting the structures of the base pairs in Chapter 8.) Thus, the base-pair edges in the major groove act as a **recognition matrix** identifiable through H bonding with a specific protein, so that it is not necessary to melt the base pairs in order to read the base sequence. Although formation of such H bonds is very important in DNA : protein recognition, other interactions also play a significant role. For example, the C-5-methyl groups unique to thymine residues are nonpolar "knobs" projecting into the major groove.

Proteins Also Recognize DNA via "Indirect Readout"

Indirect readout refers to the ability of a protein to indirectly recognize a particular nucleotide sequence through the local conformational variations that

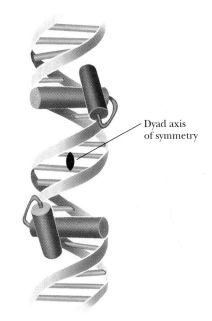

Dyad axis of symmetry

Figure 24.25 The helix-turn-helix motif. A dimeric HTH protein bound to a dyad-symmetric DNA site. (Note dyad axis of symmetry.) The recognition helix (helix 3, in red) is represented as a dark barrel situated in the major groove of each half of the dyad site. Helix 2 sits above helix 3 and aids in locking it into place. *(Adapted from Johnson, P. F., and McKnight, S. L., 1989. Eukaryotic transcriptional regulatory proteins. Annual Review of Biochemistry **58:**799, Figure 1.)*

---

[6]"Homeo" is derived from homeotic genes, a set of genes originally discovered in the fruit fly *Drosophila melanogaster,* through their involvement in the specification of body parts during development.

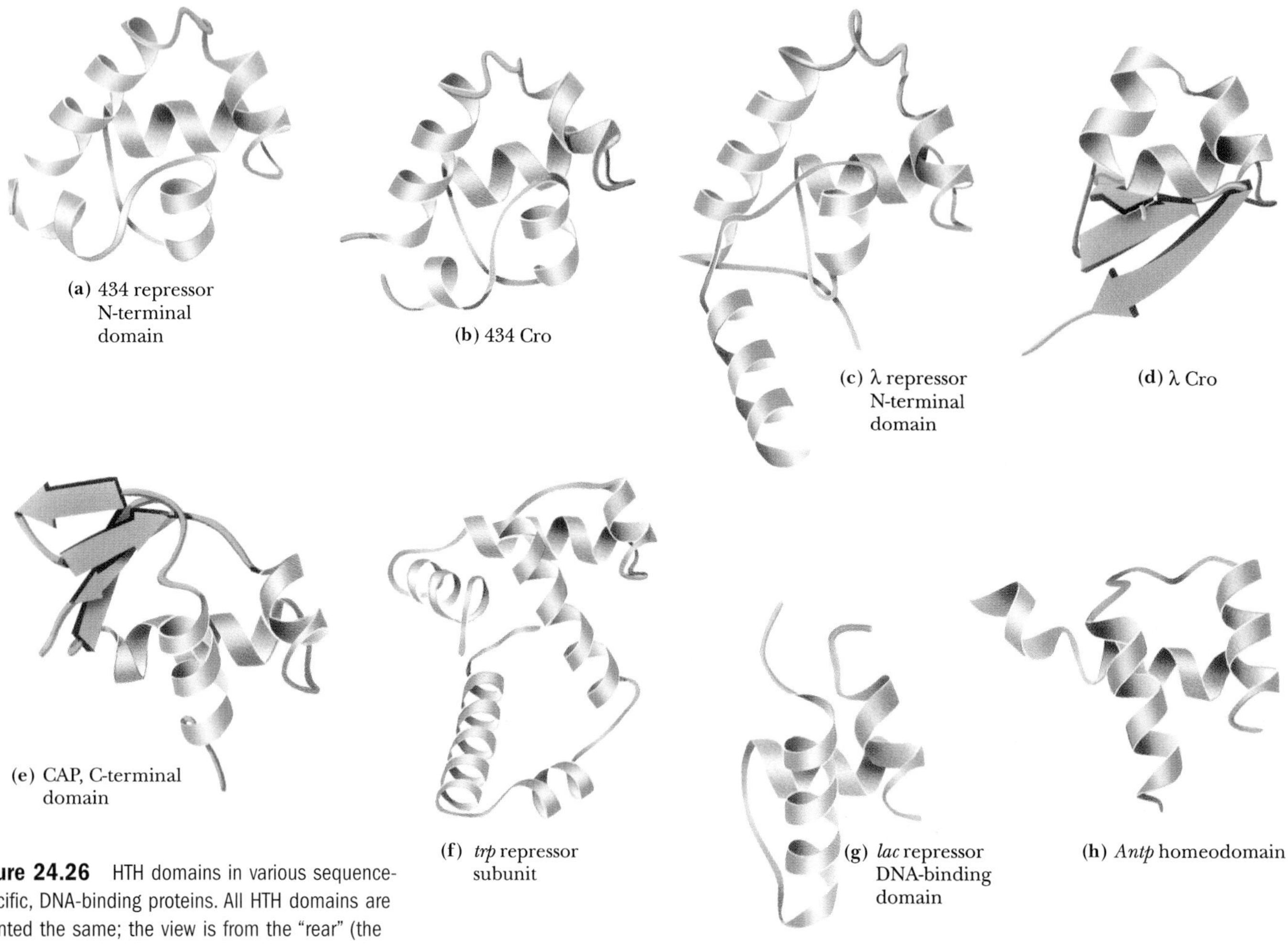

(a) 434 repressor N-terminal domain

(b) 434 Cro

(c) λ repressor N-terminal domain

(d) λ Cro

(e) CAP, C-terminal domain

(f) *trp* repressor subunit

(g) *lac* repressor DNA-binding domain

(h) *Antp* homeodomain

Figure 24.26 HTH domains in various sequence-specific, DNA-binding proteins. All HTH domains are oriented the same; the view is from the "rear" (the side opposite the DNA-binding face). The first HTH helix runs vertically down from the upper right, the turn is right center, and the "recognition" helix runs across the center from right to left "behind" the domain in this perspective. **(a)** 434 repressor N-terminal domain; **(b)** 434 *Cro*; **(c)** λ repressor N-terminal domain; **(d)** λ *Cro*; **(e)** CAP, C-terminal domain; **(f)** Trp repressor subunit; **(g)** *lac* repressor DNA-binding domain; **(h)** *Antp* homeodomain. Domains (a) through (d) are in bacteriophage proteins, (e) through (g) are bacterial representatives, and (h) is a motif found in a eukaryotic DNA-binding protein. *(Adapted from Harrison, S. C., and Aggarwal, A. K., 1990. DNA recognition by proteins with a helix-turn-helix motif.* Annual Review of Biochemistry **59**:933, Figure 2; Sauer, R. T., 1984. Protein-DNA recognition. Annual Review of Biochemistry **53**:293.)

base sequence imposes on DNA structure. Superficially, the B-form structure of DNA appears to be a uniform cylinder. Nevertheless, the conformation of DNA over a short distance along its circumference varies subtly according to local base sequence. That is, base sequences generate unique contours that proteins can recognize. Because these contours arise from the base sequence, the DNA-binding protein "indirectly reads out" the base sequence through interactions with the DNA backbone. In the *E. coli* Trp repressor : *trp* operator DNA complex, the Trp repressor engages in 30 specific hydrogen bonds to the DNA: 28 involve phosphate groups in the backbone; only 2 are to bases. Thus, some sequence-specific DNA-binding proteins are able to recognize an overall DNA conformation caused by the specific DNA sequence.

Proteins with Zn-Finger Motifs

There are many classes of zinc finger motifs. The prototype Zn-finger is a structural feature formed by a pair of Cys residues separated by 2 residues, then a run of 12 amino acids, and finally a pair of His residues separated by 3 residues ($Cys-x_2-Cys-x_{12}-His-x_3-His$). This motif may be repeated as many as 13 times over the primary structure of a Zn-finger protein. Each repeat coordinates a zinc ion via its 2 Cys and 2 His residues (Figure 24.27). The 12 or so residues separating the Cys and His coordination sites are looped out and form a distinct DNA interaction module, the so-called Zn-finger. When Zn-finger proteins associate with DNA, each Zn-finger binds in the major groove and interacts with about five nucleotides, adjacent fingers interacting with contiguous stretches of DNA. Many DNA-binding proteins with this motif have been identified. In all cases, the finger motif is repeated at least two times, with at least a 7 to 8 amino acid linker between Cys/Cys and His/His sites. Proteins with this general pattern are assigned to the **C_2H_2 class** of Zn-finger proteins to distinguish them from proteins bearing another kind of Zn-finger, the **C_x type,** which includes the C_4 and C_6 Zn-finger proteins. The C_x proteins have a variable number of Cys residues available for Zn **chelation.** For example, the vertebrate steroid receptors have two sets of Cys residues, one with four conserved cysteines (C_4) and the other with five (C_5).

chelation from the Greek *chele,* claw; binding of a metal ion to two or more nonmetallic atoms of the same molecule

Proteins with the Basic Region–Leucine Zipper (*bZIP*) Motif

The basic region–leucine zipper (*bZIP*) is a structural motif characterizing the third major class of sequence-specific, DNA-binding proteins. This motif was first recognized by Steve McKnight in **C/EBP,** a heat-stable, DNA-binding protein isolated from rat liver nuclei that binds to both CCAAT promoter elements and certain enhancer core elements.[7] The DNA-binding domain of C/EBP was localized to the C-terminal region of the protein. This region shows a notable absence of Pro residues, suggesting it might be arrayed in an α-helix. Within this region are two clusters of basic residues, A and B. Further along is a 28-residue sequence. When this latter region is displayed end-to-end down the axis of a hypothetical α-helix, beginning at Leu[315], an amphipathic cylinder is generated, similar to the one shown in Figure 6.53. One side of this amphipathic helix consists principally of hydrophobic residues (particularly leucines), whereas the other side has an array of negatively and positively charged side chains (Asp, Glu, Arg, and Lys), as well as many uncharged polar side chains (glutamines, threonines, and serines).

The Zipper Motif: Intersubunit Interaction of Leucine Side Chains

The leucine zipper motif arises from the periodic repetition of leucine residues within this helical region. The periodicity causes the Leu side chains to protrude from the same side of the helical cylinder, where they can enter into hydrophobic interactions with a similar set of Leu side chains extending from a matching helix in a second polypeptide. These hydrophobic interactions establish a stable noncovalent linkage, fostering dimerization of the two polypeptides (as you will note in Figure 24.29). The leucine zipper is not a DNA-binding domain. Instead, it functions in protein dimerization. Leucine zippers have been found in other mammalian transcriptional regulatory proteins, including *Myc, Fos,* and *Jun.*

The Basic Region of *bZIP* Proteins

The actual DNA contact surface of *bZIP* proteins is contributed by a 16-residue segment that ends exactly 7 residues before the first Leu residue of the Leu zipper. This DNA contact region is rich in basic residues and hence is referred to as the **basic region.** Two *bZIP* polypeptides join via a Leu zipper to form a Y-shaped molecule in which the stem of the Y corresponds to a coiled pair of α-helices held by the leucine zipper. The arms of the Y are the respective basic regions of each polypeptide; they act as a linked set of DNA contact surfaces

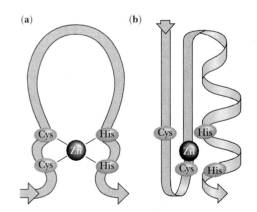

Figure 24.27 The Zn-finger motif of the C_2H_2 type showing **(a)** the coordination of Cys and His residues to Zn and **(b)** the secondary structure. *(Adapted from Evans, R. M., and Hollenberg, S. M., 1988. Zinc fingers: Gilt by association.* Cell **52:**1, Figure 1.)

[7]The acronym C/EBP designates this protein as a "*C*CAAT and *e*nhancer *b*inding *p*rotein."

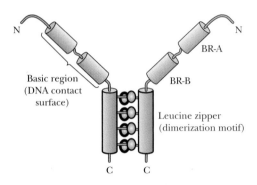

Figure 24.28 Model for a dimeric *bZIP* protein. Two *bZIP* polypeptides dimerize to form a Y-shaped molecule. The stem of the Y is the Leu zipper, and it holds the two polypeptides together. Each arm of the Y is the basic region from one polypeptide. Each arm is composed of two α-helical segments: BR-A and BR-B (basic regions A and B).

(Figure 24.28). The dimer interacts with a DNA target site by situating the fork of the Y at the center of the dyad-symmetric DNA sequence. The two arms of the Y can then track along the major groove of the DNA in opposite directions, reading the specific recognition sequence (Figure 24.29). An interesting aspect of *bZIP* proteins is that the two polypeptides need not be identical (see Figure 24.29). Heterodimers can form provided both polypeptides possess a leucine zipper region. An important consequence of heterodimer formation is that the DNA target site need not be a palindromic sequence. The respective basic regions of the two different *bZIP* polypeptides (for example, *Fos* and *Jun*) can track along the major groove reading two different base sequences. Heterodimer formation expands enormously the DNA recognition and regulatory possibilities of this set of proteins.

24.5 Post-Transcriptional Processing of mRNA in Eukaryotic Cells

Transcription and translation are concomitant processes in prokaryotes, but in eukaryotes, the two processes are spatially and temporally disconnected (see Chapter 8). *Transcription occurs on DNA in the nucleus and translation occurs on ribosomes in the cytoplasm.* Consequently, transcripts must move from the nucleus to the cytosol to be translated. On the way, these transcripts undergo **processing:** alterations that convert the newly synthesized RNAs, or *primary transcripts,* into mature messenger RNAs. Also, unlike prokaryotes, in which many mRNAs encode more than one polypeptide (that is, they are polycistronic), eukaryotic mRNAs encode only one polypeptide (that is, they are exclusively monocistronic).

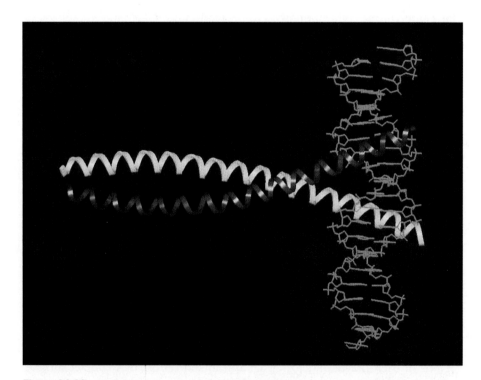

Figure 24.29 Model for the heterodimeric *bZIP* transcription factor *c-Fos:c-Jun* bound to a DNA oligomer containing the AP-1 consensus target sequence TGACTCA. *(Adapted from Glover, J. N. M., and Harrison, S. C., 1995. Crystal structure of the heterodimeric bZIP transcription factor c-Fos:c-Jun bound to DNA. Nature* **373**:257–261.)

See page 20

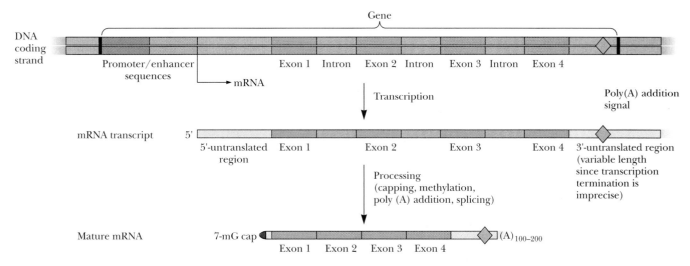

Figure 24.30 The organization of split eukaryotic genes.

Eukaryotic Genes Are Split Genes

Most genes in higher eukaryotes are split into coding regions, called **exons,**[8] and noncoding regions, called **introns** (Figure 24.30; see also Chapter 8). Introns are the intervening nucleotide sequences that are removed from the primary transcript when it is processed into a mature RNA. Gene expression in eukaryotes entails not only transcription but also the *processing of primary transcripts* to yield the mature RNA molecules we classify as mRNAs, tRNAs, rRNAs, and so forth.

The Organization of Split Genes

Split genes occur in an incredible variety of interruptions and sizes. The yeast **actin gene** is a simple example, having only a single 309-bp intron that separates the nucleotides encoding the first 3 amino acids from those encoding the remaining 350 or so amino acids in the protein. The chicken **ovalbumin gene** is composed of 8 exons and 7 introns. The two **vitellogenin genes** of the African clawed toad *Xenopus laevis* are both spread over more than 21 kbp of DNA; their primary transcripts consist of just 6 kb of message that is punctuated by 33 introns. The chicken **pro α-2 collagen gene** has a length of about 40 kbp; the coding regions constitute only 5 kb distributed over 51 exons within the primary transcript. The exons are quite small, ranging from 45 to 249 bases in size.

Clearly, the mechanism by which introns are removed and multiple exons are spliced together to generate a continuous, translatable mRNA must be both precise and complex. If one base too many or too few is excised during splicing, the coding sequence in the mRNA will be disrupted. The mammalian **DHFR (dihydrofolate reductase) gene** is split into 6 exons spread over more than 31 kbp of DNA. The 6 exons are spliced together to give a 6-kb mRNA (Figure 24.31). Note that, in three different mammalian species, the size and position of the exons are essentially the same but that the lengths of the corresponding introns vary considerably. Indeed, the lengths of introns in

---

[8]Although the term "exon" is commonly used to refer to the protein-coding regions of an interrupted or split gene, a more precise definition would specify exons as sequences that are represented in mature RNA molecules. This definition encompasses not only protein-coding genes but also the genes for various RNAs (such as tRNAs or rRNAs) from which intervening sequences must be excised in order to generate the mature gene product.

Figure 24.31 The organization of the mammalian DHFR gene in three representative species. Note that the exons are much shorter than the introns. Note also that the exon pattern is more highly conserved than the intron pattern.

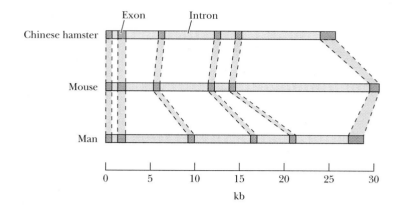

vertebrate genes range from a minimum of about 60 bases to more than 10,000 bp. Many introns have nonsense codons in all three reading frames, and thus are untranslatable. Introns are found in the genes of mitochondria and chloroplasts as well as in nuclear genes. Although introns have been observed in archaebacteria and even bacteriophage T4, none are known in the genomes of eubacteria.

Post-Transcriptional Processing of Messenger RNA Precursors

Capping and Methylation of Eukaryotic mRNAs

The protein-coding genes of eukaryotes are transcribed by RNA polymerase II to form primary transcripts or **pre-mRNAs** that serve as precursors to mRNA. As a population, these RNA molecules are very large and their nucleotide sequences are very heterogeneous because they represent the transcripts of many different genes—hence the designation **heterogeneous nuclear RNA, or hnRNA.** Shortly after transcription of hnRNA is initiated, the 5'-end of the growing transcript is capped by addition of a guanylyl residue. This reaction is catalyzed by the nuclear enzyme **guanylyl transferase** using GTP as substrate (Figure 24.32). The **cap structure** is methylated at the 7-position of the G residue. Additional methylations may occur at the 2'-O positions of the two nucleosides following the 7-methyl-G cap and at the 6-amino group of a first base adenine (Figure 24.33).

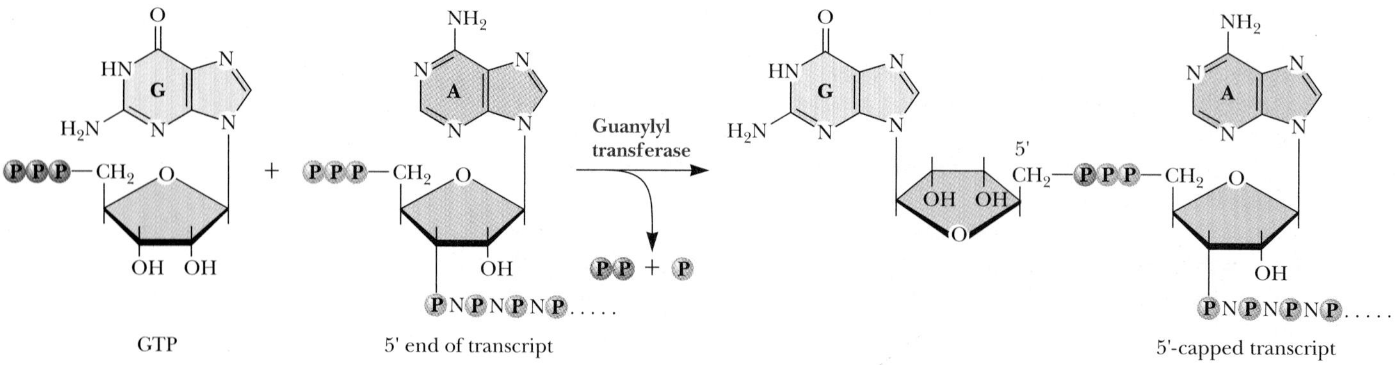

Figure 24.32 The capping of eukaryotic pre-mRNAs. Guanylyl transferase catalyzes the addition of a guanylyl residue (G_p) derived from GTP to the 5'-end of the growing transcript, which has a 5'-triphosphate group already there. In the process, pyrophosphate (*pp*) is liberated from GTP and the terminal phosphate (p) is removed from the transcript. Gppp + pppApNpNpNp ... → GpppApNpNpNp ... + *pp* + p (A is often the initial nucleotide in the primary transcript).

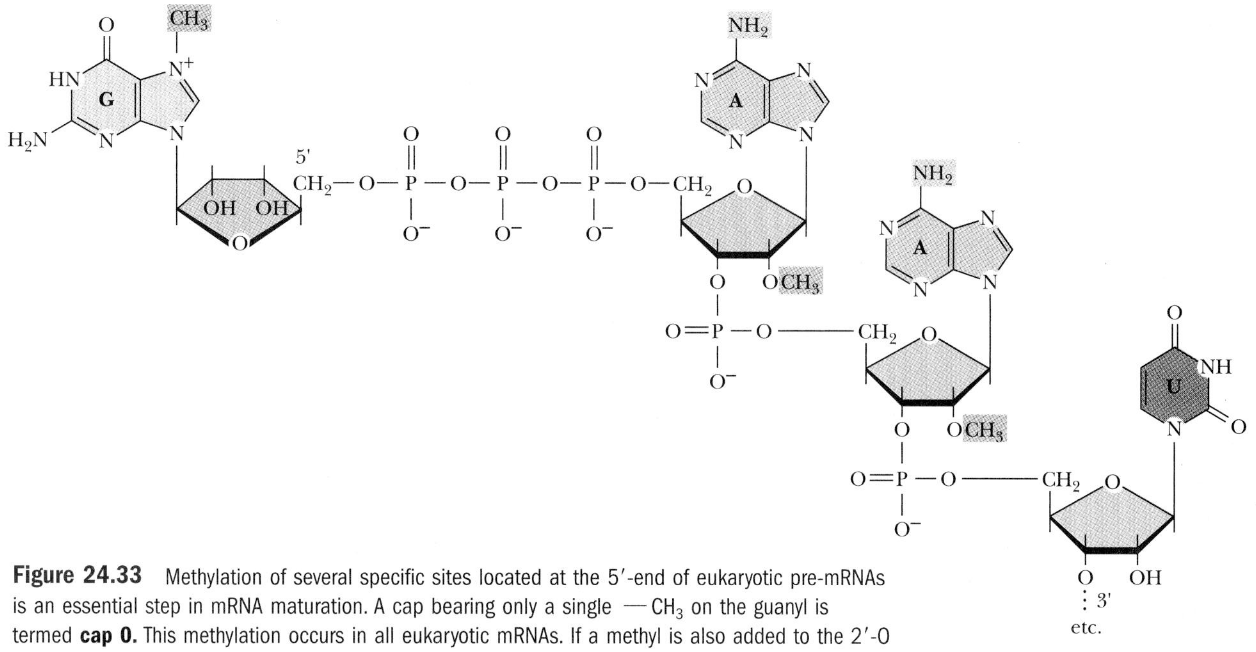

Figure 24.33 Methylation of several specific sites located at the 5′-end of eukaryotic pre-mRNAs is an essential step in mRNA maturation. A cap bearing only a single —CH_3 on the guanyl is termed **cap 0.** This methylation occurs in all eukaryotic mRNAs. If a methyl is also added to the 2′-O position of the first nucleoside after the cap, a **cap 1** structure is generated. This is the predominant cap form in all multicellular eukaryotes. Some species add a third —CH_3 to the 2′-O position of the second nucleoside after the cap, giving a **cap 2** structure. Also, if the first base after the cap is an adenine, it may be methylated on its 6-NH_2. In addition, approximately 0.1% of the adenine bases throughout the mRNA of higher eukaryotes carry methylation on their 6-NH_2 groups.

3′-Polyadenylylation of Eukaryotic mRNAs

Transcription by RNA polymerase II typically continues past the 3′-end of the mature messenger RNA. Primary transcripts show heterogeneity in sequence at their 3′-ends, indicating that the precise point where termination occurs is nonspecific. However, termination does not normally occur until RNA polymerase II has transcribed past a consensus AAUAAA sequence known as the **polyadenylylation signal.**

Most eukaryotic mRNAs have 100 to 200 adenine residues attached at their 3′-end, the **poly(A) tail.** (Histone mRNAs are the only common mRNAs that lack poly(A) tails.) These A residues are not encoded in the DNA but are added post-transcriptionally by the enzyme **poly(A) polymerase,** using ATP as substrate. The consensus AAUAAA is not itself the poly(A) addition site; instead it defines the position where poly(A) addition occurs (Figure 24.34). The consensus AAUAAA is found 10 to 35 nucleotides upstream from where the nascent primary transcript is cleaved by an endonuclease to generate a new 3′-OH end. This end is where the poly(A) tail is added. The processing events of mRNA capping and poly(A) addition take place before splicing of the primary transcript creates the mature mRNA. Interestingly, both the guanylyl transferase that adds the 5′-cap structure and the enzymes that process the 3′-end of the transcript and add the poly(A) tail are anchored to RNA pol II via interactions with its RPB1 CTD.

Nuclear Pre-mRNA Splicing

Within the nucleus, hnRNA forms **ribonucleoprotein particles (RNPs)** through association with a characteristic set of nuclear proteins. These proteins interact with the nascent RNA chain as it is synthesized, maintaining the hnRNA in an untangled, accessible conformation. The substrate for splicing, that is, intron

Figure 24.34 Poly(A) addition to the 3′-ends of transcripts occurs 10 to 35 nucleotides downstream from a consensus AAUAAA sequence, defined as the *polyadenylylation signal.* CPSF (*cleavage and polyadenylylation specificity factor*) binds to this signal sequence and mediates looping of the 3′-end of the transcript through interactions with a G/U-rich sequence even further downstream. Cleavage factors (*CFs*) then bind and bring about the endonucleolytic cleavage of the transcript to create a new 3′-end 10–35 nucleotides downstream from the polyadenylation signal. Poly(A) polymerase (PAP) then successively adds 200–250 adenylyl residues to the new 3′-end. (RNA pol II is also a significant part of the polyadenylylation complex at the 3′-end of the transcript, but for simplicity in illustration, its presence is not shown in the lower part of the figure.)

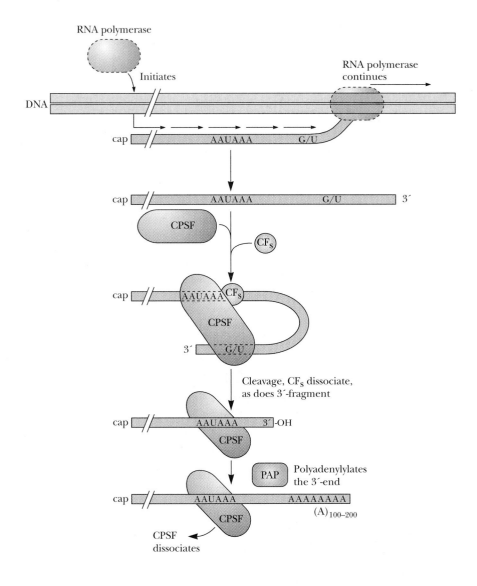

5′-Splice Site Consensus

A G : G U A A G U

Exon   Intron

3′-Splice Site Consensus

Py Py Py Py Py Py Py Py – C A G : G ––

Intron   Exon

Vertebrate genes have introns beginning with GU and ending with AG. (Rarely, introns begin with a GC dinucleotide.) Note that for the 5′ consensus AG : GUAAGU, the colon indicates the exon : intron junction/cleavage site. Note also the 3′ consensus (Py series) NCAG : G, the colon designating the 3′ intron : exon junction/cleavage site.

Figure 24.35 Consensus sequences at the splice sites in vertebrate genes.

excision and exon ligation, is the capped, polyadenylylated RNA in the form of an RNP complex. Splicing occurs exclusively in the nucleus. The mature mRNA that results is then exported to the cytoplasm to be translated. Splicing requires precise cleavage at the 5′- and 3′-ends of introns and the accurate joining of the two ends. Consensus sequences define the exon/intron junctions in eukaryotic mRNA precursors, as indicated from an analysis of the splice sites in vertebrate genes (Figure 24.35). Note that the sequences GU and AG are found at the 5′- and 3′-ends, respectively, of introns in pre-mRNAs from higher eukaryotes. In addition to the splice junctions, a conserved sequence within the intron, the **branch site,** is also essential to pre-mRNA splicing. The site lies 18 to 40 nucleotides upstream from the 3′-splice site and is represented in higher eukaryotes by the consensus sequence YNYRAY, where Y is any pyrimidine, R is any purine, and N is any nucleotide.

The Splicing Reaction: Lariat Formation

The mechanism for splicing nuclear mRNA precursors is shown in Figure 24.36. A covalently closed loop of RNA, the **lariat,** is formed by attachment of the 5′-phosphate group of the intron's invariant 5′-G to the 2′-OH at the invariant branch site A to form a 2′-5′ phosphodiester bond. Note that lariat formation creates an unusual branched nucleic acid. The lariat structure is excised when the 3'-OH of the consensus G at the 3'-end of the 5' exon (Exon 1, Figure 24.36) covalently joins with the 5'-phosphate at the 5'-end of the 3' exon (Exon 2). The reactions that occur are transesterification reactions where an OH group reacts with a phosphoester bond, displacing an —OH to form a new phosphoester link. Because the reactions lead to no net change in the number of phosphodiester linkages, no energy input (for example, as ATP) is needed. The lariat product is unstable; the 2'-5' phosphodiester branch is quickly cleaved to give a linear excised intron that is rapidly degraded in the nucleus.

Splicing Depends on snRNPs

The hnRNA (pre-mRNA) substrate is not the only RNP complex involved in the splicing process. Splicing also depends on a unique set of small nuclear ribonucleoprotein particles, so-called **snRNPs** (pronounced "snurps"). In higher eukaryotes, each snRNP consists of a small RNA molecule 100 to 200 nucleotides long and a set of about 10 different proteins. Some of the different

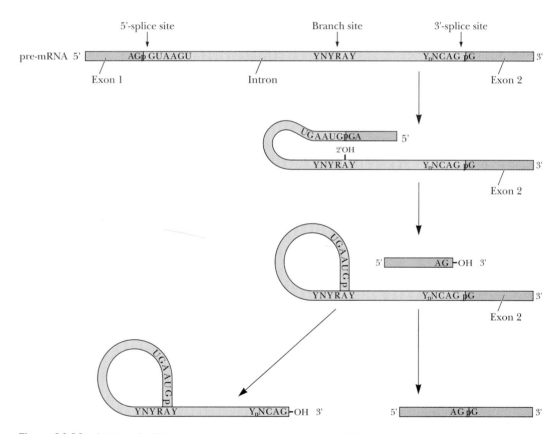

Figure 24.36 Splicing of mRNA precursors. A representative precursor mRNA is depicted. Exon 1 and Exon 2 indicate two exons separated by an intervening sequence (or intron) with consensus 5'-, 3'-, and branch sites. The fate of the phosphates at the 5'- and 3'-splice sites can be followed by tracing the fate of the respective p's. The products of the splicing reaction, the lariat form of the excised intron and the united exons, are shown at the bottom of the figure. The lariat intermediate is generated when the invariant G at the 5'-end of the intron attaches via its 5'-phosphate to the 2'-OH of the invariant A within the branch site. The consensus guanosine residue at the 3'-end of Exon 1 (the 5'-splice site) then reacts with the 5'-phosphate at the 3'-splice site (the 5'-end of Exon 2), ligating the two exons and releasing the lariat structure. Although the reaction is shown here in a stepwise fashion, 5' cleavage, lariat formation, and exon ligation/lariat excision are believed to occur in a concerted fashion. (Adapted from Sharp, P. A., 1987. Science **235**:766, Figure 1.)

Table 24.5	The snRNPs Found in Spliceosomes	
snRNP	**Length (nt)**	**Splicing Target**
U1	165	5′ splice
U2	189	Branch
U4	145	5′ splice, recruitment
U5	115	of branch point to
U6	106	5′-splice site

proteins form a "core" set common to all snRNPs, whereas others are unique to a specific snRNP. The major snRNP species are very abundant, present at greater than 100,000 copies per nucleus. The RNAs of snRNPs are typically rich in uridine, hence the classification of particular snRNPs as U1, U2, and so on. (The prominent snRNPs are given in Table 24.5.) U1 snRNA folds into a secondary structure that leaves the 11 nucleotides at its 5′-end single-stranded. The 5′-end of U1 snRNA is complementary to the consensus sequence at the 5′-splice junction of the pre-mRNA (Figure 24.37), as is a region at the 5′-end of U6 snRNA. U2 snRNA is complementary to the consensus branch site sequence.

snRNPs Form the Spliceosome

Splicing occurs when the various snRNPs come together with the pre-mRNA to form a multicomponent complex called the **spliceosome.** The spliceosome is a large complex, roughly equivalent to a ribosome in size, and its assembly requires ATP. Assembly of the spliceosome begins with the binding of U1 snRNP at the 5′-splice site of the pre-mRNA (see Figure 24.38). The branch-point sequence (UACUAAC in yeast), binds U2 snRNP, and then the triple snRNP complex of U4/U6 · U5 replaces U1 at the 5′-splice site. The substitution of base-pairing interactions between U1 and the pre-mRNA 5′-splice site by base-pairing between U6 and the 5′-splice site is just one of the many RNA rearrangements that accompany the splicing reaction. Base-pairings between U6 and U2 RNA bring the 5′-splice site and the branch-point RNA sequences into proximity. Interactions between U2 and U6 lead to release of U4 snRNP. The spliceosome is now activated for catalysis: a transesterification reaction involving the 2′-O of the invariant A residue in the branch-point sequence displaces the 5′-exon from the intron, creating the lariat intermediate. The free

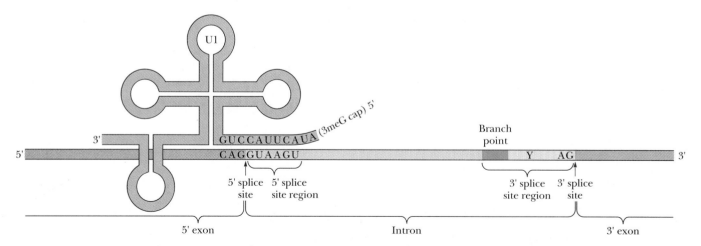

Figure 24.37 Mammalian U1 snRNA can be arranged in a secondary structure where its 5′-end is single-stranded and can base-pair with the consensus 5′-splice site of the intron. *(Adapted from Rosbash, M., and Seraphin, B., 1991. Trends in Biochemical Sciences **16**:187, Figure 1.)*

Figure 24.38 Events in spliceosome assembly. U1 snRNP binds at the 5'-splice site, followed by ▶
the association of U2 snRNP with the UACUAA*C branch-point sequence. The triple U4/U6 · U5
snRNP complex replaces U1 at the 5'-splice site and directs the juxtaposition of the branch-point
sequence with the 5'-splice site, whereupon U4 snRNP is released. Lariat formation occurs, freeing
the 3'-end of the 5'-exon to join with the 5'-end of the 3'-exon, followed by exon ligation. U2, U5,
and U6 snRNPs dissociate from the lariat following exon ligation. Spliceosome assembly,
rearrangement, and disassembly require ATP as well as various RNA-binding proteins (not shown).
*(Adapted from Staley, J. P., and Guthrie, C., 1998. Mechanical devices of the spliceosome: Motors, clocks,
springs, and things. Cell **92**:315–326, Figure 2.)*

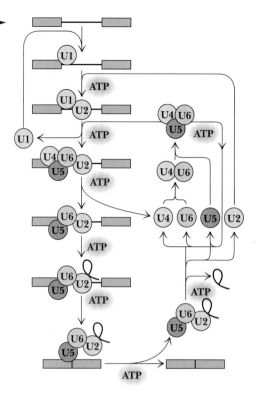

3'-O of the 5'-exon now triggers a second transesterification reaction through
attack on the P atom at the 3'-exon splice site. This second reaction joins the
two exons and releases the intron as a lariat structure. In addition to the
snRNPs, a number of proteins with RNA-annealing functions as well as proteins with ATP-dependent RNA-unwinding activity participate in spliceosome
function. The spliceosome is thus a dynamic structure that uses the pre-mRNA
as a template for assembly, carries out its transesterification reactions, and then
disassembles when the splicing reaction is over. Thomas Cech's discovery of
catalytic RNA (see ribozymes, Chapter 10) raises the possibility that the transesterification reactions are catalyzed not by snRNP proteins but by the snRNAs
themselves.

Alternative RNA Splicing

In one mode of splicing, every intron is removed and every exon is incorporated into the mature RNA without exception. This type of splicing, termed
constitutive splicing, results in a single form of mature mRNA from the primary transcript. However, many eukaryotic genes can give rise to multiple forms
of mature RNA transcripts. The mechanisms for production of multiple transcripts from a single gene include use of different promoters, selection of different polyadenylylation sites, **alternative splicing** of the primary transcript, or
even a combination of the three.

   Different transcripts from a single gene make possible a set of related
polypeptides, or **protein isoforms,** each with slightly altered functional capability. Such variation serves as a useful mechanism for increasing the apparent
coding capacity of the genome. Further, alternative splicing offers another level
at which regulation of gene expression can operate. For example, mRNAs
unique to particular cells, tissues, or developmental stages could be formed
from a single gene by choosing different 5'- or 3'-splice sites or by omitting entire exons. Translation of these mature mRNAs produces cell-specific protein
isoforms that display properties tailored to the needs of the particular cell.
Such regulated expression of distinct protein isoforms is a fundamental characteristic of eukaryotic cell differentiation and development.

The Fast Skeletal Muscle Troponin T Gene:
An Example of Alternative Splicing

In addition to many other instances, alternative splicing is a prevalent mechanism for generating protein isoforms from the genes encoding muscle proteins
(see Chapter 13), allowing distinctive isoforms aptly suited to the function of
each muscle. An impressive manifestation of alternative splicing is seen in the
expression possibilities for the rat **fast skeletal muscle troponin T gene** (Figure 24.39). This gene consists of 18 exons, 11 of which are found in all mature mRNAs (exons 1 through 3, 9 through 15, and 18) and thus are **constitutive.** Five exons, those numbered 4 through 8, are **combinatorial** in that they

Fast skeletal troponin T gene and spliced mRNAs

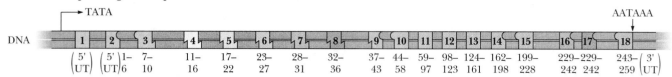

Figure 24.39 Organization of the fast skeletal muscle troponin T gene and the 64 possible mRNAs that can be generated from it. Exons are constitutive (yellow), combinatorial (green), or mutually exclusive (blue or orange). Exon 1 is composed of 5′-untranslated (UT) sequences, and Exon 18 includes the polyadenylylation site (AATAAA) and 3′-UT sequences. The TATA box indicates the transcription start site. The amino acid residues encoded by each exon are indicated below. Many exon:intron junctions fall between codons. The "sawtooth" exon boundaries indicate that the splice site falls between the first and second nucleotides of a codon; the "concave/convex" exon boundaries indicate that the splice site falls between the second and third nucleotides of a codon; and flush boundaries between codons signify that the splice site falls between intact codons. Each mRNA includes all constitutive exons, one of the 32 possible combinations of Exons 4 to 8, and either Exon 16 or 17. (Adapted from Breitbart, R. E., Andreadis, A., and Nadal-Ginard, B., 1987. Annual Review of Biochemistry **56**:467, Figure 2.)

may be individually included or excluded, in any combination, in the mature mRNA. Two exons, 16 and 17, are mutually exclusive: one or the other is always present, but never both. Sixty-four different mature mRNAs can be generated from the primary transcript of this gene by alternative splicing. Because each exon represents a cassette of genetic information encoding a subsegment of protein, alternative splicing is a versatile way to introduce functional variation within a common protein theme.

PROBLEMS

1. The 5′-end of an mRNA has the sequence

   … AGAUCCGUAUGGCGAUCUCGACGAAGACUCCUA GGGAAUCC …

   What is the nucleotide sequence of the DNA template strand from which it was transcribed? If this mRNA is translated beginning with the first AUG codon in its sequence, what is the N-terminal amino acid sequence of the protein it encodes? (See Table 25.1 for the genetic code.)

2. Describe the sequence of events involved in the initiation of transcription by *E. coli* RNA polymerase. Include in your description those features a gene must have for proper recognition and transcription by RNA polymerase.

3. RNA polymerase has two binding sites for ribonucleoside triphosphates, the initiation site and the elongation site. The initiation site has a greater K_m for NTPs than the elongation site. Suggest what possible significance this fact might have for the control of transcription in cells.

4. How does transcription in eukaryotes differ from transcription in prokaryotes?

5. DNA-binding proteins may recognize specific DNA regions either by reading the base sequence or by "indirect readout." How do these two modes of protein:DNA recognition differ?

6. The metallothionein promoter is illustrated in Figure 24.22. How long is this promoter, in nanometers? How many turns of B-DNA are found in this length of DNA? How many nucleosomes (approximately) would be bound to this much DNA? (Consult Chapter 8 to review the properties of nucleosomes.)

7. Describe why the ability of *bZIP* proteins to form heterodimers increases the repertoire of genes whose transcription might be responsive to regulation by these proteins.

8. Suppose exon 17 were deleted from the fast skeletal muscle troponin T gene (see Figure 24.39). How many different mRNAs could now be generated by alternative splicing? Suppose that exon 7 in a wild-type troponin T gene were duplicated. How many different mRNAs might be generated from a transcript of this new gene by alternative splicing?

FURTHER READING

Bailey, C. H., Bartsch, D., and Kandel, E. R., 1996. Toward a molecular definition of long-term memory storage. *Proceedings of the National Academy of Sciences, U.S.A.* **93**:13445–13452.

Barberis, A., et al., 1995. Contact with a component of the polymerase II holoenzyme suffices for gene activation. *Cell* **81**:359–368.

Bell, A., et al., 1998. The establishment of active chromatin domains. *Cold Spring Harbor Symposium on Quantitative Biology* **63**:509–514.

Berg, O. G., and von Hippel, P. H., 1988. Selection of DNA binding sites by regulatory proteins. *Trends in Biochemical Sciences* **13**:207–211. A discussion of the quantitative binding aspects of DNA:protein interactions.

Bjorklund, S., and Kim, Y. J., 1996. Mediator of transcriptional regulation. *Trends in Biochemical Sciences* **21**:335–337.

Bjorklund, S., et al., 1999. Global transcription regulators of eukaryotes. *Cell* **96**:759–767.

Breitbart, R. E., Andreadis, A., and Nadal-Ginard, B., 1987. Alternative splicing: A ubiquitous mechanism for the generation of multiple protein isoforms from single genes. *Annual Review of Biochemistry* **56**:467–495.

Brown, C. E., et al., 2000. The many HATs of transcriptional coactivators. *Trends in Biochemical Sciences* **25**:15–19.

Burley, S., 1998. X-ray crystallographic studies of eukaryotic transcription factors. *Cold Spring Harbor Symposium on Quantitative Biology* **63**:33–40.

Burley, S. K., and Roeder, R. G., 1996. Biochemistry and structural biology of transcription factor IID (TFIID). *Annual Review of Biochemistry* **65**:769–799.

Carey, M., and Smale, S. T., 2000. *Transcriptional Regulation in Eukaryotes: Concepts, Strategies, and Techniques.* New York: Cold Spring Harbor Laboratory Press.

Chan, C., Lonetto, M. A., and Gross, C. A., 1996. Sigma domain structure: One down, one to go. *Current Biology* **4**:1235–1238.

Conaway, R. C., and Conaway, J. W., 1999. Transcription elongation and human disease. *Annual Review of Biochemistry* **68**:301–319.

Cramer, D. et al., 2000. Architecture of RNA polymerase II and implications for the transcription mechanism. *Science* **288**:640–649.

Das, A., 1993. Control of transcription termination by DNA-binding proteins. *Annual Review of Biochemistry* **62**:893–930.

Dover, S. L., et al., 1997. Activation of prokaryotic transcription through arbitrary protein–protein contacts. *Nature* **386**:627–630.

Dreyfuss, G., et al., 1993. hnRNP proteins and the biogenesis of mRNA. *Annual Review of Biochemistry* **62**:289–321.

Farrell, S., et al., 1996. Gene activation by recruitment of RNA polymerase II holoenzyme. *Genes and Development* **10**:2359–2367.

Foulkes, N. S., and Sassone-Corsi, P., 1996. Transcription factors coupled to the cAMP-signalling pathway. *Biochimica et Biophysica Acta* **1288**:F101–F121.

Jacob, F., and Monod, J., 1961. Genetic regulatory mechanisms in the synthesis of proteins. *Journal of Molecular Biology* **3**:318–356. The classic paper presenting the operon hypothesis.

Johnson, P. F., and McKnight, S. L., 1989. Eukaryotic transcriptional regulatory proteins. *Annual Review of Biochemistry* **58**:799–839. A review of the structure and function of eukaryotic DNA-binding proteins that activate transcription.

Katonaga, J. T., 1998. Eukaryotic transcription: An interlaced network of transcription factors and chromatin-modifying machines. *Cell* **92**:306–313.

Kim, J. L., Nikolov, D. B., and Burley, S. K., 1993. Co-crystal structure of TBP recognizing the groove of a TATA element. *Nature* **365**:520–527.

Kim, Y., Geiger, J. H., Hahn, S., and Sigler, P. B., 1993. Crystal structure of a yeast TBP/TATA–box complex. *Nature* **365**:512–519.

Kornberg, R. D., 1998. Mechanism and regulation of yeast RNA polymerase II transcription. *Cold Spring Harbor Symposium on Quantitative Biology* **63**:229–233.

Kornberg, R. D., 1999. Eukaryotic transcriptional control. *Trends in Biochemical Sciences* **24**:M46-M49.

Kornberg, R. D., and Lorch, Y., 1999. Twenty-five years of the nucleosome, fundamental particle of the eukaryotic chromosome. *Cell* **98**:285–294.

Kramer, A., 1996. The structure and function of proteins involved in mammalian pre-mRNA splicing. *Annual Review of Biochemistry* **65**:367–409.

Kuo. M.-H., et al., 1998. Histone acetyltransferase activity of yeast Gcn5p is required for the activation of target genes *in vivo*. *Genes and Development* **12**:627–639.

Lee, T. I., et al., 2000. Redundant roles for TFIID and SAGA complexes in global transcription. *Nature* **405**:701–704.

Losick, R., 1998. Summary: Three decades after sigma. *Cold Spring Harbor Symposium on Quantitative Biology* **63**:653–668.

Malhtra, A., et al., 1996. Crystal structure of a σ_{70} subunit fragment from *E. coli* RNA polymerase. *Cell* **87**:127–136.

Matthews, K. S., 1992. DNA looping. *Microbiological Reviews* **56**:123–136. A review of DNA looping as a general mechanism in the regulation of gene expression.

Mizzen, C., et al., 1998. Signaling to chromatin through histone modifications: How clear is the signal? *Cold Spring Harbor Symposium on Quantitative Biology* **63**:469–481.

Montminty, M., 1997. Transcriptional regulation by cAMP. *Annual Review of Biochemistry* **66**:807–822.

Ng, H. H., and Bird, A., 2000. Histone deacylases: Silencers for hire. *Trends in Biochemical Sciences* **25**:121–126.

Pabo, C. O., and Sauer, R. T., 1992. Transcription factors: Structural families and principles of DNA recognition. *Annual Review of Biochemistry* **61**:1053–1095.

Patikoglou, G., and Burley, S. K., 1997. Eukaryotic transcription factor–DNA complexes. *Annual Review of Biophysics and Biomolecular Structure* **26**:289–325.

Platt, T., 1998. RNA structure in transcription elongation, termination, and antitermination. In *RNA Structure and Function*, Simons, R. W., and Grunberg-Monago, M., eds. Cold Spring Harbor, N.Y.: Cold Spring Harbor Press.

Polyakov, A., Severinova, E., and Darst, S. A., 1995. Three-dimensional structure of *E. coli* core RNA polymerase: Promoter binding and elongation conformation of the enzyme. *Cell* **83**:365–373.

Reinberg, D., et al., 1998. The RNA polymerase II general transcription factors: Past, present, and future. *Cold Spring Harbor Symposium on Quantitative Biology* **63**:83–103.

Roberts, C. W., and Roberts, J. W., 1996. Base-specific recognition of the nontemplate strand of promoter DNA by *E. coli* RNA polymerase. *Cell* **86**:495–501.

Roeder, R. G., 1996. The role of general initiation factors in transcription by RNA polymerase II. *Trends in Biochemical Sciences* **21**:327–335.

Rosbash, M., and Seraphin, B., 1991. Who's on first? The U1 snRNP59–splice site interaction and splicing. *Trends in Biochemical Sciences* **16**:187–190.

Schleif, R., 1992. DNA looping. *Annual Review of Biochemistry* **61**:199–223. An excellent review of DNA looping as a regulatory mechanism in gene expression.

Schnitzler, G. R., et al., 1998. A model for chromatin remodeling by the SW1/SNF family. *Cold Spring Harbor Symposium on Quantitative Biology* **63**:535–542.

Shadel, G. S., and Clayton, D. A., 1993. Mitochondrial transcription initiation: Variation and conservation. *Journal of Biological Chemistry* **268**:16083–16086.

Sharp, P. A., 1987. Splicing of messenger RNA precursors. *Science* **235**:766–771.

Steitz, T. A., 1990. Structural studies of protein–nucleic acid interaction: The sources of sequence-specific binding. *Quarterly Reviews of Biophysics* **23**:205–280. A comprehensive account of the structural properties of DNA-binding proteins and their interaction with DNA.

Struhl, K., 1998. Histone acetylation and transcriptional regulatory mechanisms. *Genes and Development* **12**:599–606.

Struhl, K., 1999. Fundamentally different logic of gene regulation in prokaryotes and eukaryotes. *Cell* **98**:1–4.

Struhl, K., et al., 1998. Activation and repression mechanisms in yeast. *Cold Spring Harbor Symposium on Quantitative Biology* **63**:413–421.

Suka, N., et al., 1998. The regulation of gene activity by histones and the histone deacylase RPD3. *Cold Spring Harbor Symposium on Quantitative Biology* **63**:391–399.

Thomas, M. J., et al., 1998. Transcription fidelity and proofreading by RNA polymerase II. *Cell* **93**:627–637.

Tully, T., 1997. Regulation of gene expression and its role in long-term memory and synaptic plasticity. *Proceedings of the National Academy of Sciences, U.S.A.* **94**:4239–4241.

Uptain, S. M., Kane, C. M., and Chamberlin, M. J., 1997. Basic mechanisms of transcription elongation and its regulation. *Annual Review of Biochemistry* **66**:117–172.

Utley, R. T., et al., 1998. Transcriptional activators direct histone acetyltransferase complexes to promoters. *Nature* **394**:498–502.

Vinson, C. R., Sigler, P. B., and McKnight, S. L., 1989. Scissors-grip model for DNA recognition by a family of leucine zipper proteins. *Science* **246**:911–916. A model for the interaction of a *bZIP* protein with its DNA site.

Wiener, A. M., 1993. mRNA splicing and autocatalytic introns: Distant cousins or the products of chemical determinism? *Cell* **72**:161–164. This article considers the possibility that small nuclear RNAs (such as U2 and U6) are the catalysts mediating mRNA splicing.

Winston, F., and Sudarsanam, P., 1998. The SAGA of Spt proteins and transcriptional analysis in yeast: Past, present, and future. *Cold Spring Harbor Symposium on Quantitative Biology* **63**:553–560.

Wu, C., et al., 1998. ATP-dependent remodeling of chromatin. *Cold Spring Harbor Symposium on Quantitative Biology* **63**:525–534.

Wu, J., and Grunstein, M., 2000. 25 years after the nucleosome model: Chromatin modifications. *Trends in Biochemical Sciences* **25**:619–623.

Zaman, Z., et al., 1998. Gene transcription by recruitment. *Cold Spring Harbor Symposium on Quantitative Biology* **63**:167–171.

Zlatanova, J., et al., 2000. Linker histone binding and displacement: Versatile mechanism for transcriptional regulation. *The FASEB Journal* **14**:1697–1704.

Protein Synthesis and Degradation

The Maya encoded their history in hieroglyphs carved on stelae and temples like these ruins in Tikal, Guatemala. Only recently have we been able to translate their meaning. (George Holton/ Photo Researchers. Inc.)

We are a spectacular, splendid manifestation of life. We have language and can build metaphors as skillfully and precisely as ribosomes make proteins. We have affection. We have genes for usefulness, and usefulness is about as close to a "common goal" of nature as I can guess at. And finally, and perhaps best of all, we have music.

LEWIS THOMAS (1913–1994), "The Youngest and Brightest Thing Around," in *The Medusa and the Snail*

Outline

W e turn now to the problem of how the sequence of nucleotides in an mRNA molecule is translated into the specific amino acid sequence of a protein. The problem raises both informational and mechanical questions. First, what is the genetic code that allows the information specified in a sequence of bases to be translated into the amino acid sequence of a polypeptide? That is, how is the 4-letter language of nucleic acids translated into the 20-letter language of proteins? Implicit in this question is a mechanistic problem: it is easy to see how base pairing establishes a one-to-one correspondence that allows the template-directed synthesis of polynucleotide chains in the processes of replication and transcription. However, there is no obvious chemical affinity between the purine and pyrimidine bases and the 20 different amino acids. Nor is there any structural complementarity or stereochemical connection between polynucleotides and amino acids that might guide the translation of information.

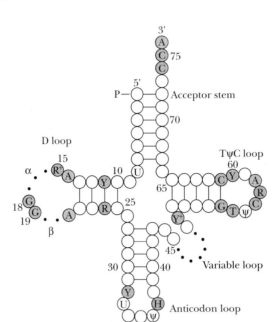

Figure 25.1 The general structure of tRNA molecules. Circles represent nucleotides in the tRNA sequence. The numbers given indicate the standardized numbering system for tRNAs (which differ in total number of nucleotides). Dots indicate places where the number of nucleotides may vary in different tRNA species. All tRNAs have the invariant 3-base sequence CCA at their 3′-ends. Recall from Chapter 8 that tRNA molecules often have modified or unusual bases.

Francis Crick reasoned that **adapter molecules** must bridge this information gap. These adapter molecules must interact specifically with both nucleic acids (mRNAs) and amino acids. At least 20 different adapter molecules would be needed, at least one for each amino acid. The various adapter molecules would be able to read the genetic code in an mRNA template and align the amino acids according to the template's directions so that they could be polymerized into a unique polypeptide. Transfer RNAs (tRNAs; Figure 25.1) are the adapter molecules (see Chapter 8). Amino acids are attached to the 3′-OH at the 3′-CCA end of tRNAs as aminoacyl esters. The formation of these aminoacyl-tRNAs, so-called "charged tRNAs," is catalyzed by specific **amino-acyl-tRNA synthetases.** There is one of these enzymes for each of the 20 amino acids, and each aminoacyl-tRNA synthetase loads its amino acid only onto those tRNAs intended to carry it. In turn, these tRNAs specifically recognize unique sequences of bases in the mRNA through complementary base pairing.

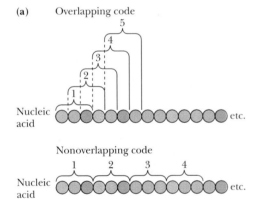

(a) Overlapping code

Nonoverlapping code

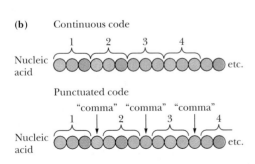

(b) Continuous code

Punctuated code

Figure 25.2 **(a)** An overlapping versus a nonoverlapping code. **(b)** A continuous versus a punctuated code.

25.1 The Genetic Code

Once it was realized that the sequence of bases in a gene specified the sequence of amino acids in a protein, various possibilities for the genetic code came under consideration. How many bases were necessary to specify each amino acid? Was the code overlapping or nonoverlapping (Figure 25.2)? Was the code punctuated or continuous? Mathematical considerations favored a triplet of bases as the minimal code word, or **codon,** for each amino acid: a doublet code based on pairs of the four possible bases, A, C, G, and U, has $4^2 = 16$ unique arrangements, an insufficient number to encode the 20 amino acids. A triplet code of four bases has $4^3 = 64$ possible code words, more than enough for the task.

The General Nature of the Genetic Code

The genetic code is a triplet code read continuously from a fixed starting point in each mRNA. It is defined by the following characteristics:

1. A group of three bases codes for one amino acid.
2. The code is not overlapping.
3. The base sequence is read from a fixed starting point without punctuation. That is, the mRNA sequences contain no "commas" signifying appropriate groupings of triplets. If the reading frame is displaced by one base, it remains shifted throughout the subsequent message; no "commas" are present to restore the "correct" frame.
4. The code is **degenerate,** meaning that, in most cases, each amino acid can be coded by any of several triplets. Recall that a triplet code yields 64 codons for 20 amino acids. Most codons (61 of 64) code for some amino acid.

The Genetic Code Translated

The complete translation of the genetic code is presented in Table 25.1. Codons, like other nucleotide sequences, are read $5' \rightarrow 3'$. Codons represent triplets of bases in mRNA or, replacing U with T, triplets along the nontranscribed (nontemplate) strand of DNA.

Table 25.1 The Genetic Code

First Position (5'-end)	Second Position				Third Position (3'-end)
	U	C	A	G	
U	UUU Phe	UCU Ser	UAU Tyr	UGU Cys	U
	UUC Phe	UCC Ser	UAC Tyr	UGC Cys	C
	UUA Leu	UCA Ser	UAA Stop	UGA Stop	A
	UUG Leu	UCG Ser	UAG Stop	UGG Trp	G
C	CUU Leu	CCU Pro	CAU His	CGU Arg	U
	CUC Leu	CCC Pro	CAC His	CGC Arg	C
	CUA Leu	CCA Pro	CAA Gln	CGA Arg	A
	CUG Leu	CCG Pro	CAG Gln	CGG Arg	G
A	AUU Ile	ACU Thr	AAU Asn	AGU Ser	U
	AUC Ile	ACC Thr	AAC Asn	AGC Ser	C
	AUA Ile	ACA Thr	AAA Lys	AGA Arg	A
	AUG Met*	ACG Thr	AAG Lys	AGG Arg	G
G	GUU Val	GCU Ala	GAU Asp	GGU Gly	U
	GUC Val	GCC Ala	GAC Asp	GGC Gly	C
	GUA Val	GCA Ala	GAA Glu	GGA Gly	A
	GUG Val	GCG Ala	GAG Glu	GGG Gly	G

* AUG signals translation initiation as well as coding for Met residues.

Third-Base Degeneracy Is Color-Coded

Third-Base Relationship	Third Bases with Same Meaning	Number of Codons
Third base irrelevant	U, C, A, G	32 (8 families)
Purines	A or G	12 (6 pairs)
Pyrimidines	U or C	14 (7 pairs)
Three out of four	U, C, A	3 (AUX = Ile)
Unique definitions	G only	2 (AUG = Met) (UGG = Trp)
Unique definition	A only	1 (UGA = Stop)

Several noteworthy features characterize the genetic code:

1. *All the codons have meaning.* Sixty-one of the 64 codons specify particular amino acids. The remaining three—UAA, UAG, and UGA—specify no amino acid and thus are **nonsense codons.** Nonsense codons serve as **termination codons**—they are "stop" signals indicating that the end of the protein has been reached.

2. *The genetic code is unambiguous.* Each of the 61 "sense" codons encodes only one amino acid.

3. *The genetic code is degenerate.* With the exception of Met and Trp, every amino acid is coded by more than one codon. Several—Arg, Leu, and Ser—are represented by six different codons. Codons coding for the same amino acid are called **synonymous codons.**

4. *Codons representing the same amino acid or chemically similar amino acids tend to be similar in sequence.* Often the third base in a codon is irrelevant, so that, for example, all four codons in the GGX family specify Gly, and the UCX family specifies Ser (see Table 25.1). This feature is known as **third-base degeneracy.** Note also that codons with a pyrimidine as second base likely encode amino acids with hydrophobic side chains, and codons with a purine in the second-base position typically specify polar or charged amino acids. The two negatively charged amino acids, Asp and Glu, are encoded by GAX codons; GA-pyrimidine gives Asp and GA-purine specifies Glu. The consequence of these similarities is that mutations are less likely to be harmful because single-base changes in a codon will result either in no change or in a substitution with an amino acid similar to the original amino acid. The degeneracy of the code is evolution's buffer against mutational disruption.

5. *The genetic code is "universal."* Although certain minor exceptions in codon usage occur (see *A Deeper Look,* below), the more striking feature of the code is its universality: codon assignments are virtually the same through-

A DEEPER LOOK

Natural Variations in the Standard Genetic Code

The genomes of some lower eukaryotes, prokaryotes, and mitochondria show some exceptions to the standard genetic code (see Table 25.1) in codon assignments. The phenomenon is more common in mitochondria. For example, the termination codon UGA codes for tryptophan in mitochondria from various animals, protozoans, and fungi. AUA, normally an Ile codon, codes for methionine in some animal and fungal mitochondrial genomes, and AGA (an Arg codon) is a termination codon in vertebrate mitochondria but is a Ser codon in fruit fly mitochondria. Mitochondria in several species of yeast use the CUX codons to specify Thr instead of Leu. Some yeast and algal mitochondria use CGG, normally an Arg codon, as a stop codon.

Less common are genomic codon variations within the genomes of prokaryotic and eukaryotic cells. Among the lower eukaryotes, certain ciliated protozoans (*Tetrahymena* and *Paramecium*) use UAA and UGA as glutamine codons rather than as stop codons. Instances in prokaryotes include use of the stop codon UGA to specify Trp by *Mycoplasma.* Perhaps most interesting is the use of some UGA codons by both prokaryotes and eukaryotes (including humans)

to specify **selenocysteine,** an analog of cysteine in which the sulfur atom is replaced by a selenium atom. Indeed, the identification of selenocysteine residues in proteins from bacteria, archaea, and eukaryotes has led some people to nominate selenocysteine as the twenty-first amino acid! Selenocysteine formation requires a novel selenocysteine-specific tRNA known as tRNA^{Sec}. This tRNA^{Sec} is loaded with a Ser residue by seryl-tRNA synthetase, the aminoacyl-tRNA synthetase for serine. Then, in an ATP-dependent process, the Ser-O is replaced by Se. Translation of certain selected UGA codons by selenocysteinyl-tRNA^{Sec} requires additional proteins and formation of a stable stem-loop secondary structure in the mRNA next to the particular UGA codon.

$$H-Se-CH_2-\underset{\underset{^+NH_3}{|}}{\overset{\overset{H}{|}}{C}}-C\underset{O^-}{\overset{O}{<}}$$

Selenocysteine

Adapted from Fox, T. D., 1987. Natural variation in the genetic code. *Annual Review of Genetics* **21:**67–91; Knight, R. D., et al., 1999. Selection, history, and chemistry: Three faces of the genetic code. *Trends in Biochemical Sciences* **24:**241–247; and Low, S. C., and Berry, M. J., 1996. Knowing when not to stop. *Trends in Biochemical Sciences* **21:**203–208.

out all organisms—archaea, eubacteria, and eukaryotes. This conformity means that all extant organisms use the same genetic code, providing strong evidence that they all evolved from a common primordial ancestor.

<table>
<tr><td>**25.2**</td><td>**The "Second" Genetic Code: Aminoacyl-tRNA Synthetase Recognition of the Proper Substrates**</td></tr>
</table>

Codon recognition is achieved by aminoacyl-tRNAs. In order for accurate translation to occur, the appropriate aminoacyl-tRNA must "read" the codon through base pairing via its **anticodon loop** (see Chapter 8). Once an aminoacyl-tRNA has been synthesized, the amino acid part makes no contribution to accurate translation of the mRNA. Thus, a **second genetic code** must exist, the code by which each aminoacyl-tRNA synthetase discriminates between the 20 amino acids and the many tRNAs and uniquely picks out its proper substrates— one specific amino acid and the tRNA(s) appropriate to it—from among the more than 400 possible combinations. The appropriate tRNA(s) are those having anticodons that can base-pair with the codon(s) specifying the particular amino acid. Clearly, the proper amino acids must be loaded onto the various tRNAs so that the mRNA is translated with fidelity. Although the primary genetic code is key to understanding the central dogma of molecular biology on how DNA encodes proteins, the second genetic code is just as crucial to the fidelity of information transfer.

Cells have 20 different aminoacyl-tRNA synthetases, one for each amino acid. Each of these enzymes catalyzes ATP-dependent attachment of its specific amino acid to the 3′-end of its **cognate tRNA molecules** (Figure 25.3). The aminoacyl-tRNA synthetase reaction serves two purposes:

cognate kindred; in this sense, cognate refers to those tRNAs having anticodons that can read one or more of the codons that specify one particular amino acid

1. It activates the amino acid so that it will readily react to form a peptide bond.
2. It bridges the information gap between amino acids and codons.

The underlying mechanisms of molecular recognition used by each amino-acyl-tRNA synthetase to bring the proper amino acid to its cognate tRNA are the embodiment of the second genetic code.

The Two Classes of Aminoacyl-tRNA Synthetases

Despite their common enzymatic function, aminoacyl-tRNA synthetases are a diverse group of proteins in terms of size, amino acid sequence, and oligomeric structure. In higher eukaryotes, at least some aminoacyl-tRNA synthetases assemble into large multiprotein complexes. The aminoacyl-tRNA synthetases fall into two fundamental classes on the basis of similar amino acid sequence motifs, oligomeric state, and acylation function (Table 25.2): class I and class II. Class I aminoacyl-tRNA synthetases first add the amino acid to the 2′-OH of the terminal adenylate residue of tRNA before shifting it to the 3′-OH; class II enzymes add it directly to the 3′-OH (see Figure 25.3). Evidently, the catalytic domains of these enzymes evolved from two different ancestral predecessors. Aminoacyl-tRNA synthetases are ranked among the oldest proteins because different forms of these enzymes were present very early in evolution. Class I and class II aminoacyl-tRNA synthetases interact with the tRNA 3′-terminal CCA and acceptor stem in a mirror-symmetric fashion with respect to each other (Figure 25.4). Class I enzymes bind to the tRNA acceptor stem helix from the minor-groove side, whereas class II enzymes bind it from the major-groove side.

Both class I and class II aminoacyl-tRNA synthetases can be approximated as two-domain structures, as can their L-shaped tRNA substrates, which have

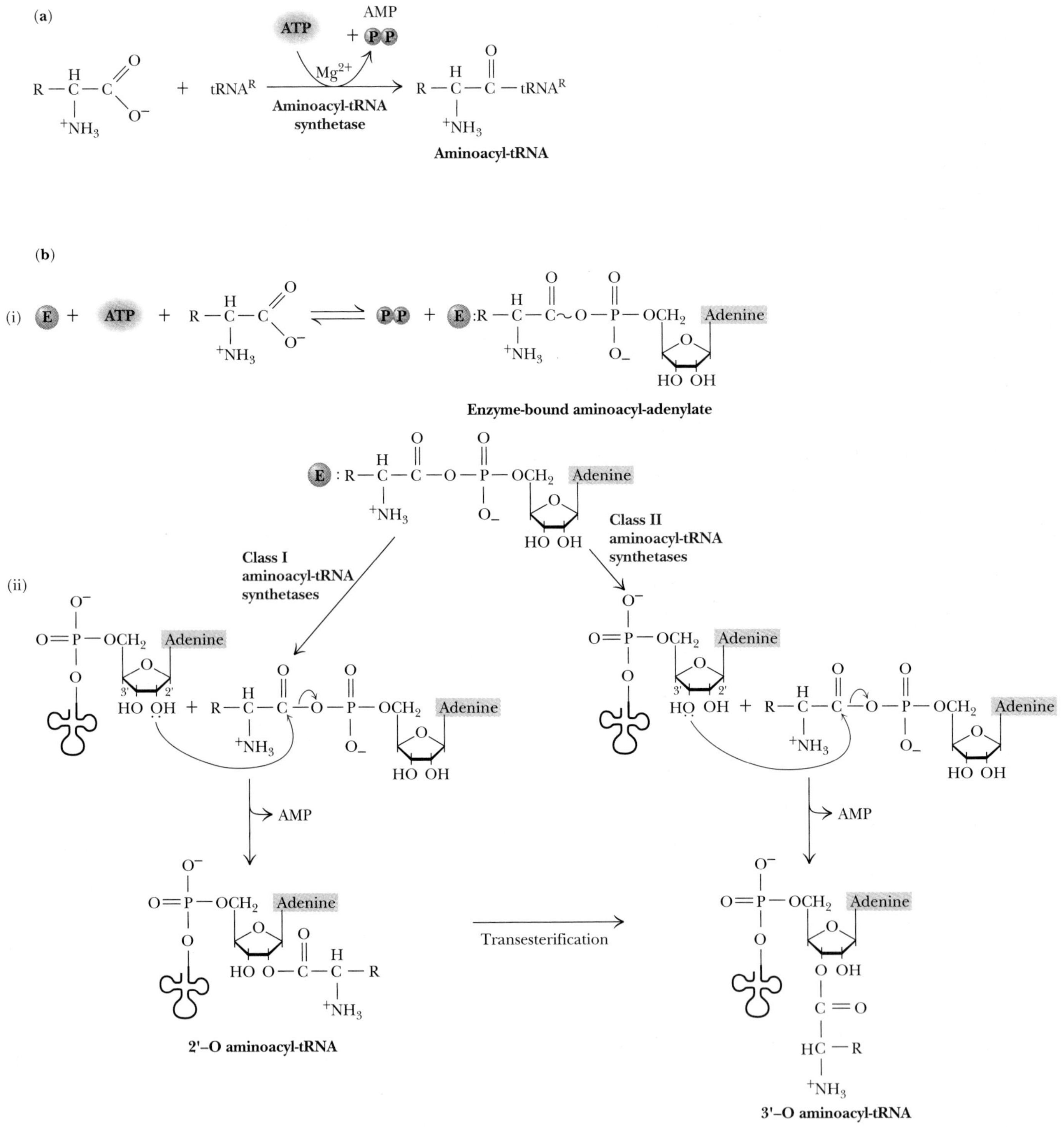

Figure 25.3 The aminoacyl-tRNA synthetase reaction. **(a)** The overall reaction. Ever-present pyrophosphatases in cells quickly hydrolyze the PP$_i$ produced in the aminoacyl-tRNA synthetase reaction, rendering aminoacyl-tRNA synthesis thermodynamically favorable and essentially irreversible. **(b)** The overall reaction commonly proceeds in two steps: (i) formation of an aminoacyl-adenylate, and (ii) transfer of the activated amino acid moiety of the mixed anhydride to either the 2′-OH (class I aminoacyl-tRNA synthetases) or 3′-OH (class II aminoacyl-tRNA synthetases) of the ribose on the terminal adenylic acid at the 3′-CCA terminus common to all tRNAs. Those aminoacyl-tRNAs formed as 2′-aminoacyl esters undergo a transesterification that moves the aminoacyl function to the 3′-O of tRNA. Only the 3′-esters are substrates for protein synthesis.

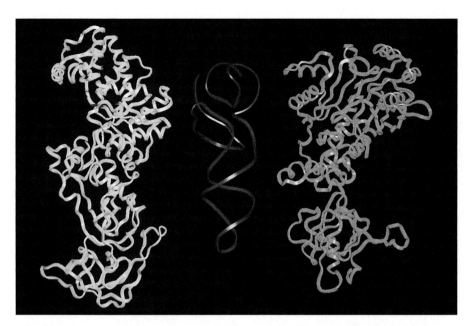

Figure 25.4 Mirror-symmetric interactions of class I versus class II aminoacyl-tRNA synthetases with their tRNA substrates. The two different classes of aminoacyl-tRNA synthetases bind to opposite faces of tRNA molecules. On the left is a space-filling model of the class I glutaminyl-tRNA^{Gln} synthetase. Class I synthetases bind to the side of their tRNA substrates shown as closest in this figure (the model tRNA structure is tRNA^{Phe} for purposes of illustration). On the right is a space-filling model of the class II aspartyl-tRNA^{Asp} synthetase; this class of synthetase binds to the side of tRNA closest to it here. *(Adapted from Arnez, J. G., and Moras, D., 1997. Structural and functional considerations of the aminoacylation reaction.* Trends in Biochemical Sciences **22**:211–216, Figure 5.)

 See *Interactive Biochemistry CD-ROM and Workbook,* pages 89, 94, and 95

the acceptor stem/CCA-3′-OH at one end and the anticodon stem-loop at the other (see Figures 8.55 and 25.5). This L-shaped tertiary structure of tRNAs separates the 3′-CCA acceptor end from the anticodon loop by a distance of 7.6 nm. The two domains of tRNAs have distinct functions: The 3′-CCA end is the site of aminoacylation, and the anticodon-containing domain interacts with the mRNA template. The two domains of tRNAs interact with the separate domains in the synthetases. One of the two major aminoacyl-tRNA synthetase domains is the catalytic domain (which defines the difference between class I and class II enzymes); this domain interacts with the tRNA 3′-CCA end. The other major domain in aminoacyl-tRNA synthetases is highly variable and interacts with parts of the tRNA beyond the acceptor-TΨC stem-loop domain, including, in some cases, the anticodon.

Table 25.2 The Two Classes of Aminoacyl-tRNA Synthetases

Class I	Class II
Arg	Ala
Cys	Asn
Gln	Asp
Glu	Gly
Ile	His
Leu	Lys
Met	Phe
Trp	Pro
Tyr	Ser
Val	Thr

Selective tRNA Recognition by Aminoacyl-tRNA Synthetases

Aside from the need to uniquely recognize their cognate amino acids, aminoacyl-tRNA synthetases must be able to discriminate between the various tRNAs. The structural features that permit the synthetases to recognize and amino-acylate their cognate tRNA(s) are *not* universal. That is, a common set of rules does not govern tRNA recognition by these enzymes. Most surprising is the fact that the recognition features are not limited to the anticodon and, in some instances, do not even include the anticodon. For most tRNAs, a set of sequence elements is recognized by its specific aminoacyl-tRNA synthetase, rather than a single distinctive nucleotide or base pair. These elements include one or more of the following: (a) at least one base in the anticodon; (b) one or more of

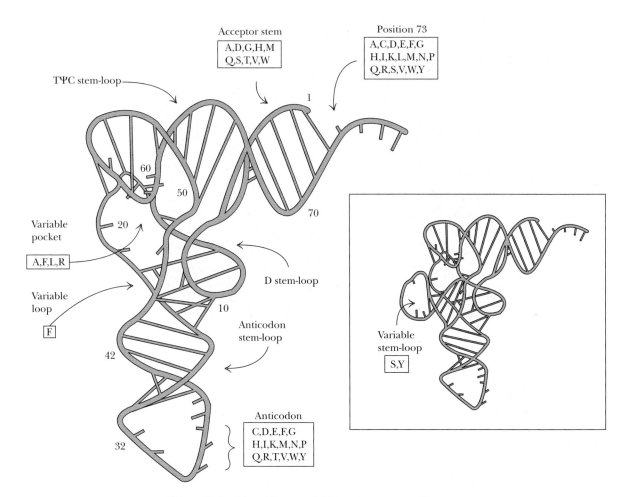

Figure 25.5 Ribbon diagram of tRNA tertiary structure. Numbers represent the consensus nucleotide sequence (see Figure 25.1). The locations of nucleotides recognized by the various aminoacyl-tRNA synthetases are indicated; shown within the boxes are one-letter designations of the amino acids whose respective aminoacyl-tRNA synthetases interact at the discriminator base (position 73), acceptor stem, variable pocket and/or loop, or anticodon. The inset shows additional recognition sites in those tRNAs having a variable loop that forms a stem-loop structure. *(Adapted from Saks, M. E., Sampson, J. R., and Abelson, J. N., 1994. The transfer RNA problem: A search for rules.* Science **263***:191–197, Figure 2.)*

the three base pairs in the acceptor stem; and (c) the base at canonical position 73 (the unpaired base preceding the CCA end), referred to as the **discriminator base** because this base is invariant in the tRNAs for a particular amino acid. Figure 25.5 presents a ribbon diagram of a tRNA molecule showing the common location of nucleotides that contribute to specific recognition by the respective aminoacyl-tRNA synthetases for each of the 20 amino acids. Interestingly, the same set of tRNA features that serves as positive determinants for binding and aminoacylation of the tRNA by its cognate aminoacyl-tRNA synthetase may act as negative determinants that prohibit binding and aminoacylation by other (noncognate) aminoacyl-tRNA synthetases. Because no common set of rules exists, the second genetic code is an **operational code** based on aminoacyl-tRNA synthetase recognition of varying sequence and structural features in the different tRNA molecules during the operation of aminoacyl-tRNA synthesis. Some examples of this code are given in Figure 25.6.

Figure 25.6 Major identity elements in four tRNA species. Each base in the tRNA is represented ▶ by a circle. Numbered filled circles indicate positions of identity elements within the tRNA that are recognized by its specific aminoacyl-tRNA synthetase. *(Adapted from Schulman, L. H., and Abelson, J., 1988. Recent excitement in understanding transfer RNA identity. Science* **240:***1591–1592.)*

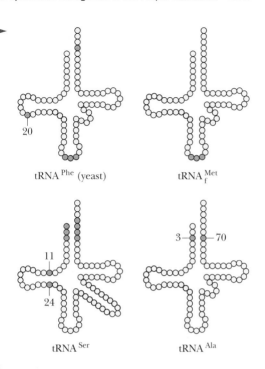

tRNA ^{Phe} (yeast)   tRNA $_f^{Met}$

tRNA ^{Ser} tRNA ^{Ala}

tRNA Recognition Sites in *E. coli* Glutaminyl-tRNA ^{Gln} Synthetase

E. coli glutaminyl-tRNA ^{Gln} synthetase, a class I enzyme, provides a good illustration of aminoacyl-tRNA synthetase : cognate tRNA interactions. This glutaminyl-tRNA ^{Gln} synthetase shares a continuous interaction with its cognate tRNA that extends from the anticodon to the acceptor stem along the entire inside of the L-shaped tRNA (Figure 25.7). Specific recognition elements include enzyme contacts with the discriminator base, acceptor stem, and anticodon, particularly the central U in the CUG anticodon. The carboxylate group of Asp ^{235} makes sequence-specific H bonds in the tRNA minor groove with the 2-NH $_2$ group of G3 in the base pair G3 : C70 of the acceptor stem. A mutant glutaminyl-tRNA ^{Gln} synthetase with Asn substituted for Asp at position 235 shows relaxed specificity; that is, it now will acylate noncognate tRNAs with Gln.

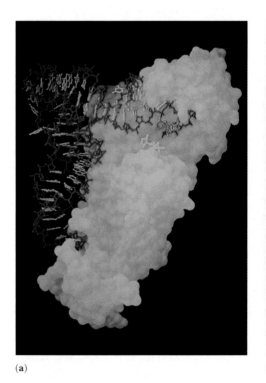

(a)

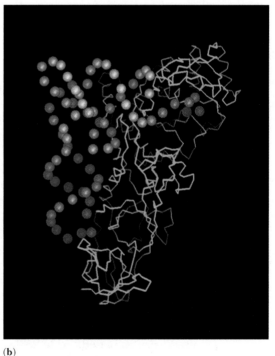

(b)

Figure 25.7 **(a)** Representation of the solvent-accessible surface of *E. coli* glutaminyl-tRNA ^{Gln} synthetase complexed with tRNA ^{Gln} and ATP, derived from analysis of the crystal structure of the complex. The protein is colored blue. The sugar–phosphate backbone of the tRNA is red; its bases are yellow. The protein : tRNA contact region extends along one side of the entire length of this extended protein. The acceptor stem of the tRNA and the ATP (green) fit into a cleft at the top of the protein in this view. The enzyme also interacts extensively with the anticodon (lower tip of tRNA ^{Gln}). **(b)** Diagram showing the structure of tRNA ^{Gln}, as represented by its phosphorus atoms (purple spheres), in complex with *E. coli* glutaminyl-tRNA ^{Gln} synthetase, as represented in the terms of its C $_\alpha$ atoms (blue). *(a, adapted from Rould, M. A., et al., 1989. Structure of E. coli glutaminyl-tRNA synthetase complexed with tRNA ^{Gln} and ATP at 2.8 Å resolution. Science* **246:***1135. Photo courtesy of Thomas A. Steitz of Yale University.)*

See pages 89, 94, and 95

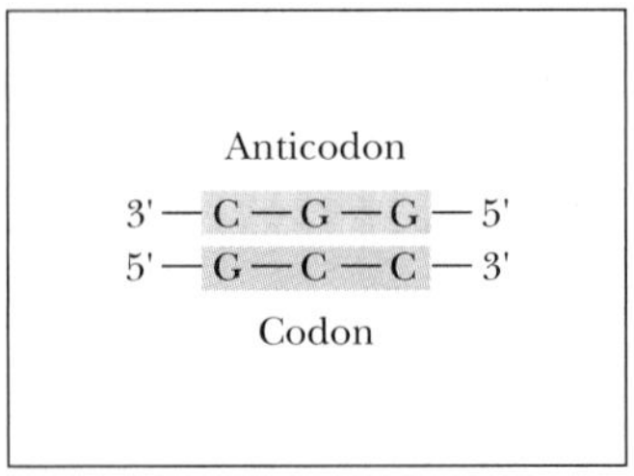

Figure 25.8 Codon–anticodon pairing. Complementary trinucleotide sequence elements align in antiparallel fashion.

Table 25.3 Base Pairing Possibilities at the Third Position of the Codon

Base on the Anticodon	Bases Recognized on the Condon
U	A, G
C	G
A	U
G	U, C
I	U, C, A

Source: Adapted from Crick, F. H. C., 1966. Codon–anticodon pairing: The wobble hypothesis. *Journal of Molecular Biology* **19**:548–555.

25.3 Codon–Anticodon Pairing and Codon Usage

Protein synthesis depends on the codon-directed binding of the proper aminoacyl-tRNAs so that the right amino acids are sequentially aligned according to the specifications of the mRNA undergoing translation. This alignment is achieved via codon–anticodon pairing in antiparallel orientation (Figure 25.8). However, considerable degeneracy exists in the genetic code at the third position. Conceivably, this degeneracy could be handled in either of two ways: (a) codon–anticodon recognition could be highly specific so that a complementary anticodon is required for each codon, or (b) fewer than 61 anticodons could be used for the "sense" codons if certain allowances were made in the base-pairing rules. Then, some anticodons could recognize more than one codon. As early as 1965, it was known that poly (U) bound *all* the Phe-tRNA^{Phe} even though UUC is also a Phe codon. The phenylalanine-specific tRNAs could recognize both UUU and UUC. Also, one particular yeast tRNA^{Ala} was able to bind to three codons: GCU, GCC, and GCA.

Francis Crick considered these results and tested alternative base-pairing possibilities by model building. He hypothesized that the first two bases of the codon and the last two bases of the anticodon form canonical Watson–Crick A:U or G:C base pairs but that pairing between the third base of the codon and the first base of the anticodon follows less stringent rules. That is, a certain amount of play, or **wobble,** might occur in base pairing at this position. Thus, the third base of the codon is sometimes referred to as the **wobble position.**

Crick's investigations suggested a set of rules for pairing between the third base of the codon and the first base of the anticodon (Table 25.3). The wobble rules indicate that a first-base anticodon U could recognize either an A or G in the codon third-base position, a first-base anticodon G could recognize either U or C in the third-base position of the codon, and a first-base anticodon I[1] could interact with U, C, or A in the codon third position.[2]

The wobble rules also predict that four-codon families (like Pro or Thr), where any of the four bases may be in the third position, require at least two different tRNAs. However, all members of the set of tRNAs specific for a particular amino acid—termed **isoacceptor tRNAs**—are served by one aminoacyl-tRNA synthetase.

Codon Usage

Because more than one codon exists for most amino acids, the possibility for variation in codon usage arises. Indeed, variation in codon usage accommodates the fact that the DNA of different organisms varies in relative A:T/G:C content. Nevertheless, even in organisms of average base composition, codon usage may be biased. Table 25.4 gives some examples from *E. coli* and humans reflecting the nonrandom usage of codons. Of over 109,000 Leu codons tabulated here from various human genes, CUG was used over 48,000 times, CUC over 23,000 times, but UUA just 6000 times.

[1]I is inosine (6-OH purine).
[2]Thus, the first base of the anticodon indicates whether the tRNA can read one, two, or three different codons: anticodons beginning with A or C read only one codon, those beginning with G or U read two, while anticodons beginning with I can read three codons.

Table 25.4 Representative Examples of Codon Usage in *E. coli* and Human Genes

The results are expressed as frequency of occurrence of a codon per 1000 codons tabulated in 1562 *E. coli* genes and 2681 human genes, respectively. (Because *E. coli* and human proteins differ somewhat in amino acid composition, the frequencies for a particular amino acid do not correspond exactly between the two species.)

Amino Acid	Codon	*E. coli* Gene Frequency/1000	Human Gene Frequency/1000
Leu	CUA	3.2	6.1
	CUC	9.9	20.1
	CUG	54.6	42.1
	CUU	10.2	10.8
	UUA	10.9	5.4
	UUG	11.5	11.1
Pro	CCA	8.2	15.4
	CCC	4.3	20.6
	CCG	23.8	6.8
	CCU	6.6	16.1
Ala	GCA	15.6	14.4
	GCC	34.4	29.7
	GCG	32.9	7.2
	GCU	13.4	18.9
Lys	AAA	36.5	21.9
	AAG	12.0	35.2
Glu	GAA	43.5	26.4
	GAG	19.2	41.6

Adapted from Wada, K., et al., 1992. Codon usage tabulated from Genbank genetic sequence data. *Nucleic Acids Research* **20**:2111–2118.

The occurrence of codons in *E. coli* mRNAs correlates well with the relative abundance of the tRNAs that read them. Preferred codons are represented by the most abundant isoacceptor tRNAs. Further, mRNAs for proteins that are synthesized in abundance tend to employ preferred codons. Rare tRNAs correspond to rarely used codons, and messages containing such codons might experience delays in translation.

Nonsense Suppression

Mutations that alter a sense codon to one of the three nonsense codons—UAA, UAG, or UGA—result in premature termination of protein synthesis and the release of truncated (incomplete) polypeptides. Geneticists found that second mutations elsewhere in the genome were able to *suppress* the effects of nonsense mutations so that the organism survived. The molecular basis for such *intergenic suppression* was a mystery until it was realized that **suppressors** were mutations in tRNA genes that altered the anticodon so that the mutant tRNA could now read a particular "stop" codon and insert an amino acid. For example, alteration of the anticodon of a tRNA^Tyr from GUA to CUA allows this tRNA to read the *amber* stop codon, UAG, and insert Tyr. (The nonsense codons are named *amber* [UAG], *ochre* [UAA], and *opal* [UGA]). **Suppressor tRNAs** are typically generated from minor tRNA species within a set of isoacceptor tRNAs, so their recruitment to a new role via mutation does not involve loss of an essential tRNA; that is, the mutation is not particularly deleterious to the organism. Suppressor tRNAs don't necessarily carry the same amino acid as the tRNA produced by the unmutated tRNA gene.

25.4 Protein Synthesis: Ribosome Structure and Assembly

Protein biosynthesis is achieved by the process of **translation.** Translation converts the language of genetic information embodied in the base sequence of a messenger RNA molecule into the amino acid sequence of a polypeptide chain. During translation, proteins are synthesized on ribosomes by linking amino acids together in the specific linear order stipulated by the sequence of codons in an mRNA. Ribosomes are the agents of protein synthesis.

Ribosomes are compact ribonucleoprotein particles found in the cytosol of all cells as well as in the matrix of mitochondria and the stroma of chloroplasts. The general structure of ribosomes is described in Chapter 8; here we consider their structure in light of their function in synthesizing proteins. Ribosomes are mechano-chemical systems that move along mRNA templates, orchestrating the interactions between successive codons and the corresponding anticodons presented by aminoacyl-tRNAs. As they align successive amino acids via codon–anticodon recognition, ribosomes also catalyze the formation of peptide bonds between adjacent amino acid residues.

The Composition of Prokaryotic Ribosomes

Escherichia coli ribosomes are representative of the structural organization of the prokaryotic versions of these supramolecular protein-synthesizing machines (Table 25.5; see also Figure 8.23). The *E. coli* ribosome is a roughly globular particle with a diameter of 25 nm, a sedimentation coefficient of 70S, and a mass of about 2520 kD. It consists of two unequal subunits that dissociate from each other at Mg^{2+} concentrations below 1 mM. The smaller, or **30S,** subunit is composed of 21 different proteins and a single rRNA, **16S ribosomal RNA (rRNA).** The larger **50S** subunit consists of 31 different proteins and two rRNAs: **23S rRNA** and **5S rRNA.** Ribosomes are roughly two-thirds RNA and one-third protein by mass. An *E. coli* cell contains around 20,000 ribosomes, constituting about 20% of the dry cell mass.

Ribosomal Proteins

There is one copy of each ribosomal protein per 70S ribosome, excepting protein L7/L12 (L7 and L12 have identical amino acid sequences and differ only in the degree of N-terminal acetylation). Only one protein is common to both the small and large subunit: S20 = L26. The largest ribosomal protein is S1 (557 residues, 61.2 kD); the smallest is L34 (46 residues, 5.4 kD). The sequences of ribosomal proteins share little similarity. These proteins are typically rich in the cationic amino acids Lys and Arg and have few aromatic amino acid

Table 25.5 Structural Organization of *E. coli* Ribosomes

	Ribosome	Small Subunit	Large Subunit
Sedimentation coefficient	70S	30S	50S
Mass (kD)	2520	930	1590
Major RNAs		16S = 1542 bases	23S = 2904 bases
Minor RNAs			5S = 120 bases
RNA mass (kD)	1664	560	1104
RNA proportion	66%	60%	70%
Protein number		21 polypeptides	31 polypeptides
Protein mass (kD)	857	370	487
Protein proportion	34%	40%	30%

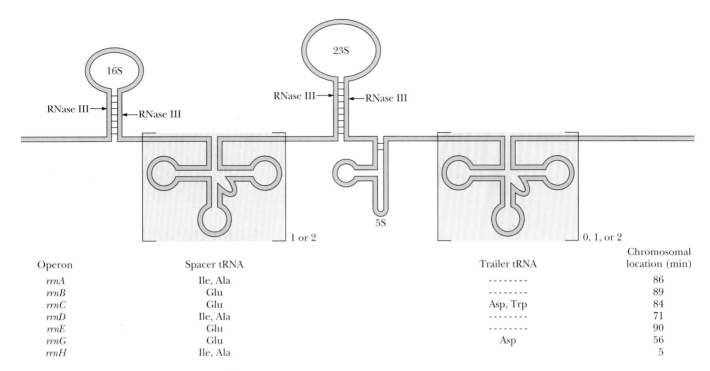

Operon	Spacer tRNA	Trailer tRNA	Chromosomal location (min)
rrnA	Ile, Ala	--------	86
rrnB	Glu	--------	89
rrnC	Glu	Asp, Trp	84
rrnD	Ile, Ala	--------	71
rrnE	Glu	--------	90
rrnG	Glu	Asp	56
rrnH	Ile, Ala		5

Figure 25.9 The seven ribosomal RNA operons in *E. coli*. These operons, or gene clusters, are transcribed to give a precursor RNA that is subsequently cleaved by RNase III and other nucleases, at the sites indicated, to generate 23S, 16S, and 5S rRNA molecules, as well as several tRNAs that are unique to each operon. Numerals to the right of the brackets indicate the number of species of tRNA encoded by each transcript.

residues, properties appropriate to proteins intended to interact strongly with polyanionic RNAs.

rRNAs

The three *E. coli* rRNA molecules—23S, 16S, and 5S—are derived from a single 30S rRNA precursor transcript that also includes several tRNAs (Figure 25.9). Ribosomal RNAs show extensive potential for intrachain hydrogen bonding and assume secondary structures reminiscent of tRNAs although substantially more complex (see Figures 8.56–8.58). About two-thirds of rRNA is double-helical. Double-helical regions are punctuated by short, single-stranded stretches, generating hairpin conformations that dominate the molecule. The three-dimensional structures of both the 30S and 50S ribosomal subunits show that the general shapes of the ribosomal subunits are determined by the confomation of the rRNA molecules within them. Figure 25.10 gives an indication of the three-dimensional structure of 16S rRNA within the 30S subunit, revealing that the size and shape of this structure is essentially that of the rRNA. Apparently, ribosomal proteins serve no function other than to brace and stabilize the rRNA conformations within the ribosomal subunits.

Self-Assembly of Ribosomes

Ribosomal subunit self-assembly is one of the paradigms for the spontaneous formation of supramolecular complexes from their macromolecular components. If the individual proteins and rRNAs composing ribosomal subunits are

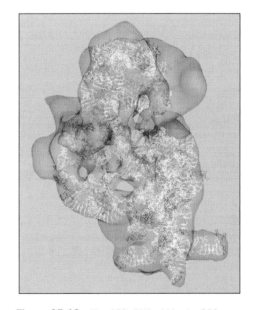

Figure 25.10 The 16S rRNA within the 30S ribosomal subunit. This view is of the "solvent side" of the 30S subunit (the side opposite to the 50S subunit-binding side—see Figure 25.13). *(Adapted from Mueller, F., and Brimacombe, R., 1997. A new model for the three-dimensional folding of Escherichia coli 16S ribosomal RNA: I. Fitting the RNA to a 3D electron microscopic map at 20 Å. Journal of Molecular Biology **271**:524-544, Figure 2. Figure courtesy of Florian Mueller and Richard Brimacombe, Max-Planck-Institute for Molecular Genetics, Berlin.)*

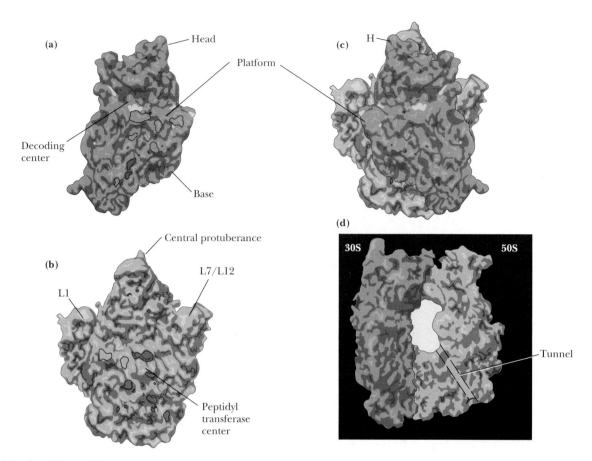

Figure 25.11 Structure of the *E. coli* ribosomal subunits and 70S ribosome, as deduced by X-ray crystallography. Prominent structural features are labeled. **(a)** and **(b)** present views of the 30S (a) and 50S (b) subunits. These views show the sides of these two that form the interface between them when they come together to form a 70S subunit **(c)**. **(d)** is a side view of the 70S ribosome; the white area represents the region where mRNA and tRNAs are bound and peptide bond formation occurs. The tunnel through the 50S subunit that the growing peptide chain transits is shown as a dashed line. The approximate dimensions of the 30S subunit are $5.5 \times 22 \times 22$ nm; the 50S subunit dimensions are $15 \times 20 \times 20$ nm. *(Adapted from Figures 2 and 3 in Cate, J. H., et al., 1999. X-ray crystal structures of 70S ribosomal functional complexes. Science **285**:2095–2104.)*

mixed together *in vitro* under appropriate conditions of pH and ionic strength, spontaneous self-assembly into functionally competent subunits takes place without the intervention of any additional factors or chaperones. The rRNA acts as a scaffold upon which the various ribosomal proteins convene. Ribosomal proteins bind in a specified order.

Ribosomal Architecture

Ribosomal subunits have a characteristic three-dimensional architecture that has been revealed by image reconstructions from cryoelectron microscopy, X-ray crystallography, and X-ray and neutron solution scattering. Such analyses have led to the images depicted in Figure 25.11. The 30S, or small, subunit features a "head" and a "base" from which a "platform" projects. A cleft is defined by the spatial relationship between the head, base, and platform (Figure 25.11a). The mRNA passes across this cleft. The platform represents the central domain of the 30S subunit; it contains one-third of the 16S rRNA. This central domain binds mRNA and the anticodon stem-loop end of aminoacyl-tRNAs, providing the framework for decoding the genetic information in mRNA by mediating codon–anticodon recognition. As such, this central domain of the 30S subunit serves as the **decoding center.** This center is composed only of 16S rRNA; no ribosomal proteins are involved in decoding the message.

The 50S, or large, subunit is a mittlike globular structure with three distinctive projections: a "central protuberance," the "stalk" containing protein L1, and a winglike ridge known as the "L7/L12 region" (Figure 25.11b). The large subunit binds the aminoacyl-acceptor ends of the tRNAs and is responsible for catalyzing formation of the peptide bonds between successive amino acids in the polypeptide chain. This catalytic center, the **peptidyl transferase,** is located at the bottom of a deep cleft. From it, a 10-nm-long tunnel passes outward through the back of the large subunit.

The small and large subunits associate with each other in the manner shown in Figure 25.11(c) and (d). The contacts between the 30S and 50S subunits are rather limited, and the subunit interface contains mostly rRNA, with relatively little contribution from ribosomal proteins. The decoding center in the 30S subunit is aligned somewhat with the peptidyl transferase and the tunnel in the large subunit, and the growing peptidyl chain is threaded through this tunnel as protein synthesis proceeds. Even though the ribosomal proteins are arranged peripherally around the rRNAs in ribosomes, rRNA occupies 30 to 40% of the ribosomal subunit surface areas.

Eukaryotic Ribosomes

Eukaryotic cells have ribosomes in their mitochondria (and chloroplasts) as well as in the cytosol. The mitochondrial and chloroplastic ribosomes resemble prokaryotic ribosomes in size, overall organization, structure, and function, a fact reflecting the prokaryotic origins of these organelles. Whereas eukaryotic cytosolic ribosomes retain many of the structural and functional properties of their prokaryotic counterparts, they are larger and considerably more complex. Further, higher eukaryotes have more complex ribosomes than lower eukaryotes. For example, the yeast cytosolic ribosomes have major rRNAs of 3392 (large subunit) and 1799 nucleotides (small subunit); the major rRNAs of mammalian cytosolic ribosomes are 4718 and 1874 nucleotides, respectively. Table 25.6 lists the properties of cytosolic ribosomes in a representative mammal, the rat. Comparison of base sequences and secondary structures of rRNAs from different organisms suggests that evolution has worked to conserve the secondary structure of these molecules, although not necessarily the nucleotide sequences creating such structure. That is, the retention of a base pair at a particular location seems more important than whether the base pair is G:C or A:U.

25.5 The Mechanics of Protein Synthesis

Like chemical polymerization processes, protein biosynthesis in all cells is characterized by three distinct phases: initiation, elongation, and termination. At each stage, the energy driving the assembly process is provided by GTP hydrolysis, and specific soluble protein factors participate in the events.

Table 25.6 Structural Organization of Mammalian (Rat Liver) Cytosolic Ribosomes

	Ribosome	Small Subunit	Large Subunit
Sedimentation coefficient	80S	40S	60S
Mass (kD)	4220	1400	2820
Major RNAs		18S = 1874 bases	28S = 4718 bases
Minor RNAs			5.8S = 160 bases
			5S = 120 bases
RNA mass (kD)	2520	700	1820
RNA proportion	60%	50%	65%
Protein number		33 polypeptides	49 polypeptides
Protein mass (kD)	1700	700	1000
Protein proportion	40%	50%	35%

Initiation involves binding of mRNA by the small ribosomal subunit, followed by association of a particular **initiator aminoacyl-tRNA** that recognizes the first codon. This codon often lies within the first 30 nucleotides or so of mRNA spanned by the small subunit. The large ribosomal subunit then joins the initiation complex, preparing it for the elongation stage.

Elongation includes the synthesis of all peptide bonds from the first to the last. The ribosome remains associated with the mRNA throughout elongation, moving along it and translating its message into an amino acid sequence. This is accomplished via a repetitive cycle of events in which successive aminoacyl-tRNAs add to the ribosome : mRNA complex as directed by codon binding, and the polypeptide chain grows by one amino acid at a time.

Three tRNA molecules may be associated with the ribosome : mRNA complex at any moment. Each lies in a distinct site (Figure 25.12). The **A,** or **acceptor, site** is the attachment site for an incoming aminoacyl-tRNA. The **P,** or **peptidyl, site** is occupied by peptidyl-tRNA, the tRNA carrying the growing polypeptide chain. The elongation reaction transfers the peptide chain from the peptidyl-tRNA in the P site to the aminoacyl-tRNA in the A site. This transfer occurs through covalent attachment of the α-amino group of the aminoacyl-tRNA to the α-carboxyl group of the peptidyl-tRNA, forming a new peptide bond. The new, longer peptidyl-tRNA now moves from the A site into the P site as the ribosome moves one codon further along the mRNA. The A site,

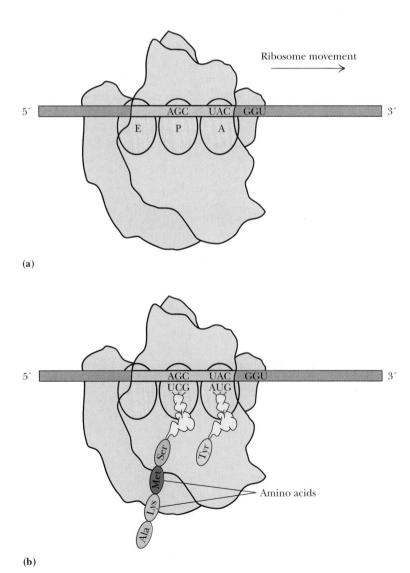

(a)

(b)

Figure 25.12 The basic steps in protein synthesis. The ribosome has three distinct binding sites for tRNA: the A, or acceptor, site; the P, or peptidyl, site; and the E, or exit, site. **(a)** Both ribosomal subunits contribute to each tRNA-binding site. The sites are aligned with codons in the mRNA. **(b)** Prior to peptide bond formation, an aminoacyl-tRNA is present in the A site and polypeptidyl-tRNA is in the P site. (continued)

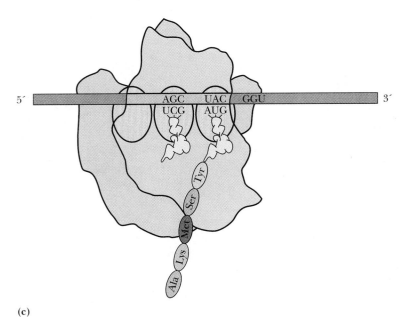

(c)

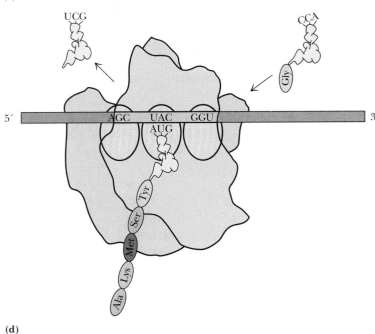

(d)

Figure 25.12 (continued) **(c)** Peptide bond formation involves transfer of the polypeptide chain to the amino group of the amino acid carried by the tRNA in the A site. **(d)** The ribosome then translocates one codon further along the mRNA and the uncharged tRNA exits via the E site. Translocation places the polypeptidyl-tRNA in the P site and aligns a new codon within the A site, ready to accept the next incoming aminoacyl-tRNA.

left vacant by this translocation, can accept the next incoming aminoacyl-tRNA. The **E,** or **exit, site** is transiently occupied by the "unloaded," or deacylated, tRNA which has lost its peptidyl chain through the peptidyl transferase reaction. These events are summarized in Figure 25.12. The contributions made to each of the three tRNA-binding sites by each ribosomal subunit are shown in Figure 25.13.

Termination is triggered when the ribosome reaches a "stop" codon on the mRNA. At this point, the polypeptide chain is released, and the ribosomal subunits dissociate from the mRNA.

Protein synthesis proceeds rapidly. In vigorously growing bacteria, about 20 amino acid residues are added to a growing polypeptide chain each second, so an average protein molecule of about 300 amino acid residues is synthesized in only 15 seconds. Eukaryotic protein synthesis is only about 10% as fast. Protein synthesis is also highly accurate: an inappropriate amino acid is incorporated only once in every 10^4 codons. We focus first on protein synthesis in *E. coli*, the system about which we know the most.

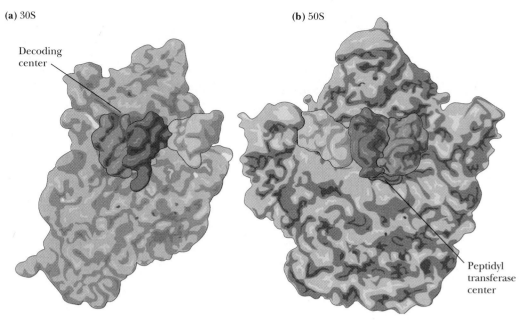

(a) 30S   **(b)** 50S

Decoding center

Peptidyl transferase center

Figure 25.13 The three tRNA-binding sites on ribosomes. The view shows the ribosomal surfaces that form the interface between the 30S **(a)** and 50S **(b)** subunits in a 70S ribosome (as though a 70S ribosome has been "opened" like a book to expose facing "pages"). The A, P, and E sites are occupied by green, blue, and yellow tRNAs, respectively. The decoding center on the 30S subunit (a) lies behind the top of the tRNAs in the A and P sites (which is where the anticodon ends of the tRNAs are located). The peptidyl transferase center on the 50S subunit (b) lies at the lower tips (acceptor ends) of the A-and P-site tRNAs. *(Adapted from Figure 5 in Cate, J. H., et al., 1999. X-ray crystal structures of 70S ribosome functional complexes. Science* **285:***2095–2104.)*

Peptide Chain Initiation in Prokaryotes

The components required for peptide chain initiation include (a) mRNA, (b) 30S and 50S ribosomal subunits, (c) a set of proteins known as **initiation factors,** (d) GTP, and (e) a specific charged initiator tRNA, **f-Met-tRNA$_i^{fMet}$.** A discussion of the properties of these components and their interaction follows.

Initiator tRNA

tRNA$_i^{fMet}$ is a particular tRNA for reading an AUG (or GUG, or even UUG) codon that signals the start site, or N-terminus, of a polypeptide chain; the $_i$ signifies "initiation." This tRNA$_i^{fMet}$ does not read internal AUG codons, so it does not participate in chain elongation. Instead, that role is filled by another methionine-specific tRNA, referred to as tRNA^{Met}, which cannot replace tRNA$_i^{fMet}$ in peptide chain initiation. (However, both of these tRNAs are loaded with Met by the same methionyl-tRNA synthetase.) The structure of *E. coli* tRNA$_i^{fMet}$ has several distinguishing features (Figure 25.14). Collectively, these features identify this tRNA as essential to initiation and inappropriate for chain elongation.

The synthesis of all *E. coli* polypeptides begins with the incorporation of a modified methionine residue, *N*-formyl-Met, as N-terminal amino acid. However, in about half of the *E. coli* proteins, this Met residue is removed once the growing polypeptide is 10 or so residues long; as a consequence, many mature proteins in *E. coli* lack N-terminal Met.

The methionine contributed in peptide chain initiation by tRNA$_i^{fMet}$ is unique in that its amino group has been formylated. This reaction is catalyzed by a specific enzyme, **methionyl-tRNA$_i^{fMet}$ formyl transferase** (Figure 25.15). Note that the addition of the formyl group to the α-amino group of Met creates an N-terminal block resembling a peptidyl grouping. That is, the initiating Met is transformed into a minimal analog of a peptidyl chain.

mRNA Recognition and Alignment

In order for the mRNA to be translated accurately, its sequence of codons must be brought into proper register with the translational apparatus of the ribo-

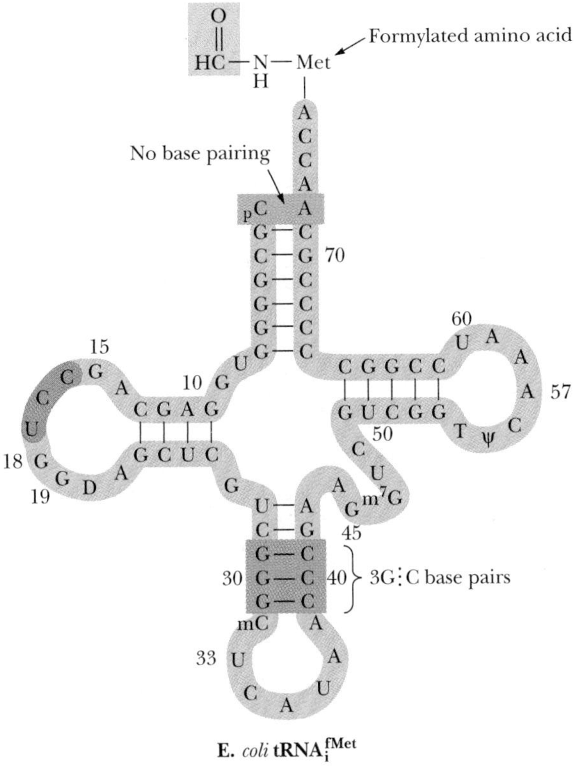

Figure 25.14 The structure of E. coli N-formyl-methionyl-tRNA$_i^{fMet}$. The features distinguishing it from noninitiator tRNAs are highlighted.

some. Recognition of translation initiation sequences on mRNAs involves the 16S rRNA component of the 30S ribosomal subunit. Base pairing between a pyrimidine-rich sequence at the 3′-end of 16S rRNA and complementary purine-rich tracts at the 5′-end of prokaryotic mRNAs positions the 30S ribosomal subunit in proper alignment with an initiation codon on the mRNA. The purine-rich mRNA sequence, the **ribosome-binding site,** is often called the **Shine–Dalgarno sequence** in honor of its discoverers. Figure 25.16 shows various Shine-Dalgarno sequences found in prokaryotic mRNAs, along with the complementary 3′-tract on E. coli 16S rRNA. The 3′-end of 16S rRNA resides in the "head" region of the 30S small subunit.

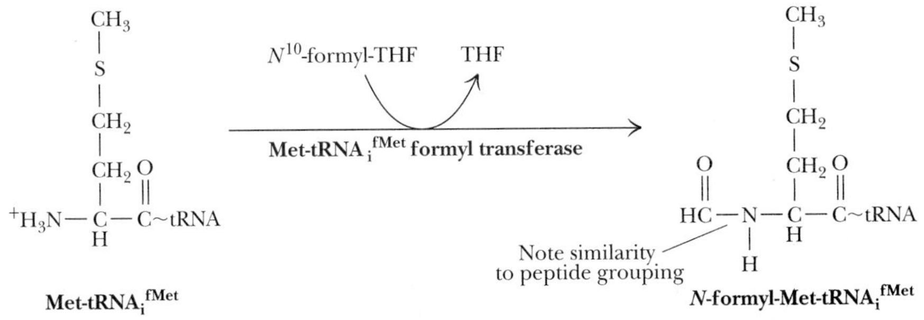

Figure 25.15 Methionyl-tRNA$_i^{fMet}$ formyl transferase catalyzes the transformylation of methionyl-tRNA$_i^{fMet}$ using N^{10}-formyl-THF as formyl donor. The tRNA for reading Met codons within a protein (tRNA^{Met}) is not a substrate for this transformylase.

Initiation
codon

araB	– U U U G G A U G G A G U G A A A C G A U G G C G A U U –
galE	– A G C C U A A U G G A G C G A A U U A U G A G A G U U –
lacI	– C A A U U C A G G G U G G U G A U U G U G A A A C C A –
lacZ	– U U C A C A C A G G A A A C A G C U A U G A C C A U G –
Q β phage replicase	– U A A C U A A G G A U G A A A U G C A U G U C U A A G –
ϕX174 phage A protein	– A A U C U U G G A G G C U U U U U U A U G G U U C G U –
R17 phage coat protein	– U C A A C C G G G G U U U G A A G C A U G G C U U C U –
ribosomal protein S12	– A A A A C C A G G A G C U A U U U A A U G G C A A C A –
ribosomal protein L10	– C U A C C A G G A G C A A A G C U A A U G G C U U U A –
trpE	– C A A A A U U A G A G A A U A A C A A U G C A A A C A –
trpL leader	– G U A A A A A G G G U A U C G A C A A U G A A A G C A –
3'-end of 16S rRNA	3' _HO_A U U C C U C C A C U A G – 5'

Figure 25.16 Various Shine–Dalgarno sequences recognized by *E. coli* ribosomes. These sequences lie about 10 nucleotides upstream from their respective AUG initiation codon and are complementary to the UCCU core sequence element of *E. coli* 16S rRNA. G:U as well as canonical G:C and A:U base pairs are involved here.

Initiation Factors

Initiation involves interaction of the **initiation factors (IFs)** with GTP, *N*-formyl-Met- tRNA$_i^{fMet}$, mRNA, and the 30S subunit to give a **30S initiation complex** to which the 50S subunit then adds to form a **70S initiation complex.** The initiation factors are soluble proteins required for assembly of proper initiation complexes. Their properties are summarized in Table 25.7.

Events in Initiation

Initiation begins when a 30S subunit : (IF-3 : IF-1) complex binds mRNA and a complex of IF-2, GTP, and f-Met-tRNA$_i^{fMet}$. The sequence of events is summarized in Figure 25.17. Although IF-3 is absolutely essential for mRNA binding by the 30S subunit, it is not involved in locating the proper translation initiation site on the message. The presence of IF-3 on 30S subunits also prevents them from reassociating with 50S subunits. IF-3 must dissociate before the 50S subunit will associate with the mRNA : 30S subunit complex.

IF-2 delivers the initiator f-Met-tRNA$_i^{fMet}$ in a GTP-dependent process. Apparently, the 30S subunit is aligned with the mRNA such that the initiation codon is situated within the "30S part" of the P site. Upon binding, f-Met-tRNA$_i^{fMet}$ enters this 30S portion of the P site. GTP hydrolysis is necessary to form an active 70S ribosome. GTP hydrolysis is triggered when the 50S subunit joins and is accompanied by IF-1 and IF-2 release. The A site of the *70S initiation complex* is ready to accept an incoming aminoacyl-tRNA; the 70S ribosome is poised to begin chain elongation.

Peptide Chain Elongation

The requirements for peptide chain elongation are (a) an mRNA : 70S ribosome : peptidyl-tRNA complex (peptidyl-tRNA in the P site), (b) aminoacyl-

Table 25.7 Properties of *E. coli* Initiation Factors

Factor	Mass (kD)	Molecules/ Ribosome	Function
IF-1	9	0.15	Assists IF-3 function
IF-2	97		Binds initiator tRNA and GTP
IF-3	23	0.25	Binds to 30S subunits and directs mRNA binding

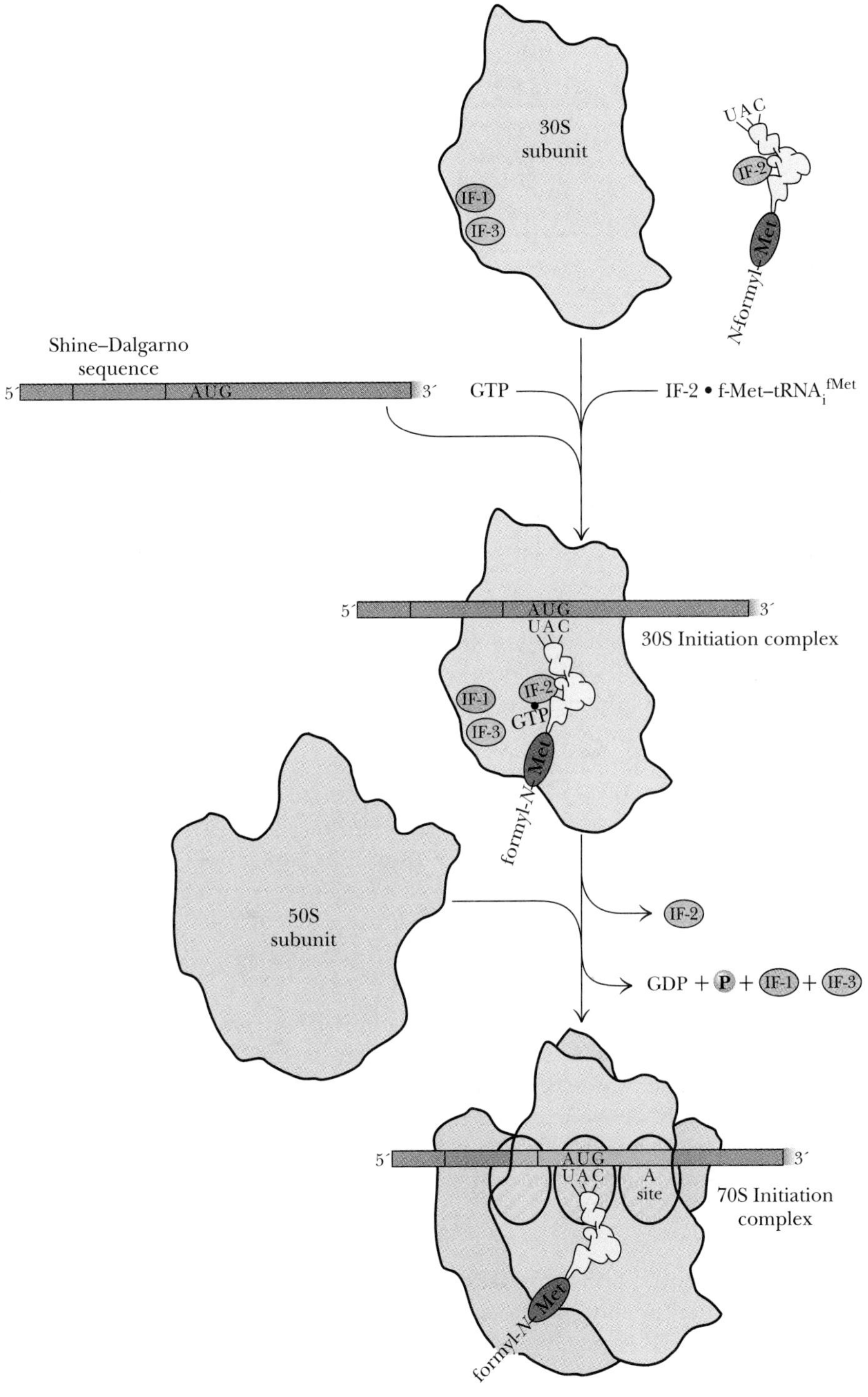

Figure 25.17 The sequence of events in peptide chain initiation.

tRNAs, (c) a set of proteins known as **elongation factors,** and (d) GTP. Chain elongation can be divided into three principal steps:

1. Codon-directed binding of the incoming aminoacyl-tRNA at the A site
2. Peptide bond formation: transfer of the peptidyl chain from the tRNA bearing it to the —NH$_2$ group of the new amino acid
3. Translocation of the "one-residue-longer" peptidyl-tRNA to the P site to make room for the next aminoacyl-tRNA at the A site. These shifts are coupled with movement of the ribosome one codon further along the mRNA.

Table 25.8 **Properties of *E. coli* Elongation Factors**

Factor	Mass (kD)	Molecules/ Cell	Function
EF-Tu	43	70,000	Binds aminoacyl-tRNA in presence of GTP
EF-Ts	74	10,000	Displaces GDP from EF-Tu
EF-G	77	20,000	Binds GTP, promotes translocation of ribosome along mRNA

The Elongation Cycle

The properties of the soluble proteins essential to peptide chain elongation are summarized in Table 25.8. These proteins are present in large quantities, reflecting the great importance of protein synthesis to cell vitality. For example, **elongation factor Tu (EF-Tu)** is the most abundant protein in *E. coli*, accounting for 5% of total cellular protein.

Aminoacyl-tRNA Binding

EF-Tu binds aminoacyl-tRNA and GTP. There is only one EF-Tu species serving all the different aminoacyl-tRNAs, and aminoacyl-tRNAs are accessible to the A site of active 70S ribosomes only in the form of aminoacyl-tRNA : EF-Tu : GTP complexes. Once correct base pairing between codon and anticodon has been established, the GTP is hydrolyzed to GDP and P_i, the aminoacyl-tRNA enters the A site, and the EF-Tu molecule is released as a EF-Tu : GDP complex (Figure 25.18).

Elongation factor Ts (EF-Ts) promotes the recycling of EF-Tu by mediating the displacement of GDP from EF-Tu and its replacement by GTP. EF-Ts accomplishes its job through entry into a transient complex with EF-Tu. GTP then displaces EF-Ts from EF-Tu (Figure 25.18).

The Decoding Center: A 16S rRNA Function

Analysis of the structures of the 70S ribosome : tRNA complexes and isolated 30S subunits has revealed the decoding center in the 30S subunit. This decoding center, where anticodon loops of the A-site and P-site tRNAs and the codons of the mRNA are matched up, is a property of 16S rRNA. No ribosomal proteins segments are found in this region; only the 16S rRNA conformation determines how codon : anticodon interactions occur (Figure 25.19).

Peptidyl Transfer

Peptidyl transfer, or **transpeptidation,** is the central reaction of protein synthesis, the actual peptide bond-forming step. No energy input (for example, in the form of ATP) is needed; the ester bond linking the peptidyl moiety to

Figure 25.19 The decoding center of the 30S ribosomal subunit is composed only of 16S rRNA. (*Left*) The 30S subunit, as viewed from the 50S subunit. The circle (cyan) shows the latch structure of the 30S subunit that encircles and encloses the mRNA. The mRNA enters along the path indicated by the arrow and follows a groove along this face of the subunit. The location of the decoding center is indicated by the red circle. (*Right*) Enlarged view of the decoding center. Helix H44 of the 16S rRNA is at the bottom, in the olive-shaded region. Two codons of the mRNA (the blue-shaded region) are immediately above H44. Three anticodon loops are shown above the mRNA codons: The A-site tRNA anticodon loop is in green; the P-site tRNA anticodon loop is in purple; and the E-site tRNA anticodon loop is in gray. *(Adapted from Figure 3 in Schluenzen, F., et al., 2000. Structure of the functionally activated small ribosomal subunit at 3.3 Å resolution. Cell **102**:615–623. Figure courtesy of Francois Franceschi, Max-Planck-Institut für Molekulare Genetik, Berlin.)*

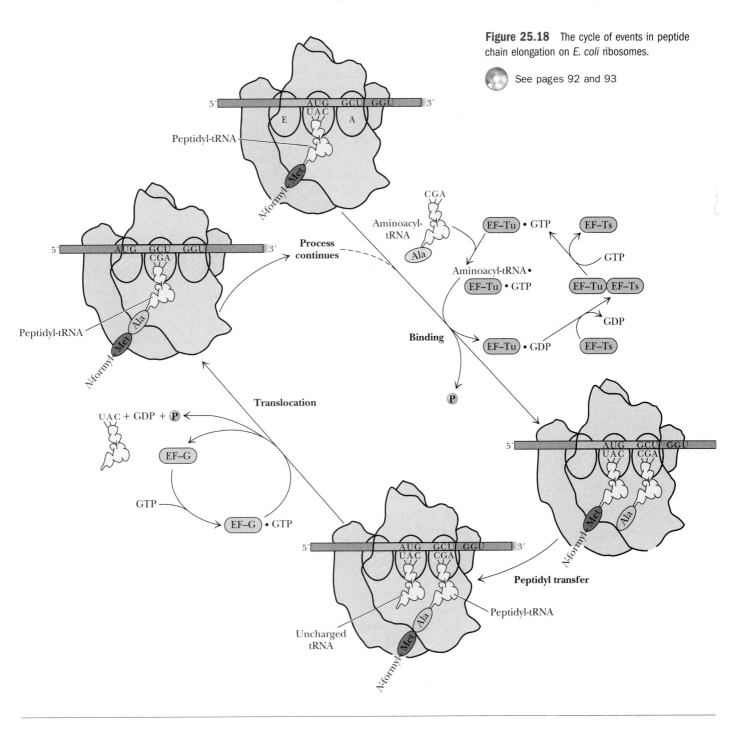

See pages 92 and 93

Figure 25.18 The cycle of events in peptide chain elongation on *E. coli* ribosomes.

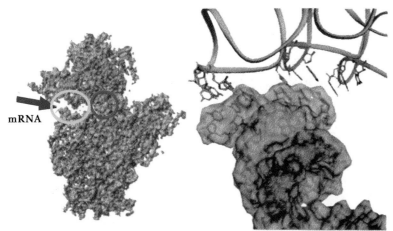

tRNA is intrinsically reactive. As noted earlier, *peptidyl transferase,* the activity catalyzing peptide bond formation, is associated with the 50S ribosomal subunit. Indeed, this reaction is a property of the 23S rRNA in the 50S subunit.

23S rRNA Is the Peptidyl Transferase Enzyme

The reaction catalyzed by the **peptidyl transferase center** of 23S rRNA is depicted in Figure 25.20. The lone electron pair on a ring N atom of the purine A^{2451} of 23S rRNA serves as a general base, abstracting a proton from the α-amino group on the A-site aminoacyl-tRNA, thus facilitating nucleophilic attack by this amino group on the carbonyl-C of the P-site peptidyl-tRNA, leading to peptide bond formation and transfer of the peptidyl chain to the aminoacyl-tRNA. The peptidyl transferase region of 23S rRNA is shown in Figure 25.21. Nucleotide sequences in this region of 23S rRNA are among the most highly conserved in all biology.

Translocation

Three things remain to be accomplished in order to return the active 70S ribosome:mRNA complex to the starting point in the elongation cycle:

1. The deacylated tRNA must be removed from the P site.
2. The peptidyl-tRNA must be moved (translocated) from the A site to the P site.
3. The ribosome must move one codon down the mRNA so that the next codon is positioned in the A site.

The precise events in translocation are still being resolved, but several distinct steps are clear. Within the 70S ribosome, the anticodon ends of both the A-site and P-site tRNAs interact with the **decoding center** of the 30S subunit, and mRNA codon:tRNA anticodon recognition takes place. In contrast, the acceptor ends (the aminoacylated ends) of both A-site and P-site tRNAs interact with the **peptidyl transferase** center of the 50S subunit. Because the growing peptidyl chain doesn't move during peptidyl transfer, the acceptor end of the A-site aminoacyl-tRNA must move into the P site as its aminoacyl function picks up the peptidyl chain. At the same time, the acceptor end of the deacylated P-site tRNA is shunted into the E site (see Figure 25.13). Then, the mRNA and the anticodon ends of tRNAs move together with respect to the 30S subunit so that the mRNA is passively dragged one codon further through the ribosome. With this movement, the anticodon end of the now one-residue-longer peptidyl-tRNA goes from the A site of the 30S subunit to the P site. Concomitantly, the anticodon end of the deacylated tRNA is moved into the E site. These movements are catalyzed by the translocation protein **elongation factor G (EF-G),** which couples the energy of GTP hydrolysis to movement. Note that translocation of the mRNA relative to the 30S subunit will deliver the next codon to the 30S A site.

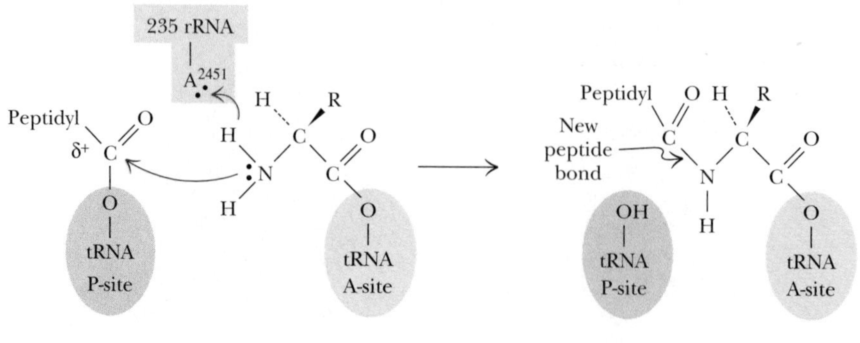

Figure 25.20 Peptide bond formation in protein synthesis. Nucleophilic attack by the α-amino group of the A-site aminoacyl-tRNA on the carbonyl-C of the P-site peptidyl-tRNA is facilitated when 23S rRNA purine A^{2451} abstracts a proton.

Figure 25.21 The peptidyl transferase center of 23S rRNA. 23S rRNA has a highly conserved secondary structure (see Figure 8.58). Only sequences implicated in the peptidyl transferase function are presented here. This region corresponds to region V of 23S rRNA. Numbers indicate base positions in the 23S rRNA nucleotide sequence. Green dots symbolize bases in the central region that are not conserved in 23S rRNAs from different sources. Purine A^{2451}, the catalytic purine in the peptidyl transferase reaction, is highlighted. 23S rRNA sites involved in interactions with peptidyl-tRNA and aminoacyl-tRNA, the substrates of protein synthesis, are indicated. Residue G2252 forms close contacts with the CCA-end of the P-site (peptidyl) tRNA, and residue G2553 forms close contacts with the CCA-end of the A-site (aminoacyl) tRNA. The loops in which these residues are found are labeled P loop and A loop, respectively. The results indicate that the tertiary structure of 23S rRNA brings the P-loop–bound peptidyl-tRNA into an interaction with base U2585 in the central region. *(Adapted from Pace, N. R., 1992. New horizons for RNA catalysis. Science* **256**:*1402–1403; Porse, B., et al., 1997. The donor substrate site within the peptidyl transferase loop of 23S rRNA and its putative interactions with the CCA-end of N-blocked aminoacyl-tRNA. Journal of Molecular Biology* **266**:*472-483; and Green, R., et al., 1998. Ribosome-catalyzed peptide-bond formation with an A-site substrate covalently linked to 23S ribosomal RNA. Science* **280**:*286–289.)*

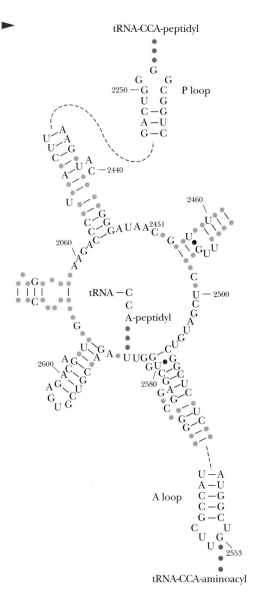

EF-G binds to the ribosome as an EF-G:GTP complex. GTP hydrolysis is essential not only for translocation but also for subsequent EF-G dissociation. Because EF-G and EF-Tu compete for a common binding site on the ribosome (the **factor-binding center**) adjacent to the A site, EF-G release is a prerequisite for return of the 70S ribosome:mRNA to the beginning point in the elongation cycle.

In this simple model of peptidyl transfer and translocation, the ends of both tRNAs move relative to the two ribosomal subunits in two discrete steps, the acceptor ends moving first and then the anticodon ends. Further, the readjustments needed to reposition the ribosomal subunits relative to the mRNA and to one another imply that the 30S and 50S subunits must move relative to one another. *This model provides a convincing explanation for why ribosomes are universally organized into a two-subunit structure: the small and large subunits must move relative to each other, as opposed to moving as a unit, in order to carry out the process of translation.*

GTP Hydrolysis Fuels the Conformational Changes That Drive Ribosomal Functions

Two GTPs are hydrolyzed for each amino acid residue incorporated into peptide during chain elongation, one upon EF-Tu–mediated binding of aa-tRNA and one more in translocation. The role of GTP (with EF-Tu as well as EF-G) is mechanical, analogous to the role of ATP in driving muscle contraction (see Chapter 13). GTP binding induces conformational changes in ribosomal components that actively engage these components in the mechanics of protein synthesis; subsequent GTP hydrolysis followed by GDP and P_i release relax the system back to the initial conformational state so that another turn in the cycle can take place. The energy expenditure for protein synthesis is at least four high-energy phosphoric anhydride bonds per amino acid. In addition to the two provided by GTP, two from ATP are expended in amino acid activation via aminoacyl-tRNA synthesis (see Figure 25.3).

Peptide Chain Termination

The elongation cycle of polypeptide synthesis continues until the 70S ribosome encounters a "stop" codon. At this point, polypeptidyl-tRNA occupies the P site and the arrival of a "stop" or nonsense codon in the A site signals that the end

A DEEPER LOOK

Molecular Mimicry: The Structures of EF-Tu : Aminoacyl-tRNA and EF-G

EF-Tu and EF-G compete for binding to ribosomes. EF-Tu has the unique capacity to recognize and bind any aminoacyl-tRNA and deliver it to the ribosome in a GTP-dependent reaction. EF-G catalyzes GTP-dependent translocation. The ternary structure of the *Thermus aquaticus* EF-Tu : Phe-tRNA^{Phe} complexed with GMPPNP (a nonhydrolyzable analog of GTP) is remarkably similar to the structure of EF-G : GDP (see figure). EF-Tu is a three-domain protein (shown in red, green, and light blue here); a space-filling representation of its bound tRNA is shown in magenta. EF-G is a six-domain protein, five of which correspond to the overall EF-Tu : tRNA structure. Domains 1 and 2 in EF-G correspond to the EF-Tu protein domains colored red and green; domains 3, 4,

and 5 of EF-G show a striking structural resemblance to the tRNA component in EF-Tu : tRNA and are also colored magenta. (EF-G has an extra domain, colored dark blue here.) Thus, parts of the EF-G structure mimic the structure of a tRNA molecule.

One view of early evolution suggests that RNA was the primordial macromolecule, fulfilling all biological functions, including those of catalysis and information storage that are now assumed for the most part by proteins and DNA. The mimicry of EF-Tu : tRNA by EF-G may represent a fossil of early macromolecular evolution when the proteins first began to take over some functions of RNA by mimicking shapes known to work as RNAs.

See pages 92 and 93

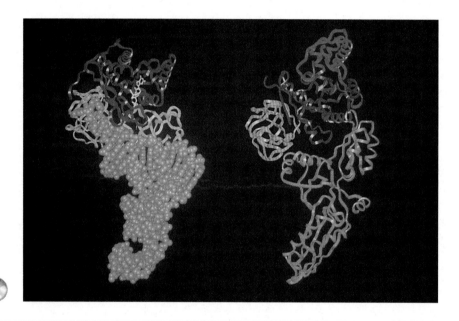

Adapted from Nyborg, J., et al., 1996. Structure of the ternary complex of EF-Tu: Macromolecular mimicry in translation. *Trends in Biochemical Sciences* **21**:81–82.

of the polypeptide chain has been reached (Figure 25.22). These nonsense codons are not "read" by any "terminator tRNAs" but instead are recognized by specific proteins known as **release factors,** so named because they promote polypeptide release from the ribosome. The release factors bind at the A site. **RF-1** recognizes UAA and UAG, while **RF-2** recognizes UAA and UGA. Like EF-G, RF-1 and RF-2 are molecular mimics of the EF-Tu : tRNA complex. As such, they interact well with the ribosomal A-site structure. Further, these release factors "read" the nonsense codons through specific tripeptide sequences that serve as the RF protein equivalent of the tRNA anticodon loop. There is about one molecule each of RF-1 and RF-2 per 50 ribosomes. Ribosomal binding of RF-1 or RF-2 is competitive with EF-G. The binding of RF-1 or RF-2 is promoted by a third release factor, **RF-3.** RF-3 function requires GTP.

The presence of release factors with a nonsense codon in the A site creates a **70S ribosome : RF-1** (or **RF-2) : RF-3-GTP : termination signal** complex that transforms the ribosomal peptidyl transferase into a hydrolase. That is, instead of catalyzing the transfer of the polypeptidyl chain from a polypeptidyl-

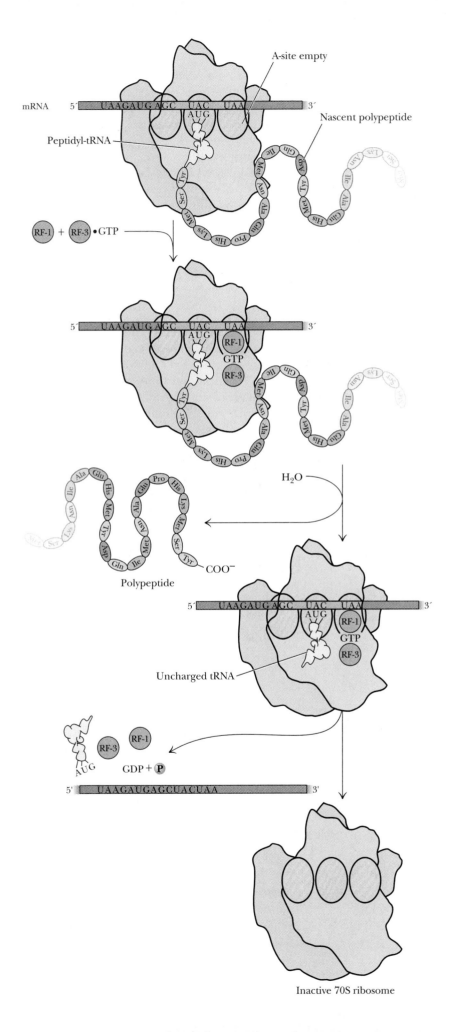

Figure 25.22 The events in peptide chain termination.

827

tRNA to an acceptor aminoacyl-tRNA, the peptidyl transferase hydrolyzes the ester bond linking the polypeptidyl chain to its tRNA carrier. In actuality, peptidyl transferase transfers the polypeptidyl chain to a water molecule instead of an aminoacyl-tRNA. GTP hydrolysis now drives conformational events leading to the dissociation of the uncharged tRNA and expulsion of the release factors from the ribosome (see Figure 25.22).

We can now recount the central role played by GTP in protein synthesis. IF-2, EF-Tu, EF-G, and RF-3 are all GTP-binding proteins, and all are part of the G protein superfamily (whose name is derived from the heterotrimeric G proteins that function in transmembrane signalling pathways, as in Figure 10.41). IF-2, EF-Tu, EF-G, and RF-3 interact with the same site on the 50S subunit, the *factor-binding center*, in the 50S cleft. This factor-binding center activates the GTPase activity of these factors, once they become bound.

The Ribosome Life Cycle

Ribosomal subunits cycle rapidly through protein synthesis. In actively growing bacteria, 80% of the ribosomes are engaged in protein synthesis at any one moment. Once a polypeptide chain is synthesized and the nascent polypeptide chain is released, the 70S ribosome dissociates from the mRNA and separates into free 30S and 50S subunits (Figure 25.23). Intact 70S ribosomes are inactive in protein synthesis because only free 30S subunits can interact with the initiation factors. Binding of initiation factor IF-3 by 30S subunits and interaction of 30S subunits with 50S subunits are mutually exclusive: 30S subunits with bound initiation factors associate with mRNA, but 50S subunit addition requires IF-3 release from the 30S subunit.

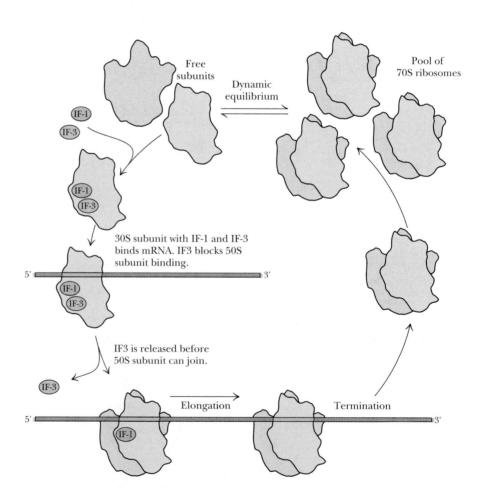

Figure 25.23 The ribosome life cycle. Note that IF-3 is released prior to 50S addition.

Polyribosomes Are the Active Structures of Protein Synthesis

Active protein-synthesizing units consist of an mRNA with several ribosomes attached to it. Such structures are **polyribosomes,** or, simply, **polysomes** (Figure 25.24). All protein synthesis occurs on polysomes. In the polysome, each ribosome is traversing the mRNA and independently translating it into polypeptide. The further a ribosome has moved along the mRNA, the greater the length of its associated polypeptide product. In prokaryotes, as many as 10 ribosomes may be found in a polysome. Ultimately, as many as 300 ribosomes may translate an mRNA, so as many as 300 enzyme molecules may be produced from a single transcript. Eukaryotic polysomes typically contain fewer than 10 ribosomes.

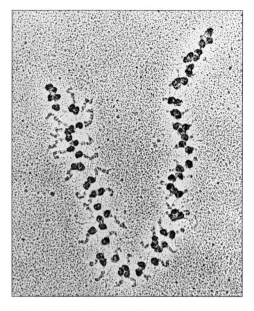

Figure 25.24 Electron micrograph of polysomes: multiple ribosomes translating the same mRNA. *(From Francke, C., et al., 1982. Electron microscopic visualization of a discrete class of giant translation units in salivary gland cells of* Chironomus tentans. *EMBO Journal **1**:59–62. Photo courtesy of Oscar L. Miller, Jr., University of Virginia.)*

25.6 Protein Synthesis in Eukaryotic Cells

Eukaryotic mRNAs are characterized by two post-transcriptional modifications: the $5'$-^{7}**methyl-GTP cap** and the **poly(A) tail** (Figure 25.25). The ^{7}methyl-GTP cap is essential for ribosomal binding of mRNAs in eukaryotes and also enhances the stability of these mRNAs by preventing their degradation by $5'$-exonucleases. The poly(A) tail enhances both the stability and translational efficiency of eukaryotic mRNAs. The Shine–Dalgarno sequences found at the $5'$-end of prokaryotic mRNAs are absent in eukaryotic mRNAs.

Peptide Chain Initiation in Eukaryotes

The events in eukaryotic peptide chain initiation are summarized in Figure 25.26, and the properties of **eukaryotic initiation factors,** symbolized as **eIFs,** are presented in Table 25.9. As might be expected, eukaryotic protein synthesis is considerably more complex than prokaryotic protein synthesis. The eukaryotic initiator tRNA is a unique tRNA functioning only in initiation. Like the prokaryotic initiator tRNA, the eukaryotic version carries only Met. However, unlike prokaryotic f-Met-tRNA$_i^{fMet}$, the Met on this tRNA is not formylated.

Eukaryotic initiation can be divided into three fundamental stages:

Stage 1: Formation of the **43S preinitiation complex** (Figure 25.26, stage 1). Initiation factors eIF-1A and eIF-3 bind to a 40S ribosomal subunit. Then, Met-

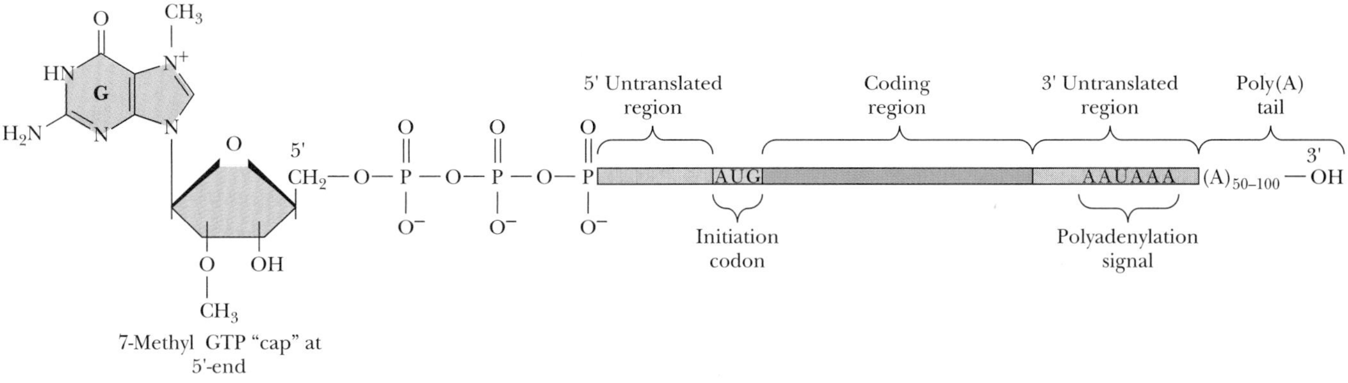

Figure 25.25 The characteristic structure of eukaryotic mRNAs. Untranslated regions ranging between 40 and 150 bases in length occur at both the $5'$- and $3'$-ends of the mature mRNA. An initiation codon at the $5'$-end, invariably AUG, signals the translation start site.

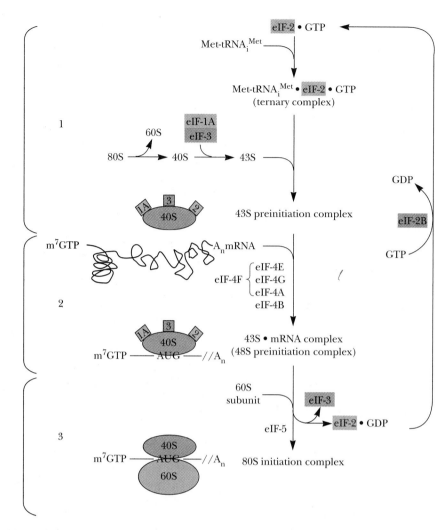

Figure 25.26　The three stages in the initiation of translation in eukaryotic cells. See Table 25.9 for a description of the functions of the eukaryotic initiation factors (eIFs). *(Adapted from Pain, V. M., 1996. Initiation of protein synthesis in eukaryotic cells.* European Journal of Biochemistry ***236:**747–771, Figure 1; and Gingras, A.-C., et al., 1999. eIF-4 initiation factors: Effectors of mRNA recruitment to ribosomes and regulators of translation. *Annual Review of Biochemistry **68:**913–963, Figure 1.)*

tRNA$_i^{Met}$ (in the form of an eIF-2:GTP:Met-tRNA$_i^{Met}$ ternary complex) is delivered to the eIF-1A:eIF-3:40S subunit complex. (Unlike in prokaryotes, binding of Met-tRNA$_i^{Met}$ by eukaryotic ribosomes occurs in the absence of mRNA, so Met-tRNA$_i^{Met}$ binding is not codon-directed.)

Stage 2:　Formation of the **48S initiation complex** (Figure 25.26, stage 2). This stage involves binding of the 43S preinitiation complex to mRNA and migration of the 40S ribosomal subunit to the correct AUG initiation codon. Binding of mRNA by the 43S preinitation complex requires a set of proteins termed the **eIF-4 group.** Collectively, these proteins recognize the 5′-terminal cap and 3′-terminal poly(A) tail of an mRNA, unwind any secondary structure in the mRNA, and transfer the mRNA to the 43S preinitiation complex. The eIF-4 group includes eIF-4B and eIF-4F. eIF-4F is a trimeric complex consisting of eIF-4A (an ATP-dependent RNA helicase), eIF-4E (which binds the 5′-terminal ^{7}methyl-GTP of mRNAs), and eIF-4G. Because eIF-4G interacts with **Pab1p,** the *p*oly(A)-*b*inding *p*rotein which binds to the poly(A) tract on mRNAs

■ **Table 25.9** **Properties of Eukaryotic Translation Initiation Factors**

Factor	Subunit	Size (kD)	Function
eIF-1		15	Enhances initiation complex formation
eIF-1A		17	Stabilizes Met-tRNA$_i$ binding to 40S ribosomes
eIF-2		125	GTP-dependent Met-tRNA$_i$ binding to 40S ribosomes
	α	36	Regulated by phosphorylation
	β	50	Binds Met-tRNA$_i$
	γ	55	Binds GTP, Met-tRNA$_i$
eIF-2B		270	Promotes guanine nucleotide exchange on eIF-2
eIF-2C		94	Stabilizes ternary complex in presence of RNA
eIF-3		550	Promotes Met-tRNA$_i$ and mRNA binding
eIF-4F		243	Binds to mRNA caps and poly(A) tails; consists of eIF-4A, eIF-4E, and eIF-4G; RNA helicase activity unwinds mRNA 2° structure
eIF-4A		46	Binds RNA; ATP-dependent RNA helicase; promotes mRNA binding to 40S ribosomes
eIF-4E		24	Binds to 5′-terminal ^{7}methyl-GTP cap on mRNA
eIF-4G		173	Binds to Pab1p
eIF-4B		80	Binds mRNA; promotes RNA helicase activity and mRNA binding to 40S ribosomes
eIF-5		49	Promotes GTPase of eIF-2, ejection of eIF
eIF-5B		175	Ribosome-dependent GTPase activity; mediates 40S and 60S joining
eIF-6			Dissociates 80S; binds to 60S

Adapted from Clark, B. F. C., et al., eds., 1996. Prokaryotic and eukaryotic translation factors. *Biochimie* **78**:1119–1122; Dever, T. E., 1999. Translation initiation: Adept at adapting. *Trends in Biochemical Sciences* **24**:398–403; and Gingras, A.-C., et al., 1999. eIF-4 initiation factors: Effectors of mRNA recruitment to ribosomes and regulators of translation. *Annual Review of Biochemistry* **68**:913–963.

(Figure 25.27), eIF-4G serves as the bridge between the cap-binding eIF-4E, the poly(A) tail of the mRNA, and the 40S subunit (through interaction with eIF-3). These interactions between the 5′-terminal ^{7}methyl-GTP cap and the 3′-poly(A) tail initiate scanning of the 40S subunit in search of an AUG codon.

eIF-4E, the mRNA cap-binding protein, represents a key regulatory element in eukaryotic translation. eIF-4F binding to the cap structure is necessary for association of eIF-4B and formation of the 48S preinitiation complex. Translation is inhibited when the eIF-4E subunit of eIF-4F binds with **4E-BP** (the eIF-

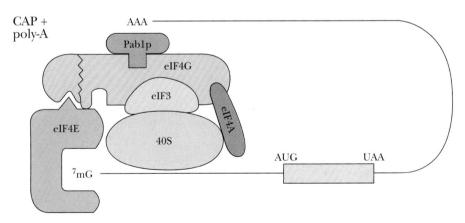

Figure 25.27 Initiation factor eIF-4G serves as a multipurpose adapter to engage the ^{7}methyl-G cap : eIF-4E complex, the Pab1p : poly(A) tract, and the 40S ribosomal subunit in eukaryotic translation initiation. *(Adapted from Hentze, M. W., 1997. eIF-4G: A multipurpose ribosome adapter?* Science ***275**:500–501.)*

4E binding protein). Growth factors stimulate protein synthesis by causing the phosphorylation of 4E-BP, which prevents its binding to eIF-4E.

Stage 3: Formation of the **80S initiation complex.** When the 48S preinitiation complex stops at an AUG codon, GTP hydrolysis in the eIF-2:Met-tRNA$_i^{Met}$ ternary complex causes ejection of the initiation factors bound to the 40S ribosomal subunit. eIF-5, in conjuction with eIF-5B, acts here by stimulating the GTPase activity of eIF-2. Ejection of the eIFs is followed by 60S subunit association to form the 80S initiation complex, whereupon translation begins (Figure 25.26, stage 3). eIF-2:GDP is recycled to eIF-2:GTP by eIF-2B (eIF-2B is a *guanine nucleotide exchange factor*).

Regulation of Eukaryotic Peptide Chain Initiation

Regulation of gene expression can be exerted post-transcriptionally through control of mRNA translation. Phosphorylation/dephosphorylation of translational components is a dominant mechanism for control of protein synthesis. Thus far, seven proteins—eIF-2α, eIF-2B, eIF-4E, eIF-4G, 40S ribosomal protein S6, and the two eukaryotic elongation factors, eEF-1 and eEF-2 (see later)— have been identified as targets of regulatory controls. Modification of some factors affects the rate of mRNA translation; modification of others affects which mRNAs are selected for translation. Peptide chain initiation, the initial phase of the synthetic process, is the optimal place for such control. Phosphorylation of S6 facilitates initiation of protein synthesis, resulting in a shift of the ribosomal population from inactive ribosomes to actively translating polysomes. S6 phosphorylation is stimulated by serum growth factors (see Chapter 26). The action of eIF-4F, the mRNA cap-binding complex, is promoted by phosphorylation. On the other hand, phosphorylation of other translational components inhibits protein synthesis. For example, the α-subunit of eIF-2 can be reversibly phosphorylated at a specific Ser residue by an eIF-2α kinase/phosphatase system (Figure 25.28). Phosphorylation of eIF-2α inhibits peptide chain initiation. Reversible phosphorylation of eIF-2 is an important control governing globin synthesis in reticulocytes. If heme for hemoglobin synthesis becomes limiting in these cells, eIF-2α is phosphorylated so globin mRNA is not translated and chains are not synthesized. Availability of heme reverses the inhibition through phosphatase-mediated removal of the phosphate group from the Ser residue.

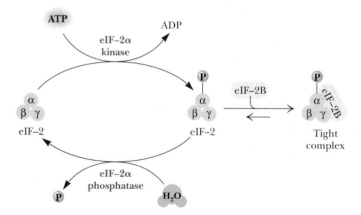

Figure 25.28 Control of eIF-2 functions through reversible phosphorylation of a Ser residue on its α-subunit. The phosphorylated form of eIF-2 (eIF-2-P) enters a tight complex with eIF-2B and is unavailable for initiation.

Peptide Chain Elongation in Eukaryotes

Eukaryotic peptide elongation occurs in very similar fashion to the process in prokaryotes. An incoming aminoacyl-tRNA enters the ribosomal A site while peptidyl-tRNA occupies the P site. Peptidyl transfer then occurs, followed by translocation of the ribosome one codon further along the mRNA. Two elongation factors, EF-1 and EF-2, mediate the elongation steps. EF-1 consists of two components: EF-1A and EF-1B. EF-1A is the eukaryotic counterpart of EF-Tu; it serves as the aminoacyl-tRNA binding factor and requires GTP. EF-1B is the eukaryotic equivalent of prokaryotic EF-Ts; it catalyzes the exchange of bound GDP on EF-1:GDP for GTP so active EF-1:GTP can be regenerated. EF-2 is the eukaryotic translocation factor. EF-2 (like its prokaryotic kin, EF-G) binds GTP, and GTP hydrolysis accompanies translocation.

Eukaryotic Peptide Chain Termination

Whereas prokaryotic termination involves three different release factors (RFs), just one RF is sufficient for eukaryotic termination. Eukaryotic RF binding to the ribosomal A site is GTP-dependent, and RF:GTP binds at this site when it is occupied by a termination codon. Then, hydrolysis of the peptidyl-tRNA ester bond, hydrolysis of GTP, release of nascent polypeptide and deacylated tRNA, and ribosome dissociation from mRNA ensue.

Inhibitors of Protein Synthesis

Protein synthesis inhibitors have served two major, and perhaps complementary, purposes. First, they have been very useful scientifically in elucidating the biochemical mechanisms of protein synthesis. Second, some of these inhibitors affect prokaryotic but not eukaryotic protein synthesis and thus are medically important antibiotics. Table 25.10 is a partial list of these inhibitors and their modes of action. The structures of some of these compounds are given in Figure 25.29.

Table 25.10 Some Protein Synthesis Inhibitors

Inhibitor	System Inhibited	Mode of Action
Initiation		
Aurintricarboxylic acid	Prokaryotic	Prevents IF binding to 30S subunit
Kasugamycin	Prokaryotic	Inhibits f-Met-tRNA$_i$ ^{fMet} binding
Streptomycin	Prokaryotic	Prevents formation of initiation complexes
Elongation: Aminoacyl-tRNA Binding		
Tetracycline	Prokaryotic	Inhibits aminoacyl-tRNA binding at A site
Streptomycin	Prokaryotic	Codon misreading, insertion of improper amino acid
Elongation: Peptide Bond Formation		
Sparsomycin	Prokaryotic	Peptidyl transferase inhibitor
Chloramphenicol	Prokaryotic	Binds to 50S subunit, blocks peptidyl transferase activity
Erythromycin	Prokaryotic	Binds to 50S subunit, blocks peptidyl transferase activity
Cycloheximide	Eukaryotic	Inhibits translocation of peptidyl-tRNA
Elongation: Translocation		
Fusidic acid	Both	Inhibits EF-G:GDP dissociation from ribosome
Thiostrepton	Prokaryotic	Inhibits ribosome-dependent EF-Tu and EF-G GTPase
Diphtheria toxin	Eukaryotic	Inactivates eEF-2 through ADP-ribosylation
Premature Termination		
Puromycin	Both	Aminoacyl-tRNA analog, which acts as a peptidyl acceptor, aborting further peptide elongation
Ribosome Inactivation		
Ricin	Eukaryotic	Catalytic inactivation of 28S rRNA

(Text continued on page 836)

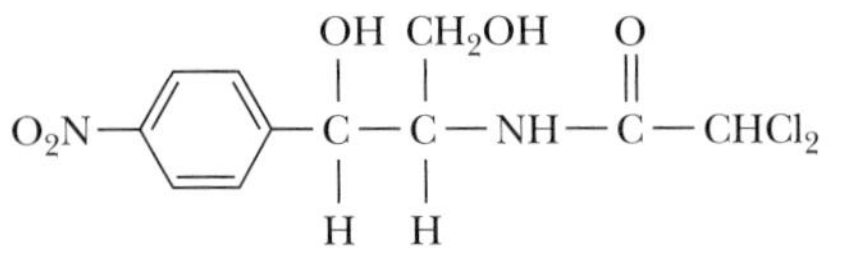

Chloramphenicol

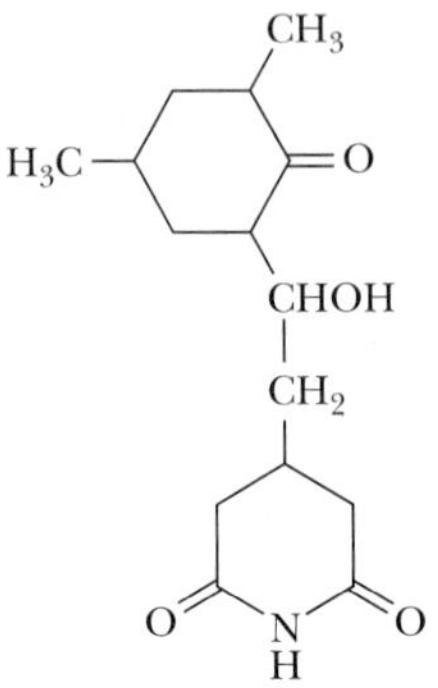

Cycloheximide

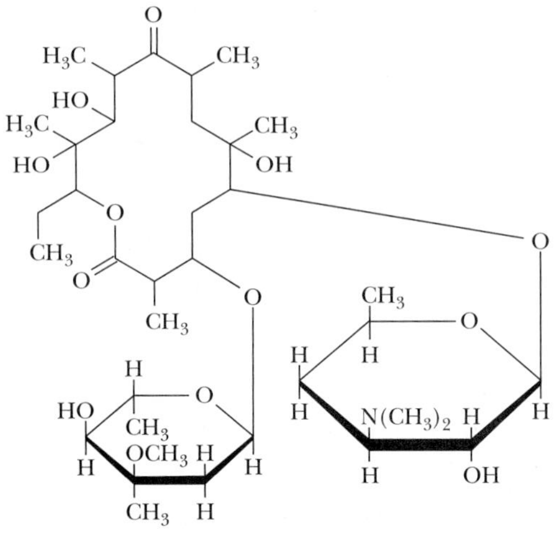

Erythromycin

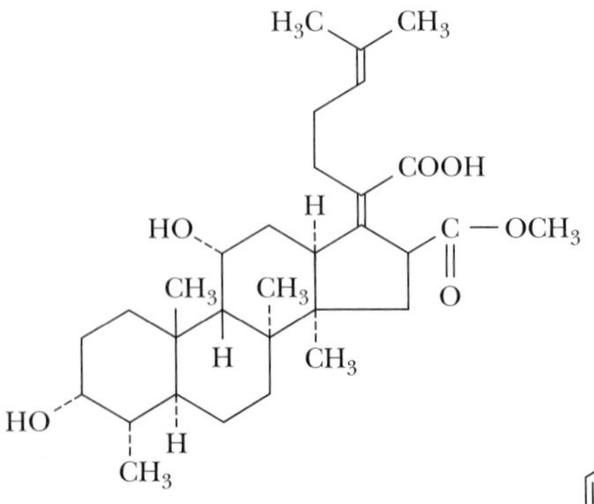

Fusidic acid

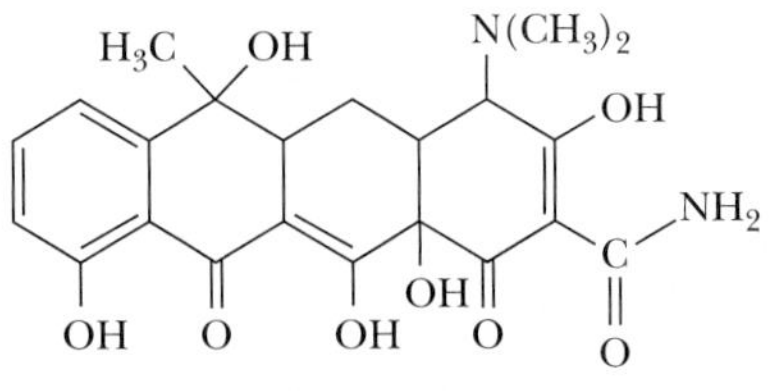

Tetracycline

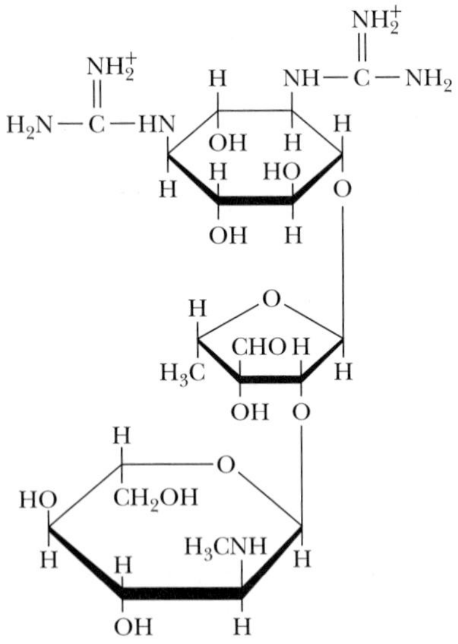

Streptomycin

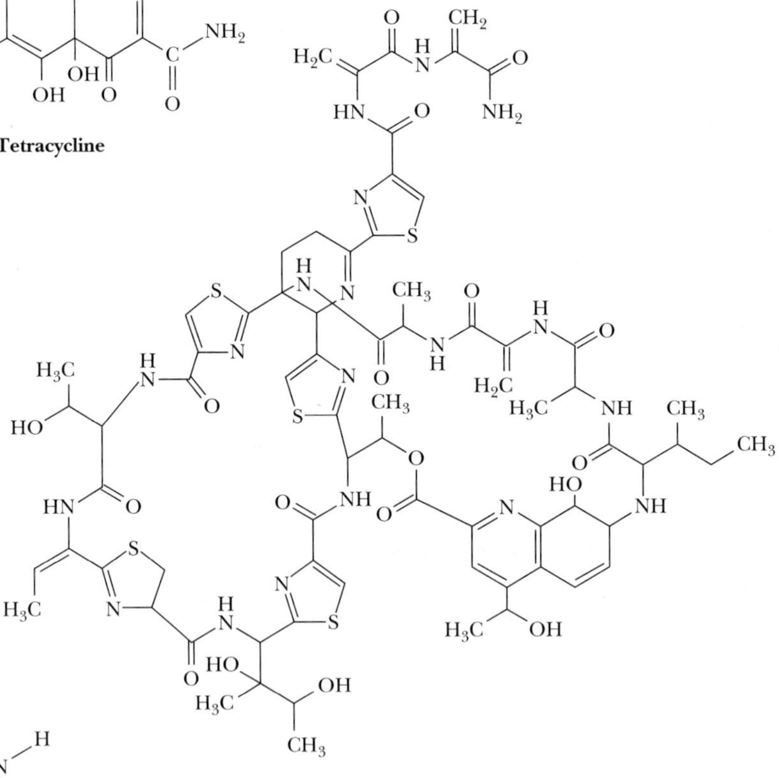

Thiostrepton

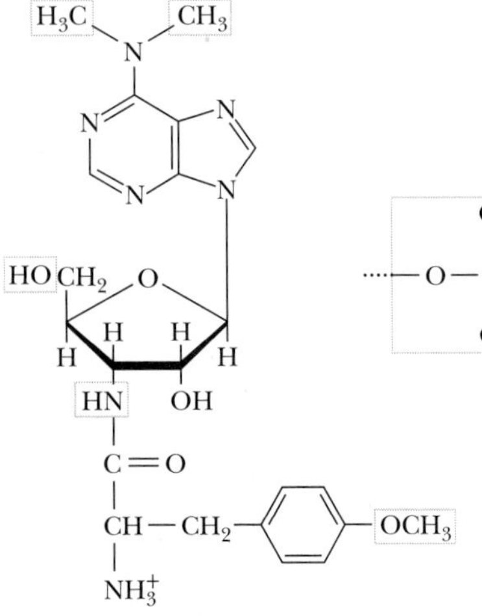

Puromycin

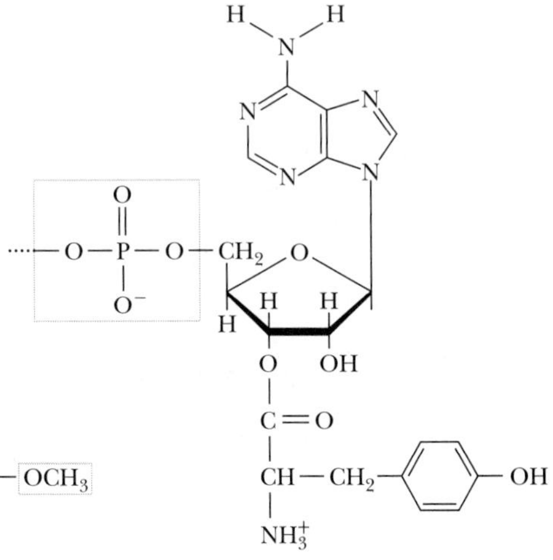

Tyrosyl-tRNA

Figure 25.29 The structures of various antibiotics that act as protein synthesis inhibitors. Puromycin mimics the structure of aminoacyl-tRNA in that it resembles the 3′-terminus of a Tyr-tRNA.

HUMAN BIOCHEMISTRY

Diphtheria Toxin

Diphtheria arises from infection by *Corynebacterium diphtheriae* bacteria carrying bacteriophage *corynephage β*. Diphtheria toxin is a phage-encoded enzyme secreted by these bacteria that is capable of inactivating a number of GTP-dependent enzymes through covalent attachment of an ADP-ribosyl moiety derived from NAD^+. That is, diphtheria toxin is an NAD^+-dependent ADP-ribosylase. One target of diphtheria toxin is the eukaryotic translocation factor, EF-2. This protein has a modified His residue known as **diphthamide.** Diphthamide is generated post-translationally on EF-2; its biological function is unknown. (EF-G of prokaryotes lacks this unusual modification and is not susceptible to diphtheria toxin.) Diphtheria toxin specifically ADP-ribosylates an imidazole-N within the diphthamide moiety of EF-2 (see figure). ADP-ribosylated EF-2 retains the ability to bind GTP but is unable to function in protein synthesis. Because diphtheria toxin is an enzyme and can act catalytically to modify many molecules of its target protein, just a few micrograms suffice to cause death.

Diphtheria toxin catalyzes the NAD^+-dependent ADP-ribosylation of selected proteins. (*right*) ADP-ribosylation of the diphthamide moiety of eukaryotic EF-2. (Diphthamide = 2-[3-carboxamido-3-(trimethylammonio)propyl]histidine.)

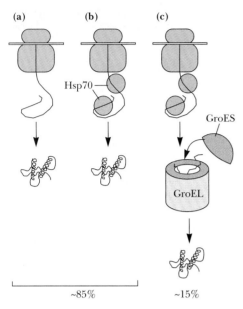

Figure 25.30 Protein folding pathways.
(a) Chaperone-independent folding. The protein folds as it is synthesized on the ribosome (green) or shortly thereafter. **(b)** Hsp70-assisted protein folding. Hsp70 (gray) binds to nascent polypeptide chains as they are synthesized and assists their folding. **(c)** Folding assisted by Hsp70 and chaperonin complexes. The chaperonin complex in *E. coli* is GroEL–GroES. The chaperonin complex in eukaryotic cells is known as TRiC (for TCP-1 ring complex) or CCT (cytosolic chaperonin containing TCP-1). The majority of proteins fold by pathways (a) or (b). *(Adapted from Netzer, W. J., and Hartl, F. U., 1998. Protein folding in the cytosol: Chaperonin-dependent and -independent mechanisms.* Trends in Biochemical Sciences **23**:68–73, Figure 2.)

25.7 Protein Folding

The information for folding each protein into its unique three-dimensional architecture resides within its amino acid sequence (primary structure). Proteins begin to fold even as they are being synthesized on ribosomes (Figure 25.30a). Nevertheless, proteins may be assisted in folding by a family of helper proteins known as **molecular chaperones** (see Chapter 5). Chaperones also serve to shepherd proteins to their ultimate cellular destinations.

Nascent polypeptide chains exiting the large ribosomal subunit are met by a ribosome-associated chaperone (**TF** or **trigger factor** in *E. coli;* **NAC [nascent chain-associated complex]** in eukaryotes). These proteins mediate transfer of the emerging nascent polypeptide chain to the Hsp70 class of chaperones. (Many proteins do not require this step for proper folding.)

The principal chaperones are the **Hsp70** and **Hsp60** classes of proteins. In Hsp70-assisted folding, proteins of the Hsp70 class bind to nascent polypeptide chains while they are still on ribosomes (Figure 25.30b). Hsp70 (known as **DnaK** in *E. coli*) recognizes exposed, extended regions of polypeptides that are rich in hydrophobic residues. By binding to these regions, Hsp70 prevents nonproductive associations and keeps the polypeptide in an unfolded (or partially folded) state until productive folding interactions can occur. Completion of folding requires release of the protein from Hsp70; release is energy-dependent and is driven by ATP hydrolysis.

A limited number of proteins requires the Hsp60 class of chaperones, which are known as **chaperonins,** for the completion of folding. Chaperonins sequester partially folded molecules from one another (and from extraneous interactions) but still allow folding to proceed. The principal chaperonin in *E. coli* is the **GroEL–GroES** complex. GroEL is made of two stacked seven-membered rings of 60-kD subunits that form a cylindrical α_{14} oligomer 15 nm high and 14 nm wide (Figure 25.31). GroEL has a 5-nm central cavity that is the site of ATP-dependent protein folding. GroES consists of a single seven-membered ring of 10-kD subunits that sits like a dome on GroEL (Figure 25.31). (The eukaryotic analog of GroEL, **CCT** (also called **TriC**) is an eight- or nine-membered double-ring structure of 55-kD subunits. (**Prefoldin** can serve as a co-chaperone for CCT, much as GroES does for GroEL.) A partially folded protein molecule is bound within the central cavity of the 14-subunit complex, where its folding is facilitated in an ATP-dependent fashion. Many cycles of binding of the folding protein to the surface of the central cavity, ATP hydrolysis, release of the protein, and rebinding to the surface take place, with folding steps occurring in the brief intervals when the protein is free from the cavity surface (Figure 25.31c). The folding of rhodanese, a 33-kD protein, requires the hydrolysis of about 130 equivalents of ATP. Once the protein has achieved its fully folded state, it is released from GroEL.

25.8 Post-Translational Processing of Proteins

Aside from these folding events, release of the completed polypeptide from the ribosome is not necessarily the final step in the covalent construction of a protein. Many proteins must undergo covalent alterations before they become functional. In the course of these so-called *post-translational* modifications, the primary structure of a protein can be altered and novel derivations can be introduced into its amino acid side chains. Hundreds of different amino acid variations have been described in proteins, virtually all arising post-translationally. The list of such modifications is very large; some are rather com-

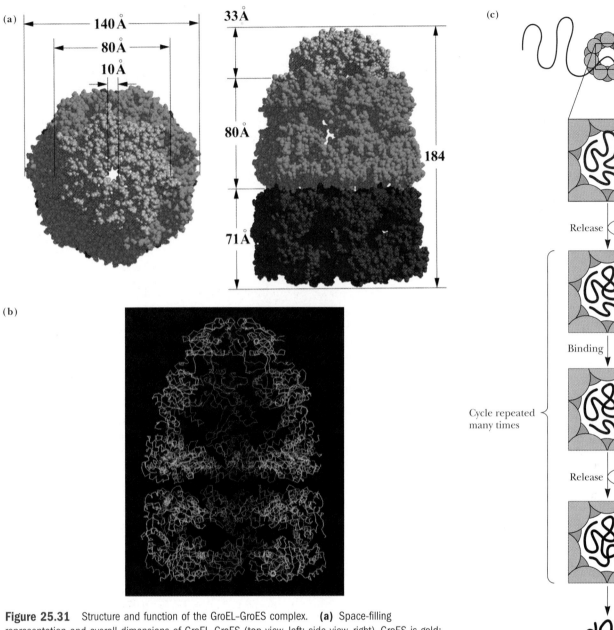

Figure 25.31 Structure and function of the GroEL–GroES complex. **(a)** Space-filling representation and overall dimensions of GroEL–GroES (top view, left; side view, right). GroES is gold; the top GroEL ring is green and the bottom GroEL ring is red. **(b)** Section through the center of the complex to reveal the central cavity. The GroEL–GroES structure is shown as a Cα carbon trace. ADP molecules bound to GroEL are shown as space-filling models. **(c)** Successive cycles of protein binding and release with the chaperonin central cavity, ATP-dependent release, and progressive folding ends in release of the protein from the complex in its fully folded form. *(Figure parts [a] and [b] adapted from Figure 1 in Xu, Z., Horwich, A. L., and Sigler, P. B., 1997. Nature **388:**741–750. Molecular graphics courtesy of Paul B. Sigler, Yale University.)*

monplace, whereas others are peculiar to a single protein. A survey of some of the more prominent chemical groups conjugated to proteins, such as carbohydrates and phosphates, is given in Chapter 5.

Proteolytic Cleavage of the Polypeptide Chain

Proteolytic cleavage, as the most prevalent form of protein post-translational modification, merits special attention. The very occurrence of proteolysis as a processing mechanism seems strange: Why join a number of amino acids in sequence and then eliminate some of them? Three reasons can be cited. First, diversity can be introduced where none exists. For example, a simple form of proteolysis, enzymatic removal of N-terminal Met residues, occurs in many proteins. **Met-aminopeptidase,** by removing the invariant Met initiating all polypeptide chains, introduces diversity at N-termini. Second, proteolysis serves as an activation mechanism so that expression of the biological activity of a protein can be delayed until appropriate. A number of metabolically active proteins, including digestive enzymes and hormones, are synthesized as larger inactive precursors termed **pro-proteins** that are activated through proteolysis (see **zymogens,** Chapter 10). The N-terminal pro-sequence on such proteins may act as an intramolecular chaperone to ensure correct folding of the active site. Third, proteolysis is involved in the targeting of proteins to their proper destinations in the cell, a process known as **protein translocation.**

Protein Translocation

Proteins targeted for service in membranous organelles or for export from the cell are synthesized in precursor form carrying an N-terminal stretch of amino acid residues, or **leader peptide,** that serves as a **signal sequence.** In effect, signal sequences serve as "zip codes" for sorting and dispatching proteins to their proper compartments. Thus, the information specifying the correct cellular localization of a protein is found within its structural gene. Once the protein is routed to its destination, the signal sequence is proteolytically clipped from the protein.

In addition to the bacterial plasma membrane, a number of eukaryotic membranes are competent in protein translocation, including the membranes of the endoplasmic reticulum (ER), nucleus, mitochondria, chloroplasts, and peroxisomes. Several common features characterize protein translocation systems:

1. Proteins to be translocated are made as preproteins containing contiguous blocks of amino acid sequence that act as sorting signals.
2. Membranes involved in translocation have specific protein receptors exposed on their cytosolic faces.
3. **Translocons,** complex structures consisting of several proteins with different functions, catalyze movement of the proteins across the membrane, and metabolic energy in the form of ATP, GTP, or a membrane potential is essential.
4. Preproteins are maintained in a loosely folded, translocation-competent conformation through interaction with molecular chaperones.

Prokaryotic Protein Translocation

Gram-negative bacteria typically have four compartments: cytoplasm, plasma (or inner) membrane, periplasmic space (or periplasm), and outer membrane. Most proteins destined for any location other than the cytoplasm are synthesized with amino-terminal leader sequences 16 to 26 amino acid residues long. These leader sequences, or *signal sequences,* consist of a basic N-terminal region, a central domain of 7 to 13 hydrophobic residues, and a nonhelical C-terminal region (Figure 25.32). The conserved features of the last part of the leader, the C-terminal region, include a helix-breaking Gly or Pro residue and amino acids with small side chains located one and three residues before the proteolytic cleavage site. Unlike the basic N-terminal and nonpolar central regions, the C-terminal features are not essential for translocation but instead serve as recognition signals for the **leader peptidase,** which removes the leader se-

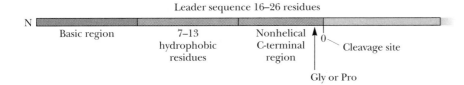

Leader sequence 16–26 residues

N — Basic region | 7–13 hydrophobic residues | Nonhelical C-terminal region | 0 — Cleavage site

Gly or Pro

Figure 25.32 General features of the N-terminal signal sequences on *E. coli* proteins destined for translocation: a basic N-terminal region, a central apolar domain, and a nonhelical C-terminal region.

quence. The exact amino acid sequence of the leader peptide is unimportant. Nonpolar residues in the center and a few Lys residues at the amino terminus are sufficient for successful translocation. The functions of leader peptides are to retard the folding of the preprotein so that molecular chaperones have a chance to interact with it and to provide recognition signals for the translocation machinery and leader peptidase.

Eukaryotic Protein Sorting and Translocation

Eukaryotic cells are characterized by many membrane-bounded compartments. In general, signal sequences targeting proteins to their appropriate compartments are located at the N-terminus as *cleavable presequences,* although many proteins have N-terminal localization signals that are not cleaved and others have internal targeting sequences that may or may not be cleaved. Proteolytic removal of the leader sequences is also catalyzed by specialized proteases, but removal is not essential to translocation. No sequence similarity is found among the targeting signals for each compartment. Thus, the targeting information resides in more generalized features of the leader sequences such as charge distribution, relative polarity, and secondary structure.

The Synthesis of Secretory Proteins and Many Membrane Proteins Is Coupled to Translocation Across the ER Membrane

The signals recognized by the endoplasmic reticulum translocation system are virtually indistinguishable from bacterial signal sequences; indeed, the two are interchangeable *in vitro.* In higher eukaryotes, translation and translocation of many proteins destined for processing via the ER are tightly coupled. That is, translocation across the ER occurs as the protein is being translated on the ribosome. As the N-terminal signal sequence of a preprotein undergoing synthesis emerges from the ribosome, it is detected by the **signal recognition particle (SRP;** Figure 25.33). SRP is a 325-kD nucleoprotein assembly that contains six polypeptides and a 300-nucleotide **7S RNA.** SRP binding of the signal sequence halts further protein synthesis on the ribosome. This prevents release of the growing protein into the cytosol before it reaches the ER and its destined translocation. The SRP-ribosome complex then moves to the cytosolic face of the ER, where it binds to the **docking protein** (also known as the **SRP receptor**), an $\alpha\beta$ heterodimeric protein. The α-subunit is anchored to the membrane by the transmembrane β-subunit; both subunits have GTPase activity. Docking is followed by dissociation of SRP in a GTP-dependent process. After targeting to the membrane, the ribosome resumes protein synthesis, delivering its growing polypeptide to the **translocon,** a complex, multifunctional entity that includes, among other proteins, the SRP receptor and the **Sec61 complex,** a heterotrimeric complex of membrane proteins. The 53-kD α-subunit of Sec61p has 10 membrane-spanning segments, whereas the β- and γ-subunits are single TMS proteins. Sec61p forms a transmembrane channel through which the nascent polypeptide is transported into the ER lumen (see Figure 25.33). The pore size of Sec61p is about 2 nm.

Soon after it enters the ER lumen, the signal peptide is clipped off by membrane-bound **signal peptidase.** Other modifying enzymes within the lumen

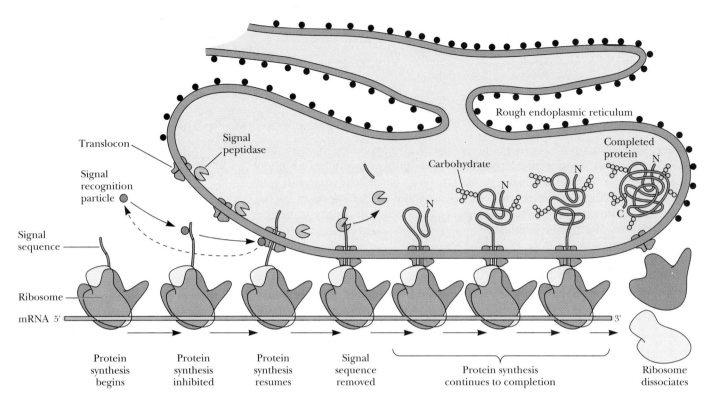

Figure 25.33 Synthesis of a eukaryotic secretory protein and its translocation via the endoplasmic reticulum.

introduce additional post-translational alterations into the polypeptide, such as glycosylation with specific carbohydrate residues. ER-processed proteins destined for secretion from the cell or inclusion in vesicles such as lysosomes end up contained within the soluble phase of the ER lumen, but polypeptides destined to become membrane proteins carry **stop-transfer** sequences within their mature domains. These stop-transfer sequences are typically a 20-residue stretch of hydrophobic amino acids that arrest passage of the protein across the ER membrane. Proteins with stop-transfer sequences remain embedded in the ER membrane with their C-termini on the cytosolic face of the ER. Such membrane proteins arrive at their intended destinations via subsequent processing of the ER. To prevent secretion of inappropriate proteins, fragmented or misfolded secretory proteins are passed from the ER back into the cytosol via Sec61p. Thus, Sec61p also serves as channel for aberrant secretory proteins to be returned to the cytosol so that they can be destroyed by the proteasome degradation apparatus (see Section 25.9).

Mitochondrial Protein Import

Most mitochondrial proteins are encoded by the nuclear genome and synthesized on cytosolic ribosomes. Mitochondria consist of four principal subcompartments: the outer membrane, the intermembrane space, the inner membrane, and the matrix. Thus, mitochondrial proteins must not only find mitochondria, they must gain access to the proper subcompartment, and once there they must attain a functionally active conformation. In principle, comparable considerations apply to protein import to chloroplasts, organelles with five principal subcompartments (outer membrane, intermembrane space, inner/thylakoid membrane, stroma, and thylakoid lumen; see Chapter 18).

Signal sequences on nuclear-encoded proteins destined for the mitochondria are N-terminal cleavable presequences 10 to 70 residues long. These mitochondrial presequences lack contiguous hydrophobic regions. Instead, they have positively charged and hydroxyl-amino acid residues spread along their entire length. These sequences form **amphipathic α-helices** (Figure 25.34) with basic residues on one side of the helix and uncharged and hydrophobic residues on the other; that is, mitochondrial presequences are positively charged amphiphatic sequences. In general, mitochondrial targeting sequences share no sequence homology. Once synthesized, mitochondrial pre-proteins are retained in an unfolded state with their target sequences exposed, through association with heat shock proteins of the Hsp70 family. Import involves binding of a preprotein to a receptor protein on the mitochondrial outer membrane and subsequent uptake via the mitochondrial protein translocation apparatus.

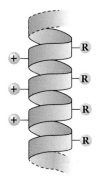

Figure 25.34 Structure of an amphipathic α-helix having basic (+) residues on one side and uncharged and hydrophobic (R) residues on the other.

25.9 Protein Degradation

Cellular proteins are in a dynamic state of turnover, with the relative rates of protein synthesis and protein degradation ultimately determining the amount of protein present at any point in time. In many instances, transcriptional regulation determines the concentrations of specific proteins expressed within cells, with protein degradation playing a minor role. In other instances, the amounts of key enzymes and regulatory proteins, such as cyclins and transcription factors, are controlled via selective protein degradation. In addition, abnormal proteins arising from biosynthetic errors or postsynthetic damage must be destroyed to prevent the deleterious consequences of their buildup. The elimination of proteins typically follows first-order kinetics, with half-lives ($t_{1/2}$) of different proteins ranging from several minutes to many days. A single, random proteolytic break introduced into the polypeptide backbone of a protein is believed sufficient to trigger its rapid disappearance because no partially degraded proteins are normally observed in cells.

Protein degradation poses a real hazard to cellular processes. To control this hazard, protein degradation is compartmentalized, either in macromolecular structures known as **proteasomes** or in degradative organelles such as lysosomes. Protein degradation within lysosomes is largely nonselective; selection occurs during lysosomal uptake. Proteasomes are found in eukaryotic as well as prokaryotic cells. The proteasome is a functionally and structurally sophisticated counterpart to the ribosome. Regulation of protein levels via degradation is an essential cellular mechanism. Regulation by degradation is both rapid and irreversible.

The Ubiquitin Pathway for Protein Degradation in Eukaryotes

Ubiquitination is the most common mechanism to label a protein for proteasome degradation in eukaryotes. **Ubiquitin** is a highly conserved, 76-residue (8.5 kD) polypeptide widespread in eukaryotes. Proteins are condemned to degradation through ligation to ubiquitin. Three proteins in addition to ubiquitin are involved in the ligation process: E_1, E_2, and E_3 (Figure 25.35). E_1 is the **ubiquitin-activating enzyme** (105-kD dimer). It becomes attached via a thioester bond to the C-terminal Gly residue of ubiquitin through ATP-driven formation of an activated ubiquitin-adenylate intermediate. Ubiquitin is then transferred from E_1 to an SH group on E_2, the **ubiquitin-carrier protein.** (E_2 is actually a family of at least seven different small proteins, several of which are heat shock proteins.) In protein degradation, E_2-S ~ ubiquitin transfers ubiquitin to free amino groups on proteins selected by E_3 (180 kD), the **ubiquitin-protein ligase.** Upon binding a protein substrate, E_3 catalyzes the transfer of

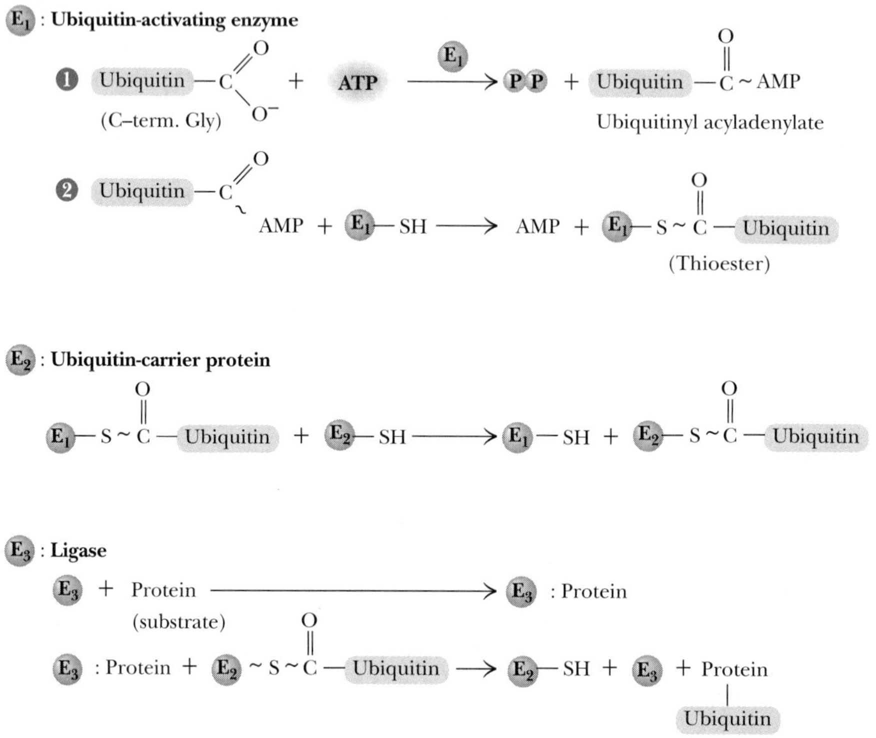

Figure 25.35 Enzymatic reactions in the ligation of ubiquitin to proteins. Ubiquitin is attached to selected proteins via isopeptide bonds formed between the ubiquitin carboxy-terminus and free amino groups (α-NH$_2$ terminus, Lys ε-NH$_2$ side chains) on the protein.

ubiquitin from E$_2$-S $\sim$ ubiquitin to free amino groups (usually Lys ε-NH$_2$) on the protein. More than one ubiquitin may be attached to a protein substrate, and tandemly linked chains of ubiquitin also occur via *isopeptide bonds* between the C-terminus of one ubiquitin and the ε-amino of Lys^{48} in another.

E$_3$ plays a central role in recognizing and selecting proteins for degradation. E$_3$ selects proteins by the nature of the N-terminal amino acid. Proteins must have a free α-amino terminus to be susceptible. Proteins with either Met, Ser, Ala, Thr, Val, Gly, or Cys at the amino terminus are resistant to the ubiquitin-mediated degradation pathway. However, proteins having Arg, Lys, His, Phe, Tyr, Trp, Leu, Asn, Gln, Asp, or Glu N-termini have half-lives of only 2 to 30 minutes.

Interestingly, proteins with acidic N-termini (Asp or Glu) show a tRNA requirement for degradation (Figure 25.36). Transfer of Arg from Arg-tRNA to the N-terminus of these proteins alters their N-terminus from acidic to basic, rendering the protein susceptible to E$_3$. It is also interesting that Met is less likely to be cleaved from the N-terminus if the next amino acid in the chain is one particularly susceptible to ubiquitin-mediated degradation.

Most proteins with susceptible N-terminal residues are *not* normal intracellular proteins but tend to be secreted proteins in which the susceptible residue has been exposed by action of a signal peptidase. Perhaps part of the function of the N-terminal recognition system is to recognize and remove from the cytosol any invading "foreign" or secreted proteins.

Other proteins targeted for ubiquitin ligation and proteasome degradation contain **PEST sequences**—short, highly conserved sequence elements rich in proline (P), glutamate (E), serine (S), and threonine (T) residues.

Proteasomes

Proteasomes are large oligomeric structures enclosing a central cavity where proteolysis takes place. The 20S proteasome from the archaebacterium

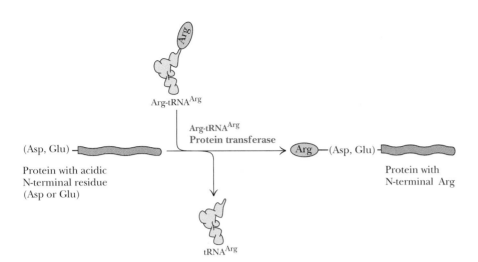

Figure 25.36 Proteins with acidic N-termini show a tRNA requirement for degradation. Arginyl-tRNA^{Arg} : protein transferase catalyzes the transfer of Arg to the free α-NH$_2$ of proteins with Asp or Glu N-terminal residues. Arg-tRNA^{Arg} : protein transferase serves as part of the protein degradation recognition system.

Thermoplasma acidophilum is a 700-kD barrel-shaped structure composed of two different kinds of polypeptide chains, α and β, arranged to form four stacked rings of $\alpha_7\beta_7\beta_7\alpha_7$ subunit organization. The barrel is about 15 nm in height and 11 nm in diameter and contains a three-part central cavity (Figure 25.37a). The proteolytic sites of the 20S proteasome are found within this cavity. Access to the cavity is controlled through a 1.3-nm opening formed by the outer α_7 rings. These rings are believed to unfold proteins destined for degradation and transport them into the central cavity. The β-subunits possess the proteolytic

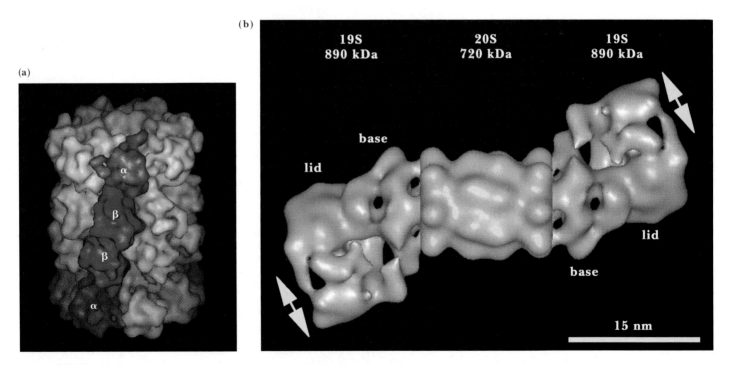

Figure 25.37 Model for the structure of the 26S proteasome. **(a)** The *Thermoplasma acidophilum* 20S proteasome core structure. **(b)** Composite model of the 26S proteasome. Arrows signify wagging motions observed in 19S cap structures. The 20S proteasome core is shown in yellow; the 19S regulator (19S cap) structures are in blue. *(Adapted from Figures 3 and 5 in Voges, D., Zwickl, P., and Baumeister, W. 1999. The 26S proteasome: A molecular machine designed for controlled proteolysis.* Annual Review of Biochemistry **68:**1015–1068. *Images courtesy of Peter Zwickl and W. Baumeister, Mat-Planck-Institut für Biochemie, Martinsried, Germany.)*

activity. The products of proteasome degradation are oligopeptides 7 to 9 residues long.

Eukaryotic cells contain two forms of proteasomes: the **20S proteasome,** and its larger counterpart, the **26S proteasome.** The eukaryotic 26S proteasome is a 45-nm-long structure composed of a 20S proteasome plus two additional substructures known as **19S regulators** (also called **19S caps** or **PA700** [for *p*roteasome *a*ctivator-*700* kD]) (see Figure 25.37b). Overall, the 26S proteasome (~ 2.5 mD) has two copies each of 32–34 distinct subunits, 14 in the 20S core and 18–20 in the cap structures. Unlike the archaeal 20S proteasome, the eukaryotic 20S core structure contains seven different kinds of α-subunits and seven different kinds of β-subunits. Interestingly, only three of the seven different β-subunits have protease active sites. The 26S proteasome forms when the 19S regulators dock to the two outer α_7 rings of the 20S proteasome cylinder. Many of the 19S regulator subunits have ATPase activity. Replacement of certain 19S regulator subunits with others changes the specificity of the proteasome. The 19S regulators cause the proteolytic function of the 20S proteasome to become ATP-dependent and specific for ubiquitinylated proteins as substrates. That is, these 19S caps act as regulatory complexes for the recognition and selection of ubiquitinylated proteins for degradation by the 20S proteasome core (Figure 25.38). The 19S regulators also carry out the unfolding and transport of ubiquitinylated protein substrates into the proteolytic central cavity.

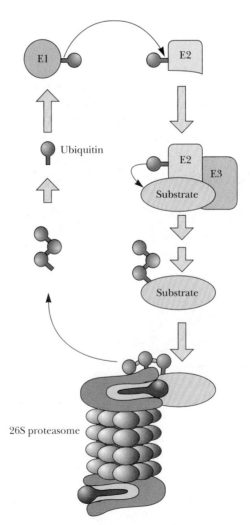

26S proteasome

Figure 25.38 Diagram of the ubiquitin-proteasome degradation pathway. Pink "lollipop" structures symbolize ubiquitin molecules. *(Adapted from Hilt, W., and Wolf, D. H., 1996. Proteasomes: Destruction as a program.* Trends in Biochemical Sciences *21:96–102, Figure 1.)*

A DEEPER LOOK

Protein Triage: A Model for Protein Quality Control

Triage is a medical term for the sorting of patients according to their need for (and their likelihood to benefit from) medical treatment. Sue Wickner, Michael Maurizi, and Susan Gottesman have pointed out that cells control the quality of their proteins through a system of triage based on the chaperones and the ubiquitination–proteasome degradation pathway. These systems recognize non-native proteins (proteins that are only partially folded, misfolded, incorrectly modified, damaged, or in an inappropriate compartment). Depending on the severity of its damage, a non-native protein is directed to chaperones for refolding or targeted for destruction by a proteasome (see figure).

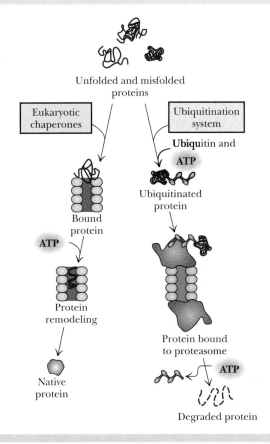

Adapted from Figures 2 and 3 in Wickner, S., Maurizi, M. R., and Gottesman, S., 1999. Posttranslational quality control: Folding, refolding, and degrading proteins. *Science* **286**:1888–1893.

PROBLEMS

1. The following sequence represents part of the nucleotide sequence of a cloned cDNA:

    … CAATACGAAGCAATCCCGCGACTAGACCTTAAC …

    Can you reach an unambiguous conclusion from these data about the partial amino acid sequence of the protein encoded by this cDNA?

2. A random (AG) copolymer was synthesized using a mixture of 5 parts adenine nucleotide to 1 part guanine nucleotide as substrate. If this random copolymer is used as an mRNA in a cell-free protein synthesis system, which amino acids will be incorporated into the polypeptide product? What will be the relative abundances of these amino acids in the product?

3. Review the evidence establishing that aminoacyl-tRNA synthetases bridge the information gap between amino acids and codons. Indicate the various levels of specificity possessed by

aminoacyl-tRNA synthetases that are essential for high-fidelity translation of messenger RNA molecules.

4. Point out why Crick's wobble hypothesis would allow fewer than 61 anticodons to be used to translate the 61 sense codons. How might "wobble" tend to accelerate the rate of translation?

5. How many codons can mutate to become nonsense codons through a single base change? Which amino acids do they encode?

6. Nonsense suppression occurs when a suppressor mutant arises that reads a nonsense codon and inserts an amino acid, as if the nonsense codon were actually a sense codon. Which amino acids do you think are most likely to be incorporated by nonsense suppressor mutants?

7. Why do you suppose eukaryotic protein synthesis is only 10% as fast as prokaryotic protein synthesis?

8. If the tunnel through the large ribosomal subunit is 10 nm long, how many amino acid residues might be contained

within it? Assume that the growing polypeptide chain is in an extended (ε) (β-sheet–like) conformation.

9. Eukaryotic ribosomes are larger and more complex than prokaryotic ribosomes. What advantages and disadvantages might this greater ribosomal complexity bring to a eukaryotic cell?

10. What ideas can you suggest to explain why ribosomes invariably exist as two-subunit structures instead of a larger, single-subunit entity?

11. How do prokaryotic cells determine whether a particular methionyl-tRNA^{Met} is intended to initiate protein synthesis or to deliver a Met residue for internal incorporation into a polypeptide chain? How do the Met codons for these two different purposes differ? How do eukaryotic cells handle these problems?

12. What is the Shine–Dalgarno sequence? What does it do? The efficiency of protein synthesis initiation may vary by as much as 100-fold for different mRNAs. How might the Shine–Dalgarno sequence be responsible for this difference?

13. In the protein synthesis elongation events described on pages 822–825, which of the following seems the most apt account of the peptidyl transfer reaction: (a) The peptidyl-tRNA delivers its peptide chain to the newly arrived aminoacyl-tRNA situated in the A site, or (b) the aminoacyl end of the aminoacyl-tRNA moves toward the P site to accept the peptidyl chain? Which of these two scenarios makes more sense to you? Why?

14. Why might you suspect that the elongation factors EF-Tu and EF-Ts are evolutionarily related to the G proteins of membrane signal transduction pathways described in Chapter 10?

15. Human rhodanese (33 kD) consists of 296 amino acid residues. Approximately how many ATP equivalents are consumed in the synthesis of the rhodanese polypeptide chain from its constituent amino acids *and* the folding of this chain into an active tertiary structure?

16. A single proteolytic break in a polypeptide chain of a native protein is often sufficient to initiate its total degradation. What does this fact suggest to you regarding the structural consequences of proteolytic nicks in proteins?

FURTHER READING

Arnez, J. G., and Moras, D., 1997. Structural and functional considerations of the aminoacylation reaction. *Trends in Biochemical Sciences* **22**:211–216.

Baku, B., and Horwich, A. L., 1998. The Hsp70 and Hsp60 chaperone machines. *Cell* **92**:351–366.

Ban, N., et al., 2000. The complete atomic structure of the large ribosomal subunit at 2.4 Å resolution. *Science* **289**:905–920.

Baumeister, W., et al., 1998. The proteasome: Paradigm of a self-compartmentalizing protease. *Cell* **92**:367–380.

Bibi, E., 1998. The role of the ribosome-translocon complex in translation and assembly of polytopic membrane proteins. *Trends in Biochemical Sciences* **23**:51–58.

Bukau, B., et al., 2000. Getting newly synthesized proteins into shape. *Cell* **101**:119–122.

Carter, A. P., et al., 2000. Functional insights from the structure of the 30S ribosomal subunit and its interactions with antibiotics. *Nature* **407**:340–348.

Caskey, C. T., 1973. "Inhibitors of protein synthesis." In *Metabolic Inhibitors*, vol. 4. Hochster, R. M., Kates, M., and Quastel, J. H., eds. New York: Academic Press.

Cate, J. H., et al., 1999. X-ray crystal structure of 70S functional ribosomal complexes. *Science* **285**:2095–2104.

Cech, T. R., 2000. The ribosome is a ribozyme. *Science* **289**:878–879.

Cedergren, R., and Miramontes, P., 1996. The puzzling origin of the genetic code. *Trends in Biochemical Sciences* **21**:199–200.

Clemons, W. M., Jr., 1999. Structure of a bacterial 30S ribosomal subunit at 5.5 Å resolution. *Nature* **400**:833–840.

Crick, F. H. C., 1966. Codon–anticodon pairing: The wobble hypothesis. *Journal of Molecular Biology* **19**:548–555. Crick's original paper on wobble interactions between tRNAs and mRNA.

Crick, F. H. C., et al., 1961. General nature of the genetic code for proteins. *Nature* **192**:1227–1232. This study foresaw the nature of the genetic code as later substantiated by biochemical results.

Dever, T. E., 1999. Translation initiation: Adept at adapting. *Trends in Biochemical Sciences* **24**:398–403.

Ellis, R. J., 1998. Steric chaperones. *Trends in Biochemical Sciences* **23**:43–45.

Ferrel, K., et al., 2000. Regulatory subunit interactions of the 26S proteasome: A complex problem. *Trends in Biochemical Sciences* **25**:83–88.

Gabashvili, I. S., et al., 2000. Solution structure of the *E.coli* 70S ribosome at 11.5Å resolution. *Cell* **100**:537–549.

Garrett, R., 1999. Mechanics of the ribosome. *Nature* **400**:811–812.

Gesteland, R. F., Cech, T. R., and Atkins, J. F., eds. 1999. *The RNA World*, 2nd ed. Cold Spring Harbor, N.Y.: Cold Spring Harbor Laboratory Press.

Gingras, A.-C., et al., 1999. eIF-4 initiation factors: Effectors of mRNA recruitment to ribosomes and regulators of translation. *Annual Review of Biochemistry* **68**:913–963.

Green, R., and Noller, H. F., 1997. Ribosomes and translation. *Annual Review of Biochemistry* **66**:679–716.

Green, R., Switzer, C., and Noller, H. F., 1998. Ribosome-catalyzed peptide-bond formation with an A-site substrate covalently linked to 23S ribosomal RNA. *Science* **280**:286–289.

Groll, M., et al., 1997. Structure of 20S proteasome from yeast at 2.4Å resolution. *Nature* **386**:463–471.

Haas, A. L., 1997. Introduction: Evolving roles for ubiquitin in cellular regulation. *The FASEB Journal* **11**:1053–1054.

Haas, A. L., and Siepman, T. J., 1997. Pathways of ubiquitin conjugation. *The FASEB Journal* **11**:1257–1268.

Hartl, F. U., 1996. Molecular chaperones in cellular protein folding. *Nature* **381**:571–580.

Hentze, M. W., 1997. eIF-4G: A multipurpose ribosome adapter? *Science* **275**:500–501.

Hershey, J. W. B., Mathews, M. B., and Sonenberg, N., eds., 1996. *Translational Control*. Cold Spring Harbor, N.Y.: Cold Spring Harbor Laboratory Press.

Hochstrasser, M., 1996. Ubiquitin-dependent protein degradation. *Annual Review of Genetics* **30**:405–439.

Huttenhofer, A., and Bock, A., 1998. RNA structures involved in seleno-protein synthesis. In *RNA Structure and Function*. Simons, R. W., and Grunberg-Monago, M., eds. Cold Spring Harbor, N.Y.: Cold Spring Harbor Laboratory Press.

Ibba, M., and Söll, D., 1999. Quality control mechanisms during translation. *Science* **286**:1893–1897.

Ibba, M., Curnow, A. W., and Söll, D., 1997. Aminoacyl-tRNA synthesis: Divergent routes to a common goal. *Trends in Biochemical Sciences* **22**:39–42.

Khaitovich, P., et al., 1999. Characterization of functionally active subribosomal particles from *Thermus aquaticus*. *Proceedings of the National Academy of Sciences, U.S.A.* **96**:85–90.

Knight, R. D., et al., 1999. Selection, history, and chemistry: Three faces of the genetic code. *Trends in Biochemical Sciences* **24**:241–247.

Low, S. C., and Berry, M. J., 1996. Knowing when not to stop: Selenocysteine incorporation in eukaryotes. *Trends in Biochemical Sciences* **21**:203–208.

Lowe, J., et al., 1995. Crystal structure of the 20S proteasome from the archaeon *T. acidophilum* at 3.4 Å resolution. *Science* **268**:533–539.

Lupas, A., et al., 1997. Self-compartmentalizing proteases. *Trends in Biochemical Sciences* **22**:399–404.

Marcotrigiano, J., et al., 1997. Cocrystal structure of the messenger RNA 5′ cap-binding protein (eIF-4E) bound to 7-methyl-GTP. *Cell* **89**:951–961.

Martin, J., and Hartl, F. U., 1997. Chaperone-assisted protein folding. *Current Opinion in Structural Biology* **7**:41–52.

Matlack, K. E. S., Mothes, W., and Rapoport, T. A., 1998. Protein translocation: Tunnel vision. *Cell* **92**:381–390.

Mueller, F., and Brimacombe, R., 1997. A new model for the three-dimensional folding of *Escherichia coli* 16S ribosomal RNA: I. Fitting the RNA to a 3D electron microscopic map at 20 Å. II. The RNA-protein interaction. III. The topography of the functional center (with van Heel, H. S., and Rinke-Appel, J.). *Journal of Molecular Biology* **271**:524–587.

Muth, G. W., et al., 2000. A single adenosine with a neutral pK_a in the ribosomal peptidyl transferase center. *Science* **289**:947–950.

Nakamura, Y., et al., 2000. Mimicry grasps reality in transcription termination. *Cell* **101**:349–352.

Netzer, W. J., and Hartl, F. U., 1998. Protein folding in the cytosol: Chaperonin-dependent and -independent mechanisms. *Trends in Biochemical Sciences* **23**:68–73.

Neuport, W., 1997. Protein import into mitochondria. *Annual Review of Biochemistry* **66**:863–917.

Nirenberg, M. W., and Leder, P., 1964. RNA codewords and protein synthesis. *Science* **145**:1399–1407. The use of simple trinucleotides and a ribosome-binding assay to decipher the genetic code.

Nissen, P., et al., 1995. Crystal structure of the ternary complex of Phe-tRNA^{Phe}, EF-Tu, and a GTP analog. *Science* **270**:1464–1472.

Nissen, P., et al., 2000. The structural basis of ribosome activity in peptide bond synthesis. *Science* **289**:920–930.

Nyborg, J., et al., 1996. Structure of the ternary complex of EF-Tu: Macromolecular mimicry in translation. *Trends in Biochemical Sciences* **21**:81–82.

Pagano, M., 1997. Cell cycle regulation by the ubiquitin pathway. *The FASEB Journal* **11**:1067–1075.

Pestova, T. V., et al., 2000. The joining of ribosomal subunits requires eIF-5B. *Nature* **403**:332–335.

Pickart, C. M., 1997. Targeting of substrates to the 26S proteasome. *The FASEB Journal* **11**:1055–1066.

Rapoport, T. A., Jungnickel, B., and Kytay, U., 1996. Protein transport across the eukaryotic endoplasmic reticulum and bacterial inner membranes. *Annual Review of Biochemistry* **65**:271–303.

Rechsteiner, M., and Rogers, W. S., 1996. PEST sequences and regulation by proteolysis. *Trends in Biochemical Sciences* **21**:267–271.

Rhoads, R. E., 1999. Signal transduction pathways that regulate eukaryotic protein synthesis. *Journal of Biological Chemistry* **274**:30337–30340.

Rould, M. A., et al., 1989. Structure of *E. coli* glutaminyl-tRNA synthetase complexed with tRNA^{Gln} and ATP at 2.8Å resolution. *Science* **246**:1135–1142.

Sachs, A. B., and Varani, G., 2000. Eukaryotic translation initiation: There are two sides (at least) to every story. *Nature Structural Biology* **7**:356–361.

Samuel, C. E., 1993. The eIF-2α protein kinases, regulators of translation in eukaryotes from yeast to humans. *Journal of Biological Chemistry* **268**:7603–7606.

Schimmel, P., 1995. An operational RNA code for amino acids and variations in critical nucleotide sequences in evolution. *Journal of Molecular Evolution* **40**:531–536.

Schimmel, P., and Schmidt, E., 1995. Making connections: RNA-dependent amino acid recognition. *Trends in Biochemical Sciences* **20**:1–2.

Schluenzen, F., et al., 2000. Structure of the functionally activated small ribosomal subunit at 3.3 Å resolution. *Cell* **102**:615–623.

Schmidt, M., and Kloetzel, P.-M., 1997. Biogenesis of 20S proteasomes: The complex maturation pathway of a complex enzyme. *The FASEB Journal* **11**:1235–1243.

Stark, H., et al., 1997. Arrangement of tRNAs in pre- and posttranslational ribosomes revealed by electron cryomicroscopy. *Cell* **88**:19–28.

Stark, H., et al., 1997. Visualization of elongation factor Tu on the *Escherichia coli* ribosome. *Nature* **389**:403–406.

Svergun, D. I., et al., 1997. Solution scattering analysis of the 70S *Escherichia coli* ribosome by contrast variation. I. Invariants and validation of electron microscopy models. II. A model of the ribosome and its RNA at 3.5 nm resolution. *Journal of Molecular Biology* **271**:588–618.

Tarun, S. Z., Jr., et al., 1997. Translation factor eIF-4G mediates *in vitro* poly(A) tail-dependent translation. *Proceedings of the National Academy of Sciences, U.S.A.* **94**:9046–9051.

Varshavsky, A., 1997. The ubiquitin system. *Trends in Biochemical Sciences* **22**:383–387.

Voges, D., et al., 1999. The 26S proteasome: A molecular machine for controlled proteolysis. *Annual Review of Biochemistry* **68**:1015–1068.

Wickner, S., Maurizi, M. R., and Gottesman, S., 1999. Posttranslational quality control: Folding, refolding, and degrading proteins. *Science* **286**:1888–1893.

Wilkinson, K. D., 1997. Regulation of ubiquitin-dependent processes by deubiquitinating enzymes. *The FASEB Journal* **11**:1245–1256.

Wilson, K., and Noller, H. F., 1998. Molecular movement inside the translational engine. *Cell* **92**:337–349.

Wimberly, B. T., et al., 2000. Structure of the 30S ribosomal subunit. *Nature* **407**:327–339.

Xu, Z., Horwich, A. L., and Sigler, P. B., 1997. The crystal structure of the asymmetric GroEL-GroES-(ADP)$_7$ chaperonin complex. *Nature* **388**:741–750.

Signal Transduction: The Relay of Metabolic and Environmental Information

How little we know, how much to discover,
What chemical forces flow from lover to lover.
How little we understand what touches off that tingle,
That sudden explosion when two tingles intermingle.

Who cares to define what chemistry this is?
Who cares with your lips on mine
How ignorant bliss is,
So long as you kiss me and the world around us
    shatters?
How little it matters how little we know.

"How Little We Know" by P. SPRINGER and C. LEIGH as recorded by Frank Sinatra, April 5, 1956, Capitol Records, Inc.

Outline

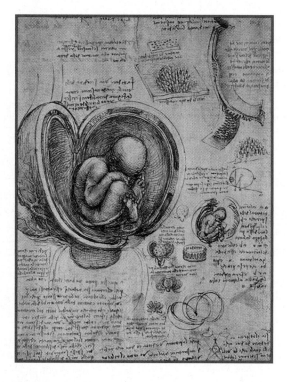

Drawing of a human fetus in utero, by Leonardo da Vinci. Human sexuality and embryonic development represent two hormonally regulated processes of universal interest. (Florentine Royal Collection. Windsor, England A.K.G., Berlin/Superstock, International)

Higher life-forms must have molecular mechanisms for detecting environmental information as well as mechanisms that allow for communication at the cell and tissue levels. Sensory systems detect and integrate physical and chemical information from the environment and pass this information along by the process of **neurotransmission.** Control and coordination of processes at the cell and tissue levels are achieved not only by neurotransmission but also by chemical signals in the form of **hormones** that are secreted by one set of cells to direct the activity of other cells. In this chapter, we will address these mechanisms of information transfer, beginning with the molecular basis of hormone action and then moving to excitable membranes, neurotransmission, and sensory systems.

Hormones are secreted by certain cells, usually located in glands, and travel, either by simple diffusion or circulation in the bloodstream, to specific target cells. As we shall see, some hormones bind to specialized receptors on the plasma membrane and induce responses within the cell without themselves entering the target cell (Figure 26.1). Other hormones actually enter the target cell and interact with specific receptors there. By these mechanisms, hormones regulate the metabolic processes of various organs and tissues; facilitate and control growth, differentiation, reproductive activities, and learning and memory; and help the organism cope with changing conditions and stresses in its environment.

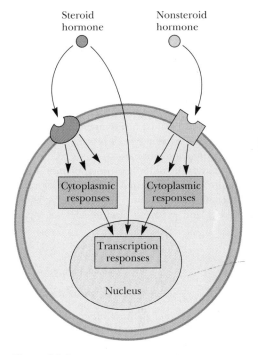

Figure 26.1 Nonsteroid hormones bind exclusively to plasma membrane receptors, which mediate the cellular responses to the hormone. Steroid hormones exert their effects either by binding to plasma membrane receptors or by diffusing to the nucleus, where they modulate transcriptional events.

26.1 Hormones

Many different chemical species act as hormones. **Steroid hormones,** all derived from cholesterol, regulate metabolism, salt and water balances, inflammatory processes, and sexual function. Several hormones are **amino acid derivatives.** Among these are epinephrine and norepinephrine, which regulate smooth-muscle contraction and relaxation, blood pressure, cardiac rate, and the processes of lipolysis and glycogenolysis (Chapter 19); and the thyroid hormones, which stimulate metabolism. **Peptide hormones** are a large group of hormones that appear to regulate processes in all body tissues, including the release of yet other hormones.

Hormones and other signal molecules in biological systems bind with very high affinities to their receptors, displaying K_D values in the range of 10^{-12} to 10^{-6} M. The concentrations of hormones are maintained at levels equivalent to or slightly above these K_D values. Once hormonal effects have been induced, the hormone is usually rapidly metabolized.

Steroid Hormones

The steroid hormones include the **glucocorticoids** (cortisol and corticosterone), the **mineralocorticoids** (aldosterone), **vitamin D,** and the **sex hormones** (progesterone and testosterone, for example) (Figure 26.2; see Chapter 20 for the details of their synthesis). The steroid hormones exert their effects in two ways. First, by entering cells and migrating to the nucleus, steroid hormones act as transcription regulators, modulating gene expression. These

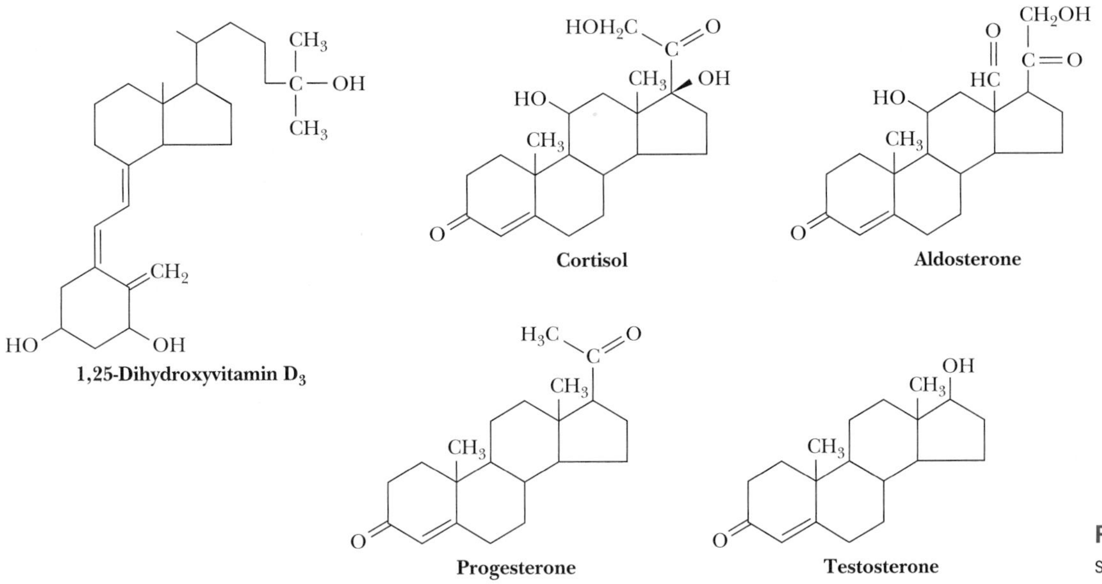

1,25-Dihydroxyvitamin D₃

Cortisol

Aldosterone

Progesterone

Testosterone

Figure 26.2 Structures of some steroid hormones.

HUMAN BIOCHEMISTRY

The Acrosome Reaction

Steroid hormones affect ion channels in the **acrosome reaction,** which must occur before human sperm can fertilize an egg. The **acrosome** is an organelle that surrounds the head of a sperm (see figure) and lies just inside and juxtaposed with the plasma membrane. The acrosome itself is essentially a large vesicle of hydrolytic and proteolytic enzymes. In the acrosome reaction, influx of Ca^{2+} ions causes the outer acrosomal membrane to fuse with the plasma membrane. These fused membrane segments separate from each other and diffuse away, freeing the acrosomal enzymes to attack the egg and exposing binding sites on the inner acrosomal mem-

brane that are thought to interact with the egg in the fertilization process.

This acrosome reaction is induced by **progesterone,** a female hormone secreted by the cumulus oophorus, a collection of ovarian follicle cells surrounding the egg. Intracellular Ca^{2+} levels increase within seconds of treating human sperm with progesterone. These effects must occur via binding of the steroid to the sperm plasma membrane. A far longer time would be required for progesterone to act through an enhancement of transcription.

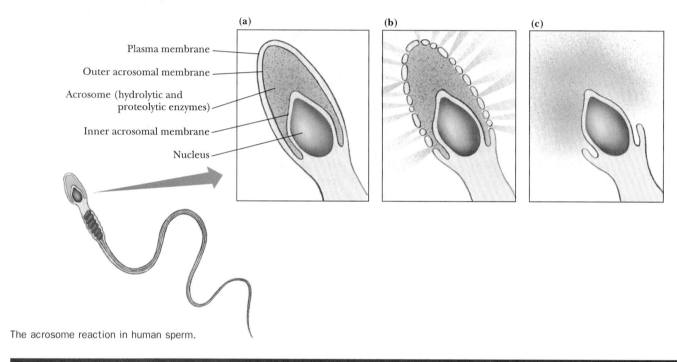

The acrosome reaction in human sperm.

effects of the steroid hormones occur on time scales of hours and involve synthesis of new proteins. Considerable evidence has accumulated, however, that steroids can also act at the cell membrane, directly regulating ligand-gated ion channels and perhaps other processes. These latter processes take place very rapidly, on time scales of seconds and minutes.

The Polypeptide Hormones

The largest class of hormones in vertebrate organisms is that of the **polypeptide hormones** (Table 26.1). One of the first polypeptide hormones to be discovered, **insulin,** was described by Frederick G. Banting and Charles Best in 1921. Insulin, which is typical of the **secreted polypeptide hormones,** is a secretion of the pancreas that controls glucose utilization and promotes the synthesis of proteins, fatty acids, and glycogen. It is discussed in detail in Chapters 4, 10, and 19.

Many other polypeptide hormones are produced and processed in a manner similar to that of insulin. Three unifying features of their synthesis and

Table 26.1 Some Representative Polypeptide Hormones

Hormone	Amino Acid Residues	Source	Target Cells	Function
Bradykinin	9	Kidney, other tissues	Blood vessels	Causes vasodilation
Calcitonin	33	Thyroid gland	Bone	Regulates plasma Ca^{2+} and phosphate
Gastrin	17	Gastrointestinal tract	GI tract, gallbladder, pancreas	Regulates digestion
Glucagon	29	Pancreas	Primarily liver	Regulates metabolism and blood glucose
Growth hormone (GH)	191	Anterior pituitary	Many: bone, fat, liver	Stimulates skeleton and muscle growth
Insulin	A, 21 B, 30	Pancreas	Primarily liver, muscle, and fat	Regulates metabolism and blood glucose
Prolactin	197	Anterior pituitary	Breast	Stimulates milk

Adapted from Rhoades, R., and Pflanzer, R., 1992. *Human Physiology,* 2nd ed. Philadelphia: Saunders College Publishing.

cellular processing should be noted. First, all secreted polypeptide hormones are originally synthesized with a signal sequence which facilitates their eventual direction to secretory granules and thence to the extracellular milieu. Second, peptide hormones are usually synthesized from mRNA as inactive precursors, termed **preprohormones,** which become activated by proteolysis. Third, a single polypeptide precursor or preprohormone may produce several different peptide hormones by suitable proteolytic processing.

An impressive example of the production of many hormone products from a single precursor is the case of **prepro-opiomelanocortin,** a 250-residue precursor peptide synthesized in the pituitary gland. A cascade of proteolytic steps produces, as the name implies, a natural *opi*ate substance (**endorphin**) and several other hormones (Figure 26.3). Endorphins and other opiate hormones are produced by the body in response to systemic stress. These substances probably contribute to the "runner's high" that marathon runners describe.

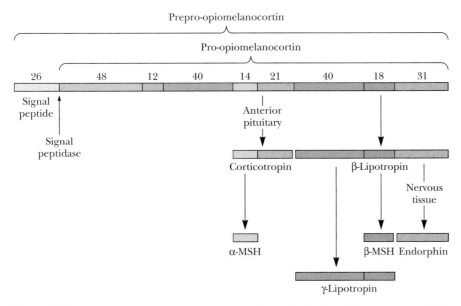

Figure 26.3 The conversion of prepro-opiomelanocortin to a family of peptide hormones, including corticotropin, β- and γ-lipotropin, α- and β-MSH, and endorphin.

26.2 Signal Transduction Pathways

Hormonal regulation depends upon the transduction of the hormonal signal across the plasma membrane to specific intracellular sites, particularly the nucleus. Signal transduction pathways consist of a step-wise progression of signaling stages: receptor → transducer → effector. The **receptor** perceives the signal, **transducers** relay the signal, and the **effectors** convert the signal into an intracellular response. Often, effector action involves a series of steps, each of which is mediated by an enzyme, and each of these enzymes can be considered as an amplifier in a pathway connecting the hormonal signal to its intracellular targets. Such enzymes are typically protein kinases (and protein phosphatases), and many steps in these signaling pathways involve phosphorylation (or dephosphorylation) of serine, threonine, or tyrosine residues on target proteins. The complexity of signal transduction is thus manifested in estimates that the human genome contains about 1000 protein kinase and protein phosphatase genes. Although most of these kinases—and other signaling proteins and molecules—have not been identified, the principles that govern these pathways are now apparent (see also Chapter 10).

Many protein kinases and phosphatases have relatively broad specificities, but they interact with their substrates via specialized recognition domains on the involved proteins. There are also cellular strategies for localization of these signaling molecules. As we shall see later, modular proteins having one or more protein–protein or lipid–protein recognition domains recruit kinases and other proteins into heteromultimeric **scaffolds** that mediate those intracellular events the hormone is intended to elicit.

Many Signaling Pathways Involve Enzyme Cascades

Signaling pathways must operate with speed and precision, facilitating the accurate relay of intracellular signals to specific targets. But how does this happen? **Enzyme cascades** are one answer to this question. Enzymes can produce (or modify) a large number of molecules rapidly and specifically. The hormonal activation of glycogen phosphorylase is mediated by cAMP and involves a cascade of three enzymes—adenylyl cyclase, cAMP-dependent protein kinase (PKA), and phosphorylase kinase (Figure 10.39, page 358). As indicated in the previous section, the cascade acts like a series of amplifiers, dramatically increasing the magnitude of the intracellular response available from a very small amount of hormone.

Connections—Complete Pathways from Membrane to Nucleus

The complete pathway from hormone binding at the plasma membrane to modulation of transcription in the nucleus is understood for a few signaling pathways. Figure 26.4 shows a complete signal transduction pathway that connects receptor tyrosine kinases, the Ras GTPase, cytoplasmic Raf, and two other protein kinases with transcription factors that alter gene expression in the nucleus. This pathway represents just one component of a complex signaling network that involves many other proteins and signaling factors. The presumed existence of about a thousand protein kinases and the growing list of signaling factors portend a complex and interwoven network of signaling interactions in nearly all cells.[1]

---

[1]The American Association for the Advancement of Science oversees a consortium of researchers who have established an outstanding Web site—the Signal Transduction Knowledge Environment (STKE). This site is an up-to-date and ongoing compilation of information about cell signaling and signal transduction. The URL is http://www.stke.org.

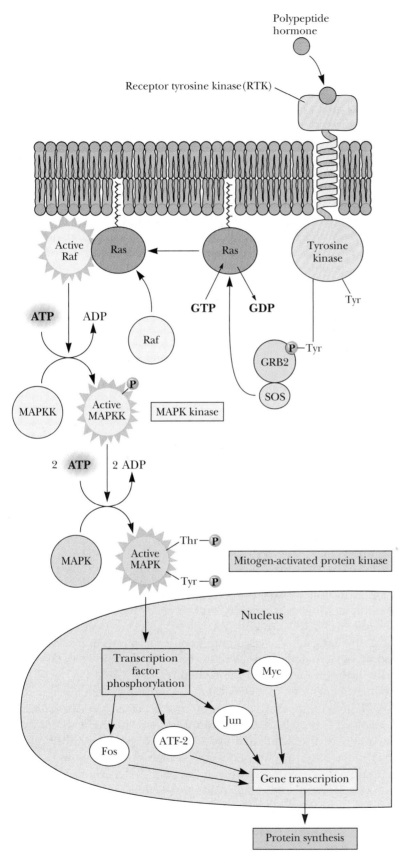

Figure 26.4 A complete signal transduction pathway that connects a hormone receptor with transcription events in the nucleus. A number of similar pathways have been characterized.

26.3 Signal-Transducing Receptors Transmit the Hormonal Message

Very often in life, the message is more important than the messenger, and this is certainly true for hormones. The structure and chemical properties of a hormone are only important for specific binding of the hormone to its appropriate receptor. Of much greater interest and importance, however, is the metabolic information carried by the hormonal signal. The information implicit in the hormonal signal is interpreted by the cell, and an intricate pattern of cellular responses ensues.

Steroid hormones may either bind to plasma membrane receptors or exert their effects within target cells, entering the cell and migrating to their sites of action via specific cytoplasmic receptor proteins. The nonsteroid hormones, which act by binding to outward-facing plasma membrane receptors, activate unique **signal transduction pathways** that mobilize various **second messengers**—cyclic nucleotides, Ca^{2+} ions, and other substances that activate or inhibit enzymes or cascades of enzymes in very specific ways. These hormonally activated processes are the focus of this chapter.

All receptors that mediate transmembrane signaling processes fit into one of three **receptor superfamilies:**

1. The **G-protein–coupled receptors** (see Section 26.4) are integral membrane proteins with an extracellular recognition site for ligands, and an intracellular recognition site for a **GTP-binding protein** (see following discussion).
2. The **single-transmembrane segment (1-TMS) catalytic receptors** are proteins with only a single transmembrane segment and substantial globular domains on both the extracellular and intracellular faces of the membrane. The extracellular domain is the ligand recognition site and the intracellular catalytic domain is either a **tyrosine kinase** or a **guanylyl cyclase.**
3. **Oligomeric ion channels** consist of associations of protein subunits, each of which contains several transmembrane segments. These oligomeric structures are **ligand-gated ion channels.** Binding of the specific ligand typically opens the ion channel. The ligands for these ion channels are neurotransmitters.

The G-Protein–Coupled Receptors

The G-protein–coupled receptors (GPCRs) have primary and secondary structures similar to that of bacteriorhodopsin (Chapter 6), with seven apparent transmembrane α-helical segments, and they are thus known as 7-*trans*membrane segment (7-TMS) proteins. Rhodopsin and the β-adrenergic receptor, for which epinephrine is a ligand, are good examples (Figure 26.5). The site for binding of cationic catecholamines to the adrenergic receptors is located within the hydrophobic core of the receptor.

Binding of hormone to a GPCR induces a conformation change that activates a GTP-binding protein, also known as a G protein (discussed in Section 26.4). Activated G proteins transduce a variety of cellular effects, including activation of adenylyl and guanylyl cyclases (which produce cAMP and cGMP from ATP), activation of phospholipases (which produce second messengers from phospholipids), and activation of Ca^{2+} and K^+ channels (which leads to elevation of cell $[Ca^{2+}]$ and $[K^+]$. (All of these effects are described in Section 26.4.)

The Single-TMS Receptors

Receptor proteins that span the plasma membrane with a single helical transmembrane segment possess an external ligand recognition site and an internal domain with enzyme activity—either **receptor tyrosine kinase** or **receptor**

(a) β₂-**adrenergic receptor**

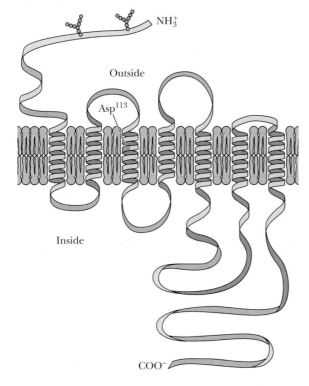

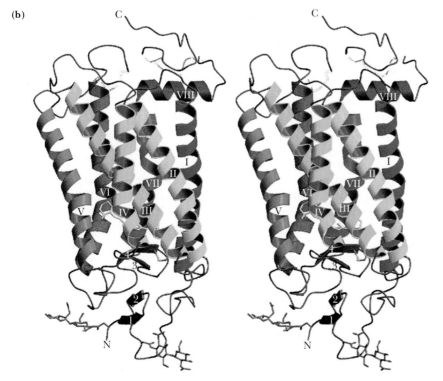

Figure 26.5 **(a)** The arrangement of the β_2-adrenergic receptor in the membrane. Substitution of Asp[113] in the third hydrophobic domain of the β-adrenergic receptor with an Asn or Gln by site-directed mutagenesis results in a dramatic decrease in affinity of the receptor for both agonists and antagonists. Significantly, this Asp residue is conserved in all other G-protein–coupled receptors that bind biogenic amines, but is absent in receptors whose ligands are not amines. Asp[113] appears to be the counter-ion for the amine moiety of adrenergic ligands. **(b)** Stereo image of rhodopsin, viewed parallel to the plane of the membrane. The seven transmembrane α-helices are indicated by roman numerals. *(Figure 2 from Palczewski, K., et al., 2000. Crystal structure of rhodopsin: A G-protein–coupled receptor. Science **289**: 739–745. Image kindly provided by K. Palczewski, University of Washington.)*

Class I receptor (EGF receptor) Class II receptor (insulin receptor) Class III receptor (FGF receptor)

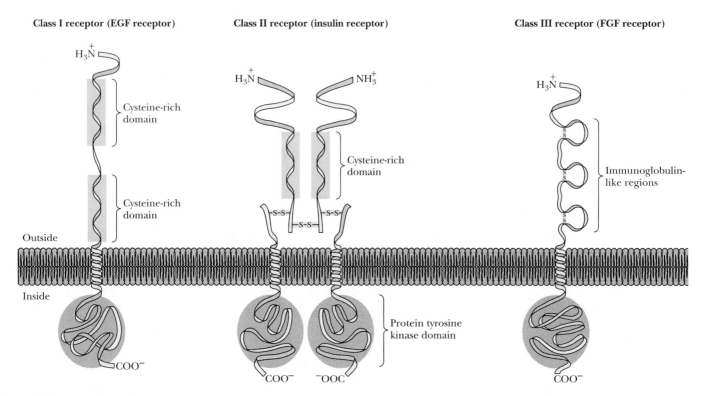

Figure 26.6 The three classes of receptor tyrosine kinases. Class I receptors are monomeric and contain a pair of Cys-rich repeat sequences. The insulin receptor, a typical Class II receptor, is a glycoprotein composed of two kinds of subunits in an $\alpha_2\beta_2$ tetramer. The α- and β-subunits are synthesized as a single peptide chain, together with an N-terminal signal sequence. Subsequent proteolytic processing yields the separate α- and β-subunits. The β-subunits of 620 residues each are integral transmembrane proteins, with only a single transmembrane α-helix; the amino terminus is outside the cell and the carboxyl terminus inside. The α-subunits of 735 residues each are extracellular proteins that are linked to the β-subunits and to each other by disulfide bonds. The insulin-binding domain is located in a cysteine-rich region on the α-subunits. Class III receptors contain multiple immunoglobulin-like domains. Shown here is fibroblast growth factor (FGF) receptor, which has three immunoglobulin-like domains. *(Adapted from Ullrich A., and Schlessinger, J., 1990. Cell* **61:**203–212.)

guanylyl cyclase. Each of these enzyme activities is manifested in two different cellular forms. Thus, guanylyl cyclase activity is found both in membrane-bound receptors and in soluble, cytoplasmic proteins. Tyrosine kinase activity, on the other hand, is exhibited by two different types of membrane proteins: the receptor tyrosine kinases are integral transmembrane proteins, whereas the non-receptor tyrosine kinases are peripheral, lipid-anchored proteins.

Receptor Tyrosine Kinases

The binding of polypeptide hormones and growth factors to receptor tyrosine kinases activates the tyrosine kinase activity of these proteins. These catalytic receptors are composed of three domains (Figure 26.6): a glycosylated extracellular hormone-binding domain; a transmembrane domain consisting of a single transmembrane α-helix; and an intracellular domain that includes both a tyrosine kinase domain that mediates the biological response to the hormone or growth factor via its catalytic activity, and a regulatory domain that contains multiple autophosphorylation sites.

There are three classes of receptor tyrosine kinases (see Figure 26.6). Class I, exemplified by the **epidermal growth factor (EGF) receptor,** has an extracellular domain containing two Cys-rich repeat sequences. Class II, typified by the **insulin receptor,** has an $\alpha_2\beta_2$ tetrameric structure with transmembrane β-subunits and a Cys-rich domain in the extracellular α-subunit. Class III receptors, such as the **platelet-derived growth factor (PDGF) receptor,** have five (or sometimes three) immunoglobulin-like extracellular domains.

Receptor Tyrosine Kinases Are Membrane-Associated Allosteric Enzymes

Given that the extracellular and intracellular domains of receptor tyrosine kinases are joined only by a single transmembrane helical segment, how does ex-

HUMAN BIOCHEMISTRY

Apoptosis: The Programmed Suicide of Cells

According to a Japanese saying, "Once we are in the land of the living, we will eventually die." True as this may be for human beings, it is also true of the cells of our bodies. In the ongoing cycle of cell division and differentiation, many surplus and harmful cells are produced. These cells must be removed or killed to maintain the integrity and homeostasis of the organism. The magnitude of such cell death is often surprising; for example, more than 95% of thymocytes die during maturation of the thymus! The cell death that occurs during embryogenesis, metamorphosis, and normal cell turnover is termed "programmed cell death" or **apoptosis** (where the second "p" is silent).

Apoptosis can be initiated in a variety of ways. One of these involves "death factors"—proteins such as **Fas ligand** and **tumor necrosis factor,** or **TNF,** that are members of the **cytokine** family of proteins. These factors bind to specific plasma membrane receptors **Fas** and **TNF-receptor,** respectively, prompting trimeric associations of the receptor proteins in the membrane. These receptor proteins (see figure) possess intracellular **death domains**—polypeptide motifs that adopt an unusual structure consisting of six antiparallel, amphipathic α-helices. These receptor death domains form oligomeric associations with other similar death domains on adaptor proteins (known as **FADD** or **MORT1**). This prompts special **death-effector domains** on FADD/MORT1 to bind to cysteine proteases known as **caspases** (so-named with a "c" for cysteine and "asp" indicating that these proteases have an active site Cys and cleave after an Asp residue in their protein substrates). Sequential activation in a cascade of caspases, together with other events, triggers apoptosis.

Inappropriate apoptosis underlies the cause of certain human diseases, including neurodegenerative disorders and cancer. An unusual caspase cleavage appears to play a role in Huntington's disease (HD), the cause of the death of folk singer Woody Guthrie. The mutation responsible for HD is an expansion of a CAG trinucleotide repeat at the 5′-end of the gene *Hdh,* which encodes an essential protein **(huntingtin)** of unknown function. The CAG extension results in an extended polyglutamine domain on the

N-terminus of the protein. Huntington's disease is manifested if this Gln repeat region exceeds 35 residues. Downstream of the poly-Gln region is a cluster of five DXXD motifs that are cleaved specifically by caspase 3. The longer the poly-Gln region, the greater is the activity of caspase 3 in cleaving the DXXD motifs. The released poly-Gln domains are cytotoxic, provoking the death of affected neurons, progressive loss of control of voluntary movements, general loss of neural function, and eventual death.

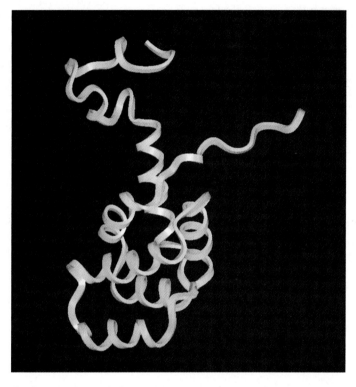

The Fas "death domain."

See *Interactive Biochemistry*
CD-ROM and Workbook, page 108

tracellular hormone binding activate intracellular tyrosine kinase activity? How is the signal transduced? As shown in Figure 26.7, signal tranduction occurs by hormone-induced oligomeric association of receptors. Binding of hormone triggers a conformational change in the extracellular domain, which induces oligomeric association. Oligomeric association allows adjacent cytoplasmic domains to interact, leading to phosphorylation of the cytoplasmic domains and stimulation of cytoplasmic tyrosine kinase activity. By virtue of these ligand-induced conformation changes and oligomeric interactions, receptor tyrosine kinases are called **membrane-associated allosteric enzymes.**

Receptor Tyrosine Kinases Phosphorylate a Variety of Cellular Target Proteins

Receptor tyrosine kinases catalyze the phosphorylation of numerous cellular target proteins, producing coordinated changes in cell behavior, including

Subclass I

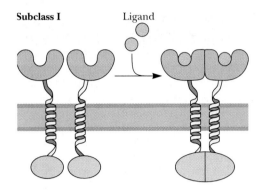

Subclass II

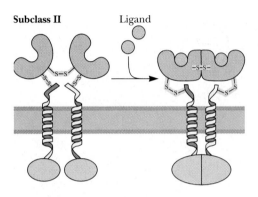

Subclass III

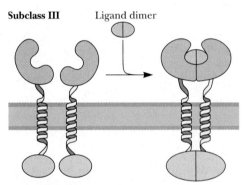

Figure 26.7 Ligand (hormone)-stimulated oligomeric association of receptor tyrosine kinases.

alterations in membrane transport of ions and amino acids, the transcription of genes, and the synthesis of proteins. Many individual phosphorylation targets have been characterized, including the γ-isozymes of phospholipase C and phosphatidylinositol-3-kinase.

Membrane-Bound Guanylyl Cyclases Are Single-TMS Receptors

Another cellular second messenger, **guanosine 3′,5′-cyclic monophosphate (cGMP),** is formed from GTP by **guanylyl cyclase,** an enzyme found in several different forms in different cellular locations. **Membrane-bound guanylyl cyclases** constitute a second class of single-TMS receptors with an extracellular hormone-binding domain; a single, α-helical transmembrane segment; and an intracellular catalytic domain (Figure 26.8). A variety of peptides act to stimulate the membrane-bound guanylyl cyclases, including **atrial natriuretic peptide (ANP),** which regulates body fluid homeostasis and cardiovascular function; the **heat-stable enterotoxins** from *E. coli*; and a series of peptides secreted by mammalian ova (eggs), which stimulate sperm motility and act as sperm chemoattractant signals. The binding of these peptides to an extracellular site on the guanylyl cyclase in the sperm plasma membrane induces a conformational change that activates the intracellular catalytic site for cyclase activity. Activation may involve oligomerization of receptors in the membrane, as for the RTKs discussed previously.

Nonreceptor Tyrosine Kinases

The first tyrosine kinases to be discovered were associated with **viral transforming proteins.** These proteins, produced by oncogenic viruses, enable the virus to transform animal cells, that is, to convert them to the cancerous state. A prime example is the tyrosine kinase expressed by the **src gene** of **Rous** or **avian sarcoma virus.** The protein product of this gene is **pp60$^{\text{v-src}}$** (*p*hospho-*p*rotein, *60* kD, *v*iral origin, *s*arcoma-causing). The v-src gene was derived from the avian proto-oncogenic gene c-src during the original formation of the virus. The cellular proto-oncogene homolog of pp60$^{\text{v-src}}$ is referred to as pp60$^{\text{c-src}}$, a 526-residue peripheral membrane protein. It undergoes two posttranslational modifications: (a) the amino group of the NH_2-terminal glycine is modified by

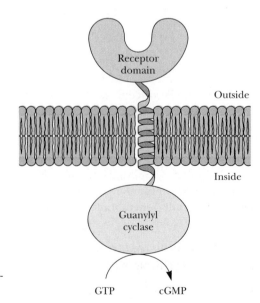

Figure 26.8 The structure of membrane-bound guanylyl cyclases.

(a)

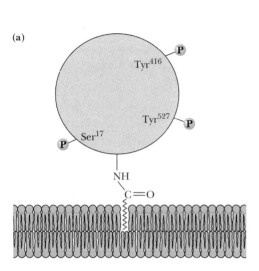

(b)

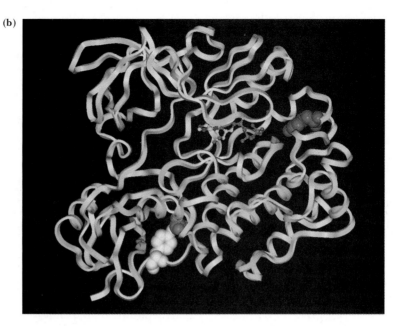

Figure 26.9 **(a)** The soluble tyrosine kinase pp60$^{v\text{-}src}$ is anchored to the plasma membrane via an N-terminal myristyl group. **(b)** The structure of protein tyrosine kinase pp60$^{c\text{-}src}$, showing AMP-PNP in the active site (ball-and-stick), Tyr^{416} (red), and Tyr^{527} (yellow). Tyr^{527} is phosphorylated (purple).

the covalent attachment of a **myristyl** group (this modification is required for membrane association of the kinase; see Figure 26.9a), and (b) Ser^{17}, Tyr^{416}, and Tyr^{527} are phosphorylated. The phosphorylation at Tyr^{416}, which increases kinase activity two- to threefold, appears to be an autophosphorylation. On the other hand, phosphorylation at Tyr^{527} is inhibitory and is catalyzed by another kinase known as CSK. The significance of nonreceptor tyrosine kinase activity to cell growth and transformation is only partially understood, but 1% of all cellular proteins (many of which are also kinases) are phosphorylated by these kinases.

Soluble Guanylyl Cyclases Are Receptors for Nitric Oxide

Nitric oxide, or **NO·,** a reactive free radical, acts as a neurotransmitter and as a second messenger, activating soluble guanylyl cyclase more than 400-fold.

A DEEPER LOOK

Nitric Oxide, Nitroglycerin, and Alfred Nobel

NO· is the active agent released by **nitroglycerin** (see figure), a powerful drug that ameliorates the symptoms of heart attacks and **angina pectoris** (chest pain due to coronary artery disease) by causing the dilation of coronary arteries. Nitroglycerin is also the active agent in dynamite. Ironically, Alfred Nobel, the inventor of dynamite who also endowed the Nobel Prizes, suffered himself from angina pectoris. In a letter to a friend in 1885, Nobel wrote, "It sounds like the irony of fate that I should be ordered by my doctor to take nitroglycerin internally."

$$
\begin{array}{ccc}
CH_2 & CH & CH_2 \\
| & | & | \\
O & O & O \\
| & | & | \\
NO_2 & NO_2 & NO_2
\end{array}
$$

The structure of nitroglycerin, a potent vasodilator.

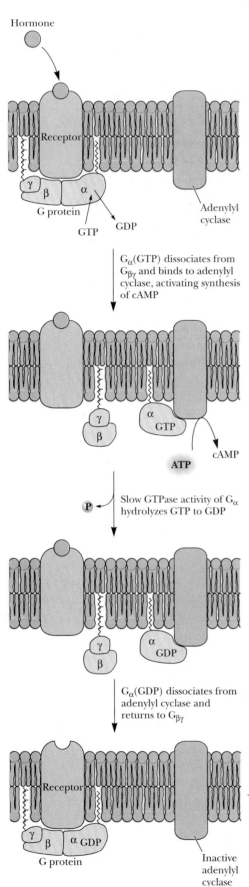

The cGMP thus produced also acts as a second messenger, inducing relaxation of vascular smooth muscle and mediating penile erection. As a dissolved gas, NO· is capable of rapid diffusion across membranes in the absence of any apparent carrier mechanism. This property makes NO· a particularly attractive second messenger because NO· generated in one cell can exert its effects quickly in many neighboring cells. NO· has a very short cellular half-life (1 to 5 sec) and is rapidly degraded by nonenzymatic pathways.

26.4 Transduction of Receptor Signals

Receptor signals are *transduced* in one of three ways, to initiate actions inside the cell:

1. Exchange of GDP for GTP by GTP-binding proteins (G proteins), which leads to generation of *second messengers,* including cAMP, phospholipid breakdown products, and Ca^{2+}.
2. Receptor-mediated activation of phosphorylation cascades that in turn trigger activation of various enzymes. This is the action of the receptor tyrosine kinases described in Section 26.3. Protein kinases and protein phosphatases acting as effectors will be discussed in Section 26.5.
3. Conformation changes that open ion channels or recruit proteins into nuclear transcription complexes. Ion channels are discussed in Section 26.7 and the formation of nuclear transcription complexes was described in Chapter 24.

GPCR Signals Are Transduced by G Proteins

The signals of G-protein–coupled receptors (GPCRs) are transduced by GTP-binding proteins, known more commonly as G proteins. The large G proteins are heterotrimers consisting of α-(45 to 47 kD), β-(35 kD), and γ-(7 to 9 kD) subunits. The α-subunit binds GDP or GTP and has an intrinsic, slow GTPase activity. The $G_{\alpha\beta\gamma}$ complex in the unactivated state has GDP at the nucleotide site (Figure 26.10). The binding of hormone to receptor stimulates a rapid exchange of GTP for GDP on G_{α}. The binding of GTP causes G_{α} to dissociate from $G_{\beta\gamma}$ and to associate with an effector protein such as adenylyl cyclase. Binding of G_{α}(GTP) activates adenylyl cyclase. The adenylyl cyclase actively synthesizes cAMP as long as G_{α}(GTP) remains bound to it. However, the intrinsic GTPase activity of G_{α} eventually hydrolyzes GTP to GDP, leading to the dissociation of G_{α}(GDP) from adenylyl cyclase and the reassociation with the $G_{\beta\gamma}$ dimer, regenerating the inactive heterotrimeric $G_{\alpha\beta\gamma}$ complex.

Two stages of amplification occur in the G-protein–mediated hormone response. First, a single hormone–receptor complex can activate many G proteins before the hormone dissociates from the receptor. Second, and more obvious, the G_{α}-activated adenylyl cyclase synthesizes many cAMP molecules. Thus, the binding of hormone to a very small number of membrane receptors stim-

◄ **Figure 26.10** Activation of adenylyl cyclase by heterotrimeric G proteins. Binding of hormone to its receptor causes a conformational change that induces the receptor to catalyze a replacement of GDP by GTP on G_{α}. The G_{α} (GTP) complex dissociates from $G_{\beta\gamma}$ and binds to adenylyl cyclase, stimulating synthesis of cAMP. Bound GTP is slowly hydrolyzed to GDP by the intrinsic GTPase activity of G_{α}. G_{α}(GDP) dissociates from adenylyl cyclase and reassociates with $G_{\beta\gamma}$. G_{α} and G_{γ} are lipid-anchored proteins. Adenylyl cyclase is an integral membrane protein consisting of 12 transmembrane α-helical segments.

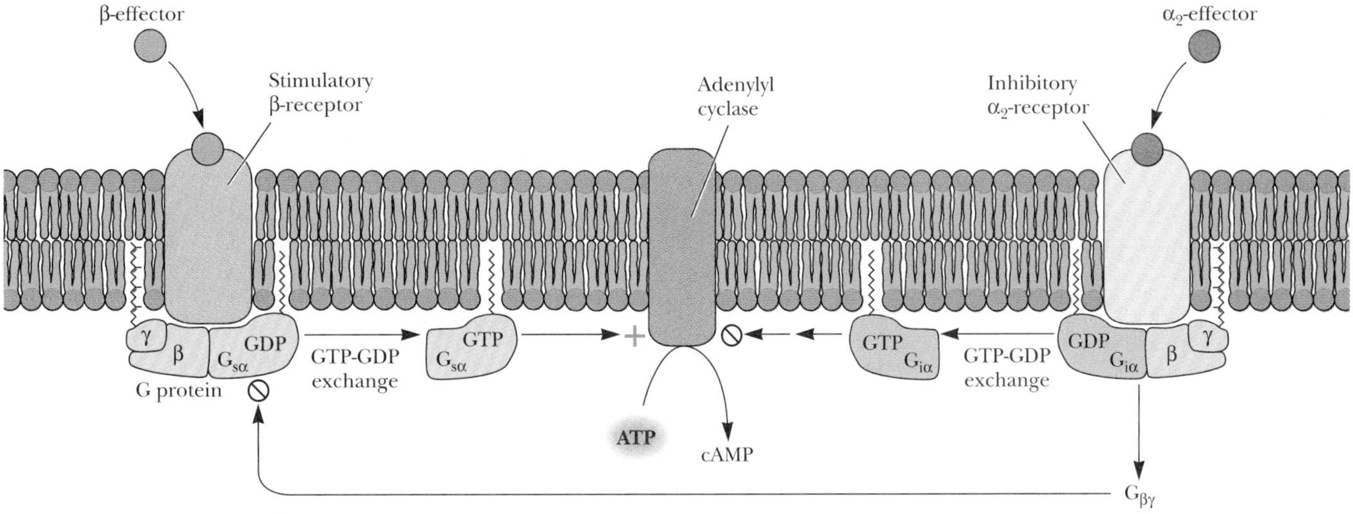

Figure 26.11 Adenylyl cyclase activity is modulated by the interplay of stimulatory (G_s) and inhibitory (G_i) G proteins. Binding of hormones to β_1- and β_2-adrenergic receptors activates adenylyl cyclase via G_s, whereas hormone binding to α_2 receptors leads to the inhibition of adenylyl cyclase. Inhibition may occur by direct inhibition of cyclase activity by $G_{i\alpha}$ or by binding of $G_{i\beta\gamma}$ to $G_{s\alpha}$.

See page 109

ulates a large increase in the concentration of cAMP within the cell. The hormone receptor, G protein, and cyclase constitute a complete hormone **signal transduction unit** (Figure 26.11).

Hormone-receptor–mediated processes regulated by G proteins may be stimulatory or inhibitory. Each hormone receptor interacts specifically with either a stimulatory G protein, denoted **G_s**, or an inhibitory G protein, denoted **G_i**.

Synthesis and Degradation of Cyclic AMP

Cyclic AMP (denoted cAMP) was identified in 1956 by Earl Sutherland, who termed cAMP a **second messenger,** since it is the intracellular response provoked by binding of hormone (the first messenger) to its receptor. Since Sutherland's discovery of cAMP, many other second messengers have been identified (Table 26.2). The concentrations of second messengers in cells are carefully regulated. Synthesis or release of a second messenger is followed quickly by

Table 26.2   Intracellular Second Messengers[a]

Messenger	Source	Effect
cAMP	Adenylyl cyclase	Activates protein kinases
cGMP	Guanylyl cyclase	Activates protein kinases, regulates ion channels, regulates phosphodiesterases
Ca^{2+}	Ion channels in ER and plasma membrane	Activates protein kinases, activates Ca^{2+}-modulated proteins
IP_3	PLC action on PI	Activates Ca^{2+} channels
DAG	PLC action on PI	Activates protein kinase C
Phosphatidic acid	Membrane component and product of PLD	Activates Ca^{2+} channels, inhibits adenylyl cyclase
Ceramide	PLC action on sphingomyelin	Activates protein kinases
Nitric oxide (NO)	NO synthase	Activates guanylyl cyclase, relaxes smooth muscle
Cyclic ADP-ribose	cADP-ribose synthase	Activates Ca^{2+} channels

[a] IP_3 is inositol-1,4,5-trisphosphate; PLC is phospholipase C; PLD is phospholipase D; PI is phosphatidylinositol; DAG is diacylglycerol.

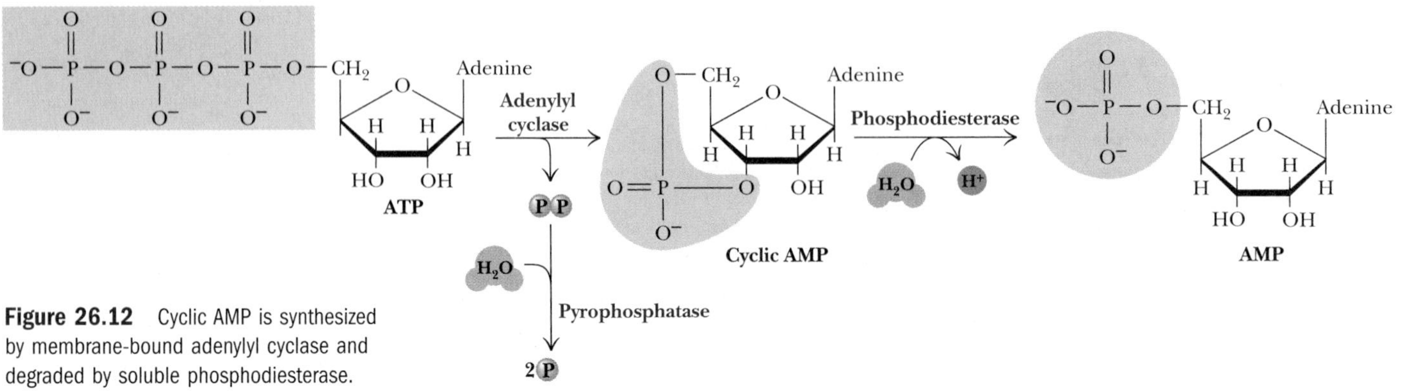

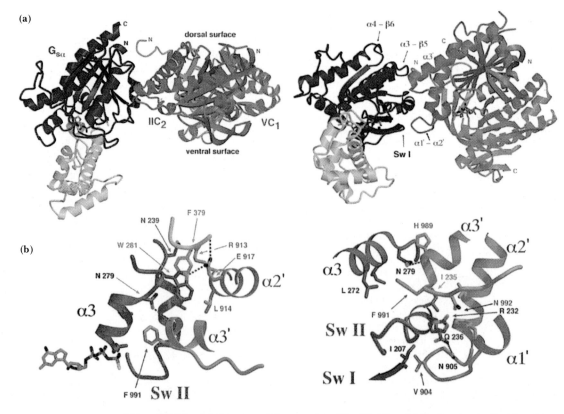

Figure 26.12 Cyclic AMP is synthesized by membrane-bound adenylyl cyclase and degraded by soluble phosphodiesterase.

degradation or removal from the cytosol. Following its synthesis by adenylyl cyclase, cAMP is broken down to 5'-AMP by phosphodiesterase (Figure 26.12).

Adenylyl cyclase (AC) is an integral membrane enzyme. Its catalytic domain, on the cytoplasmic face of the plasma membrane, includes two subdomains denoted VC_1 and IIC_2. Binding of the α-subunit of G_s (denoted $G_{s\alpha}$) activates the AC catalytic domain. Alfred Gilman, Stephen Sprang, and coworkers have determined the structure of a complex of $G_{s\alpha}$ (with bound GTP) with the cytoplasmic domains (VC_1 and IIC_2) of adenylyl cyclase (Figure 26.13). The $G_{s\alpha}$ complex binds to a cleft at one corner of the C_2 domain, and the surface of $G_{s\alpha}$-GTP that contacts adenylyl cyclase is the same surface that binds the $G_{\beta\gamma}$ dimer. The catalytic site, where ATP is converted to cyclic AMP, is far removed from the bound G protein.

See page 103

Figure 26.13 **(a)** Two views of the complex of the VC_1-IIC_2 catalytic domain of adenylyl cyclase and $G_{s\alpha}$. **(b)** Details of the $G_{s\alpha}$ complex in the same orientation as the structures above. SW I and SW II are "switch regions," whose conformations differ greatly depending on whether GTP or GDP is bound. *(Courtesy of Alfred Gilman, University of Texas Southwestern Medical Center)*

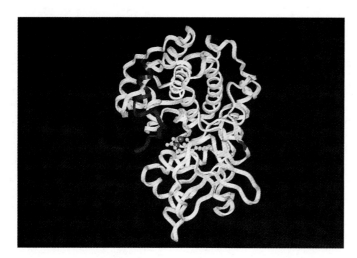

Figure 26.14 Cyclic AMP–dependent protein kinase is shown complexed with a pseudosubstrate peptide (red). This complex also includes ATP (yellow) and two Mn^{2+} ions (violet) bound at the active site.

 See page 104

cAMP Activates Protein Kinase A

All second messengers exert their cellular effects by binding to one or more target molecules. Cyclic AMP produced by adenylyl cyclase activates a protein kinase, which is thus known as cAMP-dependent protein kinase; protein kinase A, as this enzyme is also known, activates many other cellular proteins by phosphorylation. The activation of protein kinase A by cAMP and regulation of the enzyme by intrasteric control was described in detail in Chapter 10 (page 349). The structure of protein kinase A has served as a paradigm for understanding many related protein kinases (Figure 26.14).

Ras and the Small GTP-Binding Proteins

GTP-binding proteins are implicated in growth control mechanisms in higher organisms. Certain tumor virus genomes contain genes encoding 21-kD proteins that bind GTP and show regions of homology with other G proteins. The first of these genes to be identified was found in *rat sarcoma virus*, and was dubbed the ***ras* gene.** Genes implicated in tumor formation are known as **oncogenes;** often they are mutated versions of normal, noncancerous genes involved in growth regulation, so-called **proto-oncogenes.** The normal, cellular Ras protein is a GTP-binding protein that functions similarly to other G proteins described earlier, activating metabolic processes when GTP is bound and becoming inactive when GTP is hydrolyzed to GDP. The GTPase activity of the normal Ras p21 is very low, as appropriate for a G protein that regulates long-term effects like growth and differentiation. A specific **GTPase-activating protein (GAP)** increases the GTPase activity of the Ras protein. Mutant (oncogenic) Ras proteins have severely impaired GTPase activity, which apparently causes serious alterations of cellular growth and metabolism in tumor cells. The conformations of Ras proteins (Figure 26.15) in complexes with GDP are different from the corresponding complexes with GTP analogs like GMP-PNP (a nonhydrolyzable analog of GTP in which the β-P and γ-P are linked by N rather than by O). Two regions of the Ras structure change conformation upon GTP hydrolysis. These conformation changes mediate the interactions of Ras with other proteins, termed **effectors.**

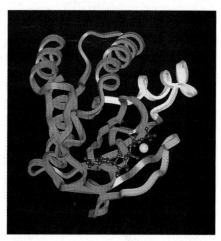

(a)

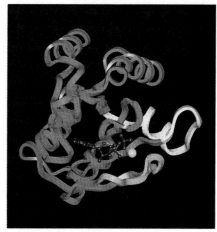

(b)

Figure 26.15 The structure of Ras complexed with **(a)** GDP and **(b)** GMP-PNP. The Ras p21-GMP-PNP complex is the active conformation of this protein.

 See page 66

A DEEPER LOOK

RGSs and GAPs: Switches That Turn Off G Proteins

Nature has made Ras and $G_{s\alpha}$ very poor enzymes by design. For example, Ras hydrolyzes GTP with a rate constant of only 0.02 min^{-1}. These G proteins are active only in the GTP-bound state, and downstream targets will dissociate upon GTP hydrolysis. If Ras and $G_{s\alpha}$ were efficient enzymes, the GTP-bound state would be short-lived, and G protein–mediated signaling would be ineffective.

But how can G proteins be switched off if they are inherently poor GTPases? The answer is provided by **GTPase-activating proteins** and **regulators of G protein signaling (RGS),** which cause dramatic increases in GTPase activity when bound to G proteins. The figure on the left shows Ras (white) with a fragment of a GAP (blue)

bound to it. GAP stands for GTPase-Activating Protein. GAPs increase the GTPase activity of Ras by a factor of 10^5. The figure on the right shows RGS (blue) bound to $G_{i\alpha}$ (yellow). RGS proteins accelerate $G_{s\alpha}$-catalyzed GTP hydrolysis by nearly a hundred-fold. In both Ras and $G_{s\alpha}$, GTPase activity and the conversion from GTP-bound to GDP-bound forms of the protein involve conformation changes in the switch regions, which are a portions of the G-protein structure that surround the GTP/GDP binding site. GAPs and RGS increase the GTPase activities of their respective G proteins by binding near the active site and stabilizing the transition state of the GTP hydrolysis reaction.

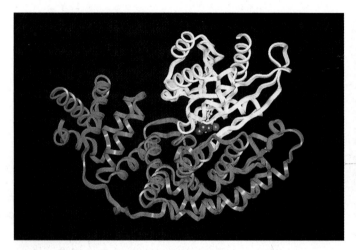

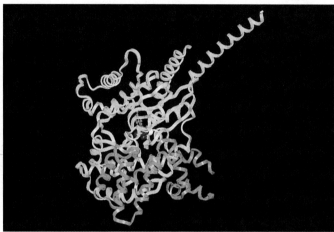

See page 110

G Proteins Are Universal Signal Transducers

A given G protein can be activated by several different hormone-receptor complexes. For example, either glucagon or epinephrine, binding to their distinctive receptor proteins, can activate the same species of G protein in liver cells. The effects are additive, and stimulation by glucagon and epinephrine together leads to higher cytoplasmic concentrations of cAMP than activation by either hormone alone.

G proteins are a universal means of signal transduction in higher organisms, activating many hormone-receptor–initiated cellular processes in addition to adenylyl cyclase. Such processes include, but are not limited to, activation of phospholipases C and A_2, and the opening or closing of transmembrane channels for K^+, Na^+, and Ca^{2+} in brain, muscle, heart, and other organs (Table 26.3). G proteins are integral components of sensory pathways such as vision and olfaction. More than 100 different G protein–coupled receptors and at least 21 distinct G proteins are known. At least a dozen different G protein effectors have been identified, including a variety of enzymes and ion channels.

Specific Phospholipases Release Second Messengers

A diverse array of second messengers is generated by the breakdown of membrane phospholipids. The binding of certain hormones and growth factors to their respective receptors triggers a sequence of events that can lead to the ac-

HUMAN BIOCHEMISTRY

Cancer, Oncogenes, and Tumor Suppressor Genes

The disease state known as **cancer** is the uncontrolled growth and proliferation of one or more cell types in the body. Control of cell growth and division is an incredibly complex process, involving the signal-transducing proteins (and small molecules) described in this chapter and many others like them. The genes that give rise to these growth-controlling proteins are of two distinct types:

1. **Oncogenes:** These genes code for proteins that are capable of stimulating cell growth and division. In normal tissues and organisms, such growth-stimulating proteins are regulated, so that growth is appropriately limited. However, mutations in these genes may result in loss of growth regulation, leading to uncontrolled cell proliferation and tumor development. These mutant genes are known as oncogenes because they induce the oncogenic state—cancer. The normal precursors of these genes are termed **proto-oncogenes** and are essential for normal cell growth and differentiation. Oncogenes are dominant because a mutation of only one of the cell's two copies of that gene can lead to tumor formation. Table A lists a few of the known proto-oncogenes (over 60 are now known).

2. **Tumor suppressor genes:** These genes code for proteins whose normal function is to turn off cell growth. A mutation in one of these growth-limiting genes may result in a protein product that has lost its growth-limiting ability. The normal forms of such genes have been shown to suppress tumor growth and are known as tumor suppressor genes. Because both cellular copies of a tumor suppressor gene must be mutated to foil its growth-limiting action, these genes are recessive in nature. Table B presents several recognized tumor suppressor genes.

Careful molecular analysis of cancerous tissue has shown that tumor development may result from mutations in several proto-oncogenes or tumor suppressor genes. The implication is that there is redundancy in cellular growth regulation. Many (if not all) tumors are the result of either interactions among two or more oncogene products or simultaneous mutations in a proto-oncogene and both copies of a tumor suppressor gene. Cells have thus evolved with overlapping growth-control mechanisms. When one is compromised by mutation, others take over.

Table A A Representative List of Proto-Oncogenes Implicated in Human Tumors

Proto-Oncogene	Neoplasm(s)
Abl	Chronic myelogenous leukemia
ErbB-1	Squamous cell carcinoma; astrocytoma
ErbB-2 (Neu)	Adenocarcinoma of breast, ovary, and stomach
Myc	Burkitt's lymphoma; carcinoma of lung, breast, and cervix
H-Ras	Carcinoma of colon, lung, and pancreas; melanoma
N-Ras	Carcinoma of genitourinary tract and thyroid; melanoma
Src	Carcinoma of colon
Jun	Several
Fos	

Adapted from Bishop, J. M., 1991. Molecular themes in oncogenesis. *Cell* **64**:235–248.

Table B Representative Tumor Suppressor Genes Implicated in Human Tumors

Tumor Suppressor Gene	Neoplasm(s)
RB1	Retinoblastoma; osteosarcoma; carcinoma of breast, bladder, and lung
p53	Astrocytoma; carcinoma of breast, colon, and lung; osteosarcoma
WT1	Wilms tumor
DCC	Carcinoma of colon
NF1	Neurofibromatosis type 1
FAP	Carcinoma of colon
MEN-1	Tumors of parathyroid, pancreas, pituitary, and adrenal cortex

Adapted from Bishop, J. M., 1991. Molecular themes in oncogenesis. *Cell* **64**:235–248.

tivation of **specific phospholipases.** The action of these phospholipases on membrane lipids produces the second messengers shown in Figure 26.16.

Inositol Phospholipid Breakdown Yields Inositol-1,4,5-Trisphosphate and Diacylglycerol

The breakdown of **phosphatidylinositol (PI)** and its derivatives by **phospholipase C** produces a family of second messengers. In the best-understood pathway, successive phosphorylations of PI produce **phosphatidylinositol-4-P (PIP)**

Table 26.3 **G Proteins and Their Physiological Effects**

G Protein	Location	Stimulus	Effector	Effect
G_s	Liver	Epinephrine, glucagon	Adenylyl cyclase	Glycogen breakdown
G_s	Adipose tissue	Epinephrine, glucagon	Adenylyl cyclase	Fat breakdown
G_s	Kidney	Antidiuretic hormone	Adenylyl cyclase	Conservation of water
G_s	Ovarian follicle	Luteinizing hormone	Adenylyl cyclase	Increased estrogen and progesterone synthesis
G_i	Heart muscle	Acetylcholine	Potassium channel	Decreased heart rate and pumping force
G_i/G_o	Brain neurons	Enkephalins, endorphins, opioids	Adenylyl cyclase, potassium channels, calcium channels	Changes in neuron electrical activity
G_q	Smooth muscle cells in blood vessels	Angiotensin	Phospholipase C	Muscle contraction, blood pressure elevation
G_{olf}	Neuroepithelial cells in the nose	Odorant molecules	Adenylyl cyclase	Odorant detection
Transducin (G_t)	Retinal rod and cone cells	Light	cGMP phosphodiesterase	Light detection
GPA1	Baker's yeast	Pheromones	Unknown	Mating

Adapted from Hepler, J., and Gilman. A., 1992. G proteins. *Trends in Biochemical Sciences* **17**:383–387.

 See page 109

and **phosphatidylinositol-4,5-bisphosphate** (**PIP₂**). Four isozymes of phospholipase C (denoted α, β, γ, and δ) hydrolyze PI, PIP, and PIP₂. Hydrolysis of PIP₂ by phospholipase C yields the second messenger **inositol-1,4,5-trisphosphate** (**IP₃**), as well as another second messenger, **diacylglycerol** (**DAG**) (Figure 26.17). IP₃ is water-soluble and diffuses to intracellular organelles where release of Ca^{2+} is activated. DAG, on the other hand, is lipophilic and remains in the plasma membrane where it activates a Ca^{2+}-dependent protein kinase known as **protein kinase C** (see following discussion).

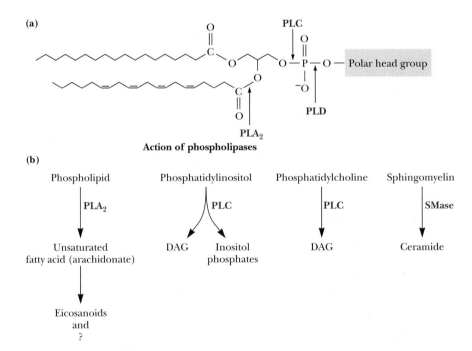

Figure 26.16 **(a)** The general action of phospholipase A₂ (PLA₂), phospholipase C (PLC), and phospholipase D (PLD). **(b)** The synthesis of second messengers from phospholipids by the action of phospholipases and sphingomyelinase.

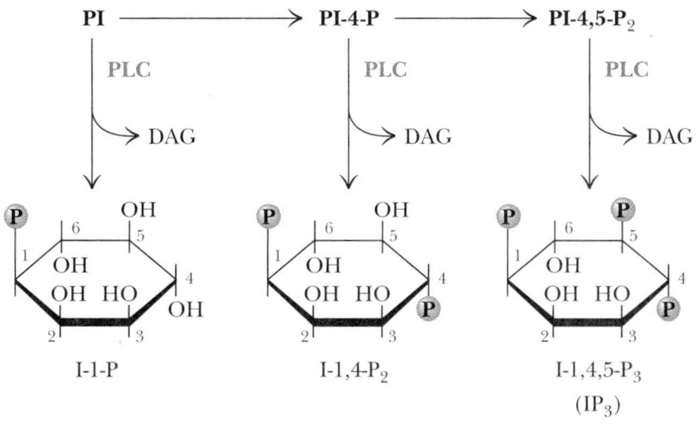

PI ⟶ PI-4-P ⟶ PI-4,5-P$_2$

PLC PLC PLC

DAG DAG DAG

I-1-P I-1,4-P$_2$ I-1,4,5-P$_3$
(IP$_3$)

Figure 26.17 The family of second messengers produced by phosphorylation and breakdown of phosphatidylinositol. PLC action instigates a bifurcating pathway culminating in two distinct and independent second messengers: DAG and IP$_3$.

Activation of Phospholipase C Is Mediated by G Proteins or by Tyrosine Kinases

Phospholipase C-β, -γ, and -δ are all Ca^{2+}-dependent, but the different phospholipase C isozymes are activated by different intracellular events. Phospholipase C-β is stimulated by G proteins (Figure 26.18). On the other hand, phospholipase C-γ is activated by **receptor tyrosine kinases** (Figure 26.19). The primary structures of phospholipase C-β and -γ are shown in Figure 26.20. The X and Y domains of phospholipase C-β and -γ are highly homologous, and both of these domains are required for phospholipase C activation. The other domains of these isozymes confer specificity for G-protein activation or tyrosine kinase activation.

Phosphatidylcholine, Sphingomyelin, and Glycosphingolipids Also Generate Second Messengers

In addition to PI, other phospholipids serve as sources of second messengers. The breakdown of phosphatidylcholine by phospholipases yields a variety of second messengers, including diacylglycerol, phosphatidic acid, and prostaglandins. The action of **sphingomyelinase** on sphingomyelin produces **ceramide,** which stimulates **ceramide-activated protein kinase.** Similarly, gangliosides (such as ganglioside G$_{M3}$ [see Chapter 6]) and their breakdown products modulate the activity of protein kinases and G-protein–coupled receptors.

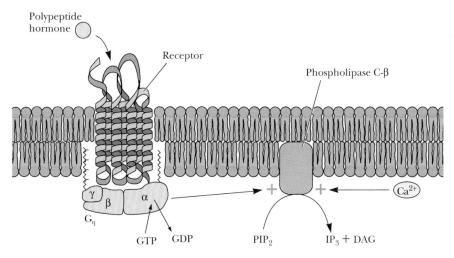

Figure 26.18 Phospholipase C-β is activated specifically by G$_q$, a GTP-binding protein, and also by Ca^{2+}.

Figure 26.19   Phospholipase C-γ is activated by receptor tyrosine kinases and by Ca^{2+}.

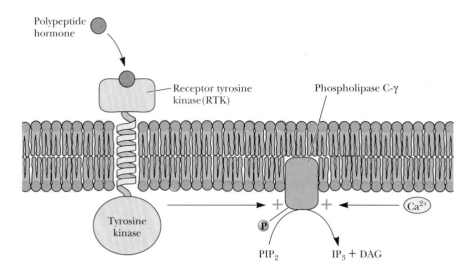

Polypeptide hormone

Receptor tyrosine kinase(RTK)

Phospholipase C-γ

Tyrosine kinase

Ca^{2+}

PIP₂

IP₃ + DAG

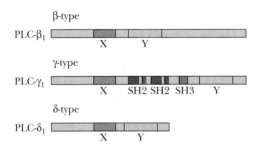

β-type

PLC-β₁

X   Y

γ-type

PLC-γ₁

X   SH2 SH2 SH3   Y

δ-type

PLC-δ₁

X   Y

Figure 26.20   The amino acid sequences of phospholipase C isozymes β, γ, and δ share two homologous domains, denoted X and Y. The sequence of γ isozyme contains src homology domains, denoted SH2 and SH3. SH2 domains (approximately 100 residues in length) interact with phosphotyrosine-containing proteins (such as RTKs), whereas SH3 domains mediate interactions with cytoskeletal proteins. *(Adapted from Dennis, E., Rhee, S., Gillah, M., and Hannun, E., 1991. Role of phospholipases in generating lipid second messengers in signal transduction. FASEB Journal* **5**:2068–2077.)

Calcium As a Second Messenger

Calcium ion is an important intracellular signal. Binding of certain hormones and signal molecules to plasma membrane receptors can cause transient increases in cytosolic Ca^{2+} levels, which in turn can activate a wide variety of enzymatic processes, including smooth-muscle contraction, exocytosis, and glycogen metabolism. (Most of these activation processes depend on special Ca^{2+}-binding proteins discussed in the following section.) Cytosolic $[Ca^{2+}]$ can be increased in two ways (Figure 26.21). As mentioned briefly earlier, cAMP can activate the opening of plasma membrane Ca^{2+} channels, allowing extracellular Ca^{2+} to stream in. On the other hand, cells also contain intracellular reservoirs of Ca^{2+} within the endoplasmic reticulum and **calciosomes,** small membrane vesicles that are similar in some ways to muscle sarcoplasmic reticulum. These special intracellular Ca^{2+} stores are not released by cAMP. They respond to IP₃, a second messenger derived from phosphatidylinositol (PI).

Intracellular Calcium-Binding Proteins

Given the central importance of Ca^{2+} as an intracellular messenger, it should not be surprising that complex mechanisms exist in cells to manage and control Ca^{2+}. Ca^{2+} signals generated by cAMP, IP₃, and other agents are translated into the desired intracellular responses by **calcium-binding proteins,** which in turn regulate many cellular processes (Figure 26.22). One of these, protein kinase C, is described in Section 26.5. The other important Ca^{2+}-binding proteins can, for the most part, be divided into two groups on the basis of struc-

HUMAN BIOCHEMISTRY

PI Metabolism and the Pharmacology of Li⁺

An intriguing aspect of the phosphoinositide story is the specific action of lithium ion, Li⁺, on several steps of PI metabolism. Lithium salts have been used in the treatment of manic-depressive illnesses for more than 30 years, but the mechanism of lithium's therapeutic effects had been unclear. Recently, however, several reactions in the phosphatidylinositol degradation pathway have been

shown to be sensitive to Li⁺ ion. Li⁺ levels similar to those employed in treatment of manic illness thus lead to the accumulation of several key intermediates. This story is far from complete, and many new insights into phosphoinositide metabolism and the effects of Li⁺ can be anticipated.

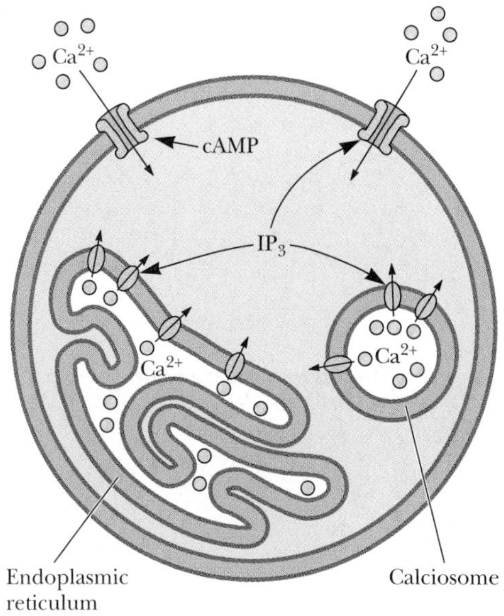

Figure 26.21 Cytosolic $[Ca^{2+}]$ increases occur via the opening of Ca^{2+} channels in the ▶ membranes of calciosomes, the endoplasmic reticulum, and the plasma membrane.

ture and function: (a) the **calcium-modulated proteins,** including **calmodulin, parvalbumin, troponin C,** and many others, all of which have in common a structural feature called the **EF hand** (see Figure 5.22), and (b) the **annexin proteins,** a family of homologous proteins that interact with membranes and phospholipids in a Ca^{2+}-dependent manner.

More than 170 calcium-modulated proteins are known (Table 26.4). All possess a characteristic peptide domain consisting of a short α-helix, a loop of 12 amino acids, and a second α-helix. Robert Kretsinger at the University of Virginia initially discovered this pattern in parvalbumin, a protein first identified in the carp fish and later in neurons possessing a high firing rate and a high oxidative metabolism. Kretsinger lettered the six helices of parvalbumin A through F. He noticed that the E and F helices, joined by a loop, resembled

▼ **Figure 26.22** IP_3-mediated signal transduction pathways. Increased $[Ca^+]$ activates protein kinases, which phosphorylate target proteins. Ca^{2+}/CaM represents calci-calmodulin (Ca^{2+} complexed with the regulatory protein calmodulin).

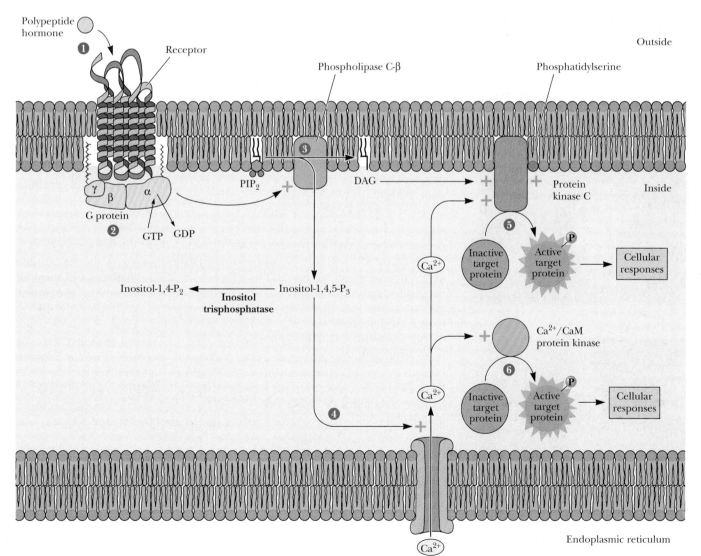

Table 26.4 **Some Calcium-Modulated Proteins**

Protein	Function
α-Actinin	Cross-linking of cytoskeletal F-actin
Calcineurin B	Protein Ser/Thr phosphatase
Calmodulin	Modulates activity of Ca^{2+}-dependent proteins
β- and γ-Crystallins	Ca^{2+}-modulated processes in eye lens
Inositol phospholipid-specific phospholipase C	Second messenger release and cell signaling
Myeloperoxidase	Inflammatory action of neutrophils
Parvalbumin	Acceleration of muscle relaxation, Ca^{2+} sequestration
S-100	Cell cycle progression, cell differentiation, cytoskeleton-membrane interactions
Troponin C	Activation of muscle contraction

Adapted from Heizmann, C. W., ed., 1991. *Novel Calcium Binding Proteins—Fundamentals and Clinical Implications.* New York: Springer-Verlag.

(a)

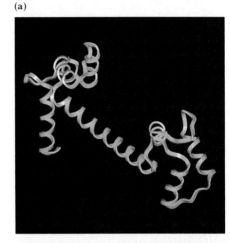

(b)

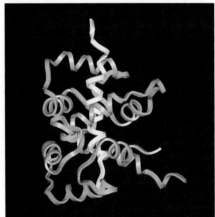

(c)

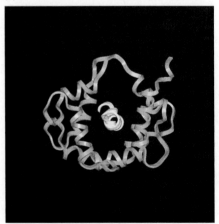

the thumb and forefinger of a right hand (see Figure 5.22), and named this structure the EF hand, a name in common use today to identify the helix-loop-helix motif in calcium-binding proteins. In the EF hand, Ca^{2+} is coordinated by six carboxyl oxygens contributed by a glutamate and three aspartates, by a carbonyl oxygen from a peptide bond, and by the oxygen of a coordinated water molecule. The EF hand was subsequently identified in calmodulin, troponin C, and calbindin-9K (Figure 26.23). Most of the known EF-hand proteins possess two or more (as many as eight) EF-hand domains, usually arranged so that two EF-hand domains can directly contact each other.

Calmodulin Target Proteins Possess a Basic Amphiphilic Helix

The conformations of EF-hand proteins change dramatically upon binding of Ca^{2+} ions. This change promotes binding of the EF-hand protein with its target protein(s). For example, calmodulin (CaM), a 148-residue protein found in many cell types, modulates the activities of a large number of target proteins, including Ca^{2+}-ATPases, protein kinases, phosphodiesterases, and NAD^+ kinase. CaM binds to these and to many other proteins with extremely high affinities (K_D values typically in the high picomolar to low nanomolar range). All CaM target proteins possess a **basic amphiphilic alpha helix (Baa helix),** to which CaM binds specifically and with high affinity. Viewed end-on, in the so-called **helical wheel** representation (Figure 26.24), a Baa helix has mostly hydrophobic residues on one face; basic residues are collected on the opposite face. However, the Baa helices of CaM target proteins, although conforming to the model, show extreme variability in sequence. How does CaM, itself a highly conserved protein, accommodate such a variety of sequence and structure? Each globular domain consists of a large hydrophobic surface flanked by regions of highly negative electrostatic potential—a surface suitable for interacting with a Baa helix. The long central helix joining the two globular regions behaves as a long, flexible tether. When the target protein is bound, the two globular domains fold together, forming a single binding site for target pep-

◀ **Figure 26.23** **(a)** Structure of uncomplexed calmodulin (gold). Calmodulin, with four Ca^{2+}-binding domains, forms a dumbbell-shaped structure with two globular domains joined by an extended, central helix. Each globular domain juxtaposes two Ca^{2+}-binding EF hand domains. An intriguing feature of these EF-hand domains is their nearly identical three-dimensional structure, despite a relatively low degree of sequence homology (only 25% in some cases). **(b, c)** Complex of calmodulin (gold) with a peptide from myosin light chain kinase (white); (b) side view; (c) top view.

 See page 105

(a)

(b)

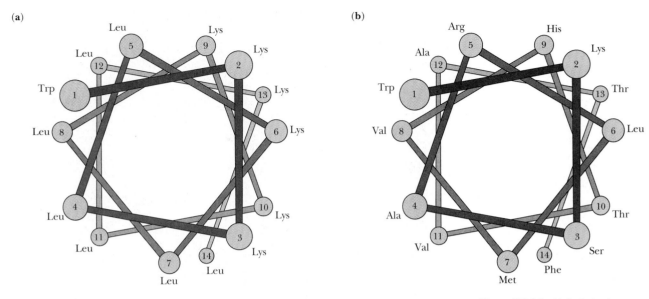

Figure 26.24 Helical wheel representations of **(a)** a model peptide, Ac-WKKLLKLLKKLLKL-CONH$_2$, and **(b)** the calmodulin-binding domain of spectrin, an erythrocyte protein. Positively charged and polar residues are indicated in green and hydrophobic residues are orange. *(Adapted from O'Neil, K., and DeGrado, W., 1990. How calmodulin binds its targets: Sequence independent recognition of amphiphilic α-helices.* Trends in Biochemical Sciences **15**:59–64.)

 See pages 20–24

tides (see Figure 26.23b). The flexible nature of the tethering helix allows the two globular domains to adjust their orientation synergistically for maximal binding of the target protein or peptide.

26.5 Effectors Convert the Signals to Actions in the Cell

Transduction of the hormonal signal leads to activation of **effectors**—usually protein kinases and protein phosphatases—that elicit a variety of actions which regulate discrete cellular functions. Of the thousands of mammalian kinases and phosphatases, the structures and functions of a few are representative.

Protein Kinase A

Most protein kinases share a common catalytic core first characterized in protein kinase A (PKA), the enzyme that phosphorylates phosphorylase kinase (see *A Deeper Look*, page 349, and Figure 10.39, page 358). The active site of the catalytic subunit of PKA in a ternary complex with MnAMP-PNP and a pseudosubstrate inhibitor peptide (Figure 26.25) includes a glycine-rich β-strand that acts as a flap over the triphosphate moiety of the bound nucleotide. A conserved residue, Asp166, is the catalytic base that deprotonates the Ser/Thr-OH during phosphorylation, and Lys168 stabilizes the transition state of the reaction. Three Glu residues on the enzyme are involved in recognition of the pseudosubstrate inhibitor peptide.

Protein Kinase C

The enzymes called "protein kinase C" are actually a family of similar enzymes—isozymes—that encompass three subclasses. The "conventional PKCs" α, βI, βII, and γ are regulated by Ca^{2+}, diacylglycerol (DAG), and phosphatidylserine (PS). Because Ca^{2+} levels increase in the cell in response to IP$_3$, the activation of conventional PKCs depends upon both of the second messengers released by the hydrolysis of PIP$_2$. The "novel PKCs" δ, ε, θ, and η are Ca^{2+}-independent but are regulated by DAG and PS. PKCs ζ, ι, and λ are termed "atypical" and are activated by PS alone. These various cofactor requirements

 See page 113

Figure 26.25 The structure of the catalytic subunit of PKA in a ternary complex with MnAMP-PNP and a pseudosubstrate inhibitor peptide. A glycine-rich β-strand acts as a flap over the triphosphate moiety of the bound nucleotide. The glycine-rich flap that covers the ATP-binding site is shown in magenta, AMP-PNP is bound in the ATP site, Asp-166 is shown in blue, and Lys-168 is shown in yellow.

 See page 104

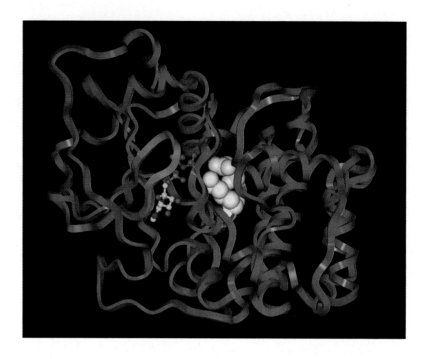

are imparted by subdomains represented in the conventional PKC polypeptide sequence. Conventional PKCs are comprised (Figure 26.26) of four conserved domains (C1–C4) and five variable regions (V1–V5). Domain C1 is a *pseudo-substrate sequence* that regulates the kinase by *intrasteric control* (see *A Deeper Look,* page 349), C2 is a Ca^{2+}-binding domain, C3 is the ATP-binding domain, and C4 binds peptide substrates.

PKCs phosphorylate serine and threonine residues on a wide range of protein substrates. A role for conventional and novel protein kinase Cs in cellular growth and division is demonstrated by their strong activation by phorbol esters (Figure 26.27). These compounds, from the seeds of *Croton tiglium,* are **tumor promoters,** agents that do not themselves cause tumorigenesis but that potentiate the effects of carcinogens. The phorbol esters mimic DAG, bind to the regulatory pseudosubstrate domain of the enzyme, and activate protein kinase C.

Protein Tyrosine Kinase pp60^c-src

The structure of protein tyrosine kinase $pp60^{c\text{-}src}$ (see also Figure 26.9b) consists of an N-terminal "unique domain," an SH2 domain, an SH3 domain, and a kinase domain that includes a small lobe comprised mainly of a twisted β-sheet and a large lobe that is predominantly α-helical (Figure 26.28). (SH2 and

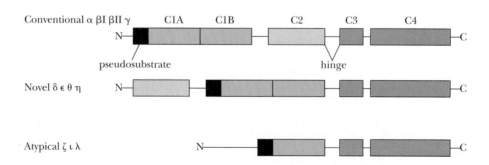

Figure 26.26 The primary structures of the PKC isozymes. Conserved domains C1–C4 are indicated.

Figure 26.27 The structure of a phorbol ester. Long-chain fatty acids predominate at the 12-position, whereas acetate is usually found at the 13-position.

12-*O*-Tetradecanoylphorbol-13-acetate

SH3 domains are discussed in Section 26.6.) Phosphorylation of Tyr^{527} in the SH2 domain inhibits tyrosine kinase activity by drawing an "activation loop" into the active site, blocking ATP and/or substrate binding. Dephosphorylation of Tyr^{527} induces a conformation change that removes the activation loop from the active site, permitting autophosphorylation of Tyr^{416}, which stimulates tyrosine kinase activity.

Protein Tyrosine Phosphatase SHP-2

The human phosphatase SHP-2 is a cytosolic nonreceptor tyrosine phosphatase. It comprises two SH2 domains, a catalytic phosphatase domain, and a C-terminal tail domain. The SH2 domains enable the enzyme to bind to its target substrates, and they also regulate the phosphatase activity. The catalytic domain of SHP-2 consists of nine α-helices and a 10-stranded mixed β-sheet that wraps around one of the helices (Figure 26.29). The other eight helices pack together on the opposite side of the β-sheet.

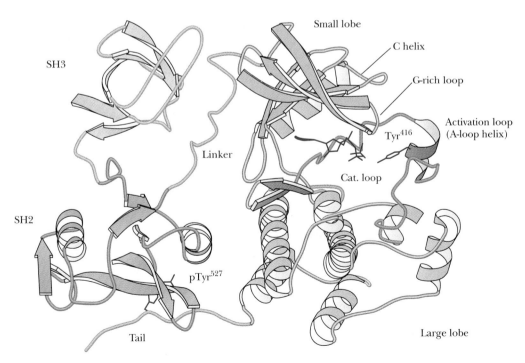

Figure 26.28 A ribbon diagram showing the structure of protein tyrosine kinase pp60[c-src] in its inactive state bound to AMP-PNP. *(Image kindly provided by Stephen C. Harrison)*

Figure 26.29 **(a)** A ribbon diagram showing the structure of protein tyrosine phosphatase SHP-2 in its autoinhibited, closed conformation. The N- and C-terminal SH2 domains are yellow and green, respectively. The catalytic domain is blue, and interdomain linkers (residues 104–111 and 217–220) are white. **(b)** A wireframe model of SHP-2. *(Image kindly provided by Steven E. Shoelson)*

 See page 88

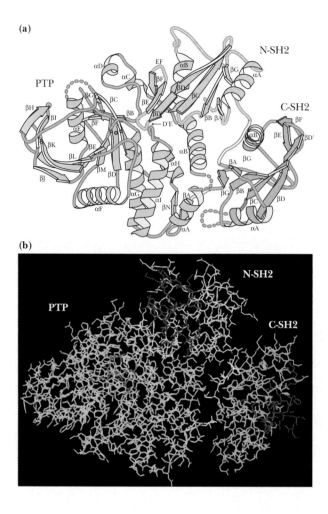

The N-terminal SH2 domain regulates phosphatase activity by binding to the phosphatase domain and directly blocking the active site. When a target peptide containing a phosphotyrosine group binds to the SH2 domain, a conformation change causes this domain to dissociate from the catalytic domain, exposing the active site and allowing peptide substrate to bind. Binding of phosphotyrosine-containing peptide to the second SH2 domain provides additional activation. Target peptides with two phosphotyrosines (one to bind to each SH2 domain) thus provide maximal activation of the phosphatase activity.

26.6 Protein Modules in Signal Transduction

Signal transduction within cells occurs via protein–protein and protein–phospholipid interactions based on protein modules. Proteins with two (or more) such modules associate simultaneously with two (or more) binding partners, leading to assembly of functional complexes, either at an activated cell-surface receptor or free in the cytoplasm. The **SH2 domain,** which binds with high affinity to peptide motifs containing a phosphotyrosine, is a good example (Figure 26.30). The binding of a particular SH2 domain to a particular phosphotyrosine motif depends on the particular sequence of residues that are

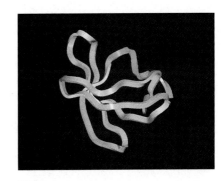

SH2 Domain

Grb2 —(SH3)—[SH2]—(SH3)

Shc —◇ PTB ◇———————————[SH2]

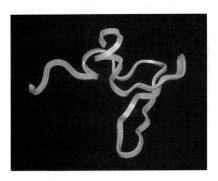

SH3 Domain

p47*phox* —[PX]——————(SH3)—(SH3)—[Pro]

Nck —(SH3)—(SH3)—(SH3)————————[SH2]

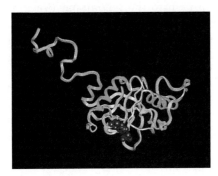

PTB Domain

IRS-1 ⬡PH⬡—◇ PTB ◇—————————

FE65 ⬠WW⬠—◇ PTB ◇—◇ PTB ◇

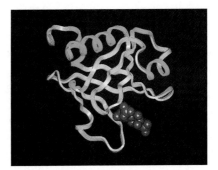

WW Domain

Human Nedd4 —[C2]—⬠WW⬠—⬠WW⬠—⬠WW⬠—⬠WW⬠—[HECT]—

Dystrophin —[Actin binding]—([Spectrin])₂₄—⬠WW⬠—[Cys]—

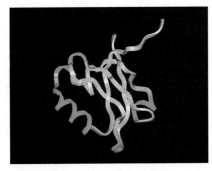

PDZ Domain

Tyrosine Phosphatase ———[PDZ]—[B4.1]—([PDZ])₅—[PTP]—

Grip —[PDZ]—[PDZ]—[PDZ]—[PDZ]—[PDZ]—[PDZ]———[PDZ]

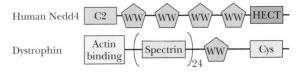

PH Domain

Sos ——————————[Dbl-H]—⬡PH⬡—[Ras-GEF]—[Pro]

Ras-GAP —[SH2]—(SH3)—[SH2]—⬡PH⬡—[GAP]——

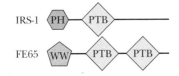

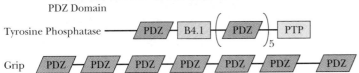

Figure 26.30 Six of the protein modules that are found in cell-signaling proteins. Shown for each is a molecular graphic image of the module, together with primary structures of several proteins in which they are found. *(WW domain coordinates provided by Harmut Oschkinat, Forschungsinstitut für Molekulare Pharmakologie, and Marius Sudol, Mount Sinai School of Medicine)*

 See pages 97–102

C-terminal to the phosphotyrosine. The SH2 domain itself is a module of about 100 residues that consists of a small β-sheet flanked by α-helices.

SH2 domains are the prototype for a growing number of protein modules that play a role in cell signaling. **PTB** modules also recognize and bind to phosphotyrosine motifs, but by recognizing sequence motifs N-terminal to the phosphotyrosines. **SH3** and **WW** modules bind proline-rich target sequences, and **PDZ** modules bind to the terminal four or five residues of a target protein. The **PH** (pleckstrin homology) module, with a fold similar to the PTB module, functions quite differently, associating with specific phosphoinositides and directing target proteins to the plasma membrane. Figure 26.30 illustrates structures of these modules, together with examples of proteins that contain them.

Protein Scaffolds Localize Signaling Molecules

Membrane-bound receptors can amplify their signaling by means of adaptor or scaffolding proteins that provide multiple docking sites for signaling modules on other proteins. Such docking proteins typically possess an N-terminal sequence that targets the docking protein to the membrane (for example, a PH domain or a myristoylation site) and a PTB domain that enables the docking protein to bind to a phosphorylated tyrosine on a receptor, as well as additional modules and phosphorylation sites that facilitate the binding of target proteins. A typical case is **IRS-1** (Insulin Receptor Substrate-1), a substrate of the insulin receptor. IRS-1 has an N-terminal PH domain followed by a PTB domain and 18 potential tyrosine phosphorylation sites. The PH and PTB domains direct IRS-1 to the membrane, facilitating both tyrosine phosphorylation of IRS-1 by the insulin receptor tyrosine kinase and subsequent mediation of additional cell signaling events. Scaffolding proteins can thus assemble sequential components of a signaling pathway.

26.7	**Excitable Membranes, Neurotransmission, and Sensory Systems**

The survival of higher organisms is predicated on the ability to respond rapidly to sensory input such as sights, sounds, and smells. The responses to such stimuli may include muscle movements and many forms of intercellular communication. Hormones can move through an organism only at speeds determined by the circulatory system. In most higher organisms, a faster mode of communication is crucial. Nerve impulses, which can be propagated at speeds up to 100 m/sec, provide a means of intercellular signaling that is fast enough to encompass sensory recognition, movement, and other physiological functions and behaviors in higher animals. The generation and transmission of nerve impulses in vertebrates is mediated by an incredibly complicated neural network that connects every part of the organism with the brain—itself an interconnected array of as many as 10^{12} cells.

Despite their complexity and diversity, the nervous systems of higher organisms all possess common features and common mechanisms. Physical or chemical stimuli are recognized by specialized **receptor proteins** in the membranes of **excitable cells.** Conformational changes in the receptor protein result in a change in enzyme activity or a change in the permeability of the membrane. These changes are then propagated throughout the cell or from cell to cell in specific and reversible ways to carry information through the organism. This section describes the characteristics of excitable cells and the mechanisms by which these cells carry information at high speeds through an organism.

Figure 26.31 The structure of a mammalian motor neuron. The nucleus and most other organelles are contained in the cell body. One long axon and many shorter dendrites project from the body. The dendrites receive signals from other neurons and conduct them to the cell body. The axon transmits signals from this cell to other cells via the synaptic knobs. Glial cells called Schwann cells envelop the axon in layers of an insulating myelin membrane. Although glial cells lie in proximity to neurons in most cases, no specific connections (such as gap junctions, for example) connect glial cells and neurons. However, gap junctions can exist between adjacent glial cells.

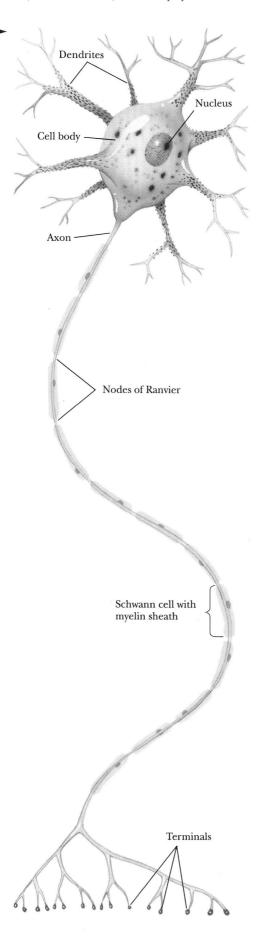

The Cells of Nervous Systems

Neurons and **neuroglia** (or **glial cells**) are cell types unique to nervous systems. The reception and transmission of nerve impulses are carried out by neurons (Figure 26.31), whereas glial cells serve protective and supportive functions. ("Neuroglia" could be translated as "nerve glue.") Glial cells differ from neurons in several ways: they do not possess axons or synapses and they retain the ability to divide throughout their life spans. Glial cells outnumber neurons by at least 10 to 1 in most animals.

Neurons are distinguished from other cell types by their long cytoplasmic extensions or projections, called **processes.** Most neurons consist of three distinct regions (see Figure 26.31): the **cell body** (called the **soma**), the **axon,** and the **dendrites.** The axon ends in small structures called **synaptic terminals, synaptic knobs,** or **synaptic bulbs.** Dendrites are short, highly branched structures emanating from the cell body that receive neural impulses and transmit them to the cell body. The space between a synaptic knob on one neuron and a dendrite ending of an adjacent neuron is the **synapse** or **synaptic cleft.**

Three kinds of neurons are found in higher organisms: sensory neurons, interneurons, and motor neurons. **Sensory neurons** acquire sensory signals, either directly or from specific receptor cells, and pass this information along to either **interneurons** or **motor neurons.** Interneurons simply pass signals from one neuron to another, whereas motor neurons pass signals from other neurons to muscle cells, thereby inducing muscle movement (motor activity).

Ion Gradients: Source of Electrical Potentials in Neurons

The impulses that are carried along axons, as signals pass from neuron to neuron, are electrical in nature. *These electrical signals occur as transient changes in the electrical potential differences (voltages) across the membranes of neurons (and other cells).* Such potentials are generated by ion gradients. The cytoplasm of a neuron at rest is low in Na^+ and Cl^- and high in K^+, relative to the extracellular fluid (Figure 26.32). These gradients are generated by the Na^+,K^+-ATPase (see Chapter 6). A resting neuron exhibits a potential difference of approximately -60 mV (that is, negative inside).

The Action Potential

Nerve impulses, also called **action potentials,** are transient changes in the membrane potential that move rapidly along nerve cells. Action potentials are created when the membrane is locally **depolarized** by approximately 20 mV—from the resting value of about -60 mV to a new value of approximately -40 mV. This small change is enough to have a dramatic effect on specific proteins in the axon membrane called **voltage-gated ion channels.** These proteins are ion channels that are specific either for Na^+ or K^+. These ion channels are normally closed at the resting potential of -60 mV. When the potential difference rises to -40 mV, the "gates" of the Na^+ channels are opened and Na^+ ions begin to flow into the cell. As Na^+ enters the cell, the membrane potential

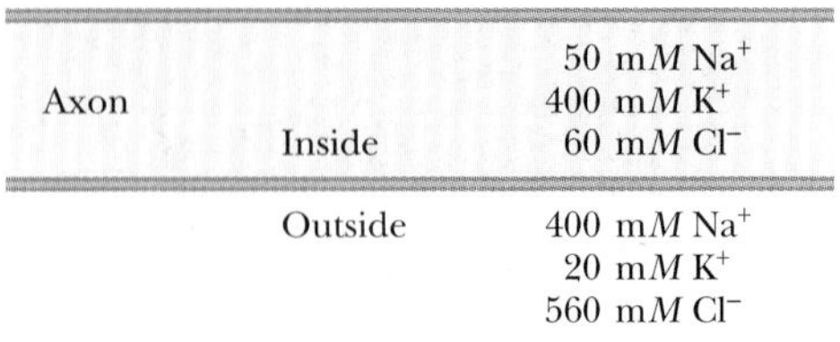

Axon		50 mM Na⁺
		400 mM K⁺
	Inside	60 mM Cl⁻
	Outside	400 mM Na⁺
		20 mM K⁺
		560 mM Cl⁻

Figure 26.32 The concentrations of Na⁺, K⁺, and Cl⁻ ions inside and outside a typical resting mammalian axon.

continues to increase, and additional Na⁺ channels are opened (Figure 26.33). The potential rises to more than +30 mV. At this point, Na⁺ influx slows and stops. As the Na⁺ channels close, K⁺ channels begin to open and K⁺ ions stream out of the cell, returning the membrane potential to negative values. The potential eventually overshoots its resting value a bit. At this point, K⁺ channels close and the resting potential is eventually restored by the action of the Na⁺,K⁺-ATPase and the other channels. These transient increases and decreases, first in Na⁺ permeability and then in K⁺ permeability, were first observed by Alan Hodgkin and Andrew Huxley. For this and related work, Hodgkin and Huxley, along with J. C. Eccles, won the Nobel Prize in physiology or medicine in 1963.

The Action Potential Is Mediated by the Flow of Na⁺ and K⁺ Ions

These changes in potential in one part of the axon are rapidly passed along the axonal membrane (Figure 26.34). The sodium ions that rush into the cell in one localized region actually diffuse farther along the axon, raising the Na⁺ concentration, depolarizing the membrane, and causing Na⁺ gates to open in that adjacent region of the axon. In this way, the action potential moves down the axon in wavelike fashion. This simple process has two very dramatic properties:

1. Action potentials propagate very rapidly—up to and sometimes exceeding 100 meters per second.
2. The action potential is not attenuated (diminished in intensity) as a function of distance transmitted.

The input of energy all the way along an axon—in the form of ion gradients maintained by Na⁺,K⁺-ATPase—ensures that the shape and intensity of the action potential is maintained over long distances. The action potential has an all-or-none character. There are no gradations of amplitude; a given neuron is either at rest (with a polarized membrane) or is conducting a nerve impulse (with a reversed polarization). Because nerve impulses display no varia-

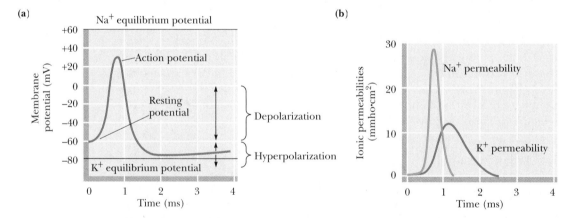

(a)

(b)

Figure 26.33 The time dependence of an action potential, compared with the ionic permeabilities of Na⁺ and K⁺. **(a)** The rapid rise in membrane potential from −60 mV to slightly more than +30 mV (a) is referred to as a "depolarization." This depolarization is caused **(b)** by a sudden increase in the permeability of Na⁺. As the Na⁺ permeability decreases, K⁺ permeability is increased and the membrane potential drops, eventually falling below the resting potential—a state of "hyperpolarization"—followed by a slow return to the resting potential. *(Adapted from Hodgkin, A., and Huxley, A., 1952. A quantitative description of membrane current and its application to conduction and excitation in nerve. Journal of Physiology* **117**:500–544.)

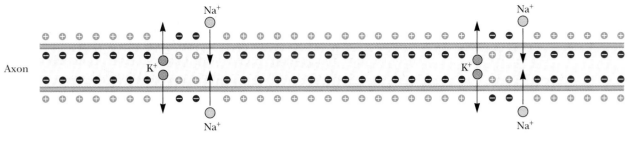

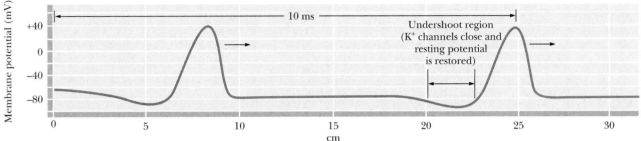

Figure 26.34 The propagation of action potentials along an axon. Figure 26.33 shows the time dependence of an action potential at a discrete point on the axon. This figure shows how the membrane potential varies along the axon as an action potential is propagated. (For this reason, the shape of the action potential is the apparent reverse of that shown in Figure 26.33.) At the leading edge of the action potential, membrane depolarization causes Na$^+$ channels to open briefly. As the potential moves along the axon, the Na$^+$ channels close and K$^+$ channels open, leading to a drop in potential and the onset of hyperpolarization. When the resting potential is restored, another action potential can be initiated.

tion in amplitude, the size of the action potential is not important in processing signals in the nervous system. Instead, it is the number of action-potential firings and the frequency of firing that carry specific information.

The Voltage-Gated Sodium and Potassium Channels

The action potential is a delicately orchestrated interplay between the Na$^+$,K$^+$-ATPase and the voltage-gated Na$^+$ and K$^+$ channels, initiated by a stimulus at the postsynaptic membrane. The density and distribution of Na$^+$ channels along the axon are different for myelinated and unmyelinated axons (Figure 26.35). In unmyelinated axons, Na$^+$ channels are uniformly distributed,

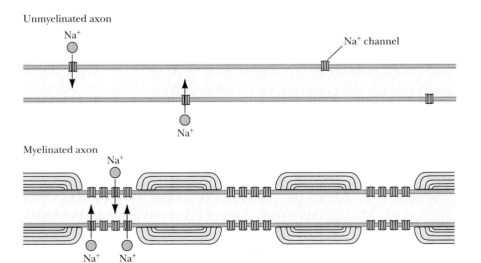

Figure 26.35 Na$^+$ channels are infrequently and randomly distributed in unmyelinated nerve. In myelinated axons, Na$^+$ channels are clustered in large numbers in the nodes of Ranvier, between the regions surrounded by myelin sheath structures.

although they are few in number—approximately 20 channels per μm^2. On the other hand, in myelinated axons, Na$^+$ channels are **clustered** at the nodes of Ranvier. In these latter regions, they occur with a density of approximately 10,000 per μm^2. Elucidation of these distributions of Na$^+$ channels was made possible by the use of several Na$^+$-channel toxins (Figure 26.36).

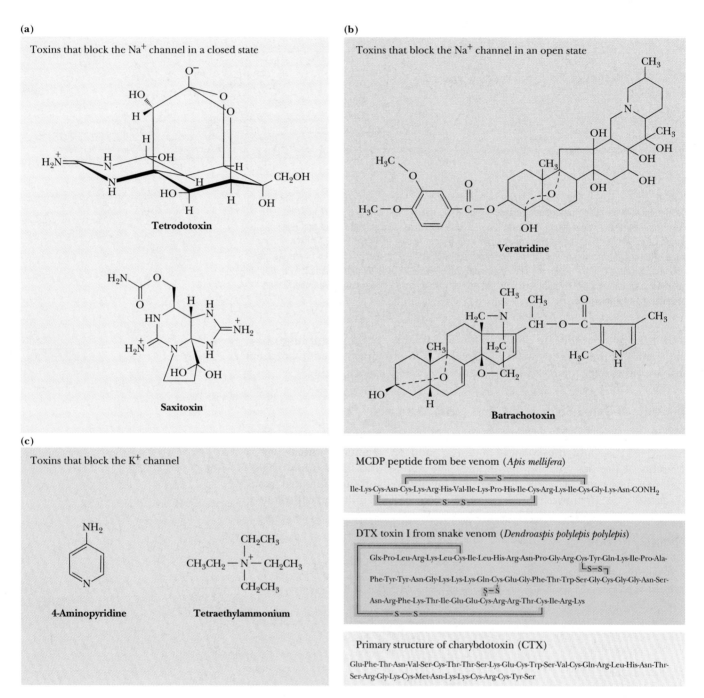

Figure 26.36 Effectors of Na$^+$ channels include **(a)** tetrodotoxin and saxitoxin, which block the Na$^+$ channel in a closed state, and **(b)** veratridine and batrachotoxin, which block the Na$^+$ channel in an open state. K$^+$-channel blockers include **(c)** 4-aminopyridine, tetraethylammonium ion, mast cell degranulating peptide (MCDP), dendrotoxin (DTX), and charybdotoxin (CTX). (See *A Deeper Look* boxes on page 882.)

The Na^+ channel in mammalian brain is a heterotrimer consisting of α- (260 kD), β_1-(36 kD), and β_2-(33 kD) subunits (Figure 26.37). All three subunits are exposed to the extracellular surface and are heavily glycosylated. The α-subunit contains the binding site for toxins and has four domains of 300 to 400 amino acids each (Figure 26.38), with approximately 50% identity or conservation in their amino acid sequences. Each domain contains six regions of probable α-helical structure which are long enough to be membrane-spanning segments.

Just as for the voltage-gated sodium channels (see *A Deeper Look: Tetrodotoxin and Other Na^+ Channel Toxins*, page 882), the high-affinity binding of several specific K^+-channel blockers has aided in the identification and characterization of voltage-gated K^+ channels (see *A Deeper Look: Potassium Channel Toxins*). The K^+ channel from rat synaptosomal membranes consists of an α-subunit of 76 to 80 kD and a β-subunit of 38 kD. Phosphorylation of the α-subunit leads to activation of the K^+ channel.

Cell–Cell Communication at the Synapse

How are neuronal signals passed from one neuron to the next? Neurons are juxtaposed at the synapse. The space between the two neurons is called the **synaptic cleft.** The number of synapses in which any given neuron is involved varies greatly. There may be as few as one synapse per postsynaptic cell (in the midbrain) to many thousands per cell. Typically, 10,000 synaptic knobs may impinge on a single spinal motor neuron, with 8000 on the dendrites and 2000 on the soma or cell body. The ratio of synapses to neurons in the human forebrain is approximately 40,000 to 1!

Synapses are actually quite specialized structures of several different types. A minority of synapses in mammals, termed electrical synapses, are characterized by a very small gap—approximately 2 nm—between the presynaptic cell

Na^+ channel

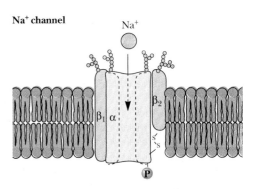

Figure 26.37 The Na^+ channel comprises three subunits, denoted α, β_1, and β_2. A disulfide bridge links α and β_2 as shown. All three subunits are glycosylated, and the α subunit can be phosphorylated on the cytoplasmic surface.

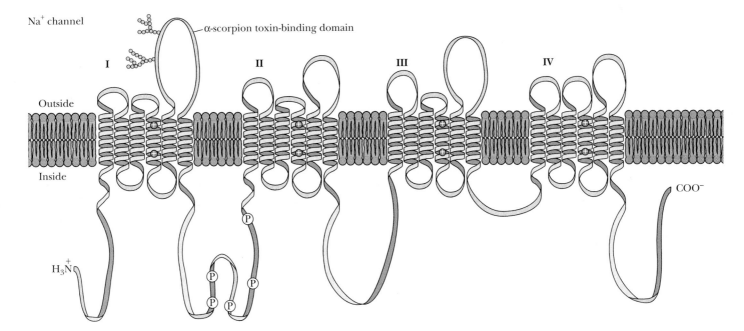

Figure 26.38 Model for the arrangement of the Na^+ channel α-subunit in the plasma membrane. The α-subunit consists of four domains (I through IV), each of which contains six transmembrane α-helices, designated S1 through S6. Phosphorylation sites (P) and location of positive charges on helix S4 are indicated.

A DEEPER LOOK

Tetrodotoxin and Other Na⁺ Channel Toxins

Tetrodotoxin and **saxitoxin** are highly specific blockers of Na⁺ channels and bind with very high affinity ($K_D < 1$ nM). This unique specificity and affinity have made possible the use of radioactive forms of these toxins to purify Na⁺ channels and map their distribution on axons. Tetrodotoxin is found in the skin and several internal organs of puffer fish, also known as blowfish or swellfish, members of the family *Tetraodontidae*, which react to danger by inflating themselves with water or air to nearly spherical (and often comical) shapes (see figure). Although tetrodotoxin poisoning can easily be fatal, puffer fish are delicacies in Japan, where they are served in a dish called *fugu*. The puffer fish must be cleaned and prepared by specially trained chefs.

Saxitoxin is made by *Gonyaulax catenella* and *G. tamarensis*, two species of marine dinoflagellates or plankton that are responsible for "red tides" that cause massive fish kills. Saxitoxin is concentrated by certain species of mussels, scallops, and other shellfish that are exposed to red tides. Consumption of these shellfish by animals, including humans, can be fatal. In addition to these toxins, which prevent the Na⁺ channel from opening, there are equally poisonous agents that block the Na⁺ channel in an open state, permitting Na⁺ to stream into the cell without control, destroying the Na⁺ gradients. Included in this group of toxins (see Figure 26.36) are **veratridine** from *Schoenocaulon officinalis*, a member of the lily family, and **batrachotoxin**, a compound found in skin secretions of a Colombian frog, *Phyllobates aurotaenia*. These skin secretions have traditionally been used as arrow poisons.

Tetrodotoxin is found in puffer fish, which are prepared and served in Japan as *fugu*. The puffer fish on the left is unexpanded; the one on the right is inflated. *(left, Zig Lesczynski/ Animals, Animals; right, Tim Rock/Animals, Animals)*

(which delivers the signal) and the postsynaptic cell (which receives the signal). At electrical synapses, the arrival of an action potential on the presynaptic membrane leads directly to depolarization of the postsynaptic membrane, initiating a new action potential in the postsynaptic cell. However, most synaptic clefts are much wider—on the order of 20 to 50 nm. In these, an action potential in the presynaptic membrane causes secretion of a chemical substance—called a neurotransmitter—by the presynaptic cell. This substance

A DEEPER LOOK

Potassium Channel Toxins

K⁺-channel blockers include **4-aminopyridine, tetraethylammonium ion,** and several peptide toxins, including the **dendrotoxins (DTX), mast cell degranulating peptide (MCDP),** and **charybdotoxin (CTX).** (see Figure 26.36) Dendrotoxin I is a 60-residue peptide from *Dendroaspis polylepsis*, the dangerous black mamba snake of sub-Saharan Africa. MCDP, a bee venom toxin that has a degranulating action on mast cells, is a potent convulsant. It is a 22-residue peptide with two disulfide bonds, one proline, and a C-terminal amide. Charybdotoxin is a minor component of the venom of the scorpion, *Leiurus quinquestriatus*. It is a 37-residue peptide with 8 positively charged residues (3 arginines, 4 lysines, and a histidine). All these agents bind with high affinity to membranes containing voltage-activated K⁺ channels.

binds to receptors on the postsynaptic cell, initiating a new action potential. Synapses of this type are thus **chemical synapses.**

Different synapses utilize specific neurotransmitters. The **cholinergic synapse,** a paradigm for chemical transmission mechanisms at synapses, employs acetylcholine as a neurotransmitter. Other important neurotransmitters and receptors fall into one of several major classes, including amino acids (and their derivatives), catecholamines, peptides, and gaseous neurotransmitters. Table 26.5 lists some, but not all, of the known neurotransmitters.

The Cholinergic Synapses

In **cholinergic synapses,** small **synaptic vesicles** inside the synaptic knobs contain large amounts of acetylcholine (approximately 10,000 molecules per vesicle; Figure 26.39). When the membrane of the synaptic knob is stimulated by an arriving action potential, special **voltage-gated Ca^{2+} channels** open and Ca^{2+} ions stream into the synaptic knob, causing the acetylcholine-containing vesicles to attach to and fuse with the knob membrane. The vesicles open, spilling acetylcholine into the synaptic cleft. The binding of acetylcholine to specific **acetylcholine receptors** in the postsynaptic membrane causes the opening of ion channels and the creation of a new action potential in the postsynaptic neuron.

A variety of toxins can alter or affect this process. The anaerobic bacterium *Clostridium botulinum,* which causes botulism poisoning, produces several toxic proteins that strongly inhibit acetylcholine release. The black widow spider, *Lactrodectus mactans,* produces a venom protein, **α-latrotoxin,** that stimulates abnormal release of acetylcholine at the neuromuscular junction. The bite of

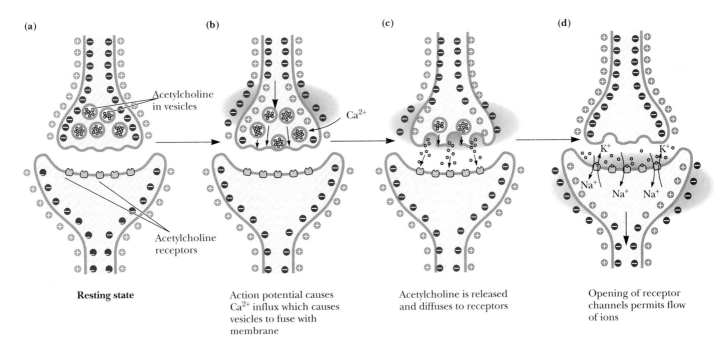

| (a) | (b) | (c) | (d) |

Acetylcholine in vesicles

Ca^{2+}

Acetylcholine receptors

K^+ K^+ Na^+ Na^+ Na^+

Resting state

Action potential causes Ca^{2+} influx which causes vesicles to fuse with membrane

Acetylcholine is released and diffuses to receptors

Opening of receptor channels permits flow of ions

Figure 26.39 Cell–cell communication at the synapse **(a)** is mediated by neurotransmitters such as acetylcholine, produced from choline by cholineacetyltransferase. The arrival of an action potential at the synaptic knob **(b)** opens Ca^{2+} channels in the presynaptic membrane. Influx of Ca^{2+} induces the fusion of acetylcholine-containing vesicles with the plasma membrane and release of acetylcholine into the synaptic cleft **(c).** Binding of acetylcholine to receptors in the postsynaptic membrane opens Na^+ channels **(d).** The influx of Na^+ depolarizes the postsynaptic membrane, generating a new action potential.

Table 26.5   Families of Neurotransmitters

Cholinergic Agents
Acetylcholine

Catecholamines
Norepinephrine (noradrenaline)
Epinephrine (adrenaline)
L-Dopa
Dopamine
Octopamine

Amino Acids (and Derivatives)
γ-Aminobutyric acid (GABA)
Alanine
Aspartate
Cystathione
Glutamate
Glycine
Histamine
Proline
Serotonin
Taurine
Tyrosine

Peptide Neurotransmitters
Cholecystokinin
Enkephalins and endorphins
Gastrin
Gonadotropin
Neurotensin
Oxytocin
Secretin
Somatostatin
Substance P
Thyrotropin releasing factor
Vasopressin
Vasoactive intestinal peptide (VIP)

Gaseous Neurotransmitters
Carbon monoxide (CO)
Nitric oxide (NO)

the black widow causes pain, nausea, and mild paralysis of the diaphragm but is rarely fatal.

Two Classes of Acetylcholine Receptors

Two different acetylcholine receptors are found in postsynaptic membranes. They were originally distinguished by their responses to **muscarine,** a toxic alkaloid in toadstools, and **nicotine** (Figure 26.40). The **nicotinic receptors** are cation channels in postsynaptic membranes and the **muscarinic receptors** are transmembrane proteins that interact with G proteins. The receptors in sympathetic ganglia and those in motor endplates of skeletal muscle are nicotinic receptors. Nicotine locks the ion channels of these receptors in their open conformation. Muscarinic receptors are found in smooth muscle and in glands.

The nicotinic acetylcholine receptor is a transmembrane glycoprotein consisting of four different subunits, α(54 kD), β(56 kD), γ(58 kD), and δ(60 kD), with a quaternary structure of $\alpha_2\beta\gamma\delta$. Each α-subunit possesses a binding site for acetylcholine.

The Nicotinic Acetylcholine Receptor Is a Ligand-Gated Ion Channel

The nicotinic acetylcholine receptor functions as a **ligand-gated ion channel.** On the basis of its structure, it is also an **oligomeric ion channel.** When acetylcholine (the ligand) binds to this receptor, a conformational change opens the channel, which is equally permeable to Na^+ and K^+. Na^+ rushes in while K^+ streams out, but because the Na^+ gradient across this membrane is steeper than that of K^+, the Na^+ influx greatly exceeds the K^+ efflux. The influx of Na^+ depolarizes the postsynaptic membrane, initiating an action potential in the adjacent membrane. After a few milliseconds, the channel closes, even though acetylcholine remains bound to the receptor. At this point the channel will remain closed until the concentration of acetylcholine in the synaptic cleft drops to about 10 nM.

Nicotiana tabacum

Figure 26.40   Two types of acetylcholine receptors are known. Nicotinic acetylcholine receptors are locked in their open conformation by nicotine. Obtained from tobacco plants, nicotine is named for Jean Nicot, French ambassador to Portugal, who sent tobacco seeds to France in 1550 for cultivation. Muscarinic acetylcholine receptors are stimulated by muscarine, obtained from the intensely poisonous mushroom *Amanita muscaria.*

Nicotine

Amanita muscaria

Muscarine

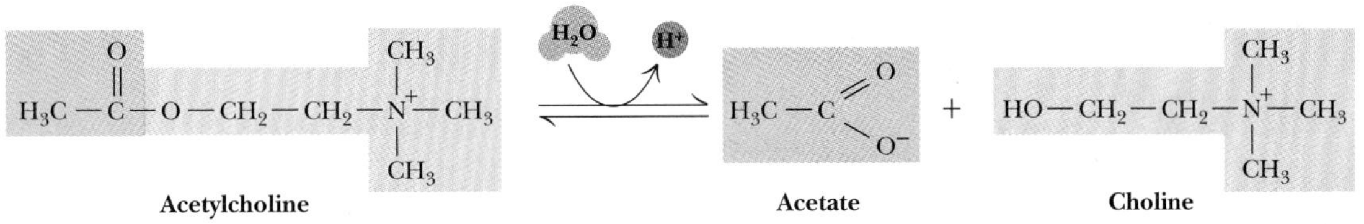

Figure 26.41 Acetylcholine is degraded to acetate and choline by acetylcholinesterase, a serine protease.

Acetylcholinesterase Degrades Acetylcholine in the Synaptic Cleft

Following every synaptic signal transmission, the synapse must be readied for the arrival of another action potential. Several things must happen very quickly. First, the acetylcholine left in the synaptic cleft must be rapidly degraded to resensitize the acetylcholine receptor and to restore the excitability of the postsynaptic membrane. This reaction is catalyzed by **acetylcholinesterase** (Figure 26.41).

When [acetylcholine] has decreased to low levels, acetylcholine dissociates from the receptor, which thereby regains its ability to open in a ligand-dependent manner. Second, the synaptic vesicles must be reformed from the presynaptic membrane by endocytosis (Figure 26.42) and then must be restocked with acetylcholine. This occurs through the action of an ATP-driven H^+ pump and an **acetylcholine transport protein.** The H^+ pump in this case is a member of the family of **V-type ATPases.** It uses the free energy of ATP hydrolysis to create an H^+ gradient across the vesicle membrane. This gradient is used by the acetylcholine transport protein to drive acetylcholine into the vesicle, as shown in Figure 26.42.

Antagonists of the nicotinic acetylcholine receptor are particularly potent neurotoxins. These agents, which bind to the receptor and prevent opening of the ion channel, include **d-tubocurarine,** the active agent in the South American arrow poison **curare,** and several small proteins from poisonous snakes. These latter agents include **cobratoxin** from cobra venom, and **α-bungarotoxin,** from *Bungarus multicinctus,* a snake common in Taiwan (Figure 26.43).

Muscarinic Receptor Function Is Mediated by G Proteins

Muscarinic receptors are 70-kD glycoproteins and are members of the 7-transmembrane segment (7-TMS) family of receptors. Activation of muscarinic receptors (by binding of acetylcholine) results in several effects, including the inhibition of adenylyl cyclase, the stimulation of phospholipase C, and the opening of K^+ channels. As shown in Figure 26.44, all of these effects of muscarinic receptors are mediated by G proteins. Many antagonists for muscarinic acetylcholine receptors are known, including **atropine** from *Atropa belladonna,* the deadly nightshade plant, whose berries are sweet and tasty but highly poisonous (see Figure 26.43).

Both the nicotinic and muscarinic acetylcholine receptors are sensitive to certain agents that inactivate acetylcholinesterase itself. Acetylcholinesterase is a serine esterase similar to trypsin and chymotrypsin (see Chapter 11). The reactive serine at the active site of such enzymes is a vulnerable target for organophosphorus inhibitors (Figure 26.45). **DIFP** and related agents form stable covalent complexes with the active-site serine, irreversibly blocking the enzyme. **Malathion** and **parathion** are commonly used insecticides, and **sarin** and **tabun** are nerve gases used in chemical warfare. All these agents effectively block nerve impulses, stop breathing, and cause death by suffocation.

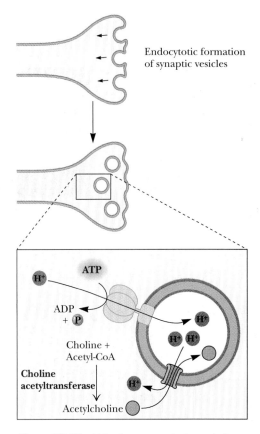

Figure 26.42 Following a synaptic transmission event, acetylcholine is repackaged in vesicles in a multistep process. Synaptic vesicles are formed by endocytosis, and acetylcholine is synthesized by choline acetyltransferase. A proton gradient is established across the vesicle membrane by an H^+-transport ATPase, and a proton–acetylcholine transport protein transports acetylcholine into the synaptic vesicles, exchanging acetylcholine for protons in an electrically neutral antiport process.

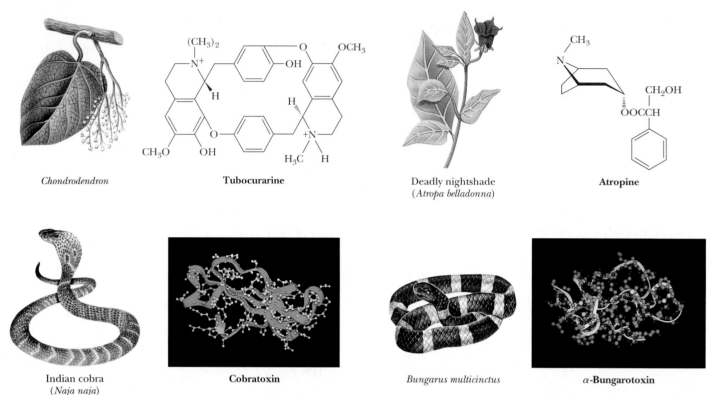

Chondrodendron

Tubocurarine

Deadly nightshade
(*Atropa belladonna*)

Atropine

Indian cobra
(*Naja naja*)

Cobratoxin

Bungarus multicinctus

α-Bungarotoxin

Figure 26.43   Tubocurarine, obtained from the plant *Chondrodendron tomentosum*, is the active agent in "tube curare," named for the bamboo tubes in which it is kept by South American tribal hunters. Atropine is produced by *Atropa belladonna*, the poisonous deadly nightshade. The species name, which means "beautiful woman," is derived from the use of atropine in years past by Italian women to dilate their pupils. Atropine is still used for pupil dilation in eye exams by ophthalmologists. Cobratoxin and α-bungarotoxin are produced by the cobra (*Naja naja*) and the banded krait snake (*Bungarus multicinctus*), respectively.

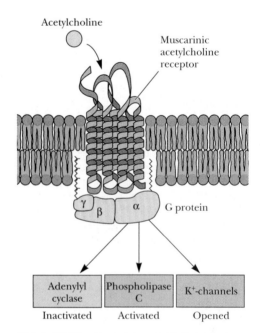

Figure 26.44   Muscarinic acetylcholine receptors are typical G-protein–coupled receptor proteins. Binding of acetylcholine to these receptors activates G proteins, which inactivate adenylyl cyclase, activate phospholipase C, and open K$^+$ channels.

Milder inhibitors of the acetylcholinesterase reaction are useful therapeutic agents. **Physostigmine,** an alkaloid found in calabar beans, and **neostigmine,** a synthetic analog, contain carbamoyl ester groups (see Figure 26.45). Reaction with the active-site serine of acetylcholinesterase leaves an intermediate that is hydrolyzed only very slowly, effectively inhibiting the enzyme. Physostigmine and neostigmine have been used to treat **myasthenia gravis,** a chronic disorder that causes muscle weakness and eventual paralysis. Myasthenia gravis is an autoimmune disease in which individuals produce antibodies that bind to their own acetylcholine receptors, blocking the response to acetylcholine. By blocking acetylcholinesterase (thus allowing acetylcholine levels in the synaptic cleft to remain high), physostigmine and neostigmine can suppress the symptoms of the disease.

Other Neurotransmitters and Synaptic Junctions

Synaptic junctions that use amino acids, catecholamines, and peptides (see Table 26.5) appear to operate much the way the cholinergic synapses do. Presynaptic vesicles release their contents into the synaptic cleft, where the neurotransmitter substance can bind to specific receptors on the postsynaptic membrane to induce a conformational change and elicit a particular response. Some of these neurotransmitters are **excitatory** in nature and stimulate postsynaptic neurons to transmit impulses, whereas others are **inhibitory** and prevent the postsynaptic neuron from carrying other signals. Just as acetylcholine acts both

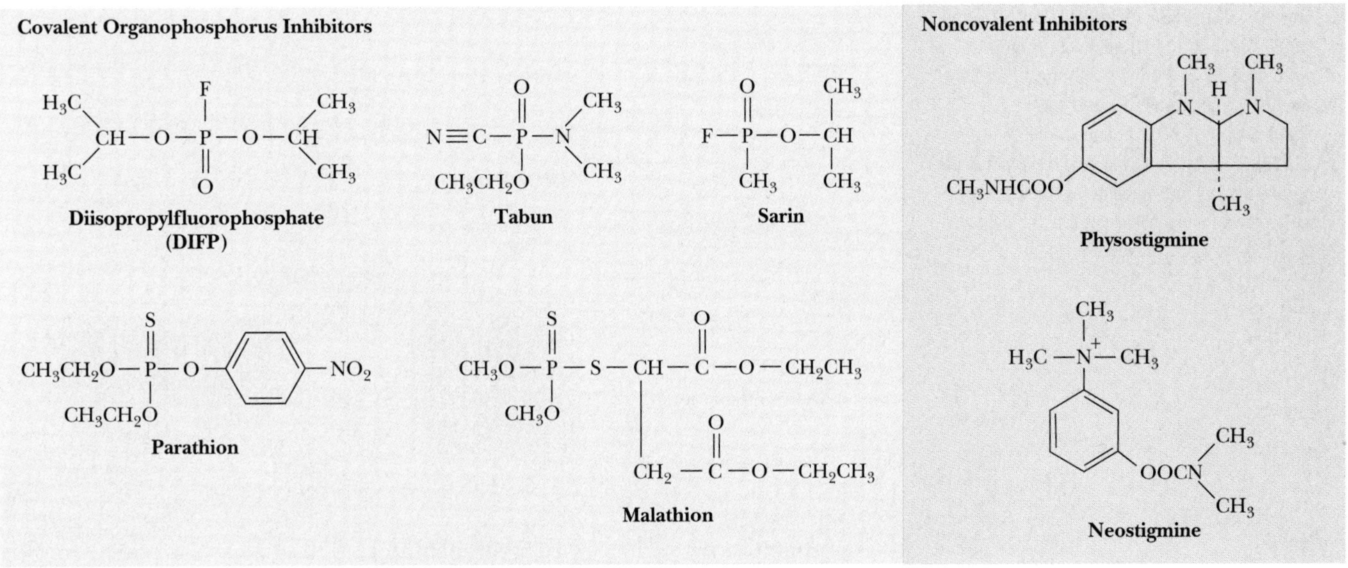

Covalent Organophosphorus Inhibitors

Noncovalent Inhibitors

Figure 26.45 Covalent inhibitors (blue) of acetylcholinesterase include DIFP, the nerve gases tabun and sarin, and the insecticides parathion and malathion. Milder, noncovalent (pink) inhibitors of acetylcholinesterase include physostigmine and neostigmine.

on nicotinic and muscarinic receptors, so most of the known neurotransmitters act on several (and in some cases, many) different kinds of receptors.

Glutamate and Aspartate: Excitatory Amino Acid Neurotransmitters

The common amino acids glutamate and aspartate act as neurotransmitters. Like acetylcholine, glutamate and aspartate are excitatory and stimulate receptors on the postsynaptic membrane to transmit a nerve impulse. No enzymes that degrade glutamate exist in the extracellular space, so glutamate must be cleared by high-affinity presynaptic and glial transporters—a process called **reuptake.**

There are at least five known subclasses of glutamate receptors. The best understood of these excitatory receptors is the *N*-methyl-D-aspartate (NMDA) receptor, a ligand-gated channel that, when open, allows Ca^{2+} and Na^+ to flow into the cell and K^+ to flow out of the cell. **Phencyclidine (PCP)** is a specific antagonist of the NMDA receptor (Figure 26.46). Phencyclidine was once used as an anesthetic agent, but legitimate human use was quickly discontinued when it was found to be responsible for bizarre psychotic reactions and behavior in its users. Since this time, PCP has been used illegally as a hallucinogenic drug under the street name of **angel dust.** Sadly, it has caused many serious, long-term psychological problems in its users.

γ-Aminobutyric Acid and Glycine: Inhibitory Neurotransmitters

Certain neurotransmitters, acting through their conjugate postsynaptic receptors, inhibit the postsynaptic neuron from propagating nerve impulses from other neurons. Two such inhibitory neurotransmitters are **γ-aminobutyric acid (GABA)** and **glycine.** These agents make postsynaptic membranes permeable to chloride ions and cause a net influx of Cl^-, which in turn causes **hyperpolarization** of the postsynaptic membrane (making the membrane potential more negative). Hyperpolarization of a neuron effectively raises the threshold for the onset of action potentials in that neuron, making the neuron resistant

Figure 26.46 The NMDA receptor is an Na$^+$ and Ca^{2+} channel, which is regulated by Zn^{2+} and glycine, stimulated by N-methyl-D-aspartate, and inhibited by phencyclidine (PCP) and the anticonvulsant drug MK-801. *(Adapted from Young, A., and Fagg, G., 1990. Excitatory amino acid receptors in the brain: Membrane binding and receptor autoradiographic approaches,* Trends in Pharmacological Sciences **11:**126–133.)

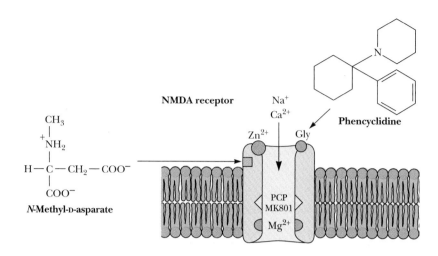

N-Methyl-D-asparate

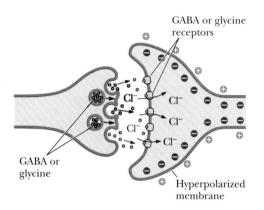

Figure 26.47 GABA (γ-aminobutyric acid) and glycine are inhibitory neurotransmitters that activate chloride channels. Influx of Cl$^-$ causes a hyperpolarization of the postsynaptic membrane.

to stimulation by excitatory neurotransmitters. These effects are mediated by the **GABA and glycine receptors,** which are **ligand-gated chloride channels** (Figure 26.47). GABA is derived by a decarboxylation of glutamate (Figure 26.48) and appears to operate mainly in the brain, whereas glycine acts primarily in the spinal cord. The glycine receptor has a specific affinity for the convulsive alkaloid **strychnine** (Figure 26.49). The effects of ethanol on the brain arise in part from the opening of GABA receptor Cl$^-$ channels.

The Catecholamine Neurotransmitters

Epinephrine, norepinephrine, dopamine, and **L-dopa** are collectively known as the catecholamine neurotransmitters. These compounds are synthesized from tyrosine (Figure 26.50), both in sympathetic neurons and in the adrenal glands. They function as neurotransmitters in the brain and as hormones in the circulatory system. However, these two pools operate independently, thanks to the **blood–brain barrier,** which permits only very hydrophobic species in the circulatory system to cross over into the brain. Hydroxylation of tyrosine (by **tyrosine hydroxylase**) to form **3,4-di**hydr*oxy*phenyl*al*anine (L-dopa) is the rate-limiting step in this pathway. Dopamine, a crucial catecholamine involved in several neurological diseases, is synthesized from L-dopa by a pyridoxal phosphate-dependent enzyme, **dopa decarboxylase.** Subsequent hydroxylation and methylation produce norepinephrine and epinephrine (see Figure 26.50). The methyl group in the final reaction is supplied by *S*-adenosylmethionine.

Each of these catecholamine neurotransmitters plays a unique role in synaptic transmission. The neurotransmitter in junctions between sympathetic nerves and smooth muscle is norepinephrine. On the other hand, dopamine is involved in other processes. Either excessive brain production of dopamine or hypersensitivity of dopamine receptors is responsible for psychotic symptoms and schizophrenia, whereas lowered production of dopamine and the loss of dopamine neurons are important factors in Parkinson's disease.

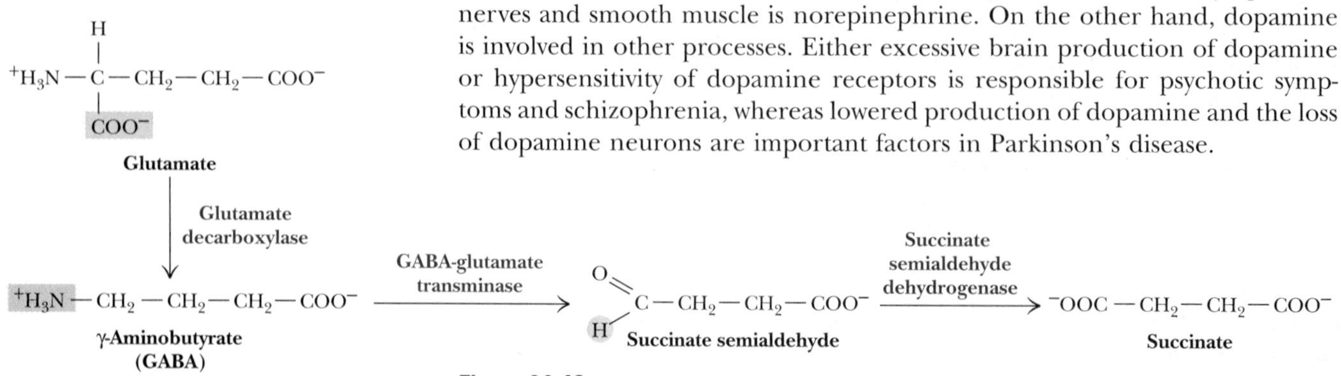

H
|
$^+$H$_3$N — C — CH$_2$ — CH$_2$ — COO$^-$
|
COO$^-$

Glutamate

↓ Glutamate decarboxylase

$^+$H$_3$N — CH$_2$ — CH$_2$ — CH$_2$ — COO$^-$ → (GABA-glutamate transaminase) → Succinate semialdehyde → (Succinate semialdehyde dehydrogenase) → $^-$OOC — CH$_2$ — CH$_2$ — COO$^-$

γ-Aminobutyrate (GABA)

Succinate semialdehyde

Succinate

Figure 26.48 Glutamate is converted to GABA by glutamate decarboxylase. GABA is degraded by the action of GABA-glutamate transaminase and succinate semialdehyde dehydrogenase to produce succinate.

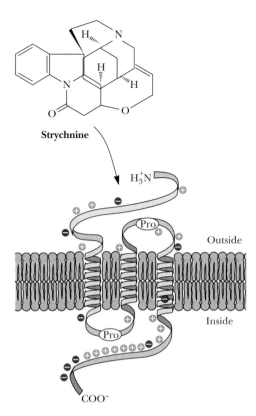

Figure 26.49 Glycine receptors are distinguished by their unique affinity for strychnine. A model for the arrangement of the 48-kD subunit of the glycine receptor in the postsynaptic membrane is shown.

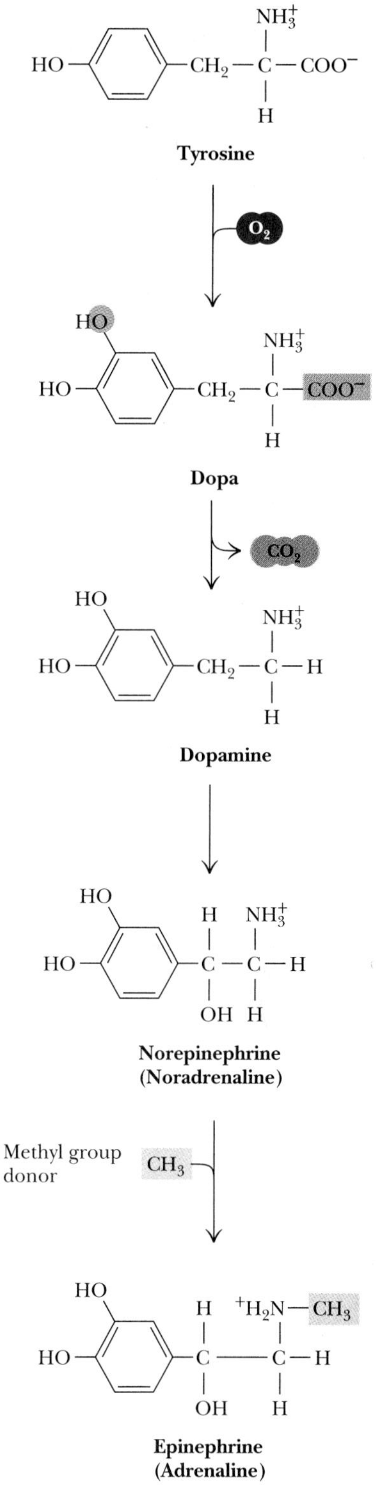

The Peptide Neurotransmitters

Many relatively small peptides have been shown to possess neurotransmitter activity (see Table 26.5). One of the challenges of this field is that the known neuropeptides may represent a very small subset of the neuropeptides that exist. Another challenge arises from the small *in vivo* concentrations of these agents and the small number of receptors that are present in neural tissue. Physiological roles for most of these peptides are complex. For example, the **endorphins** and **enkephalins** are natural opioid substances and potent pain relievers. The **endothelins** are a family of homologous regulatory peptides, synthesized by certain endothelial and epithelial cells, that act on nearby smooth muscle and connective tissue cells. They induce or affect smooth-muscle contraction; vasoconstriction; heart, lung, and kidney function; and mitogenesis and tissue remodeling. **Vasoactive intestinal peptide (VIP)** produces a G-protein/adenylyl cyclase–mediated increase in cAMP, which in turn triggers a variety of protein phosphorylation cascades, one of which leads to conversion of phosphorylase *b* to phosphorylase *a*, stimulating glycogenolysis. Moreover, VIP has synergistic effects with other neurotransmitters, such as norepinephrine. In addition to increasing cAMP levels through β-adrenergic receptors, norepinephrine acting at α1-adrenergic receptors markedly stimulates the increases in cAMP elicited by VIP. Many other effects have also been observed. For example, injection of VIP increases rapid eye movement (REM) sleep and decreases waking time in rats. VIP receptors exist in regions of the central nervous system involved in sleep modulation.

Figure 26.50 The pathway for the synthesis of catecholamine neurotransmitters. L-Dopa, dopamine, noradrenaline, and adrenaline are synthesized sequentially from tyrosine.

HUMAN BIOCHEMISTRY

The Biochemistry of Neurological Disorders

Defects in catecholamine processing are responsible for the symptoms of many neurological disorders, including clinical depression (which involves norepinephrine) and Parkinsonism (involving dopamine). Once these neurotransmitters have bound to and elicited responses from postsynaptic membranes, they must be efficiently cleared from the synaptic cleft (see figure, part a). Clearing can occur by several mechanisms. Norepinephrine and dopamine transport or reuptake proteins exist both in the presynaptic membrane and in nearby glial-cell membranes. On the other hand, catecholamine neurotransmitters can be metabolized and inactivated by two enzymes: **catechol-O-methyl-transferase** in the synaptic cleft and **monoamine oxidase** in the mitochondria (see figure, part b). Catecholamines transported back into the presynaptic neuron are accumulated in synaptic vesicles by a reuptake

mechanism. Clinical depression has been treated by two different strategies. **Monoamine oxidase inhibitors** act as antidepressants by increasing levels of catecholamines in the brain. Another class of antidepressants, the **tricyclics,** such as desipramine (see figure, part c), act on several classes of neurotransmitter reuptake transporters and facilitate more prolonged stimulation of postsynaptic receptors. **Prozac** is a more specific reuptake inhibitor and acts only on serotonin reuptake transporters.

Parkinsonism is characterized by degeneration of dopaminergic neurons as well as by consequent overproduction of postsynaptic dopamine receptors. In recent years, Parkinson's patients have been treated with dopamine agonists such as bromocriptine (see figure, part d) to counter the degeneration of dopamine neurons.

(a)

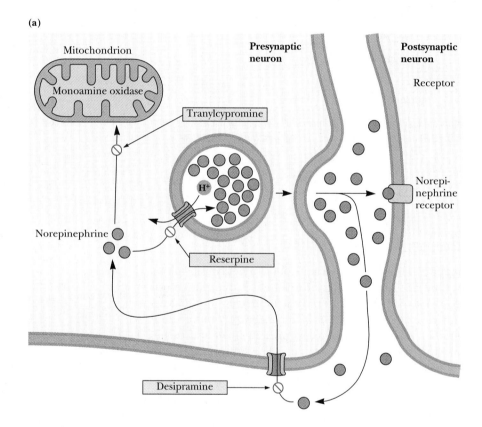

(a) The pathway for reuptake and vesicular repackaging of the catecholamine neurotransmitters. The sites of action of desipramine, tranylcypromine, and reserpine are indicated.

Catecholaminergic neurons are involved in many other interesting pharmacological phenomena. For example, **reserpine** (see figure, part e), an alkaloid from a climbing shrub of India, is a powerful sedative that depletes the level of brain monoamines by inhibiting incorporation of monoamines into synaptic vesicles.

Cocaine (see figure, part f), a highly addictive drug, binds with high affinity and specificity to reuptake transporters for the monoamine neurotransmitters in presynaptic membranes. Thus, at least one of the pharmacological effects of cocaine is to prolong the synaptic effects of these neurotransmitters.

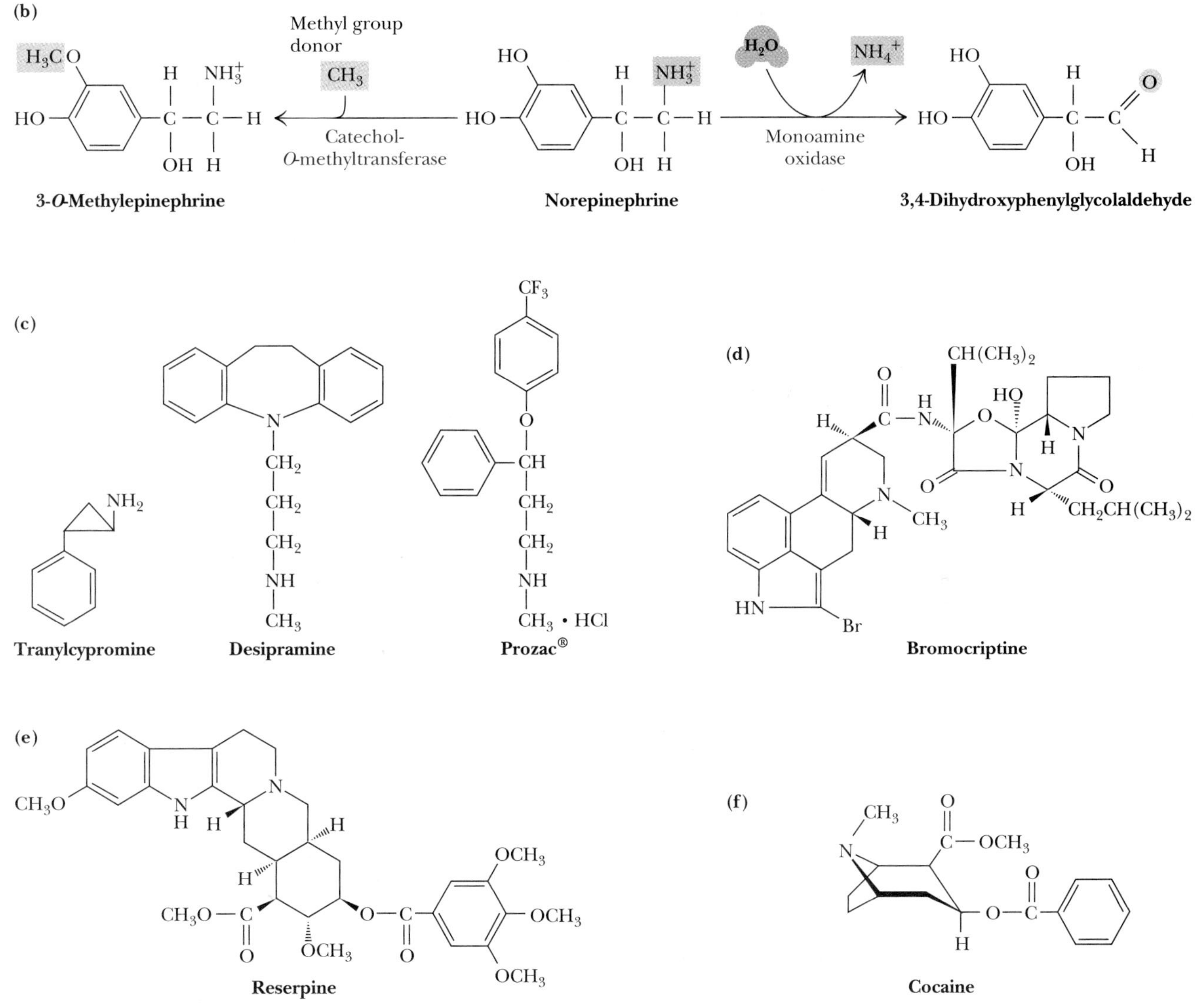

(b)

3-O-Methylepinephrine   Norepinephrine   3,4-Dihydroxyphenylglycolaldehyde

(c)

Tranylcypromine   Desipramine   Prozac®

(d)

Bromocriptine

(e)

Reserpine

(f)

Cocaine

(b) Norepinephrine can be degraded in the synaptic cleft by catechol-O-methyltransferase or in the mitochondria of presynaptic neurons by monoamine oxidase. **(c)** The structures of tranylcypromine, desipramine, and Prozac. **(d)** The structure of bromocriptine. **(e)** The structure of reserpine. **(f)** The structure of cocaine.

PROBLEMS

1. Compare and contrast the features and physiological advantages of each of the major classes of hormones, including the steroid hormones, polypeptide hormones, and the amino acid–derived hormones.

2. Compare and contrast the features and physiological advantages of each of the known classes of second messengers.

3. Nitric oxide may be merely the first of a new class of gaseous second messenger/neurotransmitter molecules. Based on your knowledge of the molecular action of nitric oxide, suggest another gaseous molecule that might act as a second messenger, and propose a molecular function for it.

4. Herbimycin A is an antibiotic that inhibits tyrosine kinase activity by binding to SH groups of cysteine in the src gene tyrosine kinase and other similar tyrosine kinases. What effect might it have on normal rat kidney cells that have been transformed by Rous sarcoma virus? Can you think of other effects you might expect for this interesting antibiotic?

5. Monoclonal antibodies that recognize phosphotyrosine are commercially available. How could such an antibody be used in studies of cell-signaling pathways and mechanisms?

6. Explain and comment on this statement: The main function of hormone receptors is that of signal amplification.

7. Synaptic vesicles are approximately 40 nm in outside diameter, and each vesicle contains about 10,000 acetylcholine molecules. Calculate the concentration of acetylcholine in a synaptic vesicle.

8. GTPγS is a nonhydrolyzable analog of GTP. Experiments with giant squid synapses reveal that injection of GTPγS into the presynaptic end (terminal) of the neuron inhibits neurotransmitter release (slowly and irreversibly). The calcium signals produced by presynaptic action potentials and the number of synaptic vesicles docking on the presynaptic membrane are unchanged by GTPγS. Propose a model for neurotransmitter release that accounts for all of these observations.

FURTHER READING

Armstrong, C., 1998. The vision of the pore. *Science* **280**:56–57.

Asaoka, Y., et al., 1992. Protein kinase C, calcium and phospholipid degradation. *Trends in Biochemical Sciences* **17**:414–417.

Avruch, J., Zhang, X.-F., and Kyriakis, J. M., 1994. Raf meets ras: Completing the framework of a signal transduction pathway. *Trends in Biochemical Sciences* **19**:279–283.

Bajjalieh, S. M., and Scheller, R. H., 1995. The biochemistry of neurotransmitter secretion. *Journal of Biological Chemistry* **270**:1971–1974.

Barford, D., Das, A. K., and Egloff, M.-P., 1998. The structure and mechanism of protein phosphatases: Insights into catalysis and regulation. *Annual Review of Biochemistry* **27**:133–164.

Bell, R., and Burns, D., 1991. Lipid activation of protein kinase C. *Journal of Biological Chemistry* **266**:4661–4664.

Bork, P., Schultz, J., and Ponting, C. P., 1997. Cytoplasmic signalling domains: The next generation. *Trends in Biochemical Sciences* **22**:296–298.

Bourne, H. R., 1997. Pieces of the true grail: A G protein finds its target. *Science* **278**:1898–1899.

Bredt, D. S., and Snyder, S. H., 1994. Nitric oxide: A physiologic messenger molecule. *Annual Review of Biochemistry* **63**:175–195.

Cadena, D., and Gill, G., 1992. Receptor tyrosine kinases. *FASEB Journal* **6**:2332–2337.

Camps, M., Nichols, A., and Arkinstall, S., 2000. Dual specificity phosphatases: A gene family for control of MAP kinase function. *FASEB Journal* **14**:6–16.

Casey, P. J., and Seabra, M. C., 1996. Protein prenyltransferases. *Journal of Biological Chemistry* **271**:5289–5292.

Catterall, W. A., 1995. Structure and function of voltage-gated ion channels. *Annual Review of Biochemistry* **64**:493–531.

Clapham, D. E., 1996. The G-protein nanomachine. *Nature* **379**:297–299.

Clapham, D. E., 1997. Some like it hot: Spicing up ion channels. *Nature* **389**:783–784.

Cohen, P., 1992. Signal integration at the level of protein kinases, protein phosphatases and their substrates. *Trends in Biochemical Sciences* **17**:408–413.

Cohen, P., 2000. The regulation of protein function by multisite phosphorylation—a 25 year update. *Trends in Biochemical Sciences* **25**:596–601.

Cohen, P., and Cohen, P. T. W., 1989. Protein phosphatases come of age. *Journal of Biological Chemistry* **264**:21435–21438.

Coleman, D. E., and Sprang, S. R., 1996. How G proteins work: A continuing story. *Trends in Biochemical Sciences* **21**:41–44.

Dennis, E. A., 1997. The growing phospholipase A_2 superfamily of signal transduction enzymes. *Trends in Biochemical Sciences* **22**:1–2.

Doyle, D. A., Cabral, J. M., Pfuetzner, R. A., et al., 1998. The structure of the potassium channel: Molecular basis of K^+ conduction and selectivity. *Science* **280**:69–77.

English, J., Pearson, G., Wilsbacher, J., et al., 1999. New insights into the control of MAP kinase pathways. *Experimental Cell Research* **253**:255–270.

Exton, J., 1990. Signaling through phosphatidylcholine breakdown. *Journal of Biological Chemistry* **265**:1–4.

Feig, L. A., Urano, T., and Cantor, S., 1996. Evidence for a *ras/ral* signaling cascade. *Trends in Biochemical Sciences* **21**:438–441.

Ferrell, J. E., 1997. How responses get more switchlike as you move down a protein kinase cascade. *Trends in Biochemical Sciences* **22**:288–289.

Fesik, S. W., 2000. Insights into programmed cell death through structural biology. *Cell* **103**:273–282.

Garbers, D., 1989. Guanylate cyclase, a cell surface receptor. *Journal of Biological Chemistry* **264**:9103–9106.

Gudermann, T., Schönberg, T., and Schultz, G., 1997. Functional and structural complexity of signal transduction via G-protein–coupled receptors. *Annual Review of Neuroscience* **20**:399–427.

Hannun, Y. A., and Obeid, L. M., 1995. Ceramide: An intracellular signal for apoptosis. *Trends in Biochemical Sciences* **20**:73–77.

Hepler, J., and Gilman, A., 1992. G proteins. *Trends in Biochemical Sciences* **17**:383–387.

Hof, P., Pluskey, S., Dhe-Paganon, S., et al., 1998. Crystal structure of the tyrosine phosphatase SHP-2. *Cell* **92**:441–450.

Hunter, T., 2000. Signaling—2000 and beyond. *Cell* **100**:113–127.

Jackson, H., and Parks, T., 1989. Spider toxins: Recent applications in neurobiology. *Annual Review of Neurosciences* **12**:405–414.

James, P., Vorherr, T., and Carafoli, E., 1995. Calmodulin-binding domains: Just two faced or multifaceted? *Trends in Biochemical Sciences* **20**:38–42.

Jordan, J. D., Landau, E. M., and Iyengar, R., 2000. Signaling networks: The origins of cellular multitasking. *Cell* **103**:193–200.

Kay, B. K., Williamson, M.P., and Sudol, M., 2000. The importance of being proline: The interaction of proline-rich motifs in signaling proteins with their cognate domains. *FASEB Journal* **14**: 231–241.

Kemp, B., and Pearson, R., 1991. Intrasteric regulation of protein kinases and phosphatases. *Biochimica et Biophysica Acta* **1094**:67–76.

Keyse, S. M., 2000. Protein phosphatases and the regulation of mitogen-activated protein kinase signalling. *Current Opinions in Cell Biology* **12**: 186–192.

Knowles, R., and Moncada, S., 1992. Nitric oxide as a signal in blood vessels. *Trends in Biochemical Sciences* **17**:399–402.

Koch, C., et al., 1991. SH2 and SH3 domains: Elements that control interactions of cytoplasmic signaling proteins. *Science* **252**:668–674.

Linder, M., and Gilman, A., 1992. G proteins. *Scientific American* **267**: 56–65.

Liu, J., Wu, J., Oliver, C., et al., 2000. Mutations of the serine phosphorylated in the protein phosphatase-1-binding motif in the skeletal muscle glycogen-targeting subunit. *Biochemical Journal* **346**:77–82.

Luttrell, L. M., Daaka, Y., and Lefkowitz, R. J., 1999. Regulation of tyrosine kinase cascades by G-protein–coupled receptors. *Current Opinions in Cell Biology* **11**:177–183.

Marshall, M. S., 1993. The effector interactions of p21^{ras}. *Trends in Biochemical Sciences* **18**:250–254.

Meskiene, I., and Hirt, H., 2000. MAP kinase pathways: Molecular plug-and-play chips for the cell. *Plant Molecular Biology* **42**:791–806.

Milligan, G., Parenti, M., and Magee, A. I., 1995. The dynamic role of palmitoylation in signal transduction. *Trends in Biochemical Sciences* **20**: 181–186.

Millward, T. A., Zolnierowicz, S., and Hemmings, B. A., 1999. Regulation of protein kinase cascades by protein phosphatase 2A. *Trends in Biochemical Sciences* **24**:186–191.

Mustelin, T., and Burn, P., 1993. Regulation of Src family tyrosine kinases in lymphocytes. *Trends in Biochemical Sciences* **18**:215–220.

Nicholson, D. W., and Thornberry, N. A., 1997. Caspases: Killer proteases. *Trends in Biochemical Sciences* **22**:299–306.

Odorizzi, G., Babst, M., and Emr, S. D., 2000. Phosphoinositide signaling and the regulation of membrane trafficking in yeast. *Trends in Biochemical Sciences* **25**:229–235.

Omer, C. A., and Kohl, N. E., 1997. CA$_1$A$_2$X-competitive inhibitors of farnesyltransferase as anticancer agents. *Trends in Pharmacological Sciences* **18**:437–444.

Palczewski, K., Kumasaka, T., Hori, T., et al., 2000. Crystal structure of rhodopsin: A G protein–coupled receptor. *Science* **289**: 739–745.

Parekh, D. B., Ziegler, W., and Parker, P. J., 2000. Multiple pathways control protein kinase C phosphorylation. *The EMBO Journal* **19**:496–503.

Pawson, T., and Scott, J. D., 1997. Signalling through scaffold, anchoring and adaptor proteins. *Science* **278**:2075–2080.

Plotkin, M., 1993. *Tales of a Shaman's Apprentice.* New York: Viking Penguin.

Prescott, S. M., 1997. A thematic series on phospholipases. *Journal of Biological Chemistry* **272**:15043.

Putney, J. W., 1998. Calcium signaling: Up, down, up, down. . . . What's the point? *Science* **279**:191–192.

Salveson, G. S., and Dixit, V. M., 1997. Caspases: Intracellular signaling by proteolysis. *Cell* **91**:443–446.

Sato, C., Ueno, Y., Asai, K., Takahashi, K., Sato, M., Engel, A., and Fujiyoshi, Y., 2001. The voltage-sensitive sodium channel is a bell-shaped molecule with several cavities. *Nature* **409**: 1047–1051.

Schlessinger, J., 2000. Cell signaling by receptor tyrosine kinases. *Cell* **103**:211–225.

Schlichting, I., et al., 1990. Time-resolved X-ray crystallographic study of the conformational change in Ha-*Ras* p21 protein on GTP hydrolysis. *Nature* **345**:309–314.

Sprang, S. R., 1997. GAP into the breach. *Science* **277**:329–330.

Sprang, S. R., 1997. G protein mechanisms: Insights from structural analysis. *Annual Review of Biochemistry* **66**:639–678.

Stamler, J., Singel, D., and Loscalzo, J., 1992. Biochemistry of nitric oxide and its redox-active forms. *Science* **258**:1898–1902.

Sternweiss, P. C., and Smrcka, A. V., 1992. Regulation of phospholipase C by G proteins. *Trends in Biochemical Sciences* **17**:502–506.

Strader, C. D., Fong, T. M., Tota, M. R., and Underwood, D., 1994. Structure and function of G-protein–coupled receptors. *Annual Review of Biochemistry* **63**:101–132.

Traylor, T., and Sharma, V., 1992. Why NO? *Biochemistry* **31**:2847–2849.

Virshup, D. M., 2000. Protein phosphatase 2A: A panoply of enzymes. *Current Opinions in Cell Biology* **12**:180–185.

Wickelgren, I., 1997. Biologists catch their first detailed look at NO enzyme. *Science* **278**:389.

Wittinghofer, A., and Pai, E., 1991. The structure of Ras protein: A model for a universal molecular switch. *Trends in Biochemical Sciences* **16**:382–387.

Xu, W., Doshi, A., Lei, M., et al., 1999. Crystal structures of c-Src reveal features of its autoinhibitory mechanism. *Molecular Cell* **3**:629–638.

ABBREVIATED ANSWERS TO PROBLEMS

For detailed answers to the end-of-chapter problems as well as additional problems to solve, see the *Student Study Guide* by David Jemiolo and Steven Theg that accompanies this textbook.

Chapter 1

1. Since bacteria (compared with humans) have simple nutritional requirements, their cells obviously contain enzyme systems that allow them to convert rudimentary precursors (even inorganic substances such as NH_4^+, NO_3^-, N_2, and CO_2) into the complex biomolecules—proteins, nucleic acids, polysaccharides, and complex lipids. On the other hand, animals have an assortment of different cell types designed for specific physiological functions; these cells possess a correspondingly greater repertoire of complex biomolecules to accomplish their intricate physiology.
2. Consult Figures 1.20 and 1.21 to confirm your answer.
3. a. Laid end to end, 250 *E. coli* cells would span the head of a pin.
   b. The volume of an *E. coli* cell is about 10^{-15} L.
   c. The surface area of an *E. coli* cell is about 6.0×10^{-12} m^2. Its surface-to-volume ratio is 6.0×10^6 m^{-1}.
   d. 600,000 molecules.
   e. 1.7 n*M*.
   f. Since we can calculate the volume of one ribosome to be 4.2×10^{-24} m^3 (or 4.2×10^{-21} L), 15,000 ribosomes would occupy 6.3×10^{-17} L, or 6.3% of the total cell volume.
   g. Since the *E. coli* chromosome contains 4600 kilobase pairs (4.6×10^6 bp) of DNA, its total length would be 1.6 mm—approximately 800 times the length of an *E. coli* cell.
   h. This DNA would encode 4300 different proteins, each 360 amino acids long.
4. a. The volume of a single mitochondrion is about 4.2×10^{-16} L (about 40% the volume calculated for an *E. coli* cell in problem 3).

   b. A mitochondrion would contain on average less than 8 molecules of oxaloacetate.
5. a. Laid end to end, 25 liver cells would span the head of a pin.
   b. The volume of a liver cell is about 8×10^{-12} liters (8000 times the volume of an *E. coli* cell).
   c. The surface area of a liver cell is 2.4×10^{-9} m^2; its surface-to-volume ratio is 3×10^5 m^{-1}, or about 0.05 (1/20) that of an *E. coli* cell. Cells with lower surface-to-volume ratios are limited in their exchange of materials with the environment.
   d. The number of base pairs in the DNA of a liver cell is 6×10^9 bp, which would amount to a total DNA length of 2 m (or 6 feet of DNA!) contained within a cell that is only 20 μm on a side.
   e. Maximal information content of liver-cell DNA = 3×10^9, which, expressed in proteins 400 amino acids in length, could encode 2.5×10^6 proteins.
6. The amino acid side chains of proteins provide a range of shapes, polarity, and chemical features that allow a protein to be tailored to fit almost any possible molecular surface in a complementary way.
7. Biopolymers may be informational molecules because they are constructed of different monomeric units ("letters") joined head to tail in a particular order ("words, sentences"). Polysaccharides are often linear polymers composed of only one (or two repeating) monosaccharide unit(s), and thus display little information content. Polysaccharides with a variety of monosaccharide units may convey information, through specific recognition by other biomolecules. Also, most monosaccharide units are typically capable of forming branched polysaccharide structures that are potentially very rich in information content (as in cell surface molecules that act as the unique labels displayed by different cell types in multicellular organisms).
8. Molecular recognition is based on structural complementarity. If complementary interactions involved covalent bonds (strong forces), stable structures would

be formed that would be less responsive to the continually changing dynamic interactions that characterize living processes.

9. Slight changes in temperature, pH, ionic concentrations, and so forth may be sufficient to disrupt weak forces (H bonds, ionic bonds, van der Waals interactions, hydrophobic interactions).

10. Living systems are maintained by a continuous flow of matter and energy through them. Despite the ongoing transformations of matter and energy by these highly organized, dynamic systems, no overt changes seem to occur in them: they are in a *steady state.*

Chapter 2

1. a. 3.3; b. 9.85; c. 5.7; d. 12.5; e. 4.4; f. 6.97.
2. a. 1.26 mM; b. 0.25 μM; c. 4×10^{-12} M; d. 2×10^{-4} M; e. 3.16×10^{-10} M; f. 1.26×10^{-7} M $(0.126 \ \mu M)$.
3. a. $[H^+] = 2.51 \times 10^{-5} \ M$;
   b. $K_a = 3.13 \times 10^{-8}$; $pK_a = 7.5$.
4. a. pH = 2.38; b. pH = 4.23.
5. Combine 187 mL of 0.1 M acetic acid with 813 mL of 0.1 M sodium acetate.
6. $[HPO_4^{2-}]/[H_2PO_4^-] = 0.398$.
7. a. Fraction of H_3PO_4: @pH 2 = 0.58; @pH 3 = 0.12; negligible @pH 6.
   b. Fraction of $H_2PO_4^-$: @pH 2 = 0.41; @pH 3 = 0.88; @pH 6 = 0.94; @pH 8 = 0.14; negligible @pH 10.
   c. Fraction of HPO_4^{2-}: negligible @pH 2 and 3; @pH 6 = 0.06; @pH 8 = 0.86; @pH 10 $\cong$ 1.0; @pH 13 = 0.2.
   d. Fraction of PO_4^{3-}: negligible at any pH < 10; @pH 13 = 0.8.
8. a. pH = 7.02; $[H_2PO_4^-]$ = 0.0200 M; $[HPO_4^{2-}]$ = 0.0133M.
   b. pH = 7.38; $[H_2PO_4^-]$ = 0.0133 M; $[HPO_4^{2-}]$ = 0.0200 M.

Chapter 3

1. K_{eq} = 613 M; ΔG° = −15.9 kJ/mol.
2. ΔG° = 1.69 kJ/mol at 20°C; ΔG° = −5.80 kJ/mol at 30°C. ΔS° = 0.75 kJ/mol · K.
3. ΔG = −24.8 kJ/mol.
4. State functions are quantities that depend on the state of the system and not on the path or process taken to reach that state. Volume, pressure, and temperature are state functions. Heat and all forms of work, such as mechanical work and electrical work, are not state functions.
5. $\Delta G^{\circ\prime} = \Delta G^\circ - 39.9 \ n$ (in kJ/mol), where n is the number of H^+ produced in any process. So $\Delta G^\circ = \Delta G^{\circ\prime} + 39.9 \ n = -30.5$ kJ/mol $+ 39.5(1)$ kJ/mol. ΔG° = 9.4 kJ/mol at 1 $M [H^+]$.

6. a. K_{eq}(AC) = (0.02×1000) = 20.
   b. ΔG°(AB) = 10.1 kJ/mol.
   ΔG°(BC) = −17.8 kJ/mol.
   ΔG°(AC) = −7.7 kJ/mol.
   K_{eq} = 20.
7. See *Student Study Guide* for resonance structures.
8. $K_{eq} = [Cr][P_i]/[CrP][H_2O]$.
   $K_{eq} = 3.89 \times 10^7$.
9. CrP in the amount of 135.3 moles would be required per day to provide 5860 kJ energy. This corresponds to 17,730 g of CrP per day. With a body content of 20 g CrP, each molecule would recycle 886 times per day.

   Similarly, 422 moles of glucose-6-P, or 109,720 g of glucose-6-P, would be required. Each molecule would recycle 5410 times/day.
10. ΔG = −46.1 kJ/mol.
11. The hexokinase reaction is a sum of the reactions for hydrolysis of ATP and phosphorylation of glucose:

$$ATP + H_2O \rightleftharpoons ADP + P_i$$
$$Glucose + P_i \rightleftharpoons G6P + H_2O$$
$$\overline{Glucose + ATP \rightleftharpoons G6P + ADP}$$

The free energy change for the hexokinase reaction can thus be obtained by summing the free energy changes for the first two reactions above.

$\Delta G^{\circ\prime}$ for hexokinase =
$\quad$ −30.5 kJ/mol + 13.9 kJ/mol = − 16.6 kJ/mol

12. Comparing the acetyl group of acetoacetyl-CoA and the methyl group of acetyl-CoA, it is reasonable to suggest that the acetyl group is more electron-withdrawing in nature. For this reason it tends to destabilize the thiol ester of acetoacetyl-CoA, and the free energy of hydrolysis of acetoacetyl-CoA should be somewhat larger than that of acetyl-CoA. In fact, $\Delta G^{\circ\prime}$ = −43.9 kJ/mol, compared with −31.5 kJ/mol for acetyl-CoA.

13. Carbamoyl phosphate should have a somewhat larger free energy of hydrolysis than acetyl phosphate, at least in part because of greater opportunities for resonance stabilization in the products. In fact, the free energy of hydrolysis of carbamoyl phosphate is −51.5 kJ/mol, compared with −43.3 kJ/mol for acetyl phosphate.

Chapter 4

1. Structures for glycine, aspartate, leucine, isoleucine, methionine, and threonine are presented in Figure 4.3.
2. Asparagine = Asn = N.
   Arginine = Arg = R.
   Cysteine = Cys = C.
   Lysine = Lys = K.
   Proline = Pro = P.
   Tyrosine = Tyr = Y.
   Tryptophan = Trp = W.

3.

Alanine dissociation:

$$H_3N^+-\overset{COOH}{\underset{CH_3}{C}}-H \rightleftharpoons H_3N^+-\overset{COO^-}{\underset{CH_3}{C}}-H \rightleftharpoons H_2N-\overset{COO^-}{\underset{CH_3}{C}}-H$$

Glutamate dissociation:

$$H_3N^+-\overset{COOH}{\underset{\underset{COOH}{CH_2}}{\underset{CH_2}{C}}}-H \rightleftharpoons H_3N^+-\overset{COO^-}{\underset{\underset{COOH}{CH_2}}{\underset{CH_2}{C}}}-H \rightleftharpoons H_3N^+-\overset{COO^-}{\underset{\underset{COO^-}{CH_2}}{\underset{CH_2}{C}}}-H \rightleftharpoons H_2N-\overset{COO^-}{\underset{\underset{COO^-}{CH_2}}{\underset{CH_2}{C}}}-H$$

Histidine dissociation:

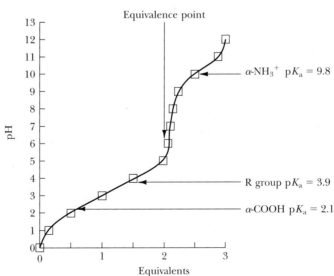

Lysine dissociation:

$$H_3N^+-\overset{COOH}{\underset{\underset{NH_3^+}{(CH_2)_4}}{C}}-H \rightleftharpoons H_3N^+-\overset{COO^-}{\underset{\underset{NH_3^+}{(CH_2)_4}}{C}}-H \rightleftharpoons H_2N-\overset{COO^-}{\underset{\underset{NH_3^+}{(CH_2)_4}}{C}}-H \rightleftharpoons H_2N-\overset{COO^-}{\underset{\underset{NH_2}{(CH_2)_4}}{C}}-H$$

Phenylalanine dissociation:

4. The proximity of the α-carboxyl group lowers the pK_a of the α-amino group.

5.

6. pH = pK_a + log (2/1) = 4.3 + 0.3 = 4.6.
   The γ-carboxyl group of glutamic acid is 2/3 dissociated at pH = 4.6.

7. pH = pK_a + log (1/4) = 10.5 + (-0.6) = 9.9.
8. a. The pH of a 0.3 M leucine hydrochloride solution is approximately 1.46.

b. The pH of a 0.3 *M* sodium leucinate solution is approximately 11.5.

c. The pH of a 0.3 *M* solution of isoelectric leucine is approximately 6.05.

9. Cystine (disulfide-linked cysteine) has two chiral carbons, the two α-carbons of the cysteine moieties. Since each chiral center can exist in two forms, there are four stereoisomers of cystine. However, it is not possible to distinguish the difference between L-cysteine/D-cysteine and D-cysteine/L-cysteine dimers. So three distinct isomers are formed:

10. Basic amino acids should adhere strongly to Dowex-50, a sulfonated resin, whereas acidic amino acids should bind least strongly, with neutral and hydrophobic amino acids being intermediate in behavior. Therefore, the order of elution should be: aspartate, valine, isoleucine, histidine, and finally arginine. (The pK_a values for valine are slightly lower than those for isoleucine—thus the elution time of valine should be somewhat faster than that of isoleucine.)

11. L-threonine is (2*S*, 3*R*)-threonine.
D-threonine is (2*R*, 3*S*)-threonine.
L-allothreonine is (2*S*, 3*S*)-threonine.
D-allothreonine is (2*R*, 3*R*)-threonine.

12. Nitrate reductase is a dimer (2 Mo/240,000 M_r).

13. Phe-Asp-Tyr-Met-Leu-Met-Lys.

14. Tyr-Asn-Trp-Met-(Glu-Leu)-Lys. Parentheses indicate that the relative positions of Glu and Leu cannot be assigned from the information provided.

15. Ser-Glu-Tyr-Arg-Lys-Lys-Phe-Met-Asn-Pro.

Chapter 5

1. The central rod domain of keratin is composed of distorted α-helices, with 3.6 residues per turn, but a pitch of 0.51 nm, compared with 0.54 nm for a true α-helix.

(0.51 nm/turn) (312 residues)/(3.6 residues/turn)
$$= 44.2 \text{ nm} = 442 \text{ Å}$$

For an α-helix, the length would be:

(0.54 nm/turn)(312 residues)/(3.6 residues/turn)
$$= 46.8 \text{ nm} = 468 \text{ Å}$$

The distance between residues is 0.347 nm for antiparallel β-sheets and 0.325 nm for parallel β-sheets. So 312 residues of antiparallel β-sheets amount to 1083 Å and 312 residues of parallel β-sheet amount to 1014 Å.

2. The collagen helix has 3.3 residues per turn and 0.29 nm per residue, or 0.96 nm/turn. Then:

(4 in/year)(2.54 cm/in)(10^7 nm/cm)/(0.96 nm/turn)
$$= 1.06 \times 10^8 \text{ turns/year}$$

(1.06×10^8 turns/year)(1 year/365 days)(1 day/24 hours)
(1 hour/60 minutes) = 201 turns/minute

3. **Asp:** The ionizable carboxyl can participate in ionic and hydrogen bonds. Hydrophobic and van der Waals interactions are negligible.

Leu: The leucine side chain does not participate in hydrogen bonds or ionic bonds, but it will participate in hydrophobic and van der Waals interactions.

Tyr: The phenolic hydroxyl of tyrosine, with a relatively high pK_a, will participate in ionic bonds only at high pH, but can both donate and accept hydrogen bonds. Uncharged tyrosine is capable of hydrophobic interactions. The relatively large size of the tyrosine side chain will permit substantial van der Waals interactions.

His: The imidazole side chain of histidine can act both as acceptor and donor of hydrogen bonds and, when protonated, can participate in ionic bonds. Van der Waals interactions are expected, but hydrophobic interactions are less likely in most cases.

4. As an imido acid, proline has a secondary nitrogen with only one hydrogen. In a peptide bond, this nitrogen possesses no hydrogens and thus cannot function as a hydrogen bond donor in α-helices. On the other hand, proline stabilizes the *cis*-configuration of a peptide bond and is thus well suited to β-turns, which require the *cis*-configuration.

5. For a right-handed crossover, moving in the N-terminal to C-terminal direction, the crossover moves in a clockwise direction when viewed from the C-terminal side toward the N-terminal side. The reverse is true for a left-handed crossover; that is, movement from N-terminus to C-terminus is accompanied by counterclockwise rotation.

6. The Ramachandran plot reveals allowable values of ϕ and ψ for α-helix and β-sheet formation. The plots consider steric hindrance and will be somewhat specific for individual amino acids. For example, peptide bonds containing glycine can adopt a much wider range of ϕ and ψ angles than can peptide bonds containing tryptophan.

7. Hydrophobic interactions frequently play a major role in subunit–subunit interactions. The surfaces that participate in subunit–subunit interactions in the B_4 tetramer are likely to possess larger numbers of hydrophobic residues than the corresponding surfaces of protein A.

8. The length is given by: (53 residues) × (0.15 nm run/residue) = 7.95 nm. The number of turns in the helix is given by: (53 residues)/(3.6 residues/turn) = 14.7 turns. There are 49 hydrogen bonds in this helix.

Chapter 6

1. (a) Since the question specifically asks for triacylglycerols that contain stearic acid AND arachidonic acid, we can discount the triacylglycerols that contain only stearic or only arachidonic acid. In this case, there are six possibilities:

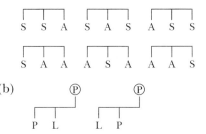

(b)

Only the structure on the left is likely to be found in a biological membrane.

2. PL/protein molar ratio in purple patches of *H. halobium* is 10.8.

3. See *Student Study Guide* for plots of sucrose solution density vs. percent by volume.

4. $r = (4Dt)^{1/2}$.

According to this equation, a phospholipid with $D = 1 \times 10^{-8}$ cm^2 will move approximately 200 nm in 10 milliseconds.

5. Fibronectin: For $t = 10$ msec, $r = 1.67 \times 10^{-7}$ cm $= 1.67$ nm. Rhodopsin: $r = 110$ nm.

All else being equal, the value of D is roughly proportional to $(M_r)^{-1/3}$. Molecular weights of rhodopsin and fibronectin are 40,000 and 460,000, respectively. The ratio of diffusion coefficients is thus expected to be:

$$(40{,}000)^{-1/3}/(460{,}000)^{-1/3} = 2.3$$

On the other hand, the values given for rhodopsin and fibronectin give an actual ratio of 4286. The explanation is that fibronectin is anchored in the membrane via interactions with cytoskeletal proteins, and its diffusion is severely restricted compared with that of rhodopsin.

6. a. Divalent cations increase T_m.
   b. Cholesterol broadens the phase transition without significantly changing T_m.
   c. Distearoylphosphatidylserine should increase T_m, due to increased chain length and also to the favorable interactions between the more negative PS headgroup and the more positive PC headgroups.

7. $\Delta G = RT \ln ([C_2]/[C_1])$.
   $\Delta G = +4.0$ kJ/mol.

8. $\Delta G = RT \ln ([C_{out}]/[C_{in}]) + Z\mathscr{F}\Delta\psi$.
   $\Delta G = +4.08$ kJ/mol $- 2.89$ kJ/mol.
   $\Delta G = +1.19$ kJ/mol.
   The unfavorable concentration gradient thus overcomes the favorable electrical potential and the outward movement of Na$^+$ is not thermodynamically favored.

9. One could solve this problem by going to the trouble of plotting the data in v vs. [S], $1/v$ vs. $1/$[S], or [S]$/v$ vs. [S] plots, but it is simpler to examine the value of [S]$/v$ at each value of [S]. The Hanes-Woolf plot makes clear that [S]$/v$ should be constant for all [S] for the case of passive diffusion. In the present case, [S]$/v$ is a constant value of 0.0588 (L/min)$^{-1}$. It is thus easy to recognize that this problem describes a system that permits passive diffusion of histidine.

10. This is a two-part problem. First calculate the energy available from ATP hydrolysis under the stated conditions, then use the answer to calculate the maximal internal fructose concentration against which fructose can be transported by coupling to ATP hydrolysis. Using a value of -30.5 kJ/mol for the $\Delta G^{\circ\prime}$ of ATP and the indicated concentrations of ATP, ADP, and P$_i$, one finds that the ΔG for ATP hydrolysis under these conditions (and at 298 K) is -52.0 kJ/mol. Putting the value of $+52.0$ kJ/mol into Equation 6.1 and solving for C_2 yields a value for the maximum possible internal fructose concentration of 1300 M! Thus, ATP hydrolysis could (theoretically) drive fructose transport against internal fructose concentrations up to this value. (In fact, this value is vastly in excess of the limit of fructose solubility.)

11. Monensin is a mobile carrier ionophore, whereas cecropin is a channel-forming ionophore. The transport rates for mobile carrier ionophores are quite sensitive to temperature (in the vicinity of the membrane phase transition), but the transport rates for channel-forming ionophores, by contrast, are relatively insensitive to temperature. The phase transition temperature for DPPC is 41.4°C. Therefore, the transport rates for monensin in DPPC should increase significantly between 35°C and 50°C, but the rates observed for cecropin *a* in DPPC membranes at 50°C should be about the same as they are at 35°C.

12. Each of the transport systems described can be inhibited (with varying degrees of specificity). Inhibition of the rhamnose transport system by one or more of these agents would be consistent with involvement of one of these transport systems with rhamnose transport. Thus, non-hydrolyzable ATP analogs should inhibit ATP-dependent transport systems, ouabain should specifically block Na$^+$ (and K$^+$) transport, uncouplers should inhibit proton-gradient–dependent systems, and fluoride should inhibit the PTS system (via inhibition of enolase).

Chapter 7

1. See structures in *Student Study Guide.*
2. The systematic name for stachyose (Figure 7.19) is β-D-fructofuranosyl-O-α-D-galactopyranosyl-(1 → 6)-O-α-D-galactopyranosyl-(1 → 6)-α-D-glucopyranoside.
3. The systematic name for trehalose is α-D-glucopyranosyl-(1 → 1)-α-D-glucopyranoside.
   Trehalose is not a reducing sugar. Both anomeric carbons are occupied in the disaccharide linkage.
4. See structures in *Student Study Guide.*
5. A sample that is 0.69 g α-D-glucose/mL and 0.31 g β-D-glucose/mL will produce a specific rotation of 83°.
6. A 0.2-g sample of amylopectin corresponds to 0.2 g/162 g/mole or 1.23×10^{-3} mole glucose residues. 50 μmole is 0.04 of the total sample or 4% of the residues. Methylation of such a sample should yield 1,2,3,6-tetramethylglucose for the glucose residues on the reducing ends of the sample. The amylopectin sample contains 2.4×10^{18} reducing ends.

Chapter 8

1. See Figure 8.16.
2. $f_A = 0.304$; $f_G = 0.195$; $f_C = 0.182$; $f_T = 0.318$.
3. 5'-TAGTGACAGTTGCGAT-3'.
4. 5'-ATCGCAACTGTCACTA-3'.
5. 5'-TACGGTCTAAGCTGA-3'.
6. There are two possibilities, a and b. (E = *EcoR1* site; B = *BamH1* site.) Note that b is the reverse of a.

a.
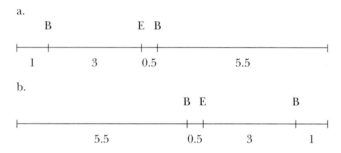

b.

```
              B  E            B
├────┼───────────┼──┼──────────┼
       5.5        0.5    3     1
```

7. See diagram at top of next column.
8. Original nucleotide: 5'-dAGACTTGACGCT
9. $\Delta Z = 0.32$ nm; $P = 3.36$ nm/turn; 10.5 base pairs per turn; $\Delta\phi = 34.3°$; c (true repeat) = 6.72 nm.
10. 27.3 nm; 122 base pairs.
11. 4325 nm (4.33 μm).
12. 6×10^9 bp/200 bp = 3×10^7 nucleosomes.
    The length of B-DNA 6×10^9 bp long = (0.34 nm) $(6 \times 10^9) = 2.04 \times 10^9$ nm (over 2 meters!). The

A	G	C	T
			──
──			
	──		
	──		
──			
		──	
			──
	──		
──			

length of 3×10^7 nucleosomes = (6 nm)(3×10^7) = 18×10^7 nm (0.18 meter).
13. From Figure 8.52: similarly shaded regions indicate complementary sequences joined via intrastrand hydrogen bonds (see figure at bottom of page).
14. Increasing order of T_m: yeast < human < salmon < wheat < *E. coli.*

Chapter 9

1. Linear and circular DNA molecules consisting of one or more copies of just the genomic DNA fragment; linear and circular DNA molecules consisting of one or more copies of just the vector DNA; linear and circular DNA molecules containing one or more copies of both the genomic DNA fragment and plasmid DNA.

2. -GAATTCCCGGGGATCCTCTAGAGTCGACCTGCAGGCATGC-

GAATTC	GGATCC	GTCGAC	GCATGC
EcoRI	*BamHI*	*SalI*	*SphI*
CCCGGG	TCTAGA	CTGCAG	
SmaI	*XbaI*	*PstI*	

3. a. AAGCTTGAGCTCGAGATCTAGATCGAT

   HindIII   *XhoI*   *XbaI*
   SacI   *BglII*   *ClaI*

   b.
   Vector: *HindIII:* 5'-A........-gap-.........CGAT-3': *ClaI*
                        3'-TTCGA....-gap-......... TA-5'
   Fragment: *HindIII:* 5'-AGCTT(NNNN-etc-NNNN)AT-3'
                        3'-A(NNNN-etc-NNNN)TAGC-5'
4. N = 3480.
5. N = 10.4 million.
6. 5'-ATGCCGTAGTCGATCAT and
   5'-ATGCTATCTGTCCTATG.

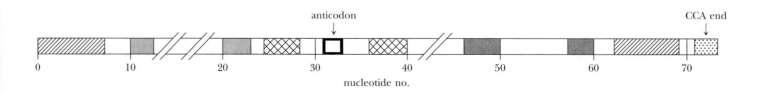

7. Thr-Met-Ile-Thr-Asn-Ser-Pro-*Asp-Pro-Phe-Ile-His-Arg-Arg-Ala-Gly-Ile-Pro-Lys-Arg-Arg-Pro* . . .

   The junction between β-galactosidase and the insert amino acid sequence is between Pro and Asp, so the first amino acid encoded by the insert is Asp. (The polylinker itself codes for Asp just at the *BamHI* site, but, in constructing the fusion, this Asp and all of this downstream section of polylinker DNA is displaced to a position after the end of the insert.)

8. $5'$-(G)AATTCNGGNATGCAYCCNGGNAAR$_\text{C}^\text{T}$T$_\text{N}^\text{Y}$-GCN$\underline{\text{AGY}}$T-GGTTYGTNGGGAATTCN-

   (*Note:* The underlined triplet AGY represents the middle Ser residue. Ser codons are either AGY or TCN [where Y = pyrimidine and N = any base]; AGY was selected here so that the mutagenesis of this codon to a Cys codon (TGY) would involve only an A → T change in the nucleotide sequence.)

   Since the middle Ser residue lies nearer to the $3'$-end of this *EcoRI* fragment, the mutant primer for PCR amplification should encompass this end. That is, it should be the primer for the $3' \to 5'$ strand of the *EcoRI* fragment:

   $$5'\text{–NNNGAATTCCCN}_c\text{ACR}_c\text{AACCA}\underline{\text{R}_c\text{CAN}}_c\text{GC-}3'$$

   where the mutated Ser → Cys triplet is underlined, NNN = several extra bases at the $5'$-end of the primer to place the *EcoRI* site internal, N$_c$ = the nucleotide complementary to the nucleotide at this position in the $5' \to 3'$ strand, and R$_c$ = the pyrimidine complementary to the purine at this position in the $5' \to 3'$ strand.

Chapter 10

1. $v/V_{max} = 0.8$.

2. $v = 91 \, \mu\text{mol/mL} \cdot \text{sec}$.

3. $K_m = 3 \times 10^{-4} \, M$.

4. From double-reciprocal Lineweaver–Burk plots: $V_{max} = 51 \, \mu\text{mol/mL} \cdot \text{sec}$ and $K_m = 3.2 \, \text{m}M$. Inhibitor (2) shows competitive inhibition with a $K_I = 2.13 \, \text{m}M$. Inhibitor (3) shows noncompetitive inhibition with a $K_I = 4 \, \text{m}M$.

5. Top left: (1) Competitive inhibition (I competes with S for binding to E). (2) I binds to and forms a complex with S.

   Top right: (1) Pure noncompetitive inhibition. (2) Random, single-displacement bisubstrate reaction, where *A* doesn't affect *B* binding and vice versa. (Other possibilities include [3] Irreversible inhibition of *E* by *I*; [4] $1/v$ versus $1/[S]$ plot at two different concentrations of enzyme, *E*.)

   Bottom: (1) Mixed, noncompetitive inhibition. (2) Ordered, single-displacement bisubstrate reaction.

6. Clancy must drink 694 mL of wine, or about one 750-mL bottle.

7. a. As $[P]$ rises, the rate of P formation shows an apparent decline, as enzyme-catalyzed conversion of $P \to S$ becomes more likely.

   b. Availability of substrates and cofactors.

   c. Changes in [enzyme] due to enzyme synthesis and degradation.

   d. Covalent modification.

   e. Allosteric regulation.

   f. Specialized controls, such as zymogen activation, isozyme variability, and modulator protein influences.

8. Proteolytic enzymes have the potential to degrade the proteins of the cell in which they are synthesized. Synthesis of these enzymes as zymogens is a way of delaying expression of their activity to the appropriate time and place.

9. Monod, Wyman, Changeux allosteric K system:

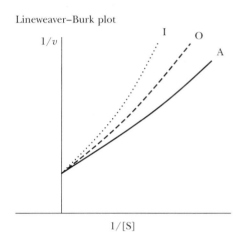

Monod, Wyman, Changeux allosteric V system:

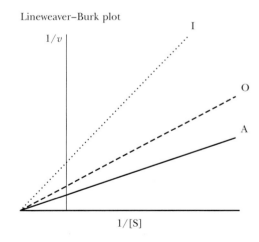

10. If L_0 is large, i.e., $T_0 \gg R_0$, activator, A, will show positive homotropic binding. If R_0 is very small, i.e., $R_0 \gg T_0$, inhibitor, I, will show positive homotropic binding.

11. More glycogen phosphorylase will be in the glycogen phosphorylase *a* (more active) form, but caffeine promotes the less active T conformation of glycogen phosphorylase.

12. a. By definition, when $[P_i] = K_{0.5}$, $v = 0.5 V_{max}$.
 b. In the presence of AMP, at $[P_i] = K_{0.5}$, $v = 0.85 V_{max}$ (from Figure 10.36c).
 c. In the presence of ATP, at $[P_i] = K_{0.5}$, $v = 0.12 V_{max}$ (from Figure 10.36b).

Chapter 11

1. a. Nucleophilic attack by an imidazole nitrogen of His[57] on the —CH_2— carbon of the chloromethyl group of TPCK covalently inactivates chymotrypsin. (See *Student Study Guide* for structures.)
 b. TPCK is specific for chymotrypsin because the phenyl ring of the phenylalanine residue interacts effectively with the binding pocket of the chymotrypsin active site. This positions the chloromethyl group to react with His[57].
 c. Replacement of the phenylalanine residue of TPCK with arginine or lysine produces reagents that are specific for trypsin.
2. a. The structure proposed by Craik et al., 1987 (*Science* **237**:905–907) is shown below. (If you look up this reference, note that the letters A and B of the figure legend for Figure 3 of this article actually refer to parts B and A of the figure, respectively. Reverse either the letters in the figure or the letters in the figure legend and it will make sense.)

 b. In the proposed model (shown above), Asn[102] of the mutant enzyme can serve only as a hydrogen-bond donor to His[57]. It is unable to act as a hydrogen-bond acceptor, as aspartate does in native trypsin. As a result, His[57] is unable to act as a general base in transfer-

ring a proton from Ser[195]. This presumably accounts for the diminished activity of the mutant trypsin.

3. a. The usual explanation for the inhibitory properties of pepstatin is that the central amino acid, statine, mimics the tetrahedral amide hydrate transition state of a good pepsin substrate with its unique hydroxyl group.
 b. Pepsin and other aspartic proteases prefer to cleave peptide chains between a pair of hydrophobic residues, whereas HIV-1 protease preferentially cleaves a Tyr-Pro amide bond. Since pepstatin more closely fits the profile of a pepsin substrate, we would surmise that it is a better inhibitor of pepsin than of HIV-1 protease. In fact, pepstatin is a potent inhibitor of pepsin ($K_I < 1$ nM) but only a moderately good inhibitor of HIV-1 protease, with a K_I of about 1 μM.
4. This problem is solved best by using the equation derived in problem 6:

$$\frac{k_e}{k_u} = e^{(\Delta G_u^{\ddagger} - \Delta G_e^{\ddagger})/RT}$$

Using this equation, we can show that the difference in activation energies for the uncatalyzed and catalyzed hydrolysis reactions ($\Delta G_u - \Delta G_e$) is 92 kJ/mol.
5. Trypsin catalyzes the conversion of chymotrypsinogen to π-chymotrypsin, and chymotrypsin itself catalyzes the conversion of π-chymotrypsin to α-chymotrypsin.
6. The enzyme-catalyzed rate is given by:

$$v = k_e[ES] = k_e'[EX^{\ddagger}]$$

$$K_e^{\ddagger} = \frac{[EX^{\ddagger}]}{[ES]}$$

$$\Delta G_e^{\ddagger} = -RT \ln K_e^{\ddagger}$$

$$K_e^{\ddagger} = e^{-\Delta G_e^{\ddagger}/RT}$$

$$[EX^{\ddagger}] = K_e^{\ddagger}[ES] = e^{-\Delta G_e^{\ddagger}/RT}[ES]$$

So $k_e[ES] = k_e' e^{-\Delta G_e^{\ddagger}/RT}[ES]$

or $k_e = k_e' e^{-\Delta G_e^{\ddagger}/RT}$

Similarly, for the uncatalyzed reaction:

$$k_u = k_u' e^{-\Delta G_u^{\ddagger}/RT}$$

Assuming that $k_u' \cong k_e'$

Then $\dfrac{k_e}{k_u} = \dfrac{e^{-\Delta G_e^{\ddagger}/RT}}{e^{-\Delta G_u^{\ddagger}/RT}}$

$$\frac{k_e}{k_u} = e^{(\Delta G_u^{\ddagger} - \Delta G_e^{\ddagger})/RT}$$

Chapter 12

1. Five liters $\times$ 0.75 mM = 3.75 mmol. Five palmitate-binding sites per albumin $\times$ 3.75 mmol = 18.75 total mmol of palmitate bound. The molecular weight of palmitate is 257 g/mol; $257 \times 18.75 \times 10^{-3} = 4.82$ g.

2. 0.24 = fraction of protein, 0.76 = fraction of lipid.
3. The concentration of albumin in serum is 0.75 mM. $K_D = 10^{-6}$ M = [albumin:drug X]/([albumin]$_{free}$ [drug X]$_{free}$). Also, [drug X]$_{bound}$ = [albumin:drug X] and [drug X]$_{bound}$ + [drug X]$_{free}$ = 10 μM. (The effect of drug X binding on [albumin]$_{free}$ can be ignored, since [albumin] $\gg$ [drug X]$_{total}$.) So, 10^{-6} M = $(0.75 \times 10^{-3}$ $M)(10^{-5}$ M − [drug X]$_{bound}$)/[drug X]$_{bound}$. Thus, [drug X]$_{free}$ = 0.0133 μM.
4. If half of 200 mg/dL cholesterol is associated with LDLs, LDLs contribute 100 mg/dL. If half the lipid (by weight) in LDLs is cholesterol, the person's blood contains 200 mg/dL of LDL lipids. Since 200 mg/dL = 2000 mg/L lipids and 60% of the weight of LDLs is lipids, this person has 3.33 g/L LDLs or 16.7 g LDLs in 5 L of blood.
5. For n = 2.8, Y_{lungs} = 0.98 and $Y_{capillaries}$ = 0.77.
   For n = 1.0, Y_{lungs} = 0.79 and $Y_{capillaries}$ = 0.61.
   Thus, with an n of 2.8 and a P_{50} of 26 torr, hemoglobin becomes almost fully saturated with O_2 in the lungs and drops to 77% saturation in resting tissue, a change of 21%. If n of 1.0 and a P_{50} of 26 torr, hemoglobin would become only 79% saturated with O_2 in the lungs and would drop to 61% in resting tissue, a change of 18%. The difference in hemoglobin O_2 saturation conditions between the values for n = 2.8 and 1.0 (21% − 18% or 3% saturation) seems small, but note that the potential for O_2 delivery (98% saturation vs. 79% saturation) is large and becomes crucial when pO_2 in actively metabolizing tissue falls below 40 torr.
6. Over time, stored erythrocytes will metabolize 2,3-BPG via the pathway of glycolysis. If [BPG] drops, hemoglobin may bind O_2 with such great affinity that it will not be released to the tissues (see Figure 12.24). The patient receiving a transfusion of [BPG]-depleted blood may actually suffocate.

Chapter 13

1. The pronghorn antelope is truly a remarkable animal, with numerous specially evolved anatomical and molecular features. These include a large windpipe (to draw in more oxygen and exhale more carbon dioxide), lungs that are three times the size of those of comparable animals (such as goats), and lung alveoli with five times the surface area, so that oxygen can diffuse more rapidly into the capillaries. The blood contains larger numbers of red blood cells and thus more hemoglobin. The skeletal and heart muscles are likewise adapted for speed and endurance. The heart is three times the size of that of comparable animals and pumps a proportionally larger volume of blood per contraction. Significantly, the muscles contain much larger numbers of energy-producing mitochondria, and the muscle fibers themselves are shorter and thus designed for faster contractions. All these characteristics enable the pronghorn antelope to run at a speed nearly twice the top speed of a thoroughbred racehorse and to sustain such speed for up to an hour.

2. Refer to Figure 13.20. Step 5, in which the myosin head conformation change occurs, is the step that should be blocked by β,γ-methylene-ATP, since hydrolysis of ATP should occur in this step and β,γ-methylene-ATP is nonhydrolyzable.
3. Phosphocreatine is synthesized from creatine (via creatine kinase) primarily in muscle mitochondria (where ATP is readily generated) and then transported to the sarcoplasm, where it can act as an ATP buffer. The creatine kinase reaction in the sarcoplasm yields the product creatine, which is transported back into the mitochondria to complete the cycle. Like many mitochondrial proteins, the expression of mitochondrial creatine kinase is directed by mitochondrial DNA, whereas sarcoplasmic creatine kinase is derived from information encoded in nuclear DNA.
4. Note in step 4 of Figure 13.20 that it is ATP that stimulates dissociation of myosin heads from the actin filaments—the dissociation of the crossbridge complex. When ATP levels decline (as happens rapidly after death), large numbers of myosin heads are unable to dissociate from actin, and the muscle becomes stiff and unable to relax.
5. The skeletal muscles of the average adult male have a cross-sectional area of approximately 5500 cm^2. The gluteus maximus muscles represent approximately 300 cm^2 of this total. Assuming 4 kg of maximal tension per square centimeter of cross-sectional area, one calculates a total tension for the gluteus maximus of 1200 kg (as stated in the problem). The same calculation shows that the total tension that could be developed by all the muscles in the body is 22,000 kg (or 24.2 tons)!

Chapter 14

1. 6.5×10^{12} (6.5 trillion) people.
2. Consult Table 14.2.
3. See Section 14.2.
4. Consult Figure 14.4.
5. The answer is found in an examination of the energetics of metabolism. Free energy from catabolism is captured in the form of ATP. The "packet" of energy necessary to cause the phosphorylation of ADP by P_i is a finite quantity, about 40 to 50 kJ/mol under cellular conditions. If the potential energy in a molecule (say, glucose) undergoing catabolism is released stepwise over many steps, the number of ATP molecules that can be formed in the process is potentially greater. Similarly, the amount of energy that can be delivered to an anabolic reaction by ATP is limited by the free energy available from its hydrolysis. That is, metabolism consists of many reactions in order to tailor the energy transaction at any step to the amount of free energy that accompanies ATP

synthesis or hydrolysis. Another reason for many steps is the need for metabolic control. Cells disassemble or assemble molecules one chemical step at a time so that the overall process can be carefully regulated.

6. Consult *Corresponding Pathways of Catabolism and Anabolism Differ in Important Ways*, p. 445. See also Figure 14.6.

7. See *The ATP Cycle*, p. 446; *NAD$^+$ Collects Electrons Released in Catabolism*, p. 447; and *NADPH Provides the Reducing Power for Anabolic Process*, p. 448.

8. See Chapter 1, pp. 22–25, and *Student Study Guide*.

9. A wide variety of mutations of mitochondrial branched-chain α-keto acid dehydrogenases (BCKD) have been identified in patients with MSUD. Some of these mutations produce BCKD with low enzyme activity, but others appear to shorten the biological half-life of the enzyme. Some mutant BCKD are degraded much more rapidly in the cell than are normal BCKD. The result is that, although such mutant BCKD molecules show normal activity, each cell contains fewer enzyme molecules and displays a lower total enzyme activity. Studies by Louis Elsas at Emory University have led to a proposal that thiamine pyrophosphate stabilizes a conformation of BCKD that is more resistant to cellular degradation. For this reason, treatment of some MSUD patients with thiamine (100–500 mg/day) increases the biological half-life of the BCKD enzyme molecules and thus increases the total mitochondrial BCKD activity.

Chapter 15

1. a. Phosphoglucoisomerase, fructose bisphosphate aldolase, triose phosphate isomerase, glyceraldehyde-3-P dehydrogenase, phosphoglycerate mutase, enolase.

   b. Hexokinase/glucokinase, phosphofructokinase, phosphoglycerate kinase, pyruvate kinase, lactate dehydrogenase.

   c. Hexokinase and phosphofructokinase.

   d. Phosphoglycerate kinase and pyruvate kinase.

   e. According to Equation (3.12), reactions in which the number of reactant molecules differs from the number of product molecules exhibit a strong dependence on concentration. That is, such reactions are extremely sensitive to changes in concentration. Using this criterion, we can predict from Table 15.1 that the free energy changes of the fructose bisphosphate aldolase and glyceraldehyde-3-P dehydrogenase reactions will be strongly influenced by changes in concentration.

   f. Reactions that occur with ΔG near zero operate at or near equilibrium. See Table 15.1 and Figure 15.22.

2. The carboxyl carbon of pyruvate derives from carbons 3 and 4 of glucose. The keto carbon of pyruvate derives from carbons 2 and 5 of glucose. The methyl carbon of pyruvate is obtained from carbons 1 and 6 of glucose.

3. Increased [ATP] or [citrate] inhibits glycolysis. Increased [AMP], [fructose-1,6-bisphosphate], [fructose-2,6-bisphosphate], or [glucose-6-P] stimulates glycolysis.

4. See *Student Study Guide* for discussion.

5. The relevant reactions of galactose metabolism are shown in Figure 15.24. See *Student Study Guide* for mechanisms.

6. Ignoring the possibility that ^{32}P might be incorporated into ATP, the only reaction of glycolysis that utilizes P_i is glyceraldehyde-3-P dehydrogenase, which converts glyceraldehyde-3-P to 1,3-bisphosphoglycerate. $^{32}P_i$ would label the phosphate at carbon 1 of 1,3-bisphosphoglycerate. The label will be lost in the next reaction, and no other glycolytic intermediates will be directly labeled. (Once the label is incorporated into ATP, it will also show up in glucose-6-P, fructose-6-P, and fructose-1,6-bisphosphate.)

7. The sucrose phosphorylase reaction produces glucose-6-P directly, bypassing the hexokinase reaction. Direct production offers the obvious advantage of saving a molecule of ATP.

8. All of the kinases involved in glycolysis, as well as enolase, are activated by Mg^{2+} ion. A Mg^{2+} deficiency could lead to reduced activity for some or all of these enzymes. However, other systemic effects of a Mg^{2+} deficiency might cause even more serious problems.

9. a. 7.52.

   b. 0.0266 mM.

10. +1.08 kJ/mol.

11. a. −31.4 kJ/mol.

    b. K_{eq} = 318,542.

    c. [Pyr]/[PEP] = 39,818.

12. a. −13.8 kJ/mol.

    b. K_{eq} = 262.

    c. [FBP]/[F6P] = 653.

Chapter 16

1. Glutamate enters the TCA cycle via a transamination to form α-ketoglutarate. The γ-carbon of glutamate entering the cycle is equivalent to the methyl carbon of an entering acetate. Thus no radioactivity from such a label would be lost in the first or second cycle, but in each subsequent cycle, 50% of the total label would be lost.

2. NAD$^+$ activates pyruvate dehydrogenase and isocitrate dehydrogenase and thus would increase TCA cycle activity. If NAD$^+$ increases at the expense of NADH, the resulting decrease in NADH would likewise activate the cycle by stimulating citrate synthase and α-ketoglutarate dehydrogenase. ATP inhibits pyruvate dehydrogenase, citrate synthase, and isocitrate dehydrogenase; reducing the ATP concentration would thus activate the cycle. Isocitrate is not a regulator of the cycle, but in-

creasing its concentration would mimic an increase in acetate flux through the cycle and increase overall cycle activity.

3. For most enzymes that are regulated by phosphorylation, the covalent binding of a phosphate group at a distant site induces a conformation change at the active site that either activates or inhibits the enzyme activity. On the other hand, X-ray crystallographic studies reveal that the phosphorylated and unphosphorylated forms of isocitrate dehydrogenase share identical structures with only small (and probably insignificant) conformational changes at Ser^{113}, the locus of phosphorylation. What phosphorylation *does* do is block isocitrate binding (with no effect on the binding affinity of $NADP^+$). As shown in Figure 2 of the paper cited in the problem (Barford, D., 1991. Molecular mechanisms for the control of enzymic activity by protein phosphorylation. *Biochimica et Biophysica Acta* **1133**:55–62), the γ-carboxyl group of bound isocitrate forms a hydrogen bond with the hydroxyl groups of Ser^{113}. Phosphorylation apparently prevents isocitrate binding by a combination of a loss of the crucial H bond between substrate and enzyme, and by repulsive electrostatic and steric effects.

conveniently assayed by measuring the decrease in absorbance in a solution of the flavoenzyme and succinate.

7. The central (C-3) carbon of citrate is reduced, and an adjacent carbon is oxidized in the aconitase reaction. The carbon bearing the hydroxyl group is obviously oxidized by isocitrate dehydrogenase. In the α-ketoglutarate dehydrogenase reaction, the departing carbon atom and the carbon adjacent to it are both oxidized. Both of the $—CH_2—$ carbons of succinate are oxidized in the succinate dehydrogenase reaction. Of these four molecules, all but citrate undergo a net oxidation in the TCA cycle.

8. Several TCA metabolite analogs are known, including malonate, an analog of succinate, and 3-nitro-2-S-hydroxypropionate, the anion of which is a transition-state analog for the fumarase reaction.

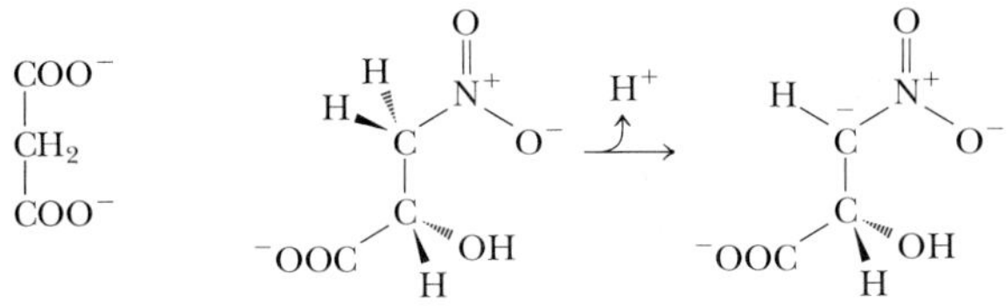

Malonate (succinate analog)   3-Nitro-2-S-hydroxypropionate   Transition-state analog for fumarase

9.

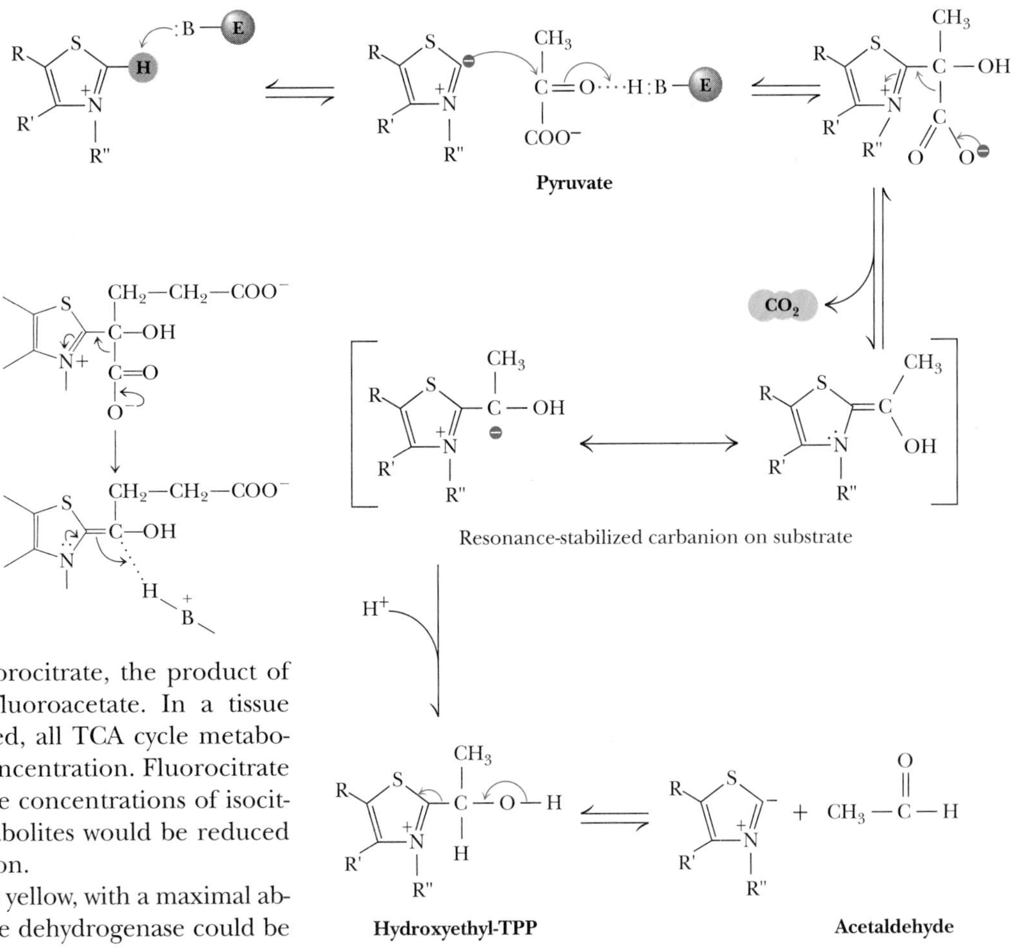

Pyruvate

Resonance-stabilized carbanion on substrate

4.

TPP   *α*-KG

Covalent TPP intermediate

Hydroxyethyl-TPP   **Acetaldehyde**

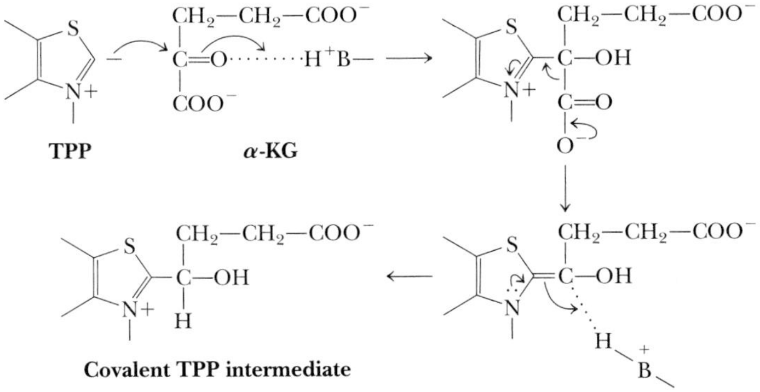

5. Aconitase is inhibited by fluorocitrate, the product of citrate synthase action on fluoroacetate. In a tissue where inhibition has occurred, all TCA cycle metabolites should be reduced in concentration. Fluorocitrate would replace citrate, and the concentrations of isocitrate and all subsequent metabolites would be reduced because of aconitase inhibition.

6. $FADH_2$ is colorless, but FAD is yellow, with a maximal absorbance at 450 nm. Succinate dehydrogenase could be

10. $[\text{isocitrate}]/[\text{citrate}] = 0.1$.   When $[\text{isocitrate}] = 0.03$ mM $[\text{citrate}] = 0.3$ mM.

Chapter 17

1. The cytochrome couple is the acceptor, and the donor is the (bound) $FAD/FADH_2$ couple, because the cytochrome couple has a higher (more positive) reduction potential.

$$\Delta \mathscr{E}_o' = 0.254\text{V} - 0.02\text{V*}$$

*This is a typical value for enzyme-bound [FAD].

$$\Delta G = -n\mathscr{F}\Delta \mathscr{E}_o' = -45.2 \text{ kJ/mol}$$

2. $\Delta \mathscr{E}_o' = -0.03$ V; $\Delta G = 5790$ J/mol.
3. This situation is analogous to that described in Section 3.3. The net result of the reduction of NAD^+ is that a proton is consumed. The effect on the calculation of free energy change is similar to that described in Equation (3.22). Adding an appropriate term to Equation (17.13) yields:

$$\mathscr{E} = \mathscr{E}_o' - RT \ln [H^+] + (RT/n\mathscr{F}) \ln ([\text{ox}]/[\text{red}])$$

4. Cyanide acts primarily via binding to cytochrome a_3, and the amount of cytochrome a_3 in the body is much lower than the amount of hemoglobin. Nitrite anion is an effective antidote for cyanide poisoning because of its unique ability to oxidize ferrohemoglobin to ferrihemoglobin, a form of hemoglobin that competes very effectively with cytochrome a_3 for cyanide. The amount of ferrohemoglobin needed to neutralize an otherwise lethal dose of cyanide is small compared with the total amount of hemoglobin in the body. Even though a small amount of hemoglobin is sacrificed by sequestering the cyanide in this manner, a "lethal dose" of cyanide can be neutralized in this manner without adversely affecting oxygen transport.
5. You should advise the wealthy investor that she should decline this request for financial backing. Uncouplers can indeed produce dramatic weight loss, but can also cause death. Dinitrophenol was actually marketed for weight loss at one time, but sales were discontinued when the potentially fatal nature of such "therapy" was fully appreciated. Weight loss is best achieved by simply making sure that the number of calories of food eaten is less than the number of calories metabolized. A person who metabolizes about 2000 kJ (about 500 Cal) more than he or she consumes every day will lose about a pound in about eight days.
6. The calculation in Section 17.8 assumes the same conditions as in this problem (1 pH unit gradient across the membrane and an electrical potential difference of 0.18 V). The transport of 1 mole of H^+ across such a membrane generates 23.3 kJ of energy. For three moles of H^+, the energy yield is 69.9 kJ. Then, from Equation (3.12), we have:

$$69,900 \text{ J/mol} = 30,500 \text{ J/mol} + RT \ln ([\text{ATP}]/[\text{ADP}][\text{P}_i])$$
$$39,400 \text{ J/mol} = RT \ln ([\text{ATP}]/[\text{ADP}][\text{P}_i])$$
$$[\text{ATP}]/[\text{ADP}][\text{P}_i] = 4.36 \times 10^6 \text{ at } 37°\text{C}$$

In the absence of the proton gradient, this same ratio is 7.25×10^{-6} at equilibrium!

7. The succinate/fumarate redox couple has the highest (i.e., most positive) reduction potential of any of the couples in glycolysis and the TCA cycle. Thus oxidation of succinate by NAD^+ would be very unfavorable in the thermodynamic sense.

Oxidation of succinate by NAD^+:

$$\Delta \mathscr{E}_o' = -0.35\text{V}; \Delta G^{\circ\prime} = 67,500 \text{ J/mol}$$

On the other hand, oxidation of succinate is quite feasible using enzyme-bound FAD.

Oxidation of succinate by [FAD]:
$\Delta \mathscr{E}_o' \approx 0$, $\Delta G^{\circ\prime} \approx 0$, depending on the exact reduction potential for bound FAD.

8. a. -73.3 kJ/mol.
   b. $K_{eq} = 7.1 \times 10^{12}$.
   c. $[\text{ATP}]/[\text{ADP}] = 19.7$.
9. a. -151.5 kJ/mol.
   b. $K_{eq} = 3.54 \times 10^{26}$.
   c. $[\text{ATP}]/[\text{ADP}] = 8.8$.
10. a. -219 kJ/mol.
    b. $K_{eq} = 2.63 \times 10^{38}$.
    c. $[\text{ATP}]/[\text{ADP}] = 36$.
11. a. -217 kJ/mol.
    b. $K_{eq} = 1.08 \times 10^{38}$.
    c. $[\text{ATP}]/[\text{ADP}] = 3479$.
12. a. -26 kJ/mol.
    b. $K_{eq} = 3.68 \times 10^4$.
    c. $[\text{NAD}^+]/[\text{NADH}] = 1.48$.
13. a. $K_{eq} = 4.8 \times 10^3$.
    b. $+48$ kJ/mol.
    c. $[\text{ATP}]/[\text{ADP}] = 2.3$.

Chapter 18

1. Efficiency $= \Delta G^{\circ\prime}/(\text{light energy input}) = n\mathscr{F}\Delta \mathscr{E}_o'/(Nhc/\lambda)$ $= 0.564$.
2. $\Delta G^{\circ\prime} = -n\mathscr{F}\Delta \mathscr{E}_o'$, so $\Delta G^{\circ\prime}/-n\mathscr{F} = \Delta \mathscr{E}_o'$. $n = 4$ for $2H_2O \longrightarrow 4e^- + 4H^+ + O_2$. $\Delta \mathscr{E}_o' = 0.0648\text{V} = (\mathscr{E}_o' (\text{primary oxidant}) - \mathscr{E}_o'(\frac{1}{2}O_2/H_2O))$. $\mathscr{E}_o' (\text{primary oxidant}) = +0.88\text{V}$.
3. ΔG for ATP synthesis $= 50,336$ J/mol. $\Delta \mathscr{E}_o' = 0.26\text{V}$.
4. $CO_2 + 3 H_2O + 2 \text{ NADPH} + 2 H^+ + 2 \text{ ATP} + \text{RuBP} \longrightarrow \text{glucose} + 2 \text{ NADP}^+ + 2 \text{ ADP} + 2 \text{ P}_i$.
5. Efficiency of noncyclic photophosphorylation $= 2 \text{ } h\nu / \text{ATP}$; efficiency of cyclic photophosphorylation $= 1.5 \text{ } h\nu/\text{ATP}$.
6. Considering the reactions involving water separately:
   1. ATP synthesis:

$$18 \text{ ADP} + 18 \text{ P}_i \longrightarrow 18 \text{ ATP} + 18 \text{ H}_2\text{O}$$

2. NADP$^+$ reduction and the photolysis of water:

$$12\ H_2O + 12\ NADP^+ \longrightarrow 12\ NADPH + 12\ H^+ + 6\ O_2$$

3. Overall reaction for hexose synthesis:

$$6\ CO_2 + 12\ NADPH + 12\ H^+ + 18\ ATP + 12\ H_2O \longrightarrow$$
$$glucose + 12\ NADP^+ + 6\ O_2 + 18\ ADP + 18\ P_i$$

$$\text{\textit{Net:}}\ 6\ CO_2 + 6\ H_2O \longrightarrow glucose + 6\ O_2$$

Of the 12 waters consumed in O_2 production in reaction 2 and the 12 waters consumed in the reactions of the Calvin–Benson cycle (reaction 3), 18 are restored by H_2O release in phosphoric anhydride bond formation in reaction 1.

Chapter 19

1. The reactions that contribute to the equation on page 583 are:

$$2\ Pyruvate + 2\ H^+ + 2\ H_2O \longrightarrow glucose + O_2$$
$$2\ NADH + 2\ H^+ + O_2 \longrightarrow 2\ NAD^+ + 2\ H_2O$$
$$4\ ATP + 4\ H_2O \longrightarrow 4\ ADP + 4\ P_i$$
$$2\ GTP + 2\ H_2O \longrightarrow 2\ GDP + 2\ P_i$$

Summing these 4 reactions produces the equation on page 583.

2. This problem essentially involves consideration of the three unique steps of gluconeogenesis. The conversion of PEP to pyruvate was shown in Chapter 15 to have a $\Delta G^{\circ\prime}$ of -31.7 kJ/mol. For the conversion of pyruvate to PEP, we need only add the conversion of a GTP to GDP (equivalent to ATP to ADP) to the reverse reaction. Thus:

Pyruvate $\longrightarrow$ PEP: $\Delta G^{\circ\prime} = +31.7$ kJ/mol $- 30.5$ kJ/mol
$$= 1.2\ \text{kJ/mol}$$

Then, using Equation (3.12):

$\Delta G = 1.2$ kJ/mol $+$ RT ln ([PEP][ADP]^2[P$_i$]/[Py][ATP]^2)
ΔG (in erythrocytes) $= -31$ kJ/mol

In the case of the fructose-1,6-bisphosphatase reaction, $\Delta G^{\circ\prime} = -16.3$ kJ/mol (see Equation [15.6]).

$\Delta G = -16.3$ kJ/mol $+$ RT ln ([F6P][P$_i$]/[F1,6BP])
ΔG (in erythrocytes) $= -36.2$ kJ/mol
For the glucose-6-phosphatase reaction, $\Delta G^{\circ\prime} = -13.8$ kJ/mol (see Table 3.3). $\Delta G = -13.9$ kJ/mol $+$ RT ln ([Glu][P$_i$]/[G6P]) $= -21.1$ kJ/mol. From these ΔG values and those in Table 15.1, ΔG for gluconeogenesis $= -103.4$ kJ/mol. $\Delta G^{\circ\prime} = -37.9$ kJ/mol.

3. Inhibition by 25 μM fructose-2,6-bisphosphate is approximately 94% at 25 μM fructose-1,6-bisphosphate and approximately 44% at 100 μM fructose-1,6-bisphosphate.

4. The hydrolysis of UDP-glucose to UDP and glucose is characterized by a $\Delta G^{\circ\prime}$ of -31.9 kJ/mol. This is more than sufficient to overcome the energetic cost of synthesizing a new glycosidic bond in a glycogen molecule. The net $\Delta G^{\circ\prime}$ for the glycogen synthase reaction is -13.3 kJ/mol.

5. According to Table 20.1, a 70-kg person possesses 1920 kJ of muscle glycogen. Without knowing how much of this is in fast-twitch muscle, we can simply use the fast-twitch data from *A Deeper Look* (page 592) to calculate a rate of energy consumption. The plot on page 592 shows that glycogen supplies are exhausted after 60 minutes of heavy exercise. Ignoring the curvature of the plot, 1920 kJ of energy consumed in 60 minutes corresponds to an energy consumption rate of 533 J/sec.

6. The reactions of the pentose phosphate pathway can combine in a variety of ways, depending on the metabolic needs of the organism. If the primary need is for ribose-5-P and the oxidative steps are bypassed, the pathway takes the form shown on the top of the following page. C-6 of glucose becomes C-5 of ribose for all five glucose molecules that pass through this scheme. There are three possibilities. The C-2 of one glucose becomes the C-3 of pyruvate, and the C-4 of the same molecule of glucose becomes C-1 of a different molecule of pyruvate. For another glucose through the pathway, C-2 becomes C-3 of pyruvate and C-4 becomes C-1 of another pyruvate. For a third glucose through the pathway, both C-2 and C-4 carbons become C-1 of pyruvate molecules.

7. The "products" of the pentose phosphate pathway as shown in Figure 19.19 are fructose-6-P and glyceraldehyde-3-P. C-2 of glucose-6-P is distributed into both C-1 and C-3 of fructose-6-P. On the other hand, C-4 of glucose-6-P is distributed into C-3 and C-4 of fructose-6-P and also into C-1 of glyceraldehyde-3-P.

8. Glycogen molecules do not have any free reducing ends, regardless of the size of the molecule. If branching occurs every 8 residues and each arm of the branch has 8 residues (or 16 per branch point), a glycogen molecule with 8000 residues would have about 500 ends. If branching occurs every 12 residues, a glycogen molecule with 8000 residues would have about 334 ends.

9. (a) Increased fructose-1,6-bisphosphate would activate pyruvate kinase, stimulating glycolysis. (b) Increased blood glucose would decrease gluconeogenesis and increase glycogen synthesis. (c) Increased blood insulin inhibits gluconeogenesis and stimulates glycogen synthesis. (d) Increased blood glucagon inhibits glycogen synthesis and stimulates glycogen breakdown. (e) Since ATP inhibits both phosphofructokinase and pyruvate kinase, and since its level reflects the energy status of the cell, a decrease in tissue ATP would have the effect of stimulating glycolysis. (f) Increasing AMP would have the same effect as decreasing ATP—stimulation of glycolysis and inhibition of gluconeogenesis. (g) Fructose-6-phosphate is not a regulatory molecule and decreases in its concentration would not markedly affect either

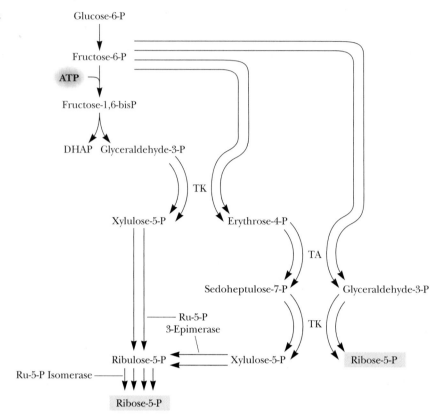

glycolysis or gluconeogenesis (ignoring any effects due to decreased [G6P] as a consequence of decreased [F6P].

10. At 298 K, assuming roughly equal concentrations of glycogen molecules of different lengths, the glucose-1-P concentration would be about 0.286 mM.

Chapter 20

1. Assuming that all fatty acid chains in the triacylglycerol are palmitic acid, the fatty acid content of the triacylglycerol is 95% of the total weight. On the basis of this assumption, one can calculate that 30 lb of triacylglycerol will yield 118.7 L of water.

2. 11-*cis*-Heptadeceonic acid is metabolized by means of seven cycles of β-oxidation, leaving a propionyl-CoA as the final product. However, the fifth cycle bypasses the acyl-CoA dehydrogenase reaction, because a *cis*-double bond is already present at the proper position. Thus, β-oxidation produces 7 NADH (=17.5 ATP), 6 FADH$_2$ (=9 ATP), and 7 acetyl-CoA (=70 ATP), for a total of 96.5 ATP. Propionyl-CoA is converted to succinyl-CoA (with expenditure of 1 ATP), which can be converted to oxaloacetate in the TCA cycle (with production of 1 GTP, 1 FADH$_2$ and 1 NADH). Oxaloacetate can be converted to pyruvate (with no net ATP formed or consumed), and pyruvate can be metabolized in the TCA cycle (producing 1 GTP, 1 FADH$_2$, and 4 NADH). The net for these conversions of propionate is 16.5 ATP. Together with the results of β-oxidation, the total ATP yield for the oxidation of one molecule of 11-*cis*-heptadecenoic acid is 113 ATP.

3. Instead of invoking hydroxylation and β-oxidation, the best strategy for oxidation of phytanic acid is α-hydroxylation, which places a hydroxyl group at C-2. This facilitates oxidative α-decarboxylation, and the resulting acid can react with CoA to form a CoA ester. This product then undergoes six cycles of β-oxidation. In addition to CO$_2$, the products of this pathway are three molecules of acetyl-CoA, three molecules of propionyl-CoA and one molecule of 2-propionyl-CoA.

4. Although acetate units cannot be used for net carbohydrate synthesis, oxaloacetate can enter the gluconeogenesis pathway in the PEP carboxykinase reaction. (For this purpose, it must be converted to malate for transport to the cytosol.) Acetate labeled at the carboxyl carbon will first label (equally) the C-3 and C-4 positions of newly formed glucose. Acetate labeled at the methyl carbon will label (equally) the C-1, C-2, C-5, and C-6 positions of newly formed glucose.

5. Fat is capable of storing more energy (37 kJ/g) than carbohydrate (16 kJ/g). Ten lb of fat contain $10 \times 454 \times 37 = 167,980$ kJ of energy. This same amount of energy would require $167,980/16 = 10,499$ g or 23 lb of stored carbohydrate.

6. The enzyme methylmalonyl-CoA mutase, which catalyzes the third step in the conversion of propionyl-CoA to succinyl-CoA, is B$_{12}$-dependent. If a deficiency in this vitamin occurs, and if large amounts of odd-carbon fatty acids were ingested in the diet, L-methymalonyl-CoA could accumulate.

7. a. Myristic acid:

$$CH_3(CH_2)_{12}CO\text{-}CoA + 94\,P_i + 94\,ADP + 20\,O_2 \longrightarrow$$
$$94\,ATP + 14\,CO_2 + 113\,H_2O + CoA$$

b. Stearic acid:

$$CH_3(CH_2)_{16}CO\text{-}CoA + 122\ P_i + 122\ ADP + 26\ O_2 \longrightarrow$$
$$122\ ATP + 18\ CO_2 + 147\ H_2O + CoA$$

c. α-Linolenic acid:

$$C_{17}H_{29}CO\text{-}CoA + 116.5\ P_i + 116.5\ ADP + 24.5\ O_2 \longrightarrow$$
$$116.5\ ATP + 18\ CO_2 + 138.5\ H_2O + CoA$$

d. Arachidonic acid:

$$C_{19}H_{31}CO\text{-}CoA + 128\ P_i + 128\ ADP + 27\ O_2 \longrightarrow$$
$$128\ ATP + 20\ CO_2 + 152\ H_2O + CoA$$

8. During the hydration step, the elements of water are added across the double bond. Also, the proton transferred to the acetyl-CoA carbanion in the thiolase reaction is derived from the solvent, so each acetyl-CoA released by the enzyme would probably contain two tritiums. Seven tritiated acetyl-CoAs would thus derive from each molecule of palmitoyl-CoA metabolized, each with two tritiums at C-2.

9. A carnitine deficiency would presumably result in defective or limited transport of fatty acids into the mitochondrial matrix and reduced rates of fatty acid oxidation.

10. Carbons C-1 and C-6 of glucose become the methyl carbons of acetyl-CoA that is the substrate for fatty acid synthesis. Carbons C-2 and C-5 of glucose become the carboxyl carbon of acetyl-CoA for fatty acid synthesis. Only citrate that is immediately exported to the cytosol provides glucose carbons for fatty acid synthesis. Citrate that enters the TCA cycle does not immediately provide carbon for fatty acid synthesis.

11. A suitable model, based on the evidence presented in this chapter, would be that the fundamental regulatory mechanism in ACC is a polymerization-dependent conformation change in the protein. All other effectors—palmitoyl-CoA, citrate, and phosphorylation-dephosphorylation—may function primarily by shifting the inactive protomer-active polymer equilibrium. Polymerization may bring domains of the protomer (i.e., bicarbonate-, acetyl-CoA-, and biotin-binding domains) closer together, or may bring these domains on separate protomers close to each other. See *Student Study Guide* for further details.

12. The pantothenic acid group may function, at least to some extent, as a flexible "arm" to carry acyl groups between the malonyl transferase and ketoacyl-ACP synthase active sites. The pantothenic acid moiety is approximately 1.9 nm in length, setting an absolute upper-limit distance between these active sites of 3.8 nm. However, on the basis of modeling considerations, it seems likely that the distance between these sites is smaller than this upper-limit value.

13. Ethanolamine + glycerol + 2 fatty acyl-CoA + 2 ATP + CTP + $H_2O \longrightarrow$ Phosphatidylethanolamine + 2 ADP + 2 CoA + CMP + PP$_i$ + P$_i$

14. The conversion of acetyl-CoA to lanosterol can be written as:

$$18\ Acetyl\text{-}CoA + 13\ NADPH + 13\ H^+ + 18\ ATP + 0.5\ O_2 \longrightarrow$$
$$Lanosterol + 18\ CoA + 13\ NADP^+ + 18\ ADP + 6\ P_i + 6\ PP_i + CO_2$$

The conversion of lanosterol to cholesterol is complicated; however, in terms of carbon counting, three carbons are lost in the conversion to cholesterol. This process might be viewed as 1.5 acetate groups for the purpose of completing the balanced equation.

15. The numbers 1–4 in the cholesterol structure indicate the carbon positions of mevalonate as shown (note that the numbering shown here is not based on the systematic numbering of mevalonate):

Mevalonate   Cholesterol

16. As shown in Figures 20.34 and 20.37, the syntheses (in eukaryotes) of phosphatidylcholine, phosphatidylethanolamine, phosphatidylinositol, and phosphatidylglycerol are dependent upon CTP. A CTP deficiency would be likely to affect all these synthetic pathways.

Chapter 21

1. a. [ATP] increase will favor adenylylation; the value of n will be greater than 6 ($n > 6$).

b. An increase in P_{IIA}/P_{IID} will favor deadenylylation; the value of n will be less than 6 ($n < 6$).

c. An increase in the $[\alpha KG]/[Gln]$ ratio will favor deadenylylation; $n < 6$.

d. $[P_i]$ decrease will favor adenylylation; $n > 6$.

2. Two ATP are consumed in the carbamoyl-P synthetase-I reaction, and 2 phosphoric anhydride bonds (equal to 2 ATP equivalents) are expended in the argininosuccinate synthetase reaction. Thus, 4 ATP equivalents are consumed in the urea cycle, as 1 urea and 1

fumarate are formed from 1 CO_2, 1 NH_3, and 1 aspartate.

3. Protein catabolism to generate carbon skeletons for energy production releases the amino groups of amino acids as excess nitrogen, which is excreted in the urine, principally as urea.

4. From Figure 21.20: ^{14}C-labeled carbon atoms derived from ^{14}C-2 of PEP are shaded.

5. 1. 2 aspartate → *transamination* → 2 oxaloacetate.
   2. 2 oxaloacetate + 2 GTP → *PEP carboxykinase* →
      2 PEP + 2 CO_2 + 2 GDP.
   3. 2 PEP + 2 H_2O → *enolase* → 2 2-PG.
   4. 2 2-PG → *phosphoglyceromutase* → 2 3-PG.
   5. 2 3-PG + 2 ATP → *3-P glycerate kinase* → 2 1,3 bisPG +
      2 ADP.
   6. 2 1,3 bisPG + 2 NADH + 2 H^+ → *G3P dehydrogenase* →
      2 G3P + 2 NAD^+ + 2 P_i.
   7. 1 G3P → *triose-P isomerase* → 1 DHAP.
   8. G3P + DHAP → *aldolase* → Fructose-1,6-bisP.
   9. F-1,6-bisP + H_2O → *FBPase* → F6P + P_i.
   10. F6P → *phosphoglucoisomerase* → G6P.
   11. G6P + H_2O → *glucose phosphatase* → glucose + P_i.

   Net: 2 aspartate + 2 GTP + 2 ATP + 2 NADH +
   6 H^+ + 4 H_2O → glucose + 2 CO_2 + 2 GDP +
   2 ADP + 4P_i + 2 NAD^+

   (Note that 4 of the 6 H^+ are necessary to balance the charge on the 4 carboxylate groups of the 2 OAA.)

   (As a consequence of reaction [1], 2 α-keto acids [e.g., α-ketoglutarate] will receive amino groups to become 2 α-amino acids [e.g., glutamate].)

6. See Figure 21.25 (purines) and Figure 21.34 (pyrimidines).

7. Assume ribose-5-P is available.
   Purine synthesis: Two ATP equivalents in the ribose-5-P pyrophosphokinase reaction, 1 in the GAR synthetase reaction, 1 in the FGAM synthetase reaction, 1 in the AIR carboxylase reaction, 1 in the CAIR synthetase reaction, and 1 in the SACAIR synthetase reaction yields IMP, the precursor common to ATP and GTP. *Net:* 7 ATP equivalents.

   a. *ATP:* 1 GTP (an ATP equivalent) is consumed in converting IMP to AMP; 2 more ATP equivalents are needed to convert AMP to ATP. Overall, ATP synthesis from ribose-5-P onward requires 10 ATP equivalents.

   b. *GTP:* 2 high-energy phosphoric anhydride bonds from ATP, but 1 NADH is produced in converting

IMP to GMP; 2 more ATP equivalents are needed to convert GMP to GTP. Overall, GTP synthesis from ribose-5-P onward requires 8 ATP equivalents. *Pyrimidine synthesis:* Starting from HCO_3^- and Gln, 2 ATP equivalents are consumed by CPS-II, and an NADH equivalent is produced in forming orotate. OMP synthesis from ribose-5-P plus orotate requires conversion of ribose-5-P to PRPP at a cost of 2 ATP equivalents. Thus, the net ATP investment in UMP synthesis is just 1 ATP.

   c. Formation of UTP from UMP requires 2 ATP equivalents. Net ATP equivalents in UTP biosynthesis = 3.

   d. CTP biosynthesis from UTP by CTP synthetase consumes 1 ATP equivalent. Overall ATP investment in CTP synthesis = 4 ATP equivalents.

8. a. See Figure 21.29.
   b. See Figure 21.38.
   c. See Figure 21.38.

9. a. Azaserine inhibits glutamine-dependent enzymes, as in steps 2 and 5 of IMP synthesis (glutamine : PRPP amido-transferase and FGAM synthetase), as well as GMP synthetase (step 2, Figure 21.28), and CTP synthetase (Figure 21.37).

   b. Methotrexate, an analog of folic acid, antagonizes THF-dependent processes, such as steps 4 and 10 (GAR transformylase and AICAR transformylase) in purine biosynthesis (Figure 21.26), and the thymidylate synthase reaction (Figure 21.48) of pyrimidine metabolism.

   c. Sulfonamides are analogs of *p*-aminobenzoic acid (PABA). Like methotrexate, sulfonamides antagonize THF formation. Thus, sulfonamides affect nucleotide biosynthesis at the same sites as methotrexate, but only in organisms such as prokaryotes that synthesize their THF from simple precursors such as PABA.

   d. Allopurinol is an inhibitor of xanthine oxidase (Figure 21.23).

10. 5-Fluorouracil inhibits the thymidylate synthase reaction (see *Human Biochemistry: Fluoro-Substituted Pyrimidine Analogs As Therapeutic Agents*, p. 702).

11. See Figure 21.45.

12. Ribose, as ribose-5-P, is released during nucleotide catabolism (as in Figure 21.31). Ribose-5-P is catabolized via the pentose phosphate pathway and glycolysis to form pyruvate, which enters the citric acid cycle. Three ribose-5-P (rearranged to give 1 ribose-5-P and 2 xylulose-5-P) give a net consumption of 2 ATP and a net production of 8 ATP and 5 NADH (= 15 ATP), when converted to 5 pyruvate via 2 fructose-6-P and 1 glyceraldehyde-3-P (consult Figure 19.19 and the pathway of glycolysis). If each pyruvate is worth 15 ATP equivalents (as in a prokaryotic cell), the overall yield of ATP from 3 ribose-5-P is 75 ATP + 23 ATP − 2 ATP = 96 ATP. Net yield per ribose-5-P is thus 32 ATP equivalents.

Chapter 22

1. Energy charge = 0.945.
2. Eight mM ATP will last 1.12 sec. Since the equilibrium constant for creatine-P + ATP $\rightarrow$ Cr + ADP is 175, [ATP] must be less than 1750 [ADP] for the reaction creatine-P + ADP $\rightarrow$ Cr + ATP to proceed to the right when [Cr-P] = 40 mM and [Cr] = 4 mM.
3. From Equation (17.13), $\mathscr{E}$(NADP$^+$/NADPH) = -0.350V; $\mathscr{E}$ (NAD$^+$/NADH) = -0.281V; thus, $\Delta\mathscr{E}$ = 0.069V, and ΔG = $-13,316$ J/mol. If an ATP "costs" 50 kJ/mol, this reaction can produce about 0.27 ATP equivalents at these concentrations of NAD$^+$, NADH, NADP$^+$, and NADPH.
4. Assume that the K_{eq} for ATP + AMP $\rightarrow$ 2 ADP = 1.2 (legend, Figure 22.2). When [ATP] = 7.2 mM, [ADP] = 0.737 mM, and [AMP] = 0.063 mM. When [ATP] decreases by 10% to 6.48 mM, [ADP] + [AMP] = 1.52 mM, and thus [ADP] = 1.30 mM and [AMP] = 0.22 mM. A 10% decrease in [ATP] has resulted in a 0.22/0.063 = 3.5-fold increase in [AMP].
5. Inhibition of acetyl-CoA carboxylase will lower the concentration of malonyl-CoA. Since malonyl-CoA inhibits uptake of fatty acids (Figure 20.31), decreases in [malonyl-CoA] favor fatty acid synthesis.

Chapter 23

1. Since DNA replication is semi-conservative, the density of the DNA will be (1.724 + 1.710)/2 = 1.717 g/mL. If replication is dispersive, the density of the DNA will also be 1.717 g/mL. To distinguish between these possibilities, heat the dsDNA obtained after one generation in ^{15}N to 100°C to separate the strands and then examine the strands by density gradient ultracentrifugation. If replication is indeed semi-conservative, two bands of ssDNA will be observed, the "heavy" one containing ^{15}N atoms, the "light" one containing ^{14}N atoms. If replication is dispersive, only a single band of intermediate density would be observed.
2. The 5′-exonuclease activity of DNA polymerase I removes mispaired segments of DNA sequence that lie in the path of the advancing polymerase. Its biological role is to remove mispaired bases during DNA repair. The 3′-exonuclease activity acts as a proofreader to see whether the base just added by the polymerase activity is properly base-paired with the template. If not (that is, if it is an improper base with respect to the template), the 3′-exonuclease removes it, and the polymerase activity can try once more to insert the proper base. An *E. coli* strain lacking DNA polymerase I 3′-exonuclease activity would show a high rate of spontaneous mutation.

3. Assume that the polymerization rate achieved by each half of the DNA polymerase III homodimer is 750 nucleotides per sec. The entire *E. coli* genome consists of 4.64×10^6 bp. At a rate of 750 bp/sec per DNA pol III homodimer (one at each replication fork), DNA replication would take almost 3100 sec (51.7 min; 0.86 hr). When *E. coli* is dividing at a rate of once every 20 min, *E. coli* cells must be replicating DNA at the rate of 4.64×10^6 bp per 20 min (2.32×10^5 bp/min or 3867 bp/sec). To achieve this rate of replication would require initiation of DNA replication once every 20 min at *ori*, and a minimum of 3933/1500 = 2.57 replication bubbles per *E. coli* chromosome, or 5.14 replication forks.
4. If there are only 10 molecules of DNA polymerase III (as DNA pol III homodimers) per *E. coli* cell and *E. coli* growing at its maximum rate has about 5 replication forks per chromosome, it seems likely that DNA polymerase III availability is sufficient to sustain growth at this rate.
5. Okazaki fragments are 1000–2000 nucleotides in length. Since DNA replication in *E. coli* generates Okazaki fragments whose total length must be 4.64×10^6 nucleotides, a total of 2300–4700 Okazaki fragments must be synthesized. Consider in comparison a human cell carrying out DNA replication. The haploid human genome is 3×10^9 bp, but most cells are diploid (6×10^9 bp). Collectively, the Okazaki fragments must total 6×10^9 nucleotides in length, distributed over 3 to 6 million separate fragments.
6. DNA gyrases are ATP-dependent topoisomerases that introduce negative supercoils into DNA. DNA gyrases introduce supercoils into double-helical DNA by breaking the sugar–phosphate backbone of its DNA strands and then religating them (see Figure 8.44). In contrast, helicases are ATP-dependent enzymes that disrupt the hydrogen bonds between base pairs that hold double-helical DNA together. Helicases move along dsDNA, leaving ssDNA in their wake.
7. A diploid human cell contains 6×10^9 bp of DNA. One origin of replication every 300 kbp gives 20,000 origins of replication in 6×10^9 bp. If DNA replication proceeds at a rate of 100 bp/sec at each of 40,000 replication forks (2 per origin), 6×10^9 bp could be replicated in 1500 sec (25 min). To provide 2 molecules of DNA polymerase per replication fork, a cell would require 80,000 molecules of this enzyme.
8. For purposes of illustration, consider how the heteroduplex bacteriophage chromosome in Figure 23.18 might have arisen (see figure on following page):

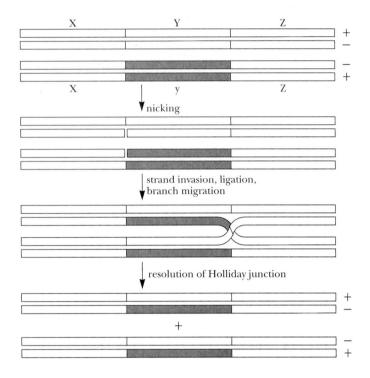

9. The mismatch repair system of *E. coli* relies on DNA methylation to distinguish which DNA strand is "correct" and which is "mismatched"; the unmethylated strand (the newly synthesized one that has not had sufficient time to become methylated is, by definition, the mismatched one). Homologous recombination involves DNA duplexes at similar stages of methylation, neither of which would be interpreted as "mismatched," and thus the mismatch repair system does not act.

10. B-DNA normally has a helical twist of about 10 bp/turn, so that the rotation per residue (base pair), $\Delta\phi$, is 36°. If the DNA is unwound to 18.6 bp per turn, $\Delta\phi$ becomes 19.4°. Thus, the change in $\Delta\phi$ is 16.6°.

11.

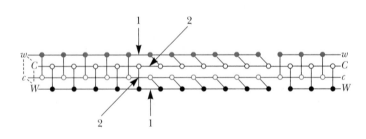

Consider two DNA duplexes, *WC* and *wc*, respectively, displayed as "ladders." The gap at the right denotes the initial cleavage, in this case of the *W* and *w* strands. Cleavage by resolvase at the arrows labeled 1 and ligation of like strands (*W* with *w* and *C* with *c*) will yield patch recombinants; cleavage at the arrows labeled 2 and ligation of like strands gives splice recombinants. See also the scissors in Figure 23.23b.

12. The DNA is: GCTA
                        CGAT

   a. HNO$_2$ causes deamination of C and A. Deamination of C yields U, which pairs the way T does, giving:

   GTTA and ACTA and (rarely) ATTA
   CAAT   TGAT   TAAT

   Deamination of A yields I, which pairs as G does, giving:

   GCTG and GCCA and (rarely) GCCG
   CGAC   CGGT   CGGC

   b. Bromouracil usually replaces T, and pairs the way C does:

   GCCA and GCTG and (rarely) GCCG
   CGGT   CGAC   CGGC

   Less often, bromouracil replaces C, and mimics T in its base-pairing:

   GTTA and ACAT and (rarely) ATTA
   CAAT   TGTA   TAAT

   c. 2-Aminopurine replaces A and normally base-pairs with T, but may also pair with C:

   GCTG and CGGT and (rarely) GCCG
   CGAC   GCCA   CGGC

13. Transposons are mobile genetic elements that can move from place to place in the genome. Insertion of a transposon within or near a gene may disrupt the gene or inactivate its expression.

14. The PrP^{SC} form of the prion protein accumulates in individuals with inheritable (genetic) diseases such as Creutzfeldt-Jakob disease. Further, prion-related diseases may be acquired by ingestion of PrP^{SG}, indicating the infectious properties of this form of the protein.

Chapter 24

1. 5'-AGAUCCGUAUGGCGAUCUCGACGAAGACUCCUAGGGAAUCC...mRNA

   3'-TCTAGGCATACCGCTAGAGCTGCTTCTGAGGATCCCTTAGG...5' = DNA template strand

The first AUG codon is underlined:

5'-AGAUCCGU<u>AUG</u>GCGAUCUC-
            GACGAAGACUCCUAGGGAAUCC ...

Reading from the AUG codon, the amino acid sequence of the protein is:

(N-term)Met-Ala-Ile-Ser-Thr-Lys-Thr-Pro-Arg-Glu-Ser- . . .

2. See Figure 24.2 for a summary of the events in transcription initiation by *E. coli* RNA polymerase. A gene must have a promoter region for proper recognition and transcription by RNA polymerase. The promoter region (where RNA polymerase binds) typically consists of 40 bp to the 5′-side of the transcription start site. The promoter is characterized by two consensus sequence elements: a hexameric TTGACA consensus element in the −35 region and a Pribnow box (consensus sequence TATAAT) in the −10 region (see Figure 24.3).

3. The fact that the initiator site on RNA polymerase for nucleotide binding has a higher K_m for NTP than the elongation site ensures that initiation of mRNA synthesis will not begin unless the available concentration of NTP is sufficient for complete synthesis of an mRNA.

4. In contrast to prokaryotes (which have only a single kind of RNA polymerase), eukaryotes possess three distinct RNA polymerases—I, II, and III—acting on three distinct sets of genes (see Section 24.3). The subunit organization of these eukaryotic RNA polymerases is more complex than that of their prokaryotic counterparts. The three sets of genes are recognized by their respective RNA polymerases because they possess distinctive categories of promoters. In turn, all three polymerases interact with their promoters via particular transcription factors that recognize promoter elements within their respective genes. Protein-coding eukaryotic genes, in analogy with prokaryotic genes, have two consensus sequence elements within their promoters, a TATA box (consensus sequence TATAAA) located in the −25 region, and *Inr*, an initiator element that encompasses the transcription start site. In addition, eukaryotic promoters often contain additional short, conserved sequence modules such as enhancers and response elements for appropriate regulation of transcription. Transcription termination differs as well as prokaryotes and eukaryotes. In prokaryotes, two transcription basic termination mechanisms occur: *rho* protein-dependent termination and DNA-encoded termination sites. Transcription in eukaryotes tends to be imprecise, occurring downstream from consensus AAUAAA sequences known as poly(A)$^+$ addition sites. An important aspect to eukaryotic transcription is post-transcriptional processing of mRNA (Section 24.5).

5. When DNA-binding proteins "read" specific base sequences in DNA, they do so by recognizing a specific matrix of H-bond donors and acceptors displayed within the major groove of B-DNA by the edges of the bases. When DNA-binding proteins recognize a specific DNA region by "indirect readout," the protein is discerning local conformational variations in the cylindrical surface of the DNA double helix. These conformational variations are a consequence of unique sequence information contributed by the base pairs that make up this region of the DNA.

6. The metallothionein promoter is about 265 bp long. Since each bp of B-DNA contributes 0.34 nm to the length of DNA, this promoter is about 90 nm long, covering 26.5 turns of B-DNA (10 bp/turn). Each nucleosome spans about 146 bp, so about 2 nucleosomes would be bound to this promoter.

7. If cells contain a variety of proteins that share a leucine zipper dimerization motif but differ in their DNA recognition helices, heterodimeric *bZIP* proteins can form that will contain two different DNA contact *basic regions* (consider this possibility in light of Figure 24.29). These heterodimers are no longer restricted to binding sites on DNA that are dyad-symmetric; the repertoire of genes with which they can interact will thus be dramatically increased.

8. Exon 17 of the fast skeletal muscle troponin T gene is one of the *mutually exclusive* exons (see Figure 24.39). Deletion of this exon would cut by half the alternative splicing possibilities, so that only 32 different mature mRNAs would be available from this gene. If exon 7, a *combinatorial* exon, were duplicated, it would likely occur as a tandem duplication. The different mature mRNAs could contain 0, 1, or 2 copies of exon 7. The 32 combinatorial possibilities shown in the center of Figure 24.39 represent those with 0 or 1 copy of exon 7. Those with 2 copies are 16 in number: *456778; 56778; 46778; 6778; 45778; 4778; 5778; 778; 4577; 577; 5677; 45677; 4677; 477; 677;* and *77*. Given the *mutually exclusive* exons 17 and 18, 96 different mature mRNAs would be possible.

Chapter 25

1. The cDNA sequence as presented:

    CAATACGAAGCAATCCCGCGACTAGACCTTAAC. . .

    represents six potential reading frames, the three inherent in the sequence as written and the three implicit in the complementary DNA strand, which is written 5′ → 3′ as:

    . . .GTTAAGGTCTAGTCGCGGGATTGCTTCGTATTG

    Two of the three reading frames of the cDNA sequence given contain stop codons. The third reading frame is a so-called open reading frame (a stretch of coding sequence devoid of stop codons):

    <u>CAA</u>TAC<u>GAA</u>GCA<u>ATC</u>CCG<u>CGA</u>CTA<u>GAC</u>CTT<u>AAC</u>. . .
    (alternate codons underlined)

    The amino acid sequence it encodes is:

    Gln-Tyr-Glu-Ala-Ile-Pro-Arg-Leu-Asp-Leu-Asn. . .

    Two of the three reading frames of the complementary DNA sequence also contain stop codons. The third may be an open reading frame (see top of following page):

. . .GTT<u>A</u>AGG<u>TCT</u>AGT<u>CGC</u>GGG<u>A</u>TTGCT<u>T</u><u>CG</u>TAT<u>T</u>G
(alternate codons underlined)

An unambiguous conclusion about the partial amino acid sequence of this cDNA cannot be reached.

2. A random (AG) copolymer would contain varying amounts of the following codons:

AAA AAG AGA GAA AGG GAG GGA GGG, codons for Lys Lys Arg Glu Arg Glu Gly Gly, respectively. Therefore, the random (AG) copolymer would direct the synthesis of a polypeptide consisting of Lys, Arg, Glu, and Gly. The relative frequencies of the various codons are a function of the probability that a base will occur in a codon. For example, if the A/G ratio is 5/1, the ratio of AAA/AAG is $(5 \times 5 \times 5)/(5 \times 5 \times 1) = 5/1$. If the 3A codon is assigned a value of 100, then the 2A1G codon has a frequency of 20. From this analysis, the relative abundances of these amino acids in the polypeptide should be:

Lys = 120; Arg = 24; Glu = 24; Gly = 4.8
(Normalized to Lys = 100: Arg = 20; Glu = 20; Gly = 4)

3. Review the information in Section 25.2, noting in particular the two levels of specificity exhibited by aminoacyl-tRNA synthetases (1: at the level at ATP $\rightleftharpoons$ PP$_i$ exchange/aminoacyl adenylate synthesis in the presence of amino acid and the absence of tRNA; and 2: at the level of loading the aminoacyl group on an acceptor tRNA).

4. The wobble rules state that a first-base anticodon U can recognize either an A or G in the codon third-base position; first-base anticodon G can recognize either U or C in the codon third-base position; and first-base anticodon I can recognize either U, C, or A in the codon third-base position. Thus, codons with third-base A or G, which are degenerate for a particular amino acid (Table 25.3 reveals that all codons with third-base purines are degenerate, except those for Met and Trp) could be served by single tRNA species with first-base anticodon U. More emphatically, codons with third-base pyrimidines (C or U) are always degenerate and could be served by single tRNA species with first-base anticodon G. Wobble involving first-base anticodon I further minimizes the number of tRNAs needed to translate the 61 sense codons. Wobble tends to accelerate the rate of translation because the non-canonical base pairs formed between bases in the third position of codons and bases occupying the first-base wobble position of anticodons are less stable. As a consequence, the codon : anticodon interaction is more transient.

5. The stop codons are UAA, UAG, and UGA.
Sense codons that are a single-base change from UAA include: CAA (Pro), AAA (Lys), GAA (Glu), UUA (Leu), UCA (Ser), UAU (Tyr), and UAC (Tyr).
Sense codons that are a single-base change from UAG include: CAG (Gln), AAG (Lys), GAG (Glu), UCG (Ser), UUG (Leu), UGG (Trp), UAU (Tyr), and UAC (Tyr).

Sense codons that are a single-base change from UGA include: CGA (Arg), AGA (Arg), GGA (Gly), UUA (Leu), UCA (Ser), UGU (Cys), UGC (Cys), and UGG (Trp). That is, 19 of the 61 sense codons are just a single base change from a nonsense codon.

6. The list of amino acids in problem 5 is a good place to start in considering the answer to this question. Those amino acid codons in which the codon base (the wobble position) is but a single base change from a nonsense codon are the more likely among this list, because pairing is less stringent at this position. These include UAU (Tyr) and UAC (Tyr) for nonsense codons UAA and UAG; and UGU (Cys), UGC (Cys), and UGG (Trp) for nonsense codon UGA.

7. The more obvious answer to this question is that eukaryotic ribosomes are larger, more complex, and hence slower than prokaryotic ribosomes. In addition, initiation of translation requires a greater number of initiation factors in eukaryotes than in prokaryotes. It is also worth noting that eukaryotic cells, in contrast to prokaryotic cells, are typically under less selective pressure to multiply rapidly.

8. Each amino acid in a polypeptide chain with an ϵ or extended conformation contributes about 0.35 nm to its length. 10 nm $\div$ 0.35 nm per amino acid = 28.6 amino acids.

9. Larger, more complex ribosomes offer greater advantages in terms of their potential to respond to the input of regulatory influences, to have greater accuracy in translation, and to enter into interactions with subcellular structures. Their larger size may slow the rate of translation, which in some instances may be a disadvantage.

10. The universal organization of ribosomes as two-subunit structures in all cells—archae, eubacteria, and eukaryotes—suggests that such an organization is fundamental to ribosome function. One possibility is that translocation along mRNA, aminoacyl-tRNA binding, peptidyl transfer, and/or tRNA release are processes that require repetitious uncoupling of physical interactions between the large and small ribosomal subunits.

11. Prokaryotic cells rely on N-formyl-Met-tRNA$_f^{Met}$ to initiate protein synthesis. The tRNA$_i^{fMet}$ molecule has a number of distinctive features, not found in noninitiator tRNAs, that earmark it for its role in translation initiation (see Figure 25.16). Further, N-formyl-Met-tRNA$_i^{fMet}$ interacts only with initiator codons (AUG or, less commonly, GUG). A second tRNA^{Met}, designated tRNA$_m^{Met}$, serves to deliver methionyl residues as directed by internal AUG codons. Both tRNA$_i^{fMet}$ and tRNA$_m^{Met}$ are loaded with methionine by the same methionyl-tRNA synthetase, and UAG is the Met codon, both in initiation and elongation. AUG initiation codons are distinctive in that they are situated about 10 nucleotides downstream from the Shine–Dalgarno sequence at the 5′-end of mRNAs; this sequence determines the trans-

lation start site (see Figure 25.16). Eukaryotic cells also have two tRNA^{Met} species, one of which is a unique tRNA$_i^{Met}$ that functions only in translation initiation. Eukaryotic mRNAs lack a counterpart to the prokaryotic Shine–Dalgarno sequence; apparently the eukaryotic small ribosomal subunit binds to the 5′-end of a eukaryotic mRNA and scans along it until it encounters an AUG codon. This first AUG codon defines the eukaryotic translation start site.

12. The Shine–Dalgarno sequence is a purine-rich sequence element near the 5′-end of prokaryotic mRNAs (Figure 25.16). It base-pairs with a complementary pyrimidine-rich region near the 3′-end of 16S rRNA, the rRNA component of the prokaryotic 30S ribosomal subunit. Base pairing between the Shine–Dalgarno sequence and 16S rRNA brings the translation start site of the mRNA into the A site on the prokaryotic ribosome. Since the nucleotide sequence of the Shine–Dalgarno element varies somewhat from mRNA to mRNA, while the pyrimidine-rich Shine–Dalgarno–binding sequence of 16S rRNA is invariant, different mRNAs vary in their affinity for binding to 30S ribosomal subunits. Those that bind with highest affinity are more likely to be translated.

13. The most apt account is (b). Polypeptide chains typically contain hundreds of amino acid residues. Such chains, attached as a peptidyl group to a tRNA in the P site, would show significant inertia to movement, compared with an aminoacyl-tRNA in the A site.

14. Elongation factors EF-Tu and EF-Ts interact in a manner analogous to the GTP-binding G proteins of signal transduction pathways (see pp. 358–359). EF-Tu binds GTP and in the GTP-bound form delivers an aminoacyl-tRNA to the ribosome, whereupon the GTP is hydrolyzed to yield EF-Tu : GDP and P_i. EF-Ts mediates an exchange of the bound GDP on EF-Tu:GDP with free GTP, regenerating EF-Tu : GTP for another cycle of aminoacyl-tRNA delivery. The α-subunit of the heterotrimeric GTP-binding G proteins also binds GTP. G_α has an intrinsic GTPase activity and the GTP is eventually hydrolyzed to form G_α:GDP and P_i. The β- and γ-subunits of G proteins facilitate the exchange of bound GDP on G_α for GTP, in analogy with EF-Ts for EF-Tu. The amino acid sequences of EF-Tu and G proteins reveal that they share a common ancestry.

15. Human rhodanese has 296 amino acid residues. Its synthesis would require the involvement of 4 ATP equivalents per residue or 1184 ATP equivalents. Folding of rhodanese by the GroEL complex consumes another 130 ATP equivalents. The total number of ATP equivalents expended in the synthesis and folding of rhodanese is approximately 1314.

16. Since a single proteolytic nick in a protein can doom it to total degradation, nicked proteins are clearly not tolerated by cells and are quickly degraded. Cells are virtually devoid of partially degraded protein fragments, which would be the obvious intermediates in protein degradation. The absence of such intermediates and the rapid disappearance of nicked proteins from cells indicates that protein degradation is a rigorously selective, efficient cellular process.

Chapter 26

1. Polypeptide hormones constitute a larger and structurally more diverse group of hormones than either the steroid or the amino-acid–derived hormones, and it thus might be concluded that the specificity of polypeptide hormone–receptor interactions, at least in certain cases, should be extremely high. The steroid hormones may act either by binding to receptors in the plasma membrane or by entering the cell and acting directly with proteins controlling gene expression, whereas polypeptides and amino-acid–derived hormones act exclusively at the membrane surface. Amino-acid–derived hormones can be rapidly interconverted in enzyme-catalyzed reactions that provide rapid responses to changing environmental stresses and conditions. See *Student Study Guide* for additional information.

2. The cyclic nucleotides are highly specific in their action, since cyclic nucleotides play no metabolic roles in animals. Ca^{2+} ion has an advantage over many second messengers because it can be very rapidly "produced" by simple diffusion processes, with no enzymatic activity required. IP_3 and DAG, both released by the metabolism of phosphatidylinositol, form a novel pair of effectors that can act either separately or synergistically to produce a variety of physiological effects. DAG and phosphatidic acid share the unique property that they can be prepared from several different lipid precursors. Nitric oxide, a gaseous second messenger, requires no transport or translocation mechanisms and can diffuse rapidly to its target sites.

3. Nitric oxide functions primarily by binding to the heme prosthetic group of soluble guanylyl cyclase, activating the enzyme. An agent that could bind in place of NO—but that does not activate guanylyl cyclase—could reverse the physiological effects of nitric oxide. Interestingly, carbon monoxide, which has long been known to bind effectively to heme groups, appears to function in this way. Solomon Snyder and his colleagues at Johns Hopkins University have shown that administration of CO to cells that have been stimulated with nitric oxide causes attenuation of the NO-induced effects.

4. Herbimycin, whose structure is shown in the figure below, reverses the transformation of cells by Rous sarcoma virus, presumably as a direct result of its inactivation of the viral tyrosine kinase. The manifestations of transformation (on rat kidney cells, for example) include rounded cell morphology, increased glucose uptake and glycolytic activity, and the ability to grow without being anchored to a physical support (termed

"anchorage-independent growth"). Herbimycin reverses all these phenotypic changes. On the basis of these observations, one might predict that herbimycin might also inactivate tyrosine kinases that bear homology to the viral pp60$^{v\text{-}src}$ tyrosine kinase. This inactivation has in fact been observed, and herbimycin is used as a diagnostic tool for implicating tyrosine kinases in cell-signaling pathways.

5. The identification of phosphorylated tyrosine residues on cellular proteins is difficult. Quantities of phosphorylated proteins are generally extremely small, and to distinguish tyrosine phosphorylation from serine/threonine phosphorylation is tedious and laborious. On the other hand, monoclonal antibodies that recognize phosphotyrosine groups on protein provide a sensitive means of detecting and characterizing proteins with phosphorylated tyrosines, using, for example, Western blot methodology.

6. Hormones act at extremely low concentration, but many of the metabolic consequences of hormonal activation (release of cyclic nucleotides, Ca^{2+} ions, DAG, etc., and the subsequent alterations of metabolic pathways) occur at and involve higher concentrations of the affected molecular species. As we have seen in this chapter, most of the known hormone receptors mediate hormonal signals by activating enzymes (adenylyl cyclase, phospholipases, protein kinases, and phosphatases). One activated enzyme can produce many thousands of product molecules before it is inactivated by cellular regulation pathways.

7. Vesicles with an outside diameter of 40 nm have an inside diameter of approximately 36 nm and an inside radius of 18 nm. The data correspond to a volume of 2.44×10^{-20} liter. Then 10,000 molecules/6.02×10^{23} molecules/mole $= 1.66 \times 10^{-20}$ mole. The concentration of acetylcholine in the vesicle is thus 1.66 moles/2.44 liter or 0.68 M.

8. The evidence outlined in this problem points to a role for cAMP in fusion of synaptic vesicles with the presynaptic membrane and the release of neurotransmitters. GTPγS may activate an inhibitory G protein, releasing $G_{\alpha i}$(GTPγS), which inhibits adenylyl cyclase and prevents the formation of requisite cAMP.

INDEX

B = box; D = definition; F = figure; G = molecular graphic; S = structure; T = table.

Kinetics
chemical, 324
of enzyme-catalyzed reactions, 325, 338
kmef group, 396
Knockout mice, 748**B**
Knoop, Franz, 607
Kornberg, Arthur, 729
Krebs, Edwin, 346, 358
Krebs, Sir Hans, 445
Krebs cycle. *See* Tricarboxylic acid cycle
Kretsinger, Robert, 869
Kringle module, 141**F**, 142**F**
Kunkel, Louis, 426**B**
Kuru, 740**B**
K_w, 38

L-selectins, 230**B**
lac Operon, 770, 771**F**
lac Repressor(s), 109, 771, 772**F**
Lactam, 243, 243**S**
Lactase, 218, 491**B**
Lactate, 487, 583
Lactate dehydrogenase, 347, 348**F**, 487, 487**F**
Lactic acid fermentation, 486**F**, 487
Lactic acidosis, 593**B**
Lactim, 243, 243**S**
Lactobacillus, 491**B**
Lactose, 217**S**, 491**B**, 770, 770**S**
β-D-Lactose, 218
Lactose intolerance, 491**B**
Lactose synthase, 491**B**
Lactrodectus mactans, 883
α-Lactrotoxin, 883
Laetrile, 220, 220**S**
Lagging strand, 728**D**, 729
LamB protein, 181, 181**S**
Lambda receptor, 181, 181**S**
Lamellae, 22, 553
Langdon, Robert, 644
Langston, J. William, 529**B**
Lanolin, 165
Lanosterol, 166**S**, 167, 647, 648**F**
Lariat, 794
Lathyrism, 133**B**
LDL receptors, 391, 393**S**
Leader peptidase, 838
Leader peptide, 838
Leading strand, 728
Leber's hereditary optic neuropathy, 507**B**
Lecithin, 159
Lecithin:cholesterol acyltransferase, 391
Lehninger, Albert, 607
Leiurus quinquestriatus, 882**B**
Leloir, Louis, 489, 588
Leloir pathway, 489, 489**F**
Lepidoteran insects, 310
Leptin, 717**B**
Leucine side chains, 789
Leucopterin, 464**B**
Leukocytes, 230**B**
Leukosialin, 234, 234**F**
Leukotrienes, 639, 641**F**
Levinthal, Cyrus, 141
Levinthal's paradox, 141

Levorotatory, 80
Lewis, Carl, 653
Li, Jia, 410**B**
Ligand-gated chloride channels, 888, 888**F**
Ligand-gated ion channel(s), 854, 884
Ligands, 15**D**, 16**F**
and enzyme interactions
computer modeling of, 339**B**
Ligation, 295
Light chain(s), 424, 425**F**
Light-driven ATP synthesis, 566
Light-driven electron flow, 562
Light energy absorption, 556, 557**F**
Light-harvesting complexes, 559
Light-harvesting pigments, 556, 557**S**
Lignin(s), 453, 674**F**
Limonene, 166**S**
Lineweaver-Burk double-reciprocal plot(s), 332, 333**F**, 334, 335**F**, 339, 340**F**
Link protein, 240
Linkers, 296, 296**F**
Linoleic acid, 155**S**, 156, 629, 630**F**
α-Linoleic acid, 155**S**, 156
1-Linoleoyl-2-palmitoylphosphatidylcholine, 159
Lipases, 636**F**
Lipid(s), 153**D**, 172**F**, 452
amphipathic, 153
biosynthesis of, 633
classification of, 154
dietary
fatty acids in, 156**B**
diffusion of, 176
metabolism of, 603
orientation in membranes, 177
phase separation of, 176, 177**F**
Lipid-anchored proteins, 178
Lipid bilayer(s), 172**F**, 173, 173**F**
hydrocarbon chain orientation in, 175
Lipoatropic diabetes, 717**B**
Lipoic acid, 462, 463**F**
Lipopolysaccharide, 230, 231**F**
Lipoprotein(s), 107, 388, 388**T**, 389**S**, 390**S**, 391**F**
defects in metabolism, 393**B**
Lipoprotein lipase, 390
Liposomes, 173
Lithium ion
pharmacology of, 868**B**
Liver, 717
Liver alcohol dehydrogenase, 146, 147**F**, 372, 372**F**
Living systems
organization of, 3
Lock and key hypothesis, 343
Lohman, Timothy, 420
Long-term memory, 785**B**
Longitudinal tubules, 421
Loops
of tRNA molecules, 284, 284**F**, 285
Lovastatin, 650**B**
Low density lipoprotein receptor, 234, 391, 393

Low density lipoproteins, 389
Lycopene, 166**S**
Lymphocyte homing receptor, 238
Lymphocytes
B-cell, 393
T-cell, 393
Lynen, F., 607
Lysine, 670
Lysogeny, 26
Lysolecithin, 161**B**
Lysosomal acid lipases, 391
Lysosomes, 24

M disk, 422
M line, 422
Macromolecule(s), 3, 8, 9**F**
directionality of, 11, 12**F**
as informational, 11, 12**F**
loss of structural order, 17
synthesis and growth of, 706
Macrophages, 393
Mad cow disease, 740**B**
Magnetic resonance imaging, 174
Maize, 747
Major groove, 269, 269**F**
Major transplantation antigens H2, 180**B**
Malate, 498**F**, 513**F**, 574
L-Malate, 505
Malate–aspartate shuttle, 545, 546**F**, 547
Malate dehydrogenase, 498**F**, 505, 505**F**, 506**S**. 506**G**, 509**F**
Malate synthase, 516**B**
Malathion, 885, 887**S**
Male baldness, 652**B**
4-Maleylacetoacetate, 679**B**
Malic enzyme, 510, 614, 614**F**
Maloney murine leukemia virus, 315, 316**F**
Malonyl-CoA, 620, 622, 625, 631
Malonyl transacylase, 625
Malonyl transferase, 625
Maltase, 219
Maltoporin, 181, 181**S**, 183**F**
Maltose, 217**S**, 218
Mandrill, 3**F**
Mannan, 225**S**
Mannitol, 213
Mannose, 489
MAP kinase, 696
Marfan's syndrome, 133**B**
Margarine, 155**B**
Markers
genetic, 307**B**
Mass spectrometry, 99
Mast cell degranulating peptide, 882**B**
Mathis, Diane, 478**B**
Matrix
mitochondrial, 521
Maurizi, Michael, 845**B**
McClintock, Barbara, 747, 748
McCully, Kilmer, 672**B**
McGwire, Mark, 654**B**
McIntosh, Richard, 417
McKnight, Steve, 789
MCM proteins, 734
Mechano-chemical coupling, 417

A	adenine (or the amino acid alanine)		$\mathscr{F}$	Faraday's constant
Ab	antibody		F_{AB}	antibody molecule fragment that binds antigen
Ac-CoA	acetyl-coenzyme A		FAD	flavin adenine dinucleotide
ACh	acetylcholine		$FADH_2$	reduced flavin adenine dinucleotide
ACP	acyl carrier protein		FBP	fructose-1,6-bisphosphate
ADH	alcohol dehydrogenase		FBPase	fructose bisphosphatase
ADP	adenosine diphosphate		Fd	ferredoxin
Ag	antigen		fMet	N-formyl-methionine
AIDS	acquired immunodeficiency syndrome		FMN	flavin mononucleotide
ALA	δ-aminolevulinic acid		F1P	fructose-1-phosphate
Ala	alanine		F6P	fructose-6-phosphate
AMP	adenosine monophosphate		G	guanine or Gibbs free energy (or the amino acid glycine)
Arg	arginine		GABA	γ-aminobutyric acid
Asn	asparagine		Gal	galactose
Asp	aspartate		GDP	guanosine diphosphate
ATCase	aspartate transcarbamoylase		GLC	gas-liquid chromatography
atm	atmosphere		Glc	glucose
ATP	adenosine triphosphate		Gln	glutamine
BChl	bacteriochlorophyll		Glu	glutamate
bp	base pair		Gly	glycine
BPG	bisphosphoglycerate		GMP	guanosine monophosphate
BPheo	bacteriopheophytin		G1P	glucose-1-phosphate
C	cytosine (or the amino acid cysteine)		G3P	glyceraldehyde-3-phosphate
cal	calorie		G6P	glucose-6-phosphate
CaM	calmodulin		G6PD	glucose-6-phosphate dehydrogenase
cAMP	cyclic 3′,5′-adenosine monophosphate		GS	glutamine synthetase
CAP	catabolite activator protein		GSH	glutathione (reduced glutathione)
cDNA	complementary DNA		GSSG	glutathione disulfide (oxidized glutathione)
CDP	cytidine diphosphate		GTP	guanosine triphosphate
CDR	complementarity-determining region		H	histidine
Chl	chlorophyll		h	hour
CM	carboxymethyl		h	Planck's constant
CMP	cytidine monophosphate		Hb	hemoglobin
CoA or CoASH	coenzyme A		HDL	high-density lipoprotein
CoQ	coenzyme Q		HGPRT	hypoxanthine-guanine phosphoribosyltransferase
cpm	counts per minute		HIV	human immunodeficiency virus
CTP	cytidine triphosphate		HMG-CoA	hydroxymethylglutaryl-coenzyme A
Cys	cysteine		hnRNA	heterogeneous nuclear RNA
D	dalton (or the amino acid aspartate)		HPLC	high-pressure (or high-performance) liquid chromatography
d	deoxy		HX	hypoxanthine
dd	dideoxy		Hyl	hydroxylysine
DAG	diacylglycerol		Hyp	hydroxyproline
DEAE	diethylaminoethyl		I	inosine (or the amino acid isoleucine)
DHAP	dihydroxyacetone phosphate		IDL	intermediate density lipoprotein
DHF	dihydrofolate		IF	initiation factor
DHFR	dihydrofolate reductase		IgG	immunoglobulin G
DNA	deoxyribonucleic acid		Ile	isoleucine
DNP	dinitrophenol		IMP	inosine monophosphate
Dopa	dihydroxyphenylalanine		IP_3	inositol-1, 4, 5-trisphosphate
E	glutamate		IPTG	isopropylthiogalactoside
$\mathscr{E}$	reduction potential		IR	infrared
E4P	erythrose-4-phosphate		ITP	inosine triphosphate
EF	elongation factor		J	joule
EGF	epidermal growth factor		K	lysine
EPR	electron paramagnetic reasonance		k	kilo (10^3)
ER	endoplasmic reticulum		K_m	Michaelis constant
F	phenylalanine		kb	kilobases